## FOURTH EDITION

# INTRODUCTORY ELECTRONIC DEVICES AND CIRCUITS
# Conventional Flow Version

## Robert T. Paynter

*St. Louis Community College*

Prentice Hall
Upper Saddle River, New Jersey ■ Columbus, Ohio

**Library of Congress Cataloging-in-Publication Data**

Paynter, Robert T.
     Introductory electronic devices and circuits  :  conventional flow
version / Robert T. Paynter. — 4th ed.
          p.          cm.
     Includes index.
     ISBN 0-13-243502-0
     1. Solid state electronics.   2. Electronic apparatus and
appliances.   3. Electronic circuits.   4. Transistor amplifiers.
     I. Title
     TK7871.85.P39        1997
     621.3815—dc20                                                          96-33852
                                                                              CIP

Cover art: © Gregory Macnicol/Photo Researchers, Inc.
Editors: Dave Garza/Linda Ludewig
Developmental Editor: Carol Hinklin Robison
Production Editor: Louise N. Sette
Copy Editor: Marianne L'Abbate
Design Coordinator: Jill E. Bonar
Text Designer: Anne Flanagan
Cover Designer: Brian Deep
Production Manager: Laura Messerly
Marketing Manager: Debbie Yarnell

This book was set in Times Roman by The Clarinda Company and was printed and bound by R. R.
Donnelley & Sons Company. The cover was printed by Phoenix Color Corp.

 © 1997, 1994, 1991, 1989 by Prentice-Hall, Inc.
Simon & Schuster/A Viacom Company
Upper Saddle River, New Jersey 07458

Printed in the United States of America

10   9   8   7   6   5   4   3   2   1

ISBN: 0-13-243502-0

Prentice-Hall International (UK) Limited, *London*
Prentice-Hall of Australia Pty. Limited, *Sydney*
Prentice-Hall of Canada, Inc., *Toronto*
Prentice-Hall Hispanoamericana, S. A., *Mexico*
Prentice-Hall of India Private Limited, *New Delhi*
Prentice-Hall of Japan, Inc., *Tokyo*
Simon & Schuster Asia Pte. Ltd., *Singapore*
Editora Prentice-Hall do Brasil, Ltda., *Rio de Janeiro*

■ *This book is dedicated to the folks who stayed the course:*

*Wayne Newcomb (a.k.a. Big Bad Bob Wayne)*
*John Gerber & the rest of the gang in River Falls*
*Craig and Linda Kessler*
*Ron and Tori Welling*
*Bill Muckler*
*Bob Colbert*
*Mike Powers*
*My parents, Bill and Mary Paynter*
*Fred and Shirley Matthews*

*and most of all*

*Susan Matthews Paynter*

# Preface: To the Instructor

The fourth edition of *Introductory Electronic Devices and Circuits* is, for me, a return to writing. As you probably know, I was unable to work on the third edition of the text. The authors of that edition (John Clemons, Fred Evangelisti, Fred Kerr, and Charles Klingensmith) did an admirable job under the circumstances, but now it is time for my book to come home. Even so, you may find traces of their work in this edition (credit for which is hereby given).

From the start, my goal has been to produce a text that students can really *use* in their studies. To this end, many of the learning aids developed throughout the previous editions have been retained in this edition:

① **Chapter outlines** provide a handy overview of the chapter organization.

② **Performance-based objectives** enable the students to measure their progress by telling them what they are expected to be able to *do* as a result of their studies.

③ **Objective identifiers** in the margins cross-reference the objectives with the chapter material. This helps the students to quickly locate the material that will enable them to fulfill a given objective.

④ **Margin notes** include a running glossary of new terms, notes that highlight the differences between theory and practice, and reminders of principles covered in earlier chapters.

⑤ **In-chapter practice problems** are included in the examples to provide the students with an immediate opportunity to apply the principles being demonstrated.

⑥ **Summary illustrations** provide a convenient summary of circuit operating principles and applications. Many provide comparisons between a variety of related circuits.

Examples of the learning aids are shown on the following pages.

The following have also been retained from previous editions:

- **Section Review** questions at the end of each section.

- An **Equation Summary** and **Key Term List** at the end of each chapter, along with **Answers to the Example Practice Problems.**

- An extensive set of practice problems at the end of most chapters. In addition to standard practice problems, the problem sets include:

  **Troubleshooting Practice Problems**

  **"The Brain Drain"** (Challengers)

  **Suggested Computer Applications Problems**

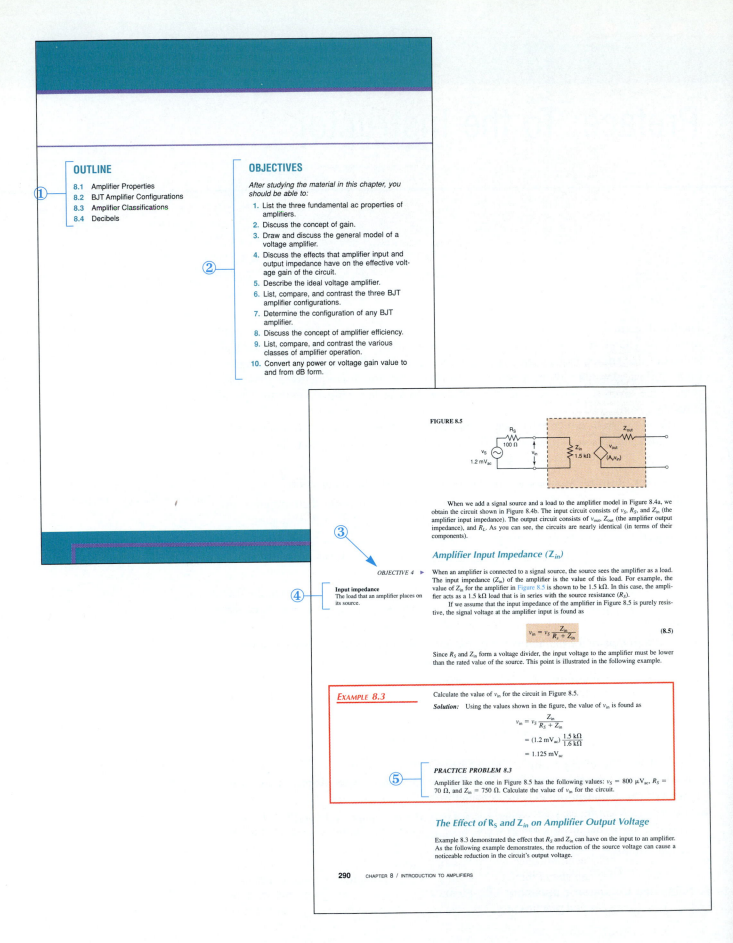

## OBJECTIVES

*After studying the material in this chapter, you should be able to:*

1. List the three fundamental ac properties of amplifiers.
2. Discuss the concept of gain.
3. Draw and discuss the general model of a voltage amplifier.
② 4. Discuss the effects that amplifier input and output impedance have on the effective voltage gain of the circuit.
5. Describe the ideal voltage amplifier.
6. List, compare, and contrast the three BJT amplifier configurations.
7. Determine the configuration of any BJT amplifier.
8. Discuss the concept of amplifier efficiency.
9. List, compare, and contrast the various classes of amplifier operation.
10. Convert any power or voltage gain value to and from dB form.

---

**FIGURE 8.5**

③

When we add a signal source and a load to the amplifier model in Figure 8.4a, we obtain the circuit shown in Figure 8.4b. The input circuit consists of $v_S$, $R_S$, and $Z_{in}$ (the amplifier input impedance). The output circuit consists of $v_{out}$, $Z_{out}$ (the amplifier output impedance), and $R_L$. As you can see, the circuits are nearly identical (in terms of their components).

### Amplifier Input Impedance ($Z_{in}$)

OBJECTIVE 4 ▶ When an amplifier is connected to a signal source, the source sees the amplifier as a load. The input impedance ($Z_{in}$) of the amplifier is the value of this load. For example, the value of $Z_{in}$ for the amplifier in Figure 8.5 is shown to be 1.5 kΩ. In this case, the amplifier acts as a 1.5 kΩ load that is in series with the source resistance ($R_S$).

④ **Input impedance**
The load that an amplifier places on its source.

If we assume that the input impedance of the amplifier in Figure 8.5 is purely resistive, the signal voltage at the amplifier input is found as

$$v_{in} = v_S \frac{Z_{in}}{R_s + Z_{in}} \tag{8.5}$$

Since $R_S$ and $Z_{in}$ form a voltage divider, the input voltage to the amplifier must be lower than the rated value of the source. This point is illustrated in the following example.

---

*EXAMPLE 8.3*

Calculate the value of $v_{in}$ for the circuit in Figure 8.5.

**Solution:** Using the values shown in the figure, the value of $v_{in}$ is found as

$$v_{in} = v_S \frac{Z_{in}}{R_S + Z_{in}}$$

$$= (1.2 \text{ mV}_{ac}) \frac{1.5 \text{ k}\Omega}{1.6 \text{ k}\Omega}$$

$$= 1.125 \text{ mV}_{ac}$$

**PRACTICE PROBLEM 8.3**

⑤ Amplifier like the one in Figure 8.5 has the following values: $v_S = 800$ μV$_{ac}$, $R_S = 70$ Ω, and $Z_{in} = 750$ Ω. Calculate the value of $v_{in}$ for the circuit.

---

### The Effect of $R_S$ and $Z_{in}$ on Amplifier Output Voltage

Example 8.3 demonstrated the effect that $R_S$ and $Z_{in}$ can have on the input to an amplifier. As the following example demonstrates, the reduction of the source voltage can cause a noticeable reduction in the circuit's output voltage.

**290** CHAPTER 8 / INTRODUCTION TO AMPLIFIERS

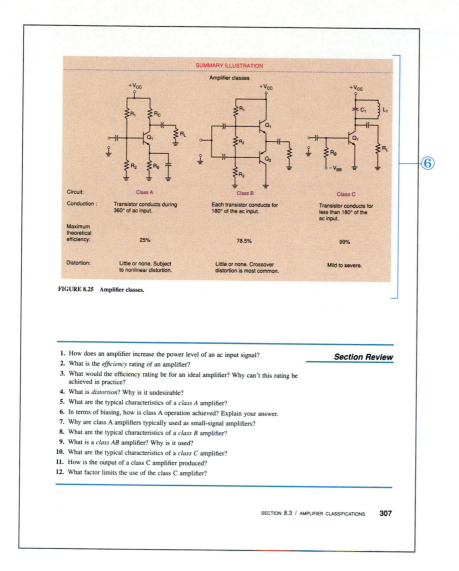

SUMMARY ILLUSTRATION

Amplifier classes

| Circuit: | Class A | Class B | Class C |
|---|---|---|---|
| Conduction : | Transistor conducts during 360° of ac input. | Each transistor conducts for 180° of the ac input. | Transistor conducts for less than 180° of the ac input. |
| Maximum theoretical efficiency: | 25% | 78.5% | 99% |
| Distortion: | Little or none. Subject to nonlinear distortion. | Little or none. Crossover distortion is most common. | Mild to severe. |

**FIGURE 8.25** Amplifier classes.

**Section Review**

1. How does an amplifier increase the power level of an ac input signal?
2. What is the *efficiency* rating of an amplifier?
3. What would the efficiency rating be for an ideal amplifier? Why can't this rating be achieved in practice?
4. What is *distortion*? Why is it undesirable?
5. What are the typical characteristics of a *class A* amplifier?
6. In terms of biasing, how is class A operation achieved? Explain your answer.
7. Why are class A amplifiers typically used as small-signal amplifiers?
8. What are the typical characteristics of a *class B* amplifier?
9. What is a *class AB* amplifier? Why is it used?
10. What are the typical characteristics of a *class C* amplifier?
11. How is the output of a class C amplifier produced?
12. What factor limits the use of the class C amplifier?

# NEW TO THIS EDITION

The fourth edition of *Introductory Electronic Devices and Circuits* incorporates many changes I had planned to make in the third edition. Some of these changes include providing the students with some new learning aids, new coverage of several topics, and improved coverage of many others. Here are some of the cornerstones of the fourth edition:

## Highlighted Figure References

From the start, one of my goals has been to provide students with the easiest possible access to any information they may need. For example, the objective identifiers were developed to help students quickly locate the information that is relevant to a given objective. In this edition, the easy access to information has been taken one step further.

I have found that many students spend quite a bit of time searching the text for information about a given figure. In this edition, it will be easier for students to locate this information. This has been accomplished by highlighting *the first reference in the text* to each figure. Your students can spend more time reading about a given figure because they will need less time to locate the discussion. The only exceptions are those figures that are contained in the examples and practice problems (since it is obvious where they are covered).

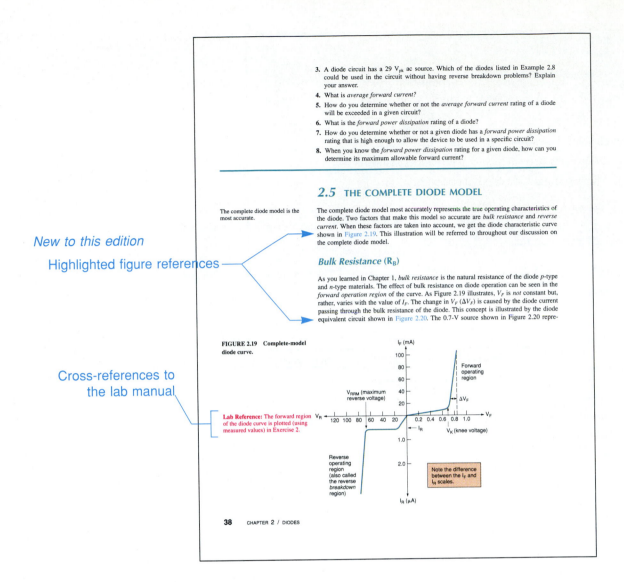

The boxed textbook page contains:

3. A diode circuit has a 29 $V_{pk}$ ac source. Which of the diodes listed in Example 2.8 could be used in the circuit without having reverse breakdown problems? Explain your answer.

4. What is *average forward current*?

5. How do you determine whether or not the *average forward current* rating of a diode will be exceeded in a given circuit?

6. What is the *forward power dissipation* rating of a diode?

7. How do you determine whether or not a given diode has a *forward power dissipation* rating that is high enough to allow the device to be used in a specific circuit?

8. When you know the *forward power dissipation* rating for a given diode, how can you determine its maximum allowable forward current?

## 2.5 THE COMPLETE DIODE MODEL

The complete diode model is the most accurate.

The complete diode model most accurately represents the true operating characteristics of the diode. Two factors that make this model so accurate are *bulk resistance* and *reverse current*. When these factors are taken into account, we get the diode characteristic curve shown in Figure 2.19. This illustration will be referred to throughout our discussion on the complete diode model.

### Bulk Resistance (R_B)

As you learned in Chapter 1, *bulk resistance* is the natural resistance of the diode *p*-type and *n*-type materials. The effect of bulk resistance on diode operation can be seen in the *forward operation region* of the curve. As Figure 2.19 illustrates, $V_F$ is *not* constant but, rather, varies with the value of $I_F$. The change in $V_F$ ($\Delta V_F$) is caused by the diode current passing through the bulk resistance of the diode. This concept is illustrated by the diode equivalent circuit shown in Figure 2.20. The 0.7-V source shown in Figure 2.20 repre-

FIGURE 2.19  Complete-model diode curve.

Lab Reference: The forward region of the diode curve is plotted (using measured values) in Exercise 2.

38  CHAPTER 2 / DIODES

---

Callout labels (in the margin, pointing to the page above):

*New to this edition*

Highlighted figure references

Cross-references to the lab manual

---

## Highlighted Lab References

Lab manuals have always contained references to the text. Now, you have a text that contains references to the lab manual. When a given circuit or procedure is covered in the lab manual, a *margin note* identifies the related lab exercise. This will help you to plan your lab schedule so that it more closely follows your progress through the text.

## Improved Use of Color

The functional use of color is enhanced in this edition to help your students distinguish between the parts that make up the various circuits and graphs. Color is also used to help your students distinguish the various learning aids and margin notes.

## Changes in Content

The approach to many topics has been improved or completely modified. Here are some of the changes you will see:

**Chapter 2.**  Several of the sections have been repositioned to provide a more logical flow of material.

**Chapter 8.**  This chapter has been completely rewritten to provide a more complete and realistic *overview* of amplifier operation. Your students are introduced to

- the overall purpose served by amplifiers
- the characteristics of the ideal amplifier
- the differences between ideal and practical amplifiers
- amplifier configurations and classes
- decibels

As it is now written, it is easier for your students to apply the information provided to amplifiers other than those containing BJTs.

**Chapter 9.**  Current gain is now approached as a *circuit characteristic* rather than one of the transistor only. Your students are shown how amplifier input circuits and loads affect current gain. The result is a more realistic approach to the entire subject of gain.

**Chapter 15.**  *Negative feedback* is approached more as an op-amp topic (rather than a discrete circuit topic). To reflect the change in approach, the material has been moved to this chapter.

**Chapter 16.**  The section on *Other Op-Amp Circuits* has returned to complete this chapter on op-amp circuits. The coverage of *instrumentation amplifiers* has also been expanded and improved.

**Chapter 20.**  This chapter more closely resembles its original form. The discussions of thyristor spec sheets and ratings that were omitted from the third edition have returned.

**Chapter 21.**  The section on *Switching Voltage Regulators* has been expanded and improved. It now contains in-depth coverage of

- the operating principles of switching regulators
- the advantages and disadvantages of using switching and linear voltage regulators
- the various switching regulator configurations
- practical IC switching regulators

I sincerely believe these changes will help you to provide your students with a more thorough understanding of electronic devices and circuits.

## ACKNOWLEDGMENTS

I am genuinely grateful to Prentice Hall for giving me the opportunity to develop this fourth edition of *Introductory Electronic Devices and Circuits*. It is, I believe, the best work that the reviewers, editors, and I have produced to date.

A project of this size could not have been completed without help from a variety of capable and concerned individuals. First and foremost, I would like to acknowledge the efforts of **Toby Boydell**, Seva Electronics (formerly of Conestoga College, Ontario). Toby has played a major role in this edition. He has provided quality input at every phase of development. He also provided the *solutions manual* for this edition and the answers found in Appendix G of the text. I would also like to thank the following professionals for the quality input they provided in their reviews of the text: Frank Brattain, Ivy Tech

State College; Gary Cardinale, DeVry Institute of Technology–Woodbridge; Robert Diffenderfer, DeVry Institute of Technology–Kansas City; Dr. Victor Gerez, Montana State University; Bob Griffin, Tarrant County Junior College; Dr. J. Jan Jellema, Eastern Michigan University; Dean Johnson, Northwest Technical College–Moorehead; Daniel Landiss, St. Louis Community College at Forest Park; Don Lovelace, Ashville-Buncombe Technical Community College; Leon Nicely, ITT Technical Institute–Ft. Lauderdale; and Malcolm Skipper, Midlands Technical College.

I would like to thank the staff at Prentice Hall for their "behind-the-scenes" work on this edition. These people deserve special recognition:

**Carol Robison**, my developmental editor, for making writing enjoyable again. She implemented the highest standards of quality control, while still managing to laugh at the worst of my jokes.

**Marianne L'Abbate**, my copy editor, for the most thorough and accurate editing I've ever seen.

**Louise Sette**, my production editor, for putting up with my delays and keeping things on track (despite my best efforts to derail them).

and

**Dave Garza**, Editor-in-Chief, for keeping the faith.

Finally, a special thanks goes out to my family and friends (especially Dick Arnoldy, Rich Reeves, and Bob Eversole) for their constant support, and to Susan for helping to put it all together . . . again.

*Bob Paynter*

# To the Student

## "WHY AM I LEARNING THIS?"

Have you ever found yourself asking this question? If you have, then take a moment to read further.

I believe that any subject is easier to learn when you know *why* you are learning it. For this reason, we're going to take a moment to discuss several things:

1. What *electronic devices* are
2. Why the study of electronic devices is important
3. How this area of study relates to the other areas of electronics
4. How you can get the most out of your study of electronic devices

One of the components that you have already studied is the *resistor*. In your basic dc electronics course, you were taught just about everything there is to know about resistors. Why? Because resistors are used in virtually every type of electronic circuit and/or system. If you take a moment to flip through the book, you'll see that there are very few circuits that do not contain at least one resistor.

A thorough understanding of resistors is necessary if you are going to be successful in understanding the area of electronic devices. *Each area of electronics is learned because it contributes to understanding the next.* If you want to understand electronic devices, you must first understand basic dc electronics. And, if you want to understand the areas of electronics that are studied after devices, you must come to understand the material in this book. Just as the knowledge of resistors is fundamental to this course, so the knowledge of devices is fundamental to the courses that will follow. This is *why* you are studying devices at this point in your education.

What *are* electronic devices? They are components with *dynamic* resistance characteristics. That is, they are components whose resistance is determined by the voltage applied to them or by the current drawn through them. Thus, some are *current-controlled resistances* while others are *voltage-controlled resistances*.

Electronic devices are somewhat complex components that are used in virtually every type of electronic system. They are used extensively in *communications systems* (such as televisions, radios, and VCRs), *digital systems* (such as PCs and calculators), and *industrial systems* (such as robotic and process control systems).

As you can see, the study of electronic devices is critical if your knowledge is to advance beyond the point where it is now. The next question is:

# "WHAT CAN I DO TO GET THE MOST OUT OF THIS COURSE?"

There are several steps that you can take to ensure that you will be successful in learning this area of electronics. The first is to realize that *learning electronics requires active participation on your part.* If you are going to learn electronics, you must take an active role in your education. It's like learning how to ride a bicycle. If you want to learn how to ride a bike, you have to hop on and take a few spills. You can't learn how to ride a bike just by "reading the book." The same can be said about learning electronics. You must become actively involved in the learning process.

How do you get involved in the learning process? Here are some habits worth developing:

1. *Attend class on a regular basis.*
2. *Take part in classroom problem solving sessions.* This means getting out your calculator and solving the problems along with the class.
3. *Do all the assigned homework.* Circuit analysis is a skill. As with any skill you gain competency only through practice.
4. *Take part in classroom discussions.* More often than not, classroom discussion can serve to clarify points that may be confusing otherwise.
5. *Become an active participant in the textbook discussions.*

Being an *active participant* in the textbook discussions means that you must do more than simply "read the book." When you are studying new material, there are several things that you need to do:

1. *Learn the terminology.* You are taught new terms because you need to know what they mean and how and when to use them. When you come across a new term in the text, take time to commit the new term to memory. How do you know when a new term is being introduced? Throughout this text, new terms are identified in the margins. When you see a new term and its definition in the margin, stop and learn the term before going on to the next section.
2. *Use your calculator to work through the examples.* When you come across an example, get your calculator out and try the example for yourself. When you do this, you develop the skill necessary to solve the problems on your own.
3. *Solve the example practice problems.* Most of the examples in this book are followed by a practice problem that is identical in nature to the example. When you see these problems, try them. Then you can check your solutions by looking up the answers at the end of the chapter.
4. *Use the chapter objectives to measure your learning.* Each chapter begins with an extensive list of performance-based objectives. *These objectives tell you what you should be able to do as a result of learning the material.*

This book contains *objective identifiers* that are located in the margins of the text. For example, if you look at page 83, you'll see "objective 2" printed in the margin. This identifier tells you that this is the point where you are taught the skill mentioned in objective 2 (see the objective list on page 75). These objective identifiers can be used to help you with your studies. If you don't know how to perform the action called for in a specific objective, just flip through the chapter until you see the appropriate identifier. At that point in the chapter, you will find the information you need.

## ONE FINAL NOTE

There is a lot of work involved in being an active learner. However, the extra effort will pay off in the end. Your understanding of electronic devices will be better as a result of your efforts, and it will also make learning the remaining areas of electronics much easier. I wish you the best of success.

*Bob Paynter*

# Contents

# 1

## FUNDAMENTAL SOLID-STATE PRINCIPLES   1

# 2

## DIODES                                                    20

# 3

# COMMON DIODE APPLICATIONS
## Basic Power Supply Circuits                                          74

# 4

# COMMON DIODE APPLICATIONS
## Clippers, Clampers, Multipliers,
## and Displays                                                        136

# 5

# SPECIAL APPLICATIONS DIODES        174

# 6

# BIPOLAR JUNCTION TRANSISTORS        202

# 7

# dc BIASING CIRCUITS        238

# 8

# INTRODUCTION TO AMPLIFIERS                   284

# 9

# COMMON-EMITTER AMPLIFIERS                   322

# 10

# OTHER BJT AMPLIFIERS 376

# 11

# POWER AMPLIFIERS 412

# 12

# FIELD-EFFECT TRANSISTORS 472

# 13

# MOSFETs 536

# 14

# AMPLIFIER FREQUENCY RESPONSE 566

# 15

# OPERATIONAL AMPLIFIERS 620

# 16

# ADDITIONAL OP-AMP APPLICATIONS 686

# 17

# TUNED AMPLIFIERS 724

# 18

# OSCILLATORS 786

# 19

# SOLID-STATE SWITCHING CIRCUITS     816

# 20

# THYRISTORS AND OPTOELECTRONIC DEVICES     872

# 21

## DISCRETE AND INTEGRATED VOLTAGE REGULATORS     924

# Introductory
# Electronic Devices
# and Circuits
## Conventional Flow Version

# 1

# FUNDAMENTAL SOLID-STATE PRINCIPLES

The circuit shown, if constructed with vacuum tubes, would have been approximately 365 cubic meters in size.

## OUTLINE

## OBJECTIVES

*After studying the material in this chapter, you should be able to:*

1. Describe the makeup of the atom and state the relationship between the number of valence electrons and its conductivity.
2. State the rules that govern the association between electrons and orbital shells.
3. Describe the relationship between conduction and temperature.
4. Contrast trivalent and pentavalent elements.
5. List the similarities and differences between *n*-type and *p*-type semiconductors.
6. Discuss diffusion current.
7. Explain how the depletion layer is formed around a *pn* junction.
8. Explain the source of barrier potential and list the barrier potential values for silicon and germanium.
9. Compare the relationship between depletion layer width, junction resistance, and junction current.
10. Define *bias*.
11. Explain the different methods of forward and reverse biasing a *pn* junction.
12. Explain why silicon is used more commonly than germanium in the production of solid-state devices.

# LITTLE DID THEY KNOW . . .

Electronic systems, such as radios, televisions, and computers, were originally constructed using *vacuum tubes*. Vacuum tubes were generally used to increase the strength of ac signals (*amplify*) and to convert ac energy to dc energy (*rectify*). Although they were able to perform these critical operations very well, vacuum tubes had several characteristic problems. They were large and fragile, and they wasted a tremendous amount of power through heat loss.

In the 1940s, a team of scientists working for Bell Labs developed the *transistor*, the first *solid-state* device capable of amplifying an ac signal. The term *solid-state* was coined because the transistor was solid, rather than hollow like the vacuum tube it was designed to replace. The transistor was also smaller and more rugged, and wasted much less power than did its vacuum-tube counterpart. Since the development of the transistor, solid-state components have replaced vacuum tubes in almost every application.

**Semiconductor**
Neither an insulator nor a conductor.

Solid-state components are made from elements that are classified as *semiconductor* elements. A **semiconductor** element is one that is neither a conductor nor an insulator but, rather, lies halfway between the two. Under certain circumstances, the resistive properties of a semiconductor can be varied between those of a conductor and those of an insulator. As you will see later, it is *this* characteristic of semiconductor elements that makes them useful as amplifiers and rectifiers.

## 1.1 ATOMIC THEORY

Before we get started, let's take a look at the reason for reviewing atomic theory. *Some coverage of atomic theory is essential to understanding the characteristics of semiconductors.* However, all you need to be concerned with is *what* happens on the atomic level,

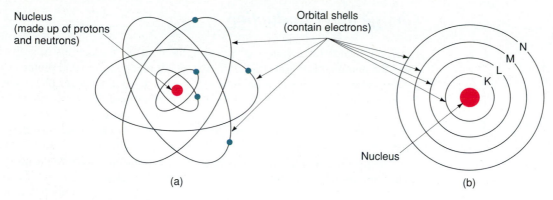

Nucleus
(made up of protons
and neutrons)

Orbital shells
(contain electrons)

Nucleus

(a)

(b)

**FIGURE 1.1    Bohr model of the atom.**

not *why* it happens. As you read through this section, realize that the material is being presented to give you a *basic* understanding of what happens on the atomic level. If you simply *accept* these principles to be true, you will be much more comfortable with the principles of semiconductors.

The atom has been shown to contain three basic particles: the *protons* and *neutrons*   ◀   *OBJECTIVE 1*
that make up the nucleus (core) of the atom, and *electrons* that orbit about the nucleus.
The basic model of the atom, called the *Bohr model*, is illustrated in Figure 1.1. The
orbital paths, or *shells*, are identified using the letters $K$ through $Q$. The innermost shell
is the $K$ shell, followed by the $L$ shell, and so on. The outermost shell for a given atom is
called the **valence shell**. The valence shell of an atom is critical because it determines the       **Valence shell**
conductivity of the atom.                                                                             The outermost shell that determines
                                                                                                     the conductivity of an atom.
The valence shell of an atom can contain up to eight electrons. The conductivity of the atom depends on the number of electrons that are in the valence shell. When
an atom has one valence electron, it is a nearly perfect conductor. When an atom has
eight valence electrons, the valence shell is said to be *complete*, and the atom is an
insulator. Therefore, *conductivity decreases with an increase in the number of valence
electrons*.

Semiconductors are atoms that contain four valence electrons. Because the number
of valence electrons in a semiconductor is halfway between one (for a conductor) and
eight (for an insulator), a semiconductor atom is neither a good conductor nor a good
insulator.

Three of the most commonly used semiconductor materials are *silicon* (Si), *germanium* (Ge), and *carbon* (C). These atoms are all represented in Figure 1.2. Note that all
these elements contain four valence electrons. Of the semiconductors shown, silicon and
germanium are used in the production of solid-state components. Carbon is used mainly
in the production of resistors and potentiometers.

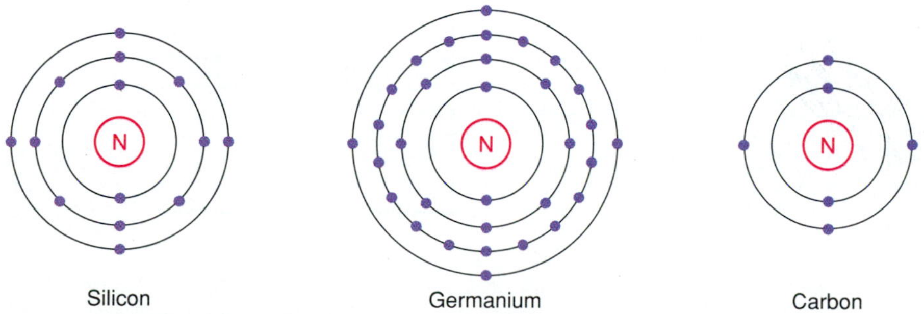

Silicon

Germanium

Carbon

**FIGURE 1.2    Semiconductor atoms.**

## Charge and Conduction

When no outside force causes conduction, the number of electrons in a given atom will equal the number of protons. Since the charges of electrons (−) and protons (+) are equal and opposite, the *net charge* on the atom is zero. If an atom *loses* one valence electron, the atom will then contain less electrons than protons, and the net charge on the atom will be positive. If an atom with an incomplete valence shell *gains* one valence electron, the atom will have more electrons than protons, and the net charge on the atom will be *negative*.

OBJECTIVE 2 ▶

Some fundamental laws regarding the relationship between electrons and orbital shells have been shown to be true. These are the following:

1. *Electrons travel in an orbital shell. They cannot orbit the nucleus in the space that exists between any two orbital shells.*

2. *Each orbital shell relates to a specific energy range. Thus, all the electrons traveling in a given orbital shell will contain the same relative amount of energy.* Note that the energy levels for the shells increase as you move away from the nucleus of the atom. Thus, the valence electrons will always have the highest energy levels in a given atom.

3. *For an electron to "jump" from one shell to another, it must absorb enough energy to make up the difference between its initial energy level and the energy level of the shell to which it is jumping.*

4. *If an electron absorbs enough energy to jump from one shell to another, it will eventually give up the energy it absorbed and return to a lower-energy shell.*

The first three of these principles are illustrated in Figure 1.3. The space between any two orbital shells is referred to as the **energy gap**. The electrons will travel through the energy gap when going from one shell to another, but they cannot continually orbit the nucleus of the atom in one of the energy gaps.

As Figure 1.3 shows, each orbital shell is related to a specific energy level. For an electron to jump from one orbital shell to another, it must absorb enough energy to make up the difference between the shells. For example, in Figure 1.3, the valence shell, or **band**, is shown to have an energy level of approximately 0.7 **electron-volt (eV)**. The **conduction band** is shown to have an energy level of 1.8 eV. Thus, for an electron to jump from the valence band to the conduction band, it would have to absorb an amount of energy equal to

$$1.8eV - 0.7eV = 1.1eV$$

**Energy gap**
The space between orbital shells.

**Band**
Another name for an orbital shell.

**eV (electron-volt)**
The energy absorbed by an electron when it is subjected to a 1-V difference of potential.

**Conduction band**
The band outside the valence shell.

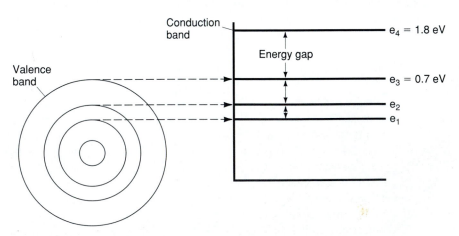

**FIGURE 1.3    Silicon energy gaps and levels.**

For conductors, semiconductors, and insulators, the valence to conduction-band energy gaps are approximately 0.4, 1.1, and 1.8 eV, respectively. The higher this energy gap, the harder it is to have conduction because of the fact that more energy must be absorbed for an electron to enter the conduction band.

When an electron absorbs enough energy to jump from the valence band to the conduction band, the electron is said to be in the *excited* state. An excited electron will eventually give up the energy it absorbed and return to its original energy level. The energy given up by the electron is in the form of *light* or *heat*.

## Covalent Bonding

**Covalent bonding** is the method by which atoms complete their valence shells by "sharing" valence electrons with other atoms. The covalent bonding of a group of silicon atoms is represented in Figure 1.4. To help illustrate the bonding process, each atom has been represented by an octagon (eight-sided figure) containing a square. The octagon represents the valence shell of the atom, and the square is used to identify the electrons that belong to the particular atom. As you can see, the center atom has eight valence electrons, four that belong to the center atom, plus one that belongs to each of the four surrounding atoms. These atoms, in turn, share electrons with four surrounding atoms, and so on. The results of this bonding are as follows:

**Covalent bonding**
A means of holding atoms together by sharing valence electrons.

1. The atoms are held together, forming a solid substance.
2. The atoms are all electrically stable because their valence shells are complete.
3. The completed valence shells cause the silicon to act as an *insulator*. Thus, pure (**intrinsic**) silicon is a very poor conductor. The same principle holds true for intrinsic germanium.

**Intrinsic**
Another way of saying *pure*.

When semiconductor atoms bond together in a set pattern like the one shown in Figure 1.4, the resulting material is called a *crystal*. A crystal is a smooth glassy solid. You have probably seen quartz crystals at some time. Silicon and germanium commonly crystallize in the same manner. At room temperature, a silicon crystal has fewer free electrons than a germanium crystal. This means less leakage current at room temperature, which is one reason why silicon is used more extensively than germanium in today's electronics industry.

Carbon can crystallize in the same fashion as silicon and germanium, but carbon crystals are too expensive to use in solid-state component production. This is because carbon crystals are *diamonds*.

## Conduction

When sufficient energy is added to a valence electron, it will jump from the valence band to the conduction band. The result of this action is that a *gap* is left in the covalent bond. This gap is referred to as a **hole**. It would follow that, *for every conduction-band electron, there must exist a valence-band hole*. The term used to describe this condition is **electron-hole pair**. The basic concept of the electron-hole pair is illustrated in Figure 1.5.

Within a few microseconds of becoming a free electron, an electron will give up its energy and fall into one of the holes in the covalent bond. This process is known as **recombination**. The time from when an electron jumps to the conduction band (becomes a free electron) until recombination occurs is called the **lifetime** of the electron-hole pair.

**Hole**
A gap in a covalent bond.

**Electron-hole pair**
An electron and its matching valence band hole.

**Recombination**
When a free electron returns to the valence shell.

**Lifetime**
The time between electron-hole pair generation and recombination.

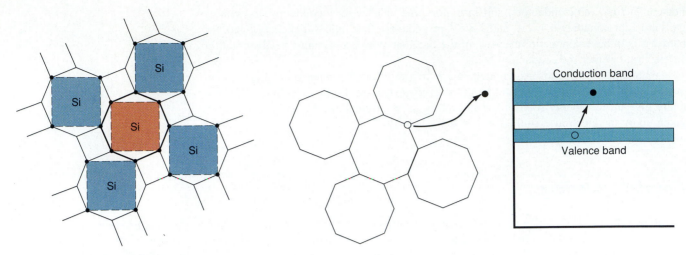

**FIGURE 1.4    Silicon covalent bonding.**    **FIGURE 1.5    Generation of an electron-hole pair.**

## Conduction Versus Temperature

*OBJECTIVE 3* ▶ At room temperature, *thermal energy* (heat) causes the constant creation of electron-hole pairs, with their subsequent recombination. Thus, a semiconductor will have some number of free electrons even when *no voltage is applied* to the element. As you increase the temperature, more electrons in the element will absorb enough energy to break free of their covalent bonds, and the number of free electrons will increase.

On the other hand, if you *decrease* the temperature, there is less thermal energy to release electrons from their covalent bonds, and the number of free electrons decreases. This situation will continue until the temperature reaches *absolute zero*. Absolute zero is, by definition, the temperature at which there is no thermal energy and occurs at −273.16 °C (−459.69 °F). Since there is no thermal energy at this temperature, there are no free electrons. There is no energy for the electrons to absorb so they cannot jump to the conduction band.

The important relationship here is the fact that *conductivity in a semiconductor is directly proportional to temperature.* This is why current increases in a circuit as the circuit warms up.

## Section Review

1. What are the three particles that make up the atom?
2. What is the relationship between the number of valence electrons and the conductivity of a given element?
3. How many valence electrons are there in a conductor? An insulator? A semiconductor?
4. What three semiconductor elements are most commonly used in electronics?
5. What are the relationships between electrons and orbital shells that were listed in this section?
6. What is an *energy gap*? What are the energy gap values for insulators, semiconductors, and conductors?
7. What forms of energy are given off by an electron that is falling into the valence band from the conduction band?
8. What is *covalent bonding*?

9. What are the effects of covalent bonding on intrinsic semiconductor materials?
10. What is an *electron-hole pair*?
11. What is *recombination*?
12. What is the typical lifetime of an electron-hole pair?
13. What is the relationship between temperature and conductivity?
14. Why does current stop at absolute zero in a semiconductor?

## 1.2 DOPING

◀ OBJECTIVE 4

As you have been shown, *intrinsic* (pure) silicon and germanium are poor conductors. This is due partially to the number of valence electrons, the covalent bonding, and the relatively large energy gap. Because of their poor conductivity, intrinsic silicon and germanium are of little use.

**Doping** is the process of *adding impurity atoms to intrinsic silicon or germanium to improve the conductivity of the semiconductor*. The term *impurity* is used to describe the doping elements because the silicon or germanium is no longer pure once the doping has occurred. Since a doped semiconductor is no longer pure, it is called an **extrinsic** semiconductor.

Two element types are used for doping: **trivalent** and **pentavalent**. A trivalent element is one that has three valence electrons. A pentavalent element is one that has five valence electrons. When trivalent atoms are added to intrinsic semiconductors, the resulting material is called a *p-type* material. When pentavalent impurity atoms are used, the resulting material is called an *n-type* material. The most commonly used doping elements are listed in Table 1.1.

**Doping**
Adding impurity elements to intrinsic semiconductors.

**Extrinsic**
Another word for *impure*.

**Trivalent**
Elements with three valence shell electrons.

**Pentavalent**
Elements with five valence shell electrons.

### n-Type Materials

When pentavalent impurities are added to silicon or germanium, the result is an excess of electrons *in the covalent bonds*. This point is illustrated in Figure 1.6. As you can see, the pentavalent arsenic atom is surrounded by four silicon atoms. The silicon atoms will bond with the arsenic atom, each sharing one arsenic electron. However, the fifth arsenic electron is not bound to any of the surrounding silicon atoms. Because of this, the fifth arsenic electron requires little energy to break free and enter the conduction band. If literally millions of arsenic atoms are added to pure silicon or germanium, there will be millions of electrons that are not a part of the covalent bonding. All these electrons can then be made to flow through the material with little difficulty.

One point that needs to be made and understood is this: Even though there are millions of electrons that are not part of the covalent bonding, the material is still electrically neutral. This is because each arsenic atom has the same number of protons as electrons, just like the silicon or germanium atoms. Since the overall number of protons and electrons in the material is equal, the net charge on the material is zero.

Why there are excess conduction band electrons in *n*-type materials.

*n*-type materials are electrically neutral.

**TABLE 1.1  Commonly Used Doping Elements**

| Trivalent Impurities | Pentavalent Impurities |
|---|---|
| Aluminum (Al) | Phosphorus (P) |
| Gallium (Ga) | Arsenic (As) |
| Boron (B) | Antimony (Sb) |
| Indium (In) | Bismuth (Bi) |

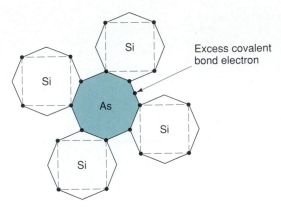

FIGURE 1.6   *n*-type material.

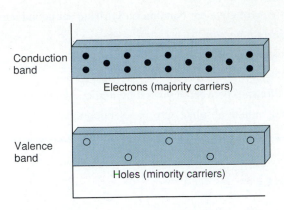

FIGURE 1.7   Energy diagram (*n*-type material).

**n-type material**
Silicon or germanium that has
pentavalent impurities.

Because there are more conduction-band electrons than valence-band holes in an **n-type material**, the electrons are called *majority carriers*, and the valence band holes are called *minority carriers*. The relationship between majority and minority carriers in an *n*-type material can be better understood by taking a look at Figure 1.7. The valence band is shown to contain *some* holes. These holes are caused by thermal energy excitation of electrons, as was discussed earlier. With the excess of conduction band electrons, however, the lifetime of an electron-hole pair is shortened significantly. This is due to the fact that the hole is filled almost immediately by one of the excess electrons in the conduction band.

Since the only holes that exist in the covalent bonding are those caused by thermal energy, the number of holes is far less than the number of conduction-band electrons. This is where the terms *majority* and *minority* come from.

Note that the term *n-type* implies an excess of electrons. As you will see, a *p-type* material is one with an excess of holes and relatively few free electrons.

## p-type Materials

**p-type material**
Silicon or germanium that has
trivalent impurities.

When intrinsic silicon or germanium is doped with a trivalent element, the resulting material is called a **p-type material**. The use of a trivalent element causes the existence of a hole in the covalent bonding structure. This point is illustrated in Figure 1.8. As you can see, the aluminum atom is shown to be surrounded by four silicon atoms. This time, however, there is a gap in the covalent bond caused by the lack of a fourth valence electron in the aluminum atom. Now, instead of an excess of electrons, we have an excess of holes. This situation is represented in Figure 1.9. The p-type material is shown to have an

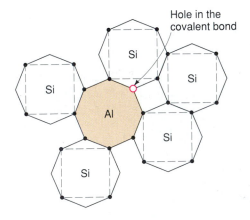

FIGURE 1.8   *p*-type material.

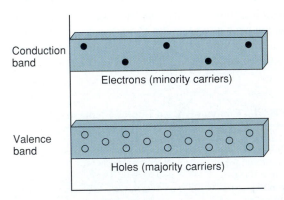

FIGURE 1.9   Energy diagram (*p*-type material).

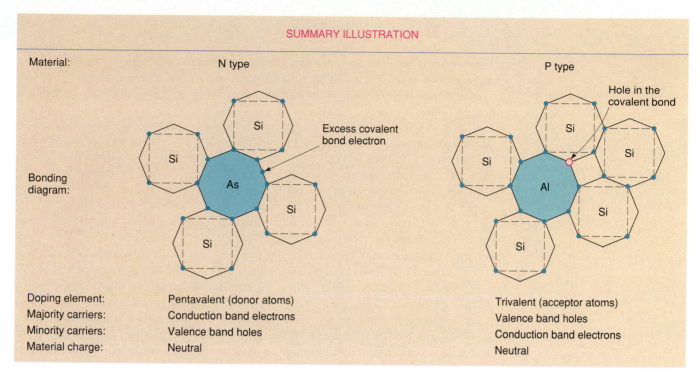

Material:  N type  P type

Bonding diagram:

Si · Si · As · Si · Si — Excess covalent bond electron

Si · Si · Al · Si · Si — Hole in the covalent bond

| Doping element: | Pentavalent (donor atoms) | Trivalent (acceptor atoms) |
| Majority carriers: | Conduction band electrons | Valence band holes |
| Minority carriers: | Valence band holes | Conduction band electrons |
| Material charge: | Neutral | Neutral |

**FIGURE 1.10**

excess of holes in the valence band. At the same time, there are some electrons in the conduction band. Again, the free electrons in the conduction band are there because of thermal energy. Since there are many more valence-band holes than conduction-band electrons, the holes are the majority carriers and the electrons are the minority carriers.

Again, you must remember that the number of electrons in the *p*-type material is equal to the number of protons. Even though there are gaps in the covalent bonds, the proton-electron balance still exists. Therefore, the net charge on the *p*-type material is zero.

*p*-type materials are electrically neutral.

## Summary

Figure 1.10 is what we call a *summary illustration*. The figure is provided to give you a quick summary form of the material presented in this section. As you will see, summary illustrations will be used extensively throughout the text.

◄ *OBJECTIVE 5*

Whenever you come to a summary illustration, take time to study the illustration and the relationships it contains. This will help you to remember the points that were made in the section.

Note that the terms *donor atom* and *acceptor atom* are used in Figure 1.10 in the description of the pentavalent and trivalent doping elements. The meanings of these two terms will be made clear in the next section.

1. What is doping? Why is it necessary?
2. What is an *impurity element*?
3. What are *trivalent* and *pentavalent* elements?
4. Despite their respective characteristics, *n*-type and *p*-type materials are still electrically neutral. Why?
5. In what ways are *n*-type and *p*-type materials similar? In what ways are they different?

*Section Review*

# *1.3* THE *pn* JUNCTION

Alone, there is little use for either an *n*-type material or a *p*-type material. However, these materials become useful when *joined* together to form a *pn junction*. Figure 1.11 illustrates the relative conditions of the *n* and *p* materials prior to joining and at the moment they are joined together.

Figure 1.11a shows the individual materials. As you can see, the *n*-type material is shown (on top) as containing an excess of electrons (solid circles), and the *p*-type material is shown as having an excess of holes (open circles). The energy diagrams illustrate the relationship between the energy levels of the two materials. Note that the valence bands of the two materials are at slightly different energy levels, as are the conduction bands. This is due to differences in the atomic makeup of the two materials.

OBJECTIVE 6 ▶ When the two materials are joined together (Figure 1.11b), the conduction and valence bands of the materials overlap. This allows free electrons from the *n*-type material to *diffuse* (wander) to the *p*-type material. This action and its results are illustrated in Figure 1.12. When a free electron wanders from the *n*-type material across the junction, it will become trapped in one of the valence-band holes in the *p*-type material. As a result, there is one net *positive* charge in the *n*-type material and one net *negative* charge in the *p*-type material. This may seem confusing at first, but it really isn't that strange. Take a look at Figure 1.13. Under normal circumstances, the covalent bond shown in the *n*-type material would have a total of 21 valence electrons (4 per silicon atom + 5 for the pentavalent atom = 21) and 21 protons that offset the negative valence charges.* This balance between valence electrons and protons results in a net charge of zero. When an electron wanders into the *p*-type material as shown, there are two results:

1.  The bond shown in the *n*-type material has lost an electron. Now the bond has only 20 electrons, but *still* has 21 offsetting protons. As a result, the overall charge is positive (+1).

*To simplify things, we consider only *valence* electrons and their "matching" protons. The actual number of total electrons and protons is not important because the electrical activity is determined by the valence shell.

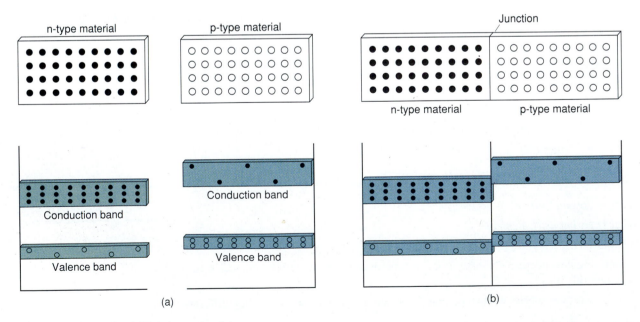

FIGURE 1.11   *pn*-junction initial energy levels.

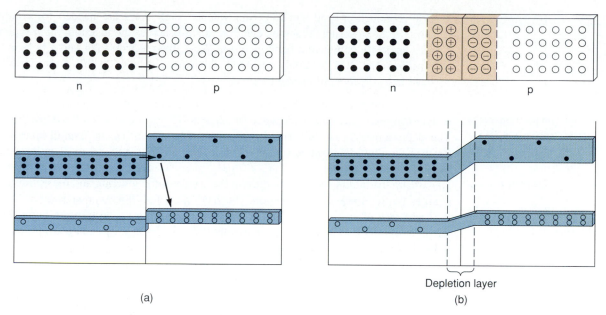

FIGURE 1.12    The forming of the depletion layer.

**2.** The covalent bond shown in the *p*-type material originally had 19 valence electrons that were offset by 19 protons, for a net charge of zero. However, when the electron from the *n*-type material falls into the hole, the bond has 20 electrons that are offset by only 19 protons. The net charge on the bond is therefore negative (−1).

Now it is time to consider what happens on a larger scale. When looking at the large-scale picture, there are several things to remember:

**1.** Each electron that diffuses across the junction leaves one positively charged bond in the *n*-type material and produces one negatively charged bond in the *p*-type material.

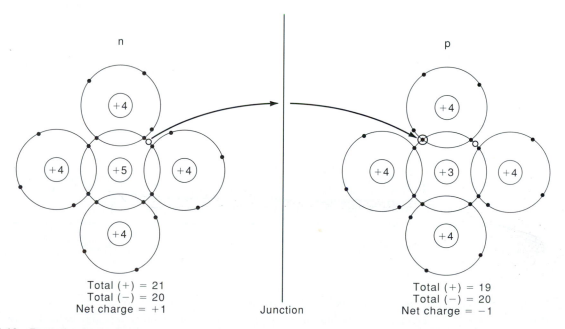

FIGURE 1.13    Depletion layer charges.

**2.** Both conduction-band electrons and valence shell holes are needed for conduction through the materials. When an electron diffuses across the junction, the *n*-type material has lost a conduction-band electron. When the electron falls into a hole in the *p*-type material, that material has lost a valence-band hole. At this point, both bonds have been *depleted* of charge carriers.

**Depletion layer**
The area around a *pn* junction that is depleted of free carriers. Note that the depletion layer is also referred to as the *depletion region*.

Since the action just described happens on a large scale, the junction ends up with a layer (on both sides) that is depleted of carriers. This layer is called the **depletion layer**. The depletion layer is represented in Figure 1.12b. Note that the overall charge of the layer is shown to be positive on the *n* side of the junction and negative on the *p* side of the junction.

**Barrier potential**
The natural potential across a *pn* junction.

With the buildup of (−) charges on the *p* side of the junction and the buildup of (+) charges on the *n* side of the junction, you have a natural *difference of potential* between the two sides of the junction. This potential is referred to as the **barrier potential**. The barrier potential for silicon is approximately 0.7 volt. For germanium, it is approximately 0.3 volt.

## Summary

When an *n*-type material is joined with a *p*-type material:

**1.** A small amount of *diffusion* occurs across the junction. The amount of diffusion is limited by the difference between the conduction-band energy levels of the two materials.

**2.** When electrons diffuse into the *p* region, they give up their energy and "fall" into the holes in the valence-band covalent bonds.

**3.** Since the pentavalent atoms (near the junction) in the *n* region have lost an electron, they have an overall *positive charge*.

**4.** Since the trivalent atoms (near the junction) in the *p* region have gained an electron, they have an overall *negative charge*.

**5.** The difference in charges on the two sides of the junction is called the *barrier potential*. The barrier potential is approximately equal to 0.7 V for silicon and 0.3 V for germanium.

**Donor atoms**
Another name for *pentavalent atoms*.

**Acceptor atoms**
Another name for *trivalent atoms*.

Note that, since the pentavalent atoms in the *n* material are giving up electrons, they are often referred to as **donor atoms**. The trivalent atoms, on the other hand, are accepting electrons from the pentavalent atoms and thus are referred to as **acceptor atoms**.

## Section Review

**1.** What is the overall charge on an *n-type* covalent bond that has just given up a conduction-band electron?

**2.** What is the overall charge on a *p-type* covalent bond that has just accepted an extra valence-band electron?

**3.** Discuss the forming of the *depletion layer*.

**4.** What is *barrier potential*? What causes it?

**5.** What is the barrier potential for silicon? For germanium?

## 1.4 BIAS

Depletion layer width and device current are inversely proportional.

A *pn* junction becomes useful when we are able to control the width of the depletion layer. By controlling the width of the depletion layer, we are able to control the resistance of the *pn* junction and thus the amount of current that can pass through the device. The relationship between the width of the depletion layer and the junction current is summarized as follows:

| Depletion Layer Width | Junction Resistance | Junction Current |
|---|---|---|
| Minimum | Minimum | Maximum |
| Maximum | Maximum | Minimum |

**Bias** is a potential applied to a *pn* junction to obtain a desired mode of operation. This potential is used to control the width of the depletion layer. The two types of bias are *forward bias* and *reverse bias*. A forward-biased *pn* junction will have a minimum depletion layer width and the indicated resistance-current characteristics. A reverse-biased *pn* junction will have a maximum depletion layer width and the indicated resistance-current characteristics. In this section, we will take a look at the two bias types.

◀ *OBJECTIVE 10*

**Bias**
A potential applied to a *pn* junction to obtain a desired mode of operation.

## Forward Bias

A *pn* junction is **forward biased** when the applied potential causes the *n*-type material to be more *negative* than the *p*-type material. When forward biased, a *pn* junction will allow current to pass with little opposition. The effects of forward bias are illustrated in Figure 1.14. Figure 1.14a shows the *pn* junction connected to a voltage source (V) and an open switch (SW₁). The energy diagram is included to show the initial energy states of the junction materials.

**Forward bias**
A potential used to reduce the resistance of a *pn* junction.

When SW₁ is closed, a negative potential is applied to the *n*-type material and a positive potential is applied to the *p*-type material. These potentials cause the following to occur:

1. The conduction band electrons in the *n*-type material are pushed toward the junction by the negative terminal potential.

2. The valence band holes in the *p*-type material are pushed toward the junction by the positive terminal potential.

The effects of *forward* bias on a *pn* junction.

*Assuming that V is greater than the barrier potential of the junction*, the electrons in the *n*-type material will gain enough energy to break through the depletion layer. When this happens, the electrons will be free to recombine with the holes in the *p*-type material and conduction will occur. Conduction through a *pn* junction is illustrated in Figure 1.14c. Once a *pn* junction begins to conduct, it provides a slight opposition to current. This opposition to current is referred to as **bulk resistance**. Bulk resistance (which is illustrated in Figure 1.14d) is the combined resistance of the *n*-type and *p*-type materials, as follows:

**Bulk resistance**
The natural resistance of a forward-biased diode.

$$r_b = r_p + r_n$$

The value of $r_b$ is typically in the range of 25 Ω or less. The exact value of $r_b$ for a given junction depends on the dimensions of the *n*-type and *p*-type materials, the amount of doping used to produce the materials, and the operating temperature.

Since the value of $r_b$ is extremely low, very little voltage is dropped across this resistance. For this reason, the voltage drop across $r_b$ is usually ignored in circuit calculations.

When a forward-biased *pn* junction begins to conduct, the **forward voltage ($V_F$)** across the junction is slightly greater than the barrier potential for the device. The values of $V_F$ are approximated as

**Forward voltage**
The voltage across a forward-biased *pn* junction.

$$V_F \cong 0.7 \quad \text{(for silicon)}$$

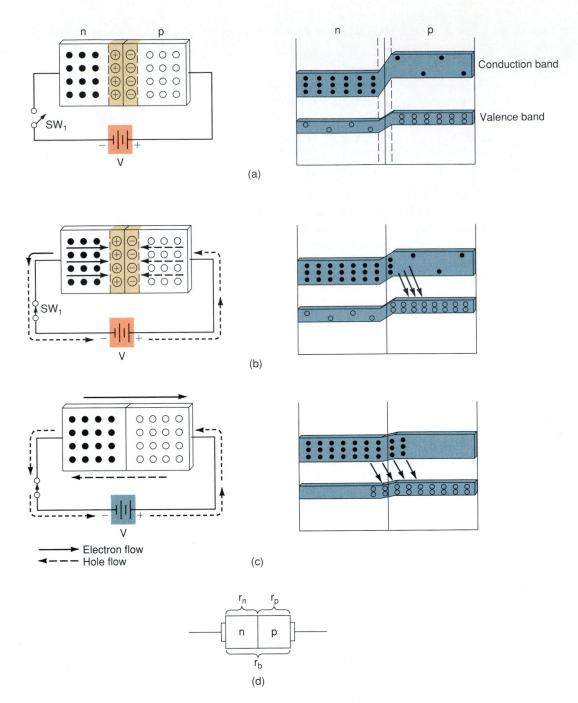

n           p             n         p

Conduction band

Valence band

(a)

SW$_1$

V

(b)

SW$_1$

V

→ Electron flow
◄ - - - Hole flow

V

(c)

$r_n$    $r_p$

n   p

$r_b$

(d)

**FIGURE 1.14**   **The effects of forward bias.**

and

$$V_F \cong 0.3 \qquad \text{(for germanium)}$$

Any difference between the measured and the approximated value of $V_F$ is a result of the current through the bulk resistance of the device. This point is discussed further in Chapter 2. The importance of the value of $V_F$ will become evident in our discussions on diode circuits in Chapter 3.

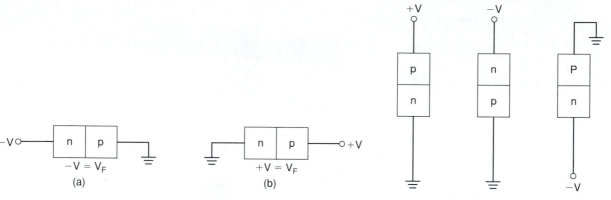

FIGURE 1.15   Forward biasing a *pn* junction.

FIGURE 1.16   Some forward-biased *pn* junctions.

It should be noted that there are two ways a *pn* junction can be forward biased:

◄ *OBJECTIVE 11*

1. By applying a potential to the *n*-type material that is more negative than the *p*-type material potential

How a *pn* junction is forward biased.

2. By applying a potential to the *p*-type material that is more positive than the *n*-type material potential

These two biasing methods are illustrated in Figure 1.15. In both cases, the fixed potential is shown as ground. Figure 1.15a shows the *n*-type material of the junction being driven negative by an amount that is sufficient to cause conduction. Assuming that the junction is made of silicon and $V \geq 0.7$ V, the junction will conduct. This is due to the fact that the *n*-type material is 0.7 V more negative than the *p*-type material. The junction shown in Figure 1.15b will conduct for the same reason. If the *p*-type material is biased at +0.7 V with respect to the *n*-type material, the diode will conduct. Thus, we can cause a junction to conduct by driving the *n*-type material more negative than the *p*-type material or by driving the *p*-type material more positive than the *n*-type material. Both methods of forward biasing a *pn* junction are used in practice.

All the junctions shown in Figure 1.16 are forward biased. See if you can relate the components shown to the forward-bias conditions discussed.

## Reverse Bias

A *pn* junction is **reverse biased** when the applied potential causes the *n*-type material to be more *positive* than the *p*-type material. When a *pn* junction is reverse biased, the depletion layer becomes wider, and junction current is reduced to almost zero. Reverse bias and its effects are illustrated in Figure 1.17. Figure 1.17a shows the junction in the forward-biased condition. As you have already been shown, the junction is allowing current to pass with little opposition. If the forward-biasing potential returns to zero, the depletion layer re-forms as electrons diffuse across the junction. This is illustrated in Figure 1.17b.

**Reverse bias**
A potential that causes a *pn* junction to have a high resistance.

If we apply a voltage with the polarity shown in Figure 1.17c, the electrons in the *n*-type material will head toward the positive terminal of the source. At the same time, the holes in the *p*-type material will move toward the negative source terminal. The electrons moving away from the *n* side of the junction will further deplete the material of free carriers. The depletion layer has effectively been widened. The same principle holds true for the *p* side of the material. Since there are fewer holes near the junction, the depletion region has grown. *The overall effect of the widening of the depletion layer is that the*

The effects of *reverse* bias on a *pn* junction.

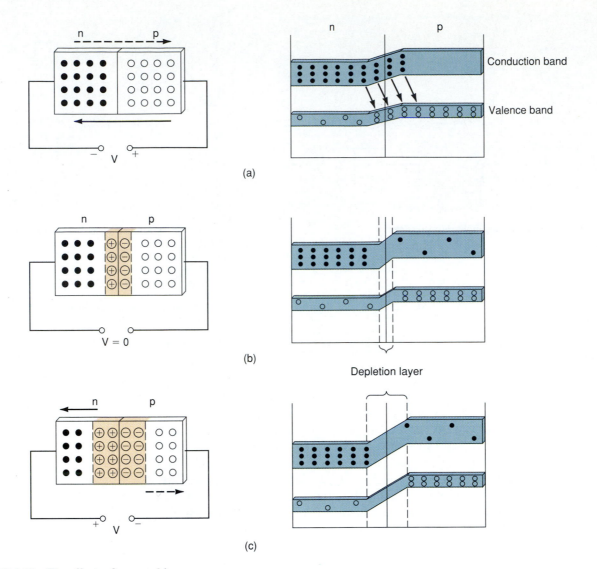

**FIGURE 1.17    The effects of reverse bias.**

*resistance of the junction has been drastically increased, and conduction drops to near zero.*

Figure 1.17 helps to demonstrate another point about reverse bias. During the time that the depletion layer is forming and growing, there is *still* majority carrier current in both materials. This current, called *diffusion current*, lasts only as long as it takes the depletion layer to reach its maximum width. This is undesirable in high-frequency circuits. Special diodes (which are discussed in Chapter 5) have been developed for these type of circuits.

How a *pn* junction is reverse biased.

Just as there are two ways to forward bias a junction, there are two ways to reverse bias a junction:

1. By applying a potential to the *n*-type material that is more positive than the *p*-type material potential

2. By applying a potential to the *p*-type material that is more negative than the *n*-type material potential

All the junctions shown in Figure 1.18 are reverse biased. Again, try to relate the diagrams to the statements made above.

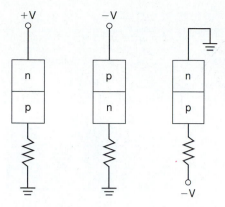

**FIGURE 1.18   Some reverse-biased *pn* junctions.**

## The Bottom Line

Bias is a potential applied to a *pn* junction that determines the operating characteristics of the device. Bias polarities and effects are summarized as follows:

| Bias Type | Junction Polarities | Junction Resistance |
|-----------|---------------------|---------------------|
| Forward | *n*-type material is more (−) than *p*-type material | Extremely *low* |
| Reverse | *p*-type material is more (−) than *n*-type material | Extremely *high* |

The voltage drop across a forward-biased *pn* junction will be approximately equal to 0.7 V for silicon and 0.3 V for germanium. When a *pn* junction is reverse biased, it acts essentially as an *open* circuit. Therefore, the voltage across a reverse-biased *pn* junction equals the applied voltage. These points are summarized in Figure 1.19.

## A Final Note

Up to this point, we have been mentioning both silicon and germanium in our discussions of semiconductor materials and junctions. Of the two, silicon is almost always used, for several reasons:

◄  *OBJECTIVE 12*

1. Silicon is more tolerant of heat.
2. Germanium oxide is water soluble, making it more difficult to process than silicon.
3. At room temperature, germanium produces more leakage current than does silicon.

Since it is the most commonly used semiconductor, we will limit future discussions to silicon. Just remember that germanium-based circuits work the same way as silicon-based circuits. The primary difference is in the value of $V_F$ for germanium versus silicon.

1. What are the resistance and current characteristics of a *pn* junction when its depletion layer is at its *maximum* width? *Minimum* width?
2. What purpose is served by the use of *bias*?
3. What is the depletion layer width of a forward-biased *pn* junction?

**Section Review**

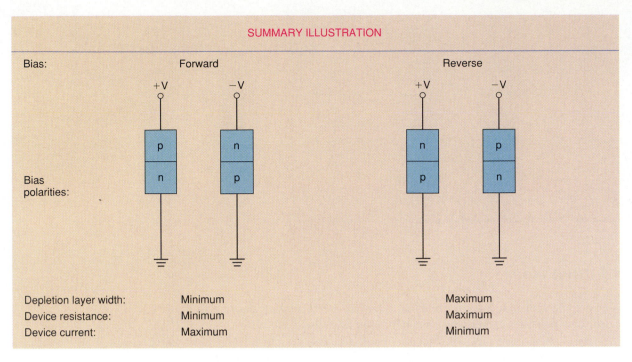

**FIGURE 1.19**

4. What is the junction resistance of a forward-biased *pn* junction?

5. What are the approximate values of $V_F$ for forward-biased silicon and germanium *pn* junctions?

6. List the commonly used methods for forward biasing a *pn* junction.

7. What is the depletion layer width of a reverse-biased *pn* junction?

8. What is the junction resistance of a reverse-biased *pn* junction?

9. List the commonly used methods for reverse biasing a *pn* junction.

10. What does the voltage across a reverse-biased *pn* junction equal?

11. Why is silicon used more commonly than germanium in the production of solid-state components?

## Key Terms

The following terms were introduced and defined in this chapter:

| | | |
|---|---|---|
| acceptor atom | electron-volt (eV) | *n*-type material |
| band | energy gap | pentavalent |
| barrier potential | extrinsic | *p*-type material |
| bias | forward bias | recombination |
| bulk resistance | forward voltage | reverse bias |
| conduction band | germanium | reverse voltage |
| covalent bonding | hole | semiconductor |
| depletion layer | intrinsic | silicon |
| diffusion current | lifetime | transient current |
| donor atom | majority carrier | trivalent |
| doping | minority carrier | valence shell |
| electron-hole pair | | |

# 2

# DIODES

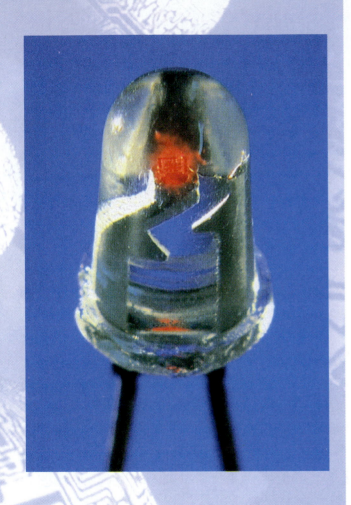

Light-emitting diodes (LEDs) have been developed that incorporate two *pn* junctions inside one lens. This allows the device to produce one color when biased with one polarity, a second color when biased in the opposite polarity, and a third color when the bias is rapidly alternated between polarities.

## OUTLINE

## OBJECTIVES

*After studying the material in this chapter, you should be able to:*

1. Identify the terminals of a *pn*-junction diode, given the schematic symbol of the component.

2. Analyze the schematic diagram of a simple diode circuit and determine:

   a. Whether or not the diode is conducting.

   b. The direction of current through a conducting diode.

3. List the three diode models and the applications for each.

4. List the main parameters of the *pn*-junction diode and explain how each limits the use of the component.

5. Determine the suitability of a given diode for a given application, using diode spec sheets and/or selector guides.

6. Identify the schematic symbol of the zener diode and determine the direction of current through the device.

7. Discuss the basic operating principles of the zener diode.

8. List the main zener diode parameters and explain how each limits the use of the component.

9. Discuss the basic operating principles of the light-emitting diode (LED).

10. Calculate the value of the *current-limiting* resistor needed for an LED in a given circuit.

11. Determine whether a given *pn*-junction diode, zener diode, or LED is good or faulty.

# WHICH CAME FIRST?

In later chapters, you will learn about a type of electronic device called a *transistor*. While the transistor is similar in construction to a *pn*-junction diode, it is actually a more complex component whose operation is based on more complex concepts.

It would seem to the casual observer that the *pn*-junction diode was developed before the transistor. After all, most complex devices are developed as outgrowths of similar, but simpler, devices. This, however, is not the case.

The *pn*-junction diode was actually developed almost six years *after* the development of the first transistor. In fact, the transistor had already been in commercial use for two years when Bell Laboratories announced the development of the *pn*-junction diode! In this instance, the chicken definitely came before the egg.

The diode is the most basic of the solid-state components. There are many diode types, each with its own operating characteristics and applications. The various diode types are easily identified by name, circuit application, and schematic symbol. It should be noted that the term *diode,* used by itself, refers to the basic *pn*-junction diode. All other diode types have other identifying names, such as *zener diode, light-emitting diode,* and so on.

**Diode**
A one-way conductor.

A **diode** is a *two-electrode* (two-terminal) device that acts as a *one-way conductor.* The most basic type of diode is the *pn-junction diode,* which is nothing more than a *pn* junction with a lead connected to each of the semiconductor materials. When forward biased, this type of diode will conduct. When reverse biased, diode conduction will drop to nearly zero.

In this chapter, we will look at the three most commonly used types of diodes: the *pn-junction diode*; the *zener diode*; and the *light-emitting diode,* or *LED.* Many other types of diodes are covered in Chapter 5.

## 2.1  INTRODUCTION TO THE *pn*-JUNCTION DIODE

OBJECTIVE 1 ▶

**Cathode**
The *n*-type terminal of a diode.

**Anode**
The *p*-type terminal of a diode.

The schematic symbol for the *pn*-junction diode is shown in Figure 2.1. The *n*-type material is called the **cathode**, and the *p*-type material is called the **anode**.

Recall that a *pn* junction will conduct when the *n*-type material (cathode) is more negative than the *p*-type material (anode). Relating this characteristic to the schematic symbol of the diode, we can make the following statement: A diode will conduct when the two following conditions are met:

1. The arrow points to the more negative of the diode potentials.
2. The voltage differential between the anode and the cathode exceeds the barrier voltage of approximately 0.3 V for a germanium diode and 0.7 V for a silicon diode.

OBJECTIVE 2 ▶

This point is illustrated in Figure 2.2, which shows several forward-biased (conducting) diodes. Note that the arrow in the diode schematic symbol points to the more negative potential in each case. Since the arrow in the schematic symbol points to the more negative potential when the diode is conducting, *diode forward current will be in the direction of the arrow*, as is shown in Figure 2.2.

A *pn*-junction diode is reverse biased when the *n*-type material (cathode) is more positive than the *p*-type material (anode). This causes the depletion region to widen and

**FIGURE 2.1  *pn*-junction diode schematic symbol.**

Anode (A)  *p* type — ▶| — Cathode (k)  *n* type

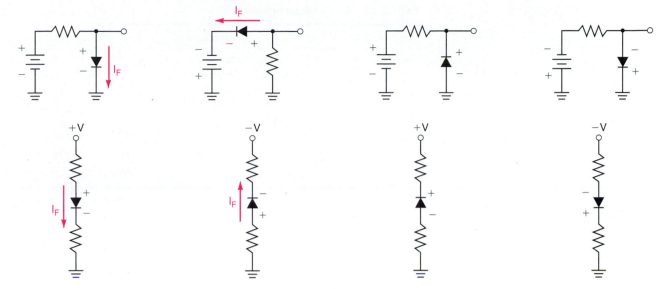

FIGURE 2.2   Forward-biased diodes.

FIGURE 2.3   Reverse-biased diodes.

prevent current. Relating this characteristic to the schematic symbol for the diode, we can make the following statement: A diode will not conduct when the arrow points to the more positive of the diode potentials. This point is illustrated in Figure 2.3, which shows several reverse-biased (nonconducting) diodes. Note that the arrow in the diode schematic symbol points to the more positive potential in each case.

## Diode Models

In this chapter, you will be introduced to three diode *models*. A **model** is a representation of a component or circuit that contains one or more of the characteristics of that component or circuit. For example, the dc model of a capacitor may represent the component as an open circuit, since a capacitor blocks dc. At the same time, the ac model of a capacitor may represent the component as a variable impedance, since the impedance of a capacitor varies inversely with frequency.

Component models are usually used to represent the component under specific circumstances or in specific applications. As you will see, the diode model that you use depends on what you are trying to do.

The first diode model that we will cover is called the *ideal diode model*. This diode model represents the diode as a simple switch that is either *closed* (conducting) or *open* (nonconducting). This model is used only in the initial stages of **troubleshooting**, as will be explained in Section 2.2.

The *practical diode model* is a bit more complex than the ideal diode model. The practical diode model includes the diode characteristics that must be considered when mathematically analyzing a diode circuit and when determining whether or not a given diode can be used in a given circuit. The practical diode model will be covered in Section 2.3.

The *complete diode model* is the most complex of the diode models. It includes the diode characteristics that are considered only under specific conditions such as *circuit development* (or *engineering*), high-frequency analysis, and so on. As you will see, the characteristics that are included in the complete diode model are not usually considered on a daily basis by the average technician. We will look at the complete diode model in Section 2.5.

The three diode models and their applications are summarized in Table 2.1.

**Model**
A representation of a component or circuit.

**Troubleshooting**
The process of locating faults in electronic equipment.

◄   *OBJECTIVE 3*

**TABLE 2.1  Diode Model Summary**

| Diode Model | Application(s) |
|---|---|
| **Ideal** | • Circuit troubleshooting |
| **Practical** | • Mathematical circuit analysis<br>• Determining whether or not a given diode<br>  can be used in a given circuit |
| **Complete** | • Circuit development (engineering)<br>• Uncommon special-condition circumstances |

*Section Review*

1. What is a diode?
2. What type of bias causes a diode to conduct?
3. What type of bias causes a diode to act as an insulator?
4. Draw the symbol for a diode and label the terminals.
5. What polarity is required to forward bias a diode?
6. When analyzing a schematic diagram, how do you know whether or not a diode is conducting?
7. When analyzing a schematic diagram, how do you determine the direction of diode current? Explain your answer.
8. When is the *ideal diode model* used?
9. When is the *practical diode model* used?
10. When is the *complete diode model* used?

## 2.2  THE IDEAL DIODE

The ideal diode acts as a switch.

The *ideal* diode has the characteristics of an open switch when it is reverse biased and those of a closed switch when forward biased. You may recall that a switch has the following characteristics:

| Condition | Characteristics |
|---|---|
| **Open** | • Infinite resistance and thus no current<br>• Full applied voltage dropped across the component<br>  terminals |
| **Closed** | • No resistance and thus maximum current<br>• No voltage dropped across the component<br>  terminals |

Based on the characteristics of a switch, we can make the following statements about the ideal diode:

Charactistics of the ideal reverse-biased diode.

1. When *reverse biased* (open switch):
   a. The diode will have infinite resistance.
   b. The diode will not pass current.
   c. The diode will drop the entire applied voltage across its terminals.

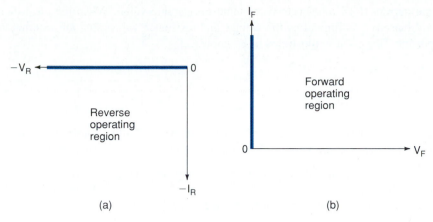

FIGURE 2.4  Characteristics of the ideal diode.

**2.** When *forward biased* (closed switch):

    **a.** The diode will have no resistance.

    **b.** The diode will have no control over the current through it.

    **c.** The diode will have no voltage drop across its terminals.

Characteristics of the ideal forward-biased diode.

These characteristics of the ideal diode are illustrated in Figure 2.4. Figure 2.4a shows the reverse-bias characteristics of the ideal diode. Note that as **reverse voltage** ($V_R$) increases, **reverse current** ($I_R$) remains at zero. This implies that the reverse-biased diode is an *open circuit* (just like an open switch), since there is no current through the device regardless of the value of the applied voltage. *Since the reverse-biased diode is acting as an open, the full applied voltage is dropped across the terminals of the device.* This point is illustrated in Example 2.1.

**Reverse voltage ($V_R$)**
The voltage across a reverse-biased diode.

**Reverse current ($I_R$)**
The current through a reverse-biased diode.

---

Determine the values of $V_{D1}$, $I_T$, and $V_{R1}$ for the circuit shown in Figure 2.5a.

*EXAMPLE 2.1*

*Solution:*  Because the arrow in the schematic symbol is pointing toward the positive terminal of the source, we know that the diode is reverse biased. Therefore:

a. The full applied voltage is dropped across $D_1$.

$$V_{D1} = V_S = 5\ \text{V}$$

b. $D_1$ will not allow conduction. Therefore, $I_T = 0$ A.
c. Since there is no current through $R_1$, there is no voltage drop across the component ($V_R = 0$ V).

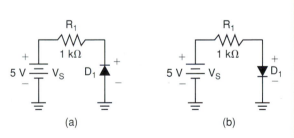

FIGURE 2.5

*PRACTICE PROBLEM 2.1*

A series circuit consists of a 12-V source, a 470-$\Omega$ resistor ($R_1$), a 330-$\Omega$ resistor ($R_2$), and a diode. If the diode is reverse biased, what is the value of $V_{R1}$? What is the value of $V_{R2}$?

---

Figure 2.4b shows the forward-bias characteristics of the ideal diode. Note that the forward voltage ($V_F$) across the diode is assumed to be 0 V for this diode model, while

What determines the value of diode forward current?

forward current ($I_F$) is shown to be at some measurable value. When the ideal diode is forward biased, $I_F$ is limited by the voltage and resistance values that are external to the component. This point is illustrated in Example 2.2.

---

**EXAMPLE 2.2**

Determine the values of $V_{D1}$, and $V_{R1}$, and $I_T$ for the circuit shown in Figure 2.5b.

*Solution:* Because the arrow in the schematic symbol is pointing toward the negative terminal of the source, we know that the diode is forward biased. Therefore:

a. $V_{D1} = 0$ V, leaving the total applied voltage to be dropped across $R_1$.
b. $V_R = V_S = 5$ V.
c. $I_T$ is determined by the source voltage and $R_1$. By formula,

$$I_T = \frac{V_R}{R_1} = \frac{5 \text{ V}}{1 \text{ k}\Omega} = 5 \text{ mA}$$

**PRACTICE PROBLEM 2.2**

A series circuit consists of a 12-V source, a 470-$\Omega$ resistor, 330-$\Omega$ resistor, and a diode. If the diode is forward biased, what is the value of $I_T$ for the circuit?

---

If you compare the values found in Examples 2.1 and 2.2 to the switch characteristics described earlier, you will see that the diode acted as an ideal switch in both cases. The forward and reverse characteristics of the ideal diode are summarized in Figure 2.6.

## When Do We Use the Ideal Diode Model?

Normally, the ideal model of the diode is used in the initial stages of circuit troubleshooting. When troubleshooting most diode circuits, your initial concern is only

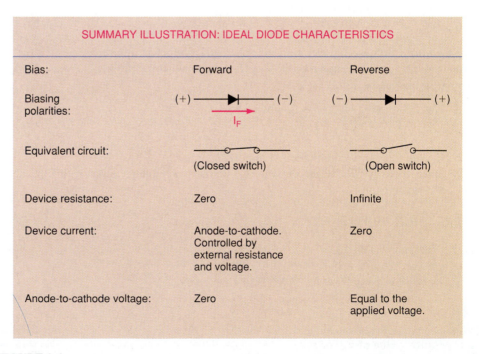

| SUMMARY ILLUSTRATION: IDEAL DIODE CHARACTERISTICS | | |
|---|---|---|
| Bias: | Forward | Reverse |
| Biasing polarities: | (+) ▸ (−)  $I_F$ | (−) ▸ (+) |
| Equivalent circuit: | (Closed switch) | (Open switch) |
| Device resistance: | Zero | Infinite |
| Device current: | Anode-to-cathode. Controlled by external resistance and voltage. | Zero |
| Anode-to-cathode voltage: | Zero | Equal to the applied voltage. |

**FIGURE 2.6**

whether or not a given diode is acting as a one-way conductor. If it is, the component is assumed to be good. If not, it is faulty and must be replaced.

## One Final Note

It was stated earlier in the chapter that the ideal model of the diode is used only for a non-detailed analysis of a diode circuit. Then we used this model to mathematically analyze the circuits in Figure 2.5. Why? We did this only to illustrate how the ideal model of the diode is treated as a perfect switch: either *on* (closed) with no resistance or *off* (open) with infinite resistance. In the next section, you will be shown how the *practical diode model* is used to calculate current and voltage values.

1. What are the forward characteristics of the ideal diode model?
2. What are the reverse characteristics of the ideal diode model?
3. When do we use the ideal diode model?

## 2.3 THE PRACTICAL DIODE MODEL

In our discussion of the ideal diode, we did not consider many of the diode characteristics that must be dealt with by working technicians on a regular basis. One of these characteristics, *forward voltage*, is normally considered in the mathematical analysis of a diode circuit. Many other practical diode characteristics are used when determining whether or not one diode may be used in place of another or in a specific circuit. These characteristics include *peak reverse voltage, average forward current*, and *forward power dissipation*.

In this section, we will take a look at *forward voltage* $(V_F)$ and the effect that it has on the mathematical analysis of basic diode circuits. The other characteristics listed are covered in detail in Section 2.4.

Whenever the term *diode* is used, we will assume that it is a silicon type. As we learned in Chapter 1, one difference between a silicon *pn* junction and a germanium *pn* junction is the difference in their values of $V_F$ when forward biased. Therefore, to work a given example as you would for a circuit containing a germanium diode, just reduce the forward voltage drop to approximately 0.3 V from 0.7 V.

## Forward Voltage $(V_F)$

In Chapter 1, we established the fact that there is a slight voltage developed across a forward-biased *pn* junction. The effect of this $V_F$ on the diode characteristic curve is illustrated in Figure 2.7.

Figure 2.7a is a composite of the ideal diode characteristic graphs that were shown in Figure 2.4. Note that the point where $I_F$ suddenly increases is labeled $V_k$ in the figure. This label is commonly used to identify what is called the *knee voltage* in a voltage-versus-current graph. The term **knee voltage** $(V_k)$ is often used to describe the point in a voltage-versus-current graph where current suddenly increases or decreases. As you can see, the ideal diode model assumes a knee voltage of 0 V.

In Figure 2.7b, we see the characteristic curve for the practical diode model. The only difference between this curve and the one shown for the ideal diode model is the value of $V_F$. In the practical diode model, the value of $V_k$ is shown to be equal to the approximated value of $V_F$ for a silicon *pn* junction, 0.7 V. In an actual circuit, the $V_F$ may fall between 0.7 V and 1.1 V, depending on the current through the device.

**Knee voltage $(V_k)$**
The voltage at which device current suddenly increases or decreases.

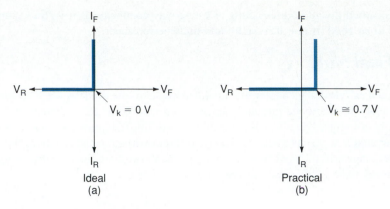

**FIGURE 2.7    Diode characteristic curves.**

Using the curve shown in Figure 2.7b, we can make the following statements about the forward operating characteristics of the practical diode:

1.  Diode current remains at zero until the knee voltage is reached.
2.  Once the applied voltage reaches the value of $V_k$, the diode turns *on* and forward conduction occurs.
3.  As long as the diode is conducting, the value of $V_F$ is approximately equal to $V_k$. In other words, $V_F$ is assumed to be approximately 0.7 V, regardless of the value of $I_F$.

Figure 2.8 will help you to see the difference between the forward operating characteristics of the ideal diode model and the practical diode model. Note the addition of the battery in the equivalent circuit for the practical diode model. This battery is used to represent the 0.7-V value of $V_F$ for the component.

## The Effect of $V_F$ on Circuit Analysis

So, how does including the value of $V_F$ change the analysis of a diode circuit? To answer this question, let's take a look at the circuit shown in Figure 2.9. According to Kirchhoff's voltage law, the sum of the component voltages in the circuit must equal the applied voltage. By formula,

$$V_S = V_F + V_R$$

If we substitute the value of $V_k$ (0.7 V) for $V_F$ and rearrange the equation to solve for $V_R$, we get

$$V_R = V_S - 0.7 \text{ V} \tag{2.1}$$

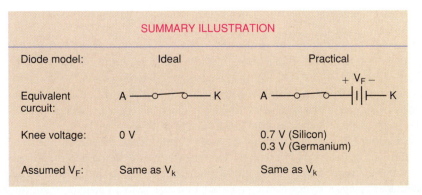

**FIGURE 2.8**

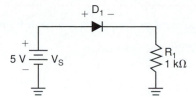

**FIGURE 2.9**

According to Ohm's law,

$$I_T = \frac{V_R}{R_1}$$

Substituting equation (2.1) in place of $V_R$ in the above equation, we get

$$I_T = \frac{V_S - 0.7\ V}{R_1} \qquad\qquad (2.2)$$

Thus, for the circuit shown,

$$V_R = V_S - 0.7\ V = 4.3\ V$$

and

$$I_T = \frac{V_S - 0.7\ V}{R_1} = \frac{5\ V - 0.7\ V}{1\ k\Omega} = 4.3\ mA$$

If you compare these values with those obtained for the same circuit in Example 2.2, you will see how including the value of $V_F$ in the circuit analysis changes the results. The two sets of values are summarized as follows:

| Value | Ideal | Practical |
|-------|-------|-----------|
| $V_F$ | 0 V   | 0.7 V     |
| $V_R$ | 5 V   | 4.3 V     |
| $I_T$ | 5 mA  | 4.3 mA    |

Examples 2.3 and 2.4 further demonstrate the use of $V_F$ in circuit calculations.

---

Determine the voltage across $R_1$ in Figure 2.10.

***EXAMPLE 2.3***

**Solution:** The voltage across the diode is assumed to be 0.7 V. Thus, the voltage across the resistor will be equal to the difference between the source voltage and the value of $V_F$. By formula,

$$V_R = V_S - 0.7\ V = 5.3\ V$$

**FIGURE 2.10**

---

EXAMPLE **2.4**

Determine the total circuit current in the circuit shown in Figure 2.10.

*Solution:* The total circuit current is found as

$$I_T = \frac{V_S - 0.7\ \text{V}}{R_1}$$

$$= \frac{6\ \text{V} - 0.7\ \text{V}}{10\ \text{k}\Omega}$$

$$= 530\ \mu\text{A}$$

### PRACTICE PROBLEM 2.4

A circuit like the one shown in Figure 2.10 has a 5-V source and a 510-$\Omega$ resistor. Determine the value of $I_T$ for the circuit.

## Percentage of Error

**Lab Reference:** Percentage of error calculations appear throughout the lab manual, beginning in Exercise 3.

In most circuit analysis problems, a calculated value is considered to be accurate enough if it is within $\pm 10\%$ of the actual measured value. The percentage of error of a given calculation is determined using

$$\% \text{ of error} = \frac{|X - X'|}{X} \times 100 \tag{2.3}$$

where $X$ = the actual measured value
$X'$ = the calculated value

How accurate must circuit calculation be?

Using the ideal diode model in circuit calculations can introduce a percentage of error in the results that is not acceptable (that is, not within $\pm 10\%$). For example, we used the ideal and practical diode models to determine the value of $V_R$ for the circuit shown in Figure 2.9. The percentage of error introduced by using the ideal diode model in this case would be found as

$$\% \text{ of error} = \frac{|4.3\ V - 5\ V|}{4.3\ V} \times 100$$

$$= 16.28\%$$

As you can see, the percentage of error is greater than 10% and therefore is not acceptable. This is why we use the practical diode model in circuit analysis problems.

If there is more than one resistor in a simple diode circuit, the *total* resistance ($R_T$) must be used in determining the value of $I_T$, as was the case in all the circuits you studied in basic electronics. This point is illustrated in Example 2.5.

Determine the value of $I_T$ for the circuit shown in Figure 2.11.

EXAMPLE 2.5

*Solution:* For the circuit shown, we can calculate the total circuit current using

$$I_T = \frac{V_S - 0.7\text{ V}}{R_T}$$

$$= \frac{5\text{ V} - 0.7\text{ V}}{3.4\text{ k}\Omega}$$

$$= 1.26\text{ mA}$$

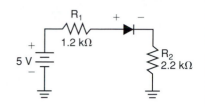

**FIGURE 2.11**

Note that we used $R_T$ in place of $R_1$ in equation (2.2) to solve the circuit in Example 2.5.

Just as you must consider the total resistance in the analysis of a diode circuit, you must consider the *sum of the diode voltage drops* if there are several *series-connected* diodes in a circuit. This point is illustrated in Example 2.6.

Determine the value of $I_T$ for the circuit shown in Figure 2.12.

EXAMPLE 2.6

*Solution:* With two diodes in the circuit, the total value of $V_F$ is assumed to be 1.4 V. Using this value in the place of 0.7 V in equation (2.2) allows us to accurately determine the value of $I_T$ as follows:

$$I_T = \frac{V_S - 1.4\text{ V}}{R_1}$$

$$= \frac{4\text{ V} - 1.4\text{ V}}{5.1\text{ k}\Omega}$$

$$= 509.8\ \mu\text{A}$$

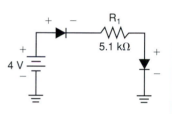

**FIGURE 2.12**

**PRACTICE PROBLEM 2.6**

A series circuit consists of two forward-biased diodes, a 470-$\Omega$ resistor, a 330-$\Omega$ resistor, and a 6-$V$ source. What is the value of $I_T$ for the circuit?

Now, let's do one more example to tie everything together.

Determine the value of $I_T$ for the circuit shown in Figure 2.13 *using the ideal diode model*. Then recalculate the value *using the practical diode model*. What is the percentage of error introduced by using the ideal diode model?

EXAMPLE 2.7

*Solution:* The ideal diode model assumes that $V_F = 0$ V. Therefore, the total applied voltage is dropped across the two resistors, and $I_T$ is found as

$$I_T = \frac{V_S}{R_T}$$

$$= \frac{10\text{ V}}{3.3\text{ k}\Omega}$$

$$= 3.03\text{ mA}$$

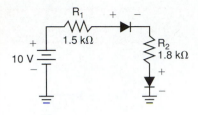

**FIGURE 2.13**

The practical diode model assumes that $V_F = 0.7$ V for each diode. Therefore, the value of $I_T$ is found as

$$I_T = \frac{V_S - 1.4\text{ V}}{R_T}$$

$$= \frac{10\text{ V} - 1.4\text{ V}}{3.3\text{ k}\Omega}$$

$$= 2.61\text{ mA}$$

The percentage of error between the two calculations is found as

$$\%\text{ of error} = \frac{|2.61\text{ mA} - 3.03\text{ mA}|}{2.61\text{ mA}} \times 100 = 16.1\%$$

Note that the value of $I_T$ found using the *practical diode model* was used in the denominator of the fraction in the percentage of error calculation. This value was used because the practical value is always assumed to be closer than the ideal to the actual value of $I_T$.

### PRACTICE PROBLEM 2.7

Refer to Practice Problem 2.6. Recalculate the value of total circuit current using the *ideal diode model*. Then determine the percentage of error introduced by using this diode model.

## One Final Note

It can be argued that it isn't always necessary to include the 0.7 V drop across a diode in the mathematical analysis of a diode circuit. For example, if we had a diode and resistor in series with a 100-V source, ignoring the 0.7 V diode drop would cause a percentage of error of less than 1%, which is well within the acceptable limits for accuracy.

While this argument is valid, there are two other considerations. First, many diode circuits contain more than one diode. When this is the case, the percentage of error introduced by using the ideal diode model becomes much larger. Second, we are always interested in getting the most accurate results possible (within reason) in the mathematical analysis of any circuit. For these reasons, we will always use the practical diode model in the mathematical analysis of any diode circuit. Remember, the only difference in using this model for analysis purposes is that you must take the value of $V_F$ for each diode into account. Otherwise, the component is assumed to work just like the ideal model.

In the next section, we will look at the factors that must be considered when determining whether or not a specific diode can be used in a given situation. While these factors are not normally considered in voltage and current calculations, they are very important when it comes to actually working on a circuit.

## Section Review

1. What diode characteristic must be considered in the mathematical analysis of a diode circuit?
2. Which diode characteristics are normally considered when replacing one diode with another?
3. What is the assumed value of a knee voltage for a silicon diode?
4. List the characteristics of the practical forward-biased diode.
5. How does including the value of $V_F$ affect the accuracy of circuit calculations?

**6.** How accurate must a calculation be to be considered acceptable?

**7.** A circuit voltage is calculated to be 10 V. The actual value of this voltage is 12.2 V. What is the percentage of error in the calculation?

---

# 2.4 OTHER PRACTICAL CONSIDERATIONS

Assume that you have just finished troubleshooting a circuit. While troubleshooting, you found that the diode in the circuit is faulty and must be replaced. Now you discover that you have a wide variety of diodes in stock, but none of them has the same part number as the one that needs replacing. How can you determine whether or not a specific diode can be used in place of the faulty one? Several diode characteristics must be considered when determining whether or not a specific diode can be used in a given circuit. These characteristics are *peak reverse voltage, average forward current,* and *forward power dissipation.*

◄ *OBJECTIVE 4*

## Peak Reverse Voltage ($V_{RRM}$)

Any insulator will conduct if the applied voltage is high enough to cause the insulator to break down. For a reverse-biased diode, the *maximum* reverse voltage that *won't* force the diode to conduct is called the **peak reverse voltage ($V_{RRM}$)**. When $V_{RRM}$ is exceeded, the depletion layer will break down, and the diode will conduct in the reverse direction. Typical values of $V_{RRM}$ range from a few volts (for zener diodes) to thousands of volts.

**Peak reverse voltage ($V_{RRM}$)**
The *maximum* reverse voltage allowable for a diode.

The effect that $V_{RRM}$ has on the diode characteristic curve is illustrated in Figure 2.14. Note that the value of reverse current ($I_R$) is shown to be zero until the value of $V_{RRM}$ ($-70$ V, in this case) is exceeded. When $V_R > V_{RRM}$, the value of $I_R$ increases rapidly as the depletion layer breaks down. Normally, when a *pn* junction is forced to conduct in the reverse direction, the device is destroyed. A *zener diode*, on the other hand, is designed to work in the reverse direction without harming the diode. This point is discussed in detail later in this chapter.

The current that occurs when $V_R > V_{RRM}$ is called **avalanche current**. This name comes from the fact that one free electron bumps other electrons in the diode, causing them to break free from their covalent bonds, which then rapidly causes even more electrons to be broken free, and so on. The result is that the diode is destroyed by excessive current and the heat it produces.

**Avalanche current**
The current that occurs when $V_{RRM}$ is reached. Avalanche current can destroy a *pn* junction diode.

**FIGURE 2.14**

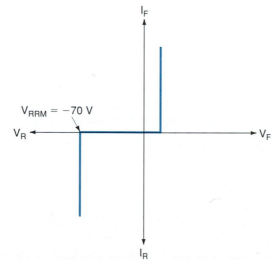

**Parameter**
A limit.

*A Practical Consideration:* Most reverse breakdown voltage ratings for *pn*-junction diodes are multiples of 50 or 100 V. Common values are 50, 100, 150, 200, and so on.

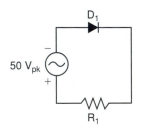

**FIGURE 2.15**

Peak reverse voltage is a very important **parameter** (limit). When you are considering whether or not to use a given diode for some application, you must be sure that the reverse voltage in the circuit will never meet or exceed the value of $V_{RRM}$ for the diode. For example, consider the circuit shown in Figure 2.15. The ac signal being applied to the circuit has a peak value of 50 V. When the ac signal source has the polarity shown, the diode is reverse biased and the full applied voltage is dropped across its terminals. Thus, $D_1$ must have a $V_{RRM}$ rating that is *greater than* 50 V. A diode is normally chosen that is capable of handling at least 20% over the maximum calculated reverse voltage. This takes care of fluctuations in the applied voltage and will not force the diode to run at its reverse voltage limit. At any time, the applied voltage can fluctuate, and if the diode is already running at its voltage limit, it could be destroyed.

We calculate the needed diode $V_{RRM}$ by:

$$\text{Calculated maximum reverse voltage} \times 1.20$$

So in this case, the minimum acceptable value of $V_{RRM}$ is 50 V $\times$ 1.20 = 60 $V_{pk}$.

One point should be made at this time: As long as the $V_{RRM}$ rating is *greater than* the maximum reverse bias in the circuit, we do not really care what its value is. For example, let's say that the diode in Figure 2.15 needed to be replaced. We could use a diode with a $V_{RRM}$ rating of 100, 200, or even 1000 V. This point is illustrated in Example 2.8.

---

## EXAMPLE 2.8

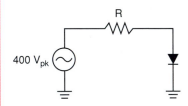

**FIGURE 2.16**

The diode in the circuit shown in Figure 2.16 is faulty and must be replaced. When checking the parts bin, you see that you have the following diodes in stock:

| Diode Part Number | Peak Reverse Voltage |
| --- | --- |
| 1N4001 | 50 |
| 1N4002 | 100 |
| 1N4003 | 200 |
| 1N4004 | 400 |
| 1N4005 | 600 |
| 1N4006 | 800 |
| 1N4007 | 1000 |

Which of these diodes could be used to replace the faulty diode?

***Solution:*** The $V_{RRM}$ rating of a diode *must be greater than* the maximum value of reverse bias that occurs in a circuit. Since the peak value of the source is 400 $V_{pk}$, a full 400 V will be dropped across the diode terminals whenever the device is reverse biased. Therefore, the $V_{RRM}$ rating of the replacement part must be 20% *greater than* 400 V. The only diodes we can use are the 1N4005, 1N4006, or 1N4007.

### PRACTICE PROBLEM 2.8

A circuit like the one in Figure 2.16 has a 175-$V_{pk}$ signal source. Which of the diodes listed in Example 2.8 could be used in the circuit?

The value of $V_{RRM}$ for a given *pn*-junction diode can be obtained from the *specification sheet* (or data sheet) for the component. A **specification sheet** (spec sheet) lists all the important parameters and operating characteristics of a device or circuit. The specification sheet for a given component can be obtained from the manufacturer or from most electronics parts stores (usually free of charge). You will be shown in Section 2.6 how to find the value of $V_{RRM}$ (and many other important parameters) on a specification sheet.

Peak reverse voltage is only one of several parameters that must be considered before trying to use a specific diode in a given application. Another parameter that must be considered is the *average forward current* rating of the diode.

**Specification sheet**
A listing of all the important parameters and operating characteristics of a device or circuit.

## Average Forward Current ($I_0$)

The **average forward current** rating of a diode is the *maximum allowable value of dc forward current*. For example, the 1N4001 diode has an average forward current rating of 1 A. This means that the dc forward current through the diode must never be allowed to exceed 1 A. If the dc forward current through the diode is allowed to exceed 1 A, the diode will be destroyed from excessive heat. The average forward current rating is a parameter found on the spec sheet for the component.

**Average forward current**
The maximum allowable value of dc forward current for a diode.

When you are considering whether or not to use a diode in a specific circuit, you must determine what the average value of forward current ($I_F$) in the circuit will be. Then you must make sure that the diode you want to use has an average forward current rating that is at least 20% greater than the value of $I_F$ for the circuit. This point is illustrated in Example 2.9.

In Chapter 3, you will be shown how to determine the *dc equivalent current* for an ac source. When dealing with ac circuits, you will need to determine the *average* (or *dc equivalent*) current for the circuit. Then this dc equivalent current value must be compared to the average forward current rating of the diode to see if it can be used in the circuit.

---

Determine the *minimum* average forward current rating that would be required for the diode in Figure 2.17.

*Solution:* The forward current in the circuit is calculated (as always) using the practical model of the diode. Thus,

$$I_F = \frac{V_S - 0.7 \text{ V}}{R_1}$$

$$= \frac{50 \text{ V} - 0.7 \text{ V}}{200 \ \Omega}$$

$$= 246.5 \text{ mA}$$

Thus, any diode used in the circuit would have to have an average forward current rating that is *greater than* 246.5 mA. In practice, you would use a diode whose average forward current rating is at least 20% *greater than* the calculated value of $I_F$, or 246.5 mA $\times$ 1.20 = 295.8 mA.

### PRACTICE PROBLEM 2.9

A circuit like the one shown in Figure 2.17 has a 100 $V_{dc}$ source. If the resistor has a value of 51 $\Omega$, what is the minimum allowable average forward current rating for the diode in the circuit?

**EXAMPLE 2.9**

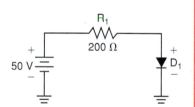

**FIGURE 2.17**

*A Practical Consideration:* Let's say that you calculate the dc forward current in a circuit to be 1 A. If you use a diode that has an average forward current rating of 1 A in the circuit, you will be pushing the diode to its limit. This will shorten the lifetime of the diode. If, on the other hand, you use a diode with an average forward current rating of 10 A, the diode will not be pushed to its limit and will last much longer.

# Forward Power Dissipation ($P_{D(MAX)}$)

Many diodes have a **forward power dissipation** rating. This rating indicates the *maximum possible power dissipation of the device when it is forward biased.*

Recall from your study of basic electronics that power is found as

$$P = IV$$

where $P$ = the power dissipated by a component
$\quad\quad I$ = the device current
$\quad\quad V$ = the voltage across the device

Using the basic power equation and the values of $V_F$ and $I_F$ for a diode, you can determine the required forward power dissipation rating for a replacement component, as is illustrated in Example 2.10.

Some diode specification sheets contain a forward power dissipation rating *instead* of an average forward current rating. When this is the case, the maximum forward current rating for the diode can be found using

$$I_{F(MAX)} = \frac{P_{D(MAX)}}{V_F} \tag{2.4}$$

where $I_{F(MAX)}$ = the maximum allowable forward current
$\quad\quad P_{D(MAX)}$ = the forward power dissipation rating of the diode
$\quad\quad\quad V_F$ = the forward voltage across the diode, assumed to be 0.7 V for a silicon *pn*-junction diode

This equation is simply a variation on the standard power dissipation equation. Its use is illustrated in Example 2.11.

---

## EXAMPLE 2.10

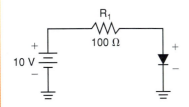

**FIGURE 2.18**

Calculate the *minimum* forward power dissipation rating for any diode that would be used in the circuit shown in Figure 2.18.

**Solution:** First, we have to calculate the total circuit current. This current is found as

$$I_F = \frac{10\text{ V} - 0.7\text{ V}}{100} = 93\text{ mA}$$

Using $I_F = 93$ mA and $V_F = 0.7$ V, the power dissipation of the diode is found as

$$P = I_F V_F = (93\text{ mA})(0.7\text{ V}) = 65.1\text{ mW}$$

To add a safety margin, take 65.1 mW $\times$ 1.20 = 78.12 mW. Thus, any diode used in the circuit would have to have a forward power dissipation rating that is greater than 78.12 mW.

### PRACTICE PROBLEM 2.10

A circuit like the one shown in Figure 2.18 has a 20 $V_{dc}$ source and a 68-$\Omega$ series resistor. What is the minimum required forward power dissipation rating for any diode used in the circuit?

EXAMPLE *2.11*

A diode has a forward power dissipation rating of 500 mW. What is the maximum allowable value of forward current for the device?

**Solution:** The value of $I_{F(MAX)}$ is found as

$$I_{F(MAX)} = \frac{P_{D(MAX)}}{V_F}$$

$$= \frac{500 \text{ mW}}{0.7 \text{ V}}$$

$$= 714.29 \text{ mA}$$

In a circumstance such as this, we would normally restrict the circuit current to 80% of the maximum allowable diode current to provide a safety margin. To do this, calculate the maximum current allowed in our circuit as follows:

$$I_{F(MAX)} \times 0.80 = \text{maximum circuit current}$$

Therefore,

$$I_{T(MAX)} = 714.29 \text{ mA} \times 0.80 = 571.43 \text{ mA}$$

The diode that we selected can be used in any circuit that has a maximum current that is less than or equal to 571.43 mA.

**PRACTICE PROBLEM 2.11**

Show that the power dissipation rating of 500 mW in Example 2.11 will be exceeded if forward current is 750 mA.

## Summary

When you are trying to replace one diode with another, three main parameters must be considered. Before substituting one diode for another, ask yourself:

1. Is the $V_{RRM}$ rating of the replacement diode greater than the maximum reverse voltage in the circuit?
2. Is the *average forward current* rating of the replacement diode greater than the average (dc) value of $I_F$ in the circuit?
3. Is the *forward power dissipation* rating of the replacement diode high enough?

If the answer to any of these questions is *no*, then you cannot use the diode in the circuit.

## One Final Note

In this section, we have considered the more practical aspects of diode operation that you will have to deal with on a regular basis. There are more diode characteristics, many of which are considered in *circuit development*, in *high-frequency circuit analysis*, and under some other special circumstances. These characteristics are included in the *complete diode model*.

*Section Review*

1. What is *peak reverse voltage*?
2. Why do you need to consider the value of $V_{RRM}$ for a diode before attempting to use the diode in a specific circuit?

3. A diode circuit has a 29 $V_{pk}$ ac source. Which of the diodes listed in Example 2.8 could be used in the circuit without having reverse breakdown problems? Explain your answer.

4. What is *average forward current?*

5. How do you determine whether or not the *average forward current* rating of a diode will be exceeded in a given circuit?

6. What is the *forward power dissipation* rating of a diode?

7. How do you determine whether or not a given diode has a *forward power dissipation* rating that is high enough to allow the device to be used in a specific circuit?

8. When you know the *forward power dissipation* rating for a given diode, how can you determine its maximum allowable forward current?

## 2.5 THE COMPLETE DIODE MODEL

The complete diode model is the most accurate.

The complete diode model most accurately represents the true operating characteristics of the diode. Two factors that make this model so accurate are *bulk resistance* and *reverse current*. When these factors are taken into account, we get the diode characteristic curve shown in Figure 2.19. This illustration will be referred to throughout our discussion on the complete diode model.

### Bulk Resistance ($R_B$)

As you learned in Chapter 1, *bulk resistance* is the natural resistance of the diode p-type and n-type materials. The effect of bulk resistance on diode operation can be seen in the *forward operation region* of the curve. As Figure 2.19 illustrates, $V_F$ is *not* constant but, rather, varies with the value of $I_F$. The change in $V_F$ ($\Delta V_F$) is caused by the diode current passing through the bulk resistance of the diode. This concept is illustrated by the diode equivalent circuit shown in Figure 2.20. The 0.7-V source shown in Figure 2.20 repre-

**FIGURE 2.19    Complete-model diode curve.**

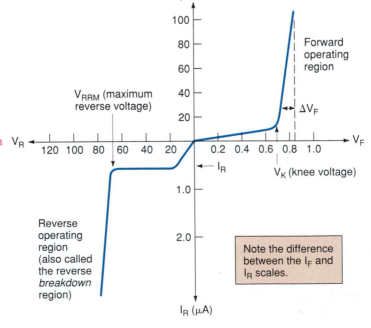

**Lab Reference:** The forward region of the diode curve is plotted (using measured values) in Exercise 2.

FIGURE 2.20   Diode equivalent
circuit.

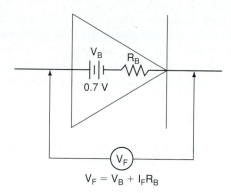

$$V_F = V_B + I_F R_B$$

sents the barrier potential of the diode. The bulk resistance of the diode is represented by
the series resistor, $R_B$. The diode current, $I_F$, passing through the bulk resistance will
develop a voltage equal to $I_F R_B$. The total voltage across the diode is the sum of $I_F R_B$ and
the barrier potential. By formula,

$$V_F = 0.7 \text{ V} + I_F R_B \qquad \text{(for silicon)} \qquad (2.5)$$

Note that, as $I_F$ increases, so does $I_F R_B$. Therefore, the total voltage across a diode varies
directly with the value of $I_F$. This is shown in Example 2.12.

---

**EXAMPLE 2.12**

Determine the voltage across the diode in Figure 2.20 for values of $I_F = 1$ mA and
$I_F = 5$ mA. Assume that the bulk resistance of the diode is 5 $\Omega$ and that the diode is
silicon.

*Solution:*   Because the diode is silicon, we use equation (2.5) to solve for the total
voltage drop across the diode. For $I_F = 1$ mA,

$$
\begin{aligned}
V_F &= 0.7 + (1 \text{ mA})(5 \text{ }\Omega) \\
&= 0.7 + 0.005 \text{ V} \\
&= 0.705 \text{ V} \qquad (705 \text{ mV})
\end{aligned}
$$

For $I_F = 5$ mA,

$$
\begin{aligned}
V_F &= 0.7 + (5 \text{ mA})(5 \text{ }\Omega) \\
&= 0.7 + 0.025 \text{ V} \\
&= 0.725 \text{ V} \qquad (725 \text{ mV})
\end{aligned}
$$

**PRACTICE PROBLEM 2.12**

A silicon diode has a bulk resistance of 8 $\Omega$ and a forward current of 12 mA. What is
the actual value of $V_F$ for the device?

---

## Effect of Bulk Resistance on Circuit Measurements

In our discussion on practical diode circuit analysis, we assumed that the value of $V_F$ for
a silicon diode is 0.7 V. While this assumed value of $V_F$ will work very well for the analysis of a circuit, you will find that measured values of $V_F$ will generally vary between 0.7
and 1.1 V.

A diode used in a low-current circuit will have a very small amount of voltage developed across its bulk resistance. Because of this, the voltage across such a diode will tend to be closer to 0.7 V. At the same time, a diode used in a high-current circuit will have a relatively large amount of voltage developed across its bulk resistance. This will cause it to have a value of $V_F$ that is closer to 1.1 V.

For routine circuit measurements, using the approximate value of 0.7 V is acceptable because the *exact* value of $V_F$ isn't critical. However, in *circuit development* (or *engineering*) applications, the circuit designer may need to very accurately predict the value of $V_F$ in a circuit. This is when equation (2.5) and other factors of the complete diode model are most commonly used.

## Reverse Current ($I_R$)

Ideally, when a diode is reverse biased, the depletion layer reaches its maximum width, and conduction through the diode stops. In reality, this is not the case. A *very small* amount of minority carrier current passes through the diode when it is reverse biased. This current is referred to as *reverse current* and is illustrated in the *reverse operating region* of the diode curve (Figure 2.19). Reverse current is made up of two independent currents: *reverse saturation current, $I_S$,* and *surface-leakage current, $I_{SL}$.* By formula,

$$I_R = I_S + I_{SL} \qquad \text{(2.6)}$$

where $I_R$ = the diode reverse current
$\quad I_S$ = the reverse saturation current
$\quad I_{SL}$ = the surface-leakage current

**Reverse saturation current ($I_S$)**
A current caused by thermal activity in a reverse-biased diode. $I_S$ is temperature dependent.

**Reverse saturation current** is a current caused by thermal activity in the two diode materials. *This current is affected by temperature,* but not by the amount of reverse bias applied to the diode. $I_S$ accounts for the major portion of reverse current. Since $I_S$ is not related to the amount of reverse bias, $I_R$ remains relatively constant across a range of reverse voltages. This can be seen by referring to Figure 2.19.

**Surface-leakage current ($I_{SL}$)**
A current along the surface of a reverse-biased diode. $I_{SL}$ is $V_R$ dependent.

**Surface-leakage current** is a current that is present on the surface of the diode. This current will increase with an increase in reverse bias. However, since its value is much lower than that of $I_S$, there is no noticeable change in $I_R$ when $I_{SL}$ changes.

The range of $I_R$ values.

The total value of $I_R$ for most diodes is typically in the microampere range or less. However, as temperature increases, $I_S$, and consequently $I_R$, increase. This point is discussed in detail later in this section.

## Effect of Reverse Current on Circuit Measurements

Until now, we have assumed that the value of $I_R$ is zero. As you know, a small amount of reverse current occurs in a diode circuit. This reverse current will cause a slight voltage to be developed across any resistance in the circuit. An example of this is shown in Figure 2.21a. We have an applied voltage of 25 V and a diode reverse leakage current of 20 μA. We can calculate the values of the component voltages as follows:
The voltage across the resistor is found as

$$20 \ \mu A \times 10 \ k\Omega = 200 \ mV$$

**Lab Reference:** This method of measuring diode reverse resistance is used in Exercise 2.

Now, the voltage dropped across the diode is found as

$$25 \ V - 200 \ mV = 24.8 \ V$$

Incidentally, the reverse resistance of the diode can be found as

$$\frac{24.8 \ V}{20 \ \mu A} = 1.24 \ M\Omega$$

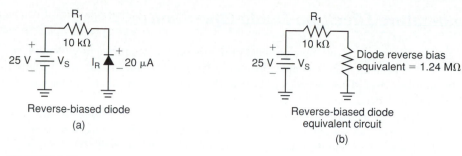

(a) Reverse-biased diode

(b) Reverse-biased diode equivalent circuit

**FIGURE 2.21**

The circuit can then be redrawn with a 1.24-MΩ resistor in place of the diode to give the reverse-bias equivalent circuit, which is shown in Figure 2.21b.

While the voltage drop across the 10-kΩ resistor would not be significant in most cases, it can be very important in any circuit where the normal value of $I_F$ is extremely small.

## Diode Capacitance

Refer to Figure 2.22. An insulator placed between two closely spaced conductors forms a capacitor. When a diode is *reverse biased*, it forms a depletion layer (insulator) between the two semiconductor materials (conductors). Therefore, a reverse-biased diode forms a small capacitor. Under normal circumstances, the capacitance associated with a reverse-biased diode can be ignored. However, when the *high-frequency* operation of a diode is analyzed, the junction capacitance can become extremely important. The effects of junction capacitance on high-frequency operation are discussed in detail in Chapter 14.

Incidentally, there is a type of diode designed to make use of its junction capacitance. This diode, called a *varactor*, is discussed in detail in Chapter 5.

*A reverse-biased pn junction has capacitance.*

## Diffusion Current

Refer to the forward operating region in Figure 2.19. The 0.7-V point on the diode curve is labeled *knee voltage*, $V_k$. Below the knee voltage, you will notice that $I_F$ does *not* instantly drop to the zero point. The reason for this is simple. When $V_F$ goes below the barrier potential of the diode, the device starts to form a depletion layer. However, until the diode is actually reverse biased, the depletion layer will not reach its maximum width, and thus will not reach its maximum resistance. In other words, as long as there is *some* forward voltage, the depletion layer will not be at its maximum width, and some amount of $I_F$ will occur. This small amount of $I_F$ is called **diffusion current**. Note that when a diode is switching from forward bias to reverse bias, the diffusion current will last only as long as it takes the depletion layer to form, generally a few milliseconds or less.

**Diffusion current**
The $I_F$ below the knee voltage.

**FIGURE 2.22   Diode capacitance.**

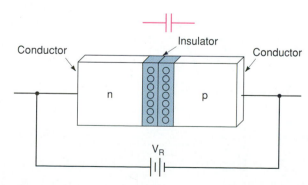

## Temperature Effects on Diode Operation

Temperature has a significant effect on most of the diode characteristics discussed in this section. The reason for this is simple: Increased temperature means increased thermal activity and decreased diode resistance, as was shown in Chapter 1. This holds true for both forward and reverse diode operation.

Temperature effects on $I_F$ and $V_F$

The effects of increased temperature on forward diode operation are illustrated in Figure 2.23. As you can see, there are two forward operation curves. One represents the diode operation at 25°C, and the other represents the diode operation at 100°C. From the graph, we can draw two conclusions regarding forward diode operation and temperature:

1. *As temperature increases,* $I_F$ *will increase for a specified value of* $V_F$. This is illustrated in Figure 2.23 by the two points labeled $I_1$ and $I_2$. As you can see, both of these points fall on the $V_F = 0.7$-V line. However, $I_2$ is greater than $I_1$. This is due to the increased thermal activity in the diode. As temperature has increased from 25° to 100°C, the value of $I_F$ has increased from 5 mA ($I_1$) to 25 mA ($I_2$).

2. *As temperature increases,* $V_F$ *will decrease for a specified value of* $I_F$. This is illustrated by the $V_1$ and $V_2$ points on the curve. Note that both of these points correspond to a value of $I_F = 20$ mA. As temperature has increased, $V_F$ has decreased from 0.75 V ($V_1$) to 0.68 V ($V_2$).

*In practice, a rise in temperature will usually be followed by both a slight increase in* $I_F$ *and a slight decrease in* $V_F$.

Temperature effects on $I_R$

The effect of temperature on the reverse operation of a diode is basically the same as the effect on $I_F$. The effect of temperature on $I_R$ is illustrated in Figure 2.24. You may recall that $I_R$ is equal to the sum of reverse saturation current ($I_S$) and surface-leakage current ($I_{SL}$). Since $I_S$ is normally much greater than $I_{SL}$, it is safe to assume that $I_R$ is approximately equal to $I_S$.

$I_S$ will *double* with every 10°C rise in temperature. Since $I_R$ is approximately equal to $I_S$, we can say that $I_R$ will double for every 10°C rise in temperature. You can see this in Figure 2.24, where $I_R$ doubled from 5 to 10 μA when the temperature increased to 35°C and then doubled again to 20 μA when the temperature increased to 45°C. A handy formula for determining the value of $I_R$ at a given temperature is

$$I'_R = I_R(2^x)$$ (2.7)

**FIGURE 2.23   Temperature effects on forward operation.**

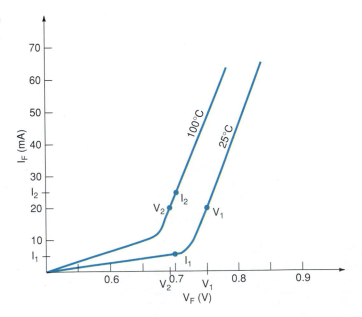

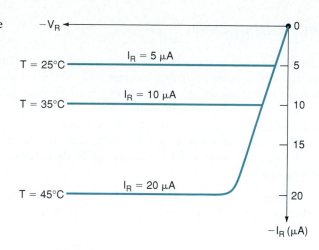

**FIGURE 2.24** Temperature effects on reverse operation.

where $I'_R$ = the value of $I_R$ at the specified temperature
$I_R$ = the value of $I_R$ at 25°C
$x = (T - 25°C)/10$   ($T$ is the temperature of interest)
Example 2.13 illustrates this relationship.

---

*EXAMPLE 2.13*

A given silicon diode is rated at a reverse current of 0.5 μA at 25°C. What is the value of $I_R$ at 50°C? 75°C? 100°C?

**Solution:**   For 50°C,

$$x = \frac{50 - 25}{10} = 2.5$$

and

$$I'_R = (0.5 \ \mu A)(2^{2.5}) = (0.5 \ \mu A)(5.66) = 2.83 \ \mu A$$

For 75°C,

$$x = \frac{75 - 25}{10} = 5$$

and

$$I'_R = (0.5 \ \mu A)(2^{5}) = (0.5 \ \mu A)(32) = 16 \ \mu A$$

For 100°C,

$$x = \frac{100 - 25}{10} = 7.5$$

and

$$I'_R = (0.5 \ \mu A)(2^{7.5}) = (0.5 \ \mu A)(181) = 90.5 \ \mu A$$

**PRACTICE PROBLEM 2.13**

A diode has a reverse current rating of 2μA at 25°C. What is the value of $I_R$ for the device at $T = 60°C$?

## The Bottom Line

When the *complete* diode model is used.

In most cases, you will not have to deal with most of the characteristics discussed in this section unless you are designing a circuit with very low variation tolerances or very high operating frequencies. Even in most practical circuit designs, such as rectifiers and amplifiers, knowing the diode's parameters is more important than knowing the *exact* values of diode voltage, current, resistance, and capacitance.

When you are involved in practical circuit analysis and troubleshooting, you need consider only the more practical aspects of diode operation. These aspects are summarized as follows:

A summary of *practical* diode characteristics.

1. When the forward voltage across a diode reaches the barrier potential, the diode will start to conduct. From that point on, the value of $V_F$ will be approximately equal to the barrier potential.

2. As $V_F$ decreases below the barrier potential, the diode will start to turn off and begin to act like an open switch. This "open-switch" characteristic will continue for all reverse voltage values up to peak reverse voltage, $V_{RRM}$.

   a. As long as $V_R \leq V_{RRM}$, the total applied voltage will be dropped across the diode.

   b. When $V_R > V_{RRM}$, the diode may break down and conduct in the reverse direction. When this happens, you will usually end up replacing the diode.

3. The forward current through a diode is limited by the applied voltage and the resistance values that are external to the diode. For all practical purposes, reverse current through a diode is zero.

## Section Review

1. What is *bulk resistance*?
2. What effect does bulk resistance have on circuit voltage measurements?
3. What is the relationship between $I_F$ and $V_F$?
4. What is reverse current?
5. Which component of reverse current is affected by temperature?
6. Which component of reverse current is affected by the amount of reverse bias?
7. Which component of reverse current makes up a majority of that reverse current?
8. What effect will $I_R$ have on the measured value of voltage across any series resistance?
9. Describe the reverse-biased diode as a capacitor.
10. What is diffusion current?
11. What effect will an increase in temperature have on $I_F$ and $V_F$?
12. What effect will an increase in temperature have on $I_R$?

## 2.6 DIODE SPECIFICATION SHEETS

The *specification sheet*, or *spec sheet*, for any component lists the *parameters* and *operating characteristics* of the device. The parameters and the limits of the device are normally shown under the categories of minimum, maximum, and typical.

Why parameters are important.

Diode operating characteristics and parameters are important for several reasons:

1. They indicate whether or not a given diode can be used for specific application.
2. They establish the operating limits of any circuit designed to use the diode.

You have already been introduced to most of the commonly used diode characteristics and parameters. In this section, we will take a look at how these characteristics and parameters may be found on the spec sheet of a given diode.

## Spec Sheet Organization

Diode spec sheets are commonly divided into two sections: *maximum ratings* and *electrical characteristics*. This can be seen in Figure 2.25, which contains the spec sheet for the 1N4001–1N4007 (or 1N400X) series diodes.

The **maximum ratings** table contains the diode parameters that must not be exceeded under any circumstances. If any of these parameters are exceeded, you will more than likely have to replace the diode.

The **electrical characteristics** table contains the guaranteed operating characteristics of the device. As long as the *maximum ratings* limits are observed, the diode is guaranteed to work within the limits shown in the *electrical characteristics* table.

Confused? Let's take a look at the *maximum reverse current* rating in the *electrical characteristics* table. The 1N400X series of diodes is guaranteed to have a maximum reverse current ($I_R$) of 10 $\mu$A at 25°C. However, the table also shows that the typical $I_R$ would be 0.05 $\mu$A. This assumes that you do not exceed the *reverse breakdown voltage* rating, which is listed on the spec sheet of the device. If you exceed any diode parameter, the *electrical characteristics* values of the device cannot be guaranteed.

**Maximum ratings**
Diode parameters that must not be exceeded under any circumstances.

**Electrical characteristics**
The guaranteed operating characteristics of the device.

## Diode Maximum Ratings

As stated earlier, many of the parameters listed in the *maximum ratings* table have already been covered in this chapter. Table 2.2 summarizes the parameters listed in Figure 2.25 and their meanings.

A few notes:

1. **Peak repetitive reverse voltage** ($V_{RRM}$) indicates the maximum allowable reverse voltage that can be applied to the device. The 1N4001 has a $V_{RRM}$ rating of 50 V. If you apply a reverse voltage to the device that is *greater than* 50 V, the device *may* go into reverse breakdown. Note that this rating is also called *working peak reverse voltage* or *dc blocking voltage*.

2. The reverse voltage ratings are the only ratings that distinguish the diodes in the 1N400X series from each other. In other words, the diodes listed have the same maximum ratings and electrical characteristics *outside of their reverse voltage characteristics*. This is typical for a given series of diodes.

3. The *average forward current* ($I_0$) rating shows that the maximum allowable average forward current *decreases* as temperature *increases*. The basis for this relationship is simple. The current through a diode generates *heat*. The maximum allowable current through a diode depends on *how much heat the component can dissipate*. When the air that surrounds a diode is hot, the diode cannot dissipate as much heat. Therefore, the limit on device current decreases.

4. The *peak surge current* ($I_{FSM}$) indicates the diode's ability to handle a *surge* (short duration, extremely high value) of current. We will cover surge current and one of its primary sources in Chapter 3.

**Peak repetitive reverse voltage** ($V_{RRM}$)
The maximum allowable reverse voltage that can be applied to a diode.

## Diode Identification

Normally, when repairing a piece of equipment, you have a schematic to guide you. However, you are sometimes required to work on equipment that you are unfamiliar with, and you may not have schematics available. Occasionally, the numbers or cathode band has

**MOTOROLA Semiconductors**

BOX 20912 • PHOENIX, ARIZONA 85036

## Designers Data Sheet

### "SURMETIC"▲ RECTIFIERS

. . . subminiature size, axial lead mounted rectifiers for general-purpose low-power applications.

**Designers Data for "Worst Case" Conditions**

The Designers▲ Data Sheets permit the design of most circuits entirely from the information presented. Limit curves — representing boundaries on device characteristics — are given to facilitate "worst case" design.

## 1N4001 thru 1N4007

### LEAD MOUNTED SILICON RECTIFIERS

50-1000 VOLTS
DIFFUSED JUNCTION

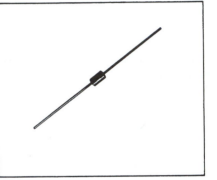

### *MAXIMUM RATINGS

| Rating | Symbol | 1N4001 | 1N4002 | 1N4003 | 1N4004 | 1N4005 | 1N4006 | 1N4007 | Unit |
|---|---|---|---|---|---|---|---|---|---|
| Peak Repetitive Reverse Voltage<br>Working Peak Reverse Voltage<br>DC Blocking Voltage | $V_{RRM}$<br>$V_{RWM}$<br>$V_R$ | 50 | 100 | 200 | 400 | 600 | 800 | 1000 | Volts |
| Non-Repetitive Peak Reverse Voltage<br>(halfwave, single phase, 60 Hz) | $V_{RSM}$ | 60 | 120 | 240 | 480 | 720 | 1000 | 1200 | Volts |
| RMS Reverse Voltage | $V_{R(RMS)}$ | 35 | 70 | 140 | 280 | 420 | 560 | 700 | Volts |
| Average Rectified Forward Current<br>(single phase, resistive load,<br>60 Hz, see Figure 8, $T_A$ = 75°C) | $I_O$ | 1.0 | | | | | | | Amp |
| Non-Repetitive Peak Surge Current<br>(surge applied at rated load<br>conditions, see Figure 2) | $I_{FSM}$ | 30 (for 1 cycle) | | | | | | | Amp |
| Operating and Storage Junction<br>Temperature Range | $T_J, T_{stg}$ | −65 to +175 | | | | | | | °C |

### *ELECTRICAL CHARACTERISTICS

| Characteristic and Conditions | Symbol | Typ | Max | Unit |
|---|---|---|---|---|
| Maximum Instantaneous Forward Voltage Drop<br>($i_F$ = 1.0 Amp, $T_J$ = 25°C) Figure 1 | $v_F$ | 0.93 | 1.1 | Volts |
| Maximum Full-Cycle Average Forward Voltage Drop<br>($I_O$ = 1.0 Amp, $T_L$ = 75°C, 1 inch leads) | $V_{F(AV)}$ | — | 0.8 | Volts |
| Maximum Reverse Current (rated dc voltage)<br>    $T_J$ = 25°C<br>    $T_J$ = 100°C | $I_R$ | 0.05<br>1.0 | 10<br>50 | μA |
| Maximum Full-Cycle Average Reverse Current<br>($I_O$ = 1.0 Amp, $T_L$ = 75°C, 1 inch leads | $I_{R(AV)}$ | — | 30 | μA |

*Indicates JEDEC Registered Data.

**MECHANICAL CHARACTERISTICS**

**CASE:** Void free, Transfer Molded
**MAXIMUM LEAD TEMPERATURE FOR SOLDERING PURPOSES:** 350°C, 3/8" from case for 10 seconds at 5 lbs. tension
**FINISH:** All external surfaces are corrosion-resistant, leads are readily solderable
**POLARITY:** Cathode indicated by color band
**WEIGHT:** 0.40 Grams (approximately)

▲Trademark of Motorola Inc.

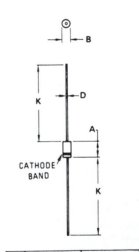

CATHODE BAND

| DIM | MILLIMETERS | | INCHES | |
|---|---|---|---|---|
| | MIN | MAX | MIN | MAX |
| A | 5.97 | 6.60 | 0.235 | 0.260 |
| B | 2.79 | 3.05 | 0.110 | 0.120 |
| D | 0.76 | 0.86 | 0.030 | 0.034 |
| K | 27.94 | — | 1.100 | — |

**CASE 59-04**
Does Not Conform to DO-41 Outline.

© MOTOROLA INC., 1975

DS 6015 R3

FIGURE 2.25    The Motorola 1N400X series specifications. (Copyright of Motorola. Used by permission.)

**TABLE 2.2   Diode Parameters**

| Rating | Discussion | |
|---|---|---|
| Peak repetitive reverse voltage, $V_{RRM}$ | This is the maximum allowable $V_R$ for the diode. This holds true for both dc and *peak* ac voltages. | |
| RMS reverse voltage, $V_{R(rms)}$ | This rating is found by converting the peak repetitive voltage rating to an rms value. You may recall that $V_{rms} = 0.707\ V_{pk}$. Using this formula, you can convert any peak maximum rating to an rms maximum rating. | |
| Average half-wave rectified forward current, $I_0$ | This rating tells you the maximum *average* forward current the diode can handle. For the 1N4001, this value is 1 A when the temperature is 75°C. If the temperature increases to 100°C, the maximum allowable value drops to 750 mA. *This does not mean that diode forward current will decrease when temperature increases!* It means that, at higher temperatures, the *limit* on forward current decreases. | The meaning of the term *half-wave rectified* will be explained in Chapter 3. |
| Peak surge current, $I_{FSM}$ | This is the maximum *surge* value of $I_F$ that the diode can handle. Surge current is nonrepetitive, meaning that it does not happen at regular intervals. With the average diode, this rating can be exceeded *once*. After that, you will have to replace the diode. | |
| Operating and storage temperature range, $T_J$, $T_{stg}$ | This one is self-explanatory. The diode can be used and stored at any temperature between −65° and 175°C. | |

worn off the components, so identification of diode polarity is difficult, if not impossible. There are several standards for diode identification that will normally hold true. The most common types of diodes are shown in Figure 2.26. The anode and cathode are labeled for identification. When working with an unknown diode, it is best to find the specifications in the data manual. However, if the numbers are not legible or a data manual is unavailable, the diode configurations shown in Figure 2.26 may help you to identify the component terminals.

## Diode Parameters and Device Substitution

It was stated earlier in the chapter that *average forward current* and *peak reverse voltage* ratings are two of the primary concerns when you are substituting one diode for another.   ◄   *OBJECTIVE 5*

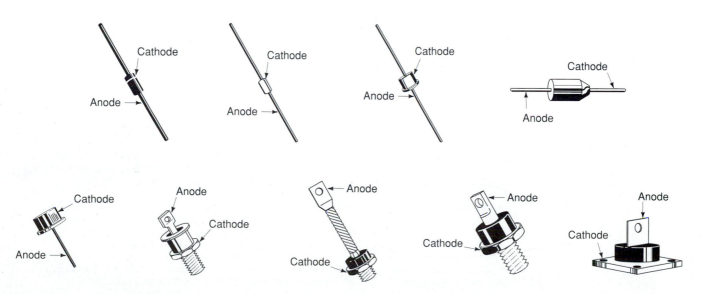

**FIGURE 2.26   Common types of diodes. (Copyright of Motorola. Used by permission.)**

$T_J$ stands for junction temperature.

To make component substitutions easier, diode data books sometimes contain *selector guides*. These guides group diodes by *average forward current* (or *forward power dissipation*) and *peak reverse voltage*. An example of a selector guide is shown in Figure 2.27.

The selector guide allows you to select a diode based on circuit requirements. For example, let's say that you need a diode with an average forward current rating of 1.5 A and a peak reverse voltage of 100 V. You would locate the column that corresponds to 1.5 A and cross-match it with the row that has a $V_{RRM}$ of 100 V. The block that corresponds to both of these values contains the number 1N5392. Thus, the 1N5392 would be suitable for the desired application.

## Electrical Characteristics

As stated earlier, the values provided under *electrical characteristics* indicate the guaranteed operating characteristics of the diode. Table 2.3 gives a brief explanation of the electrical characteristics listed in Figure 2.25.

A few notes:

1. The *maximum reverse current* rating of the 1N400X series is 10 μA when $T_J = 25°C$. If we were to use these values in equation (2.7), we would calculate the maximum reverse current at 100°C to be

$$I_R = (10 \text{ μA})(2^{7.5}) = 1.81 \text{ mA}$$

However, the rated value of $I_{R(MAX)}$ at 100°C is 50 μA. You might wonder *which one is correct*? The fact is, equation (2.7) gives you a *worst-case* value; that is, the highest that it would be under any circumstances. When the *rated* value of $I_R$ at a given temperature is less than the one calculated with equation (2.7), you assume that the rated value is the maximum.

2. Many of the characteristics list both *typical* and *maximum* values. When analyzing the operation of the component in a circuit, the value listed as *typical* would be used. When determining circuit tolerances for circuit development purposes, the *maximum* values would be used. When only one value is given, it is used for *all* circuit analyses.

## Finding Data on Spec Sheets

When you are looking for data on a spec sheet, follow a three-step procedure:

1. Determine whether the data you are looking for is a *maximum rating* (parameter) or an *electrical characteristic* (guaranteed minimum performance).

**TABLE 2.3   Diode Electrical Characteristics**

| | |
|---|---|
| Maximum forward voltage drop, $V_F$ | This is the maximum value that $V_F$ will ever reach. All $V_F$ values are guaranteed to be at least 1.1 V or less. Note that this value was determined at a temperature of 25°C. On the average, $V_F$ will decrease by about 1.8 mV for every 1°C rise in temperature. |
| Maximum full-cycle average forward voltage drop, $V_{F(av)}$ | This is the maximum *average* forward voltage ($V_F$). For the 1N400X, this value is 0.8 V. Note that this parameter is also temperature dependent and was measured at 75°C. |
| Maximum reverse current, $I_R$ | These ratings, 10 μA and 50 μA, were given for 25° and 100°C, respectively. The spec sheet lists this parameter *"at rated dc voltages."* This means that the rating is valid for all dc values of $V_R$ at or below the 50-V peak repetitive $V_R$ rating. |
| Maximum full-cycle average reverse current, $I_{R(av)}$ | This is the maximum average value of $I_R$. Note that this rating is at a temperature of 75°C. At all temperatures below 75°C, the average value of $I_R$ will be less than 30 μA. |

| $V_{RRM}$ (Volts) | $I_O$, AVERAGE RECTIFIED FORWARD CURRENT (Amperes) | | | | | |
|---|---|---|---|---|---|---|
| | 1.0 | 1.5 | 3.0 | | | 6.0 |
| | 59-03 (DO-41) Plastic  | 59-04 Plastic  | 60-01 Metal  | 267-03 Plastic  | 267-02 Plastic 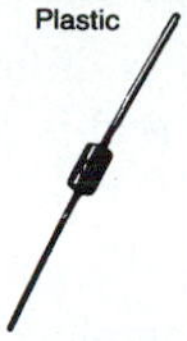 | 194-04 Plastic  |
| 50 | †1N4001 | **1N5391 | 1N4719 | **MR500 | 1N5400 | MR750 |
| 100 | †1N4002 | **1N5392 | 1N4720 | **MR501 | 1N5401 | MR751 |
| 200 | †1N4003 | 1N5393 *MR5059 | 1N4721 | **MR502 | 1N5402 | MR752 |
| 400 | †1N4004 | 1N5395 *MR5060 | 1N4722 | **MR504 | 1N5404 | MR754 |
| 600 | †1N4005 | 1N5397 *MR5061 | 1N4723 | **MR506 | 1N5406 | MR756 |
| 800 | †1N4006 | 1N5398 | 1N4724 | MR508 | | MR758 |
| 1000 | †1N4007 | 1N5399 | 1N4725 | MR510 | | MR760 |
| $I_{FSM}$ (Amps) | 30 | 50 | 300 | 100 | 200 | 400 |
| $T_A$ @ Rated $I_O$ (°C) | 75 | $T_L$ = 70 | 75 | 95 | $T_L$ = 105 | 60 |
| $T_C$ @ Rated $I_O$ (°C) | | | | | | |
| $T_J$ (Max) (°C) | 175 | 175 | 175 | 175 | 175 | 175 |

† Package Size: 0.120" Max Diameter by 0.260" Max Length.

* 1N5059 series equivalent Avalanche Rectifiers.

** Avalanche versions available, consult factory.

| $V_{RRM}$ (Volts) | $I_O$, AVERAGE RECTIFIED FORWARD CURRENT (Amperes) | | | | | | | |
|---|---|---|---|---|---|---|---|---|
| | 12 | 20 | 24 | 25 | 30 | | 40 | 50 |
| | 245A-02 (DO-203AA) Metal  | | 339-02 Plastic Note 1  | 193-04 Plastic Note 2  | 43-02 (DO-21) Metal  | | 42A-01 (DO-203AB) Metal  | 43-04 Metal  |
| 50 | MR1120 1N1199,A,B | MR2000 | MR2400 | MR2500 | 1N3491 | 1N3659 | 1N1183A | MR5005 |
| 100 | MR1121 1N1200,A,B | MR2001 | MR2401 | MR2501 | 1N3492 | 1N3660 | 1N1184A | MR5010 |
| 200 | MR1122 1N1202,A,B | MR2002 | MR2402 | MR2502 | 1N3493 | 1N3661 | 1N1186A | MR5020 |
| 400 | MR1124 1N1204,A,B | MR2004 | MR2404 | MR2504 | 1N3495 | 1N3663 | 1N1188A | MR5040 |
| 600 | MR1126 1N1206,A,B | MR2006 | MR2406 | MR2506 | | Note 3 | 1N1190A | Note 3 |
| 800 | MR1128 | MR2008 | | MR2508 | | Note 3 | Note 3 | Note 3 |
| 1000 | MR1130 | MR2010 | | MR2510 | | Note 3 | Note 3 | Note 3 |
| $I_{FSM}$ (Amps) | 300 | 400 | 400 | 400 | 300 | | 400 | 800 | 600 |
| $T_A$ @ Rated $I_O$ (°C) | | | | | | | | |
| $T_C$ @ Rated $I_O$ (°C) | 150 | 150 | 125 | 150 | 130 | | 100 | 150 | 150 |
| $T_J$ (Max) (°C) | 190 | 175 | 175 | 175 | 175 | | 175 | 190 | 195 |

Note 1. Meets mounting configuration of TO-220 outline.
Note 2. Request Data Sheet for Mounting Information.
Note 3. Available on special order.

**FIGURE 2.27** (Copyright of Motorola. Used by permission.)

2. Look in the appropriate table for the desired data. If it isn't there, try the alternative-search technique covered in step 3. If you still can't find the data, contact the manufacturers of the component.

3. Sometimes it is difficult to locate a particular parameter or characteristic because of the wording used in the spec sheet. When you can't locate a particular bit of information, try this:

   a. Determine the unit that the desired parameter or characteristic would be measured in. For example, *average forward current* would typically be measured in *mA* or *A*, while *reverse current* would typically be measured in μA.

   b. Search the *unit* column in the spec sheet tables to find the appropriate unit of measure.

   c. If you find the unit of measure, look at the data name to see if it is the value you are looking for.

The use of step three can be illustrated using Figure 2.25. Let's say that we are interested in the temperature range of the 1N400X series of diodes. We know that temperature is measured in °C. Looking under the *unit* column, we see that only one parameter is measured in °C. Looking to the left side of that measurement, we see the parameter shown is *operating and storage junction temperature*. This is the parameter that we are looking for.

While the spec sheet for the 1N400X series of diodes is relatively short and simple, the spec sheets for many other types of devices are relatively complex. When dealing with these more complex spec sheets, you will find the steps listed under step three very useful for quickly finding the information that you need.

## One Final Note

Not all operating parameters and characteristics used to describe diodes are listed on the 1N400X spec sheet. However, the most critical of the operating parameters and characteristics have been covered. Some other parameters you will often see that were not included in this sheet are *junction capacitance* (rated in *pF*), *forward power dissipation* (rated in *mW* or *W*), and *maximum switching frequency* (rated in *kHz* or *MHz*). Also, as mentioned earlier, not all spec sheets use the same terminology to describe various parameters and characteristics. However, the wording is usually close enough that you will be able to find the information you need.

**Section Review**

1. Why are parameters important?
2. What is a *maximum rating*? Give an example.
3. What is an *electrical characteristic*? Give an example.
4. What is the difference between a *maximum rating* and an *electrical characteristic*?
5. Why does the limit on forward current decrease when temperature increases?
6. What is the procedure for locating information on a diode specification sheet?

## 2.7 ZENER DIODES

**Zener diodes**
Diodes that are designed to work in the reverse operating region.

The **zener diode** is a special type of diode that is designed to work in the reverse breakdown region of the diode characteristic curve. Recall that a normal diode operated in this region will usually be destroyed by excessive current and the heat it produces. This is not the case for the zener diode.

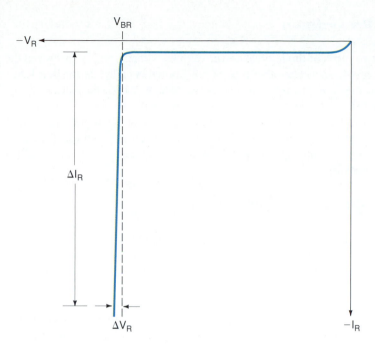

FIGURE 2.28   Reverse breakdown characteristics.

**Lab Reference:** The reverse operating curve of a zener diode is plotted (using measured values) in Exercise 3.

The value of a component designed to operate in the reverse breakdown region can be seen by taking a look at Figure 2.28. This figure shows the reverse breakdown region of the diode characteristic curve. As the curve illustrates, two things happen when the **reverse breakdown voltage, $V_{BR}$,** is reached:

**Reverse breakdown voltage ($V_{BR}$)**
The $V_R$ that causes a diode to conduct in the reverse direction.

1.  The diode current increases drastically.

2.  The reverse voltage across the diode, $V_R$, remains relatively constant.

In other words, *a diode operated in this region will have a relatively constant voltage across it, regardless of the value of current through the device.* The zener diode is used specifically for this purpose: to maintain a relatively constant voltage, regardless of variations in diode current. This makes the zener diode a good *voltage regulator*. A **voltage regulator** is a circuit designed to maintain a constant voltage, regardless of variations in load current or input voltage. Voltage regulators are discussed in detail in Chapter 3.

**Voltage regulator**
A circuit designed to maintain a constant voltage regardless of variations in load current.

Since the zener diode is used for its reverse operating characteristics, it should not surprise you that the current through a zener is normally against the arrow, as shown in Figure 2.29. Since the diode is normally reverse biased, the cathode is more *positive* than the anode when the diode is conducting. Thus, the arrow points in the opposite direction of current.

◄  *OBJECTIVE 6*

Will a zener ever conduct in the same direction as a conventional diode? Is it ever operated in the forward operating region? Rarely. Even though a zener diode has forward operating characteristics similar to those of a *pn*-junction diode, the value of the zener lies in its application as a voltage regulator. This application is lost in a forward-biased zener, so it is rarely used in the forward operating region.

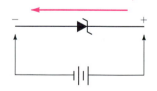

FIGURE 2.29   Zener current.

When a zener is operating in the reverse operating region, the voltage across the device will be *nearly* constant and equal to the **zener voltage ($V_Z$)** rating of the device. Zener diodes have a range of $V_Z$ ratings from about 1.8 V to several hundreds of volts, with power dissipation ratings from 500 mW to 50 W. *Note that the zener rating always tells you the approximate voltage across the device when it is operating in the reverse breakdown region.*

**Zener voltage ($V_Z$)**
The approximate voltage across a zener when operated in reverse breakdown.

## Diode Breakdown

OBJECTIVE 7 ▶

There are two types of diode breakdown. We have already briefly discussed the first type, *avalanche breakdown*. The other type of breakdown is called **zener breakdown**.

Zener breakdown occurs at much lower values of $V_R$ than does avalanche breakdown. The heavy doping of the zener diode causes the device to have a much narrower depletion layer. As a result, it takes very little $V_R$ to cause the diode to go into breakdown, typically 5 V or less. Note that the zener diodes with low $V_Z$ ratings ($< 5$ V) experience zener breakdown, while the ones with higher $V_Z$ ratings ($> 5$ V) usually experience avalanche breakdown.

**Zener breakdown**
A type of reverse breakdown that occurs at low values of $V_R$.

## Zener Operating Characteristics

A zener diode will maintain a near-constant reverse voltage for a *range* of reverse current values. These values are identified in Figure 2.30. The minimum current required to maintain voltage regulation (constant voltage) is the *zener knee current, $I_{ZK}$*. When the zener is used as a voltage regulator, the current through the diode must never be allowed to drop below this value.

$I_{ZK}$. The minimum value of $I_Z$ required to maintain voltage regulation.

The *maximum zener current, $I_{ZM}$*, is the maximum amount of current the diode can handle without being damaged or destroyed. The *zener test current, $I_{ZT}$*, is the current level at which the $V_Z$ rating of the diode is measured. For example, if a given zener diode has values of $V_Z = 9.1$ V and $I_{ZT} = 20$ mA, this means that the diode has a reverse voltage of 9.1 V when the test current is 20 mA. At other current values, the value of $V_Z$ will vary *slightly* above or below the rated value.

$I_{ZM}$. The maximum allowable value of $I_Z$.

$I_{ZT}$. The zener test current.

$I_{ZT}$ is a *test* current value and thus is not a critical value for circuit analysis. However, it is an important value in circuit design. *If you need to have a zener voltage that is as close to $V_Z$ as possible, you need to design the circuit to have a current value equal to $I_{ZT}$.* The further your circuit current is from $I_{ZT}$, the further $V_R$ will be from the ideal $V_Z$ value. $I_{ZK}$ and $I_{ZM}$ are very important for both circuit analysis and component substitution. This point will be demonstrated in our discussion on voltage regulators in Chapter 3.

**Zener impedance, $Z_z$**, is the zener diode's opposition to a *change in current*. This is evidenced by the fact that $Z_Z$ is measured at a specific change in zener current. Figure 2.31 helps to illustrate this point. The spec sheet for the 1N746–1N759 series zener diodes lists the following test conditions for measuring $Z_Z$:

**Zener impedance ($Z_Z$)**
The zener diode's opposition to a change in current.

$$I_{ZT} = 20 \text{ mA}, \qquad I_{zt} = 2 \text{ mA}$$

**FIGURE 2.30    Zener reverse current values.**

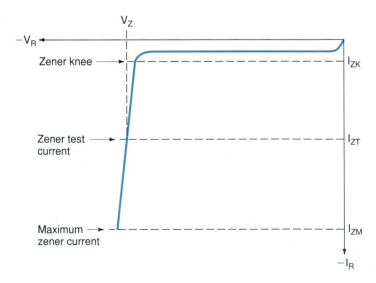

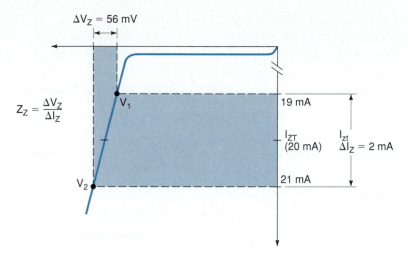

**FIGURE 2.31** Determining zener impedance.

This means that $Z_Z$ is measured while *varying* zener current by 2 mA around the value of $I_{ZT}$ (20 mA). This variation in zener current is illustrated in Figure 2.31. Note that the 2-mA variation in $I_Z$ causes a 56-mV variation in $V_Z$. From this information, $Z_Z$ is determined using

$$Z_Z = \frac{\Delta V_Z}{\Delta I_Z} \bigg| \Delta V_Z = \text{the } \textit{change} \text{ in } V_Z \qquad \textbf{(2.8)}$$

**Lab Reference:** *Zener impedance* is measured in Exercise 3.

as follows:

$$Z_Z = \frac{56 \text{ mV}}{2 \text{ mA}}$$
$$= 28 \text{ } \Omega$$

    **Static reverse current, $I_R$,** is the reverse current through the diode when $V_R$ is less than $V_Z$. In other words, this is the reverse leakage current through the diode when it is off. For the 1N746, this value is 10 µA at 25°C and 30 µA at 150°C.

**Static reverse current, ($I_R$)**
The reverse current through the diode when $V_R < V_Z$.

## Zener Equivalent Circuits

There are basically two equivalent circuits for the zener diode. Both are shown in Figure 2.32. The *ideal* model simply considers the zener to be a voltage source equal to $V_Z$. When placed in a circuit, this voltage source *opposes* the applied circuit-voltage.

The *ideal* and *practical* zener models.

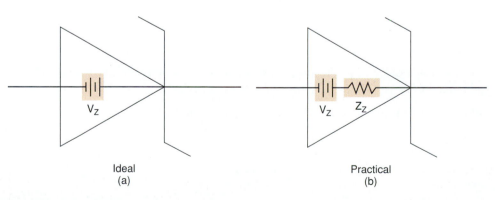

Ideal
(a)

Practical
(b)

**FIGURE 2.32** Zener equivalent circuits.

The *practical* model of the zener includes a series resistor, labeled $Z_Z$. This model of the zener is used mainly for predicting the response of the diode to a change in circuit current. This is another point that is demonstrated in our discussion on voltage regulators.

## Section Review

1. What is the primary difference between zener diodes and *pn*-junction diodes?
2. Why can the zener diode be used as a voltage regulator?
3. How do you determine the direction of zener current in a schematic diagram?
4. What is the significance of the zener voltage rating?
5. Name and define the following symbols: $V_Z$, $Z_z$, $I_{ZK}$, $I_{ZM}$, $I_{ZT}$, $I_{zt}$, and $I_R$.
6. Which rated zener currents would limit the total current in a zener diode circuit?

## 2.8 ZENER DIODE SPECIFICATION SHEETS

Zener diode spec sheets are somewhat different from *pn*-junction diode spec sheets. This can be seen by looking at Figure 2.33, which shows the spec sheet for the 1N746–1N759 series zener diodes. We will refer to this figure throughout this section.

### Maximum Ratings

◄ OBJECTIVE 8

The primary parameters for the zener diode are *dc power dissipation* and the *power derating factor*. While *operating and storage junction temperature range* is also listed in the maximum ratings section, the average technician would not be concerned with this parameter under normal circumstances.

**dc power dissipation rating**
The maximum allowable value of $P_D$ for a zener diode that is operating in the reverse breakdown.

The **dc power dissipation rating** is extremely important for zener diodes. You see, most zener diode spec sheets do not list the values of $I_{ZM}$ for the various diodes. When this is the case, you must calculate the value of $I_{ZM}$ using the zener voltage rating (listed under "Electrical Characteristics") and the dc power dissipation rating (listed under "Maximum Ratings"). When you know the zener voltage and dc power dissipation ratings for a given diode, $I_{ZM}$ can be determined using

$$I_{ZM} = \frac{P_{D(MAX)}}{V_Z} \qquad (2.9)$$

Example 2.14 demonstrates the use of this equation.

## EXAMPLE 2.14

A 1N754 zener diode has a dc power dissipation rating of 500 mW and a zener voltage rating of 6.8 V. What is the value of $I_{ZM}$ for the device?

**Solution:** Using equation (2.9), the value of $I_{ZM}$ is found as

$$I_{ZM} = \frac{P_{D(MAX)}}{V_Z}$$

$$= \frac{500 \text{ mW}}{6.8 \text{ V}}$$

$$= 73.5 \text{ mA}$$

**1.5KE6.8, A thru 1.5KE250, A See Page 4-59**

**1N746 thru 1N759**  
**1N957A thru 1N986A**  
**1N4370 thru 1N4372**

**GLASS ZENER DIODES 500 MILLIWATTS 2.4–110 VOLTS**

## Designers Data Sheet

### 500-MILLIWATT HERMETICALLY SEALED GLASS SILICON ZENER DIODES

- Complete Voltage Range — 2.4 to 110 Volts
- DO-35 Package — Smaller than Conventional DO-7 Package
- Double Slug Type Construction
- Metallurgically Bonded Construction
- Oxide Passivated Die

**Designer's Data for "Worst Case" Conditions**

The Designer's Data sheets permit the design of most circuits entirely from the information presented. Limit curves — representing boundaries on device characteristics — are given to facilitate "worst case" design.

### MAXIMUM RATINGS

| Rating | Symbol | Value | Unit |
|---|---|---|---|
| DC Power Dissipation @ $T_L \le 50°C$, Lead Length = 3/8" | $P_D$ | | |
| *JEDEC Registration | | 400 | mW |
| *Derate above $T_L = 50°C$ | | 3.2 | mW/°C |
| Motorola Device Ratings | | 500 | mW |
| Derate above $T_L = 50°C$ | | 3.33 | mW/°C |
| Operating and Storage Junction Temperature Range | $T_J$, $T_{stg}$ | | °C |
| *JEDEC Registration | | −65 to +175 | |
| Motorola Device Ratings | | −65 to +200 | |

*Indicates JEDEC Registered Data.

### MECHANICAL CHARACTERISTICS

**MAXIMUM LEAD TEMPERATURE FOR SOLDERING PURPOSES:** 230°C, 1/16" from case for 10 seconds.

**FINISH:** All external surfaces are corrosion resistant with readily solderable leads.

**POLARITY:** Cathode indicated by color band. When operated in zener mode, cathode will be positive with respect to anode.

**MOUNTING POSITION:** Any

STEADY STATE POWER DERATING

**ELECTRICAL CHARACTERISTICS** ($T_A = 25°C$, $V_F = 1.5$ V max at 200 mA for all types)

| Type Number (Note 1) | Nominal Zener Voltage $V_Z$ @ $I_{ZT}$ (Note 2) Volts | Test Current $I_{ZT}$ mA | Maximum Zener Impedance $Z_{ZT}$ @ $I_{ZT}$ (Note 3) Ohms | *Maximum DC Zener Current $I_{ZM}$ (Note 4) mA | Maximum Reverse Leakage Current $T_A = 25°C$ $I_R$ @ $V_R = 1$ V μA | $T_A = 150°C$ $I_R$ @ $V_R = 1$ V μA |
|---|---|---|---|---|---|---|
| 1N4370 | 2.4 | 20 | 30 | 150 | 100 | 200 |
| 1N4371 | 2.7 | 20 | 30 | 135 | 75 | 150 |
| 1N4372 | 3.0 | 20 | 29 | 120 | 50 | 100 |
| 1N746 | 3.3 | 20 | 28 | 110 | 10 | 30 |
| 1N747 | 3.6 | 20 | 24 | 100 | 10 | 30 |
| 1N748 | 3.9 | 20 | 23 | 95 | 10 | 30 |
| 1N749 | 4.3 | 20 | 22 | 85 | 2 | 30 |
| 1N750 | 4.7 | 20 | 19 | 75 | 2 | 30 |
| 1N751 | 5.1 | 20 | 17 | 70 | 1 | 30 |
| 1N752 | 5.6 | 20 | 11 | 65 | 0.1 | 20 |
| 1N753 | 6.2 | 20 | 7 | 60 | 0.1 | 20 |
| 1N754 | 6.8 | 20 | 5 | 55 | 0.1 | 20 |
| 1N755 | 7.5 | 20 | 6 | 50 | 0.1 | 20 |
| 1N756 | 8.2 | 20 | 8 | 45 | 0.1 | 20 |
| 1N757 | 9.1 | 20 | 10 | 40 | 0.1 | 20 |
| 1N758 | 10 | 20 | 17 | 35 | 0.1 | 20 |
| 1N759 | 12 | 20 | 30 | 30 | 0.1 | 20 |

| Type Number (Note 1) | Nominal Zener Voltage $V_Z$ (Note 2) Volts | Test Current $I_{ZT}$ mA | Maximum Zener Impedance $Z_{ZT}$ @ $I_{ZT}$ (Note 3) Ohms | $Z_{ZK}$ @ $I_{ZK}$ Ohms | $I_{ZK}$ mA | *Maximum DC Zener Current $I_{ZM}$ (Note 4) mA | Maximum Reverse Current $I_R$ Maximum μA | Test Voltage Vdc $V_R$ 5% | $V_R$ 10% |
|---|---|---|---|---|---|---|---|---|---|
| 1N957A | 6.8 | 18.5 | 4.5 | 700 | 1.0 | 47 | 150 | 5.2 | 4.9 |
| 1N958A | 7.5 | 16.5 | 5.5 | 700 | 0.5 | 42 | 50 | 5.7 | 5.4 |
| 1N959A | 8.2 | 15 | 6.5 | 700 | 0.5 | 38 | 25 | 6.2 | 5.9 |
| 1N960A | 9.1 | 14 | 7.5 | 700 | 0.5 | 35 | 10 | 6.9 | 6.6 |
| 1N961A | 10 | 12.5 | 8.5 | 700 | 0.25 | 32 | 5 | 7.6 | 7.2 |
| 1N962A | 11 | 11.5 | 9.5 | 700 | 0.25 | 28 | 5 | 8.4 | 8.0 |
| 1N963A | 12 | 10.5 | 11.5 | 700 | 0.25 | 26 | 5 | 9.1 | 8.6 |
| 1N964A | 13 | 9.5 | 13 | 700 | 0.25 | 24 | 5 | 9.9 | 9.4 |
| 1N965A | 15 | 8.5 | 16 | 700 | 0.25 | 21 | 5 | 11.4 | 10.8 |
| 1N966A | 16 | 7.8 | 17 | 700 | 0.25 | 20 | 5 | 12.2 | 11.5 |
| 1N967A | 18 | 7.0 | 21 | 750 | 0.25 | 17 | 5 | 13.7 | 13.0 |
| 1N968A | 20 | 6.2 | 25 | 750 | 0.25 | 15 | 5 | 15.2 | 14.4 |
| 1N969A | 22 | 5.6 | 29 | 750 | 0.25 | 14 | 5 | 16.7 | 15.8 |
| 1N970A | 24 | 5.2 | 33 | 750 | 0.25 | 13 | 5 | 18.2 | 17.3 |
| 1N971A | 27 | 4.6 | 41 | 750 | 0.25 | 11 | 5 | 20.6 | 19.4 |
| 1N972A | 30 | 4.2 | 49 | 1000 | 0.25 | 10 | 5 | 22.8 | 21.6 |
| 1N973A | 33 | 3.8 | 58 | 1000 | 0.25 | 9.2 | 5 | 25.1 | 23.8 |
| 1N974A | 36 | 3.4 | 70 | 1000 | 0.25 | 8.5 | 5 | 27.4 | 25.9 |
| 1N975A | 39 | 3.2 | 80 | 1000 | 0.25 | 7.8 | 5 | 29.7 | 28.1 |
| 1N976A | 43 | 3.0 | 93 | 1000 | 0.25 | 7.0 | 5 | 32.7 | 31.0 |
| 1N977A | 47 | 2.7 | 105 | 1500 | 0.25 | 6.4 | 5 | 35.8 | 33.8 |
| 1N978A | 51 | 2.5 | 125 | 1500 | 0.25 | 5.9 | 5 | 38.8 | 36.7 |
| 1N979A | 56 | 2.2 | 150 | 2000 | 0.25 | 5.4 | 5 | 42.6 | 40.3 |
| 1N980A | 62 | 2.0 | 185 | 2000 | 0.25 | 4.9 | 5 | 47.1 | 44.6 |
| 1N981A | 68 | 1.8 | 230 | 2000 | 0.25 | 4.5 | 5 | 51.7 | 49.0 |
| 1N982A | 75 | 1.7 | 270 | 2000 | 0.25 | 4.0 | 5 | 56.0 | 54.0 |
| 1N983A | 82 | 1.5 | 330 | 3000 | 0.25 | 3.7 | 5 | 62.2 | 59.0 |
| 1N984A | 91 | 1.4 | 400 | 3000 | 0.25 | 3.3 | 5 | 69.2 | 65.5 |
| 1N985A | 100 | 1.3 | 500 | 3000 | 0.25 | 3.0 | 5 | 76 | 72 |
| 1N986A | 110 | 1.1 | 750 | 4000 | 0.25 | 2.7 | 5 | 83.6 | 79.2 |

NOTES:
1. PACKAGE CONTOUR OPTIONAL WITHIN A AND B. HEAT SLUGS, IF ANY, SHALL BE INCLUDED WITHIN THIS CYLINDER, BUT NOT SUBJECT TO THE MINIMUM LIMIT OF B.
2. LEAD DIAMETER NOT CONTROLLED IN ZONE F TO ALLOW FOR FLASH, LEAD FINISH BUILDUP AND MINOR IRREGULARITIES OTHER THAN HEAT SLUGS.
3. POLARITY DENOTED BY CATHODE BAND.
4. DIMENSIONING AND TOLERANCING PER ANSI Y14.5, 1973.

| DIM | MILLIMETERS MIN | MAX | INCHES MIN | MAX |
|---|---|---|---|---|
| A | 3.05 | 5.08 | 0.120 | 0.200 |
| B | 1.52 | 2.29 | 0.060 | 0.090 |
| D | 0.46 | 0.56 | 0.018 | 0.022 |
| F | | 1.27 | | 0.050 |
| K | 25.40 | 38.10 | 1.000 | 1.500 |

All JEDEC dimensions and notes apply.

**CASE 299-02 DO-204AH GLASS**

**FIGURE 2.33** (Copyright of Motorola. Used by permission.)

*Remember, the value of* $I_{ZM}$ *is important because it determines the maximum current the diode can tolerate!* If the total current through this diode exceeds 73.5 mA, you'll end up replacing the diode.

**PRACTICE PROBLEM 2.14**

A zener diode has a dc power dissipation rating of 1 W and a zener-voltage rating of 27 V. What is the value of $I_{ZM}$ for the device?

---

**Power derating factor**
The rate at which $P_{D(MAX)}$ must be decreased per 1°C rise above a specified temperature.

The **power derating factor** (listed under "Maximum Ratings") tells you how much the dc power dissipation rating *decreases* when the operating temperature increases above a specified value. For example, the spec sheet shown in Figure 2.33 shows a derating factor of *3.33 mW/°C for temperatures above 50°C.* This means that you must *decrease* the dc power dissipation rating by 3.33 mW for every 1°C rise in operating temperature above 50°C. The following example illustrates the use of the derating factor.

---

**EXAMPLE 2.15**

*Why are there two sets of power ratings on the spec sheet?* The (Joint Electronic Device Engineering Council (JEDEC)) registered values are for military applications. Military specs call for large tolerances. By rating the device at 100 mW below its actual capability, the large tolerance required by the military is built into the specifications. For our purposes, we will use the manufacturer's parameter values.

A 1N746 is operated at a temperature of 75°C. What is the maximum dc power dissipation for the device at this temperature?

*Solution:* The first step is to determine the total derating value. This value is found as

$$(3.33 \text{ mW/°C})(75°C - 50°C) = 83.25 \text{ mW}$$

We subtract this derating value from the dc power dissipation rating to get the dc power dissipation limit:

$$P_D = 500 \text{ mW} - 83.25 \text{ mW} = 416.75 \text{ mW}$$

Thus, at 75°C and 1N746 has a power dissipation limit of 416.75 mW.

**PRACTICE PROBLEM 2.15**

Determine the power dissipation rating of the 1N746 at 125°C. Assume that the dc power dissipation rating of the device is 500 mW and the derating factor is 3.33 mW/°C above 50°C.

---

## Electrical Characteristics

The electrical characteristics portion of the zener spec sheet shows the zener voltage, current, and impedance ratings for the entire group of diodes.

Take a look at the component numbers. You'll see that many of the device numbers end with the letter *A*. When a letter follows the part number, the letter indicates the *tolerance of the ratings.* In this case, the letter *A* indicates that there is a ±5% tolerance in the ratings. Thus, the 1N980A (rated $V_Z = 62$ V) could actually have a zener voltage that falls between

$$62 \text{ V} - 3.1 \text{ V} = 58.9 \text{ V} \ (minimum)$$

and

$$62 \text{ V} + 3.1 \text{ V} = 65.1 \text{ V} \ (maximum)$$

When no letter follows the part number, the tolerance of the component is ±10%.

If you look at the list of **nominal zener voltage** (rated zener voltages), you will notice something interesting: standard zener voltage values follow the same basic progression as standard resistor values. The nominal zener voltages listed were measured at the zener test current ($I_{ZT}$).

**Nominal zener voltage**
The rated value of $V_Z$ for a given zener diode.

The *maximum zener impedance* ($Z_{ZM}$) rating is self-explanatory. Note that the listed maximum value of $Z_Z$ would be used in any calculation that involves zener impedance. This point will be illustrated in Chapter 3 when we cover zener voltage regulators.

The *maximum dc zener current* ($I_{ZM}$) column needs a bit of explaining. First, you will notice that there are two sets of numbers. Those on the left correspond to the JEDEC registered value of 400 mW for dc power dissipation. Those on the right correspond to the Motorola device rating of 500 mW for dc power dissipation. Therefore, we would be interested in the values that appear in the right-hand column. Second, if you use equation (2.9) to calculate the values of $I_{ZM}$, you'll find that you get answers that are around 10% higher than the values shown. For example, using the 500-mW Motorola dc power dissipation rating, the value of $I_{ZM}$ for the 1N758 is found as

$$I_{ZM} = \frac{P_{D(MAX)}}{V_Z}$$

$$= \frac{500 \text{ mW}}{10 \text{ V}}$$

$$= 50 \text{ mA}$$

This value is approximately 10% higher than the $I_{ZM} = 45$ mA rating shown. The rated values of $I_{ZM}$ are generally reduced by about 10% to keep you from driving the components to their absolute limits. This is similar to the situation we encountered with the *average forward current* rating of the *pn*-junction diode.

As stated earlier, the *maximum reverse leakage current* (also known as *static reverse current*) is the reverse leakage current when the device is biased off ($V_R < V_Z$). As you can see, this reverse current rating is temperature dependent, just like the reverse current rating of the *pn*-junction diode.

## Zener Diode Selector Guides

The selector guides for zener diodes are very similar to those used for the *pn*-junction diodes. A zener diode selector guide is shown in Figure 2.34.

As you can see, the critical parameters for device substitution are *nominal zener voltage* (vertical listing) and *dc power dissipation* (horizontal listing). Examples 2.16 and 2.17 show how the selector guide is used.

---

We need a substitute component for a zener diode that is faulty. The substitute component must have a nominal zener voltage of 12 V and must be capable of dissipating 1.2 W. Which of the zeners listed in Figure 2.34 can be used?

**Solution:** First, the zener diode must have a rating of 12 V. Therefore, our diode is listed in the row that corresponds to $V_Z = 12$ V. Two diodes are listed in the 12 V row that have power dissipation ratings greater than 1.2 W. These are the 1N5927A (1.5 W) and the 1N5349A (5 W). Either of these diodes could be used as a substitute component.

### PRACTICE PROBLEM 2.16

We need a zener diode with a nominal zener voltage of 75 V and a power dissipation capability of 4 W. Which zener diode(s) shown in Figure 2.34 can be used in this case?

*EXAMPLE 2.16*

*A Practical Consideration:* When we were dealing with substituting one *pn*-junction diode for another, we were concerned only with whether or not the current and peak reverse voltage ratings were high enough to survive in the circuit. When dealing with zener diodes, the power rating of the substitute diode may be higher than needed, but the $V_Z$ rating of the substitute must *equal* that of the component it is replacing. We *cannot* use a diode with a higher (or lower) $V_Z$ rating than the original component.

| Nominal Zener Voltage (*Note 1) | 500 mW Cathode = Polarity Mark (*Notes 4,11) Glass Case 362-01 | 500 mW Cathode = Polarity Mark (*Notes 9,11) | 1 Watt Cathode = Polarity Mark (*Note 6) Glass Case 59-04 (DO-41) | 1 Watt Cathode = Polarity Mark (*Notes 6,12) Glass Case 362B-01 | 1 Watt Cathode to Case (*Note 7) Metal Case 52-03 (DO-13) | 1.5 Watt Cathode = Polarity Mark (*Note 8) Sumetic 30 Case 59-03 (DO-41) | 5 Watt Cathode = Polarity Mark (*Note 8) Sumetic 40 Case 17-02 |
|---|---|---|---|---|---|---|---|
| 1.8 | | | | | | | |
| 2.0 | | | | | | | |
| 2.2 | | | | | | | |
| 2.4 | MLL4370 | MLL5221A | | | | | |
| 2.5 | | MLL5222A | | | | | |
| 2.7 | MLL4371 | MLL5223A | | | | | |
| 2.8 | | MLL5224A | | | | | |
| 3.0 | MLL4372 | MLL4225A | | | | | |
| 3.3 | MLL746 | MLL5226A | 1N4728 | MLL4728 | 1N3821 | 1N5913A | 1N5333A |
| 3.6 | MLL747 | MLL5227A | 1N4729 | MLL4729 | 1N3822 | 1N5914A | 1N5334A |
| 3.9 | MLL748 | MLL5228A | 1N4730 | MLL4730 | 1N3823 | 1N5915A | 1N5335A |
| 4.3 | MLL749 | MLL5229A | 1N4731 | MLL4731 | 1N3824 | 1N5916A | 1N5336A |
| 4.7 | MLL750 | MLL5230A | 1N4732 | MLL4732 | 1N3825 | 1N5917A | 1N5337A |
| 5.1 | MLL751 | MLL5231A | 1N4733 | MLL4733 | 1N3826 | 1N5918A | 1N5338A |
| 5.6 | MLL752 | MLL5232A | 1N4734 | MLL4734 | 1N3827 | 1N5919A | 1N5339A |
| 6.0 | | MLL5233A | | | | | |
| 6.2 | MLL753 | MLL5234A | 1N4735 | MLL4735 | 1N3828 | 1N5920A | 1N5341A |
| 6.8 | MLL754 / MLL957A | MLL5235A | 1N4736 | MLL4736 | 1N3829 / 1N3016A | 1N5921A | 1N5342A |
| 7.5 | MLL755 / MLL958A | MLL5236A | 1N4737 | MLL4737 | 1N3830 / 1N3017A | 1N5922A | 1N5343A |
| 8.2 | MLL756 / MLL959A | MLL5237A | 1N4738 | MLL4738 | 1N3018A | 1N5923A | 1N5344A |
| 8.7 | | MLL5238A | | | | | 1N5345A |
| 9.1 | MLL757 / MLL960A | MLL5239A | 1N4739 | MLL4739 | 1N3019A | 1N5924A | 1N5346A |
| 10 | MLL758 / MLL961A | MLL5240A | 1N4740 | MLL4740 | 1N3020A | 1N5925A | 1N5347A |
| 11 | MLL962A | MLL5241A | 1N4741 | MLL4741 | 1N3021A | 1N5926A | 1N5348A |
| 12 | MLL759 / MLL963A | MLL5242A | 1N4742 | MLL4742 | 1N3022A | 1N5927A | 1N5349A |
| 13 | MLL964A | MLL5243A | 1N4743 | MLL4743 | 1N3023A | 1N5928A | 1N5350A |
| 14 | | MLL5244A | | | | | 1N5351A |
| 15 | MLL965A | MLL5245A | 1N4744 | MLL4744 | 1N3024A | 1N5929A | 1N5352A |
| 16 | MLL966A | MLL5246A | 1N4745 | MLL4745 | 1N3025A | 1N5930A | 1N5353A |
| 17 | | MLL5247A | | | | | 1N5354A |
| 18 | MLL967A | MLL5248A | 1N4746 | MLL4746 | 1N3026A | 1N5931A | 1N5355A |
| 19 | | MLL5249A | | | | | 1N5356A |
| 20 | MLL968A | MLL5250A | 1N4747 | MLL4747 | 1N3027A | 1N5932A | 1N5357A |
| 22 | MLL969A | MLL5251A | 1N4748 | MLL4748 | 1N3028A | 1N5933A | 1N5358A |
| 24 | MLL970A | MLL5252A | 1N4749 | MLL4749 | 1N3029A | 1N5934A | 1N5359A |
| 25 | | MLL5253A | | | | | 1N5360A |
| 27 | MLL971A | MLL5254A | 1N4750 | MLL4750 | 1N3030A | 1N5935A | 1N5361A |
| 28 | | MLL5255A | | | | | 1N5362A |
| 30 | MLL972A | MLL5256A | 1N4751 | MLL4751 | 1N3031A | 1N5936A | 1N5363A |
| 33 | MLL973A | MLL5257A | 1N4752 | MLL4752 | 1N3032A | 1N5937A | 1N5364A |
| 36 | MLL974A | MLL5258A | 1N4753 | MLL4753 | 1N3033A | 1N5938A | 1N5365A |
| 39 | MLL975A | MLL5259A | 1N4754 | MLL4754 | 1N3034A | 1N5939A | 1N5366A |
| 43 | MLL976A | MLL5260A | 1N4755 | MLL4755 | 1N3035A | 1N5940A | 1N5367A |
| 47 | MLL977A | MLL5261A | 1N4756 | MLL4756 | 1N3036A | 1N5941A | 1N5368A |
| 51 | MLL978A | MLL5262A | 1N4757 | MLL4751 | 1N3037A | 1N5942A | 1N5369A |
| 56 | MLL979A | MLL5263A | 1N4758 | MLL4758 | 1N3038A | 1N5943A | 1N5370A |
| 60 | | MLL5264A | | | | | 1N5371A |
| 62 | MLL980A | MLL5265A | 1N4759 | MLL4759 | 1N3039A | 1N5944A | 1N5372A |
| 68 | MLL981A | MLL5266A | 1N4760 | MLL4760 | 1N3040A | 1N5945A | 1N5373A |
| 75 | MLL982A | MLL5267A | 1N4761 | MLL4761 | 1N3041A | 1N5946A | 1N5374A |
| 82 | MLL983A | MLL5268A | 1N4762 | MLL4762 | 1N3042A | 1N5947A | 1N5375A |
| 87 | | MLL5269A | | | | | 1N5376A |
| 91 | MLL984A | MLL5270A | 1N4763 | MLL4763 | 1N3043A | 1N5958A | 1N5377A |
| 100 | MLL985A | | 1N4764 | MLL4764 | 1N3044A | 1N5949A | 1N5378A |
| 110 | MLL986A | | | | 1N3045A | 1N5950A | 1N5379A |
| 120 | | | | | 1N3046A | 1N5951A | 1N5380A |
| 130 | | | | | 1N3047A | 1N5952A | 1N5831A |
| 150 | | | | | 1N3048A | 1N5953A | 1N5383A |
| 160 | | | | | 1N3049A | 1N5954A | 1N5384A |
| 170 | | | | | | | 1N5385A |
| 175 | | | | | | | |
| 180 | | | | | 1N3050A | 1N5955A | 1N5386A |
| 200 | | | | | 1N3051A | 1N5956A | 1N5388A |

**FIGURE 2.34    Zener selector guide. (Copyright of Motorola. Used by permission.)**

When we know the zener voltage and current requirements of a component and need a substitute, we have to do a little calculating to find the right substitute part. This point is illustrated in the following example.

**EXAMPLE 2.17**

We need to replace a faulty zener diode. The substitute component must have a nominal zener voltage of 20 V and must be able to handle the power generated by a maximum current of 150 mA. Which diode(s) listed in Figure 2.34 can be used in this application?

*Solution:*    First, we need to determine the power dissipation requirements of the substitute diode. The dc power requirement is found as

$$P_{D(\text{MAX})} = I_{ZM}V_Z$$
$$= (150 \text{ mA})(20 \text{ V})$$
$$= 3 \text{ W}$$

Therefore, the substitute component must be able to dissipate *at least* 3 W. Checking the 20 V/5 W location on the selector guide, we see that the only component we can use is the 1N5357A.

**PRACTICE PROBLEM 2.17**

We need a zener diode with a nominal zener voltage of 6.8 V that can handle a maximum zener current of 175 mA. Which diode(s) listed in Figure 2.34 can be used for this application?

**Section Review**

1. When a spec sheet does not list the value of $I_{ZM}$, how do you determine its value?
2. What is a *power derating factor*? How is it used?
3. Explain how you would determine whether or not one zener diode can be used in place of another.

# 2.9 LIGHT-EMITTING DIODES (LEDs)

LEDs are diodes that will emit light when biased properly. The schematic symbol for the LED is shown in Figure 2.35. Although LEDs are available in various colors, such as infrared (which is not visible), red, green, yellow, orange, and blue, the schematic symbol is the same for all colors. There is nothing in the symbol to indicate the color of a particular LED. A drawing showing the internal components of a typical LED appears in Figure 2.36.

◄ *OBJECTIVE 9*

Since LEDs have clear (or semiclear) cases, there is normally no label on the case to identify the leads. The leads are normally identified in one of three ways (as shown in Figure 2.37):

1. The leads may have different lengths, as shown in Figure 2.37a. When this scheme is used, the longer of the two leads is usually the *anode*.
2. One of the leads may be flattened, as shown in Figure 2.37b. The flattened lead is usually the *cathode*.
3. One side of the case may be flattened, as shown in Figure 2.37c. The lead closest to the flattened side is usually the *cathode*.

## LED Characteristics

LEDs have characteristic curves that are very similar to those for the *pn*-junction diodes. However, they tend to have *higher* forward voltage ($V_F$) values and *lower* reverse breakdown voltage ($V_{BR}$) ratings. The typical ranges for these values are as follows:

1. Forward voltage: $+1.2$ to $+4.3$ V (typical)
2. Reverse breakdown voltage: $-3$ to $-10$ V (typical)

The $V_F$ and $V_{BR}$ ratings of the LED mean that it typically drops more voltage than the *pn*-junction diodes when forward biased and will tend to break down at lower reverse voltages. The color of the LED is determined by the type of material used in the manufacture of the *pn* junction. Since different materials are used, the forward voltage is not the same for all LEDs.

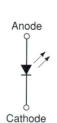

**FIGURE 2.35   The LED schematic symbol.**

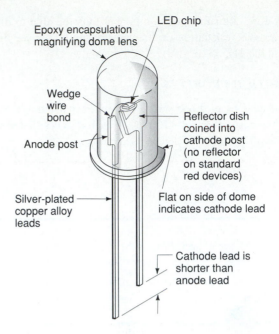

**FIGURE 2.36** Construction features of T-1 3/4 plastic LED lamp. (Courtesy of Hewlett-Packard Company. Reproduced with permission.)

Epoxy encapsulation magnifying dome lens

LED chip

Wedge wire bond

Reflector dish coined into cathode post (no reflector on standard red devices)

Anode post

Silver-plated copper alloy leads

Flat on side of dome indicates cathode lead

Cathode lead is shorter than anode lead

The maximum forward current ratings of LEDs typically range between 2 and 50 mA. Because LEDs tend to have low forward current ratings, they require the use of a series *current-limiting resistor*.

## Current-Limiting Resistors (R$_S$)

**Current-limiting resistor ($R_S$)**
A resistor in series with an LED to limit the value of $I_F$ through the component.

When used in any practical application, the LED will have a series **current-limiting resistor**, as shown in Figure 2.38. The resistor ensures that the maximum current rating of the LED will not be exceeded by the circuit. As the illustration shows, the value of the limiting resistor, $R_S$, is determined using the following equation:

$$R_S = \frac{V_{out(pk)} - V_F}{I_F}$$

(2.10)

where     $V_{out(pk)}$ = the peak output voltage of the driving circuit
$V_F$ = the *minimum* value of $V_F$ for the LED
$I_F$ = the desired value of $I_F$ for the LED

Note that $I_F$ must always be less than the rated maximum LED current.

OBJECTIVE 10 ▶     Example 2.18 demonstrates the use of equation (2.10) in determining the needed value for a current-limiting resistor.

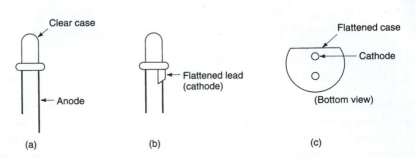

Clear case

Anode

(a)

Flattened lead (cathode)

(b)

Flattened case

Cathode

(Bottom view)

(c)

**FIGURE 2.37**

**FIGURE 2.38   An LED needs a current-limiting resistor.**

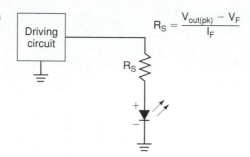

$$R_S = \frac{V_{out(pk)} - V_F}{I_F}$$

**EXAMPLE 2.18**

The driving circuit shown in Figure 2.38 has a peak output voltage of 8 V. The LED has ratings of $V_F$ = 1.8 to 2.0 V and the desired $I_F$ is 16 mA. What value of current-limiting resistor is needed in the circuit?

**Solution:**   Using the peak voltage, *minimum* $V_F$, and $I_F$, the value of $R_S$ is found as

$$R_S = \frac{V_{out(pk)} - V_F}{I_F}$$

$$= \frac{8 \text{ V} - 1.8 \text{ V}}{16 \text{ mA}}$$

$$= 387.5 \text{ } \Omega$$

The smallest standard-value resistor that has a value *greater than* 387.5 $\Omega$ is the 390-$\Omega$ resistor. This is the component we would use in this circuit.

**PRACTICE PROBLEM 2.18**

An LED with a forward voltage rating of 1.4 to 1.8 V and a desired forward current rating of 12 mA is driven by a source with a peak voltage of 14 V. What is the smallest standard resistor value that can be used as $R_S$? (*Note:* A listing of the standard resistor values appears at the end of Appendix A.)

## *Multicolor LEDs*

LEDs have been developed that will emit one color of light when the supply voltage is one polarity, a second color when the polarity is reversed, and a third when the bias polarity is rapidly alternated. One commonly used schematic symbol for these **multicolor LEDs** is shown in Figure 2.39. Multicolor LEDs contain two *pn* junctions, one for each color that the LED will emit. Because a diode junction can emit light only when forward biased, the two diode junctions are connected in reverse parallel, meaning that the anode of one diode junction is connected to the cathode of the other. When a voltage of either polarity is applied to the LED, one of the diodes will be forward biased and emit the color of that diode.

**Multicolor LED**
An LED that emits different colors when the polarity of the supply voltage changes.

If a positive potential is applied to the LED (as shown in Figure 2.39a), the *pn* junction on the *left* will light. Note that the device current passes through the left *pn* junction. If the polarity of the voltage source is reversed (as shown in Figure 2.39b), the *pn* junction on the right will light. Note that the direction of diode current has reversed and is now passing through the right *pn* junction.

Multicolor LEDs are typically *red* when biased in one direction and *green* when biased in the other. If a multicolor LED is switched fast enough between the two polarities, the LED will appear to produce a *third* color. For example, a red/green LED will

FIGURE 2.39   Multicolor LED.

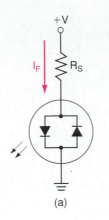
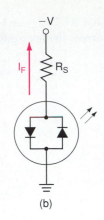

(a)                                                (b)

appear to produce a *yellow* light when rapidly switched back and forth between biasing polarities.

---

**Section Review**

1. How are the leads on an LED usually identified?
2. How do the electrical characteristics of LEDs differ from those of *pn*-junction diodes?
3. Why do LEDs need series current-limiting resistors?

---

## *2.10* DIODE TESTING

We have seen the characteristics of the *pn*-junction diode, the zener diode, and the LED. Now we will cover the testing procedure for each of these devices.

Why diodes fail.

Most diode failures are caused by excessive current, component age, or surpassing the $V_{RRM}$ rating. When a diode is damaged or destroyed by excessive current, the problem is easy to diagnose. In most cases, the diode will crack or fall apart completely. Burned connection points on a printed circuit board are also symptoms of excessive current through a diode. Figure 2.40 shows a printed circuit board after a diode failed from excessive current. Note the damage to the copper trace due to excessive heat.

Not all diodes will show physical signs of failure. However, some simple tests will tell you whether or not a given diode is faulty.

### *Testing pn Junction Diodes*

OBJECTIVE 11 ▶

*A Practical Consideration:* Some meters provide an output voltage that is less than $V_F$ for resistance measurements. These meters will give an open indication, even on a good diode. You need to check your meter documentation before using it for diode testing.

The *pn* junction diode can be tested using a digital or analog ohmmeter. The reason for using the ohmmeter is that the internal battery will supply the voltage necessary to bias the *pn* junction. This allows the diode to be checked in both the forward and reverse bias conditions. When using the analog meter, the resistance range should be set to the R × 1000 (or an equivalent) scale to limit the current supplied by the meter. If the meter is set to a lower scale, excessive current supplied by the meter could destroy the diode.

When connected as in Figure 2.41a, the diode will be forward biased, and you will obtain a low resistance reading, typically less than 1 kΩ. When connected as in Figure 2.41b, the diode will be reverse biased and will have a very high resistance reading, typically in the megohm range. *When you test a given diode, if the forward resistance is very high or the reverse resistance is very low, the diode must be replaced.*

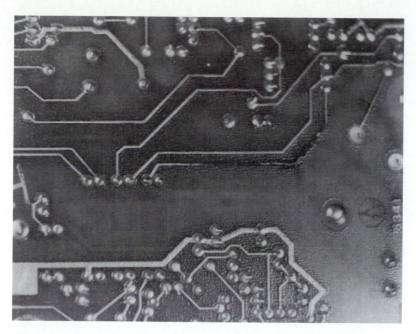

FIGURE 2.40

Many *digital multimeters* (DMM) have a diode check position on the range selector switch. This limits the current and provides sufficient voltage for checking a diode. When forward biased, the meter reading will show the approximate forward barrier voltage drop of the diode. Using the measured value of $V_F$, you can determine whether the diode is made of silicon (0.7 to 0.8 V) or germanuim (0.3 to 0.4 V). When the diode is reverse biased, the DMM will read a very high resistance.

There is an important point that needs to be made: Not all volt ohmmeters have a positive common lead when set for resistance measurements. Some meters will supply current from the common lead, while others will supply current from the "ohm" lead. Check the documentation on your meter to be sure that you know which lead is positive before testing any diodes. Otherwise, you may think that you have a faulty diode when you do not.

The ohmmeter can also be used to identify the anode and the cathode of an unmarked diode. Diodes are usually marked to indicate which end of the component is the cathode. However, with age and heat, these markings may fade. When the ohmmeter is connected to the diode so that it is forward biased, the positive meter lead is connected to the anode and the negative lead is connected to the cathode.

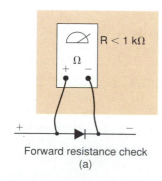

Forward resistance check
(a)

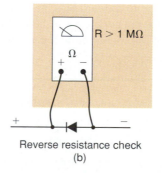

Reverse resistance check
(b)

**Lab Reference:** Diode testing with an ohmmeter is demonstrated in Exercise 2.

FIGURE 2.41   **Diode Testing.**

FIGURE 2.42

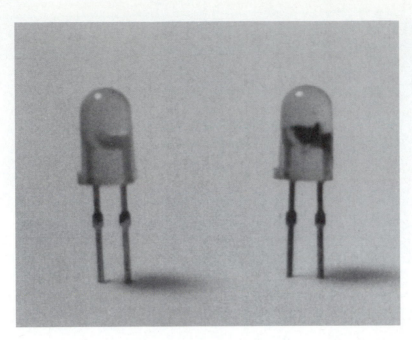

## Testing Zener Diodes

A zener diode cannot be tested in the same manner as a *pn*-junction diode simply because the zener is *designed* to conduct in *both* directions. Because of its design, you cannot test a zener diode with an ohmmeter.

The simplest test of a zener diode is to check the voltage across its terminals while it is in the circuit. If the voltage across the zener is within tolerance, the zener diode is good. If the voltage across the zener is out of tolerance, there is a strong possibility that the zener is faulty. When you suspect that a zener is faulty, the simplest method of dealing with it is to replace the diode.

## Testing LEDs

Normally you will not need to test an LED to see if it is defective. The usual cause of a faulty LED is excessive forward current, which destroys the *pn* junction. When this occurs, the discoloration of the LED due to the burnt junction is easy to recognize. In Figure 2.42, the LED on the left is normal, and the LED on the right has been damaged by excessive current.

A DMM can often be used to check the polarity of an LED if no familiar markings are on the case. With the selector in the diode check or low ohms position, a typical DMM supplies approximately 2.5 V at a very low current to the probes. Normally, this voltage is enough to forward bias the *pn* junction, causing the LED to glow dimly. When this happens, the negative lead of the DMM is connected to the cathode of the LED and the positive lead is connected to the anode.

## Section Review

1. What are the common causes of diode failure?
2. What steps are involved in testing a *pn*-junction diode with an ohmmeter?
3. What precautions should be taken before using an ohmmeter to test a *pn*-junction diode?
4. How do you test a zener diode?

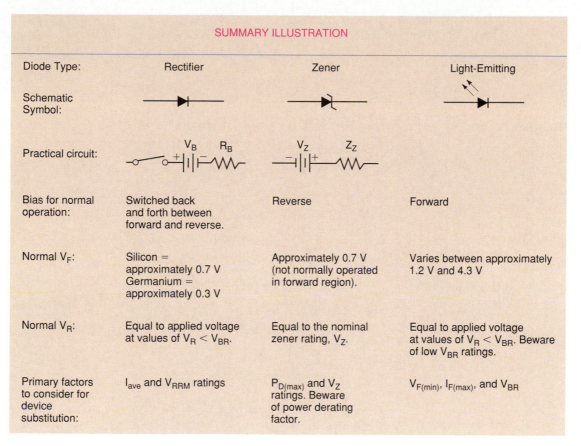

**SUMMARY ILLUSTRATION**

| Diode Type: | Rectifier | Zener | Light-Emitting |
|---|---|---|---|
| Schematic Symbol: | | | |
| Practical circuit: | | | |
| Bias for normal operation: | Switched back and forth between forward and reverse. | Reverse | Forward |
| Normal $V_F$: | Silicon = approximately 0.7 V Germanium = approximately 0.3 V | Approximately 0.7 V (not normally operated in forward region). | Varies between approximately 1.2 V and 4.3 V |
| Normal $V_R$: | Equal to applied voltage at values of $V_R < V_{BR}$. | Equal to the nominal zener rating, $V_Z$. | Equal to applied voltage at values of $V_R < V_{BR}$. Beware of low $V_{BR}$ ratings. |
| Primary factors to consider for device substitution: | $I_{ave}$ and $V_{RRM}$ ratings | $P_{D(max)}$ and $V_Z$ ratings. Beware of power derating factor. | $V_{F(min)}$, $I_{F(max)}$, and $V_{BR}$ |

**FIGURE 2.43**

## Chapter Summary

You have been shown quite a lot in this chapter. The summary illustration in Figure 2.43 will help you to remember the primary points about each diode covered. If you have difficulty remembering any of the points listed, review the appropriate section of the chapter.

## Key Terms

The following terms were introduced and defined in this chapter:

anode
avalanche current
average forward current
bulk resistance
cathode
complete diode model
current-limiting resistor
dc blocking voltage
dc power dissipation
    rating
diffusion current
diode
electrical characteristics

forward power
    dissipation
ideal diode model
junction capacitance
knee voltage
light-emitting diode
    (LED)
maximum ratings
maximum zener current
model
multicolor LED
nominal zener voltage
parameter

peak repetitive reverse
    voltage
peak reverse voltage
peak surge current
power derating factor
reverse breakdown
    voltage
reverse current
reverse saturation current
reverse voltage
selector guides
specification (spec) sheet
static reverse current

surface-leakage current     zener breakdown     zener knee current
troubleshooting     zener diode     zener test current
voltage regulator     zener impedence     zener voltage

## Equation Summary

| Equation Number | Equation | Section Number |
|---|---|---|
| (2.1) | $V_R = V_S - 0.7\text{V}$ | 2.3 |
| (2.2) | $I_T = \dfrac{V_S - 0.7\text{V}}{R_1}$ | 2.3 |
| (2.3) | $\% \text{ of error} = \dfrac{|X - X'|}{X} \times 100$ | 2.3 |
| (2.4) | $I_{F(\text{max})} = \dfrac{P_{D(\text{max})}}{V_F}$ | 2.4 |
| (2.5) | $V_F = 0.7\text{ V} + I_F R_B$ | 2.5 |
| (2.6) | $I_R = I_S + I_{SL}$ | 2.5 |
| (2.7) | $I'_R = I_R (2^X)$ | 2.5 |
| (2.8) | $Z_Z = \dfrac{\Delta V_Z}{\Delta I_Z} \mid \Delta V_Z = \text{the change in } V_Z$ | 2.7 |
| (2.9) | $I_{ZM} = \dfrac{P_{D(\text{max})}}{V_Z}$ | 2.8 |
| (2.10) | $R_S = \dfrac{V_{\text{out(pk)}} - V_F}{I_F}$ | 2.9 |

## Answers to the Example Practice Problems

**2.1.** The diode acts as an open, and an open drops the full applied voltage. Therefore, $V_{R1} = V_{R2} = 0$ V.

**2.2.** Using the ideal diode model, $I_T = 15$ mA.

**2.4.** $I_T = 8.43$ mA.

**2.6.** $I_T = 5.75$ mA.

**2.7.** $I_T$ (ideal) $= 7.5$ mA: % of error $= 30.4\%$.

**2.8.** Any diode with $V_{RRM} > 210$ V. Therefore, you could use any diode from 1N4004–07.

**2.9.** The value of $I_T$ for the circuit is 1.95 A. The rating for the diode would have to be 20% greater than this value, or 2.34A.

**2.10.** $P_{D(\text{min})} = 198.7$ mW $\times 1.20 = 238.4$ mW

**2.11.** $P_D = (750$ mA$) (0.7$ V$) = 525$ mW. This exceeds the 500-mW limit.

**2.12.** 0.796 V (or 796 mV).

**2.13.** x = 3.5, $I'_R = 22.63\mu$A

**2.14.** $I_{ZM} = 37$ mA.

**2.15.** $P_D$ (at 125°C) $= 250.25$ mW.

**2.16.** The 1N5374A.

**2.17.** $P_D$ for the device is 1.19 W. Therefore, any diode in the 6.8-V row with a $P_D$ of 1.5 W or higher could be used. (The 1N5921A or 1N5342A could be used.)

**2.18.** $R_S = 1.1$ kΩ minimum standard value; the calculated value is 1050 Ω.

FIGURE 2.44

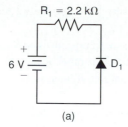

(a)

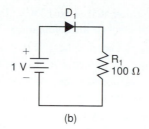

(b)

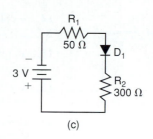

(c)

## §2.1

**Practice Problems**

1. Draw a circuit containing a dc voltage source, a resistor, and a forward-biased diode.

2. Add an arrow to the circuit you drew in Problem 1 to indicate the direction of diode current.

3. Draw a circuit containing a dc voltage source, a resistor, and a reverse-biased diode.

4. For each of the circuits shown in Figure 2.44, determine the direction (if any) of diode forward current.

5. For each of the circuits shown in Figure 2.45, determine the direction (if any) of diode forward current.

## §2.2

6. Using the *ideal diode model*, determine the voltage drop across each of the diodes in Figure 2.44.

7. Using the *ideal diode model*, determine the voltage drop across each of the components in Figure 2.45a.

## §2.3

8. Using the *practical diode model*, determine the values of $V_{D1}$, $V_{R1}$, and $I_T$ for the circuit shown in Figure 2.44a.

9. Using the *practical diode model*, determine the values of $V_{D1}$, $V_{R1}$, and $I_T$ for the circuit shown in Figure 2.44b.

10. Using the *practical diode model*, determine the values of $V_{D1}$, $V_{R1}$, $V_{R2}$, and $I_2$ for the circuit shown in Figure 2.44c.

11. Determine the values of $V_{D1}$, $V_{R1}$, $I_1$, $V_{D2}$, $V_{R2}$, and $I_2$ for the circuit shown in Figure 2.45a.*

12. Determine the values of $V_{D1}$, $V_{D2}$, $V_{R1}$, and $I_T$ for the circuit shown in Figure 2.45b.

13. Determine the values of $V_{D1}$, $V_{D2}$, $V_{R1}$, $V_{R2}$, and $I_T$ for the circuit shown in Figure 2.45c.

*From now on, the practical diode model will be assumed unless indicated otherwise.

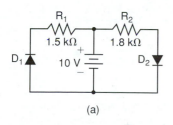

(a)

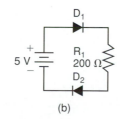

(b)

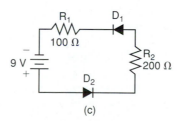

(c)

FIGURE 2.45

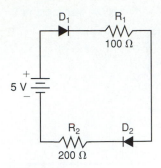

FIGURE 2.46

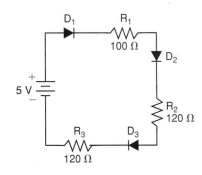

FIGURE 2.47

14. A voltage is calculated to be 12.8 V. The measured voltage is 13.2 V. What is the percentage of error in the calculation?

15. A current is calculated to be 750 $\mu$A. The measured current is 880 $\mu$A. Determine whether or not the percentage of error in the calculation is acceptable.

16. A voltage is calculated to be 144 mV. The measured voltage is 160 mV. Determine whether or not the percentage of error in the calculation is acceptable.

17. Calculate the value of $V_{R2}$ for the circuit shown in Figure 2.46. When measured, $V_{R2}$ is found to be 2.54 V. What is the percentage of error between your calculated value of $V_{R2}$ and the measured value?

18. Calculate the value of $V_{R2}$ for the circuit shown in Figure 2.47. When measured, $V_{R2}$ is found to be 970 mV. What is the percentage of error between your calculated value of $V_{R2}$ and the measured value?

### §2.4

19. What is the minimum required peak reverse voltage rating for the diode in Figure 2.48a?

20. What is the minimum required peak reverse voltage rating for the diode in Figure 2.48b? (*Hint*: Don't forget how to work with voltage dividers!)

21. It was stated in section 2.4 that practical $V_{RRM}$ ratings are usually multiples of 50 or 100 V. With this in mind, what would be the minimum acceptable $V_{RRM}$ rating for the diode in Figure 2.48c?

22. What is the minimum acceptable average forward current rating for the diode shown in Figure 2.49?

23. What is the minimum acceptable average forward power dissipation rating for the diode shown in Figure 2.49?

24. A diode has a $P_{D(MAX)}$ rating of 1.2 W. What is the maximum allowable value of forward current for the device?

25. A diode has a $P_{D(MAX)}$ rating of 750 mW. What is the maximum allowable value of forward current for the device?

### §2.5

26. A small-signal diode (silicon) has a forward current of 10 mA and a bulk resistance $(R_B)$ of 5 $\Omega$. What is the actual value of $V_F$ for the device?

27. A small-signal diode (silicon) has a forward current of 8.2 mA and a bulk resistance $(R_B)$ of 12 $\Omega$. What is the actual value of $V_F$ for the device?

28. A diode (silicon) has a bulk resistance of 20 $\Omega$. At what value of $I_F$ will the value of $V_F$ actually *equal* 0.8 V?

29. Refer to Figure 2.48a. The diode in the circuit has a maximum rated value of $I_R =$ 10 $\mu$A at 25°C. Assuming that $I_R$ reaches its maximum value at each negative peak of the input cycle, what value of voltage will be measured across $R_1$ when $I_R$ peaks? (Assume that $T = 25°C$.)

30. The circuit described in Problem 29 is operated at a temperature of 75°C. What is the voltage across $R_1$ when $I_R$ peaks?

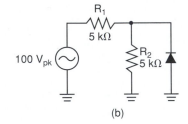

(a)

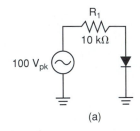

(b)

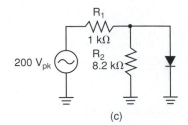

(c)

FIGURE 2.48

**31.** A 150-$\Omega$ resistor is in series with a diode that has a rating of $I_R = 20$ μA at 25°C. What reverse voltage will be measured across the resistor if the diode is reversed biased and $T = 120$°C? (Assume that $I_R$ reaches its maximum possible value.)

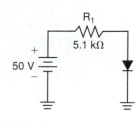

**FIGURE 2.49**

### §2.6

**32.** Refer to the diode spec sheet shown in Figure 2.50. In terms of maximum reverse voltage ratings, which of the diodes listed could be used in the circuit shown in Figure 2.48a?

**33.** Refer to the spec sheet shown in Figure 2.50. What is the maximum value of $I_R$ for the 1N5398 at $T = 150$°C?

**34.** Refer to the spec sheet in Figure 2.50. What is the surge current rating for the 1N5391?

**35.** Refer to Figure 2.27. A circuit has an average forward current of 24.5 A and a 225-$V_{pk}$ source. Which of the diodes listed has the *minimum* acceptable ratings for use in this circuit?

**36.** Refer to Figure 2.27. A circuit has an average forward current of 3.6 A and a 170-$V_{pk}$ source. Which diode has the *minimum* acceptable ratings for use in this circuit?

**37.** Refer to Figure 2.27. A circuit has an average forward power dissipation (for the diode) of 2.8 W and a 470-$V_{pk}$ source. Which diode has the *minimum* acceptable ratings for use in this circuit?

### §2.7

**38.** A zener diode spec sheet lists values of $I_{ZT} = 20$ mA and $I_{zt} = 1$ mA. If the measured change in $V_Z$ (at $I_{zt}$) is 25 mV, what is the value of zener impedance for the device?

**39.** Refer to Figure 2.51. In each circuit, determine whether or not the biasing voltage has the correct polarity for *normal* zener operation.

**40.** For each of the properly biased zener diodes in Figure 2.51, draw an arrow indicating the direction of zener current.

### §2.8

**41.** A 6.8-V zener diode has a $P_{D(MAX)}$ rating of 1 W. What is the value of $I_{ZM}$ for the device?

**42.** A 24-V zener diode has a $P_{D(MAX)}$ rating of 10 W. What is the value of $I_{ZM}$ for the device?

**43.** A zener diode with a $P_{D(MAX)}$ rating of 5 W has a derating factor of 8 mW/°C above 50°C. What is the maximum allowable value of $P_D$ for the device if it is operating at 120°C?

**44.** The MLL4678 zener diode has a $P_{D(MAX)}$ rating of 250 mW and a derating factor of 1.67 mW/°C above 50°C. What is the maximum allowable value of $P_D$ for the device if it is operating at 150°C?

**45.** Refer to Figure 2.34. Which of the diodes could be used in place of a 28-V zener diode that has a maximum power dissipation of 1.8 W?

**46.** Refer to Figure 2.34. Which of the diodes listed could be used in place of a 6.8-V zener diode that has a maximum power dissipation of 1.2 W?

**47.** Refer to Figure 2.34. Which of the diodes listed could be used in place of a 12-V zener diode that has a maximum operating current of 150 mA?

### §2.9

**48.** An LED has a range of $V_F = 1.5$ to 1.8 V and $I_{F(desired)} = 18$ mA. If the LED is driven by a 20-$V_{pk}$ source, what standard value of current-limiting resistor is needed to protect the LED?

**1N5391
thru
1N5399**

## Designers Data Sheet

### "SURMETIC" RECTIFIERS

. . . subminiature size, axial lead-mounted rectifiers for general-purpose, low-power applications.

**Designers Data for "Worst Case" Conditions**

The Designers Data Sheets permit the design of most circuits entirely from the information presented. Limits curves—representing boundaries on device characteristics—are given to facilitate "worst-case" design.

### LEAD-MOUNTED SILICON RECTIFIERS

**50–1000 VOLTS
DIFFUSED JUNCTION**

### *MAXIMUM RATINGS

| Rating | Symbol | 1N5391 | 1N5392 | 1N5393 | 1N5395 | 1N5397 | 1N5398 | 1N5399 | Unit |
|---|---|---|---|---|---|---|---|---|---|
| Peak Repetitive Reverse Voltage<br>Working Peak Reverse Voltage<br>DC Blocking Voltage | $V_{RRM}$<br>$V_{RWM}$<br>$V_R$ | 50 | 100 | 200 | 400 | 600 | 800 | 1000 | Volts |
| Nonrepetitive Peak Reverse Voltage<br>(Halfwave, Single Phase, 60 Hz) | $V_{RSM}$ | 100 | 200 | 300 | 525 | 800 | 1000 | 1200 | Volts |
| RMS Reverse Voltage | $V_{R(RMS)}$ | 35 | 70 | 140 | 280 | 420 | 560 | 700 | Volts |
| Average Rectified Forward Current<br>(Single Phase, Resistive Load,<br>60 Hz, $T_L$ = 70°C,<br>1/2" From Body) | $I_O$ | ← 1.5 → | | | | | | | Amp |
| Nonrepetitive Peak Surge Current<br>(Surge Applied at Rated Load<br>Conditions, See Figure 2) | $I_{FSM}$ | ← 50 (for 1 cycle) → | | | | | | | Amp |
| Storage Temperature Range | $T_{stg}$ | ← −65 to +175 → | | | | | | | °C |
| Operating Temperature Range | $T_L$ | ← −65 to +170 → | | | | | | | °C |
| DC Blocking Voltage Temperature | $T_L$ | ← 150 → | | | | | | | °C |

### *ELECTRICAL CHARACTERISTICS

| Characteristic and Conditions | Symbol | Typ | Max | Unit |
|---|---|---|---|---|
| Maximum Instantaneous Forward Voltage Drop<br>($i_F$ = 4.7 Amp Peak, $T_L$ = 170°C,<br>1/2 Inch Leads) | $v_F$ | — | 1.4 | Volts |
| Maximum Reverse Current (Rated dc Voltage)<br>($T_L$ = 150°C) | $I_R$ | 250 | 300 | µA |
| Maximum Full-Cycle Average Reverse Current (1)<br>($I_O$ = 1.5 Amp, $T_L$ = 70°C, 1/2 Inch Leads) | $I_{R(AV)}$ | — | 300 | µA |

*Indicates JEDEC Registered Data.

NOTE 1: Measured in a single-phase, halfwave circuit such as shown in Figure 6.25 of EIA RS-282, November 1963. Operated at rated load conditions $I_O$ = 1.5 A, $V_r$ = $V_{RWM}$, $T_L$ = 70°C.

### MECHANICAL CHARACTERISTICS

**CASE:** Transfer molded plastic

**MAXIMUM LEAD TEMPERATURE FOR SOLDERING PURPOSES:** 240°C,
1/8" from case for 10 seconds at 5 lbs. tension

**FINISH:** All external surfaces are corrosion-resistant, leads are readily solderable

**POLARITY:** Cathode indicated by color band

**WEIGHT:** 0.40 grams (approximately)

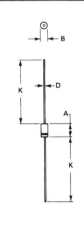

NOTES:
1. ALL RULES AND NOTES ASSOCIATED WITH JEDEC DO-41 OUTLINE SHALL APPLY.
2. POLARITY DENOTED BY CATHODE BAND.
3. LEAD DIAMETER NOT CONTROLLED WITHIN "F" DIMENSION.

| DIM | MILLIMETERS | | INCHES | |
|---|---|---|---|---|
| | MIN | MAX | MIN | MAX |
| A | 5.97 | 6.60 | 0.235 | 0.260 |
| B | 2.79 | 3.05 | 0.110 | 0.120 |
| D | 0.76 | 0.86 | 0.030 | 0.034 |
| K | 27.94 | — | 1.100 | — |

**CASE 59-04
PLASTIC**

**FIGURE 2.50** (Copyright of Motorola. Used by permission.)

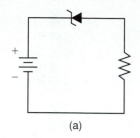 (a)

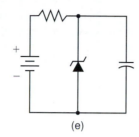 (b)

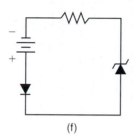 (c)

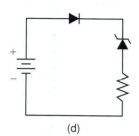 (d)

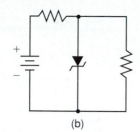 (e)

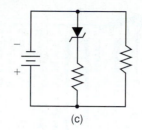 (f)

**FIGURE 2.51**

**49.** An LED has a range of $V_F$ = 1.6 to 2.0 V and $I_{F(desired)}$ = 20 mA. Determine the minimum standard resistor value that could be used as a current-limiting resistor if the LED is driven by a 32-$V_{pk}$ source.

**50.** The following table lists the results of testing several diodes. In each case, determine whether the diode is good, open, or shorted.

| | Forward Resistance | Reverse Resistance |
|---|---|---|
| a. | 1200 MΩ | 1200 MΩ |
| b. | 15 Ω | 3500 MΩ |
| c. | 75 Ω | 175 MΩ |
| d. | 30 Ω | 50 Ω |

**51.** Refer to Figure 2.48a. When the output from the source is positive, the peak voltage across $R_1$ is approximately 100 V. When the output from the source is negative, the peak voltage across $R_1$ is approximately 0 V. Is the diode good, open, or shorted? Explain your answer.

**52.** Refer to Figure 2.48a. The voltage across $R_1$ is always equal to the peak source voltage. Is the diode good, open, or shorted? Explain your answer.

**53.** Refer to Figure 2.48a. The voltage across $R_1$ is 0 V, regardless of the value of the source voltage. Is the diode good, open, or shorted? Explain your answer.

**54.** The spec sheet for the MLL755 zener diode is shown in Figure 2.52. What is the value of $I_{ZM}$ for the device when it is operated at 150°C?

**55.** The MLL756 zener diode *cannot* be used in the circuit shown in Figure 2.53. Why not? (*Note:* The temperature range shown is the normal operating temperature for the circuit.)

## 500 MILLIWATT HERMETICALLY SEALED GLASS SILICON ZENER DIODES

- Complete Voltage Range — 2.4 to 110 Volts
- Leadless Package for Surface Mount Technology
- Double Slug Type Construction
- Metallurgically Bonded Construction
- Nitride Passivated Die
- Available in 8 mm Tape and Reel
- T1 Cathode Facing Sprocket Holes
- T2 Anode Facing Sprocket Holes

**MLL746 thru MLL759**

**MLL957A thru MLL986A**

**MLL4370 thru MLL4372**

**LEADLESS GLASS ZENER DIODES**
500 MILLIWATTS
2.4-110 VOLTS

### MAXIMUM RATINGS

| Rating | Symbol | Value | Unit |
|---|---|---|---|
| DC Power Dissipation @ $T_A \le 50°C$ Derate above $T_A = 50°C$ | $P_D$ | 500 3.3 | mW mW/°C |
| Operating and Storage Junction Temperature Range | $T_J, T_{stg}$ | −65 to +200 | °C |

### MECHANICAL CHARACTERISTICS

**CASE:** Double slug type, hermetically sealed glass.

**MAXIMUM LEAD TEMPERATURE FOR SOLDERING PURPOSES:** 230°C, for 10 seconds

**FINISH:** All external surfaces are corrosion resistant and readily solderable.

**POLARITY:** Cathode indicated by color band. When operated in zener mode, cathode will be positive with respect to anode

**MOUNTING POSITION:** Any

STEADY STATE POWER DERATING

CASE 362-01 GLASS

| DIM | MILLIMETERS MIN | MILLIMETERS MAX | INCHES MIN | INCHES MAX |
|---|---|---|---|---|
| A | 3.30 | 3.70 | 0.130 | 0.146 |
| B | 1.60 | 1.70 | 0.063 | 0.067 |
| R | 2.49 | 2.59 | 0.098 | 0.102 |
| U | 0.41 | 0.55 | 0.016 | 0.022 |

## ELECTRICAL CHARACTERISTICS ($T_A$ = 25°C, $V_F$ = 1.5 V Max @ 200 mA for all types)

| Type Number (Note 1) | Nominal Zener Voltage $V_Z$ @ $I_{ZT}$ (Notes 1,2,3) Volts | Test Current $I_{ZT}$ (Note 2) mA | Maximum Zener Impedance $Z_{ZT}$ @ $I_{ZT}$ (Note 4) Ohms | Maximum DC Zener Current $I_{ZM}$ mA | | Maximum Reverse Leakage Current | |
|---|---|---|---|---|---|---|---|
| | | | | | | $T_A$ = 25°C $I_R$ @ $V_R$ = 1 V $\mu$A | $T_A$ = 150°C $I_R$ @ $V_R$ = 1 V $\mu$A |
| MLL4370 | 2.4 | 20 | 30 | 150 | 190 | 100 | 200 |
| MLL4371 | 2.7 | 20 | 30 | 135 | 165 | 75 | 150 |
| MLL4372 | 3.0 | 20 | 29 | 120 | 150 | 50 | 100 |
| MLL746 | 3.3 | 20 | 28 | 110 | 135 | 10 | 30 |
| MLL747 | 3.6 | 20 | 24 | 100 | 125 | 10 | 30 |
| MLL748 | 3.9 | 20 | 23 | 95 | 115 | 10 | 30 |
| MLL749 | 4.3 | 20 | 22 | 85 | 105 | 2 | 30 |
| MLL750 | 4.7 | 20 | 19 | 75 | 95 | 2 | 30 |
| MLL751 | 5.1 | 20 | 17 | 70 | 85 | 1 | 20 |
| MLL752 | 5.6 | 20 | 11 | 65 | 80 | 1 | 20 |
| MLL753 | 6.2 | 20 | 7 | 60 | 70 | 0.1 | 20 |
| MLL754 | 6.8 | 20 | 5 | 55 | 65 | 0.1 | 20 |
| MLL755 | 7.5 | 20 | 6 | 50 | 60 | 0.1 | 20 |
| MLL756 | 8.2 | 20 | 8 | 45 | 55 | 0.1 | 20 |
| MLL757 | 9.1 | 20 | 10 | 40 | 50 | 0.1 | 20 |
| MLL758 | 10 | 20 | 17 | 35 | 45 | 0.1 | 20 |
| MLL759 | 12 | 20 | 30 | 30 | 35 | 0.1 | 20 |

**FIGURE 2.52** (Copyright of Motorola. Used by permission.)

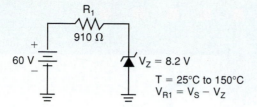

$R_1$

$910\ \Omega$

$60\ V$

$V_Z = 8.2\ V$

$T = 25°C\ to\ 150°C$

$V_{R1} = V_S - V_Z$

**FIGURE 2.53**

56. Write a program to determine the total voltage drop across a forward-biased *pn*-junction diode. The program must take into account the type of semiconductor material used, and the values of $V_k$, $I_F$, and $R_B$.

57. Write a program to determine the total voltage drop across a zener diode when $V_Z$, $Z_Z$, and $I_Z$ are known.

58. Write a program that will determine whether a given *pn*-junction diode is good, open, or shorted when provided with forward and reverse resistance readings.

59. Write a program that will determine the value of $I_{ZM}$ at specified values of $V_Z$, $P_{D(MAX)}$, power derating factor, and temperature.

*Suggested Computer Applications Problems*

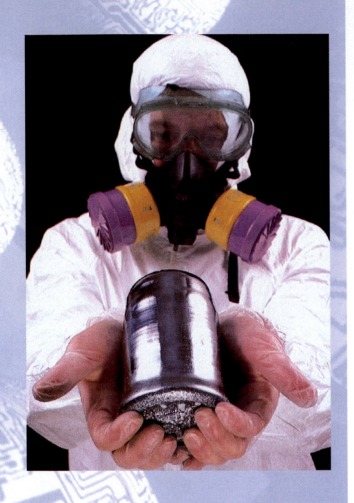

# 3

# COMMON DIODE APPLICATIONS

## Basic Power Supply Circuits

 rom that point on, research centered around the use of silicon rather than germanium."

# OUTLINE

# OBJECTIVES

*After studying the material in this chapter, you should be able to:*

1. Briefly describe the purpose served by a power supply and the function of each circuit it contains.

2. Calculate the *peak* and dc (*average*) load voltage and current values for a *positive* half-wave rectifier.

3. Describe the operation of the full-wave rectifier.

4. Calculate the values of *peak* and dc (*average*) load voltage and current for any full-wave rectifier.

5. Describe the operation of the bridge rectifier.

6. Calculate the *peak* and dc (*average*) load voltage and current values for a bridge rectifier.

7. Discuss the effects that *filtering* has on the output of a rectifier.

8. Describe the operation of the basic capacitive filter.

9. Discuss the reason that full-wave rectifiers are preferred over half-wave rectifiers.

10. Calculate the values of $I_Z$, $I_L$, and/or $I_T$ for a zener voltage regulator, given the needed circuit values and diode ratings.

11. Calculate the values of $V_{dc}$, $I_L$, and $V_{r(out)}$ for a basic power supply.

12. List the faults that commonly occur in a basic power supply and the symptoms of each.

# THE ROLE OF POWER SUPPLIES IN SEMICONDUCTOR DEVELOPMENT

Power supplies are the most commonly used circuits in electronics. Virtually every electronic system requires the use of a power supply to convert the ac line voltage to the dc voltages needed for the system's internal operation.

In addition to being the most commonly used type of circuit, the power supply also played a major role in the development of today's electronic devices. Early power supplies used vacuum tubes to *rectify* ac, that is, to convert ac to pulsating dc. These vacuum tubes wasted a tremendous amount of power.

Early semiconductor research was centered around the use of *germanium*, a semiconductor material that cannot withstand any significant amount of current and heat. With the development of the commercial *pn*-junction diode in 1954, researchers turned to the problem of developing rectifier diodes, diodes that could withstand large current values and the heat produced by those currents.

In 1955, *silicon pn*-junction rectifier diodes had been developed that could handle current values up to 2 amperes. From that point on, research centered around the use of silicon rather than germanium. At this point in time, silicon is used for most semiconductor applications. Germanium is rarely used in the production of semiconductor devices.

---

I t would take several volumes to discuss every diode application in modern electronics. In this chapter, we will concentrate on the most common diode application, the power supply. In Chapter 4, we will look at several additional diode applications.

OBJECTIVE 1 ▶

**Power supply**
Converts ac to dc.

The **power supply** of an electronic system is used to convert the ac energy provided by the wall outlet to dc energy. The power cord of any electronic system supplies the line power to the system power supply, which then provides all internal dc voltages needed for proper circuit operation.

There are two basic types of power supplies. The *linear* power supply is the original and simpler of the two, and it is the focus of this chapter. The *switching* power supply is a newer and more complex circuit that is used in an increasing number of electronic systems. Switching supplies are covered in Chapter 21.

The basic power supply can be broken down into four circuit groups, as shown in Figure 3.1. The incoming ac line voltage is usually applied to a *transformer*. This transformer may either step up or step down the line voltage, depending on the needs of the power supply. The alternating voltage out of the transformer is then applied to a rectifier. A **rectifier** is a diode circuit that converts the ac to what is called *pulsating dc*. This pulsating dc is then applied to a *filter*, which reduces the variations in dc voltage. The **filter** is usually made up of passive components, such as resistors, capacitors, and inductors. The final stage is the **voltage regulator**. A voltage regulator is used to maintain a constant output voltage.

**Rectifier**
A circuit that converts ac to pulsating dc.

**Filter**
A circuit that reduces the variations in the output of a rectifier.

**Voltage regulator**
A circuit used to maintain a constant output voltage.

At one time, voltage regulators were designed using zener diodes as the regulating element. However, the development of the IC (integrated-circuit) voltage regulator has led to the replacement of zener diodes as regulating elements. IC voltage regulators are far more efficient than zener diodes, so power supply design currently emphasizes their use. At the same time, zener regulators are easier to understand and they serve as valuable educational circuits. We will therefore concentrate on the zener regulator in this chapter. The IC voltage regulator is covered in detail in Chapter 21.

## 3.1 TRANSFORMERS

Transformers are not considered solid-state devices, but they *do* play an integral role in the operation of most power supplies. Therefore, we will begin our discussion on power supply operation by reviewing the basics of transformer operation.

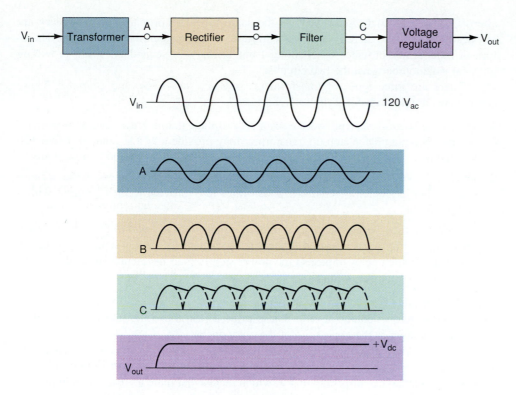

FIGURE 3.1   Basic power supply block diagram and waveforms.

The basic schematic symbol for the transformer is shown in Figure 3.2a. The component consists of two windings, called the *primary* and the *secondary*. The input to the transformer is applied to the primary and the output is taken from the secondary.

Transformers are made up of inductors that are in close proximity to each other, yet are not physically connected. An alternating voltage applied to the primary induces an

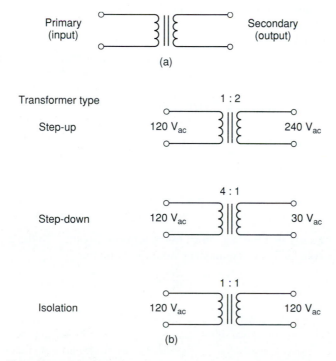

FIGURE 3.2   Transformer symbols.

alternating voltage in the secondary. At the same time, the primary and secondary are physically isolated, so there is no actual current transfer between the two circuits. Therefore, a transformer provides ac coupling from primary to secondary, while providing physical isolation between the two circuits.

There are three types of transformers: *step-up, step-down,* and *isolation.* These components are described as follows:

1. The *step-up transformer* provides a secondary voltage that is *greater than* the primary voltage. For example, a step-up transformer may provide a 240 $V_{ac}$ output with a 120 $V_{ac}$ input.

2. The *step-down transformer* provides a secondary voltage that is *less than* the primary voltage. For example, a step-down transformer may provide a 30 $V_{ac}$ output with a 120 $V_{ac}$ input.

3. An *isolation transformer* provides an output voltage that is *equal to* the input voltage. This type of transformer is used to electrically isolate the power supply from the ac power line, which helps to protect the power supply (and the technician who is working on it) from the line voltage.

Typical schematics for each of the above are provided in Figure 3.2b. As you can see, each of the symbols contains a ratio that is printed just above the transformer symbol. This ratio is known as the *turns ratio* of the component.

The turns ratio of a transformer is *the ratio of the number of turns in the primary to the number of turns in the secondary.* For example, the step-down transformer in Figure 3.2b is shown to have a turns ratio of 4:1, which means that there are four turns in the primary for each turn in the secondary.

The turns ratio of a transformer is equal to the voltage ratio of the component. By formula,

$$\frac{N_2}{N_1} = \frac{V_2}{V_1} \qquad \text{(3.1)}$$

where $N_2$ = the number of turns in the secondary
$N_1$ = the number of turns in the primary
$V_2$ = the secondary voltage ($V_{ac}$)
$V_1$ = the primary voltage ($V_{ac}$)

Thus, the primary voltage of the step-down transformer in Figure 3.2b is four times as great as the secondary voltage.

## Calculating Secondary Voltage

When the turns ratio and primary voltage of a transformer are known, the secondary voltage can be found as

$$V_2 = \frac{N_2}{N_1}V_1 \qquad \text{(3.2)}$$

For example, let's say that the step-down transformer in Figure 3.2b has a 120 $V_{ac}$ input. The secondary voltage for the component would be found as

$$V_2 = \frac{N_2}{N_1}V_1$$

$$= \frac{1}{4}(120 \text{ V}_{ac})$$

$$= 30 \text{ V}_{ac}$$

As you can see, the primary voltage (120 $V_{ac}$) is four times the secondary voltage (30 $V_{ac}$).

## Calculating Secondary Current

Ideally, transformers are 100 percent efficient. This means that the ideal transformer transfers 100 percent of its input power to the secondary. By formula,

$$P_2 = P_1$$

Since power equals the product of voltage and current,

$$V_2 I_2 = V_1 I_1$$

and

$$\frac{I_1}{I_2} = \frac{V_2}{V_1} \qquad\qquad (3.3)$$

As you can see, the current ratio is the inverse of the voltage ratio. This means that

1. For a step-down transformer, $I_2 > I_1$.
2. For a step-up transformer, $I_2 < I_1$.

In other words, current varies (from primary to secondary) in the opposite way that voltage varies. If voltage increases, current decreases, and vice versa.

Since the voltage ratio of a transformer is equal to its turns ratio, equation (3.3) can be rewritten as

$$\frac{I_1}{I_2} = \frac{N_2}{N_1}$$

or

$$I_2 = \frac{N_1}{N_2} I_1 \qquad\qquad (3.4)$$

The following example demonstrates a practical application of this relationship.

---

The fuse in Figure 3.3 is used to limit the current in the primary of the transformer. Assuming that the fuse limits the value of $I_1$ to 1 A, what is the limit on the value of the secondary current?

EXAMPLE 3.1

**Solution:** The maximum secondary current is found using the limit on $I_1$ and the turns ratio of the transformer, as follows:

$$I_2 = \frac{N_1}{N_2} I_1$$

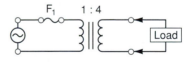

**FIGURE 3.3**

$$= \frac{1}{4}\,(1\text{ A})$$

$$= 250\text{ mA}$$

If the secondary current tries to exceed the 250 mA limit, the primary current will exceed its limit and blow the fuse.

### PRACTICE PROBLEM 3.1

A circuit like the one in Figure 3.3 has a turns ratio of 1:12 and a fuse that limits the primary current to 250 mA. Calculate the maximum allowable value of $I_2$.

**FIGURE 3.4**

The current relationships developed here have been based on the idea that a transformer is 100 percent efficient. In reality, there are a number of losses within a transformer that cause the secondary power to be somewhat lower than the primary power. However, the difference between primary power and secondary power is small enough to have little effect on the relationships covered in this section.

### Transformer Input/Output Phase Relationships

Some transformers provide a 180° phase shift from input to output; others do not. The input/output phase relationship of a transformer is not an important factor in the analysis of a power supply. However, you should be aware that a transformer with a 180° phase shift is identified as follows: In the schematic symbol, there are two dots: one on the top side of the primary and one on the bottom side of the secondary. When you see these dots, you are working with a transformer whose output voltage is 180° out of phase with its input voltage, as shown in Figure 3.4.

### Transformer Ratings

Some manufacturers' catalogs rate transformers by their turns ratios, while others list them by *secondary voltage ratings*. For example, a transformer may be listed as a 40 $V_{ac}$ transformer. When this rating is used, *it indicates the ac secondary voltage produced by a 120 $V_{ac}$ input to the primary*. In other words, it gives you the rms output from the transformer when it is supplied by a standard 120 $V_{ac}$ line input. Throughout this chapter, we will use both of these methods of rating transformers.

---

**Section Review**

1. What are names of the transformer input and output circuits?
2. List and describe the three types of transformers.
3. What is the *turns ratio* of a transformer?
4. Describe the relationship between the turns ratio of a transformer and its input and output voltages.
5. Describe the relationship between the transformer primary and secondary power.
6. Describe the relationship between the voltage ratio of a transformer and its current ratio.
7. Describe the two means by which transformers are normally listed in parts catalogs.

---

## 3.2 HALF-WAVE RECTIFIERS

There are three basic types of rectifier circuits: the *half-wave, full-wave,* and *bridge* rectifiers. Of the three, the bridge rectifier is the most commonly used, followed by the full-wave rectifier. Our discussion on rectifiers begins with the half-wave rectifier simply because it is the easiest to understand.

What does a half-wave rectifier do?     The half-wave rectifier is made up of a *diode* and a *resistor,* as shown in Figure 3.5. The half-wave rectifier is used to eliminate either the negative alternation of the input or

FIGURE 3.5    Ideal half-wave rectifier operation.

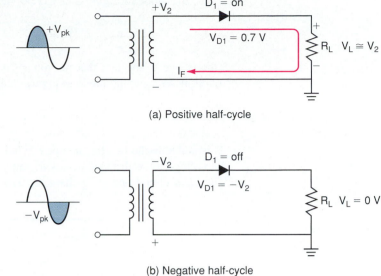

(a) Positive half-cycle

(b) Negative half-cycle

the positive alternation of the input. As you will see, the diode direction determines which half-cycle will be eliminated.

Because the half-wave rectifier configuration requires the least number of parts, it is the cheapest to produce. However, it is the least efficient of the three different types of rectifiers. Because of this, it is normally used for noncritical, low-current applications.

## Basic Circuit Operation

The negative half-cycle of the input to the rectifier in Figure 3.5 is eliminated by the one-way conduction of the diode. Figure 3.6 details the operation of the circuit for one complete cycle of the input signal. During the positive half-cycle of the input, $D_1$ is forward biased and provides a path for current. This allows a voltage ($V_L$) to be developed across $R_L$ that is approximately equal to the voltage across the secondary of the transformer ($V_2$).

When the polarity of the input signal reverses, $D_1$ is reverse biased, preventing conduction in the circuit. With no current through $R_L$, no voltage is developed across the load. In this case, the output voltage remains at approximately 0 V, and the voltage across the diode ($V_D$) is approximately equal to $V_2$.

If all this seems confusing, remember the *ideal diode model*. This diode model was shown in Chapter 2 to represent the component as either an *open* (reverse-biased) or *closed* (forward-biased) switch. When forward biased, this ideal switch drops no voltage. When reverse biased, this ideal switch drops all the applied voltage.

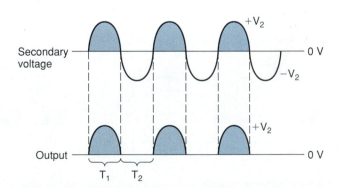

**Lab Reference:** These waveforms are observed in Exercise 4.

FIGURE 3.6    Input and output waveforms.

The diode in Figure 3.5a can be viewed as a closed switch. Therefore:

$$V_L \cong V_2 \qquad \text{(forward operation)} \qquad \textbf{(3.5)}$$

This is due to the fact that no voltage is dropped across a closed switch. By the same token, if we view the diode in Figure 3.5b as an *open* switch, we see that

$$V_{D1} = V_2 \qquad \text{(reverse operation)} \qquad \textbf{(3.6)}$$

This is due to the fact that an open switch drops all the applied voltage. This would leave no voltage to be dropped across $R_L$, and $V_L$ would equal 0 V, as is shown in the figure. The *ideal* circuit operating characteristics are summarized as follows:

| Diode Condition | $V_{D1}$ | $V_L$ |
|---|---|---|
| Forward biased | 0 V | Equal to $V_2$ |
| Reverse biased | Equal to $V_2$ | 0 V |

Using these relationships, it is easy to understand the input/output waveforms shown in Figure 3.6. During $T_1$, the diode is forward biased, and the output ($V_L$) is approximately equal to the input ($V_2$). During $T_2$, the diode is reverse biased, and the output drops to 0 V. This is the way half-wave rectifiers work.

## Negative Half-Wave Rectifiers

A *Practical Consideration:* When we talk about the *positive* and *negative* half-cycles of the input, we are referring to the polarity of the transformer secondary voltage measured from the top of the secondary to the bottom of the secondary.

Determining the output polarity of a half-wave rectifier.

Figure 3.7 shows a half-wave rectifier with the diode direction reversed. In this circuit, the diode will conduct on the *negative* half-cycle of the input, and equation (3.5) will apply. The diode will be reverse biased on the *positive* half-cycle of the input, and equation (3.6) will apply. As a result, the positive half-cycle of the input is eliminated. Note that the operating principles for this circuit are *exactly* the same as those for the *positive* half-wave rectifier shown in Figure 3.5. The only difference is that the polarity of the output has been reversed.

As you can see, the *direction* of the diode determines whether the output from the rectifier is positive or negative. For circuit recognition, the following statements will generally hold true:

1. When the diode points toward the load ($R_L$), the output from the rectifier will be *positive*.
2. When the diode points toward the transformer, the output from the rectifier will be *negative*.

These two statements will also hold true for the full-wave rectifier. The points made so far about half-wave rectifiers are summarized in Figure 3.8.

**Lab Reference:** The effects of diode direction are demonstrated in Exercise 4.

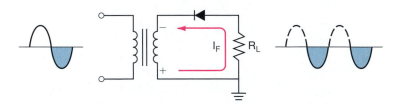

**FIGURE 3.7    Negative half-wave rectifier.**

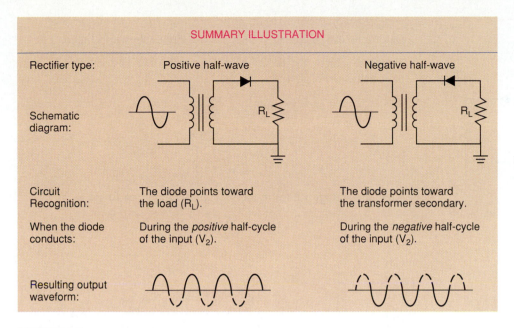

| Rectifier type: | Positive half-wave | Negative half-wave |
|---|---|---|
| Schematic diagram: | | |
| Circuit Recognition: | The diode points toward the load ($R_L$). | The diode points toward the transformer secondary. |
| When the diode conducts: | During the *positive* half-cycle of the input ($V_2$). | During the *negative* half-cycle of the input ($V_2$). |
| Resulting output waveform: | | |

**FIGURE 3.8**

## Calculating Load Voltage and Current Values

In our discussion of the *ideal* half-wave rectifier, we ignored the value of $V_F$ for the diode. When we take this value into account, the *peak load voltage*, $V_{L(pk)}$ is found as

◄ *OBJECTIVE 2*

$$V_{L(pk)} = V_{2(pk)} - V_F \qquad (3.7)$$

You shouldn't have any difficulty with this equation. It is simply a variation on equation (2.1). $V_{2(pk)}$ is the *peak secondary voltage* of the transformer, found as

Equation (2.1):
$V_R = V_S - 0.7$ V

$$V_{2(pk)} = \frac{N_2}{N_1} V_{1(pk)} \qquad (3.8)$$

where       $\dfrac{N_2}{N_1}$ = the ratio of transformer secondary turns to primary turns

$V_{1(pk)}$ = the peak transformer primary voltage

   *A word of caution:* Equation (3.8) assumes that the input to the transformer is given as a *peak* value. More often than not, source voltages are given as rms values. When this is the case, the source voltage can be converted to a peak value as follows:

$$V_{pk} = \frac{V_{rms}}{0.707} \qquad (3.9)$$

Example 3.2 illustrates the procedure for calculating the peak output voltage from a positive half-wave rectifier.

   Here is another practical situation: Most transformers are rated for a specific rms output voltage. For example, a 25 $V_{ac}$ transformer would have an rms output of 25 V when supplied from a 120 V wall outlet. When a transformer has an output voltage rating, simply divide the rated output voltage by 0.707 to obtain the value of $V_{2(pk)}$. This is shown in Example 3.3.

*Note:* You will be shown later in this section how to calculate the output values for a *negative* half-wave rectifier.

## EXAMPLE 3.2

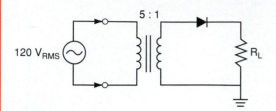

**FIGURE 3.9**

Determine the peak load voltage for the circuit shown in Figure 3.9.

**Solution:** First, the ac input to the transformer is converted to a peak value, as follows:

$$V_{1(pk)} = \frac{V_{1(rms)}}{0.707}$$

$$= \frac{120\ V_{ac}}{0.707}$$

$$= 169.7\ V_{pk}$$

Now, the voltage values in the secondary circuit are found as

$$V_{2(pk)} = \frac{N_2}{N_1}V_{1(pk)}$$

$$= \frac{1}{5}(169.7\ V_{(pk)})$$

$$= 33.94\ V_{pk}$$

and

$$V_{L(pk)} = V_{2(pk)} - V_F$$
$$= 33.94\ V_{pk} - 0.7\ V$$
$$= 33.24\ V_{pk}$$

### PRACTICE PROBLEM 3.2

A half-wave rectifier has values of $N_1 = 10$, $N_2 = 1$, and $V_{1(pk)} = 180\ V_{pk}$. What is the peak load voltage for the circuit?

## EXAMPLE 3.3

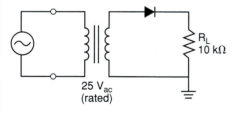

**FIGURE 3.10**

Determine the peak load voltage for the circuit shown in Figure 3.10.

**Solution:** The transformer is shown to have a 25 $V_{ac}$ rating. This value of $V_2$ is converted to peak form as follows:

$$V_{2(pk)} = \frac{V_{2(rms)}}{0.707}$$

$$= \frac{25\ V_{rms}}{0.707}$$

$$= 35.36\ V_{pk}$$

Now the value of $V_{L(pk)}$ is found as

$$V_{L(pk)} = V_{2(pk)} - 0.7\ V$$
$$= 35.36\ V_{pk} - 0.7\ V$$
$$= 34.66\ V_{pk}$$

### PRACTICE PROBLEM 3.3

A 24 $V_{ac}$ transformer is being used in a positive half-wave rectifier. What is the peak load voltage for the circuit?

Once the peak load voltage is determined, the peak load current is found as

$$I_{L(pk)} = \frac{V_{L(pk)}}{R_L}$$
(3.10)

Example 3.4 demonstrates the calculation of peak load current.

EXAMPLE 3.4

What is the peak load current for the circuit shown in Figure 3.11?

**Solution:**  The input voltage is given as an rms value. This value is converted to a *peak* value as follows:

$$V_{1(pk)} = \frac{V_{1(rms)}}{0.707}$$

$$= \frac{200\ V_{ac}}{0.707}$$

$$= 282.9\ V_{pk}$$

Now, the value of $V_{2(pk)}$ is found as

$$V_{2(pk)} = \frac{N_2}{N_1}V_{1(pk)}$$

$$= \frac{1}{5}(282.9\ V_{pk})$$

$$= 56.6\ V_{pk}$$

Finally, the load voltage and current values are found as:

$$V_{L(pk)} = V_{2(pk)} - V_F$$
$$= 56.6\ V_{pk} - 0.7\ V$$
$$= 55.9\ V_{pk}$$

and

$$I_{L(pk)} = \frac{V_{L(pk)}}{R_L}$$

$$= \frac{55.9\ V_{pk}}{10\ k\Omega}$$

$$= 5.59\ mA_{pk}$$

*Question and Answer:* At this point, you may be wondering why we are so concerned with *peak* values. Wouldn't it just be easier to leave everything in *rms* form since ac voltmeters measure *rms*? There are two reasons for dealing with peak voltages: First, we measure peak voltages when using an oscilloscope. Second (and perhaps most important) is the fact that *average* (dc) voltage and current values are easier to determine when peak values are known.

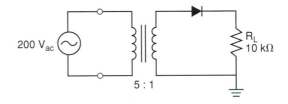

**FIGURE 3.11**

**PRACTICE PROBLEM 3.4**

A circuit like the one shown in Figure 3.11 has values of $N_1 = 12$, $N_2 = 1$, $V_1 = 150\ V_{ac}$, and $R_L = 8.2\ k\Omega$. What is the peak load current for the circuit?

## Average Load Voltage and Current

The **average load voltage**, $V_{ave}$, from an ac circuit indicates *the reading you would get if the voltage was measured with a dc voltmeter*. In other words, $V_{ave}$ is the *dc equivalent* of an ac signal. In most cases, $V_{ave}$ and $V_{dc}$ will be used to describe the same value. Since

**Average load voltage**
The dc equivalent of an ac signal. $V_{ave}$ is measured with a dc voltmeter.

rectifiers are used to convert ac to dc, $V_{ave}$ is a very important value. For a half-wave rectifier, $V_{ave}$ is found as

$$V_{ave} = \frac{V_{pk}}{\pi} \qquad \text{(half-wave rectified)} \qquad \textbf{(3.11)}$$

Another form of this equation is

$$V_{ave} = 0.318(V_{pk}) \qquad \text{(half-wave rectified)} \qquad \textbf{(3.12)}$$

where $0.318 \cong 1/\pi$. Either of these equations can be used to determine the *dc equivalent* load voltage for a *half-wave* rectifier. Example 3.5 demonstrates the process for determining the value of $V_{ave}$ for a half-wave rectifier.

---

## EXAMPLE 3.5

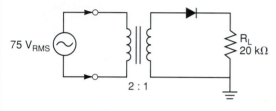

**FIGURE 3.12**

**Lab Reference:** These values are calculated and measured as part of Exercise 4.

Determine the value of $V_{ave}$ for the circuit shown in Figure 3.12.

***Solution:***

$$V_{1(pk)} = \frac{V_{1(rms)}}{0.707}$$

$$= \frac{75\ V_{ac}}{0.707}$$

$$= 106.1\ V_{pk}$$

$$V_{2(pk)} = \frac{N_2}{N_1}V_{1(pk)}$$

$$= \frac{1}{2}(106.1\ V_{pk})$$

$$= 53.04\ V_{pk}$$

$$V_{L(pk)} = 53.04\ V_{pk} - 0.7\ V$$

$$= 52.34\ V_{pk}$$

$$V_{ave} = \frac{V_{pk}}{\pi}$$

$$= \frac{52.34\ V_{pk}}{\pi}$$

$$= 16.66\ V_{dc}$$

***PRACTICE PROBLEM 3.5***

A half-wave rectifier like the one in Figure 3.12 has values of $N_1 = 14$, $N_2 = 1$, and $V_1 = 150\ V_{ac}$. What is the dc load voltage for the circuit?

---

**Average load current**
The dc equivalent current of an ac signal. $I_{ave}$ is measured with a dc ammeter.

Just as we can convert a *peak* voltage to *average* voltage, we can also convert a *peak* current to an *average* current. The value of the **average load current** for an ac waveform is the value that would be measured with a *dc ammeter*. Thus, the value of $I_{ave}$ for a waveform gives us an *equivalent dc current*. This is another very important value that we need to know for any circuit used to convert ac to dc. The value of $I_{ave}$ can be calculated in one of two ways:

1. We can determine the value of $V_{ave}$ and then use Ohm's law as follows:

$$I_{ave} = \frac{V_{ave}}{R_L}$$

2. We can convert $I_{pk}$ to average form using the same basic equations, (3.11) and (3.12), that we used to convert $V_{pk}$ to $V_{ave}$. The current forms of these equations are

$$I_{ave} = \frac{I_{pk}}{\pi} \qquad \text{(half-wave rectified)} \qquad \textbf{(3.13)}$$

and

$$I_{ave} = 0.318(I_{pk}) \qquad \text{(half-wave rectified)} \qquad \textbf{(3.14)}$$

Example 3.6 demonstrates the first method for determining the value of $I_{ave}$.

*Question and Answer:* Where did equations (3.11) through (3.14) come from? These are the equations that are taught in an ac circuits course for converting *peak* values to *average* values. You may want to review your basic electronics text for a complete explanation of the equations.

---

Determine the value of $I_{ave}$ for the circuit shown in Figure 3.12.

**EXAMPLE 3.6**

**Solution:** In Example 3.5, we determined the value of $V_{ave}$ to be 16.66 $V_{dc}$. Using this value and the value of $R_L$, the value of $I_{ave}$ is found to be

$$I_{ave} = \frac{V_{ave}}{R_L}$$

$$= \frac{16.66 \, V_{ave}}{20 \, k\Omega}$$

$$= 833 \, \mu A_{dc}$$

**PRACTICE PROBLEM 3.6**

A half-wave rectifier has an average output voltage that is equal to 24 $V_{dc}$. The load resistance is 2.2 $k\Omega$. What is the value of the dc load current for the circuit?

---

Example 3.7 demonstrates the second method of determining the value of $I_{ave}$.

---

Determine the dc load current for the rectifier shown in Figure 3.13.

**EXAMPLE 3.7**

**Solution:** The transformer has a 24 $V_{ac}$ rating. Thus, the peak secondary voltage is found as

$$V_{2(pk)} = \frac{24 \, V_{ac}}{0.707}$$

$$= 33.9 \, V_{pk}$$

The peak load voltage is now found as

$$V_{L(pk)} = V_{2(pk)} - 0.7 \, V$$
$$= 33.2 \, V_{pk}$$

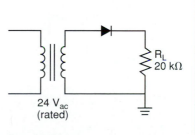

24 $V_{ac}$
(rated)

$R_L$
20 $k\Omega$

**FIGURE 3.13**

The peak load current is found as

$$I_{L(pk)} = \frac{V_{L(pk)}}{R_L}$$
$$= \frac{33.2\ V_{pk}}{20\ k\Omega}$$
$$= 1.66\ mA_{pk}$$

Finally,

$$I_{ave} = \frac{I_{pk}}{\pi}$$
$$= \frac{1.66\ mA_{pk}}{\pi}$$
$$= 529.13\ \mu A_{dc}$$

**PRACTICE PROBLEM 3.7**

A half-wave rectifier is fed by a 48 $V_{ac}$ transformer. If the load resistance for the circuit is 12 k$\Omega$, what is the dc load current for the circuit?

## Negative Half-Wave Rectifiers

The analysis of a *negative* half-wave rectifier is nearly identical to that for a positive half-wave rectifier. The only difference is that all the voltage polarities will be reversed.

You can use a simple method for performing the mathematical analysis of a negative half-wave rectifier:

1. Analyze the circuit as if it were a positive half-wave rectifier.
2. After completing your calculations, change all your voltage polarity signs from positive to negative.

This method for analyzing a negative half-wave rectifier is demonstrated in Example 3.8.

**EXAMPLE 3.8**

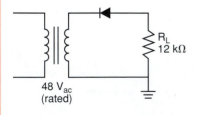

**FIGURE 3.14**

Determine the dc output voltage for the circuit shown in Figure 3.14.

***Solution:*** We'll start by solving the circuit as if it were a positive half-wave rectifier. First,

$$V_{2(pk)} = \frac{48\ V_{ac}}{0.707}$$
$$= 67.9\ V_{pk}$$

and

$$V_{L(pk)} = V_{2(pk)} - 0.7\ V$$
$$= 67.2\ V_{pk}$$

Finally,

$$V_{ave} = \frac{V_{pk}}{\pi}$$
$$= 21.39\ V_{dc}$$

Now we simply convert all the positive voltage values to negative voltage values. Thus, for the circuit shown in Figure 3.14,

$$V_{2(pk)} = -67.9\ V_{pk}$$
$$V_{L(pk)} = -67.2\ V_{pk}$$
$$V_{ave} = -21.39\ V_{dc}$$

**PRACTICE PROBLEM 3.8**

A negative half-wave rectifier is fed by a 36 $V_{ac}$ transformer. Using the method illustrated in this example, calculate the values of $V_{2(pk)}$, $V_{L(pk)}$, and the dc output voltage.

As you can see, there isn't really a whole lot of difference between the analysis of a negative half-wave rectifier and that of a positive half-wave rectifier.

## Component Substitution

The value of $I_{ave}$ is important for another reason. You may recall from Chapter 2 that the *maximum dc forward current* that can be drawn through a diode is equal to the *average forward current* ($I_0$) rating of the device. When working with rectifiers, you may need at some point to substitute one diode for another. When this is the case, you must make sure that the value of $I_{ave}$ for the diode in the circuit is less than the $I_0$ rating of the substitute component. This point will be demonstrated in Section 3.5.

Why knowing the value of $I_{ave}$ is important.

## Peak Inverse Voltage (PIV)

The maximum amount of reverse bias that a diode will be exposed to in a rectifier is called the **peak inverse voltage** or **PIV** of the rectifier. For the half-wave rectifier, the value of PIV is found as

$$PIV = V_{2(pk)} \qquad \text{(half-wave rectifier)} \qquad (3.15)$$

**Peak inverse voltage (PIV)**
The maximum reverse bias that a diode will be exposed to in a rectifier.

The basis for this equation can be seen by referring to Figure 3.5. When the diode is reverse biased (Figure 3.5b), there is no voltage dropped across the load. Therefore, all of $V_2$ is dropped across the diode in the rectifier.

The PIV of a given rectifier is important because it determines the *minimum allowable value* of $V_{RRM}$ for any diode used in the circuit. This point is demonstrated in Section 3.5.

Why PIV is important.

1. Briefly explain the forward and reverse operation of a half-wave rectifier.

*Section Review*

2. Describe the difference between the output waveforms of a positive and a negative half-wave rectifier.

3. How can you tell the output polarity of a half-wave rectifier?

4. List, in order, the steps you would take to calculate the dc output voltage from a rectifier if you were given the turns ratio of the transformer and the rms primary voltage.

5. List, in order, the steps you would take to calculate the dc output voltage from a rectifier if you knew the rated rms output voltage.

6. You have calculated the value of $V_{ave}$ for a half-wave rectifier. What piece of test equipment would you use to measure it?

7. You have calculated the value of $I_{ave}$ for a half-wave rectifier. What piece of test equipment would you use to measure it?

8. Describe the two methods for analyzing a *negative* half-wave rectifier.

9. Why is PIV important?

10. How do you determine the value of PIV for a half-wave rectifier?

## 3.3 FULL-WAVE RECTIFIERS

The full-wave rectifier consists of *two diodes* and a *resistor*, as shown in Figure 3.15a. The result of this change in circuit construction is illustrated in Figure 3.15b.

In Figure 3.15b, the output from the full-wave rectifier is compared with that of a half-wave rectifier. Note that the full-wave rectifier has two positive half-cycles out for every one produced by the half-wave rectifier.

**Center-tapped transformer**
Transformer with an output lead connected to the center of the secondary winding.

The transformer shown in Figure 3.15 is a **center-tapped transformer**. This type of transformer has a lead connected to the center of the secondary winding. The voltage from the center tap to each of the outer winding terminals is equal to one-half of the secondary voltage. For example, let's say we have a 24 V center-tapped transformer. The voltage from the center tap to each of the outer winding terminals is 12 V.

As you will see, the operation of the center-tapped transformer plays a major role in the operation of the full-wave rectifier. For this reason, the full-wave rectifier cannot be "line operated"; that is, it cannot be connected directly to the ac power input like the half-wave rectifier can.

### Basic Circuit Operation

*OBJECTIVE 3* ▶

Figure 3.16 shows the operation of the full-wave rectifier during one complete cycle of the input signal. During the positive half-cycle of the input, $D_1$ is forward biased and $D_2$ is reverse biased. Note the direction of current through the load ($R_L$). Using the ideal operating characteristics of the diode, $V_L$ can be found as

$$V_{L(pk)} = \frac{V_{2(pk)}}{2} \tag{3.16}$$

Why the load voltage is only half the value of the transformer secondary voltage.

$V_L$ is equal to half of the secondary voltage because the transformer is center tapped. The voltage from one end of a center-tapped transformer to the center tap is *always* one-half of the total secondary voltage.

**FIGURE 3.15   Full-wave rectifier.**

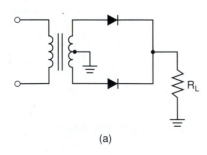

(a)

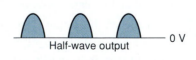
Half-wave output          0 V

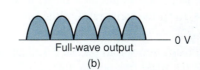
Full-wave output          0 V

(b)

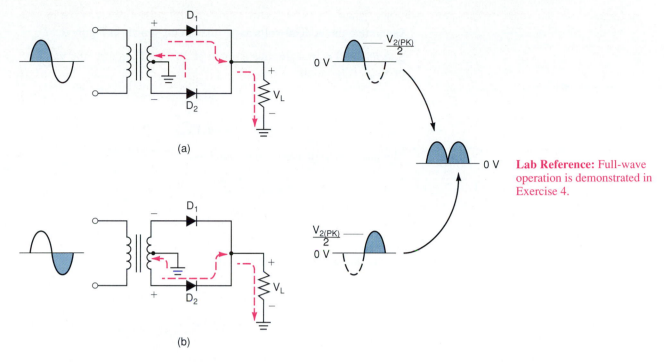

**FIGURE 3.16** Full-wave operation.

**Lab Reference:** Full-wave operation is demonstrated in Exercise 4.

When the polarity of the input reverses, $D_2$ is forward biased, and $D_1$ is reverse biased. Note that the direction of current through the load *has not changed*, even though the polarity of the transformer secondary has. Thus, another positive half-cycle is produced across the load. This gives the output waveform shown in Figure 3.15b.

## Calculating Load Voltage and Current Values

Using the *practical diode model*, the peak load voltage for a full-wave rectifier is found as ◀ *OBJECTIVE 4*

$$V_{L(pk)} = \frac{V_{2(pk)}}{2} - 0.7 \text{ V} \qquad (3.17)$$

The full-wave rectifier will produce twice as many output pulses (per input cycle) as the half-wave rectifier. In other words, for every output pulse produced by a half-wave rectifier, two will be produced by a full-wave rectifier. For this reason, the *average load voltage* for the full-wave rectifier is found as

The full-wave rectifier has twice the output frequency of the half-wave rectifier.

$$V_{ave} = \frac{2V_{L(pk)}}{\pi} \qquad (3.18)$$

or

$$V_{ave} = 0.636 \, V_{L(pk)} \qquad (3.19)$$

where $0.636 \cong 2/\pi$. Note that the value 0.636 used in equation (3.19) is twice the value of 0.318 that was used in equation (3.12) to find the value of $V_{ave}$ for a half-wave rectifier. The procedure for determining the dc load voltage for a full-wave rectifier is illustrated in Example 3.9.

## EXAMPLE 3.9

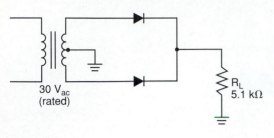

**FIGURE 3.17**

**Lab Reference:** These values are calculated and measured as part of Exercise 4.

Determine the dc load voltage for the circuit shown in Figure 3.17.

**Solution:** The transformer is rated at 30 $V_{ac}$. Therefore, the value of $V_{2(pk)}$ is found as

$$V_{2(pk)} = \frac{30 \; V_{ac}}{0.707}$$
$$= 42.4 \; V_{pk}$$

The peak load voltage is now found as

$$V_{L(pk)} = \frac{V_{2(pk)}}{2} - 0.7 \; V$$
$$= 21.2 \; V_{pk} - 0.7 \; V$$
$$= 20.5 \; V_{pk}$$

Finally, the dc load voltage is found as

$$V_{ave} = \frac{2 \; V_{L(pk)}}{\pi}$$
$$= \frac{41 \; V_{pk}}{\pi}$$
$$= 13.05 \; V_{dc}$$

### PRACTICE PROBLEM 3.9

A full-wave rectifier is fed by a 24 $V_{ac}$ center-tapped transformer. What is the dc load voltage for the circuit?

Once the peak and average load voltage values are known, it is easy to determine the values of $I_{L(pk)}$ and $I_{ave}$. Just use the known values and Ohm's law, as demonstrated in Example 3.10.

## EXAMPLE 3.10

Determine the values $I_{L(pk)}$ and $I_{ave}$ for the circuit shown in Figure 3.17.

**Solution:** In Example 3.9, we calculated the peak and average output voltages for the circuit. Using these calculated values and the value of $R_L$ shown in the circuit, we determine the circuit current values as follows:

$$I_{L(pk)} = \frac{V_{L(pk)}}{R_L}$$
$$= \frac{20.5 \; V_{pk}}{5.1 \; k\Omega}$$
$$= 4.02 \; mA_{pk}$$

and

$$I_{ave} = \frac{V_{ave}}{R_L}$$

$$= \frac{13.05 \ V_{dc}}{5.1 \ k\Omega}$$

$$= 2.56 \ \text{mA dc}$$

**PRACTICE PROBLEM 3.10**

The circuit described in Practice Problem 3.9 has a 2.2-k$\Omega$ load. What are the peak and dc load current values for the circuit?

## Negative Full-Wave Rectifiers

If we reverse the directions of the diodes in the positive full-wave rectifier, we will have a *negative* full-wave rectifier. The negative full-wave rectifier and its output waveform are shown in Figure 3.18. As you can see, the main differences between the positive and negative full-wave rectifiers are the direction that the diodes are pointing and the polarity of the output voltage.

The method you were shown for analyzing a negative half-wave rectifier will also work for analyzing a negative full-wave rectifier: simply change the voltage polarity signs from positive to negative. We will not go through these analysis procedures again, but you should have no problem making the transition from the negative half-wave to the negative full-wave rectifier. If you don't remember how to analyze a negative rectifier, refer back to Section 3.2.

## Peak Inverse Voltage

When one of the diodes in a full-wave rectifier is reverse biased, the voltage across that diode will be approximately equal to $V_2$. This point is illustrated in Figure 3.19. The 24 $V_{pk}$ across the primary develops peak voltages of $+12$ V and $-12$ V across the secondary (when measured from end to center tap). Note that $V_{2(pk)}$ equals the difference between these two voltages: 24 V. With the polarities shown, $D_1$ is conducting, and $D_2$ is reverse biased. If we assume $D_1$ to be ideal, the voltage drop across the component will equal 0 V. Thus, the cathode of $D_1$ will also be at $+12$ V. Since this point is connected directly to the cathode of $D_2$, its cathode is also at $+12$ V. With $-12$ V applied to the anode of $D_2$, the total voltage across the diode is 24 V.

The peak load voltage supplied by the full-wave rectifier is equal to one-half the secondary voltage, $V_2$. Therefore, the reverse voltage across either diode will be twice the peak load voltage. By formula,

$$PIV = 2 \ V_{L(pk)} \tag{3.20}$$

**FIGURE 3.18**

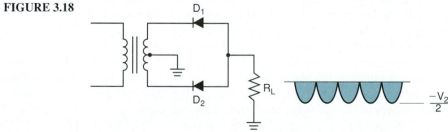

**FIGURE 3.19  Full-wave rectifier PIV.**

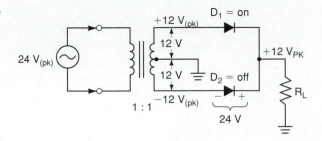

Since the peak load voltage is half the secondary voltage, we can also find the value of PIV as

$$\text{PIV} = V_{2(\text{pk})} \tag{3.21}$$

You may recall that equation (3.21) is the same as the equation we used for the half-wave rectifier, equation (3.15).

When calculating the PIV in a practical diode circuit, the 0.7 V drop across the conducting diode should be taken into consideration. In Figure 3.19, $D_1$ is on and will have a voltage drop of 0.7 V. Because $D_1$ is in series with $D_2$, the PIV across $D_2$ will be reduced by the voltage drop across $D_1$. The equation

*A Practical Consideration:* It may seem confusing to have three different equations for PIV. However, you should keep two points in mind:

1. Equations (3.20) and (3.21) are actually saying the same thing.

2. The 0.7 V drop included in equation (3.22) may become important when you are dealing with low-voltage rectifiers, like those typically found in personal computers.

$$\text{PIV} = V_{2(\text{pk})} - 0.7 \text{ V} \tag{3.22}$$

gives a more accurate PIV voltage, but in most applications the 0.7 V will not make a difference in the choice of a diode.

## Full-Wave Versus Half-Wave Rectifiers

There are quite a few similarities between full-wave and half-wave rectifiers. Figure 3.20 summarizes the relationships that you have been shown for the half-wave and full-wave rectifiers.

It would seem (at first) that the only similarity between the half-wave and the full-wave rectifiers is the method of finding the PIV values for the two circuits. However, there is another similarity that may not be as obvious. Let's assume for a moment that both of the rectifiers shown in Figure 3.20 are fed by 24 $V_{ac}$ transformers. If you were to calculate the dc output voltages for the two circuits, you would get the following values:

$$V_{ave} = 10.58 \ V_{dc} \qquad \text{(for the half-wave rectifier)}$$
$$V_{ave} = 10.36 \ V_{dc} \qquad \text{(for the full-wave rectifier)}$$

As you can see, the two circuits will produce nearly identical dc output voltages *for identical values of transformer secondary voltage.* In fact, if the values of $R_L$ for the two circuits are equal, the dc output current values for the circuits will also be nearly identical.

So why do we bother with the full-wave rectifier when we can get the same dc output values with the half-wave rectifier? There are a couple of reasons. First, *if the peak load voltages for the two circuits are equal, the full-wave rectifier will have twice the dc load voltage and power efficiency that the half-wave rectifier has.* The second reason deals with the operation of filters. As you will be shown in Section 3.6, the full-wave rectifier has twice the output frequency of the half-wave rectifier, which has an impact on the filtering of the rectifier output. When we add filters to the half-wave and full-wave rectifiers, the advantages of the full-wave rectifier will become clear.

FIGURE 3.20

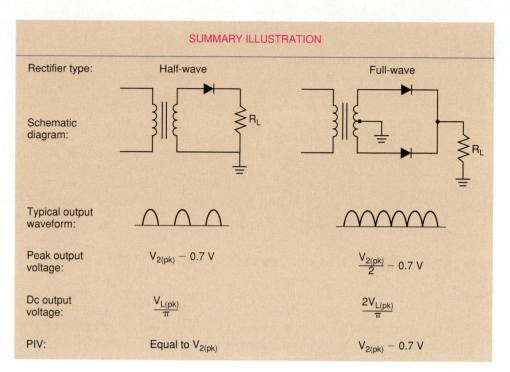

SUMMARY ILLUSTRATION

| Rectifier type: | Half-wave | Full-wave |
|---|---|---|
| Peak output voltage: | $V_{2(pk)} - 0.7\ V$ | $\dfrac{V_{2(pk)}}{2} - 0.7\ V$ |
| Dc output voltage: | $\dfrac{V_{L(pk)}}{\pi}$ | $\dfrac{2V_{L(pk)}}{\pi}$ |
| PIV: | Equal to $V_{2(pk)}$ | $V_{2(pk)} - 0.7\ V$ |

Section Review

1. Briefly explain the operation of a full-wave rectifier.
2. How do you determine the PIV across each diode in a full-wave rectifier?
3. Briefly discuss the similarities and differences between the half-wave rectifier and the full-wave rectifier.

## 3.4 FULL-WAVE BRIDGE RECTIFIERS

The bridge rectifier is the most commonly used full-wave rectifier circuit for several reasons:

Why bridge rectifiers are preferred.

1. It does not require the use of a center-tapped transformer, and therefore can be coupled directly to the ac power line, if desired.
2. Using a transformer with the same secondary voltage produces a peak output voltage that is nearly double the voltage of the full-wave center-tapped rectifier. This results in a higher dc voltage from the supply.

The bridge rectifier is shown in Figure 3.21. As you can see, the circuit consists of *four diodes* and a *resistor*.

FIGURE 3.21   Bridge rectifier.

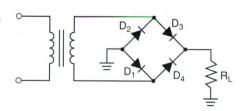

## Basic Circuit Operation

OBJECTIVE 5 ► The full-wave rectifier produces its output by alternating circuit conduction between the two diodes. When one is *on* (conducting), the other is *off* (not conducting), and vice versa. The bridge rectifier works basically the same way. The main difference is that the bridge rectifier alternates conduction between two diode *pairs*. When $D_1$ and $D_3$ (Figure 3.21) are on, $D_2$ and $D_4$ are off, and vice versa. This circuit operation is illustrated in Figure 3.22. During the positive half-cycle of the input, $V_2$ will have the polarities shown, causing $D_1$ and $D_3$ to conduct. Note the direction of current through the load resistor and the polarity of the resulting load voltage. During the negative half-cycle of the input, $V_2$ will reverse polarity. $D_2$ and $D_4$ will now conduct rather than $D_1$ and $D_3$. However, the current direction through the load has not changed, nor has the resulting polarity of the load voltage.

## Calculating Load Voltage and Current Values

OBJECTIVE 6 ► Recall that the full-wave rectifier has an output voltage equal to one-half the secondary voltage (assuming that the conducting diode is ideal). This relationship was expressed in equation (3.17). As you know, the output voltage in a full-wave rectifier is reduced to one-half the secondary voltage by the center tap on the transformer secondary. The center-tapped transformer is essential for the full-wave rectifier to work, but it cuts the output voltage in half.

The bridge rectifier does not require the use of a center-tapped transformer. Assuming the diodes in the bridge to be ideal, the rectifier will have a peak output voltage of

$$V_{L(\text{pk})} = V_{2(\text{pk})} \qquad \text{(ideal)} \tag{3.23}$$

Refer to Figure 3.22. If you consider the diodes to be ideal, the cathode and anode voltage will be equal for each diode. If you view the conducting diodes as being shorted connections to the transformer secondary, you can see that the voltage across the load resistor is equal to the voltage across the secondary.

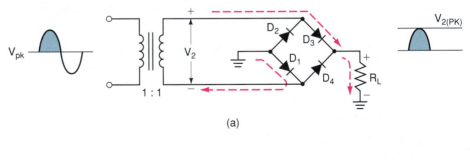

(a)

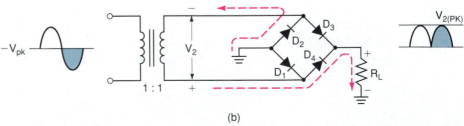

(b)

**FIGURE 3.22    Bridge rectifier operation.**

When calculating circuit output values, you will get more accurate results if you take the voltage drops across the two conducting diodes into account. To include these values, use the following equation when calculating peak load voltage:

$$V_{L(\text{pk})} = V_{2(\text{pk})} - 1.4 \text{ V} \qquad\qquad (3.24)$$

The 1.4 V value represents the sum of the diode voltage drops.

The rest of the load voltage and current values are found using the same equations as those used for the full-wave rectifier. This is illustrated in Example 3.11.

---

Determine the dc load voltage and current values for the circuit shown in Figure 3.23.

**EXAMPLE 3.11**

*Solution:* With the 12 $V_{\text{ac}}$ rated transformer, the peak secondary voltage is found as

$$V_{2(\text{pk})} = \frac{12 \ V_{\text{ac}}}{0.707}$$
$$= 16.97 \ V_{\text{pk}}$$

The peak load voltage is now found as

$$V_{L(\text{pk})} = V_{2(\text{pk})} - 1.4 \text{ V}$$
$$= 15.57 \ V_{\text{pk}}$$

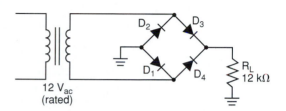

**FIGURE 3.23**

**Lab Reference:** These voltages are calculated and measured as part of Exercise 4.

The dc load voltage is found as

$$V_{\text{ave}} = \frac{2V_{L(\text{pk})}}{\pi}$$
$$= \frac{31.14 \ V_{\text{pk}}}{\pi}$$
$$= 9.91 \ V_{\text{dc}}$$

Finally, the dc load current is found as

$$I_{\text{ave}} = \frac{V_{\text{ave}}}{R_L}$$
$$= \frac{9.91 \ V_{\text{dc}}}{12 \text{ k}\Omega}$$
$$= 825.8 \ \mu A_{\text{dc}}$$

**PRACTICE PROBLEM 3.11**

A bridge rectifier is fed by an 18 $V_{\text{ac}}$ transformer. Determine the dc load voltage and current for the circuit when it has a 1.2 k$\Omega$ load.

---

It is easy to remember how the individual diodes are placed in a bridge rectifier circuit. All the diodes point toward the load. The corner opposite the load is the ground reference. The ac voltage is applied at the two remaining points of the square.

## Bridge Versus Full-Wave Rectifiers

Let's analyze one more full-wave rectifier. This will give us some values for comparing the outputs from the full-wave and bridge rectifiers.

EXAMPLE 3.12

A full-wave rectifier has a 12 $V_{ac}$ transformer and a 12 k$\Omega$ load. Determine the dc load voltage and current values for the circuit.

***Solution:*** The peak secondary voltage for the circuit is found as

$$V_{2(pk)} = \frac{12\ V_{ac}}{0.707}$$

$$= 16.97\ V_{pk}$$

The peak load voltage is now found as

$$V_{L(pk)} = \frac{V_{2(pk)}}{2} - 0.7\ V$$

$$= 7.79\ V_{pk}$$

The dc load voltage is found as

$$V_{ave} = \frac{2\ V_{L(pk)}}{\pi}$$

$$= 4.96\ V_{dc}$$

Finally, the dc load current is found as

$$I_{ave} = \frac{V_{ave}}{R_L}$$

$$= 413.3\ \mu A$$

Now let's compare the results from Examples 3.11 and 3.12. The only difference between these two circuits is that one is a full-wave rectifier and one is a bridge rectifier. For convenience, the results are summarized as follows:

| Value | Bridge Rectifier | Full-Wave Rectifier |
|---|---|---|
| Peak load voltage | 15.57 $V_{pk}$ | 7.79 $V_{pk}$ |
| DC load voltage | 9.91 $V_{dc}$ | 4.96 $V_{dc}$ |
| DC load current | 825.8 $\mu A$ | 413.3 $\mu A$ |

The primary advantages of using a bridge rectifier instead of a full-wave rectifier.

As you can see, the bridge rectifier has output values that are twice as high as a comparable full-wave rectifier. This higher output is the primary advantage of using a bridge rectifier instead of a full-wave rectifier. The power efficiency of the bridge rectifier is also higher than that of the full-wave rectifier.

## Peak Inverse Voltage

Using the ideal diode model, the PIV of each diode in the bridge rectifier is equal to $V_2$. This is the same voltage that was applied to the diodes in the full-wave center-tapped rectifier. Figure 3.24 helps to illustrate this point. In Figure 3.24a, two things have been done:

1. The conducting diodes ($D_1$ and $D_3$) have been replaced by straight wires. Assuming that the diodes are ideal, they will have the same resistance as wire; therefore, the replacement is valid.

2. The positive side of the secondary has been labeled A and the negative side has been labeled B.

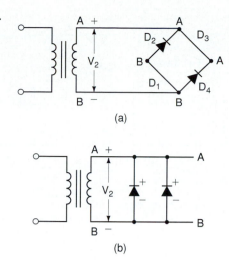

**FIGURE 3.24    Bridge rectifier PIV.**

(a)

(b)

Connecting the common A points along a straight line and doing the same with the B points gives us the circuit shown in Figure 3.24b. With this equivalent circuit, you can see that the two reverse-biased diodes and the secondary of the transformer are all in *parallel*. Since parallel voltages are equal, the PIV across each diode is equal to $V_2$. The same situation will exist for $D_1$ and $D_3$ when they are reverse-biased.

## *Putting It All Together*

The three commonly used rectifiers are the *half-wave rectifier*, *full-wave (center-tapped) rectifier*, and the *bridge rectifier*. The output characteristics of these circuits are summarized in Figure 3.25.

As you know, the *half-wave rectifier* is the simplest of the three circuits. For each input cycle, the half-wave rectifier produces a single half-cycle output. The polarity of the half-cycle output depends on the direction of the diode in the circuit. The half-wave rectifier is normally used in conjunction with a transformer. However, the circuit can be directly coupled to the ac line input.

The *full-wave (center-tapped) rectifier* uses two diodes in conjunction with a center-tapped transformer to convert an ac input to a pulsating dc output. For each input cycle, this rectifier produces two output half-cycles (as shown in Figure 3.25). As is the case with the half-wave rectifier, the output polarity of the full-wave rectifier is determined by the direction of the diodes in the circuit. Since a center-tapped transformer is required for this rectifier to operate, it *cannot* be directly coupled to the ac line input.

The *bridge rectifier* produces an output that is similar to that of a center-tapped full-wave rectifier. However, the diode configuration in this rectifier eliminates the need for a center-tapped transformer. As a result, the bridge rectifier has two distinct advantages over its full-wave counterpart:

1. The peak output (for a given peak input) is twice the value of the output produced by a center-tapped full-wave rectifier.

2. The bridge rectifier can be directly coupled to an ac line input, just like the half-wave rectifier.

Figure 3.25 summarizes the output relationships for the rectifiers we have discussed. Whenever you need to quickly review the relationships listed, you can refer to this illustration. For more detailed information, refer to the appropriate section of this chapter.

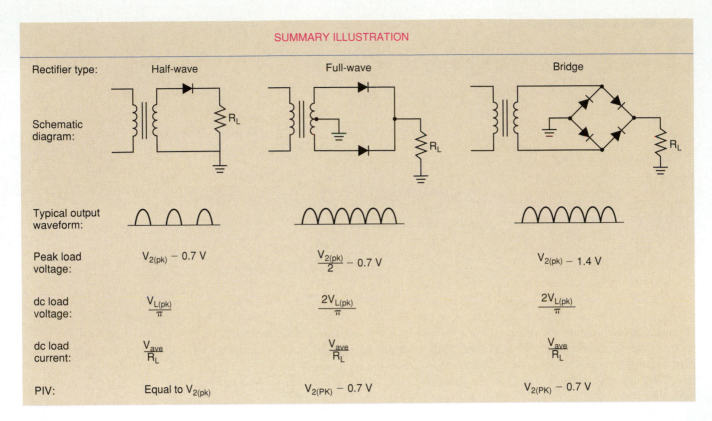

The table in the figure:

| Rectifier type: | Half-wave | Full-wave | Bridge |
|---|---|---|---|
| Schematic diagram: | | | |
| Typical output waveform: | | | |
| Peak load voltage: | $V_{2(pk)} - 0.7 \text{ V}$ | $\dfrac{V_{2(pk)}}{2} - 0.7 \text{ V}$ | $V_{2(pk)} - 1.4 \text{ V}$ |
| dc load voltage: | $\dfrac{V_{L(pk)}}{\pi}$ | $\dfrac{2V_{L(pk)}}{\pi}$ | $\dfrac{2V_{L(pk)}}{\pi}$ |
| dc load current: | $\dfrac{V_{ave}}{R_L}$ | $\dfrac{V_{ave}}{R_L}$ | $\dfrac{V_{ave}}{R_L}$ |
| PIV: | Equal to $V_{2(pk)}$ | $V_{2(PK)} - 0.7 \text{ V}$ | $V_{2(PK)} - 0.7 \text{ V}$ |

**FIGURE 3.25**

---

## *Section Review*

1. Describe the operation of the bridge rectifier.
2. List the advantages that the bridge rectifier has over the full-wave rectifier.
3. Describe the method for determining the PIV for a diode in a bridge rectifier.

---

# *3.5* WORKING WITH RECTIFIERS

Several practical factors involving rectifiers have been ignored up to this point. Most of these factors explain the differences between *theory* and *practice*.

## *Effects of Bulk Resistance and Reverse Current on Circuit Measurements*

In Chapter 2, we discussed the effects of bulk resistance and reverse current on circuit voltage measurements. In that chapter, the following points were made:

1. The current through the bulk resistance of a diode can affect the actual value of $V_F$ for the device.
2. When a diode is in series with a resistor, reverse current can cause a voltage to be developed across the resistor while the diode is biased *off*.

While the effects of bulk resistance and reverse current are not normally very drastic, they do serve to explain a few things:

1. Rectifiers tend to be very high current circuits. Thus, the effects of $I_F R_B$ on the value of $V_F$ can be fairly significant. Many rectifier diodes will have a value of $V_F$ that is closer to 1 V than to 0.7 V.

2. It is fairly common for *rectifier diodes* (those with high current capabilities) to have high reverse current ratings. For example, the MBR320-360 series of high-power rectifier diodes has a reverse current rating of 20 mA (maximum) at $T = 100°C$. This value of $I_R$ could have a major impact on the operation of a power supply. To keep the amount of reverse current to a minimum, power supplies must be kept as cool as possible.

*A Practical Consideration:* Forward voltage values in high-current rectifiers are typically as high as 1.4 V.

Many power rectifiers are mounted to a heat sink to keep the diodes cool when used in high-current circuits. The heat sink draws the heat away from the diode, keeping the *pn* junction cooler. There are many configurations for heat sinks, from flat aluminum stock, to those that have elaborate cooling fins. Many times, heat sinks have forced air flowing through and around them to keep them from becoming too hot. Many power diodes, such as the one shown in Figure 3.26, are threaded on one end. This allows the diode to be directly mounted onto a heat sink.

Reverse current is one of the reasons that many power supplies are fan cooled.

Integrated-circuit bridge rectifiers, like the one shown in Figure 3.27, are also designed to be mounted against a heat sink. There is a hole in the middle of the component to allow a mounting bolt to pass through. The bottom of the bridge may be aluminum, which is electrically isolated from all diode junctions and serves as an excellent thermal conductor to transfer the heat. (IC bridge rectifiers are discussed further at the end of this section.)

## Transformer Rating Tolerance

The tolerance of a transformer's output rating can have a major effect on circuit voltage measurements. Depending on the quality of the transformer, the output voltage may (at times) be above or below the rated value by as much as 20 percent! This can have a major impact on any voltage measured in the circuit.

Later in this chapter, you will learn about *zener voltage regulators*. As you will be shown, voltage regulators provide varying degrees of **line regulation**; that is, *they are capable (up to a point) of maintaining a constant load voltage despite changes in the rectifier output voltage.* The bottom line here is simple: Voltage regulators make it possible to have a stable dc output voltage despite the variation in output voltage of the components in the rectifier. Thus, transformer rating tolerances may cause a measured rectifier voltage to be off by a considerable margin, but they will have little effect on dc load voltage when a voltage regulator is used. Again, the effect of voltage regulation on the operation of a power supply will be covered in detail later in this chapter.

**Line regulation**
The ability of a voltage regulator to maintain a stable dc output voltage despite variations in rectifier output voltage.

## Power Rectifiers

Many power supplies require the use of rectifier diodes that have extremely high forward current and/or power dissipation ratings. The characteristics of these **power rectifiers** are illustrated in Figure 3.26, which contains the spec sheet for the MUR50 series of power rectifiers.

**Power rectifiers**
Diodes with extremely high forward current and/or power dissipation ratings.

The spec sheet shown illustrates the primary differences between power rectifiers and small-signal diodes. These differences are as follows:

1. Power rectifiers have extremely high forward current ratings. The MUR50 series diodes are capable of handling an average forward current of up to 50 A!

2. Power rectifiers (as stated earlier) tend to have values of $V_F$ that are greater than 0.7 V. The typical values of $V_F$ for the MUR50 diodes are between 0.8 V and 1.15 V.

*Question and Answer:* The spec sheet for the MUR50 series calls these components *ultrafast* power rectifiers. Ultrafast rectifiers can be switched on and off at a very high rate of speed. This is required when working with switching power supplies, which switch at a high frequency, typically from 20 to over 100 kHz.

**MUR5005**
**MUR5010**
**MUR5015**
**MUR5020**

### SWITCHMODE POWER RECTIFIERS

... designed for use in switching power supplies, inverters and as free wheeling diodes, these state-of-the-art devices have the following features:

● Ultrafast 50 Nanosecond Recovery Time

● Low Forward Voltage Drop

● Hermetically Sealed Metal DO-203AB Package

**ULTRAFAST RECTIFIERS**

**50 AMPERES**
**50 to 200 VOLTS**

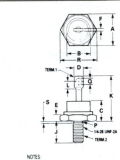

### MAXIMUM RATINGS

| Rating | Symbol | MUR | | | | Unit |
|---|---|---|---|---|---|---|
| | | 5005 | 5010 | 5015 | 5020 | |
| Peak Repetitive Reverse Voltage<br>Working Peak Reverse Voltage<br>DC Blocking Voltage | $V_{RRM}$<br>$V_{RWM}$<br>$V_R$ | 50 | 100 | 150 | 200 | Volts |
| Nonrepetitive Peak Reverse Voltage | $V_{RSM}$ | 55 | 110 | 165 | 220 | Volts |
| Average Forward Current<br>$T_C = 125°C$ | $I_{F(AV)}$ | 50 | | | | Amps |
| Nonrepetitive Peak Surge Forward Current (half cycle, 60 Hz, Sinusoidal Waveform) | $I_{FSM}$ | 600 | | | | Amps |
| Operating Junction and Storage Temperature | $T_J, T_{stg}$ | −55 to +175 | | | | °C |

### THERMAL CHARACTERISTICS

| Rating | Symbol | All Devices | Unit |
|---|---|---|---|
| Thermal Resistance, Junction to Case | $R_{\theta JC}$ | 1.0 | °C/W |

### ELECTRICAL CHARACTERISTICS

| | | | |
|---|---|---|---|
| Maximum Instantaneous Forward Voltage Drop<br>($i_F$ = 50 Amp, $T_J$ = 25°C)<br>($i_F$ = 50 Amp, $T_J$ = 125°C)<br>($i_F$ = 100 Amp, $T_J$ = 125°C) | $v_F$ | 1.15<br>0.95<br>1.10 | Volts |
| Maximum Reverse Current @ DC Voltage<br>($T_J$ = 25°C)<br>($T_J$ = 125°C) | $I_R$ | 10<br>1.0 | µA<br>mA |
| Maximum Reverse Recovery Time<br>($I_F$ = 1.0 Amp, di/dt = 50 Amp/µs, $V_R$ = 30 V, $T_J$ = 25°C) | $t_{rr}$ | 50 | ns |

NOTES:
1. DIM "P" IS DIA.
2. CHAMFER OR UNDERCUT ON ONE OR BOTH ENDS OF HEXAGONAL BASE IS OPTIONAL.
3. ANGULAR ORIENTATION AND CONTOUR OF TERMINAL ONE IS OPTIONAL.
4. THREADS ARE PLATED.
5. DIMENSIONING AND TOLERANCING PER ANSI Y14.5, 1973.

| DIM | MILLIMETERS | | INCHES | |
|---|---|---|---|---|
| | MIN | MAX | MIN | MAX |
| A | 16.94 | 17.45 | 0.669 | 0.687 |
| B | — | 16.94 | — | 0.667 |
| C | — | 11.43 | — | 0.450 |
| D | — | 9.53 | — | 0.375 |
| E | 2.92 | 5.08 | 0.115 | 0.200 |
| F | — | 2.03 | — | 0.080 |
| J | 10.72 | 11.51 | 0.422 | 0.453 |
| K | — | 25.40 | — | 1.000 |
| L | 3.86 | — | 0.156 | — |
| P | 5.59 | 6.32 | 0.220 | 0.249 |
| Q | 3.56 | 4.45 | 0.140 | 0.175 |
| R | — | 20.16 | — | 0.794 |
| S | — | 2.26 | — | 0.089 |

**CASE 257-01**
**DO-203AB**
**METAL**

### MECHANICAL CHARACTERISTICS

**CASE:** Welded, hermetically sealed
**FINISH:** All external surface corrosion resistant and terminal leads are readily solderable
**POLARITY:** Cathode to Case
**MOUNTING POSITIONS:** Any
**MOUNTING TORQUE:** 25 in-lb max

**FIGURE 3.26** (Copyright of Motorola. Used by permission.)

## SEMICONDUCTOR
TECHNICAL DATA

# MDA2500 Series

---

## RECTIFIER ASSEMBLY

. . . utilizing individual void-free molded rectifiers, interconnected and mounted on an electrically isolated aluminum heat sink by a high thermal-conductive epoxy resin.

- 400 Ampere Surge Capability
- Electrically Isolated Base
- UL Recognized
- 1800 Volt Heat Sink Isolation

## SINGLE-PHASE FULL-WAVE BRIDGE

### 25 AMPERES
### 50-600 VOLTS

---

## MAXIMUM RATINGS

| Rating (Per Diode) | Symbol | MDA 2500 | 2501 | 2502 | 2504 | 2506 | 2508 | 2510 | Unit |
|---|---|---|---|---|---|---|---|---|---|
| Peak Repetitive Reverse Voltage Working Peak Reverse Voltage DC Blocking Voltage | $V_{RRM}$ $V_{RWM}$ $V_R$ | 50 | 100 | 200 | 400 | 600 | 800 | 1000 | Volts |
| DC Output Voltage   Resistive Load   Capacitive Load | Vdc | 30 50 | 62 100 | 124 200 | 250 400 | 380 600 | 500 600 | 620 1000 | Volts |
| Sine Wave RMS Input Voltage | $V_R$(RMS) | 35 | 70 | 140 | 280 | 420 | 560 | 700 | Volts |
| Average Rectified Forward Current   (Single phase bridge resistive   load, 60 Hz, $T_C = 55°C$) | $I_O$ | | | | 25 | | | | Amp |
| Nonrepetitive Peak Surge Current   (Surge applied at rated load   conditions) | $I_{FSM}$ | | | | 400 | | | | Amp |
| Operating and Storage Junction Temperature Range | $T_J$, $T_{stg}$ | | | | $-65$ to $+175$ | | | | °C |

## THERMAL CHARACTERISTICS

| Characteristic | Symbol | Typ | Max | Unit |
|---|---|---|---|---|
| Thermal Resistance, Junction to Case   Each Die   Total Bridge | $R_{\theta JC}$ | 4.5 2.0 | 6.0 2.8 | °C/W |

## ELECTRICAL CHARACTERISTICS ($T_C = 25°C$ unless otherwise noted)

| Characteristic | Symbol | Min | Typ | Max | Unit |
|---|---|---|---|---|---|
| Instantaneous Forward Voltage (Per Diode)   ($i_F = 40$ A)* | $v_F$ | — | 0.95 | 1.05 | Volts |
| Reverse Current (Per Diode)   (Rated $V_R$) | $I_R$ | — | — | 10 | µA |

## MECHANICAL CHARACTERISTICS

**CASE:** Plastic case with an electrically isolated aluminum base.
**POLARITY:** Terminal designation embossed on case:
    +DC output
    −DC output
    AC not marked
**MOUNTING POSITION:** Bolt down. Highest heat transfer efficiency accomplished through the surface opposite the terminals. Use silicone heat sink compound on mounting surface for maximum heat transfer.
**WEIGHT:** 25 grams (approx.)
**TERMINALS:** Suitable for fast-on connections. Readily solderable, corrosion resistant. Soldering recommended for applications greater than 15 amperes.
**MOUNTING TORQUE:** 20 in-lb max

*Pulse Width = 100 ms, Duty Cycle ≤ 2%.

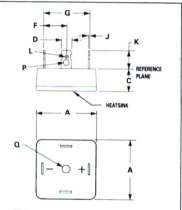

NOTES:
1. DIMENSION "Q" SHALL BE MEASURED ON HEATSINK SIDE OF PACKAGE.
2. DIMENSIONS "F" AND "G" SHALL BE MEASURED AT THE REFERENCE PLANE.

| DIM | MILLIMETERS MIN | MAX | INCHES MIN | MAX |
|---|---|---|---|---|
| A | 25.65 | 26.16 | 1.010 | 1.030 |
| C | 12.44 | 13.97 | 0.490 | 0.550 |
| D | 6.10 | 6.60 | 0.240 | 0.260 |
| F | 10.01 | 10.49 | 0.394 | 0.413 |
| G | 19.99 | 21.01 | 0.787 | 0.827 |
| J | 0.71 | 0.86 | 0.028 | 0.034 |
| K | 9.52 | 11.43 | 0.375 | 0.450 |
| L | 1.52 | 2.06 | 0.060 | 0.081 |
| P | 2.79 | 2.92 | 0.110 | 0.115 |
| Q | 4.42 | 4.67 | 0.174 | 0.184 |

**CASE 309A-03**

**FIGURE 3.27** (Copyright of Motorola. Used by permission.)

3. Power diodes tend to have relatively high reverse current ratings. For the MUR50 series, $I_R$ ranges from 10 μA at 25°C to 1 mA at 125°C.

4. Power diodes can handle much higher nonrepetitive surge currents. The MUR50 series can handle surges up to 600 A.

These ratings agree with the statements that were made earlier about the effects of bulk resistance and reverse current on circuit measurements.

## Integrated Rectifiers

**Integrated circuit**
An entire circuit that is constructed on a single piece of semiconductor material.

Advances in semiconductor device manufacturing have made it possible to construct entire circuits on a single piece of semiconductor material. A circuit that is constructed entirely on a single piece of semiconductor material is called an **integrated circuit**.

There are a number of advantages to using integrated rectifiers. Among them are:

1. Reduced cost (it takes fewer components to construct a power supply).

2. Troubleshooting is made easier.

3. Integrated rectifiers are more compact, so space is saved on the circuit board.

4. All the diodes in the bridge will be the same temperature, thus having the same $V_F$ and leakage current.

5. The bridge can easily be mounted to a heat sink.

The spec sheet for the MDA2500 series IC bridge rectifiers is shown in Figure 3.27. The single casing shown in the spec sheet contains the circuit shown above it: a bridge rectifier. This bridge rectifier is capable of handling an average forward current of 25 A and can handle surges as high as 400 A. The $V_{RRM}$ ratings for the device are comparable to those for any rectifier diode, and the reverse current rating is as low as most small-signal diodes.

If you look below the casing illustration, you will see a drawing that identifies the component leads. Figure 3.28 uses this symbol to show you how the single component would be wired to replace the circuit shown in Figure 3.23.

When using an integrated-circuit bridge, the same mathematical calculations are used as when working with individual diodes. If you look at the $V_F$ rating for the MDA2500, you'll see that values of 0.95 to 1.05 V *per diode* are typical. Thus, you must still consider the effect of *two* diode voltage drops in the mathematical analysis of any circuit that uses the component.

## Section Review

1. What effect does bulk resistance have on the measured value of $V_F$ for a rectifier diode?

2. What effect can reverse current have on measured load voltages?

3. How does cooling a power supply reduce the effects of reverse current?

4. What is the typical tolerance range for transformers?

**FIGURE 3.28**

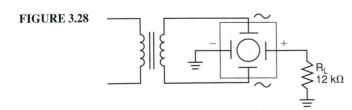

5. What is *line regulation*?

6. What are the primary differences between power rectifiers and small-signal diodes?

7. What advantages of using IC rectifiers were listed in this section?

# 3.6 FILTERS

The third circuit in a power supply is the *filter* (see Figure 3.1). Filters are used in power supplies to *reduce the variations in the rectifier output signal*. Since our goal is to produce a *constant* dc output voltage, it is necessary to remove as much of the rectifier output variation as possible.

Filters reduce the variations in the rectifier output.

The overall result of using a filter is illustrated in Figure 3.29. Here, we see the output from the half-wave rectifier, both before and after filtering. Note that there are still voltage variations after filtering; however, the *amount* of variation has been greatly reduced.

◄ *OBJECTIVE 7*

The remaining voltage variation in the output of the filter is called **ripple voltage**, $V_r$. As you will see, the amount of ripple voltage left by a given filter depends on the rectifier used, the filter component values, and the load resistance.

**Ripple voltage**
The remaining variation in the output from a filter.

Power supplies are designed to produce as little ripple voltage as possible. Too much ripple in the output can have different adverse effects, depending on the application of the power supply. In an audio amplifier, excessive power supply ripple will produce an annoying hum at a 60 or 120 Hz rate, depending on the type of rectifier used. In video circuits, excessive ripple will cause video "hum" bars in the picture. In digital circuits, it is possible to cause erroneous outputs from logic gates. Therefore, it is important for the filter circuit to remove as much ripple as possible.

## Basic Capacitor Filter

The capacitor filter is the most basic filter type and the most commonly used. This filter is simply a capacitor connected in parallel with the load resistance, as shown in Figure 3.30. The filtering action is based on the charge/discharge action of the capacitor. During the positive half-cycle of the input, $D_1$ will conduct and the capacitor will charge rapidly (Figure 3.30a). As the input starts to go negative, $D_1$ will turn off, and the capacitor will slowly discharge through the load resistance (Figure 3.30b). As the input from the rectifier drops below the charged voltage of the capacitor, the capacitor acts as the voltage source for the load. It is the difference between the charge and discharge times of the capacitor that reduces the variations in the rectifier output voltage.

◄ *OBJECTIVE 8*

The difference between the charge and discharge times of the capacitor is caused by two distinct *RC time constants* in the circuit. You may recall from your study of basic

**FIGURE 3.29** The effects of filtering on the output of a half-wave rectifier.

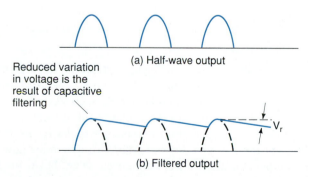

(a) Half-wave output

Reduced variation in voltage is the result of capacitive filtering

$V_r$

(b) Filtered output

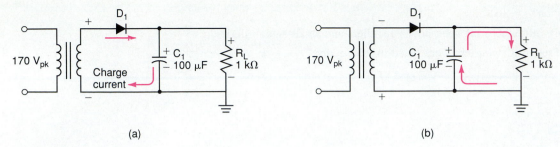

(a)                                        (b)

**FIGURE 3.30**   The basic capacitive filter.

**Lab Reference:** A similar circuit is analyzed as part of Exercise 5.

electronics that a capacitor will charge (or discharge) in *five* time constants. One time constant (represented by the Greek letter *tau*) is found as

$$\tau = RC \qquad (3.25)$$

where $R$ and $C$ are the total circuit resistance and capacitance, respectively. Since it takes five time constants for a capacitor to charge or discharge fully, this time period can be found as

$$T = 5(RC) \qquad (3.26)$$

Now refer to Figure 3.30. The capacitor charges through the diode. For the sake of discussion, let's assume that $D_1$ has a forward resistance of 5 $\Omega$. The time constant for the circuit would be found as

$$\tau = RC$$
$$= (5 \ \Omega)(100 \ \mu F)$$
$$= 500 \ \mu s$$

and the total capacitor charge time would be found as

$$T = 5(RC)$$
$$= 5(500 \ \mu s)$$
$$= 2.5 \ ms$$

Thus, the capacitor would charge to the peak input voltage in 2.5 ms. The discharge path for the capacitor is through the resistor (Figure 3.30b). For this circuit, the time constant would be found as

$$\tau = RC$$
$$= (1 \ k\Omega)(100 \ \mu F)$$
$$= 100 \ ms$$

and the total capacitor discharge time would be found as

$$T = 5(RC)$$
$$= 5(100 \ ms)$$
$$= 500 \ ms$$

Therefore, the capacitor in Figure 3.30 would have a charge time of 2.5 ms and a discharge time of 500 ms. This is why it charges almost instantly, yet barely has time to discharge before another charging voltage is provided by the rectifier.

You have seen that the values of load resistance and capacitance in the filter determine the time required for the capacitor to discharge. It is important to understand that the capacitor size and the load resistance determine how much ripple will be present in the output. Ideally, we want a pure dc value, but if the load resistance or the value of the filter capacitor is decreased, ripple voltage will increase. In Figure 3.31, this rela-

**FIGURE 3.31**

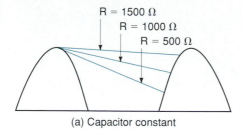

R = 1500 Ω
R = 1000 Ω
R = 500 Ω

(a) Capacitor constant

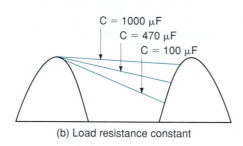

C = 1000 μF
C = 470 μF
C = 100 μF

(b) Load resistance constant

**Lab Reference:** The effects of changing capacitance on filtering are demonstrated in Exercise 5.

tionship is demonstrated using the output from a half-wave filtered power supply. In Figure 3.31a, the capacitance value is held constant while the load resistance is varied. As the value of the load resistance decreases, the capacitor will discharge faster, causing more ripple. In Figure 3.31b, the load resistance is held constant as the filter capacitor values are changed. As the value of the filter capacitor decreases, the time constant of the circuit also decreases. This causes the capacitor to discharge faster, increasing the amount of ripple.

Ripple in a power supply can be minimized by having a large value of filter capacitance combined with a high resistance load. However, these two values are limited by other considerations. Since the load resistance limits the amount of output current, its value must be limited. If the load resistance is too high, output current from the rectifier will be reduced to the point where the circuit is useless. The value of $C$ is limited by three factors:

What limits the value of a filter capacitor?

1. The maximum allowable charge time for the component.
2. The amount of *surge current*, $I_{surge}$, that the rectifier diodes can withstand.
3. The cost of "larger than needed" filter capacitors.

The capacitor is not only involved in the discharge action, it is also involved in the charging action. If you make the value of $C$ too high, your discharge time will be greatly increased, but so will the charge time. The second factor that limits the value of $C$ is *surge current*. We'll take a look at surge current and its causes and effects now.

## Surge Current

When you first turn a power supply on, the filter capacitor has no accumulated charge to oppose $V_2$. For the first instant, the discharged capacitor acts as a short circuit, as shown in Figure 3.32. As you can see, the diode current is initially limited only by the resistance of the transformer secondary and the bulk resistance of the diode. Since these resistances are usually very low, the initial current will be extremely high. This high initial current is referred to as **surge current**, and it is calculated as follows:

**Surge current**
The high initial current in a power supply.

$$I_{surge} = \frac{V_{2(pk)}}{R_w + R_B}$$

(3.27)

FIGURE 3.32

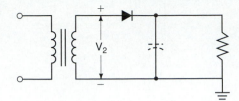

where $V_{2(pk)}$ = the *peak* secondary voltage

$R_w$ = the resistance of the secondary windings

$R_B$ = the total diode bulk resistance

Example 3.13 demonstrates the calculation of surge current for a filtered rectifier.

---

**EXAMPLE 3.13**

If the circuit shown in Figure 3.30 has values of $R_w = 0.8\ \Omega$ and $R_B = 5\ \Omega$, what is the initial surge current for the circuit?

***Solution:*** The peak secondary voltage ($V_2$) is found as

$$V_{2(pk)} = \frac{N_2}{N_1} V_{1(pk)}$$

$$= \frac{1}{2} \times 170\ V_{pk}$$

$$= 85\ V_{pk}$$

Now the surge current is found as

$$I_{surge} = \frac{V_{2(pk)}}{R_w + R_B}$$

$$= \frac{85\ V_{pk}}{0.8\ \Omega + 5\ \Omega}$$

$$= 14.655\ A$$

**PRACTICE PROBLEM 3.13**

If the circuit in Figure 3.30 has values of $R_w = 0.5\ \Omega$ and $R_B = 8\ \Omega$, what is the initial surge current for the circuit?

---

The surge current value found in Example 3.13 may seem to be extremely high, but it probably won't cause any problem. As you may recall, most rectifier diodes have relatively high surge current ratings. An example of this is the 1N400X series of diodes. Referring to the spec sheet in Figure 2.25, the nonrepetitive surge current ($I_{FSM}$) is rated at 30 A. The 14.655 A drawn from Example 3.13 is well below the $I_{FSM}$ rating of the 1N400X series diodes.

When the amount of surge current produced by a circuit is more than the rectifier diodes can handle, the problem can be resolved by using a series *current-limiting* resistor, shown in Figure 3.33 as $R_{surge}$. The current-limiting resistor will usually be a low-resistance, high-wattage component. The resistor does help lower the surge current, but it also lowers the output voltage from the circuit. This is because the output voltage from the rectifier is divided between the series resistor and the load resistance.

**FIGURE 3.33**

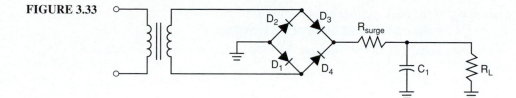

Surge current can also be limited by using a smaller value of filter capacitor. Smaller-value capacitors have higher reactance values and will charge in a shorter period of time. The relationship between capacitance, current, and time is as follows:

$$C = \frac{I(t)}{\Delta V_C} \qquad (3.28)$$

where    $C$ = the capacitance, in farads
   $I$ = the dc (average) charge/discharge current
   $t$ = the charge/discharge time
   $\Delta V_C$ = the *change* in capacitor voltage during charge/discharge

Equation (3.28) can be rearranged to produce the following:

$$t = \frac{C(\Delta V_C)}{I} \qquad (3.29)$$

As equation (3.29) shows, the time required for a capacitor to charge to a specified value is directly proportional to the value of the capacitor. Thus, a lower-value capacitor will charge faster and eliminate the surge current faster.

## Filter Output Voltages

Ideally, the filter capacitor will charge to the peak output voltage from the rectifier and will not discharge at all. However, as previous illustrations have shown, this is not the case. Figure 3.34 relates the various voltage values contained in the filter output. The dc output voltage ($V_{dc}$) is shown to equal the peak voltage ($V_{pk}$) minus one-half of the peak-to-peak value of the ripple voltage. By formula,

$$V_{dc} = V_{pk} - \frac{V_r}{2} \qquad (3.30)$$

**FIGURE 3.34**

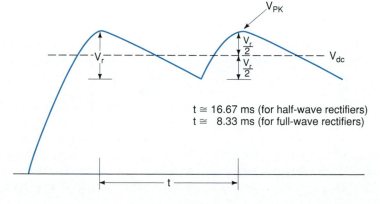

$t \cong 16.67$ ms (for half-wave rectifiers)
$t \cong 8.33$ ms (for full-wave rectifiers)

**Lab Reference:** The effects of filtering on dc load voltage are demonstrated in Exercise 5.

where $V_{pk}$ = the peak rectifier output voltage

$V_r$ = the peak-to-peak value of ripple voltage

The ripple voltage from the filter can be found using a variation of equation (3.28), as follows:

$$V_r = \frac{I_L t}{C} \qquad \textbf{(3.31)}$$

where $I_L$ = the dc load current

$t$ = the time between charging peaks

$C$ = the capacitance, in farads

The value of $t$ is shown in Figure 3.34. It is found by taking the reciprocal of the ripple frequency, which is 60 Hz for a half-wave power supply and 120 Hz for a full-wave supply. Because the time for the half-wave supply is double the time for the full-wave supply, the half-wave will have much higher ripple voltage for the same value of capacitor and load current. This point is demonstrated in Example 3.14.

---

## EXAMPLE 3.14

Determine the ripple output from each of the circuits shown in Figure 3.35. Assume that the load current is 20 mA for each circuit.

**Solution:** The half-wave rectifier (Figure 3.35a) has a time of 16.67 ms between charging pulses. Therefore,

$$V_r = \frac{I_L t}{C}$$

$$= \frac{(20 \text{ mA})(16.67 \text{ ms})}{500 \text{ } \mu\text{F}} \qquad \left( \text{where } t = \frac{1}{f} \right)$$

$$= 666.8 \text{ mV}_{pp}$$

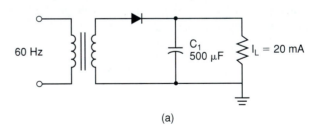

(a)

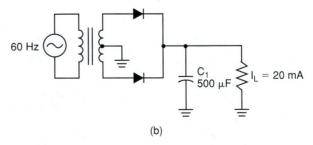

(b)

**FIGURE 3.35**

For the full-wave rectifier (Figure 3.35b), $t = 8.33$ ms. Therefore,

$$V_r = \frac{I_L t}{C}$$

$$= \frac{(20 \text{ mA})(8.33 \text{ ms})}{500 \text{ }\mu\text{F}}$$

$$= 333.2 \text{ mVpp}$$

As Example 3.14 shows, the full-wave rectifier has exactly one-half the ripple output produced by the half-wave rectifier. This is due to the shortened time period between capacitor charging pulses. Figure 3.36 shows the capacitor discharge comparison between the half-wave and the full-wave power supplies. Although the discharge slopes for the capacitor/load combinations are the same for both supplies, the time between the voltage pulses accounts for the difference in the ripple voltage between the two.

An oscilloscope can be used to measure the ripple voltage that is riding on a dc level. Because the ripple voltage is normally very small compared to the dc voltage, we must set the oscilloscope for ac coupling to block the dc component of the ripple, allowing us to expand the range of the scope so that we can measure very small ac voltages.

**Lab Reference:** An oscilloscope is used to measure $V_r$ in Exercise 5.

You were told earlier in the chapter that the primary advantage of using full-wave rectifiers in place of half-wave rectifiers would be seen when we discussed filtering. As Example 3.14 has shown, a full-wave rectifier will have half the ripple output voltage of a comparable half-wave rectifier. Since our goal is to have a steady dc voltage that has as little ripple voltage as possible, the full-wave rectifier gets us much closer to our goal than does the half-wave rectifier.

◄ *OBJECTIVE 9*

Why full-wave rectifiers are preferred over half-wave rectifiers.

## Filter Effects on Rectifier Analysis

Refer back to Figure 3.34. As the figure shows, you can find the value of $V_{dc}$ for a filtered rectifier by subtracting $V_r/2$ from the peak rectifier output voltage. There is only one problem. To determine the value of $V_r$, we have to know the value of the dc load current ($I_L$). And to determine the value of $I_L$, we have to know the value of $V_{dc}$.

What we have here is referred to as a *loop*. We have to know the value of $V_{dc}$ to find the value of $V_r$, but we have to know the value of $V_r$ to find the value of $V_{dc}$.

So, what is the solution? If you look closely at Figure 3.34, you'll see that the final value of $V_{dc}$ is *very close* to the value of $V_{pk}$. Thus, we can start our mathematical analysis of the circuit by making the following assumption:

$$V_{dc} \cong V_{L(pk)}$$

Then, using the *assumed* value of $V_{dc}$, we can calculate the approximate value of $I_L$. From there, we can find the approximate value of $V_r$ and an even closer value of $V_{dc}$. As Exam-

**FIGURE 3.36  Full-wave versus half-wave ripple.**

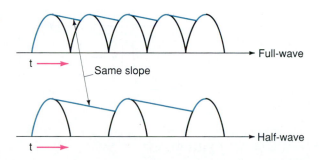

ple 3.15 illustrates, the final value of $V_{dc}$ will prove to be very close to the value that was originally assumed.

<hr />

**EXAMPLE 3.15**

Determine the value of $V_{dc}$ for the circuit shown in Figure 3.37.

**Solution:**   The transformer is rated at 24 $V_{ac}$, so

$$V_{2(pk)} = \frac{24\ V_{rms}}{0.707}$$

$$= 33.95\ V_{pk}$$

$$V_{L(pk)} = \frac{V_{2(pk)}}{2} - 0.7\ V$$

$$= 16.28\ V_{pk}$$

Now we will *assume* that

$$V_{dc} = V_{L(pk)} = 16.28\ V$$

Using this assumed value of $V_{dc}$, the dc load current is found as

$$I_L = \frac{V_{dc}}{R_L}$$

$$= 13.6\ \text{mA}$$

Now the value of the ripple voltage is found as

$$V_r = \frac{I_L t}{C}$$

$$= \frac{(13.6\ \text{mA})(8.33\ \text{ms})}{470\ \mu F}$$

$$= 241\ \text{mV}_{pp}$$

Finally, the calculated value of $V_{dc}$ is found as

$$V_{dc} = V_{pk} - \frac{V_r}{2}$$

$$= 16.27\ V_{pk} - 120.5\ \text{mV}$$

$$= 16.15\ V_{dc}$$

**PRACTICE PROBLEM 3.15**

A full-wave rectifier has a 24 $V_{ac}$ transformer, a 330 $\mu F$ filter capacitor, and a 1.5 k$\Omega$ load. Calculate the values of $V_r$ and $V_{dc}$ for the circuit.

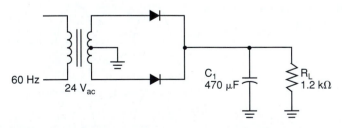

**FIGURE 3.37**

We only had a percentage of error between our assumed value of $V_{dc}$ and our calculated value of $V_{dc}$ of 0.62%. This shows that the initial assumption of $V_{dc} \cong V_{L(pk)}$ for the circuit is valid.

## Filter Effects on Diode PIV

For the full-wave and bridge rectifiers, the filter will not really have any significant effect on the peak inverse voltage across each diode. However, when filtered, the half-wave rectifier diode will have a peak inverse voltage equal to twice the secondary voltage. By formula,

$$\text{PIV} = 2\,V_{2(pk)} \qquad \text{(half-wave, filtered)} \qquad \textbf{(3.32)}$$

This point is illustrated in Figure 3.38. When the diode is conducting, $C_1$ will charge to $V_{2(pk)}$ (ignoring the voltage drop across the diode). This is shown in Figure 3.38a. When the polarity of the secondary voltage reverses (Figure 3.38b), $D_1$ turns off. For an instant, the reverse voltage across the diode is equal to the sum of the secondary voltage and the capacitor voltage. Since these two voltages are equal, the peak inverse voltage across the diode equals $2V_{2(pk)}$. Note that this value will be reduced as the capacitor discharges through the load resistance. However, the diode must have a breakdown voltage rating that is greater than $2V_{2(pk)}$.

## Other Filter Types

There are several other types of filters, as shown in Figure 3.39. Each filter type shown makes use of the reactance properties of the capacitors and/or inductors. In each filter, the series impedance is designed to be very high at the ripple frequency, while the shunt impedance is designed to be very low. Therefore, whatever ripple is not dropped by the series component is greatly reduced by the shunt component.

The inductive filters provide an added benefit. Since an inductor opposes a rapid change in current, inductive filters do not have the surge current problems that the capacitive filters do. At the same time, capacitive filters are more commonly used because of cost and size factors.

Inductive filters provide protection against surge current problems.

**FIGURE 3.38   The effects of filtering on diode PIV.**

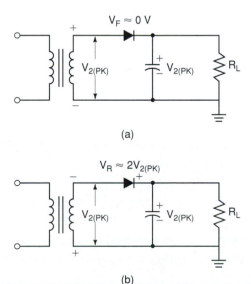

(a)

(b)

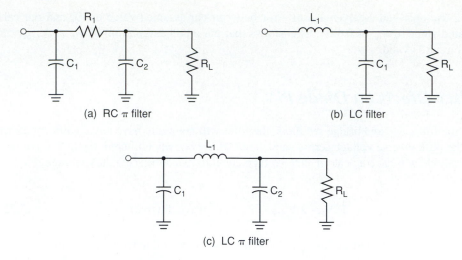

(a) RC π filter          (b) LC filter

(c) LC π filter

**FIGURE 3.39   Some other filter circuits.**

High-quality filters use capacitors to oppose any changes in voltage and inductors to oppose any changes in current. For example, refer to the LC π-filter shown in Figure 3.39c. The inductor aids in keeping the series current constant, while the capacitors short any voltage changes to ground. The result is a very stable output in terms of both current and voltage. Therefore, while the LC filters are more costly to produce, they are often used in more sophisticated circuits that require extremely low power supply variations.

## One Final Note

The use of a filter will greatly reduce the variations in the output from a rectifier. At the same time, the *ideal* power supply would provide a stable dc output voltage that had no ripple voltage at all. While there will always be *some* ripple voltage at the output of a power supply, the use of a voltage regulator will reduce the filter output ripple even further. This point will be demonstrated in Section 3.7.

**Section Review**

1. What are filters used for?
2. What is *ripple voltage*?
3. Describe the operation of the basic capacitive filter in terms of *charge/discharge time constants.*
4. What limits the value of a filter capacitor?
5. What causes *surge current* in a power supply?
6. What are the *three* devices shown in this section that limit surge current? How do they limit $I_{\text{surge}}$?
7. Describe the relationship between $V_r$ and $V_{\text{dc}}$ for a filtered rectifier.
8. Why are full-wave rectifiers preferred over half-wave rectifiers?
9. Describe the process used to calculate the values of $V_r$ and $V_{\text{dc}}$ for a filtered rectifier.
10. Explain the effect that filtering has on the PIV rating of a half-wave rectifier.

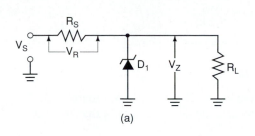

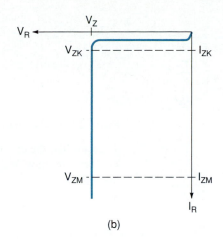

(a)

(b)

**FIGURE 3.40** The basic zener voltage regulator.

## 3.7 ZENER VOLTAGE REGULATORS

The final circuit in the basic power supply is the **voltage regulator** (as shown in Figure 3.1). There are many types of voltage regulators. Many of these circuits contain a number of transistors and/or integrated-circuits (ICs). Some of these regulators are discussed in Chapter 21. At this point, we are going to concentrate on the simple *zener regulator*, which is shown in Figure 3.40a. You should recall that a zener diode operated in the reverse breakdown region will have a constant voltage across it as long as zener current remains between the knee current ($I_{ZK}$) and the maximum current rating ($I_{ZM}$). These operating characteristics are illustrated in Figure 3.40b.

Since the load resistance is in parallel with the zener diode, load voltage remains constant as long as the zener voltage remains constant. If the zener current leaves the allowable range, zener voltage will change and so will the load voltage. Therefore, *the key to keeping the load voltage constant is to keep the zener current within its specified range*; that is, between $I_{ZK}$ and $I_{ZM}$.

**Voltage regulator**
A circuit that maintains a constant load voltage despite variations in load current demand and input voltage.

### Total Circuit Current

For a zener regulator like the one shown in Figure 3.40a, the total current that is drawn from the source is found as

◄ *OBJECTIVE 10*

$$I_T = \frac{V_S - V_Z}{R_S} \qquad (3.33)$$

where $I_T$ = the total current drawn from the filtered rectifier
  $V_S$ = the source voltage
  $V_Z$ = the nominal (rated) zener voltage
  $R_S$ = the series resistor

Example 3.16 demonstrates the use of this equation.

---

Determine the total circuit current for Figure 3.41.

**Solution:** With an applied voltage of 20 V and a zener voltage of 9.1 V, the total circuit current is found as

*EXAMPLE 3.16*

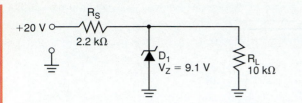

**FIGURE 3.41**

$$I_T = \frac{V_S - V_Z}{R_S}$$

$$= \frac{10.9\ \text{V}}{2.2\ \text{k}\Omega}$$

$$= 4.95\ \text{mA}$$

*PRACTICE PROBLEM 3.16*

A circuit like the one shown in Figure 3.41 has a 15 $V_{dc}$ source, a 1.5 k$\Omega$ series resistor, and a 12 V zener diode. What is the value of $I_T$ for the circuit?

## The Basis of Equation (3.33)

In the regulator shown in Figure 3.40a, the total circuit current passes through the series resistor, $R_S$. Therefore,

$$I_T = I_S$$

According to Ohm's law, we can find the total circuit current as

$$I_T = \frac{V_R}{R_S}$$

Since $V_R$ is equal to the difference between the source voltage ($V_S$) and the zener voltage ($V_Z$), the above equation can be written as

$$I_T = \frac{V_S - V_Z}{R_S}$$

## Load Current

Since the load resistance ($R_L$) is in parallel with the zener diode, the voltage across the load will always equal $V_Z$. Thus, the load current can be found as

$$I_L = \frac{V_Z}{R_L} \tag{3.34}$$

Example 3.17 demonstrates the use of this equation.

*EXAMPLE 3.17*

Determine the load current for the circuit shown in Figure 3.41.

**Solution:**  With a 9.1 V zener and a 10 k$\Omega$ resistor, the total load current is found as

$$I_L = \frac{V_Z}{R_L}$$

$$= \frac{9.1\ \text{V}}{10\ \text{k}\Omega}$$

$$= 910\ \mu\text{A}$$

*PRACTICE PROBLEM 3.17*

The zener voltage regulator described in Practice Problem 3.16 has a 12 k$\Omega$ load resistance. What is the value of load current for the circuit?

## Zener Current

Since the zener diode and the load resistance are in parallel, the sum of $I_Z$ and $I_L$ equals the total circuit current. Thus,

$$I_Z = I_T - I_L \qquad \text{(3.35)}$$

where $I_Z$ = the total zener current
$I_T$ = the total circuit current, found using equation (3.33)
$I_L$ = the load current, found using equation (3.34)

Example 3.18 demonstrates the procedure for determining the value of $I_Z$.

---

Determine the load current and zener current for the circuit shown in Figure 3.41.

**Solution:** The load current can be found using the value of $V_Z$ for load voltage since the load and the zener diode are in parallel. Therefore,

$$I_L = \frac{V_Z}{R_L}$$

$$= \frac{9.1 \text{ V}}{10 \text{ k}\Omega}$$

$$= 910 \text{ } \mu\text{A}$$

In Example 3.16, we determined $I_T$ to be 4.95 mA. Therefore,

$$I_Z = I_T - I_L$$
$$= 4.95 \text{ mA} - 910 \text{ } \mu\text{A}$$
$$= 4.04 \text{ mA}$$

*EXAMPLE 3.18*

**PRACTICE PROBLEM 3.18**

Determine the value of the zener current for the circuit described in Practice Problems 3.16 and 3.17.

---

## Load Variations

It was stated earlier that the zener regulator will maintain a constant output voltage as long as zener current stays between $I_{ZK}$ and $I_{ZM}$. The possible effects of load resistance variations on zener current are illustrated in Figure 3.42. If load resistance is reduced to 0 Ω (Figure 3.42a), all circuit current will pass through the load. In this case, zener current will drop below $I_{ZK}$, and the diode will stop regulating the output voltage. If the load resistance is increased to ∞ Ω (Figure 3.42b), total circuit current will pass through the zener diode. In this case, the diode will be destroyed unless $R_S$ is large enough to prevent $I_{ZM}$ from being reached.

What are the practical limits on the value of $R_L$? The minimum value of $R_L$ is determined by the zener voltage and the value of $I_{zk}$. To maintain zener regulation, the minimum zener current must be equal to $I_{ZK}$. Therefore,

$$I_{L(\text{max})} = I_T - I_{ZK}$$

Since $I_{L(\text{max})}$ occurs when the load resistance is at a minimum,

$$R_{L(\text{min})} = \frac{V_Z}{I_{L(\text{max})}} \qquad \text{(3.36)}$$

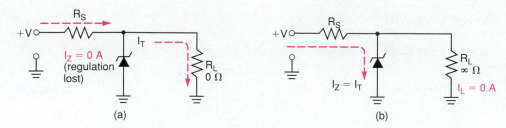

**FIGURE 3.42** The effects of load variation on the operation of a zener voltage regulator.

This relationship is illustrated in Example 3.19.

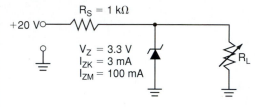

**FIGURE 3.43**

## EXAMPLE 3.19

The zener diode in Figure 3.43 has values of $I_{ZK} = 3$ mA and $I_{ZM} = 100$ mA. What is the minimum allowable value of $R_L$?

*Solution:* First, the total circuit current is found as

$$I_T = \frac{V_S - V_Z}{R_S}$$

$$= \frac{20\,V - 3.3\,V}{1\,k\Omega}$$

$$= \frac{16.7}{1\,k\Omega}$$

$$= 16.7\,mA$$

Now $I_{L(max)}$ is found as

$$I_{L(max)} = I_T - I_{ZK}$$
$$= 16.7\,mA - 3\,mA$$
$$= 13.7\,mA$$

and

$$R_{L(min)} = \frac{V_Z}{I_{L(max)}}$$

$$= \frac{3.3\,V}{13.7\,mA}$$

$$= 241\,\Omega$$

### PRACTICE PROBLEM 3.19

If the diode shown in Figure 3.43 has values of $V_Z = 5.1$ V and $I_{ZK} = 5$ mA, what is the minimum allowable value of $R_L$ for the circuit?

For the circuit shown in Figure 3.43, the zener will maintain a constant output voltage as long as $R_L$ does not go below 241 $\Omega$ . If $R_L$ goes below this value, $I_L$ will increase above its maximum allowable value, and $I_Z$ will go below $I_{ZK}$. The diode will then stop regulating the output voltage. Note that a load that draws *maximum* current is referred to as a **full load**.

**Full load**
A minimum load resistance that draws maximum current.

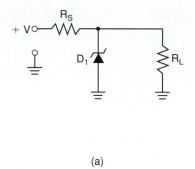

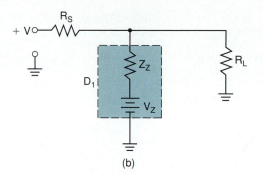

(a)                                        (b)

**FIGURE 3.44**

## Load Regulation

You have been shown that the zener regulator can maintain a constant load voltage for a range of load current values. The ability of a regulator to maintain a constant load voltage under varying load current demands is called **load regulation**. Load regulation will be discussed in detail in Chapter 21.

**Load regulation**
The ability of a regulator to maintain a constant output voltage despite changes in load current demand.

## Zener Reduction of Ripple Voltage

The zener regulator provides an added bonus: It reduces the amount of ripple voltage present at the filter output. The effect that the regulator has on ripple voltage is easy to understand when you consider the equivalent circuit of the diode. The basic zener regulator and its equivalent circuit are shown in Figure 3.44. You may recall from Chapter 2 that the zener impedance, $Z_Z$, is an ac value. In other words, it must be considered in any analysis involving a *change* in current or voltage. Since ripple voltage is a changing quantity, it will be affected by $Z_Z$.

To the ripple waveform, there is a voltage divider present in the regulator. This voltage divider is made up of the series resistance ($R_S$) and the parallel combination of $Z_Z$ and the load. The ripple output from the regulator can be found as

$$V_{r(\text{out})} = \frac{(Z_Z \| R_L)}{(Z_Z \| R_L) + R_S} V_r \qquad (3.37)$$

where    $V_{r(\text{out})}$ = the ripple present at the regulator output
    $(Z_Z \| R_L)$ = the parallel combination of $Z_Z$ and the load resistance
        $R_S$ = the regulator series resistance
        $V_r$ = the peak-to-peak ripple voltage present at the regulator input

Example 3.20 demonstrates the use of this equation.

---

The filtered output from a full-wave rectifier has a peak-to-peak ripple voltage of 1.5 V. If this signal is applied to the circuit shown in Figure 3.45, what will the ripple at the load equal?

***Solution:*** The zener diode in the circuit is shown to have values of $R_L$ = 120 Ω and $Z_Z$ = 5 Ω. With a 51 Ω series resistor and a

*EXAMPLE 3.20*

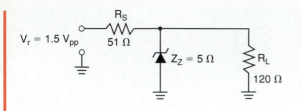

**FIGURE 3.45**

ripple input of 1.5 $V_{pp}$, the ripple at the output of the regulator is found as

$$V_{r(out)} = \frac{(Z_Z \| R_L)}{(Z_Z \| R_L) + R_S} V_r$$

$$= \frac{4.8\ \Omega}{4.8\ \Omega + 51\ \Omega}(1.5\ V_{pp})$$

$$= (0.086)(1.5\ V)$$

$$= 129\ mV_{pp}$$

*PRACTICE PROBLEM 3.20*

A zener regulator has a 91 Ω series resistance, a 200 Ω load resistance, and a zener impedance that equals 25 Ω. If the input ripple to the circuit is 1.2 $V_{pp}$, what is the amount of load ripple?

As the example shows, a voltage regulator will substantially reduce any ripple present at the output of a power supply filter.

At this point, you have been shown the basic operating principles of the voltage regulator and the other circuits that make up the basic dc power supply. Now, we're going to put them all together and analyze the operation of a complete dc power supply.

## Section Review

1. Why must zener current in a voltage regulator be kept within its specified limits?
2. What is a *full load*?
3. How does a zener voltage regulator reduce the ripple voltage from a filter?

# 3.8 PUTTING IT ALL TOGETHER

We have discussed the operation of transformers, rectifiers, filters, and zener regulators in detail. Now it is time to put them all together into a basic working power supply. In this section, we will analyze the basic power supply shown in Figure 3.46.

The power supply shown contains a transformer, bridge rectifier, capacitive filter, and zener diode voltage regulator. The transformer converts the incoming line voltage to a lower secondary voltage. The bridge rectifier converts the transformer secondary ac voltage into a *positive* pulsating dc voltage. This pulsating dc voltage is applied to the

**FIGURE 3.46**

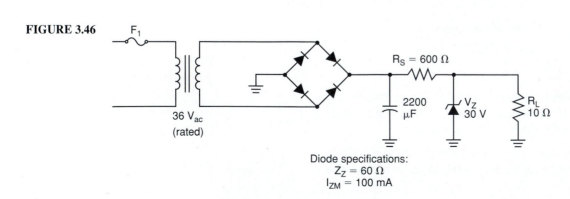

Diode specifications:
$Z_Z = 60\ \Omega$
$I_{ZM} = 100\ mA$

capacitive filter, which reduces the variations in the rectifier dc output voltage. Finally, the zener voltage regulator performs two functions:

1. It reduces the ripple (variations) in the output voltage.

2. It ensures that the dc output voltage from the power supply ($V_{dc}$) will remain relatively constant despite variations in load current demand. Thus, the combination of the four circuits has converted an ac line voltage to a steady dc supply voltage that will remain constant when load current demands change.

Our goal in analyzing a basic power supply is to determine the values of dc output voltage ($V_{dc}$), ripple voltage ($V_r$), and load current ($I_L$). The procedure for determining these values is as follows:

◄ *OBJECTIVE 11*

1. Determine the rms value of the transformer secondary voltage.

2. Determine the value of $V_{2(pk)}$.

3. Determine the value of $V_{pk}$ at the rectifier output.

4. Determine the total current through the series resistor. This current value (designated as $I_R$) will be used when calculating the value of ripple voltage.

5. Determine the value of ripple voltage from the filter.

6. Find $V_{dc}$ at the output. This value will equal the $V_Z$ rating of the zener diode under normal circumstance.

7. Using the rated value of $Z_Z$, approximate the final ripple output voltage.

8. Using $V_Z$ and $R_L$, determine the value of load current.

Example 3.21 illustrates the process for analyzing the schematic of a basic power supply.

---

Determine the values of $V_{dc}$, $V_{r(out)}$, and $I_L$ for the power supply shown in Figure 3.46.

*EXAMPLE 3.21*

*Solution:* First, we must convert the rated value of the transformer secondary voltage to a peak value, as follows:

$$V_{2(pk)} = \frac{36 \ V_{ac}}{0.707}$$
$$= 51 \ V_{pk}$$

Now we determine the value of peak voltage at the filter input.

$$V_{pk} = V_{2(pk)} - 1.4 \ V$$
$$= 49.6 \ V_{pk}$$

Next, we assume that we have a dc source voltage of 49.6 V. Using 49.6 V as our value for $V_S$, we determine the value of the current through the series resistor as follows:

$$I_R = \frac{V_S - V_Z}{R_S}$$
$$= \frac{49.6 \ V - 30 \ V}{75 \ \Omega}$$
$$= 261.3 \ mA$$

We now use the value of $I_R = 261.3$ mA to determine the value of $V_r$.

$$V_r = \frac{I_R t}{C}$$
$$= \frac{(261.3 \ mA)(8.33 \ ms)}{2200 \ \mu F}$$
$$= 989 \ mV_{pp}$$

As stated earlier, the dc output voltage from the power supply will equal the value of $V_Z$. By formula,

$$V_{dc} = V_Z = 30 \; V_{dc}$$

and

$$I_L = \frac{V_Z}{R_L}$$

$$= \frac{30 \; V}{300 \; \Omega}$$

$$= 100 \; mA$$

Finally, the value of $V_{r(out)}$ is found as

$$V_{r(out)} = \frac{(Z_Z \parallel R_L)}{(Z_Z \parallel R_L) + R_S} V_r$$

$$= \frac{50 \; \Omega}{125 \; \Omega} (989 \; mV_{pp})$$

$$= 396 \; mV_{PP}$$

### PRACTICE PROBLEM 3.21

A power supply like the one in Figure 3.46 has the following values: $V_2 = 24 \; V_{ac}$ (rated), $C = 470 \; \mu F$, $R_S = 500 \; \Omega$, $V_Z = 10 \; V$, $Z_Z = 20 \; \Omega$, and $R_L = 5.1 \; k\Omega$. Determine the values of $V_{dc}$, $I_L$, and $V_{r(out)}$ for the circuit.

## 3.9 POWER SUPPLY TROUBLESHOOTING

OBJECTIVE 12 ▶ Power supply faults may occur in the transformer, rectifier diodes, filter, or voltage regulator. When a fault develops in the power supply, the type of symptom and a few simple tests will tell you where the fault is located. In this section, we will take a look at the common power supply fault symptoms and the tests used to isolate the faulty component.

### Primary Fuse

Every power supply contains a fuse that is located in the primary circuit of the supply. This fuse is normally placed in series with the primary of the transformer, as shown in Figure 3.47.

*A Word of Caution:* When you begin to work in the field, you may meet some technicians who will use a wire to "defeat" (bypass) the primary fuse in a faulty power supply. This allows the power supply to continue to operate under circumstances that would normally cause a system shutdown. *Never defeat the fuse in a power supply.* You not only risk starting a fire, but you could also be injured or killed if you should accidentally come into contact with any of the current-carrying components in the circuit!

If a fault in the power supply causes an extremely high amount of current to be drawn from the transformer, the primary fuse will *blow* (open). When this happens, the ac line voltage from the wall outlet is prevented from reaching the power supply. This will stop the excessive current protecting the power supply and the technician who is working on it.

Many power supplies use *slow-blow* fuses. Slow-blow fuses are designed to handle the high surge current produced when a power supply is first turned on. At the same time, if the current demand on a slow-blow fuse is too high for a long enough period of time (usually around 1 second), the slow-blow fuse will open to protect the circuit.

When a power supply fuse needs to be replaced, *you must use the same value and type of fuse*. Never, under any circumstances, use a fuse that has a *higher* current rating than the one you are replacing. Using a higher-value fuse will defeat the purpose of the fuse (protecting the circuit from excessive current) and may even create a fire hazard.

As you will see throughout this section, many power supply faults will cause the primary fuse to blow, while many others will not. When a fault causes the primary fuse to blow, simply replace the fuse after the fault is diagnosed and corrected.

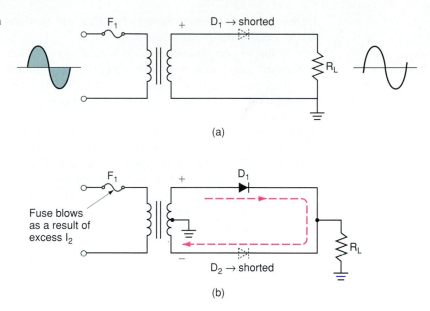

**FIGURE 3.47** The effects of a shorted rectifier diode.

(a)

Fuse blows as a result of excess $I_2$

(b)

## Transformer Faults

The transformer in a power supply can develop one of several possible faults:

1. A shorted primary or secondary winding

2. An open primary or secondary winding

3. A short between the primary or secondary winding and the transformer frame

Transformer fault symptoms.

In most cases, a shorted primary or secondary winding will cause the fuse to blow. If the fuse does not blow, the dc output from the power supply will be extremely low and the transformer itself will get extremely hot.

When the primary or the secondary winding of the transformer opens, the output from the power supply will drop to zero. In this case, the primary fuse will not blow. If you believe that either transformer winding is open, a simple resistance check of the winding will verify your suspicions. If either winding reads a very high resistance, the winding is open. For a 120 to 12 V transformer, the typical values of resistance will be approximately 50 $\Omega$ on the primary and 3 to 7 $\Omega$ on the secondary.

If either winding shorts to the transformer casing, the result will be a blown fuse. This fault is isolated by checking the resistance from the winding leads to the transformer casing. A low resistance measurement indicates that a winding-to-case short circuit exists.

With any of the problems above, the repair procedure is simple. You must replace the transformer.

## Rectifier Faults

The *half-wave rectifier* is the easiest rectifier to troubleshoot. If the diode in the rectifier *shorts*, the output from the rectifier will be a sine wave that is identical to $V_2$. This is illustrated in Figure 3.47a. Since the diode is shorted, it acts as a straight piece of wire. Therefore, neither half-cycle of the input ($V_2$) is eliminated. The output will be a replica of $V_2$.

If the diode in the half-wave rectifier opens, the output from the circuit will drop to zero. If the rectifier diode is either shorted or open, simply replace the diode.

In a *full-wave rectifier*, a shorted diode will cause the power supply fuse to blow. Figure 3.47b shows the effects of a shorted $D_2$ on circuit operation. Note that the diode has been replaced by a straight wire. When $D_1$ is forward biased by $V_2$, the transformer

*A Practical Consideration:* If one diode in a rectifier goes bad, it will usually damage or destroy one or more of the other rectifier diodes in the process. If you determine the fault to be in the rectifier of a power supply, you should replace *all* the diodes.

secondary will be shorted through $D_1$. This will produce excessive current in the secondary, causing the primary fuse to blow.

**Lab Reference:** Several rectifier faults are simulated in Exercise 4.

If you suspect that either diode in a full-wave rectifier has shorted, simply measure the forward and reverse resistance of both diodes. In each case, the diode should have a very large reverse resistance and a very small forward resistance. If the reverse resistance of either diode is extremely low, replace the diode.

If a diode in a full-wave rectifier opens, the output from the rectifier will resemble the output from a half-wave rectifier. The output ripple voltage will approximately double, and the ripple frequency will go from 120 to 60 Hz. In this case, measure the diode resistances and look for a large value of forward resistance. When you have such a reading, replace that diode, or if possible, both diodes.

The symptoms for shorted and open diodes in the *bridge rectifier* are the same as those for full-wave rectifiers. In the case of the bridge rectifier, you simply have more diodes that need to be tested.

## Filter Faults

When working with filter capacitors, technician safety is very important. The hazard of electrical shock or heat burns is always present. Because capacitors store an electrical charge, they can retain their voltage even after the power switch has been turned off or the ac plug has been disconnected. As a safety precaution, capacitors should be discharged by shorting across the terminals with a 50 to 100 $\Omega$ resistor before desoldering them or taking any measurements. Besides being a danger to the technician, test equipment can be damaged if the capacitors remained charged. Many power supply circuits have bleeder resistors connected directly across the capacitor terminals so that, if the load is disconnected from the power supply, the capacitors still discharge. Otherwise, the capacitor could stay charged until the voltage bleeds off internally through leakage.

When the *filter capacitor shorts*, the primary fuse will blow. The reason for this is illustrated in Figure 3.48. When the filter capacitor shorts, it shorts out the load resistance. This has the same effect as wiring the two sides of the bridge together (Figure 3.48a). If

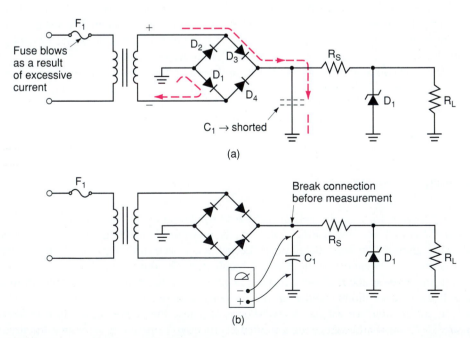

**FIGURE 3.48    Shorted *C* effects and testing.**

you trace from the high side of the bridge to the low side, you will see that the only resistance across the secondary of the transformer is the forward resistance of the two *on* diodes. This effectively shorts out the transformer secondary, causing excessive secondary current and a blown fuse in the primary.

When checking for a shorted filter capacitor, simply measure the resistance of the power supply output, as shown in Figure 3.48b. If the capacitor is shorted, you will measure a very low resistance. Be sure that you disconnect the bridge from the line before measuring the resistance. This will prevent you from forward biasing a rectifier diode with the meter, which will give you a faulty reading.

When electrolytic capacitors get old, the electrolyte tends to dry out. This process is accelerated when the capacitor is exposed to high temperatures, such as those in a power supply. As the electrolyte dries out, the capacitor tends to lose capacity. For example, a 1000-μF capacitor may drop to 700 μF. This reduced capacitance causes the ripple voltage to increase. As the capacitor continues to dry out, it eventually becomes an *open*. There are several ways to check the quality of a capacitor. The best way is to check its value with a capacitance checker. Another way of checking for reduced capacitance is to place an identical good capacitor in parallel with the suspected bad one. If the ripple decreases, the suspect capacitor is probably bad and should be replaced. An *open* filter capacitor will display similar symptoms and must also be replaced.

*Caution: Be very careful when replacing electrolytic capacitors. If you connect an electrolytic capacitor backward in a high-current circuit (such as a power supply), it will either explode or become hot enough to cause serious burns.*

## Zener Regulator Faults

If the *zener diode shorts*, the symptoms will be the same as those for a shorted filter capacitor. When the output symptoms indicate a short, simply connect an ohmmeter to check for a shorted filter capacitor. With the ohmmeter connected, disconnect one end of the zener from the circuit. If the problem remains, the filter capacitor is the cause. If the problem disappears, the zener diode is the cause.

If the *zener diode opens*, the peak output voltage and ripple will both increase. At the same time, voltage regulation will be lost. Thus, you will see a significant amount of variation in the output when the load current demands change.

## Secondary Fuse

Many power supply circuits have fuses installed in the secondary dc outputs. One purpose for multiple fuses in the secondary is that, if one section of a system fails, causing a blown fuse, the rest of the system will remain functional. It is also done because some circuits may require different voltages in different areas of the system. Multiple fuses also allow different areas of a system to be fused at different current levels. An example is a videocassette recorder, which normally has a separate fuse for each voltage that the power supply delivers. By using this method, if one individual circuit has a problem and the fuse blows, the remaining parts of the system still operate. The separate secondary fuses allow certain areas to operate when other areas are disabled.

## Troubleshooting Applications

It is easy to tell you about common power supply faults and their causes. However, it is not always easy to initially apply these fault/cause relationships to actual circuit problems. In this section, we will go through several example troubleshooting cases to show how a little thought and some simple testing can make power supply troubleshooting relatively easy.

# APPLICATION 1

*SYMPTOMS* When the power supply in Figure 3.49 is first turned on, the output is okay. After a few minutes, the dc output voltage drops to almost zero and then the primary fuse blows.

*OBSERVATION* Some components will operate as they should when cool and will develop the symptoms of a short when they get hot. Electrolytic capacitors are extremely susceptible to this type of problem. For this reason, the capacitor is suspected of being the cause of the problem.

*TESTING* To diagnose the possible capacitor problem, the capacitor must be heated in one way or another. This is due to the fact that the problem is heat related. A soldering iron connected to one of the capacitor leads will heat the component enough for testing purposes.

An ohmmeter is connected in the circuit as shown in Figure 3.48b. When first connected, the ohmmeter reads a relatively high resistance. When the hot soldering iron is connected to the capacitor lead, the component starts to get hot. After a moment, the resistance reading on the meter drops to a very low value.

*CONCLUSION* Since the capacitor resistance dropped when temperature increased, the component is *leaky* (partially shorted) and must be replaced.

There is an alternative method to test a capacitor suspected of being leaky. Replace the primary fuse and turn on the power supply. When the output voltage first drops, spray the capacitor with a specially made aerosol coolant. Such coolants are available at most electronic parts stores. If the output voltage goes back to normal when the capacitor is cooled, the capacitor is leaky and must be replaced.

# APPLICATION 2

*SYMPTOMS* When the power supply in Figure 3.49 is turned on, the transformer becomes extremely hot and the dc output voltage is nearly zero. After a moment, the primary fuse blows.

*OBSERVATION* The symptoms listed are classic for a shorted transformer primary or secondary. For this reason, the transformer is tested immediately.

*TESTING* The typical power supply transformer will have a primary resistance of nearly 50 $\Omega$ and a secondary resistance of 10 $\Omega$ or less. When tested, the transformer gives resistance readings of $R_p = 5$ $\Omega$ and $R_S = 8$ $\Omega$.

*CONCLUSION* Since the primary resistance is extremely low, the transformer primary must be shorted. Replacing the transformer corrects the problem.

Some transformers have the value of secondary resistance printed on the side of the component. To determine the approximate value of primary resistance, multiply the secondary resistance by the turns ratio. For example, a transformer with a secondary

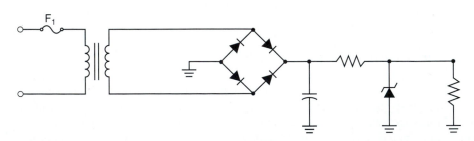

**FIGURE 3.49**

resistance of 1.5 $\Omega$ and a turns ratio of 20:1 will have a primary resistance of approximately 1.5 $\Omega \times 20 = 30\ \Omega$.

# APPLICATION 3

*SYMPTOMS*   The power supply shown in Figure 3.49 has a dc output voltage that is approximately one-half of its rated value and a large amount of ripple voltage.

*OBSERVATION*   These symptoms can be caused only by an open filter capacitor.

*TESTING*   None is required.

*CONCLUSION*   The capacitor filter must be replaced.

# APPLICATION 4

*SYMPTOMS*   The electronic clock and the channel tuner function correctly in a videotape recorder. However, the cassette drive motor does not operate when the play function is selected.

*OBSERVATION*   It appears that there is power to certain sections of the VCR and not to others.

*TESTING*   The dc voltage is checked at each of the secondary fuses. When the 12 V motor drive circuit is checked, the voltmeter reads 0 V. All other secondary fused outputs read the correct voltages.

*CONCLUSION*   No voltage reaches the cassette drive motor because of a blown secondary fuse. The motor driver circuit should be checked and the bad component replaced. After replacing the bad component and the fuse in the 12 V motor drive circuit, the VCR should be operational.

## What the Primary Fuse Tells You

If you look back through this section, you'll notice that there is a common thread to troubleshooting the circuits in the power supply: the fuse in the transformer primary. The condition of the primary fuse indicates whether the problem in the power supply is a *short* or an *open*. If the fuse is blown, a short exists somewhere in the power supply. If it is not blown, an open exists somewhere in the supply. This, of course, assumes that there is a problem in the supply to begin with.

## One Final Note

Most commercially used power supplies are far more complex than the ones we have seen in this chapter. The main difference lies in the voltage regulator circuitry, which usually contains one or more transistors and/or IC voltage regulators. However, the basic principles covered in this chapter apply to all power supplies. While the exact circuitry used may vary from one power supply to another, the basic principles of operation do not.

You will need to learn a great deal about transistors and IC voltage regulators before you will be ready to take on some of the more complex power supply circuits. However, by the time you get to Chapter 21, you will be more than ready to deal with the complex circuits covered.

## Section Review

1. List the common transformer faults and their symptoms.
2. What should you do when you diagnose a rectifier diode fault?
3. List the common filter faults and their symptoms.
4. List the common voltage regulator faults and their symptoms.
5. What does the condition of the primary fuse tell you?

## ■ Key Terms

The following terms were introduced and defined in this chapter:

average load current
average load voltage
bridge rectifier
capacitive filter
center-tapped transformer
filter
full load
full-wave rectifier
half-wave rectifier
inductive filter

integrated circuit (IC)
integrated rectifier
linear power supply
line regulation
load regulation
peak inverse voltage (PIV)
peak load current
peak load voltage
peak secondary voltage

power rectifier
power supply
rectifier
ripple voltage
surge current
switching power supply
transformer secondary
  rating
voltage regulator

## Equation Summary

| Equation Number | Equation | Section Number |
|---|---|---|
| (3.1) | $\dfrac{N_2}{N_1} = \dfrac{V_2}{V_1}$ | 3.1 |
| (3.2) | $V_2 = \dfrac{N_2}{N_1} V_1$ | 3.1 |
| (3.3) | $\dfrac{I_1}{I_2} = \dfrac{V_2}{V_1}$ | 3.1 |
| (3.4) | $I_2 = \dfrac{N_1}{N_2} I_1$ | 3.1 |
| (3.5) | $V_L \cong V_2$ (forward operation) | 3.2 |
| (3.6) | $V_{D1} = V_2$ (reverse operation) | 3.2 |
| (3.7) | $V_{L(pk)} = V_{2(pk)} - V_F$ | 3.2 |
| (3.8) | $V_{2(pk)} = \dfrac{N_2}{N_1} V_{1(pk)}$ | 3.2 |
| (3.9) | $V_{pk} = \dfrac{V_{rms}}{0.707}$ | 3.2 |
| (3.10) | $I_{L(pk)} = \dfrac{V_{L(pk)}}{R_L}$ | 3.2 |
| (3.11) | $V_{ave} = \dfrac{V_{pk}}{\pi}$ (half-wave rectified) | 3.2 |

| Equation Number | Equation | Section Number | Equation Summary |
|:---|:---:|:---:|:---|
| **(3.12)** | $V_{ave} = 0.318(V_{pk})$    (half-wave rectified) | 3.2 | |
| **(3.13)** | $I_{ave} = \dfrac{I_{pk}}{\pi}$    (half-wave rectified) | 3.2 | |
| **(3.14)** | $I_{ave} = 0.318(I_{pk})$    (half-wave rectified) | 3.2 | |
| **(3.15)** | $PIV = V_{2(pk)}$    (half-wave rectifier) | 3.2 | |
| **(3.16)** | $V_{L(pk)} = \dfrac{V_{2(pk)}}{2}$ | 3.3 | |
| **(3.17)** | $V_{L(pk)} = \dfrac{V_{2(pk)}}{2} - 0.7\ V$ | 3.3 | |
| **(3.18)** | $V_{ave} = \dfrac{2V_{L(pk)}}{\pi}$ | 3.3 | |
| **(3.19)** | $V_{ave} = 0.636\ V_{L(pk)}$ | 3.3 | |
| **(3.20)** | $PIV = 2\ V_{L(pk)}$ | 3.3 | |
| **(3.21)** | $PIV = V_{2(pk)}$ | 3.3 | |
| **(3.22)** | $PIV = V_{2\ (pk)} - 0.7\ V$ | 3.3 | |
| **(3.23)** | $V_{L(pk)} = V_{2(pk)}$    (ideal) | 3.4 | |
| **(3.24)** | $V_{L(pk)} = V_{2(pk)} - 1.4\ V$ | 3.4 | |
| **(3.25)** | $\tau = RC$ | 3.6 | |
| **(3.26)** | $T = 5(RC)$ | 3.6 | |
| **(3.27)** | $I_{surge} = \dfrac{V_{2(pk)}}{R_w + R_B}$ | 3.6 | |
| **(3.28)** | $C = \dfrac{I(t)}{\Delta V_C}$ | 3.6 | |
| **(3.29)** | $t = \dfrac{CV}{I}$ | 3.6 | |
| **(3.30)** | $V_{dc} = V_{pk} - \dfrac{V_r}{2}$ | 3.6 | |
| **(3.31)** | $V_r = \dfrac{I_L t}{C}$ | 3.6 | |
| **(3.32)** | $PIV = 2\ V_{2(pk)}$    (half-wave, filtered) | 3.6 | |
| **(3.33)** | $I_T = \dfrac{V_S - V_Z}{R_S}$ | 3.7 | |
| **(3.34)** | $I_L = \dfrac{V_Z}{R_L}$ | 3.7 | |

| Equation Summary | Equation Number | Equation | Section Number |
|---|---|---|---|
| | (3.35) | $I_Z = I_T - I_L$ | 3.7 |
| | (3.36) | $R_{L(min)} = \dfrac{V_Z}{I_{L(max)}}$ | 3.7 |
| | (3.37) | $V_{r(out)} = \dfrac{(Z_Z \| R_L)}{(Z_Z \| R_L) + R_S} V_r$ | 3.7 |

**Answers to the Example Practice Problems**

**3.1.** $I_2 = 20.83$ mA (maximum)

**3.2.** $V_{L(pk)} = 17.3\ V_{pk}$

**3.3.** $V_{L(pk)} = 33.25\ V_{pk}$

**3.4.** $I_{L(pk)} = 2.07\ mA_{pk}$

**3.5.** $V_{ave} = 4.60\ V_{dc}$

**3.6.** $I_{ave} = 10.91$ mA

**3.7.** $I_{ave} = 1.78$ mA

**3.8.** $V_{2(pk)} = -50.92$ V, $V_{L(pk)} = -50.22\ V_{pk}$, $V_{ave} = -15.99\ V_{dc}$

**3.9.** $V_{ave} = 10.36\ V_{dc}$

**3.10.** $I_{L(pk)} = 7.4$ mA, $I_{ave} = 4.71\ mA_{dc}$

**3.11.** $V_{ave} = 15.32\ V_{dc}$, $I_{ave} = 12.77$ mA

**3.13.** $I_{surge} = 10A$

**3.15.** $V_r = 274\ mV_{pp}$, $V_{dc} = 16.14$ V

**3.16.** $I_T = 2$ mA

**3.17.** $I_L = 1$ mA

**3.18.** $I_Z = 1$ mA

**3.19.** $R_{L(min)} = 515.2\ \Omega$

**3.20.** $V_{r(out)} = 236\ mV_{pp}$

**3.21.** $V_{dc} = 10$ V, $I_L = 1.96$ mA, $V_{r(out)} = 30.73\ mV_{pp}$

**Practice Problems**

*§3.1*

1. Determine the transformer secondary $V_{pk}$ in Figure 3.50.
2. Determine the transformer secondary $V_{rms}$ in Figure 3.51.
3. If the transformer in Figure 3.57a has 250 mA of current in the primary, what will be the value of the secondary current?
4. What is $V_{pk}$ across the transformer secondary in Figure 3.56 (page 132)?
5. What is the $V_{rms}$ across the transformer secondary in Figure 3.57a (page 132)?
6. If a transformer has 40 $V_{pk}$ across the primary and 320 $V_{pk}$ across the secondary, what is the turns ratio?

*§3.2*

7. Determine the peak load voltage for the circuit shown in Figure 3.50.
8. Determine the peak load voltage for the circuit shown in Figure 3.51.
9. Determine the peak load voltage for the circuit shown in Figure 3.52.
10. Determine the peak load current for the circuit shown in Figure 3.52.

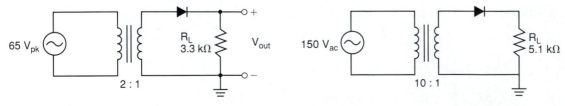

**FIGURE 3.50**  **FIGURE 3.51**

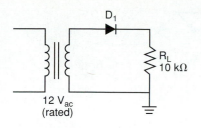

FIGURE 3.52

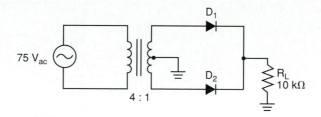

FIGURE 3.53

11. Determine the average (dc) load voltage for the circuit shown in Figure 3.50.

12. Determine the average (dc) load voltage for the circuit shown in Figure 3.51.

13. Determine the average (dc) load voltage for the circuit shown in Figure 3.52.

14. Assume that the diode in Figure 3.50 is reversed. Determine the new values of $V_{L(pk)}$, $V_{ave}$, and $I_{ave}$ for the circuit.

15. Assume that the diode in Figure 3.51 is reversed. Determine the new values of $V_{L(pk)}$, $V_{ave}$, and $I_{ave}$ for the circuit.

16. Assume that the diode in Figure 3.52 is reversed. Determine the new values of $V_{L(pk)}$, $V_{ave}$, and $I_{ave}$ for the circuit.

17. Determine the PIV for the diode in Figure 3.51.

18. A negative half-wave rectifier with a 12 k$\Omega$ load is driven by a 20 $V_{ac}$ transformer. Draw the schematic for the circuit and determine the following values: PIV, $V_{L(pk)}$, $V_{ave}$, and $I_{ave}$.

§3.3

19. Determine the values of $V_{L(pk)}$, $V_{ave}$, and $I_{ave}$ for the circuit shown in Figure 3.53.

20. Determine the values of $V_{L(pk)}$, $V_{ave}$, and $I_{ave}$ for the circuit shown in Figure 3.54.

21. Determine the values of $V_{L(pk)}$, $V_{ave}$, and $I_{ave}$ for the circuit shown in Figure 3.55.

22. Determine the PIV of the circuit shown in Figure 3.53.

23. Determine the PIV of the circuit shown in Figure 3.54.

24. Determine the PIV of the circuit shown in Figure 3.55.

25. Assume that the diodes in Figure 3.53 are both reversed. Determine the new values of $V_{L(pk)}$, $V_{ave}$, and $I_{ave}$ for the circuit.

26. A negative full-wave rectifier with a 910 $\Omega$ load is driven by a 16 $V_{ac}$ transformer. Draw the schematic diagram of the circuit and determine the following values: $V_{L(pk)}$, PIV, $V_{ave}$, and $I_{ave}$.

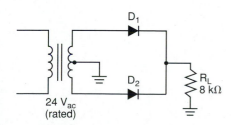

FIGURE 3.54

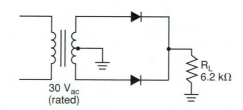

FIGURE 3.55

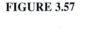

FIGURE 3.56

150 V$_{ac}$      4 : 1      R$_L$ 10 kΩ

*§3.4*

27. Determine the peak and average output voltage and current values for the circuit shown in Figure 3.56.

28. Repeat Problem 27. Assume that the transformer is a 16 V$_{ac}$ transformer.

29. What is the minimum allowable value of $V_{RRM}$ for each of the diodes in Problem 28?

30. A bridge rectifier with a 1.2 kΩ load is driven by a 48 V$_{ac}$ transformer. Draw the schematic diagram for the circuit and calculate the dc load voltage and current values. Also, determine the PIV for each diode in the circuit.

*§3.6*

31. The circuit in Figure 3.57a has values of $R_W = 1\ \Omega$ and $R_B = 6\ \Omega$. What is the value of surge current for the circuit?

32. A half-wave rectifier has values of $R_W = 2\Omega$, $R_B = 12\Omega$, and $V_2 = 36\ V_{ac}$. What is the value of surge current for the circuit?

33. What are the values of $V_{dc}$ and $V_r$ for the circuit shown in Figure 3.57a?

34. What are the values of $V_{dc}$ and $V_r$ for the circuit shown in Figure 3.57b?

35. The circuit shown in Figure 3.58 has the following values: $V_2 = 18\ V_{ac}$ (rated), $C = 470\ \mu F$, and $R_L = 820\ \Omega$. Determine the values of $V_r$, $V_{dc}$, and $I_L$ for the circuit.

36. The circuit shown in Figure 3.58 has the following values: $V_2 = 24\ V_{ac}$ (rated), $C = 1200\ \mu F$, and $R_L = 200\ \Omega$. Determine the values of $V_r$, $V_{dc}$, and $I_L$ for the circuit.

37. What is the PIV for the circuit described in problem 36?

38. What is the PIV for the circuit shown in Figure 3.57a?

FIGURE 3.57

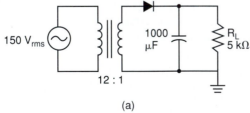

150 V$_{rms}$      12 : 1      1000 μF      R$_L$ 5 kΩ

(a)

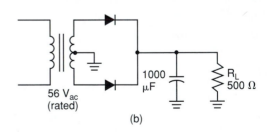

56 V$_{ac}$ (rated)      1000 μF      R$_L$ 500 Ω

(b)

FIGURE 3.58

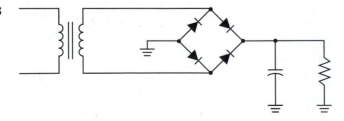

**FIGURE 3.59**

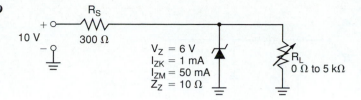

**FIGURE 3.60**

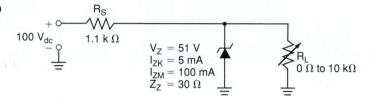

## §3.7

**39.** For the circuit shown in Figure 3.59, determine the total circuit current.

**40.** For the circuit shown in Figure 3.59, determine the value of $I_L$ for $R_L = 2\ k\Omega$.

**41.** For the circuit shown in Figure 3.59, determine the value of $I_Z$ for $R_L = 2\ k\Omega$.

**42.** For the circuit shown in Figure 3.59, determine the value of $I_Z$ for $R_L = 3\ k\Omega$.

**43.** For the circuit shown in Figure 3.59, determine the minimum allowable value of $R_L$.

**44.** For the circuit shown in Figure 3.60, determine the total circuit current.

**45.** For the circuit shown in Figure 3.60, determine the value of $I_Z$ for $R_L = 5\ k\Omega$.

**46.** For the circuit shown in Figure 3.60, determine the minimum allowable value of $R_L$.

**47.** Determine the values of $I_T$, $I_L$, and $I_Z$ for the circuit shown in Figure 3.61.

**48.** The 12-V input to Figure 3.61 has 770 mV$_{pp}$ of ripple voltage. What is the output ripple voltage for the circuit?

## §3.8

**49.** Calculate the dc output voltage and current values for the circuit shown in Figure 3.62.

**50.** Calculate the output ripple voltage for the circuit shown in Figure 3.62.

**FIGURE 3.61**

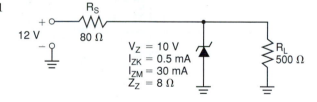

**FIGURE 3.62**

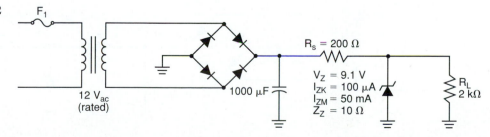

FIGURE 3.63

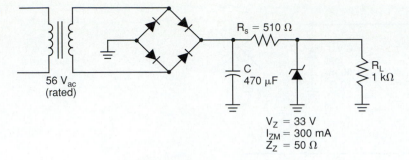

51. Calculate the dc output values and output ripple voltage for the circuit shown in Figure 3.63.

## Troubleshooting Practice Problems

52. The circuit shown in Figure 3.64a has an output signal identical to $V_2$. Discuss the possible cause(s) of the problem.

53. The circuit shown in Figure 3.64b has the output waveform shown. Discuss the possible cause(s) of the problem.

54. The circuit in Figure 3.65 has an output that equals $0\ V_{dc}$. The primary fuse is not blown. Discuss the possible cause(s) of the problem.

55. The load resistance in Figure 3.65 opens. Will this cause the zener diode to be destroyed? Use circuit calculations to explain your answer.

## The Brain Drain

56. Refer to Figure 2.34 (p. 59). Assume that the zener diode in Figure 3.65 has opened. Which of the diodes in Figure 2.34 could be used as a replacement component?

57. The circuit shown in Figure 3.66 has measured output values of $V_{dc} = 12$ V and $V_r = 1.22\ V_{pp}$. Determine whether or not there is a problem in the circuit. (*Remember: Transformer tolerances can be as high as $\pm 20$ percent.*)

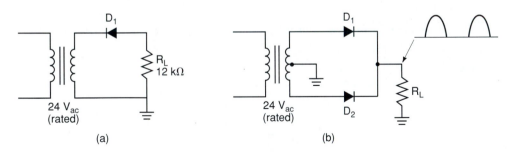

(a)                    (b)

**FIGURE 3.64**

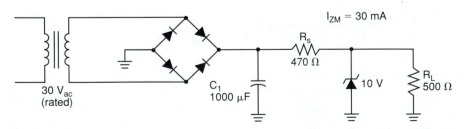

**FIGURE 3.65**

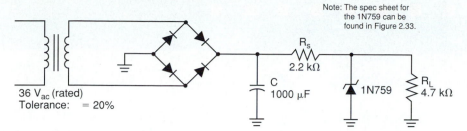

Note: The spec sheet for the 1N759 can be found in Figure 2.33.

$R_s$
2.2 kΩ

C
1000 μF

1N759

$R_L$
4.7 kΩ

36 $V_{ac}$ (rated)
Tolerance: = 20%

**FIGURE 3.66**

58. The rectifier diodes in Figure 3.67 must be replaced. Using Figure 2.27 (p. 49), determine which, if any, of the 1N4001-7 series diodes can be used as a replacement component in the circuit.

59. Write a program designed to calculate the values of $V_{out(pk)}$, $V_{ave}$, and $I_{ave}$ when provided with the transformer secondary rating and the load resistance for a half-wave rectifier.

*Suggested Computer Applications Problems*

60. Write a program that will determine the values of $V_{out(pk)}$, $V_{ave}$, and $I_{ave}$ for a full-wave rectifier when provided with the transformer secondary rating and the load resistance.

61. Modify the program in Problem 60 to take into account the effects of a filter capacitor. The program should request the value of $C$ (along with the values from Problem 60) and should provide the values of dc output voltage and ripple voltage.

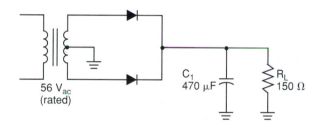

$C_1$
470 μF

$R_L$
150 Ω

56 $V_{ac}$
(rated)

**FIGURE 3.67**

# 4

# COMMON DIODE APPLICATIONS

## Clippers, Clampers, Multipliers, and Displays

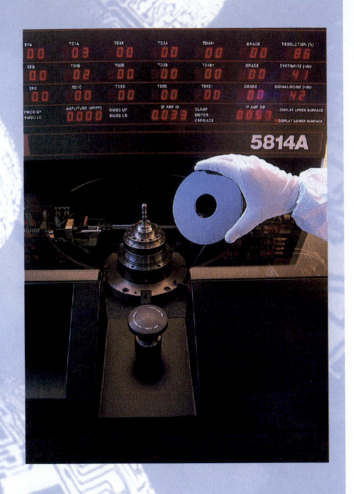

**M**ultisegment displays are used to display alphanumeric symbols; that is, they are used to display numbers, letters, and punctuation marks.

## OUTLINE

## OBJECTIVES

*After studying the material in this chapter, you should be able to:*

1. State the purposes served by clippers, clampers, and voltage multipliers.

2. Discuss the operation of *series clippers.*

3. Discuss and analyze the operation of *shunt clippers.*

4. Discuss and analyze the operation of *biased shunt clippers.*

5. Describe the effects of *negative clampers* and *positive clampers* on an input waveform.

6. Describe the circuit operation of a clamper.

7. Describe and analyze the operation of the *half-wave voltage doubler.*

8. Describe and analyze the operation of the *full-wave voltage doubler.*

9. Discuss the use of the LED as a power-level indicator.

10. Discuss the use of the LED in multisegment displays.

11. Describe the common fault symptoms that occur in clippers and clampers.

12. Describe the procedure for troubleshooting voltage multipliers.

**Limiters** (or **clippers**)
Circuits used to eliminate a portion of an ac signal.

**dc restorers** (or **clampers**)
Circuits used to change the dc reference of an ac signal.

**Voltage multipliers**
Circuits used to produce a dc output voltage that is some multiple of an ac peak input voltage.

**Multisegment display**
An LED circuit used to display **alphanumeric symbols** (numbers, letters, and punctuation marks).

**D**iodes are among the primary components in power supplies, as we saw in Chapter 3, but they have many other common applications as well. In this chapter, we will take a look at several diode circuits and applications.

The first type of circuit that we will cover is the **limiter,** or **clipper.** Clippers are used to clip off or eliminate a portion of an ac signal. As you will see, there is a wide variety of clipper circuits.

The second circuit we will cover is the **dc restorer,** or **clamper.** Clampers are used to restore or change the dc reference of an ac signal. As you will see, it is very common to have an ac signal that is referenced at some dc voltage. For example, you may have a 12 $V_{pp}$ ac signal that varies equally above and below 2 V dc. A clamper could be used to change this dc reference voltage.

The third type of circuit we will discuss is the **voltage multiplier.** A voltage multiplier will produce a dc voltage that is some multiple of an ac peak voltage.

Finally, we will look at the most common LED application: the **multisegment display.** Multisegment displays are used to display **alphanumeric symbols**; that is, they are used to display numbers, letters, and punctuation marks.

## *4.1* CLIPPERS

You have already been introduced to the basic operating principles of the clipper. The half-wave rectifier is basically a clipper that eliminates one of the alternations of an ac signal.

*Clippers* are classifed as being either *series clippers* or *shunt clippers.*

There are two types of clippers: the *series* clipper and the *shunt* clipper. Each of these can be *positive* or *negative.* All these clipper types are shown in Figure 4.1. The *series clipper* contains a diode that is *in series with the load.* The *shunt clipper,* on the other hand, contains a diode that is *in parallel with the load.*

### *Series Clippers*

**Series clipper**
A clipper that has an output when the diode is forward biased (conduction).

The **series clipper** (shown in Figure 4.2) has the same circuit operating characteristics as the half-wave rectifier. In fact, if you compare the circuit shown in Figure 4.2 to those shown in Figure 3.8, you will see that the half-wave rectifier is nothing more than a series clipper.

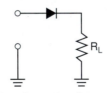

Negative series clipper                    Positive series clipper

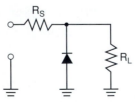

Negative shunt clipper                    Positive shunt clipper

**FIGURE 4.1**

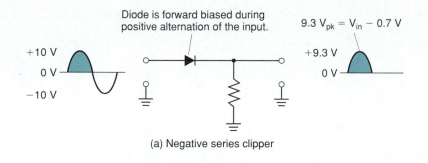

(a) Negative series clipper

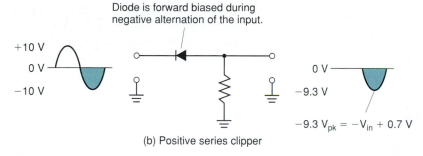

(b) Positive series clipper

**FIGURE 4.2**  **Series clipper operation.**

Since the operation of the series clipper is identical to that of the half-wave recti-  ◄ *OBJECTIVE 2*
fier, we will not go into any great detail on series clipper operation at this point. As a
review, the following points are made about series clippers:

1. When the diode in a negative series clipper is *forward biased* by the input signal, it
   conducts, and the load voltage is found as

$$V_L = V_{in} - 0.7 \text{ V} \qquad (4.1)$$

2. When the diode in the negative series clipper is *reverse biased* by the input signal, it
   does not conduct. Therefore,

$$V_{D1} = V_{in} \qquad (4.2)$$

and

$$V_L = 0 \text{ V} \qquad (4.3)$$

An example of a negative series clipper and its associated waveforms are shown in
Figure 4.2a.

3. The *positive* series clipper operates in the same fashion. The only differences are the
   following:

   a. The output voltage polarities are reversed.

   b. The current directions through the circuit are reversed.

   The positive series circuit and output waveform are shown in Figure 4.2b.

For a complete review of series clippers, refer to the coverage of half-wave rectifiers in Chapter 3.

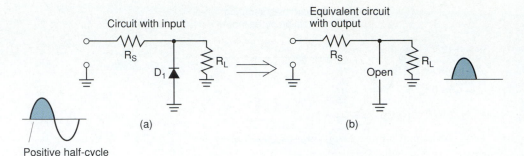

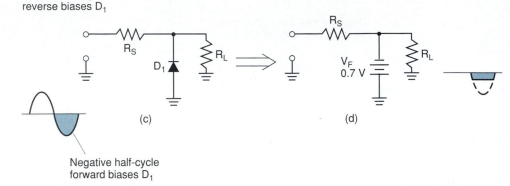

Positive half-cycle
reverse biases $D_1$

Negative half-cycle
forward biases $D_1$

**FIGURE 4.3    Shunt clipper operation.**

## Shunt Clippers

Lab Reference: The operation
illustrated in Figure 4.3 is
demonstrated in Exercise 6.

OBJECTIVE 3 ►

**Shunt clipper**
A clipper that has an output when
the diode is reverse biased (not
conducting).

*Note*: For the time being, we will
concentrate on the *negative shunt
clipper*. The *positive shunt clipper*
will be discussed later in this
section.

The operation of the **shunt clipper** is exactly the opposite of that of the series clipper. The series clipper has an output when the diode is forward biased and no output when the diode is reverse biased. The *shunt clipper* has an output when the diode is *reverse biased* and shorts the input signal to ground when the diode is forward biased. This operation is illustrated in Figure 4.3.

The circuit shown in Figure 4.3 is a *negative shunt clipper*. When the diode in the negative shunt clipper is reverse biased, it is effectively removed from the circuit. This is shown in Figure 4.3b. With the diode reversed biased, the resistors form a voltage divider, and the load voltage can be found by

$$V_L = \frac{R_L}{R_L + R_S} V_{in} \qquad (4.4)$$

While the output signal will resemble the positive alternation of the input, the peak output voltage will be somewhat less than the peak input voltage. This point is illustrated in Example 4.1.

---

**EXAMPLE 4.1**

The negative shunt clipper shown in Figure 4.4 has a peak input voltage of $+12 \ V_{pk}$. What is the peak load voltage for the circuit?

*Solution:*   When the input is *positive*, the diode is reverse biased and does not conduct. Therefore, the peak load voltage is found as

---

$$V_L = \frac{R_L}{R_L + R_S} V_{in}$$

$$= \frac{5.1 \text{ k}\Omega}{6.1 \text{ k}\Omega} (+12 \text{ }V_{pk})$$

$$= 10 \text{ }V_{pk}$$

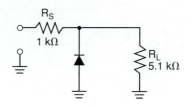

**FIGURE 4.4**

### PRACTICE PROBLEM 4.1

A negative shunt clipper has values of $R_L = 510 \text{ }\Omega$ and $R_S = 100 \text{ }\Omega$. If the input voltage is $+15 \text{ }V_{pk}$, what is the peak load voltage?

During the negative alternation of the input signal, the diode in the negative shunt clipper is *forward biased* and the voltage across the diode equals $V_F$. A representation of this, along with the resulting waveform, is shown in Figure 4.3d. Since the load is in parallel with the diode, load voltage also equals the forward voltage across the diode. By formula,

$$V_L = -V_F \qquad (4.5)$$

*Just a Reminder:* A diode won't conduct until its cathode is 0.7 V *more negative* than its anode. Thus, the diode in the negative shunt clipper won't conduct until the cathode is at −0.7 V. That is why $V_F$ is given as a negative value in equation (4.5).

Since the output side of $R_S$ is held to approximately −0.7 V when the diode is forward biased, the voltage across $R_S$ (which is designated $V_R$) is equal to the *difference between* $V_{in}$ and the value $V_F$. By formula,

$$V_R = -V_{in} + 0.7 \text{ V} \qquad (4.6)$$

when the diode is forward biased. Example 4.2 illustrates the circuit conditions that exist when the diode in a negative shunt clipper is forward biased.

---

The circuit described in Example 4.1 has a −12 $V_{pk}$ input voltage. Determine the values of $V_L$ and $V_R$ for the circuit.

**EXAMPLE 4.2**

*Solution:* Since the diode is forward biased, the load voltage is equal to the value of $V_F$ for the diode. Thus,

$$V_L = -0.7 \text{ V}$$

The voltage across the series resistor ($R_S$) is now found as

$$V_R = -V_{in} + 0.7 \text{ V}$$
$$= -12 \text{ }V_{pk} + 0.7 \text{ V}$$
$$= -11.3 \text{ }V_{pk}$$

### PRACTICE PROBLEM 4.2

The circuit described in Practice Problem 4.1 has a −15 $V_{pk}$ input. Determine the values of $V_L$ and $V_R$ for the circuit.

---

So far we have worked with the *negative shunt clipper*. As Example 4.3 illustrates, the *positive shunt clipper* works according to the same basic principles and is analyzed in the same fashion.

EXAMPLE 4.3

The *positive shunt clipper* shown in Figure 4.5 has the input wave-form shown. Determine the value for $V_L$ for each of the input alternations.

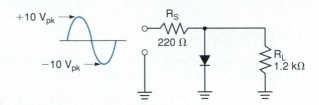

**FIGURE 4.5**

**Solution:** When the input is *positive*, the diode is forward biased. Thus,

$$V_L = 0.7 \text{ V}$$

and,

$$V_R = V_{in} - 0.7 \text{ V}$$
$$= 10 \; V_{pk} - 0.7 \text{ V}$$
$$= 9.3 \; V_{pk}$$

When the input to the circuit is *negative,* the diode is reverse biased and effectively removed from the circuit. Thus,

$$V_L = \frac{R_L}{R_L + R_S} V_{in}$$
$$= \frac{1.2 \text{ k}\Omega}{1.42 \text{ k}\Omega} (-10 \; V_{pk})$$
$$= -8.45 \; V_{pk}$$

***PRACTICE PROBLEM 4.3***

A *positive shunt clipper* with values of $R_S = 100 \; \Omega$ and $R_L = 1.1$ k$\Omega$ has a $\pm 12 \; V_{pk}$ input signal. Determine the value of $V_L$ for each alternation of the input. Also, determine the value of $V_R$ when the diode is forward biased.

The diode in a *positive* shunt clipper conducts when the input is *positive*. Thus, for the positive shunt clipper, equations (4.5) and (4.6) are modified as follows:

$$V_L = V_F$$
$$V_R = V_{in} - 0.7 \text{ V}$$

Using the values calculated in Example 4.3, we can draw the output waveform for the circuit. The waveform (shown in Figure 4.6) is clipped at 0.7 V and peaks at $-8.45 \; V_{pk}$.

## The Purpose Served by R$_S$

$R_S$ is included in the shunt clipper as a *current-limiting* resistor. Consider what would happen if the input signal forward biased the diode and $R_S$ was not in the circuit. This situation is illustrated in Figure 4.7.

Without $R_S$ in the circuit, the diode will short the signal source to ground during the positive alternation of the input signal. This will probably result in one of the following:

$V_{out} = 0.7$ V

$V_{out} = -8.45 \; V_{pk}$

**FIGURE 4.6**

FIGURE 4.7

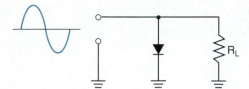

1. The diode being destroyed by excessive forward current
2. One or more components in the signal source being destroyed by the excessive current demand of the clipper

For example, assume that the circuit input signal in Figure 4.7 has a value of $+V_{pk} = +12$ V. When the input signal is at its positive peak, we have the diode shorting the $+12$ $V_{pk}$ to ground. The resulting high current will damage either the diode or the signal source.

In any practical situation, the value of $R_S$ will be *much lower than* the value of $R_L$. When this is the case, the load voltage (when the diode is reverse biased) will be approximately equal to the value of $V_{in}$. This point is illustrated in Example 4.4.

---

Determine the peak load voltage for the circuit shown in Figure 4.8. Assume that the diode is reverse biased.

*EXAMPLE **4.4***

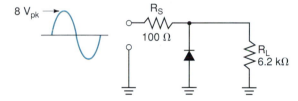

**FIGURE 4.8**

***Solution:*** The peak load voltage is found using equation (4.4) as follows:

$$V_L = \frac{R_L}{R_L + R_S} V_{in}$$

$$= \frac{6.2 \text{ k}\Omega}{6.3 \text{ k}\Omega} (8 \ V_{pk})$$

$$= 7.87 \ V_{pk}$$

With the relationship of $R_S$ and $R_L$ in this circuit, we could have assumed that $V_{in}$ and $V_L$ were approximately equal in any practical analysis situation.

---

## Biased Clippers

A **biased clipper** is a shunt clipper that contains a dc voltage source in series with the diode. This allows the circuit to clip voltages other than the normal diode $V_F$ of 0.7 V. Figure 4.9 shows examples of a positive- and negative-biased clipper. The bias voltage, or $V_B$, is in series with the shunt diode, causing it to have a reference above or below ground. The point at which the diode will clip the waveform is equal to the sum of $V_F$ and $V_B$.

◄ *OBJECTIVE 4*

**Biased clipper**
A shunt clipper that uses a dc voltage source to bias the diode.

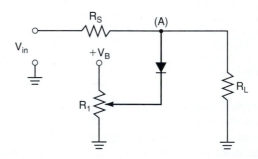

(a) Positive-biased clipper

(b) Negative-biased clipper

**FIGURE 4.9**

**Lab Reference:** The operation of a biased clipper is demonstrated in Exercise 6.

**Positive-biased clipper**
Clips the input signal at $V_B + 0.7$ V.

**Negative-biased clipper**
Clips the input signal at $V_B - 0.7$ V.

The **positive-biased clipper** (Figure 4.9a) will clip the input signal at $V_B + 0.7$ V. The actual value at which the circuit will clip the input signal depends on the value of the biasing voltage ($V_B$). If $V_B$ is 2 V, the input signal will be clipped at 2.7 V. If the value of $V_B$ is 5 V, the input signal will be clipped at 5.7 V, and so on.

The **negative-biased clipper** (Figure 4.9b) works in the same fashion, but it will clip the input signal at $-V_B - 0.7$ V. If $V_B$ is $-2$ V, the input signal will be clipped at $-2.7$ V. If the value of $V_B$ is $-5$ V, the input signal will be clipped at $-5.7$ V, and so on.

In practice, a potentiometer is used to provide an adjustable value of $V_B$, as shown in Figure 4.10. In this circuit, the biasing voltage ($+V_B$) is connected to the diode via the potentiometer ($R_1$). $R_1$ is adjusted in this circuit to provide the desired clipping limit at point $A$ in the circuit. Note that the clipping voltage will never be higher than $V_B + 0.7$ V or as low as ground due to the fact that there will be *some* voltage dropped across $R_S$ whenever the diode is conducting. By reversing the direction of the diode and the polarity of $V_B$, the circuit in Figure 4.10 can be modified to work as a negative-biased clipper.

## One Final Note

As you can see, there are many different clipper configurations. These configurations, which are used for a variety of applications, are summarized in Figure 4.11. What are they used for? This is what we will be discussing in the next section.

**FIGURE 4.10**

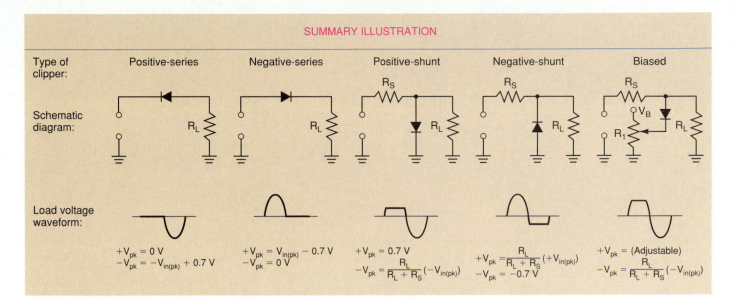

| Type of clipper: | Positive-series | Negative-series | Positive-shunt | Negative-shunt | Biased |
|---|---|---|---|---|---|

Schematic diagram:

Load voltage waveform:

Positive-series:
$+V_{pk} = 0 \text{ V}$
$-V_{pk} = -V_{in(pk)} + 0.7 \text{ V}$

Negative-series:
$+V_{pk} = V_{in(pk)} - 0.7 \text{ V}$
$-V_{pk} = 0 \text{ V}$

Positive-shunt:
$+V_{pk} = 0.7 \text{ V}$
$-V_{pk} = \dfrac{R_L}{R_L + R_S}(-V_{in(pk)})$

Negative-shunt:
$+V_{pk} = \dfrac{R_L}{R_L + R_S}(+V_{in(pk)})$
$-V_{pk} = -0.7 \text{ V}$

Biased:
$+V_{pk} = \text{(Adjustable)}$
$-V_{pk} = \dfrac{R_L}{R_L + R_S}(-V_{in(pk)})$

**FIGURE 4.11**

1. What purpose is served by a *clipper*?
2. What is another name for a clipper?
3. What purpose is served by a *clamper*?
4. What is another name for a clamper?
5. Discuss the differences between series and shunt clippers.
6. What purpose is served by $R_S$ in the shunt clipper?
7. Describe the operation of a biased clipper.

# 4.2 CLIPPER APPLICATIONS

Clippers are used in a wide variety of electronic systems. They are generally used to perform one of several functions:

1. Altering the shape of a waveform
2. Circuit transient protection
3. Detection

You have already seen one example of the first function in the half-wave rectifier. This circuit alters the shape of an ac signal, changing it to pulsating dc. You will see another application of this type in this section, along with several transient protection circuits.

## Transient Protection

A **transient** is an abrupt current or voltage spike that has an extremely short duration. A *current surge* would be one type of transient.

    Transients can do a large amount of damage to circuits whose inputs must stay within certain voltage or current limits. For example, many digital circuits have inputs that can handle only voltages that fall within a specified range. For these circuits, a volt-

**Transient**
An abrupt current or voltage spike.

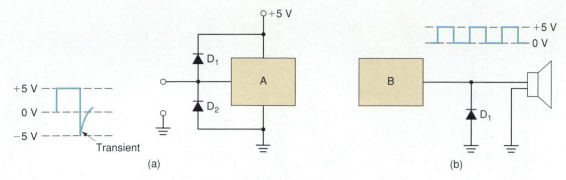

+5 V
0 V
-5 V
Transient

(a)

+5 V
0 V

(b)

**FIGURE 4.12**

age transient that goes outside the specified voltage range cannot be allowed to reach the circuit. A clipper could be used in this case to prevent such a transient from reaching the digital circuit. This point is illustrated in Figure 4.12a.

The block labeled A represents a digital circuit whose input voltage must not be allowed to go outside the range of 0 to +5 V. The clipper will protect the circuit from any voltage transients outside this range. For example, let's assume that the transient shown in the figure occurs on the input line. The transient will forward bias $D_2$, causing the diode to conduct. With $D_2$ conducting, the transient will be shorted to ground, protecting the input to circuit A. If the input square wave should go above 5 V, $D_1$, which is located between the +5-V supply and the input, would be forward biased. This would short any input greater than 5 V back to the +5-V supply.

These examples assume that we are using ideal diode representations. In an actual circuit, the input could go down to −0.7 V before $D_2$ would conduct and go up to 5.7 V before $D_1$ would conduct. This is because of the $V_F$ of the diode.

*A Practical Consideration*: Let's say that circuit B in Figure 4.12b goes bad. When it is replaced, the new circuit works fine until it tries to drive the speaker. After driving the speaker for a few seconds, the new circuit B also goes bad. The problem in this case is an open diode. If the diode is open, the counter emf produced by the speaker isn't shorted to ground and destroys each new circuit that is used to replace the previous one.

The circuit in Figure 4.12b includes a clipper designed to protect the *output* of circuit B from transients produced when a square wave is used to drive a speaker (a common practice with computer sound effects). A speaker is essentially a big coil (inductor). When a square wave is used to drive the coil in a speaker, that coil produces a *counter emf*. This counter emf is produced each time that the output voltage makes the transition from +5 to 0 V. The counter emf will be equal in magnitude to the original voltage, but opposite in polarity. Thus, each time circuit B tries to drive the output line to 0 V, the speaker will try to force the line to continue to −5 V. This −5 V counter emf could destroy the driving circuit. However, the clipping diode ($D_1$) will short the counter emf produced by the speaker coil to ground before it can harm the driving circuit.

## The AM Detector

**AM detector**
A diode clipper that converts a varying amplitude ac input to a varying dc level.

Another common application for the diode clipper can be found in a typical AM receiver. In this case, the clipper is part of a circuit called an **AM detector**. A simple detector is shown in Figure 4.13.

The purpose of the detector is to produce an output voltage representative of the peak variations in the input signal. This output is produced by the capacitor, which charges to the average of the input signal.

As shown, the input signal to the circuit has an average value of 0 V. This is due to the fact that each positive peak has an equal and opposite negative peak. The clipping diode solves this problem by eliminating the negative portion of the input waveform. With the negative portion eliminated, the capacitor charges and discharges at the rate of input variation. This provides a signal at the load that is a reproduction of the peak variations in the input.

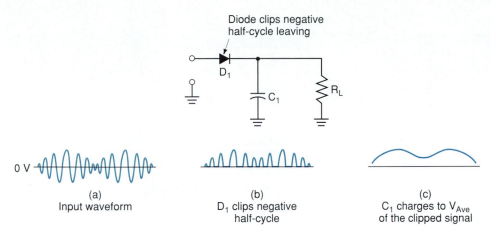

Diode clips negative
half-cycle leaving

$D_1$

$C_1$    $R_L$

(a)
Input waveform

(b)
$D_1$ clips negative
half-cycle

(c)
$C_1$ charges to $V_{Ave}$
of the clipped signal

0 V

**FIGURE 4.13    $C_1$ charges to $V_{ave}$ of the clipped signal.**

## One Final Note

The applications covered in this section were intended only to give you an idea of clipper applications. To cover every possible circuit would require several more chapters. Even though we have covered only a few clipper circuits, you should now have a good idea of the purposes they serve.

1. What is a *transient*?
2. Why must digital circuit inputs be protected from voltage transients?
3. Describe the operation of the AM detector.

## *4.3* CLAMPERS (dc RESTORERS)

A clamper is a circuit designed to shift a waveform either above or below a given reference voltage without distorting the waveform. There are two types of clampers: the *positive clamper* and the *negative clamper*. The input/output characteristics of these two circuits are illustrated in Figure 4.14.

Figure 4.14a shows a typical ac sine wave. As you can see, the sine wave has peak values of $\pm 10$ V for an overall value of 20 $V_{PP}$. We will use this waveform to help illustrate the operation of the two clamper circuits.

A **positive clamper** will shift its input waveform so that the *negative* peak output voltage of the waveform will be approximately equal to the dc reference voltage of the clamper. For example, let's say that we have a positive clamper with a dc reference voltage of 0 V. (You will be shown later where this *dc reference voltage* comes from.) If we were to apply the waveform in Figure 4.14a to this clamper, we would get the waveform shown in Figure 4.14b. As you can see, the waveform now has peak values of $+ 20$ V and 0 V (the negative peak) for an overall value of 20 $V_{PP}$. In this case, the positive clamper has shifted the entire waveform so that the negative peak voltage is approximately equal to the clamper's dc reference voltage.

The **negative clamper** will shift its input waveform so that the *positive* peak output voltage of the waveform will be approximately equal to the dc reference voltage of the clamper. Now let's assume that we have a *negative* clamper with a dc ref-

◀ *OBJECTIVE 5*

**Positive clamper**
A circuit that shifts an entire input signal *above* a dc reference voltage.

**Negative clamper**
A circuit that shifts an entire input signal *below* a dc reference voltage.

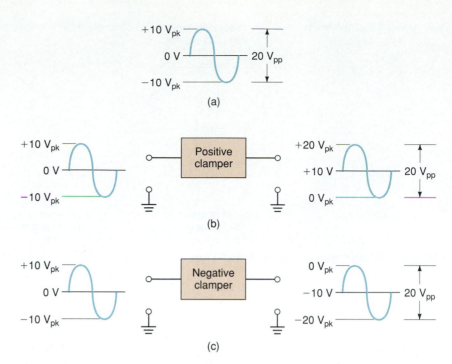

**FIGURE 4.14**

erence voltage of 0 V. If we apply the original waveform to this circuit, we get the output waveform shown in Figure 4.14c. In this case, the positive peak voltage of the input has been shifted so that it is approximately equal to the clamper's dc reference voltage.

## $V_{ave}$ *Shift*

You were shown in Chapter 3 how to determine the dc average ($V_{ave}$) value of a *rectified* sine wave. When we are dealing with a sine wave that is *not* rectified, the value of $V_{ave}$ falls halfway between the positive and negative peak voltage values. For example, if we were to measure the waveform in Figure 4.14a with a dc voltmeter, we would get a reading of 0 V. This is because the dc average of the waveform (0 V) falls halfway between +10 and −10 $V_{pk}$.

The effect of clamper operation on the value of $V_{ave}$ for a waveform.

When a waveform is shifted by a clamper, the value of $V_{ave}$ for the waveform may not equal 0 V any more. For example, consider the output waveform shown in Figure 4.14b. This waveform has peak values of +20 and 0 $V_{pk}$. The dc average, which falls halfway between the peak values, is +10 V. Thus, if we were to measure this output waveform with a dc voltmeter, we could get a reading of 10 $V_{dc}$. By the same token, if we were to measure the output waveform shown in Figure 4.14c with a dc voltmeter, we would get a reading of −10 $V_{dc}$.

Clampers do not affect the peak-to-peak and rms values of a waveform.

An interesting point can be made at this time. While the value of $V_{ave}$ for a waveform will normally change when the waveform goes through the clamper, the *ac values* of a waveform will not. For example, all three waveforms in Figure 4.14 have a peak-to-peak voltage of 20 $V_{PP}$. If we were to convert this value to rms, we would get a value of 7.07 $V_{ac}$. If we were to measure the three waveforms with an ac voltmeter, they would all get a reading of approximately 7.07 $V_{ac}$. Thus, while the clamper changes the peak and average (dc) values of a waveform, it does not change the peak-to-peak or rms value of the original signal.

**FIGURE 4.15** Clamper *charge* and *discharge* time constants.

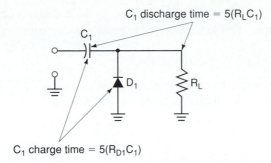

## Clamper Operation

The schematic diagram of the positive clamper is shown in Figure 4.15. As you can see, this clamper is very similar to the shunt clipper, with the exception of the added capacitor.

The clamper works on the basis of **switching time constants.** When the diode is forward biased, it provides a charging path for the capacitor. For this circuit, the charging time constant is found as

$$\tau = R_{D1}C_1$$

and the approximate total charge time is found as

$$T_C = 5(R_{D1}C_1) \tag{4.7}$$

where $R_{D1}$ = the bulk resistance of the diode.

When the diode is reverse biased, the capacitor will start to discharge through the resistor. Thus, the discharge time constant for the circuit is found as

$$\tau = R_L C_1$$

and the approximate total discharge time is found as

$$T_D = 5(R_L C_1) \tag{4.8}$$

Note the different resistance values that appear in equations (4.7) and (4.8). As Example 4.5 shows, the difference between the capacitor charge and discharge times is significant.

**Switching time constants**
A term used to describe a condition when a capacitor has different charge and discharge times.

---

Determine the capacitor charge and discharge times for the circuit shown in Figure 4.15. Assume that the forward resistance of the diode is 10 Ω, the value of $R_L$ is 10 kΩ, and $C_1 = 1$ μF.

*Solution:* The capacitor charges through the diode. Therefore, the charge time is found as

$$\begin{aligned} T_C &= 5(R_{D1}C_1) \\ &= 5(10 \text{ Ω} \times 1 \text{ μF}) \\ &= 50 \text{ μs} \end{aligned}$$

The capacitor discharges through the resistor. Thus, the discharge time is found as

$$\begin{aligned} T_D &= 5(R_L C_1) \\ &= 5(10 \text{ kΩ} \times 1 \text{ μF}) \\ &= 50 \text{ ms} \end{aligned}$$

*EXAMPLE 4.5*

As you can see, it takes a lot longer for the capacitor to discharge than it does for it to charge. This is the basis of clamper circuit operation.

### PRACTICE PROBLEM 4.5

A clamper has values of $R_{D1} = 8\Omega$, $C_1 = 4.7 \mu F$, and $R_L = 1.2 k\Omega$. Determine the charge and discharge times for the capacitor.

*OBJECTIVE 6* ▶ Using the values obtained in Example 4.5, we will discuss the overall operation of the clamper. For ease of discussion, we're going to take a look at the circuit's response to a *square-wave* input. (When designed properly, clampers work for a variety of input waveforms.) We will also assume that the diodes are *ideal* components. The response of a clamper to a square-wave input is illustrated in Figure 4.16.

Figure 4.16a shows the input waveform. When the input goes to its positive peak (+5 V), $D_1$ is forward biased. This provides a low-resistance current path for charging $C_1$, as shown in Figure 4.16b. Assuming that the charge time of the capacitor ($5R_{D1}C_1$) is very *short*, $C_1$ will charge to the full value of the peak input voltage, 5 V. With the full applied voltage dropped across $C_1$, the load voltage has a positive peak value of 0 V.

When the input waveform begins to go negative, the output side of the capacitor (which was at 0 V) also begins to go negative. This turns $D_1$ off, and the capacitor is forced to discharge through the resistor, as shown in Figure 4.16c. Assuming that the discharge time of the capacitor ($5R_LC_1$) is very *long*, $C_1$ will lose very little of its charge. With the input voltage and $V_C$ having the values and polarities shown in Figure 4.16c, the load voltage has a negative peak value of $-10$ V. Thus, with the input shown, the clamper provides an output square wave that varies between 0 V and $-10$ V, as shown in Figure 4.16d.

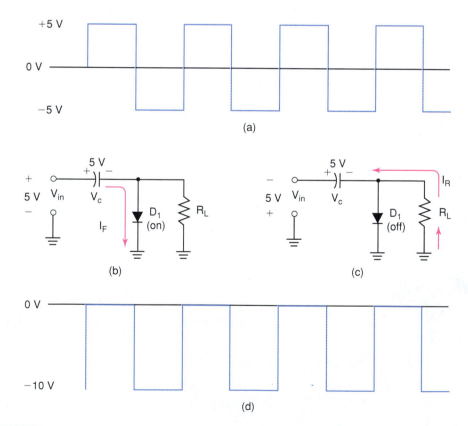

**FIGURE 4.16**

Figure 4.17 shows what occurs when the size of the capacitor is reduced and the input frequency is held constant. The time constant is no longer large enough to keep the capacitor fully charged.

Remember that the clamper illustrated here uses the ideal diode model. The diode appears as either an open or a short. In reality, the $V_F$ of the diode (0.7 V) would affect the output waveform. The capacitor could charge only up to the difference between $V_{in(pk)}$ and the $V_F$ of the diode. This would cause the output waveform to be shifted by approximately −0.7 V.

## Negative Clampers Versus Positive Clampers

As was the case with clippers, *the difference between the negative clamper and the positive clamper is simply the direction of the diode.* This point is illustrated in Figure 4.18. Figure 4.18a shows a negative clamper and its effect on a 40 $V_{PP}$ sine-wave input. By reversing the diode, we get the circuit operation shown in Figure 4.18b. Here the diode direction is reversed, making it a positive clamper. Since the diode is returned to ground (0 V), the circuit shifts the input waveform until the negative peak voltage of the waveform is approximately equal to 0 V.

It is important to remember that, when reversing the diode, the capacitor (if polarized) must also be reversed. The positive side of a polarized capacitor must always point toward the most positive voltage. In a negative clamper, the positive terminal of the capacitor will be connected to the signal source. In a positive clamper, the positive side of the capacitor will point toward the load.

There is a quick and easy memory trick for determining what type of clamper you are dealing with: *If the diode is pointing up (away from ground), the circuit is a positive clamper. If the diode is pointing down (toward ground), the circuit is a negative clamper.*

## Biased Clampers

**Biased clampers** allow us to shift a waveform so that it falls above or below a dc reference other than 0 V. Several biased clampers are shown in Figure 4.19.

The circuit in Figure 4.19a is easy to analyze using the procedure we have just established. The diode is pointing down, so the circuit is a negative clamper. What is the

**Biased clamper**
Allows a waveform to be shifted above or below a dc reference other than 0 V.

**Lab Reference:** Both positive and negative clampers are analyzed in Exercise 7.

5 V
0

(a) Clamper output with long discharge time constant

5 V
0

(b) Clamper output with short discharge time constant

**FIGURE 4.17**

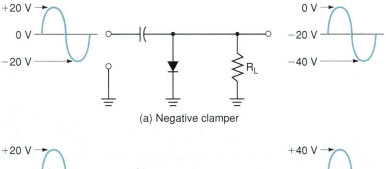

(a) Negative clamper

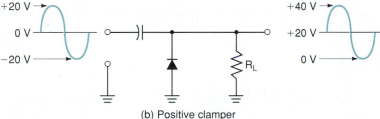

(b) Positive clamper

**FIGURE 4.18**

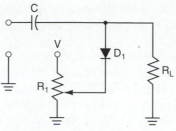

(a) Negative-biased clamper

(b) Positive-biased clamper

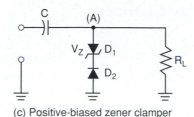

(c) Positive-biased zener clamper

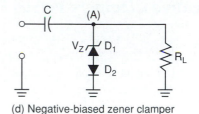

(d) Negative-biased zener clamper

**FIGURE 4.19**

**Lab Reference:** The operation of a zener clamper is demonstrated in Exercise 7.

dc reference voltage for the circuit? That depends on the value of the dc voltage source ($V$) and the setting of $R_1$. If $V$ is negative, the diode potential will fall somewhere between $-V$ and approximately 0 V. For example, let's say that $V$ is $-20$ V, and $R_1$ is set so that the diode potential is $-10$ V. With this setting, the clamper will shift the input waveform so that its positive peak voltage is approximately $-10$ V.

The circuit shown in Figure 4.19b is a positive-biased clamper. It works using the same concepts as the negative-biased clamper. The output waveform is now shifted so that the negative peak voltage of the waveform is approximately the same positive voltage set at the anode of $D_1$. By having $R_1$ adjustable, we can adjust the dc reference to any voltage between $V$ and ground.

The circuits in Figures 4.19c and 4.19d require a bit more explaining. The circuits shown here are referred to as **zener clampers**. The zener diodes are used to determine the dc reference voltages of the circuits. For example, if a 10 V zener is used, the dc reference voltage of Figure 4.19c will be approximately $-10$ V. If a 10 V zener is used in Figure 4.19d, the dc reference voltage of the circuit will be approximately $+10$ V. If 20 V zeners are used, the dc references of the above circuits will be $+20$ V or $-20$ V.

Why are the *pn*-junction diodes included? The operation of the clamper depends on the *one-way* action of the diode. As you learned in Chapter 2, a zener will conduct in both the zener breakdown region and the forward-bias region. For example, if point A in Figure 4.19d is more positive than the value of $V_Z$, the zener will conduct, as will the forward-biased *pn*-junction diode. On the other hand, if point A is negative, the zener diode will try to conduct in its forward operating region. However, this same negative potential will reverse bias the *pn*-junction diode, blocking conduction. Thus, the *pn*-junction diode ensures that the zener diode will conduct only in its reverse breakdown operating region.

It is easy to tell the difference between a positive or negative clamper circuit that uses a zener diode as a voltage reference. The direction of the *pn*-junction diode determines the polarity of the clamper because the diode determines the polarity of the charge on the capacitor.

**Zener clampers**
Clampers that use zener diodes to determine the dc reference voltages.

## *One Final Note*

Zener clampers are somewhat limited; that is, they come in only two varieties:

1. *Negative* clampers with *positive* dc reference voltages
2. *Positive* clampers with *negative* dc reference voltages

You cannot have a *positive* clamper with a *positive* dc reference voltage, for example. This is due to the fact that the *pn*-junction diode and the zener diode must be connected either as common-cathode (Figure 4.19c) or common-anode (Figure 4.19d) components. If $D_2$ in Figure 4.19c is turned around and point A is positive, nothing will stop $D_1$ from going into forward conduction. This will defeat the whole purpose of including $D_2$ in the circuit. Thus, $D_2$ in Figure 4.19c cannot be reversed to produce a negative clamper. By the same token, $D_2$ in Figure 4.19d cannot be reversed to produce a positive clamper.

## Summary

The clamper shifts a waveform to a predetermined dc reference voltage. The dc reference voltage for a clamper is determined by the potential to which the diode in the circuit is referenced.

When the diode in a clamper is pointing up (away from the ground), the circuit is a positive clamper. The positive clamper will shift the input waveform so that its negative peak voltage is approximately equal to the dc reference voltage of the circuit. When the diode in a clamper is pointing down (toward ground), the circuit is a negative clamper. The negative clamper will shift the input waveform so that its positive peak voltage is approximately equal to the dc reference voltage of the circuit.

1. Describe the difference between *positive clampers* and *negative clampers* in terms of input/output relationships.
2. What effect does a clamper have on the value of $V_{ave}$ for a given input waveform?
3. What piece of test equipment would you use to measure the value of $V_{ave}$ for a clamper output signal?
4. What effect does a clamper have on the rms voltage of a sine-wave input?
5. Explain the operation of a clamper in terms of switching time constants.
6. What determines the dc reference voltage of a clamper?
7. How can you tell whether a given clamper is a *positive* or *negative* clamper?
8. Why can't you have a positive zener clamper with a positive dc reference voltage?
9. Describe the purpose served by biased clampers.

## 4.4 VOLTAGE MULTIPLIERS

Voltage multipliers are circuits that provide a *dc output* that is a multiple of the *peak input voltage*. For example, a *voltage doubler* provides a dc output voltage that is *twice* the peak input voltage, and so on.

While voltage multipliers provide an output voltage that is much greater than the peak input voltage, they are not power generators. When a voltage multiplier increases the peak input voltage by a given factor, the peak input current is *decreased* by approximately the same factor. Thus, the actual output power from a voltage multiplier is *never* greater than the input power.

Because the current in a voltage multiplier is reduced every time the voltage is increased, the device will normally end up with very low current capability. Because of this, voltage multipliers are usually used in high-voltage, low-current applications. One typical application is to supply the high-voltage, low-current signal required to operate the cathode-ray tube (a.k.a. the picture tube) in a television.

*A Practical Consideration*: Since the diodes in a voltage multiplier will dissipate some power, the output power for the circuit will actually be *less than* the input power.

In this section, we will take a look at several types of voltage multipliers. For ease of discussion, the multiplier circuits are shown as being driven by a simple voltage source, $V_S$. It should be noted that multiplier circuits, in practice, are generally used in power supply applications. Thus, they are usually connected to the secondary of a power supply transformer.

## Half-Wave Voltage Doublers

OBJECTIVE 7 ▶

**Half-wave voltage doubler**
Provides a dc output voltage that is approximately twice its peak input voltage. The name is derived from the fact that the output capacitor is charged during one-half of the input cycle.

The **half-wave voltage doubler** is made up of two diodes and two capacitors, normally electrolytics. The circuit is shown in Figure 4.20, along with its voltage source and load resistance.

The operation of the half-wave doubler is easier to understand if we assume that the diodes are *ideal* components. During the negative alternation of the input (shown in Figure 4.21a), $D_1$ is forward biased and $D_2$ is reverse biased by the input signal polarity. If we represent $D_1$ as a *short* and $D_2$ as an *open,* we get the equivalent circuit shown in the figure. As you can see, $C_1$ will charge until its plate-to-plate voltage is equal to the source voltage. At the same time, $C_2$ will be in the process of discharging through the load resistance. (The source of this charge on $C_2$ will be explained in a moment.)

When the input polarity reverses, we have the circuit conditions shown in Figure 4.21b. Since $D_1$ is *off,* it is represented as an *open* in the equivalent circuit. Also, $D_2$ (which is *on*) is represented as a *short.* Using the equivalent circuit, it is easy to see that $C_1$ (which is charged to the peak value of $V_S$) and the source voltage ($V_S$) now act as *series-aiding* voltage sources. Thus, $C_2$ will charge to the sum of the series peak voltages, $2 V_{S(pk)}$.

When $V_S$ returns to its original polarity, $D_2$ is again turned off. With $D_2$ off, the only discharge path for $C_2$ is through the load resistance. Normally, the time constant of this circuit will be such that $C_2$ has little time to lose any of its charge before the input reverses polarity again. In other words, during the negative alternation of the input, $C_2$ will discharge *slightly.* Then, during the positive alternation $D_2$ is turned on and $C_2$ recharged until its plate-to-plate voltage again equals $2 V_{S(pk)}$.

The time constant formed by the output capacitor and the load resistance prevents the capacitor from discharging rapidly.

Since $C_2$ barely discharges between input cycles, the output waveform of the half-wave voltage doubler closely resembles that of a filtered half-wave rectifier. Typical input and output waveforms for a half-wave voltage doubler are shown in Figure 4.22. As the figure shows, the circuit will have a dc output voltage and ripple voltage that closely resemble the output from a filtered rectifier. The dc output voltage is approximated as

$$V_{dc} \cong 2 V_{S(pk)} \tag{4.9}$$

The output ripple from a given multiplier is calculated in the same manner as for a filtered half-wave rectifier. As such, the amount of ripple depends primarily on the values of the capacitors and the current demand of the load. High capacitor values and low current demands reduce the amount of ripple present at the output of a given multiplier.

**FIGURE 4.20**

**Lab Reference:** Half-wave voltage doubler operation is demonstrated in Exercise 8.

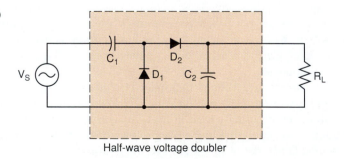

Half-wave voltage doubler

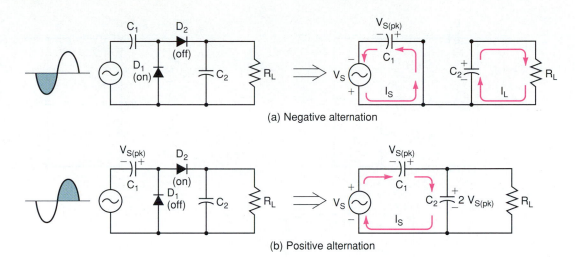

(a) Negative alternation

(b) Positive alternation

**FIGURE 4.21**

If the directions of the diodes in Figure 4.20 are reversed, along with the polarity of the two capacitors, we will have a negative voltage doubler. The circuit operates in the same way as the positive doubler except that $C_1$ will charge to a negative voltage, which will add to the incoming negative peak voltage. The circuit output will be a negative voltage at twice the value of $V_{in(pk)}$.

*Reversing the directions of the diodes and capacitors reverses the output polarity from the half-wave doubler.*

## Full-Wave Voltage Doublers

The **full-wave voltage doubler** closely resembles the half-wave doubler. It contains two diodes and two capacitors ($C_1$ and $C_2$), as shown in Figure 4.23a. In the figure, the voltage source and load resistance are both shown, along with an added filter capacitor, $C_3$. This filter capacitor is used to reduce the ripple output from the voltage doubler.

◄ *OBJECTIVE 8*

During the positive half-cycle of the input, $D_1$ is forward biased and $D_2$ is reverse biased by the voltage source. This gives us the equivalent circuit shown in Figure 4.23b. Again, we have idealized the diodes to simplify the circuit. Using the equivalent circuit, you can see that $C_1$ will charge to the value of $V_{S(pk)}$.

When the input polarity is reversed, $D_1$ is reverse biased and $D_2$ is forward biased. This gives us the equivalent circuit shown in Figure 4.23c. As this circuit shows, $C_2$ now charges to the value of $V_{S(pk)}$. Since $C_1$ and $C_2$ are in series, the total voltage across the two components is found as

*A Practical Consideration:* Sometimes a voltage multiplier will have a resistor in parallel with the output capacitor. This resistor, called a *bleeder resistor*, is used to provide a discharge path for the capacitor if the load opens.

$$V_{dc} = 2 V_{S(pk)}$$

With the added filter capacitor ($C_3$), there will be very little ripple voltage at the output of the full-wave voltage doubler under normal circumstances. This is one of the

**FIGURE 4.22**

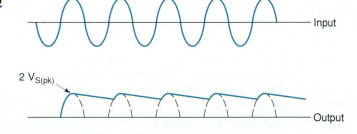

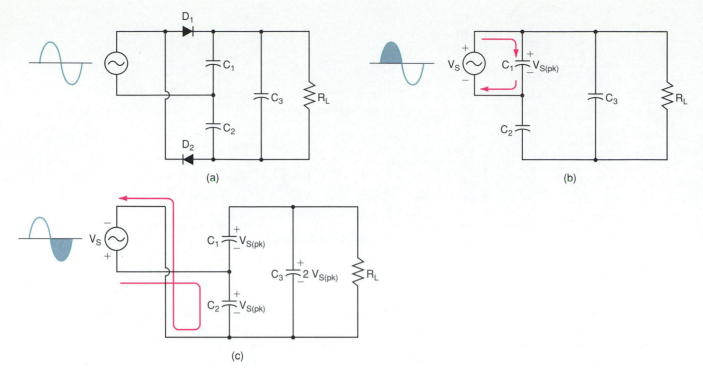

**FIGURE 4.23** Full-wave voltage doubler operation.

**Lab Reference:** Full-wave voltage doubler operation is demonstrated in Exercise 8.

advantages of using the full-wave voltage doubler in place of the half-wave doubler. Another advantage of this circuit is that it can be used to produce a *dual-polarity power supply*. This application will be shown later in this section.

## Voltage Triplers and Quadruplers

*Voltage triplers* and *voltage quadruplers* are both variations on the basic half-wave voltage doubler. The schematic diagram for the voltage tripler is shown in Figure 4.24.

**Voltage tripler**
Produces a dc output voltage that is three times the peak input voltage.

As you can see, the **voltage tripler** is very similar to the half-wave voltage doubler shown in Figure 4.21. In fact, the $D_1$, $D_2$, $C_1$, and $C_2$ circuitry forms a half-wave doubler. The added components ($D_3$, $C_3$, and $C_4$) form the rest of the voltage tripler.

There are two keys to the operation of this circuit. The first is the fact that the half-wave voltage doubler works *exactly* as described earlier in this section. The second is the fact that $D_3$ will conduct whenever $D_1$ conducts.

When $V_S$ is negative, as shown in Figure 4.24a, $D_1$ and $D_3$ both conduct, allowing $C_1$ and $C_3$ to charge up to $V_{S(pk)}$. Figure 4.24b shows what happens when $V_S$ goes positive. $D_2$ is turned on, allowing $C_2$ to charge up to $2\,V_{S(pk)}$. The voltages across $C_2$ and $C_3$ now add up to $3\,V_{S(pk)}$.

$C_4$ is a filter capacitor that is added to reduce the ripple in the dc output voltage. Since this capacitor is in parallel with the $C_2/C_3$ series circuit, it will charge to a plate-to-plate voltage equal to $3\,V_{S(pk)}$. Note that the value of $C_4$ is chosen so that the time constant formed with the load resistance will be extremely long. Thus, $C_4$ will barely discharge between charging cycles.

**Voltage quadrupler**
Produces a dc output voltage that is four times the peak input voltage.

The **voltage quadrupler** will provide a dc output that is four times the peak input voltage. As Figure 4.25 shows, the voltage quadrupler is simply made up of parallel half-wave voltage doublers. The half-wave doubler that is made up of $D_1$, $D_2$, $C_1$, and $C_2$ will charge $C_2$ until its plate-to-plate voltage is equal to $2\,V_{S(pk)}$. The half-wave doubler formed by the other components will charge $C_4$ until its plate-to-plate voltage is also

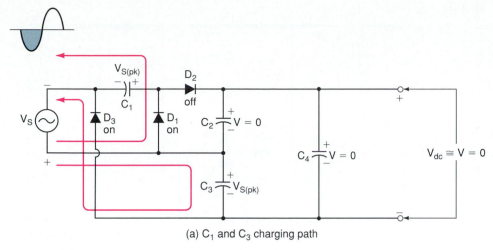

(a) $C_1$ and $C_3$ charging path

**Lab Reference:** Voltage tripler operation is demonstrated in Exercise 8.

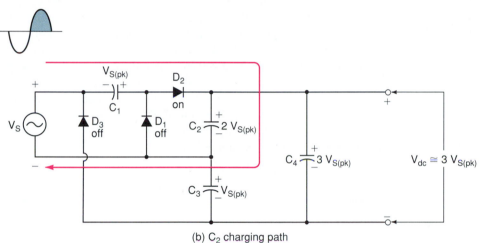

(b) $C_2$ charging path

**FIGURE 4.24    Voltage tripler operation.**

equal to $2 V_{S(pk)}$. The series combination of $C_2$ and $C_4$ will charge the filter capacitor ($C_5$) until its plate-to-plate voltage is equal to $4 V_{S(pk)}$. Again, the value of $C_5$ is chosen so that it will retain most of its charge between charging cycles.

## A Dual-Polarity Power Supply

One application for the voltage multiplier can be seen in a basic dual-polarity dc power supply. A **dual-polarity power supply** is one that provides both positive and negative dc output voltages. One such supply is shown in Figure 4.26.

**Dual-polarity power supply**
A dc supply that provides both positive and negative dc output voltages.

**FIGURE 4.25**

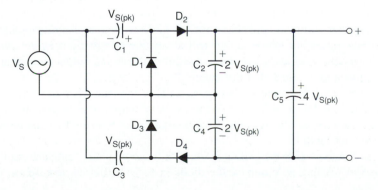

FIGURE 4.26

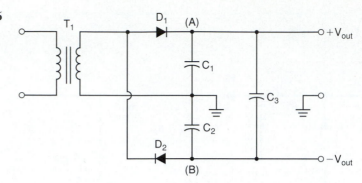

**Lab Reference:** The operation of a dual-polarity circuit is demonstrated in Exercise 8.

*A Practical Consideration:* When working with voltage multipliers, you should be extremely careful because of the high voltages that may be present within the circuit. All capacitors should be discharged before any component is removed from the circuit. When replacing capacitors in voltage multipliers, the replacement capacitors must have the same or higher voltage ratings as the originals. Using a smaller value capacitor (or reversing the capacitor polarity) can result in circuit damage and/or injury to you.

The transformer in the figure is connected to a full-wave voltage doubler. The common point for the capacitors ($C_1$ and $C_2$) is used as the power supply ground, meaning that all circuit voltages are measured with respect to this point. Thus, point A will be *positive* with respect to ground and point B will be *negative* with respect to ground. Note that the two dc output voltages will be approximately equal in magnitude to $V_{2(pk)}$. For example, if $V_{2(pk)}$ is 12 $V_{pk}$, the power supply will have outputs approximately equal to +12 $V_{dc}$ and $-12$ $V_{dc}$. The filter capacitor ($C_3$) is added to reduce the ripple in the dc output voltages.

A dual-voltage supply such as this is used only in low-current, noncritical applications where exact voltages are not a concern. A practical, dual-polarity power supply is more complicated and includes more filtering and some type of voltage regulation for each polarity. You will see one such power supply in Chapter 21.

## Section Review

1. Describe the operation of the half-wave voltage doubler.
2. Describe the operation of the full-wave voltage doubler.
3. Why are full-wave voltage doublers preferred over half-wave voltage doublers?
4. What is the similarity between the voltage tripler and the half-wave voltage doubler?
5. What is the similarity between the voltage quadrupler and the half-wave voltage doubler?

## 4.5 LED APPLICATIONS

The most common application for the LED is as an *indicator* that can replace a neon or incandescent lamp. The power indicator that lights on the front of your stereo whenever the system is on is probably an LED.

OBJECTIVE 9 ▶

Another common application for the LED is as a *level indicator* in switching circuits. Given a circuit whose output will always be at either a *high dc voltage level* or a *low dc voltage level,* the LED can be used to indicate the output *state* at any given time. This application is illustrated in Figure 4.27.

The circuit in Figure 4.27a uses an LED to indicate when the output from the driving circuit is at the +10 V level. With a +10 V output, the diode is forward biased and lights. When the output from the driving circuit is at 0 V, the LED is not forward biased and does not light.

In Figure 4.27b, the LED is used to indicate when the output from the driving circuit is 0 V. With a 0 V output, the diode is forward biased and lights. When the out-

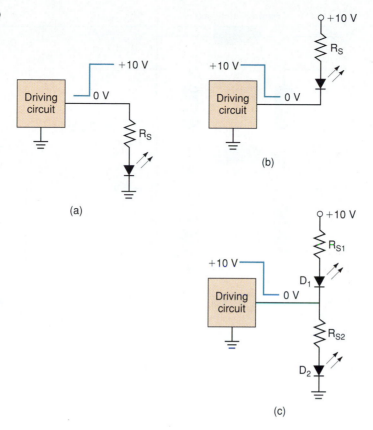

FIGURE 4.27  An LED as a level indicator.

(a)

(b)

(c)

put from the driving circuit is +10 V, the diode has +10 V on both sides of it and is not forward biased.

The circuit in Figure 4.27c is a combination of the first two circuits. $D_2$ lights when the output from the driving circuit is at +10 V, and $D_1$ lights when the output from the driving circuit is 0 V.

## Multisegment Displays

LEDs are used as the primary components in *multisegment displays*. Multisegment displays are used to display alphanumeric characters. One common multisegment display is the **seven-segment display** shown in Figure 4.28. The seven-segment display uses seven individual LEDs that are arranged in a figure 8 configuration. By lighting a combination of individual LEDs, any number from 0 to 9, along with several letters, can be displayed. For example, if LEDs b, c, f, and g are lit, the display will show the number 4.

The seven-segment display shown in Figure 4.28 is called a **common-cathode display.** This term means that the cathodes of the LEDs are tied together. Because of this, the display can function with only one ground connection. The individual LEDs are lit by applying a *high* voltage at the appropriate pins. Normally, this high voltage is within the range of +5 to +10 V. Note that each LED in the seven-segment display must have its own series current-limiting resistor.

Another type of display is the **common-anode display.** This type of display has a single +V input that is common to all the LEDs. Then individual LEDs are lit by applying a *ground* to the appropriate cathode.

Variations on the basic seven-segment display include those with added LEDs and built-in decoders. Figure 4.29a shows a display that allows any letter or number to be produced. Figure 4.29b represents a 5 × 7 dot matrix read-out that can display the digits 0 to 9, the alphabet, and other symbols by turning on a combination of the individual LEDs.

◄ *OBJECTIVE 10*

**Seven-segment display**
A device made up of seven LEDs shaped in a figure 8 that is used to display numbers.

**Common-cathode display**
A display in which all LED cathodes are tied to a single pin.

**Common-anode display**
A display in which all LED anodes are tied to a single pin.

Internal configuration

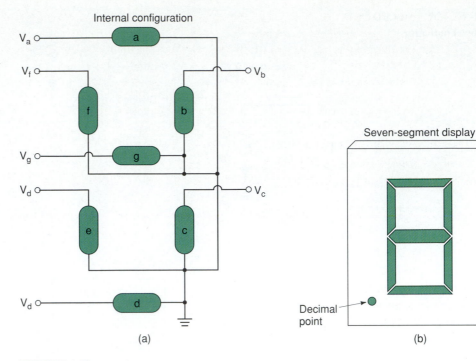

(a)

Seven-segment display

Decimal
point

(b)

**FIGURE 4.28**

## Liquid-Crystal Displays (LCDs)

**Liquid-crystal display (LCD)**
A display consisting of segments
that reflect (or do not reflect)
ambient light.

As a technician, you will probably encounter several other types of displays when working in the field. The most common of these is the **liquid-crystal display (LCD)**. The LCD is a display consisting of segments that can be made to reflect (or not reflect) ambient light.

LCDs are low-power devices that operate at voltages typically lower than 10 V. This makes the LCD ideal for use in portable battery-operated devices, such as wrist watches, cellular telephone displays, and laptop computers.

The segments in an LCD are usually arranged in the same pattern as an LED display like any of those shown in this section. When a voltage is applied across a given segment in a liquid-crystal display, that segment does not pass or reflect light, giving it a dark appearance. When the voltage is removed, the segment becomes clear, allowing the

**FIGURE 4.29**

(a) 16-Segment
alphanumeric display

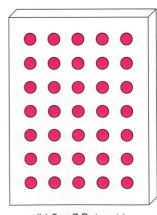

(b) 5 × 7 Dot matrix
display

background material to reflect the light in the room. Characters and numbers are produced in an LCD by applying a voltage to the appropriate segments.

If an LCD is designed to reflect ambient (external) light, it is usable only in well lighted areas. LCDs designed for use in low-light areas contain an LED (or some other source of light) that makes the display appear "on" when activated.

**Section Review**

1. Discuss the use of an LED as a power-level indicator.
2. Discuss the use of LEDs in multisegment displays.
3. Describe the LCD and its operation.

## 4.6 DIODE CIRCUIT TROUBLESHOOTING

Now that we have covered the operating principles of clippers, clampers, voltage multipliers, and multisegment displays, we will take a look at some of the fault symptoms that commonly occur in these circuits.

### Clipper Faults

The faults that normally occur in series clippers are the same as those of the half-wave rectifier that were discussed in Chapter 3. The fault–symptom relationships shown in Table 4.1 are those for the basic shunt clipper (shown in Figure 4.30) and the biased clipper.

◄ *OBJECTIVE 11*

### Clamper Faults

The fault–symptom relationships shown in Table 4.2 are those for the basic clamper, shown in Figure 4.31a.

The *biased clamper* (Figure 4.31b) may develop any of the fault–symptom combinations listed in Table 4.2. In addition, it can develop any of the fault–symptom combinations given in Table 4.3.

The *zener clamper* (Figure 4.31c) may develop any of the capacitor fault–symptom combinations listed in Table 4.2. Also, if either diode *opens*, the results will be the same as those shown in Table 4.2 for an open diode. If either diode in the zener clamper *shorts*, the results are slightly different from those listed in Table 4.2. The results of either diode shorting in the zener clamper are shown in Table 4.4.

As always, there is a strong possibility that both diodes in the zener clamper will be destroyed if either diode shorts. You may recall from our discussion on rectifier circuits that it is not uncommon for a shorting diode to destroy any series diodes in the process.

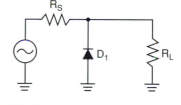

**FIGURE 4.30**

**TABLE 4.1  Shunt Clipper Faults**

| Fault | Symptom(s) |
|---|---|
| $R_S$ open | Since the source is isolated from the load, the load voltage and current both drop to zero. |
| $D_1$ open | Clipping action is lost, so the output waveform is nearly identical in shape to the input waveform. |
| $D_1$ shorted | The symptoms are the same as those for $R_S$ open because the full applied voltage is dropped across $R_S$. The shorted diode shorts out the load. |

*A Practical Consideration:* Carbon resistors do not internally short. If a short exists, the short is normally caused by a solder bridge or by accidentally using the wrong value of resistor. A visual inspection will find either of these two problems.

**TABLE 4.2  Clamper Faults**

Faults common to most clampers.

| Fault | Symptom(s) |
|---|---|
| $C_1$ open | The voltage source is isolated from the rest of the circuit, so load voltage and current drop to zero. |
| $C_1$ shorted | The circuit resembles a shunt clipper, which lacks the current-limiting resistor. Either the diode will be destroyed from excessive current or the signal source will be shorted to ground and destroyed. |
| $C_1$ leaky | The capacitor attempts to charge but will not be able to hold the charge for any period of time. This causes the dc reference of the output signal to constantly change. The waveform itself may also be extremely distorted. |
| $D_1$ open | All clamping action is lost and the output waveform is centered around 0 V. |
| $D_1$ shorted | The voltage source is shorted to ground by the short $RC$ time constant of the capacitor and the shorted diode. The output from the voltage source is loaded down, and the voltage source itself may be damaged. |

Faults common to most clampers.

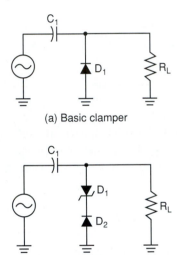
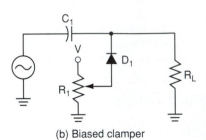

(a) Basic clamper

(b) Biased clamper

(c) Zener clamper

**FIGURE 4.31**

**TABLE 4.3  Additional Biased-Clamper Faults**

Additional faults that may develop in *biased clampers*.

| Fault | Symptom(s) |
|---|---|
| $V_B$ missing | If the dc biasing voltage is missing (drops to 0 V), the clamper works like an unbiased clamper. If the dc voltage source in an electronic system goes out, many other circuits in the system (including the clamper signal source) may also stop working. |
| $R_1$ open | If $R_1$ opens, the results are the same as $D_1$ opening. All clamping action is lost, and the output waveform is centered around 0 V. |

**TABLE 4.4  Additional Zener Clamper Faults**

Additional faults that may develop in the *zener clamper*.

| Fault | Symptom(s) |
|---|---|
| $D_1$ shorted | If $D_1$ shorts, the clamper acts as an unbiased clamper. As you can see, shorting $D_1$ in Figure 4.31c gives us the exact same circuit as the one shown in Figure 4.31a. |
| $D_2$ shorted | The zener diode turns on during both alternations of the input waveform. The result is an extremely distorted output waveform. |

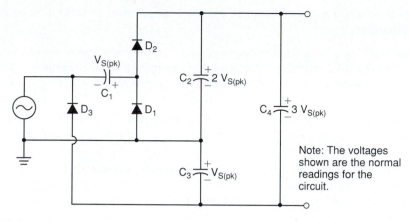

**FIGURE 4.32**

The best thing to do whenever you find a shorted diode is to replace all diodes in series with the component.

## Voltage-Multiplier Faults

Of all the circuits covered in this chapter, voltage multipliers are by far the most difficult to troubleshoot. The maze of diodes and capacitors that make up the voltage multiplier can develop many faults, each having symptoms that may closely resemble those of some other fault.

◄ *OBJECTIVE 12*

The best approach to troubleshooting the voltage multiplier is to start by reading the voltage across the output filter capacitor. For example, if we were to troubleshoot the voltage tripler shown in Figure 4.32, we would start by reading the voltage across $C_4$.

The value of the voltage across the filter capacitor may help to isolate the area where the fault is located. For example, let's say that we are testing the voltage tripler shown in Figure 4.32, and the voltage across $C_4$ is close to 2 $V_{S(pk)}$. This would indicate that $C_3$ is not charging since the voltage across this component is usually equal to $V_S$. We would then test $D_3$ and $C_3$, the two components that lie in the $C_4$ charging path.

If the voltage across $C_4$ were approximately equal to $V_{S(pk)}$, we would then check the voltage across $C_2$ since the voltage across this component is normally equal to 2 $V_{S(pk)}$. If the voltage across the component is not correct, we would check the voltage across $C_1$. If the voltage across $C_1$ is correct, we would test $D_2$ and $C_2$. If the voltage across $C_1$ is not correct, we would test $D_1$ and $C_1$.

When testing voltage multipliers, it helps if you have a quick reference showing you the normal capacitor voltage readings for the various multiplier circuits. Such a reference is supplied in Table 4.5.

*A Practical Consideration*: Don't expect the capacitor voltages to be exactly equal to the peak values we have assumed in our dicussions. The diodes will drop some voltage, and the capacitors will actually be charging and discharging continually. However, the average capacitor voltages should be close to the appropriate multiples of $V_{S(pk)}$ that you were shown earlier.

**TABLE 4.5   Normal Capacitor Voltage Readings for Various Multiplier Circuits**

| Capacitor | Half-Wave Doubler | Full-Wave Doubler | Tripler | Quadrupler |
|---|---|---|---|---|
| $C_1$ | $V_{S(pk)}$ | $V_{S(pk)}$ | $V_{S(pk)}$ | $V_{S(pk)}$ |
| $C_2$ | 2 $V_{S(pk)}$ | $V_{S(pk)}$ | 2 $V_{S(pk)}$ | 2 $V_{S(pk)}$ |
| $C_3$ | | 2 $V_{S(pk)}$ | $V_{S(pk)}$ | $V_{S(pk)}$ |
| $C_4$ | | | 3 $V_{S(pk)}$ | 2 $V_{S(pk)}$ |
| $C_5$ | | | | 4 $V_{S(pk)}$ |

When testing a voltage multiplier, use the listing shown in Table 4.5 and the following procedure:

1. *Measure the voltage across the output capacitor. The reading may direct you to test one of the parallel charging capacitors.* For example, if the voltage across $C_4$ (Figure 4.32) is short by $V_S$, you would check $C_3$. If it is short by $2\ V_{S(pk)}$, you would check $C_2$.

2. Measure the voltage across any capacitor that lies in the charge path of an uncharged capacitor. For example, if $C_2$ (Figure 4.32) is not charged, you would measure the voltage across $C_1$.

3. Continue the process until you find the uncharged capacitor that is closest to the source. Then test all the components in the charge path for that capacitor. For example, if $C_2$ (Figure 4.32) is not charged and the voltage across $C_1$ is normal, check $D_2$ and $C_2$. If the voltages across $C_1$ and $C_2$ are both wrong, check $D_1$ and $C_1$.

If none of the capacitor charges are correct, and there is no problem with either $C_1$ or $D_1$, the voltage source is the problem.

## Display Faults

**Decoder–driver**
An IC that is used to drive a multisegment display.

Multisegment displays are driven by ICs called **decoder–drivers.** The most common symptom of a fault in the driver-display circuit is the failure of one or more segments to light.

The testing of a driver-display circuit is simple. When a segment should be *on,* check the driving pin for that segment. If the voltage there is correct (a positive voltage for a *common-cathode* display and 0 V for a *common-anode* display), the problem is the display. If the voltage is not there, the problem is likely the decoder–driver. This assumes that the inputs to the decoder–driver are correct.

Some multisegment displays have the decoder–driver circuit built within the unit. This eliminates the need for external decoder and driver circuits and for current-limiting resistors. When testing this type of component, the inputs are checked and compared to the output of the display. If the output of the display is incorrect, the entire unit should be replaced.

## Section Review

1. Describe the diode faults that can occur in the zener clamper and the symptoms of each.

2. Why should you replace both diodes in a zener clamper if one of them shorts?

3. Describe the process used to troubleshoot a voltage multiplier.

## ■ Key Terms

The following terms were introduced and defined in this chapter:

alphanumeric symbols
AM detector
biased clamper
biased clipper
clamper
clipper
common-anode display
common-cathode display

dc reference voltage
dc restorer
decoder–driver
dual-polarity power
   supply
full-wave voltage doubler
half-wave voltage doubler
limiter

liquid-crystal display
   (LCD)
multisegment display
negative-biased clipper
negative clamper
negative shunt clipper
positive-biased clipper
positive clamper

## Equation Summary

| Equation Number | Equation | | Section Number |
|---|---|---|---|
| (4.1) | $V_L = V_{in} - 0.7\text{ V}$ | Positive series clipper | 4.1 |
| (4.2) | $V_{D1} = V_{in}$ | Negative series clipper | 4.1 |
| (4.3) | $V_L = 0\text{ V}$ | Negative series clipper | 4.1 |
| (4.4) | $V_L = \dfrac{R_L}{R_L + R_S} V_{in}$ | Negative shunt clipper | 4.1 |
| (4.5) | $V_L = -V_F$ | Positive shunt clipper | 4.1 |
| (4.6) | $V_R = V_{in} + 0.7\text{ V}$ | Positive shunt clipper | 4.1 |
| (4.7) | $T_C = 5(R_{D1}C_1)$ | | 4.3 |
| (4.8) | $T_D = 5(R_L C_1)$ | | 4.3 |
| (4.9) | $V_{dc} \cong 2 V_{S(pk)}$ | | 4.4 |

### Answers to the Example Practice Problems

**4.1.** 12.54 $V_{pk}$

**4.2.** $V_L = -0.7\text{ V}$, $V_R = -14.3\ V_{pk}$

**4.3.** Positive alternation: $V_L = 0.7\text{ V}$, $V_R = 11.3\ V_{pk}$; negative alternation:
$V_L = -11\ V_{pk}$

**4.5.** $T_C = 188\ \mu s$, $T_D = 28.2\ ms$

### Practice Problems

**§4.1**

1. Determine the positive peak load voltage from the circuit in Figure 4.33a.
2. Draw the output waveform for the circuit in Figure 4.33a. Label the peak voltage values on the waveform.
3. Determine the negative peak load voltage for the circuit shown in Figure 4.33b.
4. Draw the output waveform for the circuit in Figure 4.33b. Label the peak voltage values on the waveform.
5. Determine the peak voltage values for the output from Figure 4.33c. Then draw the waveform and include the voltage values in the drawing.
6. Repeat Problem 5 for the circuit shown in Figure 4.33d.
7. The potentiometer in Figure 4.34a is set so that the anode of $D_1$ is at $-2$ V. If the value of $V_S$ is 14 $V_{pp}$, what are the peak load voltages for the circuit?
8. The potentiometer in Figure 4.34a is set so that the anode voltage of $D_1$ is $-8$ V. Assuming that $V_S = 24\ V_{pp}$, determine the peak load voltages for the circuit and draw the output waveform.

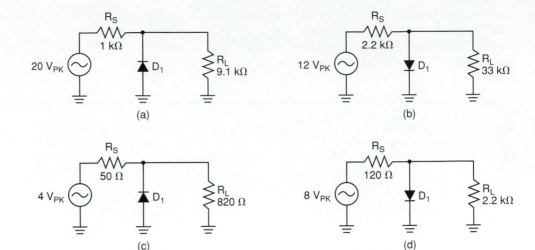

FIGURE 4.33

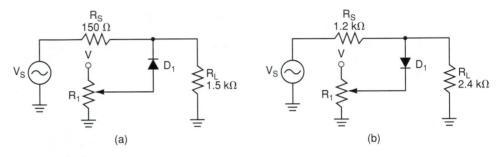

FIGURE 4.34

9. The potentiometer in Figure 4.34b is set so that the cathode voltage of $D_1$ is +4 V. Assuming that $V_S = 22$ $V_{pp}$, determine the peak load voltages for the circuit and draw the output waveform.

10. The potentiometer in Figure 4.34b is set so that the cathode voltage of $D_1$ is +2 V. Assuming that $V_S = 4$ $V_{pp}$, determine the peak load voltages for the circuit and draw the output waveform. *(Be careful on this one!)*

§4.3

11. A clamper with a 24 $V_{pp}$ input shifts the waveform so that its peak voltages are 0 and −24 V. Determine the dc average of the output waveform.

12. A clamper with a 14 $V_{pp}$ input shifts the waveform so that its peak voltages are 0 and +14 V. Determine the dc average of the output waveform.

13. The circuit in Figure 4.35a has values of $R_{D1} = 24$ Ω, $R_L = 2.2$ kΩ, and $C_1 = 4.7$ μF. Determine the charge and discharge times for the capacitor.

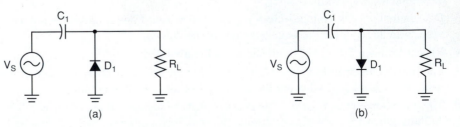

FIGURE 4.35

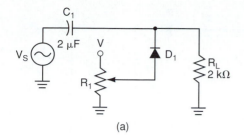

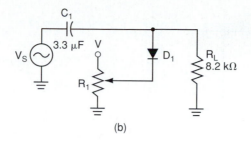

**FIGURE 4.36**

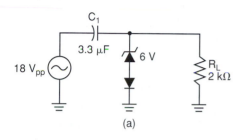

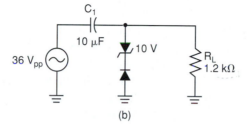

**FIGURE 4.37**

**14.** The circuit in Figure 4.35a has a 14 $V_{pp}$ input signal. Draw the output waveform and determine its peak voltage values.

**15.** The circuit shown in Figure 4.35a has values of $R_{D1} = 8\ \Omega$, $R_L = 1.2\ k\Omega$, and $C = 33\ \mu F$. Determine the charge and discharge times for the capacitor.

**16.** The circuit shown in Figure 4.35b has values of $R_{D1} = 14\ \Omega$, $R_L = 1.5\ k\Omega$, and $C = 1\ \mu F$. Determine the charge and discharge times for the capacitor.

**17.** The input to the circuit in Figure 4.35b is a 12 $V_{pp}$ signal. Determine the peak load voltages for the circuit and draw the output waveform.

**18.** The potentiometer in Figure 4.36a is set so that the voltage at the anode of $D_1$ is +3 V. The value of $V_S$ is 9 $V_{pp}$. Draw the output waveform and determine its peak voltage values.

**19.** The potentiometer in Figure 4.36b is set so that the cathode voltage of $D_1$ is +6 V. The value of $V_S$ is 30 $V_{pp}$. Draw the output waveform for the circuit and determine its peak voltage values.

**20.** Draw the output waveform for the circuit in Figure 4.37a and determine its peak voltage values.

**21.** Draw the output waveform for the circuit in Figure 4.37b and determine its peak voltage values.

**§4.4**

**22.** The circuit in Figure 4.38 has a 15 $V_{pk}$ input signal. Determine the values of $V_{C1}$ and $V_{C2}$ for the circuit. Assume that the diodes are ideal components.

**FIGURE 4.38**

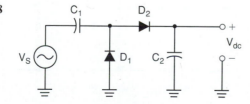

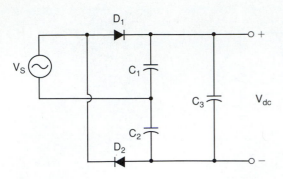

FIGURE 4.39

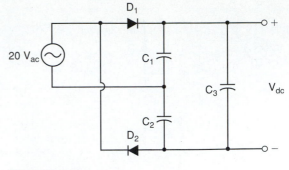

FIGURE 4.40

23. The circuit in Figure 4.38 has a 48 $V_{pk}$ input signal. Determine the values of $V_{C1}$ and $V_{C2}$ for the circuit. Assume that the diodes are ideal components.

24. The circuit in Figure 4.39 has a 24 $V_{pk}$ input signal. Determine the values of $V_{C1}$, $V_{C2}$, and $V_{C3}$ for the circuit. Assume that the diodes are ideal components.

25. The circuit in Figure 4.39 has a 25 $V_{ac}$ input signal. Determine the values of $V_{C1}$, $V_{C2}$, and $V_{C3}$ for the circuit. Assume that the diodes are ideal components. (*Be careful on this one!*)

26. Determine the dc output voltage for the circuit shown in Figure 4.40. Assume that the diodes are ideal components.

27. Determine the dc output voltage for the circuit shown in Figure 4.41. Assume that the diodes are ideal components.

28. The circuit shown in Figure 4.42 has a 20 $V_{pk}$ input signal. Determine the values of $V_{C1}$, $V_{C2}$, $V_{C3}$, and $V_{C4}$ for the circuit. Assume that the diodes are ideal components.

29. The circuit shown in Figure 4.42 has a 15 $V_{ac}$ input signal. Determine the values of $V_{C1}$, $V_{C2}$, $V_{C3}$, and $V_{C4}$ for the circuit. Assume that the diodes are ideal components.

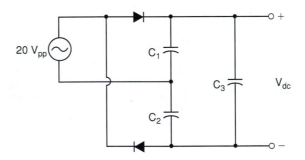

FIGURE 4.41

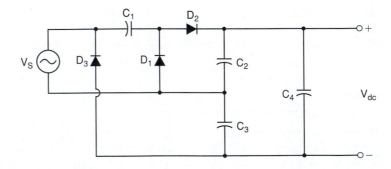

FIGURE 4.42

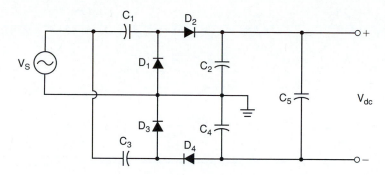

**FIGURE 4.43**

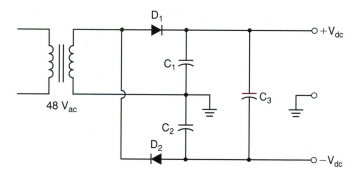

**FIGURE 4.44**

30. The circuit shown in Figure 4.43 has an 18 $V_{pk}$ input signal. Determine the values of $V_{C1}$, $V_{C2}$, $V_{C3}$, $V_{C4}$, and $V_{C5}$ for the circuit. Assume that the diodes are ideal components.

31. The circuit shown in Figure 4.43 has a 36 $V_{ac}$ input signal. Determine the values of $V_{C1}$, $V_{C2}$, $V_{C3}$, $V_{C4}$, and $V_{C5}$ for the circuit. Assume that the diodes are ideal components.

32. Determine the dc output voltage values for the dual-polarity power supply shown in Figure 4.44. Assume that the diodes are ideal components.

33. Assume that the transformer in Figure 4.44 is a 36 $V_{ac}$ transformer and that the diodes in the circuit are ideal components. Determine the output dc voltage values for the circuit.

***Troubleshooting Practice Problems***

34. The circuit in Figure 4.45a has the input/output waveforms shown. Determine whether or not there is a problem in the circuit. If there is, discuss the possible cause(s) of the problem.

35. The circuit in Figure 4.45b has the input/output waveforms shown. Determine whether or not there is a problem in the circuit. If there is, discuss the possible cause(s) of the problem.

36. The circuit in Figure 4.46a has the input/output waveforms shown. Determine whether or not there is a problem in the circuit. If there is, discuss the possible cause(s) of the problem.

37. The circuit in Figure 4.46b has the input/output waveforms shown. Determine whether or not there is a problem in the circuit. If there is, discuss the possible cause(s) of the problem.

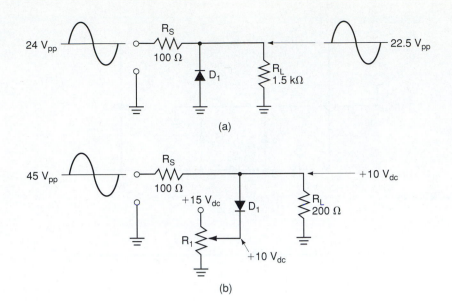

(a)

(b)

**FIGURE 4.45**

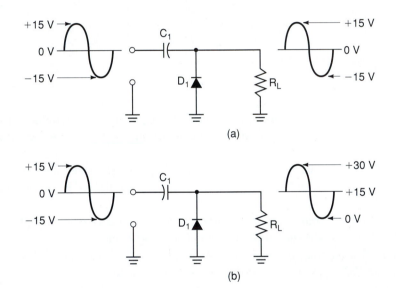

(a)

(b)

**FIGURE 4.46**

**38.** The circuit in Figure 4.47a has the input/output waveforms shown. Determine whether or not there is a problem in the circuit. If there is, discuss the possible cause(s) of the problem.

**39.** The circuit in Figure 4.47b has the input/output waveforms shown. Determine whether or not there is a problem in the circuit. If there is, discuss the possible cause(s) of the problem.

**40.** The circuit in Figure 4.48 has the capacitor voltages shown. Determine whether or not there is a problem in the circuit. If there is, list the steps you would take to find the faulty component.

**41.** The circuit in Figure 4.49 has the capacitor voltages shown. Determine whether or not there is a problem in the circuit. If there is, list the steps you would take to find the faulty component.

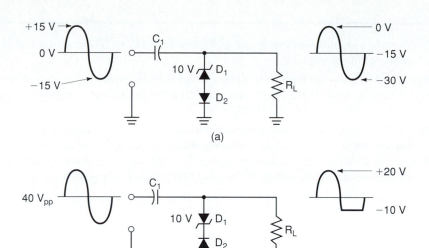

(a)

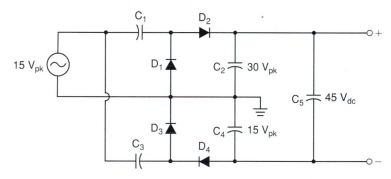

(b)

**FIGURE 4.47**

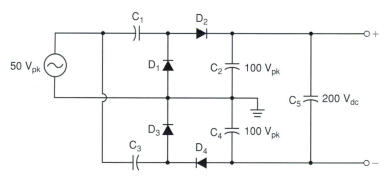

**FIGURE 4.48**

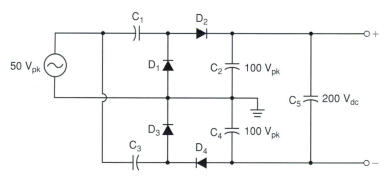

**FIGURE 4.49**

42. Explain the output waveform shown in Figure 4.50. (*Note:* There are no faulty components in the circuit.)

43. Figure 4.51 shows the basic half-wave voltage doubler with the diodes and capacitors reversed. Analyze the circuit and show that it will provide a *negative* dc output that is approximately twice the peak input voltage.

44. If you look closely at the voltage multipliers discussed in this chapter, you may notice that there is a relationship between the circuit components and the value of the circuit multiplier. If you can determine the relationship, you will be able to determine the dc output voltage for the multipler shown in Figure 4.52.

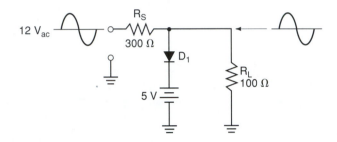

**FIGURE 4.50**

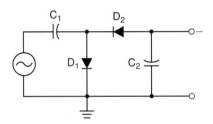

**FIGURE 4.51**

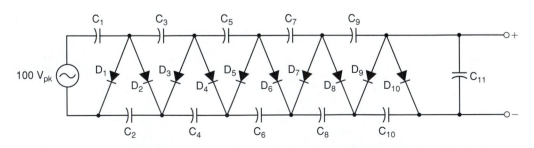

**FIGURE 4.52**

**45.** Write a program that will determine the peak output voltages for a biased shunt clipper. The program should allow the user to input the value of $V_{S(pk)}$ and the value of the diode biasing voltage. It should also be able to solve both negative and positive clippers.

**46.** Modify the program in Problem 45 to accept the value of $V_S$ in peak, peak-to-peak, or rms form.

**47.** Write a program that will determine the peak output voltages and dc output voltage for either a positive or negative unbiased clamper. The program should provide for $V_S$ inputs in any standard voltage form (peak, rms, and so on).

# 5

# SPECIAL APPLICATIONS DIODES

**S**everal of the devices discussed in this chapter are used in ultrahigh-frequency communications applications.

# OUTLINE

# OBJECTIVES

*After studying the material in this chapter, you should be able to:*

1. State the purpose served by a *varactor.*
2. Discuss the relationship between varactor bias and junction capacitance.
3. Describe the method by which a varactor can be used as the tuning component in a parallel (or series) *LC* circuit.
4. Describe a *surge* and the danger it presents to an electronic system.
5. List the characteristics that every surge-protection circuit needs.
6. Describe the differences between *constant-current* diodes and *pn*-junction diodes.
7. Describe the operation of the *tunnel* diode.
8. Describe the construction and operation of the Schottky diode.
9. Describe the construction and operation of the PIN diode.
10. Compare and contrast the forward operation of the PIN and *pn*-junction diodes.

In the first four chapters, we discussed the operation, common circuit applications, and troubleshooting of *pn*-junction diodes, zener diodes, and LEDs. In this chapter, we are going to take a relatively brief look at several other types of diodes. The diodes covered in this chapter are called *special applications diodes* because they are used to perform functions other than those performed by the *pn*-junction diode, zener diode, and LED. The diodes covered in this chapter are rarely used for any purposes other than those we will be discussing.

The diodes we will be discussing have very little in common with each other. For this reason, each section in this chapter will read as a *stand-alone* section. This means that you do not have to cover the sections in this chapter in any particular order. You can choose those sections you wish to cover in any order.

## 5.1 VARACTOR DIODES

*OBJECTIVE 1* ▶

**Varactor**
A diode that has junction capacitance when reverse biased.

*OBJECTIVE 2* ▶

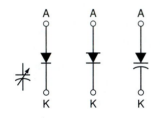

**FIGURE 5.1    Varactor schematic symbols.**

The **varactor** is a type of *pn*-junction diode that has relatively high junction capacitance when reverse biased. The capacitance of the junction is controlled by the amount of reverse voltage applied to the device. This makes the component very useful as a *voltage-controlled capacitor*. Note that the varactor is also referred to as a *varicap, tuning diode,* or *epicap*. The commonly used schematic symbols for the varactor are shown in Figure 5.1.

The use of a varactor as a voltage-controlled capacitor is easy to understand when you consider the reverse bias characteristics of the device. When a *pn*-junction is reverse biased, the depletion layer acts as an insulator between the *p*-type and *n*-type materials, as shown in Figure 5.2a. As you know, a capacitor is made up of an insulator (called the *dielectric*) that separates two conductors (called the *plates*). If we view the *p*-type and *n*-type materials in the varactor as the plates and the depletion layer as the dielectric, it is easy to view the reverse-biased component as a capacitor.

The capacitance of a reverse-biased varactor junction is found as follows:

$$C_T = \epsilon \frac{A}{W_d} \tag{5.1}$$

where $C_T$ = the total junction capacitance
       $\epsilon$ = the permittivity of the semiconductor material
       $A$ = the cross-sectional area of the junction
       $W_d$ = the *width* of the depletion layer

Equation (5.1) is fairly useless when it comes to any practical applications. Let's face it: it would take a lot of intricate mathematical analyses to determine all the values needed to solve the equation. However, it is introduced here because it does serve to illustrate one important concept; that is, the value of $C_T$ is *inversely proportional* to the width of the depletion layer. Since the width of the depletion layer varies directly with the amount of

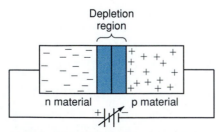

(a) *pn* junction with small reverse bias voltage applied

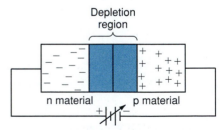

(b) *pn* junction with increased reverse bias voltage applied

**FIGURE 5.2**

**FIGURE 5.3   Varactor bias versus capacitance curve.**

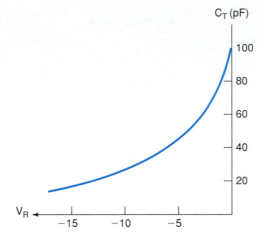

reverse bias (as shown in Figure 5.2b), we can say that *the capacitance of the varactor varies inversely with the amount of reverse bias that is applied to the device*. As $V_R$ increases, $C_T$ decreases, and vice versa. This relationship between $V_R$ and $C_T$ is further illustrated in Figure 5.3. As the graph shows, you increase or decrease the junction capacitance by simply changing the amount of reverse bias applied to the device. This makes the varactor ideal for use in circuits that require voltage-controlled tuning. One such application is introduced later in this section and again in Chapter 17.

## Varactor Specifications

The spec sheet for a given varactor will provide the information you need for any circuit analysis. Figure 5.4 shows the spec sheet and operating curves for the Motorola MV209 series varactors. We will use this spec sheet for our discussion on the commonly used varactor parameters and characteristics.

The *maximum ratings* for the varactor are the same as those used for the *pn*-junction diode and the zener diode. The same holds true for the *reverse breakdown voltage* and *reverse leakage current* ratings that appear in the *electrical characteristics* section of the spec sheet. Since you are already familiar with these ratings, we will not discuss them here. If you need to review any of them, refer back to the appropriate discussions in Chapter 2.

The **diode capacitance temperature coefficient** (**$TC_C$**) rating of a varactor tells you how much the component's capacitance will change for each 1°C rise in temperature above 25°C. The $TC_C$ rating for the MV209 series varactors is 300 ppm/°C. This means that the varactor capacitance will *increase* by 300 *parts per million* for each 1°C rise in temperature above 25°C.

What is meant by *parts per million*? It means that the capacitance will increase by *300 millionths (0.0003) of its normal value*. If a capacitor's normal value was 1 pF, the change would be found as

$$\Delta C = (0.0003)(1 \text{ pF}) = 0.0003 \text{ pF}$$

Since the normal capacitance rating of the MV209 is typically 29 pF, the increase in capacitance per 1°C rise in temperature would be found as

$$\Delta C = (0.0003)(29 \text{ pF}) = 0.0087 \text{ pF}$$

Thus, the MV209 varactor would have to experience a temperature increase of nearly 115°C to produce an increase in capacitance of 1 pF. A graph that shows the relationship between the ambient temperature and the actual capacitance of the varactor is shown in Figure 5.4. This graph shows that the capacitance of the varactor is relatively stable with

Varactor maximum ratings are the same as those for *pn*-junction diodes and zener diodes.

**Diode capacitance temperature coefficient (*$TC_C$*)**
The amount by which varactor capacitance changes when temperature changes.

How to determine the $\Delta C$ for a temperature change of 1°C.

Temperature has little effect on the capacitance rating of most varactors.

## SILICON EPICAP DIODE

. . . designed for general frequency control and tuning applications; providing solid-state reliability in replacement of mechanical tuning methods.
- High Q with Guaranteed Minimum Values at VHF Frequencies
- Controlled and Uniform Tuning Ratio
- Available in Surface Mount Package

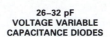

### MMBV109L MV209

CASE 182-02, STYLE 1
(TO-226AC)

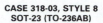
2 o———|◄|———o 1
Cathode        Anode

CASE 318-03, STYLE 8
SOT-23 (TO-236AB)

3 o———|◄|———o 1
Cathode        Anode

**26–32 pF
VOLTAGE VARIABLE
CAPACITANCE DIODES**

### MAXIMUM RATINGS

| Rating | Symbol | MV209 | MMBV209,L | Unit |
|---|---|---|---|---|
| | | Value | | |
| Reverse Voltage | $V_R$ | 30 | | Volts |
| Forward Current | $I_F$ | 200 | | mA |
| Forward Power Dissipation @ $T_A = 25°C$<br>Derate above 25°C | $P_D$ | 280<br>2.8 | 200<br>2.0 | mW<br>mW/°C |
| Junction Temperature | $T_J$ | + 125 | | °C |
| Storage Temperature Range | $T_{stg}$ | − 55 to + 150 | | °C |

### DEVICE MARKING

MMBV109L = 4A

### ELECTRICAL CHARACTERISTICS ($T_A = 25°C$ unless otherwise noted.)

| Characteristic | Symbol | Min | Typ | Max | Unit |
|---|---|---|---|---|---|
| Reverse Breakdown Voltage<br>($I_R = 10\ \mu Adc$) | $V_{(BR)R}$ | 30 | — | — | Vdc |
| Reverse Voltage Leakage Current<br>($V_R = 25\ Vdc$) | $I_R$ | — | — | 0.1 | $\mu Adc$ |
| Diode Capacitance Temperature Coefficient<br>($V_R = 3.0\ Vdc, f = 1.0\ MHz$) | $TC_C$ | — | 300 | — | ppm/°C |

| Device | $C_t$, Diode Capacitance<br>$V_R = 3.0\ Vdc, f = 1.0\ MHz$<br>pF | | | Q, Figure of Merit<br>$V_R = 3.0\ Vdc$<br>f = 50 MHz<br>(Note 1) | $C_R$, Capacitance Ratio<br>$C_3/C_{25}$<br>f = 1.0 MHz<br>(Note 2) | |
|---|---|---|---|---|---|---|
| | Min | Nom | Max | Min | Min | Max |
| MMBV109L, MV209 | 26 | 29 | 32 | 200 | 5.0 | 6.5 |

FIGURE 1 — DIODE CAPACITANCE

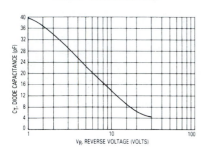

FIGURE 2 — FIGURE OF MERIT

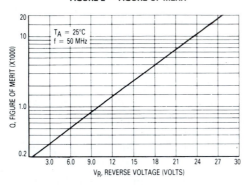

FIGURE 3 — LEAKAGE CURRENT

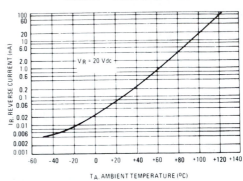

FIGURE 4 — DIODE CAPACITANCE

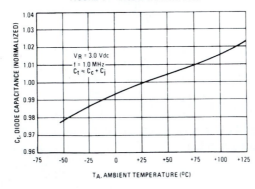

### NOTES ON TESTING AND SPECIFICATIONS

1. Q is calculated by taking the G and C readings of an admittance bridge, such as Boonton Electronics Model 33AS8, at the specified frequency and substituting in the following equation:

$$Q = \frac{2\pi f C}{G}$$

2. $C_R$ is the ratio of $C_t$ measured at 3.0 Vdc divided by $C_t$ measured at 25 Vdc.

**FIGURE 5.4    Spec sheet and operating curves for the Motorola MV209 series varactors. (Copyright of Motorola. Used by permission.)**

a varying temperature. There are also circuits available that compensate for the effect of temperature variations if capacitance stability is critical.

The **diode capacitance ($C_t$)** rating of a varactor is self-explanatory. For the MV209 series varactors, the $C_t$ rating is 26 to 32 pF when $V_R$ is 3 $V_{dc}$. Note that the *nominal* value (29 pF) is the *rated* value of the component. In other words, the MV209 would be rated as a 29 pF varactor diode, even though the actual value could fall anywhere within the specified range at $V_R = 3 V_{dc}$.

The **capacitance ratio ($C_R$)** rating of the varactor tells you how much the junction capacitance varies over the given range of voltages. For the MV209 series, the $C_R$ rating ranges from 5.0 to 6.5 for $V_R = 3$ to 25 $V_{dc}$. This means that the capacitance at $V_R = 3 V_{dc}$ will be 5.0 to 6.5 times as high as it is at $V_R = 25 V_{dc}$.

The *higher* the value of $C_R$, the wider the range of capacitance values for a given varactor. For example, let's say that we have two varactors, each having a $C_t$ rating of 51 pF when $V_R = 5 V_{dc}$. Let's also assume that $D_1$ has a rating of $C_R = 1.5$ for $V_R = 5$ to 10 $V_{dc}$, and $D_2$ has a rating of $C_R = 4$ for $V_R = 5$ to 10 $V_{dc}$. To find the capacitance of each diode at $V_R = 10 V_{dc}$, we would divide the value of $C_t$ by the value of $C_R$. Thus,

$$C_{D1} = \frac{51 \text{ pF}}{1.5} = 34 \text{ pF} \qquad \text{(when } V_R = 10 \, V_{dc})$$

and

$$C_{D2} = \frac{51 \text{ pF}}{4} = 12.75 \text{ pF} \qquad \text{(when } V_R = 10 \, V_{dc})$$

Thus, $D_1$ would have a range of capacitance from 34 to 51 pF, and $D_2$ would have a range from 12.75 to 51 pF.

The $C_R$ rating of a varactor diode is important when designing circuits. A varactor with a *high* $C_R$ rating could be used in a *coarse tuning* circuit, while a varactor with a *low* $C_R$ rating could be used in a *fine tuning* circuit. Tuning will be discussed further when we take a look at varactor applications later in this section.

While the $C_R$ rating of a varactor can be used to determine its capacitance *range*, it is not used to determine the specific diode capacitance at a specific value of $V_R$. For example, we would not use the $C_R$ rating of the MV209 to determine its capacitance at $V_R = 6 V_{dc}$. Instead, you need to use the *capacitance versus reverse voltage* curve that is shown in Figure 5.4. Using this curve, we could determine the value of capacitance at $V_R = 6 V_{dc}$ to be approximately 23 pF.

The *figure of merit* rating is the $Q$ of the junction capacitance. You may recall that the $Q$ of a capacitor is the *ratio of energy stored in the capacitor to the energy lost through leakage current*. In other words, it is the ratio of power returned to the circuit by the capacitor to power dissipated by the capacitor. For the MV209, the $Q$ rating is 200 (minimum). This means that the power returned by the capacitor to the circuit (when it discharges) will be at least 200 times the power that it dissipated.

## Varactor Applications

Varactors are used almost exclusively in *tuned circuits*, which are discussed in detail in Chapter 17. A tuned *LC* circuit that contains a varactor is shown in Figure 5.5. Note that the capacitance of the varactor is in *parallel* with the inductor. Thus, the varactor and the inductor form a *parallel LC circuit*, or *LC tank circuit*.

Before we analyze the operation of the varactor in the circuit, a few points should be made. First, note the direction of the varactor. Since the varactor is pointing toward the positive source voltage, we know that it is reverse biased. Therefore, it is acting as a voltage-variable capacitance. *For normal operation, a varactor diode is operated in its reverse operating region.* A forward-biased varactor diode would serve no special purpose because it has the same forward characteristics as a standard *pn*-junction diode. Second, $R_W$ in the circuit is the *winding resistance* of the inductor. This winding resistance is

**Diode capacitance ($C_t$)**
The rated value (or range) of C for a varactor at a specific value of $V_R$.

**Capacitance ratio ($C_R$)**
The factor by which C changes from one specified value of $V_R$ to another.

The importance of the varactor $C_R$ rating.

You can use $C_R$ to find the range of $C_t$ values. However, you must use the *capacitance versus reverse voltage* graph to find the value of C at various values of $V_R$.

◀ *OBJECTIVE 3*

For normal operation, the varactor is reverse biased.

**FIGURE 5.5** *LC* tank circuit.

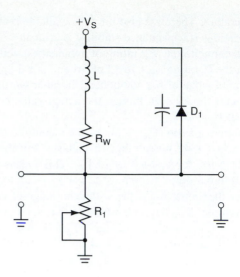

in series with the potentiometer, $R_1$. Thus, $R_W$ and $R_1$ form a voltage divider used to determine the amount of reverse bias across $D_1$, and therefore its capacitance. By adjusting the setting of $R_1$, we can vary the diode capacitance. This, in turn, varies the *resonant frequency* of the *LC* circuit.

The resonant frequency of the *LC* tank circuit is found using

$$f_r = \frac{1}{2\pi \sqrt{LC}}$$
(5.2)

If the amount of varactor reverse bias is *decreased*, the value of $C$ for the component *increases*. The increase in $C$ will cause the resonant frequency of the circuit to *decrease*. Thus, *a decrease in reverse bias causes a decrease in resonant frequency*. By the same token, an increase in varactor reverse bias causes an increase in the value of $f_r$. This point is illustrated in Example 5.1.

---

**EXAMPLE 5.1**

The *LC* tank circuit in Figure 5.5 has a 1-mH inductor. The varactor has the following specifications: $C_t = 100$ pF when $V_R = 5$ $V_{dc}$, and $C_R = 2.5$ for $V_R = 5$ to 10 $V_{dc}$. Determine the resonant frequency for the circuit at $V_R = 5$ $V_{dc}$ and $V_R = 10$ $V_{dc}$.

*Solution:*  When the varactor reverse bias is 5 $V_{dc}$, the value of $C$ is 100 pF. For this condition, the value of $f_r$ is found as

$$f_r = \frac{1}{2\pi \sqrt{LC}}$$

$$= \frac{1}{2\pi \sqrt{(1 \text{ mH})(100 \text{ pF})}}$$

$$= 503.29 \text{ kHz}$$

The value of $C$ at $V_R = 10$ $V_{dc}$ is found by dividing the $C_t$ rating by the value of $C_R$, as follows:

$$C = \frac{100 \text{ pF}}{2.5}$$

$$= 40 \text{ pF}$$

Now, the value of $f_r$ at $V_R = 10\ V_{dc}$ is found as

$$f_r = \frac{1}{2\pi \sqrt{LC}}$$

$$= \frac{1}{2\pi \sqrt{(1\ mH)(40\ pF)}}$$

$$= 795.77\ kHz$$

Thus, when the varactor reverse bias increased from 5 to 10 $V_{dc}$, the value of $f_r$ increased from 503.29 kHz to 795.77 kHz.

### PRACTICE PROBLEM 5.1

A circuit like the one shown in Figure 5.5 has a 3.3-mH inductor and a varactor with the following specifications: $C_t = 51\ pF$ at $V_R = 4\ V_{dc}$ and $C_R = 1.8$ for $V_R = 4$ to $10\ V_{dc}$. Determine the value of $f_r$ for $V_R = 4\ V_{dc}$ and $V_R = 10\ V_{dc}$.

You were told in our discussion on varactor parameters that high-$C_R$ varactors are used in *coarse tuning* circuits, while low-$C_R$ varactors are used in *fine tuning* circuits. This point can be illustrated with the aid of Example 5.2.

**EXAMPLE 5.2**

The diode in Figure 5.5 is replaced with one that has ratings of $C_t = 100\ pF$ when $V_R = 5\ V_{dc}$ and $C_R = 1.02$ for $V_R = 5$ to $10\ V_{dc}$. Determine the frequency range of the circuit for $V_R = 5$ to $10\ V_{dc}$.

**Solution:**  In Example 5.1 we determined the value of $f_r$ at $V_R = 5\ V_{dc}$ to be 503.29 kHz. This value has not changed. For the new circuit, the value of $C$ at $V_R = 10\ V_{dc}$ is found as

$$C = \frac{100\ pF}{1.02}$$

$$= 98\ pF$$

The value of $f_r$ at $V_R = 10\ V_{dc}$ is now found as

$$f_r = \frac{1}{2\pi \sqrt{LC}}$$

$$= \frac{1}{2\pi \sqrt{(1\ mH)(98\ pF)}}$$

$$= 508.40\ kHz$$

### PRACTICE PROBLEM 5.2

The varactor diode in Figure 5.5 has been replaced with a diode that has the ratings of $C_t = 100\ pF$ when $V_R = 5\ V_{dc}$, and $C_R = 1.70$ for $V_R = 5$ to $10\ V_{dc}$. Determine the frequency range of the circuit for $V_R = 5$ to $10\ V_{dc}$.

Table 5.1 summarizes the values found in Examples 5.1 and 5.2. As you can see, the range of frequencies for the circuit in Example 5.2 is much smaller than that for the circuit in Example 5.1.

**TABLE 5.1  Results from Examples 5.1 and 5.2**

| Example | $f_r$ at $V_R = 5\ V_{dc}$ | $f_r$ at $V_R = 10\ V_{dc}$ | $\Delta f_r$ |
|---------|---------------------------|-----------------------------|--------------|
| 5.1 | 503.29 kHz | 795.77 kHz | 292.48 kHz |
| 5.2 | 503.29 kHz | 508.4 kHz | 5.11 kHz |

*Coarse tuning* versus *fine tuning*.

The circuit in Example 5.1 would be a *coarse tuning* circuit; that is, it would be used to vary the value of $f_r$ over a wide range of values. The circuit in Example 5.2, on the other hand, would be a *fine tuning* circuit. It would be used to select a frequency within a smaller range. It would be used to center in on an *exact* frequency, while the coarse tuning circuit would be used to obtain an *approximate* frequency value.

## Section Review

1. A varactor acts as what type of capacitance?
2. What is the relationship between the amount of reverse bias applied to a varactor and its capacitance?
3. What is the *diode capacitance temperature coefficient* rating? What is its unit of measure?
4. What is the relationship between varactor capacitance and temperature?
5. What is the *capacitance ratio* of a varactor?
6. Why is the $C_R$ rating of a varactor important?
7. What is the $Q$ of a capacitor?
8. Explain the operation of the circuit shown in Figure 5.5.
9. What is the relationship between varactor reverse bias and the resonant frequency of its tuned circuit?
10. What is the difference between *coarse tuning* and *fine tuning*?
11. What is the relationship between the $C_R$ rating of a varactor and the type of tuning provided?

## 5.2  TRANSIENT SUPPRESSORS AND CONSTANT-CURRENT DIODES

In this section we will discuss two diodes that are very similar in operation to the standard zener diode. The first is the **transient suppressor**. Transient suppressors are zener diodes that have extremely high *surge*-handling capabilities. These diodes are used to protect voltage-sensitive circuits from surges that can occur under a variety of circumstances. The second diode we will discuss is the **constant-current diode**. The constant-current diode is an extremely high impedance diode that will maintain a constant device current over a wide range of forward operating voltages. Thus, the constant-current diode can be viewed as a "current version" of the zener diode. Note that the constant-current diode is not actually a zener diode, despite the similarities between the two. Rather, it is a variation on the *pn*-junction diode.

**Transient suppressors**
Zener diodes that have extremely high surge-handling capabilities.

**Constant-current diodes**
Diodes that maintain a constant device current over a wide range of forward operating voltages.

### Transient Suppressors

In Chapter 4, you were introduced to the idea of using a shunt clipper to protect a circuit from a *surge*, or *transient*. The *transient suppressor* uses a configuration similar to a shunt

**FIGURE 5.6**

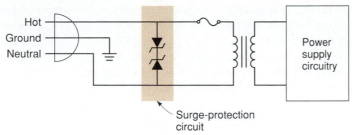

(a) Surge protection between hot and neutral lines

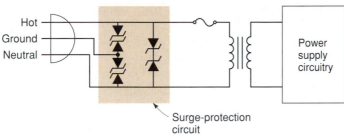

(b) Full surge protection to ground

clipper to protect the input of a power supply from any ac line surges. A basic surge-protection circuit is shown in Figure 5.6a. The circuit shown would protect the power supply from surges that occur between the *hot* and *neutral* lines. A more complete surge-protection circuit is shown in Figure 5.6b. As you can see, a transient suppressor is placed between *each pair* of incoming lines. This protects the power supply from a surge on any one incoming line.

You may recall that a surge is an abrupt high-voltage condition that lasts for a very brief time, usually in the microsecond or millisecond range. Left unchecked, a surge in the ac power lines can cause serious damage to the power supply of any electronic system, such as a television or personal computer. Transients and surges are caused by various conditions. The most common come from electric motors, air-conditioning and heating units, arcing switches, and lightning. The circuits shown in Figure 5.6 would protect the power supply from such surges by shorting out any voltages greater than the $V_Z$ ratings of the diodes.

◄ *OBJECTIVE 4*

Transient suppressors have many other uses besides conditioning the ac voltage that goes into a power supply. They are used in many telecommunication, automotive, and consumer devices. Because of the wide range of applications, transient suppressors are available in a wide range of voltages.

For a surge-protection circuit to operate properly, it must have several characteristics:

◄ *OBJECTIVE 5*

**1.** The diodes used must have extremely high power dissipation ratings. This is due to the fact that most ac power line surges contain a relatively high amount of power, generally in the hundreds of watts or higher.

**2.** The diodes must be able to turn on very rapidly. If the diodes in the surge-protection circuit are too slow, the power supply could be damaged before they have a chance to turn on.

These requirements for the surge-protection circuit are easily fulfilled by using transient suppressors.

Transient suppressors have the same general operating characteristics as those of the standard zener diode. In fact, the schematic symbol for the transient suppressor is identical to that of a standard zener diode. The main difference between the transient suppressor and the standard zener diode is the suppressor's *surge-handling capability*. The

The transient suppressor can dissipate power values in the kilowatt range for very brief periods.

transient suppressor is designed to dissipate extremely high amounts of power for a very limited time. For example, the Motorola 1N5908 through 1N6389 series of transient suppressors can dissipate up to 1.5 kW for a period of slightly less than 10 ms. This amount of power, even for a short period of time, would destroy any standard zener diode.

You might think that the time limit on the power rating of a transient suppressor could be a problem, but it isn't. Remember, surges generally last only a few milliseconds. Thus, the transient suppressor can generally handle any surge that occurs.

## Transient Suppressor Specifications

The *maximum ratings* and *electrical characteristics* shown in Figure 5.7 are those for the 1N5908 through 1N6389 series transient suppressors. We will use the values shown for our discussion on the typical suppressor specifications.

**Peak power dissipation rating ($P_{pk}$)**
Indicates the amount of surge power that the suppressor can dissipate.

The maximum ratings are fairly standard, with the exception of the **peak power dissipation rating ($P_{pk}$).** This rating indicates the surge-handling capability of the suppressors. Several notes (not shown) on the specification sheet indicate that this rating has some conditions:

**MAXIMUM RATINGS**

| Rating | Symbol | Value | Units |
|---|---|---|---|
| Peak Power Dissipation (1) @ $T_L < 25°C$ | $P_{PK}$ | 1500 | Watts |
| Steady State Power Dissipation @ $T_L < 75°C$. Lead Length = 3/8″ Derated above $T_L = 75°C$ | $P_D$ | 5.0 50 | Watts mW/°C |
| Forward Surge Current (2) @ $T_A = 25°C$ | $I_{FSM}$ | 200 | Amps |
| Operating and Storage Temperature Range | $T_J$, $T_{stg}$ | −65 to +175 | °C |
| Lead Temperature not less than 1/16″ from the case for 10 seconds. 230°C | | | |

**MECHANICAL CHARACTERISTICS**

**CASE:** Void-free, transfer-molded, thermosetting plastic

**FINISH:** All external surfaces are corrosion resistant and leads are readily solderable and weldable

**POLARITY:** Cathode indicated by polarity band. When operated in zener mode, will be positive with respect to anode

**MOUNTING POSITION:** Any

**ELECTRICAL CHARACTERISTIC** ($T_A = 25°C$ unless otherwise noted) $V_F$* = 3.5V max, $I_F$** = 100A) (C suffix denotes standard back to back versions Test both polarities)

| JEDEC Device | Device | Breakdown Voltage $V_{BR}$ Volts Min | @ $I_T$ (mA) | Maximum Reverse Stand-Off Voltage $V_{RWM}$*** (Volts) | Maximum Reverse Leakage @ $V_{RWM}$ $I_R$ (µA) | Maximum Reverse Surge Current $I_{RSM†}$ (Amps) | Maximum Reverse Voltage @ $I_{RSM†}$ (Clamping Voltage) $V_{RSM}$(Volts) | Clamping Voltage Peak Pulse Current @ $I_{pp1†} = 1.0A$ $V_{C1}$ (Volts max) | Peak Pulse Current @ $I_{pp2†} = 10A$ $V_{C2}$ (Volts max) |
|---|---|---|---|---|---|---|---|---|---|
| 1N6373 | ICTE-5/MPTE-5 | 6.0 | 1.0 | 5.0 | 300 | 160 | 9.4 | 7.1 | 7.5 |
| — | ICTE-5C/MPTE-5C | 6.0 | 1.0 | 5.0 | 300 | 160 | 9.4 | 8.1 | 8.3 |
| 1N6374 | ICTE-8/MPTE-8 | 9.4 | 1.0 | 8.0 | 25 | 100 | 15.0 | 11.3 | 11.5 |
| 1N6382 | ICTE-8C/MPTE-8C | 9.4 | 1.0 | 8.0 | 25 | 100 | 15.0 | 11.4 | 11.6 |
| 1N6375 | ICTE-10/MPTE-10 | 11.7 | 1.0 | 10 | 2.0 | 90 | 16.7 | 13.7 | 14.1 |
| 1N6383 | ICTE-10C/MPTE-10C | 11.7 | 1.0 | 10 | 2.0 | 90 | 16.7 | 14.1 | 14.5 |
| 1N6376 | ICTE-12/MPTE-12 | 14.1 | 1.0 | 12 | 2.0 | 70 | 21.2 | 16.1 | 16.5 |
| 1N6384 | ICTE-12C/MPTE-12C | 14.1 | 1.0 | 12 | 2.0 | 70 | 21.2 | 16.7 | 17.1 |

**FIGURE 5.7   1N5908–1N6389 series transient suppressor ratings and characteristics. (Copyright of Motorola. Used by permission.)**

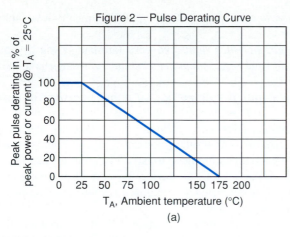

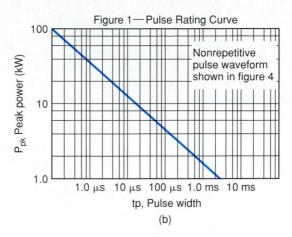

**FIGURE 5.8** **1N5908–1N6389 series transient suppressor derating curves. (Copyright of Motorola. Used by permission.)**

1. The operating temperature is 25°C.

2. The surge has a duration of approximately 8 ms.

3. The maximum number of surges is four per minute.

These conditions indicate that the $P_{pk}$ rating of the suppressor is:

1. Temperature dependent

2. Time dependent

3. Frequency dependent

The temperature dependency of the $P_{pk}$ rating is illustrated in Figure 5.8a. As you can see, the $P_{pk}$ rating is derated (by percent) at temperatures above 25°C. Example 5.3 illustrates the use of this type of power derating curve.

---

A 1N5908 surge suppressor has a $P_{pk}$ of 1500 W at 25°C. It is being used in a circuit that has an ambient temperature of 100°C. What is the value of $P_{pk}$ at this temperature?

**EXAMPLE 5.3**

**Solution:** The point where the curve intersects the $T = 100$°C line corresponds to a derating percentage of 50%. Thus, $P_{pk}$ will be 50% of its maximum value at this temperature, or 750 W.

***PRACTICE PROBLEM 5.3***

What is the $P_{pk}$ rating of the 1N5908 at $T = 150$°C?

---

You may be wondering why we bothered with the temperature curve when the spec sheet for the 1N5908 series suppressors lists a derating value. *The derating value listed in the maximum ratings table applies to the $P_D$ rating of the components, not their $P_{pk}$ rating.* To derate the $P_{pk}$ rating, you must use the power versus temperature curve.

The time dependency of the $P_{pk}$ rating is illustrated in Figure 5.8b. As you can see, $P_{pk}$ and surge duration (*pulse width*) *are inversely proportional*. As the duration of the surge increases, the power-dissipating capability of the suppressors decreases.

The final maximum rating is the *forward surge current* rating. As you can see, the forward surge current rating for the transient suppressor is extremely high. This is due to

The derating value given on the suppressor spec sheet does *not* apply to the $P_{pk}$ rating.

the combination of their high power dissipation ratings and the typically low value of $V_F$ (approximately 0.7 V).

Two of the *electrical characteristics* should be very familiar to you by now. The *reverse breakdown voltage* ($V_{BR}$) and *reverse leakage current* ($I_R$) ratings have been used throughout our study of diodes.

The **maximum reverse stand-off voltage** ($V_{RWM}$) rating is extremely important. This rating indicates the *maximum allowable peak or dc reverse voltage for any circuit using the component*. For example, refer back to Figure 5.6. Assuming that the circuit shown is driven by a 120 $V_{ac}$ line, the peak primary voltage would be approximately 170 $V_{pk}$. Thus, the diodes used in the circuit would need to have $V_{RWM}$ ratings that are *greater than* 170 V. Otherwise, the diodes could turn on during the normal operating cycles of the power supply. *Note that $V_{RWM}$ is the primary rating that is considered when attempting to substitute one suppressor for another.*

The *maximum reverse surge current* ($I_{RSM}$) rating is self-explanatory. It is the maximum surge current that the suppressor can handle for the rated period of time (in this case, approximately 8 ms). Note that the *maximum reverse voltage $V_{RSM}$* rating *decreases* as $I_{RSM}$ *increases*. This is due to the fact that the product of reverse voltage and reverse current must be *less than* the value of $P_{pk}$. (Do you remember your basic power formulas?)

The **clamping voltage** ($V_{RSM}$) is the maximum rated value of $V_Z$. When turned on, the reverse voltage across the transient suppressor will be less than or equal to this value.

The **maximum temperature coefficient** rating indicates the percentage by which $V_{BR}$ will change per 1°C increase in temperature above 25°C. At this point, you shouldn't have any trouble with this type of rating.

## Selector Guides

Selector guides for transient suppressors contain all the ratings that we have discussed in this section. The selector guides for a given series of transient suppressors can be obtained from the series manufacturer.

## Back-to-Back Suppressors

If you look closely at the ratings for the 1N6374 and the 1N6382, you'll see that their ratings are identical. You may be wondering, then, what the difference is between these two suppressors.

The 1N6382 is a **back-to-back suppressor**. This type of suppressor, whose schematic symbol is shown in Figure 5.9, actually contains two transient suppressors that are connected *internally*, like the two diodes shown in Figure 5.6. The obvious advantages to using this type of component are reduced circuit manufacturing costs and simpler circuitry. Note that the back-to-back suppressors have no $V_F$ rating since they are designed to break down at the rated value of $V_{BR}$ in both directions.

How can you tell whether a given suppressor is a single diode or a back-to-back suppressor? The component listings will contain some type of indicator that will be identified clearly in the *Notes* listing on the spec sheet. For the 1N5908 series sheet shown in Figure 5.7, the use of the suffix *C* in the ICTE number indicates that the device is a back-to-back suppressor.

## Constant-Current Diodes

Constant-current diodes are drastically different from any of the diodes you have seen so far. This can be seen by taking a look at Figure 5.10, which shows the forward operating curves for the Motorola 1N5283 through 1N5314 series **current regulator diodes**.

---

**Maximum reverse stand-off voltage ($V_{RWM}$)**
The maximum allowable peak or dc reverse voltage that will not turn on a transient suppressor.

*Maximum reverse voltage* ($V_{RSM}$), or *clamping voltage*, is the maximum allowable value of reverse voltage when diode current equals $I_{RSM}$.

**Clamping voltage**
The maximum value of $V_Z$ for a transient suppressor.

**Maximum temperature coefficient**
The percentage of change in $V_{BR}$ per 1°C rise in operating temperature.

**Back-to-back suppressor**
A single package containing two transient suppressors that are connected as shown in Figure 5.9.

(anode 1) ———▶◀——— (anode 2)

**FIGURE 5.9  Back-to-back suppressor.**

**Current regulator diode**
Another name for the constant-current diode.

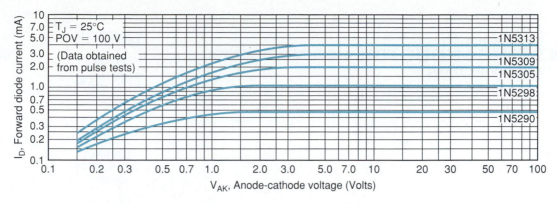

**FIGURE 5.10   (Copyright of Motorola. Used by permission.)**

When a conventional *pn*-junction diode is forward biased, $V_F$ is approximately 0.7 V and $I_F$ is determined by the components in the diode circuit. This does not hold true for the constant-current diode. As the forward curve in Figure 5.10 shows, the value of $V_F$ for a constant-current diode can have a wide range of values, in this case, anywhere from 0.1 to 100 V. At the same time, the value of $I_F$ for the constant-current diode is limited by the diode itself, not the components in the diode circuit. For example, the 1N5290 curve shows that the value of $I_F$ for the device will increase as $V_F$ increases from 0.15 V to approximately 1 V. At that point, the diode regulates the value of $I_F$ to around 500 μA (0.5 mA). The value of $I_F$ through the 1N5290 is held at this value for any value of $V_F$ between 1 and 100 V.

◄ *OBJECTIVE 6*

Because the operation of the constant-current diode is so radically different from that of any other diode, it has its own schematic symbol. This symbol is shown in Figure 5.11.

## Constant-Current Diode Specifications

Figure 5.12 shows the spec sheet for the 1N5283 through 1N5314 series constant-current diodes. The only new *maximum rating* is the **peak operating voltage (POV)** rating. The POV is the *maximum allowable value of $V_F$*. As you can see, the constant-current diode can have values of $V_F$ other than 0.7 V, in this case, anywhere up to 100 V maximum.

**Peak operating voltage (POV)**
The maximum allowable value of $V_F$.

For reasons that you'll see in a moment, we will start our coverage of the *electrical characteristics* with the last rating given, the **maximum limiting voltage, $V_L$**. The $V_L$ rating indicates *the voltage at which the diode begins to regulate current*. For the 1N5290, the value of $V_L$ is 1.05 V. This is approximately equal to the value of $V_L$ that we obtained from the current versus voltage graph in Figure 5.10.

**Maximum limiting voltage ($V_L$)**
The voltage at which the diode starts to limit current.

The **regulator current ($I_P$)** rating is the *regulated value of forward current for values of $V_F$ that are between $V_L$ and POV*. As long as the forward voltage is kept between the rated values of $V_L$ and POV, the current through the diode will be maintained at the value of $I_P$. Note that the value of $I_P$ for the 1N5290 is shown to be 423 μA (0.423 mA). This value is approximately equal to the value we obtained from the current versus voltage graph in Figure 5.10.

**Regulator current ($I_P$)**
The regulated value of forward current for forward voltages that are between $V_L$ and POV.

The *minimum dynamic impedance ($Z_T$)* rating of the constant-current diode demonstrates another big difference between this diode and the *pn*-junction diode. You may recall that the bulk resistance of a *pn*-junction diode is typically very small, around 10 Ω or less. The impedance of the constant-current diode is extremely high, typically in the high kilohm to low megohm range. The *minimum knee impedance ($Z_k$)* is even higher!

**FIGURE 5.11   Constant-current diode schematic symbol.**

(anode) ⎯⎯⎯⎯○┤⎯⎯⎯⎯ (cathode)

$\longrightarrow$
$I_F$

**1N5283
thru
1N5314**

**CURRENT
REGULATOR
DIODES**

## CURRENT REGULATOR DIODES

Field-effect current regulator diodes are circuit elements that provide a current essentially independent of voltage. These diodes are especially designed for maximum impedance over the operating range. These devices may be used in parallel to obtain higher currents.

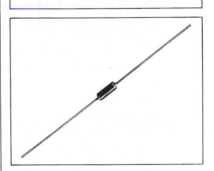

## MAXIMUM RATINGS

| Rating | Symbol | Value | Unit |
|--------|--------|-------|------|
| Peak Operating Voltage ($T_J = -55°C$ to $+200°C$) | POV | 100 | Volts |
| Steady State Power Dissipation @ $T_L = 75°C$ Derate above $T_L = 75°C$ Lead Length = 3/8″ (Forward or Reverse Bias) | $P_D$ | 600 4.8 | mW mW/°C |
| Operating and Storage Junction Temperature Range | $T_J$, $T_{stg}$ | −55 to +200 | °C |

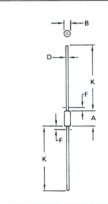

| DIM | MILLIMETERS | | INCHES | |
|-----|-----|-----|-----|-----|
| | MIN | MAX | MIN | MAX |
| A | 5.84 | 7.62 | 0.230 | 0.300 |
| B | 2.16 | 2.72 | 0.085 | 0.107 |
| D | 0.46 | 0.56 | 0.018 | 0.022 |
| F | – | 1.27 | – | 0.050 |
| K | 25.40 | 38.10 | 1.000 | 1.500 |

All JEDEC dimensions and notes apply

**CASE 51-02
DO-204AA
GLASS**

NOTES:
1. PACKAGE CONTOUR OPTIONAL WITHIN DIA B AND LENGTH A. HEAT SLUGS, IF ANY, SHALL BE INCLUDED WITHIN THIS CYLINDER, BUT SHALL NOT BE SUBJECT TO THE MIN LIMIT OF DIA B.
2. LEAD DIA NOT CONTROLLED IN ZONES F, TO ALLOW FOR FLASH, LEAD FINISH BUILDUP, AND MINOR IRREGULARITIES OTHER THAN HEAT SLUGS.

**FIGURE 5.12** **Specification sheet for the 1N5283–1N5314 series current regulator diodes. (Copyright of Motorola. Used by permission.)**

| Type No. | Regulator Current Ip (mA) @ V$_T$ = 25 V | | | Minimum Dynamic Impedance @ V$_T$ = 25 V Z$_T$ (MΩ) | Minimum Knee Impedance @ V$_K$ = 6.0 V Z$_K$ (MΩ) | Maximum Limiting Voltage @ I$_L$ = 0.8 Ip (min) V$_L$ (Volts) |
|---|---|---|---|---|---|---|
| | nom | min | max | | | |
| 1N5283 | 0.22 | 0.198 | 0.242 | 25.0 | 2.75 | 1.00 |
| 1N5284 | 0.24 | 0.216 | 0.264 | 19.0 | 2.35 | 1.00 |
| 1N5285 | 0.27 | 0.243 | 0.297 | 14.0 | 1.95 | 1.00 |
| 1N5286 | 0.30 | 0.270 | 0.330 | 9.0 | 1.60 | 1.00 |
| 1N5287 | 0.33 | 0.297 | 0.363 | 6.6 | 1.35 | 1.00 |
| 1N5288 | 0.39 | 0.351 | 0.429 | 4.10 | 1.00 | 1.05 |
| 1N5289 | 0.43 | 0.387 | 0.473 | 3.30 | 0.870 | 1.05 |
| 1N5290 | 0.47 | 0.423 | 0.517 | 2.70 | 0.750 | 1.05 |
| 1N5291 | 0.56 | 0.504 | 0.616 | 1.90 | 0.560 | 1.10 |
| 1N5292 | 0.62 | 0.558 | 0.682 | 1.55 | 0.470 | 1.13 |
| 1N5293 | 0.68 | 0.612 | 0.748 | 1.35 | 0.400 | 1.15 |
| 1N5294 | 0.75 | 0.675 | 0.825 | 1.15 | 0.335 | 1.20 |
| 1N5295 | 0.82 | 0.738 | 0.902 | 1.00 | 0.290 | 1.25 |
| 1N5296 | 0.91 | 0.819 | 1.001 | 0.880 | 0.240 | 1.29 |
| 1N5297 | 1.00 | 0.900 | 1.100 | 0.800 | 0.205 | 1.35 |
| 1N5298 | 1.10 | 0.990 | 1.210 | 0.700 | 0.180 | 1.40 |
| 1N5299 | 1.20 | 1.08 | 1.32 | 0.640 | 0.155 | 1.45 |
| 1N5300 | 1.30 | 1.17 | 1.43 | 0.580 | 0.135 | 1.50 |
| 1N5301 | 1.40 | 1.26 | 1.54 | 0.540 | 0.115 | 1.55 |
| 1N5302 | 1.50 | 1.35 | 1.65 | 0.510 | 0.105 | 1.60 |
| 1N5303 | 1.60 | 1.44 | 1.76 | 0.475 | 0.092 | 1.65 |
| 1N5304 | 1.80 | 1.62 | 1.98 | 0.420 | 0.074 | 1.75 |
| 1N5305 | 2.00 | 1.80 | 2.20 | 0.395 | 0.061 | 1.85 |
| 1N5306 | 2.20 | 1.98 | 2.42 | 0.370 | 0.052 | 1.95 |
| 1N5307 | 2.40 | 2.16 | 2.64 | 0.345 | 0.044 | 2.00 |
| 1N5308 | 2.70 | 2.43 | 2.97 | 0.320 | 0.035 | 2.15 |
| 1N5309 | 3.00 | 2.70 | 3.30 | 0.300 | 0.029 | 2.25 |
| 1N5310 | 3.30 | 2.97 | 3.63 | 0.280 | 0.024 | 2.35 |
| 1N5311 | 3.60 | 3.24 | 3.96 | 0.265 | 0.020 | 2.50 |
| 1N5312 | 3.90 | 3.51 | 4.29 | 0.255 | 0.017 | 2.60 |
| 1N5313 | 4.30 | 3.87 | 4.73 | 0.245 | 0.014 | 2.75 |
| 1N5314 | 4.70 | 4.23 | 5.17 | 0.235 | 0.012 | 2.90 |

**FIGURE 5.12** (continued)

## Constant-Current Diode Applications

Most of the circuits that would use constant-current diodes are relatively complex and thus are not covered here. However, we *can* (at this point) cover some of the *types* of applications that these diodes would typically be used for.

The **series current regulator** is used to maintain a constant input current to a circuit. The basic configuration for a series current regulator is shown in Figure 5.13a. The block labeled A is the circuit whose input current must be regulated. With the constant-current diode placed in series with the input to A, its input current will be maintained at the value of $I_P$ for the diode.

**Series current regulator**
Maintains a constant circuit input current over a wide range of input voltages.

Two important points need to be made about the series current regulator shown:

1. The constant-current diode will maintain a constant value of $I_P$, thus eliminating the sinusoidal voltage that would otherwise appear at the input to circuit A. In other

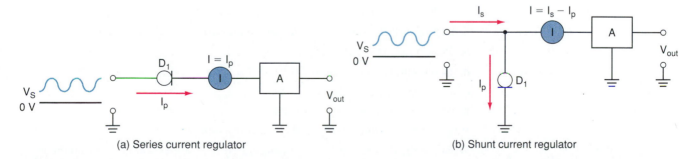

(a) Series current regulator          (b) Shunt current regulator

**FIGURE 5.13** Basic current regulator configurations.

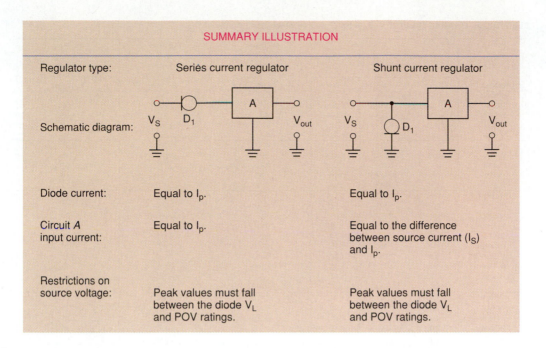

SUMMARY ILLUSTRATION

| Regulator type: | Series current regulator | Shunt current regulator |
|---|---|---|
| Schematic diagram: | $V_S$  $D_1$  A  $V_{out}$ | $V_S$  $D_1$  A  $V_{out}$ |
| Diode current: | Equal to $I_p$. | Equal to $I_p$. |
| Circuit *A* input current: | Equal to $I_p$. | Equal to the difference between source current ($I_S$) and $I_p$. |
| Restrictions on source voltage: | Peak values must fall between the diode $V_L$ and POV ratings. | Peak values must fall between the diode $V_L$ and POV ratings. |

**FIGURE 5.14**

words, there will be a constant voltage at the input to A as long as the input resistance of A does not change.

2. For the circuit to operate properly, the voltage across the diode must fall between its $V_L$ and POV ratings.

The shunt current regulator shown in Figure 5.13b provides circuit A with an input current equal to the *difference between* the source current ($I_S$) and the regulated current ($I_P$). Unlike the series current regulator, the voltage at the input of A would be approximately equal to $V_S$. Thus, the shunt current regulator would be used to provide circuit A with an input voltage equal to $V_S$ and an input current that is less than $I_S$.

Figure 5.14 summarizes the operation of the series and shunt current regulators.

## Section Review

1. What is a *transient suppressor*?
2. What is a *constant-current diode*?
3. What is a *surge*? How are ac power line surges commonly generated?
4. Why is a surge in the ac power line hazardous?
5. What are the required characteristics for a surge-protection circuit?
6. How does a *transient suppressor* differ from a standard zener diode?
7. Define each of the following transient suppressor ratings:
   a. Peak power dissipation ($P_{pk}$)
   b. Maximum reverse stand-off voltage ($V_{RWM}$)
   c. Clamping voltage ($V_{RSM}$)
   d. Maximum temperature coefficient
8. Why don't back-to-back suppressors have $V_F$ ratings?
9. What are the differences between constant-current diodes and *pn*-junction diodes?
10. Draw the schematic symbol for the constant-current diode and label its terminals.

11. Define each of the following constant-current diode specifications:
    a. Peak operating voltage (POV)
    b. Maximum limiting voltage ($V_L$)
    c. Regulator current ($I_P$)
12. What purpose is served by the series current regulator?
13. Describe the operation of the series current regulator.
14. What purpose is served by the shunt current regulator?
15. Describe the operation of the shunt current regulator.

# 5.3 TUNNEL DIODES

**Tunnel diodes** are components used in the **ultrahigh frequency (UHF)** and microwave-frequency range. They have many applications in high-frequency communication electronics. Applications using tunnel diodes include amplifiers, oscillators, modulators, and demodulators. Because of the way that they are manufactured, they exhibit a unique characteristic curve, unlike any of the diodes that we have already studied. The schematic symbol and operating curve for the tunnel diode are shown in Figure 5.15. The operating curve is a result of the extremely heavy doping used in the manufacturing of the tunnel diode. In fact, the tunnel diode is doped approximately 1000 times as heavily as standard *pn*-junction diodes.

In the forward operating region of the tunnel diode, we are interested in the area between the **peak voltage** ($V_{pk}$) and the **valley voltage** ($V_V$). At $V_F = V_{pk}$, forward current is called *peak current* ($I_{pk}$). Note that $I_{pk}$ is the maximum value that $I_F$ will reach under normal circumstances. As $V_F$ is increased to the value of $V_V$, $I_F$ decreases to its minimum value, called *valley current* ($I_V$). As you can see, *forward voltage and current are inversely proportional when the diode is operated between the values of $V_{pk}$ and $V_V$.* The term used to describe a device whose current and voltage are inversely proportional is **negative resistance**. Thus, the region of operation between $V_{pk}$ and $V_V$ is called the *negative resistance region*.

Tunnel diodes are operated almost exclusively in the negative resistance region. One common use of the tunnel diode is as the active component in an oscillator.

## Tunnel Diode Oscillator

An **oscillator** is a circuit that is used to convert dc to an ac signal. The ac signal created by the tunnel diode oscillator could be used for many applications that require a high-frequency sine wave. Oscillators are discussed in detail in Chapter 18; however, the tunnel diode oscillator can be understood using basic diode principles. The basic tunnel diode oscillator, or

**Tunnel diode**
A heavily doped diode used in high-frequency communications circuits.

**Ultrahigh frequency (UHF)**
The band of frequencies between 300 MHz and 3 GHz.

◄ *OBJECTIVE 7*

**Peak** and **valley ratings**
The peak and valley voltage and current values for a tunnel diode can be obtained from the component spec sheet. When operated between $V_{pk}$ and $V_V$, tunnel diode forward voltage and current are inversely proportional.

**Negative resistance**
Any device whose current and voltage are inversely proportional.

**Oscillator**
An ac signal generator.

**FIGURE 5.15   Tunnel diode symbol and characteristic curve.**

 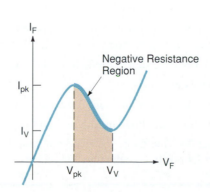

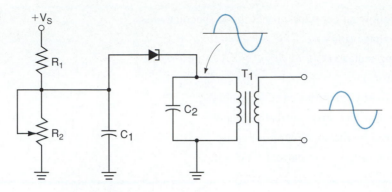

**FIGURE 5.16** Tunnel diode oscillator.

**Negative resistance oscillator**
Another name for the tunnel diode oscillator.

The tank circuit in the negative resistance oscillator is formed by the primary of the transformer and $C_2$.

**negative resistance oscillator**, is shown in Figure 5.16. Note that the circuit is shown to have a dc input voltage (labeled $+V_S$) and an ac output signal. The output frequency of the circuit will be approximately equal to the resonant frequency of the *LC* tank circuit.

The negative resistance oscillator in Figure 5.16 uses a tunnel diode to sustain oscillations; that is, to produce the alternations in the circuit output. Note that $R_1$ and $R_2$ in the circuit are used to dc bias the tunnel diode. The biasing of the diode is set (by design) so that the following conditions are met:

1. When the waveform produced by the tank circuit is at its *maximum* (positive peak) value, the *difference* between this peak voltage and the biasing voltage is equal to the $V_V$ rating of the tunnel diode. Thus, at the positive peak of the tank circuit output, the voltage across the tunnel diode is equal to $V_V$.

2. When the waveform produced by the tank circuit is at its *minimum* (negative peak) value, the difference between this peak voltage and the biasing voltage is equal to the $V_{pk}$ rating of the tunnel diode. Thus, at the negative peak of the tank circuit output, the voltage across the tunnel diode is equal to $V_{pk}$.

You may recall that the current through a tunnel diode is maximum when the voltage across the component is equal to $V_{pk}$. Thus, when the tank circuit waveform is at its negative peak, the diode conduction is maximum, and power is returned to the tank circuit. When the tank circuit is at its positive peak, the voltage across the diode is equal to $V_V$, and the diode current is at its *minimum* value, $I_V$.

You may be wondering how the oscillations start in the first place. When power is first applied to the circuit, the diode will momentarily conduct. The pulse of current that passes through the diode will produce the first half-cycle in the tank circuit.

Negative resistance oscillators cannot be used efficiently at low output frequencies.

The negative resistance oscillator has one major drawback. While the circuit works very well at high frequencies (in the upper megahertz range and higher), it cannot be used efficiently at lower frequencies. Lower-frequency oscillators are generally made with *transistors*.

---

## Section Review

1. What is a tunnel diode?
2. Describe the negative resistance operating region of the tunnel diode. Include the relationship between diode forward current and forward voltage.
3. Why does the tunnel diode have the operating curve shown in Figure 5.15?
4. What is an *oscillator*?
5. What determines the output frequency of a negative resistance oscillator?
6. How does the diode in Figure 5.16 keep the tank circuit oscillations from dying out?
7. What does the diode forward voltage in Figure 5.16 equal when the tank circuit signal is at its positive peak value? Its negative peak value?

# 5.4 OTHER DIODES

The diodes covered in this section are used less often than the other diodes we have discussed. Except for this fact, they are not necessarily related to each other in characteristics or applications.

## Schottky Diodes

The **Schottky diode** has very little junction capacitance. Because of this, it can be operated at much higher frequencies than the typical *pn*-junction diode. The reduced junction capacitance also results in a much faster *switching time*. For this reason, Schottky devices are used more and more in digital switching applications.

The Schottky diode is often referred to by any of the following names: *Schottky barrier diode*, *hot-carrier diode*, and *surface-barrier diode*. The schematic symbol and characteristic curve for this component are shown in Figure 5.17. As the characteristic curve indicates, the Schottky diode has lower $V_F$ and $V_{BR}$ values than those of the *pn*-junction diode. Typically, the Schottky diode will have a $V_F$ of approximately 0.3 V and a $V_{BR}$ of less than −50 V. These are much lower than the typical *pn*-junction ratings of $V_F = 0.7$ V and $V_{BR} = -150$ V.

The low junction capacitance and high-switching-speed capability of the Schottky diode are the result of the construction of the component. Schottky diodes have a junction that uses metal in place of the *p*-type material, as shown in Figure 5.18. As a result, the component has no depletion layer to dissolve or rebuild.

By forming a junction with a semiconductor and metal, you still have a junction, but now there is very little junction capacitance. With very little junction capacitance, the Schottky diode can be switched back and forth very rapidly between forward and reverse operation. While many components are capable of switching at this frequency rate, most of them are *low-current* devices. The Schottky diode is a relatively high-current device, capable of switching rapidly while providing forward currents in the neighborhood of 50 A. In sinusoidal and *low-current* switching circuits, the Schottky diode is capable of operating at frequencies of 20 GHz and more.

Schottky diodes are also used in the manufacture of integrated circuit chips to decrease the **propagation delay** time of the internal circuits. Shorter propagation time increases the maximum operating speed of integrated circuits. This relationship is discussed further in Chapter 19.

**Schottky diode**
A high-speed diode with very little junction capacitance.

◀ *OBJECTIVE 8*

**Propagation delay**
The time required for a signal to get from the input to the output in an integrated circuit.

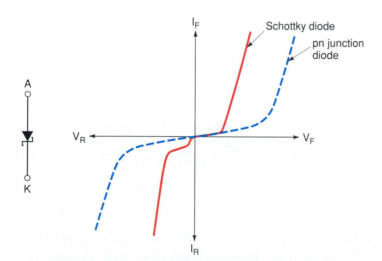

**FIGURE 5.17** Schottky diode symbol and characteristic curve.

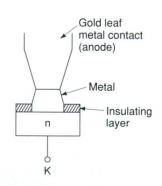

**FIGURE 5.18** Schottky diode construction.

# PIN Diodes

*OBJECTIVE 9* ▶

The PIN diode is made up of *three* semiconductor materials. The construction of the PIN diode is illustrated in Figure 5.19. The center material is made up of *intrinsic* (pure) silicon. The *p*- and *n*-type materials are very heavily doped and therefore have very low resistances.

**When reverse biased, the PIN diode acts as a capacitor.**

When reverse biased, the PIN diode acts as a capacitor. The reason for this is illustrated in Figure 5.20. You may recall that an intrinsic semiconductor acts as an insulator. Thus, the intrinsic material in the PIN diode can be viewed as the dielectric of a capacitor.

By comparison to the intrinsic material, the heavily doped *p*- and *n*-type materials can be viewed as *conductors*. Therefore, we have a dielectric (the intrinsic material) sandwiched between two conductors (the *p*- and *n*-type materials). This forms the PIN diode capacitor.

**The capacitance of a reverse-biased PIN diode remains relatively constant over a wide range of reverse voltages.**

The capacitance of a reverse-biased PIN diode will be relatively constant over a wide range of $V_R$ values. Figure 5.21 shows the *capacitance versus $V_R$* curve for the Motorola MMBV3700 PIN diode. Note that the capacitance of the device remains at approximately 0.65 pF for values of $V_R$ between $-10$ and $-50$ V. At values of $V_R$ between 0 and $-10$ V, the MMBV3700 has the same capacitance characteristics as the varactor. This is typical for PIN diodes.

**When forward biased, the PIN diode acts as a *current-controlled* resistor.**

When forward biased, the intrinsic material is forced into conduction. As the number of free carriers in the intrinsic material increases, the resistance of the material decreases. Thus, when forward biased, the PIN diode acts as a *current-controlled resistance*. This point is illustrated in Figure 5.22. Note that the *series resistance* (diode resistance) *decreases* as forward current increases. This is due to the increase in the number of free carriers that are in the intrinsic material.

*OBJECTIVE 10* ▶

Figure 5.23 shows the forward operating curve for the PIN diode. As you can see, the current versus voltage curve gradually increases, starting at the $V_F = 0.75$ V point. If you compare this forward operating curve to the *pn*-junction diode curve, you'll see two major differences:

1. The *pn*-junction diode curve shows conduction to start at nearly 0 V, while the PIN diode *starts* conducting (in this case) at 0.75 V (750 mV).

**FIGURE 5.19    PIN diode.**

anode —— [ p | i | n ] —— cathode

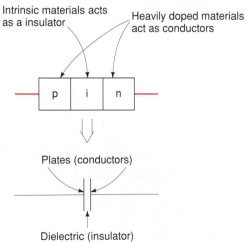

**FIGURE 5.20    PIN diode capacitance.**

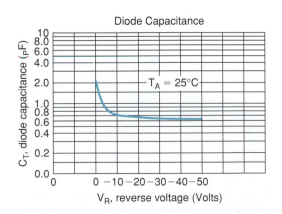

**FIGURE 5.21    MMBV3700 capacitance curve. (Copyright of Motorola. Used by permission.)**

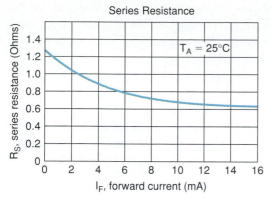

FIGURE 5.22 (Copyright of Motorola. Used by permission.)

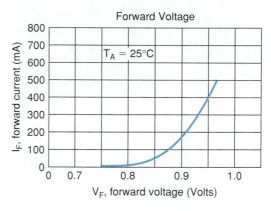

FIGURE 5.23 (Copyright of Motorola. Used by permission.)

2. The *pn*-junction diode has a definite turning point in the curve (called the *knee volt-age*), while the PIN diode shows no definite knee voltage.

PIN diode operation versus *pn*-junction diode operation.

These two differences are caused by the construction of the PIN diode. For the PIN diode to conduct, $V_F$ must overcome the resistance of the insulating intrinsic material. For the MMBV3700, $V_F$ must be at least 750 mV before the intrinsic material will allow conduction.

The lack of a knee voltage, or turning point in the curve, is due to the fact that the PIN diode does not have a *pn* junction. Without a *pn* junction, the device does not have any sudden *turn-on* point. Rather, conduction increases at a gradual rate.

As was stated earlier, the PIN diode is used primarily in UHF and microwave applications. They are also used as rf switches in many amateur radio systems. The low reverse capacitance and current-controlled resistance make it ideal for high-frequency communication circuits.

The most common applications of PIN diodes are switches or **modulators**. If the $V_F$ value is below 0.75 V, the diode has virtually no current leakage because the intrinsic material acts as an insulator. When forward biased above the 0.75 V threshold, the device has a low-value resistance and will pass a high-frequency signal with minimal reduction in amplitude, thus functioning as an ideal switch.

**Modulator**
A circuit that combines two signals of different frequencies into a single signal.

## Step-Recovery Diodes

The **step-recovery diode** is an ultrafast diode. Like the PIN diode, the step-recovery diode's characteristics are due to the unusual method of doping used. In the case of the step-recovery diode, the *p*- and *n*-type materials are doped much more heavily at the ends of the component than they are at the junction. This point is illustrated by the graph in Figure 5.24, which shows the doping level increasing as the distance from the junction increases.

**Step-recovery diode**
A heavily doped diode with an ultrafast switching time.

FIGURE 5.24

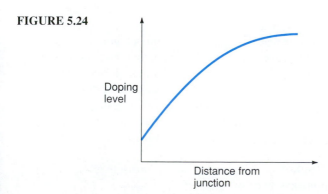

The unusual doping of the step-recovery diode affects the time required for the device to switch from *off* to *on*, and vice versa. The typical switching time for a step-recovery diode is in the low-picosecond range. This makes them ideal for switching applications in the VHF frequency range and above.

## One Final Note

There are many more types of diodes, too many to cover adequately in one chapter. However, the diodes covered in this chapter have been selected to give you a basis for understanding some of the other diode types.

**Section Review**

1. What are the commonly used names for the Schottky diode?
2. Why can the Schottky diode be operated at much higher frequencies than the typical *pn*-junction diode?
3. How does the construction of the Schottky diode differ from that of the *pn*-junction diode?
4. Describe the construction of the PIN diode.
5. Explain the reverse characteristics of the PIN diode.
6. Explain the forward characteristics of the PIN diode.
7. Why is the forward operating curve of the PIN diode so different from that of a *pn*-junction diode?
8. Describe the doping of the step-recovery diode.
9. What is the result of the step-recovery diode doping scheme?

■ **Key Terms**

The following terms were introduced and defined in this chapter:

back-to-back suppressor
capacitance ratio ($C_R$)
clamping voltage
coarse tuning
constant-current diode
current-controlled
   resistance
current regulator diode
diode capacitance ($C_t$)
diode capacitance
   temperature coefficient
   ($TC_c$)
epicap
fine tuning
flywheel effect
hot-carrier diode
maximum limiting
   voltage ($V_L$)

maximum reverse stand-
   off voltage
maximum reverse surge
   current
maximum temperature
   coefficient
modulator
negative resistance
negative resistance
   oscillator
oscillator
peak operating voltage
   (POV)
peak power dissipation
   rating ($P_{pk}$)
peak voltage
PIN diode
propagation delay

regulator current ($I_P$)
Schottky barrier diode
Schottky diode
series current regulator
shunt current regulator
step-recovery diode
surface-carrier diode
transient suppressor
tunnel diode
tunnel diode oscillator
ultrahigh frequency (UHF)
valley current
valley voltage
varactor
varicap

| Equation Number | Equation | Section Number | Equation Summary |
|---|---|---|---|
| (5.1) | $$C_T = \epsilon \frac{A}{W_d}$$ | 5.1 | |
| (5.2) | $$f_r = \frac{1}{2\pi \sqrt{LC}}$$ | 5.1 | |

## §5.1

**Practice Problems**

1. A varactor has ratings of $C_t = 50$ pF and $TC_C = 500$ ppm/°C. Determine the change in capacitance for each 1°C rise in temperature.

2. A varactor has ratings of $C_t = 48$ pF at 25°C and $TC_C = 800$ ppm/°C. Determine the change in capacitance for each 1°C rise in temperature.

3. The varactor in Figure 5.25 has the following values: $C_t = 48$ pF at $V_R = 3\ V_{dc}$, and $C_R = 4.8$ for $V_R = 3$ to $12\ V_{dc}$. Determine the resonant frequency for the circuit at $V_R = 3\ V_{dc}$ and $V_R = 12\ V_{dc}$.

4. The varactor in Figure 5.26 has the following values: $C_t = 68$ pF at $V_R = 4\ V_{dc}$, and $C_R = 1.12$ for $V_R = 4$ to $10\ V_{dc}$. Determine the resonant frequency for the circuit at $V_R = 4\ V_{dc}$ and $V_R = 10\ V_{dc}$.

## §5.2

5. Figure 5.27 shows the surge power curves for the Motorola MPZ-16 series transient suppressors. How much power can these diodes handle if the surge duration is 8 ms and $T_C = 35°C$?

6. Refer to Figure 5.27. Determine the maximum power dissipation for the MPZ-16 series diodes when the surge duration is 15 ms and $T_C = 35°C$.

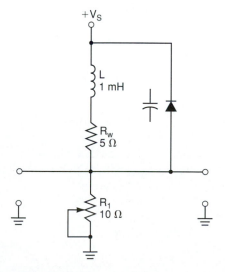

**FIGURE 5.25**          **FIGURE 5.26**

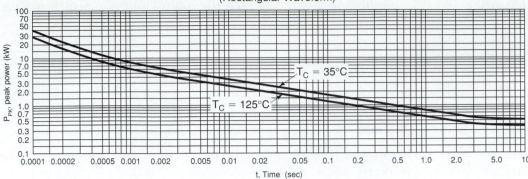

Figure 1—Maximum Non-Repetitive Surge Power
(Rectangular Waveform)

**FIGURE 5.27**

## The Brain Drain

7. The varactor in Figure 5.28a has the specifications and characteristics curve shown in Figure 5.28b. What value of $V_R$ is required to set the resonant frequency of the tank circuit to 100 kHz?

8. The 1N6303 transient suppressor has the characteristics shown in the selector guide in Figure 5.29. Determine the dynamic zener impedance ($Z_Z$) of this suppressor. (*Hint:* Review the method used to determine $Z_Z$ that was shown in Chapter 2.)

9. The circuit shown in Figure 5.30 cannot tolerate an input voltage surge that is greater than 30 percent above the rated peak value of $V_{in}$. Which of the suppressors listed in Figure 5.29 would be best suited for use in this circuit?

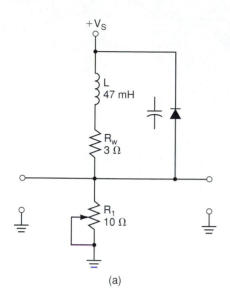

(a)

# BB 139
## VHF/FM VARACTOR DIODE
### DIFFUSED SILICON PLANAR

* $C_3/C_{25}$  5.0–6.5
* MATCHED SETS (Note 2)

**ABSOLUTE MAXIMUM RATINGS** (Note 1)

**Temperatures**

| | |
|---|---|
| Storage Temperature Range | −55°C to +150°C |
| Maximum Junction Operating Temperature | +150°C |
| Lead Temperature | +260°C |

**Maximum Voltage**

| | | |
|---|---|---|
| WIV | Working Inverse Voltage | 30 V |

**DO-35 OUTLINE**

NOTES:
Copper clad steel leads, tin plated
Gold plated leads available
Hermetically sealed glass package
Package weight is 0.14 gram

**ELECTRICAL CHARACTERISTICS** (25°C Ambient Temperature unless otherwise noted)

| SYMBOL | CHARACTERISTIC | MIN | TYP | MAX | UNITS | TEST CONDITIONS |
|---|---|---|---|---|---|---|
| BV | Breakdown Voltage | 30 | | | V | $I_R = 100 \mu A$ |
| $I_R$ | Reverse Current | | 10<br>0.1 | 50<br>0.5 | nA<br>μA | $V_R = 28$ V<br>$V_R = 28$ V, $T_A = 60°c$ |
| C | Capacitance | 4.3 | 29<br>5.1 | 6.0 | pF<br>pF | $V_R = 3.0$ V.1 = 1 MHz<br>$V_R = 25$ V, 1 = 1 MHz |
| $C_3/C_{25}$ | Capacitance Ratio | 5.0 | 5.7 | 6.5 | | $V_R = 3$ V / 25 V, 1 = 1 MHz |
| Q | Figure of Merit | 150 | | | | $V_R = 3.0$ V, 1 = 100 MHz |
| $R_S$ | Series Resistance | | 0.35 | | Ω | C510 pF, f = 600 MHz |
| $L_S$ | Series Inductance | | 2.5 | | nH | 1.5 mm from case |
| $f_o$ | Series Resonant Frequency | | 1.4 | | GHz | $V_R = 25$ V |

NOTES
1  These ratings are limiting values above which the serviceability of the diode may be impaired.
2  The capacitance difference between any two diodes in one set is less than 3% over the reverse voltage range of 0.6 V to 28 V.

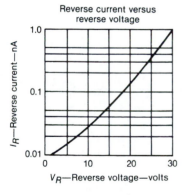

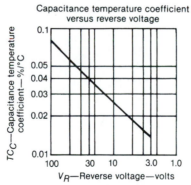

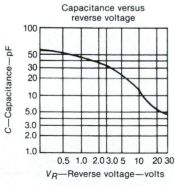

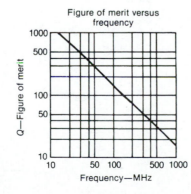

**FIGURE 5.28   (Copyright of Motorola. Used by permission.)**

**CASE 41-11**

**PEAK POWER DISSIPATION @ 1.0 ms = 1500 WATTS**

| Breakdown Voltage | | Device Type | | I$_{RSM}$ Maximum Reverse Surge Current Amp | V$_{RSM}$ Maximum Reverse Voltage @ I$_{RSM}$ Volts | Case |
|---|---|---|---|---|---|---|
| V(BR) Volts Nom | @I$_T$ mA | | | | | |
| 6.0 | 1.0 | 1N5908 | | 120 | 8.5 | 41-11 |
| 6.8 | 10 | 1N6267 | 1.5KE6.8 | 139 | 10.8 | |
| 7.5 | 10 | 1N6268 | 1.5KE7.5 | 128 | 11.7 | |
| 8.2 | 10 | 1N6269 | 1.5KE8.2 | 120 | 12.5 | |
| 9.1 | 1.0 | 1N6270 | 1.5KE9.1 | 109 | 13.8 | |
| 10 | 1.0 | 1N6271 | 1.5KE10 | 100 | 15.0 | |
| 11 | 1.0 | 1N6272 | 1.5KE11 | 93 | 16.2 | |
| 12 | 1.0 | 1N6273 | 1.5KE12 | 87 | 17.3 | |
| 13 | 1.0 | 1N6274 | 1.5KE13 | 79 | 19.0 | |
| 15 | 1.0 | 1N6275 | 1.5KE15 | 68 | 22.0 | |
| 16 | 1.0 | 1N6276 | 1.5KE16 | 64 | 23.5 | |
| 18 | 1.0 | 1N6277 | 1.5KE18 | 56.5 | 26.5 | |
| 20 | 1.0 | 1N6278 | 1.5KE20 | 51.5 | 29.1 | |
| 22 | 1.0 | 1N6279 | 1.5KE22 | 47.0 | 31.9 | |
| 24 | 1.0 | 1N6280 | 1.5KE24 | 43.0 | 34.7 | |
| 27 | 1.0 | 1N6281 | 1.5KE27 | 38.5 | 39.1 | |
| 30 | 1.0 | 1N6282 | 1.5KE30 | 34.5 | 43.5 | |
| 33 | 1.0 | 1N6283 | 1.5KE33 | 31.5 | 47.7 | |
| 36 | 1.0 | 1N6284 | 1.5KE36 | 29.0 | 52 | |
| 39 | 1.0 | 1N6285 | 1.5KE39 | 26.5 | 56.4 | |
| 43 | 1.0 | 1N6286 | 1.5KE43 | 24 | 61.9 | |
| 47 | 1.0 | 1N6287 | 1.5KE47 | 22.2 | 67.8 | |
| 51 | 1.0 | 1N6288 | 1.5KE51 | 20.4 | 73.5 | |
| 56 | 1.0 | 1N6289 | 1.5KE56 | 18.6 | 80.5 | |
| 62 | 1.0 | 1N6290 | 1.5KE62 | 16.9 | 89 | |
| 68 | 1.0 | 1N6291 | 1.5KE68 | 15.3 | 98 | |
| 75 | 1.0 | 1N6292 | 1.5KE75 | 13.9 | 108 | |
| 82 | 1.0 | 1N6293 | 1.5KE82 | 12.7 | 118 | |
| 91 | 1.0 | 1N6294 | 1.5KE91 | 11.4 | 131 | |
| 100 | 1.0 | 1N6295 | 1.5KE100 | 10.4 | 144 | |
| 110 | 1.0 | 1N6296 | 1.5KE110 | 9.5 | 158 | |
| 120 | 1.0 | 1N6297 | 1.5KE120 | 8.7 | 173 | |
| 130 | 1.0 | 1N6298 | 1.5KE130 | 8.0 | 187 | |
| 150 | 1.0 | 1N6299 | 1.5KE150 | 7.0 | 215 | |
| 160 | 1.0 | 1N6300 | 1.5KE160 | 6.5 | 230 | |
| 170 | 1.0 | 1N6301 | 1.5KE170 | 6.2 | 244 | |
| 180 | 1.0 | 1N6302 | 1.5KE180 | 5.8 | 258 | |
| 200 | 1.0 | 1N6303 | 1.5KE200 | 5.2 | 287 | |
| 220 | 1.0 | | 1.5KE220 | 4.3 | 344 | |
| 250 | 1.0 | | 1.5KE250 | 5.0 | 360 | |

Breakdown Voltage for Standard is ± 10% tolerance; ± 5% version is available by adding "A", i.e., 1N6267A. 1.5KE6.8A. Clipper (back to back) versions are available by ordering the 1.5KE series with a "C" or "CA" suffix, i.e., 1.5KE6.8C or 1.5KE6.8CA.

**FIGURE 5.29**   (Copyright of Motorola. Used by permission.)

**FIGURE 5.30**

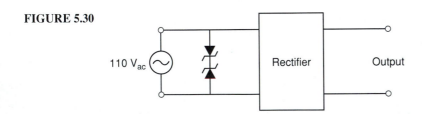

# 6

# BIPOLAR JUNCTION TRANSISTORS

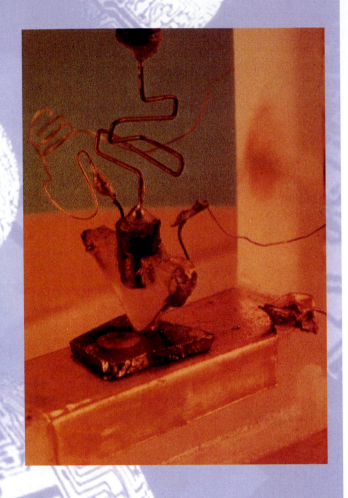

In the 1940s, a team of scientists working at Bell Labs developed the transistor, the first solid-state device capable of amplifying an ac signal. The term *solid-state* was coined because the transistor was solid, rather than hollow like the vacuum tube it was designed to replace.

# OUTLINE

# OBJECTIVES

*After studying the material in this chapter, you should be able to:*

1. Name and identify (by schematic symbol) each terminal of the bipolar junction transistor and explain its relationship to the other terminals of the transistor.

2. Describe the construction of a bipolar junction transistor (BJT) and the differences between the *npn* and *pnp* transistors.

3. Describe the characteristics of a bipolar junction transistor in the *cutoff*, *saturation*, and the *active* regions of operation.

4. Discuss the transistor as a current-controlled device, and state the relationship among the three terminal currents.

5. Define *beta* and *alpha*, and use both in the calculations of circuit currents.

6. Calculate the maximum allowable base current for a transistor, given the maximum allowable value of collector current and the maximum beta rating of the device.

7. List and describe the nine transistor circuit voltages and discuss the importance of the transistor voltage ratings.

8. Describe the characteristic curves of a BJT.

9. Describe the relationship among *beta*, *temperature*, and *dc collector current*.

10. Describe the five ohmmeter checks used to test for a faulty BJT.

11. Explain the difference between *discrete* and *integrated* transistors.

12. Describe the characteristics of *high-current*, *high-voltage*, and *high-power* transistors.

13. Describe the basic construction of *surface-mount components* and list the advantages they have over other ICs.

# A COMMON BELIEF

Popular belief holds that the *bipolar junction transistor*, or BJT, was developed by Schockley, Brattain, and Bardeen in 1948. However, this is not entirely accurate.

The transistor developed by the Bell Laboratories team in 1948 was a *point-contact transistor*. This device consisted of a thin germanium wafer connected to two extremely thin wires. The wires were spaced only a few thousandths of an inch apart. A current introduced to one of the wires was amplified by the germanium wafer, and the larger output current was taken from the other wire.

The BJT wasn't actually developed until late 1951. The component was developed by Dr. Schockley and another Bell Laboratories team. The first time the component was used in any type of commercial venture was in October 1952, when the Bell System employed transistor circuits in the telephone switching circuits in Englewood, N.J.

---

**Transistor**
A three-terminal device whose output current, voltage, and/or power are controlled by its input current.

**Amplifier**
A circuit used to increase the strength of an ac signal.

The main building block of modern electronic systems is the *transistor*. The **transistor** is a *three-terminal device whose output current, voltage, and/or power are controlled by its input current*. In communications systems, it is used as the primary component in the **amplifier,** a circuit that is used to increase the strength of an ac signal. In digital computer electronics, the transistor is used as a *high-speed electronic switch* capable of switching between two operating states (open and closed) at a rate of several billions of times per second.

There are two basic transistor types: the *bipolar junction transistor* (BJT) and the *field-effect transistor* (FET). As you will see, these two transistor types differ in their operating characteristics and their internal construction. You should note that the single term *transistor* is used to identify the BJT. The field-effect transistor is simply referred to as a FET.

In this chapter we will take a look at the bipolar junction transistor and its basic operating principles. FETs are discussed in detail in Chapters 12 and 13.

## 6.1 INTRODUCTION TO BIPOLAR JUNCTION TRANSISTORS (BJTs)

OBJECTIVE 1 ▶

The bipolar junction transistor, or *BJT*, is a *three-terminal* component. The three terminals are called the *emitter*, the *collector*, and the *base*. The emitter and collector terminals are made up of the same type of semiconductor material (either *p*-type or *n*-type) while the base is made of the other type of material. The construction of the BJT is illustrated in Figure 6.1.

As you can see, there are two types of BJTs. The first type, called the **npn transistor**, has *n*-type emitter and collector terminals and a *p*-type base. The **pnp transistor** is constructed in the opposite manner: this transistor has *p*-type emitter and collector terminals and an *n*-type base.

**npn transistor**
A BJT with *n*-type emitter and collector materials and a *p*-type base.

**pnp transistor**
A BJT with *p*-type emitter and collector materials and an *n*-type base.

### BJT Schematic Symbols

Figure 6.1 also shows the schematic symbols for the *npn* and *pnp* transistors. The arrow on the schematic symbol is important for three reasons:

1. It identifies the component terminals. *The arrow is always drawn on the emitter terminal*. The terminal opposite the emitter is the *collector*, and the center terminal is the *base*.

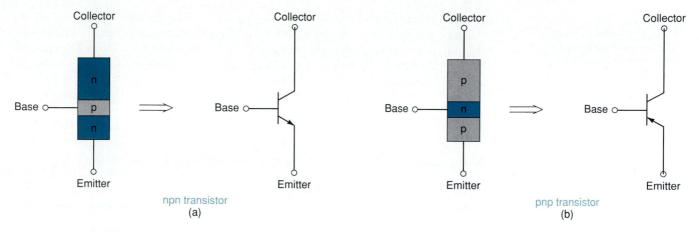

npn transistor
(a)

pnp transistor
(b)

**FIGURE 6.1** BJT construction and schematic symbols.

**2.** *The arrow always points toward the* n-*type material.* If the arrow points toward the *base*, the transistor is a *pnp* type. If it points toward the emitter, the transistor is an *npn* type.

**3.** *The arrow indicates the direction of the emitter current.* As with the *pn*-junction diode, the direction of current is indicated by the arrow. As you will see, knowing the direction of the emitter current tells you the directions of the other terminal currents.

## Transistor Currents

The terminal currents of a transistor are illustrated in Figure 6.2. The emitter, collector, and base currents of the transistor are identified as $I_E$, $I_C$, and $I_B$, respectively. Under normal circumstances, $I_E$ has the greatest value of the three, followed by $I_C$. The base current ($I_B$) normally has a much lower value than either of the other currents. Note that the current directions for the *npn* transistor are the opposite of those for the *pnp* transistor.

The transistor is a *current-controlled device*; that is, the values of the collector and emitter currents are determined primarily by the value of the base *current*. Under normal circumstances, the values of $I_C$ and $I_E$ vary directly with the value of $I_B$. An increase or decrease in the value of $I_B$ causes a similar change in the values of $I_C$ and $I_E$. This relationship is discussed in detail in Section 6.2.

The value of $I_C$ for a given transistor is normally some multiple of the value of $I_B$. The factor by which current increases from base to collector is referred to as the

**FIGURE 6.2** Transistor terminal currents.

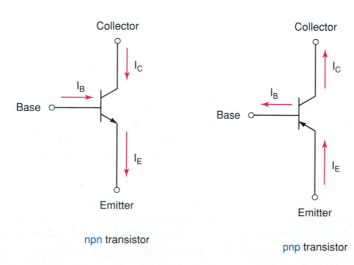

npn transistor

pnp transistor

**Current gain (β)**
The factor by which current increases from the base of a transistor to its collector.

**forward current gain** of the device and is represented using the Greek letter *beta (β)*. To determine the amount of the collector current for a transistor, you simply multiply the base current by the beta rating of the component. For example, let's say that the transistor in Figure 6.2a has values of $I_B = 50$ μA and $\beta = 120$. The value of $I_C$ is for the device is found as

$$I_C = \beta I_B$$
$$= (120)(50 \text{ μA})$$
$$= 6 \text{ mA}$$

The significance of this relationship is demonstrated throughout this chapter.

---

## Section Review

1. What is the primary application for the transistor in communications electronics?
2. What is the primary application for the transistor in digital electronics?
3. What are the two basic types of transistors?
4. What are the terminals of a bipolar junction transistor (BJT) called?
5. What are the two types of BJTs? How do they differ from each other?
6. What is indicated by the arrow on the BJT schematic symbol?
7. Draw and label the schematic symbols of an *npn* and a *pnp* transistor.

---

## 6.2 TRANSISTOR CONSTRUCTION AND OPERATION

OBJECTIVE 2 ▶

The two junctions in a BJT are referred to as the *base-emitter junction* and the *collector-base junction*.

The transistor is made up of three separate semiconductor materials. The three materials are joined together in such a way as to form two *pn* junctions, as shown in Figure 6.3.

The point at which the emitter and base are joined forms a single pn junction called the *base-emitter* junction. The *collector-base* junction is the point where the base and collector meet. The two junctions are normally operated in one of three biasing combinations, as follows:

| Base-Emitter Junction | Collector-Base Junction | Operating Region |
|---|---|---|
| Reverse biased | Reverse biased | Cutoff |
| Forward biased | Reverse biased | Active |
| Forward biased | Forward biased | Saturation |

The transistor operating regions are called *cutoff*, *active*, and *saturation*.

When both junctions are reverse biased, the transistor is said to be in *cutoff*. When the base-emitter junction is forward biased and the collector-base junction is reverse biased, the transistor is said to be operating in the *active* region. When both junctions are forward biased, the transistor is said to be operating in the *saturation* region. These "regions" refer to areas on a characteristic curve that will be discussed later in this chapter.

In our discussion of the transistor operating regions, we will be concentrating on the *npn* transistor. Note that all the principles covered apply equally to the *pnp* transistor. The differences between these two transistor types are found in the current directions and the biasing voltage polarities, as discussed later in this chapter.

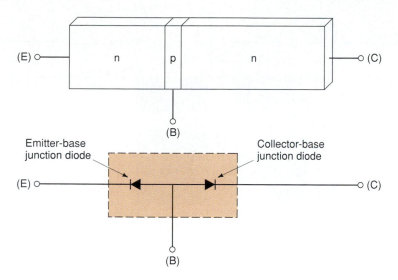

**FIGURE 6.3   BJT construction.**

## Zero Biasing

Figure 6.3 shows the *npn* transistor at room temperature with no biasing potential applied. You may recall from Chapter 1 that an unbiased *pn* junction will form a depletion layer at room temperature due to recombination of free carriers produced by thermal energy. The depletion layers that form at room temperature are shown in Figure 6.4

Since both junctions have a depletion layer, they are both *reverse biased* at room temperature. Note that the depletion layers extend farther into the base region than either of the other two. This is due to the fact that the base region has a lower doping level. With a lower doping level, there are fewer free carriers for recombination, so the depletion layer extends farther into this region.

*Zero bias* describes the biasing of the BJT at room temperature with no potentials applied.

## Cutoff

Normally, two biasing potentials are applied to the transistor. These potentials are shown in Figure 6.5. Note that both biasing sources in this figure have polarities that will reverse bias their respective junctions.

With the polarities shown, the two depletion layers will extend well into the emitter, base, and collector regions. With the larger depletion layers, only an extremely small

◄ *OBJECTIVE 3*

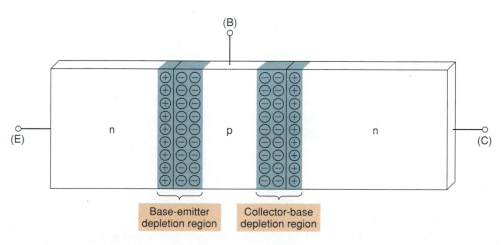

**FIGURE 6.4   Zero biasing.**

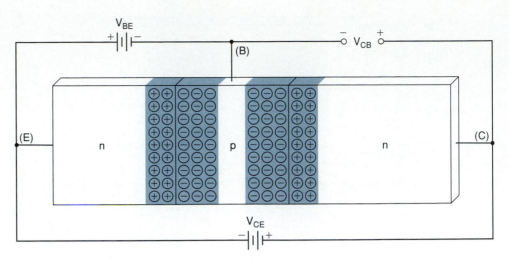

**FIGURE 6.5   Cutoff.**

**Cutoff**
A BJT operating state where $I_C$ is nearly zero.

amount of reverse current passes from the emitter to the collector, and the transistor is said to be in **cutoff**. For example, a 2N3904 transistor with a collector-emitter voltage ($V_{CE}$) of 40 V and a reverse base-emitter voltage ($V_{BE}$) of only 3 V will have only 50 nA of collector current ($I_C$). This is extremely small when compared to the current of 200 mA of $I_C$ that the component is capable of handling when the base-emitter junction is forward biased.

## Saturation

**Saturation**
A BJT operating region where $I_C$ reaches its maximum value.

The opposite of cutoff is **saturation**. Saturation is the condition where *further increases in $I_B$ will not cause increases in $I_C$*. When a transistor is saturated, $I_C$ has reached its maximum possible value, as determined by the collector supply voltage ($V_{CC}$) and the total resistance in the collector-emitter circuit. This point is illustrated in Figure 6.6.

Assume for a moment that $V_{CE}$ for the transistor is 0 V (an *ideal* situation). If this is the case, $I_C$ will depend completely on the values of $V_{CC}$, $R_C$ (the collector resistor), and $R_E$ (the emitter resistor). By Ohm's law, the maximum value of $I_C$ would be found as

$$I_C = \frac{V_{CC}}{R_C + R_E}$$

Now, let's say that $I_B$ is increased to the point where $I_C$ will reach its maximum value and will be able to increase no further. Then, further increases in $I_B$ will *not* increase $I_C$, and the relationship $I_C = \beta I_B$ will no longer hold true.

If $I_B$ is increased beyond the point where $I_C$ can increase, both of the transistor junctions will become forward biased. This point is illustrated in Figure 6.7. $V_{CE}$ for the transistor is shown to be approximately 0.3 V, which is typical for a saturated transistor. With the 0.7-V value of $V_{BE}$, the collector-base junction is biased to the difference between the two, 0.4 V. Note that this voltage indicates that the collector-base junction of the transistor is *forward* biased. When this occurs, the emitter, collector, and base currents are in the direction shown in the illustration.

## Active Operation

**Active region**
The BJT operating region between saturation (maximum $I_C$) and cutoff (minimum $I_C$).

A transistor is said to be operating in the **active region** when the base-emitter junction is forward biased and the collector-base junction is reverse biased. Generally, the transistor is said to be in active operation when it is between cutoff and saturation. The biasing for active operation is illustrated in Figure 6.8.

FIGURE 6.6　Saturation circuit conditions.

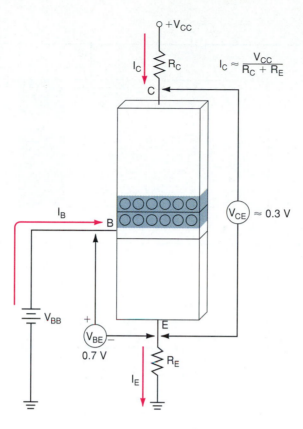

The operation of the transistor in this region is easiest to understand by considering just the base-emitter voltage ($V_{BE}$). When $V_{BE}$ is great enough to overcome the barrier potential of the junction, current is generated in the emitter and base regions.

If the base and emitter regions of the transistor acted as a normal diode, all emitter current would come from the base. However, because the base region is very lightly doped, the resistance of the base material is greater than the resistance of the reverse-biased collector-base junction. Thus, the vast majority of the emitter current is drawn from the collector circuit through the reverse-biased collector-base junction.

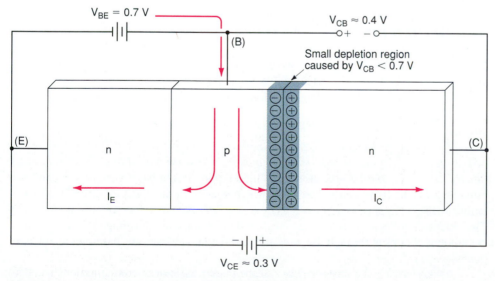

FIGURE 6.7　Saturation currents in an *npn* transistor.

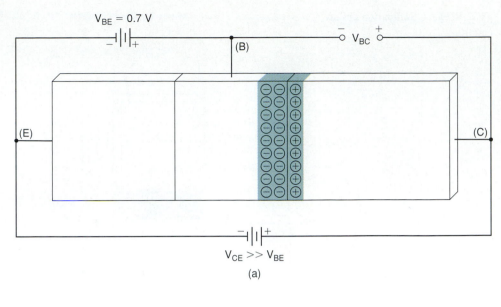

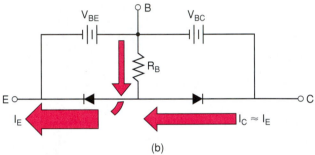

**FIGURE 6.8   Active operation.**

The idea of current through a reverse-biased *pn* junction should not seem that strange to you. After all, the zener diode is designed to allow current through a reverse-biased junction. The collector-base junction of the transistor is also designed to allow a reverse current. It can handle a large amount of current when reverse biased without doing damage to the junction.

## The Bottom Line

When a transistor is in *cutoff*, both junctions are reverse biased, and the current through all three terminals will be nearly zero. When *saturated*, $I_C$ is at its maximum possible value. In this case, both transistor junctions will be forward biased, depending on the value of $I_B$. In either case, $I_C$ is limited by $V_{CC}$ and the resistance in the collector-emitter circuit.

The region between cutoff and saturation is the active region. When a transistor is operating in this region, *its base-emitter junction is forward biased and its collector-base junction is reverse biased*. Because of the light doping of the base, very little recombination occurs in the base region, and most of $I_E$ comes from the collector circuit. The base-emitter voltage ($V_{BE}$) is approximately equal to 0.7 V, the barrier potential of the junction. The values of collector-base voltage ($V_{CB}$) and collector-emitter voltage ($V_{CE}$) depend on the amount of current through the transistor and on the values of the external circuit components.

As a reference, the characteristics of the three transistor operating regions are summarized in Figure 6.9.

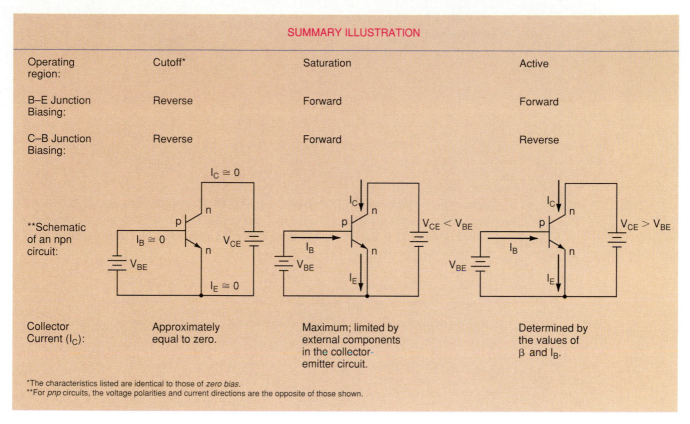

FIGURE 6.9

The image shows a summary illustration table with the following content:

| Operating region: | Cutoff* | Saturation | Active |
|---|---|---|---|
| B–E Junction Biasing: | Reverse | Forward | Forward |
| C–B Junction Biasing: | Reverse | Forward | Reverse |
| **Schematic of an npn circuit: | $I_C \cong 0$, $I_B \cong 0$, $V_{CE}$, $V_{BE}$, $I_E \cong 0$ | $I_C$, $I_B$, $V_{BE}$, $I_E$, $V_{CE} < V_{BE}$ | $I_C$, $I_B$, $V_{BE}$, $I_E$, $V_{CE} > V_{BE}$ |
| Collector Current ($I_C$): | Approximately equal to zero. | Maximum; limited by external components in the collector-emitter circuit. | Determined by the values of $\beta$ and $I_B$. |

*The characteristics listed are identical to those of *zero bias*.
**For *pnp* circuits, the voltage polarities and current directions are the opposite of those shown.

**Section Review**

1. How are the two junctions of a transistor biased when the component is in:
   a. Cutoff?
   b. The active region?
   c. Saturation?
2. What is the value of $I_C$ when a transistor is in cutoff?
3. What controls the value of $I_C$ when a transistor is saturated?
4. Describe the basic operation of a transistor biased for active-region operation.

# 6.3 TRANSISTOR CURRENTS, VOLTAGES, AND RATINGS

There are several transistor current and voltage ratings. Some of these ratings are parameters and some of them are typical electrical characteristics. In this section, we will take a look at several transistor current and voltage relationships, component ratings, and what they mean in everyday transistor applications.

## Transistor Currents

As you known, the transistor is a current-controlled device. In many applications, the base current is varied to produce variations in $I_C$ and $I_E$. Because of the construction of the component, *a small change in $I_B$ produces a large change in the other terminal cur-*

◄ *OBJECTIVE 4*

FIGURE 6.10 BJT current
relationships.

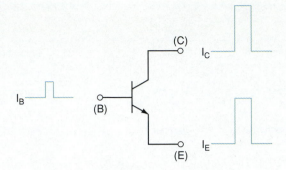

rents. This relationship is illustrated in Figure 6.10. As you can see, the small change in base current produces a large change in $I_C$ and in $I_E$. The large change in the emitter and collector currents is due to the current gain of the transistor. This point is illustrated in Example 6.1.

*EXAMPLE 6.1*

Determine the value of collector current for the transistor shown in Figure 6.11.

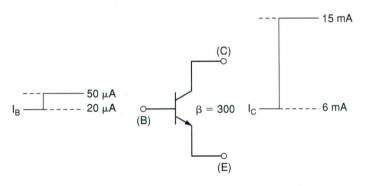

**FIGURE 6.11**

*Solution:* The base current in Figure 6.11 has an initial value of 20 μA. The beta rating of the component is 300. Using these values, the initial value of collector current is found as

$$I_C = \beta I_B$$
$$= (300)(20 \text{ μA})$$
$$= 6 \text{ mA}$$

When $I_B$ increases to 50 μA, the collector current also increases. At the new value of $I_B$, the collector current is found as

$$I_C = \beta I_B$$
$$= (300)(50 \text{ μA})$$
$$= 15 \text{ mA}$$

Thus, a 30 μA change in base current causes a 9 mA change in collector current.

**PRACTICE PROBLEM 6.1**

A transistor has values of $I_B = 50$ μA and $\beta = 350$. Determine the value of $I_C$ for the device.

The effect that a change in base current has on the output of a transistor will be demonstrated further when we discuss the ac operation of transistors.

## The Relationship Between $I_E$, $I_C$, and $I_B$

According to Kirchhoff's current law, the current leaving a component must equal the current entering the component. With this in mind, it is easy to see that $I_E$ must equal the sum of the other two components. By formula,

$$I_E = I_B + I_C \qquad (6.1)$$

Since $I_B$ is normally *much less* than $I_C$, the collector and emitter currents are approximately equal. By formula,

$$I_C \cong I_E \qquad (6.2)$$

The current relationship shown in equations (6.1) and (6.2) are correct for both the *npn* and *pnp* transistors.

The validity of equation (6.2) can be seen by looking at the results in Example 6.1. After the increase, $I_B$ was given as 50 μA and $I_C$ was determined to be 15 mA. According to equation (6.1), $I_E$ for the device would be found as

$$I_E = I_B + I_C$$
$$= 50 \text{ μA} + 15 \text{ mA}$$
$$= 15.05 \text{ mA}$$

As you can see, the values of $I_E$ and $I_C$ are approximately equal for the transistor in Example 6.1.

*A Practical Consideration*: The higher the value of β is, the closer $I_C$ will be to the $I_E$. This can be seen by comparing your results from Practice Problem 6.1 with the results from Example 6.1.

## DC Beta

The **dc beta** rating of a transistor is the *ratio of dc collector current to dc base current.* By formula,

◄ *OBJECTIVE 5*

$$\beta = \frac{I_C}{I_B} \qquad (6.3)$$

**dc beta**
The ratio of dc collector current to dc base current.

This is an extremely important rating because *the most common transistor circuits have the input signal applied to the base and the output signal taken from the collector.* Thus, when the transistor is used in these circuits, the dc beta rating of the transistor represents the overall *dc current gain* of the transistor.

We can use equations (6.1) and (6.3) to define the other terminal currents as follows:

$$I_C = \beta I_B \qquad (6.4)$$

**Lab Reference:** The dc beta of a transistor is determined using measured current values in Exercise 9.

and

$$I_E = I_B + I_C$$
$$= I_B + \beta I_B$$

or

$$I_E = I_B(1 + \beta) \qquad (6.5)$$

As Examples 6.2, 6.3, and 6.4 illustrate, you can use beta and any one terminal current to find the other two terminal currents.

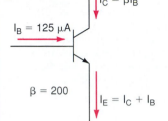

EXAMPLE 6.4

Determine the values of $I_B$ and $I_E$ for the circuit shown in Figure 6.14.

**Solution:** The base current can be found as

$$I_B = \frac{I_C}{\beta}$$

$$= \frac{50 \text{ mA}}{400}$$

$$= 125 \text{ } \mu A$$

Now the emitter current can be found as

$$I_E = I_C + I_B$$
$$= 50 \text{ mA} + 125 \text{ } \mu A$$
$$= 50.125 \text{ mA}$$

*PRACTICE PROBLEM 6.4*

The transistor in Figure 6.14 has values of $I_C = 80$ mA and $\beta = 170$. Determine the values of $I_B$ and $I_E$ for the device.

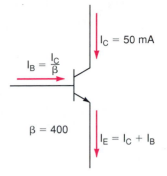

**FIGURE 6.14**

Because the beta rating of a transistor is a *ratio* of current values, it has no unit of measure. The typical range for transistor dc beta ratings is up to 300. This means that the typical transistor will have a dc collector current that can be up to 300 times as high as the dc base current when operated in the active region.

There is one point that should be made at this time: *Transistors have both dc beta ratings and ac beta ratings. Dc beta is the ratio of dc collector current to dc base current, while ac beta is the ratio of ac collector current to ac base current.* We will discuss ac beta and its applications in Chapter 9.

Transistors have both dc and ac beta ratings.

## dc Alpha

The **dc alpha** rating of a transistor is the *ratio of collector current to emitter current*. By formula,

$$\alpha = \frac{I_C}{I_E} \qquad (6.6)$$

The alpha rating of a given transistor will *always be less than unity* (1). The reason for this is illustrated in Figure 6.15.

Kirchhoff's current law states that the current leaving a point (or component) must equal the current entering the point (or component). The relationship among the three transistor terminal components was stated in equation (6.1) as

$$I_E = I_B + I_C$$

Therefore,

$$I_C = I_E - I_B$$

Since $I_C$ is always less than $I_E$ (by an amount equal to $I_B$), the fraction $I_C/I_E$ must always work out to be less than 1.

**dc alpha ($\alpha$)**
The ratio of dc collector current to dc emitter current. Also referred to as *collector current efficiency.*

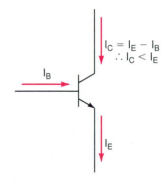

**FIGURE 6.15  Why alpha is always less than unity.**

The alpha rating of a transistor will usually be 0.9 or higher. Note that alpha (like beta) has no units since it is a ratio between two current values. Equation (6.6) can be rearranged to give us the following useful relationships:

$$I_C = \alpha I_E \qquad (6.7)$$

and

$$I_E = \frac{I_C}{\alpha} \qquad (6.8)$$

Using these relationships, we can calculate base current ($I_B$) as

$$I_B = I_E - I_C$$
$$= I_E - \alpha I_E$$

or

$$I_B = I_E (1 - \alpha) \qquad (6.9)$$

## The Relationship Between Alpha and Beta

As you will see later in this chapter, the spec sheet for a given transistor will list the value of beta for the device, but it will not list the value of alpha. This is because beta is used far more commonly in transistor circuit calculations than is alpha. This fact will become evident when we cover the dc and ac analyses of transistor circuits.

Since alpha is rarely listed on transistor spec sheets, you need to be able to determine its value using the value of beta. You can determine the value of alpha from the value of beta with the following equation:

$$\alpha = \frac{\beta}{1 + \beta} \qquad (6.10)$$

Example 6.5 illustrates the process of finding alpha when a beta value is known. It also demonstrates the validity of equation (6.10).

---

### EXAMPLE 6.5

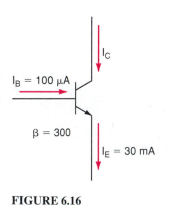

**FIGURE 6.16**

Determine the alpha rating for the transistor shown in Figure 6.16. Then determine the value of $I_C$ using both the alpha rating and the beta rating of the transistor.

*Solution:* The beta rating is given as 300. Therefore, the alpha rating is found as

$$\alpha = \frac{\beta}{1 + \beta}$$
$$= \frac{300}{301}$$
$$= 0.9967$$

The value of $I_C$ can be found as

$$I_C = \alpha I_E$$
$$= (0.9967)(30 \text{ mA})$$
$$= 29.9 \text{ mA}$$

or $I_C$ can be found as

$$I_C = \beta I_B$$
$$= (300)(100 \ \mu A)$$
$$= 30.0 \ mA$$

The two values of $I_C$ are close enough to validate both methods of calculating $I_C$.

## PRACTICE PROBLEM 6.5

A transistor has the following values: $\beta = 349$, $I_E = 350$ mA, and $I_B = 1$ mA. Determine the value of $\alpha$ for the device. Then calculate the value of $I_C$ using both $\alpha$ and $\beta$. How do the two values of $I_C$ compare?

## Maximum Current Ratings

Most transistor specifications sheets will list a maximum collector current rating for both saturation and cutoff. When the transistor is saturated, the collector current can go as high as several hundred milliamperes. High-power transistors can have current ratings as high as several amperes.

The maximum allowable base current value can be found by dividing the maximum $I_C$ value by the *maximum* dc beta rating. By formula,

$$I_{B(max)} = \frac{I_{C(max)}}{\beta_{max}} \quad \textbf{(6.11)}$$

Example 6.6 demonstrates the use of this formula.

◄ *OBJECTIVE 6*

*A Practical Consideration*: The beta rating of a transistor is usually listed as a *range* of values on the device spec sheet. Dealing with beta *ranges* is addressed later in this chapter.

---

The transistor shown in Figure 6.17 has the following ratings: $I_{C(max)} = 500$ mA and $\beta_{max} = 300$. Determine the maximum allowable value of $I_B$ for the device.

*Solution:* Using the ratings given, the value of $I_{B(max)}$ is found as

$$I_{B(max)} = \frac{I_{C(max)}}{\beta_{max}}$$
$$= \frac{500 \ mA}{300}$$
$$= 1.67 \ mA$$

If the base current is allowed to exceed 1.67 mA for the transistor shown, the collector current will exceed its maximum rating of 500 mA, and the transistor will probably be destroyed.

*EXAMPLE 6.6*

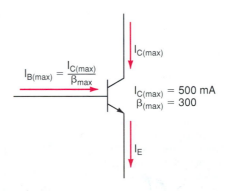

**FIGURE 6.17**

## PRACTICE PROBLEM 6.6

A transistor has ratings of $I_{C(max)} = 1$A and $\beta_{max} = 120$. Determine the value of $I_{B(max)}$ for the device.

**FIGURE 6.18**

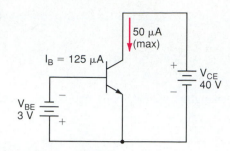

Transistors also have maximum *cutoff* current ratings. These ratings are usually in the low-nanoampere range and are specified for exact values of $V_{CE}$ and reverse $V_{BE}$. It was stated earlier that the 2N3904 has a maximum cutoff current rating of 50 nA when the reverse value of $V_{BE}$ is 3 V and the value of $V_{CE}$ is 40 V. These values are illustrated in Figure 6.18.

## Transistor Voltages

OBJECTIVE 7 ▶

Several voltages are normally involved in any discussion of transistor operation. These voltages are listed in Table 6.1. The voltages listed are identified in Figure 6.19. $V_{CC}$, $V_{EE}$, and $V_{BB}$ are identified as dc power supplies. $V_C$, $V_B$, and $V_E$ are all shown as voltages that are measured *from their respective terminals to ground*. $V_{CE}$, $V_{BE}$, and $V_{CB}$ are all shown as being measured between the terminals indicated by the subscripts. Be sure that you are able to distinguish between these voltages, as it will make future discussions of transistor operation easier to follow.

## Transistor Voltage Ratings

Most transistor specification sheets list a maximum value of *collector–base voltage*, $V_{CB}$. This rating indicates the maximum amount of reverse bias that can be applied to the collector–base junction without damaging the transistor. This rating is important because the collector–base junction is reverse biased for active region operation, as shown in Figure

**TABLE 6.1   TRANSISTOR VOLTAGES**

| Voltage Abbreviation | Definition |
|---|---|
| $V_{CC}$ | *Collector supply voltage.* This is a power supply voltage applied directly or indirectly to the collector of the transistor. |
| $V_{BB}$ | *Base supply voltage.* This is a dc voltage used to bias the base of the transistor. It may come directly from a dc voltage supply or may be applied indirectly to the base by a resistive circuit. |
| $V_{EE}$ | *Emitter supply voltage.* This, again, is a dc biasing voltage. In many cases, $V_{EE}$ is simply a *ground* connection. |
| $V_C$ | This is the dc voltage *measured from the collector terminal of the component to ground.* |
| $V_B$ | This is the dc voltage *measured from the base terminal to ground.* |
| $V_E$ | This is the dc voltage *measured from the emitter terminal to ground.* |
| $V_{CE}$ | This is the dc voltage *measured between the collector and emitter terminals of the transistor.* |
| $V_{BE}$ | This is the dc voltage *measured between the base and emitter terminals of the transistor.* |
| $V_{CB}$ | This is the dc voltage *measured between the collector and base terminals of the transistor.* |

*A Memory Trick:* There is a relatively simple way to remember the voltages listed here. When the voltage has a double subscript (such as *CC, BB,* or *EE*), it is a supply voltage. When two different subscripts are shown (such as *CE, BE,* or *CB*), the voltage is measured between the two terminals. When only one subscript is shown, the voltage is measured from that terminal to ground.

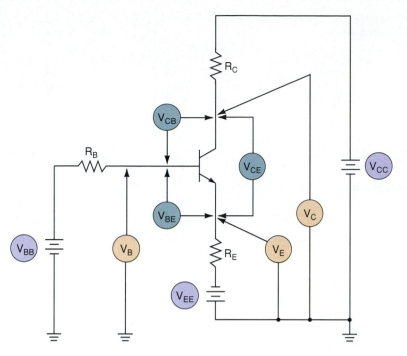

**FIGURE 6.19   BJT amplifier voltages.**

6.20. In this circuit, $V_{CE}$ is 50 V and $V_{BE}$ is 0.75 V. The value of $V_{CB}$ is equal to the difference between the other two voltages: 49.25 V. If this voltage is greater than the $V_{CB}$ rating of the transistor, the transistor will probably be destroyed.

Every transistor has three breakdown voltage ratings. These ratings indicate the *maximum reverse voltages* that the transistor can withstand. For the 2N3904, these voltage ratings are as follows:

| Rating | Value($V_{dc}$) |
|---|---|
| $BV_{CBO}$ | 60 |
| $BV_{CEO}$ | 40 |
| $BV_{EBO}$ | 6 |

The voltages listed are illustrated in Figure 6.21. If any of these reverse voltage ratings are exceeded, the transistor may not survive the experience.

There are many transistor ratings that have not been covered in this section. These ratings include *junction capacitance, maximum power dissipation, frequency limitations, operating temperature ranges,* and others. All these ratings will be covered as they are needed in future chapters.

**FIGURE 6.20   Collector–base junction biasing.**

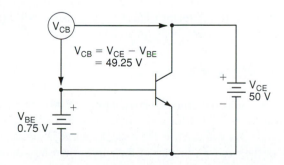

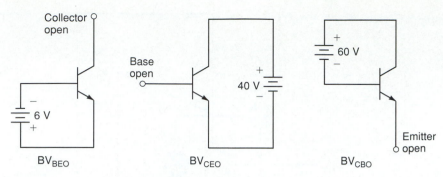

**FIGURE 6.21   BJT breakdown voltage ratings.**

1. What is meant by the term *current-controlled device*?
2. What is meant by the term *current gain*?
3. What symbol is commonly used to represent current gain?
4. Under normal circumstances, what is the relationship between:
    a. Base current and collector current?
    b. Collector current and emitter current?
5. What is *dc beta*?
6. Why doesn't beta have any units of measure?
7. What are the two types of beta ratings?
8. What is *dc alpha*?
9. What is the limit on the value of alpha?
10. Between beta and alpha, which rating is used more commonly?
11. Why do you need to be able to determine the value of alpha using the value of beta? How is this determination made?
12. What are the commonly used transistor voltage ratings?
13. What will happen if any of the reverse voltage ratings of a transistor are exceeded?
14. Define the following transistor voltages: $V_{CC}$, $V_{BB}$, $V_{EE}$, $V_C$, $V_B$, $V_E$, $V_{CE}$, $V_{BE}$, and $V_{CB}$.

# *6.4* TRANSISTOR CHARACTERISTIC CURVES

*OBJECTIVE 8* ▶ In this section, we are going to take a look at three characteristic curves that describe the operation of the transistor. We will start by looking at the collector and base curves. (There is no need for an emitter curve since its current characteristics are the same as those of the collector.) Then we will take a look at the beta curve, which shows the relationship between beta, $I_C$, and temperature.

## *Collector Curves*

**Collector curve**
Relates the values of $I_C$, $I_B$, and $V_{CE}$.

The **collector characteristic curve** relates the values of $I_C$, $I_B$, and $V_{CE}$. Each collector curve is derived for a specified value of $I_B$. Such a curve is shown in Figure 6.22.

As you can see, the collector curve is divided into three parts. The *saturation* region of the curve is the part of the curve where $V_{CE}$ is less than the *knee voltage*, $V_k$. Between

FIGURE 6.22 A
collector characteristic
curve.

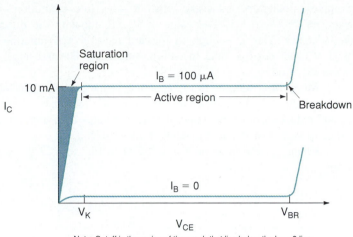

Note: Cutoff is the region of the graph that lies below the $I_B = 0$ line.

the knee voltage and the breakdown voltage ($V_{BR}$) is the *active* region. Finally, the area beyond $V_{BR}$ is the *breakdown* region. To help you understand these regions of the characteristic curve, we will discuss the operation of the transistor shown in Figure 6.23.

The *saturation* region represents the characteristics of the transistor when both junctions are forward biased. It was previously stated that the collector current in this operating region would be limited by the components external to the transistor. The value of $V_{CE}$ for the circuit in Figure 6.22 can be found as

$$V_{CE} = V_{CC} - I_C R_C \qquad (6.12)$$

The input to the base of the transistor is shown as $I_B = 100$ μA, the value of $R_C$ is 970 Ω, and beta is 100. These values give us the following output values:

$$I_C = \beta I_B$$
$$= (100)(100 \text{ μA})$$
$$= 10 \text{ mA}$$

and

$$V_{CE} = V_{CC} - I_C R_C$$
$$= 10 \text{ V} - (10 \text{ mA})(970 \text{ Ω})$$
$$= 10 \text{ V} - 9.7 \text{ V}$$
$$= 0.3 \text{ V}$$

This shows that when the $I_C$ is at 10 mA, the transistor is in saturation. We can also see that $V_{CE}$ is less than the 0.95-V value of $V_K$. Figure 6.24 shows the terminal

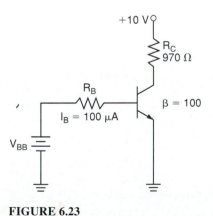

FIGURE 6.23

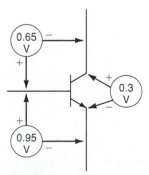

FIGURE 6.24

voltages around the transistor when it goes into saturation. $V_{BE}$ is 0.95 V and the value of $V_{CE}$ is 0.3 V. The value of $V_{CB}$ is equal to the difference between $V_{BE}$ and $V_{CE}$, 0.65 V. This is enough to forward bias the CB junction (the $p$-type material is 0.65 V more positive than the $n$-type material). The transistor is operating in the *saturation* region.

*A Practical Consideration:* $I_C$ actually *increases slightly* when $V_{CE}$ increases over its entire range. However, the change in $I_C$ is small enough to idealize to zero.

Most linear amplifiers are operated in the *active* region of operation. Figure 6.22 indicates that this region is characterized by the fact that $I_C$ is dependent on the value of $I_B$ and independent of the value of $V_{CE}$. The value of $I_C$ remains relatively constant for a range of values of $V_{CE}$ that extends from $V_K$ to $V_{BR}$.

Changing the value of $R_C$ to 400 Ω in Figure 6.23 gives us the following conditions:

$$I_C = \beta I_B$$
$$= (100)(100\ \mu A)$$
$$= 10\ mA$$

and

$$V_{CE} = V_{CC} - I_C R_C$$
$$= 10\ V - (10\ mA)(400\ \Omega)$$
$$= 6\ V$$

As you can see, changing the size of $R_C$ has not changed the value of $I_C$. However, it *has* changed the value of $V_{CE}$ from 0.3 V to 6 V. This shows that *changing the value of $V_{CE}$ will have little effect on the value of $I_C$.* It also shows that $I_C$ is found as a product of $I_B$ times beta. To change $I_C$, the value of $I_B$ or beta must change.

Figure 6.25 shows what happens when the value of $I_B$ is increased to 150 μA. With beta held constant at 100, the value of $I_C$ increases proportionately with the increase in $I_B$. The change in $I_C$ is still independent of changes in $V_{CE}$.

The *breakdown* region occurs when the value of $V_{CE}$ exceeds the breakdown voltage rating of the transistor. The value of $I_C$ increases dramatically until the transistor burns up from excessive heat caused by the increase in current.

Most collector characteristic curves are composite curves.

When several $I_B$ versus $I_C$ curves are plotted for a given transistor, a composite graph similar to the one in Figure 6.26 is created. The graph shows the collector currents produced by fixed values of $I_B$ for the transistor.

Each curve in Figure 6.26 shows the same flat response when $V_{CE}$ is in the active region, the same relative knee voltage, and the same breakdown voltage.

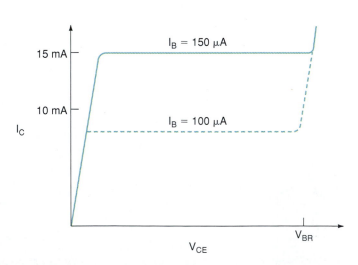

**FIGURE 6.25** **The effect of changing $I_B$ on a collector characteristic curve.**

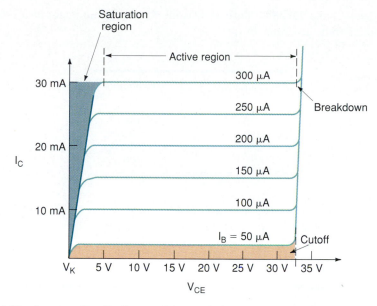

FIGURE 6.26   A composite of collector characteristic curves for a BJT.

**Lab Reference:** A composite of collector curves is developed using measured values in Exercise 9.

## Base Curves

The **base curve** of a transistor plots $I_B$ as a function of $V_{BE}$, as shown in Figure 6.27. Note that this curve closely resembles the forward operating curve of a typical *pn*-junction diode.

**Base curve**
A curve illustrating the relationship between $I_B$ and $V_{BE}$.

## Beta Curves

**Beta curves** show how the value of dc beta varies with both *temperature* and *dc collector current*. This point is illustrated in Figure 6.28. As you can see, the value of beta is greater at $T = 100°C$ than it is at 25°C. Also, beta increases (up to a point) for increases in the dc value of $I_C$. However, when $I_C$ increases beyond a certain value, beta starts to decrease.

◀   *OBJECTIVE 9*

**Beta curve**
A curve that shows the relationship between beta and temperature and/or collector current.

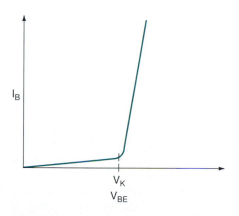

FIGURE 6.27   A base characteristic curve.

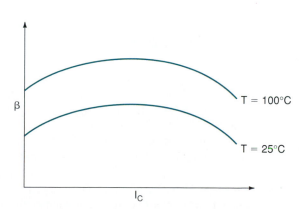

FIGURE 6.28   The relationship among beta, $I_C$, and temperature.

The spec sheet for the 2N3904 transistor lists the following minimum beta values, each measured at the indicated value of $I_C$:

| Minimum Beta | Condition |
|---|---|
| 40 | $I_C = 0.1$ mA$_{dc}$ |
| 70 | $I_C = 1.0$ mA$_{dc}$ |
| 100 | $I_C = 10$ mA$_{dc}$ |
| 60 | $I_C = 50$ mA$_{dc}$ |
| 30 | $I_C = 100$ mA$_{dc}$ |

As you can see, the value of beta increases as $I_C$ is increased to 10 mA$_{dc}$. However, as $I_C$ increases above 10 mA$_{dc}$, the value of beta begins to decrease. This goes along with the curves shown in Figure 6.28.

## Section Review

1. What does the saturation region represent on the transistor collector curve?
2. What does the active region represent on the transistor collector curve?
3. Does the value of $I_C$ depend on the value of $V_{CE}$? Explain your answer.
4. What happens when a transistor is operated in its breakdown region?
5. The base curve in Figure 6.27 indicates that the emitter-base junction of a transistor acts like what component? Explain your answer.
6. What is the relationship between beta and temperature?
7. What is the relationship between beta and dc collector current?

# 6.5 TRANSISTOR SPECIFICATION SHEETS

The spec sheet for a given transistor contains a wide variety of dc and ac operating characteristics. In this section, we will look at some of the commonly used *dc* maximum ratings and electrical characteristics that are used. In Chapter 9, we will look at the *ac* characteristics and operating curves that are typically used. For our discussion on the transistor *dc* ratings, we will use the spec sheet for the Motorola 2N3903/2N3904, which is shown in Figure 6.29.

## Maximum Ratings

*A Practical Consideration*: The "O" in the $V_{CEO}$, $V_{EBO}$, and $V_{CBO}$ ratings indicates that the third terminal is *open* when the rating is measured. For example, $V_{EBO}$ is measured with the collector terminal open. This ensures that the BJT is in cutoff when the parameter is measured.

You have already been introduced to many of the maximum ratings listed in Figure 6.29. The $V_{CEO}$, $V_{CBO}$, and $V_{EBO}$ ratings are the maximum reverse voltage ratings that we discussed in the last section. The $I_C$ rating is the maximum allowable continuous value of $I_C$, 200 mA$_{dc}$ in this case.

The *total device dissipation* ($P_D$) rating of the transistor is the same type of rating that is used for the zener diode and *pn*-junction diode. As the spec sheet shows, the 2N3904 has a $P_D$ rating of 625 mW when the *ambient* (or *room*) temperature ($T_A$) is 25°C. If the *case* temperature ($T_C$) is held to 25°C, the device $P_D$ rating increases to 1.5 W. Note that the case temperature can be held to 25°C by fan cooling or through the use of a heat sink. As always, both ratings must be derated as temperature increases.

## MAXIMUM RATINGS

| Rating | Symbol | Value | Unit |
|---|---|---|---|
| Collector-Emitter Voltage | $V_{CEO}$ | 40 | Vdc |
| Collector-Base Voltge | $V_{CBO}$ | 60 | Vdc |
| Emitter-Base Voltage | $V_{EBO}$ | 6.0 | Vdc |
| Collector Current — Continuous | $I_C$ | 200 | mAdc |
| Total Device Dissipation @ $T_A$ = 25°C  Derate above 25°C | $P_D$ | 625  5.0 | mW  mW/°C |
| *Total Device Dissipation @ $T_C$ = 25°C  Derate above 25°C | $P_D$ | 1.5  12 | Watts  mW/°C |
| Operating and Storage Junction  Temperature Range | $T_J$, $T_{stg}$ | − 55 to + 150 | °C |

## *THERMAL CHARACTERISTICS

| Characteristic | Symbol | Max | Unit |
|---|---|---|---|
| Thermal Resistance, Junction to Ambient | $R_{\theta JA}$ | 200 | °C/W |
| Thermal Resistance, Junction to Case | $R_{\theta JC}$ | 83.3 | °C/W |

*Indicates Data in addition to JEDEC Requirements.

## 2N3903
## 2N3904★

**CASE 29-04, STYLE 1**
**TO-92 (TO-226AA)**

## GENERAL PURPOSE
## TRANSISTORS

**NPN SILICON**

★This is a Motorola
designated preferred device.

## ELECTRICAL CHARACTERISTICS ($T_A$ = 25°C unless otherwise noted.)

| Characteristic | Symbol | Min | Max | Unit |
|---|---|---|---|---|
| **OFF CHARACTERISTICS** | | | | |
| Collector-Emitter Breakdown Voltage(1)  ($I_C$ = 1.0 mAdc, $I_B$ = 0) | $V_{(BR)CEO}$ | 40 | — | Vdc |
| Collector-Base Breakdown Voltage  ($I_C$ = 10 μAdc, $I_E$ = 0) | $V_{(BR)CBO}$ | 60 | — | Vdc |
| Emitter-Base Breakdown Voltage  ($I_E$ = 10 μAdc, $I_C$ = 0) | $V_{(BR)EBO}$ | 6.0 | — | Vdc |
| Base Cutoff Current  ($V_{CE}$ = 30 Vdc, $V_{EB}$ = 3.0 Vdc) | $I_{BL}$ | — | 50 | nAdc |
| Collector Cutoff Current  ($V_{CE}$ = 30 Vdc, $V_{EB}$ = 3.0 Vdc) | $I_{CEX}$ | — | 50 | nAdc |
| **ON CHARACTERISTICS** | | | | |
| DC Current Gain(1)  ($I_C$ = 0.1 mAdc, $V_{CE}$ = 1.0 Vdc)     2N3903  2N3904 | $h_{FE}$ | 20  40 | —  — | — |
| ($I_C$ = 1.0 mAdc, $V_{CE}$ = 1.0 Vdc)     2N3903  2N3904 | | 35  70 | —  — | |
| ($I_C$ = 10 mAdc, $V_{CE}$ = 1.0 Vdc)     2N3903  2N3904 | | 50  100 | 150  300 | |
| ($I_C$ = 50 mAdc, $V_{CE}$ = 1.0 Vdc)     2N3903  2N3904 | | 30  60 | —  — | |
| ($I_C$ = 100 mAdc, $V_{CE}$ = 1.0 Vdc)     2N3903  2N3904 | | 15  30 | —  — | |
| Collector-Emitter Saturation Voltage(1)  ($I_C$ = 10 mAdc, $I_B$ = 1.0 mAdc)  ($I_C$ = 50 mAdc, $I_B$ = 5.0 mAdc) | $V_{CE(sat)}$ | —  — | 0.2  0.3 | Vdc |
| Base-Emitter Saturation Voltage(1)  ($I_C$ = 10 mAdc, $I_B$ = 1.0 mAdc)  ($I_C$ = 50 mAdc, $I_B$ = 5.0 mAdc) | $V_{BE(sat)}$ | 0.65  — | 0.85  0.95 | Vdc |

**FIGURE 6.29**   The 2N3903–3904 specification sheet. (Copyright of Motorola. Used by permission.)

# Thermal Characteristics

The thermal ratings of the transistor are used primarily in circuit development applications and will not be discussed here.

## Off Characteristics

The **off characteristics** describe the operation of the transistor when it is operated in *cutoff*. The first three ratings, $V_{(BR)CEO}$, $V_{(BR)CBO}$, and $V_{(BR)EBO}$, are the same ratings that appeared in the maximum ratings listing. They are repeated here for the convenience of the technician or engineer who must rapidly find the ratings.

The **collector cutoff current ($I_{CEX}$)** rating indicates the maximum value of $I_C$ when the device is in cutoff. For the 2N3904, $I_{CEX}$ is 50 nA.

The **base cutoff current ($I_{BL}$)** rating indicates the maximum amount of base current present when the emitter-base junction is in cutoff. For the 2N3904, $I_{BL}$ is a maximum of 50 nA.

As the $I_{BL}$ and $I_{CEX}$ ratings indicate, the terminal currents of a cutoff transistor are extremely low. In the case of the 2N3904, $I_B$ and $I_C$ will be no greater than 50 nA. The value of $I_E$ will therefore be no greater than the sum of the two, or 100 nA.

## On Characteristics

The **on characteristics** describe the dc operating characteristics for both the active and saturation regions of operation.

The **dc current gain ($h_{FE}$)** rating of the transistor is the value of *dc beta*. Note that the label $h_{FE}$ is normally used to represent dc beta, so we will be using this label in all our discussions from now on. The meaning of this label is discussed in Chapter 9. For now, just remember that $h_{FE}$ is the label commonly used to represent dc beta.

As you can see, the values of $h_{FE}$ are measured at different values of $I_C$. This goes along with the discussion we had earlier on *beta versus collector current*.

The **collector-emitter saturation voltage ($V_{CE(sat)}$)** rating indicates the maximum value of $V_{CE}$ when the device is in saturation. For the 2N3904, the value of $V_{CE(sat)}$ is 0.2 V when $I_C = 10$ mA$_{dc}$ and 0.3 V when $I_C = 50$ mA$_{dc}$. Thus, at the rated values of $I_C$, $V_{CE}$ will be no greater than 0.2 or 0.3 V.

The **base-emitter saturation voltage ($V_{BE(sat)}$)** rating indicates the maximum value of $V_{BE}$ when the device is in saturation. As you can see, this value is shown to be 0.85 V or 0.95 V, depending on the rated value of $I_B$.

There are many other commonly used transistor specifications. However, these specifications deal mainly with the *ac* operation of the device. We will cover these specs when we cover transistor ac operation in Chapter 9.

---

**Section Review**

1. What do the *off characteristics* of a transistor indicate?
2. What is *base cutoff current*?
3. What is *collector cutoff current*?
4. What do the *on characteristics* of a transistor indicate?
5. What label is commonly used to represent dc beta?
6. What is *collector-emitter saturation voltage*?
7. What is *base-emitter saturation voltage*?

---

## 6.6 TRANSISTOR TESTING

*OBJECTIVE 10* ▶  You may recall that a diode can be tested with an ohmmeter. The transistor can be checked in the same basic manner. However, several more measurements are needed to verify that a transistor is good. The process for testing a transistor is illustrated in Fig-

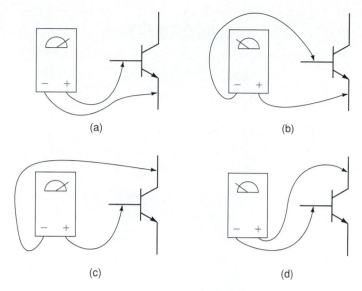

(a)                          (b)

(c)                          (d)

**Lab Reference:** BJT testing with an ohmmeter is demonstrated in Exercise 16.

**FIGURE 6.30**   **Transistor resistance checks.**

ure 6.30. The emitter-base junction is tested for a low forward resistance and a high reverse resistance. When connected to the ohmmeter, as in Figure 6.30a, the meter should read a low resistance, typically 1 kΩ or less. When connected as shown in Figure 6.30b, the resistance should be high, typically 100 kΩ or higher. If both these tests show the emitter-base junction to be good, the collector-base junction should be tested in the same manner. These tests, illustrated in Figure 6.30c and d, should yield the same results. The forward resistance should be approximately 1 kΩ or less, and the reverse resistance should be approximately 100 kΩ or higher.

The last test involves checking the collector-emitter terminals. These terminals should have an extremely high resistance, regardless of the polarity of the meter. If the transistor passes all the resistance checks, it is a good transistor. If it fails *any one check*, it is faulty and must be replaced.

*A few final points:* First, the polarities shown in Figure 6.30 are for an *npn* transistor. When testing a *pnp* transistor, the ohmmeter leads must be reversed. The results of the tests, however, will be the same. Also, before testing a transistor with a VOM, be sure to review the points made in Chapter 2 (Section 2.10) regarding the use of a VOM for testing *pn* junctions.

## Transistor Checkers

A *transistor checker* is a specialized piece of test equipment that is designed to test most *npn* and *pnp* transistors. In addition to determining whether or not a transistor is faulty, a transistor checker can be used to determine the actual value of $h_{FE}$ (dc beta) for a given transistor. A typical transistor checker is shown in Figure 6.31.

1. Describe the method used to test a transistor with an ohmmeter.

2. List the precautions that must be taken when checking a transistor with an ohmmeter. (These precautions can be found in Section 2.10.)

3. What test can be made with a transistor checker that cannot be made with a standard ohmmeter?

*Section Review*

**FIGURE 6.31    A typical transistor checker. (Courtesy of Sencore, Inc.)**

# *6.7* *RELATED TOPICS*

In this section, we will wrap up the chapter by briefly discussing some topics that do not really fit into any of the previous sections.

## pnp *Versus* npn *Transistors*

*pnp* and *npn* transistors differ only in voltage polarities and current directions.

In all our previous discussions about BJT operation, we concentrated on the *npn* transistor. The *pnp* transistor has the same basic operating characteristics as the *npn* transistor. The primary difference is the polarity of the biasing voltages and the direction of the terminal currents. These differences are illustrated in Figure 6.32. As you can see, $V_{CC}$, $V_{CE}$, and $V_{BE}$ are all *negative* voltages for the *pnp* transistor. $V_{CC}$ and $V_{BB}$ generate currents that exit the transistor through their respective terminals. The combination of $I_C$ and $I_B$ enters the transistor via the emitter terminal.

All the ratings that have been discussed apply to the *pnp* transistor as well as to the *npn* transistor. Also, all the equations used in this chapter apply to the *pnp* transistor.

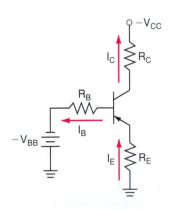

**FIGURE 6.32   *pnp* voltage polarities and current directions.**

In practice, the transistor biasing voltages are normally provided by the system's dc power supply.

## *Supply Voltages*

Throughout the chapter, $V_{CC}$ and $V_{BB}$ have been represented as batteries in all the circuit diagrams. In practice, the biasing voltages for transistor circuits are usually derived from the dc power supply of an electronic system. This point is illustrated in Figure 6.33.

In the circuit shown, the (+) and (−) outputs from the dc power supply are used as $V_{CC}$ and ground, respectively. Note that the output from the dc power supply would be used to bias a variety of other circuits as well. Figure 6.33 has been limited to one transistor circuit for the sake of simplicity.

There are a few more points that should be made: First, the value of $V_{BB}$ in Figure 6.33 is developed by the combination of $V_{CC}$, $R_B$, and the transistor itself. *In most practical circuits, $V_{BB}$ is actually drawn from the collector power supply, $V_{CC}$.* This point will be made clear throughout Chapter 7 when we cover *dc biasing circuits.* Second, *pnp tran-*

FIGURE 6.33

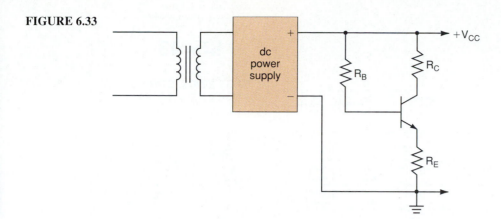

*sistors are used in systems that have negative dc power supplies*. There are some cases where *pnp* transistors are used in systems with positive dc power supplies, but they are used mainly in negative-supply systems.

## Integrated Transistors

**Integrated transistors** come in packages that house more than one transistor in integrated form. You may recall that integrated circuits contain more than one device in a single casing. Figure 6.34 shows a typical IC transistor casing. Note that the package contains four individual transistors, as is indicated by the *pin* listing (located in the lower left-hand corner of the figure).

Each transistor in an IC (like the one shown) has the same types of maximum ratings and electrical characteristics as *individual* (or **discrete**) transistors. The only real difference is the small amount of space that is taken up by the four transistors in the IC package.

## High-Voltage Transistors

**High-voltage transistors** have unusually high reverse breakdown voltage ratings. For example, the maximum ratings for the Motorola BFW43 transistor are shown in Figure 6.35. Note that the $V_{CEO}$ and $V_{CBO}$ ratings are much higher than those for the 2N3904 (see

◄ *OBJECTIVE 11*

**Integrated transistor**
An IC that contains more than one transistor.

**Discrete**
A term used to describe devices packaged in individual casings.

◄ *OBJECTIVE 12*

**High-voltage transistors**
BJTs with high reverse breakdown voltage ratings.

**FIGURE 6.34** (Copyright of Motorola. Used by permission.)

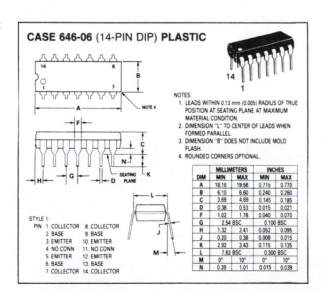

CASE 646-06 (14-PIN DIP) PLASTIC

NOTES:
1. LEADS WITHIN 0.13 mm (0.005) RADIUS OF TRUE POSITION AT SEATING PLANE AT MAXIMUM MATERIAL CONDITION.
2. DIMENSION "L" TO CENTER OF LEADS WHEN FORMED PARALLEL.
3. DIMENSION "B" DOES NOT INCLUDE MOLD FLASH.
4. ROUNDED CORNERS OPTIONAL.

| DIM | MILLIMETERS | | INCHES | |
|-----|-----|-----|-----|-----|
| | MIN | MAX | MIN | MAX |
| A | 18.16 | 19.56 | 0.715 | 0.770 |
| B | 6.10 | 6.60 | 0.240 | 0.260 |
| C | 3.69 | 4.69 | 0.145 | 0.185 |
| D | 0.38 | 0.53 | 0.015 | 0.021 |
| F | 1.02 | 1.78 | 0.040 | 0.070 |
| G | 2.54 BSC | | 0.100 BSC | |
| H | 1.32 | 2.41 | 0.052 | 0.095 |
| J | 0.20 | 0.38 | 0.008 | 0.015 |
| K | 2.92 | 3.43 | 0.115 | 0.135 |
| L | 7.62 BSC | | 0.300 BSC | |
| M | 0° | 10° | 0° | 10° |
| N | 0.39 | 1.01 | 0.015 | 0.039 |

STYLE 1:
PIN 1. COLLECTOR   8. COLLECTOR
2. BASE   9. BASE
3. EMITTER   10. EMITTER
4. NO CONN   11. NO CONN
5. EMITTER   12. EMITTER
6. BASE   13. BASE
7. COLLECTOR   14. COLLECTOR

### MAXIMUM RATINGS

| Rating | Symbol | Value | Unit |
|---|---|---|---|
| Collector-Emitter Voltage | $V_{CEO}$ | 150 | Vdc |
| Collector-Base Voltage | $V_{CBO}$ | 150 | Vdc |
| Emitter-Base Voltage | $V_{EBO}$ | 6.0 | Vdc |
| Collector Current — Continuous | $I_C$ | 0.1 | Adc |
| Total Device Dissipation @ $T_A$ = 25°C<br>Derate above 25°C | $P_D$ | 0.4<br>2.66 | Watt<br>mW/°C |
| Total Device Dissipation @ $T_C$ = 25°C<br>Derate above 25°C | $P_D$ | 1.4<br>8.0 | Watt<br>mW/°C |
| Operating and Storage Junction<br>Temperature Range | $T_J$, $T_{stg}$ | − 65 to + 200 | °C |

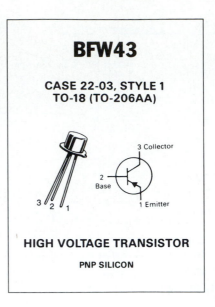

**BFW43**

**CASE 22-03, STYLE 1
TO-18 (TO-206AA)**

3 Collector

2
Base

1 Emitter

3 2 1

**HIGH VOLTAGE TRANSISTOR**

PNP SILICON

**FIGURE 6.35**   (Copyright of Motorola. Used by permission.)

Figure 6.29). High-voltage transistors are used in circuits that have extremely high supply voltages, such as television CRT (cathode ray tube) control circuits.

## High-Current Transistors

**High-current transistors**
BJTs with high maximum $I_C$ ratings.

**High-current transistors** have very high maximum $I_C$ ratings. For example, the maximum ratings for the Motorola 2N4237 transistor are shown in Figure 6.36. Note that the $I_C$ rating of the component is 1 A dc. This is significantly higher than the $I_C$ rating of the 2N3904 (Figure 6.29). Obviously, these devices would be used in applications that have high current demands, such as current regulator circuits.

### MAXIMUM RATINGS

| Rating | Symbol | 2N4237 | 2N4238 | 2N4239 | Unit |
|---|---|---|---|---|---|
| Collector-Emitter Voltage | $V_{CEO}$ | 40 | 60 | 80 | Vdc |
| Collector-Base Voltage | $V_{CBO}$ | 50 | 80 | 100 | Vdc |
| Emitter-Base Voltage | $V_{EBO}$ | | 6.0 | | Vdc |
| Base Current | $I_B$ | | 500 | | mA |
| Collector Current — Continuous | $I_C$ | | 1.0 | | Adc |
| Total Device Dissipation<br>@ $T_A$ = 25°C<br>Derate above 25°C | $P_D$ | | 1.0<br>5.3 | | Watt<br>mW/°C |
| Total Device Dissipation<br>@ $T_C$ = 25°C<br>Derate above 25°C | $P_D$ | | 6.0<br>34 | | Watts<br>mW/°C |
| Operating and Storage Junction<br>Temperature Range | $T_J$, $T_{stg}$ | | − 65 to + 200 | | °C |

**2N4237
thru
2N4239**

**CASE 79-04, STYLE 1
TO-39 (TO-205AD)**

3 Collector

2
Base

1 Emitter

3 2 1

**GENERAL PURPOSE
TRANSISTORS**

**NPN SILICON**

**FIGURE 6.36**   (Copyright of Motorola. Used by permission.)

## MOTOROLA
## ■ SEMICONDUCTOR ■
### TECHNICAL DATA

**NPN**
## 2N3902

### 3.5 AMPERE
### POWER TRANSISTORS
### NPN SILICON

**400 VOLTS**
**100 WATTS**

### HIGH VOLTAGE NPN SILICON TRANSISTORS

. . . designed for use in high-voltage inverters, converters, switching regulators and line operated amplifiers.

- High Collector-Emitter Voltage — $V_{CEX}$ = 700 Vdc
- Excellent DC Current Gain —
  $h_{FE}$ = 10 (Min) @ $I_C$ = 2.5 Adc
- Low Collector-Emitter Saturation Voltage —
  $V_{CE(sat)}$ = 0.8 Vdc (Max) @ $I_C$ = 1.0 Adc

### *MAXIMUM RATINGS

| Rating | Symbol | 2N3902 | Unit |
|---|---|---|---|
| Collector-Emitter Voltage | $V_{CEO}$ | 400 | Vdc |
| Collector-Emitter Voltage | $V_{CEX}$ | 700 | Vdc |
| Emitter-Base Voltage | $V_{EB}$ | 5.0 | Vdc |
| Collector Current — Continuous | $I_C$ | 3.5 | Adc |
| Base Current | $I_B$ | 2.0 | Adc |
| Total Device Dissipation @ $T_C$ = 75°C<br>Derate above 75°C | $P_D$ | 100<br>1.33 | Watts<br>W/°C |
| Operating Junction Temperature Range | $T_J$ | −65 to +150 | °C |
| Storage Temperature Range | $T_{stg}$ | −65 to +200 | °C |

### THERMAL CHARACTERISTICS

| Characteristic | Symbol | Max | Unit |
|---|---|---|---|
| Thermal Resistance, Junction to Case | $\theta_{JC}$ | 0.75 | °C/W |

*Indicates JEDEC Registered Data

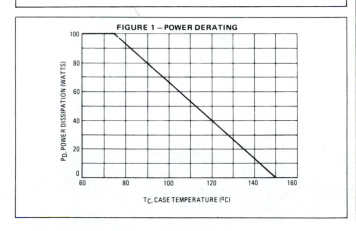

**FIGURE 1 — POWER DERATING**

$T_C$, CASE TEMPERATURE (°C)

$P_D$, POWER DISSIPATION (WATTS)

STYLE 1:
PIN 1. BASE
2. EMITTER
CASE COLLECTOR

-T- SEATING PLANE

NOTES:
1. DIMENSIONING AND TOLERANCING PER ANSI Y14.5M, 1982
2. CONTROLLING DIMENSION: INCH.
3. ALL RULES AND NOTES ASSOCIATED WITH REFERENCED TO-204AA OUTLINE SHALL APPLY.

| DIM | MILLIMETERS | | INCHES | |
|---|---|---|---|---|
| | MIN | MAX | MIN | MAX |
| A | — | 39.37 | — | 1.550 |
| B | — | 21.08 | — | 0.830 |
| C | 6.35 | 8.25 | 0.250 | 0.325 |
| D | 0.97 | 1.09 | 0.038 | 0.043 |
| E | 1.40 | 1.77 | 0.055 | 0.070 |
| F | 30.15 BSC | | 1.187 BSC | |
| G | 10.92 BSC | | 0.430 BSC | |
| H | 5.46 BSC | | 0.215 BSC | |
| J | 16.89 BSC | | 0.665 BSC | |
| K | 11.18 | 12.19 | 0.440 | 0.480 |
| Q | 3.84 | 4.19 | 0.151 | 0.165 |
| R | — | 26.67 | — | 1.050 |
| U | 4.83 | 5.33 | 0.190 | 0.210 |
| V | 3.84 | 4.19 | 0.151 | 0.165 |

**CASE 1-06**
**TO-204AA**
**(TO-3)**

FIGURE 6.37   (Copyright of Motorola. Used by permission.)

# High-Power Transistors

**High-power transistors**
BJTs with high power dissipation ratings.

**High-power transistors** are designed for use in high-power circuits, such as regulated and switching power supplies. High-power transistors, as you may have guessed, are devices that have extremely high power dissipation ratings. For example, the maximum ratings for the Motorola 2N3902 power transistor are shown in Figure 6.37.

As you can see, the $P_D$ rating of the 2N3902 is 100 W. Power dissipation ratings of this magnitude are needed in many regulated power supply applications. (Regulated power supplies are covered in Chapter 21.)

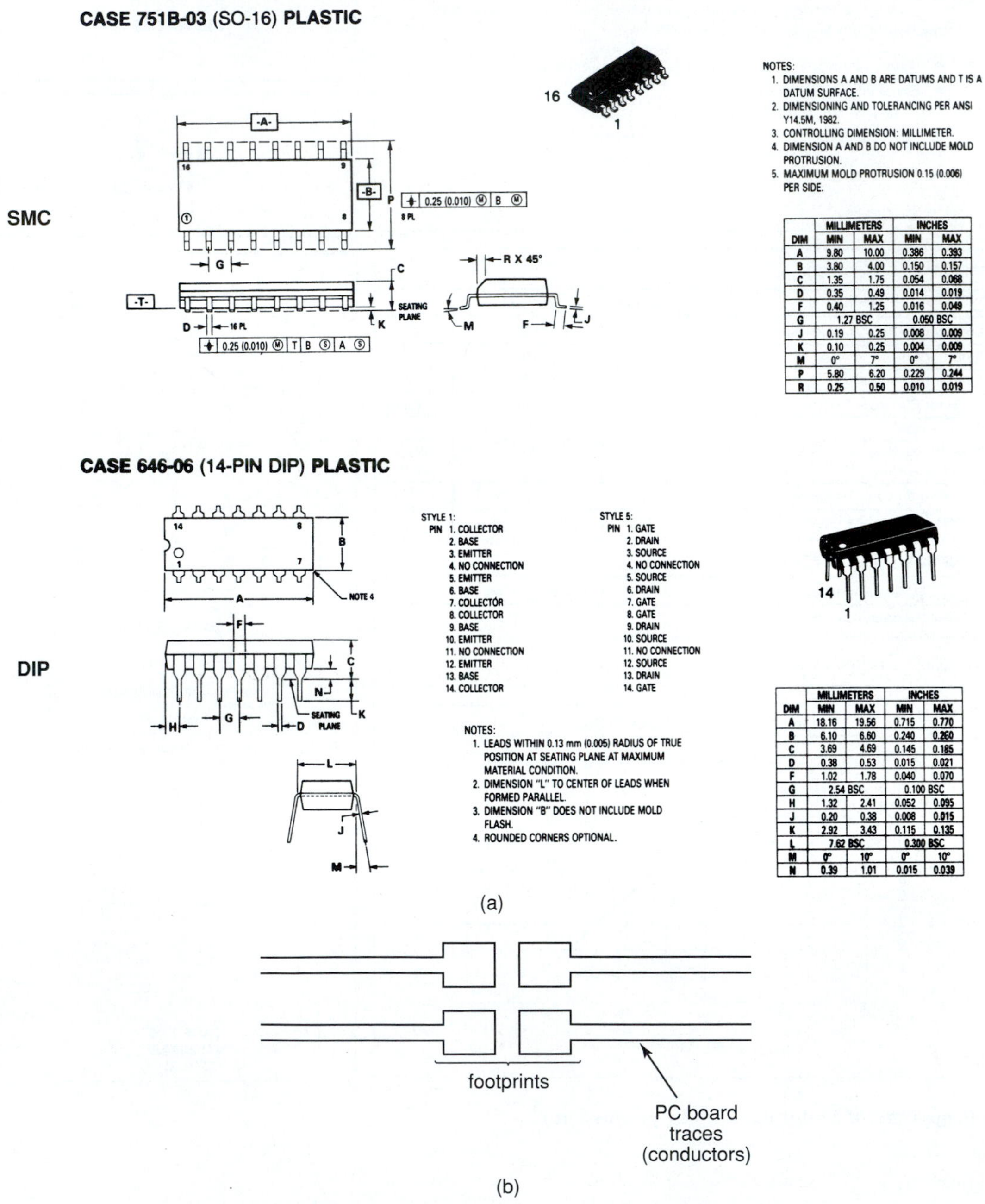

**CASE 751B-03 (SO-16) PLASTIC**

SMC

NOTES:
1. DIMENSIONS A AND B ARE DATUMS AND T IS A DATUM SURFACE.
2. DIMENSIONING AND TOLERANCING PER ANSI Y14.5M, 1982.
3. CONTROLLING DIMENSION: MILLIMETER.
4. DIMENSION A AND B DO NOT INCLUDE MOLD PROTRUSION.
5. MAXIMUM MOLD PROTRUSION 0.15 (0.006) PER SIDE.

| DIM | MILLIMETERS MIN | MILLIMETERS MAX | INCHES MIN | INCHES MAX |
|---|---|---|---|---|
| A | 9.80 | 10.00 | 0.386 | 0.393 |
| B | 3.80 | 4.00 | 0.150 | 0.157 |
| C | 1.35 | 1.75 | 0.054 | 0.068 |
| D | 0.35 | 0.49 | 0.014 | 0.019 |
| F | 0.40 | 1.25 | 0.016 | 0.049 |
| G | 1.27 BSC | | 0.050 BSC | |
| J | 0.19 | 0.25 | 0.008 | 0.009 |
| K | 0.10 | 0.25 | 0.004 | 0.009 |
| M | 0° | 7° | 0° | 7° |
| P | 5.80 | 6.20 | 0.229 | 0.244 |
| R | 0.25 | 0.50 | 0.010 | 0.019 |

**CASE 646-06 (14-PIN DIP) PLASTIC**

DIP

STYLE 1:
PIN 1. COLLECTOR
2. BASE
3. EMITTER
4. NO CONNECTION
5. EMITTER
6. BASE
7. COLLECTOR
8. COLLECTOR
9. BASE
10. EMITTER
11. NO CONNECTION
12. EMITTER
13. BASE
14. COLLECTOR

STYLE 5:
PIN 1. GATE
2. DRAIN
3. SOURCE
4. NO CONNECTION
5. SOURCE
6. DRAIN
7. GATE
8. GATE
9. DRAIN
10. SOURCE
11. NO CONNECTION
12. SOURCE
13. DRAIN
14. GATE

NOTES:
1. LEADS WITHIN 0.13 mm (0.005) RADIUS OF TRUE POSITION AT SEATING PLANE AT MAXIMUM MATERIAL CONDITION.
2. DIMENSION "L" TO CENTER OF LEADS WHEN FORMED PARALLEL.
3. DIMENSION "B" DOES NOT INCLUDE MOLD FLASH.
4. ROUNDED CORNERS OPTIONAL.

| DIM | MILLIMETERS MIN | MILLIMETERS MAX | INCHES MIN | INCHES MAX |
|---|---|---|---|---|
| A | 18.16 | 19.56 | 0.715 | 0.770 |
| B | 6.10 | 6.60 | 0.240 | 0.260 |
| C | 3.69 | 4.69 | 0.145 | 0.185 |
| D | 0.38 | 0.53 | 0.015 | 0.021 |
| F | 1.02 | 1.78 | 0.040 | 0.070 |
| G | 2.54 BSC | | 0.100 BSC | |
| H | 1.32 | 2.41 | 0.052 | 0.095 |
| J | 0.20 | 0.38 | 0.008 | 0.015 |
| K | 2.92 | 3.43 | 0.115 | 0.135 |
| L | 7.62 BSC | | 0.300 BSC | |
| M | 0° | 10° | 0° | 10° |
| N | 0.39 | 1.01 | 0.015 | 0.039 |

(a)

footprints

PC board traces (conductors)

(b)

**FIGURE 6.38** Surface-mount component dimensions. (Copyright of Motorola. Used by permission.)

## Surface-Mount Components

The primary advantage in using integrated transistors is the fact that several transistors can be housed in a single, relatively small package. This size advantage has been taken one step further with the development of *surface-mount* technology. **Surface-mount components (SMCs)** are integrated components that are much smaller and lighter than their standard IC counterparts. The term *surface-mount* is derived from the fact that these ICs are mounted directly *onto* the surface of a PC board, rather than being mounted in holes on the board or fitted into IC sockets.

Figure 6.38a shows a comparison between a standard *dual in-line* package (DIP) and an equivalent surface-mount package. The difference in their sizes can be seen by comparing the *maximum A, B,* and *C* dimensions of the two packages. The maximum size of the DIP, in millimeters, is given as $19.56 \times 6.6 \times 4.69$. For the SMC, the maximum size, in millimeters, is given as $10 \times 4 \times 1.75$. As you can see, the SMC is roughly half the size of its DIP counterpart. This means that more of these components can be mounted in a given area of a PC board.

If you look closely at the pins on the SMC in Figure 6.38a, you'll see that the pins are flattened at the bottom. This allows the component to be soldered directly onto *footprints* on the PC board. These footprints (represented in Figure 6.38b) are typically no larger than $1 \text{ mm}^2$. Because of this and the extreme sensitivity that SMCs have to temperature, great care must be taken when soldering replacement SMCs onto a PC board. *Before attempting to solder an SMC onto a PC board, always read the manufacturer's soldering guidelines.* These guidelines are often provided with the components.

◄ *OBJECTIVE 13*

**Surface-mount components (SMCs)**
ICs that are much smaller and lighter than their standard IC counterparts.

---

### Section Review

1. What are the primary differences between *pnp* and *npn* transistor circuits?
2. Which transistor is used primarily in systems with *positive* dc supply voltages?
3. Which transistor is used primarily in systems with *negative* dc supply voltages?
4. What is the difference between *integrated* and *discrete* transistors?
5. What are *high-voltage transistors*? When are they typically used?
6. What are *high-current transistors*? When are they typically used?
7. What are *high-power transistors*? When are they typically used?
8. Describe the differences between surface-mount components and standard DIP integrated circuits.

---

### Key Terms

The following terms were introduced and defined in this chapter:

active region
amplifier
amplitude
base
base current ($I_B$)
base curve
base cutoff current ($I_{BL}$)
base-emitter junction
base-emitter saturation
   voltage ($V_{BE(\text{sat})}$)
beta ($\beta$)

beta curve
bipolar junction transistor
   (BJT)
breakdown voltage
collector
collector-base junction
collector biasing voltage
   ($V_{CC}$)
collector characteristic
   curve
collector current ($I_C$)

collector cutoff current
   ($I_{CEX}$)
collector-emitter
   saturation voltage
   ($V_{CE(\text{sat})}$)
current gain ($\beta$)
cutoff
dc alpha
dc beta
dc current gain ($h_{FE}$)
discrete

emitter
emitter current ($I_E$)
emitter follower
high-current transistor
high-power transistor
high-voltage transistor

integrated transistor
*npn* transistor
off characteristics
on characteristics
*pnp* transistor

saturation
surface-mount
   components
transistor
transistor checker

| Equation Summary | Equation Number | Equation | Section Number |
|---|---|---|---|
| | (6.1) | $I_E = I_B + I_C$ | 6.3 |
| | (6.2) | $I_C \cong I_E$ | 6.3 |
| | (6.3) | $\beta = \dfrac{I_C}{I_B}$ | 6.3 |
| | (6.4) | $I_C = \beta I_B$ | 6.3 |
| | (6.5) | $I_E = I_B(1 + \beta)$ | 6.3 |
| | (6.6) | $\alpha = \dfrac{I_C}{I_E}$ | 6.3 |
| | (6.7) | $I_C = \alpha I_E$ | 6.3 |
| | (6.8) | $I_E = \dfrac{I_C}{\alpha}$ | 6.3 |
| | (6.9) | $I_B = I_E(1 - \alpha)$ | 6.3 |
| | (6.10) | $\alpha = \dfrac{\beta}{1 + \beta}$ | 6.3 |
| | (6.11) | $I_{B(\text{max})} = \dfrac{I_{C(\text{max})}}{\beta_{\text{max}}}$ | 6.3 |
| | (6.12) | $V_{CE} = V_{CC} - I_C R_C$ | 6.4 |

**Answers to the Example Practice Problems**

**6.1.** $I_C = 17.5$ mA

**6.2.** $I_C = 20$ mA, $I_E = 20.05$ mA

**6.3.** $I_B = 85.11$ μA, $I_C = 11.92$ mA

**6.4.** $I_B = 470.59$ μA, $I_E = 80.47$ mA

**6.5.** $\alpha = 0.997$, $I_C = 349$ mA (using β), $I_C = 348.95$ mA (using α)

**6.6.** $I_{B(\text{max})} = 8.33$ mA

**Practice Problems**

§6.3

1. A BJT has values of $\beta = 320$ and $I_B = 12$ μA. Determine the value of $I_C$ for the device.

2. A BJT has values of $\beta = 400$ and $I_B = 30$ μA. Determine the value of $I_C$ for the device.

**3.** A BJT has values of $\beta = 254$ and $I_B = 1.01$ mA. Determine the value of $I_C$ for the device.

**4.** A BJT has values of $\beta = 144$ and $I_B = 82$ μA. Determine the value of $I_C$ for the device.

**5.** A BJT has values of $I_B = 20$ μA and $I_C = 1.1$ mA. Determine the value of $I_E$ for the device.

**6.** A BJT has values of $I_B = 1.1$ mA and $I_C = 344$ mA. Determine the value of $I_E$ for the device.

**7.** Complete the following table.

| Beta | $I_B$ | $I_C$ |
|---|---|---|
| a. 150 | 25 μA | _____ |
| b. ____ | 75 μA | 1.5 mA |
| c. 240 | 100 μA | _____ |
| d. 325 | _____ | 20 mA |

**8.** Complete the following table.

| Beta | $I_B$ | $I_C$ |
|---|---|---|
| a. ____ | 50 μA | 12 mA |
| b. 440 | _____ | 35 mA |
| c. 175 | 45 μA | _____ |
| d. ____ | 120 μA | 84 mA |

**9.** Complete the following table.

| $I_B$ | $I_C$ | $I_E$ |
|---|---|---|
| a. 25 μA | 1 mA | _____ |
| b. _____ | 1.8 mA | 1.98 mA |
| c. 120 μA | _____ | 3 mA |
| d. _____ | 7.5 mA | 8 mA |
| e. 50 μA | _____ | 20 mA |
| f. 175 μA | 9.825 mA | _____ |

**10.** A BJT has values of $I_B = 35$ μA and $\beta = 100$. Determine the values of $I_C$ and $I_E$ for the device.

**11.** A BJT has values of $I_B = 150$ μA and $\beta = 400$. Determine the values of $I_C$ and $I_E$ for the device.

**12.** A BJT has values of $I_B = 48$ μA and $\beta = 120$. Determine the values of $I_C$ and $I_E$ for the device.

**13.** A BJT has values of $I_C = 12$ mA and $\beta = 440$. Determine the values of $I_B$ and $I_E$ for the device.

**14.** A BJT has values of $I_C = 50$ mA and $\beta = 400$. Determine the values of $I_B$ and $I_E$ for the device.

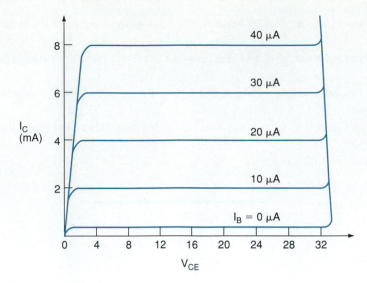

**FIGURE 6.39**

15. A BJT has values of $I_E = 65$ mA and $\beta = 380$. Determine the values of $I_B$ and $I_C$ for the device.

16. A BJT has values of $I_E = 120$ mA and $\beta = 60$. Determine the values of $I_B$ and $I_C$ for the device.

17. A BJT has a value of $\beta = 426$. Determine the value of $\alpha$ for the device.

18. A BJT has a value of $\beta = 350$. Determine the value of $\alpha$ for the device.

19. A BJT has the following parameters: $I_{C(max)} = 120$ mA and $\beta = 50$ to $120$. Determine the maximum allowable value of $I_B$ for the device.

20. A BJT has the following parameters: $I_{C(max)} = 250$ mA and $\beta = 35$ to $100$. Determine the maximum allowable value of $I_B$ for the device.

### §6.4

21. Refer to Figure 6.39. What is the breakdown voltage of the transistor represented by the collector curve?

22. Refer to Figure 6.39. What is the value of $I_C$ when $I_B$ is 40 μA?

23. Refer to Figure 6.39. What is the maximum value of $V_{CE}$ when the device is saturated?

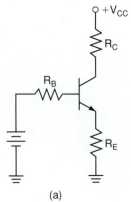

(a)

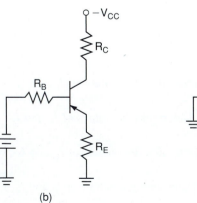

(b)

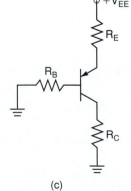

(c)

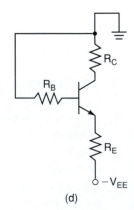

(d)

**FIGURE 6.40**

*§6.7*

**24.** For each circuit shown in Figure 6.40, identify the type of transistor and indicate the directions of the terminal currents.

## Troubleshooting Practice Problems

**25.** Following are the results of several transistor tests. In each case, determine whether or not the transistor is good.

| *Base-Emitter* | | *Collector-Base* | | *Emitter-to-Collector* |
|---|---|---|---|---|
| *Forward* | *Reverse* | *Forward* | *Reverse* | *Resistance* |
| a. 250 Ω | 300 kΩ | 150 kΩ | 140 kΩ | 1200 kΩ |
| b. 800 Ω | 115 kΩ | 800 Ω | 115 kΩ | 14 kΩ |
| c. 377 Ω | 152 kΩ | 900 Ω | 180 kΩ | 1500 kΩ |
| d. 100 kΩ | 100 kΩ | 190 Ω | 144 kΩ | 3500 kΩ |

**26.** Draw a series of diagrams to show how you would connect a VOM with a negative ground lead to test a *pnp* transistor. In each case, indicate the reading you would obtain for a good transistor.

**27.** Repeat Problem 26, showing how you would connect a VOM with a positive ground lead to test a *pnp* transistor.

## Suggested Computer Applications Problems

**28.** Write a program that will solve for $I_C$, $I_E$, and $\alpha$, given the values of $I_B$ and $h_{FE}$ (dc beta).

**29.** Write a program that will solve the table shown in Problem 9, given any two of the values.

**30.** Write a program that will accept information like that shown in Problem 25 and tell you whether or not a given transistor is good. Assume a good transistor would have a minimum reverse-to-forward resistance ratio of 100:1.

# 7

# dc BIASING
# CIRCUITS

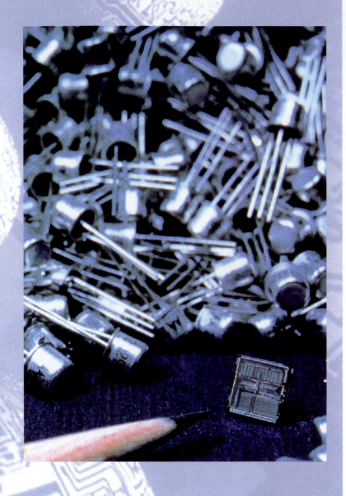

**M**anufacturing technology has developed to the point where literally hundreds of thousands of components can be produced on a single silicon wafer.

## OUTLINE

## OBJECTIVES

*After studying the material in this chapter, you should be able to:*

1. State the purpose of dc biasing circuits.
2. Plot the dc load line for an amplifier, given the values of $V_{CC}$ and the total collector-emitter circuit resistance.
3. Describe the $Q$-point of an amplifier and explain what the point represents.
4. Describe and analyze the operation of a base-bias circuit.
5. Determine if a circuit is midpoint biased, given the values of $I_C$, $V_{CE}$, and $V_{CC}$ for the circuit.
6. Describe and analyze the operation of an emitter-bias circuit.
7. Troubleshoot emitter-bias circuits.
8. Describe and analyze the operation of a voltage-divider biasing circuit.
9. Estimate the value of $I_{CQ}$ for an amplifier without detailed calculation and justify the use of this estimate.
10. Describe the troubleshooting procedure for a voltage-divider circuit.
11. Describe and analyze the operation of a collector-feedback bias circuit.
12. Describe and analyze the operation of the emitter-feedback bias circuit.

# THE WORLD OF SILICON MINIATURE COMPONENTS

Not long after the development of the transistor, the race was on to make miniature versions of almost every type of component and circuit. It was found that almost any type of circuit or component could be made using silicon. Resistors and capacitors, for example, could be produced on silicon. This made it possible for entire amplifiers (including the biasing components) to be made on a single silicon wafer.

Manufacturing technology has developed to the point where literally hundreds of thousands of components can be produced on a single silicon wafer that is much smaller than a penny. Yet the simplest of all electronic components, the *inductor*, took longer to develop in semiconductor form than any other component. Inductors were first fabricated in microminiature form in the late 1980s. Even now, these inductors are considered impractical because of the amount of chip space (or "real estate") that it takes to produce them.

*OBJECTIVE 1* ▶

The purpose of the *dc biasing circuit* is to set up the initial dc values of $I_B$, $I_C$, and $V_{CE}$.

$\mathbf{T}$he ac operation of an amplifier depends on the initial dc values of $I_B$, $I_C$, and $V_{CE}$. As $I_B$ is varied from an initial value, $I_C$ and $V_{CE}$ are varied from their initial values. This operation is illustrated in Figure 7.1. The *purpose of initial dc biasing is to set up the initial values of $I_B$, $I_C$, and $V_{CE}$.* Several bias methods are used to achieve different results, but each provides a means of setting the initial operating point of the transistor. In this chapter, we will discuss the operation and troubleshooting of the most commonly used dc biasing circuits, including base, emitter, and voltage-divider bias, as well as collector- and emitter-feedback bias circuits.

## 7.1 INTRODUCTION TO dc BIASING: THE dc LOAD LINE

**dc load line**
A graph of all $V_{CE}$ versus $I_C$ combinations.

*OBJECTIVE 2* ▶

Equation (6.12):
$V_{CE} = V_{CC} - I_C R_C$

The **dc load line** is a graph that *represents all the possible combinations of $I_C$ and $V_{CE}$ for a given amplifier*. For every possible value of $I_C$, an amplifier will have a corresponding value of $V_{CE}$. The dc load line represents all the $I_C/V_{CE}$ combinations for the circuit. A generic dc load line is shown in Figure 7.2.

The ends of the load line are labeled $I_{C(sat)}$ and $V_{CE(off)}$. The value of the $I_{C(sat)}$ point represents the *ideal* value of saturation current for the circuit. If we look at the saturated transistor as being a short circuit from emitter to collector, then $V_{CE}$ is zero, and equation (6.12) becomes

**FIGURE 7.1  Typical amplifier operation.**

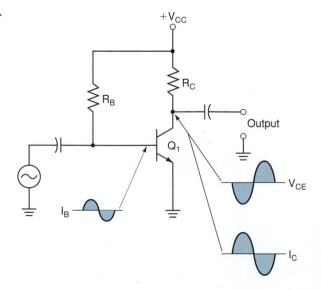

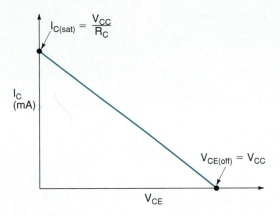

FIGURE 7.2   A generic dc load line.

$$V_{CC} = I_C R_C \qquad \text{(saturation)} \qquad \textbf{(7.1)}$$

Rearranging the formula for $I_C$, we get

$$I_{C(\text{sat})} = \frac{V_{CC}}{R_C} \qquad \text{(ideal)} \qquad \textbf{(7.2)}$$

Conversely, when the transistor is in cutoff, we can look at the component as being an open circuit from emitter to collector. Because the transistor acts as an open, there is no collector current. Thus, $I_C R_C = 0$, and equation (6.12) becomes

$$V_{CE(\text{off})} = V_{CC} \qquad \textbf{(7.3)}$$

As the following example shows, equations (7.2) and (7.3) are used to plot the end points of the dc load line.

Plot the dc load line for the circuit shown in Figure 7.3a.

*EXAMPLE **7.1***

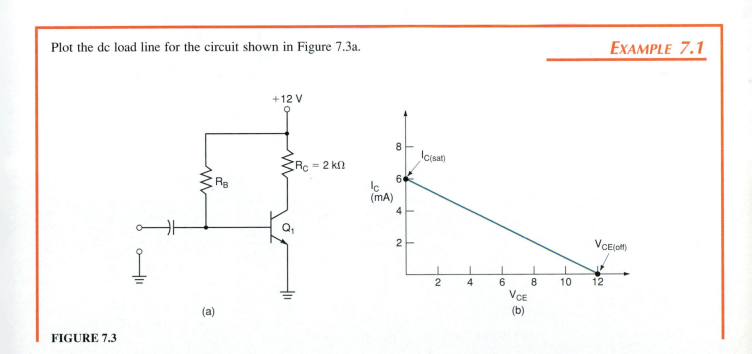

FIGURE 7.3

*Solution:* With the circuit values shown,

$$V_{CE(off)} = V_{CC}$$
$$= 12 \text{ V}$$

and

$$I_{C(sat)} = \frac{V_{CC}}{R_C}$$
$$= \frac{12\text{V}}{2 \text{ k}\Omega}$$
$$= 6 \text{ mA}$$

The load line therefore had end points of $V_{CE(off)} = 12$ V and $I_{C(sat)} = 6$ mA. The plotted line is shown in Figure 7.3b. While $R_B$ is shown in the figure, it has no effect on load line end points.

### PRACTICE PROBLEM 7.1

A circuit like the one in Figure 7.3a has values of $V_{CC} = +8$ V and $R_C = 1.1$ k$\Omega$. Plot the dc load line for the circuit.

As stated earlier, the dc load line represents all the possible combinations of $I_C$ and $V_{CE}$ for a given amplifier. This point is illustrated in the following example.

## EXAMPLE 7.2

Plot the dc load line for the circuit shown in Figure 7.4. Then use equation (6.12) to verify the load line $V_{CE}$ values for $I_C = 1$ mA, $I_C = 2$ mA, and $I_C = 5$ mA.

*Solution:* With the circuit values shown, $I_{C(sat)}$ is found as

$$I_{C(sat)} = \frac{V_{CC}}{R_C}$$
$$= \frac{10 \text{ V}}{1 \text{ k}\Omega}$$
$$= 10 \text{ mA}$$

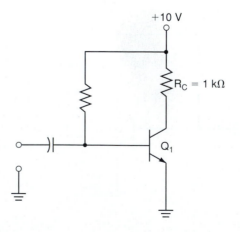

**FIGURE 7.4**

Now $V_{CE(off)}$ is found as

$$V_{CE(off)} = V_{CC}$$
$$= 10 \text{ V}$$

Using $I_{C(sat)}$ and $V_{CE(off)}$ as the end points, the dc load line for the circuit is plotted as shown in Figure 7.5a.

**FIGURE 7.5**

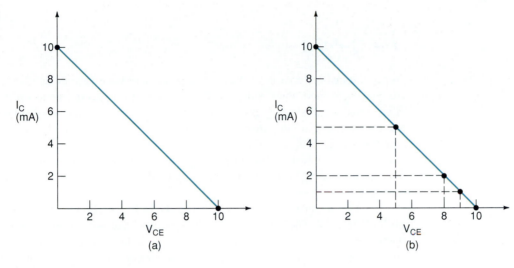

As Figure 7.5b shows, the following $I_C$ versus $V_{CE}$ combinations should be possible in this circuit:

| $I_C$ (mA) | $V_{CE}$ (V) |
|---|---|
| 1 | 9 |
| 2 | 8 |
| 5 | 5 |

These combinations (which were chosen at random) are verified by equation (6.12) as follows:

For $I_C = 1$ mA,

$$V_{CE} = V_{CC} - I_C R_C$$
$$= 10 \text{ V} - (1 \text{ mA})(1 \text{ k}\Omega)$$
$$= 10 \text{ V} - 1 \text{ V}$$
$$= 9 \text{ V}$$

For $I_C = 2$ mA,

$$V_{CE} = V_{CC} - I_C R_C$$
$$= 10 \text{ V} - (2 \text{ mA})(1 \text{ k}\Omega)$$
$$= 10 \text{ V} - 2 \text{ V}$$
$$= 8 \text{ V}$$

For $I_C = 5$ mA,

$$V_{CE} = V_{CC} - I_C R_C$$
$$= 10 \text{ V} - (5 \text{ mA})(1 \text{ k}\Omega)$$
$$= 10 \text{ V} - 5 \text{ V}$$
$$= 5 \text{ V}$$

**Lab Reference:** A dc load line is plotted and used to predict circuit values in Exercise 10.

These calculations verify the values obtained from the dc load line.

## The Q-Point

OBJECTIVE 3 ▶

**Q-point**
A point on the dc load line that indicates the values of $V_{CE}$ and $I_C$ for an amplifier at rest.

**Quiescent**
At rest.

When a transistor does not have an ac input, it will have specific dc values of $I_C$ and $V_{CE}$. As you have seen, these values will correspond to a specific point on the dc load line. This point is called the **Q-point.** The letter Q comes from the word **quiescent,** meaning *at rest.* A quiescent amplifier is one that has no ac signal applied and therefore has constant dc values of $I_C$ and $V_{CE}$.

When the dc load line of an amplifier is superimposed on the collector curves for the transistor, the Q-point value can easily be determined. This point is illustrated in Figure 7.6.

Assume that the collector curves shown in Figure 7.6 are the curves for the transistor in Figure 7.4. The load line found in Example 7.2 has been superimposed over the collector curves. The Q-point would be the point where the load line intersects the appropriate collector curve. For example, if the amplifier were operated at $I_B = 20$ μA, the Q-point would be located at the point where the dc load line intersects the $I_B = 20$ μA curve, as shown in the illustration. From the curve, we could then determine that the circuit had Q-point values of $I_C = 4$ mA and $V_{CE} = 6$ V, the coordinates that correspond to the Q-point location. For linear operation of an amplifier, it is desirable to have the Q-point centered on the load line.

When you have a centered Q-point, $V_{CE}$ is half the value of $V_{CC}$, and $I_C$ is half the value of $I_{C(sat)}$. This is illustrated in Figure 7.7. As you can see, the centered Q-point provides values of $I_C$ and $V_{CE}$ that are one-half of their maximum possible values. When a circuit is designed to have a centered Q-point, the amplifier is said to be **midpoint biased.**

**Midpoint bias**
Having a Q-point that is centered on the load line.

Midpoint biasing allows optimum ac operation of the amplifier. This point is illustrated in Figure 7.8. When an ac signal is applied to the base of the transistor, $I_C$ and $V_{CE}$

**FIGURE 7.6**

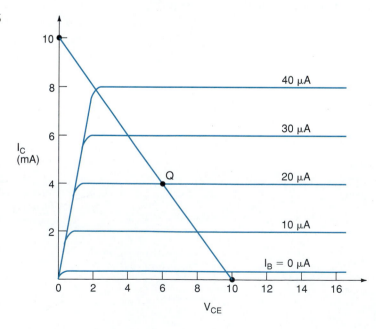

FIGURE 7.7 The Q-point for a
midpoint-biased circuit.

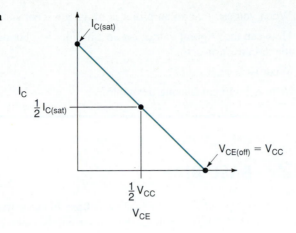

**FIGURE 7.7** The *Q*-point for a
midpoint-biased circuit.

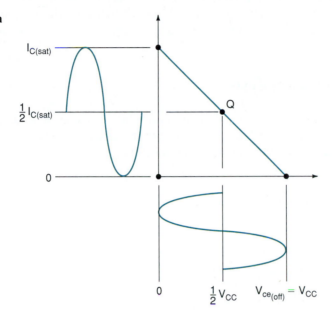

**FIGURE 7.8** Optimum
amplifier operation.

will both vary around their *Q*-point values. When the *Q*-point is centered, $I_C$ and $V_{CE}$ can both make the maximum possible transitions above and below their initial dc values. When the *Q*-point is above center on the load line, the input may cause the transistor to saturate. When this happens, part of the output sine wave is *clipped* off. If the *Q*-point is below midpoint on the load line, the input may cause the transistor to go into cutoff. This can also cause a portion of the output sine wave to be clipped.

Clipping can be caused by not designing for midpoint bias.

The ac operation of the amplifier is covered in detail in later chapters. For now, it is important that you understand the need for midpoint biasing. Having a midpoint-biased circuit will allow the best possible ac operation of the circuit.

**1.** What purpose is served by a dc biasing circuit?

*Section Review*

**2.** What is a *dc load line*?

**3.** What is represented by the point where the load line meets the $I_C$ axis on a graph?

**4.** What is represented by the point where the load line meets the $V_{CE}$ axis on a graph?

**5.** What does the *Q*-point of an amplifier represent?

**6.** What is meant by the term *quiescent*?

7. When you say that an amplifier is in its quiescent state, what does that mean?

8. How can the *Q*-point values for an amplifier be obtained from the collector curves of the circuit's transistor?

9. What is *midpoint bias*?

10. Why is midpoint biasing desirable?

## 7.2 BASE BIAS

OBJECTIVE 4 ▶

**Base bias**
Consists of a single base resistor between the base terminal and $V_{CC}$ and no emitter resistor. Also known as **fixed bias.** The name *fixed bias* stems from the fact that $I_B$ will not vary from one silicon transistor to another (or one germanium transistor to another).

**Lab Reference:** Base-bias operation is demonstrated in Exercise 10.

The simplest type of transistor biasing is **base bias**, or **fixed bias.** A base-bias circuit is shown in Figure 7.9. Current in the circuit enters from $V_{CC}$ and splits, with the larger portion entering the collector of the transistor and the smaller portion entering the base. These currents combine and exit the transistor via the emitter. As long as power is applied to the circuit, the terminal currents will be in the direction indicated in Figure 7.9.

### Circuit Analysis

The primary goal of the dc analysis of a biasing circuit is to determine the *Q*-point values of $I_C$ and $V_{CE}$. For a base-bias circuit, this analysis proceeds as follows: First, using Kirchoff's law, the value of $V_{RB}$ can be found as

$$V_{RB} = V_{CC} - V_{BE}$$

Using Ohm's law, the value of $I_B$ can be found as

$$I_B = \frac{V_{RB}}{R_B}$$

For convenience, we combine these two calculations into a single equation as follows:

$$I_B = \frac{V_{CC} - V_{BE}}{R_B} \qquad (7.4)$$

Once the value of $I_B$ is known, $I_C$ is calculated using the equation

*Remember:*
$h_{FE} = \beta_{dc}$

$$I_C = h_{FE} I_B \qquad (7.5)$$

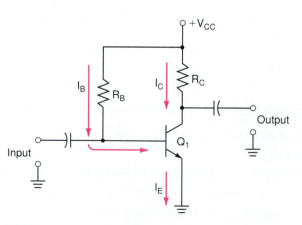

**FIGURE 7.9** Base bias.

Finally, $V_{CE}$ can be found as follows:

$$V_{CE} = V_{CC} - I_C R_C \qquad \textbf{(7.6)}$$

This is identical to equation (6.12), which gives you the $Q$-point value of $V_{CE}$. The complete dc analysis of a base-bias circuit is demonstrated in the following example.

---

Determine the $Q$-point values of $I_C$ and $V_{CE}$ for the circuit shown in Figure 7.10.

**EXAMPLE 7.3**

*Solution:*  First, $I_B$ is found as

$$I_B = \frac{V_{CC} - 0.7 \text{ V}}{R_B}$$

$$= \frac{8 \text{ V} - 0.7 \text{ V}}{360 \text{ k}\Omega}$$

$$= \frac{7.3 \text{ V}}{360 \text{ k}\Omega}$$

$$= 20.28 \text{ } \mu\text{A}$$

Next, $I_C$ is found as

$$I_C = h_{FE} I_B$$
$$= (100)(20.28 \text{ } \mu\text{A})$$
$$= 2.028 \text{ mA}$$

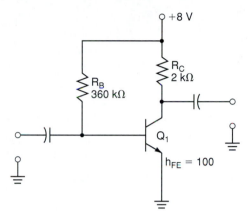

**FIGURE 7.10**

Finally, $V_{CE}$ is found as

$$V_{CE} = V_{CC} - I_C R_C$$
$$= 8 \text{ V} - (2.028 \text{ mA})(2 \text{ k}\Omega)$$
$$= 8 \text{ V} - 4.056 \text{ V}$$
$$= 3.94 \text{ V}$$

*PRACTICE PROBLEM 7.3*

A base-bias circuit like the one in Figure 7.10 has the following values: $V_{CC} = +14$ V, $R_C = 720 \text{ }\Omega$, $h_{FE} = 200$, and $R_B = 270 \text{ k}\Omega$. Determine the $Q$-point values of $I_C$ and $V_{CE}$ for the circuit.

---

Once the $Q$-point values of $I_C$ and $V_{CE}$ are known, you can determine whether or not the circuit is midpoint biased. As Examples 7.4 and 7.5 will show, there are two ways to determine whether or not a circuit is midpoint biased.

◄ *OBJECTIVE 5*

---

Construct the dc load line for the circuit in Figure 7.10, and plot the $Q$-point from the values obtained in Example 7.3. Determine whether or not the circuit is midpoint biased.

**EXAMPLE 7.4**

*Solution:*  The end points of the load line are found as

$$I_{C(\text{sat})} = \frac{V_{CC}}{R_C}$$

$$= \frac{8 \text{ V}}{2 \text{ k}\Omega}$$

$$= 4 \text{ mA}$$

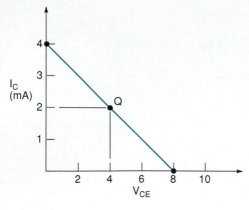

**FIGURE 7.11**

and

$$V_{CE(off)} = V_{CC}$$
$$= 8 \text{ V}$$

Using these end points, the dc load line is plotted as shown in Figure 7.11. Plotting the point that corresponds to $I_C = 2.028$ mA and $V_{CE} = 3.94$ V (from Example 7.3) gives us the $Q$-point shown in the figure. As you can see, the amplifier is very nearly midpoint biased.

### PRACTICE PROBLEM 7.4

A base bias circuit as shown in Figure 7.10 has the following values: $V_{CC} = 12$ V, $R_C = 1$k$\Omega$, $h_{FE} = 150$, and $R_B = 500$ k$\Omega$. Construct the load line and determine if the circuit is operating at midpoint.

---

**EXAMPLE 7.5**

Determine whether or not the circuit in Figure 7.10 is midpoint biased without drawing a dc load line for the circuit.

**Solution:** By definition, a circuit is midpoint biased when the $Q$-point value of $V_{CE}$ is one-half of $V_{CC}$. This relationship was illustrated in Figure 7.7.

In Example 7.3, we determined that the amplifier in Figure 7.10 has the following values:

$$V_{CC} = +8 \text{ V}$$
$$V_{CE} = 3.94 \text{ V}$$

Since the $Q$-point value of $V_{CE}$ is approximately one-half of $V_{CC}$, we can conclude that the circuit is midpoint biased without the use of a dc load line.

### PRACTICE PROBLEM 7.5

Determine whether or not the circuit described in Practice Problem 7.3 is midpoint biased. Use the methods demonstrated in Examples 7.4 *and* 7.5 to show your conclusion to be correct.

## Base-Bias Applications

Base-bias circuits are used primarily in *switching* applications. When a transistor is used as a switch, it is constantly driven back and forth between saturation and cutoff. We will discuss this application for base-bias circuits in detail in Chapter 19.

## Q-Point Shift

**Q-point shift**
A condition where a change in operating temperature indirectly causes a change in $I_C$ and $V_{CE}$.

Even though they are easy to build and analyze, base-bias circuits are rarely used in any applications that require the use of midpoint-biased amplifiers. The reason is the fact that base-bias circuits are extremely susceptible to a problem called **Q-point shift**. The term *Q-point shift* describes a condition where a change in operating temperature indirectly causes a change in the $Q$-point values of $I_C$ and $V_{CE}$.

If you look at the equation for $V_{CE}$, you can see that $V_{CE}$ will change if $I_C$ changes. If you look at the equation for $I_C$, you'll see that the value of $I_C$ will change if either $I_B$ or

$h_{FE}$ changes. As you were shown in Chapter 6, dc beta ($h_{FE}$) varies with temperature. If temperature *increases*, $h_{FE}$ will also increase.

When $h_{FE}$ *increases*, $I_C$ will *increase*. This increase in $I_C$ will cause a *decrease* in $V_{CE}$. Thus, a change in temperature will indirectly cause an increase in $I_C$ and a decrease in $V_{CE}$. When this occurs, the circuit will no longer be midpoint biased. This point is illustrated in the following example.

*A Practical Consideration:* Cooling an amplifier (decreasing its temperature) will have the opposite effect: $h_{FE}$ and $I_C$ will *decrease,* and $V_{CE}$ will *increase.*

**EXAMPLE 7.6**

The transistor in Figure 7.12 has values of $h_{FE} = 100$ when $T = 25°C$ and $h_{FE} = 150$ when $T = 100°C$. Determine the Q-point values of $I_C$ and $V_{CE}$ at both of these temperatures.

**Solution:** This is the same circuit that we analyzed in Example 7.3. At that time, we calculated the following values (when $h_{FE} = 100$):

$$I_B = 20.28 \ \mu A$$
$$I_C = 2.028 \ mA$$
$$V_{CE} = 3.94 \ V$$

When $T = 100°C$, the value of $h_{FE}$ changes to 150. Assuming that we have the same initial value of $I_B$, the new value of $I_C$ is found as

$$I_C = h_{FE}I_B$$
$$= (150)(20.28 \ \mu A)$$
$$= 3.04 \ mA$$

Using this value of $I_C$, the new value of $V_{CE}$ is found as

$$V_{CE} = V_{CC} - I_C R_C$$
$$= 8 \ V - (3.04 \ mA)(2 \ k\Omega)$$
$$= 1.92 \ V$$

As you can see, the amplifier is nowhere near midpoint biased when the temperature increases to 100°C. This is not acceptable for a linear amplifier.

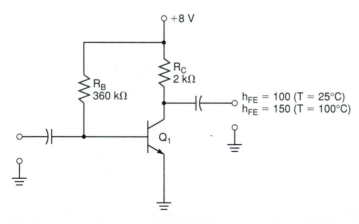

**FIGURE 7.12**

**Lab Reference:** Base-bias instability is demonstrated in Exercise 10.

***PRACTICE PROBLEM 7.6***

Refer to Practice Problem 7.3. The value of $h_{FE}$ shown in this practice problem was measured at 25°C. Determine the Q-point values for the circuit at 100°C if the value of $h_{FE}$ at this temperature is 380.

**FIGURE 7.13**

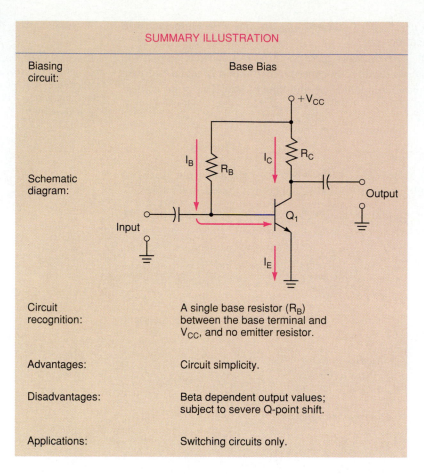

SUMMARY ILLUSTRATION

| | |
|---|---|
| Biasing circuit: | Base Bias |
| Schematic diagram: | |
| Circuit recognition: | A single base resistor ($R_B$) between the base terminal and $V_{CC}$, and no emitter resistor. |
| Advantages: | Circuit simplicity. |
| Disadvantages: | Beta dependent output values; subject to severe Q-point shift. |
| Applications: | Switching circuits only. |

**Beta-dependent circuit**
A circuit whose $Q$-point values are affected by changes in $h_{FE}$.

**Beta-independent circuit**
A circuit whose $Q$-point values are independent of changes in $h_{FE}$.

Since the $Q$-point values of $I_C$ and $V_{CE}$ for the base-bias circuit are affected by changes in $h_{FE}$, the circuit is referred to as a **beta-dependent circuit.** As we continue our discussion on dc biasing circuits, we will concentrate on several **beta-independent circuits,** that is, circuits whose $Q$-point values are independent of $h_{FE}$ and, therefore, relatively independent of changes in temperature.

## Summary

The base-bias circuit is the simplest of the dc biasing circuits. The characteristics and applications of this circuit are summarized in Figure 7.13.

## One Final Note

$I_{CQ}$
The $Q$-point value of $I_C$.

$V_{CEQ}$
The $Q$-point value of $V_{CE}$.

Throughout this section, we have referred to *the Q-point values of $I_C$ and $V_{CE}$*. To simplify further discussions, we will use the labels $I_{CQ}$ and $V_{CEQ}$ to represent the *Q-point values of $I_C$ and $V_{CE}$.*

## Section Review

1. Describe the construction and current paths in a base-bias circuit.
2. What is the goal of the dc analysis of an amplifier?
3. List the steps taken, in order, to determine the $Q$-point values of $I_C$ ($I_{CQ}$) and $V_{CE}$ ($V_{CEQ}$).

**4.** Once you have calculated the values of $I_{CQ}$ and $V_{CEQ}$ for an amplifier, how can you determine whether or not the circuit is midpoint biased?

**5.** What is *Q-point shift*?

**6.** Describe the *Q*-point shift process.

**7.** What is a *beta-dependent circuit*?

**8.** What is a *beta-independent circuit*?

---

# 7.3 EMITTER BIAS

An **emitter-bias** circuit consists of several resistors and a *dual-polarity power supply,* as shown in Figure 7.14. The current action in this circuit is basically the same as that of the base-bias circuit. The collector current and base current come from $V_{CC}$ and ground, respectively. These currents combine and exit the transistor via the emitter terminal. The emitter current then passes through the emitter resistor ($R_E$) to the negative dc supply ($V_{EE}$).

◄ *OBJECTIVE 6*

**Emitter bias**
A bias circuit that consists of a dual-polarity power supply and a grounded base resistor.

## Circuit Analysis

We will start our analysis of the emitter-bias circuit by making an assumption. For reasons that will be explained later, we will assume that the transistor base is at ground. With this in mind, we can redraw Figure 7.14 as shown in Figure 7.15a.

To analyze what is happening in the emitter circuit, it is helpful to redraw the emitter-base junction of the transistor as a simple diode (as shown in Figure 7.15b). The equation we will use to find the value of the emitter current is derived as follows: According to Kirchhoff's voltage law, the sum of the voltages in the emitter-base circuit must equal zero. Written mathematically,

$$-V_{EE} + V_{RE} + V_{BE} = 0 \text{ V}$$

or

$$V_{RE} = V_{EE} - V_{BE}$$

**FIGURE 7.14   Emitter bias.**

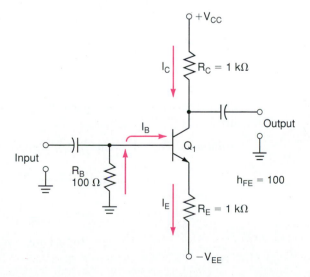

**Lab Reference:** Emitter-bias operation is demonstrated in Exercise 11.

FIGURE 7.15

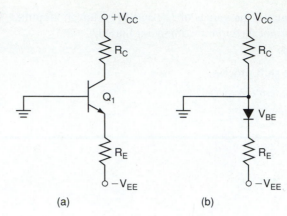

(a)                    (b)

According to Ohm's law,

$$I_E = \frac{V_{RE}}{R_E}$$

Combining the equations for $I_E$ and $V_{RE}$, we get

$$I_E = \frac{V_{EE} - V_{BE}}{R_E} \tag{7.7}$$

Assuming that $I_{CQ} \cong I_E$ (which is usually the case), the value of $I_{CQ}$ can be found as

$$I_{CQ} \cong \frac{V_{EE} - V_{BE}}{R_E} \tag{7.8}$$

**Lab Reference:** The output stability of emitter-bias is demonstrated in Exercise 11.

If you compare equation (7.8) with our previous equations for $I_C$, you will notice that beta is not involved in the collector current equation for the emitter-bias circuit. This means that the output characteristics of this amplifier type are relatively independent of beta. A change in beta for the transistor will not have a major effect on the dc output characteristics of this type of amplifier.

Since $I_{CQ}$ is independent of the value of $I_B$, the practical analysis of the emitter-bias circuit involves only two steps. First, we calculate the approximate value of $I_{CQ}$ using equation (7.8). Then we calculate the approximate value of $V_{CEQ}$ using

$$V_{CEQ} \cong V_{CC} - I_{CQ}R_C \tag{7.9}$$

## EXAMPLE 7.7

Determine the values of $I_{CQ}$ and $V_{CEQ}$ for the circuit shown in Figure 7.16.

***Solution:*** First, the value of $I_{CQ}$ is approximated as

$$I_{CQ} \cong \frac{V_{EE} - V_{BE}}{R_E}$$

$$= \frac{12\ \text{V} - 0.7\ \text{V}}{1.5\ \text{k}\Omega}$$

$$= 7.53\ \text{mA}$$

Now we can solve for $V_{CEQ}$ as

$$V_{CEQ} \cong V_{CC} - I_{CQ}R_C$$
$$= 12 \text{ V} - (7.53 \text{ mA})(750 \text{ }\Omega)$$
$$= 12 \text{ V} - 5.65 \text{ V}$$
$$= 6.35 \text{ V}$$

**PRACTICE PROBLEM 7.7**

A circuit like the one shown in Figure 7.16 has the following values: $R_C = 1.5 \text{ k}\Omega$, $R_E = 3 \text{ k}\Omega$, $V_{CC} = +15 \text{ V}$, and $V_{EE} = -15 \text{ V}$. Determine the values of $I_{CQ}$ and $V_{CEQ}$ for the circuit.

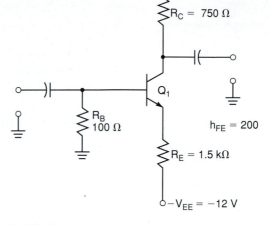

**FIGURE 7.16**

## Saturation and Cutoff

The presence of two power supplies in the emitter-bias circuit has an effect on the basic equations for $I_{C(\text{sat})}$ and $V_{CE(\text{off})}$. Normally, the two supply voltages are *equal in value*. For example, if $V_{CC}$ is $+15 \text{ V}_{\text{dc}}$, $-V_{EE}$ will be $-15 \text{ V}_{\text{dc}}$. When the two supply voltages are equal in value, we can use the following equations to find $I_{C(\text{sat})}$ and $V_{CE(\text{off})}$:

$$I_{C(\text{sat})} = \frac{2 V_{CC}}{R_C + R_E} \qquad \textbf{(7.10)}$$

and

$$V_{CE(\text{off})} = 2 V_{CC} \qquad \textbf{(7.11)}$$

Remember, these equations are valid *only* when the two supply voltages are equal in value. The following example demonstrates the process used to find $I_{C(\text{sat})}$ and $V_{CE(\text{off})}$ in this case.

Determine the value of $I_{C(\text{sat})}$ and $V_{CE(\text{off})}$ for the circuit shown in Figure 7.16.

**EXAMPLE 7.8**

**Solution:** Since the supply voltages are equal in value, $I_{C(\text{sat})}$ is found as

$$I_{C(\text{sat})} = \frac{2 V_{CC}}{R_C + R_E}$$
$$= \frac{24 \text{ V}_{\text{dc}}}{2250 \text{ }\Omega}$$
$$= 10.67 \text{ mA}$$

And the value of $V_{CE(\text{off})}$ is found as

$$V_{CE(\text{off})} = 2 V_{CC}$$
$$= 24 \text{ V}$$

If you ever come across an emitter-bias circuit that has unequal supply voltages, simply use the *difference between the two supply voltages* in place of 2 $V_{CC}$ in equations (7.10) and (7.11).

## Basis of Equations (7.10) and (7.11)

The value of $I_{C(sat)}$ is determined by two factors:

**1.** *Total* voltage across the collector–emitter circuit

**2.** *Total* resistance in the collector–emitter circuit

The total voltage across the collector–emitter circuit in the emitter-bias amplifier is equal to the difference between the two supply voltages. Since these two supply voltages are normally equal, the difference between them will be twice the value of $V_{CC}$. Thus, we can normally use 2 $V_{CC}$ in the calculation of $I_{C(sat)}$. The total resistance in the collector–emitter circuit is the sum of $R_C$ and $R_E$. Using the values of 2 $V_{CC}$ and $(R_C + R_E)$, we obtain equation (7.10).

The value of $V_{CE(off)}$ is equal to the total voltage applied to the collector–emitter circuit. In the base-bias circuit, the total voltage across the circuit was equal to $V_{CC}$. In the case of emitter bias, the total voltage is (again) equal to the difference between the supply voltages, or 2 $V_{CC}$ under normal circumstances, and thus we arrive at equation (7.11).

## Base Input Resistance

In order to discuss the base characteristics of the emitter-bias circuit, we need to take a look at the dc input resistance of the base terminal of the transistor. As you will see, this resistance value weighs heavily into our initial assumption that $V_B$ is approximately equal to 0 $V_{dc}$.

The dc input resistance of the base of a transistor is found as

$$R_{IN(base)} = h_{FE}R_E \tag{7.12}$$

This is the equation we will use whenever we need to determine the dc input resistance of the base terminal, as illustrated in the following example.

**EXAMPLE 7.9**

Determine the base input resistance for the circuit shown in Figure 7.16.

**Solution:**  The value of $h_{FE}$ is shown to be 200, and the value of $R_E$ is shown to be 1.5 kΩ. Using these values, the value of $R_{IN(base)}$ is found as

$$
\begin{aligned}
R_{IN(base)} &= h_{FE}R_E \\
&= (200)(1.5\ \text{k}\Omega) \\
&= 300\ \text{k}\Omega
\end{aligned}
$$

**PRACTICE PROBLEM 7.9**

Determine the base input resistance for the circuit described in Practice Problem 7.7. Assume that the value of $h_{FE}$ for the device is 180.

## The Basis of Equation (7.12)

The basis of equation (7.12) is relatively easy to understand when you idealize the transistor current and voltage relationships that have been established. Refer to Figure 7.17a. If we assume that the base-emitter junction of the transistor is an ideal component, then $V_{BE} = 0$ V. In this case,

$$V_B = V_E$$

As shown in the figure, $V_B$ is defined by Ohm's law as

$$V_B = I_B R_{IN(base)}$$

and $V_E$ is defined as

$$V_E = I_E R_E$$

We know that $I_C$ and $I_E$ are approximately equal. If we idealize this relationship, we can say that

$$I_E = I_C$$

or

$$I_E = h_{FE} I_B$$

Using this relationship, equation (7.12) is derived as follows:

$$V_B = V_E$$
$$I_B R_{IN(base)} = I_E R_E$$
$$I_B R_{IN(base)} = h_{FE} I_B R_E$$
$$R_{IN(base)} = h_{FE} R_E$$

## Base Voltage

It was stated earlier that $V_B$ for an emitter bias circuit is approximately 0 V. This relationship is based on the assumption that $R_{IN(base)} \gg R_B$ (which, by design, is always the case). Now, consider the circuit shown in Figure 7.17b. This is the equivalent of the emit-

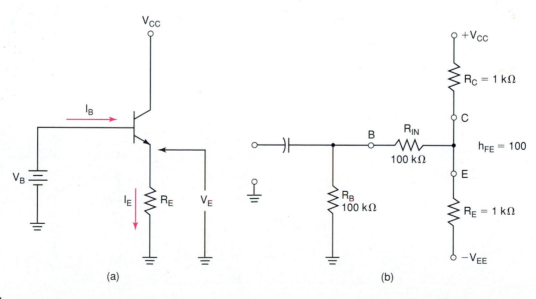

(a)          (b)

**FIGURE 7.17**

ter-bias circuit shown in Figure 7.14. For simplicity, the emitter-base diode has been ignored. Using equation (7.12), $R_{\text{IN(base)}}$ for the circuit is found as

$$R_{\text{IN(base)}} = h_{FE}R_E$$
$$= (100)(1 \text{ k}\Omega)$$
$$= 100 \text{ k}\Omega$$

Now, look at the base-emitter loop in Figure 7.17b. The circuit consists of $V_{EE}$, $R_E$, $R_{\text{IN(base)}}$, and $R_B$. Due to the extremely low value of $R_B$, almost all of $V_{EE}$ is dropped across $R_{\text{IN(base)}}$ and $R_E$. Therefore, the voltage across $R_B$ will be very nearly 0 V (the value we originally assumed for $V_B$).

## Circuit Troubleshooting

OBJECTIVE 7 ▶ When an emitter-bias circuit is operating properly (as shown in Figure 7.18), it will have the following circuit conditions:

1. $V_B$ will be approximately 0 V.
2. $V_E$ will be approximately 0.7 V less than $V_B$.
3. $V_C$ will be approximately one-half of $V_{CC}$.

When one or more of the terminal voltages shown are way out of range, there is probably something wrong with the circuit. In this section we will take a look at some of the problems that can develop in the dc emitter-bias circuit and how these problems will affect the overall operation of the circuit. In our discussion, we will refer to Figure 7.18.

$R_C$ *Open.* When the collector resistor opens, there is no path for dc current in the collector circuit. Because of this, the transistor will be in cutoff, and there will be only a

**Lab Reference:** Several emitter-bias faults are simulated in Exercise 11.

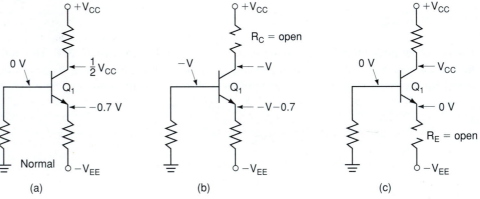

FIGURE 7.18   Emitter-bias fault symptoms.

small amount of current in the emitter circuit. This emitter current will be caused by the forward-biased emitter–base junction of the transistor. The results of this circuit operation are illustrated in Figure 7.18b. Note that the following circuit conditions exist when $R_C$ is open:

1. $V_B$ is negative.
2. $V_E$ is more negative than $V_B$, typically close to the value of $V_{EE}$.
3. $V_C$ is negative, somewhere between the values of $V_B$ and $V_E$.

The collector voltage reading ($V_C$) is caused by several things. First, with the collector circuit essentially open, the voltages at the other two terminals will be "reflected" in the collector circuit. Second, when you connect a voltmeter to the collector, the meter will provide a path for a small amount of current through its internal resistance to ground. This small amount of $I_C$ will cause a negative reading to appear at the collector terminal.

***$R_E$ Open.*** When the emitter resistor opens, no current can pass through the emitter terminal of the transistor and, again, the component will be in cutoff. However, since it is the *negative* supply that has been isolated from the circuit, the voltage readings will be different from those for an open collector resistor. The circuit conditions that correspond to an open emitter resistor are illustrated in Figure 7.18c. Note that the following conditions exist when the emitter resistor is open:

1. $V_B$ is approximately 0 V.
2. $V_E$ will be between 0 V and $V_{CC}$.
3. $V_C$ is approximately equal to $V_{CC}$.

Since there is no current in the collector circuit, no voltage is dropped across $R_C$. This causes $V_C$ to be approximately equal to $V_{CC}$. Since there is no base current, $V_B$ will still be approximately equal to 0 V.

When measuring the voltage at the emitter terminal of the transistor, you will get one of two readings, depending on the point to which you are referencing the reading. If you read from the emitter terminal to ground, you will read approximately 0 V. However, if you read across the emitter resistor to $-V_{EE}$, you may get a positive voltage reading. The reason for this can be seen by taking a look at Figure 7.19. When the meter is connected to $-V_{EE}$, the open-emitter circuit is completed through the internal resistance of the meter. In other words, the meter becomes the new $R_E$. The current through the meter develops a voltage at the emitter terminal. This voltage will register on the meter.

***$R_B$ Open.*** When the base resistor opens, there is no biasing current in the base circuit of the transistor. Because of this, the transistor is again biased in cutoff. The circuit condi-

**FIGURE 7.19  Meter input resistance can complete a circuit.**

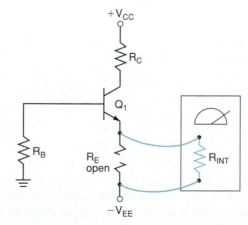

tions of an open base resistor are illustrated in Figure 7.18d. Note that the following conditions exist when the base resistor is open:

1. $V_B$ falls somewhere between 0 V and $-V_{EE}$.
2. $V_E$ is approximately equal to $-V_{EE}$.
3. $V_C$ is approximately equal to $V_{CC}$.

Since the transistor is the source of the open circuit between $V_{EE}$ and $V_{CC}$, the difference between these two potentials is dropped across the transistor. No current is present in either the emitter or collector circuits, so their respective resistors do not drop any voltage; therefore, we obtain the readings listed above.

Note that when reading the base voltage you provide a path for base current with the meter. Thus, you may read a slight negative voltage at the base of the transistor.

***$Q_1$ Open.*** This condition gives you the same readings as having an open base resistor; that is, the transistor will be in cutoff. A simple check of the transistor (as outlined in Chapter 6) or a resistance reading of $R_B$ tells you which of the two components is faulty.

***$Q_1$ Leaky.*** The term **leaky** is used to describe a transistor that is partially shorted from emitter to collector (Figure 7.18e). When a transistor is leaky, it may appear to be saturated, even when it should not be. The collector and emitter voltages may be nearly equal, and the 0.7-V drop from emitter to base will still be there. When a transistor is leaky, it must be replaced.

## *Practical Troubleshooting Considerations*

Causes of an apparent shorted carbon resistor.

When you are troubleshooting *any* dc biasing circuit, a few practical considerations should be kept in mind. These points, which can save you a great deal of time and trouble, are as follows:

1. *Standard carbon resistors do not short internally*. When a carbon-composition resistor appears to be short circuited, the short is caused by one of two conditions:
   a. Another component, wire, copper run (on a printed circuit board), or solder bridge has shorted the component. In this case, the problem is solved by removing the cause of the short.
   b. The wrong resistor value has been used in the circuit. It is not uncommon for people to misread the color code on a resistor. If a circuit calls for a 10-k$\Omega$ resistor, and you accidentally use a 100-$\Omega$ resistor, the resistor will appear as a short circuit.

   When either $R_C$ or $R_E$ is shorted, the transistor terminal connected to the shorted resistor will read the supply voltage. If $R_C$ is shorted, the collector terminal will be at $V_{CC}$. If $R_E$ is shorted, the emitter terminal will be at $-V_{EE}$. In either of these cases, the transistor will not last long.

   If $R_B$ shorts, the dc biasing circuit will not be greatly affected. However, the ac operation of the transistor will stop.

2. *An open resistor can normally be checked without any test equipment*. When a resistor goes open circuit, it usually *burns* open. Because of this, the component will normally have a single color band: *black from end to end*. In most cases, you can troubleshoot an open resistor simply by looking at it. If it has turned black or if the wire or copper run connected to the component shows signs of burning, the resistor probably needs to be replaced.

As a final note, do not overlook the obvious. If either power supply is out, an emitter-bias circuit will not work. If a transistor is placed in a circuit using the wrong connections, the circuit will not work. The bottom line is this: When you are troubleshooting,

**FIGURE 7.20**

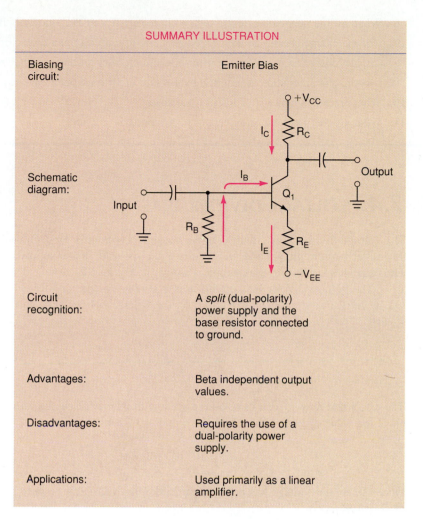

SUMMARY ILLUSTRATION

Biasing circuit:        Emitter Bias

Schematic diagram:

Circuit recognition:     A *split* (dual-polarity) power supply and the base resistor connected to ground.

Advantages:          Beta independent output values.

Disadvantages:       Requires the use of a dual-polarity power supply.

Applications:        Used primarily as a linear amplifier.

look for the obvious first. Then, if you need to, use your test equipment to troubleshoot the circuit. See what the circuit is doing as opposed to what it *should* be doing. Then, based on the comparison, check the suspected components.

## Summary

The emitter-bias circuit is a beta-independent circuit; meaning that its output characteristics ($V_{CEQ}$ and $I_{CQ}$) are not affected by variations in beta. The characteristics of the emitter-bias circuit are summarized in Figure 7.20.

**Section Review**

1. Draw an emitter-bias circuit and show the directions of the circuit currents.
2. Explain why the emitter-bias circuit is a beta-independent circuit.
3. List, in order, the steps required to determine the $Q$-point output values of an emitter-bias circuit.
4. When and why do we use the value $2 V_{CC}$ when determining the values of $I_{C(sat)}$ and $V_{CE(off)}$ for an emitter-bias circuit?
5. What adjustment must be made in equations (7.10) and (7.11) when $V_{CC}$ and $V_{EE}$ are *not* equal in value?

6. What are the normal terminal voltage characteristics of the emitter-bias circuit?

7. List the characteristics of an emitter-bias circuit with an open collector resistor.

8. List the characteristics of an emitter-bias circuit with an open emitter resistor.

9. List the characteristics of an emitter-bias circuit with an open base resistor.

10. What are the symptoms of a *leaky* transistor?

11. What are the most common causes of a "shorted" resistor?

---

# 7.4 VOLTAGE-DIVIDER BIAS

*OBJECTIVE 8* ►

**Voltage-divider bias**
A biasing circuit that contains a voltage divider in its base circuit. This type of bias is sometimes referred to as **universal bias.**

The **voltage-divider bias** circuit is by far the most commonly used. It has the stability of an emitter-bias circuit, yet does not require a dual-polarity power supply for its operation. The basic voltage-divider bias circuit is shown in Figure 7.21.

The voltage-divider bias circuit uses a simple voltage divider to set the value of $V_B$. If we ignore the effects of $R_{IN(base)}$ on the voltage divider, we can find the value of $V_B$ as

$$V_B = \frac{R_2}{R_1 + R_2} V_{CC} \tag{7.13}$$

The effect that $R_{IN(base)}$ can have on the voltage divider is discussed later in this section. Once the value of $V_B$ is established, the value of $V_E$ can be found as

$$V_E = V_B - 0.7 \text{ V} \tag{7.14}$$

Once the emitter voltage has been determined, Ohm's law can be used to find the value of $I_E$ as follows:

$$I_E = \frac{V_E}{R_E} \tag{7.15}$$

Now, assuming that $I_{CQ} = I_E$, $V_{CEQ}$ can be found as

$$V_{CEQ} = V_{CC} - I_{CQ}R_C - I_{CQ}R_E$$

**FIGURE 7.21    Voltage-divider bias.**

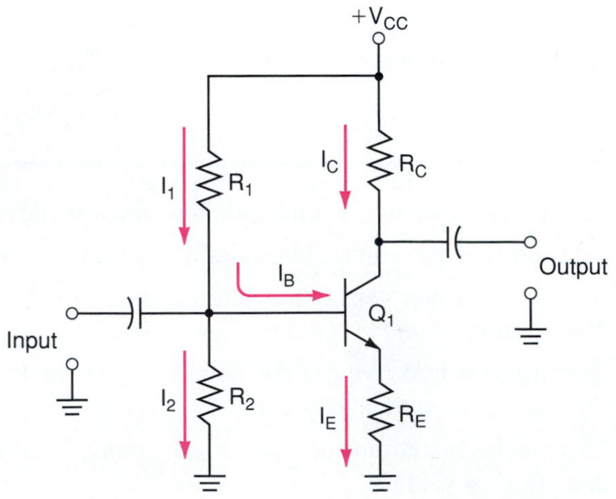

**Lab Reference:** The operation of voltage-divider bias is demonstrated in Exercise 12.

or

$$V_{CEQ} = V_{CC} - I_{CQ}(R_C + R_E) \qquad (7.16)$$

The following example illustrates the process used to determine the values of $I_{CQ}$ and $V_{CEQ}$ for the voltage-divider bias circuit.

Determine the values of $I_{CQ}$ and $V_{CEQ}$ for the circuit shown in Figure 7.22.

**EXAMPLE 7.10**

**Solution:** $R_1$ and $R_2$ set up the base voltage for the amplifier. This voltage is found as

$$V_B = \frac{R_2}{R_1 + R_2} V_{CC}$$

$$= \frac{4.7 \text{ k}\Omega}{22.7 \text{ k}\Omega} (10 \text{ V})$$

$$= (0.207)(10 \text{ V})$$

$$= 2.07 \text{ V}$$

$V_E$ is found as

$$V_E = V_B - 0.7 \text{ V}$$
$$= 2.07 \text{ V} - 0.7 \text{ V}$$
$$= 1.37 \text{ V}$$

$I_{CQ}$ is then found as

$$I_{CQ} \cong \frac{V_E}{R_E}$$

$$= \frac{1.37 \text{ V}}{1.1 \text{ k}\Omega}$$

$$= 1.25 \text{ mA}$$

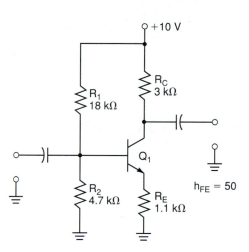

**FIGURE 7.22**

**Lab Reference:** These values are predicted and measured in Exercise 12.

Finally, $V_{CEQ}$ is found as

$$V_{CEQ} = V_{CC} - I_{CQ}(R_C + R_E)$$
$$= 10 \text{ V} - (1.25 \text{ mA})(3 \text{ k}\Omega + 1.1 \text{ }\Omega)$$
$$= 10 \text{ V} - (1.25 \text{ mA})(4.1 \text{ k}\Omega)$$
$$= 10 \text{ V} - 5.13 \text{ V}$$
$$= 4.87 \text{ V}$$

**PRACTICE PROBLEM 7.10**

A circuit like the one in Figure 7.22 has the following values: $R_C =$ 620 $\Omega$, $R_E = 180 \text{ }\Omega$, $R_1 = 12 \text{ k}\Omega$, $R_2 = 2.7 \text{ k}\Omega$, and $V_{CC} = 10 \text{ V}$. Determine the values of $I_{CQ}$ and $V_{CEQ}$ for the circuit.

Once the value of $I_E$ is known, the value of $I_B$ can be found simply by using equation (6.5), which can be rewritten as

Equation (6.5):
$I_E = I_B(1 + h_{FE})$

$$I_B = \frac{I_E}{1 + h_{FE}} \qquad (7.17)$$

## EXAMPLE 7.11

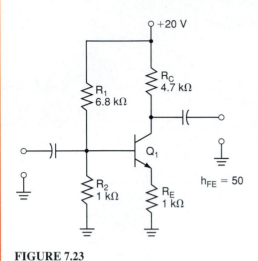

**FIGURE 7.23**

Determine the value of $I_B$ for the circuit shown in Figure 7.23.

**Solution:**  First, $V_B$ is found as

$$V_B = \frac{R_2}{R_1 + R_2} V_{CC}$$

$$= 1\frac{1\text{ k}\Omega}{7.8\text{ k}\Omega}(20\text{ V})$$

$$= 2.56\text{ V}$$

and $V_E$ is found as

$$V_E = V_B - 0.7\text{ V}$$
$$= 1.86\text{ V}$$

Now $I_E$ is found as

$$I_E = \frac{V_E}{R_E}$$

$$= \frac{1.86\text{ V}}{1\text{ k}\Omega}$$

$$= 1.86\text{ mA}$$

Finally, $I_B$ is found as

$$I_B = \frac{I_E}{1 + h_{FE}}$$

$$= \frac{1.86\text{ mA}}{51}$$

$$= 36.5\text{ }\mu\text{A}$$

### PRACTICE PROBLEM 7.11

Determine the value of $I_B$ for the circuit described in Practice Problem 7.10. Assume that the value of $h_{FE}$ for the transistor is 200.

## Which Value of $h_{FE}$ Do I Use?

When you are analyzing a voltage-divider bias circuit (or any other dc biasing circuit, for that matter), you will often have to refer to the spec sheet for the transistor to obtain its value of $h_{FE}$. The only problem is that the spec sheet may not list a single value of $h_{FE}$. Normally, a transistor spec sheet (like the one shown in Figure 7.24) will list any combination of the following values:

**1.** A *maximum* value of $h_{FE}$

**2.** A *minimum* value of $h_{FE}$

**3.** A *typical* value of $h_{FE}$

When only one value of $h_{FE}$ is listed on the spec sheet, you must use that value in any circuit analysis.

When two or more values of $h_{FE}$ are listed, you should first look to see if one of them is the *typical* value of $h_{FE}$. If a *typical* value is listed, use that value.

When the spec sheet lists a *minimum* value and a *maximum* value of $h_{FE}$, you must use the *geometric average of the two values*. The geometric average of $h_{FE}$ is found as

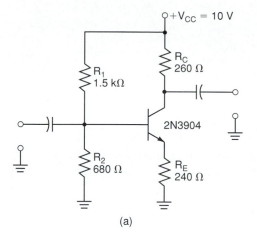

(a)

| Characteristic | | Symbol | Min | Max | Unit |
|---|---|---|---|---|---|
| **ON CHARACTERISTICS** | | | | | |
| DC Current Gain * ($I_C$ = 0.1 mAdc, $V_{CE}$ = 1 Vdc) | 2N3903 2N3904 | $h_{FE}$* | 20 40 | — — | — |
| ($I_C$ = 1.0 mAdc, $V_{CE}$ = 1 Vdc) | 2N3903 2N3904 | | 35 70 | — — | |
| ($I_C$ = 10 mAdc, $V_{CE}$ = 1 Vdc) | 2N3903 2N3904 | | 50 100 | 150 300 | |
| ($I_C$ = 50 mAdc, $V_{CE}$ = 1 Vdc) | 2N3903 2N3904 | | 30 60 | — — | |
| ($I_C$ = 100 mAdc, $V_{CE}$ = 1 Vdc) | 2N3903 2N3904 | | 15 30 | — — | |
| Collector-Emitter Saturation Voltage* ($I_C$ = 10 mAdc, $I_B$ = 1 mAdc) | | $V_{CE(sat)}$* | — | 0.2 | Vdc |
| ($I_C$ = 50 mAdc, $I_B$ = 5 mAdc) | | | — | 0.3 | |
| Base-Emitter Saturation Voltage* ($I_C$ = 10 mAdc, $I_B$ = 1 mAdc) | | $V_{BE(sat)}$* | 0.65 | 0.85 | Vdc |
| ($I_C$ = 50 mAdc, $I_B$ = 5 mAdc) | | | — | 0.95 | |
| **SMALL SIGNAL CHARACTERISTICS** | | | | | |
| High Frequency Current Gain ($I_C$ = 10 mA, $V_{CE}$ = 20 V, f = 100 mc) | 2N3903 2N3904 | $|h_{fe}|$ | 2.5 3.0 | — — | — |
| Current-Gain—Bandwidth Product ($I_C$ = 10 mA, $V_{CE}$ = 20 V, f = 100 mc) | 2N3903 2N3904 | $f_T$ | 250 300 | — — | mc |
| Output Capacitance ($V_{CB}$ = 5 Vdc, $I_E$ = 0, f = 100 kc) | | $C_{ob}$ | — | 4 | pf |
| Input Capacitance ($V_{OB}$ = 0.5 Vdc, $I_C$ = 0, f = 100 kc) | | $C_{ib}$ | — | 8 | pf |
| Small Signal Current Gain ($I_C$ = 1.0 mA, $V_{CE}$ = 10 V, f = 1 kc) | 2N3903 2N3904 | $h_{fe}$ | 50 100 | 200 400 | — |
| Voltage Feedback Ratio ($I_C$ = 1.0 mA, $V_{CE}$ = 10 V, f = 1 kc) | 2N3903 2N3904 | $h_{re}$ | 0.1 0.5 | 5.0 8.0 | X10⁻⁴ |
| Input Impedance ($I_C$ = 1.0 mA, $V_{CE}$ = 10 V, f = 1 kc) | 2N3903 2N3904 | $h_{ie}$ | 0.5 1.0 | 8 10 | Kohms |
| Output Admittance ($I_C$ = 1.0 mA, $V_{CE}$ = 10 V, f = 1 kc) | Both Types | $h_{oe}$ | 1.0 | 40 | μmhos |
| Noise Figure ($I_C$ = 100 μA, $V_{CE}$ = 5 V, $R_g$ = 1 Kohms, Noise Bandwidth = 10 cps to 15.7 kc) | 2N3903 2N3904 | NF | — — | 6 5 | db |
| **SWITCHING CHARACTERISTICS** | | | | | |
| Delay Time | $V_{CC}$ = 3 Vdc, $V_{OB}$ = 0.5 Vdc, $I_C$ = 10 mAdc, $I_{B1}$ = 1 mA | $t_d$ | — | 35 | nsec |
| Rise Time | | $t_r$ | — | 35 | nsec |
| Storage Time | $V_{CC}$ = 3 Vdc, $I_C$ = 10 mAdc, $I_{B1}$ = $I_{B2}$ = 1 mAdc   2N3903 2N3904 | $t_s$ | — — | 175 200 | nsec |
| Fall Time | | $t_f$ | — | 50 | nsec |

*Pulse Test: Pulse Width = 300 μsec, Duty Cycle = 2%     $V_{OB}$ = Base Emitter Reverse Bias

(b)

**FIGURE 7.24** Electrical characteristics listing for the 2N3904. (Copyright of Motorola. Used by permission.)

$$h_{FE(\text{ave})} = \sqrt{h_{FE(\min)} \times h_{FE(\max)}} \qquad (7.17)$$

The following example illustrates the use of $h_{FE(\text{ave})}$ in the analysis of a transistor amplifier.

---

Determine the value of $I_B$ for the circuit shown in Figure 7.24a.

***EXAMPLE 7.12***

**Solution:**   First, $V_B$ is found as

$$V_B = \frac{R_2}{R_1 + R_2}$$
$$= 3.12 \text{ V}$$

---

and $V_E$ is found as

$$V_E = V_B - 0.7 \text{ V}$$
$$= 2.42 \text{ V}$$

Now $I_E$ is found as

$$I_E = \frac{V_E}{R_E}$$
$$\cong 10 \text{ mA}$$

Assuming that $I_{CQ} = 10$ mA, we check the specification sheet shown in Figure 7.24b. As shown, beta ($h_{FE}$) has a range of 100 to 300 when $I_C = 10$ mA, thus $h_{FE(ave)}$ is found as

$$h_{FE(ave)} = \sqrt{h_{FE(min)} \times h_{FE(max)}}$$
$$= \sqrt{100 \times 300}$$
$$= 173$$

Finally, $I_B$ is found as

$$I_B = \frac{I_E}{h_{FE(ave)} + 1}$$
$$= \frac{10 \text{ mA}}{174}$$
$$= 57.47 \text{ μA}$$

### PRACTICE PROBLEM 7.12

Determine the value of $I_B$ for the circuit shown in Figure 7.24a if Q1 is a 2N3903.

## Saturation and Cutoff

The dc load line for the voltage-divider bias circuit is plotted using the end points of $I_{C(sat)}$ and $V_{CE(off)}$. When the transistor is saturated, $V_{CE}$ is approximately equal to 0 V. Thus, the collector current equals the supply voltage divided by the total resistance between $V_{CC}$ and ground. By formula,

$$I_{C(sat)} = \frac{V_{CC}}{R_C + R_E} \tag{7.19}$$

When the transistor is in cutoff, all the supply voltage is dropped across the transistor. Thus,

$$V_{CE(off)} = V_{CC} \tag{7.20}$$

Using these equations, the dc load line for Figure 7.24 is plotted as shown in Figure 7.25.

## Base Voltage: A Practical Consideration

So far, we have solved for $V_B$ using

$$V_B = V_{CC} \frac{R_2}{R_1 + R_2}$$

FIGURE 7.25 The dc load line
for the circuit in Figure 7.24a.

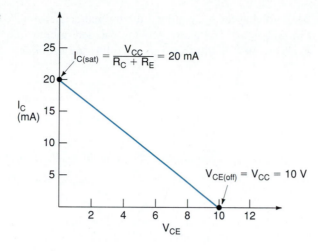

$$I_{C(sat)} = \frac{V_{CC}}{R_C + R_E} = 20 \text{ mA}$$

$$V_{CE(off)} = V_{CC} = 10 \text{ V}$$

This equation is valid for $V_B$ as long as one condition is fulfilled. For equation (7.12) to be valid, $R_2$ must be *less than or equal to one-tenth the base input resistance*. By formula

*Rule of Thumb*: If $h_{FE}R_E \geq 10R_2$, you can ignore the value of $h_{FE}R_E$.

$$R_2 \leq \frac{h_{FE}R_E}{10} \qquad (7.21)$$

Note that equation (7.12) is valid only when the condition shown in equation (7.21) is fulfilled. The reason for this can be seen by taking a look at Figure 7.26.

Figure 7.26a shows the base circuit and emitter circuit for a voltage-divider biased amplifier. Note the current arrows, which indicate that $R_2$ is *in parallel* with the input resistance of the amplifier. The value of $R_{IN(base)}$ is shown as $h_{FE}R_E$ in Figure 7.26b.

As long as $h_{FE}R_E$ is at least 10 times the value of $R_2$, the value of $h_{FE}R_E$ can be ignored in determining the value of $V_B$. However, if $h_{FE}R_E$ is *less than 10 times the value of $R_2$, it must be considered in the determination of* $V_B$.

When $h_{FE}R_E < 10R_2$, the following procedure must be used to determine the value of $V_B$:

**Lab Reference:** Equation (7.21) relates directly to voltage-divider bias stability. This relationship is demonstrated in Exercise 13.

1. Calculate the parallel equivalent resistance from the base of the transistor to ground. This resistance is found as

*Don't Forget*: The ‖ symbol means "in parallel with." Thus, $R_{eq}$ would be found by using the values of $R_2$ and $h_{FE}R_E$ in a parallel equivalent resistance equation.

$$R_{eq} = R_2 \, \| \, h_{FE}R_E \qquad (7.22)$$

2. Solve for the base voltage as follows:

$$V_B = V_{CC} \frac{R_{eq}}{R_1 + R_{eq}} \qquad (7.23)$$

FIGURE 7.26

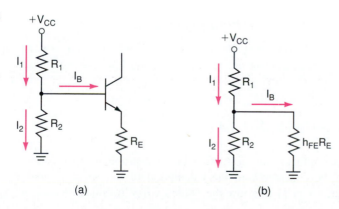

(a)

(b)

After solving for $V_B$, continue the normal sequence of calculations to find the values of $I_{CQ}$ and $V_{CEQ}$. The entire process for determining the values of $I_{CQ}$ and $V_{CEQ}$ when $h_{FE}R_E < 10R_2$ is illustrated in the following example.

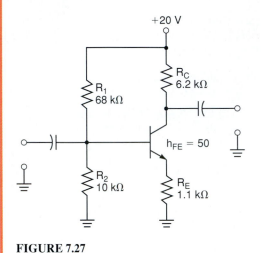

**EXAMPLE 7.13**

**FIGURE 7.27**

Determine the values of $I_{CQ}$ and $V_{CEQ}$ for the amplifier shown in Figure 7.27.

**Solution:**   The value of $h_{FE}R_E$ for the amplifier is found as

$$h_{FE}R_E = (50)(1.1 \text{ k}\Omega)$$
$$= 55 \text{ k}\Omega$$

Since this value is less than 10 times the value of $R_2$, we must use equations (7.22) and (7.23) to determine the value of $V_B$, as follows:

$$R_{eq} = R_2 \parallel h_{FE}R_E$$
$$= (10 \text{ k}\Omega) \parallel (55 \text{ k}\Omega)$$
$$= 8.46 \text{ k}\Omega$$

and

$$V_B = V_{CC} \frac{R_{eq}}{R_1 + R_{eq}}$$

$$= (20 \text{ V}) \frac{8.46 \text{ k}\Omega}{68 \text{ k}\Omega + 8.46 \text{ k}\Omega}$$

$$= 2.21 \text{ V}$$

The rest of the problem is solved in the usual fashion, as follows:

$$V_E = V_B - 0.7 \text{ V}$$
$$= 1.51 \text{ V}$$

and

$$I_{CQ} \cong I_E = \frac{V_E}{R_E}$$

$$= \frac{1.51}{1.1 \text{ k}\Omega}$$

$$= 1.37 \text{ mA}$$

Finally,

$$V_{CEQ} = V_{CC} - I_{CQ}(R_C + R_E)$$
$$= 20\text{V} - (1.37 \text{ mA})(7.3 \text{ k}\Omega)$$
$$= 9.99 \text{ V}$$

**PRACTICE PROBLEM 7.13**

Assume that the transistor in Figure 7.27 has a value of $h_{FE} = 80$. Recalculate the values of $I_{CQ}$ and $V_{CEQ}$ for the circuit.

## Estimating the Value of $I_{CQ}$

*OBJECTIVE 9* ▶   You may have noticed that we have a bit of a problem. To determine whether or not $R_{\text{IN(base)}}$ must be considered in the calculation of $V_B$, we have to know the value of $h_{FE}$. However, we need to know the value of $I_{CQ}$ to find the value of $h_{FE}$ on any spec sheet. And we cannot determine the value of $I_{CQ}$ without having first determined the value of $V_B$.

To determine the value of $h_{FE}$ when the value of $V_B$ isn't certain, we need a way of estimating the value of $I_{CQ}$ without using $V_B$. We can approximate the value of $I_{CQ}$ by making the following assumption: *The circuit was designed for midpoint bias.* Thus, the value of $I_{CQ}$ can be approximated as being one-half of $I_{C(sat)}$. So, the process for *estimating* the value of $I_{CQ}$ is as follows:

1. Calculate the value of $I_{C(sat)}$.

2. Approximate $I_{CQ}$ as being one-half the calculated value of $I_{C(sat)}$. Once you have approximated the value of $I_{CQ}$, you can use that approximated value to look up the value of $h_{FE}$ on the spec sheet of the transistor. Then use the value of $h_{FE}$ from the spec sheet to proceed with your analysis.

The validity of this method can be seen by referring to Example 7.13. Using equation (7.19), the value of $I_{C(sat)}$ for the circuit is calculated to be 2.74 mA. Half of $I_{C(sat)}$ would then be 1.37 mA, which equals the calculated value of $I_{CQ}$ in the example.

You may be wondering why we can assume that every voltage-divider bias circuit is designed for midpoint bias. The reason is simple: Voltage-divider bias circuits are used primarily in linear amplifiers. As stated earlier in the chapter, these amplifiers are almost always designed for midpoint bias to provide for the largest possible output.

## Circuit Troubleshooting

There really is not very much difference between troubleshooting a voltage-divider bias circuit and troubleshooting an emitter-bias circuit. In this section, we will cover some of the problems that may arise in a voltage-divider bias circuit. As you will see, the symptoms are essentially the same as those found in the emitter-bias circuit. ◀ *OBJECTIVE 10*

**$R_1$ Open.** When $R_1$ is open, there is no path for current in the base circuit. Therefore, no voltage is developed across $R_2$, and $V_B$ is zero. Because of the loss of $V_B$, the transistor will be biased off. Thus, there is no emitter current, and $V_E$ equals 0 V.

Without any current in the collector circuit, $I_C R_C$ equals 0 V, and $V_C$ equals $V_{CC}$. Thus, the following circuit conditions exist when $R_1$ is open:

1. $V_B$ is 0 V.
2. $V_E$ is 0 V.
3. $V_C$ equals $V_{CC}$.

These circuit conditions are illustrated in Figure 7.28a.

**$R_2$ Open.** If $R_2$ goes open circuit, the current through $R_1$ equals $I_B$. Since this current normally equals $(I_2 + I_B)$, the current through $R_1$ *decreases* significantly. This causes the voltage drop across $R_1$ to decrease, and $V_B$ increases. The increase in $V_B$ will cause the transistor to saturate. Thus, $V_C$ and $V_E$ are very close to equal in value and $I_C$ equals $I_{C(sat)}$. The existing circuit conditions are summarized as follows:

1. $V_B$ equals $V_E + 0.7$ V.
2. $V_E$ equals $R_E \times I_{C(sat)}$.
3. $V_C$ is approximately equal to $V_E$.

These circuit conditions are illustrated in Figure 7.28b.

**$R_C$ Open.** When $R_C$ opens, there is no path for collector current. In this condition, $I_E$ equals $I_B$, and the value of $V_E$ drops significantly. The value of $V_B$ is still approximately 0.7 V greater than $V_E$. In summary:

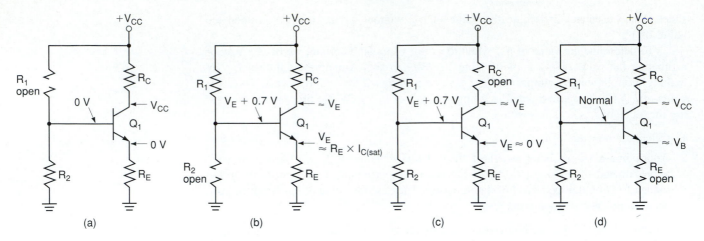

**FIGURE 7.28**  Voltage-divider bias fault system.

**1.** $V_B$ is approximately 0.7 V greater than $V_E$.

**2.** $V_E$ is slightly greater than 0 V.

**3.** $V_C$ is approximately equal to $V_E$..

In other words, all three terminal voltages are *low*. These circuit conditions are illustrated in Figure 7.28c.

**Lab Reference:** Several voltage-divider bias circuit faults are simulated in Exercise 12.

***$R_E$ Open.***    When $R_E$ opens, there is no emitter current or collector current. The base current in the circuit drops to zero, but this has little effect on the value of $V_B$ (which is equal to $I_2 R_2$).

When you read the emitter voltage with a voltmeter, the meter completes the emitter circuit, causing you to obtain a reading that is somewhat higher than normal. The collector voltage equals $V_{CC}$, since there is no current in the collector circuit. In summary:

**1.** $V_B$ is normal.

**2.** The *meter reading* of $V_E$ may be slightly higher than normal.

**3.** $V_C$ equals $V_{CC}$.

These circuit conditions are illustrated in Figure 7.28d.

## Summary

Voltage-divider bias is one of the most commonly used of the biasing circuits. It has all the advantages of emitter bias and yet does not require the use of a dual-polarity power supply.

The process for determining the $Q$-point values for the circuit is not really too complicated. However, you need to remember two points:

**1.** Start by approximating the value of $I_{CQ}$.

**2.** Determine whether or not the base input resistance will affect the calcualtion of $V_B$.

Once you have made the determination on $V_B$, the process becomes fairly routine.

The characteristics of the voltage-divider bias circuit are summarized in Figure 7.29. If you compare the characteristics shown to those of the emitter-bias circuit (Figure 7.20), you'll see that there is very little difference between these two circuits other than the power supply requirements.

**FIGURE 7.29**

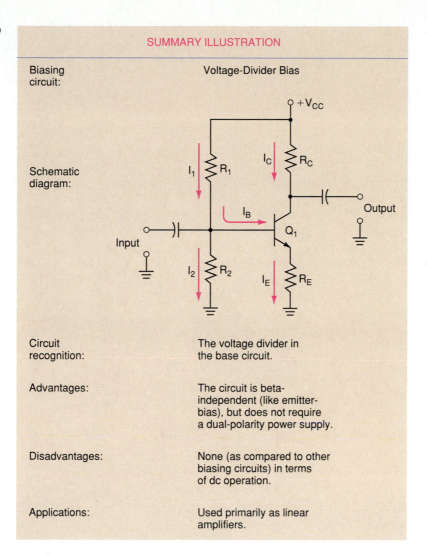

SUMMARY ILLUSTRATION

| | |
|---|---|
| Biasing circuit: | Voltage-Divider Bias |
| Schematic diagram: | |
| Circuit recognition: | The voltage divider in the base circuit. |
| Advantages: | The circuit is beta-independent (like emitter-bias), but does not require a dual-polarity power supply. |
| Disadvantages: | None (as compared to other biasing circuits) in terms of dc operation. |
| Applications: | Used primarily as linear amplifiers. |

**Section Review**

1. List, in order, the steps required to determine the values of $I_{CQ}$ and $V_{CEQ}$ for a voltage-divider bias circuit.

2. What is the procedure for determining the proper spec sheet value of $h_{FE}$ for the analysis of a circuit?

3. What additional steps must be taken in the analysis of a voltage-divider bias circuit when $h_{FE}R_E < 10R_2$?

4. Why do you need to know how to estimate the value of $I_{CQ}$?

5. How do you estimate the value of $I_{CQ}$?

6. Why is estimating the value of $I_{CQ}$ valid in most cases?

7. List the characteristic symptoms for each of the resistors in a voltage-divider bias circuit opening.

# 7.5   FEEDBACK-BIAS CIRCUITS

In this section, we will take a look at two **feedback-bias** circuits. The term *feedback* is used to describe a circuit that "feeds" a portion of the output voltage or current back to the input. For example, the **collector-feedback bias** circuit is constructed so that the col-

**Feedback bias**
A term used to describe a circuit that "feeds" a portion of the output voltage or current back to the input to control the circuit operation.

**Collector-feedback bias**
A bias circuit constructed so that $V_C$ will directly affect $V_B$.

**Emitter-feedback bias**
A bias circuit constructed so that $V_E$ will directly affect $V_B$.

lector voltage ($V_C$) has a direct effect on the base voltage ($V_B$). The **emitter-feedback bias** circuit is constructed so that the emitter voltage ($V_E$) has a direct effect on the base voltage. One reason for wiring these circuits in this way is to help reduce the effects of beta variations on the $Q$-point values of the circuits. This point will be made clear in our discussions of the two circuits.

## Collector-Feedback Bias

*OBJECTIVE 11* ▶

The collector-feedback bias circuit obtains its stability by connecting the base resistor directly to the collector of the transistor. As you can see in Figure 7.30a, this circuit configuration looks quite similar to the base-bias circuit. The circuit currents are shown in Figure 7.30b. As you can see, the path for $I_B$ includes the collector resistor, $R_C$. This was not the case for the base-bias circuit.

The analysis of the collector-feedback bias circuit begins with finding the value of $I_B$. To derive the equation for $I_B$, we start with Kirchhoff's voltage equation for the base circuit, as follows:

$$V_{CC} = V_{RC} + V_{RB} + V_{BE}$$
$$= I_C R_C + I_B R_B + V_{BE}$$

Substituting $h_{FE}I_B$ for $I_C$,

$$= I_B h_{FE} R_C + I_B R_B + V_{BE}$$
$$= I_B(h_{FE}R_C + R_B) + V_{BE}$$

By rearranging and solving for $I_B$, you have:

$$I_B = \frac{V_{CC} - V_{BE}}{R_B + h_{FE}R_C} \qquad (7.24)$$

After you have found the value of $I_B$, you proceed just as you did with the base-bias circuit, as follows:

$$I_{CQ} = h_{FE}I_B$$

and

$$V_{CEQ} = V_{CC} - I_{CQ}R_C$$

The following example illustrates the $Q$-point analysis of a collector-feedback bias circuit.

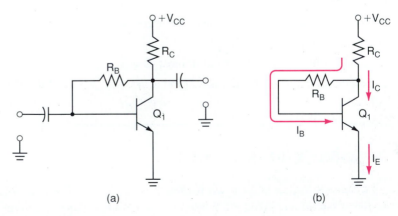

**Lab Reference:** Collector-feedback bias operation is demonstrated in Exercise 14.

(a)                    (b)

**FIGURE 7.30   Collector-feedback bias.**

Determine the values of $I_{CQ}$ and $V_{CEQ}$ for the amplifier shown in Figure 7.31.

**EXAMPLE 7.14**

**Solution:** First, the value of $I_B$ is found as

$$I_B = \frac{V_{CC} - V_{BE}}{R_B + h_{FE}R_C}$$

$$= \frac{10\text{ V} - 0.7\text{ V}}{(300\text{ k}\Omega) + (100)(1.5\text{ k}\Omega)}$$

$$= \frac{9.3\text{ V}}{450\text{ k}\Omega}$$

$$= 20.7\text{ }\mu\text{A}$$

Next, $I_{CQ}$ is found as

$$I_{CQ} = h_{FE}I_B$$
$$= 2.07\text{ mA}$$

Finally, $V_{CEQ}$ is found as

$$V_{CEQ} = V_{CC} - I_C R_C$$
$$= 10\text{ V} - (2.07\text{ mA})(1.5\text{ k}\Omega)$$
$$= 6.895\text{ V}$$

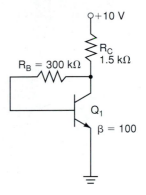

**FIGURE 7.31**

**PRACTICE PROBLEM 7.14**

A circuit like the one shown in Figure 7.31 has values of $R_C = 2\text{ k}\Omega$, $R_B = 240\text{ k}\Omega$, $h_{FE} = 120$, and $V_{CC} = +12\text{ V}$. Determine the values of $I_{CQ}$ and $V_{CEQ}$ for the circuit.

## Circuit Stability

The collector-feedback bias circuit is relatively stable against changes in beta. The key to this stability is the fact that $I_B$ and beta are *inversely related* in this circuit. As beta increases, $I_B$ decreases, and vice versa. This relationship can be seen in equation (7.24), which shows $I_B$ to vary inversely with the product of $(h_{FE}R_C)$.

Now consider the response of the amplifier in Figure 7.31 to a change in temperature. If temperature increases, $I_C$ increases. This causes beta to increase. In response to the increase in beta, $I_B$ decreases, limiting the increase in the value of $I_C$. The decrease in $I_B$ has partially offset the initial increase in $I_C$.

The relationship just described also explains the fact that a collector-feedback bias circuit *cannot saturate*. No matter how large beta becomes, the transistor will not go into saturation because the increase in beta will be offset by a decrease in $I_B$.

**Lab Reference:** Collector-feedback bias stability and fault symptoms are both demonstrated in Exercise 14.

## Circuit Troubleshooting

If either resistor in the collector-feedback bias circuit opens, the transistor will go into cutoff. In the case of this circuit, the quickest method of troubleshooting is simply to replace both resistors. If the circuit continues to malfunction, replace the transistor.

## Emitter-Feedback Bias

The collector-feedback bias circuit is designed so that the collector circuit weighs into the base current calculations. Since $I_B$ is partially controlled by the value of $h_{FE}R_C$, feedback has been obtained.

◄ *OBJECTIVE 12*

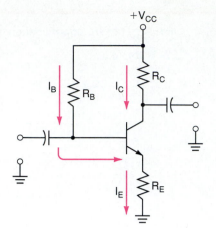

Lab Reference: Emitter-feedback bias operation is demonstrated in Exercise 15.

**FIGURE 7.32   Emitter-feedback bias.**

The emitter-feedback bias circuit works in basically the same fashion. However, in this case, it is the *emitter circuit* that affects the value of $I_B$. The basic emitter-feedback bias circuit is shown in Figure 7.32. Note that the circuit is almost identical to the base-bias circuit, with the exception of the added emitter resistor. Also note the directions of the circuit currents.

The base circuit consists of $R_E$, the base-emitter junction of the transistor, $R_B$, and $V_{CC}$. The Kirchhoff's voltage equation of this circuit is

$$V_{CC} = I_B R_B + V_{BE} + I_E R_E$$

Since $I_E = I_B(h_{FE} + 1)$, we can rewrite the above equation as

$$V_{CC} = I_B R_B + V_{BE} + I_B(h_{FE} + 1)R_E$$

Subtracting $V_{BE}$ from both sides of the equation, we get

$$V_{CC} - V_{BE} = I_B R_B + I_B(h_{FE} + 1)R_E$$

or

$$V_{CC} - V_{BE} = I_B[R_B + (h_{FE} + 1)R_E]$$

Finally, solving for $I_B$ gives us

$$I_B = \frac{V_{CC} - V_{BE}}{R_B + (h_{FE} + 1)R_E} \qquad (7.25)$$

After the value of $I_B$ is found, the value of $I_{CQ}$ is determined as

$$I_{CQ} = h_{FE} I_B$$

Finally, $V_{CEQ}$ is found as

$$V_{CEQ} = V_{CC} - I_{CQ}(R_C + R_E)$$

Note that this is the same equation that we used to determine the value of $V_{CEQ}$ for the voltage-divider bias circuit. It makes sense that these two circuits have the same $V_{CEQ}$ equation since their collector–emitter circuits are identical. Example 7.15 illustrates the process of determining the $Q$-point values for an emitter-feedback bias circuit.

Determine the values of $I_{CQ}$ and $V_{CEQ}$ for the amplifier shown in Figure 7.33.

EXAMPLE 7.15

**Solution:**   First, the value of $I_B$ is found as

$$I_B = \frac{V_{CC} - V_{BE}}{R_B + (h_{FE} + 1)R_E}$$

$$= \frac{16 \text{ V} - 0.7 \text{ V}}{680 \text{ k}\Omega + (51)(1.6 \text{ k}\Omega)}$$

$$= 20.09 \text{ μA}$$

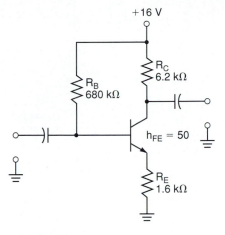

$I_{CQ}$ is found as

$$I_{CQ} = h_{FE}I_B$$
$$= (50)(20.09 \text{ μA})$$
$$= 1 \text{ mA}$$

Finally, the value of $V_{CEQ}$ is found as

$$V_{CEQ} = V_{CC} - I_{CQ}(R_C + R_E)$$
$$= 16 \text{ V} - (1 \text{ mA})(7.8 \text{ k}\Omega)$$
$$= 8.2 \text{ V}$$

**FIGURE 7.33**

**PRACTICE PROBLEM 7.15**

A circuit like the one in Figure 7.33 has values of $V_{CC} = 16$ V, $R_B = 470$ kΩ, $R_C = 1.8$ kΩ, $R_E = 910$ Ω, and $h_{FE} = 100$. Determine the values of $I_{CQ}$ and $V_{CEQ}$ for the circuit.

## Circuit Stability

$Q$-point stability in the emitter-feedback bias circuit is nearly identical to that of the collector-feedback bias circuit. If beta increases, the value of $(h_{FE} + 1)R_E$ also increases. This causes $I_B$ to decrease, as can be seen in equation (7.25). Also, like the collector-feedback bias circuit, the emitter-feedback bias circuit cannot saturate since $I_B$ and $h_{FE}$ vary inversely.

**Lab Reference:** Emitter-feedback bias stability and fault symptoms are both demonstrated in Exercise 15.

## Circuit Troubleshooting

The resistor faults that can develop in the emitter-feedback bias circuit are nearly identical to those of the voltage-divider bias circuit. The fault symptoms for $R_C$ and $R_E$ are identical between the two circuits. The $R_B$ fault symptoms are identical to those for $R_1$ in the voltage-divider bias circuit.

## Summary

The feedback bias circuits are designed so that the values of $I_B$ and $h_{FE}$ vary inversely. Thus, when $h_{FE}$ increases, $I_B$ decreases. This partially offsets the effect that the increase in $h_{FE}$ has on the value of $I_C$.

While the *collector-feedback bias circuit* is the simpler of the two, the *emitter-feedback bias circuit* has ac characteristics that make it the more commonly used. Neither of the two circuits can be driven into saturation. The characteristics of the two feedback bias circuits are summarized in Figure 7.34.

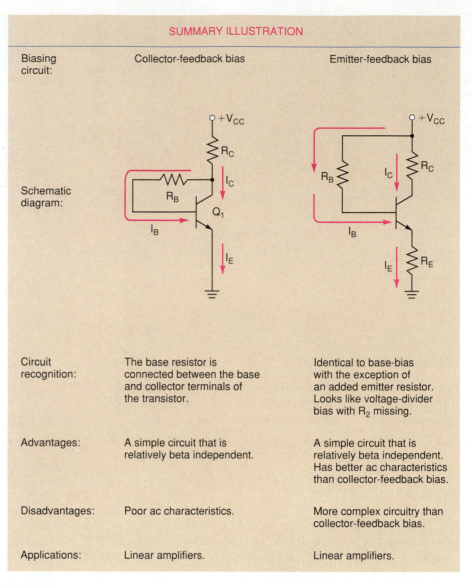

| Biasing circuit: | Collector-feedback bias | Emitter-feedback bias |
|---|---|---|
| Schematic diagram: | | |
| Circuit recognition: | The base resistor is connected between the base and collector terminals of the transistor. | Identical to base-bias with the exception of an added emitter resistor. Looks like voltage-divider bias with $R_2$ missing. |
| Advantages: | A simple circuit that is relatively beta independent. | A simple circuit that is relatively beta independent. Has better ac characteristics than collector-feedback bias. |
| Disadvantages: | Poor ac characteristics. | More complex circuitry than collector-feedback bias. |
| Applications: | Linear amplifiers. | Linear amplifiers. |

**FIGURE 7.34**

## Section Review

1. What is *feedback*? What is it used for?

2. Describe the construction and current characteristics of the *collector-feedback bias* circuit.

3. Explain why collector-feedback bias circuits are relatively stable against changes in beta.

4. Describe the construction and current characteristics of the *emitter-feedback bias* circuit.

5. List, in order, the steps required to determine the $Q$-point values of the emitter-bias circuit.

6. Explain why emitter-feedback bias circuits are relatively stable against changes in beta.

7. What are the troubleshooting similarities between emitter-feedback bias and voltage-divider bias?

## 7.6 PUTTING IT ALL TOGETHER: WHY ARE THERE SO MANY DIFFERENT TYPES OF BIAS?

In this chapter, we have analyzed the operation of various dc biasing circuits. You may be wondering why so many different circuits are used for biasing transistors. The answer lies in the fact that each of the circuits covered has its own advantages and limitations, for example:

1. *Base bias* has the advantage of circuit simplicity, but its *Q*-point values are very unstable. Thus, the use of base bias is limited to switching circuits.

2. *Emitter bias* has better *Q*-point stability than base bias but requires a dual-polarity to operate. Thus, the use of emitter bias is limited to systems that utilize a dual-polarity power supply.

3. *Voltage-divider bias* has the *Q*-point stability of emitter bias yet does not require the use of a dual-polarity power supply. In a system that contains a single-polarity supply, voltage-divider bias is used in place of emitter bias.

4. The feedback bias circuits (*collector-feedback* and *emitter-feedback*) provide a degree of *Q*-point stability, though not as much as emitter bias or voltage-divider bias. When lower *Q*-point stability is acceptable, the simpler circuitry involved in feedback bias is used.

The operation of an amplifier depends on the initial values of $I_B$, $I_C$, and $V_{CE}$. The purpose of the dc biasing circuit is to set the initial operating values for these variables.

**Chapter Summary**

The dc load line represents all possible combinations of $I_C$ and $V_{CE}$. On one end, the load line represents the values of the saturated transistor. On the other end, it represents the circuit values for the cutoff transistor.

When the *Q-point* (operating point) of the transistor falls on the halfway point on the load line, the amplifier is *midpoint biased*. This is the optimum biasing condition for any linear amplifier.

The *base-bias* circuit makes a poor linear amplifier because its output depends solely on the value of beta. Since beta varies with changes in both $I_C$ and temperature, this transistor configuration is extremely unstable. Base-bias circuits are used almost exclusively in *switching applications*.

The *emitter-bias* circuit is more stable than the base-bias circuit, but it has the drawback of requiring a split-polarity power supply. Since beta does not weigh into the output circuit calculations, the emitter-bias circuit is stable against changes in beta.

The *voltage-divider bias* circuit is by far the most commonly used. This amplifier type is extremely stable but does not need a split-polarity power supply.

The *collector-feedback bias* circuit has the simplicity of the base-bias circuit, yet it is still relatively stable against changes in beta. Although the output of this circuit type *will* vary when beta changes, the results are not as drastic as those experienced with the base-bias circuits.

The *emitter-feedback bias* circuit is a bit more complex than the collector-feedback bias circuit, but it has better ac characteristics. The emitter-feedback bias circuit is also relatively stable against changes in beta.

## Key Terms

The following terms were introduced and defined in this chapter:

| | | |
|---|---|---|
| base bias | emitter-feedback bias | Q-point |
| beta-dependent circuit | feedback bias | Q-point shift |
| beta-independent circuit | fixed bias | quiescent |
| collector-feedback bias | leaky | universal bias |
| dc load line | midpoint bias | voltage-divider bias |
| emitter bias | | |

## Equation Summary

| Equation Number | Equation | Section Number |
|---|:---:|---|
| (7.1) | $V_{CC} = I_C R_C$ (saturation) | 7.1 |
| (7.2) | $I_{C(\text{sat})} = \dfrac{V_{CC}}{R_C}$ (ideal) | 7.1 |
| (7.3) | $V_{C(\text{off})} = V_{CC}$ | 7.1 |
| (7.4) | $I_B = \dfrac{V_{CC} - V_{BE}}{R_B}$ | 7.2 |
| (7.5) | $I_{CQ} = h_{FE} I_B$ | 7.2 |
| (7.6) | $V_{CEQ} = V_{CC} - I_C R_C$ | 7.2 |
| (7.7) | $I_E = \dfrac{V_{EE} - V_{BE}}{R_E}$ | 7.3 |
| (7.8) | $I_{CQ} \cong \dfrac{V_{EE} - V_{BE}}{R_E}$ | 7.3 |
| (7.9) | $V_{CEQ} = V_{CC} - I_{CQ} R_C$ | 7.3 |
| (7.10) | $I_{C(\text{sat})} = \dfrac{2 V_{CC}}{R_C + R_E}$ | 7.3 |
| (7.11) | $V_{CE(\text{off})} = 2 V_{CC}$ | 7.3 |
| (7.12) | $R_{\text{IN(base)}} = h_{FE} R_E$ | 7.3 |
| (7.13) | $V_B = \dfrac{R_2}{R_1 + R_2} V_{CC}$ | 7.4 |
| (7.14) | $V_E = V_B - 0.7 \text{ V}$ | 7.4 |
| (7.15) | $I_E = \dfrac{V_E}{R_E}$ | 7.4 |
| (7.16) | $V_{CEQ} = V_{CC} - I_{CQ}(R_C + R_E)$ | 7.4 |
| (7.17) | $I_B = \dfrac{I_E}{1 + h_{FE}}$ | 7.4 |
| (7.18) | $h_{FE(\text{ave})} = \sqrt{h_{FE(\text{min})} \times h_{FE(\text{max})}}$ | 7.4 |

| Equation Number | Equation | Section Number |
|---|---|---|
| (7.19) | $I_{C(\text{sat})} = \dfrac{V_{CC}}{R_C + R_E}$ | 7.4 |
| (7.20) | $V_{CE(\text{off})} = V_{CC}$ | 7.4 |
| (7.21) | $R_2 \leq \dfrac{h_{FE}R_E}{10}$ | 7.4 |
| (7.22) | $R_{eq} = R_2 \parallel h_{FE}R_E$ | 7.4 |
| (7.23) | $V_B = V_{CC}\dfrac{R_{eq}}{R_1 + R_{eq}}$ | 7.4 |
| (7.24) | $I_B = \dfrac{V_{CC} - V_{BE}}{R_B + h_{FE}R_C}$ | 7.5 |
| (7.25) | $I_B = \dfrac{V_{CC} - V_{BE}}{R_B + (h_{FE} + 1)R_E}$ | 7.5 |

Answers to the
Example Practice
Problems

**7.1.** $V_{CE(\text{off})} = 8$ V, $I_{C(\text{sat})} = 7.27$ mA
**7.2.** The end points of the load line are $V_{CE(\text{off})} = 16$ V and $I_{C(\text{sat})} = 8$ mA. $V_{CE} = 12$ V at $I_C = 2$ mA, $V_{CE} = 8$ V at $I_C = 4$ mA, and $V_{CE} = 4$ V at $I_C = 6$ mA.
**7.3.** $I_C = 9.85$ mA, $V_{CE} = 6.91$ V
**7.4.** The circuit is *not* midpoint biased.
**7.5.** The circuit *is* midpoint biased.
**7.6.** $I_C = 18.72$ mA, $V_{CE} = 522.5$ mV
**7.7.** $I_{CQ} = 4.77$ mA, $V_{CEQ} = 7.85$ V
**7.8.** $I_{C(\text{sat})} = 6.67$ mA, $V_{CE(\text{off})} = 30$ V
**7.9.** $R_{\text{IN(base)}} = 540$ kΩ
**7.10.** $I_{CQ} = 6.32$ mA, $V_{CEQ} = 4.95$ V
**7.11.** $I_B = 31.44$ μA
**7.12.** $I_B = 114.15$ μA
**7.13.** $I_{CQ} = 1.48$ mA, $V_{CEQ} = 9.2$ V
**7.14.** $I_{CQ} = 2.83$ mA, $V_{CEQ} = 6.35$ V
**7.15.** $I_{CQ} = 2.72$ mA, $V_{CEQ} = 8.62$ V

### §7.1

Practice Problems

1. The circuit in Figure 7.35 has values of $V_{CC} = +8$ V and $R_C = 3.3$ kΩ. Plot the dc load line for the circuit.

2. The circuit in Figure 7.35 has values of $V_{CC} = 24$ V and $R_C = 9.1$ kΩ. Plot the dc load line for the circuit.

3. The circuit in Figure 7.35 has values of $V_{CC} = 14$ V and $R_C = 1$ kΩ. Plot the dc load line. Then use equation (7.6) to verify the load line $V_{CE}$ values for $I_C = 2$ mA, $I_C = 8$ mA, and $I_C = 10$ mA.

4. The circuit in Figure 7.35 has values of $V_{CC} = 20$ V and $R_C = 2.4$ kΩ. Plot the dc load line. Then use equation (7.6) to verify the load line $V_{CE}$ values for $I_C = 1$ mA, $I_C = 5$ mA, and $I_C = 7$ mA.

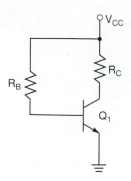

**FIGURE 7.35**

5. The circuit in Figure 7.35 has values of $V_{CC} = 8$ V and $R_C = 1$ kΩ. Plot the dc load line for the circuit. Then use your dc load line to determine the midpoint-bias values of $I_C$ and $V_{CE}$.

6. The circuit in Figure 7.35 has values of $V_{CC} = 36$ V and $R_C = 36$ kΩ. Plot the dc load line for the circuit. Then use your dc load line to determine the midpoint-bias values of $I_C$ and $V_{CE}$.

**§7.2**

7. Determine the $Q$-point values of $I_C$ and $V_{CE}$ for the circuit shown in Figure 7.36a.

8. Determine the $Q$-point values of $I_C$ and $V_{CE}$ for the circuit shown in Figure 7.36b.

9. Determine the $Q$-point values of $I_C$ and $V_{CE}$ for the circuit shown in Figure 7.36c.

10. Determine the $Q$-point values of $I_C$ and $V_{CE}$ for the circuit shown in Figure 7.36d.

11. Plot the dc load line for the circuit in Figure 7.36a. Then, using your answers from Problem 7, determine whether or not the circuit is midpoint biased.

12. Plot the dc load line for the circuit in Figure 7.36b. Then, using your answers from Problem 8, determine whether or not the circuit is midpoint biased.

13. Without the use of the dc load line, determine whether or not the circuit in Figure 7.36c is midpoint biased. Use your answers from Problem 9.

14. Without the use of a dc load line, determine whether or not the circuit in Figure 7.36d is midpoint biased. Use your answers from Problem 10.

15. The value of $h_{FE}$ shown in Figure 7.36a is measured at 25°C. At 100°C, the value of $h_{FE}$ for the transistor is 120. Calculate the values of $I_C$ and $V_{CE}$ for the circuit when it is operated at 100°C.

16. The value of $h_{FE}$ shown in Figure 7.36b is measured at 25°C. At 100°C, the value of $h_{FE}$ for the transistor is 120. Calculate the values of $I_C$ and $V_{CE}$ for the circuit when it is operated at 100°C.

**§7.3**

17. Determine the values of $I_{CQ}$ and $V_{CEQ}$ for the circuit shown in Figure 7.37.

18. For the circuit shown in Figure 7.37, *double* the values of $R_C$ and $R_E$. Now recalculate the output values for the circuit. How do these values compare with those obtained in Problem 17?

19. Calculate the values of $I_{C(sat)}$ and $V_{CE(off)}$ for the circuit shown in Figure 7.37. Use the original component values shown.

20. For the circuit shown in Figure 7.37, *double* the values of $R_C$ and $R_E$. Now recalculate the values of $I_{C(sat)}$ and $V_{CE(off)}$. How do these values compare with the ones obtained in Problem 19?

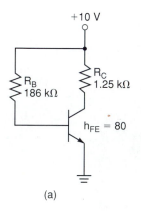

(a)

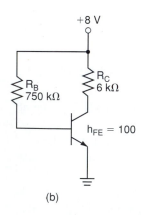

(b)

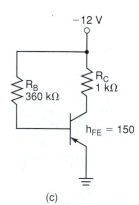

(c)

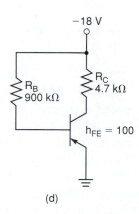

(d)

**FIGURE 7.36**

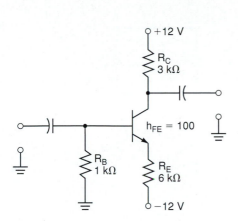

FIGURE 7.37

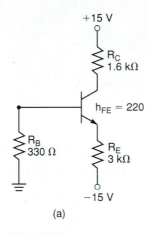

(a)

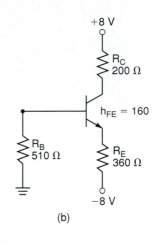

(b)

FIGURE 7.38

21. Determine the values of $I_{CQ}$, $V_{CEQ}$, $I_{C(sat)}$, and $V_{CE(off)}$ for the circuit shown in Figure 7.38a.

22. Determine the values of $I_{CQ}$, $V_{CEQ}$, $I_{C(sat)}$, and $V_{CE(off)}$ for the circuit shown in Figure 7.38b.

23. Determine the value of $R_{IN(base)}$ for the circuit shown in Figure 7.38a.

24. Determine the value of $R_{IN(base)}$ for the circuit shown in Figure 7.38b.

25. Determine the values of $I_{CQ}$, $V_{CEQ}$, and $I_B$ for the circuit shown in Figure 7.39a.       $test\ h_{FE}\ R_E$

26. Determine the values of $I_{CQ}$, $V_{CEQ}$, and $I_B$ for the circuit shown in Figure 7.39b.

27. Determine the values of $I_{CQ}$, $V_{CEQ}$, and $I_B$ for the circuit shown in Figure 7.39c.

28. Determine the values of $I_{CQ}$, $V_{CEQ}$, and $I_B$ for the circuit shown in Figure 7.39d.

29. A transistor spec sheet lists the following values of $h_{FE}$ at $I_C = 1$ mA: $h_{FE} = 50$ (min), $h_{FE} = 75$ (typical), and $h_{FE} = 120$ (max). Determine which value would be used in any amplifier analysis.

30. A transistor spec sheet lists the following values of $h_{FE}$ at $I_C = 1$ mA: $h_{FE} = 80$ (min) and $h_{FE} = 200$ (max). Determine the value of $h_{FE}$ that would be used in any circuit analysis.

31. Determine the values of $I_{C(sat)}$ and $V_{CE(off)}$ for the circuit in Figure 7.39a. Then, using your answers from Problem 25, determine whether or not the circuit is midpoint biased.

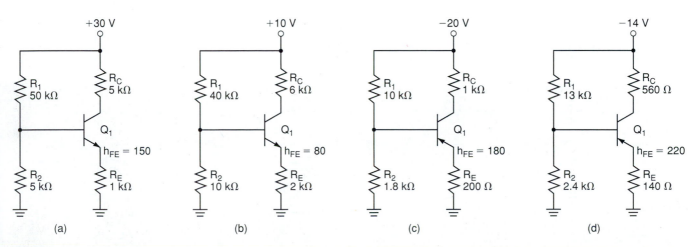

(a)         (b)         (c)         (d)

FIGURE 7.39

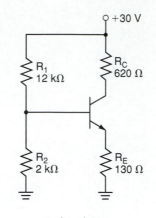

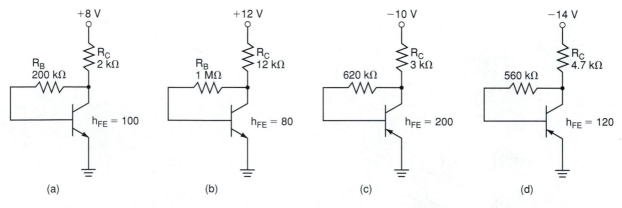

FIGURE 7.40

32. Determine the values of $I_{C(sat)}$ and $V_{CE(off)}$ for the circuit in Figure 7.39b. Then, using your answers from Problem 26, determine whether or not the circuit is mid-point biased.

33. Determine the values of $I_{C(sat)}$ and $V_{CE(off)}$ for the circuit in Figure 7.39c. Then, using your answers from Problem 27, determine whether or not the circuit is mid-point biased.

34. Determine the values of $I_{C(sat)}$ and $V_{CE(off)}$ for the circuit in Figure 7.39d. Then, using your answers from Problem 28, determine whether or not the circuit is mid-point biased.

35. Determine whether or not the circuit in Figure 7.40a is midpoint biased.

36. Determine whether or not the circuit in Figure 7.40b is midpoint biased.

§7.5

37. Determine the values of $I_{CQ}$ and $V_{CEQ}$ for the circuit shown in Figure 7.41a.

38. Determine the values of $I_{CQ}$ and $V_{CEQ}$ for the circuit shown in Figure 7.41b.

FIGURE 7.41

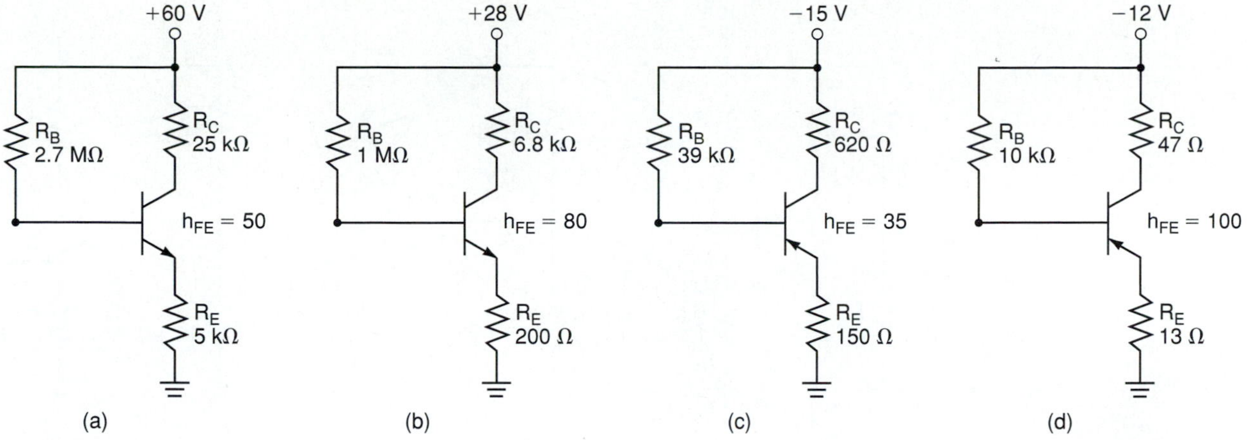

FIGURE 7.42

39. Determine the values of $I_{CQ}$ and $V_{CEQ}$ for the circuit shown in Figure 7.41c.
40. Determine the values of $I_{CQ}$ and $V_{CEQ}$ for the circuit shown in Figure 7.41d.
41. Determine the values of $I_{CQ}$ and $V_{CEQ}$ for the circuit shown in Figure 7.42a.
42. Determine the values of $I_{CQ}$ and $V_{CEQ}$ for the circuit shown in Figure 7.42b.
43. Determine the values of $I_{CQ}$ and $V_{CEQ}$ for the circuit shown in Figure 7.42c.
44. Determine the values of $I_{CQ}$ and $V_{CEQ}$ for the circuit shown in Figure 7.42d.

45. What fault is indicated in Figure 7.43a? Explain your answer.
46. What fault is indicated in Figure 7.43b? Explain your answer.
47. What fault is indicated in Figure 7.43c? Explain your answer.
48. What fault is indicated in Figure 7.44a? Explain your answer.
49. What fault is indicated in Figure 7.44b? Explain your answer.
50. What fault is indicated in Figure 7.44c? Explain your answer.

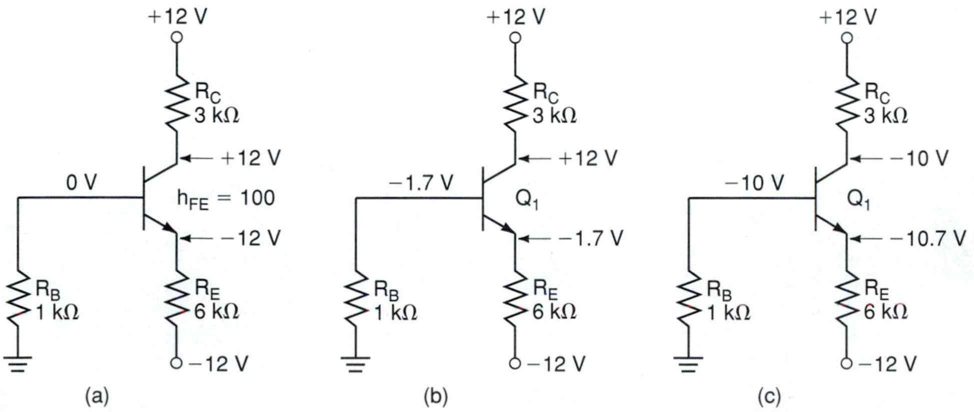

*Troubleshooting Practice Problems*

FIGURE 7.43

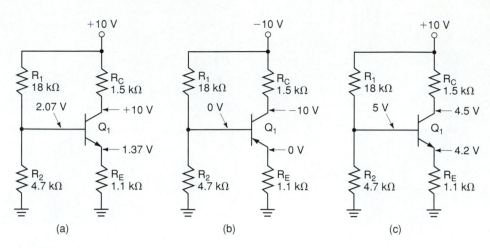

**FIGURE 7.44**

---

**The Brain Drain**

51. The transistor in Figure 7.45 has a value of $h_{FE} = 200$ for $I_C = 1$ to 8 mA. The value of $V_{CE(sat)}$ for the component is 0.3 V. Derive the collector curves for the device for $I_C = 1$ to 4 mA and $V_{CE} = 0$ V to $V_{CE(off)}$. Then plot the dc load line for the circuit on the collector curves to determine the value of $I_B$ required for mid-point-bias operation.

52. The spec sheet for the transistor in Figure 7.46 is located in Appendix A. Determine the values of $I_{CQ}$, $V_{CEQ}$, $I_B$, $V_B$, $I_{C(sat)}$, and $V_{CE(off)}$.

53. *A collector-feedback bias circuit will be midpoint biased if $R_B = h_{FE}R_C$. Using equation (7.23), prove this statement to be true. (Hint: Idealize the emitter-base junction diode in the equation.)*

54. The 2N3904 transistor cannot be used in the circuit shown in Figure 7.47. Why not? (Hint: Consider the dc load line for the amplifier and the *maximum ratings* shown in the 2N3904 spec sheet in Section 6.5.)

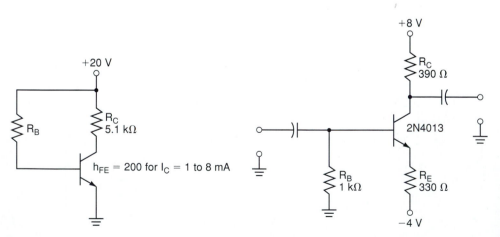

**FIGURE 7.45**          **FIGURE 7.46**

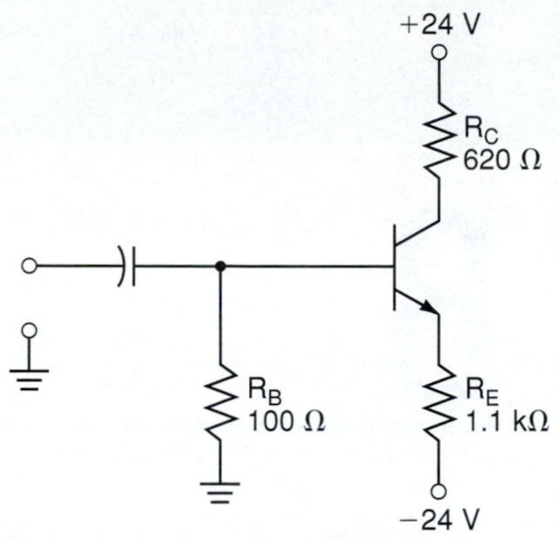

+24 V

$R_C$
620 Ω

$R_B$
100 Ω

$R_E$
1.1 kΩ

−24 V

**FIGURE 7.47**

55. Write a program that will determine whether or not a given base-bias circuit is midpoint biased, given the values of $R_C$, $R_B$, $V_{CC}$, and $h_{FE}$ for the transistor. Set up the program so that it will provide you with the values of $I_{CQ}$ and $V_{CEQ}$.

56. Write a program that will determine whether or not a given voltage-divider bias circuit is midpoint biased, given the values of $R_1$, $R_2$, $R_C$, $R_E$, and $V_{CC}$. Set up the program so that it will provide you with the values of $I_{CQ}$ and $V_{CEQ}$.

*Suggested Computer Applications Problems*

# 8

# INTRODUCTION TO AMPLIFIERS

The analysis of circuit operation generally involves the use of more than just a simple DMM.

# OUTLINE

# OBJECTIVES

*After studying the material in this chapter, you should be able to:*

1. List the three fundamental ac properties of amplifiers.
2. Discuss the concept of gain.
3. Draw and discuss the general model of a voltage amplifier.
4. Discuss the effects that amplifier input and output impedance have on the effective voltage gain of the circuit.
5. Describe the ideal voltage amplifier.
6. List, compare, and contrast the three BJT amplifier configurations.
7. Determine the configuration of any BJT amplifier.
8. Discuss the concept of amplifier efficiency.
9. List, compare, and contrast the various classes of amplifier operation.
10. Convert any power or voltage gain value to and from dB form.

When you cannot hear the output from a stereo, you turn up the volume. When the picture on your television is too dark, you increase the brightness setting. In both of these cases, you are taking a relatively weak signal and making it stronger; that is, you are increasing its power level. The process of increasing the power of an ac signal is referred to as **amplification**. The circuits used to provide amplification are referred to as **amplifiers.** Several typical amplifiers are shown in Figure 8.1.

**Amplification**
The process of increasing the power of an ac signal.

Amplifiers are some of the most widely used circuits that you will encounter. They are used extensively in audio, video, and telecommunications systems. They are also used in digital systems, biomedical systems, and so on. In fact, you would find it difficult to come up with an electronics system that doesn't contain at least one amplifier.

In the upcoming chapters, you will learn how to analyze and work with many different types of amplifiers. You'll be shown (among other things) how to calculate the values of several amplifier properties, such as *gain, input impedance,* and *output impedance.* In this chapter, we will discuss these properties and the roles they play in amplifier operation. We will also look at some other topics that relate to amplifiers in general.

## 8.1 AMPLIFIER PROPERTIES

*OBJECTIVE 1* ▶ All amplifiers have three fundamental properties: *gain, input impedance,* and *output impedance.* These properties can be combined to form a *general amplifier model* like the one shown in Figure 8.2. Note that the diamond shape in the model is used (in this case)

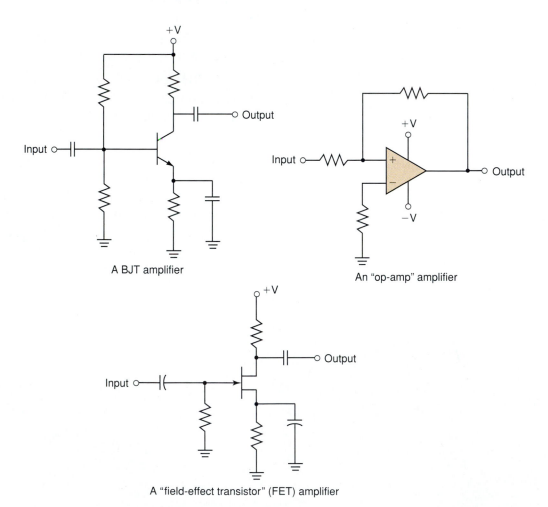

A BJT amplifier

An "op-amp" amplifier

A "field-effect transistor" (FET) amplifier

**FIGURE 8.1  Typical amplifiers.**

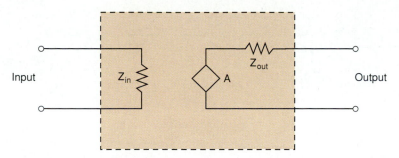

**FIGURE 8.2** The general amplifier model.

to represent the *gain* of the circuit. We'll modify this symbol slightly after we discuss the concept of gain.

## Amplifier Gain

All amplifiers exhibit a property called **gain.** The gain of an amplifier is *a multiplier that exists between the circuit's input and output.* For example, if the gain of an amplifier is 100, then the output signal will be 100 times as great as the input signal under normal operating conditions.

There are three types of gain: *voltage gain, current gain,* and *power gain.* Gain is represented using the letter *A,* as shown in Table 8.1. Note that the subscript in each symbol identifies the type of gain.

As you have been told, all amplifiers provide some degree of power gain. However, not all amplifiers are designed for this purpose. For example, a *voltage amplifier* is designed to provide a specific value of $A_v$. The fact that it also provides some value of power gain is usually a secondary consideration. The same can be said for a *current amplifier,* which is designed to provide a specific value of $A_i$. Only a *power amplifier* is designed to provide a specific value of $A_p$. The type of amplifier used in a given application depends on the type of gain desired.

In this chapter, we will focus primarily on the gain and impedance characteristics of voltage amplifiers. Current and power amplifiers will be addressed in later chapters.

◄ OBJECTIVE 2

**Gain**
A multiplier that exists between the input and output of a circuit.

**TABLE 8.1  Gain Symbols**

| Type of Gain | Symbol |
| --- | --- |
| Voltage | $A_v$ |
| Current | $A_i$ |
| Power | $A_p$ |

## Gain As a Ratio

Traditionally, *gain* is defined as *the ratio of an output value to its corresponding input value.* For example, *voltage gain* can be defined as *the ratio of ac output voltage to ac input voltage.* By formula,

Gain can be calculated as the ratio of a circuit's output to its corresponding input.

$$A_v = \frac{v_{out}}{v_{in}} \qquad (8.1)$$

where

$v_{out}$ = the ac output voltage from the amplifier
$v_{in}$ = the ac input voltage to the amplifier

The calculation of voltage gain using input and output values is demonstrated in Example 8.1.

---

The symbol shown in Figure 8.3 is a generic symbol for an amplifier. Calculate the voltage gain for the amplifier represented in the figure.

*EXAMPLE 8.1*

**Solution:**  Using the values of $v_{out}$ and $v_{in}$ shown in the figure, the voltage gain of the amplifier is found as

---

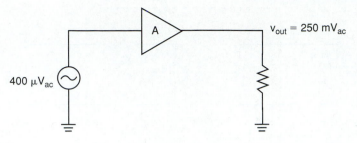

$$V_{out} = 250 \text{ mV}_{ac}$$

$$400 \; \mu V_{ac}$$

**FIGURE 8.3**

$$A_v = \frac{v_{out}}{v_{in}}$$

$$= \frac{250 \text{ mV}_{ac}}{400 \; \mu V_{ac}}$$

$$= 625$$

This result indicates that the ac output voltage for the circuit is, under normal operating circumstances, 625 times as great as the ac input voltage.

***PRACTICE PROBLEM 8.1***

An amplifier like the one represented in Figure 8.3 has the following values: $v_{out} = 72 \text{ mV}_{ac}$ and $v_{in} = 300 \; \mu V_{ac}$. Calculate the voltage gain of the circuit.

---

Since gain is a ratio, it has no units.

Note that the value of voltage gain obtained in the example has no units. This is always the case. Since each type of gain can be defined as a ratio of an output value to an input value, gain has no unit of measure.

Equation (8.1) can be somewhat misleading because it implies that the value of $A_v$ is determined by the values of $v_{out}$ and $v_{in}$. In fact, the value of $A_v$ is determined by circuit component values.

Under normal circumstances, gain is a constant. Thus, a change in the input to an amplifier usually results in a change in the corresponding output value, not in a change in gain. For example, when the value of $v_{in}$ for an amplifier changes, the new value of $v_{out}$ is found as

$$v_{out} = A_v v_{in} \tag{8.2}$$

This relationship is demonstrated in Example 8.2.

---

***EXAMPLE 8.2***

The input of the circuit in Figure 8.3 changes to $240 \; \mu V_{ac}$. Calculate the new value of $v_{out}$.

***Solution:*** The gain of the circuit was found in Example 8.1 to be 625. Using this value of $A_v$, the new value of $v_{out}$ is found as

$$v_{out} = A_v v_{in}$$
$$= (625)(240 \; \mu V_{ac})$$
$$= 150 \text{ mV}_{ac}$$

Like voltage gain, current and power gain can be defined as output-to-input ratios. These ratios are

$$A_i = \frac{I_{out}}{i_{in}}$$   (8.3)

and

$$A_p = \frac{P_{out}}{P_{in}}$$   (8.4)

Since they can be defined as ratios, current and power gain (like voltage gain) have no units. They are simply multipliers that exist between the input and output of an amplifier.

## The General Amplifier Model

Figure 8.2 showed a general model for a voltage amplifier. Now that we have defined gain, we need to modify the amplifier model, as shown in Figure 8.4a. Note that the diamond shape is actually used to represent $v_{out}$ (as a *voltage source*). The value of this voltage source is given as $A_v v_{in}$ to show that the output voltage is a function of the input voltage and the voltage gain of the circuit.

◄ *OBJECTIVE 3*

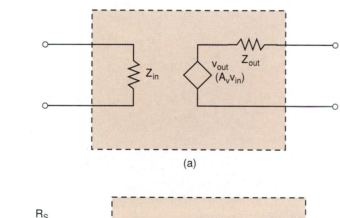

(a)

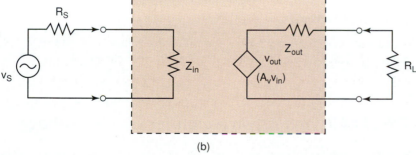

(b)

**FIGURE 8.4**

**FIGURE 8.5**

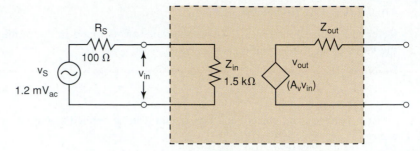

When we add a signal source and a load to the amplifier model in Figure 8.4a, we obtain the circuit shown in Figure 8.4b. The input circuit consists of $v_S$, $R_S$, and $Z_{in}$ (the amplifier input impedance). The output circuit consists of $v_{out}$, $Z_{out}$ (the amplifier output impedance), and $R_L$. As you can see, the circuits are nearly identical (in terms of their components).

## Amplifier Input Impedance ($Z_{in}$)

OBJECTIVE 4 ▶

**Input impedance**
The load that an amplifier places on its source.

When an amplifier is connected to a signal source, the source sees the amplifier as a load. The input impedance ($Z_{in}$) of the amplifier is the value of this load. For example, the value of $Z_{in}$ for the amplifier in Figure 8.5 is shown to be 1.5 kΩ. In this case, the amplifier acts as a 1.5 kΩ load that is in series with the source resistance ($R_S$).

If we assume that the input impedance of the amplifier in Figure 8.5 is purely resistive, the signal voltage at the amplifier input is found as

$$v_{in} = v_S \frac{Z_{in}}{R_s + Z_{in}}$$

(8.5)

Since $R_S$ and $Z_{in}$ form a voltage divider, the input voltage to the amplifier must be lower than the rated value of the source. This point is illustrated in the following example.

---

**EXAMPLE 8.3**

Calculate the value of $v_{in}$ for the circuit in Figure 8.5.

**Solution:** Using the values shown in the figure, the value of $v_{in}$ is found as

$$v_{in} = v_S \frac{Z_{in}}{R_S + Z_{in}}$$

$$= (1.2 \text{ mV}_{ac}) \frac{1.5 \text{ k}\Omega}{1.6 \text{ k}\Omega}$$

$$= 1.125 \text{ mV}_{ac}$$

**PRACTICE PROBLEM 8.3**

Amplifier like the one in Figure 8.5 has the following values: $v_S = 800$ μV$_{ac}$, $R_S = 70$ Ω, and $Z_{in} = 750$ Ω. Calculate the value of $v_{in}$ for the circuit.

---

## The Effect of $R_S$ and $Z_{in}$ on Amplifier Output Voltage

Example 8.3 demonstrated the effect that $R_S$ and $Z_{in}$ can have on the input to an amplifier. As the following example demonstrates, the reduction of the source voltage can cause a noticeable reduction in the circuit's output voltage.

Assume that the amplifier described in Example 8.3 has a value of $A_v = 500$. Calculate the values of $v_{out}$ for $v_{in} = 1.2 \text{ mV}_{ac}$ and $v_{in} = 1.125 \text{ mV}_{ac}$.

EXAMPLE **8.4**

**Solution:**    For $v_{in} = 1.2 \text{ mV}_{ac}$, the value of $v_{out}$ is found as

$$v_{out} = A_v v_{in}$$
$$= (500)(1.2 \text{ mV}_{ac})$$
$$= 600 \text{ mV}_{ac}$$

For $v_{in} = 1.125 \text{ mV}_{ac}$, the value of $v_{out}$ is found as

$$v_{out} = A_v v_{in}$$
$$= (500)(1.125 \text{ mV}_{ac})$$
$$= 562.5 \text{ mV}_{ac}$$

**PRACTICE PROBLEM 8.4**

Refer to Practice Problem 8.3. Using $A_v = 480$, calculate the values of $v_{out}$ for $v_{in} = v_S$ and for the value of $v_{in}$ found in the practice problem.

---

Here is what Example 8.4 has shown us: Had we ignored the effects of $R_S$ and $Z_{in}$ on the amplifier in Figure 8.5, we would have calculated an output voltage for the circuit of 600 mV$_{ac}$. However, because of the reduction in $v_S$, the actual value of $v_{out}$ is 562.5 mV$_{ac}$. Thus, $R_S$ and $Z_{in}$ cause the output of the circuit to be 37.5 mV lower than it first appears to be. As you will see, the combination of $Z_{out}$ and $R_L$ can cause an even greater reduction in the ideal output from an amplifier.

## Amplifier Output Impedance (Z$_{out}$)

When a load is connected to an amplifier, the amplifier acts as the source for that load. As with any source, there is some measurable value of source resistance, in this case, the **output impedance** of the amplifier. For example, consider the circuit shown in Figure 8.6. If we assume that the value of $Z_{out}$ for the amplifier is 200 $\Omega$, the load sees the amplifier as a voltage source with an internal resistance of 200 $\Omega$. If we assume that the output impedance of the amplifier in Figure 8.6 is purely resistive, the value of the load voltage can be found using the voltage divider equation, as follows:

**Output impedance**
The "source resistance" that an amplifier presents to its load.

$$v_L = v_{out} \frac{R_L}{Z_{out} + R_L} \tag{8.6}$$

Example 8.5 demonstrates the effect that the combination of $Z_{out}$ and $R_L$ can have on the load voltage produced by an amplifier.

**FIGURE 8.6**

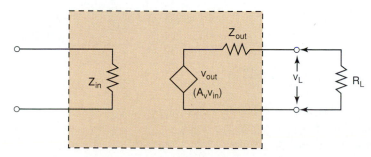

EXAMPLE 8.5

Calculate the value of the load voltage ($v_L$) for the circuit shown in Figure 8.7.

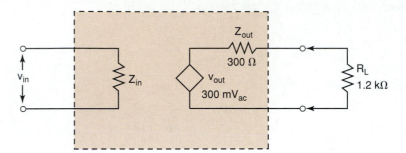

**FIGURE 8.7**

**Solution:** Using the values shown in the figure, the value of $v_L$ is found as

$$v_L = v_S \frac{R_L}{Z_{out} + R_L}$$

$$= (300 \text{ mV}_{ac}) \frac{1.2 \text{ k}\Omega}{1.5 \text{ k}\Omega}$$

$$= 240 \text{ mV}_{ac}$$

As you can see, $Z_{out}$ and $R_L$ have combined to cause a 60 mV$_{ac}$ reduction in the amplifier output voltage.

***PRACTICE PROBLEM 8.5***

A circuit like the one in Figure 8.7 has the following values: $v_{out} = 480 \text{ mV}_{ac}$, $Z_{out} = 240 \, \Omega$, and $R_L = 1.5 \text{ k}\Omega$. Calculate the value of the load voltage for the circuit.

## The Combined Effects of the Input and Output Circuits

The combination of the input and output circuits can cause a fairly significant reduction in the *effective* voltage gain of an amplifier. For example, consider the circuit shown in Figure 8.8. In order to see the combined effects of the input and output circuits, we need to calculate the value of $v_L$ for the circuit. The first step is to determine the value of $v_{in}$, as follows:

$$v_{in} = v_S \frac{Z_{in}}{R_S + Z_{in}}$$

$$= (15 \text{ mV}_{ac}) \frac{980 \, \Omega}{1 \text{ k}\Omega}$$

$$= 14.7 \text{ mV}_{ac}$$

Now, the value of $v_{out}$ is found as

$$v_{out} = A_v v_{in}$$
$$= (340)(14.7 \text{ mV}_{ac})$$
$$= 5 \text{ V}_{ac}$$

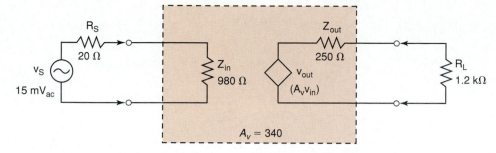

**FIGURE 8.8**

Finally, the load voltage is found as

$$v_L = v_{out} \frac{R_L}{Z_{out} + R_L}$$

$$= (5 \text{ V}_{ac}) \frac{1.2 \text{ k}\Omega}{1.45 \text{ k}\Omega}$$

$$= 4.14 \text{ V}_{ac}$$

According to these calculations, the amplifier has increased the 15 mV$_{ac}$ source voltage to a 4.14 V$_{ac}$ load voltage. We consider the *effective* voltage gain of the amplifier to equal the ratio of load voltage to source voltage. By formula,

$$A_{v(eff)} = \frac{v_L}{v_S} \qquad (8.7)$$

Therefore, the effective voltage gain of the circuit can be found as

$$A_{v(eff)} = \frac{v_L}{v_S}$$

$$= \frac{4.14 \text{ V}_{ac}}{15 \text{ mV}_{ac}}$$

$$= 276$$

As you can see, the input and output circuits have reduced the voltage gain of this amplifier from 340 to an effective value of 276.

The effects that the input and output circuits have on the voltage gain of an amplifier can be significantly reduced by:

*How to reduce the effects of the input and output circuits on amplifier voltage gain.*

**1.** *Increasing* the value of $Z_{in}$

**2.** *Decreasing* the value of $Z_{out}$

For example, consider the circuit shown in Figure 8.9. This is essentially the same circuit as the one shown in Figure 8.8. However, for the sake of discussion, we have changed the values of $Z_{in}$ and $Z_{out}$. For this circuit:

$$v_{in} = v_S \frac{Z_{in}}{R_S + Z_{in}}$$

$$= (15 \text{ mV}_{ac}) \frac{8 \text{ k}\Omega}{8.02 \text{ k}\Omega}$$

$$= 14.96 \text{ mV}_{ac}$$

$$v_{out} = A_v v_{in}$$
$$= (340)(14.96 \text{ mV}_{ac})$$
$$= 5.1 \text{ V}_{ac}$$

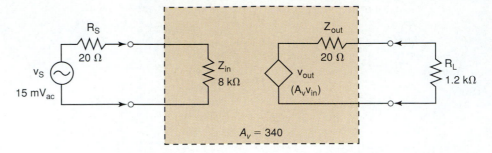

**FIGURE 8.9**

$$v_L = v_{out} \frac{R_L}{Z_{out} + R_L}$$

$$= (5.1 \text{ V}_{ac}) \frac{1.2 \text{ k}\Omega}{1.22 \text{ k}\Omega}$$

$$= 5 \text{ V}_{ac}$$

and the effective value of $A_v$ for the circuit can be found as

$$A_{v(eff)} = \frac{v_L}{v_S}$$

$$= \frac{5 \text{ V}_{ac}}{15 \text{ mV}_{ac}}$$

$$= 333.3$$

As you can see, the effective voltage gain of the circuit has increased significantly due to the changes made in the values of $Z_{in}$ and $Z_{out}$.

In most practical circuits, little can be done to change the resistance of a given signal source or load. However, the amplifier values of $Z_{in}$ and $Z_{out}$ can be affected by the choice of the active components used as well as the type of biasing circuit and component values. This point will be demonstrated further in upcoming chapters.

## The Ideal Voltage Amplifier

OBJECTIVE 5 ▶ Now that you have been introduced to the basic amplifier properties, we will look at the characteristics of the *ideal* voltage amplifier. The ideal voltage amplifier, if it could be constructed, would have the following characteristics (among others):

1. Infinite gain (if needed)
2. Infinite input impedance
3. Zero output impedance

The first of these characteristics needs little explanation. An ideal amplifier would be capable of producing any value of gain, no matter how high that value needed to be. In reality, values of $A_v$ are limited. The limit of $A_v$ depends in part of the type of active component(s) used in the circuit.

The impedance characteristics of the ideal voltage amplifier are illustrated in Figure 8.10. With infinite input impedance, there would be no current in the input circuit and, therefore, no voltage dropped across the source resistance ($R_S$). With no voltage dropped across $R_S$,

$$v_{in} = v_S \qquad \text{(for the ideal amplifier)}$$

In other words, the source voltage would not be reduced by the combined effects of $R_S$ and $Z_{in}$.

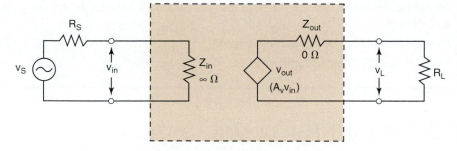

**FIGURE 8.10**

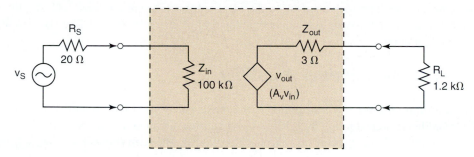

**FIGURE 8.11**

With zero output impedance, there would be no voltage divider in the output circuit of the amplifier. Therefore,

$$v_L = v_{out} \qquad \text{(for the ideal amplifier)}$$

Since there would be no reduction of voltage by either the input or output circuits, the effective voltage gain of the circuit would equal the calculated value of $A_v$.

Values of $Z_{in} = \infty\ \Omega$ and $Z_{out} = 0\ \Omega$ have not yet been achieved in practical circuits. However, it is possible to "effectively" achieve them through proper circuit design. For example, consider the circuit shown in Figure 8.11. When compared to the value of $R_S$ (20 $\Omega$), the value of $Z_{in} = 100\ k\Omega$ is, for all practical purposes, infinite. When compared to the value of $R_L$ (1.2 k$\Omega$), the value of $Z_{out} = 3\ \Omega$ is, for all practical purposes, zero. As you can see, the values of $Z_{in}$ and $Z_{out}$ would be a consideration in any circuit design since it is possible to minimize their effects on the voltage gain of the circuit.

In upcoming chapters, you will see that the ideal amplifier has other characteristics that relate to power dissipation, frequency response, and signal reproduction. At this point, however, we have established the primary characteristics of the ideal voltage amplifier.

**Section Review**

1. What is *amplification*?
2. What is *gain*? What are the three types of gain?
3. Define *voltage gain* as a ratio.
4. What determines the voltage gain of an amplifier?
5. What effect does the combination of source resistance ($R_S$) and amplifier input impedance ($Z_{in}$) have on the value of $v_{in}$ for an amplifier? What effect does it have on the value of $v_{out}$?
6. What effect does the combination of amplifier output impedance ($Z_{out}$) and load resistance ($R_L$) have on the value of $v_L$ for an amplifier?

**7.** Explain the overall effect that the input and output circuits have on the voltage gain of an amplifier.

**8.** How is the effect in Question 7 minimized?

**9.** List and explain the characteristics of the ideal voltage amplifier.

## 8.2 BJT AMPLIFIER CONFIGURATIONS

*OBJECTIVE 6* ▶ Now that we have established the gain, input impedance, and output impedance characteristics of the ideal voltage amplifier, we will take a brief look at several types of BJT amplifiers to see how they compare to the ideal amplifier. There are three BJT amplifier *configurations,* each having its unique combination of characteristics. These BJT amplifier configurations are shown in Figure 8.12.

### The Common-Emitter Amplifier

The **common-emitter (CE) amplifier** is the most widely used BJT amplifier. A typical CE amplifier is shown in Figure 8.12a. As you can see, the input is applied to the *base* of the transistor and the output is taken from the *collector.* The term *common-emitter* is used for two reasons:

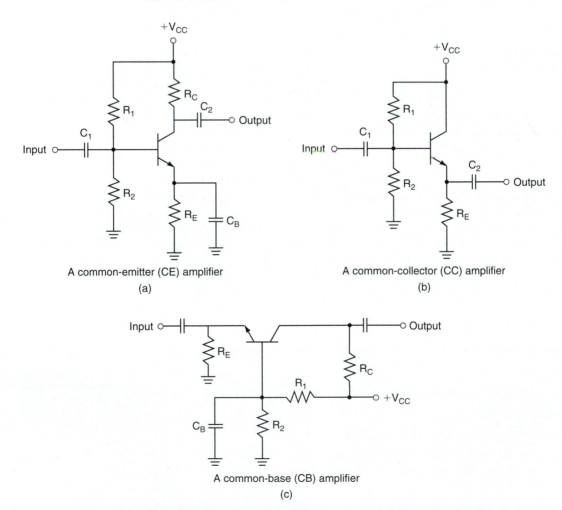

A common-emitter (CE) amplifier

(a)

A common-collector (CC) amplifier

(b)

A common-base (CB) amplifier

(c)

**FIGURE 8.12   BJT amplifier configurations.**

**TABLE 8.2   Property Ranges**

| Property | Low | Midrange | High |
|---|---|---|---|
| Gain | Less than 100 | 100–1000 | Greater than 1000 |
| Impedance | Less than 1 kΩ | 1 kΩ–10 kΩ | Greater than 10 kΩ |

1. The emitter terminal of the transistor is *common* to both the input and output circuits.

2. The emitter terminal of the transistor is normally set at 0 $V_{ac}$ and thus is at *ac ground* (or *ac common*). The ac ground is provided by the "bypass capacitor" ($C_B$) connected to the emitter terminal of the transistor. (The means by which this capacitor provides an ac ground will be discussed in Chapter 9.)

In order to discuss the characteristics of the CE amplifier (or any other), we need to establish some boundaries. For the sake of comparison, we'll classify gain and impedance values as being *low, midrange,* or *high*. These classifications are broken down as shown in Table 8.2. It should be noted that the ranges given are open to debate and should not be taken as gospel. They are used merely as a basis for comparison.

Using the ranges given in Table 8.2, we can classify the CE amplifier as an ampli-fier that typically has:

Typical CE amplifier characteristics.

1. Midrange values of voltage and current gain

2. High power gain

3. Midrange input impedance

4. Midrange output impedance

The CE amplifier is also unique among BJT amplifiers because it produces a 180° voltage phase shift from its input to its output, as shown in Figure 8.13. The basis of this voltage phase shift will be discussed in Chapter 9.

## The Common-Collector Amplifier

The **common-collector (CC) amplifier** is another widely used BJT amplifier. A typical CC amplifier is shown in Figure 8.12b. As you can see, the input is applied to the *base* of the transistor and the output is taken from the *emitter*. In this case, it is the *collector* ter-minal of the transistor that is part of both the input and output circuits and provides the ac ground (or common).

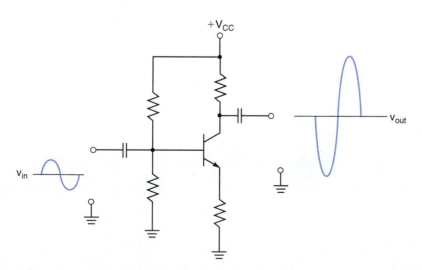

**FIGURE 8.13**

The CC amplifier typically has the following characteristics:

1. Midrange current gain
2. Extremely low voltage gain (slightly less than 1)
3. High input impedance
4. Low output impedance

The most unique characteristic here is the extremely *low voltage gain* of this configuration.

For reasons that are discussed in Chapter 10, the voltage gain of the CC amplifier is slightly less than *unity* (1). If we were to assume that $A_v = 1$ for a CC amplifier, the output waveform would be identical to the input waveform (as shown in Figure 8.14).

**Emitter follower**
Another name for the common-collector amplifier.

Since the ac signal at the emitter closely "follows" the ac voltage at the base, the CC amplifier is commonly referred to as an **emitter follower.** Note that this circuit is most commonly used for its current gain and impedance characteristics, as will be shown in Chapter 10.

## The Common-Base Amplifier

The **common-base (CB) amplifier** is the least often used BJT amplifier configuration. A typical CB amplifier is shown in Figure 8.12c. As you can see, the input is applied to the *emitter* of the transistor and the output is taken from the *collector.* The "bypass capacitor" ($C_B$) in the base circuit provides the ac ground (or common) at that terminal.

The CB amplifier typically has the following characteristics:

1. Midrange voltage gain
2. Extremely low current gain (slightly less than 1)
3. Low input impedance
4. High output impedance

If you compare the CB amplifier to the ideal voltage amplifier, you'll see immediately one of the reasons that it is rarely used. The low input impedance and high output impedance of the circuit are exact opposites of the impedance characteristics of the ideal voltage amplifier. In other words, its effective voltage gain will be nowhere near the ideal value of $A_v$.

The extremely low current gain of the CB amplifier is due to the fact that the input is applied to the emitter and the output is taken from the collector. Since collector current

**FIGURE 8.14**

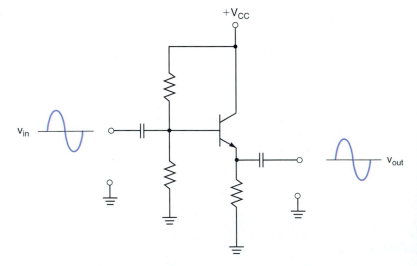

**TABLE 8.3    A Comparison of CE, CC, and CB Circuit Characteristics**

| Amplifier Type | $A_v$ | $A_i$ | $A_p$ | $Z_{in}$ | $Z_{out}$ |
|---|---|---|---|---|---|
| CE | Midrange | Midrange | High | Midrange | Midrange |
| CC | <1 | Midrange | $\cong A_i$ | High | Low |
| CB | Midrange | <1 | $\cong A_v$ | Low | High |

is always slightly less than emitter current, the value of $A_i$ for this circuit must be less than unity (1).

## Comparing the BJT Amplifier Configurations

For the sake of comparison, the gain and impedance characteristics of common-emitter, common-collector, and common-base amplifiers are listed in Table 8.3. There is an important point that should be made at this time. Note the power gain entries in Table 8.3 for the CC and CB amplifiers. These entries are based on the following relationship:

$$A_p = A_v A_i \qquad\qquad (8.8)$$

Just as power equals the product of voltage and current, power gain equals the product of $A_v$ and $A_i$. In the case of the CC amplifier, if we assume that $A_v = 1$, then power gain must equal current gain. In the case of the CB amplifier, if we assume that $A_i = 1$, then power gain must equal voltage gain.

If you compare the values given in Table 8.3, you'll see that CC and CB amplifiers are nearly opposites in terms of their gain and impedance characteristics. At the same time, the CE amplifier is sort of a "middle-of-the-road" circuit.

## Determining the Configuration of an Amplifier

There are many different ways to construct CE, CC, and CB amplifiers. The question is, When you see a BJT amplifier that you haven't seen before, how can you tell which configuration you are dealing with?    ◀ *OBJECTIVE 7*

The simplest way to determine the configuration of a given BJT amplifier is to use this technique: *Locate the input and output terminals. The third terminal is the common one.* For example, look at the circuit in Figure 8.15. The input is applied to the *emitter*

**FIGURE 8.15**

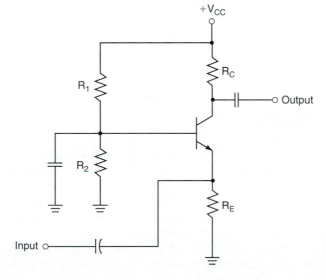

and the output is taken from the *collector*. The third terminal is the *base*. Therefore, the circuit is a *common-base amplifier*. The other two BJT amplifier configurations can be identified in the same fashion.

In Chapters 9 and 10, we will thoroughly analyze the amplifier configurations that have been introduced in this section. Among other things, you will be shown how to calculate the gain and impedance values that we have used to describe these circuits.

---

## Section Review

1. What is the basis for the term *common-emitter*?
2. What are the gain and impedance characteristics of CE amplifiers?
3. In terms of input and output ac voltages, how is the CE amplifier unique among the BJT amplifiers?
4. What is the basis for the term *common-collector*?
5. What are the gain and impedance characteristics of CC amplifiers?
6. What is another name commonly used for the CC amplifier? What is the basis of this name?
7. What is the basis for the term *common-base*?
8. What are the gain and impedance characteristics of CB amplifiers?
9. Write a brief comparison of CE, CC, and CB amplifiers.
10. How do you determine the configuration of a BJT amplifier you haven't seen before?

---

# 8.3 AMPLIFIER CLASSIFICATIONS

Some BJT amplifiers contain transistors that conduct during the entire cycle of the input signal. Others contain one or more transistors that conduct only during a portion of the input cycle. For example, consider the circuits shown in Figure 8.16. The transistor in Figure 8.16a conducts for the full 360° of the input cycle. When an amplifier contains a transistor that conducts during the entire cycle of the input, it is referred to as a **class A amplifier.** In contrast, the **class B amplifier** shown in Figure 8.16b contains two transistors that each conduct for approximately 180° of the input cycle. The **class C amplifier** shown in Figure 8.16c contains a single transistor that conducts for less than 180° of the input cycle.

In this section, we will look at each of these types of amplifiers and the ways in which they differ. First, however, we will look at some of the factors that determine which class of amplifier is used in a given application.

**Class A amplifier**
An amplifier with a single transistor that conducts during the entire input cycle.

**Class B amplifier**
An amplifier with two transistors that each conduct for approximately half of the input cycle.

**Class C amplifier**
An amplifier with one transistor that conducts for less than 180° of the input cycle.

## Amplifier Efficiency

OBJECTIVE 8 ▶

Amplifiers actually increase the power level of an ac input *by transferring power from the dc power supply to the input signal*. For example, consider the circuit shown in Figure 8.17. The input signal is shown to have a power rating of 1.5 mW. With a value of $A_p = 300$, the load power is shown to be 450 mW. The difference between $P_{in}$ and $P_{out}$ is 448.5 mW. Where did this power come from? The power was actually transferred by the amplifier from the dc power supply to the load.

The ideal amplifier would deliver 100 percent of the power it draws from the dc power supply to the load. In practice, however, this does not occur because the compo-

The *ideal* amplifier would be 100 percent efficient.

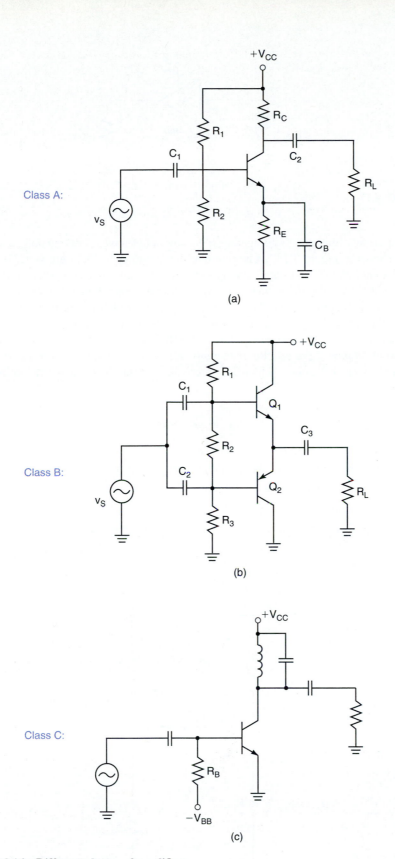

FIGURE 8.16   Different classes of amplifiers.

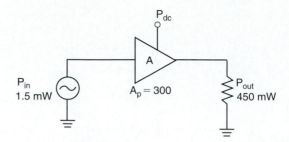

**FIGURE 8.17**

nents in the amplifier are all dissipating *some* amount of power. For example, consider the circuit shown in Figure 8.18. Assuming that the amplifier is operating normally, there is always some amount of current through each of the amplifier components. Since each component has a measurable value of current through it and a measurable voltage across it, each is dissipating power. This power dissipation (which is illustrated in the figure) reduces the amount of power available to be transferred to the load.

A *figure of merit* for any amplifier is its **efficiency.** The efficiency of an amplifier is *the amount of power drawn from the supply that is actually delivered to the load,* given as a percentage. Efficiency, which is represented by the Greek letter η (called *eta*), is found as

$$\eta = \frac{P_L}{P_{dc}} \times 100 \qquad\qquad (8.9)$$

where η = the efficiency of the amplifier, written as a percentage (%)

$P_L$ = the ac load power

$P_{dc}$ = the dc input power

Example 8.6 further illustrates the concept of amplifier efficiency.

**Efficiency (η)**
The amount of power drawn from the dc power supply that an amplifier actually delivers to its load, given as a percentage.

**FIGURE 8.18**

EXAMPLE **8.6**

An amplifier is continuously drawing 1.2 W from its dc power supply. If the ac load power for the circuit is 240 mW, what is the amplifier's efficiency rating?

*Solution:*   The efficiency rating of the amplifier is found as

$$\eta = \frac{P_L}{P_{dc}} \times 100$$

$$= \frac{240 \text{ mW}}{1.2 \text{ W}} \times 100$$

$$= 20\%$$

Thus, only 20 percent of the power drawn from the supply is actually delivered to the load. The other 80 percent is used by the amplifier itself.

**PRACTICE PROBLEM 8.6**

An amplifier is continuously drawing 3.3 W from its dc power supply. If the ac load power for the circuit is 450 mW, what is the efficiency rating of the circuit?

The higher the efficiency rating of an amplifier, the closer it comes to the ideal. As you will see, the maximum possible efficiency ratings for class A, class B, and class C amplifiers differ significantly.

## Distortion

One of the goals in amplification is to produce an output waveform that has the exact same *shape* as the input waveform. Ideally, every linear amplifier should be capable of producing a duplicate of any input waveform.

**Distortion** is defined as *any undesired change in the shape of a waveform*. The waveforms in Figure 8.19 illustrate several different types of distortion that can be produced by amplifiers. As upcoming chapters will show, each of these types of distortion is characteristic of one or more of the amplifier classes.

**Distortion**
Any undesired change in the shape of a waveform.

## Class A Amplifiers

Most of the amplifiers you will see in the field, whether they contain a BJT or some other active device, are class A amplifiers. Under normal operating conditions, a class A amplifier has:

◄   *OBJECTIVE 9*

1. An active device that conducts during the entire 360° of the input cycle
2. An output that contains little or no distortion
3. A maximum theoretical efficiency of 25 percent

Class A amplifier characteristics.

Class A operation is achieved in a BJT amplifier by midpoint biasing the transistor. As you may recall, midpoint bias allows the BJT output to vary widely around the Q-point without hitting saturation or cutoff, thus ensuring linear operation. This point is illustrated in Figure 8.20. As you can see, the output from the amplifier represented would

**FIGURE 8.19**

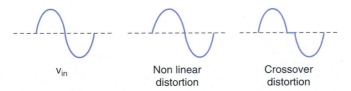

$v_{in}$

Non linear distortion

Crossover distortion

FIGURE 8.20

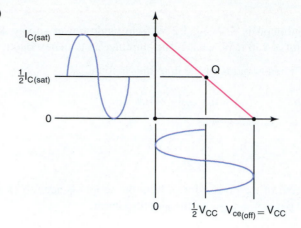

be a near-perfect reproduction of the input. Biasing the BJT above or below midpoint would result in one of two situations:

**1.** A reduction in the maximum possible amplifier output

**2.** Distortion

This point is discussed further in Chapter 11, along with the basis of the 25 percent maximum efficiency rating for the class A amplifier.

Because of their relatively poor efficiency ratings, class A amplifiers are generally used as *small-signal* (low-power) amplifiers. Small-signal amplifiers are primarily used to drive higher-power stages. One example of a higher-power stage is the audio output stage in a stereo system.

## Class B Amplifiers

The class B amplifier typically contains *two* transistors that are connected as shown in Figure 8.21. Each transistor in this type of amplifier conducts during one alternation of the ac input cycle and is in cutoff during the other alternation. With one transistor con-

**FIGURE 8.21   Class B amplifier operation.**

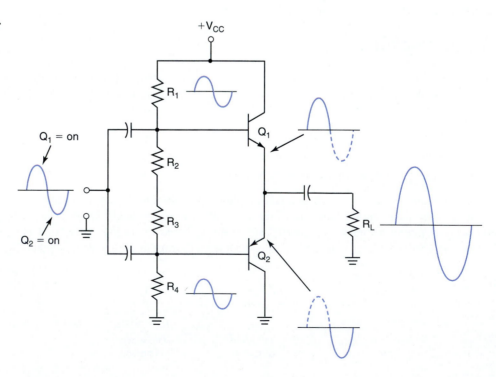

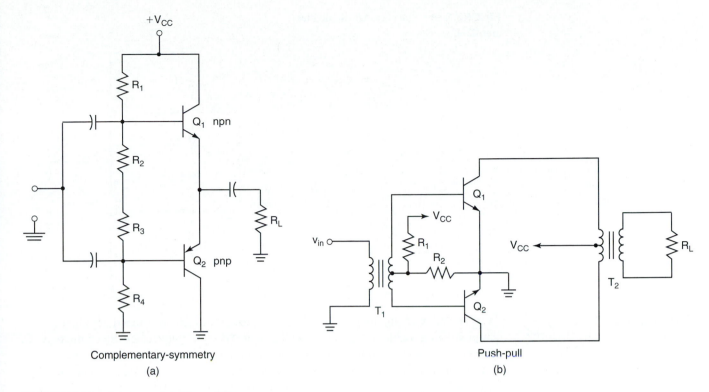

**FIGURE 8.22** **Common class B amplifiers.**

ducting during the negative alternation of the ac input and the other conducting during the positive alternation, a complete 360° output waveform is produced (as shown). Under normal operating conditions, a class B amplifier has:

Class B amplifier characteristics.

1. Two transistors that are biased at cutoff (each conducts during one alternation of the ac input cycle)
2. An output that contains little or no distortion
3. A maximum theoretical efficiency of approximately 78.5 percent

The two most common class B amplifiers are the *push-pull* amplifier and the *complementary-symmetry* amplifier shown in Figure 8.22. These amplifiers are discussed in detail in Chapter 11.

The relatively high efficiency rating of the class B amplifier makes it very useful as a higher-power amplifier. The audio output stage mentioned earlier is typically a class B amplifier.

Class AB operation.

One variation of the class B amplifier is the **class AB amplifier** shown in Figure 8.23. In the class AB amplifier, which is also known as a *diode-biased amplifier,* each transistor conducts for *slightly more than 180° of the ac input cycle.* As you will see in Chapter 11, this circuit is used to prevent a specific type of distortion that can be produced by a standard class B amplifier.

## Class C Amplifiers

The class C amplifier contains a single transistor that conducts for less than 180° of the ac input cycle. A typical BJT class C amplifier is shown in Figure 8.24.

The BJT in the class C amplifier is biased deeply into cutoff. The ac input to the amplifier causes the transistor to conduct for a brief time during the input cycle. The rest of the output waveform shown in Figure 8.24 is produced by the LC tank in the collector circuit of the amplifier.

FIGURE 8.23   The class AB (diode bias) amplifier.

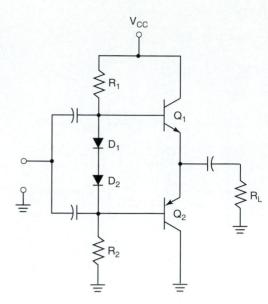

**Tuned amplifier**

An amplifier that produces a usable output over a specific range of frequencies.

The class C amplifier, by its design, is a **tuned amplifier**. A tuned amplifier is one that produces a usable output for a specific range of input frequencies. Since the class C amplifier is a tuned amplifier, coverage of this type of circuit is reserved for Chapter 17. At this point, we are interested only in establishing the class C amplifier as one that typically has:

Class C amplifier characteristics.

1. A single transistor that conducts for less than 180° of the ac input cycle
2. An output that may contain a significant amount of distortion
3. A maximum theoretical efficiency rating of approximately 99 percent

It would seem that the high efficiency rating of the class C amplifier would make it the ideal power amplifier. However, the distortion characteristics of the circuit limit its use.

## Summary

Most amplifiers fall into one of the classifications described in this section. The characteristics of class A, B, and C amplifiers are summarized in Figure 8.25.

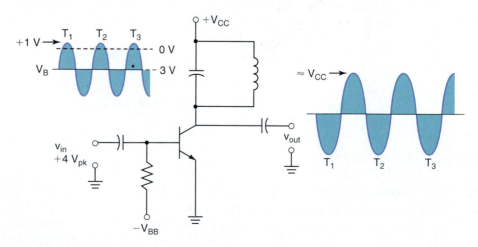

FIGURE 8.24   Class C operation.

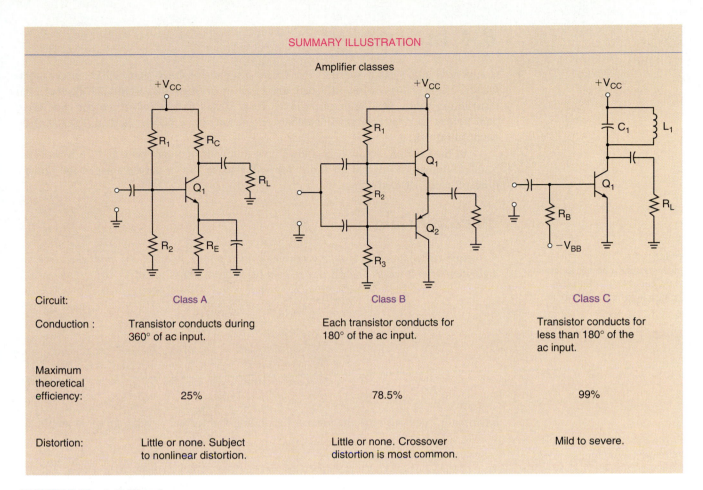

FIGURE 8.25    Amplifier classes.

1. How does an amplifier increase the power level of an ac input signal?
2. What is the *efficiency* rating of an amplifier?
3. What would the efficiency rating be for an ideal amplifier? Why can't this rating be achieved in practice?
4. What is *distortion*? Why is it undesirable?
5. What are the typical characteristics of a *class A* amplifier?
6. In terms of biasing, how is class A operation achieved? Explain your answer.
7. Why are class A amplifiers typically used as small-signal amplifiers?
8. What are the typical characteristics of a *class B* amplifier?
9. What is a *class AB* amplifier? Why is it used?
10. What are the typical characteristics of a *class C* amplifier?
11. How is the output of a class C amplifier produced?
12. What factor limits the use of the class C amplifier?

# 8.4 DECIBELS

OBJECTIVE 10 ▶ Component and system specification sheets often list voltage gain and/or power gain values. These values, when provided, are often in decibel (dB) form. The decibel is a logarithmic value. Writing numbers in dB form allows us to easily represent very large gain values as relatively small numbers. For example, a power gain of 1,000,000 is equal to 60 dB.

In this section, you will be shown how to covert numbers back and forth between standard numeric form and dB form. You will also be shown the advantages and limitations of using dB gain values.

## dB Power Gain

**Decibel**
A ratio of output power to input power, equal to 10 times the common logarithm of that ratio.

A **decibel (dB)** is *a ratio of output power to input power, equal to 10 times the common logarithm of that ratio*. The dB power gain of an amplifier is found as

$$A_{p(\text{dB})} = 10 \log A_p \tag{8.10}$$

or

$$A_{p(\text{dB})} = 10 \log \frac{P_{\text{out}}}{P_{\text{in}}} \tag{8.11}$$

Example 8.7 demonstrates the process for finding the decibel power gain of an amplifier.

---

**EXAMPLE 8.7**

A given amplifier has values of $P_{\text{in}} = 100 \ \mu\text{W}$ and $P_{\text{out}} = 2 \ \text{W}$. Calculate the dB power gain of the amplifier.

**Solution:** The dB power gain of the amplifier is found as

$$A_{p(\text{dB})} = 10 \log \frac{P_{\text{out}}}{P_{\text{in}}}$$

$$= 10 \log \frac{2 \ \text{W}}{100 \ \mu\text{W}}$$

$$= 10 \log (20,000)$$

$$= 43.01 \ \text{dB}$$

**PRACTICE PROBLEM 8.7**

An amplifier has values of $P_{\text{in}} = 420 \ \mu\text{W}$ and $P_{\text{out}} = 6 \ \text{W}$. What is the dB power gain of the amplifier?

---

When you need to convert a dB power gain value to standard numeric form, the following equation is used:

$$A_p = \log^{-1} \frac{A_{p(\text{dB})}}{10} \tag{8.12}$$

where $(\log^{-1})$ represents the *inverse log* of the fraction. Example 8.8 demonstrates the conversion of a dB power gain value to standard numeric form.

EXAMPLE 8.8

A given amplifier has a power gain of 3 dB. What is the ratio of output power to input power for the circuit?

*Solution:* The ratio of output power to input power is found as

$$A_p = \log^{-1} \frac{A_{p(dB)}}{10}$$

$$= \log^{-1} \frac{3 \text{ dB}}{10}$$

$$= 1.995$$

Thus, a gain of 3 dB represents a ratio of output power to input power that is approximately equal to 2; that is, the output power is approximately *twice* the value of the input power.

### PRACTICE POWER 8.8

An amplifier has a dB power gain of 20 dB. What is the ratio of output power to input power for the circuit?

When the input power and dB power gain of an amplifier are known, the output power is found by first converting the dB power gain to standard numeric form. Then the output power is found as the product of $A_p$ and $P_{in}$. Example 8.9 demonstrates this process.

EXAMPLE 8.9

A given amplifier has values of $P_{in} = 50$ mW and $A_{p(dB)} = 3$ dB. Calculate the amplifier output power.

*Solution:* First, the value of 3 dB is converted to standard numeric form as follows:

$$A_p = \log^{-1} \frac{A_{p(dB)}}{10}$$

$$= \log^{-1} \frac{3 \text{ dB}}{10}$$

$$= 2$$

Now, using $A_p = 2$, the value of the output power is found as

$$P_{out} = A_p P_{in}$$
$$= (2)(50 \text{ mW})$$
$$= 100 \text{ mW}$$

### PRACTICE PROBLEM 8.9

An amplifier has values of $A_{p(dB)} = 5.2$ dB and $P_{in} = 40$ μW. Calculate the output power for the circuit.

The standard numeric form of 3 dB was rounded off to 2 in Example 8.9. When working with power gain values, it is standard practice to assume that 3 dB = 2.

When output power and dB power gain are known, the input power is found as shown in Example 8.10. Note the similarity between this procedure and the one shown in Example 8.9.

EXAMPLE **8.10**

A circuit has values of $A_{p(dB)} = -3$ dB and $P_{out} = 50$ mW. Calculate the value of the circuit input power.

***Solution:*** First, we must convert the value of $-3$ dB to standard numeric form, as follows:

$$A_p = \log^{-1} \frac{A_{p(dB)}}{10}$$

$$= \log^{-1} \frac{-3 \text{ dB}}{10}$$

$$= 0.5$$

Now, the value of the input power is found as:

$$P_{in} = \frac{P_{out}}{A_p}$$

$$= \frac{50 \text{ mW}}{0.5}$$

$$= 100 \text{ mW}$$

### PRACTICE PROBLEM 8.10

An amplifier with a power gain of 12.4 dB has an output power of 2.2 W. Calculate the value of $P_{in}$ for the circuit.

If we take a moment to compare the results of Examples 8.9 and 8.10, we can come to some very important conclusions regarding the use of dB values. For convenience, the results for the two examples are listed here:

| Example | dB Power Gain | $P_{in}$ (mW) | $P_{out}$ (mW) |
|---------|---------------|---------------|----------------|
| 8.9     | 3 dB          | 50 mW         | 100 mW         |
| 8.10    | −3 dB         | 100 mW        | 50 mW          |

The conclusions that we can draw from these results are as follows:

Positive versus negative dB values.

1. *Positive* dB values represent a power *gain*, while *negative* dB values represent a power *loss*. When the gain was +3 dB, the output power was greater than the input power. When the gain was −3 dB, the output power was actually less than the input power.

2. Positive and negative decibels of equal value represent gains and losses of equal value. A +3 dB gain caused power to double while a −3 dB gain caused power to be cut in half.

The relationship described in conclusion (2) can be stated mathematically as follows:

*If x dB represents a power gain of y, then*
*−x dB represents a lower loss equal to 1/y.*

Table 8.4 lists some examples of this relationship.

The fact that equivalent (+) and (−) dB values represent gains and losses that are reciprocals is one of the reasons that dBs are commonly used. Granted, they can be a bit

**FIGURE 8.26**

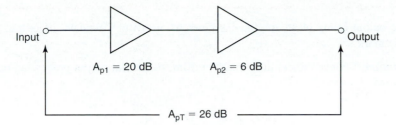

**TABLE 8.4  Some Examples of Equal Positive and Negative dB Values**

| +db Value | Gain | −dB Value | Loss |
|-----------|------|-----------|------|
| 3 | 2 | −3 | ½ |
| 6 | 4 | −6 | ¼ |
| 12 | 16 | −12 | 1/16 |
| 20 | 100 | −20 | 1/100 |

intimidating at first. However, once you get used to seeing gains represented as dB values, you will have no problem using them.

Another advantage of using dB gain values is as follows: In multistage amplifiers, the total gain of the circuit is equal to the *sum* of the individual amplifier dB gains. For example, let's say that two amplifiers are connected as shown in Figure 8.26. The total power gain for the circuit is found as 20 dB + 6 dB = 26 dB. The basis of this relationship is discussed in Chapter 9.

## The dBm Reference

On some specification sheets, you will see power values that are listed as *dBm* values. For example, a stereo may be listed as having a maximum output power of 50 dBm. This rating tells you that the maximum output power from the stereo is *50 dB above 1 mW*. Power, measured in dBm, is found as

$$P_{\text{dBm}} = 10 \log \frac{P}{1 \text{ mW}} \tag{8.13}$$

Note that dBm values represent *actual power levels*. In contrast, dB values represent power *ratios*. This point is illustrated in Examples 8.11 and 8.12.

---

An amplifier has a rating of $A_{p(\text{dB})} = 50$ dB. Calculate the output power of the amplifier.

**Solution:**  The problem cannot be solved as given. The value of 50 dB represents the *ratio* of output power to input power. Since no input power value was given, we cannot determine the value of the output power.

*EXAMPLE 8.11*

---

EXAMPLE 8.12

The output rating of an amplifier is given as 50 dBm. Calculate the output power for the circuit.

**Solution:** Since the circuit is rated in dBm, the actual output power can be found as follows:

$$A_p = \log^{-1} \frac{50 \text{ dB}}{10}$$
$$= 1 \times 10^5$$

and

$$P_{out} = A_p (1 \text{ mW})$$
$$= (1 \times 10^5)(1 \text{ mW})$$
$$= 100 \text{ W}$$

We have solved the problem by converting the dBm rating to a value of gain and then multiplying that gain by 1 mW. This is the standard approach to this problem.

**PRACTICE PROBLEM 8.12**

An amplifier has an output power rating of 32 dBm. Determine the actual output power from the circuit.

As you can see, dBm values can be used to determine actual power values, while dB values simply indicate the *ratio* of output power to input power.

## dB Voltage Gain

The dB voltage gain of an amplifier is found as

$$A_{v(dB)} = 20 \log A_v \tag{8.14}$$

or

$$A_{v(dB)} = 20 \log \frac{v_{out}}{v_{in}} \tag{8.15}$$

Equation (8.15) is derived using Ohm's law. As you know, $P = V^2/R$. Therefore,

$$P_{out} = \frac{v_{out}^2}{R_{out}} \quad \text{and} \quad P_{in} = \frac{v_{in}^2}{R_{in}}$$

If we use these equations in place of $P_{in}$ and $P_{out}$, equation (8.11) can be modified as follows:

$$\text{Gain (dB)} = 10 \log \frac{P_{out}}{P_{in}}$$

$$= 10 \log \frac{v_{out}^2/R_{out}}{v_{in}^2/R_{in}}$$

If we assume that $R_{\text{out}} = R_{\text{in}}$, then

$$\text{Gain (dB)} = 10 \log \frac{v_{\text{out}}^2}{v_{\text{in}}^2}$$

$$= 10 \log \left(\frac{v_{\text{out}}}{v_{\text{in}}}\right)^2$$

$$= 20 \log \frac{v_{\text{out}}}{v_{\text{in}}}$$

Thus, when dealing with dB voltage gain conversions, a multiplier of *20* is used (in place of the *10*). Example 8.13 demonstrates the process for calculating dB voltage gain.

---

An amplifier has values of $v_{\text{in}} = 25 \text{ mV}_{\text{ac}}$ and $v_{\text{out}} = 2 \text{ V}_{\text{ac}}$. Calculate the dB voltage gain of the circuit.

**EXAMPLE 8.13**

*Solution:* Using equation (8.14) and the values given in the problem, the dB voltage gain of the amplifier is found as

$$A_{v(\text{dB})} = 20 \log \frac{v_{\text{out}}}{v_{\text{in}}}$$

$$= 20 \log \frac{2 \text{ V}_{\text{ac}}}{25 \text{ mV}_{\text{ac}}}$$

$$= 38.1 \text{ dB}$$

**PRACTICE PROBLEM 8.13**

An amplifier with a 25 mV$_{\text{ac}}$ input has a 7.8 V$_{\text{ac}}$ output. Calculate the dB voltage gain of the circuit.

---

When you want to convert a dB voltage gain value to standard numeric form, the following equation is used:

$$A_v = \log^{-1} \frac{A_{v(\text{dB})}}{20} \qquad (8.16)$$

Again, note the change in the denominator of the equation. Example 8.14 demonstrates the use of this equation.

---

A given amplifier has a voltage gain of 6 dB. What is the ratio of output voltage to input voltage for the circuit?

**EXAMPLE 8.14**

*Solution:* Using equation (8.16), the ratio of output voltage to input voltage is found as

$$A_v = \log^{-1} \frac{A_{v(\text{dB})}}{20}$$

$$= \log^{-1} \frac{6 \text{ dB}}{20}$$

$$= 2$$

A given amplifier has a voltage gain of 12 dB. What is the ratio of output voltage to input voltage?

As is the case with dB power values, negative dB voltage values indicate a voltage *loss*. This point is illustrated in Example 8.15.

**EXAMPLE 8.15**

A circuit has a dB voltage gain of −6 dB. What is the ratio of output voltage to input voltage for the circuit?

*Solution:* Using equation (8.16), the ratio of output voltage to input voltage is found as

$$A_v = \log^{-1} \frac{A_{v(dB)}}{20}$$

$$= \log^{-1} \frac{6 \text{ dB}}{20}$$

$$= 0.5$$

**PRACTICE PROBLEM 8.15**

A circuit has a voltage gain of −12 dB. What is the ratio of output voltage to input voltage for the circuit?

Do you see a pattern here? All the rules for dB power gains apply to dB voltage gains as well. The only difference is that dB voltage gain calculations use a constant of 20 instead of 10.

## Changes in dB Gain

When the voltage gain of an amplifier changes by a given number of decibels, the power gain of the amplifier changes by the same number of decibels. By formula,

$$\Delta A_{p(dB)} = \Delta A_{v(dB)} \tag{8.17}$$

For example, if the voltage gain of an amplifier changes by −3 dB, the power gain of the amplifier also changes by −3 dB. The values of $A_p$ and $A_v$ for the circuit may or may not be equal, but they will *change* by the same number of decibels. This point will become important when we study the frequency response characteristics of amplifiers in Chapter 14.

## One Final Note on Decibels

Experience has shown that most people take a while to get used to working with dB gain values. Although the whole concept of decibels may seem a bit strange at first, rest assured that everything will "click" for you eventually. In the meantime, the following list of summary points on decibels will help. Whenever you are not sure about the meaning of a given set of dB values, refer back to this list:

Decibel (dB) characteristics.

1. Decibels are *logarithmic* representations of gain values.
2. Decibel power gain is found as $10 \log A_p$.
3. Decibel voltage gain is found as $20 \log A_v$.
4. When $A_v$ changes by a given number of decibels, $A_p$ changes by the same number of decibels.
5. You *cannot* use dB voltage and power gain values as multipliers. For example, if you want to determine $v_{out}$, given $v_{in}$ and $A_{v(dB)}$, you must convert $A_{v(dB)}$ to standard numeric form before multiplying to find $v_{out}$.

There are an extensive number of practice problems involving decibels at the end of this chapter.

---

## Section Review

1. What is a *decibel*?
2. Why are power gain values often written in dB form?
3. What is the primary restriction on using dB power gain values?
4. What is the relationship between positive and negative dB values?
5. What is the dBm reference? How does it differ from dB power gain?
6. What is the primary difference between dB voltage calculations and dB power calculations?
7. What is the relationship between changes in dB voltage gain and dB power gain?

---

## Chapter Summary

*Amplification* is the process of increasing the power level of an ac signal. The circuits used to perform this function are referred to as *amplifiers*.

Amplifiers are widely used circuits that exhibit a property called *gain*. The gain of an amplifier is a multiplier that exists between the input and output of the circuit. There are three types of gain:

1. Current gain ($A_i$)
2. Voltage gain ($A_v$)
3. Power gain ($A_p$)

The actual values of $A_i$, $A_v$, and $A_p$ for a given amplifier depend on the type of amplifier you are dealing with and the circuit component values.

The *general model* of a voltage amplifier represents the circuit as one with input impedance, output impedance, and an internal voltage source. The general model of a voltage amplifier (shown in Figure 8.4) is used to illustrate the relationship between an amplifier, its source, and its load.

The *ideal voltage amplifier* has the following characteristics:

1. Infinite gain (if needed)
2. Infinite input impedance
3. Zero output impedance
4. 100 percent efficiency
5. Little or no distortion of the input signal

Only the first three of these characteristics are actually illustrated in the general amplifier model.

There are three BJT amplifier configurations, each with its own characteristics. The characteristics of the *common-emitter (CE)*, *common-collector (CC)*, and *common-base (CB)* amplifiers were summarized in Table 8.3. Of the three, the CE amplifier is the only one that produces a 180° voltage phase shift between its input and output. The input and output voltage waveforms for the other two BJT configurations are in phase.

BJT amplifiers generally fall into one of three classes. The *class A* amplifier contains a transistor that conducts during the entire 360° of the ac input cycle. The *class B* amplifier contains two transistors that conduct during alternate half-cycles of the ac input. Together, they reproduce the entire 360° cycle of the input. The *class C* amplifier contains a single transistor that conducts for less than 180° of the input cycle. The remainder of the ac is reproduced at the output by the action of an LC tank in the collector circuit of the amplifier.

Each amplifier class has its advantages and limitations. The class A amplifier produces the lowest amount of distortion but also has the worst efficiency rating of the three. The class B amplifier has a much better maximum efficiency rating but requires the use of two transistors. Another disadvantage of the class B amplifier is that it does not provide voltage gain because the transistors are used in an emitter-follower configuration. The class C amplifier has the highest overall efficiency rating. However, this type of amplifier can severely distort its input signal.

When dealing with circuit and system specification sheets, gain values are often given in *decibel (dB)* form. The dB gain of a circuit is a *logarithmic* value. Power gain, in dB, is found as (10 log $A_p$). Voltage gain, in dB, is found as (20 log $A_v$). When dealing with dB values, they must be converted into standard numeric form before they are used in any circuit calculations.

---

## Key Terms

The following terms were introduced and defined in this chapter:

| | | |
|---|---|---|
| amplification | common-collector (CC) | efficiency (η) |
| amplifier | amplifier | emitter follower |
| amplifier input impedance | common-emitter (CE) | gain |
| amplifier output | amplifier | general amplifier model |
| impedance | current gain ($A_i$) | ideal voltage amplifier |
| class A amplifier | dBm reference | output impedance |
| class AB amplifier | dB power gain | power gain ($A_p$) |
| class B amplifier | dB voltage gain | tuned amplifier |
| class C amplifier | decibel (dB) | voltage gain ($A_v$) |
| common-base (CB) | distortion | |
| amplifier | | |

---

## Equation Summary

| Equation Number | Equation | Section Number |
|---|---|---|
| **(8.1)** | $A_v = \dfrac{v_{\text{out}}}{v_{\text{in}}}$ | 8.1 |
| **(8.2)** | $v_{\text{out}} = A_v v_{\text{in}}$ | 8.1 |
| **(8.3)** | $A_i = \dfrac{i_{\text{out}}}{i_{\text{in}}}$ | 8.1 |
| **(8.4)** | $A_p = \dfrac{P_{\text{out}}}{P_{\text{in}}}$ | 8.1 |

| Equation Number | Equation | Section Number |
|---|---|---|
| (8.5) | $v_{in} = v_S \dfrac{Z_{in}}{R_s + Z_{in}}$ | 8.1 |
| (8.6) | $v_L = v_{out} \dfrac{R_L}{Z_{out} + R_L}$ | 8.1 |
| (8.7) | $A_{v(eff)} = \dfrac{v_L}{v_S}$ | 8.1 |
| (8.8) | $A_p = A_v A_i$ | 8.2 |
| (8.9) | $\eta = \dfrac{P_L}{P_{dc}} \times 100$ | 8.3 |
| (8.10) | $A_{p(dB)} = 10 \log A_p$ | 8.4 |
| (8.11) | $A_{p(dB)} = 10 \log \dfrac{P_{out}}{P_{in}}$ | 8.4 |
| (8.12) | $A_p = \log^{-1} \dfrac{A_{p(dB)}}{10}$ | 8.4 |
| (8.13) | $P_{dBm} = 10 \log \dfrac{P}{1\,mW}$ | 8.4 |
| (8.14) | $A_{v(dB)} = 20 \log A_v$ | 8.4 |
| (8.15) | $A_{v(dB)} = 20 \log \dfrac{v_{out}}{v_{in}}$ | 8.4 |
| (8.16) | $A_v = \log^{-1} \dfrac{A_{v(dB)}}{20}$ | 8.4 |
| (8.17) | $\Delta A_{p(dB)} = \Delta A_{v(dB)}$ | 8.4 |

**8.1.** 240
**8.2.** 86.4 mV$_{ac}$
**8.3.** 731.7 μV$_{ac}$
**8.4.** $v_{out}$ = 351 mV$_{ac}$ when $v_{in}$ = 731.7 μV$_{ac}$
     $v_{out}$ = 384 mV$_{ac}$ when $v_{in}$ = 800 μV$_{ac}$
**8.5.** 413.8 mV$_{ac}$
**8.6.** 13.6%
**8.7.** 41.6 dB
**8.8.** 100
**8.9.** 132.45 μW
**8.10.** 126.6 mW
**8.12.** 1.59 W
**8.13.** 49.9 dB
**8.14.** 3.98
**8.15.** 0.25

*Answers to the*
*Example Practice*
*Problems*

# Practice Problems

## §8.1

**1.** Complete the table below.

| $v_{in}$ | $v_{out}$ | $A_v$ |
|---|---|---|
| 1.2 mV$_{ac}$ | 300 mV$_{ac}$ | _____ |
| 200 μV$_{ac}$ | 18 mV$_{ac}$ | _____ |
| 24 mV$_{ac}$ | 2.2 V$_{ac}$ | _____ |
| 800 μV$_{ac}$ | 140 mV$_{ac}$ | _____ |

**2.** Complete the table below.

| $v_{in}$ | $v_{out}$ | $A_v$ |
|---|---|---|
| 38 mV$_{ac}$ | 600 mV$_{ac}$ | _____ |
| 6 mV$_{ac}$ | 9.2 V$_{ac}$ | _____ |
| 500 μV$_{ac}$ | 88 mV$_{ac}$ | _____ |
| 48 mV$_{ac}$ | 48 mV$_{ac}$ | _____ |

**3.** For each combination of $v_{in}$ and $A_v$, calculate the value of $v_{out}$.

| $v_{in}$ | $A_v$ | $v_{out}$ |
|---|---|---|
| 240 μV$_{ac}$ | 540 | _____ |
| 1.4 mV$_{ac}$ | 300 | _____ |
| 24 mV$_{ac}$ | 440 | _____ |
| 800 μV$_{ac}$ | 720 | _____ |

**4.** Calculate the value of $v_{out}$ for each of the amplifiers shown in Figure 8.27.

**5.** An amplifier has values of $v_S = 600$ μV$_{ac}$, $Z_{in} = 1.2$ kΩ, and $R_S = 180$ Ω. Calculate the value of $v_{in}$ for the circuit.

**6.** An amplifier has values of $v_S = 18$ mV$_{ac}$, $Z_{in} = 720$ Ω, and $R_S = 60$ Ω. Calculate the value of $v_{in}$ for the circuit.

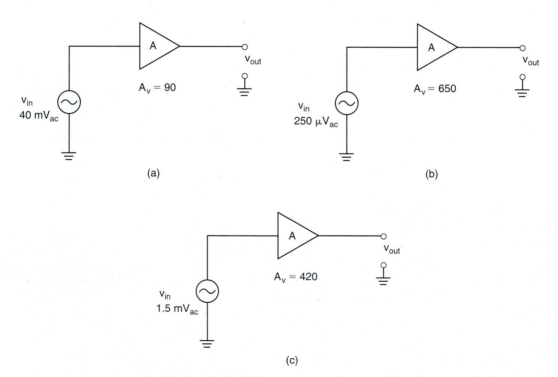

(a)

(b)

(c)

**FIGURE 8.27**

FIGURE 8.28

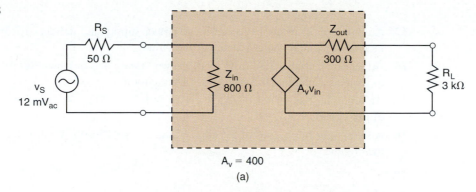

$A_v = 400$

(a)

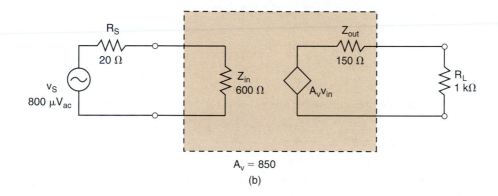

$A_v = 850$

(b)

7. An amplifier has values of $v_{out} = 8.8\ V_{ac}$, $Z_{out} = 120\ \Omega$, and $R_L = 3\ k\Omega$. Calculate the value of $v_L$ for the circuit.

8. An amplifier has values of $v_{out} = 12\ V_{ac}$, $Z_{out} = 80\ \Omega$, and $R_L = 5.2\ k\Omega$. Calculate the value of $v_L$ for the circuit.

9. Calculate the *effective* voltage gain of the amplifier in Figure 8.28a.

10. Calculate the *effective* voltage gain of the amplifier in Figure 8.28b.

11. Calculate the *effective* voltage gain of the amplifier in Figure 8.29a.

12. Calculate the *effective* voltage gain of the amplifier in Figure 8.29b.

FIGURE 8.29

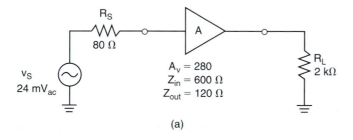

$A_v = 280$
$Z_{in} = 600\ \Omega$
$Z_{out} = 120\ \Omega$

(a)

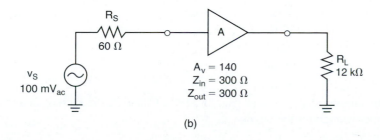

$A_v = 140$
$Z_{in} = 300\ \Omega$
$Z_{out} = 300\ \Omega$

(b)

13. An amplifier draws 8.8 W from its dc power supply and delivers 1.2 W of power to its load. Calculate the efficiency of the circuit.

14. An amplifier draws 2 W from its dc power supply and delivers 300 mW of power to its load. Calculate the efficiency of the circuit.

15. An amplifier has values of $P_{dc} = 1.4$ W and $P_L = 200$ mW. What is the efficiency of the circuit?

16. An amplifier has values of $P_{dc} = 3$ W and $P_L = 115$ mW. What is the efficiency of the circuit?

17. Complete the table below.

| $P_{in}$ | $P_{out}$ | $A_{p(dB)}$ |
|---|---|---|
| 10 mW | 5 W | _____ |
| 100 mW | 1 W | _____ |
| 1.25 mW | 4 W | _____ |
| 12 W | 6 W | _____ |

18. Complete the table below.

| $P_{in}$ | $P_{out}$ | $A_{p(dB)}$ |
|---|---|---|
| 500 μW | 1 W | _____ |
| 22 W | 1.4 W | _____ |
| 33 mW | 2.64 W | _____ |
| 120 mW | 2.2 W | _____ |

19. An amplifier has a power gain of 48 dB. What is the value of $A_p$ for the circuit?

20. An amplifier has a power gain of −18 dB. What is the value of $A_p$ for the circuit?

21. An amplifier has values of $A_{p(dB)} = 12$ dB and $P_{in} = 30$ mW. Calculate the output power for the circuit.

22. An amplifier has values of $A_{p(dB)} = 3$ dB and $P_{in} = 400$ mW. Calculate the output power for the circuit.

23. A circuit has values of $A_{p(dB)} = -12$ dB and $P_{in} = 800$ mW. Calculate the output power for the circuit.

24. A circuit has values of $A_{p(dB)} = -6$ dB and $P_{in} = 1.2$ mW. Calculate the output power for the circuit.

25. A circuit has values of $A_{p(dB)} = 5$ dB and $P_{out} = 2.2$ W. Calculate the value of $P_{in}$ for the circuit.

26. A circuit has values of $A_{p(dB)} = -12$ dB and $P_{out} = 600$ μW. Calculate the value of $P_{in}$ for the circuit.

27. The output of a circuit is rated at 40 dBm. Calculate the circuit output power.

28. The output of a circuit is rated at 22 dBm. Calculate the circuit output power.

29. The output of a circuit is rated at −1.2 dBm. Calculate the circuit output power.

30. The output of a circuit is rated at 60 dBm. Calculate the circuit output power.

31. Complete the table below.

| $v_{out}$ | $v_{in}$ | $A_{v(dB)}$ |
|---|---|---|
| 3.6 $V_{ac}$ | 120 m$V_{ac}$ | _____ |
| 800 m$V_{ac}$ | 50 μ$V_{ac}$ | _____ |
| 14 $V_{ac}$ | 200 m$V_{ac}$ | _____ |
| 300 m$V_{ac}$ | 150 m$V_{ac}$ | _____ |

32. An amplifier has a voltage gain of 22 dB. Calculate the ratio of output voltage to input voltage for the circuit

**33.** An amplifier has a voltage gain of 12 dB. Calculate the ratio of output voltage to input voltage for the circuit.

**34.** A circuit has a voltage gain of −3 dB. Calculate the ratio of output voltage to input voltage for the circuit.

**35.** A circuit has a voltage gain of −14 dB. Calculate the ratio of output voltage to input voltage for the circuit.

---

**36.** Refer to the circuit in Figure 8.30. Calculate the *effective* dB voltage gain of the circuit and the output power in dBm.

**37.** Calculate the *effective* dB power gain of the circuit in Figure 8.31.

**FIGURE 8.30**

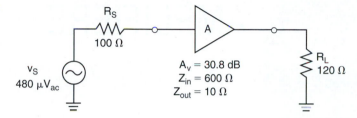

**FIGURE 8.31**

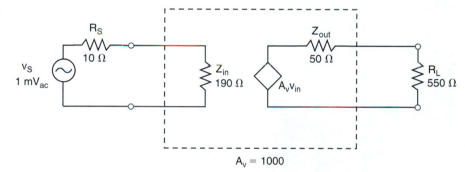

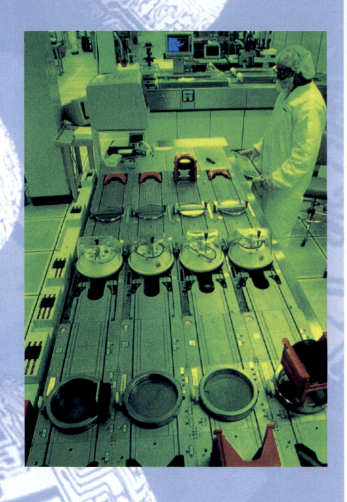

# 9

## COMMON-EMITTER AMPLIFIERS

The vignette in this chapter describes "Mr. Meticulous," the first automated transistor manufacturing machine. As you can see, we've come a long way since the days of Mr. Meticulous.

## OUTLINE

## OBJECTIVES

*After studying the material in this chapter, you should be able to:*

1. Describe *gain* and list the types of gain associated with each of the three transistor configurations.
2. Describe the input/output voltage and current phase relationships of the three transistor amplifier configurations.
3. Calculate the ac emitter resistance of a transistor in a given amplifier.
4. List the two primary roles that capacitors serve in amplifiers.
5. Derive the ac equivalent circuit for a given amplifier.
6. Explain why the voltage gain of a common-emitter amplifier may be unstable.
7. Calculate the output power for a common-emitter amplifier, given the amplifier values of $A_v$, $A_i$, and $P_{in}$.
8. Discuss the relationship between the load resistance and voltage gain of a common-emitter amplifier.
9. Calculate the values of $Z_{in}$ and $Z_{in(base)}$ for a common-emitter amplifier.
10. Discuss the effects of *swamping* on the ac characteristics of a common-emitter amplifier.
11. List and describe the four ac *h*-parameters.
12. Troubleshoot a multistage common-emitter amplifier to determine which amplifier stage is faulty.

# MODERN (?) TRANSISTOR MANUFACTURING

Here is a bit of history that you might enjoy. It has to do with the first automated transistor manufacturing machine, *Mr. Meticulous*. (I'm not kidding, folks . . . that was its *real* name.)

The development of Mr. Meticulous is accredited to R. L. Wallace, Jr. Shortly after the debut run of this machine, the *Bell Laboratories Record* (March 1955) had this to say about Mr. Meticulous:

When fashioned by human hands over any extended period of time, some transistors are produced which are substandard and useless for research purposes. But the new machine, familiarly known as "Mr. Meticulous," never gets tired, never loses his precision or accuracy. His hand never shakes and his highly organized electronic "brain" rarely has mental lapses. The machine, originated by R. L. Wallace, Jr., may someday be a pilot model for industrial machines to be used in assembly-line transistor manufacture.

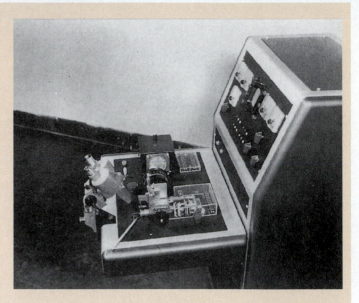

In Chapter 8, you were introduced to the basic principles of amplifier ac operation. At this point, we will turn our attention to the ac operation of *common-emitter* amplifiers. Then, in Chapter 10, we will look at both *common-collector* and *common-base* amplifiers.

## *9.1* ac CONCEPTS

You have already been introduced to many of the ac properties of amplifiers, including *gain*, *amplifier input impedance*, and *amplifier output impedance*. In this section, we will look at some ac concepts that relate specifically to the common-emitter amplifier.

### *Amplifier Gain*

OBJECTIVE 1 ▶ As you know, there are three types of gain: *current gain* ($A_i$), *voltage gain* ($A_v$), and *power gain* ($A_p$). Under normal circumstances, the common-emitter amplifier exhibits all three types of gain. In contrast, the common-collector amplifier has only current and power gain, and the common-base amplifier has only voltage and power gain. The choice of amplifier configuration often depends on the type(s) of gain desired. Since the common-emitter amplifier can be used to provide any type of gain, it is the most often used of the BJT amplifiers.

### *Input/Output Phase Relationships*

OBJECTIVE 2 ▶ For every amplifier type, *the input and output currents are in phase*. When the input current increases, the output current increases. When the input current decreases, the output current decreases.

**Common-emitter amplifier**
The only BJT amplifier with a 180° voltage phase shift from input to output.

The **common-emitter amplifier** is unique because the input and output voltages are 180° out of phase, even though the input and output currents are in phase. The cause of this input/output phase relationship is illustrated in Figure 9.1. The amplifier shown has

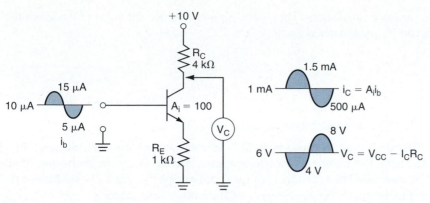

**FIGURE 9.1** Common-emitter phase relationship: current and voltage relationship.

**Lab Reference:** The phase relationship between CE input and output signals is demonstrated in Exercise 17.

a current gain ($A_i$) of 100. This means that the ac output current is 100 times the ac input current. The input current varies by 5 $\mu$A both above and below a 10 $\mu$A dc level. Multiplying the input current values by the $Ai$ of the amplifier gives us the output values shown. When the input current is at 10 $\mu$A, the output current is at $(10\ \mu A)(100) = 1$ mA. When the input current is at 15 $\mu$A, the output current is at $(15\ \mu A)(100) = 1.5$ mA, and so on. *Note that the input current and the output currents are in phase.*

To understand the voltage relationship, two points need to be considered:

1. The output of the circuit is taken from the collector and is measured with respect to ground. Therefore,

$$V_{\text{out}} = V_C \qquad (9.1)$$

2. The collector voltage is equal to the supply voltage minus the drop across the collector resistor. This relationship has been expressed already as

$$V_C = V_{CC} - I_C R_C$$

If you calculate the value of $V_C$ for each of the values of $I_C$ indicated, you will see that *$V_C$ decreases as $I_C$ increases, and vice versa.* When $I_C$ increases to 1.5 mA, $V_C$ *decreases* from 6 V to 4 V. As $I_C$ decreases from 1.5 mA to 500 $\mu$A, $V_C$ increases from 4 V to 8 V. The phase relationship is due strictly to the fact that $I_C$ and $V_C$ are inversely proportional.

Now that you have seen the relationship between $I_C$ and $V_C$, the basis of the input/output voltage phase relationship can be given as follows:

1. Input voltage and current are in phase.
2. Input current and output current are in phase. Therefore, input voltage and output current are in phase.
3. Output current is 180° out of phase with output voltage ($V_C$). Therefore, *input voltage and output voltage are 180° out of phase.*

Remember that the common-emitter amplifier is the only configuration that has this input/output voltage phase relationship. For both the common-base and common-collector amplifiers, the input and output voltages are in phase. For all three amplifier configurations, input and output currents are in phase.

## ac Emitter Resistance

You may recall that the zener diode has a *dynamic* resistance value that is considered only in ac calculations. The same holds true for the emitter-base junction of a transistor. This junction has a dynamic resistance, called **ac emitter resistance**, that is used in some gain

**ac emitter resistance ($r'_e$)**
The dynamic resistance of the transistor emitter, used in voltage gain and input impedance calculations.

OBJECTIVE 3 ▶ and impedance calculations. For a small-signal amplifier, the value of the ac emitter resistance can be *approximated* using

$$r_e' = \frac{25\ \text{mV}}{I_E} \tag{9.2}$$

where $r_e'$ = the ac emitter resistance
$I_E$ = the dc emitter current, found as $V_E/R_E$

The derivation of equation (9.2) is lengthy and involves calculus and thus is not covered here. For those readers who are interested in seeing the derivation of equation (9.2), it is included in Appendix D of the text. Example 9.1 shows how equation (9.2) can be used to determine the ac resistance of the emitter-base junction.

## EXAMPLE 9.1

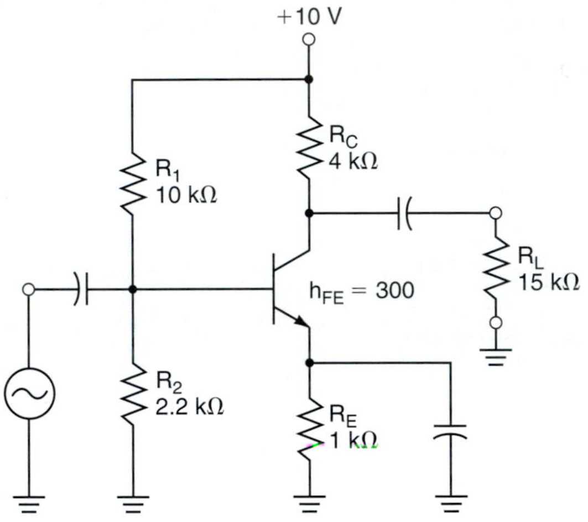

FIGURE 9.2

Determine the ac emitter resistance for the transistor in Figure 9.2.

***Solution:*** First, the value of $V_B$ is found as

$$V_B = \frac{R_2}{R_1 + R_2} V_{CC}$$
$$= 1.8\ \text{V}$$

$V_E$ is found as

$$V_E = V_B - 0.7\ \text{V}$$
$$= 1.1\ \text{V}$$

and $I_E$ is found as

$$I_E = \frac{V_E}{R_E}$$
$$= 1.1\ \text{mA}$$

Finally, the ac emitter resistance is found as

$$r_e' = \frac{25\ \text{mV}}{I_E}$$
$$= \frac{25\ \text{mV}}{1.1\ \text{mA}}$$
$$= 22.73\ \Omega$$

***PRACTICE PROBLEM 9.1***

A voltage-divider biased amplifier has values of $R_1 = 40\ \text{k}\Omega$, $R_2 = 10\ \text{k}\Omega$, $R_C = 6\ \text{k}\Omega$, $R_E = 2\ \text{k}\Omega$, $V_{CC} = +10\ \text{V}$, and $h_{FE} = 80$. Determine the ac emitter resistance of the transistor.

Since $r_e'$ is a resistance, its value can also be calculated using Ohm's law and some measured circuit values. Figure 9.3 shows the diode curve for the base-emitter junction of a transistor. If the transistor is biased at the operating point labeled $Q_1$, the change in $I_B$ causes the corresponding change in $V_{BE}$. Note that the changing values are shown as $\Delta I_B$ and $\Delta V_{BE}$. Using these values, the ac emitter resistance can be found as

$$r_e' = \frac{\Delta V_{BE}}{\Delta I_B} \tag{9.3}$$

FIGURE 9.3

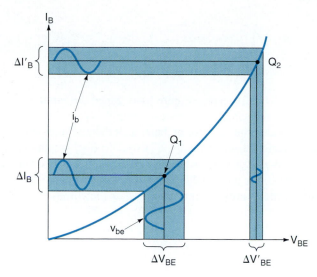

Note that the value of $\Delta V_{BE}$ produced by $\Delta I_B$ decreases if the transistor is biased at the operating point labeled $Q_2$. Thus, the value of $r_e'$ is affected by the biasing point of the amplifier.

## ac Beta

The ac current gain of an amplifier is different than the dc current gain. This has to do with the fact that dc current gain is measured with $I_B$ and $I_C$ being constant, while ac current gain is measured with *changing* (ac) current values. Figure 9.4 helps to illustrate this point. Figure 9.4a shows the method by which dc beta is determined. A set value of $I_C$ is divided by a set value of $I_B$. Since the curve of $I_B$ versus $I_C$ is not linear, the value of dc beta changes from one value of $I_C$ to another.

Figure 9.4b shows the measurement of **ac beta**. $I_B$ is varied to cause a variation of $I_C$ around the Q-point. The change in $I_C$ is then divided by the change in $I_B$ to obtain the value of ac beta. By formula,

**ac beta (or $h_{fe}$)**
The ratio of ac collector current to ac base current.

$$\beta_{ac} = \frac{\Delta I_C}{\Delta I_B} \qquad\qquad (9.4)$$

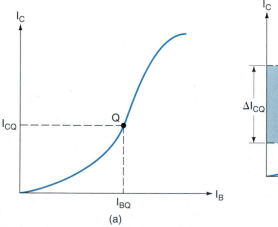

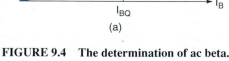

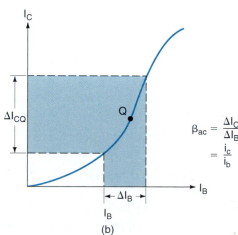

**FIGURE 9.4    The determination of ac beta.**

or

$$\beta_{ac} = \frac{i_c}{i_b} \qquad\qquad (9.5)$$

The measurement of $\beta_{ac}$ is illustrated in Figure 9.4b. Since the input and output currents for a transistor change at the same rate, you can simply divide the ac output current by the ac input current to obtain the value of ac beta, as indicated in equation (9.5).

$h_{FE}$ = dc beta
$h_{fe}$ = ac beta

Transistor specification sheets list dc beta as $h_{FE}$. To distinguish the ac value of current gain, ac beta is listed as $h_{fe}$. As is usually the case, the lowercase subscript is used to tell you that the current gain is an ac value, not a dc value.

For a common-emitter amplifier, the value of $h_{fe}$ is the current gain ($A_i$) for the *transistor.* By formula,

$$A_i = h_{fe} \qquad\qquad (9.6)$$

It should be noted that equation (9.6) defines the current gain of the *transistor*, not the amplifier. The calculation of $A_i$ for a common-emitter applifier is more involved, as will be shown later in this chapter.

The concepts and equations introduced in this section will be used constantly throughout our discussions on transistor ac operation. As you use them more and more, you will become more comfortable with them.

## Section Review

1. What are the three types of gain?
2. What types of gain are associated with each amplifier configuration?
3. Which amplifier type has an input/output voltage phase shift of 180°?
4. What is the phase relationship between the input and output voltage and current of each transistor amplifier configuration?
5. List, in order, the steps you need to take to determine the ac emitter resistance of a transistor.

# *9.2* THE ROLE OF CAPACITORS IN AMPLIFIERS

OBJECTIVE 4 ▶

Capacitors play two primary roles in common-emitter amplifiers:

Lab Reference: The effects of coupling and bypass capacitors on amplifier operation are demonstrated in Exercise 17.

1. *Coupling* capacitors pass an ac signal from one amplifier to another, while providing dc isolation between the two.
2. *Bypass* capacitors are used to "short circuit" ac signals to ground while not affecting the dc operation of the circuit.

In this section we will take a look at both of these capacitor applications.

## *Coupling Capacitors*

**Cascaded**
A term meaning *connected in series.*

In most practical applications, you will not see a single transistor amplifier. Rather, there will be a series of *cascaded* amplifiers. The term **cascaded** means *connected in series.*

When amplifiers are cascaded, the original signal is increased by the gain of each individual stage. In the end, the original signal has been provided with more gain than could have been provided by a single amplifier. Each amplifier in the series string is referred to as a **stage.** The overall series of amplifier stages is referred to as a **multistage amplifier.** Multistage amplifiers are covered in detail later in this chapter. What we are interested in at this point is the fact that *capacitors are commonly used to connect one amplifier stage to another.* A capacitor used in this application is called a *coupling capacitor.*

**Coupling capacitors** are easy to spot in a schematic diagram. They are always seen between the output of one amplifier and the input to the next. For example, take a look at the multistage amplifier shown in Figure 9.5. You will notice that a series capacitor is placed between the collector of each transistor and the base of the next. These capacitors are coupling capacitors. Any time that you see a capacitor in this position, it is a coupling capacitor. The term *coupling* means *to connect two circuits together electronically so that a signal will pass from one to the other with little or no distortion.* Coupling capacitors:

1. Pass the ac signal from one stage to the next with little or no distortion
2. Provide dc isolation between the two stages

These two functions are performed easily by the capacitor, which provides little opposition to an ac signal, while completely blocking a dc voltage and current. This blocking or isolation between stages simply means that the voltage on the collector of one stage will not affect the voltage at the base of the next stage, and vice versa. You may recall from your study of basic electronics that the reactance of a capacitor is found as

$$X_C = \frac{1}{2\pi f C} \tag{9.7}$$

where $X_C$ = the opposition to current provided by the capacitor
$f$ = the frequency of the signal applied to the capacitor
$C$ = the capacitance, in farads

This equation shows that the *opposition of a capacitor to current is inversely proportional to the frequency of the applied signal.* As frequency increases, the capacitor acts more and more like a short circuit. As frequency decreases, the capacitor acts more and more like an open circuit. At dc, the frequency of the applied signal is 0 Hz. The reactance at this frequency would be infinite. Thus, the capacitor will block dc.

The effects of the coupling capacitor on cascaded amplifiers are illustrated in Figure 9.6. Figure 9.6a is the **ac equivalent circuit** for the circuit in Figure 9.5. Note that the capacitors have been replaced by a wire connection. This represents the low opposition that the coupling capacitor presents to the ac signal. Figure 9.6b is the *dc equivalent* of Figure 9.5. In this circuit, the capacitors have been replaced by breaks in the circuit con-

**Stage**
A single amplifier in a cascaded group of amplifiers.

**Multistage amplifier**
An amplifier with a series of stages.

**Coupling capacitor**
A capacitor connected between amplifier stages to provide dc isolation between the stages while allowing the ac signal to pass without distortion.

The purposes served by coupling capacitors.

**ac equivalent circuit**
A representation of a circuit that shows how the circuit appears to an ac source.

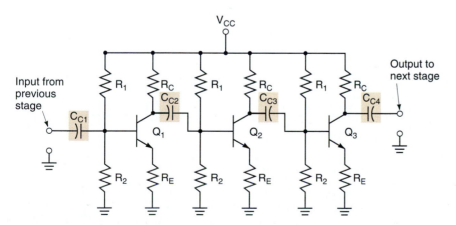

**FIGURE 9.5** **Coupling capacitors in a multistage amplifier.**

**Lab Reference:** The dc and ac characteristics of a multistage CE amplifier are observed in Exercise 18.

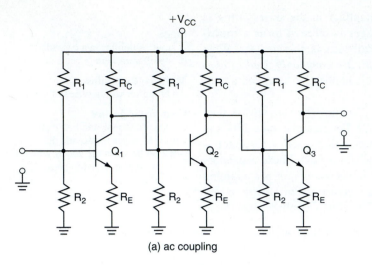

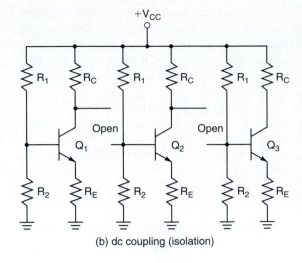

(a) ac coupling

(b) dc coupling (isolation)

**FIGURE 9.6** The effects of coupling capacitors.

nections. This represents the infinite opposition the capacitor presents to the dc voltage and current levels of the two stages.

The need for dc isolation (blocking) between amplifier stages can be seen in Figure 9.7. In this circuit, two individual amplifiers are shown, with an open switch between the two. For the second stage, $V_B$ is found as

$$V_B = \frac{R_2}{R_1 + R_2} V_{CC}$$

$$= 1.8 \text{ V}$$

*Note:* The discussion here ignores the effects of $I_{CQ}$ for the first stage to keep things as simple as possible.

This is the biasing potential for $Q_2$. Now, assume that the switch is closed. This would directly couple the output of the first stage to the input of the second. This coupling changes the picture because $R_C$ of the first stage is now in parallel with $R_1$ of the second stage. The equivalent circuit for this condition is shown in Figure 9.7b. The voltage divider at the input of stage 2 consists of the parallel combination of $R_C$ and $R_1$ in series with $R_2$. To see what this does to $V_B$ of the second stage, we first determine the total resistance of $R_C$ in parallel with $R_1$. This resistance is found as

$$R_{eq} = \frac{R_1 R_c}{R_1 + R_c}$$

$$= 2.65 \text{ k}\Omega$$

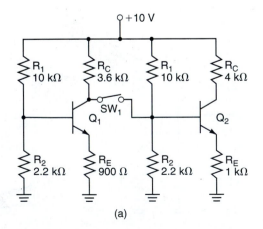

(a)

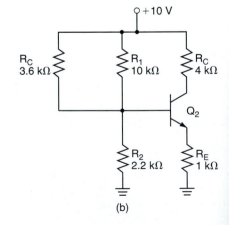

(b)

**FIGURE 9.7** Why dc isolation is important.

Now $V_B$ of the second stage is found as

$$V_B = \frac{R_2}{R_{eq} + R_2} V_{CC}$$
$$= 4.54 \text{ V}$$

With this new value of $V_B$, the biasing for the second stage has been severely affected. The 4.54 V at the base of $Q_2$ would undoubtedly cause the transistor to saturate, and the linear amplification of the circuit would be lost.

The use of the coupling capacitor allows each transistor amplifier stage to maintain its independent biasing characteristics, while allowing the ac output from one stage to pass on to the next stage. *The exact value of the coupling capacitor is not critical, provided that its reactance is extremely low at the lowest frequency used by the circuit.*

## Bypass Capacitors

**Bypass capacitors** function in the same manner as coupling capacitors, but they are used for a different reason. Bypass capacitors are connected across the emitter resistor of an amplifier to keep the resistor from affecting the ac operation of the circuit. A bypass capacitor is shown in the circuit in Figure 9.8. Note that the capacitor effectively *bypasses* the emitter resistor, thus the name.

Figure 9.9 shows how the bypass capacitor affects the ac and dc operation of the amplifier. Figure 9.9a shows the *ac equivalent* of the bypass capacitor. Note that the capacitor has been replaced by a wire, just as it is in the coupling application. In this case, the capacitor establishes an *ac ground* at the emitter of the transistor. Thus, for ac purposes, $R_E$ does not exist.

What does this have to do with anything? As you will be shown later in this chapter, the *voltage gain* of a common-emitter amplifier depends on the *total ac emitter resistance*. The lower this ac resistance, the higher the voltage gain of the amplifier. By shorting out $R_E$, the bypass capacitor ensures that the value of $R_E$ will not weigh into the voltage gain calculations for the amplifier. This increases the voltage gain of the circuit. We will examine this point more thoroughly later in this chapter.

Figure 9.9b shows the *dc equivalent* of the bypass capacitor. As before, the capacitor has been replaced with an open wire. This represents the infinite opposition that the capacitor presents to any dc voltage or current. Since the capacitor acts as an open to dc voltages and currents, it has no effect on the value of $I_E$, and therefore no effect on the value of $r'_e$.

**Bypass capacitor**
A capacitor used to establish an ac ground at a specific point in a circuit.

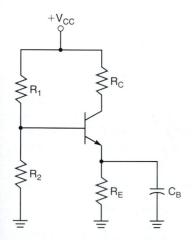

FIGURE 9.8   Bypass capacitor.

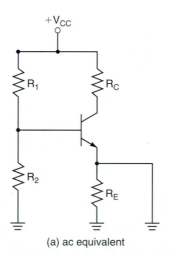

(a) ac equivalent

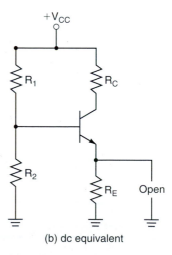

(b) dc equivalent

FIGURE 9.9   The effects of bypass capacitors.

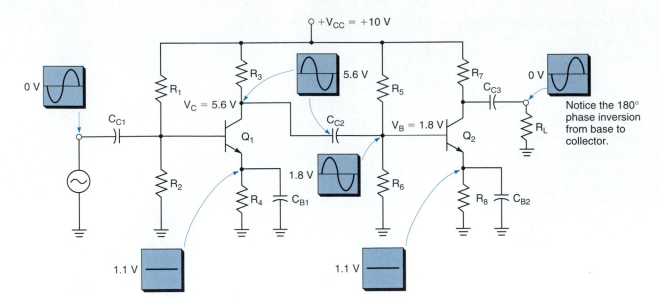

**FIGURE 9.10** Typical common-emitter amplifier signals.

Remember that the capacitor must act as a short at the lowest operating frequency. If the value of $X_C$ at the lowest frequency is not 0 Ω, the emitter will not be at ac ground. This defeats the original purpose of adding the bypass capacitor, which was to increase voltage gain.

## Amplifier Signals

Capacitor coupling is not the only type of coupling used in amplifier circuits. Another type, called *transformer coupling,* is discussed in Chapter 11.

The effects of coupling and bypass capacitors on the ac signals in a multistage amplifier can be seen by looking at Figure 9.10. The signal at the collector of $Q_1$ is shown to be a sine wave riding on a 5.6 $V_{dc}$ voltage. This dc voltage is the quiescent value of $V_C$ for the first stage. Note that this signal is applied to the coupling capacitor, $C_{C2}$. On the other side of the capacitor, you see the same ac sine wave. The only difference is that the sine wave is now riding on the dc value of $V_B$ for the second stage. Thus, the ac signal has been passed from the first stage to the next, while the dc reference of the signal has been changed from $V_C$ of the first stage to $V_B$ of the second.

Based on what we have seen, we can sum up the effects of the coupling capacitor as follows:

1. The ac signal on the input side of the capacitor equals the ac signal on the output side at the designed operating frequency.

2. The dc voltage on the input side of the capacitor equals $V_C$ of the source stage.

3. The dc voltage on the output side of the capacitor equals the value of $V_B$ for the second stage.

The effects of bypass capacitors on ac operation.

The emitters of both amplifiers show no change in $V_E$. This is caused by the bypass capacitors, which short the ac component of the emitter signal to ground. Thus, the signal present at each emitter is the pure dc value of $V_E$. Since $V_E$ remains constant, $I_E$ and $r'_e$ remain constant. This is important since $r'_e$ will weigh into all voltage gain calculations.

## Section Review

1. What two purposes are served by capacitors in multistage amplifiers?

2. What is meant by the term *coupling*?

3. Why is dc isolation between amplifier stages important?

**4.** What effect does a bypass capacitor have on the ac operation of an amplifier?

**5.** What effect does a coupling capacitor have on the ac signal that is coupled from one amplifier stage to the next?

**6.** Why don't you see an ac voltage at the emitter terminal of an amplifier with a bypass capacitor?

## 9.3 THE AMPLIFIER ac EQUIVALENT CIRCUIT

At this point, we need to derive the ac equivalent circuit for the common-emitter ampli-fier. This circuit, which is used in many ac calculations, is easily derived using this two-step process:

◄ *OBJECTIVE 5*

**1.** *Short circuit all capacitors.*

**2.** *Replace all dc sources with a ground symbol.*

How to derive an ac equivalent circuit.

The basis for the first step is easy to understand. Since the capacitors act as short circuits to the ac signals, they are replaced with wires, as they were in Figures 9.6a and 9.9a.

The second step is based on the fact that *dc sources have extremely low ac resis-tance values*. This point is illustrated in Figure 9.11. In the circuit shown, $V_{CC}$ is repre-sented as a simple dc battery. You may recall from your study of basic electronics that the internal resistance of a dc source (battery) is *ideally* 0 Ω. Thus, we can replace the dc source with a simple wire, as shown in Figure 9.11b. Note that this wire is replaced by the ground (reference) symbol in Figure 9.11c.

The process for finding the ac equivalent of a common-emitter amplifier is demon-strated in Example 9.2.

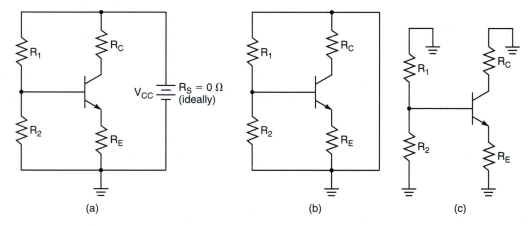

**(a)**                    **(b)**                    **(c)**

**FIGURE 9.11**   Deriving the CE ac-equivalent circuit.

Derive the ac equivalent circuit for the amplifier shown in Figure 9.12a.

*EXAMPLE 9.2*

*Solution:*   The first step is to short circuit all the capacitors ($C_C$ and $C_B$) and replace $V_{CC}$ with a ground symbol. This gives us the circuit shown in Figure 9.12b. As you can see, shorting the bypass capacitor leaves us with a straight wire across $R_E$. Thus, for the ac equivalent circuit, $R_E$ is shorted and is replaced by a wire. Figure 9.12c shows the equivalent circuit without $R_E$. Note also that $R_1$ and $R_C$ have been drawn as going

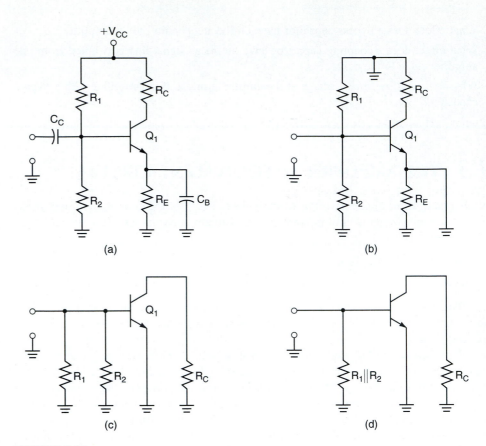

**FIGURE 9.12**

to ground. (Be sure that you can see what is happening at this point before you go on.)

As a final step, $R_1$ and $R_2$ can be combined into a single equivalent resistance. Since the ac equivalent circuit shows these two resistors as both connected to ground, they are in parallel. Figure 9.12d shows the final ac equivalent circuit for Figure 9.12a.

### PRACTICE PROBLEM 9.2

Derive the ac equivalent circuit for the amplifier shown in Figure 9.2. Include the component values in the equivalent circuit.

You need to be absolutely clear on one point: *The ac equivalent circuit is based on the ac characteristics of the amplifier and does not affect the dc analysis or troubleshooting of the amplifier.* The ac equivalent circuit is used for ac operation analysis only.

**Section Review**

1. What steps are taken to derive the ac equivalent circuit for a given amplifier?
2. What effects do the ac characteristics of an amplifier have on the dc characteristics of the circuit?

# 9.4 AMPLIFIER GAIN

As you know, there are three types of gain: *voltage*, *current*, and *power* gain. In this section, we will take a closer look at all three. The main emphasis in this section is on the common-emitter circuit. The gain characteristics of the common-base and common-collector circuits are covered in detail in Chapter 10.

*Remember*: *Gain* is a ratio; a multiplier that exists between the input and output of an amplifier.

## Voltage Gain

As you have seen, the **voltage gain** ($A_v$) of an amplifier is the factor by which the ac signal voltage increases from the input of the amplifier to the output. The value of $A_v$ for a common-emitter amplifier can be found by dividing the ac output voltage by the ac input voltage, as follows:

$$A_v = \frac{v_{out}}{v_{in}} \tag{9.8}$$

**Voltage gain ($A_v$)**
The factor by which ac signal voltage increases from the amplifier input to the amplifier output.

where   $A_v$ = the voltage gain of the amplifier
$v_{out}$ = the ac output voltage from the amplifier
$v_{in}$ = the ac input voltage to the amplifier

Example 9.3 reviews the calculaton of amplifier voltage gain.

---

An amplifier has values of $v_{out} = 12\ V_{ac}$ and $v_{in} = 60\ mV_{ac}$. What is the voltage gain of the circuit?

**EXAMPLE 9.3**

**Solution:**   The voltage gain for the circuit is found as follows:

$$A_v = \frac{v_{out}}{v_{in}}$$

$$= \frac{12\ V_{ac}}{60\ mV_{ac}}$$

$$= 200$$

*A Practical Consideration*: The value of $A_v$ can be determined using peak, peak-to-peak, or rms values. As long as $v_{out}$ and $v_{in}$ are measured in the same fashion (that is, both peak, both rms, and so on), you will obtain the same value of $A_v$.

Remember, voltage gain is a ratio and thus has no units. Thus, the value of $A_v$ for the amplifier described above is simply 200.

### PRACTICE PROBLEM 9.3

An amplifier has measured values of $v_{out} = 14.2\ V_{ac}$ and $v_{in} = 120\ mV_{ac}$. Determine the voltage gain of the circuit.

---

Equation (9.8) works well enough when the values of $v_{out}$ and $v_{in}$ are known, but there is a problem. The values of $v_{out}$ and $v_{in}$ are not usually shown on the schematic diagram for an amplifier. When this is the case, you may need to be able to determine the value of $A_v$ without knowing the values of $v_{out}$ and $v_{in}$.

The value of $A_v$ for a common-emitter amplifier is equal *to the ratio of total ac collector resistance to total ac emitter resistance*. For the circuit shown in Figure 9.13a, this ratio is found as

**Lab Reference:** The relationships given in equations (9.8) and (9.9) are demonstrated in Exercise 17.

$$A_v = \frac{r_C}{r_e'} \tag{9.9}$$

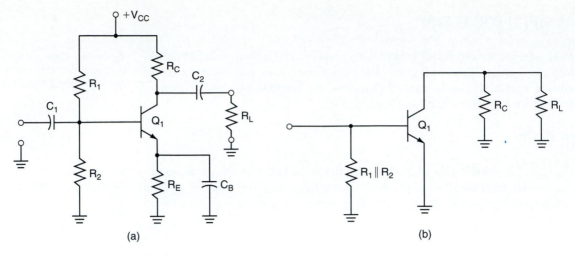

FIGURE 9.13

where $r_C$ = the total ac resistance of the collector circuit
$r'_e$ = the ac emitter resistance of the transistor

The ac equivalent for the circuit in Figure 9.13a is shown in Figure 9.13b. As you can see, the collector resistor ($R_C$) and the load ($R_L$) are in parallel. Therefore, the total ac resistance in the collector circuit is found as

$$r_C = R_C \parallel R_L \qquad (9.10)$$

Examples 9.4 and 9.5 show you how to determine the voltage gain of an amplifier by analyzing the schematic diagram.

*EXAMPLE 9.4*

Determine the voltage gain ($A_v$) of the circuit in Figure 9.14.

**Solution:** The value of $A_v$ is found by using the following steps:

Step 1: $V_B$ is found as

$$V_B = \frac{R_2}{R_1 + R_2} V_{CC}$$
$$= 2.35 \text{ V}$$

Step 2: $V_E$ is found as

$$V_E = V_B - 0.7 \text{ V}$$
$$= 1.65 \text{ V}$$

Step 3: $I_E$ is found as

$$I_E = \frac{V_E}{R_E}$$
$$= 750 \ \mu\text{A}$$

Step 4: $r'_e$ is found as

$$r'_e = \frac{25 \text{ mV}}{I_E}$$
$$= 33.3 \ \Omega$$

FIGURE 9.14

Step 5: $r_C$ is found as

$$r_C = R_C \| R_L$$
$$= 9.7 \text{ k}\Omega$$

Step 6: $A_v$ is found as

$$A_v = \frac{r_C}{r_e'}$$
$$= 291$$

Therefore, the ac output voltage from this circuit will be 291 times its ac input voltage.

### PRACTICE PROBLEM 9.4

An amplifier like the one in Figure 9.14 has the following values: $V_{CC} = 30$ V, $R_1 = 51$ k$\Omega$, $R_2 = 5.1$ k$\Omega$, $R_C = 5.1$ k$\Omega$, $R_L = 10$ k$\Omega$, $R_E = 910$ $\Omega$, and $h_{FE} = 250$. Determine the voltage gain of the amplifier.

---

Determine the value of $A_v$ for the circuit shown in Figure 9.15.

**EXAMPLE 9.5**

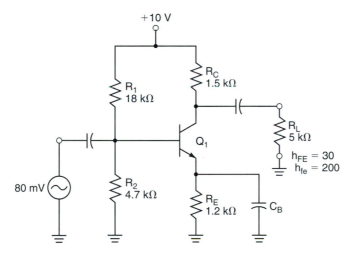

**FIGURE 9.15**

**Solution:** Since $h_{FE}R_E < 10\,R_2$, we have to consider the base input resistance in our $V_B$ calculation, as follows:

$$R_{eq} = R_2 \| h_{FE}R_E$$
$$= 4.16 \text{ k}\Omega$$

and

$$V_B = \frac{R_{eq}}{R + R_{eq}}\, V_{CC}$$
$$= 1.88 \text{ V}$$

Now

$$V_E = V_B - 0.7 \text{ V}$$
$$= 1.18 \text{ V}$$

and

$$I_E = \frac{V_E}{R_E}$$
$$= 983 \ \mu\text{A}$$

The value of $r'_e$ is now found as

$$r'_e = \frac{25 \text{ mV}}{I_E}$$
$$= 25.4 \ \Omega$$

and

$$r_C = R_L \| R_C$$
$$= 1.15 \text{ k}\Omega$$

Finally,

$$A_v = \frac{r_C}{r'_e}$$
$$= 45.3$$

### PRACTICE PROBLEM 9.5

An amplifier like the one shown in Figure 9.15 has the following values: $V_{CC} = +10$ V, $R_1 = 39$ k$\Omega$, $R_2 = 10$ k$\Omega$, $R_C = 6.2$ k$\Omega$, $R_E = 2.2$ k$\Omega$, $R_L = 10$ k$\Omega$, and $h_{FE} = 40$. Determine the value of $A_v$ for the circuit.

## The Basis for Equation (9.9)

To understand where equation (9.9) comes from, take a look at Figure 9.16. Figure 9.16a shows a common-emitter amplifier. The ac equivalent of this circuit is shown in Figure 9.16b. In Figure 9.16c, the transistor has been replaced with its own equivalent circuit. The equivalent circuit for the transistor contains three components:

1. A resistor ($r'_e$), which represents the ac emitter resistance of the transistor
2. A diode, which represents the emitter–base junction of the transistor
3. A current source, which represents the current drawn through $r_C$ by the collector of the transistor

Note that the input voltage ($v_{in}$) is applied across the diode and $r'_e$. Assuming the diode to be ideal, the ac emitter current can be found as

$$i_e = \frac{v_{in}}{r'_e} \tag{9.11}$$

Thus,

$$v_{in} = i_e r'_e \tag{9.12}$$

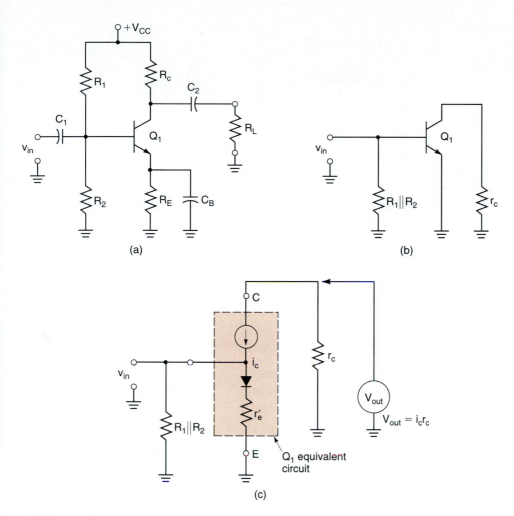

**FIGURE 9.16**

and $v_{out}$ is found as

$$v_{out} = i_c r_C \qquad \textbf{(9.13)}$$

Now we can substitute equations (9.12) and (9.13) into equation (9.8) as follows:

$$A_v = \frac{v_{out}}{v_{in}}$$

$$= \frac{i_c r_C}{i_e r_e'}$$

Since $i_e$ and $i_c$ are approximately equal, they can be dropped completely, leaving equation (9.9):

$$A_v = \frac{r_C}{r_e'}$$

## *Voltage Gain Instability*

One important consideration for the common-emitter amplifier is the *stability* of its voltage gain. An amplifier should have gain values that are stable so that the output of the circuit is predictable under all normal circumstances.

◄ *OBJECTIVE 6*

**FIGURE 9.17**

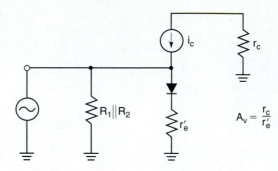

The voltage gain of a common-emitter amplifier can tend to be somewhat unstable. The reason for this can be seen in Figure 9.17. The voltage gain of an amplifier is equal to the ratio of collector resistance to the emitter resistance. For the amplifier shown in Figure 9.17, the emitter resistance is equal to $r'_e$. Like beta, $r'_e$ is somewhat dependent on temperature. Because of this, the voltage gain of the amplifier tends to be unstable. To reduce the effects of temperature on voltage gain, the *swamped* amplifier is used. Swamped amplifiers are discussed in detail in Section 9.6.

## *Calculating* $v_{out}$

When you know the voltage gain of an amplifier, determining the ac output voltage is easy. Simply use a rewritten form of equation (9.8) as follows:

$$v_{out} = A_v v_{in} \qquad (9.14)$$

---

**EXAMPLE 9.6**

The circuit used in Example 9.5 has an ac input signal of 80 mV. Determine the ac output voltage from the circuit.

**Solution:**    The output voltage is found as

$$v_{out} = A_v v_{in}$$

Using the value of $A_v$ from Example 9.5, we obtain

$$v_{out} = (45.3)\,(80\text{ mV}_{ac})$$
$$= 3.62\text{ V}_{ac}$$

**PRACTICE PROBLEM 9.6**

The amplifier described in Practice Problem 9.5 has a value of $v_{in} = 20\text{ mV}_{ac}$. Determine the value of $v_{out}$ for the amplifier.

---

## *Current Gain*

**Current gain ($A_i$)**
The factor by which ac current increases from the input of an amplifier to the output.

**Current gain ($A_i$)** is the factor by which ac current increases *from the input of an amplifier to the output*. The value of $A_i$ for a common emitter amplifier can be found by dividing the ac output current by the ac input current. By formula.

$$A_i = \frac{i_{out}}{i_{in}} \qquad (9.15)$$

where  $A_i$ = the current gain of the amplifier
$\quad\quad i_{\text{out}}$ = the ac output (load) current
$\quad\quad i_{\text{in}}$ = the ac input (source) current

As you know, the current gain of the *transistor* in a common-emitter amplifier is given on its spec sheet as $h_{fe}$. However, the current gain for the *amplifier* is always *lower* than the value of $h_{fe}$. This is due to two factors:

1. The ac input current is divided between the transistor and the biasing network.
2. The ac collector current is divided between the collector resistor and the load.

In Section 9.5, you will be shown how to calculate the value of $A_i$ for a common-emitter amplifier. For now, you need to remember only this value of $A_i$ is typically much lower than the value of the transistor current gain ($h_{fe}$).

## Power Gain

◄  OBJECTIVE 7

As you know, the **power gain (Ap)** of an amplifier is the factor by which ac signal power increases from the input of an amplifier to its output. Power gain can be found by multiplying current gain by voltage gain. By formula,

$$A_p = A_i A_v \qquad (9.16)$$

**Power gain ($A_p$)**
The factor by which ac signal power increases from the input of an amplifier to the output.

Once the power gain of an amplifier has been determined, you can calculate the output power of an amplifier provided at the load as

$$P_{\text{out}} = A_p P_{\text{in}} \qquad (9.17)$$

where  $P_{\text{out}}$ = the amplifier output power
$\quad\quad A_p$ = the power gain of the amplifier
$\quad\quad P_{\text{in}}$ = the amplifier input power

Example 9.7 demonstrates the procedure for calculating amplifier output power.

---

The amplifier shown in Figure 9.15 has values of $A_v = 45.3$ and $A_i = 20$. Determine the power gain ($A_p$) of the amplifier and the output power when $P_{\text{in}} = 80\ \mu\text{W}$.

**EXAMPLE 9.7**

**Solution:**  The power gain of the amplifier is found as

$$A_p = A_i A_v$$
$$= (20)(45.3)$$
$$= 906$$

The output power is therefore found as

$$P_{\text{out}} = A_p P_{\text{in}}$$
$$= (906)(80\ \mu\text{W})$$
$$= 72.48\ \text{mW}$$

**PRACTICE PROBLEM 9.7**

The circuit described in Practice Problem 9.5 has the following values: $A_i = 20$ and $P_{\text{in}} = 60\ \mu\text{W}$. Determine the output power for the circuit.

---

1. How is the voltage gain ($A_v$) of a common-emitter amplifier defined in terms of circuit resistance values?
2. How is the ac collector resistance of a common-emitter determined?
3. List, in order, the steps usually required to calculate the value of $A_v$ for a common-emitter amplifier.
4. Why does the voltage gain of a common-emitter amplifier tend to be unstable?

## 9.5 GAIN AND IMPEDANCE CALCULATIONS

To complete the picture of common-emitter amplifier operation, there are several gain and impedance topics that need to be addressed. These topics are discussed in this section.

### The Effects of Loading

*OBJECTIVE 8* ▶ As you know, the value of the load plays a role in the calculation of $A_v$ for a common-emitter amplifier. The lower the resistance of the load, the greater the effect. This point is illustrated in Figure 9.18. The circuits in the figure are identical, other than their values of the load resistance. As you can see, the value of $A_v$ for the circuit in Figure 9.18b is much lower than that of the circuit in Figure 9.18a. Thus, *the lower the resistance of the load, the lower the voltage gain of a common-emitter amplifier.* Conversely, the greater the resistance of the load, the greater the voltage gain of a common-emitter amplifier.

**Lab Reference:** The effects of a change in load resistance on voltage gain are demonstrated in Exercise 18.

For any amplifier, the greatest value of $A_v$ occurs when the load *opens*. With an open load, the ac collector circuit consists only of $R_C$. Therefore,

$$r_C = R_C \qquad \text{(open-load)} \qquad \textbf{(9.18)}$$

Since $R_C$ must always be greater than $(R_C \parallel R_L)$, the value of $A_v$ for a given amplifier reaches its maximum possible value when the load opens. This point is demonstrated in Example 9.8

---

**EXAMPLE 9.8**

The load in Figure 9.18a opens. Calculate the *open-load* voltage gain of the circuit.

**Solution:** With the load open,

$$\begin{aligned} r_C &= R_C \\ &= 3 \text{ k}\Omega \end{aligned}$$

and

$$\begin{aligned} A_v &= \frac{r_C}{r'_e} \\ &= \frac{3 \text{ k}\Omega}{25 \text{ }\Omega} \\ &= 120 \qquad \text{(maximum)} \end{aligned}$$

**PRACTICE PROBLEM 9.8**

Calculate the open-load voltage gain of the amplifier shown in Figure 9.15 Compare this value to the one calculated in Example 9.5.

---

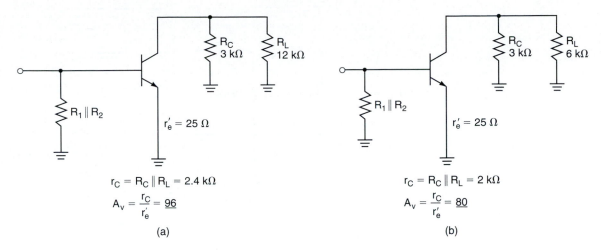

$$r_C = R_C \| R_L = 2.4 \text{ k}\Omega$$
$$A_v = \frac{r_C}{r'_e} = 96$$

(a)

$$r_C = R_C \| R_L = 2 \text{ k}\Omega$$
$$A_v = \frac{r_C}{r'_e} = 80$$

(b)

**FIGURE 9.18   The effect of a change in $R_L$ on the value of $A_v$.**

When one common-emitter is used to drive another, the input impedance of the second amplifier serves as the load resistance of the first. Thus, to calculate the $A_v$ of the first amplifier stage correctly, you must be able to calculate the input impedance of the second stage.

## Calculating Amplifier Input Impedance

The input impedance of an amplifier is determined using the ac equivalent circuit for the amplifier. As Figure 9.19 shows, the input impedance of an amplifier is the parallel impedance formed by $R_1, R_2$, and the base of the transistor. By formula,

◄ *OBJECTIVE 9*

$$Z_{in} = R_1 \| R_2 \| Z_{in(base)} \tag{9.19}$$

where     $Z_{in}$ = the input impedance to the amplifier
     $Z_{in(base)}$ = the input impedance of the transistor base

Figure 9.19 shows a typical common-emitter amplifier and its ac equivalent circuit. Note that $Z_{in(base)}$ is the input impedance *to the base of the transistor*. The amplifier input impedance ($Z_{in}$) equals the parallel combination of $Z_{in(base)}$ and the biasing resistors.

You may recall that the dc input resistance of a transistor base is found as

$$R_{in(base)} = h_{FE}R_E$$

This relationship was derived in Chapter 7. The input impedance of the base can be derived in the same manner. The only differences are as follows:

**1.** The ac emitter resistance is used in the calculation.

**2.** The value of ac beta ($h_{fe}$) is used in place of dc beta ($h_{FE}$).

Therefore, $Z_{in(base)}$ can be found as

$$Z_{in(base)} = h_{fe}r'_e \tag{9.20}$$

*A Practical Consideration:* The value of $Z_{in(base)}$ for a transistor is given as $h_{ie}$ on the component's spec sheet. The meaning and use of this rating are discussed later in this chapter.

Once the input impedance of the transistor is found, it is considered to be in parallel with the base biasing resistors. This parallel circuit gives us the total input impedance to the amplifier, as illustrated in Example 9.9.

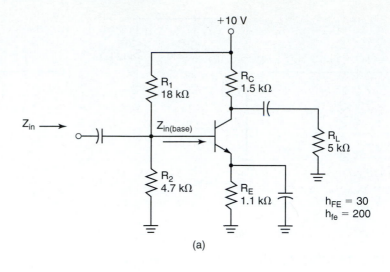

(a)

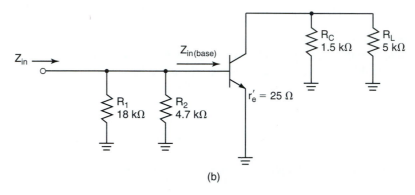

(b)

**FIGURE 9.19** A common-emitter amplifier and its ac equivalent circuit.

---

EXAMPLE 9.9

Determine the input impedance to the amplifier represented by the equivalent circuit in Figure 9.19.

**Solution:** The value of $Z_{in(base)}$ can be found as

$$Z_{in(base)} = h_{fe}r'_e$$
$$= (150)(25\ \Omega)$$
$$= 3.75\ k\Omega$$

$Z_{in}$ for the amplifier is found as

$$Z_{in} = R_1 \parallel R_2 \parallel Z_{in(base)}$$
$$= 18\ k\Omega \parallel 4.7\ k\Omega \parallel 3.75\ k\Omega$$
$$= 1.87\ k\Omega$$

**PRACTICE PROBLEM 9.9**

Determine the value of $Z_{in}$ for the amplifier described in Practice Problem 9.5. Assume that $h_{fe} = 200$ for the transistor.

## Calculating the Value of A_i

Earlier in the chapter, you were told that the overall current gain ($A_i$) of a common-emitter amplifier is always *lower* than the current gain of the transistor ($h_{fe}$). We are now ready to take a closer look at this relationship.

As you know, the current gain of a common-emitter amplifier is found as

$$A_i = \frac{i_{out}}{i_{in}}$$

where $i_{in}$ is the *ac source current* and $i_{out}$ is the *ac load current*. You also know that the current gain of the *transistor* in a common-emitter amplifier is found as

$$h_{fe} = \frac{i_c}{i_b}$$

If you refer back to Figure 9.19b, you'll see that the input of the common-emitter contains a *current divider*. This current divider is made up of $R_1$, $R_2$, and the base of the transistor. Since a portion of the ac source current passes through the biasing resistors, $i_b < i_{in}$. At the same time, the collector circuit of the amplifier forms another current divider. Since a portion of the ac collector current passes through the collector resistor, $i_{out} < i_c$. These factors combine to cause the overall current gain of the amplifier ($A_i$) to be significantly lower than the current gain of the transistor ($h_{fe}$).

In Appendix C, we derive an equation for current gain that takes into account the reduction in gain caused by the transistor input and output circuitry. This equation is:

$$A_i = h_{fe}\left(\frac{Z_{in}r_C}{Z_{in(base)}R_L}\right) \tag{9.21}$$

where

$A_i$ = the current gain of the common-emitter amplifier
$h_{fe}$ = the current gain of the transistor
$(Z_{in}r_C)/(Z_{in(base)}R_L)$ = the reduction factor introduced by the biasing and output components

Example 9.10 demonstrates the relationship between $h_{fe}$ and $A_i$.

---

Calculate the value of $A_i$ for the circuit shown in Figure 9.19b.

**EXAMPLE 9.10**

**Solution:** For the circuit shown,

$$\begin{aligned} Z_{in(base)} &= h_{fe}r'_e \\ &= (200)(25\ \Omega) \\ &= 5\ k\Omega \end{aligned}$$

$$\begin{aligned} Z_{in} &= R_1 \parallel R_2 \parallel Z_{in(base)} \\ &= 2.14\ k\Omega \end{aligned}$$

and

$$\begin{aligned} r_C &= R_C \parallel R_L \\ &= 1.15\ k\Omega \end{aligned}$$

Now, these values are used (along with $h_{fe} = 200$) in equation (9.21) as follows:

$$A_i = h_{fe}\left(\frac{Z_{in}r_C}{Z_{in(base)}R_L}\right)$$

$$= (200)\left[\frac{(2.14\ \text{k}\Omega)(1.15\ \text{k}\Omega)}{(5\ \text{k}\Omega)(5\ \text{k}\Omega)}\right]$$

$$= (200)(0.098)$$

$$= 19.68$$

As you can see, the value of $A_i$ for this circuit is significantly lower than the value of $h_{fe}$.

**PRACTICE PROBLEM 9.10**

A common-emitter amplifier has the following values: $h_{fe} = 300$, $r_C = 2.15\ \text{k}\Omega$, $R_L = 6.2\ \text{k}\Omega$, $Z_{(inbase)} = 8.2\ \text{k}\Omega$, and $Z_{in} = 3.8\ \text{k}\Omega$. Calculate the value of $A_i$ for the circuit.

## Multistage Amplifier Gain Calculations

**Lab Reference:** The voltage gain characteristics of a multistage amplifier are demonstrated in Exercise 18.

When you want to determine the overall value of $A_v$, $A_i$, and/or $A_p$ for a multistage amplifier, *you must begin by determining the appropriate gain values for the individual stages.* Once the overall gain values for the individual stages are determined, you can determine the desired overall gain value by using any (or all) of the following equations:

$$A_{vT} = (A_{v1})(A_{v2})(A_{v3}).\ .\ . \tag{9.22}$$

$$A_{iT} = (A_{i1})(A_{i2})(A_{i3})\ .\ .\ . \tag{9.23}$$

$$A_{pT} = (A_{vT})(A_{iT}) \tag{9.24}$$

Equations (9.22) and (9.23) indicate that the overall value of $A_v$ or $A_i$ is simply the product of the individual stage gain values. Equation (9.24) indicates that the overall power gain is found as the product of the overall values of $A_v$ and $A_i$.

As you recall, the exact sequence of equations used to determine the overall value of $A_v$ for a given amplifier stage is as follows:

1. Perform a basic dc analysis of the amplifier to determine the value of $I_E$.
2. Using equation (9.2), determine the value of $r'_e$ for the amplifier.
3. Using equation (9.11), determine the value of $r_C$ for the amplifier.
4. Using the values of $r'_e$ and $r_C$ found in steps 2 and 3, determine the value of $A_v$ for the stage.

When you are dealing with a two-stage amplifier, the value of $Z_{in}$ for the second stage must be used in place of $R_L$ for the $r_C$ calculation of the first stage. This point is illustrated in Example 9.11.

**EXAMPLE 9.11**

Determine the voltage gain for the first stage of the circuit in Figure 9.20.

**Solution:** Using the established procedure, $r'_e$ for the first stage is found to be 19.8 $\Omega$, and $r'_e$ for the second stage is found to be 17.4 $\Omega$. For the second stage, $h_{fe}$ is 200; therefore,

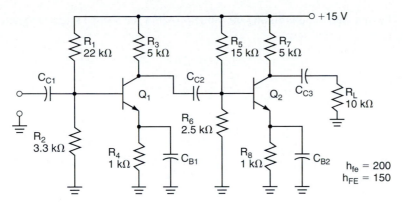

**FIGURE 9.20**

$$Z_{in(base)} = h_{fe}r'_e$$
$$= (200)(17.4\ \Omega)$$
$$= 3.48\ k\Omega$$

The input impedance for the second stage is found as

$$Z_{in} = R_5 \parallel R_6 \parallel Z_{in(base)}$$
$$= 1.33\ k\Omega$$

The input impedance of the second stage (1.33 k$\Omega$) is the load for the first stage. Therefore, $r_C$ for the first stage is found as

$$r_C = R_3 \parallel Z_{in}$$
$$= 1.05\ k\Omega$$

Finally, the value of $A_v$ for the first stage is found as

$$A_v = \frac{r_C}{r'_e}$$
$$= \frac{1.05\ k\Omega}{19.8\ \Omega}$$
$$= 53.03$$

### PRACTICE PROBLEM 9.11

Assume for a moment that the value of $h_{fe}$ for the second stage in Figure 9.20 has increased to 280. Determine the new value of $A_v$ for the first amplifier stage.

As Example 9.11 showed, you must determine the $Z_{in}$ of a load stage before you can determine the actual $A_v$ of a source stage. After determining the $Z_{in}$ of the load stage, this value is used in place of $R_L$. Then, using $r_C$, the voltage gain of the first stage is calculated.

Once the value of $A_v$ for the first stage is known, we can calculate the value of $A_v$ for the second stage and for the overall amplifier. This is illustrated in Example 9.12.

Determine the value of $A_{vT}$ for the amplifier in Figure 9.20.

**EXAMPLE 9.12**

**Solution:** We know (from Example 9.11) that the value of $A_v$ for the first stage is 53.03 and that the value of $r'_e$ for the second stage is 17.4 $\Omega$.

The first step in determining the value of $A_v$ for the second stage is to determine the value of $r_C$, as follows:

$$r_C = R_7 \parallel R_L$$
$$= 3.33 \text{ k}\Omega$$

$A_v$ for the second stage is found as

$$A_v = \frac{r_C}{r_e'}$$
$$= \frac{3.33 \text{ k}\Omega}{17.4 \text{ }\Omega}$$
$$= 191.38$$

Now that we know the values of $A_v$ for both stages, the overall voltage gain for the two-stage amplifier is found as

$$A_{vT} = (A_{v1})(A_{v2})$$
$$= (53.03)(191.38)$$
$$= 10.15 \times 10^3$$

### PRACTICE PROBLEM 9.12

Assume that the second stage of the amplifier in Figure 9.20 has the following: $h_{fe} = 240$ and $R_L = 22 \text{ k}\Omega$. Determine the value of $A_{vT}$ for the two-stage amplifier.

We mentioned earlier that *swamping* reduces the effects of changes in $r_e'$ on voltage gain. It also reduces the loading effect of the amplifier on a previous stage. We will now look at swamped amplifiers and how they overcome variations in $r_e'$ while reducing circuit loading.

---

**Section Review**

1. How does an open load affect the voltage gain of an amplifier?
2. List, in order, the steps required to determine the value of $A_v$ for a common-emitter amplifier.
3. Why is it important to be able to determine the value of $Z_{in}$ for a load stage?
4. List, in order, the steps required to determine the value of $Z_{in}$ for a common-emitter amplifier.
5. Why is the value of $A_i$ for a given amplifier always lower than the transistor's value of $h_{fe}$?
6. How do you determine the values of $A_{vT}$ and $A_{iT}$ for a multistage amplifier?
7. How do you determine the value of $A_{pT}$ for a multistage amplifier?

---

## 9.6 SWAMPED AMPLIFIERS

OBJECTIVE 10 ▶

**Swamped amplifier**
An amplifier that uses a partially bypassed emitter resistance to increase ac emitter resistance.

A **swamped amplifier** reduces variations in voltage gain by *increasing the ac resistance of the emitter circuit*. By increasing this resistance, it also increases $Z_{in(base)}$, reducing the amplifier's loading effect on a previous stage. A swamped amplifier is shown in Figure 9.21a.

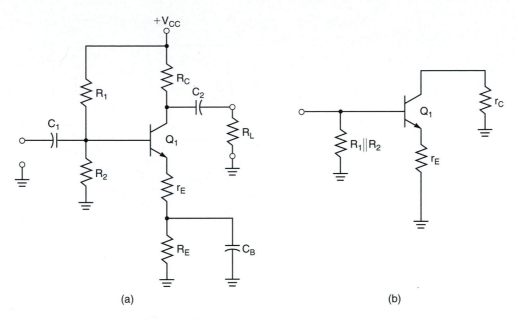

FIGURE 9.21    The swamped common-emitter amplifier and its ac equivalent circuit.

The swamped amplifier has a higher ac emitter resistance because *only part of the dc emitter resistance is bypassed*. The bypass capacitor eliminates only the value of $R_E$. The other emitter resistor, $r_E$, is part of the ac equivalent circuit, as shown in Figure 9.21b. Since the voltage gain of the amplifier is equal to the ratio of ac collector resistance to ac emitter resistance, the voltage gain for the amplifier is found as

$$A_v = \frac{r_C}{r_e' + r_E} \qquad \textbf{(9.25)}$$

The following example demonstrates the use of equation (9.25) in the calculation of $A_v$ for a swamped amplifier.

---

Determine the value of $A_v$ for the amplifier in Figure 9.22.

*Solution:*    Using the established procedure, $V_E$ is found to equal 1.37 V. The total dc resistance in the emitter circuit equals $(R_E + r_E)$. Therefore, $I_E$ is found as

$$I_E = \frac{V_E}{R_E + r_E}$$
$$= 1.14 \text{ mA}$$

Now, the value of $r_e'$ is found as

$$r_e' = \frac{25 \text{ mV}}{I_E}$$
$$= 21.9 \ \Omega$$

The value of $r_C$ is now found as

$$r_C = R_C \| R_L$$
$$= 1.3 \text{ k}\Omega$$

*EXAMPLE 9.13*

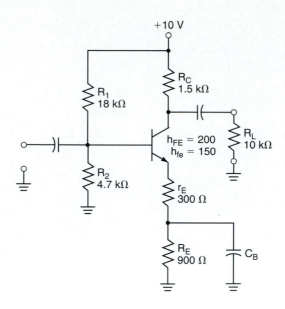

**FIGURE 9.22**

Finally, $A_v$ is found as

$$A_v = \frac{r_C}{r'_e + r_E}$$

$$= \frac{1.3\ \text{k}\Omega}{321.9\ \Omega}$$

$$= 4.04$$

### PRACTICE PROBLEM 9.13

Assume that the amplifier in Figure 9.22 has values of $R_E = 820\ \Omega$ and $r_E = 330\ \Omega$. Determine the value of $A_v$ for the circuit.

Why swamping improves gain stability.

Swamping improves the gain stability of an amplifier when $r_E \gg r'_e$. Since most of the ac emitter resistance is determined by the value of $r_E$, any change in $r'_e$ will have little effect on the overall gain of the amplifier. This point is illustrated in Example 9.14.

### EXAMPLE 9.14

Determine the change in gain for the amplifier in Example 9.13 when $r'_e$ doubles in value.

**Solution:** When $r'_e$ doubles in value, $A_v$ becomes

$$A_v = \frac{r_C}{r'_e + r_E}$$

$$= \frac{1.3\ \text{k}\Omega}{343.8\ \Omega}$$

$$= 3.78$$

The change in gain is therefore

$$4.04 - 3.78 = 0.26$$

This is a change of only 6.44% from the original value of $A_v$.

When an amplifier is not swamped, doubling the value of $r'_e$ will cause the value of $A_v$ to decrease by 50 percent. Thus, the gain of an amplifier is stabilized by swamping the emitter circuit.

## The Effect of Swamping on $Z_{in}$

The input impedance of a transistor base is shown in equation (9.20) to equal beta times the ac resistance of the emitter. This ac resistance is equal to $(r'_e + r_E)$ for the swamped amplifier. Thus,

$$Z_{in(base)} = h_{fe}(r'_e + r_E) \qquad (9.26)$$

The result is that the input impedance of the transistor base is increased by an amount equal to $h_{fe}r_E$, as shown in Example 9.15.

---

Determine the value of $Z_{in(base)}$ for the circuits shown in Figure 9.23.

*EXAMPLE 9.15*

**Solution:**   The circuits have equal values of dc emitter resistance. Therefore, following the established procedure for finding $r'_e$ gives us a value of 25 Ω for both circuits. Now, for Figure 9.23a,

$$Z_{in(base)} = h_{fe}r'_e$$
$$= 5 \text{ k}\Omega$$

For Figure 9.23b,

$$Z_{in(base)} = h_{fe}(r'_e + r_E)$$
$$= (200)(25 \text{ }\Omega + 200 \text{ }\Omega)$$
$$= 45 \text{ k}\Omega$$

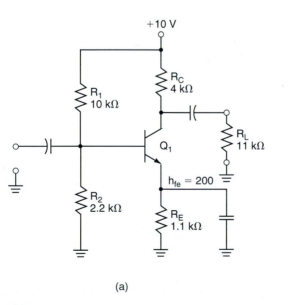

(a)

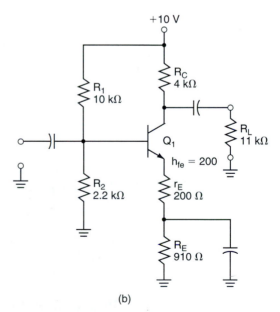

(b)

**FIGURE 9.23**

Increasing the value of $Z_{in(base)}$ *increases* the overall value of $Z_{in}$ for the amplifier, as shown in Example 9.16.

---

**EXAMPLE 9.16**

Determine the value of $Z_{in}$ for the amplifiers shown in Figure 9.23.

**Solution:** For Figure 9.23a,

$$Z_{in} = R_1 \| R_2 \| Z_{in(base)}$$
$$= 10 \text{ k}\Omega \| 2.2 \text{ k}\Omega \| 5 \text{ k}\Omega$$
$$= 1.33 \text{ k}\Omega$$

For Figure 9.23b,

$$Z_{in} = R_1 \| R_2 \| Z_{in(base)}$$
$$= 10 \text{ k}\Omega \| 2.2\text{k}\Omega \| 45 \text{ k}\Omega$$
$$= 1.73 \text{ k}\Omega$$

Thus, we see that increasing the value of $Z_{in(base)}$ increases the overall value of $Z_{in}$ for an amplifier.

**PRACTICE PROBLEM 9.16**

Determine the value of $Z_{in}$ for the amplifier shown in Figure 9.22.

---

The increased $Z_{in}$ for Figure 9.23b means that this circuit will have less effect on the $A_v$ of a source amplifier. This point is illustrated in the following example.

---

**EXAMPLE 9.17**

The amplifiers in Figure 9.23 are both driven by a source amplifier with values of $r'_e = 25 \ \Omega$ and $R_C = 8 \text{ k}\Omega$. Determine the value of $A_v$ for the source amplifier when each circuit is connected as the load.

**Solution:** When the circuit in Figure 9.23a is connected to the source amplifier, the voltage gain of the amplifier is found as

$$A_v = \frac{r_C}{r'_e}$$

$$= \frac{8 \text{ k}\Omega \| 1.33 \text{ k}\Omega}{25 \ \Omega} \qquad (r_C = R_C \| Z_{in})$$

$$= 45.6$$

When the circuit in Figure 9.23b is connected to the source amplifier, the voltage gain of the amplifier is found as

$$A_v = \frac{r_C}{r'_e}$$

$$= \frac{8 \text{ k}\Omega \| 1.73 \text{ k}\Omega}{25 \ \Omega} \qquad (r_C = R_C \| Z_{in})$$

$$= 56.9$$

Thus, the reduced loading by the circuit in Figure 9.23b increased the gain of the source amplifier.

---

## The Disadvantage of Swamping

You have been shown that a swamped amplifier is more stable against variations in $r'_e$. You have also been shown how this amplifier increases the value of $A_v$ for a source amplifier.

The main disadvantage of swamping is that the overall voltage gain of this amplifier is lower than that of a comparable standard common-emitter amplifier. While the gain of a swamped amplifier is more stable, it is lower than the gain of an amplifier that isn't swamped. This can be seen by looking at the two amplifiers in Figure 9.23 again. Note that the two circuits are identical for dc analysis purposes. Both have a total of 1.1 kΩ in their emitter circuits.

The two amplifiers differ only in their ac characteristics. Note that the total ac resistance in the emitter circuit of Figure 9.23a is 25 Ω; the value of $r'_e$. This low resistance provides the circuit with a gain of 117. The total ac resistance in the emitter circuit of Figure 9.23b is 225 Ω. This higher emitter resistance reduces the overall gain of the amplifier to 13. Thus, swamping reduces the overall gain of the amplifier. At the same time, you must remember that the swamped amplifier will have a much more stable value of $A_v$ than the standard common-emitter amplifier. Also, the increase in $A_v$ experienced by a source amplifier that is driving a swamped amplifier will partially offset the reduction in gain.

Swamping improves stability but reduces $A_v$.

## The Emitter Circuit Is the Key

When you are dealing with common-emitter amplifiers, you must be able to distinguish between *swamped* (or **gain-stabilized**) amplifiers and standard amplifiers. The key to distinguishing one common-emitter amplifier type from the other is to look at the emitter circuit. If the emitter resistance is *completely bypassed,* you are dealing with a standard common-emitter amplifier. If the emitter resistance is only partially bypassed, you are dealing with a swamped, or gain-stabilized, amplifier. The circuit recognition features of these two types of common-emitter amplifiers, along with the different equations for the two, are summarized in Figure 9.24.

**Gain-stabilized amplifier**
A term often used to describe a swamped amplifier.

---

1. How does swamping reduce the effect of variations in the value of $r'_e$?
2. How does swamping affect the value of $A_v$ for a source amplifier? Explain your answer.
3. What is the primary disadvantage of using amplifier swamping? Explain your answer.

*Section Review*

---

# 9.7 *h*-PARAMETERS

**Hybrid parameters**, or ***h*-parameters**, are transistor specifications that describe the component operating characteristics under specific circumstances. Each of the four *h*-parameters is measured under *no-load* or *full-load* conditions. These *h*-parameters are then used in circuit analysis applications.

The four *h*-parameters for a transistor in a common-emitter amplifier are as follows:

◄ *OBJECTIVE 11*

$h_{ie}$ = the base input impedance
$h_{fe}$ = the base-to-collector current gain
$h_{oe}$ = the output admittance
$h_{re}$ = the reverse voltage feedback ratio

***h*-parameters**
Transistor specifications that describe the operation of the device under full-load or no-load conditions.

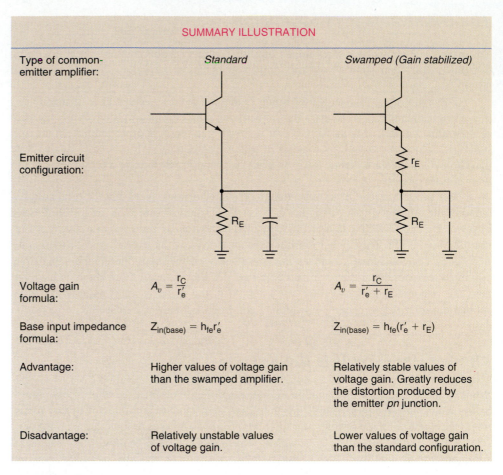

## SUMMARY ILLUSTRATION

| | Standard | Swamped (Gain stabilized) |
|---|---|---|
| Type of common-emitter amplifier: | | |
| Emitter circuit configuration: | | |
| Voltage gain formula: | $A_v = \dfrac{r_C}{r'_e}$ | $A_v = \dfrac{r_C}{r'_e + r_E}$ |
| Base input impedance formula: | $Z_{in(base)} = h_{fe}r'_e$ | $Z_{in(base)} = h_{fe}(r'_e + r_E)$ |
| Advantage: | Higher values of voltage gain than the swamped amplifier. | Relatively stable values of voltage gain. Greatly reduces the distortion produced by the emitter *pn* junction. |
| Disadvantage: | Relatively unstable values of voltage gain. | Lower values of voltage gain than the standard configuration. |

**FIGURE 9.24**

It should be noted that all these values represent *ac characteristics* of the transistor, as measured under specific circumstances. Before discussing the applications of *h*-parameters, let's take a look at the parameters and the method of measurement used for each.

### Input Impedance ($h_{ie}$)

**Input impedance ($h_{ie}$)**
The input impedance of the transistor, measured under full-load conditions.

The **input impedance** parameter, $h_{ie}$, is measured with the *output shorted*. A shorted output is a full load, so $h_{ie}$ *represents the input impedance to the transistor under full-load conditions*. The measurement of $h_{ie}$ is illustrated in Figure 9.25a. As you can see, the collector and emitter terminals are shorted. Then an ac signal is applied to the base-emitter junction of the transistor. With the input voltage applied, the base current is measured. Then $h_{ie}$ is determined as

$$h_{ie} = \frac{v_{in}}{i_b} \qquad \text{(output shorted)} \qquad \textbf{(9.27)}$$

Why $h_{ie}$ is measured under full-load conditions.

Why short the output? You may recall that any resistance in the emitter circuit is reflected back to the base. This condition was described in the equation

$$Z_{in(base)} = h_{fe}(r'_e + r_E)$$

By shorting the collector and emitter terminals, the measured value of $h_{ie}$ does not reflect any external resistance in the circuit.

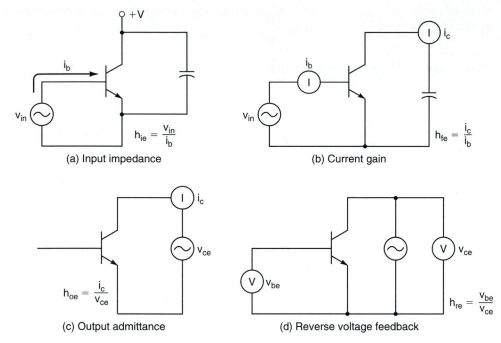

FIGURE 9.25    The measurement of *h*-parameters.

## Current Gain ( h$_{fe}$)

The base-to-collector **current gain**, $h_{fe}$, is also measured with the *output shorted*. Again, this represents a full load, so *$h_{fe}$ represents the current gain of the transistor under full-load conditions*. The measurement of $h_{fe}$ is illustrated in Figure 9.25b. With the output shorted and a signal applied to the base, both the base and collector currents are measured. Then $h_{fe}$ is determined as

$$h_{fe} = \frac{i_c}{i_b} \qquad \text{(output shorted)} \qquad (9.28)$$

**Current gain (*$h_{fe}$*)**
The ac beta of the component, measured under full-load conditions.

In this case, it is clear why the output is shorted. If the output were left open, $i_c$ would equal zero. Shorting the output gives us a measurable value of $i_c$ that can be reproduced in a practical test. In other words, anyone can achieve the same results simply by shorting the output terminals and applying the same signal to the transistor.

## Output Admittance ( h$_{oe}$)

The **output admittance**, $h_{oe}$, is measured with the *input open*. The measurement of $h_{oe}$ is illustrated in Figure 9.25c. As you can see, a signal is applied across the collector-emitter terminals. Then, with this signal applied, the value of $i_c$ is measured. The value of $h_{oe}$ is then determined as

$$h_{oe} = \frac{i_c}{v_{ce}} \qquad \text{(input open)} \qquad (9.29)$$

**Output admittance (*$h_{oe}$*)**
The admittance of the collector-emitter circuit, measured under no-load conditions. This parameter is used mainly in amplifier design. Since admittance is the reciprocal of impedance, the unit of measure is often listed as *mhos* (siemens is the preferred unit of measure).

Measuring $h_{oe}$ with the input open makes sense when you consider the effect of shorting the input. If the input to the transistor were shorted, there would be some base current ($i_b$). Since this current would come from the emitter, the value of $i_c$ would not be at its maximum potential. In other words, by not allowing $i_b$ to be generated, $i_c$ is at its absolute maximum value. Therefore, $h_{oe}$ is an accurate measurement of the *maximum* output admittance.

## Reverse Voltage Feedback Ratio ($h_{re}$)

**Reverse voltage feedback ratio ($h_{re}$)**
The ratio of $v_{be}$ to $v_{ce}$, measured under no-load conditions. This parameter is used mainly in amplifier design. Since $h_{re}$ is a ratio of two voltages, it has no unit of measure.

The **reverse voltage feedback ratio,** $h_{re}$, indicates *the amount of output voltage reflected back to the input*. This value is measured with the *input open*. The measurement of $h_{re}$ is illustrated in Figure 9.25d. A signal is applied to the collector-emitter terminals. Then, with the input open, the voltage that is fed back to the base-emitter junction is measured. The value of $h_{re}$ is then determined as

$$h_{re} = \frac{v_{be}}{v_{ce}} \qquad \text{(input open)} \tag{9.30}$$

Since the voltage at the base terminal will always be less than the voltage across the emitter–collector terminals, $h_{re}$ will always be less than 1. By measuring $h_{re}$ with the input open, you ensure that the voltage fed back to the base will always be at its maximum possible value (since maximum voltage is always developed across an open circuit).

## Circuit Calculations Involving h-Parameters

Circuit calculations involving *h*-parameters can be very simple or very complex, depending on the following:

1. What you are trying to determine
2. How exact you want your calculations to be

For our purposes, we are interested in only four *h*-parameter circuit equations. These equations are as follows:

$$A_i = h_{fe}\left(\frac{Z_{in}r_C}{h_{ie}R_L}\right) \tag{9.21}$$

$$Z_{in(base)} = h_{ie} \tag{9.31}$$

$$r_e' = \frac{h_{ie}}{h_{fe}} \tag{9.32}$$

$$A_v = \frac{h_{fe}r_C}{h_{ie}} \tag{9.33}$$

Equation (9.21) needs no explanation because it was introduced earlier in the chapter. The only change in the equation is the substitution of $h_{ie}$ for $Z_{in(base)}$, as given in equation (9.31). Equation (9.31) is relatively easy to understand if you refer back to Figure 9.12d. Here we see the ac equivalent circuit for the amplifier shown in Figure 9.12a. Looking at the ac equivalent circuit, it is easy to see that the only impedance between the base of the transistor and the emitter ground connection is the input impedance of the transistor. This input impedance is $h_{ie}$.

Equation (9.32) is derived using equations (9.20) and (9.31). You may recall from earlier discussions that

$$Z_{\text{in(base)}} = h_{fe}r'_e \qquad \text{[equation (9.20)]}$$

Substituting equation (9.31) for $Z_{\text{in(base)}}$, we obtain

$$h_{ie} = h_{fe}r'_e$$

Simply rearranging for $r'_e$ gives us equation (9.32).

You may be wondering why we would go to all this trouble to find $r'_e$ with $h$-parameters when we can simply use

$$r'_e = \frac{25 \text{ mV}}{I_E}$$

The fact of the matter is that the $25 \text{ mV}/I_E$ equation really isn't very accurate. By using the $h$-parameter equation, we are able to obtain a much more accurate value of $r'_e$ and thus can more closely calculate the value of $A_v$ for a given common-emitter amplifier.

Equation (9.33) is derived as follows:

$$A_v = \frac{r_C}{r'_e}$$

$$= \frac{1}{r'_e} r_C$$

$$= \frac{h_{fe}}{h_{ie}} r_C$$

$$= \frac{h_{fe}r_C}{h_{ie}}$$

*A Practical Consideration*: Depending on the transistor used, the given value of 25 mV in the $r'_e$ equation can actually be any value between 25 and 52 mV. This gives us

$$\frac{25 \text{ mV}}{I_E} \le r'_e \le \frac{52 \text{ mV}}{I_E}$$

This is why we use the more accurate $h$-parameter equations whenever possible.

The equations that we have discussed in this section will be used throughout our discussion on ac amplifier analysis. It should be noted that the entire subject of $h$-parameters and their derivations is far more complex than has been presented here. The subject is covered thoroughly in Appendix C for those readers who are interested. If you do not wish to get involved in $h$-parameter derivations, you may continue from this point without loss of continuity.

## Determining h-Parameter Values

The spec sheet for a given transistor will list the values of the device's $h$-parameters in the *electrical characteristics* portion of the sheet. This is illustrated in Figure 9.26, which shows the electrical characteristics portion of the spec sheet for the Motorola 2N4400–4401 series transistors.

When *minimum* and *maximum* $h$-parameter values are given, we must determine the *geometric average* of the two values. Thus, the values of $h_{ie}$ and $h_{fe}$ that we would use in the analysis of a 2N4400 circuit would be found as

$$h_{ie} = \sqrt{h_{ie(\text{min})} \times h_{ie(\text{max})}}$$

$$= \sqrt{(500 \ \Omega)(7.5 \text{ k}\Omega)}$$

$$= 1.94 \text{ k}\Omega$$

and

$$h_{fe} = \sqrt{h_{fe(\text{min})} \times h_{fe(\text{max})}}$$

$$= \sqrt{(20)(250)}$$

$$= 71$$

Example 9.18 shows how the above values would be used in the ac analysis of a 2N4400 amplifier.

*A Practical Consideration*: $h$-parameter values are often listed in an electrical characteristics section called *small-signal characteristics*.

| Characteristic | Symbol | Min | Max | Unit |
|---|---|---|---|---|
| Emitter-Base Capacitance<br>($V_{BE}$ = 0.5 Vdc, $I_C$ = 0, f = 100 kHz) | $C_{eb}$ | — | 30 | pF |
| Input Impedance<br>($I_C$ = 1.0 mAdc, $V_{CE}$ = 10 Vdc, f = 1.0 kHz)     2N4400<br>2N4401 | $h_{ie}$ | 0.5<br>1.0 | 7.5<br>15 | k ohms |
| Voltage Feedback Ratio<br>($I_C$ = 1.0 mAdc, $V_{CE}$ = 10 Vdc, f = 1.0 kHz) | $h_{re}$ | 0.1 | 8.0 | $\times 10^{-4}$ |
| Small-Signal Current Gain<br>($I_C$ = 1.0 mAdc, $V_{CE}$ = 10 Vdc, f = 1.0 kHz)     2N4400<br>2N4401 | $h_{fe}$ | 20<br>40 | 250<br>500 | — |
| Output Admittance<br>($I_C$ = 1.0 mAdc, $V_{CE}$ = 10 Vdc, f = 1.0 kHz) | $h_{oe}$ | 1.0 | 30 | $\mu$hos |

**FIGURE 9.26**   (Copyright of Motorola. Used by permission.)

---

## EXAMPLE 9.18

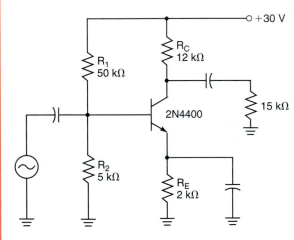

**FIGURE 9.27**

Determine the values of $Z_{in}$ and $A_v$ for the circuit shown in Figure 9.27.

**Solution:**   Using the established analysis procedure, the value of $I_C$ for the circuit is found to be approximately 1 mA. Since the values of $h_{ie}$ and $h_{fe}$ for the 2N4400 are listed at $I_C$ = 1 mA (from the spec sheet in Figure 9.26), we can use the geometric averages of $h_{ie}$ = 1.94 kΩ and $h_{fe}$ = 71 that were determined earlier. Using these values,

$$Z_{in(base)} = h_{ie}$$
$$= 1.94 \text{ k}\Omega$$

and

$$Z_{in} = R_1 \parallel R_2 \parallel Z_{in(base)}$$
$$= 1.35 \text{ k}\Omega$$

Now

$$r_C = R_C \parallel R_L$$
$$= 6.67 \text{ k}\Omega$$

and

$$A_v = \frac{h_{fe} r_C}{h_{ie}}$$

$$= \frac{(71)(6.67 \text{ k}\Omega)}{1.94 \text{ k}\Omega}$$

$$= 244$$

***PRACTICE PROBLEM 9.18***

An amplifier like the one in Figure 9.27 has values of $R_C$ = 12 kΩ, $R_L$ = 4.7 kΩ, $R_1$ = 33 kΩ, $R_2$ = 4.7 kΩ, and $I_C$ = 1 mA. At 1 mA, the transistor has h-parameter values of $h_{ie}$ = 1 kΩ to 5 kΩ and $h_{fe}$ = 70 to 350. Determine the values of $Z_{in}$ and $A_v$ for the circuit.

As Example 9.18 pointed out, the h-parameter values listed on the spec sheet of the 2N4400 were measured at $I_C$ = 1 mA. As was the case with $h_{FE}$, the ac h-parameters will

## h PARAMETERS
$V_{CE} = 10$ Vdc, f = 1.0 kHz, $T_A = 25°C$

This group of graphs illustrates the relationship between $h_{fe}$ and other "h" parameters for this series of transistors. To obtain these curves, a high-gain and a low-gain unit were selected from both the 2N4400 and 2N4401 lines, and the same units were used to develop the correspondingly numbered curves on each graph.

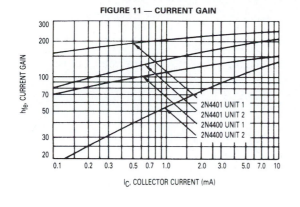

FIGURE 11 — CURRENT GAIN

2N4401 UNIT 1
2N4401 UNIT 2
2N4400 UNIT 1
2N4400 UNIT 2

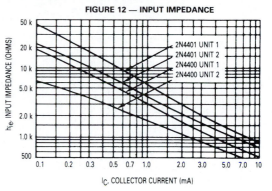

FIGURE 12 — INPUT IMPEDANCE

2N4401 UNIT 1
2N4401 UNIT 2
2N4400 UNIT 1
2N4400 UNIT 2

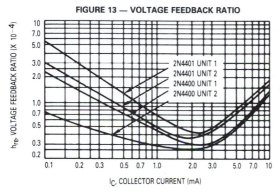

FIGURE 13 — VOLTAGE FEEDBACK RATIO

2N4401 UNIT 1
2N4401 UNIT 2
2N4400 UNIT 1
2N4400 UNIT 2

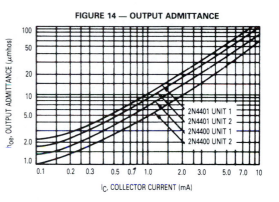

FIGURE 14 — OUTPUT ADMITTANCE

2N4401 UNIT 1
2N4401 UNIT 2
2N4400 UNIT 1
2N4400 UNIT 2

**FIGURE 9.28** (Copyright of Motorola. Used by permission.)

all vary with the value of $I_C$. This point is illustrated in Figure 9.28, which shows the *h*-parameter curves for the 2N4400–4401 series transistors.

First, a word or two about the curves. Each *h*-parameter graph shows four curves. Two of the curves are identified as being for the 2N4400, and two of them are identified as being for the 2N4401. Just as we were given *minimum* and *maximum h*-parameter values at $I_{CQ} = 1$ mA, we have minimum (Unit 2) and maximum (Unit 1) curves for the transistors.

To determine the value of a given parameter at a given value of $I_{CQ}$, use the following procedure:

**1.** Using the graph, determine the minimum and maximum *h*-parameter values at the value of $I_{CQ}$.

**2.** Use the geometric average of the two values obtained in the analysis of the amplifier.

The use of this procedure is demonstrated in Example 9.19.

---

An amplifier that uses a 2N4400 has values of $I_{CQ} = 5$ mA and $r_C = 460$ Ω. Determine the value of $A_v$ for the circuit.

***EXAMPLE 9.19***

**Solution:** From the $h_{fe}$ curve, we can approximate the limits of $h_{fe}$ to be 110 to 140 when $I_{CQ} = 5$ mA. The geometric average of these two values is found as

$$h_{fe} = \sqrt{h_{fe(min)} \times h_{fe(max)}}$$
$$= 124$$

The $h_{ie}$ graph shows the limits of $h_{ie}$ to be 600 Ω and 800 Ω when $I_{CQ}$ = 5 mA. The geometric average of these two values is found as

$$h_{ie} = \sqrt{h_{ie(min)} \times h_{ie(max)}}$$
$$= 693\ \Omega$$

Using $h_{fe}$ = 124 and $h_{ie}$ = 693 Ω, the value of $A_v$ is determined to be

$$A_v = \frac{h_{fe} r_C}{h_{ie}}$$

$$= \frac{(124)(460)}{693\ \Omega}$$

$$= 82.3$$

**PRACTICE PROBLEM 9.19**

A 2N4401 is used in a circuit with values of $I_{CQ}$ = 2 mA and $r_C$ = 1.64 kΩ. Determine the value of $A_v$ for the amplifier using the curves shown in Figure 9.28.

## A Few More Points

In this section, we have concentrated on $h_{ie}$ and $h_{fe}$. That's because these two parameters are the only ones required for "everyday" circuit analysis. The other two h-parameters, $h_{oe}$ and $h_{re}$, are used primarily in circuit development applications.

Whenever you need to determine the voltage gain of an amplifier, the required h-parameter values can be obtained easily from the spec sheet of the transistor. If a range of values is given, you can actually follow either of two procedures:

1. If you want an *approximate* value of $A_v$, you can use the geometric average of the h-parameter values listed.

2. If you are interested in the *worst-case* values of $A_v$, you can analyze the amplifier using *both* the minimum and maximum values of $h_{fe}$ and $h_{ie}$. By doing this, you will obtain the *minimum* and *maximum* values of $A_v$. These two values would be the voltage gain limits for the amplifier.

**Section Review**

1. What is $h_{ie}$? How is the value of $h_{ie}$ measured?
2. What is $h_{fe}$? How is the value of $h_{fe}$ measured?
3. What is $h_{oe}$? How is the value of $h_{oe}$ measured?
4. What is $h_{re}$? How is the value of $h_{re}$ measured?
5. Which of the four h-parameters are commonly used in circuit analysis?
6. Where are h-parameter values usually listed on a transistor spec sheet?
7. When minimum and maximum h-parameter values are listed, what value do you usually use for circuit analyses?
8. What is the procedure for determining the values of $h_{ie}$ and $h_{fe}$ at a specified value of $I_{CQ}$?
9. How is the circuit analysis procedure changed for determining the *worst-case* values of $A_v$?

# 9.8 AMPLIFIER TROUBLESHOOTING

We have already discussed the dc troubleshooting procedure for an amplifier. However, that discussion assumed that you already knew there was a problem in the amplifier. How can you tell when one out of several amplifiers is the cause of a problem? For example, take a look at Figure 9.29. If the final output from this series of amplifiers is bad, how do you know which amplifier is the source of the trouble?

◄ *OBJECTIVE 12*

When you troubleshoot a series of amplifiers, *you start at the final output stage.* When you verify that the output from the amplifier is bad, check the input. *If the amplifier's input is bad, go to the next amplifier output.* There is no reason to troubleshoot an amplifier with a bad input. An amplifier cannot possibly have a normal output if its input is not normal.

When you verify that the input to an amplifier is bad, check the output of the previous stage. If that output is good, there is a problem in the circuit that is coupling the two amplifiers. If the output from the previous stage is not normal, check its input, and so on. The complete procedure for troubleshooting a three-stage amplifier is illustrated in the flowchart in Figure 9.30. A *flowchart* is a step-by-step illustration of a problem-solving procedure.

## *Amplifier Input/Output Signals*

Figure 9.31 shows the signals you should see at the inputs and outputs of each stage of Figure 9.29. Note that the standard common-emitter amplifiers (stages 2 and 3) show no ac signal at the emitter terminal of the transistors. This is due to the fact that the ac voltage is developed across $r'_e$ in the transistors themselves. The swamped amplifier (stage 1) has a small ac signal present at the emitter terminal, but only a dc level at the bypass capacitor connection point. In all three cases, the collector signal is much larger in amplitude than the input signal at the base, and the two signals are 180° out of phase.

When troubleshooting the amplifier, you should find the amplifier stage that has the bad input/output signal condition. In other words, find the stage that has a normal input signal and a bad output signal. When you find that stage, troubleshoot the dc circuit as you were shown in Chapter 7.

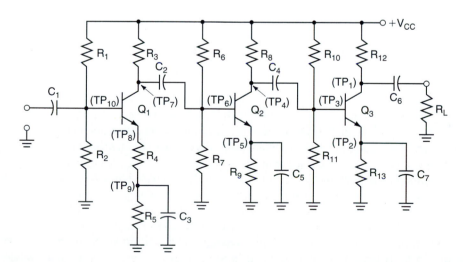

**FIGURE 9.29**

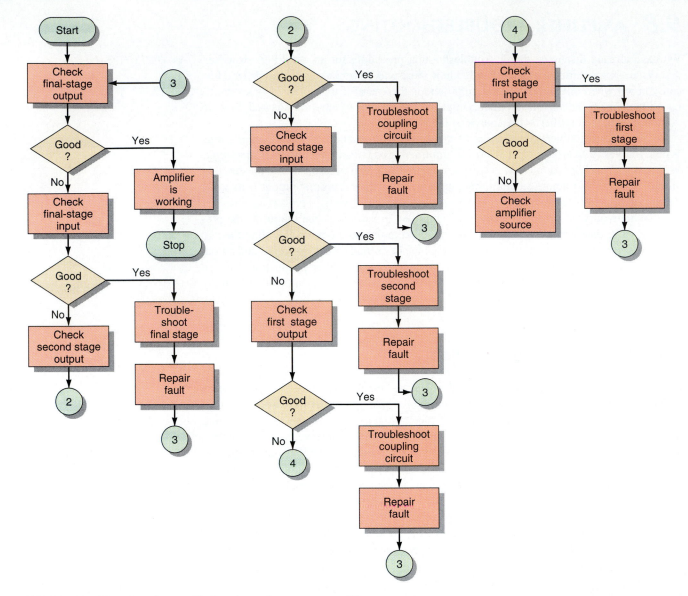

**FIGURE 9.30** Flowchart for troubleshooting a three-stage amplifier.

## Nonlinear Distortion

As you know, *distortion* is *an unwanted change in the shape of an ac signal*. One problem that commonly occurs in common-emitter amplifiers is called *nonlinear distortion*. The output waveform from an amplifier experiencing nonlinear distortion is shown in Figure 9.32. Note the difference between the shape of the negative alternation of the signal (normal) and that of its positive alternation (distorted). The change in the shape of the positive alternation is a result of nonlinear distortion.

**Nonlinear distortion** is *a type of distortion that is caused by driving the base–emitter junction of the transistor into its nonlinear operating region*. This point is illustrated with the help of Figure 9.33, which shows the base characteristic curve of a transistor.

Normally, a transistor is operated so that the base–emitter junction stays in the linear region of operation. In this region, a change in $V_{BE}$ causes a *linear* (constant rate) change in $I_B$. If the transistor is operated in the nonlinear operating region, a change in $V_{BE}$ will cause a change in $I_B$ that is not linear. For example, compare the slope of the

**Nonlinear distortion**
A type of distortion caused by driving the base–emitter junction of a transistor into its nonlinear operating region.

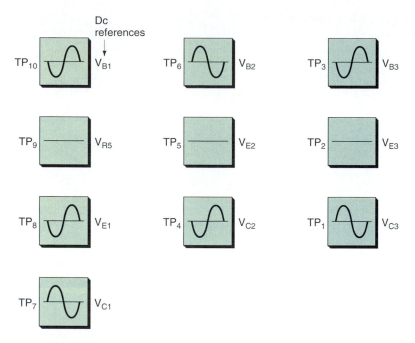

**FIGURE 9.31**   Signals for Figure 9.29.

curve at the points immediately above and below $V_K$. Since the slope of the curve changes, the rate of change in $I_B$ also changes.

The base–emitter junction of a transistor may be driven into its nonlinear operating region in one of two sets of circumstances:

1. A transistor is normally biased so that $I_B$ will have a value well within the linear region of operation. If an amplifier is poorly biased so that the $Q$-point value of $I_B$ is near the

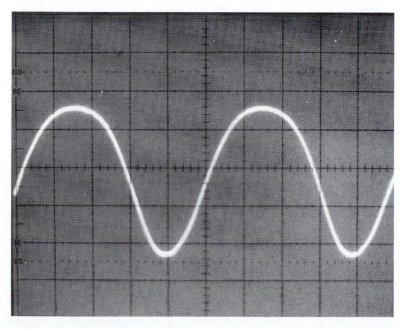

**FIGURE 9.32**

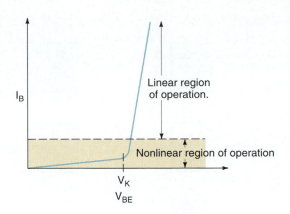

**FIGURE 9.33  A base characteristic curve due to $r_e'$.**

nonlinear region of operation, a relatively small ac input signal will drive the amplifier into the nonlinear region of operation, and nonlinear distortion will result.

2. The amplitude of the amplifier input signal may be sufficient to drive even a well-biased amplifier into nonlinear operation, causing nonlinear distortion.

If the first of these two problems (poor biasing) is present, the only solution is to redesign the amplifier for a higher value of $I_B$. When the second problem (overdriving the amplifier) is present, the solution is to reduce the amplitude of the amplifier input signal.

Nonlinear distortion can cause a variety of problems in communications electronics. In stereos, it may cause the audio to sound "grainy" or may even cause the tones of various musical instruments to change. In television video circuits, it can cause distortion in the picture on the CRT. To avoid these types of problems, amplifier design normally emphasizes avoiding nonlinear distortion. Since amplifiers are normally designed to avoid it, the most common cause of nonlinear distortion is overdriving an amplifier.

*Swamping reduces nonlinear distortion because most of the ac input signal is developed across $r_E$, which is a linear device (as compared to the nonlinearity of the emitter pn junction). This linear operation allows us to use more of the load line and still have minimal distortion.*

## A Few Pointers

When you are troubleshooting common-emitter amplifiers, remember that the input signal for a given stage will have a much lower amplitude than the output. For this reason, you should reduce the VOLTS/DIV setting on your oscilloscope when going from output to input. For example, consider a circuit with a voltage gain of 100. *Under normal circumstances, this amplifier will have an output signal that is 100 times as large as the input signal.* If you try to read these two signals using the same VOLTS/DIV setting on the oscilloscope, you will not be able to see the input signal.

Another point is this: Don't worry about seeing the 180° phase shift on your oscilloscope. If you want to, you can *external trigger* the oscilloscope on the amplifier input and then see the phase shift. However, viewing the phase shift is not really important. If the amplifier is working, the phase shift will be there. You will never run into a problem where the amplifier has the right signal but the phase shift is not there, so don't worry about it. Just look to see if the amplifier has a small-signal input and a large-signal output. If it does, it is all right. If neither signal is normal, you need to move on to the preceding stage. If the input signal is good and the output signal is bad, you have found the bad stage.

## Section Review

1. What is the procedure for troubleshooting a multistage amplifier?
2. When troubleshooting an amplifier, why do you need to pay close attention to the VOLTS/DIV setting of the oscilloscope?

There are three types of gain: *current gain, voltage gain,* and *power gain.* The types of gain exhibited by each amplifier type are listed as follows:

| Amplifier Type | Types of Gain |
|---|---|
| Common-emitter | Current/voltage/power |
| Common-collector | Current/power |
| Common-base | Voltage/power |

For every type of amplifier, the input and output currents are in phase. The common-emitter amplifier has a $180°$ voltage phase shift from input to output, while the common-base and common-collector input and output voltages are in phase.

The current gain ($A_i$) of a common-emitter amplifier is lower than the ac beta ($h_{fe}$) rating of the transistor. This is due to the current dividers that are present in the amplifier input and output circuits. The value of $h_{fe}$ is dependent on the value of $I_{CQ}$, as is the value of dc beta ($h_{FE}$).

The voltage gain ($A_v$) of the common-emitter amplifier depends on the ac values of collector and emitter resistance. In an amplifier where the external emitter resistance is *completely* bypassed, the ac emitter resistance is equal to $r'_e$, the ac resistance of the emitter–base junction. The voltage gain of a standard common-emitter amplifier is not entirely stable because $r'_e$ varies with temperature and the value of $I_{CQ}$.

To overcome the instability of $A_v$, *swamping* is used. This involves bypassing only a portion of the emitter resistance. This results in a portion of the emitter resistance still weighing into the ac model of the amplifier. The overall effect is to reduce the effects of $r'_e$, for a loss of overall voltage gain. The power gain of the common-emitter amplifier is equal to the product of $A_i$ and $A_v$.

The four *h*-parameters are current gain ($h_{fe}$), input impedance ($h_{ie}$), output admittance ($h_{oe}$), and reverse voltage transfer ratio ($h_{re}$). Of the four, only $h_{ie}$ and $h_{fe}$ are used in common circuit analyses. The values of $h_{ie}$ and $h_{fe}$ for a given transistor can be obtained from the spec sheet of the device. As with $h_{FE}$, you should use the geometric averages of $h_{fe}$ and $h_{ie}$ when minimum and maximum parameter values are listed on the spec sheet.

Troubleshooting amplifier stages involves analyzing the ac signals to determine which amplifier has a normal input and a bad output. When this amplifier is found, the dc troubleshooting procedures covered in Chapter 7 are used to locate the faulty component.

---

The following new terms were introduced and defined in this chapter:

| | | |
|---|---|---|
| ac beta (or $h_{fe}$) | gain-stabilized amplifier | output admittance ($h_{oe}$) |
| ac emitter resistance ($r'_e$) | *h*-parameters | reverse voltage feedback |
| ac equivalent circuit | input impedance ($h_{ie}$) | ratio ($h_{re}$) |
| bypass capacitor | linear | stage |
| cascaded | multistage amplifier | swamped amplifier |
| coupling capacitor | nonlinear distortion | |

| Equation Summary | Equation Number | Equation | Section Number |
|---|---|---|---|
| | (9.1) | $V_{out} = V_C$ | 9.1 |
| | (9.2) | $r_e' = \dfrac{25 \text{ mV}}{I_E}$ | 9.1 |
| | (9.3) | $r_e' = \dfrac{\Delta V_{BE}}{\Delta I_B}$ | 9.1 |
| | (9.4) | $\beta_{ac} = \dfrac{\Delta I_C}{\Delta I_B}$ | 9.1 |
| | (9.5) | $\beta_{ac} = \dfrac{i_c}{i_b}$ | 9.1 |
| | (9.6) | $A_i = h_{fe}$ (for the transistor only) | 9.1 |
| | (9.7) | $X_C = \dfrac{1}{2\pi f C}$ | 9.2 |
| | (9.8) | $A_v = \dfrac{v_{out}}{v_{in}}$ | 9.4 |
| | (9.9) | $A_v = \dfrac{r_C}{r_e'}$ | 9.4 |
| | (9.10) | $r_C = R_C \parallel R_L$ | 9.4 |
| | (9.11) | $i_e = \dfrac{v_{in}}{r_e'}$ | 9.4 |
| | (9.12) | $v_{in} = i_e r_e'$ | 9.4 |
| | (9.13) | $v_{out} = i_c r_C$ | 9.4 |
| | (9.14) | $v_{out} = A_v v_{in}$ | 9.4 |
| | (9.15) | $A_i = \dfrac{i_{out}}{i_{in}}$ | 9.4 |
| | (9.16) | $A_p = A_i A_v$ | 9.4 |
| | (9.17) | $P_{out} = A_p P_{in}$ | 9.4 |
| | (9.18) | $r_C = R_C$ (open-load) | 9.5 |
| | (9.19) | $Z_{in} = R_1 \parallel R_2 \parallel Z_{in(base)}$ | 9.5 |
| | (9.20) | $Z_{in(base)} = h_{fe} r_e'$ | 9.5 |
| | (9.21) | $A_i = h_{fe} \dfrac{Z_{in} r_C}{Z_{in(base)} R_L}$ | 9.5 |
| | (9.22) | $A_{vT} = (A_{v1})(A_{v2})(A_{v3}) \ldots$ | 9.5 |
| | (9.23) | $A_{iT} = (A_{i1})(A_{i2})(A_{i3}) \ldots$ | 9.5 |

| Equation Number | Equation | Section Number |
|---|---|---|
| **(9.24)** | $A_{pT} = (A_{vT})(A_{iT})$ | 9.5 |
| **(9.25)** | $A_v = \dfrac{r_C}{r'_e + r_E}$ | 9.6 |
| **(9.26)** | $Z_{in(base)} = h_{fe}(r'_e + r_E)$ | 9.6 |
| **(9.27)** | $h_{ie} = \dfrac{v_{in}}{i_b}$   (output shorted) | 9.7 |
| **(9.28)** | $h_{fe} = \dfrac{i_c}{i_b}$   (output shorted) | 9.7 |
| **(9.29)** | $h_{oe} = \dfrac{i_c}{v_{ce}}$   (input open) | 9.7 |
| **(9.30)** | $h_{re} = \dfrac{v_{be}}{v_{ce}}$   (input open) | 9.7 |
| **(9.31)** | $Z_{in(base)} = h_{ie}$ | 9.7 |
| **(9.32)** | $r'_e = \dfrac{h_{ie}}{h_{fe}}$ | 9.7 |
| **(9.33)** | $A_v = \dfrac{h_{fe}r_C}{h_{ie}}$ | 9.7 |

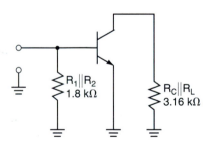

**FIGURE 9.34**

## Practice Problems

**§9.1**

1. An amplifier has an emitter current of 12 mA. Determine the value of $r'_e$ for the circuit.

2. An amplifier has an emitter current of 10 mA. Determine the value of $r'_e$ for the circuit.

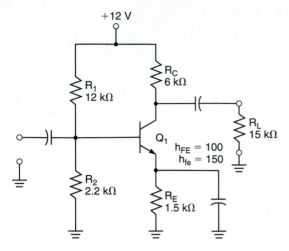

FIGURE 9.35

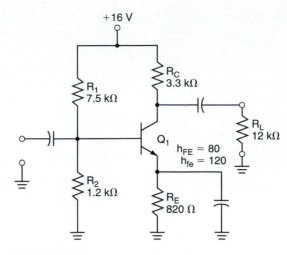

FIGURE 9.36

**3.** An amplifier has values of $V_E = 2.2$ V and $R_E = 910$ Ω. Determine the value of $r'_e$ for the circuit.

**4.** An amplifier has values of $V_E = 12$ V and $R_E = 4.7$ kΩ. Determine the value of $r'_e$ for the circuit.

**5.** An amplifier has values of $V_B = 3.2$ V and $R_E = 1.2$ kΩ. Determine the value of $r'_e$ for the circuit.

**6.** An amplifier has values of $V_B = 4.8$ V and $R_E = 3.9$ kΩ. Determine the value of $r'_e$ for the circuit.

**7.** Determine the value of $r'_e$ for the circuit shown in Figure 9.35.

**8.** Determine the value of $r'_e$ for the circuit shown in Figure 9.36.

### §9.3

**9.** Derive the ac equivalent circuit for the amplifier shown in Figure 9.35. Include all component values.

**10.** Derive the ac equivalent circuit for the amplifier shown in Figure 9.36. Include all component values.

**11.** Derive the ac equivalent circuit for the amplifier shown in Figure 9.37. Include all component values.

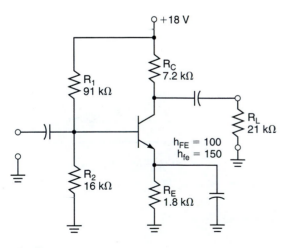

FIGURE 9.37

FIGURE 9.38

12. Derive the ac equivalent circuit for the amplifier shown in Figure 9.38. Include all component values.

## §9.4

13. An amplifier has values of $v_{in} = 120$ mV$_{ac}$ and $V_{out} = 4$ V$_{ac}$. Determine the value of $A_v$ for the circuit.

14. An amplifier has values of $v_{in} = 82$ mV$_{ac}$ and $v_{out} = 6.4$ V$_{ac}$. Determine the value of $A_v$ for the circuit.

15. An amplifier has values of $r_C = 2.2$ k$\Omega$ and $r'_e = 22.8$ $\Omega$. Determine the value of $A_v$ for the circuit.

16. An amplifier has values of $r_C = 4.7$ k$\Omega$ and $r'_e = 32$ $\Omega$. Determine the value of $A_v$ for the circuit.

17. An amplifier has values of $r_C = 2.7$ k$\Omega$ and $I_E = 1$ mA. Determine the value of $A_v$ for the circuit.

18. An amplifier has values of $r_C = 3.3$ k$\Omega$ and $I_E = 2$ mA. Determine the value of $A_v$ for the circuit.

19. Determine the value of $A_v$ for the amplifier shown in Figure 9.35.

20. Determine the value of $A_v$ for the amplifier shown in Figure 9.36.

21. Determine the value of $A_v$ for the amplifier shown in Figure 9.37.

22. Determine the value of $A_v$ for the amplifier shown in Figure 9.38.

23. The amplifier in Figure 9.37 has a 12 mV$_{ac}$ input signal. Determine the value of $v_{out}$ for the circuit.

24. The amplifier in Figure 9.36 has a 22 mV$_{ac}$ input signal. Determine the value of $v_{out}$ for the circuit.

25. Determine the value of $A_p$ for the circuit in Figure 9.35. Assume that the circuit has a value of $A_i = 14$.

26. Determine the value of $A_p$ for the circuit in Figure 9.36. Assume that the circuit has a value of $A_i = 11$.

27. Determine the value of $A_p$ for the circuit in Figure 9.37. Assume that the circuit has a value of $A_i = 31$.

28. Determine the value of $A_p$ for the circuit in Figure 9.38. Assume that the circuit has a value of $A_i = 7.5$.

29. An amplifier has values of $A_v = 110$ and $A_i = 40$. Determine the values of $A_p$ and $P_{out}$ when $P_{in} = 10$ mW.

30. An amplifier has values of $A_v = 68.8$ and $A_i = 1.44$. Determine the values of $A_p$ and $P_{out}$ when $P_{in} = 240$ mW.

## §9.5

31. Calculate the open-load voltage gain for the circuit in Figure 9.35.

32. Calculate the open-load voltage gain for the circuit in Figure 9.36.

33. Calculate the open-load voltage gain for the circuit in Figure 9.37.

34. Calculate the open-load voltage gain for the circuit in Figure 9.38.

35.. Determine the values of $Z_{in(base)}$ and $Z_{in}$ for the amplifier in Figure 9.35.

36. Determine the values of $Z_{in(base)}$ and $Z_{in}$ for the amplifier in Figure 9.36.

37. Determine the values of $Z_{in(base)}$ and $Z_{in}$ for the amplifier in Figure 9.37.

38. Determine the values of $Z_{in(base)}$ and $Z_{in}$ for the amplifier in Figure 9.38.

39. Determine whether or not the assumed value of $A_i$ in Problem 25 is correct.

40. Determine whether or not the assumed value of $A_i$ in Problem 26 is correct.

41. Determine whether or not the assumed value of $A_i$ in Problem 27 is correct.

42. Determine whether or not the assumed value of $A_i$ in Problem 28 is correct.

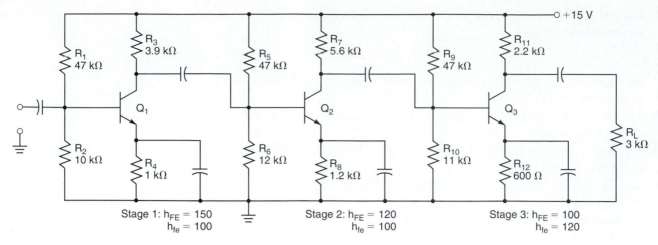

FIGURE 9.39

**43.** Determine the value of $A_v$ for the third-stage of the amplifier in Figure 9.39.

**44.** Determine the value of $A_v$ for the second-stage of the amplifier in Figure 9.39.

**45.** Determine the value of $A_v$ for the first-stage of the amplifier in Figure 9.39.

**46.** A two-stage amplifier has values of $A_{v1} = 23.8$, $A_{v2} = 122$, $A_{i1} = 24$, and $A_{i2} = 38$. Determine the values of $A_{vT}$, $A_{iT}$, and $A_{pT}$ for the circuit.

**47.** A two-stage amplifier has values of $A_{v1} = 88.6$, $A_{v2} = 90.3$, $A_{i1} = 11$, and $A_{i2} = 21$. Determine the values of $A_{vT}$, $A_{iT}$, and $A_{pT}$ for the circuit.

**48.** A two-stage amplifier has values of $A_{v1} = 24.8$, $A_{v2} = 77.1$, $A_{i1} = 30$, and $A_{i2} = 9$. Determine the values of $A_{vT}$, $A_{iT}$, and $A_{pT}$ for the circuit.

**49.** Determine the values of $A_{vT}$, $A_{iT}$, and $A_{pT}$ for the amplifier shown in Figure 9.39.

### §9.6

**50.** Determine the value of $A_v$ for the amplifier shown in Figure 9.40.

**51.** Determine the value of $A_v$ for the amplifier shown in Figure 9.41.

**52.** Determine the value of $A_v$ for the amplifier shown in Figure 9.42.

**53.** Determine the values of $Z_{in(base)}$ and $Z_{in}$ for the amplifier shown in Figure 9.40.

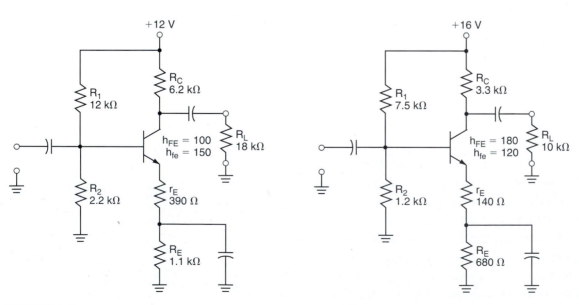

FIGURE 9.40

FIGURE 9.41

**FIGURE 9.42**

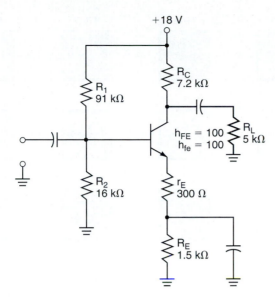

54. Determine the values of $Z_{in(base)}$ and $Z_{in}$ for the amplifier shown in Figure 9.41.

§9.7

55. An amplifier has values of $h_{fe} = 100$, $h_{ie} = 5$ k$\Omega$, and $r_C = 3.8$ k$\Omega$. Determine the values of $Z_{in(base)}$, $r'_e$, and $A_v$ for the circuit.

56. An amplifier has values of $h_{fe} = 120$, $h_{ie} = 4$ k$\Omega$, and $r_C = 3.8$ k$\Omega$. Determine the values of $Z_{in(base)}$, $r'_e$, and $A_v$ for the circuit.

57. Refer to Figure 9.43. The transistor described is used in an amplifier with values of $I_{CQ} = 1$ mA and $r_C = 2.48$ k$\Omega$. Determine the values of $Z_{in(base)}$ and $A_v$ for the circuit.

58. Refer to Figure 9.43. The transistor described is used in an amplifier with values of $I_{CQ} = 2$ mA and $r_C = 1.18$ k$\Omega$. Determine the values of $Z_{in(base)}$ and $A_v$ for the circuit.

59. Refer to Figure 9.43. The transistor described is used in an amplifier with values of $I_{CQ} = 5$ mA and $r_C = 878$ $\Omega$. Determine the values of $Z_{in(base)}$ and $A_v$ for the circuit.

60. Refer to Figure 9.43. The transistor described is used in an amplifier with values of $I_{CQ} = 10$ mA and $r_C = 1.05$ k$\Omega$. Determine the values of $Z_{in(base)}$ and $A_v$ for the circuit.

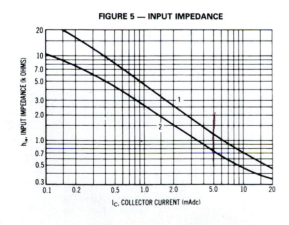

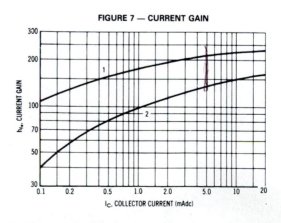

**FIGURE 9.43   Motorola 2N2222 curves. (Copyright of Motorola. Used by permission.)**

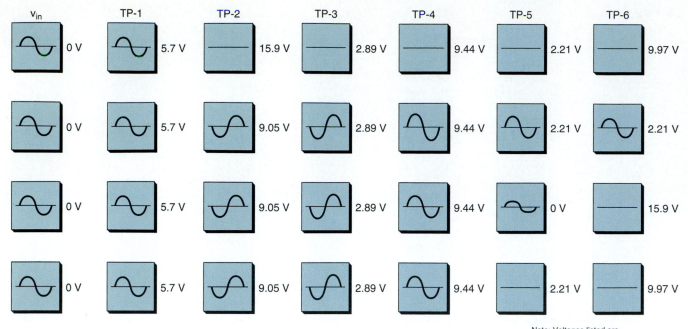

| $v_{in}$ | TP-1 | TP-2 | TP-3 | TP-4 | TP-5 | TP-6 |
|---|---|---|---|---|---|---|
| 0 V | 5.7 V | 15.9 V | 2.89 V | 9.44 V | 2.21 V | 9.97 V |
| 0 V | 5.7 V | 9.05 V | 2.89 V | 9.44 V | 2.21 V | 2.21 V |
| 0 V | 5.7 V | 9.05 V | 2.89 V | 9.44 V | 0 V | 15.9 V |
| 0 V | 5.7 V | 9.05 V | 2.89 V | 9.44 V | 2.21 V | 9.97 V |

Note: Voltages listed are DC reference voltages, as measured from the designated test point to ground.

**FIGURE 9.44**

*Troubleshooting Practice Problems*

**61.** In Figure 9.44, several sets of waveforms are shown. Each row represents a series of signal checks performed on the circuit shown in Figure 9.45. For each set of waveforms shown, answer the following questions:

**a.** Do the waveforms indicate any problem in the circuit?

**b.** If your answer to part (a) is yes, where do you think the problem is probably located? (Simply give the amplifier stage or indicate the coupling between two specific stages.)

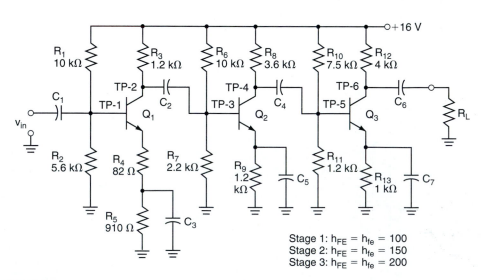

Stage 1: $h_{FE} = h_{fe} = 100$
Stage 2: $h_{FE} = h_{fe} = 150$
Stage 3: $h_{FE} = h_{fe} = 200$

**FIGURE 9.45**

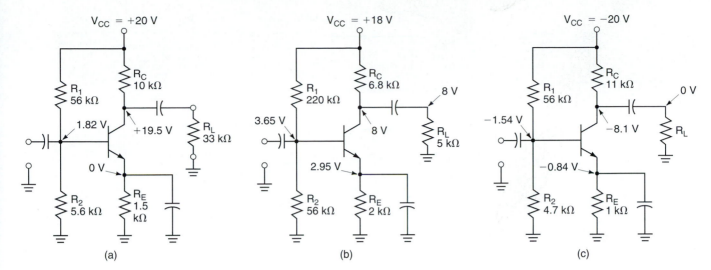

**FIGURE 9.46**

62. (Review) For each circuit shown in Figure 9.46, state whether or not a problem is indicated by the dc voltages shown, and state the possible cause(s) of the problem, if any.

63. Determine the values of $Z_{in}$, $A_v$, $A_i$, and $A_p$ for the amplifier shown in Figure 9.47.

64. The load resistor in Figure 9.47 opens. What is the resulting change in the peak-to-peak output voltage?

65. Determine the worst-case values of $A_v$ for the circuit shown in Figure 9.48.

*The Brain Drain*

66. Write a program to determine the values of $A_v$, $Z_{in(base)}$, and $Z_{in}$ for a voltage-divider-biased common-emitter amplifier, given the circuit resistor values, the load resistance, the values of $h_{ie}$ and $h_{fe}$, and the value of $V_{CC}$.

67. If you wanted to write a program for determining the input impedance and gain of a swamped amplifier, what input information would the user have to provide?

68. Write the program described in Problem 67.

*Suggested Computer Applications Problems*

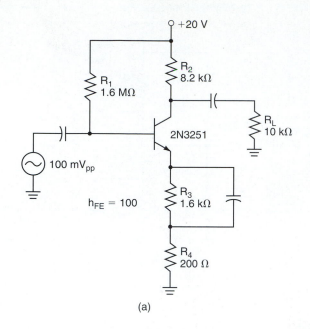

(a)

h PARAMETERS
$V_{CE} = 10$ V, f = 1 kc, $T_A = 25$ °C

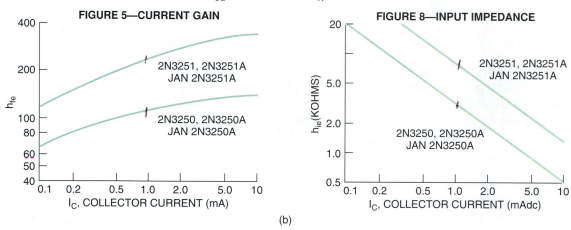

(b)

FIGURE 9.47   Operating curves. (Copyright of Motorola. Used by permission.)

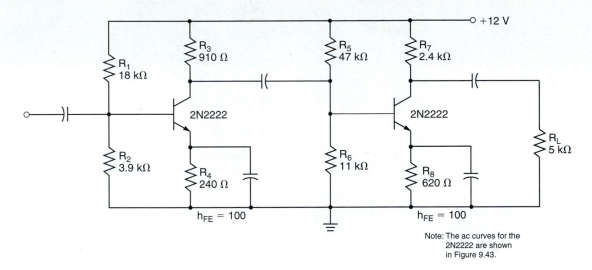

FIGURE 9.48

# 10

# OTHER BJT
# AMPLIFIERS

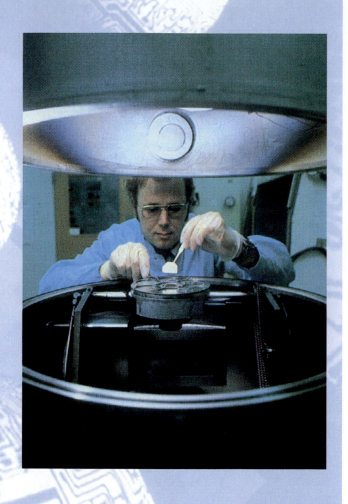

The process of producing solid-state components requires extremely high temperatures. The ovens shown here are used to heat the silicon wafers to a temperature of approximately 1400°C.

# OUTLINE

# OBJECTIVES

*After studying the material in this chapter, you should be able to:*

1. Perform a complete dc analysis of an emitter follower.

2. Perform a complete ac analysis of an emitter follower.

3. Discuss the use of emitter-feedback bias and collector-bypass capacitors in emitter-follower circuits.

4. Describe the use of the emitter follower as a current amplifier and as a buffer.

5. Perform a complete dc analysis of a Darlington emitter-follower.

6. Perform a complete ac analysis of a Darlington emitter-follower.

7. Perform a complete ac analysis of a common-base amplifier

8. Describe the use of the common-base amplifier as a voltage amplifier and a high-frequency buffer.

9. List the overall gain and impedance characteristics of all the commonly used BJT amplifiers.

# AFTER CE AMPLIFIERS: THE OTHER CONFIGURATIONS

You have just finished a long chapter that focused on the operating principles of common-emitter (CE) amplifiers. Now you are starting the *single* chapter that covers the two other types of BJT amplifiers: the common-base (CB) and the common-collector (CC) amplifiers. At this point, you may be wondering why we spent so much more time and energy on CE amplifiers than we are on the other two types combined.

There are two reasons for the extended coverage of the CE amplifier. First, the CE amplifier is used much more extensively than either CC or CB amplifiers. While CB and CC amplifiers are used in a wide variety of applications and systems, they are not used nearly as often as the CE amplifier. Second, many of the principles covered in connection with the CE amplifier apply to the CB and CC amplifiers as well. Since these principles were covered earlier, they need to be only briefly reviewed in this chapter. We will spend most of our time in this chapter discussing the ways in which CB and CC amplifiers differ from CE amplifiers.

I n Chapter 9, we concentrated on the common-emitter amplifier. Now we will take a look at the common-collector amplifier and the common-base amplifier. We will also discuss the Darlington amplifier, a two-transistor configuration commonly used in common-emitter and common-collector applications.

The **common-collector amplifier** is used to provide *current gain* and in *impedance-matching* applications. The input to this circuit is applied to the base, while the output is taken from the emitter. The voltage gain of the common-collector circuit is always less than 1, and the output voltage is in phase with the input voltage. Since the output signal "follows" the input signal, the common-collector amplifier is commonly referred to as the *emitter follower*.

The **Darlington amplifier** consists of two transistors connected so they provide dc and ac beta values equal to the product of the individual transistor beta values. The Darlington amplifier can be connected to any configuration. It is used most often in common-emitter and emitter-follower circuits.

The **common-base amplifier** is used to provide *voltage gain* for *high-frequency applications*. The common-base circuit has current gain that is always less than 1 and, again, an output voltage that is in phase with the input voltage.

**Common-collector amplifier**
A current amplifier that has no voltage gain. Also called an **emitter follower**.

**Darlington amplifier**
A two-transistor circuit used for its extremely high current gain and input impedance.

**Common-base amplifier**
A voltage amplifier that has no current gain.

## *10.1* THE EMITTER FOLLOWER (COMMON-COLLECTOR AMPLIFIER)

The *emitter follower* is a *current amplifier* with a voltage gain that is less than 1. The input is applied to the base of the transistor, while the output is taken from the emitter. The emitter follower is shown in Figure 10.1.

As you can see, there is no collector resistor in the circuit. Notice also that there is no emitter bypass capacitor. These are the two circuit recognition features for the emitter follower.

### *dc Operation*

OBJECTIVE 1 ▶ For the circuit shown in Figure 10.1, the base voltage is established by a voltage divider. This circuit works in the same way as the voltage-divider bias circuit of the common-emitter amplifier. Thus, the base voltage is found as

$$V_B = \frac{R_2}{R_1 + R_2} V_{CC} \tag{10.1}$$

**FIGURE 10.1** The emitter-follower (common-collector circuit).

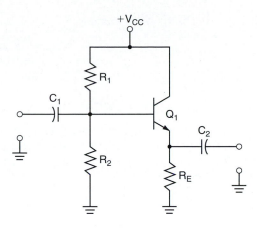

Once the base voltage is found, the emitter voltage is found as

$$V_E = V_B - 0.7 \text{ V} \qquad \textbf{(10.2)}$$

and the emitter current is found as

$$I_E = \frac{V_E}{R_E} \qquad \textbf{(10.3)}$$

**Lab Reference:** These dc values are calculated and measured in Exercise 20.

Now $V_{CEQ}$ is found as

$$V_{CEQ} = V_{CC} - V_E \qquad \textbf{(10.4)}$$

If you refer to Chapter 7, you will see that the first three formulas are identical to those we used for the analysis of common-emitter bias circuits. Example 10.1 shows how the dc analysis of an emitter follower is performed.

---

Determine the values of $V_B$, $V_E$, $I_E$, and $V_{CEQ}$ for the circuit shown in Figure 10.2.

*Solution:* The base voltage is found as

$$V_B = \frac{R_2}{R_1 + R_2} V_{CC}$$

$$= 5 \text{ V}$$

The emitter voltage is then found as

$$V_E = V_B - 0.7 \text{ V}$$
$$= 4.3 \text{ V}$$

Now $I_E$ is found as

$$I_E = \frac{V_E}{R_E}$$

$$= 860 \text{ } \mu\text{A}$$

**EXAMPLE 10.1**

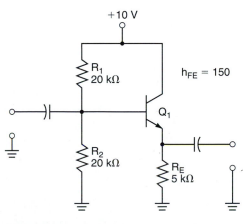

**FIGURE 10.2**

And $V_{CEQ}$ is found as

$$V_{CEQ} = V_{CC} - V_E$$
$$= 10\text{ V} - 4.3\text{ V}$$
$$= 5.7\text{ V}$$

**PRACTICE PROBLEM 10.1**

A circuit like the one in Figure 10.2 has the following values: $V_{CC} = +18$ V, $R_E = 910$ Ω, $R_1 = 16$ kΩ, $R_2 = 22$ kΩ, and $h_{FE} = 200$. Determine the values of $V_B$, $V_E$, $I_E$, and $V_{CEQ}$ for the circuit.

## The dc Load Line

As with the common-emitter circuit, the dc load line of the emitter follower is determined by the saturation and cutoff characteristics of the amplifier.

When the transistor is saturated, the *ideal* value of $V_{CE}$ is 0 V. Since $V_E = V_{CC}$, the value of $I_{C(\text{sat})}$ is found as

$$I_{C(\text{sat})} \cong \frac{V_{CC}}{R_E} \qquad (10.5)$$

As with the common-emitter circuit, the total applied voltage will be dropped across the transistor when it is in cutoff. Thus, $V_{CE(\text{off})}$ is found as

$$V_{CE(\text{off})} = V_{CC} \qquad (10.6)$$

These two equations are used to derive the dc load line of an emitter-follower amplifier, as illustrated in Example 10.2.

---

**EXAMPLE 10.2**

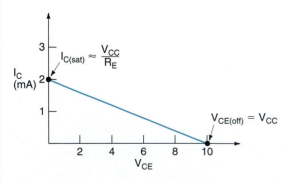

FIGURE 10.3

Derive and draw the dc load line for the circuit shown in Figure 10.2.

**Solution:** The value of $I_{C(\text{sat})}$ is found as

$$I_{C(\text{sat})} = \frac{V_{CC}}{R_E}$$
$$= 2\text{ mA}$$

and now the value of $V_{CE(\text{off})}$ is found as

$$V_{CE(\text{off})} = V_{CC}$$
$$= 10\text{ V}$$

Thus, the dc load line for Figure 10.2 is drawn as shown in Figure 10.3.

**PRACTICE PROBLEM 10.2**

Derive and draw the dc load line for the emitter follower described in Practice Problem 10.1.

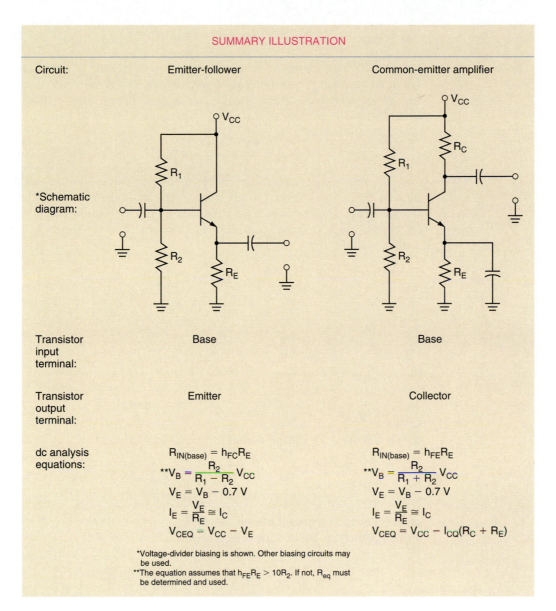

**FIGURE 10.4**

That's all there is to the dc analysis of the emitter follower. As you can see, this analysis is nearly identical to that of the common-emitter amplifier. The similarities and differences between the dc characteristics of the emitter-follower and the common-emitter amplifier are illustrated in Figure 10.4.

---

**Section Review**

1. What is an emitter follower?
2. What are the circuit recognition features of the emitter follower?
3. List, in order, the steps involved in the dc analysis of an emitter follower.
4. How do you determine the end points of the emitter-follower dc load line?

# 10.2 ac ANALYSIS OF THE EMITTER FOLLOWER

OBJECTIVE 2 ▶ Figure 10.5a shows a typical emitter follower. The key to the ac operation of this circuit lies in the fact that the load is capacitively coupled to the transistor *emitter*. Because of this, the ac equivalent of the circuit is drawn as shown in Figure 10.5b. Note that the resistor shown in the emitter circuit ($r_E$) represents the *total ac emitter resistance*. This resistance is found as

$$r_E = R_E \parallel R_L \qquad (10.7)$$

**Lab Reference:** The gain characteristics of the emitter follower are demonstrated in Exercise 20.

The low voltage gain of the emitter follower is due to the fact that the output circuit acts as a voltage divider to the input signal, as shown in Figure 10.5c. As you can see, the input is applied to both the input circuit ($R_1 \parallel R_2$) and the output circuit. The output circuit, which is made up of $r'_e$ and $r_E$, divides the input signal. Using the voltage divider equation,

$$v_{out} = v_{in} \frac{r_E}{r'_e + r_E}$$

or

$$\frac{v_{out}}{v_{in}} = \frac{r_E}{r'_e + r_E}$$

Finally, since

$$A_v = \frac{v_{out}}{v_{in}}$$

we can calculate the voltage gain of the circuit as

$$A_v = \frac{r_E}{(r'_e + r_E)} \qquad (10.8)$$

In most practical applications $r_E$ is much greater than $r'_e$. When $r_E \gg r'_e$, the following approximation can be used for voltage gain:

$$A_v \cong 1 \qquad (\text{when } r_E \gg r'_e) \qquad (10.9)$$

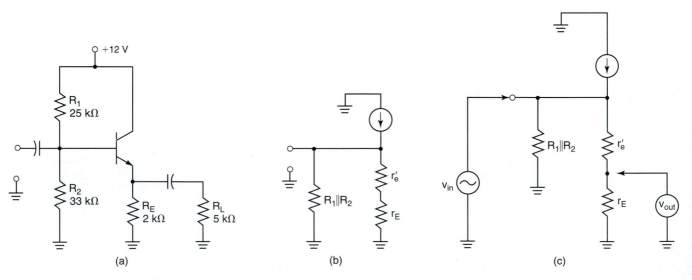

**FIGURE 10.5    A typical emitter follower and its ac equivalents.**

The greater the difference between $r_E$ and $r'_e$, the closer the amplifier voltage gain comes to 1. In practice, the voltage gain of the emitter follower is usually between 0.8 and 0.999. This is illustrated in Example 10.3.

---

Determine the value of $A_v$ for the circuit shown in Figure 10.5a.

EXAMPLE 10.3

**Solution:**   First, the value of $r_E$ is found as

$$r_E = R_E \| R_L$$
$$= 1.43 \text{ k}\Omega$$

We do not know the $h$-parameter values for the transistor, so we need to approximate the value of $r'_e$ using

$$r'_e = \frac{25 \text{ mV}}{I_E}$$

Using the established procedure, the value of $I_E$ for the circuit is found to be 3.1 mA, and the value of $r'_e$ is found as

$$r'_e = \frac{25 \text{ mV}}{I_E}$$
$$= 8.1 \text{ } \Omega$$

Using the value of $r'_e$ and the value of $r_E$, the voltage gain of the amplifier is found as

$$A_v = \frac{r_E}{r'_e + r_E}$$
$$= \frac{1.43 \text{ k}\Omega}{8.1 \text{ } \Omega + 1.43 \text{ k}\Omega}$$
$$= 0.9944$$

**PRACTICE PROBLEM 10.3**

Assume that the amplifier described in Practice Problem 10.1 has a 4-k$\Omega$ load. Determine the value of $A_v$ for the circuit.

---

## Current Gain

In Chapter 9, you were shown that the current gain ($A_i$) of a common-emitter amplifier is significantly lower than the current gain of the transistor ($h_{fe}$). This relationship was due to the current divisions that occurred in both the input and output circuits. The same principle holds true for the emitter follower. The current gain of an emitter follower is found as

$$A_i = h_{fc} \left( \frac{Z_{in} r_E}{Z_{in(base)} R_L} \right) \qquad (10.10)$$

where $h_{fc} \cong h_{fe}$. Note that the subscript $c$ merely indicates that the parameter applies to an emitter-follower (common-collector) amplifier, rather than to a common-emitter amplifier. Although there is a more exact $h$-parameter derivation for $h_{fc}$, and thus $A_i$, using $h_{fc} = h_{fe}$ in equation (10.10) will always do the trick for circuit calculations.

*A Practical Consideration:* If you look in Appendix C, you'll see that the exact equation for the value of $h_{fc}$ is

$$h_{fc} = h_{fe} + 1$$

Also shown is the equation derivation. Since $h_{fe}$ is normally much greater than 1, we can simply assume that $h_{fc} \cong h_{fe}$.

## Power Gain

As was the case with the common-emitter amplifier, the power gain ($A_p$) of an emitter follower is found as the product of current gain ($A_i$) and voltage gain ($A_v$). By formula,

$$A_p = A_i A_v \qquad \textbf{(10.11)}$$

Since the value of $A_v$ is always slightly less than 1, the power gain ($A_p$) of an emitter follower will always be slightly less than the current gain ($A_i$) of the circuit. This point is illustrated in Example 10.4.

---

**EXAMPLE 10.4**

The amplifier in Figure 10.5 (Example 10.3) has a value of $A_i = 2.7$. Determine the power gain ($A_p$) of the amplifier.

***Solution:*** In Example 10.3, we determined the voltage gain ($A_v$) of the amplifier to be 0.9944. With a current gain of $A_i = 2.7$, the power gain of the circuit is found as

$$
\begin{aligned}
A_p &= A_i A_v \\
&= (2.7)(0.9944) \\
&= 2.68
\end{aligned}
$$

As you can see, the value of $A_p$ for the emitter follower is slightly less than the value of current gain, $A_i$.

***PRACTICE PROBLEM 10.4***

The amplifier described in Practice Problem 10.1 has a 4-k$\Omega$ load and a current gain of $A_i = 24$. Determine the power gain of the amplifier.

---

## Input Impedance ($Z_{in}$)

**Lab Reference:** The input impedance of an emitter follower is measured in Exercise 20.

As with the common-emitter amplifier, the input impedance of the emitter follower is found as the parallel equivalent resistance of the base resistors and the transistor input impedance. By formula,

$$Z_{in} = R_1 \, \| \, R_2 \, \| \, Z_{in(base)} \qquad \textbf{(10.12)}$$

The base input impedance is equal to the product of the device current gain and the total ac emitter resistance. By formula,

$$Z_{in(base)} = h_{fc}(r'_e + r_E) \qquad \textbf{(10.13)}$$

where $h_{fc}$ = the transistor current gain
$\quad\;\; r'_e$ = the ac resistance of the transistor emitter
$\quad\;\; r_E$ = the parallel combination of $R_E$ and $R_L$

Example 10.5 illustrates the procedure used to determine the input impedance of an emitter follower.

EXAMPLE 10.5

Determine the input impedance of the amplifier shown in Figure 10.5a. Assume that the value of $h_{fe}$ for the transistor is listed on the spec sheet as 220.

**Solution:** In Example 10.3, we calculated values of $r_E = 1.43 \text{ k}\Omega$ and $r'_e = 8.1 \, \Omega$. Assuming that $h_{fc}$ is approximately equal to $h_{fe}$, the impedance of the transistor is found as

$$Z_{\text{in(base)}} = h_{fc}(r'_e + r_E)$$
$$= (220)(8.1 \, \Omega + 1.43 \text{ k}\Omega)$$
$$= 316.4 \text{ k}\Omega$$

Now the input impedance of the amplifier is found as

$$Z_{\text{in}} = R_1 \| R_2 \| Z_{\text{in(base)}}$$
$$= 25 \text{ k}\Omega \| 33 \text{ k}\Omega \| 316.4 \text{ k}\Omega$$
$$= 13.6 \text{ k}\Omega$$

**PRACTICE PROBLEM 10.5**

The amplifier described in Practice Problem 10.1 has a current gain of $h_{fe} = 240$ and a load resistance of $R_L = 2 \text{ k}\Omega$. Determine the input impedance of the circuit.

When the value of $r_E$ is much greater than $r'_e$, as was the case in Example 10.5, equation (10.13) can be simplified as follows:

$$Z_{\text{in(base)}} \cong h_{fc}r_E \qquad (\text{when } r_E \gg r'_e) \qquad \textbf{(10.14)}$$

The validity of this approximation can be seen by recalculating the values of $Z_{\text{in(base)}}$ and $Z_{\text{in}}$ for the circuit in Figure 10.5a. If you ignore the value of $r'_e$ in the calculations, you will end up with the same value of $Z_{\text{in}}$ that was found in Example 10.5.

## Output Impedance ($Z_{\text{out}}$)

As you may recall, **output impedance** is *the impedance that a circuit presents to its load.* When a load is connected to a circuit, the output impedance of the circuit acts as the source impedance for that load.

**Output impedance**
The impedance that a circuit presents to its load. The output impedance of a circuit is effectively the source impedance for its load.

Since the emitter follower is often used in *impedance-matching* applications, $Z_{\text{out}}$ is an important consideration. One main function of the emitter follower is to "match" a source impedance to a load impedance to improve power transfer from the source to the load. This point will be made clear in the following section.

The output impedance of an emitter follower is found as

$$Z_{\text{out}} = R_E \left\| \left( r'_e + \frac{R'_{\text{in}}}{h_{fc}} \right) \right. \qquad \textbf{(10.15)}$$

where $Z_{\text{out}}$ = the output impedance of the amplifier
$R'_{\text{in}} = R_1 \| R_2 \| R_S$
$R_S$ = the output resistance of the input voltage source

Equation (10.15) is derived in Appendix D. The use of the equation is demonstrated in Example 10.6.

## EXAMPLE 10.6

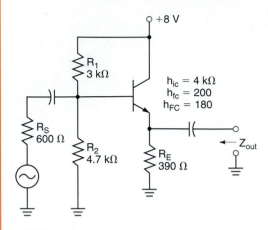

**FIGURE 10.6**

Determine the output impedance ($Z_{\text{out}}$) of the amplifier shown in Figure 10.6.

**Solution:**   Using the *h*-parameter method of analysis, the value of $r'_e$ is found as

$$r'_e = \frac{h_{ie}}{h_{fe}}$$

$$= \frac{4\text{ k}\Omega}{200}$$

$$= 20\text{ }\Omega$$

The value of $R'_{\text{in}}$ is now found as

$$R'_{\text{in}} = R_1 \| R_2 \| R_S$$

$$= 3\text{ k}\Omega \| 4.7\text{ k}\Omega \| 600\text{ }\Omega$$

$$= 452\text{ }\Omega$$

Finally, the output impedance of the amplifier is found as

$$Z_{\text{out}} = R_E \| \left( \frac{r'_e + R'_{\text{in}}}{h_{fc}} \right)$$

$$= 390\text{ }\Omega \| \left( 20\text{ }\Omega + \frac{452\text{ }\Omega}{200} \right)$$

$$= 390\text{ }\Omega \| 22.26\text{ }\Omega$$

$$= 21.06\text{ }\Omega$$

### PRACTICE PROBLEM 10.6

An emitter follower has the following values: $V_{CC} = +12$ V, $R_E = 200$ $\Omega$, $R_1 = 2.5$ k$\Omega$, $R_2 = 3.3$ k$\Omega$, $R_S = 500$ $\Omega$, $h_{FC} = 180$, $h_{fc} = 200$, and $h_{ic} = 3$ k$\Omega$. Determine the value of $Z_{\text{out}}$ for the amplifier.

Calculating the output impedance of an amplifier is not an everyday practice for most technicians. However, when we discuss the emitter follower as an impedance-matching circuit in the next section, you will see that being able to determine the output impedance of an emitter follower provides us with a useful circuit analysis tool.

## Putting It All Together

The ac analysis of an emitter follower typically involves determining the values of $A_i$, $A_v$, $A_p$, and $Z_{\text{in}}$ for the circuit. Assuming that *h*-parameters for the transistor are available, the current, voltage, and power gain values are found as

$$A_i = h_{fc} \left( \frac{Z_{\text{in}} r_E}{Z_{\text{in(base)}} R_L} \right)$$

$$A_v = \frac{r_E}{r'_e + r_E}$$

$$A_p = A_v A_i$$

and the input impedance is found as

$$Z_{\text{in}} = R_1 \| R_2 \| Z_{\text{in(base)}}$$

where

$$Z_{in(base)} = h_{fc}(r'_e + r_E)$$

The value of $Z_{in(base)}$ for an emitter follower is typically higher than that of a common-emitter amplifier. Because of this, the value of $Z_{in}$ will typically be higher for an emitter follower than for a comparable common-emitter amplifier.

The voltage gain ($A_v$) of an emitter follower is always less than 1. If you have trouble seeing this in the $A_v$ equation for the circuit, consider this: The emitter voltage of a transistor must always be slightly less than the base voltage for the transistor to be biased properly. Since $V_E$ is always slightly less than $V_B$, it would make sense that any ac emitter voltage would have to be slightly less than any ac base voltage.

The output impedance of an emitter follower is considered only in *impedance-matching* applications. When the circuit is used as an impedance-matching circuit, you need to be able to calculate the value of $Z_{out}$ for the amplifier. In any other application, this calculation would not be necessary.

1. What is the limit on the value of voltage gain ($A_v$) for an emitter follower?

2. List, in order, the steps that you would take to determine the value of $A_v$ for an emitter follower.

3. How is the value of current gain ($A_i$) for an emitter follower determined?

4. What is the limit on the value of power gain ($A_p$) for an emitter follower?

5. What effect does loading have on the voltage gain of an emitter follower?

6. List, in order, the steps you would take to determine the value of loaded voltage gain ($A_v$) for an emitter follower.

8. List, in order, the steps you would take to determine the value of amplifier input impedance ($Z_{in}$) for an emitter follower.

9. What is output impedance ($Z_{out}$)?

10. List, in order, the steps you would take to determine the value of $Z_{out}$ for an emitter follower.

10. Describe the procedure used to perform a complete ac analysis of an emitter follower.

11. When is the value of $Z_{out}$ for an emitter follower an important consideration?

# 10.3 PRACTICAL CONSIDERATIONS, APPLICATIONS, AND TROUBLESHOOTING

Two practical considerations regarding emitter followers have been ignored up to this point. These are the use of emitter-feedback bias and the need for a *collector bypass capacitor*. Each of these practical considerations will be discussed in detail in this section. We will also discuss the most common applications for the emitter follower: as a *current amplifier* and as an *impedance-matching circuit*. Finally, we will take a brief look at the troubleshooting of the emitter follower.

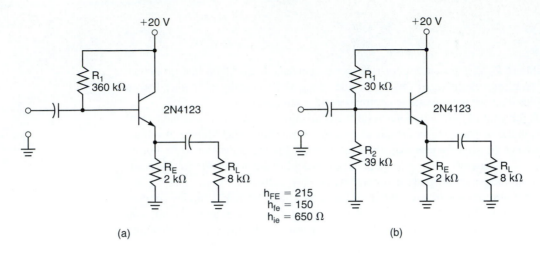

$h_{FE} = 215$
$h_{fe} = 150$
$h_{ie} = 650 \ \Omega$

(a)                                    (b)

**FIGURE 10.7**

## Using Emitter-feedback Bias to Increase Amplifier Input Impedance

OBJECTIVE 3 ▶ In many applications, such as impedance matching, one goal of emitter-follower design is to have as high an input impedance value as possible. When high input impedance is desirable, emitter-feedback bias is often used. An emitter follower with emitter-feedback bias is shown in Figure 10.7a. A comparable emitter follower with voltage-divider bias is shown in Figure 10.7b.

Why $Z_{in}$ is higher when emitter-feedback bias is used.

An emitter follower with emitter-feedback bias will have a much higher input impedance than a comparable circuit with voltage-divider bias. This point is easy to understand if you consider the equation for the input impedance of a voltage-divider biased emitter follower. As you know, this equation is

$$Z_{in} = R_1 \parallel R_2 \parallel Z_{in(base)}$$

Since the emitter-feedback bias circuit has only one base resistor, the input impedance is found as

$$Z_{in} = R_1 \parallel Z_{in(base)} \qquad (10.16)$$

As you know, the total resistance in a parallel circuit will be less than the lowest individual resistance value. Since $R_1$ in Figure 10.7b is much lower than $R_1$ in Figure 10.7a, it would make sense that the value of $Z_{in}$ for the voltage-divider bias circuit would have to be lower than the value of $Z_{in}$ for the emitter-feedback bias circuit. This point is illustrated further in Example 10.7.

---

**EXAMPLE 10.7**

Remember:

$h_{ic} = h_{ie}$

$h_{fc} \cong h_{fe}$

The amplifiers in Figure 10.7 were designed to have nearly identical output characteristics using identical transistors with the specifications shown. Determine the input impedance values for the two circuits.

**Solution:** Since the circuits are using identical transistors, emitter resistance values, and load resistance values, the value of $Z_{in(base)}$ will be the same for the two circuits. This value of $Z_{in(base)}$ is found using the following procedure:

$$r'_e = \frac{h_{ic}}{h_{fc}}$$

$$= 4.3 \ \Omega$$

Note: The values of $h_{ie}$ and $h_{fe}$ shown in Figure 10.7 were obtained from the 2N4123 operating curves.

$$r_E = R_E \, \| \, R_L$$
$$= 1.6 \ \text{k}\Omega$$

and

$$Z_{in(base)} = h_{fc}(r'_e + r_E)$$
$$= (150)(4.3 \ \Omega + 1.6 \ \text{k}\Omega)$$
$$= 240.6 \ \text{k}\Omega$$

For the emitter-feedback bias circuit, the value of $Z_{in}$ is found as

$$Z_{in} = R_1 \, \| \, Z_{in(base)}$$
$$= 360 \ \text{k}\Omega \, \| \, 240.6 \ \text{k}\Omega$$
$$= 144.2 \ \text{k}\Omega$$

For the voltage-divider biased circuit, the value of $Z_{in}$ is found as

$$Z_{in} = R_1 \, \| \, R_2 \, \| \, Z_{in(base)}$$
$$= 30 \ \text{k}\Omega \, \| \, 39 \ \text{k}\Omega \, \| \, 240.6 \ \text{k}\Omega$$
$$= 15.84 \ \text{k}\Omega$$

### PRACTICE PROBLEM 10.7

Two transistors similar to those in Figure 10.7 have values of $h_{FC} = 160$, $h_{fc} = 120$, and $h_{ic} = 540 \ \Omega$. Both circuits have values of $R_E = 1.5 \ \text{k}\Omega$ and $R_L = 2.7 \ \text{k}\Omega$. In the emitter-bias circuit, $R_1 = 300 \ \text{k}\Omega$, and in the voltage-divider circuit, $R_1 = 6.2 \ \text{k}\Omega$ and $R_2 = 9.1 \ \text{k}\Omega$. Calculate and compare the input impedance values for the two circuits.

As you can see, the value of $Z_{in}$ for an emitter-feedback biased emitter follower will be much greater than that of a comparable circuit with voltage-divider bias. This is why emitter-feedback biasing is commonly used in emitter-follower impedance-matching applications. This point will become clear later in this section.

## The Collector Bypass Capacitor

A problem can occur as a result of the normal operation of an emitter follower. This problem can be seen in the oscilloscope display shown in Figure 10.8. The upper trace in Figure 10.8 is the input signal to an emitter follower. As you can see, the output waveform (lower trace) is very distorted. The cause of this distortion can be explained with the help of Figure 10.9.

Why a collector bypass capacitor is used.

In Figure 10.9, $V_{CC}$ is represented as a simple battery with some value of internal resistance ($R_{int}$). The dc power supply is part of a loop that also contains the transistor $Q_1$, $C_{C2}$, and $R_L$.

If the value of $R_L$ is low enough, the capacitor may try to charge and discharge when an ac signal is applied to the circuit. As the capacitor charges and discharges through the dc power supply, the value of $V_{CC}$ will actually vary slightly, as shown in the following equation:

$$\Delta V = (\Delta I)(R_{int})$$

FIGURE 10.8

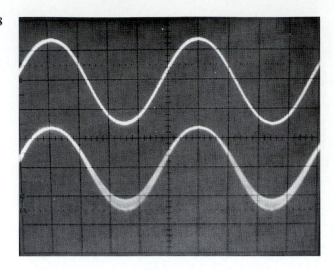

where $\Delta V$ = the change in $V_{CC}$

$\Delta I$ = the change in loop current caused by the capacitor charge/discharge action

$R_{\text{int}}$ = the internal resistance of the power supply

Thus, the charge/discharge action of the capacitor will cause a change in the value of $V_{CC}$, resulting in the distorted output waveform shown in Figure 10.8.

There are two ways to prevent the output distortion shown in Figure 10.8. The first is to use a **regulated dc power supply**. A regulated dc power supply is one that has *extremely low* internal resistance, typically less than 10 Ω. When a regulated dc power supply is used, the value of $\Delta V$ will be extremely small. In most cases, it will be too small to see in an oscilloscope display.

Another practical solution to the problem is the use of a **collector bypass capacitor** or *decoupling capacitor*. An emitter follower with a collector bypass capacitor ($C_B$) is shown in Figure 10.10.

The collector bypass capacitor is connected between $V_{CC}$ and ground and thus is in parallel with the dc power supply. When $C_{C2}$ tries to charge or discharge through the dc power supply, the change in current ($\Delta I$) is shorted around the supply by $C_B$. Thus, $V_{CC}$ does not change, and the distortion shown in Figure 10.8 is eliminated.

For $C_B$ to be effective, its value must be extremely small. A small-value $C_B$ will ensure that the dc power supply is completely bypassed. Typically, $C_B$ should be 0.01 μF or lower in value.

**Regulated dc power supply**
A dc power supply with extremely low internal resistance.

**Collector bypass capacitor**
A capacitor connected between $V_{CC}$ and ground parallel with the dc power supply. Often referred to as a *decoupling capacitor.*

Typically, a collector bypass capacitor will have a value of 0.01 μF or less.

**FIGURE 10.9**

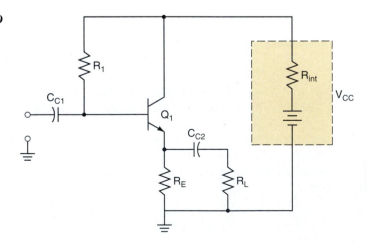

FIGURE 10.10

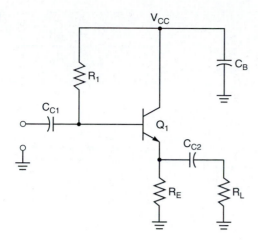

## Emitter-Follower Applications

The emitter follower is used when you need *current amplification without voltage gain.* There are many instances (especially in digital electronics) when an increase in current is required, but an increase in voltage is not. Because the emitter follower ideally has a voltage gain of 1 and relatively high current gain, it is typically used in these applications.

Consider the circuit shown in Figure 10.11. The first stage of the amplifier has specific values of $A_i$ and $A_v$. What if $A_v$ is high enough to provide the final required output voltage, but $A_i$ is not high enough? The emitter follower would be used to provide the additional current gain without increasing $A_v$ beyond its desired value.

Another application for the emitter follower is an **impedance-matching** circuit. Such an application is illustrated in Figure 10.12. The input and output impedance values shown were calculated for the circuit in Figure 10.7a.*

You may recall from your study of basic electronics that *maximum power is transferred to a load when the source and load impedances are equal.* When the source impedance equals the load impedance, the two impedances are said to be *matched.* When the impedances are matched, maximum power is delivered from the source to the load. The closer you get to a perfect match, the more power you will get at the load.

Note that the emitter follower shown in Figure 10.12 has a *relatively high input impedance* and a *relatively low output impedance.* This makes it a perfect circuit for con-

◄ *OBJECTIVE 4*

Emitter followers are used when $A_i$ is needed and $A_v$ is not.

**Impedance matching**
A circuit design technique that aids in power transfer between a source and its load.

Emitter followers may be used to provide maximum power transfer to low Z loads.

*The value of $Z_{out} = 21.06\ \Omega$ was calculated in Example 10.6 using an assumed value of $R_S = 600\ \Omega$. The value $Z_{in} = 144\ k\Omega$ was calculated in Example 10.7.

**FIGURE 10.11**

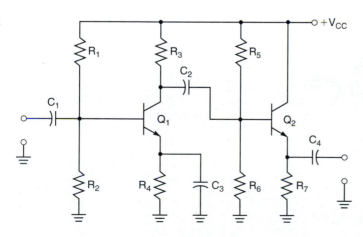

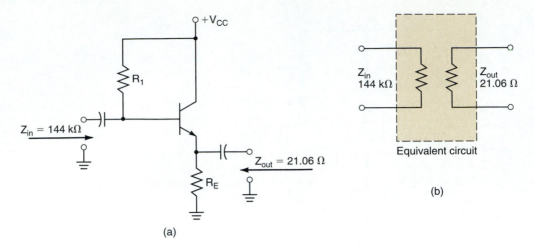

Equivalent circuit

(b)

(a)

FIGURE 10.12

**Buffer**

Any circuit that is used for impedance matching that aids in the transfer of power from a source to its load.

*necting a low-impedance load to a high-impedance source.* For example, look at the circuits shown in Figure 10.13. In Figure 10.13a, a source with 144 kΩ of internal resistance is connected to a 20-Ω load. In this case, very little of the source power would actually be delivered to the load. In fact, most of the source power would be used by its own internal resistance. To match the source to the load, the emitter follower from Figure 10.12 could be placed between the original source and load. This connection is shown in Figure 10.13b. Note that the source impedance very closely matches the input impedance of the amplifier. Because of this, there is improvement in the power transfer from the source to the amplifier input. The low output impedance of the amplifier matches the load resistance very closely. Therefore, the power transfer from the amplifier to the load is improved. The end result is that maximum power has been transferred from the original source to the original load. An emitter follower used for this purpose is usually called a **buffer**.

The specific purpose served by a given emitter follower depends on the circuit in which it is used. However, you will find that it is always used for one of the two purposes mentioned in this section: providing current gain and/or impedance matching.

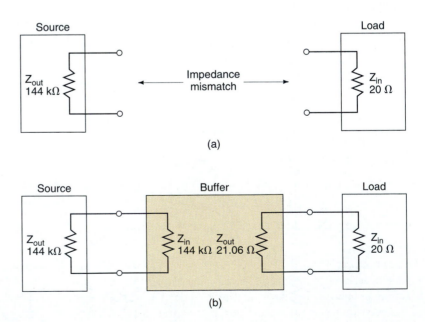

FIGURE 10.13

## Troubleshooting

Troubleshooting an emitter follower is not very different from troubleshooting a common-emitter amplifier. However, there is one important fact that you must keep in mind: *The emitter follower does not provide any voltage gain.*

**Lab Reference:** Several emitter follower faults are simulated in Exercise 20.

When you troubleshoot a common-emitter amplifier, you expect to see a much higher output voltage than input voltage. This is because the common-emitter amplifier has a relatively high amount of voltage gain. If you checked out a common-emitter circuit and saw that its ac output voltage and ac input voltage were equal, you would have good reason to believe that the circuit was faulty. However, for the emitter follower, this condition is *normal*. In fact, the output voltage will always be slightly less than the input voltage for the emitter follower.

When any component in an emitter follower goes bad, you will generally have one of two conditions: Either the transistor will saturate or it will cut off. When the output is very nearly equal to $V_{CC}$ or ground, there is reason to believe that the emitter follower is faulty. The output voltage ($V_E$) will be close to $V_{CC}$ only if the transistor is *saturated*. In this case, check to see if $R_2$ (base circuit) is open. If the output is close to ground, the transistor is in *cutoff*. In this case, check $R_1$ (base circuit) to see if it is open. If not, replace the transistor.

If no ac signal appears at the output (emitter terminal) and $V_E$ is at some value between ground and $V_{CC}$, check the input signal to the base. If a signal is there, check $R_E$ to see if it is open. If it is not, the transistor is probably faulty and should be replaced.

## Summary

The major characteristics of the emitter follower (common collector) are as follows:

1. Low voltage gain (less than 1)
2. Relatively high current gain
3. Approximately equal values of $A_p$ and $A_i$
4. High input impedance and low output impedance
5. Input and output ac voltages that are in phase

---

1. Why is emitter-feedback bias commonly used in emitter followers?
2. List, in order, the steps you would take to determine the value of amplifier input impedance for an emitter follower with emitter-feedback bias.
3. What problem is solved by the use of a collector bypass capacitor?
4. Describe the two common applications of the emitter follower.
5. What is the purpose of a *buffer*?
6. What must be kept in mind when troubleshooting an emitter follower?

***Section Review***

---

## 10.4 THE DARLINGTON AMPLIFIER

The **Darlington amplifier** is a special-case emitter follower that uses two transistors to increase the overall values of circuit current gain ($A_i$) and input impedance ($Z_{in}$). The two transistors are connected as shown in Figure 10.14. The emitter of the first is tied to the base of the second, and the transistor collector terminals are tied together. These transistor connections result in the current paths shown in the figure.

**Darlington amplifier**
A special-case emitter follower that uses two transistors to increase the overall current gain and amplifier input impedance.

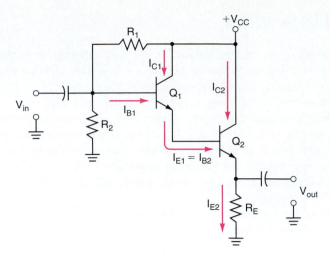

**FIGURE 10.14** The Darlington amplifier.

## dc Analysis

OBJECTIVE 5 ▶ The circuit consists of two transistors, each with a base-to-emitter voltage, $V_{BE}$. Thus, $I_{E2}$ is found by taking $2\,V_{BE}$ from the base voltage and dividing the difference by $R_E$. By formula,

$$I_{E2} = \frac{V_{B1} - 2\,V_{BE}}{R_E} \qquad (10.17)$$

Since the transistors are directly coupled,

$$I_{E1} = I_{B2} \qquad (10.18)$$

and since $I_{B2} \cong I_{E2}/h_{FC2}$,

$$I_{E1} \cong \frac{I_{E2}}{h_{FC2}} \qquad (10.19)$$

*Don't forget:*
$h_{FC} \cong h_{FE}$
$h_{fc} \cong h_{fe}$

You may recall that the dc input resistance of a transistor is found as

$$R_{IN} = h_{FC}R_E$$

For the input transistor, the emitter resistance is the input resistance of the second transistor. By formula,

$$R_{IN1} = h_{FC1}R_{IN2}$$

and since $R_{IN2} = h_{FC2}R_E$, the equation above becomes

$$R_{IN1} = h_{FC1}h_{FC2}R_E \qquad (10.20)$$

**Darlington pair**
Two transistors connected as shown in Figure 10.14.

As Example 10.8 illustrates, the **Darlington pair** has a much higher input resistance than the typical transistor configuration. The base imput resistance can easily be in the megohm range.

---

**EXAMPLE 10.8**

Determine the input resistance to the *Darlington pair* shown in Figure 10.15. Use the spec sheet in Section 7.4 of the text.

**Solution:** The dc value of $I_E$ is found as

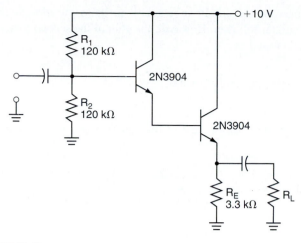

**FIGURE 10.15**

$$I_E = \frac{V_{B1} - 2 V_{BE}}{R_E}$$

$$= 1.09 \text{ mA}$$

The spec sheet for the 2N3904 lists a minimum $h_{FE}$ of 70 at $I_{CQ} = 1$ mA. Using this value as $h_{FC}$, the input resistance of the Darlington pair is found as

$$R_{IN1} = h_{FC1} h_{FC2} R_E$$
$$= (70)(70)(3.3 \text{ k}\Omega)$$
$$= 16.17 \text{ M}\Omega$$

***PRACTICE PROBLEM 10.8***

A Darlington amplifier like the one in Figure 10.15 has the following values: $h_{FC} = 240$ (each transistor) and $R_E = 390 \ \Omega$. Determine the input resistance to the Darlington pair.

## ac Analysis

For the Darlington amplifier, the input impedance is found using the same equations as for the standard emitter follower. When voltage-divider bias is used, $Z_{in}$ is found as    ◀ *OBJECTIVE 6*

$$Z_{in} = R_1 \| R_2 \| Z_{in(base)}$$

When emitter-feedback bias is used,

$$Z_{in} = R_1 \| Z_{in(base)}$$

The value of $Z_{in(base)}$ for the Darlington pair is found as

$$Z_{in(base)} = h_{ic1} + h_{fc1}(h_{ic2} + h_{fc2}r_E) \qquad\qquad \textbf{(10.21)}$$

where $h_{ic}$ = the input impedance of the identified transistor
   $h_{fc}$ = the ac current gain of the identified transistor
   $r_E$ = the parallel combination of $R_E$ and $R_L$

The derivation of equation (10.21) is covered in Appendix D.

The output impedance of the amplifier is found in the same way as the output impedance of a standard emitter follower. However, the gain of the second transistor must be considered. Thus, $Z_{out}$ is found as

$$Z_{out} = R_E \left\| \left[ r'_{e2} + \frac{r'_{e1} + (R'_{in}/h_{fc1})}{h_{fc2}} \right] \right. \tag{10.22}$$

Since $R_E$ is usually much greater than the rest of the equation, $Z_{out}$ can be approximated as

$$Z_{out} \cong r'_{e2} + \frac{r'_{e1} + (R'_{in}/h_{fc1})}{h_{fc2}} \tag{10.23}$$

where $R'_{in} = R_1 \| R_2 \| R_S$

$\quad r'_e$ = the ac emitter resistance of the identified transistor

$\quad h_{fc}$ = the ac current gain of the identified transistor

*The ac current gain of the transistors is equal to the product of the individual gains, and this product can be in the thousands.* This is due to the fact that input current is increased by the gain of the first transistor and, after being applied to the base of the second transistor, is increased again by the gain of that component. Even with the current gain reduction caused by the input and output circuits, the Darlington amplifier has extremely high overall current gain. By formula,

$$A_i = h_{fc1} h_{fc2} \left( \frac{Z_{in} r_E}{Z_{in(base)} R_L} \right) \tag{10.24}$$

As Example 10.9 illustrates, the current gain of a Darlington amplifier is typically much higher than that of a single-transistor emitter follower. As before, the voltage gain ($A_v$) of the amplifier is slightly less than 1.

---

**EXAMPLE 10.9**

Determine the circuit gain and impedance values for the amplifier in Figure 10.15. Assume that the circuit has values of $R_S = 3.3\ k\Omega$, $R_L = 1\ k\Omega$, $h_{fc1} = 120$, $h_{fc2} = 150$, $h_{ic1} = 40\ k\Omega$, and $h_{ic2} = 3\ k\Omega$.

*Solution:* The input impedance to the Darlington pair is found as

$$
\begin{aligned}
Z_{in(base)} &= h_{ic1} + h_{fc1}(h_{ic2} + h_{fc2}r_E) \\
&= 40\ k\Omega + (120)[3\ k\Omega + (150)(3.3\ k\Omega)] \\
&= 40\ k\Omega + (120)(498\ k\Omega) \\
&= 59.8\ M\Omega
\end{aligned}
$$

The amplifier input impedance is now found as

$$
\begin{aligned}
Z_{in} &= R_1 \| R_2 \| Z_{in(base)} \\
&= 59.9\ k\Omega
\end{aligned}
$$

Again, this high input impedance is characteristic of a Darlington amplifier.

To determine the value of $Z_{out}$, we need to start by determining the values of $r'_e$ for the two transistors. These values are found as

$$r'_{e1} = \frac{h_{ic1}}{h_{fc1}}$$

$$= 333\ \Omega$$

and

$$r'_{e2} = \frac{h_{ic2}}{h_{fc2}}$$

$$= 20 \ \Omega$$

The input resistance of the circuit ($R'_{in}$) is found as

$$R'_{in} = R_1 \| R_2 \| R_S$$
$$= 3.13 \text{ k}\Omega$$

Finally, the value of $Z_{out}$ is found as

$$Z_{out} \cong r'_{e2} + \frac{r'_{e1} + (R'_{in}/h_{fc1})}{h_{fc2}}$$

$$= 20 \ \Omega + \frac{333 \ \Omega + (3.13 \text{ k}\Omega/120)}{150}$$

$$= 20 \ \Omega + 2.4 \ \Omega$$

$$= 22.4 \ \Omega$$

The low output impedance is expected in this amplifier.

In order to calculate the value of the circuit current gain, we need to determine the value of $r_E$, as follows:

$$r_E = R_E \| R_L$$
$$= (3.3 \text{ k}\Omega \| 1 \text{ k}\Omega)$$
$$= 767 \ \Omega$$

Now, the current gain of the amplifier can be found as

$$A_i = h_{fc1}h_{fc2}\left(\frac{Z_{in}r_E}{Z_{in(base)}R_L}\right)$$

$$= (120)(150)\left[\frac{(59.9 \text{ k}\Omega)(767 \ \Omega)}{(59.8 \text{ M}\Omega)(50 \ \Omega)}\right]$$

$$= (18,000)(0.0154)$$

$$= 277.2$$

As with any emitter follower, the overall voltage gain of the circuit will be slightly less than 1.

### PRACTICE PROBLEM 10.9

Determine the circuit gain and impedance values for the circuit in Figure 10.15. Assume that the circuit has the following values: $R_1 = R_2 = 240$ k$\Omega$, $R_E = 510$ $\Omega$, $R_S = 6$ k$\Omega$, $R_L = 100$ $\Omega$, $h_{fc1} = 80$, $h_{fc2} = 180$, $h_{ic1} = 38$ k$\Omega$, and $h_{ic2} = 4$ k$\Omega$.

## Summary

Examples 10.8 and 10.9 demonstrate the major characteristics of the Darlington amplifier. These characteristics are as follows:

1. A voltage gain that is less than 1 ($A_v < 1$).
2. Extremely high base input impedance.
3. High current gain.

**4.** Extremely low output impedance.

**5.** Input to output voltages and currents that are in phase.

Because the characteristics of the Darlington amplifier are basically the same as those of the emitter follower, the two circuits are used for similar applications. When you need higher input impedance and current gain and/or lower output impedance than the standard emitter follower can provide, you use a Darlington amplifier.

## The "Quick" Analysis

As you can see, the detailed analysis of the Darlington amplifier is long and involved. When you are interested only in quickly determining approximate circuit values, the following approximations can be used:

$$A_i = h_{fc1} h_{fc2} \left( \frac{Z_{in} r_E}{Z_{in(base)} R_L} \right)$$

$$A_v \cong 1$$

$$Z_{in(base)} \cong h_{fc1} h_{fc2} r_E \tag{10.25}$$

$$Z_{out} \cong r'_{e2} + \frac{r'_{e1}}{h_{fc2}} \tag{10.26}$$

Equation 10.25 ignores the values of $r'_e$ for the two transistors. Had we used this equation in Example 10.9, we would have obtained a value of $Z_{in(base)} = 3.17$ M$\Omega$. Using this value of $Z_{in(base)}$, we would have obtained a value of $Z_{in} = 58.9$ k$\Omega$, a difference of only 1 percent. The final approximation ignores the value of $(R'_{in}/h_{fc1})$ in the calculation of $Z_{out}$. Had we used this equation in Example 10.9, we would have obtained a value of $Z_{out} = 22.2$ $\Omega$, rather than the 22.4 $\Omega$ value found in the example.

## One Final Note

Darlington transistors are commonly available. Like standard transistors, they have only three terminals, but they have much higher values of $h_{fe}$, $h_{ie}$, and $h_{FE}$. When you work with one of these transistors, all the relationships discussed in this section will apply.

---

**Section Review**

1. What is a *Darlington amplifier*?
2. How are the two transistors in a Darlington amplifier interconnected?
3. Why is the current gain of a Darlington amplifier higher than that of a standard emitter follower?
4. List the major characteristics of the Darlington amplifier.

---

## *10.5* THE COMMON-BASE AMPLIFIER

Common-base applications.

The *common-base amplifier* is used to provide *voltage gain with no current gain* and for *high-frequency applications*. The input to the common-base amplifier is applied to the emitter, while the output is taken from the collector. The schematic of the common-base amplifier is shown in Figure 10.16.

As you can see, there are two ways in which the common-base amplifier is commonly biased. Figure 10.16a shows the transistor in an *emitter-bias* configuration. Figure

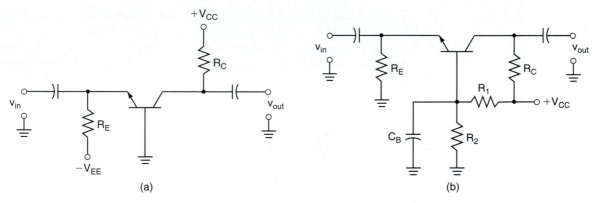

**FIGURE 10.16    Typical common-base amplifiers.**

10.16b shows the transistor in a *voltage-divider bias* configuration. The dc analysis of these two circuits is the same as for the equivalent common-emitter configurations.

**Lab Reference:** The dc characteristics of an emitter-biased CB amplifier are demonstrated in Exercise 21.

Note that the input to each amplifier is applied to the emitter, while the output is taken from the collector. The base of Figure 10.16a is coupled directly to ground. The base of Figure 10.16b is grounded (in terms of ac operation) by the bypass capacitor across $R_2$.

## ac Analysis

The ac equivalent circuit for the common-base amplifier is shown in Figure 10.17. We will use this circuit for our discussion on the ac characteristics of the common base amplifier.

◄   *OBJECTIVE 7*

The input signal is applied to the emitter, so $v_{in}$ is developed across $r_e'$. The output signal is developed across the ac collector resistance. Therefore, voltage gain is found as

$$A_v = \frac{v_{out}}{v_{in}}$$

$$= \frac{i_e r_C}{i_e r_e'}$$

and

*A Practical Consideration:* The conversion of $h_{ie}$ and $h_{fe}$ to common-base form is fairly complex (see Appendix C). In this case, you're better off using

$$A_v = \frac{r_C}{r_e'} \qquad (10.27)$$

$$r_e' = \frac{25 \text{ mV}}{I_E}$$

Since the emitter current is approximately equal to the collector current, the input and output currents are approximately equal. Therefore, current gain is usually approximated as

$$A_i \cong 1 \qquad (10.28)$$

**Lab Reference:** The gain characteristics of a common-base amplifier are demonstrated in Exercise 21.

**FIGURE 10.17    The common-base ac equivalent circuit.**

This approximation is used for simplicity. The actual current gain of a common-base amplifier ($h_{fb}$) equals the ratio of ac collector current to ac emitter current. As you may recall, this ratio is the *alpha* (ac) rating of the transistor, which is slightly less than 1.

The input impedance to the amplifier is the parallel combination of $r'_e$ and $R_E$. By formula,

$$Z_{in} = r'_e \parallel R_E \qquad (10.29)$$

Since $R_E$ is usually much greater than $r'_e$, the input impedance to the amplifier can be approximated as

$$Z_{in} \cong r'_e \qquad (10.30)$$

The output impedance of the amplifier is the parallel combination of $R_C$ and the collector resistance of the transistor ($1/h_{ob}$). By formula,

$$Z_{out} = R_C \parallel \frac{1}{h_{ob}} \qquad (10.31)$$

Since $1/h_{ob}$ is typically very high (in the hundreds of kilohms), the output impedance is approximated as

$$Z_{out} \cong R_C \qquad (10.32)$$

Example 10.10 illustrates the ac analysis of a common-base amplifier.

**$h_{ob}$**
The common-base value of output admittance. You can usually assume that $1/h_{ob}$ is much greater than any practical value of $R_C$.

---

*EXAMPLE 10.10*

**FIGURE 10.18**

Determine the gain and impedance values for the circuit shown in Figure 10.18.

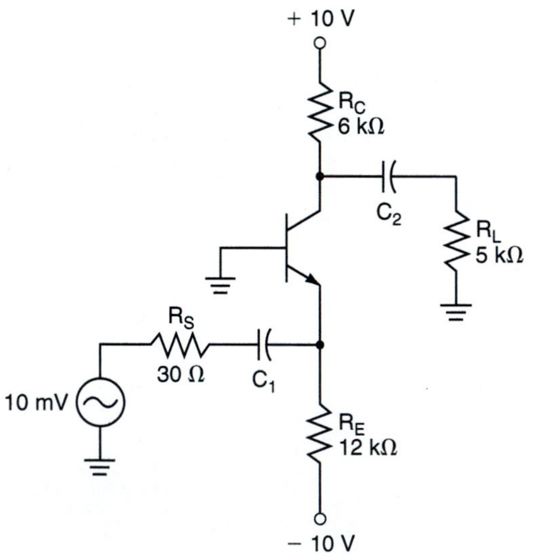

**Solution:** First, we need to determine the value of $I_E$. Since the circuit is an emitter-bias circuit, $I_E$ is found as

$$I_E = \frac{V_{EE} - V_{BE}}{R_E}$$
$$= 775 \ \mu A$$

The value of $r'_e$ can now be approximated as

$$r'_e = \frac{25 \ mV}{I_E}$$
$$= 32.3 \ \Omega$$

Since $R_E \gg$ than $r'_e$, we can approximate the value of $Z_{in}$ as

$$Z_{in} \cong r'_e$$
$$= 32.3 \ \Omega$$

The output impedance is found as

$$Z_{out} \cong R_C$$
$$= 6 \ k\Omega$$

The ac resistance of the collector circuit is found as

$$r_C = R_C \parallel R_L$$
$$= 2.73 \ k\Omega$$

and the voltage gain of the amplifier is found as

$$A_v = \frac{r_C}{r'_e}$$

$$= 84.5$$

Finally, since the output current is approximately equal to the input current, the current gain is found as

$$A_i \cong 1$$

**PRACTICE PROBLEM 10.10**

An emitter bias circuit like the one in Figure 10.18 has values of $V_{CC} = +15$ V, $V_{EE} = -15$ V, $R_E = 30$ kΩ, $R_C = 15$ kΩ, and $R_L = 3$ kΩ. Determine the approximate values of $A_v$, $A_i$, $Z_{in}$, and $Z_{out}$ for the amplifier.

## Summary

Example 10.10 demonstrated the ac analysis of a common-base amplifier. The results found in the example also serve to illustrate the ac characteristics of this type of amplifier. These characteristics are:

1. Relatively high voltage gain.
2. Current gain is less than 1 ($A_i < 1$).
3. Low input impedance.
4. High output impedance.
5. Input/output voltage and current that are in phase.

As you will see in the next section, the characteristics listed above make the common-base amplifier well suited for several applications.

**Section Review** ◄

1. Which biasing circuits are commonly used for common-base amplifiers?
2. List, in order, the steps you would take to perform a complete ac analysis of a common-base amplifier.
3. What are the gain characteristics of a common-base amplifier?
4. What are the input/output characteristics of a common-base amplifier?

## 10.6 COMMON-BASE APPLICATIONS AND TROUBLESHOOTING

As stated earlier, the common-base amplifier is used to provide *voltage gain without current gain* and for *high-frequency* applications. Of the two, the high-frequency applications are far more common.    ◄ *OBJECTIVE 8*

   Since the common-base circuit has a high voltage gain and a value of $A_i \cong 1$, the circuit can be used to provide voltage gain without increasing the value of circuit current. Consider the case where the output current from an amplifier is more than enough for the application, but the voltage needs to be boosted. The common-base amplifier would serve well in this situation because it would increase the voltage without increasing the current.

The high-frequency application for common-base amplifiers ties in closely with the impedance-matching application of the emitter follower. The common-base amplifier has the *opposite* impedance characteristics of the emitter follower, as shown in Figure 10.19a. Note that the common-base circuit typically has *low input impedance* and *high output impedance*. The emitter follower has just the opposite impedance characteristics.

How does this relate to high-frequency operation? Most high-frequency voltage sources have *very low output impedance*. When you want to connect a high-frequency, low-$Z_{out}$ source to a high-impedance load, you need a circuit to match the source impedance to the load impedance. This function is served very well by the common-base circuit, as shown in Figure 10.19b and c.

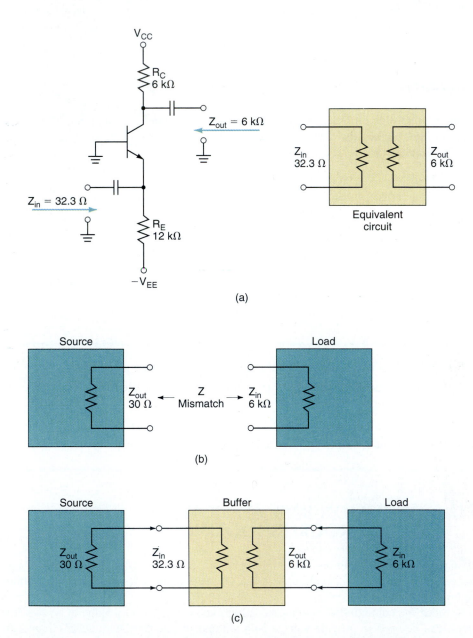

FIGURE 10.19  The common-base amplifier as an impedance-matching circuit.

# Troubleshooting

Troubleshooting the common-base amplifier is very nearly the same as troubleshooting the common-emitter circuit. The output voltage should show gain when compared to the input voltage. However, *there is no phase shift from input to output.*

When you troubleshoot a common-base circuit, check the input and output signals. If both signals are present and there is voltage gain from emitter to collector, the amplifier is good. If the input signal is normal and the output signal is not, check the amplifier using the techniques discussed in Chapter 7.

**Section Review** ◀

1. What are the two common applications for the common-base amplifier?
2. Explain the use of the common-base amplifier as an impedance-matching circuit.
3. What is the difference between troubleshooting a common-base amplifier and troubleshooting a common-emitter amplifier?

## Transistor Amplifiers: A Summary

◀ *OBJECTIVE 9*

In Chapters 6 through 9 you were introduced to a wide variety of transistor amplifiers. While the dc characteristics of the amplifiers do not vary significantly from one configuration to another, the ac characteristics of the circuits do have significant differences. As a review, the ac characteristics of the common-emitter, emitter-follower (common-collector), common-base, and Darlington amplifiers are compared in Table 10.1.

**TABLE 10.1   Amplifier Configurations and Their ac Characteristics**

|          | Common Emitter | Emitter Follower | Common Base | Darlington |
|----------|----------------|------------------|-------------|------------|
| $A_v$    | High           | Less than 1      | High        | Less than 1 |
| $A_i$    | High           | High             | Less than 1 | Extremely high |
| $A_p$    | High           | High[a]          | High[b]     | High[a]    |
| $Z_{in}$ | Low            | High             | Extremely low | Extremely high |
| $Z_{out}$| High           | Very low         | High        | Extremely low |

[a] Slightly less than the value of $A_i$.
[b] Slightly less than the value of $A_v$.

The following terms were introduced and defined in this chapter:

**Key Terms**

buffer
collector bypass capacitor

Darlington amplifier
Darlington pair

impedance matching
regulated dc power supply

| Equation Number | Equation | Section Number | Equation Summary |
|-----------------|----------|----------------|------------------|
| **(10.1)** | $V_B = \dfrac{R_2}{R_1 + R_2} V_{CC}$ | 10.1 | |
| **(10.2)** | $V_E = V_B - 0.7 \text{ V}$ | 10.1 | |

| Equation Number | Equation | Section Number |
|---|---|---|
| (10.3) | $I_E = \dfrac{V_E}{R_E}$ | 10.1 |
| (10.4) | $V_{CEQ} = V_{CC} - V_E$ | 10.1 |
| (10.5) | $I_{C(\text{sat})} \cong \dfrac{V_{CC}}{R_E}$ | 10.1 |
| (10.6) | $V_{CE(\text{off})} = V_{CC}$ | 10.1 |
| (10.7) | $r_E = R_E \parallel R_L$ | 10.2 |
| (10.8) | $A_v = \dfrac{r_E}{r'_e + r_E}$ | 10.2 |
| (10.9) | $A_v \cong 1 \mid r_E \gg r'_e$ | 10.2 |
| (10.10) | $A_i = h_{fc}\left(\dfrac{Z_{\text{in}}r_E}{Z_{\text{in(base)}}R_L}\right)$ | 10.2 |
| (10.11) | $A_p = A_i\,A_v$ | 10.2 |
| (10.12) | $Z_{\text{in}} = R_1 \parallel R_2 \parallel Z_{\text{in(base)}}$ | 10.2 |
| (10.13) | $Z_{\text{in(base)}} = h_{fc}\,(r'_e + r_E)$ | 10.2 |
| (10.14) | $Z_{\text{in(base)}} \cong h_{fc}\,r_E$ when $r_E \gg r'_e$ | 10.2 |
| (10.15) | $Z_{\text{out}} = R_E \parallel \left(r'_e + \dfrac{R'_{\text{in}}}{h_{fc}}\right)$ | 10.2 |
| (10.16) | $Z_{\text{in}} = R_1 \parallel Z_{\text{in(base)}}$ | 10.3 |
| (10.17) | $I_{E2} = \dfrac{V_{B1} - 2\,V_{BE}}{R_E}$ | 10.4 |
| (10.18) | $I_{E1} = I_{B2}$ | 10.4 |
| (10.19) | $I_{E1} \cong \dfrac{I_{E2}}{h_{fc2}}$ | 10.4 |
| (10.20) | $R_{\text{IN1}} = h_{FC1}h_{FC2}R_E$ | 10.4 |
| (10.21) | $Z_{\text{in(base)}} = h_{ic1} + h_{fc1}\,(h_{ic2} + h_{fc1}\,r_E)$ | 10.4 |
| (10.22) | $Z_{\text{out}} = R_E \parallel \left[r'_{e2} + \dfrac{r'_{e1} + (R'_{\text{in}}/h_{fc1})}{h_{fc2}}\right]$ | 10.4 |
| (10.23) | $Z_{\text{out}} \cong r'_{e2} + \dfrac{r'_{e1} + (R'_{\text{in}}/h_{fc1})}{h_{fc2}}$ | 10.4 |
| (10.24) | $A_i = h_{fc1}h_{fc2}\left(\dfrac{Z_{\text{in}}r_E}{Z_{\text{in(base)}}R_L}\right)$ | 10.4 |
| (10.25) | $Z_{\text{in(base)}} \cong h_{fc1}h_{fc2}r_E$ | 10.4 |

| Equation Number | Equation | Section Number |
|---|---|---|
| (10.26) | $Z_{\text{out}} \cong r'_{e2} + \dfrac{r'_{e1}}{h_{fc2}}$ | 10.4 |
| (10.27) | $A_v = \dfrac{r_C}{r'_e}$ | 10.5 |
| (10.28) | $A_i \cong 1$ | 10.5 |
| (10.29) | $Z_{\text{in}} = r'_e \parallel R_E$ | 10.5 |
| (10.30) | $Z_{\text{in}} \cong r'_e$ | 10.5 |
| (10.31) | $Z_{\text{out}} = R_C \parallel \dfrac{1}{h_{ob}}$ | 10.5 |
| (10.32) | $Z_{\text{out}} \cong R_C$ | 10.5 |

**10.1.** $V_B = 9.92$ V, $V_E = 9.22$ V, $I_E = 10.13$ mA, $V_{CEQ} = 8.78$ V

**10.2.** $V_{CE(\text{off})} = 18$ V, $I_{C(\text{sat})} = 19.78$ mA

**10.3.** 0.9967

**10.4.** 23.92

**10.5.** 8.73 kΩ

**10.6.** 15.54 Ω

**10.7.** 83.8 kΩ, 3.57 kΩ

**10.8.** 22.5 MΩ

**10.9.** $A_i = 858.98$, $Z_{\text{in}} \cong 111$ kΩ, $Z_{\text{out}} = 25.2$ Ω

**10.10.** $A_v = 47.71$, $A_i \cong 1$, $Z_{\text{in}} \cong 52.4$ Ω, $Z_{\text{out}} \cong 15$ kΩ

**Practice Problems**

**§10.1**

1. Determine the values of $V_B$, $V_E$, and $I_E$ for the circuit in Figure 10.20.
2. Determine the values of $V_B$, $V_E$, and $I_E$ for the circuit in Figure 10.21.
3. Plot the dc load line for the circuit in Figure 10.20.

**FIGURE 10.20**

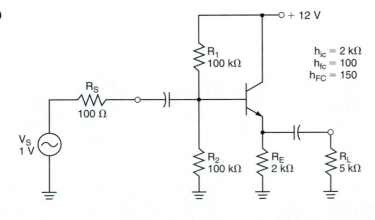

$h_{ic} = 2$ kΩ
$h_{fc} = 100$
$h_{FC} = 150$

FIGURE 10.21

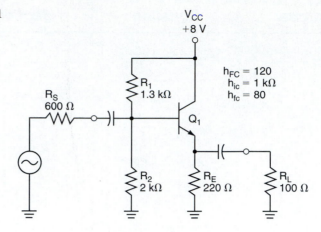

4. Plot the dc load line for the circuit in Figure 10.21.
5. Determine the value of $Z_{in}$ for the circuit in Figure 10.20.
6. Determine the value of $Z_{in}$ for the circuit in Figure 10.21.
7. Determine the values of $A_v$, $A_i$, and $A_p$ for the circuit in Figure 10.20.
8. Determine the values of $A_v$, $A_i$, and $A_p$ for the circuit in Figure 10.21.
9. Determine the value of $Z_{out}$ for the circuit in Figure 10.20.
10. Determine the value of $Z_{out}$ for the circuit in Figure 10.21.
11. Perform a basic ac analysis on the amplifier in Figure 10.22.
12. Perform a basic ac analysis on the amplifier in Figure 10.23.

### §10.2

13. Determine the value of $Z_{in}$ for the circuit in Figure 10.24.
14. Determine the value of $Z_{in}$ in the circuit in Figure 10.25.

### §10.4

$R_{IN1}$

15. Determine the values of $I_E$ and $R_{in1}$ for the amplifier in Figure 10.26
16. Determine the values of $I_E$ and $R_{in1}$ for the amplifier in Figure 10.27.
17. Determine the values of $Z_{in}$ and $Z_{out}$ for the circuit in Figure 10.26.
18. Determine the values of $Z_{in}$ and $Z_{out}$ for the circuit in Figure 10.27.
19. Determine the value of $A_i$ for the circuit in Figure 10.26.
20. Determine the value of $A_i$ for the circuit in Figure 10.27.

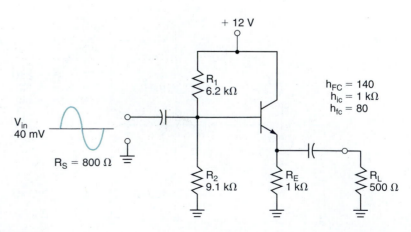

**FIGURE 10.22**

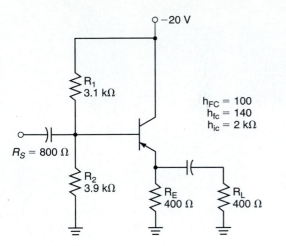

FIGURE 10.23

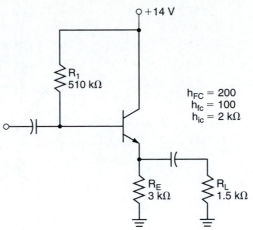

FIGURE 10.24

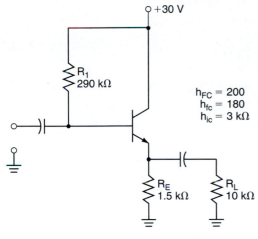

FIGURE 10.25

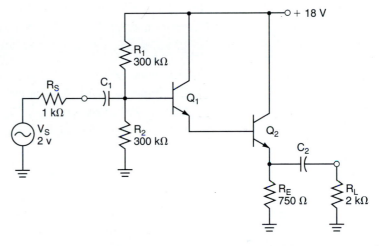

$h_{fc1} = 40$
$h_{fc2} = 160$

$h_{ic1} = 40 \text{ k}\Omega$
$h_{ic2} = 2 \text{ k}\Omega$

$h_{FC1} = 60$
$h_{FC2} = 140$

FIGURE 10.26

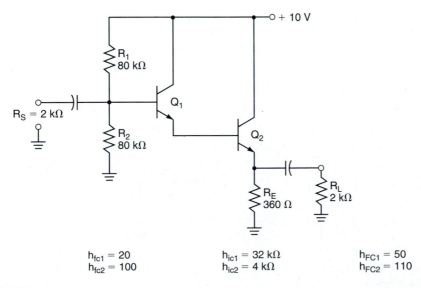

$h_{fc1} = 20$
$h_{fc2} = 100$

$h_{ic1} = 32 \text{ k}\Omega$
$h_{ic2} = 4 \text{ k}\Omega$

$h_{FC1} = 50$
$h_{FC2} = 110$

FIGURE 10.27

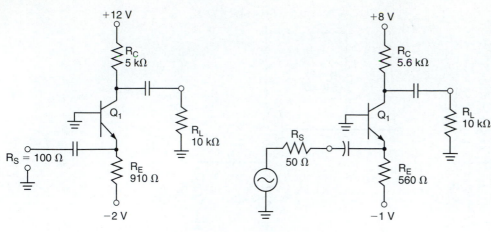

**FIGURE 10.28**    **FIGURE 10.29**

21. Perform a "quick" analysis on the amplifier in Figure 10.26.
22. Perform a "quick" analysis on the amplifier in Figure 10.27.

*§10.5*

23. Determine the value of $A_v$ for the amplifier in Figure 10.28.
24. Determine the value of $A_v$ for the amplifier in Figure 10.29.
25. Determine the values of $Z_{in}$ and $Z_{out}$ for the amplifier in Figure 10.28.
26. Determine the values of $Z_{in}$ and $Z_{out}$ for the amplifier in Figure 10.29.

**Troubleshooting Practice Problems**

27. Refer to Figure 10.21. The circuit has a constant dc emitter voltage of 0 V. Discuss the possible causes of the problem.
28. Refer to Figure 10.22. The dc emitter voltage for this circuit is approximately 1.7 V. Discuss the possible causes for this problem.
29. The circuit in Figure 10.30 has the waveforms shown in Figure 10.31. Discuss the possible causes of the problem.

**FIGURE 10.30**

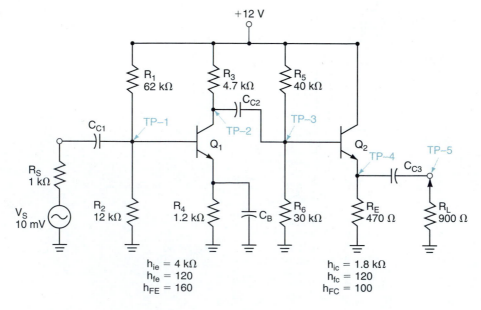

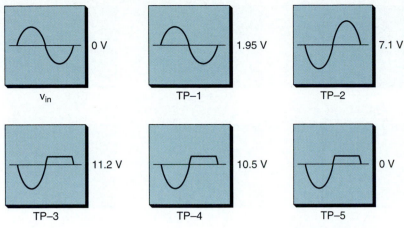

**FIGURE 10.31**

Note: Voltages listed are DC reference voltages, as measured from the designated test point to ground.

30. The circuit shown in Figure 10.28 has a constant $V_C$ of $+12 V_{dc}$. Discuss the possible causes of the problem.

31. The circuit shown in Figure 10.27 has a dc emitter voltage ($Q_2$) of approximately 0 V. Discuss the possible causes of the problem.

32. The circuit shown in Figure 10.26 has the output waveform shown in Figure 10.32. Discuss the possible causes of the problem.

33. The circuit shown in Figure 10.30 has the signals shown in Figure 10.33. Determine the possible causes of the problem.

34. The circuit in Figure 10.30 has the output waveforms shown in Figure 10.34. Determine the possible causes of the problem.

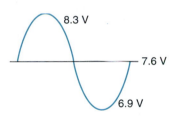

**FIGURE 10.32**

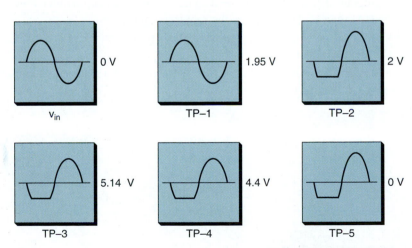

Note: Voltages listed are DC reference voltages, as measured from the designated test point to ground.

**FIGURE 10.33**

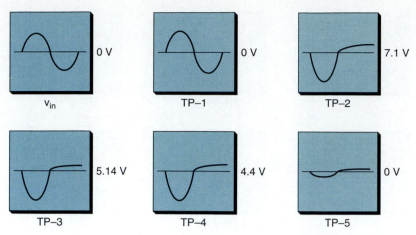

$v_{in}$    0 V

TP–1    0 V

TP–2    7.1 V

TP–3    5.14 V

TP–4    4.4 V

TP–5    0 V

Note: Voltages listed are DC reference voltages, as measured from the designated test point to ground.

**FIGURE 10.34**

*The Brain Drain*

**35.** Determine the values of $A_{vT}$, $A_{iT}$, $A_{pT}$, and $Z_{in}$ for the amplifier shown in Figure 10.30.

**36.** Determine the values of $A_i$, $Z_{in}$, and $Z_{out}$ for the amplifier in Figure 10.35 using the "quick" analysis equations. The $h_{ie}$ and $h_{fe}$ curves for the two transistors are shown in the figure.

**37.** Determine the values of $A_v$, $A_i$, $A_p$, $Z_{in}$, and $Z_{out}$ for the amplifier in Figure 10.36. The spec sheet for the Darlington transistor is located in Appendix A. Use the "quick" analysis approximations in your analysis.

*Suggested Computer Applications Problems*

**38.** Write a program that will perform the dc analysis of a standard emitter follower or a Darlington emitter follower, given the required inputs.

**39.** Write a program that will perform the ac analysis of a standard emitter follower, given the required input values.

**40.** Write a program that will perform the "quick" analysis of a Darlington emitter follower, given the required input values.

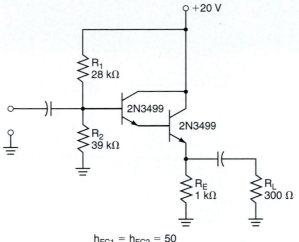

$$h_{FC1} = h_{FC2} = 50$$

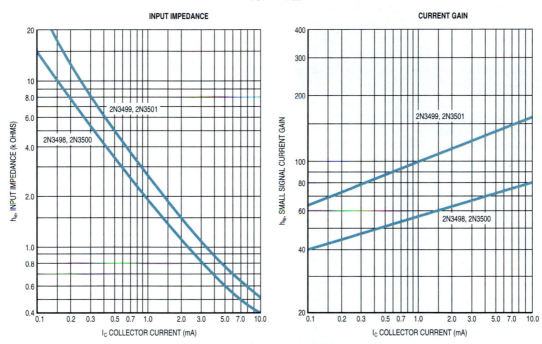

INPUT IMPEDANCE

CURRENT GAIN

2N3499, 2N3501

2N3498, 2N3500

$h_{ie}$, INPUT IMPEDANCE (k OHMS)

$I_C$ COLLECTOR CURRENT (mA)

2N3499, 2N3501

2N3498, 2N3500

$h_{fe}$, SMALL SIGNAL CURRENT GAIN

$I_C$ COLLECTOR CURRENT (mA)

**FIGURE 10.35** (Copyright of Motorola. Used by permission.)

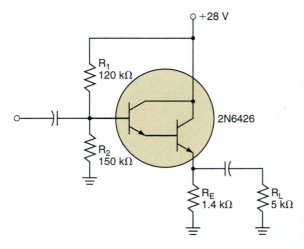

+28 V

R₁
120 kΩ

R₂
150 kΩ

2N6426

R_E
1.4 kΩ

R_L
5 kΩ

**FIGURE 10.36**

# 11

# POWER AMPLIFIERS

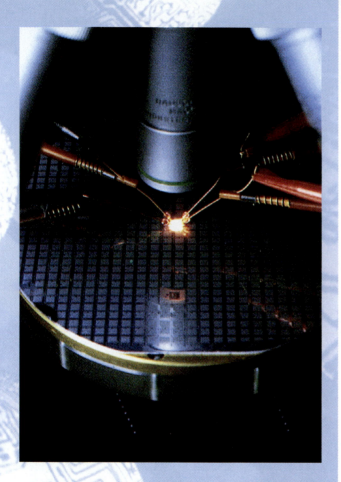

The final stage of device production is testing one of a group of devices for proper operation.

## OUTLINE

## OBJECTIVES

*After studying the material in this chapter, you should be able to:*

1. Identify the *Q*-point location on the dc load line for each of the four types of amplifiers.

2. Compare and contrast the ac load line and dc load line of an amplifier.

3. Explain the concept of amplifier compliance.

4. Explain why high amplifier efficiency ratings are desirable in power amplifiers.

5. Describe and analyze the operation of the transformer-coupled class A amplifier.

6. Describe and analyze the operation of class B amplifiers.

7. Perform all the calculations necessary to determine the values of source power and load power for a complementary-symmetry amplifier.

8. Define and describe class AB operation.

9. Explain how class AB operation eliminates crossover distortion and thermal runaway.

10. Perform the complete analysis of a class AB amplifier.

11. Calculate the transistor power dissipation requirements for a given class A, class B, or class AB amplifier.

# POWER AMPLIFIERS: WHEN ARE THEY USED?

In this chapter, you will be studying a group of amplifiers referred to as *power amplifiers*. Many of the amplifier circuits and principles covered in this chapter are used in *communications electronics*. Communications electronics is the study of the circuits and systems used to transmit information from one point to another. Television and radio transmitters and receivers, two-way transceivers, and radar systems are all examples of communications systems.

There are three basic types of power amplifiers: class A, class B, and class C power amplifiers. In this chapter, we will concentrate on the first two, the class A and class B amplifiers. These two types of circuits are used whenever the power of a signal needs to be increased. One familiar application is the audio amplifier. Audio amplifiers are circuits used to amplify signals whose frequencies lie in the range of approximately 20 Hz to 20 kHz.

$\mathbf{P}$ower amplifiers are used to deliver high values of power to low-resistance loads. The typical power amplifier will have an output power greater than 1 W with a load resistance that will range from 300 $\Omega$ (for transmission antennas) to 4 $\Omega$ (for audio speakers). Although these load values do not cover every possibility, they do illustrate the fact that power amplifiers usually drive low-resistance loads.

The ideal power amplifier would deliver 100 percent of the power it draws from the supply to the load. In practice, however, this can never occur. The reason for this is the fact that the components in the amplifier will all dissipate *some* of the power that is being drawn from the supply. As you were shown in Chapter 8, amplifier efficiency is calculated using the formula

$$\eta = \frac{\text{ac output power}}{\text{dc input power}} \times 100$$

OBJECTIVE 1 ▶ This equation indicates that lower dc input power results in higher amplifier efficiency. The dc input power to an amplifier varies with the position of the $Q$-point on the load line. The bias points for each of the four amplifier classes are shown in Figure 11.1.

In this chapter, we will look at the operation of class A, class B, and class AB amplifiers. Since class C amplifiers are tuned circuits, they will be covered in Chapter 17.

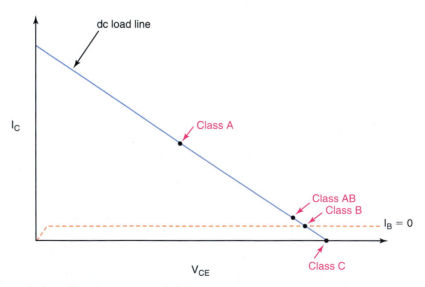

**FIGURE 11.1** *$Q$-point locations for each class of amplifier.*

# 11.1 THE ac LOAD LINE

In Chapter 7 you were introduced to the dc load line. This line is used to represent all possible combinations of $V_{CE}$ and $I_C$ for a given amplifier. The **ac load line** is used for the same basic purpose: It represents all possible ac combinations of $i_c$ and $v_{ce}$.

Up to this point, the ac load line has not been an important consideration. However, the ac load line serves as an effective tool for teaching many of the principles of class A and class B amplifier operation. For this reason, we will take a brief look at the ac load line before covering any of the common power amplifier circuits.

The ac load line for a given amplifier will *not* follow the plot of the dc load line. In other words, the two load lines will have different values of $I_{C(sat)}$ and $V_{CE(off)}$. The reason for this lies in the fact that the dc load of an amplifier is different from the ac load. For example, consider the circuit shown in Figure 11.2. In terms of dc operation, the load on the transistor is equal to $R_C$. The value of $R_L$ does not weigh into the picture because the coupling capacitor provides dc isolation between the transistor and $R_L$. The ac equivalent circuit (Figure 11.2b) shows the ac load to consist of $R_C$ in parallel with $R_L$. As you will see, this change in the load resistance has a considerable effect on the ac load line.

**ac load line**
A graph that represents all possible combinations of $i_c$ and $v_{ce}$.

◄ *OBJECTIVE 2*

## ac Saturation Current

If you take a close look at the ac equivalent circuit in Figure 11.2b, you will notice that the transistor is the ac voltage source for the ac load resistance, $r_C$. Since the ac collector current ($i_c$) passes through this resistance, the following Ohm's law relationship must hold true:

$$i_c = \frac{v_{ce}}{r_C} \qquad (11.1)$$

Now take a look at Figure 11.3a. As illustrated, the value of $i_c$ is equal to a given change in $I_C$. This is expressed as

$$i_c = I_C - I_{CQ}$$

where  $i_c$ = the *change* in collector current
  $I_C$ = the peak value of ac collector current
  $I_{CQ}$ = the *quiescent* value of $I_C$

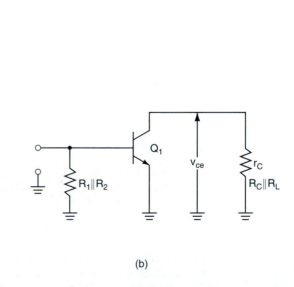

(a)                    (b)

**FIGURE 11.2**

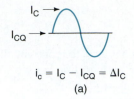
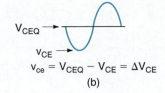

$i_c = I_C - I_{CQ} = \Delta I_C$

(a)

$v_{ce} = V_{CEQ} - V_{CE} = \Delta V_{CE}$

(b)

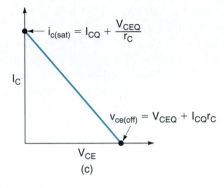

$i_{c(sat)} = I_{CQ} + \dfrac{V_{CEQ}}{r_C}$

$v_{ce(off)} = V_{CEQ} + I_{CQ}r_C$

$I_C$

$V_{CE}$

(c)

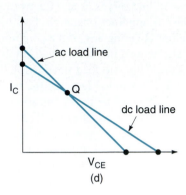

ac load line

$I_C$

Q

dc load line

$V_{CE}$

(d)

**FIGURE 11.3   ac load line.**

Figure 11.3b shows that $v_{ce}$ is equal to the change in $V_{CE}$. By formula,

$$v_{ce} = V_{CEQ} - V_{CE}$$

where   $v_{ce}$ = a *change* in collector–emitter voltage
       $V_{CE}$ = the minimum value of collector–emitter voltage
       $V_{CEQ}$ = the *quiescent* value of $V_{CE}$

If we substitute these two equations in place of $i_c$ and $v_{ce}$ in equation (11.1), we get

$$I_C - I_{CQ} = \frac{V_{CEQ} - V_{CE}}{r_C}$$

or

$$I_C - I_{CQ} = \frac{V_{CEQ}}{r_C} - \frac{V_{CE}}{r_C} \tag{11.2}$$

At saturation, $V_{CE}$ is ideally equal to 0 V. Substituting this into equation (11.2) gives us

$$i_{c(sat)} = I_{CQ} + \frac{V_{CEQ}}{r_C} \tag{11.3}$$

As Figure 11.3c shows, this is the saturation point for the ac load line.

## V_ce(off)

The equation for $v_{ce(off)}$ is derived from equation (11.2). At cutoff, the value of $I_C$ is ideally zero. Substituting this value into equation (11.2) gives us

$$v_{ce(off)} = V_{CEQ} + I_{CQ}r_C \tag{11.4}$$

As Figure 11.3c shows, this is the cutoff point for the ac load line. Note that the ac load line crosses the dc load line at the $Q$-point, as is shown in Figure 11.3d. This will always be the case for a given amplifier.

# What Does the ac Load Line Tell You?

The ac load line is used to tell you the *maximum possible output voltage swing for a given common-emitter amplifier*. In other words, the ac load line will tell you the maximum possible peak-to-peak output voltage from a given amplifier. This maximum $V_{PP}$ is referred to as the *compliance* of the amplifier.

◄ **OBJECTIVE 3**

The **compliance** of an amplifier is found by determining the maximum possible transitions of $I_C$ and $V_{CE}$ from their respective values of $I_{CQ}$ and $V_{CEQ}$. The maximum possible transitions are illustrated in Figure 11.4. As Figure 11.4a shows, the maximum possible transition for $V_{CE}$ is equal to the difference between $V_{CE(off)}$ and $V_{CEQ}$. Since this transition is equal to $I_{CQ}r_c$, the maximum peak output voltage from the amplifier is equal to $I_{CQ}r_c$. Two times this value will give us the maximum peak-to-peak transition of the output voltage, as follows:

**Compliance**
The maximum *undistorted* peak-to-peak output signal.

$$PP = 2I_{CQ}r_c \qquad (11.5)$$

where PP = the output compliance, in peak-to-peak voltage
$I_{CQ}$ = the *quiescent* value of $I_C$
$r_C$ = the ac resistance in the collector circuit

When $I_C = I_{C(sat)}$, $V_{CE}$ is ideally equal to 0 V. When $I_C = I_{CQ}$, $V_{CE}$ is at $V_{CEQ}$. These two points are illustrated in Figure 11.4b. Note that when $I_C$ makes its maximum possible transition (from $I_{CQ}$ to $I_{C(sat)}$), the output voltage changes by an amount equal to $V_{CEQ}$. Thus, the maximum peak-to-peak transition would be equal to twice this value, as follows:

$$PP = 2V_{CEQ} \qquad (11.6)$$

where    PP = the output compliance, in peak-to-peak voltage
$V_{CEQ}$ = the *quiescent* value of $V_{CE}$

Having two different equations for compliance may seem a bit confusing at first, but it really isn't so strange. Equation (11.5) sets the limit in terms of $V_{CE(off)}$. In other words, if you exceed the value obtained by the equation, the output voltage will try to exceed $V_{CE(off)}$, which it cannot do. The result of exceeding the value obtained by equa-

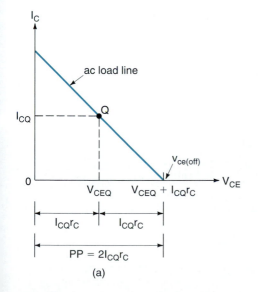

(a)

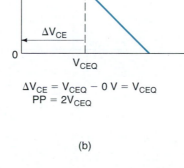

(b)

**FIGURE 11.4    Amplifier compliance.**

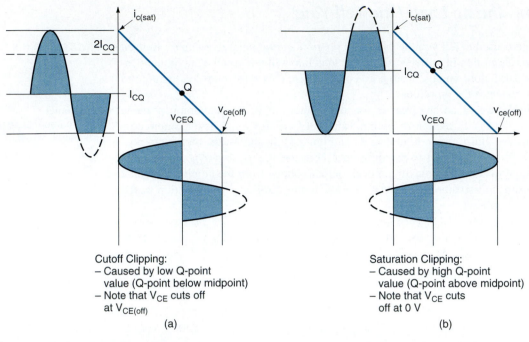

Cutoff Clipping:
– Caused by low Q-point
   value (Q-point below midpoint)
– Note that $V_{CE}$ cuts off
   at $V_{CE(off)}$

(a)

Saturation Clipping:
– Caused by high Q-point
   value (Q-point above midpoint)
– Note that $V_{CE}$ cuts
   off at 0 V

(b)

**FIGURE 11.5**  Cutoff and saturation clipping.

**Cutoff clipping**
A type of distortion caused by
driving a transistor into cutoff.

**Saturation clipping**
A type of distortion caused by
driving a transistor into saturation.

tion (11.5) is shown in Figure 11.5a. This is called **cutoff clipping** because the output voltage is clipped off at the value of $V_{CE(off)}$.

Equation (11.6) sets the limits in terms of $I_{C(sat)}$. If you exceed the value obtained by this equation, the output will experience **saturation clipping**. Saturation clipping is illustrated in Figure 11.5b.

When determining the output compliance for a given amplifier, solve both equations (11.5) and (11.6). *The lower of the two results is the compliance of the amplifier.* The compliance of the amplifier is the maximum *undistorted* output signal. The procedure for determining the compliance of an amplifier is illustrated in Example 11.1.

---

*EXAMPLE  11.1*

Determine the output compliance for the amplifier shown in Figure 11.6.

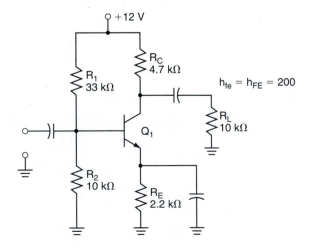

**FIGURE 11.6**

**Solution:** Using the established procedure, $I_{CQ}$ and $V_{CEQ}$ are found to be 950 μA and 5.45 V, respectively. (*Refer to Chapter 7 if you have trouble solving for these values.*) The value of $r_C$ is equal to the parallel combination of $R_C$ and $R_L$: 3.2 kΩ. Now the amplifier is solved for *both* compliance values, as follows:

$$PP = 2 V_{CEQ}$$
$$= 2(5.45 \text{ V})$$
$$= 10.9 \text{ V}_{PP}$$

and

$$PP = 2I_{CQ}r_C$$
$$= 2(950 \text{ μA})(3.2 \text{ kΩ})$$
$$= 6.08 \text{ V}_{PP}$$

Since the overall compliance is always equal to the *smaller* of the two values obtained, the PP for this amplifier is equal to 6.08 $V_{PP}$. This means that the peak-to-peak output voltage can be no more than 6.08 V. If this value is exceeded, cutoff clipping will occur.

### PRACTICE PROBLEM 11.1

A common-emitter amplifier like the one in Figure 11.6 has the following values: $V_{CC}$ = 20 V, $R_1$ = 10 kΩ, $R_2$ = 1.8 kΩ, $R_C$ = 620 Ω, $R_E$ = 200 Ω, $R_L$ = 1.2 kΩ, and $h_{FE}$ = 180. Determine the compliance (PP) of the amplifier.

The importance of the compliance values obtained in Example 11.1 is illustrated in Figure 11.7. Here, we see the dc and ac load lines, along with two possible output signals. Waveform A shows the result of trying to drive the amplifier to PP = 10.9 $V_{PP}$. As you can see, the amplifier experiences *cutoff clipping* as a result of trying to exceed the *ac cutoff point* (8.94 V). On the other hand, waveform B is not clipped because its peak-to-peak value was limited to the compliance of the circuit, 6.08 $V_{PP}$.

## A Practical Consideration

In Chapter 9 you were introduced to the concept of *nonlinear distortion*. In many cases, nonlinear distortion will occur long before cutoff clipping occurs. In other words, if the amplifier input amplitude is sufficient to cause cutoff clipping to occur, it is more than sufficient to cause nonlinear distortion in the amplifier output.

When the compliance of an amplifier is limited by cutoff clipping ($2I_{CQ}r_C$), you need to be aware that nonlinear distortion may occur before the compliance of the amplifier is reached. At the same time, if saturation clipping ($2 V_{CEQ}$) is the limiting factor, nonlinear distortion should not be a problem.

Determining the compliance of an amplifier is not an everyday task for the average technician. At the same time, you need to be aware that the *ac equivalent circuit* of an amplifier often limits the output from that amplifier. This will become an important consideration as we continue our discussion on power amplifiers.

1. What is the *ac load line*?

2. Why is the ac load line of an amplifier different from the dc load line?

3. What is *compliance*?

4. How is the compliance of an amplifier determined?

*Section Review*

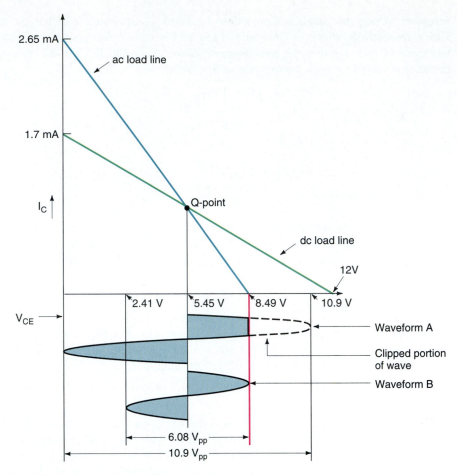

**FIGURE 11.7    Using ac load line to determine compliance.**

## 11.2 *RC*-COUPLED CLASS A AMPLIFIERS

We have already performed most of the analyses that are commonly required for *RC*-coupled class A amplifiers. The ac and dc analyses of these amplifiers were covered in Chapters 9 and 10.

These same analyses are used for the class A power amplifiers, and we will not repeat their derivations or applications here. If you have trouble with any of the standard dc or ac relationships for the common-emitter or emitter-follower amplifiers, refer to the appropriate discussion(s) to review the material. In this section, we will concentrate on the power relationships in the *RC*-coupled class A amplifier.

### *Amplifier dc Power*

The total dc power that an amplifier draws from the power supply is found as

$$P_S = V_{CC}I_{CC} \qquad (11.7)$$

As Figure 11.8 illustrates, $I_{CC}$ is equal to the sum of $I_{CQ}$ and the current through the voltage divider circuit, $I_1$. The calculation of total dc power is demonstrated in Example 11.2.

**FIGURE 11.8    Total supply current.**

Determine the total dc power that is drawn from the supply by the amplifier shown in Figure 11.9.

EXAMPLE 11.2

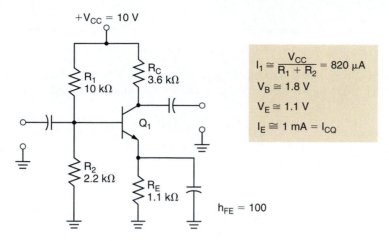

**FIGURE 11.9**

**Solution:**  The value of $I_1$ is found as

$$I_1 = \frac{V_{CC}}{R_1 + R_2}$$

$$= 820 \ \mu A$$

Using the established procedure, the value of $I_{CQ}$ is found as

$$I_{CQ} \cong 1 \ mA$$

Now the value of $I_{CC}$ is found as

$$I_{CC} = I_1 + I_{CQ}$$

$$= 1.82 \ mA$$

Finally, $P_S$ is found as

$$P_S = V_{CC}I_{CC}$$

$$= 18.2 \ mW$$

**PRACTICE PROBLEM 11.2**

Determine how much power is drawn from the dc power supply by the amplifier described in Practice Problem 11.1.

## ac Load Power

The ac load power is the amount of power that is transferred to the load. The equations and examples shown in this section demonstrate only a few of the many ways that the ac load power can be calculated.

AC load power can be calculated using the standard $V^2/R$ equation. Specifically,

$$P_L = \frac{V_L^2}{R_L} \qquad \textbf{(11.8)}$$

*A Practical Consideration:* This equation is used when $V_{out}$ is measured with an ac voltmeter.

where $P_L$ = the ac load power

$V_L$ = the *rms* load voltage

The use of this equation is demonstrated in Example 11.3.

---

**EXAMPLE 11.3**

Determine the ac load power for the circuit shown in Figure 11.10.

**Solution:** The voltmeter reads 10.6 $V_{ac}$, as indicated. Since ac voltmeters read rms voltage, the value shown can be plugged directly into equation (11.8). Thus,

$$P_L = \frac{(10.6\ V_{ac})^2}{200\ \Omega}$$

$$= 561.8\ mW$$

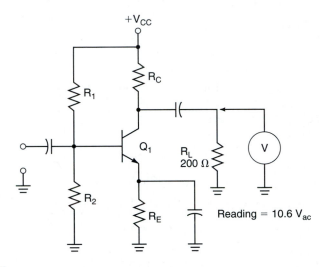

**FIGURE 11.10**

**PRACTICE PROBLEM 11.3**

An amplifier with a 1.2-k$\Omega$ load resistance has an output voltage value of 4.62 $V_{ac}$. Determine the value of load power for the amplifier.

---

A convenient form of equation (11.8) that can be used when you know the peak output voltage is as follows:

*A Practical Consideration:* This equation can be used when $V_{out}$ is measured with an oscilloscope.

$$P_L = \frac{(0.707\ V_{pk})^2}{R_L} \tag{11.9}$$

This equation simply converts $V_{pk}$ to an rms value in the numerator. The use of this equation is demonstrated in Example 11.4.

---

**EXAMPLE 11.4**

Determine the ac load power for the circuit shown in Figure 11.11.

**Solution:** The oscilloscope is shown to be measuring 20 $V_{pk}$. Using equation (11.9), the ac load power is found as

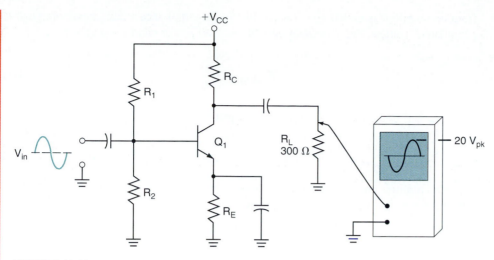

**FIGURE 11.11**

$$P_L = \frac{(0.707\ V_{pk})^2}{R_L}$$

$$= \frac{(14.14\ V)^2}{300\ \Omega}$$

$$= 666.5\ mW$$

***PRACTICE PROBLEM 11.4***

An amplifier has a 670-$\Omega$ load. Using an oscilloscope, the peak output voltage is measured at $+8\ V_{pk}$. Determine the amount of power that the amplifier is supplying to the load.

When the peak-to-peak output voltage is known, the ac load power can be found as

$$P_L = \frac{V_{PP}^2}{8\ R_L} \qquad (11.10)$$

*A Practical Consideration:* This equation can be used when $V_{out}$ is measured with an oscilloscope.

This equation is derived by substituting $V_{PP}/2$ in place of $V_{pk}$ in equation (11.9) as follows:

$$P_L = \frac{(0.707\ V_{pk})^2}{R_L}$$

$$= \frac{(0.3535\ V_{PP})^2}{R_L}$$

$$= \frac{0.125\ V_{PP}^2}{R_L}$$

Now, since $0.125 = \frac{1}{8}$, this equation is rewritten as

$$P_L = \frac{V_{PP}^2}{8\ R_L}$$

Equation (11.10) is important because it allows us to calculate the *maximum possible value of ac load power.* You may recall that the maximum possible output voltage from an amplifier is called *compliance*, PP. Since compliance is the maximum peak-to-peak output voltage, the maximum possible ac load power can be found as

$$P_{L(max)} = \frac{PP^2}{8\,R_L}$$

(11.11)

Example 11.5 demonstrates the use of this equation.

---

**EXAMPLE 11.5**

A given amplifier has a compliance of 18 $V_{PP}$. If the load value is 100 $\Omega$, what is the maximum possible value of ac load power?

*Solution:*   Using the values given, the maximum possible load power is found as

$$P_{L(max)} = \frac{PP^2}{8\,R_L}$$

$$= \frac{(18\ V_{PP})^2}{800\ \Omega}$$

$$= 405\ mW$$

**PRACTICE PROBLEM 11.5**

An amplifier has a compliance of 20 $V_{PP}$ and a load resistance of 140 $\Omega$. What is the maximum possible load power for the circuit?

---

## Amplifier Efficiency

Once the values of $P_S$ and $P_L$ have been calculated for an amplifier, we can use these values to calculate the *efficiency* of the circuit. As you were told in Chapter 8, the efficiency of an amplifier is the portion of the power drawn from the dc power supply that is actually delivered to the load, given as a percentage.

OBJECTIVE 4 ▶       The higher the efficiency of an amplifier, the better. Why? Because a high efficiency rating indicates that a very small percentage of the power drawn from the supply is used by the amplifier itself. Any power used by the amplifier must be dissipated in the form of heat. Heat in any amplifier is undesirable because it reduces the effective life of the components.

As you were shown in Chapter 8, the maximum theoretical efficiency of an *RC*-coupled class A amplifier is 25 percent. In practice, the efficiency of this type of amplifier is always much lower than 25 percent. This point is illustrated in Example 11.6.

---

**EXAMPLE 11.6**

Determine the maximum efficiency of the amplifier described in Example 11.1.

*Solution:*   In Example 11.1, we determined the compliance of the amplifier to be PP = 6.08 V. With a 10-k$\Omega$ load resistance (shown in Figure 11.6), we can find the maximum value of load power as

$$P_{L(max)} = \frac{PP^2}{8\,R_L}$$

$$= \frac{(6.08\ V_{PP})^2}{80\ k\Omega}$$

$$= 462\ \mu W$$

Using the established procedures, the following values are determined for the amplifier:

$$I_1 = 279.1\ \mu A$$
$$I_{CQ} = 950.3\ \mu A$$

and

$$I_{CC} = I_{CQ} + I_1$$
$$= 1.23\ mA$$

Now $P_S$ is found as

$$P_S = V_{CC}I_{CC}$$
$$= (12\ V)(1.23\ mA)$$
$$= 14.76\ mW$$

Finally,

$$\eta = \frac{P_L}{P_S} \times 100$$

$$= \frac{462\ \mu W}{14.76\ mW} \times 100$$

$$= 3.13\%$$

As you can see, even when the amplifier is driven to compliance its efficiency rating will be only 3.13 percent. This is considerably less than the maximum theoretical value of 25 percent.

### PRACTICE PROBLEM 11.6

The amplifier described in Practice Problem 11.1 is driven to compliance. Determine the maximum efficiency of the amplifier.

The maximum theoretical efficiency value of 25 percent for the *RC*-coupled class A amplifier is derived in Appendix D for those who wish to review the derivation.

The maximum theoretical efficiency of the *transformer-coupled* class A amplifier is 50 percent. The reason for the higher efficiency rating of this amplifier is due to its operating characteristics. We will take a look at this type of amplifier in the next section.

**Section Review**

1. List, in order, the steps required to determine the value of $I_{CC}$ for an *RC*-coupled class A amplifier.
2. How do you calculate the value of $P_L$ when $V_{out}$ is measured with an ac voltmeter?
3. How do you calculate the value of $P_L$ when $V_{out}$ is measured with an oscilloscope?
4. When is load power ($P_L$) at its maximum value?

5. What is the maximum theoretical efficiency of an *RC*-coupled class A amplifier?

6. Why is a high efficiency rating desirable for a power amplifier?

## *11.3* TRANSFORMER-COUPLED CLASS A AMPLIFIERS

*OBJECTIVE 5* ▶

**Transformer-coupled class A amplifier**
A class A amplifier that uses a transformer to couple the output signal to the load.

A **transformer-coupled class A amplifier** uses a transformer to couple the output signal from the amplifier to the load. A typical transformer-coupled amplifier is shown in Figure 11.12. The ac characteristics of the transformer shown play an important role in the overall operation of the amplifier. For this reason, we will once again review the basic transformer operating characteristics.

### Transformers

You may recall that the *turns ratio* of a transformer determines the relationship between primary and secondary values of voltage, current, and impedance. These relationships are summarized as follows:

$$\frac{N_1}{N_2} = \frac{V_1}{V_2} = \frac{I_2}{I_1} \tag{11.12}$$

and

$$\left(\frac{N_1}{N_2}\right)^2 = \frac{Z_1}{Z_2} \tag{11.13}$$

where   $N_1, N_2$ = the number of turns in the primary and secondary, respectively
   $V_1, V_2$ = the primary and secondary voltages
   $I_1, I_2$ = the primary and secondary currents
   $Z_1, Z_2$ = the primary and secondary impedances

**Step-down transformer**
One with a secondary voltage that is less than the primary voltage.

   Most transformers are classified as being either a **step-down transformer** or a *step-up* transformer. A step-down transformer is one with a secondary voltage that is less than the primary voltage. A step-down transformer will always have the following characteristics:

**FIGURE 11.12   Transformer-coupled class A amplifier.**

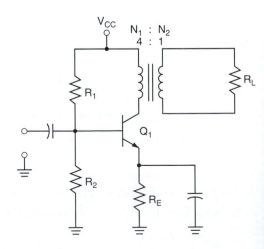

$$N_1 > N_2$$
$$V_1 > V_2$$
$$I_1 < I_2$$
$$Z_1 > Z_2$$

A **step-up transformer** is one with a secondary voltage that is greater than the primary voltage. A step-up transformer will always have the following characteristics:

$$N_1 < N_2$$
$$V_1 < V_2$$
$$I_1 > I_2$$
$$Z_1 > Z_2$$

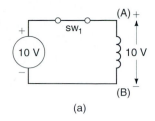

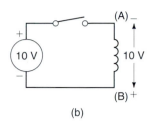

**FIGURE 11.13**

As you will see, these relationships play an important role in the ac operation of the transformer-coupled class A amplifier.

Another important characteristic of the transformer is one that is shared by all inductors, that is, the ability to produce a **counter emf,** or "*kick*" *emf*. When an inductor experiences a rapid change in supply voltage, it will produce a voltage *with a polarity that is opposite to the original voltage polarity.* For example, take a look at the circuit shown in Figure 11.13. Figure 11.13a shows an inductor with 10 V applied to it. Note that point A is shown as being *positive* and point B is shown as being *negative*. Figure 11.13b shows the same inductor with the voltage source removed. The inductor is shown as still having 10 V across the component. However, now point A is *negative* and point B is *positive*. This is the counter *emf* that is produced when there is a rapid change in the inductor supply voltage. Note that the amplitude of the counter emf may be much greater than that of the original applied voltage.

The counter emf shown in Figure 11.13b is caused by the electromagnetic field that surrounds the inductor. When this field collapses, it induces a voltage that is equal to (or greater than) the original supply voltage in value and opposite to the original voltage in polarity. *Note that this counter emf will be present only for a brief period of time.* As the field collapses into the inductor, the voltage *decreases* in value until it eventually reaches 0 V. As you will see, this characteristic also plays an important role in the operation of the transformer-coupled class A amplifier.

**Step-up transformer**
One that has a secondary voltage that is greater than the primary voltage.

**Counter emf**
A voltage produced by a transformer that has the opposite polarity of the voltage that caused it. Often referred to as *kick emf*.

We have reviewed all the transformer principles needed to understand the operation of transformer-coupled amplifiers. Now we will take a look at the dc and ac operating characteristics of these circuits.

## dc Operating Characteristics

The dc biasing of a transformer-coupled class A amplifier is very similar to any other class A amplifier with one important exception: *The value of $V_{CEQ}$ is designed to be as close as possible to $V_{CC}$.* Having a $V_{CEQ}$ that is approximately equal to $V_{CC}$ is the only acceptable situation in this amplifier type. The reason for this will become clear in our discussion of the ac characteristics of the amplifier.

$$V_{CEQ} \cong V_{CC}$$

Figure 11.14 shows a basic transformer-coupled amplifier and its dc load line. Note that the load line shown for the amplifier is very close to being a vertical line, indicating that $V_{CEQ}$ will be approximately equal to $V_{CC}$ for all the values of $I_C$ shown. The nearly vertical load line of the transformer-coupled amplifier is caused by the extremely low dc resistance of the transformer primary. You may recall that $V_{CEQ}$ for an amplifier is found as

Why the dc load line for a transformer-coupled class A amplifier is nearly vertical.

$$V_{CEQ} = V_{CC} - I_{CQ}(R_C + R_E)$$

In Figure 11.14, $R_E$ is shown to be 12 $\Omega$. Let's assume for a moment that the primary winding resistance ($R_W$) of the transformer is 10 $\Omega$. If this is the case, $V_{CEQ}$ at $I_{CQ} = 50$ mA is found as

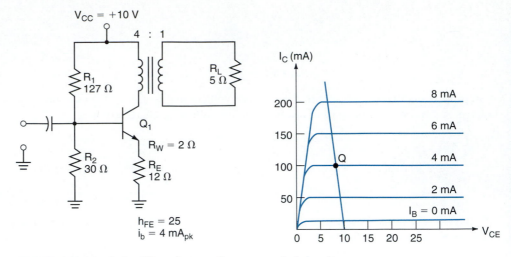

**FIGURE 11.14** dc load line of a transformer-coupled circuit.

$$V_{CEQ} = 10 \text{ V} - (50 \text{ mA})(22 \text{ }\Omega)$$
$$= 8.9 \text{ V}$$

$V_{CEQ}$ at $I_{CQ} = 200 \text{ mA}$ would be found as

$$V_{CEQ} = 10 \text{ V} - (200 \text{ mA})(22 \text{ }\Omega)$$
$$= 5.6 \text{ V}$$

Thus, the change in $V_{CEQ}$ was only 3.3 V over a range of $I_{CQ} = 50$ to 200 mA. This slight change in $V_{CEQ}$ would produce the nearly vertical dc load line shown.

You should note that the value of $R_L$ would be ignored in the dc analysis of the transformer-coupled class A amplifier. The reason for this is the fact that transformers provide dc isolation between the primary and secondary. Since the load resistance is in the secondary of the transformer, it does not affect the dc analysis of the primary circuitry.

## ac Operating Characteristics

The first step in analyzing the ac operation of the transformer-coupled amplifier is to plot the ac load line of the circuit. The process for plotting the ac load line is as follows:

Plotting the ac load line.

1. Determine the maximum possible change in $V_{CE}$.
2. Determine the corresponding change in $I_C$.
3. Plot a line that passes through the $Q$-point and the value of $I_{C(max)}$.
4. Locate the two points where the load line passes through the lines representing the minimum and maximum values of $I_B$. These two points are then used to find the maximum and minimum values of $I_C$ and $V_{CE}$.

The first step is simple. Since $V_{CE}$ cannot change by an amount greater than $(V_{CEQ} - 0 \text{ V})$, the maximum value of $\Delta V_{CE}$ is approximately equal to $V_{CEQ}$. This is illustrated in Figure 11.15. Finding the maximum corresponding change in $I_C$ takes several calculations. First, you must use equation (11.13) to find the value of $Z_1$ for the transformer. For the circuit represented in Figures 11.14 and 11.15, $Z_1$ is found as

$$Z_1 = \left(\frac{N_1}{N_2}\right)^2 Z_2$$
$$= 4^2(5 \text{ }\Omega)$$
$$= 80 \text{ }\Omega$$

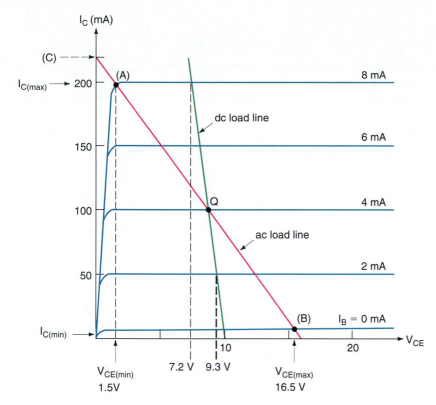

**FIGURE 11.15  ac operating characteristics.**

Now, using the values of $\Delta V_{CE}$ and $Z_1$, $\Delta I_C$ is found as

$$\Delta I_C = \frac{\Delta V_{CE}}{Z_1} \qquad\qquad \textbf{(11.14)}$$

For the circuit in Figure 11.14,

$$\Delta I_C = \frac{8.6 \text{ V}}{80 \text{ } \Omega}$$
$$= 107.5 \text{ mA}$$

The value of $I_{C(\text{max})}$ is now found using

$$I_{C(\text{max})} = I_{CQ} + \Delta I_C \qquad\qquad \textbf{(11.15)}$$

For the circuit shown,

$$I_{C(\text{max})} = 100 \text{ mA} + 107.5 \text{ mA}$$
$$= 207.5 \text{ mA}$$

The ac load line is now drawn from the point on the *y*-axis that corresponds to 207.5 mA (indicated by the dashed arrow, C, in Figure 11.15) through the *Q*-point to the *x*-axis. The completed line is shown in Figure 11.15.

Now the minimum and maximum values of $I_B$ are determined. Figure 11.14 indicates a peak ac value of $I_B$ equal to 4 mA. Thus, $I_B$ can change by 4 mA both above and below the initial value of $I_B$. With an initial value of $I_B = 4$ mA, we get the following values:

*Remember:*

$$I_{B(\text{max})} = 4 \text{ mA} + 4 \text{ mA} = 8 \text{ mA}$$

$$I_{BQ} = \frac{I_E}{h_{FE} + 1}$$

and

$$I_{B(min)} = 4\text{ mA} - 4\text{ mA} = 0\text{ mA}$$

The points where the ac load line intersects the $I_B = 8$ mA and $I_B = 0$ mA lines are indicated in Figure 11.15 as points A and B, respectively. *Note that these two points indicate the operating limits of $V_{CE}$ and $I_C$.* Point A is used to determine the values of $I_{C(max)}$ and $V_{CE(min)}$, while point B is used to determine the values of $I_{C(min)}$ and $V_{CE(max)}$. Thus, for this circuit,

$$V_{CE(max)} = 16.5\text{ V} \qquad \text{when } I_{C(min)} = 5\text{ mA}$$

$$V_{CE(min)} = 1.5\text{ V} \qquad \text{when } I_{C(max)} = 200\text{ mA}$$

At this point, you may be wondering how we managed to obtain a value of $V_{CE(max)}$ that is greater than the value of $V_{CC}$ for the circuit. The high value of $V_{CE(max)}$ is caused by the *counter emf* produced by the transformer primary. This point is illustrated by the circuits in Figure 11.16, which show the circuit conditions that correspond to points A and B in Figure 11.15. When the circuit is operating at point A, the value of $I_C$ is approximately 200 mA. With 200 mA through $R_E$, $V_E$ is equal to 2.4 V. As indicated above, the value of $V_{CE(min)}$ is 1.5 V. This means that there is a total of 3.9 V from the collector of the transistor to ground, leaving 6.1 V across the primary of the transformer. Figure 11.16b shows what happens when $I_B$ and $I_C$ drop to a minimum. At the instant that $I_{C(min)}$ (point B) is reached, the 6.1 V is still across the primary of the transformer, but the voltage polarity has reversed. This means that the bottom side of the transformer is now 6.1 V *more positive* than $V_{CC}$. Thus, the voltage from the transistor collector to ground is equal to 16.1 V. This is very close to the 16.5 V derived from the ac load line.

As you can see, a transformer-coupled amplifier will have two very important characteristics:

1. $V_{CEQ}$ will be very close to the value of $V_{CC}$.
2. The maximum output voltage will be very close to 2 $V_{CEQ}$. Thus it can approach the value of 2 $V_{CC}$.

As you will see, these characteristics are the basis for the higher efficiency rating for the transformer-coupled amplifier.

## Amplifier Efficiency

As was stated earlier, the maximum theoretical efficiency of the transformer-coupled class A amplifier is 50 percent. The derivation of this value is shown in Appendix D.

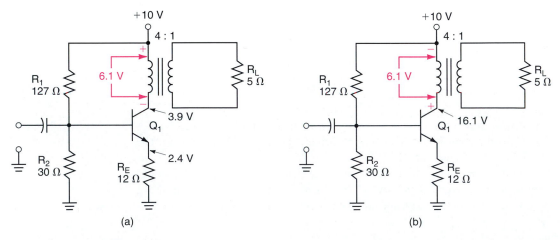

**FIGURE 11.16    Effects of counter emf.**

The actual efficiency rating of a transformer-coupled class A amplifier will generally be less than 40 percent. There are several reasons for the difference between the practical and theoretical efficiency ratings for the amplifier:

1. The derivation of the $\eta = 50$ percent value assumes that $V_{CEQ} = V_{CC}$. In practice, $V_{CEQ}$ will always be some value that is less than $V_{CC}$.

2. The transformer is subject to various power losses, as you were taught in your study of ac electronics. Among these losses are copper loss and hysteresis loss. These transformer power losses are not considered in the derivation of the $\eta = 50$ percent value.

## Calculating Maximum Load Power and Efficiency

To calculate the maximum load power for a transformer-coupled class A amplifier, start by determining the value of $I_{CQ}$ for the circuit, using the established procedure for the common-emitter circuit. Once the value of $I_{CQ}$ is known, we can approximate the value of source power as

$$P_S = V_{CC}I_{CQ} \qquad (11.16)$$

This approximation is valid because $I_{CQ} \gg I_1$ in a typical transformer-coupled class A amplifier. Once the value of $P_S$ is known, we have to calculate the maximum load power for the amplifier.

A transformer-coupled amplifier will have a maximum peak-to-peak output that is approximately equal to the difference between the values of $V_{CE(\text{max})}$ and $V_{CE(\text{min})}$ that are found using the ac load line for the circuit. Once these values have been found,

*A Practical Consideration:* The value of PP for a transformer-coupled class A amplifier will be slightly less than the $V_{CE(\text{off})}$ value found with the ac load line.

$$\text{PP} = V_{CE(\text{max})} - V_{CE(\text{min})} \qquad (11.17)$$

The value of compliance (PP) found in equation (11.17) indicates the maximum possible peak-to-peak voltage *across the primary of the transformer.* Using this value and the turns ratio of the transformer, the maximum possible peak-to-peak load voltage is found as

$$V_{\text{PP}} = \frac{N_2}{N_1} \text{PP} \qquad (11.18)$$

Once the value of peak-to-peak load voltage ($V_{\text{PP}}$) is known, the values of load power and efficiency are calculated as shown in Section 11.2. The entire process for calculating the maximum efficiency of a transformer-coupled class A amplifier is demonstrated in Example 11.7.

---

Determine the maximum efficiency of the amplifier in Figure 11.16.

**EXAMPLE 11.7**

**Solution:** Using the established procedure, the value of $I_{CQ}$ is found to be 102 mA. If you refer to Figure 11.15, you'll see that the calculated value of $I_{CQ}$ is very close to the load line value of $I_{CQ} = 100$ mA.

The value of $P_S$ is now found as

$$\begin{aligned} P_S &= V_{CC}I_{CQ} \\ &= (10 \text{ V})(102 \text{ mA}) \\ &= 1.02 \text{ W} \end{aligned}$$

The compliance of the amplifier is now found as

$$PP = V_{CE(max)} - V_{CE(min)}$$
$$= 16.5 \text{ V} - 1.5 \text{ V}$$
$$= 15 \text{ V}_{PP}$$

Note that the values of $V_{CE(max)}$ and $V_{CE(min)}$ were found using the ac load line in Figure 11.15.

Once the compliance of the amplifier is known, the maximum peak-to-peak load voltage is found as

$$V_{PP} = \frac{N_2}{N_1} PP$$

$$= \frac{1}{4} (15 \ V_{PP})$$

$$= 3.75 \ V_{PP}$$

Now the maximum load power is found using equation (11.10) as follows:

$$P_{L(max)} = \frac{V_{PP}^2}{8 R_L}$$

$$= \frac{(3.75 \ V_{PP})^2}{40 \ \Omega}$$

$$= 351.56 \text{ mW}$$

Finally, the efficiency rating of the amplifier is found as

$$\eta = \frac{P_L}{P_S} \times 100$$

$$= \frac{351.56 \text{ mW}}{1.02 \text{ W}} \times 100$$

$$= 34.46\%$$

The maximum efficiency found here would be slightly higher than the actual efficiency of the amplifier since the calculation does not take transformer losses into account.

***PRACTICE PROBLEM 11.7***

A transformer-coupled class A amplifier has the following values: $V_{CE(max)} = 22.4$ V, $V_{CE(min)} = 2.4$ V, $N_1 = 5$, $N_2 = 1$, $R_L = 4 \ \Omega$, $V_{CC} = 12$ V, and $I_{CQ} = 120$ mA. Calculate the maximum efficiency of the circuit.

## One Final Note

The transformer-coupled class A amplifier has several advantages over the *RC*-coupled circuit. Three of these are:

1. The efficiency of the transformer-coupled circuit is higher.

2. The transformer can be used for impedance matching between the amplifier and its load.

**Tuned amplifier**
A circuit designed to have a specific value of gain over a specified range of frequencies.

3. The transformer-coupled circuit can be converted easily into a **tuned amplifier,** that is, an amplifier designed to provide gain over a specified range of frequencies.

Tuned amplifiers are discussed in Chapter 17.

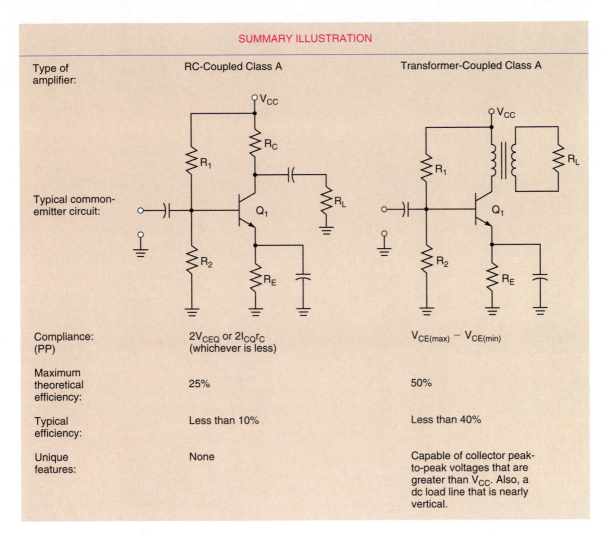

SUMMARY ILLUSTRATION

| Type of amplifier: | RC-Coupled Class A | Transformer-Coupled Class A |
|---|---|---|
| Typical common-emitter circuit: | | |
| Compliance: (PP) | $2V_{CEQ}$ or $2I_{CQ}r_C$ (whichever is less) | $V_{CE(max)} - V_{CE(min)}$ |
| Maximum theoretical efficiency: | 25% | 50% |
| Typical efficiency: | Less than 10% | Less than 40% |
| Unique features: | None | Capable of collector peak-to-peak voltages that are greater than $V_{CC}$. Also, a dc load line that is nearly vertical. |

**FIGURE 11.17**

In the next section, we will move on to the class B amplifier. A summary of the two class A amplifiers we have discussed is shown in Figure 11.17.

*Section Review*

1. What is a *transformer-coupled class A amplifier*?
2. List the characteristics of the step-down transformer.
3. List the characteristics of the step-up transformer.
4. Describe *counter emf* and its cause.
5. Describe the dc load line of a transformer-coupled amplifier.
6. What is the process for plotting the ac load line of a transformer-coupled class A amplifier?
7. What is the maximum theoretical efficiency of the transformer-coupled class A amplifier? Why is the actual efficiency always less than this value?
8. List, in order, the steps required to calculate the maximum load power and efficiency rating of a transformer-coupled class A amplifier.
9. What are the advantages of using the transformer-coupled class A amplifier?
10. What is a tuned amplifier?

# 11.4 CLASS B AMPLIFIERS

The primary disadvantage of using class A power amplifiers is the fact that their efficiency ratings are so low. As you have been shown, a majority of the power drawn from the supply by a class A amplifier is used up by the amplifier itself. This goes against the primary purpose of a power amplifier, which is to deliver the power drawn from the supply to the load.

OBJECTIVE 6 ▶

The class B amplifier was developed to improve on the low efficiency rating of the class A amplifier. The maximum theoretical efficiency rating of a class B amplifier is approximately 78.5 percent. This means that up to 78.5 percent of the power drawn from the supply can ideally be delivered to the load.

Unlike the class A amplifier, the class B amplifier consumes very little power when there is no input signal. This is a major improvement over the class A amplifier. However, because each transistor in a class B amplifier conducts for only approximately 180°, each amplifier requires two transistors to accurately reproduce an input waveform.

Figure 11.18 shows the most commonly used type of class B configuration. This circuit configuration is referred to as a **complementary-symmetry amplifier,** or **push–pull emitter follower.** The circuit recognition feature is the use of *complementary transistors* (that is, one of the transistors is an *npn* and the other is a *pnp*). The biasing circuit components may change from one amplifier to another, but complementary-symmetry amplifiers will always contain complementary transistors.

The **standard push–pull amplifier** contains two transistors of the same type with the emitters tied together. It uses a *center-tapped transformer,* or a *transistor phase splitter* on the input and a center-tapped transformer on the output. This amplifier type is shown in Figure 11.19. Note the transistor types and the transformer. This is the standard push–pull amplifier configuration.

Why is the complementary-symmetry configuration preferred over the standard push–pull? The center-tapped transformer makes the standard push–pull circuit much more expensive to construct than the complementary-symmetry amplifier. Since the complementary-symmetry amplifier is by far the most commonly used, we will concentrate on this circuit configuration in our discussion on class B operation.

**Complementary-symmetry amplifier (push–pull emitter follower)**
A class B circuit configuration using *complementary transistors* (a pair, one *npn* and one *pnp*, with matched characteristics).

**Standard push–pull amplifier**
A class B circuit that uses two identical transistors and a center-tapped transformer.

Why complementary-symmetry amplifiers are preferred.

**FIGURE 11.18   Class B complementary-symmetry amplifier.**

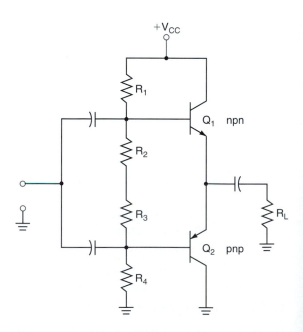

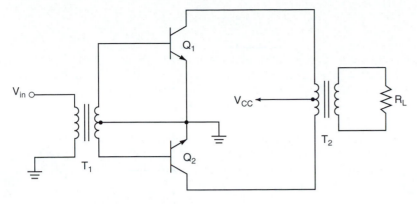

**FIGURE 11.19    Class B push–pull amplifier.**

## Class B Operation Overview

The term *push–pull* comes from the fact that the two transistors in a class B amplifier conduct on alternating half-cycles of the input. For example, consider the circuit shown in Figure 11.20. During the positive half-cycle of the input, $Q_1$ is biased *on* and $Q_2$ is biased *off*. During the negative half-cycle of the input, $Q_1$ is biased *off*, and $Q_2$ is biased *on*. The fact that *both transistors are never fully on at the same time* is the key to the high efficiency rating of the amplifier. This point will be discussed in detail later in this section.

The biasing of the two transistors is the key to its operation. When the amplifier is in its quiescent state (no input), both transistors are biased at *cutoff*. When the input goes positive, $Q_1$ is biased above cutoff, and conduction results through the transistor. During this time, $Q_2$ is still biased at cutoff. When the input goes into its negative half-cycle, $Q_1$

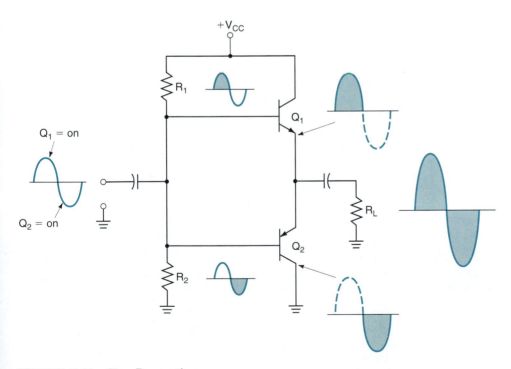

**FIGURE 11.20    Class B operation.**

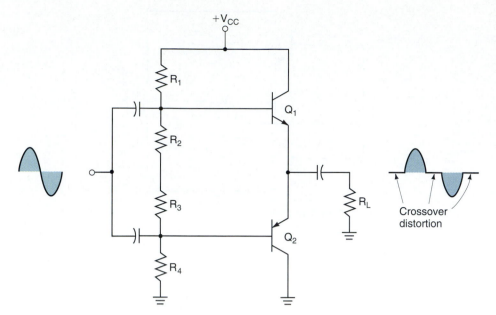

FIGURE 11.21   Crossover distortion.

**Lab Reference:** Crossover distortion can be seen (by design) in Exercise 22.

**Crossover distortion**
Distortion caused by class B transistor biasing. Crossover distortion occurs during the time neither of the transistors is conducting.

is returned to the cutoff state, and $Q_2$ is biased above cutoff. As a result, conduction through $Q_2$ starts to build, while $Q_1$ remains off.

Because of the biasing arrangement, class B amplifiers are subject to a type of distortion called **crossover distortion.** Crossover distortion is illustrated in Figure 11.21. Note the flat lines between the half-cycles of the output signal. These flat lines are the crossover distortion. Crossover distortion is caused by having a short period of time when *both transistors are off.* When both transistors are off, the output drops to 0 V. To prevent crossover distortion, both transistors will normally be biased at a level that is slightly *above* cutoff. As you will see, this type of biasing allows the amplifier to provide a linear output that contains no distortion.

## dc Operating Characteristics

**Lab Reference:** The dc characteristics of a Class B amplifier are demonstrated in Exercise 22.

The class B amplifier has a vertical dc load line. The reason for this is the fact that there are no resistors in the emitter or collector circuits of the transistors. For example, consider the circuit shown in Figure 11.22a. Assuming that both transistors are biased right at the cutoff point, the $Q$-point is established at the $I_B = 0$ line on the collector characteristic curves. This is shown in Figure 11.22b. Now assume that we could turn both transistors on at the same time. If they were both on, the following conditions would exist:

1. The voltage drops across the two transistors (from emitter to collector) would remain the same because the resistance *ratio* of the two components would not change.

2. The value of $I_C$ could be very high because there are no resistors present in the emitter or collector circuits to restrict the current. Current is limited only by the internal resistance of the transistors when they are in saturation.

Thus, the voltage across the transistors would remain fairly constant, and the current through the collector and emitter circuits would not be restricted. This gives us the vertical dc load line. Note that this line shows little change in $V_{CE}$ but *does* show a large possible increase in the value of $I_C$.

The dc load line illustrates two other points about the dc operation of the class B amplifier. First, note that

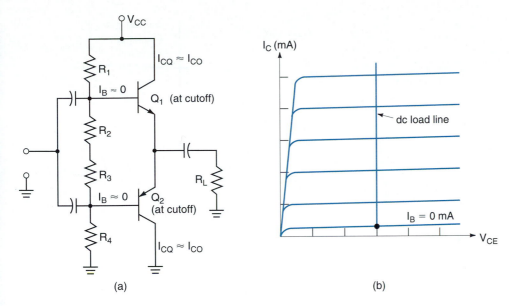

FIGURE 11.22    dc load line for a class B amplifier.

$$V_{CEQ} = \frac{V_{CC}}{2} \qquad\qquad (11.19)$$

This relationship is due to the fact that the amplifier is built using **matched transistors.** Matched transistors have the same operating characteristics, except for the fact that one is an *npn* and the other is a *pnp*. For example, the 2N3904 and the 2N3906 are matched transistors. They have the same operating parameters and specifications, except for the fact that the 2N3904 is an *npn* transistor and the 2N3906 is a *pnp* transistor. Matched transistors should be used in class B amplifiers because any difference between the operating characteristics of the individual transistors will produce output distortion or nonlinearity.

When matched transistors are used, the values of $V_{CE}$ for the two are equal when $I_C$ of one transistor is approximately equal to $I_C$ of the other. Now refer to the circuit shown in Figure 11.22. Since the transistors are wired in series, $I_{C1}$ is approximately equal to $I_{C2}$. Thus $V_{CE1}$ equals $V_{CE2}$. Since the two voltages are equal and must add up to $V_{CC}$, each equals one-half of $V_{CC}$.

Another point to be made is shown in the following equation:

$$I_{CQ} \cong 0 \qquad\qquad (11.20)$$

This approximation is valid because both transistors are biased just inside the cutoff region. If they were both biased farther into the cutoff region, the value of $I_{CQ}$ would approach the ideal value, 0 A. To help you understand this point better, take a look at Figure 11.23. This is a "close-up" of the cutoff region of the characteristic curve for the transistors shown in Figure 11.22. For discussion purposes, we will refer to the two biasing points shown as *soft cutoff* and *hard cutoff*. When the transistors are biased at the soft-cutoff point, $I_{CQ}$ will be at a higher level than that for the hard-cutoff point. As the transistor biasing is adjusted nearer the hard-cutoff point, the value of $I_{CQ}$ approaches zero. (It can never reach the ideal value of zero because there will always be *some* amount of leakage current through the transistors.)

It would seem that hard cutoff would be the ideal type of biasing to use on a class B amplifier. However, this is not the case. Biasing a class B amplifier at the hard-cutoff point causes crossover distortion. This is due to the transition time required

**Matched transistors**
Transistors that have the same operating characteristics.

Hard-cutoff causes crossover distortion.

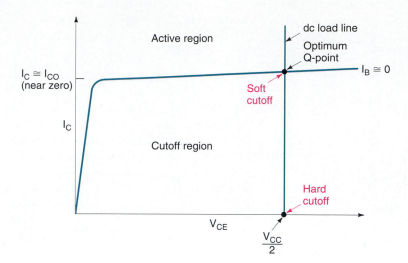

FIGURE 11.23    Magnified view of cutoff region.

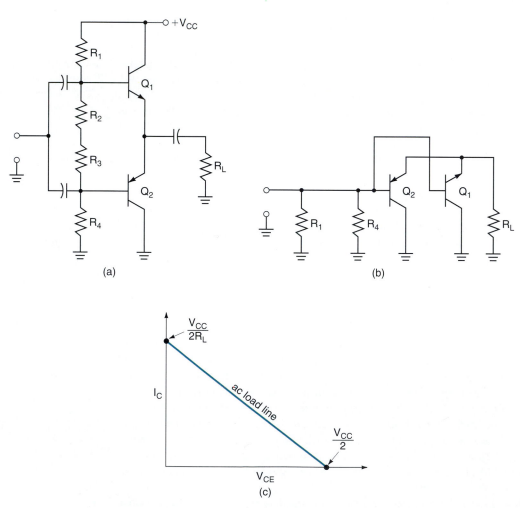

(a)

(b)

(c)

FIGURE 11.24    Class B ac characteristics.

for the transistor to come out of cutoff into the active region of operation. Biasing a transistor at soft cutoff reduces this transition time, thus reducing crossover distortion.

## ac Operating Characteristics

Before we can cover the methods used to bias class B amplifiers, it is important for you to understand the ac characteristics of this amplifier type. The ac characteristics of the class B amplifier are illustrated in Figure 11.24.

Figure 11.24b shows the ac equivalent circuit for the class B amplifier. You may recall that this circuit is obtained by grounding the dc voltage source ($V_{CC}$) and shorting the capacitors in the circuit. With this equivalent circuit, it is easy to see that the voltage across $R_L$ is equal to $V_{CE}$ of the transistors. We stated earlier that this voltage would equal one-half of $V_{CC}$. We can therefore find the value of $I_{C(sat)}$ as

**Lab Reference:** The ac operation of a Class B amplifier is demonstrated in Exercise 22.

$$I_{C(sat)} = \frac{V_{CC}}{2R_L} \qquad \textbf{(11.21)}$$

We stated earlier that the transistors in a class B amplifier are normally biased at cutoff. Thus, the normal value of $V_{CE(off)}$ is found as

$$V_{CE(off)} = \frac{V_{CC}}{2} \qquad \textbf{(11.22)}$$

These two points are used to derive the ac load line of the class B amplifier. This load line is illustrated in Figure 11.24c.

---

Determine the end-point values of the ac load line for the circuit shown in Figure 11.25.

*Solution:* The value of $I_{C(sat)}$ is found as

$$I_{C(sat)} = \frac{V_{CC}}{2R_L}$$

$$= \frac{10 \text{ V}}{2(10 \text{ }\Omega)}$$

$$= 500 \text{ mA}$$

Now, the value of $V_{CE(off)}$ is found as

$$V_{CE(off)} = \frac{V_{CC}}{2}$$

$$= \frac{10 \text{ V}}{2}$$

$$= 5 \text{ V}$$

The ac/dc load lines for the circuit are shown in Figure 11.26. Note that the load lines meet at the Q-point.

**EXAMPLE 11.8**

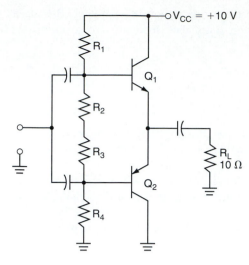

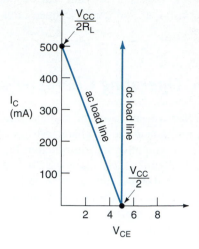

FIGURE 11.25                                    FIGURE 11.26

*PRACTICE PROBLEM 11.8*

A class B amplifier has values of $V_{CC} = +12$ V and $R_L = 2.2$ k$\Omega$. Plot the dc and ac load lines for the circuit.

## Amplifier Impedance

You may recall that the input impedance to the base of an emitter follower is found as

$$Z_{\text{in(base)}} = h_{fc}(r'_e + r_E)$$

*Don't forget:* $r'_e$ *is the ac resistance of the transistor emitter, found as*

$$r'_e = \frac{h_{ic}}{h_{fc}}$$

If you take a look at the class B amplifier, you will notice that the load resistor is connected to the emitters of the two transistors. Since the load resistor is not bypassed, its value must be considered in the calculation of $Z_{\text{in(base)}}$ for the amplifier. The transistor input impedance is therefore found as

$$Z_{\text{in(base)}} = h_{fc}(r'_e + R_L) \qquad \textbf{(11.23)}$$

The output of the class B amplifier is taken from the emitters of the transistors, so the amplifier output impedance is equal to the ac resistance of the emitter circuit. As you may recall from Chapter 10, this impedance is found as

$$Z_{\text{out}} = r'_e + \frac{R'_{\text{in}}}{h_{fc}} \qquad \textbf{(11.24)}$$

where

$$R'_{\text{in}} = R_1 \parallel R_4 \parallel R_S$$

Note that the values of $R_2$ and $R_3$ in the class B amplifier are not used in equation (11.24). This is due to the fact that they are shorted by the input coupling capacitors in the ac equivalent circuit of the amplifier.

## Amplifier Gain

Since the complementary-symmetry amplifier is basically an *emitter follower,* the current gain is found as with any emitter follower. By formula,

$$A_i = h_{fc}\left(\frac{Z_{in} r_E}{Z_{in(base)} R_L}\right)$$

Since $r_E = R_L$ for the class B amplifier, the equation for $A_i$ can be simplified to

$$A_i = h_{fc}\left(\frac{Z_{in}}{Z_{in(base)}}\right)$$

The voltage gain of the class B amplifier is found in the same manner as a standard emitter follower, as follows:

$$A_v = \frac{R_L}{R_L + r_e'}$$

As with any amplifier, the power gain is the product of $A_v$ and $A_i$. By formula,

$$A_p = A_v A_i$$

## Power Calculations

The class B amplifier has the same output power characteristic as the class A amplifier. ◄ OBJECTIVE 7
By formula,

$$P_L = \frac{V_{PP}^2}{8R_L}$$

The maximum load power is also found in the same manner as it is for the class A amplifier. By formula,

$$P_{L(max)} = \frac{PP^2}{8R_L}$$

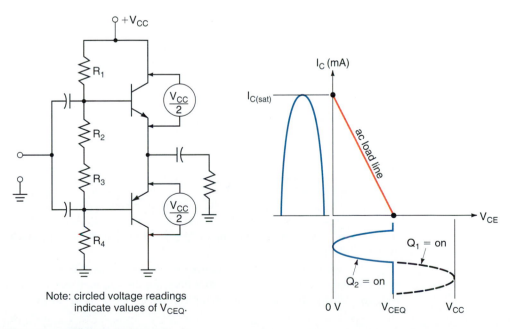

Note: circled voltage readings
indicate values of $V_{CEQ}$.

**FIGURE 11.27  Class B amplifier compliance.**

To calculate the maximum possible load power for a class B amplifier, we need to be able to determine its *compliance*. The compliance of a class B amplifier is found as

$$PP = 2V_{CEQ} \qquad \text{(11.25)}$$

Since $V_{CEQ} \cong V_{CC}/2$, equation (11.25) can be rewritten as

$$PP \cong V_{CC} \qquad \text{11.26)}$$

The compliance of a class B amplifier is illustrated in Figure 11.27. Each transistor is capable of making the full transition from $V_{CEQ}$ to approximately 0 V. This means that each transistor has a $\Delta V_{CE}$ approximately equal to $V_{CEQ}$. Since there are two transistors that conduct on alternate half-cycles of the input, the total transition of the output is approximately equal to $2V_{CEQ}$. The procedure for calculating the maximum load power for a class B amplifier is demonstrated in Example 11.9.

---

**EXAMPLE 11.9**

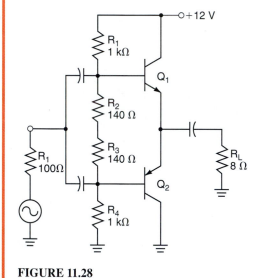

**FIGURE 11.28**

Determine the maximum load power for the circuit shown in Figure 11.28.

**Solution:**   The compliance of the amplifier is found as

$$PP \cong V_{CC}$$
$$= 12 \text{ V}$$

Now the maximum load power is found as

$$P_{L(max)} = \frac{PP^2}{8R_L}$$
$$= \frac{(12 \text{ V}_{PP})^2}{8(8 \text{ }\Omega)}$$
$$= \frac{144 \text{ V}}{64 \text{ }\Omega}$$
$$= 2.25 \text{ W}$$

***PRACTICE PROBLEM 11.9***

Determine the maximum load power for the circuit described in Practice Problem 11.8.

---

The total power that the amplifier draws from the supply is found as

$$P_S = V_{CC}I_{CC}$$

where

$$I_{CC} = I_{C1(ave)} + I_1 \qquad \text{(11.27)}$$

The equation for $P_S$ is the same one that is used for the class A amplifier. However, equation (11.27) needs some explaining.

The total current drawn from the supply is the sum of the *average* $Q_1$ collector current and the current through the amplifier base circuit, as is shown in Figure 11.29. The average value of the current through the collector of $Q_1$ is given as

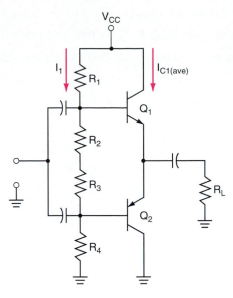

**FIGURE 11.29    Class B amplifier supply current.**

$$I_{C(\text{ave})} = \frac{I_{\text{pk}}}{\pi}$$

or

$$I_{C(\text{ave})} \cong 0.318 I_{\text{pk}} \quad \Bigg| \quad \left( 0.318 \cong \frac{1}{\pi} \right)$$

where $I_{\text{pk}}$ is the peak current through the transistor. Note that this is the standard $I_{\text{ave}}$ equation for the half-wave rectifier. Since the transistor is on for alternating half cycles, it effectively acts as a half-wave rectifier.

If you refer to Figure 11.27, you'll see that the peak current through each transistor is equal to $I_{C(\text{sat})}$ for the transistor, given as

$$I_{C(\text{sat})} = \frac{V_{CC}}{2R_L}$$

Combining these two equations,

$$I_{C1(\text{ave})} = \frac{0.159 V_{CC}}{R_L} \qquad \textbf{(11.28)}$$

Equation (11.28) uses $V_{CC}$ for the output voltage of the amplifier because it is assumed that the amplifier is driven to compliance. If the output of the amplifier is not at compliance, $V_{PP(\text{out})}$ must be substituted for $V_{CC}$. This equation for calculating $I_{C1(\text{ave})}$ is

$$I_{C1(\text{ave})} = \frac{0.159 V_{PP(\text{out})}}{R_L} \qquad \textbf{(11.29)}$$

Examples 11.10 and 11.11 demonstrate the process of determining the total power drawn from the supply and total load power.

## EXAMPLE 11.10

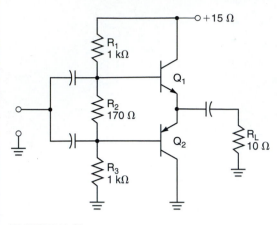

**FIGURE 11.30**

Determine the value of $P_S$ for the circuit shown in Figure 11.30.

**Solution:** Neglecting the base current from the two transistors, $I_1$ is found as

$$I_1 = \frac{V_{CC}}{R_T}$$

$$= \frac{15\text{ V}}{2.17\text{ k}\Omega}$$

$$= 6.91\text{ mA}$$

Now $I_{C(ave)}$ is found as

$$I_{C(ave)} = \frac{0.159\ V_{CC}}{R_L}$$

$$= \frac{0.159\ (15\text{ V})}{10\ \Omega}$$

$$= 238.5\text{ mA}$$

Using these two values, $I_{CC}$ is found as

$$I_{CC} = I_{C1(ave)} + I_1$$
$$= 245.4\text{ mA}$$

Finally, the total power demand on the supply is determined as

$$P_S = V_{CC}I_{CC}$$
$$= (15\text{ V})(245.4\text{ mA})$$
$$= 3.68\text{ W}$$

**PRACTICE PROBLEM 11.10**

Refer to Figure 11.28. Determine the value of $P_S$ for the circuit.

---

## EXAMPLE 11.11

Determine the maximum load power for the circuit in Figure 11.30.

**Solution:** The compliance of the amplifier is equal to $V_{CC}$, or 15 V. Using this value, the maximum load power is calculated as

$$P_{L(max)} = \frac{PP^2}{8R_L}$$

$$= \frac{(15\text{ V}_{PP})^2}{80\ \Omega}$$

$$= 2.81\text{ W}$$

**PRACTICE PROBLEM 11.11**

Refer to Figure 11.28. Determine the maximum value of $P_L$ for the circuit.

## Class B Amplifier Efficiency

It was stated earlier in the chapter that the maximum theoretical efficiency of a class B amplifier is 78.5 percent. As is the case with the class A amplifiers, any practical efficiency rating must be less than the maximum theoretical value.

The derivation of the $\eta = 78.5$ percent value is shown in Appendix D. If you take a look at the derivation, you will see it assumes that the compliance of the class B amplifier is equal to $V_{CC}$. Since both transistors in the class B amplifier still have a slight value of $V_{CE}$ when saturated, the actual compliance of the amplifier is slightly less than the value of $V_{CC}$. Thus, the class B amplifier efficiency rating never reaches the value of 78.5 percent.

Once the values of $P_S$ and $P_L$ for a given class B are known, the efficiency of the circuit is calculated in the same manner as it is for the class A amplifiers. This point is illustrated in Example 11.12.

---

**EXAMPLE 11.12**

Determine the efficiency of the amplifier used in Examples 11.10 and 11.11 (Figure 11.30).

**Solution:** The values of load power and dc supply power were calculated in Examples 11.10 and 11.11 as

$$P_L = 2.81 \text{ W} \qquad P_S = 3.68 \text{ W}$$

Using these two values, the maximum efficiency of the amplifier is found as

$$\eta = \frac{P_L}{P_S} \times 100$$

$$= \frac{2.81 \text{ W}}{3.68 \text{ W}} \times 100$$

$$= 76.36\%$$

**PRACTICE PROBLEM 11.12**

Determine the efficiency of the amplifier described in Practice Problems 11.10 and 11.11.

---

The efficiency of the class B amplifier is directly related to $V_{out}$. The highest efficiency will be at compliance. As the output signal is reduced, efficiency decreases. Example 11.13 shows how the efficiency of the class B amplifier changes when the output signal is reduced.

---

**EXAMPLE 11.13**

Determine the efficiency of the amplifier used in Example 11.10 (Figure 11.30) if the load voltage is reduced by 50 percent to 7.5 $V_{PP}$.

**Solution:** The value of $I_1$ remains at 6.91 mA (as calculated in Example 11.10) because the biasing network has not changed. Since the amplifier is no longer driven to compliance, the value of $I_{C(ave)}$ must be calculated as follows:

$$I_{C(ave)} = \frac{0.159 \, V_{PP(out)}}{R_L}$$

$$= \frac{(0.159)(7.5 \text{ V})}{10 \, \Omega}$$

$$= 119 \text{ mA}$$

Thus, the circuit has a total current of

$$6.91 \text{ mA} + 119 \text{ mA} = 125.91 \text{ mA}$$

Next, the total power drawn from the power supply is calculated. $V_{CC}$ is used for this calculation because it stays constant, despite the change in output voltage. Therefore,

$$P_S = V_{CC}I_{CC}$$
$$= 1.89 \text{ W}$$

Next, the maximum load power is found as

$$P_{L(max)} = \frac{V_{PP}^2}{8R_L}$$

$$= \frac{(7.5 \text{ V}_{PP})^2}{8(10 \text{ }\Omega)}$$

$$= 703 \text{ mW}$$

Finally, the amplifier efficiency is found as

$$\eta = \frac{P_L}{P_S} \times 100$$

$$= \frac{703 \text{ mW}}{1.89 \text{ W}} \times 100$$

$$= 37.2\%$$

**PRACTICE PROBLEM 11.13**

Determine the efficiency of the amplifier described in Example 11.13 if the load voltage is 11 $V_{PP}$.

## Summary

The class B amplifier is a two-transistor circuit that has a higher maximum efficiency rating than either of the common class A amplifiers. There are two types of class B amplifiers: the *push–pull amplifier* and the *complementary-symmetry amplifier*. Of the two, the complementary-symmetry amplifier is more commonly used for two reasons:

1. The complementary-symmetry amplifier does not require the use of transformers and thus is cheaper to produce.
2. Since the complementary-symmetry amplifier doesn't have a transformer, it is not subject to transformer losses. Thus, it will have a higher efficiency and better frequency response than a comparable push–pull amplifier.

The two transistors in a class B amplifier are biased at cutoff. When an ac signal is applied to the amplifier, the positive alternation of the ac signal will turn one transistor on, and the negative alternation will turn the other transistor on. The resulting amplifier output is an ac signal that has a peak-to-peak value approximately equal to $V_{CC}$ (when driven to compliance). The dc and ac characteristics of the complementary-symmetry amplifier are summarized in Figure 11.31.

**Section Review**

1. What advantage does the class B amplifier have over the class A amplifier?
2. What are the two common types of class B amplifiers? What are the advantages and disadvantages of each?
3. What are *complementary transistors*? What kind of amplifier requires the use of complementary transistors?
4. Describe the operating characteristics of the class B amplifier.

**FIGURE 11.31**

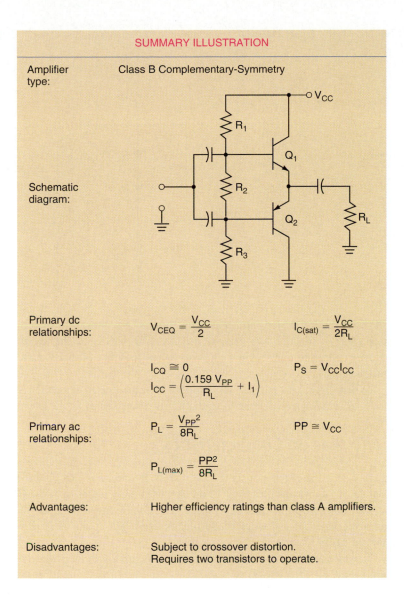

SUMMARY ILLUSTRATION

| | |
|---|---|
| Amplifier type: | Class B Complementary-Symmetry |
| Schematic diagram: | |

Primary dc relationships:

$$V_{CEQ} = \frac{V_{CC}}{2} \qquad I_{C(sat)} = \frac{V_{CC}}{2R_L}$$

$$I_{CQ} \cong 0 \qquad P_S = V_{CC}I_{CC}$$

$$I_{CC} = \left(\frac{0.159\, V_{PP}}{R_L} + I_1\right)$$

Primary ac relationships:

$$P_L = \frac{V_{PP}^2}{8R_L} \qquad PP \cong V_{CC}$$

$$P_{L(max)} = \frac{PP^2}{8R_L}$$

Advantages: Higher efficiency ratings than class A amplifiers.

Disadvantages: Subject to crossover distortion.
Requires two transistors to operate.

5. What are the typical values of $I_{CQ}$ and $V_{CEQ}$ for a class B amplifier?

6. What is the typical compliance of a class B amplifier?

7. Why is the practical efficiency rating of a class B amplifier less than the maximum theoretical value of 78.5 percent?

# 11.5 CLASS AB AMPLIFIERS (DIODE BIAS)

Up to this point, we have used voltage-divider bias for all our class B amplifiers. Problems can develop with the class B amplifier when voltage-divider bias is used:

1. Crossover distortion can occur.

2. Thermal runaway can occur.

A biasing circuit often used to eliminate the problems of crossover distortion and thermal runaway is shown in Figure 11.32. The circuit shown, called **diode bias**, uses two diodes in place of the resistor(s) between the transistor bases. The diodes

**Diode bias**
A biasing circuit that uses two diodes in place of the resistor(s) between the bases of the two transistors.

**FIGURE 11.32   Diode biasing.**

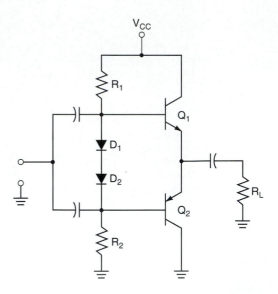

in the bias circuit, called **compensating diodes**, are chosen to match the characteristic values of $V_{BE}$ for the two transistors. As you will see, the diodes eliminate both crossover distortion and thermal runaway when they are properly matched to the amplifier transistors.

When diode bias is used, an amplifier like the one in Figure 11.32 is referred to as a *class AB amplifier*. This type of operation is discussed in detail later in this section.

## Diode Bias dc Characteristics

You may recall that the transistors in a class B amplifier are biased at cutoff, causing the value of $I_{CQ}$ for the amplifier to be approximately equal to zero. When diode bias is used, the transistors are actually biased just above cutoff; that is, there will be some measurable value of $I_{CQ}$ when diode bias is used. This point is illustrated with the help of Figure 11.33.

In order to understand the operation of the circuit shown, we must start with a few assumptions:

**1.** $V_{CEQ}$ is approximately one-half the value of $V_{CC}$, as is shown in the figure.

**2.** The current through $R_2$ causes 5.3 V to be developed across this resistor.

**Lab Reference:** The dc operation of a class AB amplifier is demonstrated in Exercise 22.

Assuming the above conditions are met, the base of $Q_2$ will be at 5.3 V and $V_E$ of $Q_2$ will be 6 V. Since $V_B$ for this *pnp* transistor is 0.7 V more negative than $V_E$, $Q_2$ will conduct.

With 1.4 V developed across the biasing diodes, $V_B$ of $Q_1$ will be 5.3 V + 1.4 V = 6.7 V. The value of $V_E$ for $Q_1$ is 6 V, as is shown in the figure. Since $V_B$ for this *npn* transistor is 0.7 V more positive than $V_E$, $Q_1$ will also conduct. Thus, both transistors in the diode-biased amplifier will conduct, and some measurable value of $I_{CQ}$ will be present, as shown in Figure 11.34.

Now let's take a look at the assumptions we made (above) and where they came from. Even though we have changed the transistor biasing circuit, the output transistors are in the same configuration as they were in the standard class B amplifier. Thus, the dc load line characteristics of the amplifier haven't changed, and $V_{CEQ}$ will always be approximately one-half of $V_{CC}$.

To understand where the assumed value of $V_{R2} = 5.3$ V came from, we have to take a look at the operation of the biasing circuit. The total voltage across the two resistors in

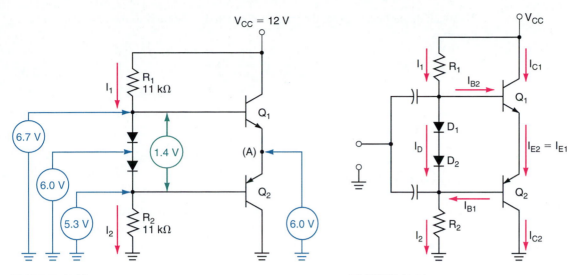

**FIGURE 11.33**                    **FIGURE 11.34**

the biasing circuit is equal to the difference between $V_{CC}$ and the 1.4-V drop across the diodes. By formula,

$$V_{R1} + V_{R2} = V_{CC} - 1.4 \text{ V}$$

Using the voltage-divider equation, we can find the value of $V_{R2}$ as

$$V_{R2} = \frac{R_2}{R_1 + R_2}(V_{CC} - 1.4 \text{ V})$$

Since $V_{B(Q2)} = V_{R2}$, the equation above can be rewritten as

$$V_{B(Q2)} = \frac{R_2}{R_1 + R_2}(V_{CC} - 1.4 \text{ V}) \qquad\qquad \textbf{(11.30)}$$

When $R_1 = R_2$ (as was the case with the circuit in Figure 11.33), the fraction in equation (11.30) will equal ½, and the equation can be rewritten as

$$V_{B(Q2)} = \frac{V_{CC}}{2} - 0.7 \text{ V}$$

or

$$V_{B(Q2)} = V_{CEQ} - 0.7 \text{ V} \qquad \text{(when } R_1 = R_2\text{)} \qquad\qquad \textbf{(11.31)}$$

*Don't Forget:*

$$V_{CEQ} \cong \frac{V_{CC}}{2}$$

Equation (11.31) is very important for several reasons: First, it points out the fact that $Q_2$ (and thus $Q_1$) will automatically be biased properly when $R_1 = R_2$. The actual values of the resistors are relatively unimportant, as long as they are equal. When $R_1 = R_2$, you can assume that the base voltages are equal to $V_{CEQ} \pm 0.7$ V. When $R_1 \neq R_2$, you must use equation (11.30) to find the value of $V_{B(Q2)}$ and then add 1.4 V to this value to find $V_{B(Q1)}$. The second reason that equation (11.31) is so important is because almost all diode-biased circuits are designed with equal-value resistors. Thus, equation (11.31) is used far more often than equation (11.30).

You may be wondering at this point why we didn't consider the values of $I_{B1}$ and $I_{B2}$ in our analysis of the biasing circuit. The reason is simple: When the transistors are properly matched, $I_{B1}$ and $I_{B2}$ will be equal in value. Therefore, the resistor currents ($I_1$ and $I_2$) will also be equal in value.

*A Practical Consideration:* The values of the biasing resistors must be low enough to allow the diodes and the transistor base–emitter junctions to conduct. Normally, the biasing resistors will be in the low-kΩ range.

When diode bias is used, the value of $I_1$ (which is needed in the calculation of $I_{CC}$) is found as

$$I_1 = \frac{V_{CC} - 1.4 \text{ V}}{R_1 + R_2} \qquad \textbf{(11.32)}$$

Again, we do not need to consider the effects of $I_{B1}$ and $I_{B2}$ since they do not affect the resistor current values.

## Class AB Operation

OBJECTIVE 8 ▶ You have been shown that $I_{CQ}$ will have some measurable value when diode bias is used. For this reason, the amplifier can no longer technically be called a *class B* amplifier. This point is illustrated with the help of Figure 11.35, which shows a typical input waveform for the circuit in Figure 11.34.

To simplify our discussion, we are going to make an assumption: A transistor will conduct until its base and emitter voltages are equal, at which time it will turn off. With this in mind, let's look at the circuit's response to the waveform shown in Figure 11.35.

$Q_1$ will conduct as long as its base voltage is more positive than 6 V. The value of $V_B$ drops to 6 V at $t_1$ and $t_3$ since the $-0.7$ V value of $v_{in}$ subtracts from the 6.7 V value of $V_{B(Q1)}$. Thus, we can assume that $Q_1$ conducts for the entire time between $t_1$ and $t_3$. The same principle applies to $Q_2$. At $t_2$ and $t_4$, $Q_2$ will turn off, since the $+0.7$ V value of $V_{in}$ adds to the value of $V_{B(Q2)}$, causing $V_B$ and $V_E$ to be equal. Thus, $Q_2$ conducts for the entire time between $t_2$ and $t_4$.

As you can see, the transistors in the diode bias circuit will conduct for *slightly more than* 180°. Because of this, the circuit is classified as a *class AB amplifier*. In class AB operation, the transistors conduct for a portion of the input cycle that is greater than 180° and less than 360°. It can be seen that both transistors will be conducting at the same time for a small portion of the wave. Figure 11.1 shows the $Q$-point location for a class AB amplifier.

While the diode bias circuit is technically classified as a class AB amplifier, most technicians simply refer to it as a class B amplifier. This is because of the strong similarities in operation between the two types of amplifiers.

## Eliminating Crossover Distortion

OBJECTIVE 9 ▶ Since both transistors in the class AB amplifier are conducting when the input signal is at 0 V, the amplifier does not have the crossover distortion problems that the class B amplifier may have. Crossover distortion occurs only when *both* transistors are in cutoff, so the problem will not normally occur in class AB amplifiers.

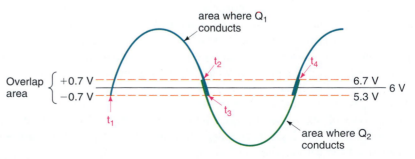

**FIGURE 11.35**

## Eliminating Thermal Runaway

For diode bias to eliminate the problem of thermal runaway, two conditions must be met:

1. The diodes and the transistor base–emitter junctions must be very nearly *perfectly* matched.

2. The diodes and the transistors must be in **thermal contact.** This means that they must be in physical contact with each other so that the operating temperature of the diodes will always equal the operating temperature of the transistors.

**Thermal contact**
Placing two or more components in physical contact with each other so that their operating temperatures will be equal.

When the diodes and transistors are matched, the forward voltage drops across the diodes ($V_{F1}$ and $V_{F2}$) will be approximately equal to the $V_{BE}$ drops of the transistors, all other factors being equal. In other words, as long as their operating temperatures are the same,

$$V_{F1} = V_{BE(Q1)}$$

and

$$V_{F2} = V_{BE(Q2)}$$

When the transistors are in thermal contact with the diodes, the components will all experience the same operating temperature. Thus, if the operating temperature of the transistors increases by a given amount, the operating temperature of the diodes will increase by the same amount, and the values of $V_F$ and $V_{BE}$ will remain approximately equal. With this in mind, let's take a look at the circuit response to an increase in temperature.

You may recall from our discussion on *temperature versus diode conduction* that an increase in diode temperature will cause a slight increase in diode current and a slight decrease in the value of $V_F$. Assuming that the transistor and diode temperature increases are equal, here's what happens:

1. The increase in diode temperature causes the 1.4 V drop across $D_1$ and $D_2$ to *decrease.* At the same time, $I_D$ *increases* (see Figure 11.34).

2. With the decrease in diode voltage, $I_1$ and $I_2$ *increase.* This causes the voltage drops across $R_1$ and $R_2$ to *increase.*

3. With the change in resistor voltage drops, the value of $V_{B(Q1)}$ *decreases* [$V_{B(Q1)} = V_{CC} - V_{R1}$], and the value of $V_{B(Q2)}$ *increases.*

4. The base voltage changes bring the values of $V_B$ closer to the value of $V_E$ for the two transistors, reducing the forward bias on the transistors.

5. The reduction in forward bias decreases $I_B$, causing a decrease in $I_C$. The decrease in $I_C$ prevents thermal runaway from occurring.

As you can see, diode bias reduces the chance of thermal runaway. The diodes can be placed in thermal contact with the power transistors in one of several ways:

1. Attached to the heat-sink tab of a power transistor

2. Attached to the heat sink on which the transistor is mounted

When replacing power transistors in a class AB amplifier, you must reattach the compensating diodes to their original location for continued thermal protection.

Some class B and AB amplifiers have two resistors added to the emitter output circuit. These resistors, shown in Figure 11.36, act as swamping resistors, reducing the effects of minor characteristic differences between the matched transistor pair. The resistors will have low resistance values, typically 0.47 to 10 Ω. Because they form a voltage divider with the load, any replacement resistor must have the same value as the original, or the result will be a distorted output.

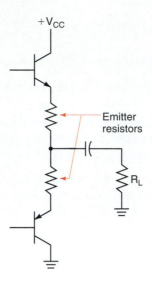

**FIGURE 11.36 Emitter output circuit.**

+V<sub>CC</sub>

Emitter resistors

$R_L$

## Class AB Amplifier Analysis

*OBJECTIVE 10* ▶ We can now use the information on dc biasing to approach the process for analyzing a class AB amplifier. For this discussion, we will use the example circuit shown in Figure 11.37a.

We will start our analysis by determining the values of $I_{C(\text{sat})}$ and $V_{CE(\text{off})}$. For this circuit, $I_{C(\text{sat})}$ is found as

$$I_{C(\text{sat})} = \frac{V_{CC}}{2\,R_L}$$

$$= 750 \text{ mA}$$

and $V_{CE(\text{off})}$ is found as

$$V_{CE(\text{off})} = \frac{V_{CC}}{2}$$

$$= 6 \text{ V}$$

Using these two values, the dc and ac load lines in Figure 11.37b are plotted.

Next, we determine the value of $I_1$. The value of $I_1$ is found as

$$I_1 = \frac{V_{CC} - 1.4 \text{ V}}{R_1 + R_2}$$

$$= \frac{12 \text{ V} - 1.4 \text{ V}}{1020 \text{ }\Omega}$$

$$= 10.4 \text{ mA}$$

The average current in the collector circuit of the amplifier is found as

$$I_{C1(\text{ave})} = \frac{0.159\, V_{CC}}{R_L}$$

$$= \frac{(0.159)(12 \text{ V})}{8 \text{ }\Omega}$$

$$= 238.5 \text{ mA}$$

Using this value and the value of $I_1$ calculated earlier, we can find the value of $I_{CC}$ as

$$I_{CC} = I_{C1(\text{ave})} + I_1$$
$$= 238.5 \text{ mA} + 10.4 \text{ mA}$$
$$= 248.9 \text{ mA}$$

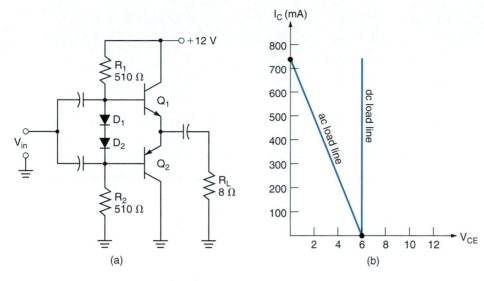

FIGURE 11.37   Class AB amplifier and its load lines.

**Lab Reference:** The ac operation of the class AB amplifier is demonstrated in Exercise 22.

Now we can calculate the total power drawn from the supply as

$$P_S = V_{CC}I_{CC}$$
$$= (12\ V)(248.9\ mA)$$
$$= 2.99\ W$$

Assuming the amplifier is being driven to compliance ($V_{PP} = V_{CC}$), we can calculate the maximum load power as

$$P_L \cong \frac{V_{PP}^2}{8\ R_L}$$

$$= \frac{12\ V^2}{8(8\ \Omega)}$$

$$= 2.25\ W$$

Finally, the maximum efficiency of the amplifier is found as

$$\eta = \frac{P_L}{P_S} \times 100$$

$$= 75.25\%$$

As you can see, the basic analysis of a class AB amplifier is actually fairly simple. In fact, with the exception of the $I_1$ calculation, it is identical to the analysis of the standard class B amplifier. When you deal with a standard class B amplifier, $I_1$ is found by dividing $V_{CC}$ by the total series base resistance. When dealing with the class AB amplifier, you must take the diode voltage drops into account.

The class AB amplifier is used far more commonly than the standard class B amplifier. For this reason, we will concentrate on the class AB amplifier from this point on. Just remember, except for the biasing circuit and the value of $I_{CQ}$, the class AB and the class B amplifiers are nearly identical. In fact, the term *class B* is commonly used to describe both amplifier types.

---

1. What are the two primary problems that can occur in standard class B amplifiers? What type of circuit is used to eliminate these problems?

2. Explain how the diode bias circuit develops the proper values of $V_B$ for the amplifier transistors.

*Section Review*

3. How do you find the values of $V_B$ in a diode bias circuit when $R_1 = R_2$?

4. How do you find the values of $V_B$ in a diode bias circuit when $R_1 \neq R_2$?

5. What is *class AB operation*?

6. How does the class AB amplifier eliminate crossover distortion?

7. How does the class AB amplifier eliminate thermal runaway?

8. List, in order, the steps you would take to perform the complete analysis of a class AB amplifier.

9. Why is the class AB amplifier often referred to as a class B amplifier?

## 11.6 TROUBLESHOOTING CLASS AB AMPLIFIERS

In this section we will concentrate on some of the faults that may develop in the diode-biased complementary-symmetry amplifier. We will also discuss some techniques you can use to isolate these faults. At the end of the section, you will be introduced to some other biasing circuits for class AB amplifiers. Although they are a bit more complex than the basic diode-bias circuit, they each serve their own purpose. Do not let the more complex circuits fool you; they may contain more components than the amplifiers discussed previously, but they work in the same basic fashion.

### Troubleshooting

**Lab Reference:** Several class AB fault symptoms are simulated in Exercise 22.

Class AB amplifiers can be more difficult to troubleshoot than class A amplifiers because of the dual-transistor configuration. Any of the standard transistor faults can develop in *either* of the two transistors. However, there are some techniques you can use that will make troubleshooting these amplifiers relatively simple.

The first step in troubleshooting a class AB amplifier is the same as with any amplifier: *You must make sure that the amplifier is the source of the trouble.* For example, assume that the two-stage amplifier shown in Figure 11.38 is not working (there is no output signal). For this circuit, there are *three* checks that must be made before you can assume there is a problem with the push–pull amplifier:

Initial tests.

1. You must verify that the $V_{CC}$ and ground connections in the push–pull amplifier are good.

**Lab Reference:** The operation of a two-stage amplifier similar to the one in Figure 11.38 is demonstrated in Exercise 23.

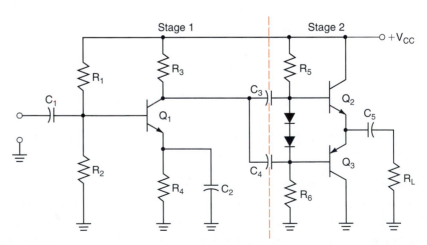

**FIGURE 11.38**

**2.** You must disconnect the load to make sure that it is not loading down the amplifier and preventing it from working.

**3.** You must be sure that the push–pull is getting the proper input from the first stage.

Picture this: You spend a couple of hours troubleshooting the push–pull amplifier, only to discover that the amplifier never had an input signal to begin with! This may sound a bit farfetched, but 99 out of 100 technicians would admit to having done something like this at one time or another. (And I would question the 100th!)

Many times the problem is not in the amplifier itself but in the supporting circuitry. By checking each of the conditions noted, the process of troubleshooting can be greatly simplified. The first two places to check when troubleshooting an amplifier are $V_{CC}$ and ground. By verifying that you have the proper $V_{CC}$ and ground connections, you have eliminated any problem that could have been caused by blown fuses, disconnected plugs, or a faulty power supply.

Next, make sure that the load is not the source of the problem. A shorted load will always prevent an amplifier from having an output. If you disconnect the load and the amplifier starts to work properly, then the load is the cause of the problem. Once the load has been verified as being okay, check that the amplifier is getting an input signal. If the input is bad, you need to continue on to the previous stage (that is, there is no reason to suspect that the push–pull is the cause of the problem). If all these tests check out, there is a problem with the amplifier.

Once you have concluded that the push–pull amplifier is the source of the problem, the first step is to *disconnect the ac signal source.* For the circuit in Figure 11.38, this means disconnecting the output of the previous stage from the input to the push–pull amplifier. There are two ways to do this, depending on the type of circuit you are dealing with:

**1.** If the input signal is fed to the push–pull by a wire, you can desolder the wire to isolate the stage.

**2.** If the input arrives by a copper run on a printed circuit board (which is usually the case), desolder and remove the coupling component that connects the two stages. For Figure 11.38, this would mean removing the coupling capacitors between the two stages, $C_3$ and $C_4$.

When the amplifier has been isolated from the signal source, you can perform some relatively simple voltage checks. Figure 11.39 shows the types of readings you should get in Figure 11.38 when the amplifier has been isolated and all the components are working properly. For example, assume that the circuit has been biased properly and that all components are good. If $V_{CC}$ is +20 V, you should get the *approximate* readings shown in Figure 11.40.

After disconnecting the signal source, perform the dc voltage checks.

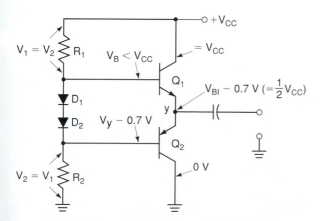

**FIGURE 11.39** Typical class AB amplifier dc voltages.

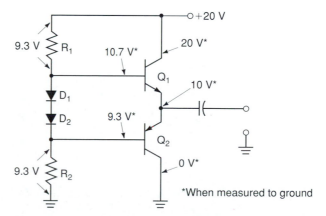

**FIGURE 11.40**

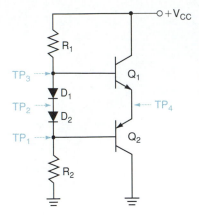

FIGURE 11.41    Commonly used dc test points.

With most class A amplifiers, the troubleshooting procedure begins with analyzing the emitter and collector voltages of the transistor and then working your way back to the input. The procedure is essentially the same for the class AB amplifier. We check the voltage at the point where the two emitters are connected, and then we check the base voltages. Table 11.1 summarizes some common faults and their symptoms. The test points referenced in the table are shown in Figure 11.41.

TABLE 11.1    Open-Component Troubleshooting

| Open Component | Symptoms |
|---|---|
| $R_2$ | The voltages at $TP_1$ and $TP_3$ will be higher than normal. The voltage at $TP_4$ will be higher than normal. |
| $D_1$ or $D_2$ | The voltages at $TP_1$, $TP_3$, and $TP_4$ will all be slightly lower than normal. |
| $R_1$ | The voltages at $TP_1$ and $TP_3$ will be at or near 0 V. The voltage at $TP_4$ will be slightly lower than normal. |
| $Q_1$ | All base circuit voltages will be normal. The voltage at $TP_4$ will be very low. |
| $Q_2$ | All base circuit voltages will be normal. The voltage at $TP_4$ will be very high. |

What to do if you suspect the transistors to be faulty.

If the dc voltages in the circuit appear to be within tolerance and the amplifier does not work, simply replace both transistors. This saves the time required for testing to see which transistor is faulty. Besides, if one of the transistors has gone bad, odds are that it has damaged the other in the process. Therefore both transistors should be replaced. Now, to help you get more comfortable with the information in this section, we will go through some troubleshooting applications.

# APPLICATION 1

The circuit shown in Figure 11.42 has no output. After checking the load, input signal, and $V_{CC}$/ground connections, the input to the circuit is disconnected. Then a voltmeter is used to obtain the readings shown. In this case, the problem is easy to spot. Since the readings on both sides of $D_2$ are equal, the diode is shorted. Note that the reading at $TP_1$ is 5.65 V, up from the 5.3 V that should be there. In this case, replacing the shorted diode (and its matching diode) solves the problem.

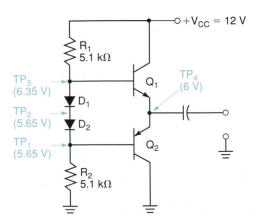

FIGURE 11.42

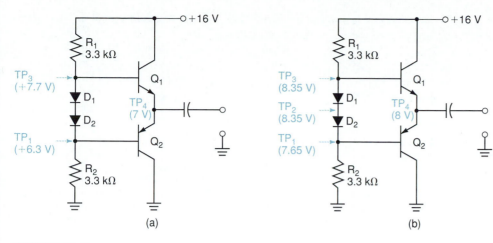

FIGURE 11.43

# APPLICATION 2

The circuit in Figure 11.43a is tested. The results are shown in the diagram. The readings in Figure 11.43a indicate that there is an open diode in the biasing circuit. A resistance check on $D_2$ indicates that it is the open diode.

When $D_2$ is replaced, the circuit readings change to those shown in Figure 11.43b. These readings indicate that $D_1$ has shorted. Replacing $D_1$ causes the circuit to operate properly again.

# APPLICATION 3

What problem is indicated by the readings in Figure 11.44? What would you do to repair the problem? (Assume that the amplifier inputs all check out and that the load is not the problem.) The answer can be found after Application 4.

FIGURE 11.44

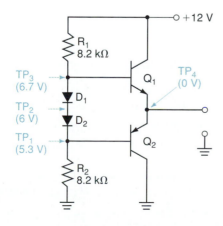

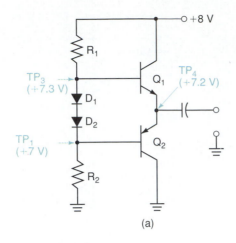 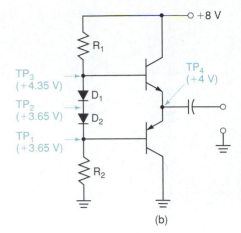

(a)                                    (b)

**FIGURE 11.45**

# APPLICATION 4

Figure 11.45a shows the initial readings in a faulty amplifier. All outside factors have been eliminated as being the cause of the problem. The readings indicate that $R_2$ is open. After replacing $R_2$, the circuit test points are checked again, giving the readings shown in Figure 11.45b. These readings indicate that $D_2$ is shorted. Replacing this component (and its matching diode) restores normal operation to the circuit.

In Application 3, all the biasing voltages are correct. The only logical assumption at this point would be that there is a problem with one of the transistors. Therefore, the next step would be to replace both transistors.

## Other Class AB Amplifiers

Next we will look at a few other types of class AB amplifiers and biasing circuits. The purpose of this discussion is *not* to teach you *everything* about these circuits, but, rather, to introduce you to some alternative circuit configurations.

Darlington complementary-symmetry amplifier.

The first circuit is the *Darlington complementary-symmetry amplifier*, shown in Figure 11.46. In this amplifier, the two transistors have been replaced by Darlington pairs. The Darlington pairs are used to increase the input impedance of the class B amplifier stage. This higher stage impedance reduces the load on the preceding amplifier stage, enabling it to have a higher voltage gain. Also, you may recall from earlier discussions that the Darlington pair has very high characteristic current gain. This circuit has much higher current gain than the standard complementary-symmetry amplifier. Because of this, it is used in applications where a high amount of load power is required. Recall that load power is calculated as

$$A_P = A_v A_i$$

Since the value of $A_i$ is much higher for a Darlington pair than it is for a single transistor, the overall gain of the Darlington class AB amplifier will be much higher than for a standard push–pull.

Note the four biasing diodes between the bases of $Q_1$ and $Q_4$. Four diodes are needed to compensate for the 1.4 V value of $V_{BE}$ for *each* Darlington pair. Since each pair

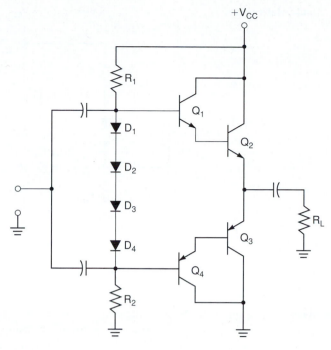

**FIGURE 11.46** Darlington class AB amplifier.

has a $V_{BE}$ of 1.4 V, there is a 2.8 V difference of potential between the bases of $Q_1$ and $Q_4$. The four diodes, when properly matched, will maintain the 2.8 V difference.

The *transistor-biased complementary-symmetry amplifier* uses transistors instead of diodes in the biasing circuit. This amplifier is shown in Figure 11.47. The amplifier is used primarily in circuits that contain integrated (IC) transistors. Note that the biasing transistors are wired in such a way as to cause them to act as diodes. The collector–base

Transistor-biased complementary-symmetry amplifier.

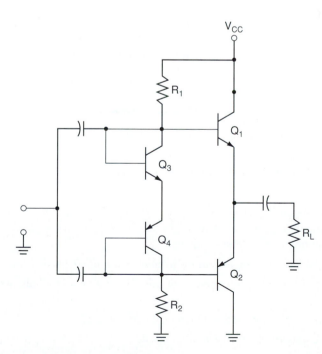

**FIGURE 11.47** Transistor-biased amplifier.

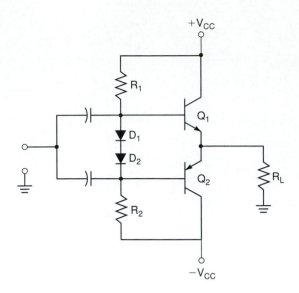

**FIGURE 11.48   Dual-polarity class AB amplifier.**

junction of each biasing transistor is shorted, leaving only an emitter–base junction in the circuit. This junction acts as a simple diode. Why go to the trouble of using transistors in the biasing circuit? Most transistor ICs contain four transistors. By using the extra transistors in an IC as the biasing diodes, you ensure that the diodes are perfectly matched to the circuit transistors ($Q_1$ and $Q_2$).

Split-supply class AB amplifier.

When the output from a class AB amplifier must be centered around 0 V (instead of around the value of $V_{CC}/2$), the *split-supply class AB amplifier* may be used. This amplifier is shown in Figure 11.48. The two power supply connections for this circuit will be equal and opposite in polarity. For example, voltage supplies of ±10 V may be used, but voltage supplies of +10 V and −5 V would not be used. With matched power supplies, each transistor will drop its own supply voltage, and the output signal will be centered at 0 V. This allows the circuit to be directly coupled to the load, as shown in Figure 11.48.

There are many different biasing configurations for class AB amplifiers. When you come up against a biasing circuit you have never seen before, just remember the basic principles of class B and class AB operation. Within reason, the operation of any class B or class AB amplifier will follow these principles.

---

## Section Review

1. What is the first step in troubleshooting a class AB amplifier (or any other amplifier, for that matter)?

2. Why are class AB amplifiers more difficult to troubleshoot than class A amplifiers?

3. What is the general approach to troubleshooting a class AB amplifier?

4. You are troubleshooting a class AB amplifier and come to the conclusion that the trouble is one of the two transistors. Briefly discuss the reasons for replacing both transistors rather than one or the other of them.

5. When is the *Darlington complementary-symmetry amplifier* used?

6. Why are four biasing diodes needed in the Darlington complementary-symmetry amplifier?

7. When is transistor biasing used in place of diode biasing?

8. When is a *split-supply class AB amplifier* used?

# 11.7 RELATED TOPICS

In this section we will take a look at a few topics that relate to power amplifiers, specifically, maximum power ratings and calculations and the use of heat sinks.

## Maximum Power Ratings

You may recall that transistors have a *maximum power dissipation rating*. When a transistor is used for a specific power amplifier application, you must make sure that the power dissipated by the transistor in the circuit does not exceed the rating of the transistor you are trying to use. From the specification sheet for the 2N3904, on page 225, the $P_{D(max)}$ value is 625 mW (assuming that no power derating is required). Thus, you could not use the 2N3904 in any power amplifier that would require its transistor to dissipate more than 625 mW.

◄ *OBJECTIVE 11*

How do you determine the amount of power a transistor will have to handle in a specific circuit? For class A amplifiers, use the equation

$$P_D = V_{CEQ}I_{CQ} \qquad \textbf{(11.33)}$$

For class B and class AB amplifiers, use the equation

$$P_D = \frac{(V_{PP})^2}{40R_L} \qquad \textbf{(11.34)}$$

to find the $P_D$ of the individual transistor where $V_{PP}$ = the *peak-to-peak* load voltage. The derivations of these equations are fairly involved, and thus are reserved for Appendix D. Examples 11.14 and 11.15 show how the equations are used to determine the transistor power requirements of class A and class AB amplifiers.

---

What is the value of $P_D$ for the transistor in Figure 11.49?

***EXAMPLE 11.14***

**Solution:** Using the established procedure, $I_{CQ}$ and $V_{CEQ}$ are found to be

$$I_{CQ} = 1 \text{ mA} \qquad V_{CEQ} = 5.3 \text{ V}$$

Using these two values, $P_D$ is found as

$$\begin{aligned} P_D &= V_{CEQ}I_{CQ} \\ &= (5.3 \text{ V})(1 \text{ mA}) \\ &= 5.3 \text{ mW} \end{aligned}$$

### PRACTICE PROBLEM 11.14

A class A amplifier has the following values: $R_E = 1.2$ kΩ, $R_C = 2.7$ kΩ, $V_{CC} = 16$ V, and $V_E = 2.4$ V. Determine the required value of $P_D$ for the transistor in the amplifier.

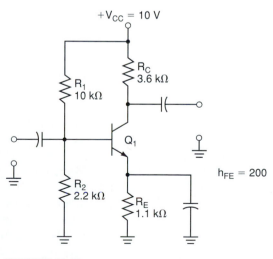

**FIGURE 11.49**

EXAMPLE *11.15*

What is the value of $P_D$ for each transistor in Figure 11.50? (Assume that the output is equal to the compliance of the amplifier.)

**Solution:** The compliance of the amplifier is 12 V. Thus $V_{PP} = 12$ V, and

$$P_D = \frac{V_{PP}^2}{40\,R_L}$$

$$= \frac{12\,V_{PP}^2}{40\,(8\ \Omega)}$$

$$= \frac{144\,V_{PP}}{320\ \Omega}$$

$$= 450\text{ mW}$$

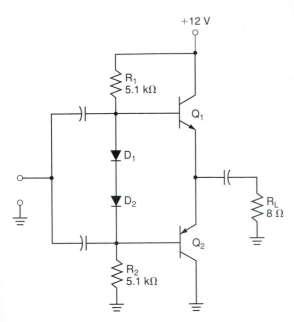

FIGURE 11.50

**PRACTICE PROBLEM 11.15**

A class AB amplifier with values of $V_{CC} = 15$ V and $R_L = 12\ \Omega$ is driven to compliance. Determine the required value of $P_D$ for each transistor in the circuit.

## Component Cooling

When large numbers of electronic components are used in an enclosed system, the power dissipated by the components can cause a significant rise in the temperature of the components. In this situation, the derating factor of the transistors becomes important.

To help decrease the loss in $P_D$ values caused by a closed system, fans are often used to reduce the temperature within the system. If you have ever worked with a computer terminal, you may have noticed that it is usually fan cooled. Fan cooling decreases component temperature and thus helps to maintain the original $P_D$ values of the components.

Another device used to reduce the case temperature of a given transistor is the heat sink. **Heat sinks** are large metallic objects that help cool transistors by increasing their *effective* surface area. Several heat sinks are shown in Figure 11.51.

When a heat sink is connected *properly* to a transistor, the case heat from the transistor is transferred to the heat sink. Because the heat sink has more surface area than the transistor casing, it is able to dissipate a much greater amount of heat. Thus, the transistor case temperature is held at a lower value, and the $P_D$ derating *value* is reduced.

When a heat sink is connected to a transistor, a **heat-sink compound** is used to aid in the transfer of heat to the sink. This compound can be purchased at any electronics parts store. The heat sink compound is applied to the transistor casing, and then the transistor is connected (attached) to the heat sink itself. The transistor must be attached to the heat sink mechanically (that is, heat-sink compound is *not* glue; it does not act as an adhesive between the transistor and the heat sink).

Power transistor cases may not be at the same voltage potential as the heat sink. In this situation, a thin insulator is inserted between the two. When used with heat-sink compound, the spacer acts as an *electrical insulator* and a *thermal conductor*. Many times these spacers will stick to an old transistor as it is removed from the circuit. If the

**Heat sink**
A large metallic object that helps to cool transistors by increasing their surface area.

**Heat-sink compound**
A compound used to aid in the transfer of heat from a transistor to a heat sink.

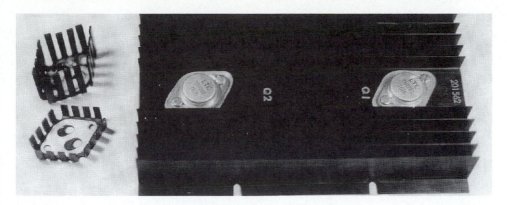

**FIGURE 11.51    Heat sinks.**

spacer is not replaced in the correct position when the new transistor is installed, the replacement transistor will short when power is applied to the circuit. Care must also be taken to ensure that the leads do not come in contact with the heat sink. The following procedure should be followed for replacing transistors that are mounted to a heat sink:

1. Remove the bad transistor from the heat sink and wipe off the old heat-sink compound (from the heat sink).

2. *Lightly* coat the new transistor with heat-sink compound. Do not use more than is necessary to build a thin coat on the component.

3. Replace insulator if necessary.

4. Connect the transistor to the heat sink. If there were any mechanical connectors between the old transistor and the heat sink, such as screws, be sure to replace them in the new circuit. Any insulating sleeves on the mounting screws must also be replaced.

5. *Be sure that the transistor leads are not touching the heat sink.*

*A Practical Consideration:* Most power transistors are bolted to the heat sink (see $Q_1$ and $Q_2$ in Figure 11.51). When properly mounted, the transistor leads will *not* be able to touch the heat sink. If the leads *are* touching the heat sink, you have inserted the transistor backward.

### Section Review

1. What is a *heat sink*?
2. What is *heat-sink compound*?
3. What is the proper procedure for replacing a transistor connected to a heat sink?
4. What precautions must be taken when performing the procedure in Question 3?

### Key Terms

The following terms were introduced and defined in this chapter:

ac load line
complementary-symmetry
  amplifier
complementary transistors
compliance
counter emf
crossover distortion
cutoff clipping

diode bias
heat sink
heat-sink compound
matched transistors
push–pull emitter follower
*RC*-coupled class A
  amplifier
saturation clipping

standard push–pull
  amplifier
step-down transformer
step-up transformer
thermal contact
transformer-coupled class
  A amplifier
tuned amplifier

| Equation Summary | Equation Number | Equation | Section Number |
|---|---|---|---|
| | **(11.1)** | $i_c = \dfrac{v_{ce}}{r_C}$ | 11.1 |
| | **(11.2)** | $I_C - I_{CQ} = \dfrac{V_{CEQ}}{r_C} - \dfrac{V_{CE}}{r_C}$ | 11.1 |
| | **(11.3)** | $i_{c(\text{sat})} = I_{CQ} + \dfrac{V_{CEQ}}{r_C}$ | 11.1 |
| | **(11.4)** | $v_{ce(\text{off})} = V_{CEQ} + I_{CQ}r_C$ | 11.1 |
| | **(11.5)** | $\text{PP} = 2I_{CQ}r_C$ | 11.1 |
| | **(11.6)** | $\text{PP} = 2V_{CEQ}$ | 11.1 |
| | **(11.7)** | $P_S = V_{CC}I_{CC}$ | 11.2 |
| | **(11.8)** | $P_L = \dfrac{V_L^2}{R_L}$ | 11.2 |
| | **(11.9)** | $P_L = \dfrac{(0.707\,V_{\text{pk}})^2}{R_L}$ | 11.2 |
| | **(11.10)** | $P_L = \dfrac{V_{\text{PP}}^2}{8\,R_L}$ | 11.2 |
| | **(11.11)** | $P_{L(\text{max})} = \dfrac{\text{PP}^2}{8\,R_L}$ | 11.2 |
| | **(11.12)** | $\dfrac{N_1}{N_2} = \dfrac{V_1}{V_2} = \dfrac{I_2}{I_1}$ | 11.3 |
| | **(11.13)** | $\left(\dfrac{N_1}{N_2}\right)^2 = \dfrac{Z_1}{Z_2}$ | 11.3 |
| | **(11.14)** | $\Delta I_C = \dfrac{\Delta V_{CE}}{Z_1}$ | 11.3 |
| | **(11.15)** | $I_{C(\text{max})} = I_{CQ} + \Delta I_C$ | 11.3 |
| | **(11.16)** | $P_S = V_{CC}I_{CQ}$ | 11.3 |
| | **(11.17)** | $\text{PP} = V_{CE(\text{max})} - V_{CE(\text{min})}$ | 11.3 |
| | **(11.18)** | $V_{\text{PP}} = \dfrac{N_2}{N_1}\,\text{PP}$ | 11.3 |
| | **(11.19)** | $V_{CEQ} = \dfrac{V_{CC}}{2}$ | 11.4 |
| | **(11.20)** | $I_{CQ} \cong 0$ | 11.4 |
| | **(11.21)** | $I_{C(\text{sat})} = \dfrac{V_{CC}}{2\,R_L}$ | 11.4 |

| Equation Number | Equation | Section Number |
|---|---|---|
| (11.22) | $V_{CE(\text{off})} = \frac{V_{CC}}{2}$ | 11.4 |
| (11.23) | $Z_{\text{in(base)}} = h_{fc}\,(r'_e + R_L)$ | 11.4 |
| (11.24) | $Z_{\text{out}} = r'_e + \frac{R_{\text{in}}}{h_{fc}}$ | 11.4 |
| (11.25) | $\text{PP} = 2V_{CEQ}$ | 11.4 |
| (11.26) | $\text{PP} \cong V_{CC}$ | 11.4 |
| (11.27) | $I_{CC} = I_{C1(\text{ave})} + I_1$ | 11.4 |
| (11.28) | $I_{C1(\text{ave})} = \frac{0.159V_{CC}}{R_L}$ | 11.4 |
| (11.29) | $I_{C1(\text{ave})} = \frac{0.159\ V_{\text{PP(out)}}}{R_L}$ | 11.4 |
| (11.30) | $V_{B(Q2)} = \frac{R_2}{R_1 + R_2}(V_{CC} - 1.4\text{ V})$ | 11.5 |
| (11.31) | $V_{B(Q2)} = V_{CEQ} - 0.7\text{ V} \mid (R_1 = R_2)$ | 11.5 |
| (11.32) | $I_1 = \frac{V_{CC} - 1.4\text{ V}}{R_1 + R_2}$ | 11.5 |
| (11.33) | $P_D = V_{CEQ}I_{CQ}$ | 11.7 |
| (11.34) | $P_D = \frac{V_{\text{PP}}^2}{40R_L}$ | 11.7 |

## Answers to the Example Practice Problems

**11.1.** 9.57 $V_{\text{PP}}$
**11.2.** 267.9 mW
**11.3.** 17.79 mW
**11.4.** 47.76 mW
**11.5.** 357.14 mW
**11.6.** 3.56%
**11.7.** 34.72%
**11.8.** The dc load line is vertical at $V_{CEQ} = 6$ V. The ac load line has endpoints of $V_{CE(\text{off})} = 6$ V and $I_{C(\text{sat})}$ 2.73 mA.
**11.9.** 8.18 mW
**11.10.** 2.93 W
**11.11.** 2.25 W
**11.12.** 76.8%
**11.13.** 55.5%
**11.14.** 16.4 mW
**11.15.** 468.75 mW

### §11.1

1. Calculate the compliance of the amplifier in Figure 11.52.
2. Calculate the compliance of the amplifier in Figure 11.53.
3. Calculate the compliance of the amplifier in Figure 11.54. Determine the type of clipping that the circuit would be most likely to experience.
4. Calculate the compliance of the amplifier in Figure 11.55. Determine the type of clipping that the circuit would be most likely to experience.

### §11.2

5. Calculate the value of $P_S$ for the amplifier in Figure 11.52.
6. Calculate the value of $P_S$ for the amplifier in Figure 11.53.
7. Calculate the value of $P_L$ for the amplifier in Figure 11.52.
8. Calculate the value of $P_L$ for the amplifier in Figure 11.53.
9. Calculate the maximum possible load power for the circuit in Figure 11.52.

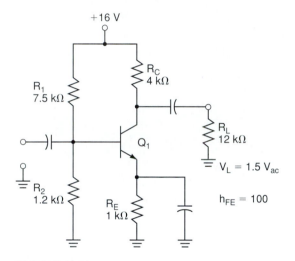

**FIGURE 11.52**

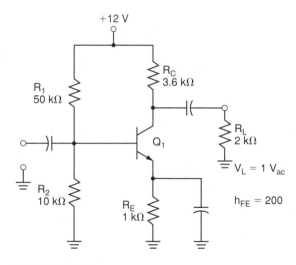

**FIGURE 11.53**

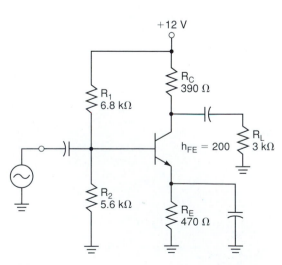

**FIGURE 11.54**

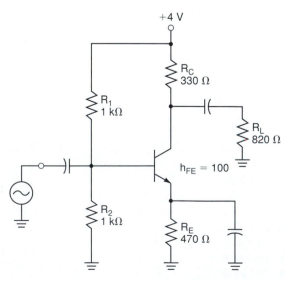

**FIGURE 11.55**

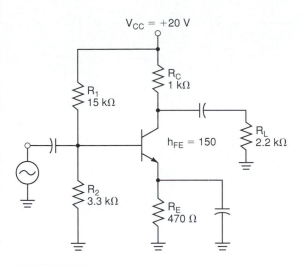

FIGURE 11.56

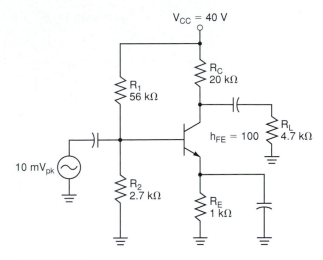

FIGURE 11.57

10. Calculate the maximum possible load power for the circuit in Figure 11.53.

11. Assuming that the amplifier in Figure 11.52 is driven to compliance, calculate the efficiency of the circuit.

12. Assuming that the amplifier in Figure 11.53 is driven to compliance, calculate the efficiency of the circuit.

13. Determine the values of $P_S$, $P_L$, and the efficiency for the amplifier in Figure 11.54. Assume that the circuit is driven to compliance.

14. Determine the values of $P_S$, $P_L$, and the amplifier in Figure 11.55. Assume that the circuit is driven to compliance.

15. Calculate the maximum efficiency for the amplifier in Figure 11.56.

16. Calculate the maximum efficiency for the amplifier in Figure 11.57.

### §11.3

17. Derive the dc and ac load lines for the circuit in Figure 11.58.

18. Derive the dc and ac load lines for the circuit in Figure 11.59.

19. Determine the maximum load power for the circuit in Figure 11.58.

20. Determine the maximum load power for the circuit in Figure 11.59.

21. Calculate the maximum efficiency of the amplifier in Figure 11.58.

22. Calculate the maximum efficiency of the amplifier in Figure 11.59.

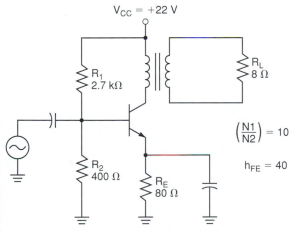

FIGURE 11.58

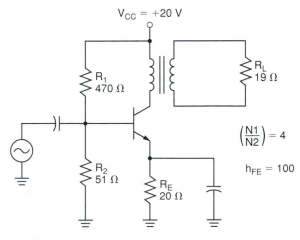

FIGURE 11.59

FIGURE 11.60

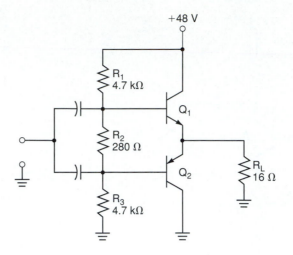

§*11.4*

23. A class B amplifier has values of $V_{CC} = +18$ V and $R_L = 3$ kΩ. Plot the ac and dc load lines for the circuit.

24. A class B amplifier has values of $V_{CC} = +24$ V and $R_L = 200$ Ω. Plot the ac and dc load lines for the circuit.

25. Calculate the maximum load power for the amplifier described in Problem 23.

26. Calculate the maximum load power for the amplifier described in Problem 24.

27. Calculate the value of $P_S$ for the amplifier in Figure 11.60.

28. Calculate the value of $P_S$ for the amplifier in Figure 11.61.

29. Calculate the value of $P_{L(max)}$ for the amplifier in Figure 11.60.

30. Calculate the value of $P_{L(max)}$ for the amplifier in Figure 11.61.

31. Calculate the maximum efficiency for the amplifier in Figure 11.60.

32. Calculate the maximum efficiency for the amplifier in Figure 11.61.

§*11.5*

33. Determine the values of $V_{CEQ}$, $V_{B(Q1)}$, and $V_{B(Q2)}$ for the class AB amplifier in Figure 11.62.

34. Determine the values of $V_{CEQ}$, $V_{B(Q1)}$, and $V_{B(Q2)}$ for the class AB amplifier in Figure 11.63.

35. Calculate the maximum efficiency of the class AB amplifier in Figure 11.62.

36. Calculate the maximum efficiency of the class AB amplifier in Figure 11.63.

**FIGURE 11.61**

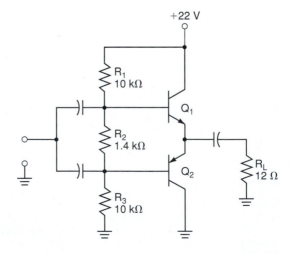

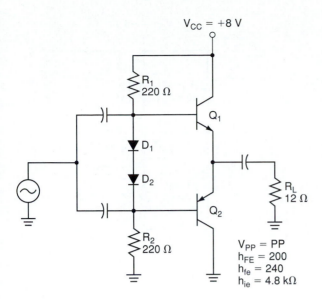

FIGURE 11.62

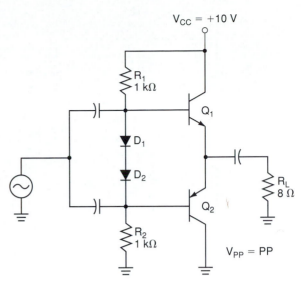

FIGURE 11.63

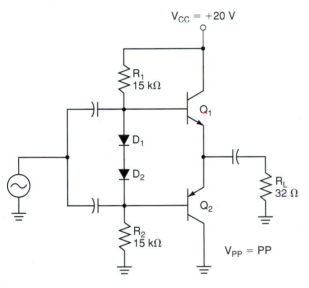

FIGURE 11.64

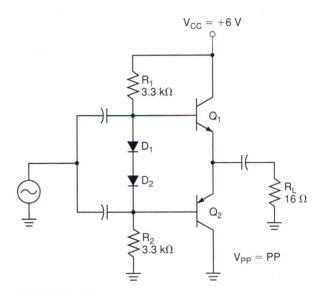

FIGURE 11.65

**37.** Calculate the maximum efficiency of the class AB amplifier in Figure 11.64.

**38.** Calculate the maximum efficiency of the class AB amplifier in Figure 11.65.

**§11.7**

**39.** Calculate the value of $P_D$ for the transistor in Figure 11.54.

**40.** Calculate the value of $P_D$ for the transistor in Figure 11.55.

**41.** Calculate the value of $P_D$ for the transistor in Figure 11.58.

---

**42.** The transformer-coupled class A amplifier in Figure 11.66 has the dc voltages shown. Discuss the possible cause(s) of the problem.

**43.** The transformer-coupled class A amplifier in Figure 11.67 has the dc voltages shown. Discuss the possible cause(s) of the problem.

*Troubleshooting Practice Problems*

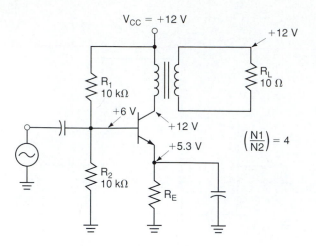

**FIGURE 11.66**

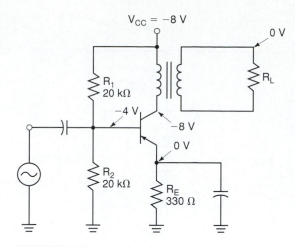

**FIGURE 11.67**

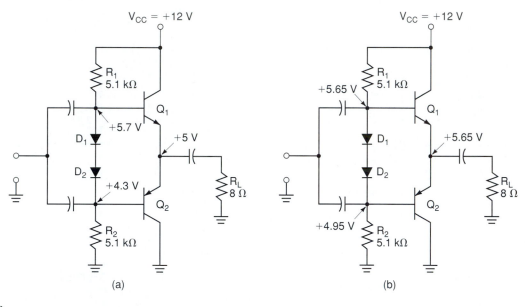

(a)                                     (b)

**FIGURE 11.68**

44. Determine the fault(s) that would cause the readings shown in Figure 11.68a.
45. Determine the fault(s) that would cause the readings shown in Figure 11.68b.

*The Brain Drain*

46. Calculate the maximum allowable input power for the amplifier in Figure 11.62.
47. Calculate the values of $A_{vT}$, $A_{iT}$, and $A_{pT}$ for the two-stage amplifier in Figure 11.69.
48. Answer the following questions for the circuit in Figure 11.69. Use circuit calculations to explain your answers.

    a. Is the amplifier driven to compliance?

    b. What type of clipping would the amplifier be most likely to experience?

49. Transistor $Q_2$ in Figure 11.69 is faulty. Can the 2N3904 be used in place of the transistor? Explain your answer using circuit calculations. The spec sheet for the 2N3904 is located on page 225 (*Hint*: The power-handling requirement of the circuit is the primary consideration in this problem.)

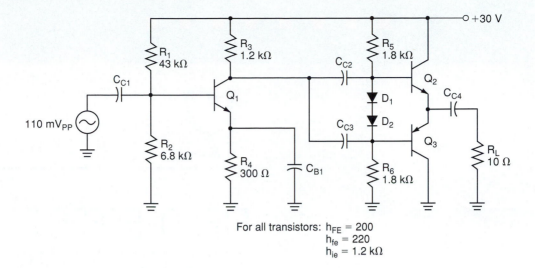

For all transistors: $h_{FE} = 200$
$h_{fe} = 220$
$h_{ie} = 1.2$ kΩ

**FIGURE 11.69**

50. Write a program that will determine the efficiency of a class AB amplifier when provided with the needed input values.

51. Write a program that will perform the complete dc analysis of a transformer-coupled class A amplifier, given the needed circuit values.

52. Write a program that will perform the complete ac analysis of a transformer-coupled class A amplifier, given the needed circuit values.

53. Write a program that will perform the complete dc and ac analysis of a class AB amplifier, given the needed values.

*Suggested Computer
Applications Problems*

# 12

# FIELD-EFFECT TRANSISTORS

The first field-effect transistor (shown here) was developed by the Bell Labs team of Ross and Dacey in 1955.

## OUTLINE

## OBJECTIVES

*After studying the material in this chapter, you should be able to:*

1. List the two types of field-effect transistors (FETs).

2. Explain the relationship between JFET channel width and drain current ($I_D$).

3. State the relationship between gate–source voltage ($V_{GS}$) and drain current ($I_D$).

4. Describe the gate input impedance characteristics of the JFET.

5. Determine the range of *Q*-point values for a given JFET biasing circuit.

6. List and explain the primary advantages and disadvantages of each of the three types of JFET biasing configurations.

7. Describe and analyze the ac operation of the common-source amplifier.

8. State the purpose of swamping a JFET amplifier.

9. Explain why the input impedance of a JFET amplifier tends to be higher than that of a comparable BJT amplifier.

10. Describe and analyze the ac operation of common-drain and common-gate amplifiers.

11. Describe the procedure used to troubleshoot a JFET amplifier.

12. List and define the commonly used JFET parameters and ratings.

13. Discuss the use of a JFET amplifier as a *buffer*.

14. Discuss the difference between FET meters and standard VOMs.

# THAT "OTHER" DEVELOPMENT

Schockley, Brattain, and Bardeen have received their share of the limelight for the development of the transistor in 1948. However, there was another team of scientists that made a major contribution to the field of solid-state electronics in 1955.

In the May 1955 volume of the *Bell Laboratories Record*, an announcement was made of the development of another type of transistor. This transistor was developed by I. M. Ross and G. C. Dacey. The *Bell Laboratories Record* said of this development:

*It would appear that it would find its main applications where considerations of size, weight, and power consumption dictate the use of a transistor, and where the required frequency response is higher than could be achieved with a simple junction transistor.*

The component developed by Ross and Dacey has had a major impact over the years, especially in the area of integrated-circuit technology. What was this "other" component? It was the *field-effect transistor.*

---

**Field-effect transistors (FETs)**
Voltage-controlled devices.

Yοu may recall that the *bipolar junction transistor* is a *current-controlled device*; that is, the output characteristics of the device are controlled by the base *current,* not the base voltage. Another type of transistor, called the **field-effect transistor**, or FET, is a *voltage-controlled device.* The output characteristics of the FET are controlled by the input voltage, not by the input current.

OBJECTIVE 1 ▶ There are two basic types of FETs: the *junction field-effect transistor,* or JFET, and the *metal oxide semiconductor FET,* or MOSFET. As you will be shown, the operation of these two components varies, as do their applications and limitations. In this chapter, we will discuss JFETs and their circuits. MOSFETs and their circuits are covered in Chapter 13.

## 12.1 INTRODUCTION TO JFETS

The physical construction of the JFET is significantly different from that of the bipolar junction transistor. You may recall that the bipolar transistor has three separate materials: either two *n*-type materials and a single *p*-type material, or two *p*-type materials and a single *n*-type material. The JFET has only *two* materials, a single *n*-type material and a single *p*-type material. The construction of the JFET is illustrated in Figure 12.1.

**FIGURE 12.1  JFET construction.**

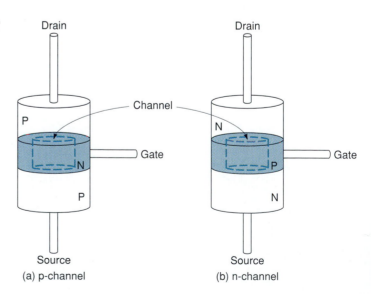

As you can see, the device has three terminals, just like the bipolar junction transistor. However, the terminals of the JFET are labeled **source**, **drain**, and **gate**. The *source* can be viewed as the equivalent of the BJT's *emitter*, the *drain* as the equivalent of the *collector*, and the *gate* as the equivalent of the *base*.

The material that connects the source to the drain is referred to as the **channel**. When this material is an *n*-type, the JFET is referred to as an *n-channel JFET*. Obviously, an FET with a *p*-type channel would be referred to as a *p-channel JFET*. Both JFET types are shown in Figure 12.1. Note that the gate material *surrounds* the channel in the same manner as a belt surrounding your waist.

The schematic symbols for the *p*-channel and *n*-channel JFETs are shown in Figure 12.2. Note that the arrow points "in" or "out" from the component. Just as with the BJT schematic symbol, the arrow points toward the *n*-type material. When the arrow is pointing *in*, it is pointing toward the channel. Thus, Figure 12.2a is the schematic symbol for the *n*-channel JFET. When the arrow is pointing *out*, it can be viewed as pointing toward the gate. If the gate is an *n*-type material, the channel must be a *p*-type material. Thus, Figure 12.2b is the schematic symbol for the *p*-channel JFET. If you think you may have trouble distinguishing the two symbols, just think "*n* . . . in.*"* This will remind you that the *n*-channel JFET has the arrow pointing *in*.

You know that the *npn* transistor is normally used with *positive* supply voltages, while the *pnp* transistor is normally used with *negative* supply voltages. The same relationship holds true for JFETs. The *n-channel* JFET is normally used with *positive* supply voltages, while the *p-channel* JFET is normally used with *negative* supply voltages. This point is illustrated in Figure 12.3. Note that the *n*-channel JFET circuit (Figure 12.3a) has a *positive* drain supply voltage ($V_{DD}$), while the *p*-channel JFET circuit (Figure 12.3b) has a *negative* supply voltage.

**Source**
The JFET equivalent of the BJT emitter.

**Drain**
The JFET equivalent of the BJT collector.

**Gate**
The JFET equivalent of the BJT base.

**Channel**
The material that connects the source and drain. The gate surrounds the channel.

JFET supply voltage polarities.

## Operation Overview

In our discussions on BJTs, we concentrated on the *npn* transistor. At the same time, it was stated that all the relationships discussed also held true for the *pnp* transistor. The only differences were the voltage polarities and the current directions. In the same manner, our discussions on JFETs will concentrate on the *n-channel JFET*. Again, all the relationships covered will also apply to the *p*-channel JFET. The only difference will be the voltage polarities and current directions.

The overall operation of the JFET is based on *varying the width of the channel to control the drain current*. This point is illustrated in Figure 12.4, which presents JFET construction in a slightly different way. In Figure 12.4a, the drain source voltage ($V_{DS}$)

◄ *OBJECTIVE 2*

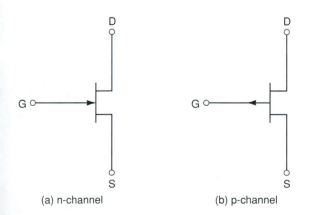

(a) n-channel        (b) p-channel

**FIGURE 12.2   JFET schematic symbols.**

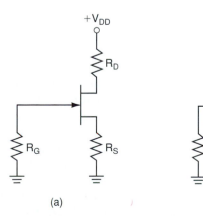

(a)            (b)

**FIGURE 12.3   JFET supply voltages.**

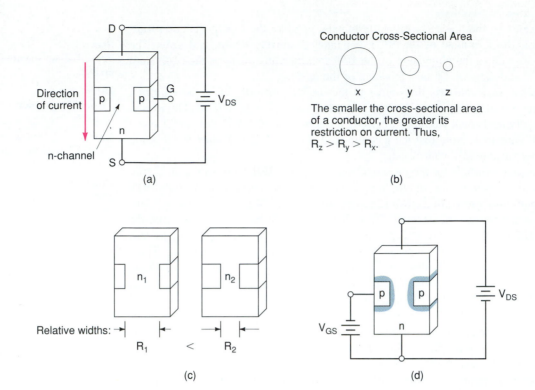

The smaller the cross-sectional area of a conductor, the greater its restriction on current. Thus, $R_z > R_y > R_x$.

**FIGURE 12.4**  **The relationship between channel width and drain current.**

is causing conduction through the JFET. You may recall that the amount of current through a conductor is *directly proportional* to the cross-sectional area (width) of the conductor. Thus, as the width of a conductor *decreases,* so does the amount of current through it at a set voltage. Now, assuming that $V_{DS}$ is a set voltage, we can decrease the amount of drain current by decreasing the width of the channel. This is shown in Figure 12.4b.

OBJECTIVE 3 ▶  How do we decrease the width of the channel? By effectively *increasing* the width of the gate. For example, in Figure 12.4c you can see two JFETs with different gate widths. Since the JFET on the right has a wider gate, it has a narrower channel. Thus, the resistance of this JFET (between the source and drain) is greater than the resistance of the JFET on the left.

How do we increase the width of the gate? By applying a *reverse bias* to the gate–source junction. In Figure 12.4d, the biasing supply is shown as $V_{GS}$. Note that the negative terminal of the supply goes to the *p*-type gate, while the positive terminal is connected to the *n*-type source. This reverse biases the gate–source junction, causing a depletion layer to form. As you know, a depletion layer will prevent current. Thus, the area of the channel is effectively *decreased,* causing the drain current to decrease.

The relationship between $V_{GS}$ and $I_D$.

There are two ways to control channel width. First, by varying the value of $V_{GS}$, we can vary the width of the channel and, in turn, vary the amount of drain current. This point is illustrated in Figure 12.5. Note that, as $V_{GS}$ increases, so does the size of the depletion layer. *As the size of the depletion layer increases, the effective size of the channel decreases.* As the effective size of the channel decreases, so does the amount of drain current. The relationship between $V_{GS}$ and $I_D$ can be summarized as follows: *As $V_{GS}$ increases (becomes more negative), $I_D$ decreases.*

The relationship between $V_{DS}$ and $I_D$.

Figure 12.5 serves to illustrate another important point. In Figure 12.5a, $V_{GS}$ is shown to be 0 V. At the same time, a small depletion layer is shown to be surrounding the gate. This small depletion layer is caused by the current through the channel. Increasing $I_D$ will cause the size of this depletion layer to increase. Thus, the second method of

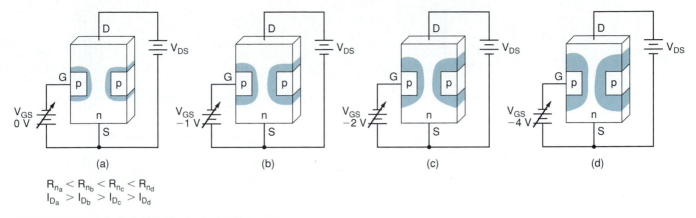

$$R_{n_a} < R_{n_b} < R_{n_c} < R_{n_d}$$
$$I_{D_a} > I_{D_b} > I_{D_c} > I_{D_d}$$

**FIGURE 12.5** The relationship between $V_{GS}$ and $I_D$.

increasing the size of the depletion layer is to hold $V_{GS}$ constant and increase the value of $V_{DS}$. Increasing $V_{DS}$ causes $I_D$ to increase, which in turn causes the size of the depletion layer to increase. This is illustrated in Figure 12.6.

At this point, there seems to be a contradiction in the theory of operation. It would seem that the increase in the size of the depletion layer would increase the resistance of the channel, preventing $I_D$ from increasing. How is it that $I_D$ is increasing if the conduction of the JFET channel is decreasing? While the channel of the component *is* becoming narrower, it is doing so very slowly. Because of this, the value of $I_D$ continues to increase. However, a point will be reached where further increases in $V_{DS}$ will be offset by proportional increases in the resistance of the channel. Thus, after a specified value of $V_{DS}$, further increases in $V_{DS}$ will *not* cause an increase in $I_D$. The value of $V_{DS}$ at which this occurs is called the **pinch-off voltage**. As $V_{DS}$ is increased, $I_D$ will increase until $V_{DS} = V_P$ (pinch-off voltage). If $V_{DS}$ is increased above the value of $V_P$, the resistance of the channel will increase at the same rate as $V_{DS}$ and $I_D$ will remain constant. This is shown in the JFET drain curve in Figure 12.7a.

Look at the portion of the curve to the left of $V_P$. As $V_{DS}$ rises from 0 V, $I_D$ shows a constant increase. During this time, the resistance of the channel is also increasing, but at a lower rate than $V_{DS}$. By the time the value of $V_P$ is reached, $V_{DS}$ and the resistance of the channel are increasing at the same rate. Thus, $I_D$ remains fairly constant for further increases in $V_{DS}$. Note that the region of operation between $V_P$ and $V_{BR}$ (breakdown voltage) is called the *constant-current* region. As long as $V_{DS}$ is kept within this range, $I_D$ will remain constant for a constant value of $V_{GS}$.

When $V_{GS}$ is equal to 0 V, you essentially have the gate and source terminals *shorted* together. When you have a shorted gate–source junction, $I_D$ reaches its maximum possible value, $I_{DSS}$. $I_{DSS}$, the **shorted gate–drain current,** is the *maximum* value of cur-

**Pinch-off voltage ($V_P$)**
With $V_{GS}$ constant, the value of $V_{DS}$ that causes maximum $I_D$. The pinch-off voltage rating of a JFET is measured at $V_{GS} = 0$ V.

The *constant-current* region.

**Lab Reference:** The relationship between $V_{DS}$ and $I_D$ is observed and graphed in Exercise 24.

**Shorted gate–drain current ($I_{DSS}$)**
The maximum possible value of $I_D$.

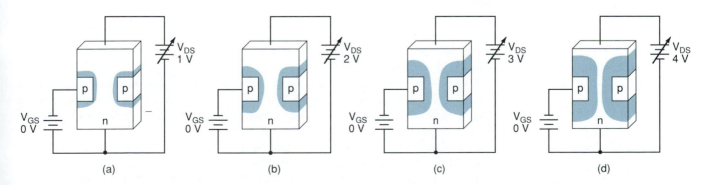

**FIGURE 12.6** The effects of varying $V_{DS}$ with constant $V_{GS}$.

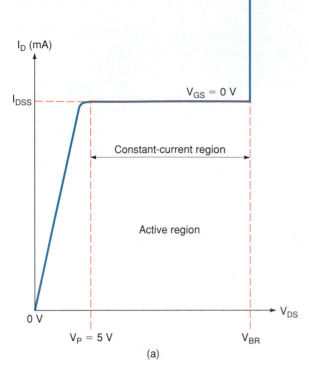

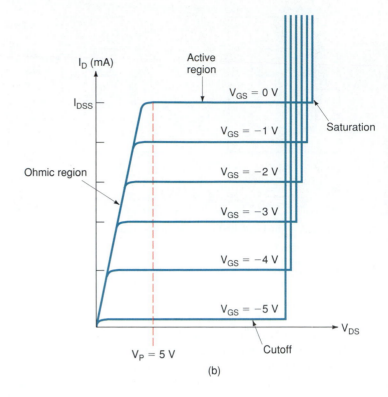

**FIGURE 12.7  JFET drain curves.**

rent that can pass through the channel of a given JFET. This value, which is given on the spec sheet for a given JFET, is measured under the following conditions:

$$V_{GS} = 0 \text{ V} \qquad V_{DS} = V_P$$

$I_D$ is always less than or equal to $I_{DSS}$.

Any current value for a given JFET must be less than or equal to the value of $I_{DSS}$. In this respect, $I_{DSS}$ can be viewed as the equivalent of $I_{C(\text{sat})}$ for a given BJT circuit. The relationship between $V_{GS}$, $V_{DS}$, and $I_{DSS}$ is also shown in Figure 12.7.

As $V_{GS}$ is made more negative than 0 V, the JFET will pinch off at a lower voltage than the $V_P$ rating for the component. Also, the constant-current value of $I_D$ will decrease from $I_{DSS}$. This is shown in Figure 12.7b.

Let's focus on the decrease in $I_D$ for a moment. As Figure 12.7b illustrates, continual increases in $V_{GS}$ cause continual decreases in the constant-current value of $I_D$. If we continue to increase $V_{GS}$, we will eventually hit a point where minimum current will pass through the JFET. The value of $V_{GS}$ that causes $I_D$ to drop to approximately *zero* is called $V_{GS(\text{off})}$, the **gate–source cutoff voltage.** Note that the value of $V_{GS(\text{off})}$ defines the limits on the value of $V_{GS}$. For conduction to occur through the device, the value of $V_{GS}$ must be somewhere between 0 V and $V_{GS(\text{off})}$.

**Gate-source cutoff voltage**
**($V_{GS(\text{off})}$)**
The value of $V_{GS}$ that causes $I_D$ to drop to zero.

$V_P$ and $V_{GS(\text{off})}$ are equal in *magnitude*.

Here's an interesting point: $V_{GS(\text{off})}$ will always have the same magnitude as $V_P$. For example, if $V_P$ is 8 V, then $V_{GS(\text{off})}$ will be $-8$ V. Since these two values are always equal and opposite, only one is usually listed on the spec sheet for a given JFET.

You must be sure to remember that there is a definite difference between $V_P$ and $V_{GS(\text{off})}$. $V_P$ is the value of $V_{DS}$ that causes the JFET to become a constant-current component. It is measured at $V_{GS} = 0$ V and will have a constant drain current of $I_D = I_{DSS}$. $V_{GS(\text{off})}$, on the other hand, is the value of $V_{GS}$ that causes $I_D$ to drop to nearly zero.

# JFET Biasing

The gate–source junction of a JFET will *always* be reverse biased. Even when the gate and source terminals are shorted, $I_D$ will cause the junction to be reverse biased, as shown in Figures 12.5 and 12.6. *The gate–source junction of a JFET must never be allowed to become forward biased* because the gate material is not constructed to handle any significant amount of current. If the junction is allowed to become forward biased, current will be generated through the gate material. Under most circumstances, this current may destroy the component.

◄ *OBJECTIVE 4*

Forward biasing a JFET gate–source junction may destroy the component.

The fact that the gate is always reverse biased leads us to another important characteristic of the device. *JFETs have an extremely high characteristic gate impedance.* This impedance is typically in the mid to high megohm (MΩ) range. For example, the spec sheet for the 2N5457 (provided in Appendix A) lists the following input characteristics:

$$\text{gate reverse current } (I_{GSS}) = 1.0 \text{ nA maximum at } 25°C$$

when

$$V_{DS} = 0 \text{ V} \qquad \text{and} \qquad V_{GS} = -15 \text{ V}$$

If we use Ohm's law on the values of $V_{GS}$ and $I_{GSS}$ listed above, we can calculate a gate resistance of 15 GΩ!

The advantage of this extremely high input impedance can be seen by taking a look at the circuit shown in Figure 12.8. Here, the JFET amplifier is driven by a high-impedance source. Because of the extremely high input impedance of the JFET gate, it draws no current from the source. Thus, it is as if there were no load on the source at all. The high input impedance of the JFET has led to its extensive use in microcomputer circuits. The low current requirements of the component make it perfect for use in ICs, where thousands of transistors must be etched onto a single piece of silicon. The low current draw helps the IC to remain relatively cool, thus allowing more components to be placed in a smaller physical area.

High input (gate) impedance is the primary advantage that FETs have over BJTS.

# Component Control

It was stated earlier that the JFET is a *voltage-controlled* device, while the BJT is a *current-controlled* device. We have now established the foundation necessary to look at this point in detail. We will start by reviewing the BJT as a current-controlled component. Consider the circuit shown in Figure 12.9. Using the formulas listed, the circuit values for Figure 12.9a can be calculated as

**FIGURE 12.8**

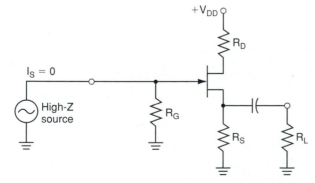

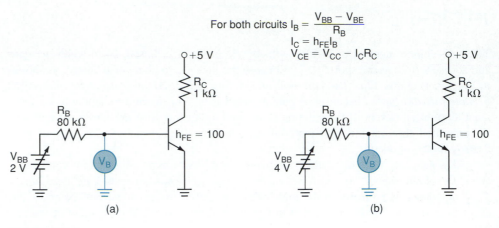

For both circuits $I_B = \dfrac{V_{BB} - V_{BE}}{R_B}$

$I_C = h_{FE}I_B$

$V_{CE} = V_{CC} - I_C R_C$

(a)                                                        (b)

**FIGURE 12.9**

$$I_B = 16.3 \ \mu A$$
$$I_C = 1.63 \ mA$$
$$V_{CE} = 3.38 \ V$$

Using the same formulas on the circuit in Figure 12.9b gives the following results:

$$I_B = 41.3 \ \mu A$$
$$I_C = 4.13 \ mA$$
$$V_{CE} = 0.88 \ V$$

Note that $V_B = V_{BE}$ for each of the circuits shown. Thus, from one circuit to another, the value of base voltage did not change. It stayed at approximately 0.7 V. Yet there were drastic changes in the output characteristics of the two amplifiers. These changes were caused by the change in $I_B$, not by any change in $V_B$. Therefore, it can be seen that the BJT is a current-controlled component. The output characteristics are determined by the input current, not the input voltage.

In contrast, the JFET has *no* gate current. It has been shown that the size of the channel is controlled by the amount of reverse bias applied to the gate–source junction. Since $V_{GS}$ controls the JFET, it is easy to see that the JFET is *not* a current-controlled device, but rather a voltage-controlled device.

Since the JFET has no gate current, there is no beta rating for the device. However, the output current ($I_D$) can be defined in terms of the circuit input as follows:

*A Practical Consideration*: Equation (12.1) will work only when

$$|V_{GS}| < |V_{GS(off)}|$$

If the above condition is not fulfilled, the equation will give you a false value of $I_D$. Just remember, if $V_{GS}$ is greater than (or equal to) $V_{GS(off)}$, $I_D$ will be zero.

$$I_D = I_{DSS}\left(1 - \frac{V_{GS}}{V_{GS(off)}}\right)^2 \qquad \textbf{(12.1)}$$

where     $I_{DSS}$ = the shorted gate–drain current rating of the device

$V_{GS}$ = the gate–source voltage

$V_{GS(off)}$ = the gate–source cutoff voltage

Example 12.1 demonstrates the use of equation (12.1).

**EXAMPLE 12.1**

Determine the value of drain current for the circuit shown in Figure 12.10.

*Solution:*   The drain current for the circuit is found as

$$I_D = I_{DSS}\left(1 - \frac{V_{GS}}{V_{GS(off)}}\right)^2$$

$$= (3\text{ mA})\left(1 - \frac{-2\text{V}}{-6\text{ V}}\right)^2$$

$$= (3\text{ mA})(0.444)$$

$$= 1.33\text{ mA}$$

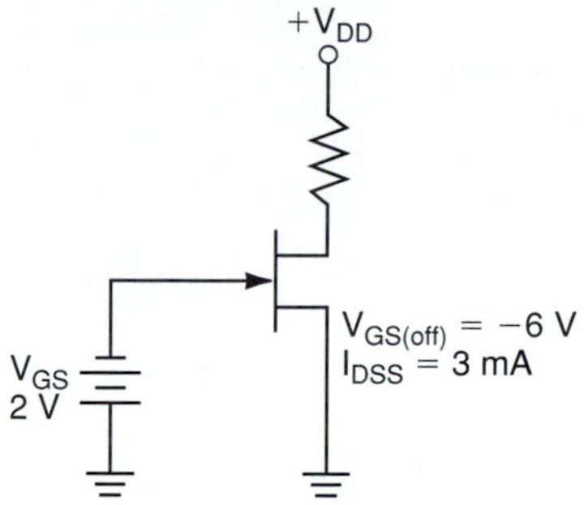

FIGURE 12.10

**PRACTICE PROBLEM 12.1**

A JFET with parameters of $I_{DSS} = 12$ mA and $V_{GS(off)} = -6$ V is used in a circuit that provides a $V_{GS}$ of $-3$ V. Determine the value of $I_D$ for the circuit.

If you take a closer look at equation (12.1), you will notice that $V_{GS}$ is the only value on the right side of the equation that will change for a specified JFET. In other words, for a specified JFET, $I_{DSS}$ and $V_{GS(off)}$ are parameters. Based on the fact that $I_{DSS}$ and $V_{GS(off)}$ are constants for a given JFET, we can assume that $I_D$ is a function of the value of $V_{GS}$. As $V_{GS}$ changes (which it will), $I_D$ changes.

We can use a series of $V_{GS}$ versus $I_D$ values to plot what is called the *transconductance curve* for a specified JFET. A **transconductance curve** is a graph of all possible combinations of $V_{GS}$ and $I_D$. The process for plotting the transconductance curve of a given JFET is as follows:

**Transconductance curve**
A plot of all possible combinations of $V_{GS}$ and $I_D$.

1. Plot a point on the x-axis that corresponds to the value of $V_{GS(off)}$.

How to plot a transconductance curve.

2. Plot a point on the y-axis that corresponds to the value of $I_{DSS}$.

3. Select two or three values of $V_{GS}$ between 0 V and $V_{GS(off)}$. For each value of $V_{GS}$ selected, determine the corresponding values of $I_D$ using equation (12.1).

4. Plot the points from step 3, and connect all the plotted points with a smooth curve.

Step 1 is based on the fact that $I_D = 0$ when $V_{GS} = V_{GS(off)}$. Thus, the point for $V_{GS(off)}$ versus $I_D$ will always fall on the x-axis of the graph. Step 2 is based on the fact that $I_D = I_{DSS}$ when $V_{GS} = 0$ V. Thus, the point for $V_{GS} = 0$ versus $I_D$ will always fall on the $I_{DSS}$ value point on the y-axis. Example 12.2 demonstrates the use of the process given for plotting the transconductance curve of a given JFET.

Plot the transconductance curve for the 2N5457. For this device, $V_{GS(off)}$ is $-6$ V and $I_{DSS}$ is 3 mA.

*EXAMPLE 12.2*

*Solution:* With the values given, we know that our end points for the curve are $(-6\text{ V}, 0\text{ mA})$ and $(0\text{ V}, 3\text{ mA})$. We now use three values of $V_{GS}$ ($-1$ V, $-3$ V, and $-5$ V) and calculate the corresponding value of $I_D$ for each.

At $V_{GS} = -1$ V,

$$I_D = (3\text{ mA})\left(1 - \frac{-1\text{ V}}{-6\text{ V}}\right)^2$$

$$= 2.08\text{ mA}$$

At $V_{GS} = -3$ V,

$$I_D = (3\text{ mA})\left(1 - \frac{-3\text{ V}}{-6\text{ V}}\right)^2$$

$$= 0.75\text{ mA} \qquad (750\ \mu\text{A})$$

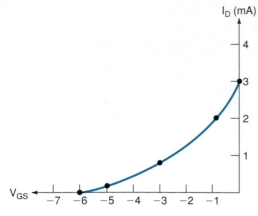

**FIGURE 12.11**

At $V_{GS} = -5$ V,

$$I_D = (3 \text{ mA})\left(1 - \frac{-5 \text{ V}}{-6 \text{ V}}\right)^2$$

$$= 0.083 \text{ mA} \qquad (83 \text{ μA})$$

We now have the following sets of $V_{GS}$ versus $I_D$ values:

| $V_{GS}$ (V) | $I_D$ (mA) |
|---|---|
| −6 | 0 |
| −5 | 0.083 |
| −3 | 0.75 |
| −1 | 2.08 |
| 0 | 3 |

The points are now plotted and connected to form the curve shown in Figure 12.11.

**Lab Reference:** A curve similar to the one in Figure 12.11 is plotted using measured values in Exercise 25.

### PRACTICE PROBLEM 12.2

A JFET has parameters of $V_{GS(off)} = -20$ V and $I_{DSS} = 12$ mA. Plot the transconductance curve for the device using $V_{GS}$ values of 0 V, −5 V, −10 V, −15 V, and −20 V.

*A Practical Consideration*: Many JFET spec sheets will list only a *maximum* value of $V_{GS(off)}$. When only a maximum value is listed:

a. Only one transconductance curve is plotted for the device.
b. It is generally expected (although not guaranteed) that the ratio of $V_{GS(off)}$ (maximum) to $V_{GS(off)}$ (minimum) will be approximately equal to the ratio of $I_{DSS}$ (maximum) to $I_{DSS}$ (minimum). For example, if the ratio of $I_{DSS}$ values is approximately 5 : 1, the ratio of $V_{GS(off)}$ values can be generally assumed to also be around 5 : 1. However, this ratio is neither guaranteed nor exact.

The transconductance curve for a given JFET will be used in both the dc and ac analyses of any amplifier using that JFET. For this reason, it is important that you be able to plot the transconductance curve for any given JFET. When you need to plot the transconductance curve for a JFET, simply obtain the values of $V_{GS(off)}$ and $I_{DSS}$ from the spec sheet for the device, and then follow the procedure outlined in this section.

One important point needs to be made at this time: Most JFET spec sheets will list more than one value of $V_{GS(off)}$ and $I_{DSS}$. For example, the spec sheet for the 2N5457 lists the following:

| | |
|---|---|
| $V_{GS(off)(min)} = -0.5$ V | $I_{DSS(min)} = 1$ mA |
| $V_{GS(off)(max)} = -6$ V | $I_{DSS(max)} = 5$ mA |

When a range of values is given, you must use the two *minimum* values to plot one curve and the two *maximum* values to plot another curve on the same graph. This procedure is demonstrated in Example 12.3.

### EXAMPLE 12.3

Using the minimum and maximum values of $V_{GS(off)}$ and $I_{DSS}$ for the 2N5457, plot the two transconductance curves for the device.

*Solution:* The maximum values of $V_{GS(off)}$ and $I_{DSS}$ are −6 V and 5 mA, respectively. Using these values and equation (12.1), we can solve for the following points:

| $V_{GS}(V)$ | $I_D(mA)$ | |
|---|---|---|
| $-6$ | 0 | |
| $-4$ | 0.556 | (556 µA) |
| $-2$ | 2.222 | |
| 0 | 5 | |

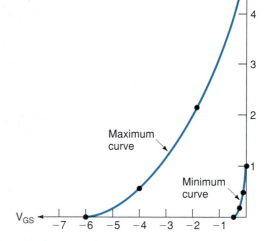

Using the minimum values of $V_{GS(off)} = -0.5$ V and $I_{DSS} = 1$ mA, we can solve for the following points with equation (12.1):

| $V_{GS}(V)$ | $I_D(mA)$ | |
|---|---|---|
| $-0.5$ | 0 | |
| $-0.4$ | 0.04 | (40 µA) |
| $-0.2$ | 0.36 | (360 µA) |
| 0 | 1 | |

The point sets listed are used to plot two separate curves. These curves are shown in Figure 12.12.

**FIGURE 12.12**

**PRACTICE PROBLEM 12.3**

The 2N5486 JFET has values of $V_{GS(off)} = -2$ to $-6$ V and $I_{DSS} = 8$ to $20$ mA. Plot the minimum and maximum transconductance curves for the device.

A few more relationships that involve the principle of transconductance must be discussed. However, these principles relate to the ac operation of the JFET. Therefore, we will cover them in our discussion on the ac characteristics of JFETs. At this point, we have the foundation needed to cover JFET biasing circuits.

## One Final Note

In this section, we have used $n$-channel JFETs in all the examples. This is due to the fact that $n$-channel JFETs are used far more commonly than $p$-channel JFETs. Just remember, all the $n$-channel JFET principles apply equally to $p$-channel JFETs. The only differences are the voltage polarities and the direction of the drain and source currents. The $n$- and $p$-channel JFETs are contrasted in Figure 12.13.

1. What is a *field-effect transistor*? What are the two types of JFETs?

*Section Review*

2. What do we call the three terminals of a JFET?

3. Describe the physical relationship between the JFET gate and channel.

4. What are the two types of JFETs?

5. Draw the schematic symbol for each type of JFET.

6. What supply voltage polarity is typically used for each type of JFET?

7. What is the relationship between channel width and drain current?

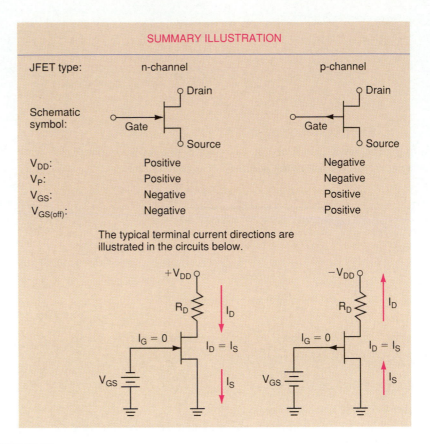

**FIGURE 12.13**

8. What is the relationship between $V_{GS}$ and channel width?

9. What is the relationship between $V_{GS}$ and drain current?

10. What is *pinch-off voltage*?

11. What effect does an increase in $V_{DS}$ have on $I_D$ when $V_{DS} < V_P$?

12. What effect does an increase in $V_{DS}$ have on $I_D$ when $V_{DS} > V_P$?

13. What is the operating region above $V_P$ called?

14. What is $I_{DSS}$?

15. What is the relationship between $I_D$ and $I_{DSS}$?

16. What is $V_{GS(off)}$?

17. What is the primary restriction on the value of $V_{GS}$?

18. Why do JFETs typically have extremely high gate input impedance?

## *12.2* JFET BIASING CIRCUITS

*OBJECTIVE 5* ▶  JFET biasing circuits are very similar to BJT biasing circuits. The main difference between JFET circuits and BJT circuits is the operation of the active components themselves. In this section, we will cover those dc operating principles and relationships that differ from the BJT circuits already covered. We will not go into lengthy explanations of the circuit operation, as these explanations were made earlier.

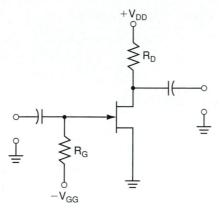

**FIGURE 12.14** Gate bias.

**Lab Reference:** The operation of gate bias is demonstrated in Exercise 26.

## Gate Bias

Do you remember *base bias*? The JFET equivalent of this circuit type is called **gate bias.** The gate-bias circuit is shown in Figure 12.14. The gate supply voltage $(-V_{GG})$ is used to ensure that the gate–source junction is reverse biased. Since there is no gate current, there is no voltage dropped across $R_G$, and the value of $V_{GS}$ is found as

$$V_{GS} = V_{GG} \tag{12.2}$$

**Gate bias**
The JFET equivalent of BJT base bias.

Using $V_{GG}$ in equation (12.1) as $V_{GS}$ allows us to calculate the value of $I_D$. Once $I_D$ is known, $V_{DS}$ for the JFET can be found as

$$V_{DS} = V_{DD} - I_D R_D \tag{12.3}$$

Example 12.4 demonstrates the complete dc analysis of a simple gate-bias circuit.

---

The JFET in Figure 12.15 has values of $V_{GS(off)} = -8$ V and $I_{DSS} = 16$ mA. Determine the values of $V_{GS}$, $I_D$, and $V_{DS}$ for the circuit.

**Solution:** Since none of $V_{GG}$ is dropped across the gate resistor, $V_{GS}$ is found as

$$V_{GS} = V_{GG}$$
$$= -5 \text{ V}$$

Using this value of $V_{GS}$ and the parameters listed above, the value of $I_D$ is found as

$$I_D = I_{DSS} \left( 1 - \frac{V_{GS}}{V_{GS(off)}} \right)^2$$
$$= 16 \text{ mA} \left( 1 - \frac{-5 \text{ V}}{-8 \text{ V}} \right)^2$$
$$= 16 \text{ mA} (0.1406)$$
$$= 2.25 \text{ mA}$$

**EXAMPLE 12.4**

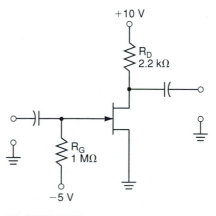

**FIGURE 12.15**

Now the value of $V_{DS}$ is found as

$$V_{DS} = V_{DD} - I_D R_D$$
$$= 10 \text{ V} - (2.25 \text{ mA})(2.2 \text{ k}\Omega)$$
$$= 5.05 \text{ V}$$

### PRACTICE PROBLEM 12.4

Assume that the JFET in Figure 12.15 has values of $V_{GS(\text{off})} = -10\text{V}$ and $I_{DSS} = 12$ mA. Determine the values of $V_{GS}$, $I_D$, and $V_{DS}$ for the circuit.

Example 12.4 illustrated the process for analyzing a gate-bias circuit. However, it assumed that $V_{GS(\text{off})}$ and $I_{DSS}$ have one specified value each. As you know, this is not usually the case.

**Lab Reference:** Gate bias instability is demonstrated in Exercise 26.

The fact that a given type of JFET can have a range of values for $V_{GS(\text{off})}$ and $I_{DSS}$ leads us to a major problem with the gate-bias circuit: *Gate bias will not provide a stable Q-point from one JFET to another.* This problem is illustrated in Example 12.5.

### EXAMPLE 12.5

Determine the range of Q-point values for the circuit shown in Figure 12.16. Assume that the JFET has ranges of $V_{GS(\text{off})} = -1$ to $-7$ V and $I_{DSS} = 2$ to 9 mA.

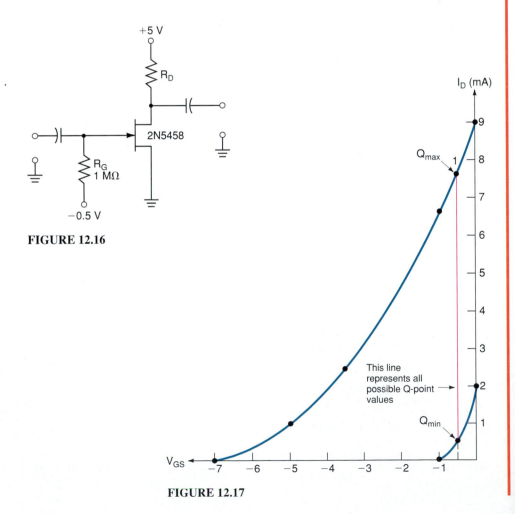

**FIGURE 12.16**

**FIGURE 12.17**

**Solution:** Using the established procedure, the two transconductance curves for the 2N5458 are plotted. These curves are shown in Figure 12.17. Since $V_{GS}$ is set to $-0.5$ V, a vertical line called a *bias line* is drawn from this point through the two transconductance curves, as shown in Figure 12.17. The point where the line intersects the maximum curve is the value of $Q_{max}$. The point where the line intersects the minimum curve is the value of $Q_{min}$. The line between the two points represents all the possible $Q$-points for a 2N5458 used in the circuit in Figure 12.16.

*A Practical Consideration*: When only a maximum curve can be plotted, the $Q$-point can fall anywhere between the intersection of the bias line and the curve and the intersection of the bias line and either axis of the graph.

As Example 12.5 shows, the circuit does not provide a stable $Q$-point. You may place one 2N5458 in the circuit and get an $I_D$ of 5 mA. Another 2N5458 may give you an $I_D$ of 2 mA, and so on. Because of the instability of the circuit, gate bias is rarely used for anything other than switching applications.

You may be wondering why the gate-bias circuit contains a gate resistor when there is no gate current. Since resistors are usually used to develop a voltage, it would seem that this resistor would be useless. After all, there is no voltage drop across the component and no current for it to limit. Actually, $R_G$ is used for ac operation purposes. Take a look at the circuits shown in Figure 12.18. Figure 12.18a shows the results of eliminating $R_G$. The ac signal produced by the first stage is coupled to the second, only to be shorted to the dc supply, $-V_{GG}$. The resistor is needed to prevent this from happening. As Figure 12.18b shows, placing a gate resistor in the second stage restores the

The purpose served by $R_G$.

**FIGURE 12.18** The purpose served by the gate resistor, $R_G$.

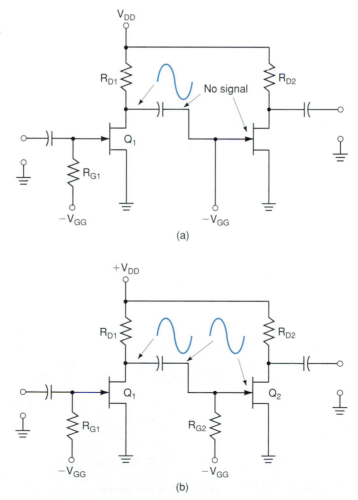

(a)

(b)

ac signal. Therefore, while the gate resistor has little to do with the dc operation of the circuit, it is a vital part of the ac operation of the circuit. As you will see, the *self-bias* circuit uses a gate resistor for the same reason.

## Self-Bias

**Self-bias**
A JFET biasing circuit that uses a source resistor to establish a negative $V_{GS}$.

**Self-bias** is a more viable type of JFET biasing. The self-bias circuit replaces the gate supply ($-V_{GG}$) with a source resistor ($R_S$). The self-bias circuit is shown in Figure 12.19. Note that the gate is grounded through $R_G$, and a resistor has been added in the source circuit. This resistor is used to produce the $-V_{GS}$ needed for the operation of the JFET. Figure 12.20 helps to illustrate this point.

In the JFET circuit, *all* the source current is supplied by the drain circuit. This is due to the fact that there is no significant gate current. Therefore, we can define source current as

$$I_S = I_D \tag{12.4}$$

With all of the drain current passing through the source resistor, the voltage across the resistor is found as

$$V_S = I_D R_S \tag{12.5}$$

**Lab Reference:** Self-bias operation is demonstrated in Exercise 26.

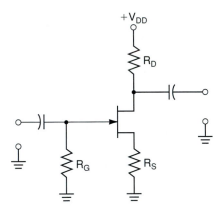

**FIGURE 12.19  Self-bias.**

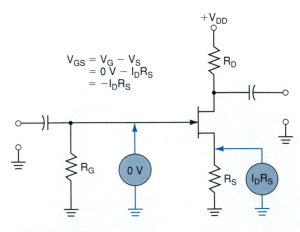

**FIGURE 12.20  $V_{GS}$ for the self-bias current.**

Since there is no significant current in the gate circuit, no voltage is developed across the gate resistor. Thus, the gate voltage (with respect to ground) is given as

$$V_G = 0 \text{ V} \tag{12.6}$$

Now we can use the relationships shown in equations (12.5) and (12.6) to show how the source resistor is used to reverse bias the gate-source junction. The voltage *from gate to source* can be expressed as

$$V_{GS} = V_G - V_S \tag{12.7}$$

Substituting the values in equations (12.5) and (12.6) in place of $V_G$ and $V_S$ gives us

$$V_{GS} = 0 \text{ V} - I_D R_S$$

which can be written as

$$V_{GS} = -I_D R_S \tag{12.8}$$

Thus, $-V_{GS}$ is developed by the current through the source resistor.

If equation (12.8) is confusing, just remember that $V_{GS}$ is the voltage measured from the gate terminal *to the source terminal*. There is no rule stating that the voltage at the gate terminal must be negative, only that the gate terminal must be more negative than the source terminal. If the source terminal is at some positive voltage and the gate is grounded, the gate terminal is *more negative* (less positive) than the source terminal. Figure 12.21 helps to illustrate this point. In all three circuits, the gate is 2 V more negative than the source. Thus, the value of $V_{GS}$ for each circuit is $-2$ V. Note that the value of $V_G$ changes from one circuit to another, as does the value of $V_S$. However, the value of $V_{GS}$ is constant for the three circuits. Make sure that you are comfortable with this point before continuing.

Equation (12.8) is the *bias line equation* for the self-bias circuit. To plot the dc bias line for a self-bias circuit, you simply follow this procedure:

1. Plot the minimum and maximum transconductance curves for the JFET used in the circuit.

How to plot the self-bias dc bias line.

2. Choose any value of $V_{GS}$ and determine the corresponding value of $I_D$ using

$$I_D = \frac{-V_{GS}}{R_S} \tag{12.9}$$

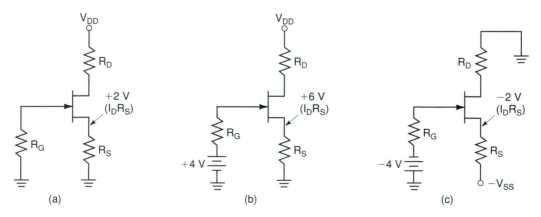

FIGURE 12.21   Three circuits with $V_{GS} = -2$ V.

3. Plot the point determined by equation (12.9), and draw a line from this point to the graph *origin* (the [0, 0] point).

4. The points where the line crosses the two transconductance curves define the limits of the *Q*-point operation of the circuit.

The procedure given here is demonstrated in Example 12.6.

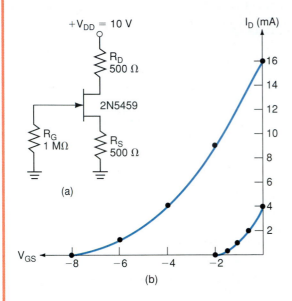

*EXAMPLE 12.6*

Determine the range of *Q*-point values for the circuit shown in Figure 12.22.

**FIGURE 12.22**

**FIGURE 12.23**

***Solution:*** The 2N5459 spec sheet lists the following values for $V_{GS(off)}$ and $I_{DSS}$:

$$V_{GS(off)} = -2 \text{ to } -8 \text{ V}$$

$$I_{DSS} = 4 \text{ to } 16 \text{ mA}$$

Using these two sets of values and equation (12.1), the transconductance curves are plotted, as is shown in Figure 12.22b.

The value of $V_{GS} = -4$ V is chosen at random to calculate $I_D$. Using equation (12.9), $I_D$ at $V_{GS} = -4$ V is found as

$$I_D = \frac{-V_{GS}}{R_S}$$

$$= 8 \text{ mA}$$

This point $(-4, 8)$ is plotted on the graph, and a line is drawn from the point to the graph origin. The plot of the line is shown in Figure 12.23. Note that the points where the bias line intersects the transconductance curves are labeled $Q_{max}$ and $Q_{min}$. Using the graph coordinates of these two points, we obtain the maximum and minimum values of $V_{GS}$ and $I_D$ shown in Figure 12.23.

### PRACTICE PROBLEM 12.6

A JFET with parameters of $V_{GS(off)} = -5$ to $-10$ V and $I_{DSS} = 5$ to 10 mA is used in a self-bias circuit with a value of $R_S = 2$ kΩ. Plot the transconductance curves and bias line for the circuit.

As Example 12.6 has shown, there is still some instability in the $Q$-point. In other words, we have not succeeded in obtaining a completely stable $Q$-point value by switching to self-bias. However, as Figure 12.24 indicates, the $Q$-point of a self-bias circuit is far more stable for a given JFET than the gate-bias circuit. In Figure 12.24, the bias line from Example 12.6 is shown as plotted in the example. A dashed line is plotted showing the same transistor in a gate-bias circuit. From these two dc bias lines, a couple of points can be made:

1. The gate-bias circuit is not subject to any change in $V_{GS}$ and thus is more stable in this respect.

2. The possible change in $I_D$ from one 2N5459 to another is much greater for the gate-bias circuit than for the self-bias circuit.

Thus, the gate-bias circuit has the more stable value of $V_{GS}$, while the self-bias circuit has the more stable value of $I_D$. Which is more important? This question is best answered by considering the ac operation of amplifiers in general. AC characteristics such as compliance, amplifier power dissipation, and load power all depend on the output voltage and current characteristics of the amplifier. You may recall that PP, $P_L$, and $P_S$ for the BJT amplifier are all defined in terms of $V_{CEQ}$ and $I_{CQ}$. For the JFET amplifier, the ac output characteristics are defined in terms of $V_{DS}$ and $I_D$. The relationship between $V_{DS}$ and $I_D$ is given as

$$V_{DS} = V_{DD} - I_D(R_D + R_S) \qquad (12.10)$$

If the ac output characteristics of the JFET amplifier are to be stable, $I_D$ must be as stable as possible. Since self-bias provides a more stable value of $I_D$ than gate bias, it is the preferred biasing method of the two.

The complete dc analysis of a self-bias circuit is actually pretty simple. The process starts with plotting the transconductance curves and the dc bias line, as was done in Example 12.6. Then, using the $Q$-point limits, the minimum and maximum values of $V_{GS}$ and $I_D$ are determined. These values can then be used to determine the minimum and maximum values of $V_{DS}$. This analysis process is demonstrated in Example 12.7.

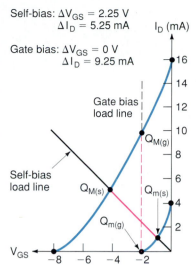

Self-bias: $\Delta V_{GS} = 2.25$ V
$\qquad\qquad \Delta I_D = 5.25$ mA

Gate bias: $\Delta V_{GS} = 0$ V
$\qquad\qquad \Delta I_D = 9.25$ mA

**FIGURE 12.24    Self-bias stability versus gate-bias stability.**

*A Practical Consideration*: Even though self-bias provides more stability, gate bias is still preferred for many switching applications. JFET switching applications are discussed in Chapter 19.

---

Determine the dc characteristics of the amplifier used in Example 12.6 (Figure 12.22).

*Solution:*   We have already determined the limits of $V_{GS}$ and $I_D$ (Figure 12.23) as

$$V_{GS} = -0.75 \text{ to } -3 \text{ V}$$
$$I_D = 1.6 \text{ to } 6 \text{ mA}$$

Using the two values of $I_D$ listed, we can determine the limits of $V_{DS}$. When $I_D = 1.6$ mA, $V_{DS}$ is found as

$$
\begin{aligned}
V_{DS} &= V_{DD} - I_D(R_D + R_S) \\
&= 10 \text{ V} - (1.6 \text{ mA})(1 \text{ k}\Omega) \\
&= 10 \text{ V} - 1.6 \text{ V} \\
&= 8.4 \text{ V}
\end{aligned}
$$

When $I_D = 6$ mA, $V_{DS}$ is found as

$$
\begin{aligned}
V_{DS} &= V_{DD} - I_D(R_D + R_S) \\
&= 10 \text{ V} - (6 \text{ mA})(1 \text{ k}\Omega) \\
&= 10 \text{ V} - 6 \text{ V} \\
&= 4 \text{ V}
\end{aligned}
$$

*EXAMPLE 12.7*

Thus, the value of $V_{DS}$ will fall between 4 and 8.4 V, depending on the particular 2N5459 used in the circuit.

**PRACTICE PROBLEM 12.7**

The circuit described in Practice Problem 12.6 has a 1-kΩ drain resistor and a 9-$V_{dc}$ source. Determine the range of $V_{DS}$ values for the circuit.

Example 12.7 illustrates one important point: Even though the self-bias circuit is more stable than gate bias, it still leaves a lot to be desired. A range of $V_{DS} = 4$ to 8.4 V would hardly be stable enough for most linear applications. We must therefore look for a circuit to provide a much greater amount of stability. *Voltage-divider* bias does the job extremely well.

## Voltage-Divider Bias

The voltage-divider bias JFET amplifier is very similar to its BJT counterpart. This biasing circuit is shown in Figure 12.25. The gate voltage for this amplifier is found in the same manner as $V_B$ in the BJT circuit. By formula,

$$V_G = V_{DD} \frac{R_2}{R_1 + R_2} \qquad (12.11)$$

Also, the difference between $V_G$ and $V_S$ is equal to $V_{GS}$. Therefore, the value of $I_D$ can be found as

$$I_D = \frac{V_S}{R_S}$$

or

$$I_D = \frac{V_G - V_{GS}}{R_S} \qquad (12.12)$$

Now we have a problem. If you refer to Examples 12.6 and 12.7 on the self-bias circuit, you will see that $V_{GS}$ changes from one transistor to another. The same holds true for the voltage-divider biased circuit. This can make it difficult to determine the exact value of $I_D$. However, plotting the dc bias line for this circuit will show you that the problem is not really a major one.

**FIGURE 12.25   Voltage-divider bias.**

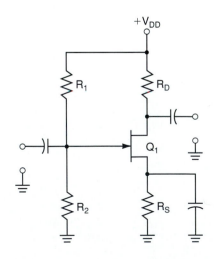

The method used to plot the dc bias line for the voltage-divider bias circuit is a bit strange—easy, but strange. The exact process is as follows:

How to plot the voltage-divider bias dc bias line.

1. As usual, plot the transconductance curves for the specific JFET.
2. Calculate the value of $V_G$ using equation (12.11).
3. *On the positive x-axis of the graph*, plot a voltage point equal to $V_G$.
4. Solve for $I_D$ using

$$I_D = \frac{V_G}{R_S} \qquad (12.13)$$

5. Locate the point on the *y*-axis that corresponds to the value found in step 4.
6. Draw a line from the $V_G$ point through the point found with equation (12.13), and continue the line to intersect *both* transconductance curves. The intersection points represent the $Q_{max}$ and $Q_{min}$ points.

Example 12.8 demonstrates the use of this procedure.

---

Plot the dc bias line for the circuit shown in Figure 12.26.

**EXAMPLE 12.8**

**Solution:** Figure 12.26 contains the 2N5459, the JFET whose transconductance curves were plotted in previous examples. Those transconductance curves can be seen in Figure 12.27. The value of $V_G$ for the circuit is calculated as

$$V_G = V_{DD} \frac{R_2}{R_1 + R_2}$$

$$= 15 \text{ V}$$

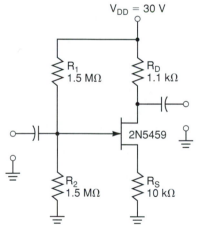

**FIGURE 12.26**

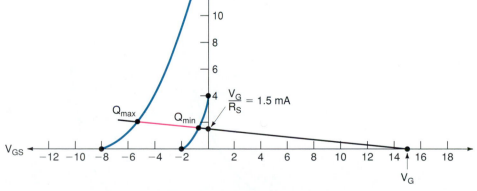

**FIGURE 12.27**  **The voltage-divider bias dc bias line.**

This point is now plotted on the *positive* x-axis, as shown in Figure 12.27. The value of $V_G$ is also used to find the y-axis intersect point, as follows:

$$I_D = \frac{V_G}{R_S}$$
$$= 1.5 \text{ mA}$$

Note that this point is marked on the y-axis of the graph. A line is then drawn through the two points and on through the two curves to make the $Q_{max}$ and $Q_{min}$ points.

### PRACTICE PROBLEM 12.8

In Practice Problem 12.3, you plotted the two transconductance curves for the 2N5486. This component is used in a voltage-divider biased amplifier with values of $V_{DD} = 36$ V, $R_1 = 10$ MΩ, $R_2 = 3.\,3$ MΩ, $R_D = 1.8$ kΩ, and $R_S = 3$ kΩ. Plot the dc bias line for the amplifier.

Figure 12.27 shows that we again have a relatively unstable value of $V_{GS}$. However, the stability of $I_D$ has improved a great deal. Figure 12.28 shows a close-up of the lower portion of Figure 12.27 to help illustrate this point. While $V_{GS}$ is varying between approximately $-5$ and $-0.5$ V, $I_D$ varies only from 2 mA to approximately 1.5 mA. Thus, this amplifier provides a much more stable value of $I_D$ than either gate bias or self-bias. For comparison, the results from Figures 12.24 and 12.28 are summarized as follows:

| Circuit Type | Change in 2N5459 Drain Current Values (mA) |
| --- | --- |
| Gate bias | 9.25 |
| Self-bias | 5.25 |
| Voltage-divider bias | 0.5 |

As you can see, the voltage-divider bias circuit is by far the most stable of the three.

It would seem that the stability of $I_D$ would be impossible given the large possible variations in $V_{GS}$ for the amplifier in Example 12.8. Since $I_D$ is a function of $V_{GS}$, as was stated in equation (12.1), it would seem that a large change in $V_{GS}$ would cause a large

**FIGURE 12.28   A close-up of Figure 12.27.**

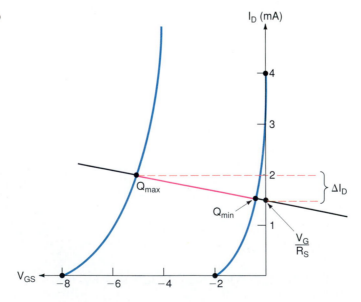

change in $I_D$. However, an analysis of the circuit would show the $Q$-point values obtained to be valid. This analysis is performed in Example 12.9.

EXAMPLE 12.9

Example 12.8 gave minimum and maximum $Q$-point values for Figure 12.26 as

$$Q_{max} = -5 \text{ V at 2 mA}$$
$$Q_{min} = -0.5 \text{ V at 1.5 mA} \qquad \text{(approximate)}$$

Verify these two $Q$-point values as being possible operating combinations of $V_{GS}$ and $I_D$.

**Solution:** We will tackle the $Q_{max}$ values of $V_{GS} = -5$ V and $I_D = 2$ mA first. Just as before, we start by determining the value of $V_G$, as follows:

$$V_G = V_{DD} \frac{R_2}{R_1 + R_2}$$
$$= 15 \text{ V}$$

Now recall that $V_S$ is equal to the difference between $V_G$ and $V_{GS}$. By formula,

$$V_S = V_G - V_{GS}$$

Since $V_{GS}$ is a *negative* value, the value of $V_S$ will end up being *higher* than the value of $V_G$, as follows:

$$V_S = V_G - V_{GS}$$
$$= 15 \text{ V} - (-5 \text{ V})$$
$$= 15 \text{ V} + 5 \text{ V}$$
$$= 20 \text{ V}$$

Therefore, $I_D$ is found as

$$I_D = \frac{V_G - V_{GS}}{R_S}$$
$$= \frac{20 \text{ V}}{10 \text{ k}\Omega}$$
$$= 2 \text{ mA}$$

Thus, the combination of $V_{GS} = -5$ V and $I_D = 2$ mA has been shown to be a possible combination for the circuit. Next, we will verify the $Q_{min}$ values of $V_{GS} = -0.5$ V and $I_D \cong 1.5$ mA. At $V_{GS} = -0.5$ V and $V_G = 15$ V,

$$I_D = \frac{V_G - V_{GS}}{R_S}$$
$$= \frac{15 \text{ V} - (-0.5 \text{ V})}{10 \text{ k}\Omega}$$
$$= \frac{15 \text{ V} + 0.5 \text{ V}}{10 \text{ k}\Omega}$$
$$= \frac{15.5 \text{ V}}{10 \text{ k}\Omega}$$
$$= 1.55 \text{ mA}$$

Again, the results obtained from plotting the bias line of Figure 12.24 have been shown to be accurate.

## PRACTICE PROBLEM 12.9

Using the technique shown in Example 12.9, verify your $Q$-point values for Practice Problem 12.8.

When you are analyzing a voltage-divider biased JFET amplifier, the procedure is the same as the one that we followed in plotting the transconductance curves and verifying the results. The only other value that must be determined is $V_{DS}$. $V_{DS}$ is determined for this circuit as it is with all the others—with equation (12.10). In Example 12.10, we calculate the range of possible $V_{DS}$ values for the circuit we have covered in this section.

---

**EXAMPLE 12.10**

Determine the minimum and maximum values of $V_{DS}$ for the circuit in Figure 12.26.

**Solution:** When $I_D = 2$ mA,

$$
\begin{aligned}
V_{DS} &= V_{DD} - I_D(R_D + R_S) \\
&= 30\text{ V} - (2\text{ mA})(11.1\text{ k}\Omega) \\
&= 30\text{ V} - 22.2\text{ V} \\
&= 7.8\text{ V}
\end{aligned}
$$

When $I_D = 1.55$ mA,

$$
\begin{aligned}
V_{DS} &= V_{DD} - I_D(R_D + R_S) \\
&= 30\text{ V} - (1.55\text{ mA})(11.1\text{ k}\Omega) \\
&= 30\text{ V} - 17.21\text{ V} \\
&= 12.79\text{ V}
\end{aligned}
$$

**PRACTICE PROBLEM 12.10**

Determine the minimum and maximum values of $V_{DS}$ for the circuit described in Practice Problem 12.8. Use the calculated values of $I_D$ from Practice Problem 12.9.

---

We now have a situation where the value of $I_D$ is much more stable than it has been with previous circuits. However, the smaller variation in $I_D$ *still* allows a rather drastic change in $V_{DS}$.

## Current-Source Bias

**Current-source bias**
A JFET biasing circuit that provides high Q-point stability by making the value of $I_D$ independent of the JFET.

The **current-source bias** circuit provides high $Q$-point stability by making the value of $I_D$ independent of the JFET. Two current-source bias circuits are shown in Figure 12.29. For both circuits shown, the JFET drain current is equal to the *collector* current of the BJT. By formula,

$$ I_D = I_C \tag{12.14} $$

Since the value of $I_C$ is independent of the variations in JFET parameters, so is the value $I_D$. This assumes, of course, that the value of $I_C$ is less than the *lowest* value of $I_{DSS}$ for the JFET. For example, let's assume that the JFET in Figure 12.29a has a value of $I_{DSS} = $ 5mA to 12 mA. As long as $I_C$ for the circuit is less than 5 mA, the value of $I_D$ for the JFET will be independent of the JFET itself. Thus, a stable $Q$-point is obtained by designing the circuit so that $I_C$ equals the desired value of $I_D$.

In Figure 12.29a, the value of $I_C$ is found as it would be for any voltage-divider biased BJT amplifier. In Figure 12.29b, the value of $I_C$ is found as it would be for any emitter bias circuit.

Even though current-source bias provides the most stable $Q$-point value of $I_D$, the complexity of the circuit makes it undesirable for most applications; that is, the improved stability (over voltage-divider bias) is usually insufficient to warrant the extra circuitry.

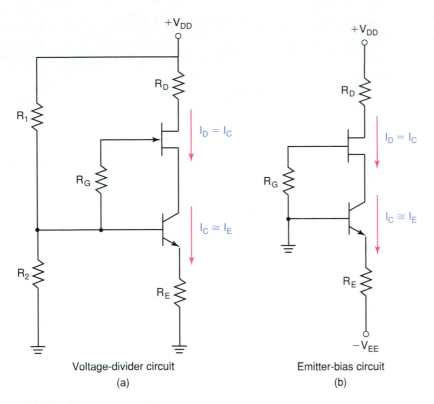

Voltage-divider circuit
(a)

Emitter-bias circuit
(b)

**FIGURE 12.29   Current-source bias.**

## Summary

Gate bias, self-bias, and voltage-divider bias are the three most commonly used JFET   ◄   *OBJECTIVE 6*
biasing circuits. Each of these circuits has its own type of dc bias line and circuit advantages and disadvantages. The characteristics of these three dc biasing circuits are summarized in Figure 12.30 (on page 498).

Figure 12.30 (on page 498)

---

1. What is the primary disadvantage of gate bias?
2. How does self-bias produce a negative value of $V_{GS}$?
3. What is the process used to plot the dc bias line for the self-bias circuit?
4. What is the process used to plot the dc bias line for the voltage-divider bias circuit?

*Section Review*

---

## 12.3   ac OPERATING CHARACTERISTICS: THE COMMON-SOURCE AMPLIFIER

In many ways, the ac operation of a JFET amplifier is similar to that of the BJT amplifier. However, some differences need to be examined in detail, such as the effects of changes in JFET transconductance.

In the next two sections, we will examine the ac operation of JFET amplifiers. We will start with the *common-source* amplifier simply because it is the most widely used of the JFET amplifiers. Then, in the next section, we will look at the operation of the *common-gate* and *source-follower* (common-drain) amplifiers.

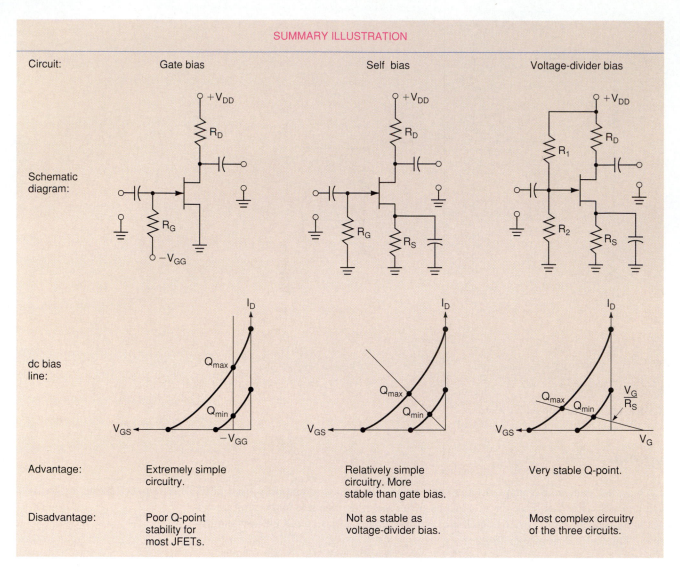

| Circuit: | Gate bias | Self bias | Voltage-divider bias |
|---|---|---|---|

| | | | |
|---|---|---|---|
| Advantage: | Extremely simple circuitry. | Relatively simple circuitry. More stable than gate bias. | Very stable Q-point. |
| Disadvantage: | Poor Q-point stability for most JFETs. | Not as stable as voltage-divider bias. | Most complex circuitry of the three circuits. |

**FIGURE 12.30**

## *Operation Overview*

*OBJECTIVE 7* ▶ There are many similarities between BJT amplifiers and JFET amplifiers. At the same time, there are differences that make JFET amplifiers more desirable in some applications and limit their use in others. One of the biggest differences is the fact that JFETs are voltage-controlled components and BJTs are current-controlled components. Another major difference is the input impedance of the JFET, as opposed to the input impedance of the BJT. As you will see, these differences play a major role in which transistor type is selected for a specific application.

In general terms, the JFET amplifier responds to an input signal in much the same way as a BJT amplifier. Figure 12.31 shows a **common-source amplifier,** which is the JFET equivalent of the *common-emitter* amplifier. Note that the input and output signals for the common-source amplifier closely resemble those of the common-emitter amplifier. Both amplifiers have a 180° signal phase shift from input to output. Both amplifiers provide *voltage gain* for the input signal.

Even though the common-source and common-emitter amplifiers serve the same basic purpose, the means by which they operate are quite different. The ac operation of

**Common-source amplifier**
The JFET equivalent of a common-emitter amplifier.

FIGURE 12.31 The common-source amplifier.

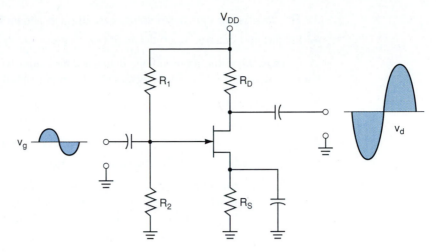

the common-source amplifier is better explained with the help of Figure 12.32. The initial dc operating values are represented as straight-line values in the blocks. These voltage values are assumed to be as follows:

$$V_G = +8 \text{ V}$$
$$V_S = +12 \text{ V}$$
$$V_D = +24 \text{ V}$$
$$V_{GS} = V_G - V_S = -4 \text{ V}$$

Here's what happens when $V_G$ is caused to increase by the input signal:

**1.** $V_G$ increases to $+10$ V.

**2.** $V_{GS}$ becomes $-2$ V ($V_{GS} = V_G - V_S = 10 \text{ V} - 12 \text{ V} = -2 \text{ V}$).

How the common-source amplifier responds to an ac input voltage.

FIGURE 12.32 Common-source amplifier ac operation.

Lab Reference: The ac operation of the CS amplifier is demonstrated in Exercise 27.

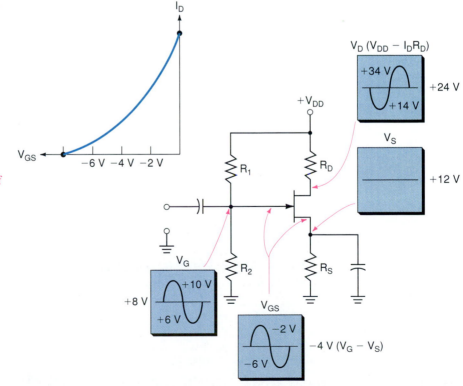

**3.** The decrease in $V_{GS}$ causes an *increase* in $I_D$, as shown in the transconductance curve.

**4.** Increasing $I_D$ causes $V_D$ to decrease ($V_D = V_{DD} - I_D R_D$).

Thus, when the input voltage increases, the output voltage decreases. Here is what happens when $V_G$ is caused to decrease by the input signal:

**1.** $V_G$ decreases to $+6$ V.

**2.** $V_{GS}$ becomes $-6$ V ($V_{GS} = V_G - V_S = 6$ V $- 12$ V $= -6$ V).

**3.** The *increase* in $V_{GS}$ causes a *decrease* in $I_D$, as shown in the transconductance curve.

**4.** Decreasing $I_D$ causes $V_D$ to increase ($V_D = V_{DD} - I_D R_D$).

Does this seem simple enough? Good. There is only one problem with the whole thing: We have seen how *a change in $V_{GS}$ causes a change in $I_D$*. When you start dealing with changing quantities of $V_{GS}$ and $I_D$, you begin to get into the whole issue of transconductance. Next, we will take a closer look at transconductance and its effects on ac operation.

## Transconductance

**Transconductance ($g_m$)**
A ratio of a change in drain current ($I_D$) to a change in $V_{GS}$, measured in microsiemens ($\mu$S) or micromhos ($\mu$mhos).

We have been plotting transconductance curves throughout this chapter as graphs of $V_{GS}$ versus $I_D$. **Transconductance** is *a ratio of a change in drain current to a change in gate–source voltage*. By formula,

$$g_m = \frac{\Delta I_D}{\Delta V_{GS}} \tag{12.15}$$

where     $g_m$ = the transconductance of the JFET at a given value of $V_{GS}$
$\Delta I_D$ = the change in $I_D$
$\Delta V_{GS}$ = the change in value of $V_{GS}$

As a rating, transconductance is measured in *microsiemens* ($\mu$S) or *micromhos* ($\mu$mhos). You may recall that conductance is measured in *siemens* or *mhos*.

We have plotted several transconductance curves throughout the chapter. The maximum-value transconductance curve for the 2N5459 will help you see one of the problems with using transconductance in ac calculations. This curve is shown in Figure 12.33. Note that two sets of points were selected on the curve, each marking a $\Delta V_{GS}$ of 1 V. Points $A_1$ and $A_2$ correspond to a change in $V_{GS}$ of 1 V, as do points $B_1$ and $B_2$. For each set of points, the corresponding change in $I_D$ is shown.

The values of $\Delta V_{GS}$ and their corresponding values of $\Delta I_D$ are summarized as follows:

| $\Delta V_{GS}$ (V) | $\Delta I_D$ (mA) |
|---|---|
| $-6$ to $-5 = 1$ | 1.25 |
| $-3$ to $-2 = 1$ | 2.75 |

Using equation (12.15), we can calculate the values of $g_m$ that correspond to the changes above as

$$g_{m(A)} = \frac{\Delta I_D}{\Delta V_{GS}}$$

$$= \frac{1.25 \text{ mA}}{1 \text{ V}}$$

$$= 1250 \ \mu\text{S}$$

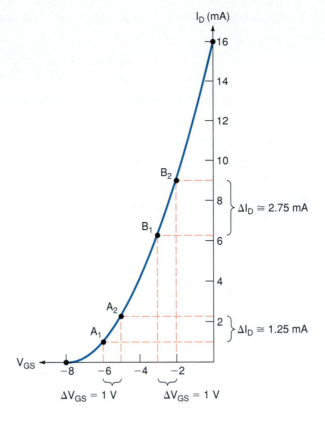

**FIGURE 12.33   The maximum-value transconductance curve for the 2N5459.**

$\Delta I_D \cong 2.75 \text{ mA}$

$\Delta I_D \cong 1.25 \text{ mA}$

$\Delta V_{GS} = 1 \text{ V}$    $\Delta V_{GS} = 1 \text{ V}$

and

$$g_{m(B)} = \frac{\Delta I_D}{\Delta V_{GS}}$$

$$= \frac{2.75 \text{ mA}}{1 \text{ V}}$$

$$= 2750 \ \mu\text{S}$$

As you can see, the value of $g_m$ is not constant for the entire transconductance curve. Figure 12.34 shows a single transconductance curve, with three possible dc bias lines. A single JFET operated at these three points would have three separate values of $g_m$. Luckily,

**FIGURE 12.34**

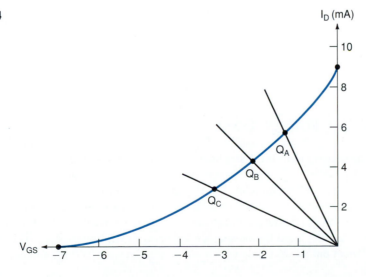

there is an equation that will tell you the value of $g_m$ at a specific value of $V_{GS}$:

$$g_m = g_{m0}\left(1 - \frac{V_{GS}}{V_{GS(\text{off})}}\right) \tag{12.16}$$

where  $g_m$ = the value of transconductance for the specific application
$\quad\quad g_{m0}$ = the maximum value of $g_m$, measured at $V_{GS} = 0$ V

Example 12.11 shows how the value of $g_m$ is calculated for a specific application.

---

**EXAMPLE 12.11**

The 2N5459 has a rating of $g_{m0} = 4000$ μS. Determine the values of $g_m$ at $V_{GS} = -3$ V and $V_{GS} = -5$ V for this device. For the 2N5459, $V_{GS(\text{off})} = -8$ V.

*Solution:*  At $V_{GS} = -3$ V,

$$g_m = g_{m0}\left(1 - \frac{V_{GS}}{V_{GS(\text{off})}}\right)$$
$$= 4000\ \mu S\left(1 - \frac{-3V}{-8V}\right)$$
$$= 4000\ \mu S\ (0.625)$$
$$= 2500\ \mu S$$

At $V_{GS} = -5$ V,

$$g_m = 4000\ \mu S\left(1 - \frac{-5V}{-8V}\right)$$
$$= 4000\ \mu S\ (0.375)$$
$$= 1500\ \mu S$$

**PRACTICE PROBLEM 12.11**

The 2N5486 has maximum values of $g_{m0} = 8000$ μS and $V_{GS(\text{off})} = -6$ V. Using these values, determine the values of $g_m$ at $V_{GS} = -2$ V and $V_{GS} = -4$ V.

---

On many specification sheets, the value of $g_{m0}$ will be listed as $y_{fs}$ or $g_{fs}$. When no value of maximum transconductance is given, you can approximate the value of $g_{m0}$ as

$$g_{m0} \cong \frac{2I_{DSS}}{V_{GS(\text{off})}} \tag{12.17}$$

While equation (12.17) will not give you an exact value of $g_{m0}$, the value obtained will usually be close enough to serve as a valid approximation.

## Amplifier Voltage Gain

The ac operation of the JFET amplifier is closely related to the value of $g_m$ at the dc value of $V_{GS}$. Consider the circuit shown in Figure 12.35. Figure 12.35a shows the basic common-source JFET amplifier. The circuit shown in Figure 12.35b shows the ac equivalent of the original circuit. Note that the biasing resistors ($R_1$ and $R_2$) have been replaced with a single parallel equivalent resistor, $r_G$. Also, the drain and load resistances ($R_D$ and $R_L$) have been replaced by the single parallel equivalent resistor, $r_D$. This is the standard ac equivalent circuit, as developed in our discussions on BJT amplifiers.

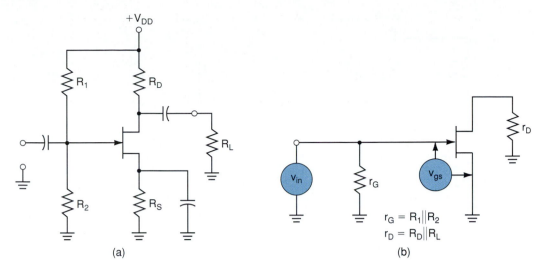

FIGURE 12.35   The common-source amplifier and its ac equivalent circuit.

For Figure 12.35b, the voltage gain is found as

$$A_v = g_m r_D \qquad\qquad (12.18)$$

Equation (12.18) points out another problem with the JFET amplifier. As you can see, the value of $A_v$ depends on the value of $g_m$. You have also been shown the following:

1. The value of $g_m$ depends on the value of $V_{GS}$.
2. The value of $V_{GS}$ can vary from one JFET to another.

This leads to the conclusion that *the voltage gain of an amplifier can vary significantly from one JFET to another.* This point is illustrated in Example 12.12.

---

Determine the maximum and minimum values of $A_v$ for the amplifier shown in Figure 12.36. Assume that the 2N5459 has values of $g_{m0} = 6000\ \mu S$ at $V_{GS(off)} = -8$ V and $g_{m0} = 2000\ \mu S$ at $V_{GS(off)} = -2$ V.

*EXAMPLE 12.12*

*Solution:*   The transconductance curves and dc bias line for the amplifier are shown in Figure 12.36b. As shown, the maximum and minimum values of $V_{GS}$ are $-5$ and $-1$ V, respectively. Using the maximum values of $V_{GS(off)}$ and $V_{GS}$, $g_m$ is found as

$$g_m = g_{m0}\left(1 - \frac{V_{GS}}{V_{GS(off)}}\right)$$

$$= 6000\ \mu S\left(1 - \frac{-5\ V}{-8\ V}\right)$$

$$= 2250\ \mu S$$

Using the minimum values of $V_{GS(off)}$ and $V_{GS}$, $g_m$ is found as

$$g_m = 2000\ \mu S\left(1 - \frac{-1\ V}{-2\ V}\right)$$

$$= 1000\ \mu S$$

The value of $r_D$ is found as

$$r_D = R_D \parallel R_L$$
$$= 7.58\ k\Omega$$

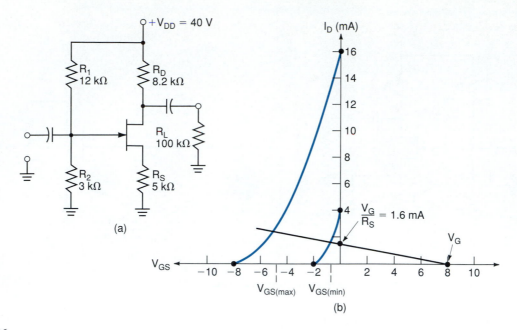

**FIGURE 12.36**

The maximum value of $A_v$ can now be found as

$$A_v = r_D g_{m(max)}$$
$$= 7.58 \text{ k}\Omega(2250 \text{ }\mu\text{S})$$
$$= 17.06$$

The minimum value of $A_v$ is found as

$$A_v = r_D g_{m(min)}$$
$$= 7.58 \text{ k}\Omega(1000 \text{ }\mu\text{S})$$
$$= 7.58$$

### PRACTICE PROBLEM 12.12

The amplifier described in Practice Problem 12.8 has a load resistance of 4.7 k$\Omega$. The 2N5486 has a $g_{m0}$ range of 4000 to 8000 $\mu$S. Using the values of $V_{GS}$ from Practice Problem 12.9, calculate the range of $A_v$ values for the amplifier.

## The Basis for Equation (12.18)

The ac equivalent circuit shown in Figure 12.35b would have an output voltage found as

$$v_{out} = i_d r_D \qquad \text{(12.19)}$$

where $v_{out}$ = the ac output voltage
   $i_d$ = the change in $I_D$ caused by the ac input signal

Now recall that $g_m$ is defined in equation (12.15) as

$$g_m = \frac{\Delta I_D}{\Delta V_{GS}}$$

Another way of writing this equation is

$$g_m = \frac{i_d}{v_{gs}}$$

(12.20)

where $v_{gs}$ is the change in $V_{GS}$ caused by the amplifier input signal, equal to that input signal. Equation (12.20) can be rewritten as

$$i_d = g_m v_{gs}$$

Substituting this equation into equation (12.19) gives us

$$v_{out} = g_m v_{gs} r_D$$

(12.21)

Now recall that voltage gain ($A_v$) is the ratio of output voltage to input voltage. By formula,

$$A_v = \frac{v_{out}}{v_{in}} \qquad (v_{in} = v_{gs})$$

We can rearrange equation (12.21) to obtain a valid voltage-gain formula for the common-source JFET amplifier:

$$\frac{v_{out}}{v_{in}} = g_m r_D$$

or

$$A_v = g_m r_D$$

This is equation (12.18).

## JFET Swamping

Just as we swamped the BJT amplifier to reduce the effects of variations in $r_e'$, we can ◄ *OBJECTIVE 8* swamp the JFET amplifier to reduce the effects of variations in $g_m$. The swamped JFET amplifier closely resembles the swamped BJT amplifier, as shown in Figure 12.37. As you

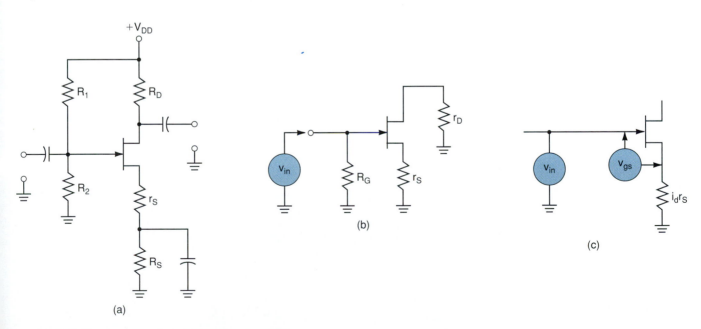

**FIGURE 12.37   A swamped common-source amplifier.**

can see, the source resistance is only partially bypassed, just as the emitter resistance is partially bypassed in the swamped BJT amplifier.

The voltage gain of the swamped JFET amplifier is found as

$$A_v = \frac{r_D}{r_S + (1/g_m)} \qquad (12.22)$$

If the value of $r_S \gg 1/g_m$, the resistor will swamp out the effects of variations in $g_m$. This point is illustrated in Example 12.13.

**EXAMPLE 12.13**

Determine the maximum and minimum gain values for the circuit in Figure 12.38.

**Solution:** Basically, the circuit in Figure 12.38 is the same as the one shown in Figure 12.36. The difference of 100 $\Omega$ in the source circuit is not enough to significantly change the values obtained in Example 12.12, so we will use those values here.

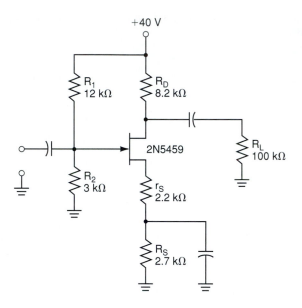

**FIGURE 12.38**

Using the values from Figure 12.38 and Example 12.12, the maximum value of $A_v$ is found as

$$A_v = \frac{r_D}{r_S + (1/g_m)}$$

$$= \frac{7.58 \text{ k}\Omega}{2.2 \text{ k}\Omega + 444 \text{ }\Omega}$$

$$= 2.87$$

The minimum value of $A_v$ can be calculated as

$$A_v = \frac{r_D}{r_S + (1/g_m)}$$

$$= \frac{7.58 \text{ k}\Omega}{2.2 \text{ k}\Omega + 1 \text{ k}\Omega}$$

$$= 2.37$$

As you can see, the stability of $A_v$ is much better for the swamped amplifier than it is for the standard common-source amplifier.

Example 12.13 demonstrates the one drawback to using swamping. If you compare the values of $A_v$ for Examples 12.12 and 12.13, you will notice that you lose quite a bit of gain when you swamp the JFET amplifier. This is the same problem that we had with swamping BJT amplifiers. *Anytime you swamp an amplifier, you improve stability but you lose some voltage gain.* Your value of $A_v$ may be much more stable, but it will also be much lower than it was without the swamping.

## The Basis for Equation (12.22)

The input voltage to the circuit in Figure 12.37b is in parallel with the gate–source circuit. Because of this, $v_{in}$ must equal the sum of $v_{gs}$ and the voltage developed across $r_S$. By formula,

$$v_{in} = v_{gs} + i_d r_S$$

This point is illustrated by Figure 12.37c. Since $i_d = v_{gs} g_m$, the equation above can be rewritten as

$$v_{in} = v_{gs} + v_{gs} g_m r_S$$

or

$$v_{in} = v_{gs}(1 + g_m r_S)$$

The voltage gain of the amplifier is found as

$$A_v = \frac{v_{out}}{v_{in}}$$

$$= \frac{g_m v_{gs} r_D}{v_{gs}(1 + g_m r_S)}$$

Eliminating $v_{gs}$ from the equation above gives us

$$A_v = \frac{g_m r_D}{1 + g_m r_S}$$

Finally, we can multiply the fraction above by 1 in the form of

$$\frac{1/g_m}{1/g_m}$$

to obtain

$$A_v = \frac{r_D}{r_S + (1/g_m)}$$

This is equation 12.22.

## Amplifier Input Impedance

Due to the extremely high input impedance of the JFET, the overall input impedance to a JFET amplifier will be higher than for a similar BJT amplifier. For the *gate-bias* and *self-bias* circuits, the amplifier input impedance is found as    ◄ *OBJECTIVE 9*

$$Z_{in} \cong R_G \qquad \qquad (12.23)$$

For the *voltage-divider biased* amplifier, $Z_{in}$ is found as

$$Z_{in} \cong R_1 \| R_2 \qquad \qquad (12.24)$$

Example 12.14 serves to illustrate the fact that a JFET amplifier will have a higher value of $Z_{in}$ than a similar BJT amplifier.

**EXAMPLE 12.14**

Determine the values of $Z_{in}$ for the amplifiers shown in Figure 12.39.

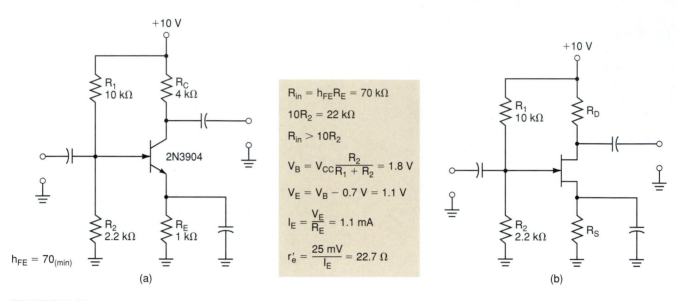

**FIGURE 12.39**

***Solution:*** Figure 12.39 shows the calculations for the BJT amplifier up to the determination of $r_e'$. The value of current gain shown is the minimum spec sheet value for $I_C = 1$ mA. Using the values shown, the input impedance to the BJT amplifier is found as

$$Z_{in} = R_1 \| R_2 \| h_{fe}(r_e')$$
$$= 844.7\ \Omega$$

For the JFET amplifier, the input impedance is found as

$$Z_{in} \cong R_1 \| R_2$$
$$= 1.8\ k\Omega$$

Thus, using the same biasing resistors, the JFET amplifier will have a much higher overall value of $Z_{in}$ than that of the BJT amplifier.

Because of the higher input impedance of the JFET amplifier, the circuit will present less of a load to the circuit driving the amplifier. Thus, in any multistage amplifier, the overall gain of an amplifier will be higher if its load is a JFET amplifier rather than an equivalent BJT amplifier. Example 12.15 helps to illustrate this point.

EXAMPLE *12.15*

The amplifier in Example 12.12 is used to drive the amplifiers described in Example 12.14. Assume that $g_m$ for the driving amplifier is fixed at 2000 $\mu$S. Determine the value of $A_v$ for the driving amplifier when each load amplifier is connected to the output.

*Solution:* When the BJT amplifier is the load, the value of $r_D$ for Figure 12.36 is found as

$$r_D = R_D \parallel Z_{in}$$
$$= 8.2 \text{ k}\Omega \parallel 844.7 \text{ }\Omega$$
$$= 765.8 \text{ }\Omega$$

With this load, the voltage gain of the amplifier is

$$A_v = g_m r_D$$
$$= (2000 \text{ }\mu\text{S})(765.8 \text{ }\Omega)$$
$$= 1.53$$

When the JFET amplifier is the load, the value of $r_D$ for Figure 12.36 is found as

$$r_D = R_D \parallel Z_{in}$$
$$= 8.2 \text{ k}\Omega \parallel 1.8 \text{ k}\Omega$$
$$= 1.476 \text{ k}\Omega$$

With this load, the voltage gain of the amplifier is

$$A_v = g_m r_D$$
$$= (2000 \text{ }\mu\text{S})(1.476 \text{ k}\Omega)$$
$$= 2.952$$

Thus, the gain of the circuit in Figure 12.36 is higher when driving the JFET amplifier than it is when driving the BJT amplifier.

## One Final Note

You may have noticed that the bulk of material in this section was related only to the *voltage-divider biased* JFET amplifier. If you derive the ac equivalent circuits for the gate-bias and self-bias circuits, you will see that these ac equivalent circuits are the same as the voltage-divider bias equivalent circuit. For this reason, these other biasing circuits were not discussed in this section.

In this section, you have been shown that the JFET common-source amplifier has extremely high input impedance when compared to the common-emitter amplifier. You have also been shown that the voltage gain of the JFET amplifier (which tends to be relatively low) changes when the value of transconductance for the JFET changes. By swamping the JFET amplifier, we can reduce the variations in voltage gain that occur when transconductance varies. However, the increased stability in voltage gain is obtained at the cost of even lower voltage gain.

**Section Review**

1. Explain the common-source amplifier circuit response to an ac input signal.
2. What is *transconductance*?
3. What are the units of transconductance?
4. How does the value of $g_m$ relate to the position of a circuit $Q$-point on the transconductance curve?

5. What is the relationship between JFET amplifier voltage gain and the value of $V_{GS}$ for the circuit?

6. What purpose is served by swamping a JFET amplifier?

7. How does the JFET amplifier input impedance compare to that of a comparable BJT amplifier? Explain your answer.

---

## *12.4* ac OPERATING CHARACTERISTICS: COMMON-DRAIN AND COMMON-GATE AMPLIFIERS

*OBJECTIVE 10* ▶ The *common-drain* and *common-gate* configurations are the JFET equivalents for the *common-collector* and *common-base* BJT amplifiers. In this section, we will take a brief look at each of these amplifiers.

### The Common-Drain (Source-Follower) Amplifier

**Source follower (common-drain amplifier)**
The JFET equivalent of the emitter follower.

The **source follower** accepts an input signal at its gate and provides an output signal at its source terminal. The input and output signals for this amplifier are *in phase,* thus the name *source follower.* The basic source follower is shown in Figure 12.40. The characteristics of the source follower are summarized as follows:

| | |
|---|---|
| Input impedance: | High |
| Output impedance: | Low |
| Gain: | $A_v < 1$ |

Because of its input and output impedance characteristics, the source follower is used primarily as an *impedance-matching* circuit; that is, it is used to couple a high-impedance source to a low-impedance load. You may recall from Chapter 10 that the *emitter follower* is also used for this purpose.

The voltage gain ($A_v$) of the source follower is found using the following equation:

$$A_v = \frac{r_S}{r_S + (1/g_m)} \tag{12.25}$$

The voltage gain of the swamped common-source amplifier is found as

$$A_v = \frac{r_D}{r_S + (1/g_m)}$$

where $r_S = R_S \parallel R_L$. The basis for equation (12.25) is easy to see if you compare the source follower in Figure 12.40 with the swamped common-source amplifier shown in Figure 12.37. As you can see, both circuits have some value of unbypassed source resistance. Thus, the voltage gains of the two circuits are calculated in the same basic fashion. The only differences are as follows:

**FIGURE 12.40   The source follower.**

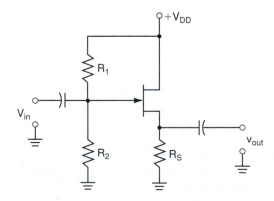

1. The unbypassed source resistance in the source follower is the parallel combination of the source resistor ($R_S$) and the circuit load resistance ($R_L$).

2. The output voltage from the source follower appears across $r_S$. Therefore, $r_S$ is used in place of $r_D$ in the swamped amplifier equation.

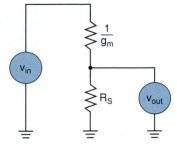

**FIGURE 12.41**

Since $r_S$ appears in both the numerator and the denominator of equation (12.25), it should be obvious that the value of $A_v$ for the source follower will always be less than 1.

The high input impedance of the amplifier is caused, again, by the high $Z_{in}$ of the JFET. As with the common-source amplifier, the total input impedance to the amplifier will equal the parallel equivalent resistance of the input circuit. When voltage-divider bias is used, $Z_{in}$ is found as

$$Z_{in} = R_1 \parallel R_2 \qquad \textbf{(12.26)}$$

When gate bias or self-bias is used,

$$Z_{in} = R_G \qquad \textbf{(12.27)}$$

Again, the high input impedance of the gate eliminates it from any $Z_{in}$ calculations.

The output impedance of the amplifier is determined with a little help from equation (12.25). If you look closely at the formula, you will notice that you essentially have a voltage-divider equation. Based on this, the circuit in Figure 12.41 is drawn. This is a resistive equivalent of the source-follower circuit. To the load, the circuit appears as two resistors connected *in parallel*. Thus, the output impedance of the amplifier is given as

$$Z_{out} = R_S \parallel \frac{1}{g_m} \qquad \textbf{(12.28)}$$

Now, how does equation (12.28) show that the value of $Z_{out}$ is always low? The normal range of $g_m$ is from 1000 μS up. For $g_m = 1000$ μS,

$$\frac{1}{g_m} = \frac{1}{0.001} = 1 \text{ k}\Omega$$

Since we used the lowest typical value of $g_m$ in the calculation above, it is safe to say that $1/g_m$ *is typically lower than* 1 kΩ. Since the total resistance of a parallel circuit must be lower than the lowest individual resistance value, it is safe to say that $Z_{out}$ *will typically be less than* 1 kΩ.

Example 12.16 demonstrates the process of analyzing a source-follower circuit. It also illustrates the fact that the source follower has a high $Z_{in}$, a low $Z_{out}$, and an overall voltage gain of less than unity.

*Don't Forget*: Output impedance is measured with the load disconnected. Thus, $r_S$ is replaced by $R_S$ in Figure 12.41 and equation (12.28).

---

Determine the maximum and minimum values of $A_v$ and $Z_{out}$ for the circuit shown in Figure 12.42. Also, determine the value of $Z_{in}$ for the circuit.

*EXAMPLE 12.16*

*Solution:* Using the dc bias line shown in Figure 12.42b, we determine the minimum and maximum values of $V_{GS}$ to be $-1$ V and $-6$ V, respectively. The 2N5459 has values stated on the spec sheet of $g_{m0} = 6000$ μS at $V_{GS(off)} = -8$ V, and $g_{m0} = 2000$ μS at $V_{GS(off)} = -2$ V. These values are used with our minimum and maximum values of $V_{GS}$ to determine the minimum and maximum values of $g_m$, as follows:

*Reference:* The data sheet for the 2N5486 JFET is shown in Figure 12.51 (page 521).

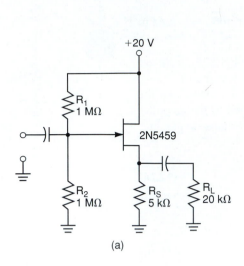

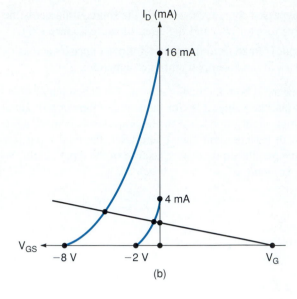

FIGURE 12.42

$$g_{m(\text{max})} = g_{m0}\left(1 - \frac{V_{GS}}{V_{GS(\text{off})}}\right)$$

$$= 6000 \ \mu S \left(1 - \frac{-6V}{-8 \ V}\right)$$

$$= 1500 \ \mu S$$

and

$$g_{m(\text{min})} = 2000 \ \mu S \left(1 - \frac{-1.0 \ V}{-2 \ V}\right)$$

$$= 1000 \ \mu S$$

The value of $r_S$ is now found as

$$r_S = R_S \parallel R_L$$
$$= 5 \ k\Omega \parallel 20 \ k\Omega$$
$$= 4 \ k\Omega$$

Using the maximum value of $g_m$, the value of $A_v$ is found as

$$A_v = \frac{r_S}{r_S + (1/g_m)}$$

$$= \frac{4 \ k\Omega}{4 \ k\Omega - 667 \ \Omega}$$

$$= 0.8571 \qquad (\text{maximum})$$

Using the minimum value of $g_m$, the value of $A_v$ is found as

$$A_v = \frac{r_S}{r_S + (1/g_m)}$$

$$= \frac{4 \ k\Omega}{4 \ k\Omega + 1 \ k\Omega}$$

$$= 0.8 \qquad (\text{minimum})$$

Using the maximum and minimum values of $g_m$, the corresponding values of $Z_{\text{out}}$ are found as

$$Z_{\text{out}} = R_S \parallel \frac{1}{g_m}$$
$$= 5\ \text{k}\Omega \parallel 1\ \text{k}\Omega$$
$$= 833\ \Omega \qquad \text{(maximum)}$$

and

$$Z_{\text{out}} = R_S \parallel \frac{1}{g_m}$$
$$= 5\ \text{k}\Omega \parallel 667\ \Omega$$
$$= 588\ \Omega \qquad \text{(minimum)}$$

Finally, the value of $Z_{\text{in}}$ is found as

$$Z_{\text{in}} = R_1 \parallel R_2$$
$$= 500\ \text{k}\Omega$$

### PRACTICE PROBLEM 12.16

The 2N5486 is used in the amplifier shown in Figure 12.42. Determine the range of $A_v$ and $Z_{\text{out}}$ values for the circuit.

---

Example 12.16 served to show us several things. First, the example demonstrated the fact that source followers have high input impedance and low output impedance values. Even when $Z_{\text{out}}$ was at its maximum value, the ratio of $Z_{\text{in}}$ to $Z_{\text{out}}$ was close to 600:1. The example also demonstrated that there is a relationship among $g_m$, $A_v$, and $Z_{\text{out}}$. When $g_m$ is at its maximum value, $A_v$ is at its maximum value and $Z_{\text{out}}$ is at its minimum value. The reverse also holds true. Based on these observations, we can state the following relationships for the source follower:

1. $A_v$ *is directly proportional to* $g_m$.
2. $Z_{\text{out}}$ *is inversely proportional to* $g_m$.

As always, the value of $Z_{\text{in}}$ is not related to the value of $g_m$.

You have been shown that the source follower is a JFET amplifier that has high input impedance, low output impedance, and a voltage gain that is less than 1. At this point, we will take a look at the common-gate amplifier.

## The Common-Gate Amplifier

The **common-gate amplifier** accepts an input signal at its source terminal and provides an output signal at its drain terminal. The basic common-gate amplifier is shown in Figure 12.43. The characteristics of the common-gate circuit are summarized as follows:

**Common-gate amplifier**
The JFET equivalent of the common-base amplifier.

| | |
|---|---|
| Input impedance: | Low |
| Output impedance: | High (as compared to $Z_{\text{in}}$) |
| Gain: | $A_v > 1$ |

As with the common-base circuit, the common-gate amplifier can be used to couple a low impedance source to a high impedance load. In this respect, the common-gate circuit (like the source follower) is used as an *impedance-matching* circuit. Also, just as with the common-base amplifier, the common-gate amplifier will have a value of $A_v$ greater than 1.

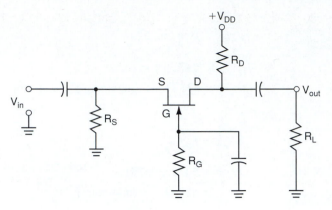

**FIGURE 12.43   The common-gate amplifier.**

Like the common-source amplifier, the common-gate amplifier has a voltage gain equal to the product of its transconductance and ac drain resistance. By formula,

$$A_v = g_m r_D$$

You saw the derivation of this equation in Section 12.3.

The input signal to the common-gate amplifier is applied to the source. Figure 12.41 showed that the impedance of the source is found as $R_S$ in parallel with $1/g_m$. Thus, the input impedance of the common-gate amplifier is found as

$$Z_{in} = R_S \parallel \frac{1}{g_m} \qquad (12.29)$$

As you can see, this is identical to the $Z_{out}$ equation (12.28) for the source follower. This makes sense, as the input impedance of the common-gate amplifier is measured at the same terminal (the source) as the output impedance of the source follower.

Figure 12.44 shows that the output impedance would be made up of the parallel combination of $R_D$ and the resistance of the JFET drain. By formula,

$$Z_{out} = R_D \parallel r_d \qquad (12.30)$$

**Output admittance ($y_{os}$)**
The admittance of the JFET drain, given on the component's spec sheet.

where $r_d$ = the resistance of the JFET drain terminal. The resistance of the JFET drain can be calculated using another JFET parameter: **output admittance, $y_{os}$.** In your study of basic electronics, you were taught that admittance is the reciprocal of impedance. We can therefore determine the value of the JFET drain resistance by taking the reciprocal of $y_{os}$. By formula,

$$r_d = \frac{1}{y_{os}} \qquad (12.31)$$

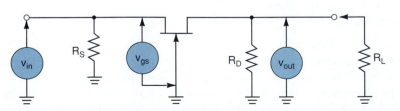

**FIGURE 12.44**

The value of $r_d$ is typically *much greater* than $R_D$. Because of this, the output impedance can normally be determined as

$$Z_{\text{out}} \cong R_D \qquad (12.32)$$

Example 12.17 illustrates this point.

**EXAMPLE 12.17**

Determine the value of $r_d$ for the JFET in Figure 12.45. Also, determine the exact and approximated values of $Z_{\text{out}}$.

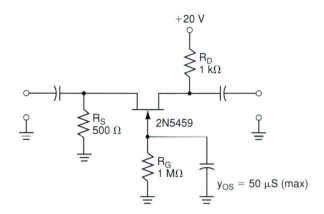

**FIGURE 12.45**

**Solution:** The spec sheet for the 2N5459 lists a maximum value of $y_{os} = 50$ μS. Using this value, the *minimum* value of $r_d$ is found as

$$r_d = \frac{1}{y_{os}}$$

$$= \frac{1}{50 \text{ μS}}$$

$$= 20 \text{ k}\Omega$$

Using this value of $r_d$, the exact value of $Z_{\text{out}}$ is found as

$$Z_{\text{out}} = R_D \parallel r_d$$
$$= 1 \text{ k}\Omega \parallel 20 \text{ k}\Omega$$
$$= 952 \text{ }\Omega$$

The approximated value of $Z_{\text{out}}$ is found as

$$Z_{\text{out}} = R_D$$
$$= 1 \text{ k}\Omega$$

In Example 12.17, the *maximum* value of $y_{os}$ was used to find the *minimum* value of $r_d$. This was done to illustrate a point. When the minimum value of $r_d$ was used, the "exact" value of $Z_{\text{out}}$ was still within 10 percent of the approximated value. If you were to use any higher value of $r_d$, the exact and approximated values of $Z_{\text{out}}$ would be even closer to each other. For example, the spec sheet for the 2N5459 lists a *typical* value of $y_{os} = 10$ μS. Using this value, $r_d = 100$ k$\Omega$. Using this value for $r_d$, $Z_{\text{out}}$ is calculated as

FIGURE 12.46

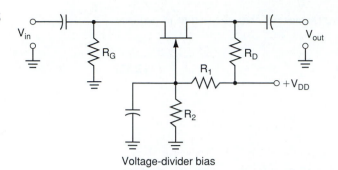

Voltage-divider bias

990 Ω. This is even closer to the approximated value of $Z_{out}$ than the value obtained in Example 12.17. We can therefore assume that equation (12.32) is valid in all cases.

In our discussion of the common-gate circuit, we have used only self-bias circuits. This is certainly not the only type of biasing used with common-gate circuits. Figure 12.46 shows how voltage-divider biasing can be used in common-gate applications.

At this point, we have covered the most basic JFET circuits. Other circuits that use JFETs will appear off and on throughout the remainder of this book and will be discussed at the appropriate points.

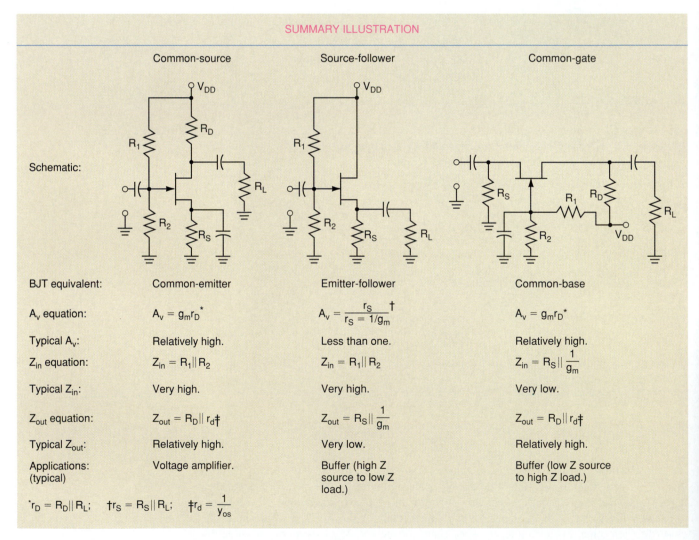

SUMMARY ILLUSTRATION

|  | Common-source | Source-follower | Common-gate |
|---|---|---|---|
| Schematic: | | | |
| BJT equivalent: | Common-emitter | Emitter-follower | Common-base |
| $A_v$ equation: | $A_v = g_m r_D$* | $A_v = \dfrac{r_S}{r_S = 1/g_m}$† | $A_v = g_m r_D$* |
| Typical $A_v$: | Relatively high. | Less than one. | Relatively high. |
| $Z_{in}$ equation: | $Z_{in} = R_1 \| R_2$ | $Z_{in} = R_1 \| R_2$ | $Z_{in} = R_S \| \dfrac{1}{g_m}$ |
| Typical $Z_{in}$: | Very high. | Very high. | Very low. |
| $Z_{out}$ equation: | $Z_{out} = R_D \| r_d$‡ | $Z_{out} = R_S \| \dfrac{1}{g_m}$ | $Z_{out} = R_D \| r_d$‡ |
| Typical $Z_{out}$: | Relatively high. | Very low. | Relatively high. |
| Applications: (typical) | Voltage amplifier. | Buffer (high Z source to low Z load.) | Buffer (low Z source to high Z load.) |

*$r_D = R_D \| R_L$;  †$r_S = R_S \| R_L$;  ‡$r_d = \dfrac{1}{y_{os}}$

FIGURE 12.47

## Summary

You have now been shown the gain and impedance characteristics of the common-source, source-follower, and common-gate amplifiers. These three amplifiers and their characteristics are summarized in Figure 12.47.

*Section Review*

1. What are the gain and impedance characteristics of the source follower?
2. List, in order, the steps taken to determine the range of $A_v$ values for a source follower.
3. What are the gain and impedance characteristics of the common-gate amplifier?
4. List, in order, the steps taken to determine the exact and approximated values of $Z_{out}$.

## 12.5 TROUBLESHOOTING JFET CIRCUITS

One of the major advantages of troubleshooting circuits that use JFETs is the fact that JFET faults are easy to diagnose. This is due to the fact that the JFET has only one junction that can go bad, but shorts and opens can result between the three terminals. As you will be shown, JFETs can be treated without removing them from their circuits. You will also see that the procedures used for troubleshooting JFET circuits are the same as those used for troubleshooting BJT amplifiers.

### JFET Faults

JFET faults are about as difficult to detect as a flat tire, because there is only *one* junction that can develop either a short or an open. To make things even easier, the shorted-junction and open-junction JFETs have distinct symptoms that can be observed in the circuit. Therefore, a minimum of effort is required to determine whether or not the JFET is the cause of an amplifier failure.

◄ *OBJECTIVE 11*

Figure 12.48a shows the normal operating characteristics in a voltage-divider biased JFET amplifier. Note that the circuit has the following operating characteristics:

Normal JFET operating characteristics.

1. The current through $R_1$ is equal to the current through $R_2$.
2. The source and drain terminal currents are equal.
3. There is no current through the gate terminal of the device.

The *shorted-junction* JFET will not have the same current characteristics. This is shown in Figure 12.48b. Since the gate–source junction is shorted, $V_{GS}$ will be approximately 0 V. As a result, $I_D \cong I_{DSS}$. Since there is some amount of gate current (as shown in the figure), $I_S$ and $I_D$ will not be equal, as is usually the case. The gate current also affects the biasing circuit, so $I_1 \neq I_2$. These are the classic symptoms of a shorted-gate. In summary:

Shorted-gate symptoms.

1. $I_1 \neq I_2$
2. $I_S \cong I_{DSS}$
3. $I_S \neq I_D$
4. $I_G > 0$

The *open-junction* JFET can be detected by observing two simple conditions:

Open-gate symptoms.

1. $V_{GS} \neq 0$ V
2. $I_D \cong I_{DSS}$

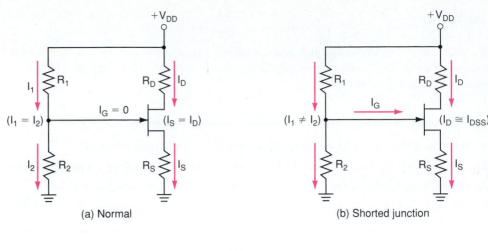

(a) Normal

(b) Shorted junction

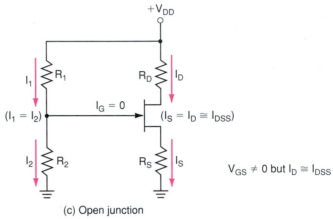

(c) Open junction

**FIGURE 12.48    JFET fault symptoms.**

You see, with the gate junction open, the depletion layer that normally restricts $I_D$ cannot form. At the same time, $V_{GS}$ will be at some measureable value. If you have a measurable value of $V_{GS}$ and $I_D$ is approximately equal to $I_{DSS}$, the gate–source junction of the JFET is open. The circuit conditions that relate to this fault are shown in Figure 12.48c.

## JFET Circuit Troubleshooting

As with any other amplifier, you have to start the troubleshooting process by making sure that the amplifier is the source of the trouble. This point cannot be stressed often enough. When you find a JFET amplifier with no output, check to make sure that the amplifier has a good input signal, and check the $V_{DD}$ and ground connections to the circuit. If all these tests have the proper results, you can assume that the amplifier is faulty.

We will restrict our discussion of amplifier faults to *open* components. As you know, most resistors do not simply develop a *shorted* condition. If a resistor is shorted, the short is usually caused by another component or a solder bridge. A visual inspection will eliminate these possibilities.

For the voltage-divider biased circuit, the common faults and their symptoms are listed in Table 12.1. Refer to Figure 12.48a to identify components.

Of the problems listed, $R_1$ going open is the most difficult to detect. If $R_1$ opens, it is like removing the component from the circuit. If you do that, you simply have a self-bias circuit. Because of this, $R_1$ could open and the unit containing the amplifier could take a long time to fail. The main effect of $R_1$ opening on circuit operation is a shift of

**TABLE 12.1  JFET Troubleshooting**

| Fault | Symptoms |
|-------|----------|
| $R_1$ open | $V_{GS}$ increases; $I_2 = 0$; $V_D$ increases; the circuit acts as a self-bias circuit. |
| $R_2$ open | JFET is destroyed by forward bias; if the JFET is replaced, the new component is also destroyed; $V_2 = V_{DD}$. |
| $R_D$ open | $V_D = 0$ V; all terminal currents drop to 0. |
| $R_S$ open | $V_S = V_{DD}$; $V_{RD} = 0$ V. |

the $Q$-point. If the amplifier develops a problem with output signal clipping, check the voltage across $R_1$. If this voltage is equal to $V_{DD}$, the component is open.

If $R_2$ opens, the gate circuit consists of $R_1$ and $V_{DD}$. At the instant $R_2$ opens, the current through $R_1$ will drop to zero, dropping the voltage across the component to zero. This places $V_{DD}$ at the gate of the JFET, which will forward bias and destroy the component. If you replace the JFET, the same thing will happen to the new component.

When $R_D$ opens, it effectively removes $V_{DD}$ from the drain–source circuit. All the supply voltage will be dropped across the open component, so the voltage from drain to ground will be 0 V. Also, the lack of supply voltage will cause $I_D$ and $I_S$ to drop to zero. At this point, the JFET may become forward biased by $V_G$, destroying the component. If you test a circuit and the value of $V_D$ indicates that $R_D$ is open, be prepared to replace the JFET.

If $R_S$ opens, source current will drop to zero. Since the resistor is open, the supply voltage is dropped across the resistor, $R_S$. With $V_{DD}$ dropped across $R_S$, $V_{GS}$ will be abnormally high, probably higher than $V_{GS(off)}$. The voltages across the gate–biasing resistors will be normal.

Now let's put all this into perspective by going through a couple of troubleshooting applications.

# APPLICATION 1

The circuit shown in Figure 12.49 has no output signal. After verifying that all inputs were valid, the input was disconnected from the circuit. Then the dc voltages shown were measured. The problem: $R_2$ is open. The first indication that $R_2$ is open is the fact that $V_{DD}$ can be measured across this component. But where did the other readings come from? Is there another problem in the circuit? Yes. When $R_2$ opened, it caused the JFET junction to become forward biased. This, in turn, caused the junction of the JFET to open.

When the function of the JFET opened, there was no depletion region to restrict the current through the channel. Thus, the current in the source–drain circuit increased as much as $R_S$, $R_D$, and the resistance of the channel would allow, in this case to about 1.62 mA. This value of $I_D$ caused 11 V to be developed across $R_S$. The difference between $V_S$ and $V_G$ is now approximately 9 V, as shown. *Note the polarity of $V_{GS}$. Normally, $V_{GS}$ is a negative voltage. The positive value of $V_{GS}$ is a good indication that the junction has opened.* For this circuit, both $R_2$ and the JFET would have to be replaced.

# APPLICATION 2

The circuit in Figure 12.50 has been verified as being faulty. After disconnecting the input, the voltages shown were measured. This problem is a simple one. The fact that $V_{DD}$ is measured across the source resistor indicates that $R_S$ is open. Note that the lack of $I_D$ causes the reading shown across the drain resistor. Also, the gate circuit is not affected by the fault, so the voltage readings across $R_1$ and $R_2$ are normal.

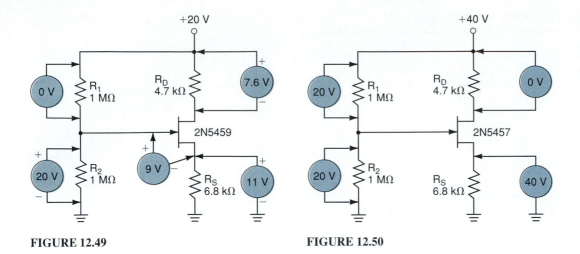

**FIGURE 12.49**

**FIGURE 12.50**

Again, we have not covered every problem that could possibly develop with a JFET amplifier. However, using a solid understanding of the component and a little common sense, you should have no difficulty in diagnosing JFET amplifier faults.

## Section Review

1. What is the primary difference between troubleshooting JFET amplifiers and BJT amplifiers?
2. List the common symptoms of a shorted-gate junction.
3. List the common symptoms of an open-gate junction.
4. Describe the procedure used to troubleshoot a JFET amplifier.

# 12.6 JFET SPECIFICATION SHEETS AND APPLICATIONS

Up to this point, we have based our calculations on several JFET parameters. Now it is time to take a look at the common JFET parameters, what they mean, and when they are used. In this section, we will also take a look at some practical JFET applications.

## JFET Specification Sheets

OBJECTIVE 12 ▶ We have used the parameters of the 2N5486 in most of the example practice problems. In this section, we will use the spec sheet for this device. The 2N5485 through 2N5486 JFET spec sheet is shown in Figure 12.51.

## Maximum Ratings

Most of the maximum ratings for these devices are self-explanatory at this point. In our BJT spec sheet discussions, we covered all the commonly used maximum voltage, current, and power ratings. The JFET maximum ratings are no different.

One maximum rating may surprise you, the **gate current**. In the section of maximum ratings in Figure 12.51, you will see the following:

**Gate current ($I_G$)**
The maximum amount of current that can be drawn through the JFET gate without damaging the device.

| Rating | Symbol | Value | Unit |
|--------|--------|-------|------|
| Gate current | $I_{G(f)}$ | 10 | mAdc |

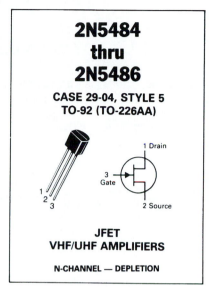

**2N5484
thru
2N5486**

**CASE 29-04, STYLE 5
TO-92 (TO-226AA)**

1 Drain

3
Gate

2 Source

**JFET
VHF/UHF AMPLIFIERS**

**N-CHANNEL — DEPLETION**

Refer to 2N4416 for graphs.

## MAXIMUM RATINGS

| Rating | Symbol | Value | Unit |
|--------|--------|-------|------|
| Drain-Gate Voltage | $V_{DG}$ | 25 | Vdc |
| Reverse Gate-Source Voltage | $V_{GSR}$ | 25 | Vdc |
| Drain Current | $I_D$ | 30 | mAdc |
| Forward Gate Current | $I_{G(f)}$ | 10 | mAdc |
| Total Device Dissipation @ $T_C$ = 25°C<br>Derate above 25°C | $P_D$ | 310<br>2.82 | mW<br>mW/°C |
| Operating and Storage Junction Temperature Range | $T_J$, $T_{stg}$ | −65 to +150 | °C |

## ELECTRICAL CHARACTERISTICS ($T_A$ = 25°C unless otherwise noted.)

| Characteristic | | Symbol | Min | Typ | Max | Unit |
|----------------|--|--------|-----|-----|-----|------|
| **OFF CHARACTERISTICS** | | | | | | |
| Gate-Source Breakdown Voltage<br>($I_G$ = −1.0 μAdc, $V_{DS}$ = 0) | | $V_{(BR)GSS}$ | −25 | — | — | Vdc |
| Gate Reverse Current<br>($V_{GS}$ = −20 Vdc, $V_{DS}$ = 0)<br>($V_{GS}$ = −20 Vdc, $V_{DS}$ − 0, $T_A$ − 100°C) | | $I_{GSS}$ | —<br>— | —<br>— | −1.0<br>−0.2 | nAdc<br>μAdc |
| Gate Source Cutoff Voltage<br>($V_{DS}$ = 15 Vdc, $I_D$ = 10 nAdc) | 2N5484<br>2N5485<br>2N5486 | $V_{GS(off)}$ | −0.3<br>−0.5<br>−2.0 | —<br>—<br>— | −3.0<br>−4.0<br>−6.0 | Vdc |
| **ON CHARACTERISTICS** | | | | | | |
| Zero-Gate-Voltage Drain Current<br>($V_{DS}$ = 15 Vdc, $V_{GS}$ = 0) | 2N5484<br>2N5485<br>2N5486 | $I_{DSS}$ | 1.0<br>4.0<br>8.0 | —<br>—<br>— | 5.0<br>10<br>20 | mAdc |
| **SMALL-SIGNAL CHARACTERISTICS** | | | | | | |
| Forward Transfer Admittance<br>($V_{DS}$ = 15 Vdc, $V_{GS}$ = 0, f = 1.0 kHz) | 2N5484<br>2N5485<br>2N5486 | $|y_{fs}|$ | 3000<br>3500<br>4000 | —<br>—<br>— | 6000<br>7000<br>8000 | μmhos |
| Input Admittance<br>($V_{DS}$ = 15 Vdc, $V_{GS}$ = 0, f = 100 MHz)<br>($V_{DS}$ = 15 Vdc, $V_{GS}$ = 0, f = 400 MHz) | 2N5484<br>2N5485, 2N5486 | Re($y_{is}$) | —<br>— | —<br>— | 100<br>1000 | μmhos |
| Output Admittance<br>($V_{DS}$ = 15 Vdc, $V_{GS}$ = 0, f = 1.0 kHz) | 2N5484<br>2N5485<br>2N5486 | $|y_{os}|$ | —<br>—<br>— | —<br>—<br>— | 50<br>60<br>75 | μmhos |
| Output Conductance<br>($V_{DS}$ = 15 Vdc, $V_{GS}$ = 0, f = 100 MHz)<br>($V_{DS}$ = 15 Vdc, $V_{GS}$ = 0, f = 400 MHz) | 2N5484<br>2N5485, 2N5486 | Re($y_{os}$) | —<br>— | —<br>— | 75<br>100 | μmhos |
| Forward Transconductance<br>($V_{DS}$ = 15 Vdc, $V_{GS}$ = 0, f = 100 MHz)<br><br>($V_{DS}$ = 15 Vdc, $V_{GS}$ = 0, f = 400 MHz) | 2N5484<br><br>2N5485<br>2N5486 | Re($y_{fs}$) | 2500<br><br>3000<br>3500 | —<br><br>—<br>— | —<br><br>—<br>— | μmhos |

**FIGURE 12.51   (Copyright of Motorola. Used by permission.)**

This rating indicates that the gate can handle a maximum forward current of 10 mA. Does this mean that the device could be operated in the forward region? No. The control of drain current depends on the amount of *reverse* bias on the JFET gate. The maximum drain current ($I_{DSS}$) is reached when the reverse bias reaches 0 V. Any increase in $V_{GS}$ above 0 V will not cause an increase in $I_D$. However, should the gate *accidentally* become forward biased, $I_G$ must be greater than 10 mA$_{dc}$ for the device to be destroyed. As long as $I_G < 10$ mA$_{dc}$, the device will be safe.

## Off Characteristics

**Gate–source breakdown voltage $V_{(BR)GSS}$**
The value of $V_{GS}$ that will cause the gate–source junction to break down.

**Gate reverse current ($I_{GSS}$)**
The maximum amount of gate current that will occur when the gate–source junction is reverse biased.

The **gate–source breakdown voltage** defines the limit on $V_{GS}$. If $V_{GS}$ is allowed to exceed this value, the junction will break down and the JFET will have to be replaced. For the 2N5484–6 series JFETs, the gate–source breakdown voltage is shown to be $-25$ V.

The **gate reverse current** rating indicates the maximum value of gate current when the gate–source junction is reverse biased. For the 2N5484–6 series JFETs, this rating is 1 nA$_{dc}$ when the ambient temperature ($T_A$) is 25°C. Note that the negative current value is used to indicate the direction of the gate current. $I_{GSS}$ is a *temperature-dependent* rating. As the spec sheet shows, the value of $I_{GSS}$ increases as temperature increases. The relationship between junction reverse current and temperature was first discussed in Chapter 2.

Since we have already discussed $V_{GS(off)}$ and $I_{DSS}$ (listed under "On Characteristics") in detail, we will not discuss them further here. Note the wide ranges of $V_{GS(off)}$ and $I_{DSS}$ values within the 2N5484–6 series. This is typical for most JFET groups.

## Small-Signal Characteristics

*Note*: The discussion here on conductance, susceptance, and admittance is very brief and is intended only as a review. You can review these principles further in the appropriate sections of your basic electronics text.

At first glance, this section of the spec sheet can be somewhat confusing. It seems that we have several $y_{fs}$ and $y_{os}$ parameters listed. Which do we use? To answer this question, we have to briefly review some basic ac principles.

*Conductance* is *the reciprocal of resistance. Susceptance* is *the reciprocal of reactance.* Combined, they make up *admittance,* which is *the reciprocal of impedance* (the combination of resistance and reactance). While resistance, reactance, and impedance are all measures of *opposition to current,* conductance, susceptance, and admittance are all measures of *the relative ease with which current will pass through a component.*

The *forward transfer admittance, input admittance,* and *output admittance* ratings all take the effects of susceptance into account. You see, the JFET has measurable input and output *capacitances.* These capacitances all have some amount of reactance, and thus some amount of susceptance. The admittance ratings include these values of susceptance. The *output conductance* and *forward transconductance* ratings, on the other hand, take only the component resistance into account.

The bottom line is this: Admittance values are based on both component resistance and component reactance. Conductance values are based solely on component resistance values. The question now is, Which ratings should be used for circuit analysis?

If you look closely at the 2N5484–6 series spec sheet, you'll see that the following ratings were measured in the 100- to 400-MHz frequency range:

- Input admittance
- Output conductance
- Forward transconductance

If you are analyzing a circuit that is operated at or near this frequency range, you should use these ratings in your analysis.

The following ratings were all measured at a frequency of approximately 1 kHz:

- Forward transfer admittance
- Output admittance

If you are analyzing a circuit that is operated at or near this frequency, you should use these ratings in your analysis.

The problem we have encountered in this section of the JFET spec sheet brings up a very interesting point. On many spec sheets, you will find multiple ratings that use the same unit of measure. When this occurs, you will always find that one or more of the conditions under which the ratings were measured differ between the two ratings. (In this case, the frequency of operation differed among the various ratings.) When this occurs, you should check to see which ratings were measured under the conditions that most closely resemble those of the circuit you are analyzing. Then use those ratings in your circuit analysis. For example, if we were analyzing a circuit that has an operating frequency of 10 kHz, we should use the *forward transfer admittance* rating for our $g_{m0}$ values and the *output admittance* to find the value of $r_d$. If we were analyzing a circuit that has an operating frequency of 70 MHz, we would use the *forward transconductance* rating to obtain our $g_{m0}$ value and the *output conductance* rating to obtain our value of $r_d$.

## JFET Capacitance Ratings

JFETs have three capacitance ratings that affect the high-frequency operation of the components. These ratings are discussed in detail in Chapter 14. At this point, we will briefly discuss several JFET circuit applications.

## JFET Applications

The high input impedance of the JFET amplifier makes it ideal for use in situations where source loading is a critical consideration. For example, consider the circuit shown in Figure 12.52. The JFET amplifier is used as a *buffer* between the low $Z_{out}$ source and the load. Because of the high input impedance of the JFET amplifier (5 M$\Omega$), the amplifier presents virtually no load to the source amplifier. This has the effect of increasing the gain

◄ *OBJECTIVE 13*

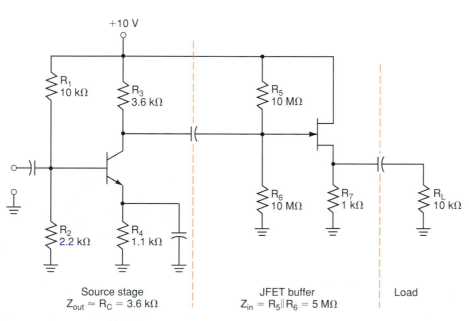

FIGURE 12.52   The JFET buffer.

of the source stage. Recall that the gain of a BJT common-emitter amplifier is found as

$$A_v = \frac{R_C \| R_L}{r_e'}$$

If the load resistor had been connected directly to the source stage, the gain of the source stage would have been found as

$$A_v = \frac{2.65 \text{ k}\Omega}{r_e'}$$

where 2.65 k$\Omega$ is the parallel combination of $R_C$ and $R_L$. With the JFET buffer, the gain of the first stage is found as

$$A_v = \frac{3.59 \text{ k}\Omega}{r_e'}$$

Regardless of the value of $r_e'$ for the first stage, the gain of that stage is increased due to the use of the buffer amplifier. The low $Z_{out}$ of the buffer would also aid in transferring power to the load. Thus, the overall circuit works much more efficiently with the buffer amplifier than without it.

OBJECTIVE 14 ▶ The high input impedance of the JFET also makes it useful in voltmeters. The FET voltmeter provides more reliable voltage readings than a standard VOM because the input to the meter is a high-impedance circuit that does not load the circuit under test. The input to an FET meter is represented (rather loosely) in Figure 12.53. Figure 12.53a is used only to show you that the input to an FET meter is applied to the gate of a reverse-biased JFET. While the actual circuitry is more complicated, the diagram illustrates the fact that the FET meter has a very high input impedance. The effect of this high input impedance on circuit measurements is illustrated in Figure 12.53b. Note that the VOM has an input impedance (typically) of 20 k$\Omega$/V. Thus, when the meter is on the 10-V scale, the input impedance is 20 k$\Omega$ × 10 V = 200 k$\Omega$. When the 200-k$\Omega$ input impedance of the meter is across $R_2$, the parallel combination of resistances reduces the voltage across the resistor. As shown, the meter gives a reading of 4.44 V. The FET meter, on the other hand, provides no loading on the circuit. Thus, the more accurate reading of 5 V is obtained.

In communications electronics, the JFET is used for a variety of functions. One of these is as the active component in an **RF amplifier** (radio-frequency amplifier). The RF amplifier is the first circuit to which a receiver input signal is applied. A JFET RF amplifier is shown in Figure 12.54. The gate and drain circuits of the amplifier are *tuned* circuits, meaning that they are designed to operate on and around specified frequencies. We will cover tuned amplifiers later in the text. The advantage of using the JFET in this cir-

**RF amplifier**
Radio-frequency amplifier; the input circuit of a radio receiver.

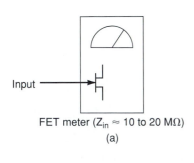

FET meter ($Z_{in} \approx 10$ to $20$ M$\Omega$)

(a)

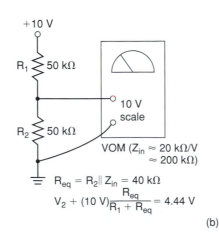

$R_{eq} = R_2 \| Z_{in} = 40$ k$\Omega$

$V_2 + (10 \text{ V})\dfrac{R_{eq}}{R_1 + R_{eq}} = 4.44$ V

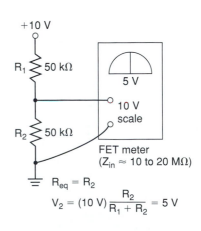

$R_{eq} = R_2$

$V_2 = (10 \text{ V})\dfrac{R_2}{R_1 + R_2} = 5$ V

(b)

**FIGURE 12.53   The FET meter.**

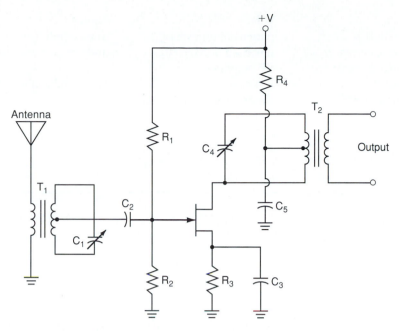

**FIGURE 12.54    A JFET RF amplifier.**

cuit is the fact that JFETs are *low-noise* components. **Noise** is any unwanted interference with a transmitted signal. Noise can be generated by either the transmitter or receiver or by factors outside the communications system. The JFET will not generate any significant amount of noise and thus is useful as an RF amplifier. Another advantage of using the JFET RF amplifier is the fact that it requires no input current to operate. The antenna of the receiver receives a very weak signal that has an extremely low amount of current. Since the JFET is a voltage-controlled component, it is well suited to respond to the low current signal provided by the antenna.

There are many other applications for JFET amplifiers. In this section, we have looked at only a few of them. The main thing to remember is that a JFET amplifier is used to serve the same basic purpose as a BJT amplifier. In applications where it is advantageous to have a high $Z_{in}$ amplifier, the JFET amplifier is preferred over the BJT amplifier.

**Noise**
Any unwanted interference with a transmitted signal.

*Section Review*

1. Define each of the following JFET ratings: gate current, gate–source breakdown voltage, gate reverse current, gate–source cutoff voltage, and zero gate voltage drain current.
2. What is the advantage of using a JFET buffer over a BJT buffer?
3. Why are FET meter voltage readings more accurate than those made by other VOMs?
4. What is an *RF amplifier*?
5. What advantages does the JFET RF amplifier have over the BJT RF amplifier?

The following terms were introduced and defined in this chapter:

*Key Terms*

| | | |
|---|---|---|
| channel | common-source amplifier | field-effect transistor |
| common-drain amplifier | current-source bias | (FET) |
| common-gate amplifier | drain | gate |

gate bias
gate current ($I_G$)
gate reverse current ($I_{GSS}$)
gate–source breakdown
    voltage ($V_{(BR)GSS}$)
gate–source cutoff
    voltage, $V_{GS(off)}$

noise
output admittance ($y_{os}$)
pinch-off voltage ($V_P$)
RF amplifier
self-bias
shorted-gate drain current
    ($I_{DSS}$)

source
source follower
transconductance ($g_m$, $g_{fs}$,
    or $y_{fs}$)
transconductance curve

| Equation Summary | Equation Number | Equation | Section Number |
|---|---|---|---|
| | **(12.1)** | $I_D = I_{DSS}\left(1 - \dfrac{V_{GS}}{V_{GS(off)}}\right)^2$ | 12.1 |
| | **(12.2)** | $V_{GS} = V_{GG}$ | 12.2 |
| | **(12.3)** | $V_{DS} = V_{DD} - I_D R_D$ | 12.2 |
| | **(12.4)** | $I_S = I_D$ | 12.2 |
| | **(12.5)** | $V_S = I_D R_S$ | 12.2 |
| | **(12.6)** | $V_G = 0 \text{ V}$ | 12.2 |
| | **(12.7)** | $V_{GS} = V_G - V_S$ | 12.2 |
| | **(12.8)** | $V_{GS} = -I_D R_S$ | 12.2 |
| | **(12.9)** | $I_D = \dfrac{-V_{GS}}{R_S}$ | 12.2 |
| | **(12.10)** | $V_{DS} = V_{DD} - I_D(R_D + R_S)$ | 12.2 |
| | **(12.11)** | $V_G = V_{DD}\dfrac{R_2}{R_1 + R_2}$ | 12.2 |
| | **(12.12)** | $I_D = \dfrac{V_G - V_{GS}}{R_S}$ | 12.2 |
| | **(12.13)** | $I_D = \dfrac{V_G}{R_S}$ | 12.2 |
| | **(12.14)** | $I_D = I_C$ | 12.2 |
| | **(12.15)** | $g_m = \dfrac{\Delta I_D}{\Delta V_{GS}}$ | 12.3 |
| | **(12.16)** | $g_m = g_{m0}\left(1 - \dfrac{V_{GS}}{V_{GS(off)}}\right)$ | 12.3 |
| | **(12.17)** | $g_{m0} \cong \dfrac{2I_{DSS}}{V_{GS(off)}}$ | 12.3 |
| | **(12.18)** | $A_v = g_m r_D$ | 12.3 |
| | **(12.19)** | $v_{out} = i_d r_D$ | 12.3 |

| Equation Number | Equation | Section Number |
|---|---|---|
| **(12.20)** | $g_m = \dfrac{i_d}{v_{gs}}$ | 12.3 |
| **(12.21)** | $v_{out} = g_m v_{gs} r_D$ | 12.3 |
| **(12.22)** | $A_v = \dfrac{r_D}{r_S + (1/g_m)}$ | 12.3 |
| **(12.23)** | $Z_{in} \cong R_G$ | 12.3 |
| **(12.24)** | $Z_{in} \cong R_1 \parallel R_2$ | 12.3 |
| **(12.25)** | $A_v = \dfrac{r_S}{r_S + (1/g_m)}$ | 12.4 |
| **(12.26)** | $Z_{in} = R_1 \parallel R_2$ | 12.4 |
| **(12.27)** | $Z_{in} = R_G$ | 12.4 |
| **(12.28)** | $Z_{out} = R_S \parallel \dfrac{1}{g_m}$ | 12.4 |
| **(12.29)** | $Z_{in} = R_S \parallel \dfrac{1}{g_m}$ | 12.4 |
| **(12.30)** | $Z_{out} = R_D \parallel r_d$ | 12.4 |
| **(12.31)** | $r_d = \dfrac{1}{y_{os}}$ | 12.4 |
| **(12.32)** | $Z_{out} \cong R_D$ | 12.4 |

**12.1.** 3 mA
**12.2.** See Figure 12.55.
**12.3.** See Figure 12.56.
**12.4.** $V_{GS} = -5$ V, $I_D = 3$ mA, $V_{DS} = 3.4$ V
**12.6.** See Figure 12.57.

*Answers to the Example Practice Problems*

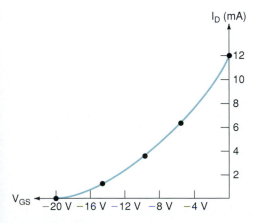

**FIGURE 12.55**

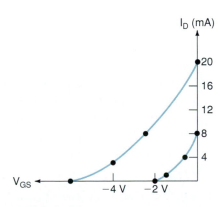

**FIGURE 12.56**

ANSWERS TO THE EXAMPLE PRACTICE PROBLEMS

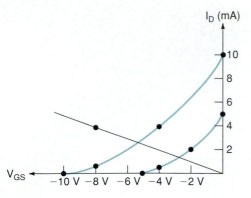

**FIGURE 12.57**

**FIGURE 12.58**

**12.7.** 1.5 to 5.25 V

**12.8.** See Figure 12.58.

**12.10.** 16.8 V (minimum), 20.4 V (maximum)

**12.11.** $g_m = 5333$ μS at $V_{GS} = -2$ V; $g_m = 2667$ μS at $V_{GS} = -4$ V

**12.12.** $A_v = 3.25$ to 5.2

**12.13.** $A_v = 3.3$ to 3.77

**12.16.** $A_v = 0.898$ to 0.9231, $Z_{out} = 313$ to 417 Ω

## *Practice Problems*

### §12.1

1. A JFET amplifier has values of $I_{DSS} = 8$ mA, $V_{GS(off)} = -12$ V, and $V_{GS} = -6$ V. Determine the value of $I_D$ for the circuit.

2. A JFET amplifier has values of $I_{DSS} = 16$ mA, $V_{GS(off)} = -5$ V, and $V_{GS} = -4$ V. Determine the value of $I_D$ for the amplifier.

3. A JFET amplifier has values of $I_{DSS} = 14$ mA, $V_{GS(off)} = -6$ V, and $V_{GS} = 0$ V. Determine the value of $I_D$ for the circuit.

4. A JFET amplifier has values of $I_{DSS} = 10$ mA, $V_{GS(off)} = -8$ V, and $V_{GS} = -5$ V. Determine the value of $I_D$ for the circuit.

5. A JFET amplifier has values of $I_{DSS} = 12$ mA, $V_{GS(off)} = -10$ V, and $V_{GS} = -4$ V. Determine the value of $I_D$ for the circuit.

6. The JFET amplifier described in Problem 5 has a value of $V_{GS} = -14$ V. Determine the value of $I_D$ for the circuit.

7. The 2N5484 has values of $V_{GS(off)} = -0.3$ to $-3$ V and $I_{DSS} = 1.0$ to 5.0 mA. Plot the minimum and maximum transconductance curves for the device.

8. The 2N5485 has values of $V_{GS(off)} = -0.5$ to $-4.0$ V and $I_{DSS} = 4.0$ to 10 mA. Plot the minimum and maximum transconductance curves for the device.

9. The 2N5457 has values of $V_{GS(off)} = -0.5$ to $-6.0$ V and $I_{DSS} = 1.0$ to 5.0 mA. Plot the minimum and maximum transconductance curves for the device.

10. The 2N5458 has values of $V_{GS(off)} = -1.0$ to $-7.0$ V and $I_{DSS} = 2.0$ to 9.0 mA. Plot the minimum and maximum transconductance curves for the device.

11. The 2N3437 has parameters of $V_{GS(off)} = -5.0$ V (maximum) and $I_{DSS} = 0.8$ to 4.0 mA. Plot the maximum transconductance curve for the device.

12. The 2N3438 has parameters of $V_{GS(off)} = -2.5$ V (maximum) and $I_{DSS} = 0.2$ to 1.0 mA. Plot the maximum transconductance curve for the device.

### §12.2

13. Determine the values of $I_D$ and $V_{DS}$ for the amplifier in Figure 12.59a.

14. Determine the values of $I_D$ and $V_{DS}$ for the amplifier in Figure 12.59b.

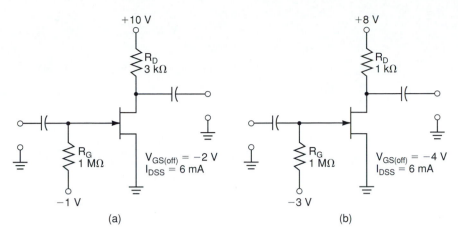

**FIGURE 12.59**

15. The 2N5486 (described in Section 12.1, Practice Problem 12.3) is used in the circuit in Figure 12.59a. Determine the range of $I_D$ values for the circuit.

16. A JFET with values of $V_{GS(off)} = -5$ to $-10$ V and $I_{DSS} = 4$ to 8 mA is used in the circuit in Figure 12.59b. Determine the range of $I_D$ values for the circuit.

17. The 2N3437 (described in Problem 11) is used in the circuit in Figure 12.60. Determine the range of $I_D$ values for the circuit. (*Hint:* See the margin note under *A Practical Consideration* on page 487 of the text.)

18. The 2N3438 (described in Problem 12) is used in the circuit in Figure 12.61. Determine the range of $I_D$ values for the circuit. (*Hint:* See the margin note under *A Practical Consideration* on page 487 of the text.)

19. Determine the ranges of $V_{GS}$, $I_D$, and $V_{DS}$ for the circuit in Figure 12.62. The 2N5484 is described in Problem 7.

20. Determine the ranges of $V_{GS}$, $I_D$, and $V_{DS}$ for the circuit in Figure 12.63. The 2N5485 is described in Problem 8.

21. Determine the ranges of $V_{GS}$, $I_D$, and $V_{DS}$ for the circuit in Figure 12.64. The 2N5458 is described in Problem 10.

22. Determine the ranges of $V_{GS}$, $I_D$, and $V_{DS}$ for the circuit in Figure 12.65. The 2N3437 is described in Problem 11.

23. Determine the ranges of $I_D$ and $V_{DS}$ for the circuit in Figure 12.66.

24. Determine the ranges of $I_D$ and $V_{DS}$ for the circuit in Figure 12.67.

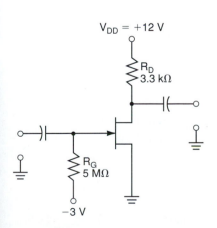

**FIGURE 12.60**

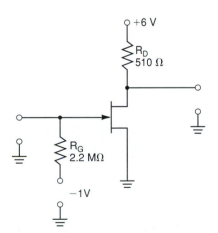

**FIGURE 12.61**

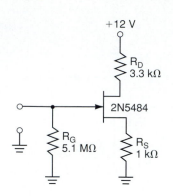

FIGURE 12.62

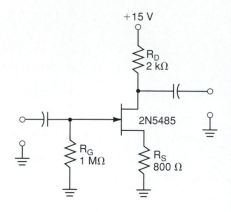

FIGURE 12.63

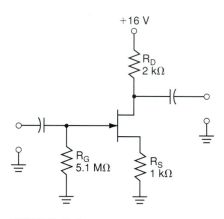

FIGURE 12.64

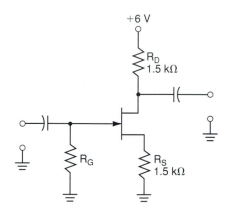

FIGURE 12.65

FIGURE 12.66

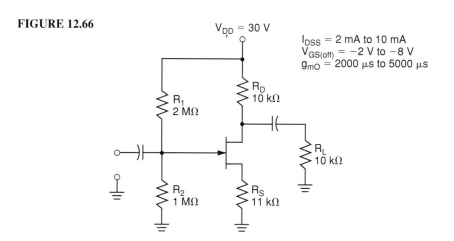

$V_{DD} = 30$ V

$I_{DSS} = 2$ mA to 10 mA
$V_{GS(off)} = -2$ V to $-8$ V
$g_{mO} = 2000$ μs to 5000 μs

25. Determine the ranges of $I_D$ and $V_{DS}$ for the circuit in Figure 12.68.
26. Determine the ranges of $I_D$ and $V_{DS}$ for the circuit in Figure 12.69.

*§12.3*

27. The 2N5484 has values of $g_{mO} = 3000$ to 6000 μS and $V_{GS(off)} = -0.3$ to $-3.0$ V. Determine the range of $g_m$ when $V_{GS} = -0.2$ V.
28. The 2N5485 has values of $g_{mO} = 3500$ to 7000 μS and $V_{GS(off)} = -0.5$ to $-4.0$ V. Determine the range of $g_m$ when $V_{GS} = -0.4$ V.

**FIGURE 12.67**

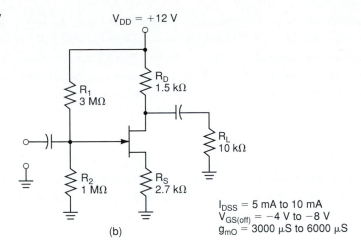

$V_{DD} = +12$ V

$R_1$ 3 MΩ

$R_D$ 1.5 kΩ

$R_L$ 10 kΩ

$R_2$ 1 MΩ

$R_S$ 2.7 kΩ

$I_{DSS} = 5$ mA to 10 mA
$V_{GS(off)} = -4$ V to $-8$ V
$g_{mO} = 3000$ μS to 6000 μS

(b)

**FIGURE 12.68**

$I_{DSS} = 5$ mA to 14 mA
$V_{GS(off)} = -2$ V to $-6$ V
$g_{mo} = 1000$ μS to 2500 μS

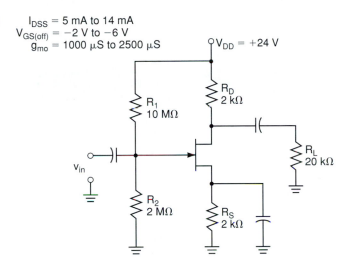

$V_{DD} = +24$ V

$R_1$ 10 MΩ

$R_D$ 2 kΩ

$R_L$ 20 kΩ

$v_{in}$

$R_2$ 2 MΩ

$R_S$ 2 kΩ

**FIGURE 12.69**

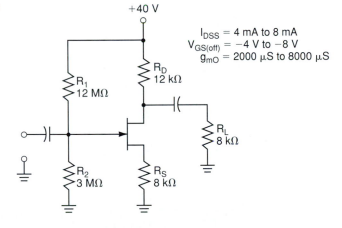

$+40$ V

$I_{DSS} = 4$ mA to 8 mA
$V_{GS(off)} = -4$ V to $-8$ V
$g_{mO} = 2000$ μS to 8000 μS

$R_1$ 12 MΩ

$R_D$ 12 kΩ

$R_L$ 8 kΩ

$R_2$ 3 MΩ

$R_S$ 8 kΩ

**29.** The 2N3437 has maximum values of $V_{GS(off)} = -5$ V and $g_{m0} = 6000$ μS. Determine the maximum values of $g_m$ at $V_G = -1$ V, $V_{GS} = -2.5$ V, and $V_{GS} = -4$ V.

**30.** The 2N3438 has maximum values of $V_{GS(off)} = 2.5$ V and $g_{m0} = 4500$ μS. Determine the maximum values of $g_m$ at $V_{GS} = -1$ V, $V_{GS} = -1.5$ V, and $V_{GS} = -2$ V.

**31.** Determine the range of $A_v$ values for the circuit in Figure 12.66.

**32.** Determine the range of $A_v$ values for the circuit in Figure 12.67.

**33.** Determine the range of $A_v$ values for the circuit in Figure 12.68.

**34.** Determine the range of $A_v$ values for the circuit in Figure 12.69.

**35.** Determine the range of $A_v$ values for the swamped amplifier shown in Figure 12.70.

**36.** Determine the range of $A_v$ values for the swamped amplifier shown in Figure 12.71.

**37.** Determine the value of $Z_{in}$ for the amplifier in Figure 12.66.

**38.** Determine the value of $Z_{in}$ for the amplifier in Figure 12.67.

### §12.4

**39.** Determine the values of $A_v$, $Z_{in}$, and $Z_{out}$ for the amplifier in Figure 12.72.

**40.** Determine the values of $A_v$, $Z_{in}$, and $Z_{out}$ for the amplifier in Figure 12.73.

**41.** Determine the ranges of $A_v$, $Z_{in}$, and $Z_{out}$ for the amplifier in Figure 12.74.

**42.** Determine the ranges of $A_v$, $Z_{in}$, and $Z_{out}$ for the amplifier in Figure 12.75.

**43.** Determine the ranges of $A_v$, $Z_{in}$, and $Z_{out}$ for the circuit in Figure 12.76.

**44.** Determine the ranges of $A_v$, $Z_{in}$, and $Z_{out}$ for the circuit in Figure 12.77.

**FIGURE 12.70**

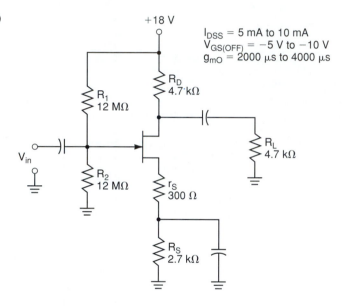

**FIGURE 12.71**

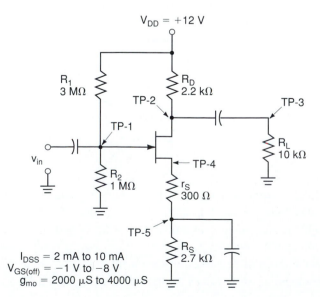

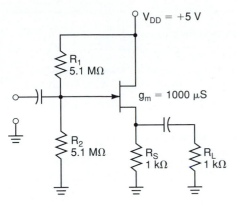

**FIGURE 12.72**

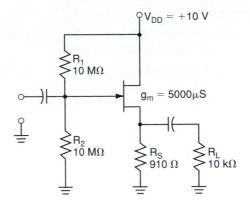

**FIGURE 12.73**

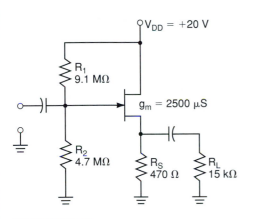

**FIGURE 12.74**

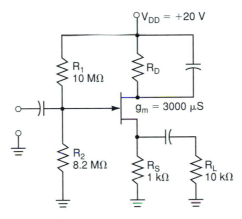

**FIGURE 12.75**

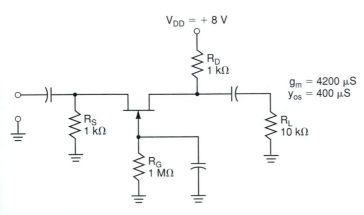

**FIGURE 12.76**

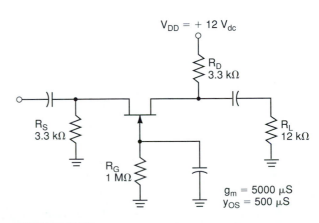

**FIGURE 12.77**

45. The circuit in Figure 12.71 has the waveforms shown in Figure 12.78. Discuss the possible causes of the problem.

46. A voltage-divider biased JFET amplifier suddenly starts to have the instability of a self-bias circuit. Changing the JFET does not solve the problem. What is wrong?

47. The circuit in Figure 12.79 has the waveforms shown. Discuss the possible causes of the problem.

48. The circuit in Figure 12.80 has the dc voltages shown. Discuss the possible causes of the problem.

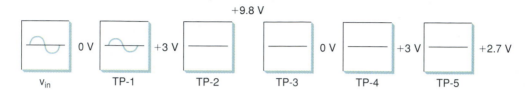

**FIGURE 12.78**

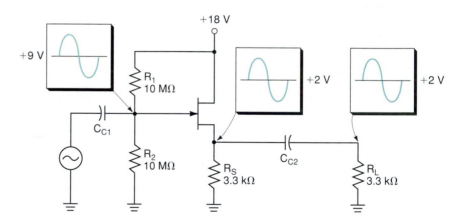

**FIGURE 12.79**

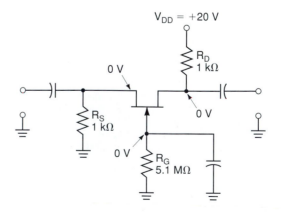

**FIGURE 12.80**

**49.** The biasing circuit shown in Figure 12.81 is a *current-source bias* circuit. Determine how this circuit obtains a stable value of $I_D$.

**50.** Can the 2N5486 be used safely in the circuit shown in Figure 12.82? Explain your answer using circuit calculations.

**51.** The 2N5486 cannot be substituted for the JFET in Figure 12.83. Why not?

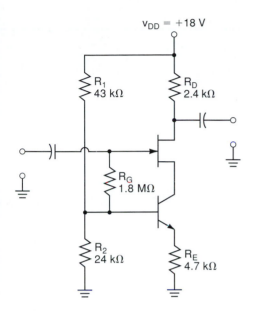

**FIGURE 12.81**

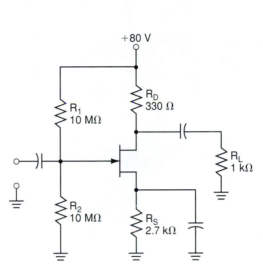

**FIGURE 12.82**

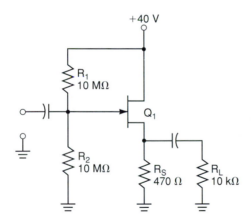

**FIGURE 12.83**

# 13

## MOSFETS

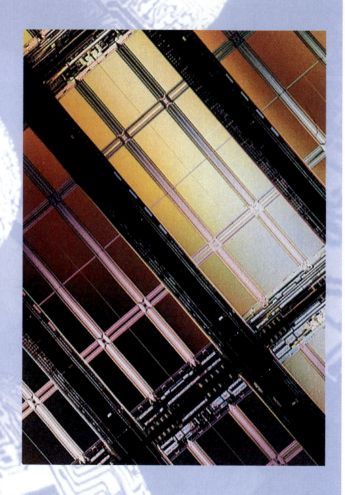

**A**dvances in MOSFET technology have lead to the development of VLSI (very large scale integrated) circuits. Many of these circuits (like the ones shown) are essentially entire electronic systems that are etched on a single piece of silicon.

# OUTLINE

# OBJECTIVES

*After studying the material in this chapter, you should be able to:*

1. Identify the two types of MOSFETs and describe the construction of each.

2. Describe the devices and precautions used to protect MOSFETs.

3. Describe the depletion-mode operation of the D-MOSFET.

4. Describe the enhancement-mode operation of the D-MOSFET.

5. Describe and analyze zero-bias operation.

6. Define threshold voltage, $V_{GS(th)}$, and discuss its significance.

7. List the FET biasing circuits that can and cannot be used with E-MOSFETs.

8. Describe the physical construction and capacitance characteristics of the dual-gate MOSFET.

9. Describe the physical construction and current-handling capabilities of the VMOS.

10. List the advantages of VMOS over BJTs.

11. List the advantages that CMOS logic circuits have over TTL logic circuits.

12. Describe the purpose served by power MOSFET drivers in digital communications.

# The MOSFET: A Major Component in Integrated Circuit Technology

Odds are that you have never heard of MOS (metal-oxide semiconductor) technology before. At the same time, you probably have one or more items with you that were made using MOSFET circuitry.

Most calculators, digital watches, and desktop computers are made using VLSI (very large scale integrated) circuits. These circuits contain up to hundreds of thousands of components that are all etched on a single piece of semiconductor material. The development of VLSI technology has made many items, such as those listed above, possible.

All electronic components dissipate heat. All electronic components must have some measurable size. These are two of the problems that have always faced the developers of integrated circuits, or ICs. The goal of IC technology has always been to produce the maximum number of components in the smallest amount of space possible. However, the need for heat-dissipation space and the actual space taken up by the components have limited the number of components that could be etched into a single piece of silicon.

The development of MOS technology has provided one solution to the problem of component spacing. Since MOSFET circuits are extremely low input current circuits, the heat-dissipation problem has been greatly reduced in these circuits. At the same time, MOSFET circuits can be made much smaller than their BJT counterparts. For these reasons, MOSFET technology has come to dominate most of the VLSI circuit market.

---

**T**he main drawback to JFET operation is the fact that the JFET gate *must* be reverse biased for the device to operate properly. Recall that the reverse bias of the gate is varied to deplete the channel of free carriers, thus controlling the effective size of the channel. This type of operation is referred to as **depletion-mode operation.** A depletion-type device would be one that uses an input voltage to *reduce* the channel size from its *zero-bias* size.

The *metal oxide semiconductor FET,* or **MOSFET**, is a device that can be operated in the **enhancement mode.** This means that the input signal can be used to *increase* the size of the channel. This also means that the device is not restricted to operating with its gate reverse biased. This is an improvement over the standard JFET. In this chapter, we will look at the various types of MOSFETs, their construction and operation, as well as their applications and fault symptoms.

**Depletion-mode operation**
Using an input voltage to reduce channel size.

**MOSFET**
An FET that can be operated in the enhancement mode.

**Enhancement-mode operation**
Using an input voltage to increase channel size.

## 13.1 MOSFET CONSTRUCTION AND HANDLING

OBJECTIVE 1 ▶

**D-MOSFETs**
Can be operated in both the depletion mode and the enhancement mode.

**E-MOSFETs**
Are restricted to enhancement-mode operation.

There are two basic types of MOSFETs: *depletion-type MOSFETs,* or *D-MOSFETs,* and *enhancement-type MOSFETs,* or *E-MOSFETs.* **D-MOSFETs** can be operated in both the depletion mode and the enhancement mode, whereas **E-MOSFETs** are restricted to operating in the enhancement mode. The primary difference between D-MOSFETs and E-MOSFETs is their physical construction. The construction difference between the two is illustrated in Figure 13.1. As you can see, the D-MOSFET has a physical channel (shaded area) between the source and drain terminals. The E-MOSFET, on the other hand, has no such channel. The E-MOSFET depends on the gate voltage to *form* a channel between the source and drain terminals. This point will be discussed in detail later in this chapter.

Note that both MOSFETs in Figure 13.1 show an *insulating layer* between the gate and the rest of the component. This insulating layer is made up of *silicon dioxide* ($SiO_2$), a glasslike insulating material. The gate terminal is made of a metal conductor. Thus, going from gate to substrate, you have *metal-oxide-semiconductor,* which is where the

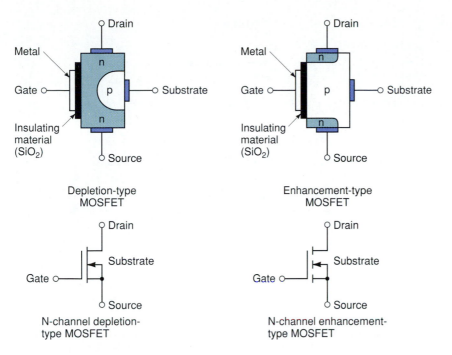

Depletion-type
MOSFET

Enhancement-type
MOSFET

N-channel depletion-
type MOSFET

N-channel enhancement-
type MOSFET

*Note*: The substrate does not have to be tied to the source. Desired operation dictates
this decision.

**FIGURE 13.1   MOSFET construction and schematic symbols.**

term *MOSFET* comes from. Since the gate is insulated from the rest of the component, the MOSFET is sometimes referred to as an *insulated-gate FET,* or *IGFET*. However, this term is rarely used in place of the term *MOSFET*.

The foundation of the MOSFET is called the **substrate.** This material is represented in the schematic symbol by the center line that is connected to the source terminal. The connection of these lines indicates that the substrate and the source are connected internally. Note that an *n*-channel MOSFET will have a *p*-material substrate, and a *p*-channel MOSFET will have an *n*-material substrate.

In the schematic symbol for the MOSFET, the arrow is placed on the substrate. As with the JFET, an arrow pointing *in* represents an *n-channel* device, while an arrow pointing *out* represents a *p-channel* device.

Figure 13.2a and b are *p*-channel depletion-type MOSFETs. Figure 13.2c and d are *p*-channel enhancement-type MOSFETs. Note that the substrate does not have to be tied to the source. It can be tied to different supply voltages, depending on the operating mode desired.

**Substrate**
The foundation of a MOSFET.

## Component Handling

The layer of $SiO_2$ that insulates the gate from the channel is extremely thin and can be destroyed easily by *static electricity*. The static electricity generated by the human body can be sufficient to ruin a MOSFET. Because of this, some precautions must be taken to protect the MOSFET.

Many MOSFET devices are now manufactured with protective diodes etched between the gate and the source, as shown in Figure 13.3. The diodes are configured so that they will conduct in either direction, provided that a predetermined voltage is reached. This voltage will be higher than any working voltage normally applied to the MOSFET. For example, if a MOSFET with a protected input is rated for a maximum $V_{GS}$

◄   *OBJECTIVE 2*

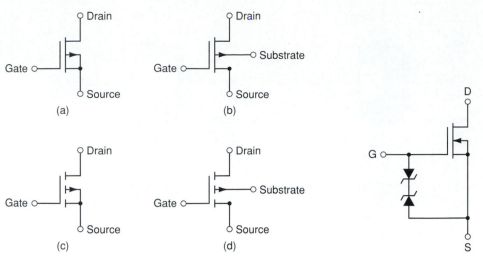

FIGURE 13.2   Substrate connections.

FIGURE 13.3   MOSFET
static protection.

of $\pm 30$ V, the zener diodes will be designed to conduct at any voltage outside the $\pm 30$ V range. This will protect the device from excessive static buildup, as well as accidentally high values of $V_{GS}$.

When MOSFETs without protected inputs are used, some basic handling precautions will keep the components from being damaged. These precautions are as follows:

MOSFET handling precautions.

Lab Reference: You will need to keep these precautions in mind while performing Exercise 28.

1. Store the devices with the leads shorted together or in conductive foam. *Never store MOSFETs in Styrofoam.* (Styrofoam is the best static electricity generator ever devised.)

2. Do not handle MOSFETs unless you need to. When you handle any MOSFET, hold the component by the case, not the leads.

3. Do not install or remove any MOSFET while power is applied to a circuit. Also, be sure that any signal source is removed from a MOSFET circuit before turning the supply voltage off or on.

If you follow the guidelines listed above, you will have no difficulty in handling MOSFET devices.

## Section Review

1. What is a *MOSFET*?

2. What is *depletion-mode operation*?

3. What is *enhancement-mode operation*?

4. In terms of operating modes, what is the difference between D-MOSFETs and E-MOSFETs?

5. In terms of physical construction, what is the difference between D-MOSFETs and E-MOSFETs?

6. What is the difference between the schematic symbol of a D-MOSFET and that of an E-MOSFET?

7. Discuss the means by which MOSFET inputs are internally protected in some devices.

8. What precautions should be observed when handling MOSFETs?

9. Why are the precautions in Question 8 necessary?

## 13.2 D-MOSFETs

As stated earlier, the D-MOSFET is capable of operating in both the depletion mode and the enhancement mode. When it is operating in the depletion mode, the characteristics of the D-MOSFET are very similar to those of the JFET. The overall operation of the D-MOSFET is illustrated in Figure 13.4.

Figure 13.4a shows the D-MOSFET operating conditions when $V_{GS} = 0$ V (that is, when the gate and source terminals are *shorted*). As stated in Chapter 12, $I_{DSS}$ is the value of $I_D$ when the source and gate terminals are shorted together. Therefore, when $V_{GS} = 0$ V, $I_D = I_{DSS}$.

When the gate–source junction is *reverse* biased ($V_{GS}$ is negative), the biasing voltage depletes the channel of free carriers. This *effectively* reduces the width of the channel, increasing its resistance. The *depletion mode* is shown in Figure 13.4b. Note that this operating state is exactly the same as the normal JFET operating state. The white area below the insulating layer in Figure 13.4b represents the depleted carrier region. The shaded area below this region is the effective channel width. Since this area is less than the area shown in Figure 13.4a, the resistance of the MOSFET channel is higher when $V_{GS}$ is negative. Thus, $I_D$ will be at some value less than $I_{DSS}$. The exact value of $I_D$ depends on the value of $I_{DSS}$ and $V_{GS}$.

Figure 13.4c shows the *enhancement mode* of operation. In this circuit, $V_{GS}$ is *positive*. When $V_{GS}$ is positive, the channel is *effectively* widened. This reduces the resistance of the channel, and $I_D$ increases above $I_{DSS}$. As you know, this type of operation is not possible with the JFET.

How does a positive $V_{GS}$ effectively *widen* the channel? You must remember that the majority carriers in a *p*-type material are valence band *holes*. The holes in the *p*-type

◄ *OBJECTIVE 3*

Depletion-mode operation is very similar to JFET operation.

◄ *OBJECTIVE 4*

How enhancement-mode operation works.

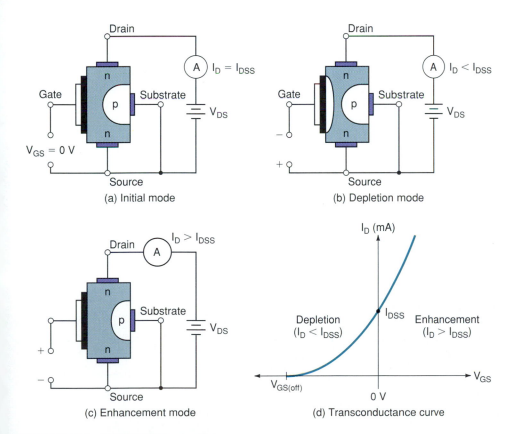

**FIGURE 13.4  D-MOSFET operation.**

substrate are *repelled* by the positive gate voltage. Left behind is a region that is depleted of valence band holes. At the same time, the conduction band electrons (minority carriers) in the *p*-type material are *attracted* toward the channel by the positive gate voltage. With the buildup of electrons near the channel, the area below the physical channel *effectively* becomes an *n*-type material. The extended *n*-type channel now allows more current to pass through the channel and $I_D > I_{DSS}$.

For enhancement-mode operation, $I_D > I_{DSS}$.

The combination of these three operating states is represented by the D-MOSFET transconductance curve in Figure 13.4d. Note that $I_D = I_{DSS}$ when $V_{GS} = 0$ V. When $V_{GS}$ is *negative*, $I_D$ decreases below the value of $I_{DSS}$. $I_D$ reaches zero when $V_{GS} = V_{GS(off)}$, just as with the JFET. When $V_{GS}$ is *positive*, $I_D$ increases above the value of $I_{DSS}$. The maximum allowable value of $I_D$ is given on the spec sheet of a given MOSFET.

You may have noticed that the transconductance curve for the D-MOSFET is very similar to the curve for a JFET. Because of this similarity, the JFET and the D-MOSFET have the same transconductance equation. This equation was given in Chapter 12 as

$$I_D = I_{DSS}\left(1 - \frac{V_{GS}}{V_{GS(off)}}\right)^2 \qquad \text{(13.1)}$$

Example 13.1 demonstrates the use of equation (13.1) in plotting the transconductance curve for a D-MOSFET.

**EXAMPLE 13.1**

A D-MOSFET has parameters of $V_{GS(off)} = -6$ V and $I_{DSS} = 1$ mA. Plot the transconductance curve for the device.

*Solution:* From the parameters, we can determine two of the points on the curve as follows:

$$V_{GS} = -6 \text{ V} \qquad I_D = 0$$
$$V_{GS} = 0 \text{ V} \qquad I_D = 1 \text{ mA}$$

These points define the conditions at $V_{GS} = V_{GS(off)}$ and at $I_D = I_{DSS}$. Now we will use equation (13.1) to determine the coordinates of several more points. Remember that the only value we change is $V_{GS}$. When $V_{GS} = -3$ V,

$$I_D = (1 \text{ mA})\left(1 - \frac{-3 \text{ V}}{-6 \text{ V}}\right)^2$$
$$= 0.25 \text{ mA} \qquad (250 \text{ }\mu\text{A})$$

When $V_{GS} = -1$ V,

$$I_D = (1 \text{ mA})\left(1 - \frac{-1 \text{ V}}{-6 \text{ V}}\right)^2$$
$$= 0.694 \text{ mA} \qquad (694 \text{ }\mu\text{A})$$

When $V_{GS} = +1$ V,

$$I_D = (1 \text{ mA})\left(1 - \frac{+1 \text{ V}}{-6 \text{ V}}\right)^2$$
$$= 1.36 \text{ mA}$$

When $V_{GS} = +3$ V,

$$I_D = (1 \text{ mA})\left(1 - \frac{+3 \text{ V}}{-6 \text{ V}}\right)^2$$
$$= 2.25 \text{ mA}$$

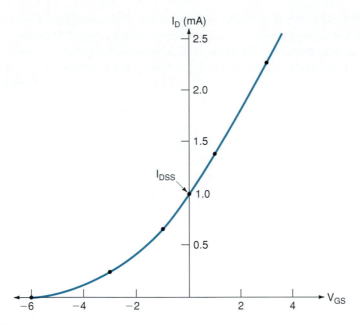

**FIGURE 13.5**

We now have the following combinations of $V_{GS}$ and $I_D$:

| $V_{GS}$ (V) | $I_D$ (mA) |
|:---:|:---:|
| −6 | 0 |
| −3 | 0.25 |
| −1 | 0.69 |
| 0 | 1.00 |
| +1 | 1.36 |
| +3 | 2.25 |

Using the values listed, we can plot the transconductance curve shown in Figure 13.5.

### PRACTICE PROBLEM 13.1

A D-MOSFET has values of $V_{GS(off)} = -4$ to $-6$ V and $I_{DSS} = 8$ to 12 mA. Plot the minimum and maximum transconductance curves for the device. Use $V_{GS}$ values of −4 V, −2 V, 0 V, 2 V, and 4 V.

As you can see, the process for plotting a D-MOSFET transconductance curve is almost the same as plotting one for a JFET. The only differences are that you must use positive values of $V_{GS}$ (as well as negative) and that you will get values of $I_D$ that are greater than $I_{DSS}$.

## D-MOSFET Drain Curves

The drain curves of a given D-MOSFET can be used to plot the dc load line of a given circuit, just as the collector curves can be used to plot the load line of a given BJT. The drain curves for a D-MOSFET are plotted using corresponding values of $V_{GS}$ and $I_D$. A composite of drain curves is shown in Figure 13.6. The curves shown were plotted using the values from Example 13.1. Also, the curves for $V_{GS} = -2$ V and $V_{GS} = 2$ V were added. Note that, above $V_P$, the value of $V_{DS}$ becomes constant for each combination of $V_{GS}$ and $I_D$. Also note that the value of $V_P$ is relative to the gate voltage, and thus changes from one curve to another. The value of $V_P$ for $V_{GS} = +3$ V is highlighted in Figure 13.6 to illustrate this point. At lower values of $V_{GS}$, $V_P$ is lower. Increasing the value of $V_{GS}$ increases the value of $V_P$.*

The actual dc load line for a D-MOSFET would depend on the circuit containing the device. The *ideal* saturation point on the load line is found as

$$I_{D(\text{sat})} = \frac{V_{DD}}{R_D} \qquad \text{(13.2)}$$

The *ideal* value of $V_{DS(\text{off})}$ is found as

$$V_{DS(\text{off})} = V_{DD} \qquad \text{(13.3)}$$

Compare these equations to the load-line equations for the BJT amplifier. The load line for the D-MOSFET is plotted as shown in Figure 13.7. Note the differences between the *actual* and *ideal* values of $I_{D(\text{sat})}$ and $V_{DS(\text{off})}$.

## Transconductance

The value of $g_m$ is found for a D-MOSFET the same way that it is for the JFET. By formula,

$$g_m = g_{m0}\left(1 - \frac{V_{GS}}{V_{GS(\text{off})}}\right) \qquad \text{(13.4)}$$

*Note that the $V_P$ rating of a MOSFET is measured at $V_{GS} = 0$ V.

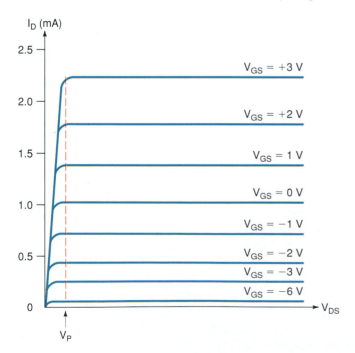

FIGURE 13.6   MOSFET drain curves.

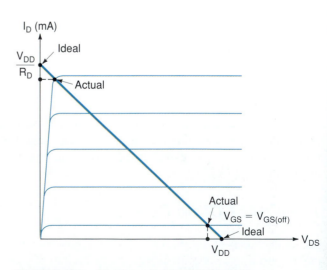

FIGURE 13.7   Plotting a D-MOSFET load line.

This equation leads to another characteristic of the D-MOSFET. Just as $I_D$ can be greater than $I_{DSS}$, $g_m$ can be greater than $g_{m0}$. Of course, this happens only when the device is operated in the enhancement mode.

## D-MOSFET Biasing Circuits

Again, the similarities between D-MOSFETs and JFETs pay off. The primary D-MOSFET biasing circuits are exactly the same as those used for JFETs. The D-MOSFET will also have the same overall characteristics as the JFET when used in the common-source, common-drain, and common-gate configurations. All the dc and ac relationships covered for the JFET will work for D-MOSFET circuits as well.

The primary difference between D-MOSFETs and JFETs is the fact that the D-MOSFET does *not* require a negative value of $V_{GS}$. In fact, one common method of biasing a D-MOSFET is to set $V_{GS}$ to 0 V. This biasing circuit configuration, called **zero bias**, is illustrated in Figure 13.8.

◄  *OBJECTIVE 5*

**Zero bias**
A D-MOSFET biasing circuit that has quiescent values of $V_{GS} = 0$ V and $I_D = I_{DSS}$.

With the JFET self-bias circuit, $I_{DQ}$ caused a voltage to develop across the source resistor, $R_S$. For the D-MOSFET zero-bias circuit, the source resistor is not necessary. With no source resistor, the value of $V_S$ is 0 V. This gives us a value of $V_{GS} = 0$ V, which biases the amplifier at $I_D = I_{DSS}$. The value of $R_D$ is selected so that $V_{DS}$ is approximately equal to one-half of $V_{DD}$. It is just that simple.

## D-MOSFET Input Impedance

The gate impedance of a D-MOSFET is extremely high. For example, one MOSFET has a maximum gate current of 10 pA when $V_{GS}$ is 35 V. This would calculate to a gate impedance of $3.5 \times 10^{12}$ Ω! With an input impedance in this range, the MOSFET presents virtually no load to a source circuit.

## D-MOSFETs Versus JFETs

Figure 13.9 summarizes many of the characteristics of JFETs and D-MOSFETs. As you can see, the components are very similar in a number of respects. The D-MOSFET has the advantages of higher input impedance and the ability to operate in the enhancement

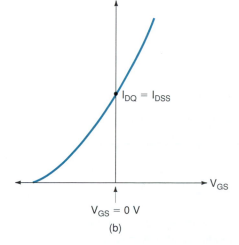

(a)          (b)

**FIGURE 13.8   Zero biasing.**

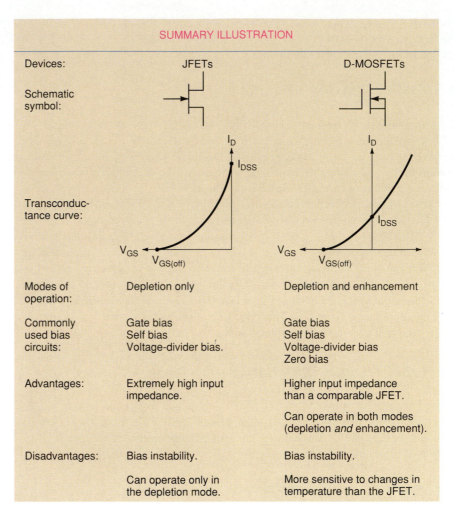

| Devices: | JFETs | D-MOSFETs |
|---|---|---|
| Schematic symbol: | | |
| Transconductance curve: | | |
| Modes of operation: | Depletion only | Depletion and enhancement |
| Commonly used bias circuits: | Gate bias<br>Self bias<br>Voltage-divider bias. | Gate bias<br>Self bias<br>Voltage-divider bias<br>Zero bias |
| Advantages: | Extremely high input impedance. | Higher input impedance than a comparable JFET.<br><br>Can operate in both modes (depletion *and* enhancement). |
| Disadvantages: | Bias instability.<br><br>Can operate only in the depletion mode. | Bias instability.<br><br>More sensitive to changes in temperature than the JFET. |

**FIGURE 13.9**

mode. It can also use zero bias, while the JFET cannot. At the same time, the D-MOS-FET is more sensitive to increases in temperature, and there are precautions that must be taken when handling many MOSFETs.

*Section Review*

1. Describe the depletion-mode relationship between $V_{GS}$ and $I_D$ for a D-MOSFET.
2. Describe the enhancement-mode relationship between $V_{GS}$ and $I_D$ for a D-MOSFET.
3. Describe the relationship between $I_D$ and $I_{DSS}$ for the D-MOSFET.
4. Describe the relationship between $g_m$ and $g_{m0}$ for the D-MOSFET.
5. Describe the quiescent conditions of the *zero-bias* circuit.
6. List the similarities and differences between the JFET and the D-MOSFET.

## *13.3* E-MOSFETs

It was stated earlier that the E-MOSFET is capable of operating only in the enhancement mode. In other words, the gate potential must be positive with respect to the source. The reason for this can be seen in Figure 13.10. When the value of $V_{GS}$ is 0 V, there is no

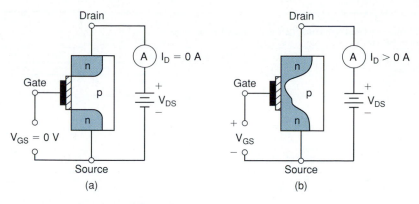

FIGURE 13.10   E-MOSFET operation.

channel connecting the source and drain materials. As a result, there can be no significant amount of drain current. This is illustrated in Figure 13.10a.

When a positive potential is applied to the gate, the *n*-channel E-MOSFET responds in the same manner as the D-MOSFET. As Figure 13.10b shows, the positive gate voltage forms a channel between the source and drain by depleting the area of valence band holes. As the holes are repelled by the positive gate voltage, that voltage is also attracting the minority carrier electrons in the *p*-type material toward the gate. This forms an *effective n*-type bridge between the source and drain, providing a path for drain current.

If the value of $V_{GS}$ is *increased,* the newly formed channel becomes wider, allowing $I_D$ to increase. If the value of $V_{GS}$ *decreases,* the channel becomes narrower, and $I_D$ decreases. This is illustrated by the E-MOSFET transconductance curve shown in Figure 13.11. As you can see, this transconductance curve is similar to those covered previously. The primary differences are:

1. All values of $V_{GS}$ that cause the device to conduct are *positive.*

2. The point at which the device turns on (or off, depending on how you look at it) is called **threshold voltage,** $V_{GS(th)}$.

Note that the value of $I_{DSS}$ for the E-MOSFET is approximately 0 A. For example, the 3N169 E-MOSFET spec sheet lists an $I_{DSS}$ of 10 nA when the ambient temperature is

*A Practical Consideration:* In this section, our discussion is limited to the *n*-channel E-MOSFET. Its p-channel counterpart operates according to the same principles. The only operating differences (as usual) are the current directions and the voltage polarities.

◄   *OBJECTIVE 6*

**Threshold voltage ($V_{GS(th)}$)**
The value of $V_{GS}$ that turns the E-MOSFET *on.*

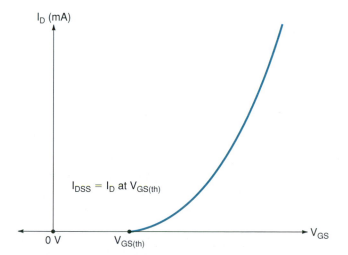

**Lab reference:** The curve for an E-MOSFET is plotted using measured values of $V_{GS}$ and $I_D$ in Exercise 28.

FIGURE 13.11   **E-MOSFET transconductance curve.**

ELECTRICAL CHARACTERISTICS (T$_A$ = 25°C unless otherwise noted)
Substrate connected to source.

| Characteristic | Figure No. | Symbol | Min | Max | Unit |
|---|---|---|---|---|---|
| **OFF CHARACTERISTICS** | | | | | |
| Drain-Source Breakdown Voltage ($I_D$ = 10 µAdc, $V_{GS}$ = 0) | — | $V_{(BR)DSS}$ | 25 | — | Vdc |
| *Gate Leakage Current ($V_{GS}$ = –35 Vdc, $V_{DS}$ = 0) | — | $I_{GSS}$ | | 10 | pAdc |
| ($V_{GS}$ = –35 Vdc, $V_{DS}$ = 0, $T_A$ = 125°C) | | | — | 100 | |
| *Zero-Gate-Voltage Drain Current ($V_{DS}$ = 10 Vdc, $V_{GS}$ = 0) | — | $I_{DSS}$ | — | 10 | nAdc |
| ($V_{DS}$ = 10 Vdc, $V_{GS}$ = 0, $T_A$ = 125°C) | | | — | 1.0 | µAdc |
| **\*ON CHARACTERISTICS** | | | | | |
| Gate-Source Threshold Voltage 3N169 3N170 3N171 ($V_{DS}$ = 10 Vdc, $I_D$ = 10 µAdc) | — | $V_{GS(th)}$ | 0.5 1.0 1.5 | 1.5 2.0 3.0 | Vdc |
| "ON" Drain Current ($V_{GS}$ = 10 Vdc, $V_{DS}$ = 10 Vdc) | 3 | $I_{D(on)}$ | 10 | — | mAdc |
| Drain-Source "ON" Voltage ($I_D$ = 10 mAdc, $V_{GS}$ = 10 Vdc) | — | $V_{DS(on)}$ | — | 2.0 | Vdc |
| **SMALL SIGNAL CHARACTERISTICS** | | | | | |
| *Drain-Source Resistance ($V_{GS}$ = 10 Vdc, $I_D$ = 0, f = 1.0 kHz) | 4 | $r_{ds(on)}$ | — | 200 | Ohms |
| Forward Transfer Admittance ($V_{DS}$ = 10 Vdc, $I_D$ = 2.0 mAdc, f = 1.0 kHz) | 1 | $|Y_{fs}|$ | 1000 | — | µmhos |
| *Reverse Transfer Capacitance ($V_{DS}$ = 0, $V_{GS}$ = 0, f = 1.0 MHz) | 2 | $C_{rss}$ | — | 1.3 | pF |
| *Input Capacitance ($V_{DS}$ = 10 Vdc, $V_{GS}$ = 0, f = 1.0 MHz) | 2 | $C_{iss}$ | — | 5.0 | pF |
| *Drain-Substrate Capacitance ($V_{D(SUB)}$ = 10 Vdc, f = 1.0 MHz) | — | $C_{d(sub)}$ | — | 5.0 | pF |
| **\*SWITCHING CHARACTERISTICS** | | | | | |
| Turn-On Delay Time | 6,10 | $t_{d(on)}$ | — | 3.0 | ns |
| Rise Time | 7,10 | $t_r$ | — | 10 | ns |
| Turn-Off Delay Time | 8,10 | $t_{d(off)}$ | — | 3.0 | ns |
| Fall Time | 9,10 | $t_f$ | — | 15 | ns |

Switching conditions: ($V_{DD}$ = 10 Vdc, $I_{D(on)}$ = 10 mAdc, $V_{GS(on)}$ = 10 Vdc, $V_{GS(off)}$ = 0, $R_G'$ = 50 Ohms)

*Indicates JEDEC Registered Data.

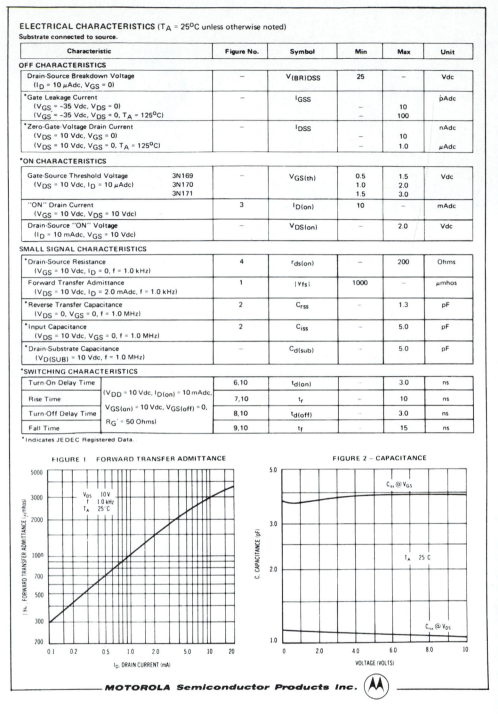

FIGURE 1 — FORWARD TRANSFER ADMITTANCE

FIGURE 2 — CAPACITANCE

**MOTOROLA Semiconductor Products Inc.**

**FIGURE 13.12** The 3N169, 70, 71 series specification sheet. (Copyright of Motorola. Used by permission.)

25°C. For all practical purposes, this value of $I_D$ is zero. Obviously, since the value of $I_{DSS}$ for the E-MOSFET is near zero, the standard transconductance formula will not work for the E-MOSFET. To determine the value of $I_D$ at a given value of $V_{GS}$, you must use the following formula:

$$I_D = k[V_{GS} - V_{GS(th)}]^2 \tag{13.5}$$

where $k$ is a *constant* for the MOSFET, found as

$$k = \frac{I_{D(\text{on})}}{[V_{GS(\text{on})} - V_{GS(\text{th})}]^2} \qquad (13.6)$$

The values used in equation (13.6) to determine the value of $k$ are obtained from the spec sheet of the E-MOSFET being used. For example, consider the spec sheet shown in Figure 13.12. This is the sheet for the 3N169, 70, 71 series E-MOSFETs. The 3N169 has the following parameters listed:

$$I_{D(\text{on})} = 10 \text{ mA}_{\text{dc}} \qquad (\text{minimum})$$
$$V_{GS(\text{th})} = 0.5 \text{ V}_{\text{dc}} \qquad (\text{minimum})$$

Now, if you look closely at the test conditions for the $I_{D(\text{on})}$ rating, you will see that this current was measured when $V_{GS} = 10$ V. Using this value and the two parameters listed above, $k$ would be found as

$$k = \frac{I_{D(\text{on})}}{[V_{GS(\text{on})} - V_{GS(\text{th})}]^2}$$
$$= \frac{10 \text{ mA}}{(10 \text{ V} - 0.5 \text{ V})^2}$$
$$= 111 \times 10^{-6}$$

The value of $k = 111 \times 10^{-6}$ would now be used in equation (13.5) to determine the value of $I_D$ for a given value of $V_{GS}$. The entire process is demonstrated in Example 13.2.

---

Determine the value of $I_D$ for the circuit shown in Figure 13.13. Use the spec sheet in Figure 13.12 to obtain the values needed to calculate the $k$ of the E-MOSFET.

*EXAMPLE 13.2*

***Solution:*** The spec sheet for the 3N171 lists the following parameters:

$$I_{D(\text{on})} = 10 \text{ mA at } V_{GS} = 10 \text{ V}$$
$$V_{GS(\text{th})} = 1.5 \text{ V}$$

Using these *minimum* values, the value of $k$ is found as

$$k = \frac{10 \text{ mA}}{(10 \text{ V} - 1.5 \text{ V})^2}$$
$$= 138 \times 10^{-6}$$

For this circuit, $V_{GS} = V_G$. The value of $V_G$ is found as

$$V_G = V_{DD} \frac{R_2}{R_1 + R_2}$$
$$= 5 \text{ V}$$

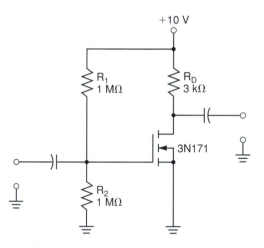

**FIGURE 13.13**

Using the value of $V_{GS} = 5$ V in equation (13.5), the value of $I_D$ for the circuit is found as

$$I_D = k[V_{GS} - V_{GS(\text{th})}]^2$$
$$= (138 \times 10^{-6})(5 \text{ V} - 1.5 \text{ V})^2$$
$$= 1.69 \text{ mA}$$

## E-MOSFET Biasing Circuits

*OBJECTIVE 7* ▶ One of the problems with the E-MOSFET is the fact that many of the biasing circuits used for JFETs and D-MOSFETs cannot be used with this device. The reason for this is the fact that $V_{GS}$ must be *positive* for an *n*-channel device. Since self-bias and zero bias produce values of $V_{GS}$ that are equal to zero or more negative than zero, neither of these circuits can be used. However, voltage-divider bias can be used, as can gate bias and *drain-feedback bias*.

In our study of BJTs, we covered a biasing circuit called *collector-feedback bias.*

**Drain-feedback bias**
The MOSFET equivalent of collector-feedback bias.

**Drain-feedback bias** is the MOSFET equivalent of this biasing method. Figure 13.14a shows a drain-feedback bias circuit. For reference purposes, a collector-feedback bias circuit is shown in Figure 13.14b.

The analysis of this circuit is relatively simple. First, refer back to the spec sheet in Figure 13.12. If you look at the $I_{D(on)}$ parameter, you will notice that it was measured under the following condition:

$$V_{GS} = V_{DS} \tag{13.7}$$

Now take a look at the drain-feedback circuit. With the superhigh impedance of the gate, there is no current in the gate circuit. Therefore, no voltage is dropped across the gate resistor, $R_G$. Since no voltage is dropped across $R_G$, the gate is at the same potential as the drain. Therefore, the circuit fulfills the condition specified in equation (13.7), and $I_D$ equals $I_{D(on)}$. The value of $V_{DS}$ for a drain-feedback bias circuit can be found as

$$V_{DS} = V_{DD} - R_D I_{D(on)} \tag{13.8}$$

Example 13.3 demonstrates the complete analysis of a drain-feedback bias circuit.

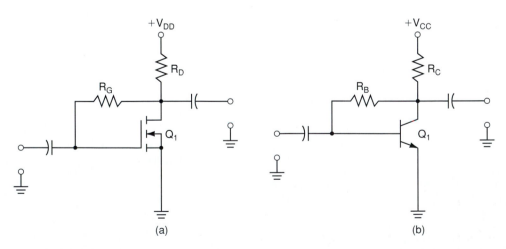

(a)                    (b)

**FIGURE 13.14   Drain-feedback bias.**

EXAMPLE 13.3

Determine the values of $I_D$ and $V_{DS}$ for the circuit shown in Figure 13.15.

**Solution:** The spec sheet for the 3N170 lists an $I_{D(on)}$ of 10 mA when $V_{GS} = V_{DS}$. Since these voltages *are* equal in the drain-feedback bias circuit, $I_D$ equals $I_{D(on)}$: 10 mA. Using this value, $V_{DS}$ (and thus $V_{GS}$) is found as

$$V_{DS} = V_{DD} - R_D I_{D(on)}$$
$$= 20\text{ V} - (1\text{ k}\Omega)(10\text{ mA})$$
$$= 10\text{ V}$$

**PRACTICE PROBLEM 13.3**

The E-MOSFET described in Practice Problem 13.2 is used in a drain-feedback bias circuit with values of $V_{DD} = 14$ V and $R_D = 510$ $\Omega$. Determine the values of $I_D$ and $V_{DS}$ for the circuit.

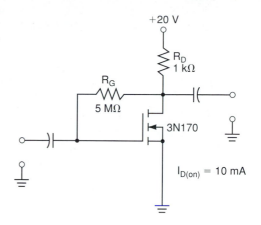

**FIGURE 13.15**

That is all there is to analyzing a drain-feedback bias circuit. Don't forget: The drain-feedback bias circuit sets up a *positive* value of $V_{GS}$. Because of this, it cannot be used with JFETs or D-MOSFETs. D-MOSFETs can be operated in the enhancement mode, but they are never biased with a value of $V_{GS}$ that is more positive than 0 V.

## Summary

So far, you have been introduced to D-MOSFETs and E-MOSFETs. A comparison of the two can be seen in Figure 13.16, which summarizes many of the important characteristics of these devices.

*Section Review*

1. How does a positive value of $V_{GS}$ develop a channel in an E-MOSFET?
2. What is *threshold voltage*, $V_{GS(th)}$?
3. When $V_{GS} = V_{GS(th)}$, what does the value of $I_D$ for an E-MOSFET equal?
4. List, in order, the steps required to determine the value of $I_D$ for an E-MOSFET at a given value of $V_{GS}$.
5. Which D-MOSFET biasing circuits can be used with E-MOSFETs?
6. Which D-MOSFET biasing circuits cannot be used with E-MOSFETs?
7. Describe the quiescent conditions for the *drain-feedback bias* circuit.

## 13.4 DUAL-GATE MOSFETs

The operation of MOSFETs is limited at high frequencies because of their high gate-to-channel capacitance. The source of this high capacitance can be seen in Figure 13.17. The metal plate used for the gate is a conductor. The silicon dioxide between

◄ *OBJECTIVE 8*

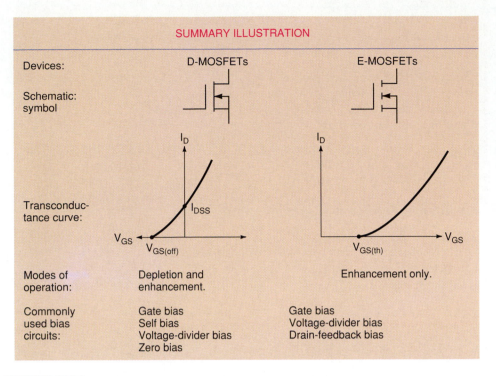

Devices:

D-MOSFETs

E-MOSFETs

Schematic: symbol

Transconductance curve:

$I_D$

$I_{DSS}$

$V_{GS}$

$V_{GS(off)}$

$I_D$

$V_{GS(th)}$

$V_{GS}$

Modes of operation:

Depletion and enhancement.

Enhancement only.

Commonly used bias circuits:

Gate bias
Self bias
Voltage-divider bias
Zero bias

Gate bias
Voltage-divider bias
Drain-feedback bias

**FIGURE 13.16**

**Dual-gate MOSFET**
A MOSFET constructed with two gates to reduce gate input capacitance.

the gate and channel is an insulator. The channel itself can be viewed as a conductor when compared to the silicon dioxide layer. Thus, the combination of the three forms a capacitor.

The **dual-gate MOSFET** uses two gate terminals to reduce the overall capacitance of the component. The physical construction and schematic symbols for the dual-gate MOSFET are shown in Figure 13.18. The reduced capacitance of the dual-gate MOSFET is a result of the way in which the component is used. Normally, the component is used so it acts as two series-connected MOSFETs. This is demonstrated in Section 13.7. When the dual-gate MOSFET is used as two series MOSFETs, the effect is similar to connecting two capacitors in series. You should recall that the total capacitance in a series connection is lower than either individual component value. Thus, by connecting the two gates in a series configuration, the overall capacitance is reduced.

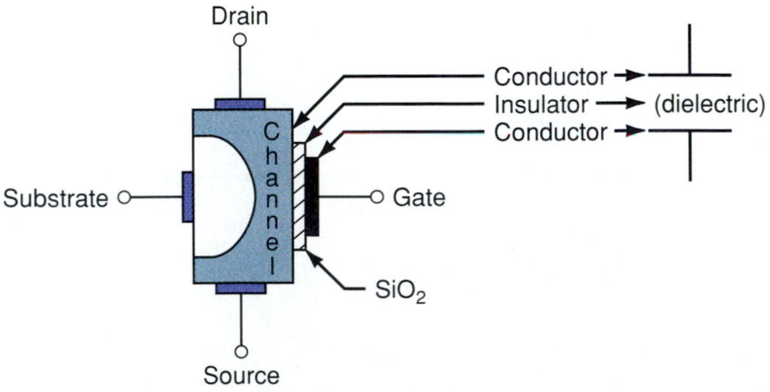

**FIGURE 13.17    MOSFET input capacitance.**

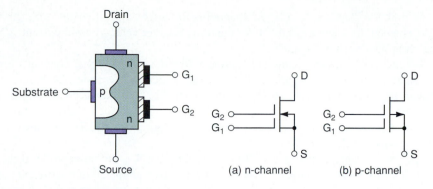

FIGURE 13.18 Dual-gate MOSFET construction and schematic symbols.

---

1. Why does the MOSFET have high input capacitance?
2. How does the dual-gate MOSFET provide reduced input capacitance?

---

## 13.5 POWER MOSFETs

Advances in engineering have produced a variety of MOSFETs that are designed specifically for high current, voltage, and/or power applications. In this section, we will take a look at two high-power MOSFETs. Some applications for these components are covered in Section 13.7.

### VMOS

The **vertical MOSFET, or VMOS,** is a component designed to handle much larger drain currents than the standard MOSFET. The current-handling capability of VMOS is a result of its physical construction, which is illustrated in Figure 13.19. As you can see, the component has materials that are labeled *p, n+,* and *n−*. The *n*-material labels indicate differences in doping levels. Also, there is no physical channel connecting the source (at top) and the drain (at bottom). Thus, VMOS is an *enhancement-type* MOSFET.

◄ *OBJECTIVE 9*

**Vertical MOSFET (VMOS)**
An E-MOSFET designed to handle high values of drain current.

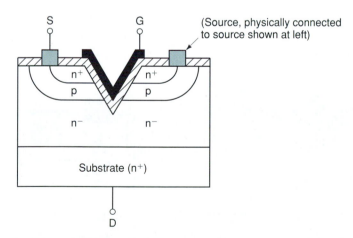

FIGURE 13.19 Vertical MOSFET (VMOS) construction.

With the V-shaped gate, a larger channel is formed by a positive gate voltage. With a larger channel, the device is capable of handling a larger value of drain current. VMOS operation is illustrated in Figure 13.20.

When a positive gate voltage is applied to the device, an *n*-type channel forms in the *p*-material region. This channel connects the source to the drain. As Figure 13.20 illustrates, the shape of the gate causes a wider channel to form than is created in the standard MOSFET. Because of this, the value of drain current is much higher for this component. It was the creation of power MOSFETs that gave MOSFETs the capability of competing successfully with BJTs.

*OBJECTIVE 10* ▶

VMOS is not susceptible to thermal runaway.

Another advantage of using VMOS is the fact that it is not susceptible to thermal runaway like its BJT counterpart. VMOS has a positive temperature coefficient, which means that the resistance of the device increases when the temperature increases. Thus, an increase in temperature will cause a decrease in drain current. This is why it can never experience thermal runaway. The positive temperature coefficient also allows VMOS to be connected in parallel to increase load power. BJTs cannot be connected this way due to differences in their $V_{BE}$ characteristics.

Two or more VMOS in parallel automatically regulate current by way of the positive temperature coefficient. Here is how it works: If one of the VMOS transistors demands more current, its heat causes an increase in temperature, which causes an increase in resistance, thus reducing current. This guarantees approximately equal amounts of current in each transistor.

Because of its higher drain current ratings and positive temperature coefficients, the VMOS device can be used in several applications where the standard MOSFET cannot be used. Some of these applications will be discussed briefly in Section 13.7.

## LDMOS

**Lateral double-diffused MOSFET (LDMOS)**
A high-power MOSFET that uses a narrow channel and a heavily doped *n*-type region to obtain high $I_D$ and low $r_{d(on)}$.

*A Practical Consideration:* VMOS and LDMOS devices have $g_{m0}$ values in the mho (or siemens) range. This means that they are capable of very high voltage gain values when used in common-source amplifiers.

Another type of power MOSFET is the **lateral double-diffused MOSFET, or LDMOS.** This type of MOSFET uses a very small channel region and a heavily doped *n*-type region $(n-)$ to obtain high drain current and low channel resistance $[r_{d(on)}]$. The basic construction of this enhancement-type MOSFET is shown in Figure 13.21.

The narrow channel (shaded area) is made up of the *p*-type material that lies between the $n-$ substrate and the $n+$ (lightly doped) source material. Since only *n*-type material lies between the channel and the drain, the effective length of the channel is extremely short. This, coupled with the *n* material in the channel-to-drain path, provides an extremely low typical value of $r_{d(on)}$. With a low channel resistance, the LDMOS device can handle very high values of current without generating any damaging amount of heat (power dissipation).

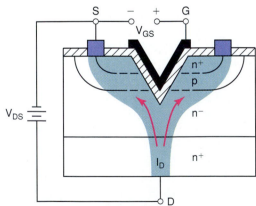

**FIGURE 13.20    VMOS operation.**

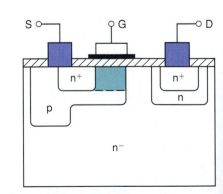

**FIGURE 13.21    LDMOS.**

The LDMOS has typical values of $r_{d(\text{on})}$ that are in the range of 2 Ω or less. With this low channel resistance, it is typically capable of handling currents as high as 20 A.

**Section Review**

1. Describe the physical construction of the VMOS.
2. How does the physical construction of the VMOS allow for a high value of $I_D$?
3. Describe the physical construction of the LDMOS.
4. How does the physical construction of the LDMOS allow for high values of $I_D$?

## 13.6 COMPLEMENTARY MOSFETs (CMOS): A MOSFET APPLICATION

The main contribution to electronics made by the MOSFET can be found in the area of *digital* (computer) electronics. In digital circuits, signals are made up of rapidly switching dc levels. An example of a digital signal is shown in Figure 13.22. This type of signal, referred to as a *rectangular wave,* is made up of two dc levels, called *logic levels.* For the waveform shown in Figure 13.22, the logic levels are 0 V and +5 V.

A group of circuits that have similar operating characteristics is referred to as a **logic family.** All the circuits in a given logic family will respond to the same logic levels, will have similar speed and power dissipation characteristics, and can be directly connected together. One such logic family is **complementary MOS, or CMOS logic.** This logic family is made up entirely of MOSFETs. The basic CMOS *inverter* is shown in Figure 13.23. An **inverter** is a digital circuit that converts one logic level to the other. When the input is at one logic level, the output will be at the other. For the circuit in Figure 13.23, a 0 V input will produce a +5 V output and a +5 V input will produce a 0 V output. The purpose served by such a circuit is beyond the scope of this book, but we will take a minute to look at its operation.

The operation of the CMOS inverter is easy to understand if we take a moment to look at each of the transistors as an individual *switch.* Figure 13.24 shows each of the MOSFETs as individual circuits. The *p*-channel MOSFET (Figure 13.24a) is an *enhancement*-type MOSFET, meaning that there is no physical channel from source to drain.

**Logic families**
Groups of digital circuits with nearly identical characteristics.

**Complementary MOS (CMOS)**
A logic family made up of MOSFETs.

**Inverter**
A logic-level converter.

**FIGURE 13.22**

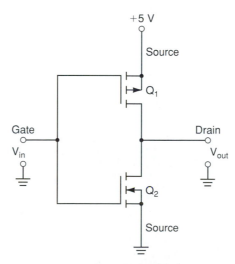

**FIGURE 13.23  CMOS inverter.**

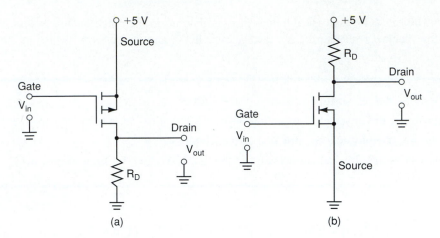

FIGURE 13.24

When the value of $V_{GS}$ for the circuit is 0 V, there is no current through the component. When $V_{GS}$ is $-5$ V, a channel is formed, and current can pass through the device. Note that the source is connected to the +5 V supply. Assuming that the input to the circuit can be only 0 or +5 V, we have the following voltage relationships for the circuit:

| $V_{in}$ (V) | $V_{SS}$ (V) | $V_{GS} = V_{in} - V_{SS}$ (V) |
|---|---|---|
| +5 | +5 | 0 |
| 0 | +5 | -5 |

Thus, when the input is +5 V, the MOSFET does not conduct, and when the input is 0 V, the MOSFET does conduct. Note that the MOSFET is said to be *on* when it is conducting and *off* when it is not. Now consider the *n*-channel MOSFET shown in Figure 13.24b. This is also an enhancement-type MOSFET. However, because it is an *n*-channel MOSFET, it has exactly the opposite characteristics of the *p*-channel MOSFET. In other words, it will conduct when $V_{GS} = +5$ V and will not conduct when the value of $V_{GS} = 0$ V. This is summarized as follows:

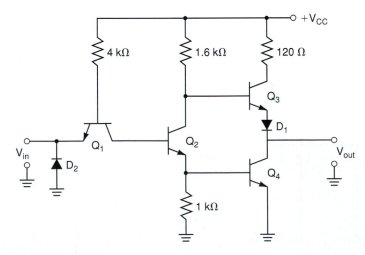

FIGURE 13.25   TTL inverter.

| $V_{in}$ (V) | $V_S$ (V) | $V_{GS} = V_{in} - V_S$ (V) |
|:---:|:---:|:---:|
| +5 | 0 | +5 |
| 0 | 0 | 0 |

Thus, when the input is +5 V, the MOSFET will conduct, and when the input is 0 V, the MOSFET will not conduct. Now let's put the two circuits together, as shown in Figure 13.23. Note that $Q_1$ serves as the drain resistor for $Q_2$, while $Q_2$ serves as the drain resistor for $Q_1$. When the input to the circuit is +5 V, $Q_1$ is off and $Q_2$ is on. Thus, the conduction path is between the output and ground through $Q_2$. When the input is at 0 V, $Q_1$ is on and $Q_2$ is off. Thus, the conduction path is between the output and $V_{DD}$ through $Q_1$. This operation is summarized as follows:

| $V_{in}$ (V) | $Q_1$ | $Q_2$ | $V_{out}$ (V) |
|:---:|:---:|:---:|:---:|
| 0 | On | Off | +5 |
| +5 | Off | On | 0 |

The relationship between $V_{in}$ and $V_{out}$ is as it should be for an inverter.

So, why is the CMOS logic circuit so popular? This question is best answered by taking a quick look at the BJT counterpart of the CMOS inverter.  ◀ *OBJECTIVE 11*

The basic TTL (a bipolar transistor logic family) inverter is shown in Figure 13.25. It is obvious that the CMOS circuit is far less complex than its TTL counterpart. This means that many more CMOS circuits can be developed in a given amount of space on an integrated circuit; that is, the CMOS circuits have a greater *packing density* than TTL circuits. In addition to the improved density, CMOS circuits have the following advantages over TTL circuits:

Why CMOS logic is so popular.

1. They draw little current from the supply and thus have better power dissipation ratings.

2. The low input current requirement of CMOS circuitry allows one CMOS output to drive an unlimited number of parallel CMOS loads; that is, you could connect any number of CMOS circuits in parallel and drive them with a single source. In contrast, the TTL circuit is usually restricted to driving no more than 10 other TTL inputs.

These advantages will become clearer to you as your study of electronics progresses in the area of digital electronics. At this point, however, you should have a pretty good idea of the benefits of using CMOS logic.

*Section Review*

1. What are *logic levels*?
2. What is a *logic family*?
3. What is an *inverter*?
4. Describe the physical construction of the CMOS inverter.
5. Describe the operation of the CMOS inverter.
6. List the advantages that CMOS logic has over TTL logic.

## 13.7 OTHER MOSFET APPLICATIONS

While CMOS logic is one of the primary MOSFET applications, there are many other applications as well. In this section, we will look at a representative sample of these MOSFET applications.

It is important to note that the D-MOSFET can be used in most applications where the JFET can be used. Thus, the full range of MOSFET applications goes well beyond those discussed in this section.

### Cascode Amplifiers

**Cascode amplifier**
A low-$C_{in}$ amplifier used in high-frequency applications.

Both JFETs and MOSFETs have relatively high input capacitances, as stated earlier. These input capacitances can adversely affect the high-frequency operation of the components. To overcome the effects of high input capacitance, the **cascode amplifier** was developed. The cascode amplifier consists of a common-source amplifier in series with a common-gate amplifier, as shown in Figure 13.26.

$Q_1$ in Figure 13.26 is a common-gate amplifier that is voltage divider biased. $Q_2$ is the common-source circuit and is self-biased. The input to the circuit is applied to $Q_2$, and the output is taken from the drain of $Q_1$. Therefore, the two MOSFETs are in series.

Since each MOSFET has a given amount of input capacitance, the series connection of the components is the same as hooking two capacitors in series. The resulting total capacitance is therefore less than either individual capacitance value. The lower overall capacitance of the cascode amplifier makes it more suitable for high-frequency operation, where values of capacitive reactance can become critical.

A simple cascode amplifier can be constructed using a dual-gate MOSFET. The dual-gate MOSFET equivalent of Figure 13.26 is shown in Figure 13.27. Gate 1 of the MOSFET is connected to act as a common-gate amplifier, while gate 2 is connected to act as the common-source amplifier. This amplifier has the same capacitance characteristics and high-frequency capabilities as those of the amplifier in Figure 13.26.

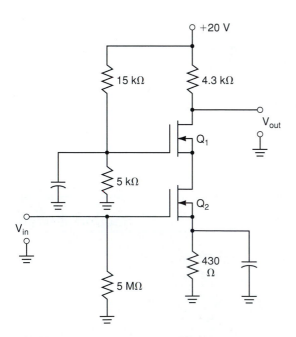

FIGURE 13.26   Cascode amplifier.

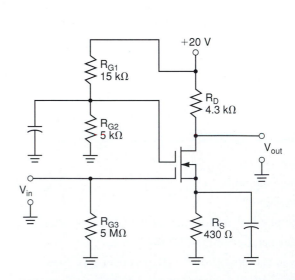

FIGURE 13.27   Dual-gate MOSFET cascode amplifier.

## RF Amplifier

The dual-gate MOSFET can also be used in an RF amplifier. Recall that the RF amplifier is the "front-end" amplifier in an AM receiver. A dual-gate MOSFET RF amplifier is shown in Figure 13.28. Note the zener diodes drawn inside the dual-gate MOSFET. These diodes are built into the MOSFET to protect the component from static electricity. As you know, the MOSFET is extremely sensitive to static electricity. The zener diodes protect the component inputs. Except for that, they have no effect on the operation of the circuit.

The RF amplifier shown is similar to the one discussed in Chapter 12. The antenna is connected to gate 1 by a tuned circuit, and the output of the amplifier is coupled to the next circuit (the mixer) by another tuned circuit. When the dual-gate MOSFET is used, almost no current is required from the antenna to cause the circuit to operate properly.

The advantage of using the dual-gate MOSFET is that it allows for the use of an *automatic gain control* (AGC) of the RF amplifier. A dc voltage is applied to gate 2 from the AGC circuit. This voltage is directly proportional to the strength of the signal received. When this AGC voltage is at a relatively high level (indicating a strong signal), the gain of the dual-gate MOSFET is decreased. When a weak signal is received, the voltage from the AGC circuitry decreases, and the gain of the dual-gate MOSFET increases. The overall effect of this operation is to prevent *fading*. If you have ever been in a car with an AM radio, you have experienced fading when going under a bridge. When the AM signal strength decreases, the radio fades. The use of AGC circuitry makes this fading minimal.

## Power MOSFET Drivers

At some point, you will probably study **digital communications.** In digital communications, information is converted (using one of several methods) into a series of digital signals. Those signals are then transmitted and received in digital form. Finally, the receiver converts the information back into its original form.

**Digital communications**
A method of transmitting and receiving information in digital form.

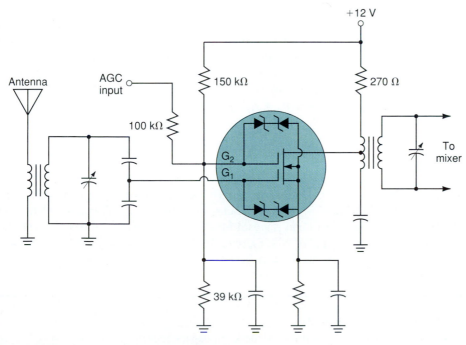

**FIGURE 13.28    Dual-gate MOSFET RF amplifier.**

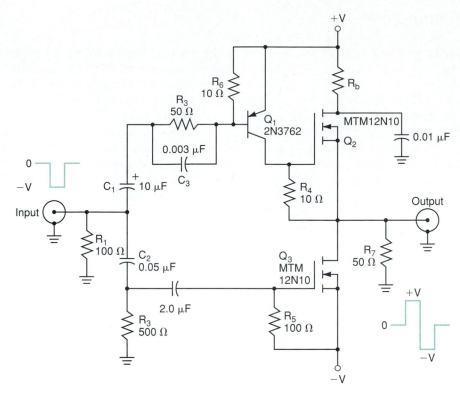

**FIGURE 13.29** (Copyright of Motorola. Used by permission.)

Many digital communications systems require the use of power amplifiers that can produce high-speed, high-current digital outputs. These signals can be produced using a power MOSFET driver circuit like the one shown in Figure 13.29.

The high-current capability of the driver is provided by the low channel resistance [$r_{d(on)}$] of the power MOSFETs. Remember, power MOSFETs typically have $r_{d(on)}$ values of 2 Ω or less. Thus, depending on the values of $+V$, $-V$, and $R_b$, the MOSFET pair can output currents as high as 20 to 35 A.

The high-speed quality of the circuit is provided by a number of factors. First, the resistance values in the driver circuit are extremely low. This allows the capacitances in the circuit to charge and discharge extremely rapidly. Also, note the parallel circuit made up of $R_3$ and $C_3$ in the base circuit of $Q_1$. The capacitor ($C_3$) is called a **speed-up capacitor.** This capacitor is used to improve the switching time (the time taken to go back and forth between saturation and cutoff) of $Q_1$. The operation of speed-up capacitors is discussed in detail in Chapter 19.

**Speed-up capacitor**
A capacitor used in the base circuit of a BJT to allow the device to switch rapidly between saturation and cutoff.

## VMOS Applications

You may remember that the Darlington pair BJT configuration is used in applications where the standard BJT will not handle the current requirements. The same relationship exists between VMOS devices and standard MOSFETs. A VMOS device may be used in almost any enhancement-type MOSFET circuit. However, it is usually used only when its high current capabilities are required.

**Section Review**

1. What is a *cascode amplifier*?
2. How does the cascode amplifier reduce input capacitance?

3. Describe the dual-gate MOSFET cascode amplifier.

4. Describe the dual-gate MOSFET RF amplifier.

5. What is *automatic gain control (AGC)*? How is it achieved in the dual-gate MOSFET RF amplifier?

6. What is *fading*?

7. What is *digital communications*?

8. What purpose does the power MOSFET driver serve in digital communications?

9. When is VMOS typically used?

---

The following terms were introduced and defined in this chapter:

*Key Terms*

| | | |
|---|---|---|
| cascode amplifier | E-MOSFET | MOSFET |
| complementary MOS (CMOS) | enhancement-mode operation | speed-up capacitor |
| depletion-mode operation | inverter | substrate |
| digital communications | lateral double-diffused | threshold voltage, $V_{GS(\text{th})}$ |
| D-MOSFET | MOSFET (LDMOS) | vertical MOSFET (VMOS) |
| drain-feedback bias | logic families | zero bias |
| dual-gate MOSFET | logic levels | |

---

| Equation Number | Equation | Section Number |
|---|---|---|
| (13.1) | $I_D = I_{DSS}\left(1 - \dfrac{V_{GS}}{V_{GS(\text{off})}}\right)^2$ | 13.2 |
| (13.2) | $I_{D(\text{sat})} = \dfrac{V_{DD}}{R_D}$ | 13.2 |
| (13.3) | $V_{DS(\text{off})} = V_{DD}$ | 13.2 |
| (13.4) | $g_m = g_{m0}\left(1 - \dfrac{V_{GS}}{V_{GS(\text{off})}}\right)$ | 13.2 |
| (13.5) | $I_D = k[V_{GS} - V_{GS(\text{th})}]^2$ | 13.3 |
| (13.6) | $k = \dfrac{I_{D(\text{on})}}{[V_{GS(\text{on})} - V_{GS(\text{th})}]^2}$ | 13.3 |
| (13.7) | $V_{GS} = V_{DS}$ | 13.3 |
| (13.8) | $V_{DS} = V_{DD} - R_D I_{D(\text{on})}$ | 13.3 |

*Equation Summary*

---

13.1.  See Figure 13.30.

13.2.  27.44 mA

13.3.  $I_D = 14$ mA, $V_{DS} = 6.86$ V

*Answers to the Example Practice Problems*

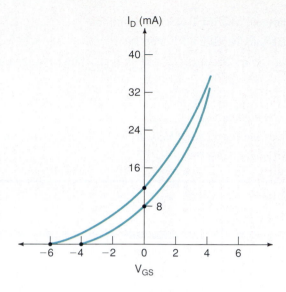

**FIGURE 13.30**

## Practice Problems

### §13.2

1. A D-MOSFET has values of $V_{GS(off)} = -4$ V and $I_{DSS} = 8$ mA. Plot the transconductance curve for the device.

2. A D-MOSFET has values of $V_{GS(off)} = -8$ V and $I_{DSS} = 12$ mA. Plot the transconductance curve for the device.

3. A D-MOSFET has values of $V_{GS(off)} = -4$ to $-6$ V and $I_{DSS} = 6$ to 9 mA. Plot the minimum and maximum transconductance curves for the device.

4. A D-MOSFET has values of $V_{GS(off)} = -4$ to $-10$ V and $I_{DSS} = 5$ to 10 mA. Plot the minimum and maximum transconductance curves for the device.

5. The D-MOSFET described in Problem 1 has a value of $g_{m0} = 2000$ μS. Determine the values of $g_m$ at $V_{GS} = -2$ V, $V_{GS} = 0$ V, and $V_{GS} = +2$ V.

6. The D-MOSFET described in Problem 2 has a value of $g_{m0} = 3000$ μS. Determine the values of $g_m$ at $V_{GS} = -5$, $V_{GS} = 0$ V, and $V_{GS} = +5$ V.

### §13.3

7. An E-MOSFET has ratings of $V_{GS(th)} = 4$ V and $I_{D(on)} = 12$ mA at $V_{GS} = 8$ V. Determine the value of $I_D$ for the device when $V_{GS} = 6$ V.

8. An E-MOSFET has ratings of $V_{GS(th)} = 2$ V and $I_{D(on)} = 10$ mA at $V_{GS} = 8$ V. Determine the value of $I_D$ for the device when $V_{GS} = +12$ V.

9. An E-MOSFET has ratings of $V_{GS(th)} = 1$ V and $I_{D(on)} = 8$ mA at $V_{GS} = 4$ V. Determine the values of $I_D$ for the device when $V_{GS} = 1$ V, $V_{GS} = 4$ V, and $V_{GS} = 5$ V.

10. An E-MOSFET has values of $V_{GS(th)} = 3$ V and $I_{D(on)} = 2$ mA at $V_{GS} = 5$ V. Determine the values of $I_D$ for the device when $V_{GS} = 3$ V, $V_{GS} = 5$ V, and $V_{GS} = 8$ V.

11. Calculate the values of $I_D$ and $V_{DS}$ for the amplifier in Figure 13.31.

12. Calculate the values of $I_D$ and $V_{DS}$ for the amplifier in Figure 13.32.

13. Calculate the values of $I_D$ and $V_{DS}$ for the amplifier in Figure 13.33.

14. Calculate the values of $I_D$ and $V_{DS}$ for the amplifier in Figure 13.34.

15. Calculate the values of $I_D$ and $V_{DS}$ for the amplifier in Figure 13.35.

16. Calculate the values of $I_D$ and $V_{DS}$ for the amplifier in Figure 13.36.

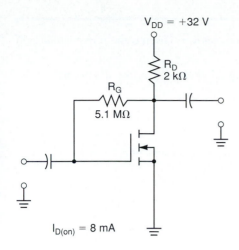

FIGURE 13.31

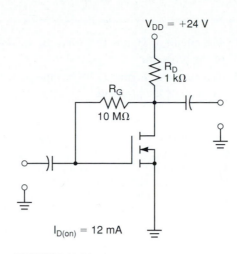

FIGURE 13.32

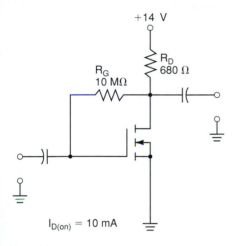

FIGURE 13.33

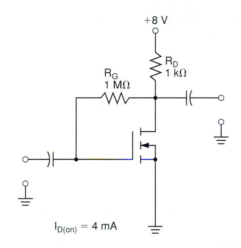

FIGURE 13.34

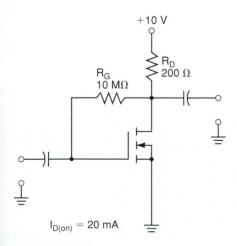

FIGURE 13.35

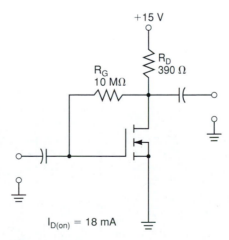

FIGURE 13.36

17. The circuit in Figure 13.37 has the waveforms shown. Discuss the possible cause(s) of the problem.

18. The circuit in Figure 13.38 has the dc voltages indicated. Discuss the possible cause(s) of the problem.

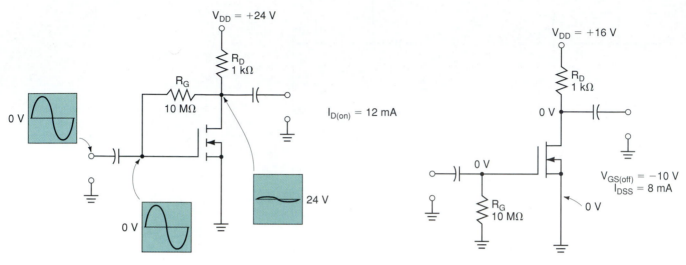

**FIGURE 13.37**                                                              **FIGURE 13.38**

## The Brain Drain

19. Determine the range of $A_v$ values for the amplifier in Figure 13.39.

20. Determine the range of $A_v$ values for the amplifier in Figure 13.40.

21. Determine the range of $I_D$ values for the amplifier in Figure 13.41.

22. The constant-current bias circuit in Figure 13.42 won't work. Why not?

23. The MOSFET shown in Figure 13.43 has a maximum gate current of 800 pA when $V_{GS} = 32$ V. Determine the input impedance of the device and the circuit. Then determine the input impedance of the circuit, assuming that the MOSFET has infinite input impedance. What is the difference between the two values of $Z_{in}$?

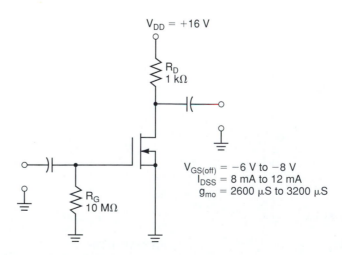

**FIGURE 13.39**

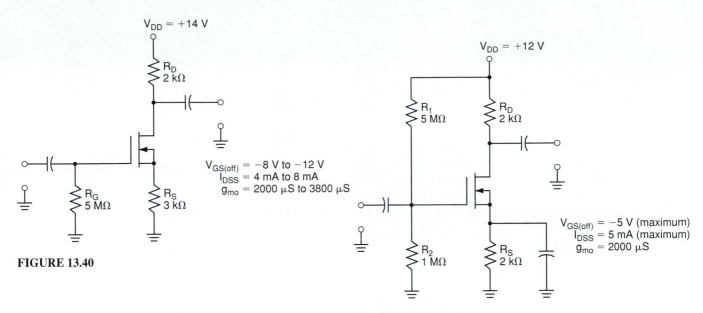

FIGURE 13.40

$V_{DD} = +14$ V

$R_D$
2 kΩ

$R_G$
5 MΩ

$R_S$
3 kΩ

$V_{GS(off)} = -8$ V to $-12$ V
$I_{DSS} = 4$ mA to 8 mA
$g_{mo} = 2000$ μS to 3800 μS

$V_{DD} = +12$ V

$R_1$
5 MΩ

$R_D$
2 kΩ

$R_2$
1 MΩ

$R_S$
2 kΩ

$V_{GS(off)} = -5$ V (maximum)
$I_{DSS} = 5$ mA (maximum)
$g_{mo} = 2000$ μS

FIGURE 13.41

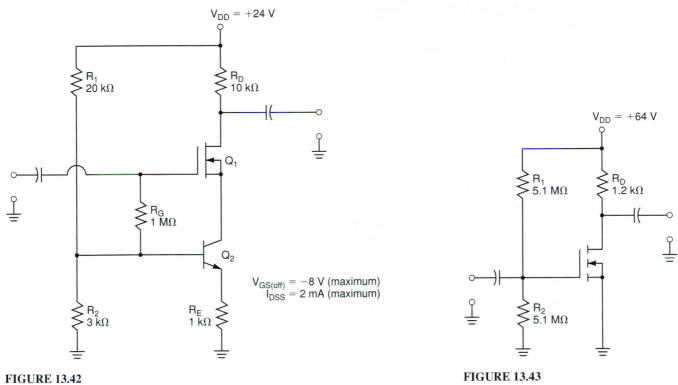

FIGURE 13.42

$V_{DD} = +24$ V

$R_1$
20 kΩ

$R_D$
10 kΩ

$Q_1$

$R_G$
1 MΩ

$Q_2$

$R_2$
3 kΩ

$R_E$
1 kΩ

$V_{GS(off)} = -8$ V (maximum)
$I_{DSS} = 2$ mA (maximum)

FIGURE 13.43

$V_{DD} = +64$ V

$R_1$
5.1 MΩ

$R_D$
1.2 kΩ

$R_2$
5.1 MΩ

**Suggested Computer Applications Problems**

24. Write a program that will perform the complete dc analysis of a zero-bias circuit when provided with the proper information.

25. Write a program that will determine the value of $I_D$ for an E-MOSFET at any given value of $V_{GS}$. The program should include the determination of the value of $k$ for the device.

26. Write a program that will perform the complete dc analysis of a drain-feedback bias circuit when provided with the required information.

# 14

# AMPLIFIER FREQUENCY RESPONSE

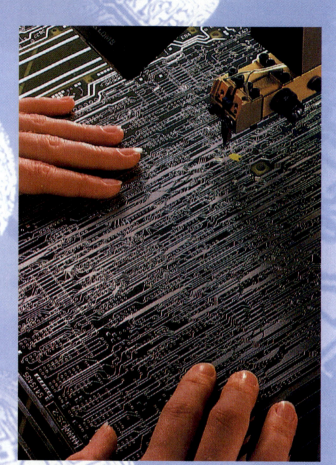

M odern production technology is used to produce circuit boards that are far more complex than their predecessors.

## OUTLINE

## OBJECTIVES

*After studying the material in this chapter, you should be able to:*

1. Define *bandwidth, cutoff frequency,* and *midpoint frequency,* and identify each on a frequency-response curve.

2. Calculate any two of the following values, given the other two: $f_1$, $f_2$, center frequency ($f_0$), or bandwidth.

3. Describe the *decade* and *octave* frequency multipliers.

4. Compare and contrast the *Bode plot* with the frequency-response curve.

5. Perform a complete low-frequency analysis of a BJT amplifier.

6. Discuss the concept of gain roll-off and calculate its effect on voltage gain at a given operating frequency.

7. Explain why BJT internal capacitances are not considered in low-frequency analyses.

8. Calculate the Miller input and output capacitance values for a BJT amplifier.

9. Perform a complete high-frequency analysis of a BJT amplifier.

10. Compare high-frequency roll-off rates to low-frequency roll-off rates.

11. Perform the low-frequency-response analysis of an FET amplifier.

12. Perform the high-frequency-response analysis of an FET amplifier.

13. Discuss and analyze the frequency response of a multistage amplifier.

# 14.1 BASIC CONCEPTS

In this chapter, we will take a look at the effects of operating frequency on BJT and FET circuit gain. As we progress through our discussions, you will see that several principles apply to all types of amplifiers. In this section, we will take a look at each of these principles.

## Bandwidth

**Bandwidth (BW)**
The range of frequencies over which gain is relatively constant.

Most amplifiers have relatively constant gain over a certain range, or *band,* of frequencies. This band of frequencies is called the **bandwidth** of the amplifier. The bandwidth for a given amplifier depends on the circuit component values and the type of active component(s) used. Later in this chapter, you will be shown how to calculate the bandwidth for any given amplifier.

When an amplifier is operated within its bandwidth, the current, voltage, and power gain values for the amplifier are calculated as shown earlier in the text. For clarity, these values of $A_i$, $A_v$, and $A_p$ are referred to as *midband gain* values. For example, the *midband power gain, $A_{p(mid)}$*, of an amplifier is the power gain of the circuit when it is operated within its bandwidth. Again, the exact value of $A_{p(mid)}$ (as well as the other gain values) will be calculated as shown earlier in the text and will vary from one circuit to another.

**Frequency-response curve**
A curve showing the relationship between amplifier gain and operating frequency.

A graphic representation of the relationship between amplifier gain and operating frequency is the **frequency-response curve.** A *simplified* frequency-response curve is shown in Figure 14.1.

As the frequency-response curve shows, the power gain of an amplifier remains relatively constant for a band of frequencies. When the operating frequency starts to go outside this range, the gain of the amplifier begins to drop off rapidly.

Two frequencies of interest, $f_1$ and $f_2$, are marked on the frequency-response curve. These are the frequencies at which power gain drops to 50 percent of $A_{p(mid)}$. For example, let's assume that the amplifier represented by Figure 14.1 has the following values: $f_1 = 10$ kHz, $f_2 = 100$ kHz, and $A_{p(mid)} = 500$. As long as the amplifier is operated between $f_1$ and $f_2$, the power gain of the amplifier will be approximately 500. If the operating frequency increases to 100 kHz or drops to 10 kHz, the power gain of the amplifier will drop to 250. If the input frequency goes far enough outside the $f_1$ or $f_2$ limits, the gain of the amplifier will continue to drop until it eventually reaches *unity gain* ($A_p = 1$). At this point, the amplifier no longer serves any useful purpose.

*A Practical Consideration:* The gain of an amplifier will start to drop off well before the operating frequency reaches $f_1$ or $f_2$. However, by the time $f_1$ or $f_2$ is reached, the value of $A_p$ will have dropped by approximately 50 percent.

The frequencies $f_1$ and $f_2$ are called the **cutoff frequencies** of an amplifier. These two frequencies are considered to be the limits of the bandwidth of the amplifier. This

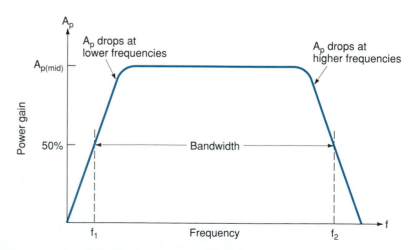

**FIGURE 14.1    A simplified frequency response curve.**

point is illustrated in Figure 14.1. Since $f_1$ and $f_2$ are the bandwidth limits, the bandwidth of an amplifier can be found as the *difference* between $f_1$ and $f_2$, as follows:

$$BW = f_2 - f_1 \qquad \textbf{(14.1)}$$

**Cutoff frequencies**
The frequencies at which $A_p$ drops to 50 percent of $A_{p(mid)}$. These frequencies are designated as $f_1$ and $f_2$.

where BW = the amplifier bandwidth, in hertz
$\quad f_2$ = the *upper* cutoff frequency
$\quad f_1$ = the *lower* cutoff frequency

The following chart shows some examples of $f_1$, $f_2$, and BW value combinations.

| $f_1$ (kHz) | $f_2$ (kHz) | BW (kHz) |
|---|---|---|
| 10 | 100 | 90 |
| 25 | 250 | 225 |
| 0 (dc) | 400 | 400 |

Note that the bandwidth in each case is found as the *difference* between $f_1$ and $f_2$. Every amplifier has upper and lower cutoff frequencies, and these frequencies are used to determine the bandwidth of the amplifier. This point is illustrated further in Example 14.1.

---

The frequency-response curve in Figure 14.2 represents the operation of a given amplifier. Determine the bandwidth of the circuit.

**EXAMPLE 14.1**

***Solution:*** The cutoff frequencies correspond to the *half-power points* on the curve, as shown. For this particular circuit, these values are approximately

$$f_1 = 13 \text{ kHz}$$

and

$$f_2 = 222 \text{ kHz}$$

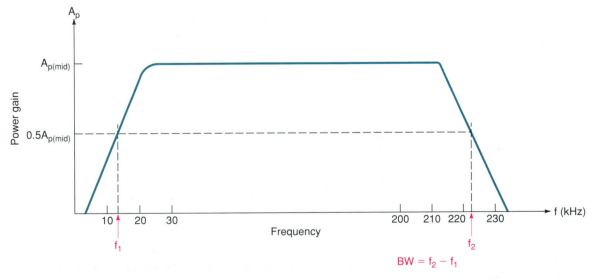

**FIGURE 14.2**

The bandwidth is found as

$$BW = f_2 - f_1$$
$$= 222 \text{ kHz} - 13 \text{ kHz}$$
$$= 209 \text{ kHz}$$

**PRACTICE PROBLEM 14.1**

Determine the bandwidth for the amplifier represented by the frequency-response curve in Figure 14.3.

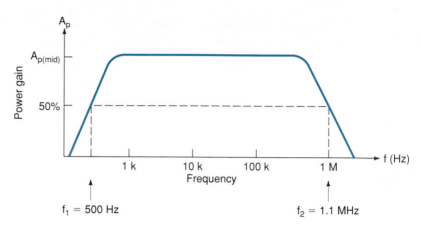

**FIGURE 14.3**

**Center frequency ($f_0$)**
The frequency that equals the geometric average of $f_1$ and $f_2$.

There are cases when it is desirable to know the **center frequency** of a given amplifier. The center frequency ($f_0$) is *the geometric average of the cutoff frequencies.* By formula,

$$f_0 = \sqrt{f_1 f_2} \qquad\qquad (14.2)$$

$A_p$ is at its maximum value when an amplifier is operated at $f_0$.

When the operating frequency of an amplifier is equal to the value of $f_0$, the power gain of the amplifier is at its maximum value. Then, as the operating frequency varies toward $f_1$ or $f_2$, the power gain begins to slowly drop off. By the time the operating frequency reaches either of the cutoff frequencies, power gain will have dropped to 50 percent of $A_{p(mid)}$. Example 14.2 demonstrates the calculation of $f_0$ for an amplifier.

**EXAMPLE 14.2**

Determine the value of $f_0$ for the amplifier represented by Figure 14.2.

***Solution:*** The values of $f_1$ and $f_2$ were found to be 13 kHz and 222 kHz, respectively. Using these two values in equation (14.2), the value of the center frequency is found as

$$f_0 = \sqrt{f_1 f_2}$$
$$= \sqrt{(13 \text{ kHz})(222 \text{ kHz})}$$
$$= 53.72 \text{ kHz}$$

**PRACTICE PROBLEM 14.2**

Determine the value of $f_0$ for the amplifier represented in Figure 14.3.

Since $f_0$ is the *geometric average* of the cutoff frequencies, the ratio of $f_0$ to $f_1$ equals the ratio of $f_2$ to $f_0$. By formula,

$$\frac{f_0}{f_1} = \frac{f_2}{f_0} \qquad (14.3)$$

Example 14.3 illustrates the fact that these ratios are equal.

**EXAMPLE 14.3**

Show that the ratio of $f_0$ to $f_1$ is equal to the ratio of $f_2$ to $f_0$ for the amplifier described in Examples 14.1 and 14.2.

**Solution:** The following values were obtained in the two examples:

$$f_1 = 13 \text{ kHz}$$
$$f_2 = 222 \text{ kHz}$$
$$f_0 = 53.72 \text{ kHz}$$

The ratio of $f_0$ to $f_1$ is found as

$$\frac{f_0}{f_1} = \frac{53.72 \text{ kHz}}{13 \text{ kHz}}$$
$$= 4.13$$

The ratio of $f_2$ to $f_0$ is found as

$$\frac{f_2}{f_0} = \frac{222 \text{ kHz}}{53.72 \text{ kHz}}$$
$$= 4.13$$

As you can see, the frequency ratios are equal.

**PRACTICE PROBLEM 14.3**

Show that equation (14.3) holds true for the circuit described in Practice Problems 14.1 and 14.2.

Equation (14.3) is important because it provides us with some insight into the relationship between $f_0$, $f_1$, and $f_2$. When an amplifier is operated at its center frequency, the input frequency can vary by the same *factor* in each direction before the power gain of the amplifier drops to 50 percent of $A_{p(\text{mid})}$. For example, let's say that an amplifier has the following values:

$$f_1 = 6 \text{ kHz}$$
$$f_2 = 150 \text{ kHz}$$
$$f_0 = 30 \text{ kHz}$$

As you can see, $f_0$ is five times the value of $f_1$. At the same time, $f_2$ is five times the value of $f_0$. Thus, if the amplifier is operated at $f_0$, the input frequency can decrease by a factor of 5 before hitting the lower cutoff frequency ($f_1$) or can increase by a factor of 5 before hitting the upper cutoff frequency ($f_2$). This is where the term *center frequency* comes from.

Equation (14.3) can be transposed to provide us with two other useful equations. Each can be used to predict the value of $f_1$ or $f_2$ when the other cutoff frequency and the center frequency are known, as follows:

$$f_1 = \frac{f_0^2}{f_2}$$

(14.4)

and

$$f_2 = \frac{f_0^2}{f_1}$$

(14.5)

Once the unknown cutoff frequency has been determined, we can predict the bandwidth of the amplifier, as Examples 14.4 and 14.5 demonstrate.

---

**EXAMPLE 14.4**

The values of $f_0$ and $f_2$ for an amplifier are measured at 60 kHz and 300 kHz, respectively. Determine the values of $f_1$ and BW for the amplifier.

**Solution:**  The value of $f_1$ is found using equation (14.4), as follows:

$$f_1 = \frac{f_0^2}{f_2}$$
$$= \frac{(60 \text{ kHz})^2}{300 \text{ kHz}}$$
$$= 12 \text{ kHz}$$

The value of BW is now found as

$$\text{BW} = f_2 - f_1$$
$$= 300 \text{ kHz} - 12 \text{ kHz}$$
$$= 288 \text{ kHz}$$

**PRACTICE PROBLEM 14.4**

An amplifier has values of $f_2 = 300$ kHz and $f_0 = 30$ kHz. Determine the bandwidth of the amplifier.

---

**EXAMPLE 14.5**

The values of $f_0$ and $f_1$ for an amplifier are measured at 40 kHz and 8 kHz, respectively. Determine the values of $f_2$ and BW for the amplifier.

**Solution:**  The value of $f_2$ is found using equation (14.5), as follows:

$$f_2 = \frac{f_0^2}{f1}$$
$$= \frac{(40 \text{ kHz})^2}{8 \text{ kHz}}$$
$$= 200 \text{ kHz}$$

The value of BW is now found as

$$BW = f_2 - f_1$$
$$= 200\ \text{kHz} - 8\ \text{kHz}$$
$$= 192\ \text{kHz}$$

**PRACTICE PROBLEM 14.5**

An amplifier has values of $f_0 = 25$ kHz and $f_1 = 5$ kHz. Determine the bandwidth of the amplifier.

You have been shown the relationship among the cutoff frequencies ($f_1$ and $f_2$), the center frequency ($f_0$), and the bandwidth (BW) of an amplifier. In later sections, you will be shown the circuit analysis equations used to determine all these values.

## Measuring $f_1$ and $f_2$

When you are working with an amplifier, the $f_1$ and $f_2$ frequencies for the amplifier can be measured easily using an oscilloscope. The procedure is as follows:

1. Set the amplifier for the maximum undistorted output signal. This is done by varying the *amplitude* of $v_{\text{in}}$ to the amplifier. As a first step, set the input frequency to approximately 100 kHz.

2. Establish that you are operating in the midband of the amplifier by varying the *frequency* of the input several kilohertz in both directions. If you are in the midband range of the amplifier, the *amplitude* of the amplifier output will *not* change significantly as you vary the input frequency.

3. If you are not in the midband frequency range, adjust $f_{\text{in}}$ until you are.

4. Adjust the *volts-div* calibration control until the waveform fills exactly *seven* major divisions (peak-to-peak).

5. To determine the value of $f_1$, decrease $f_{\text{in}}$ until the output waveform fills only *five* major divisions. At this frequency, the amplitude of the output has changed by a ratio of $5/7 \cong .707$. This ratio indicates that we are operating the circuit at its cutoff frequency.

6. Increase the frequency until the same thing happens on the high end of operation. Measure $f_2$ at this frequency.

In basic electronics, you learned that *power varies with the square of voltage.* The same principle holds true for $A_p$ and $A_v$. Thus, when $A_p$ drops to 0.5 of $A_{p(\text{mid})}$, the voltage gain will drop by the *square root* of the *0.5* factor, 0.707. As you will see, the fact that $A_v$ equals 0.707 $A_{v(\text{mid})}$ at the cutoff frequencies is important when deriving the equations for $f_1$ and $f_2$.

**Lab Reference:** This technique for measuring amplifier cutoff frequencies is demonstrated in Exercise 32.

*A Practical Consideration:* Most ac voltmeters do not have the frequency-handling capability needed to measure bandwidth. For this reason, the method covered here emphasizes the use of the oscilloscope. When using the oscilloscope to measure bandwidth, you should use a ×10 probe to reduce the effects of the scope's input capacitance on your measurements.

## Gain and Frequency Measurements

Up to this point, we have used *simplified* frequency-response curves in the various illustrations. A more practical frequency-response curve is shown in Figure 14.4. As you can see, this curve is a bit more complicated than any you have seen thus far. The primary differences are the following:

1. The *y*-axis of the graph represents a power gain *ratio,* rather than specific power gain values.

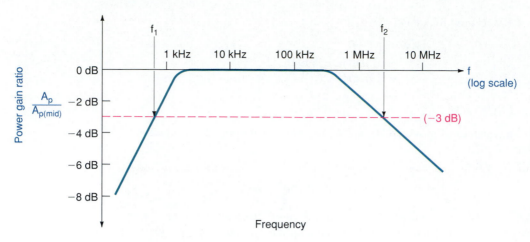

**FIGURE 14.4** **A more practical frequency-response curve.**

2. The power gain ratios are measured in *decibels* (dB).

3. The frequency units are based on a *logarithmic* scale; that is, each major division is a whole number multiple of the previous major division.

We will look at all of these differences so that you will have no problem with reading an actual frequency-response curve.

## Decibel Power Gain

A decibel (dB) is a ratio of output power to input power that is equal to 10 times the common logarithm of that ratio. One advantage of using decibels is that they allow us to represent very large power ratios as relatively small numbers. In Chapter 8, we discussed the concept of dB gain values in detail. The material presented here is intended only as a brief review.

Frequency response curves and specification sheets often list gain values that are measured in decibels. As shown in Chapter 8, the decibel power gain of an amplifier is found as

$$A_{p(\text{dB})} = 10 \log A_p \tag{14.6}$$

or

$$A_{p(\text{dB})} = 10 \log \frac{P_{\text{out}}}{P_{\text{in}}} \tag{14.7}$$

Examples 14.6 and 14.7 review the process for finding the power gain of an amplifier in decibels.

**EXAMPLE 14.6**

A given amplifier has values of $P_{\text{in}} = 100 \ \mu\text{W}$ and $P_{\text{out}} = 2 \ \text{W}$. What is the dB power gain of the amplifier?

*Solution:* The dB power gain of the amplifier is found as

$$A_{p(\mathrm{dB})} = 10 \log \frac{P_{\mathrm{out}}}{P_{\mathrm{in}}}$$

$$= 10 \log \frac{2\ \mathrm{W}}{100\ \mu\mathrm{W}}$$

$$= 10 \log (20{,}000)$$

$$= 10(4.301)$$

$$= 43.01\ \mathrm{dB}$$

**PRACTICE PROBLEM 14.6**

An amplifier has values of $P_{\mathrm{in}} = 420\ \mu\mathrm{W}$ and $P_{\mathrm{out}} = 6\ \mathrm{W}$. What is the dB power gain of the amplifier?

For multistage amplifiers, the total gain of the circuit is the *sum* of the gains in decibels.

For the multistage amplifier represented in Figure 14.5, prove that the total power gain is equal to $A_{p(\mathrm{dB})} + A_{p2(\mathrm{dB})} + A_{p3(\mathrm{dB})}$.

**EXAMPLE 14.7**

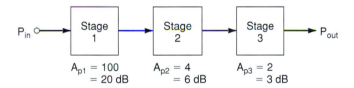

**FIGURE 14.5**

**Solution:** The gain of each stage is shown as a ratio value and as a decibel value. As you recall, the total power gain of a multistage amplifier is equal to the product of the individual power ratio values. By formula,

$$A_{p(\mathrm{total})} = (A_{p1})(A_{p2})(A_{p3})$$
$$= (100)(4)(2)$$
$$= 800$$

If we convert the ratio value of (800) to decibels, we get the following:

$$A_{p(\mathrm{dB})} = 10 \log A_p$$
$$= 10 \log 800$$
$$= 29\ \mathrm{dB}$$

Now, if we simply add the decibel gains of the three stages, we get the following:

$$A_{p(\mathrm{total})} = A_{p1(\mathrm{dB})} + A_{p2(\mathrm{dB})} + A_{p3(\mathrm{dB})}$$
$$= 20\ \mathrm{dB} + 6\ \mathrm{dB} + 3\ \mathrm{dB}$$
$$= 29\ \mathrm{dB}$$

Since both methods of determining total dB power gain yielded the same results, we have shown that you can determine total multistage gain by simply adding the individual dB power gain values.

Now, let's relate what you have learned about dB gain values to the frequency-response curve in Figure 14.4. Note that the y-axis is labeled as the *ratio* of power gain at a given frequency to midband power gain. When the circuit represented is operated in the midband range,

$$A_p = A_{p(\text{mid})}$$

and

$$\frac{A_p}{A_{p(\text{mid})}} = 1$$

If we convert a power ratio of 1 to dB form, we obtain

$$A_{p(\text{dB})} = 10 \log 1$$
$$= 0 \text{ dB}$$

Thus, the midband power ratio shown in Figure 14.4 is labeled as 0 dB. This means that the power gain of the amplifier is equal to $A_{p(\text{mid})}$ throughout the midband frequency range. This is what we have been saying all along.

When $(A_p/A_{p(\text{mid})}) = -3$ dB, the power gain of the amplifier has dropped to 50 percent of $A_{p(\text{mid})}$. Therefore, the output power has dropped to approximately 50 percent of its midband value. Based on this relationship, we can redefine $f_1$ and $f_2$ (the cutoff frequencies) as follows: *The cutoff frequencies of an amplifier are those frequencies at which the power gain of the amplifier drops by 3 dB*. For this reason, the cutoff frequencies of an amplifier are often called the *upper and lower 3 dB points*.

## Frequency Units and Terminology

OBJECTIVE 3 ▶

The frequency response curve in Figure 14.4 utilizes a *logarithmic scale* to measure frequency values. A logarithmic scale is one that uses a *geometric* progression of units. In a geometric progression, the value of each division is *a whole number multiple of the value for the previous division*. For example, in Figure 14.4, the value of each division on the horizontal scale is *ten times* the value of the previous division.

When the multiplier rate for a frequency scale is 10, the scale is referred to as a **decade scale.** Note that a **decade** is a frequency multiplier equal to 10. The scale in Figure 14.4 is a decade scale.

Another commonly used frequency multiplier is the **octave.** One octave is a frequency multiplier of *two*. Thus, when you increase frequency by one octave, you *double* the original frequency. The following table shows a series of *fundamental* (starting) frequencies and the decade and octave frequencies for each:

**Decade**
A frequency multiplier equal to ten (10).

**Octave**
A frequency multiplier equal to two (2).

| Fundamental | Octave | Decade |
|---|---|---|
| 100 HZ | 200Hz | 1 kHz |
| 500 Hz | 1 kHz | 5 kHz |
| 1.5 MHz | 3 MHz | 15 MHz |
| $f_x$ | $2f_x$ | $10f_x$ |

The terms *octave* and *decade* are used extensively in describing the rate at which gain varies with frequency. For example, an amplifier may be said to have a *roll-off rate of 20 dB/decade*. This means that the gain of the amplifier decreases at a rate of 20 dB for each decade that the frequency goes beyond the limit of either $f_1$ or $f_2$. Roll-off rates are discussed in more detail later in this chapter.

## Bode Plots

The **Bode plot** (pronounced "bō-dē" plot) is a variation on the basic frequency-response curve. A Bode plot is shown in Figure 14.6. As you can see, the Bode plot differs from the frequency-response curves you have seen so far in one respect: The value of $\Delta A_{p(mid)}$ is assumed to be zero until the cutoff frequencies are reached. Then the power gain of the amplifier is assumed to drop at a set rate of 20 dB/decade. For comparison, the standard frequency-response curve (labeled FRC) is included in the figure. Note that the power gain of the amplifier is shown to be down by 3 dB at each of the cutoff frequencies on the FRC, while the power gain is still shown to be at its midband value on the Bode plot.

Bode plots are typically used in place of other types of frequency-response curves because they are easier to read. Since Bode plots are the most commonly used frequency-response graphs, we will use them throughout the remainder of this chapter. Just remember, while the Bode plot shows $\Delta A_{p(dB)}$ to be zero at the cutoff frequencies, the actual value is $-3$ dB.

At this point, we have looked at the overall relationship between gain and frequency. We have also discussed the basic units used to measure gain and frequency. Now we will move on to specific types of circuits, their respective calculations, and the reasons for their response characteristics.

◄ *OBJECTIVE 4*

**Bode plot**
A frequency-response curve that assumes $\Delta A_{p(mid)}$ is zero until the cutoff frequency is reached.

---

1. What is *bandwidth*?

2. Draw a frequency-response curve and identify the following: $f_1$, $f_2$, bandwidth, and midpoint frequency.

3. What is the numeric relationship among the values of $f_1$, $f_0$, and $f_2$ for a given amplifier?

4. What is the procedure for measuring $f_1$ and $f_2$ with an oscilloscope?

5. Which oscilloscope probe should be used for high-frequency measurements? Why?

6. In what form are gain values typically measured?

7. What type of frequency scale is normally used in frequency-response curves?

8. What are the advantages of using decibel gain values?

9. Why are the cutoff frequencies referred to as the *upper* and *lower 3-dB points*?

### Section Review

**Lab Reference:** A Bode plot similar to the one in Figure 14.6 is plotted using measured values in Exercise 32.

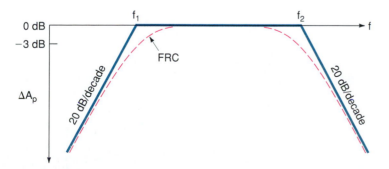

**FIGURE 14.6   Bode plot.**

10. What is a *decade*? What is an *octave*?

11. Why are decade and octave frequency scales used?

12. Compare and contrast the *Bode plot* with a frequency-response curve.

## 14.2 BJT AMPLIFIER FREQUENCY RESPONSE

The cutoff frequencies of an amplifier are determined by the combination of amplifier capacitance and resistance values. In this section, you will be shown how the capacitance and resistance values in a BJT amplifier determine the frequency-response characteristics of the circuit.

### Low-Frequency Response

*OBJECTIVE 5* ▶ We'll start our discussion on BJT amplifier low-frequency response by taking a look at its base circuit. For reference, this circuit is shown in Figure 14.7. As the figure shows, the voltage applied to the base of the transistor is equal to the voltage across $R_{in}$. Since $R_{in}$ is in series with the coupling capacitor, the two components form a voltage divider. For this divider circuit, the voltage across $R_{in}$ is found as

$$v_b = v_{in} \frac{R_{in}}{\sqrt{X_C^2 + R_{in}^2}} \tag{14.8}$$

Now, let's take a look at what happens to the fraction in equation (14.8) when $X_C = R_{in}$. When this happens, the fraction simplifies as follows:

$$\frac{R_{in}}{\sqrt{X_C^2 + R_{in}^2}} = \frac{R_{in}}{\sqrt{R_{in}^2 + R_{in}^2}}$$

$$= \frac{R_{in}}{\sqrt{2\,R_{in}^2}}$$

$$= \frac{R_{in}}{\sqrt{2}\,R_{in}}$$

$$= \frac{1}{\sqrt{2}}$$

$$= 0.707 \quad \text{(approximately)}$$

If we replace the fraction in equation (14.8) with the 0.707 value obtained, we get the following relationship:

$$v_b = 0.707 v_{in} \quad \text{(when } X_C = R_{in}) \tag{14.9}$$

**FIGURE 14.7 The base circuit of a BJT amplifier.**

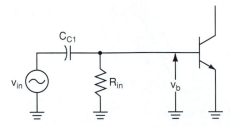

**FIGURE 14.8**

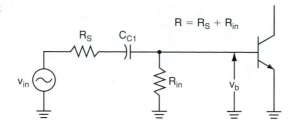

Thus, for the circuit in Figure 14.7, $v_b$ will drop to $0.707v_{in}$ when $X_C = R_{in}$. When this happens, the gain of the amplifier is *effectively* reduced to $0.707A_{v(mid)}$. As you remember, $A_v = 0.707A_{v(mid)}$ at the half-power points. Thus, the lower half-power point ($f_1$) for Figure 14.7 will occur at the frequency that causes $X_C$ to equal $R_{in}$.

At what frequency does this occur? If we swap the values of $f$ and $X_C$ in the basic capacitive reactance equation, we obtain the following equation:

$$f = \frac{1}{2\pi X_C C} \tag{14.10}$$

Since $X_C$ is equal to the total ac resistance in the base circuit ($R$) at $f_1$, we can rewrite this equation as

$$f_{1B} = \frac{1}{2\pi R C} \tag{14.11}$$

where $f_{1B}$ = the lower cutoff frequency of the base circuit
     $R$ = the *total* ac resistance in the base circuit
     $C$ = the value of the base coupling capacitor

In equation (14.11), $R$ is shown as being the *total* ac resistance in the base circuit. For Figure 14.7, the total ac resistance was shown to be $R_{in}$. However, if you want a more accurate value of $f_{1B}$, you must take the resistance of the source into account. This resistance is shown in Figure 14.8. For the circuit shown, the half-power point will occur at the frequency where $X_C$ is equal to the *series* total of $R_S$ and $R_{in}$. By formula,

$$f_{1B} = \frac{1}{2\pi(R_S + R_{in})C_{C1}} \tag{14.12}$$

where  $R_S$ = the resistance of the ac signal source
      $R_{in} = R_1 \parallel R_2 \parallel h_{ie}$

This is the equation that is used to determine the lower cutoff frequency of the base circuit, as Example 14.8 demonstrates.

---

Determine the value of $f_{1B}$ for the circuit shown in Figure 14.9.

***EXAMPLE 14.8***

**Solution:**  First, the value of $R_{in}$ is found as

$$R_{in} = R_1 \parallel R_2 \parallel h_{ie}$$
$$= 18\ \text{k}\Omega \parallel 4.7\ \text{k}\Omega \parallel 4.4\ \text{k}\Omega$$
$$= 2018\ \Omega$$

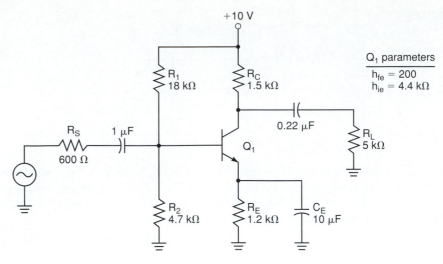

**FIGURE 14.9**

Now, using this value in equation (14.12), we obtain the value of $f_{1B}$ as follows:

$$f_{1B} = \frac{1}{2\pi(R_S + R_{in})C}$$

$$= \frac{1}{2\pi(2618\ \Omega)(1\ \mu F)}$$

$$= \frac{1}{16.45 \times 10^{-3}}$$

$$= 60.79\ \text{Hz}$$

Thus, at 60.79 Hz, the base circuit of Figure 14.9 would cause the power gain of the amplifier to drop by 3 dB.

### PRACTICE PROBLEM 14.8

A BJT amplifier has values of $R_1 = 56\ \text{k}\Omega$, $R_2 = 5.6\ \text{k}\Omega$, $R_S = 1\ \text{k}\Omega$, $C_{C1} = 1\ \mu F$, and $h_{ie} = 5\ \text{k}\Omega$. Determine the value of $f_{1B}$ for the circuit.

The collector circuit of the BJT amplifier works according to the same principle as the base circuit. When $X_C$ of the output coupling capacitor is equal to the total ac resistance in the collector circuit $(R_C + R_L)$, the output voltage will drop to 0.707 times its midband value. Again, the value of $A_v$ has then been effectively reduced to $0.707A_{v(\text{mid})}$.

It should come as no surprise that the equation for the cutoff frequency of the collector circuit is very similar to that for the base circuit. The actual equation is as follows:

$$f_{1C} = \frac{1}{2\pi(R_C + R_L)C} \tag{14.13}$$

where
$f_{1C}$ = the lower cutoff frequency of the collector circuit
$(R_C + R_L)$ = the sum of the resistances in the collector circuit
$C$ = the value of the output coupling capacitor

Example 14.9 demonstrates the use of this equation.

Determine the value of $f_{1C}$ for the circuit in Figure 14.9.

EXAMPLE 14.9

**Solution:** Using the values shown in the figure, the value of $f_{1C}$ is found as

$$f_{1C} = \frac{1}{2\pi(R_C + R_L)C}$$

$$= \frac{1}{2\pi(6.5 \text{ k}\Omega)(0.22 \text{ }\mu\text{F})}$$

$$= \frac{1}{8.984 \times 10^{-3}}$$

$$= 111.3 \text{ Hz}$$

**PRACTICE PROBLEM 14.9**

The circuit described in Practice Problem 14.8 has values of $R_C = 3.6$ k$\Omega$. $R_L = 1.5$ k$\Omega$, and $C_{C2} = 2.2$ $\mu$F. Determine the value of $f_{1C}$ for the circuit.

In Examples 14.8 and 14.9, we got values of $f_{1B} = 60.79$ Hz and $f_{1C} = 111.3$ Hz for the amplifier in Figure 14.9. Based on these two values, what would the cutoff frequency for the amplifier be? If you said 111.3 Hz, you were right. An amplifier will cut off at the frequency closest to $f_o$. Thus, for low-frequency response, the amplifier will cut off at the *highest* calculated value of $f_1$.

Before deriving the equation for the emitter cutoff frequency, we need to refer back to some relationships established in Chapter 10. These were

$$Z_{out} = R_E \| \left( r_e' + \frac{R'_{in}}{h_{fe}} \right) \qquad (14.14)$$

where

$$R'_{in} = R_1 \| R_2 \| R_S$$

In equation (14.14), the output impedance of the amplifier is considered to be the parallel combination of $R_E$ and $(r_e' + R'_{in}/h_{fe})$. Under normal circumstances, the value of $R_E$ is much greater than the value of $(r_e' + R'_{in}/h_{fe})$. Therefore, we can approximate the total ac resistance in the emitter circuit as

$$R_{out} \cong r_e' + \frac{R'_{in}}{h_{fe}} \qquad (14.15)$$

Using this value, the value of $f_{1E}$ is found as

$$f_{1E} = \frac{1}{2\pi R_{out} C_E} \qquad (14.16)$$

The following example demonstrates the use of this equation.

EXAMPLE 14.10

Determine the value of $f_{1E}$ for the circuit in Figure 14.9.

**Solution:** First, the value of $R_{out}$ is found as

$$R_{out} \cong r_e' + \frac{R_{in}'}{h_{fe}}$$

$$= 22\ \Omega + \frac{4.7\ \text{k}\Omega \parallel 18\ \text{k}\Omega \parallel 600\ \Omega}{200}$$

$$= 22\ \Omega + 2.58\ \Omega$$

$$= 24.58\ \Omega$$

Now the value of $f_{1E}$ can be found as

$$f_{1E} = \frac{1}{2\pi R_{out} C_E}$$

$$= \frac{1}{2\pi (24.58\ \Omega)(10\ \mu\text{F})}$$

$$= \frac{1}{1.544 \times 10^{-3}}$$

$$= 647.5\ \text{Hz}$$

**PRACTICE PROBLEM 14.10**

The amplifier described in Practice Problems 14.8 and 14.9 has values of $h_{fe} = 200$, $R_E = 1.3\ \text{k}\Omega$, and $C_E = 100\ \mu\text{F}$. Determine the value of $f_{1E}$ for the circuit.

At this point, what would you say is the value of $f_1$ for the circuit in Figure 14.9? Would you say 647.5 Hz? Good!

## Gain Roll-Off

OBJECTIVE 6 ▶

**Roll-off rate**
The rate of gain reduction when a circuit is operated beyond its cutoff frequencies.

The term **roll-off rate** is used to describe the rate at which the voltage gain of an amplifier (in dB) drops off after $f_1$ or $f_2$ has been passed. The *low-frequency* roll-off for the common-emitter amplifier is calculated using the following equation:

$$\Delta A_{v(\text{dB})} = 20 \log \frac{1}{\sqrt{1 + (f_1/f)^2}} \qquad (14.17)$$

where $f$ = the frequency of operation
$f_1$ = the lower cutoff frequency of the amplifier

The derivation of equation (14.17) is shown in Appendix D.

Before we get involved in any calculations, one point needs to be made about equation (14.17): *Equation (14.17) is used to determine the change in voltage gain.* For example, if equation (14.17) yields a result of $-3$ dB at a given frequency, $f$, this means that $A_{v(\text{dB})}$ is 3 dB lower than its midband value at that frequency. The following example serves to illustrate this point.

A given amplifier has values of $A_{v(mid)} = 45$ dB and $f_1 = 2$ kHz. What is the gain of the amplifier when it is operated at 500 Hz?

**Solution:**   The *change in $A_v$* is found using equation (14.17), as follows:

$$\Delta A_{v(dB)} = 20 \log \frac{1}{\sqrt{1 + (f_1/f)^2}}$$

$$= 20 \log \frac{1}{\sqrt{1 + (2000/500)^2}}$$

$$= 20 \log \frac{1}{\sqrt{1 + 16}}$$

$$= 20 \log(0.2425)$$

$$= -12.3 \text{ dB}$$

Now, the value of $A_{v(dB)}$ at 500 Hz is found as

$$A_{v(dB)} = A_{v(mid)} + \Delta A_{v(dB)}$$
$$= 45 \text{ dB} + (-12.3 \text{ dB})$$
$$= 32.7 \text{ dB}$$

### PRACTICE PROBLEM 14.11

An amplifier with a gain of 23 dB has a lower cutoff frequency of 8 kHz. If the circuit is operated at 4 kHz, what is the gain of the amplifier?

Equation (14.17) tells us one very important fact about the low-frequency roll-off of an amplifier: *The values of R and C in the amplifier have nothing to do with the roll-off rate of a given terminal circuit.* For each terminal circuit (base versus emitter versus collector), the roll-off rate is strictly a function of the difference between $f$ and $f_1$. The $RC$ circuit determines *when* the roll-off begins. However, once it has begun, the values of $R$ and $C$ are no longer significant. We can therefore conclude that *all RC circuits will roll off at the same rate*.

Values of R and C do not affect roll-off rates.

What *is* the low-frequency roll-off rate? We can determine this by setting $f$ to several multiples of $f_1$ and solving the equation. For example, when $f = f_1$, $\Delta A_{v(dB)}$ is found as

$$\Delta A_{v(dB)} = 20 \log \frac{1}{\sqrt{1 + 1^2}} \quad \left| \quad \frac{f_1}{f} = 1 \right.$$
$$= -3 \text{ dB}$$

When $f = 0.1f_1$,

$$\Delta A_{v(dB)} = 20 \log \frac{1}{\sqrt{1 + 10^2}} \quad \left| \quad \frac{f_1}{f} = 10 \right.$$
$$= -20 \text{ dB}$$

When $f = 0.01f_1$,

$$\Delta A_{v(dB)} = 20 \log \frac{1}{\sqrt{1 + 100^2}} \quad \left| \quad \frac{f_1}{f} = 100 \right.$$
$$= -40 \text{ dB}$$

SECTION 14.2 / BJT AMPLIFIER FREQUENCY RESPONSE     **583**

When $f = 0.001 f_1$,

$$\Delta A_{v(dB)} = 20 \log \frac{1}{\sqrt{1 + 1000^2}} \quad \bigg| \quad \frac{f_1}{f} = 1000$$

$$= -60 \text{ dB}$$

What is the roll-off rate of an *RC* circuit?

Do you see a pattern here? Everytime we decrease the value of $f$ by a factor of 10 (a *decade*), the gain of the amplifier drops another 20 dB. Therefore, we can state that the low-frequency roll-off for voltage gain is *20 dB/decade*. The Bode plot that corresponds to this roll-off rate is shown in Figure 14.10.

The Bode plot in Figure 14.10 is best explained in a series of steps, as follows:

**Lab Reference:** Gain roll-off is demonstrated in Exercise 32.

1. At $f_1$, the change in voltage gain is shown to be 0 dB.

2. At $f = 0.1 f_1$ (one decade down), the voltage gain of the amplifier has dropped by 20 dB to $-20$ dB.

3. At $f = 0.01 f_1$ (next decade down), the voltage gain of the amplifier has dropped by another 20 dB. The total value of $\Delta A_{v(dB)}$ is $-40$ dB.

4. At $f = 0.001 f_1$ (next decade down), the voltage gain of the amplifier has dropped by another 20 dB. The total value of $\Delta A_{v(dB)}$ is now $-60$ dB.

If the process were to continue, the next several decades would have total $\Delta A_{v(dB)}$ values of $-80$ dB, $-100$ dB, and so on. At each decade interval, the voltage gain of the circuit would be 20 dB down from the previous decade interval.

Let's relate this to the circuit in Figure 14.9. We determined that the cutoff frequency for the emitter circuit ($f_{1E}$) was 647.5 Hz. For the sake of discussion, we will make two assumptions:

1. The cutoff frequency is 600 Hz. This will simplify our frequency scale.

2. The emitter circuit is the *only* circuit that will cut off. In other words, we will neglect the effects of the base and collector circuits. (You will be shown very shortly what happens when these circuits are brought into the picture.)

Figure 14.11 shows the Bode plot for the emitter circuit in Figure 14.9. The cutoff frequency is 600 Hz. One decade down from this frequency is 60 Hz. At this frequency, the voltage gain of the amplifier has dropped by 20 dB, giving us a total $\Delta A_{v(dB)}$ of $-20$ dB, and so on.

**FIGURE 14.10**

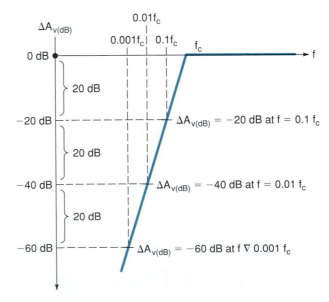

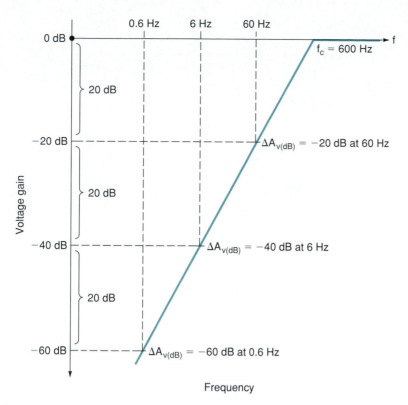

**FIGURE 14.11**   The Bode plot for the emitter circuit of Figure 14.9.

If we had wished to plot the roll-off curve for the base circuit in Figure 14.9, the roll-off rate of 20 dB per decade would not have been very useful. Since $f_{1B}$ was about 60 Hz to begin with, one or two decades down does not leave us with a whole lot of frequency! It helps, then, to be able to determine the voltage gain roll-off at *octave* frequency rates. The following table shows the $\Delta A_{v(dB)}$ values associated with octave frequency intervals. The values shown were determined by substituting the value of $(f_1/f)$ into equation (14.17) and solving for $\Delta A_{v(dB)}$.

| $f$ | $f_1/f$ | $A_{v(dB)}$ (*approximate*) |
|---|---|---|
| $1/2\,f_1$ | 2 | $-6$ |
| $1/4\,f_1$ | 4 | $-12$ |
| $1/8\,f_1$ | 8 | $-18$ |

The pattern here is easy to see. The voltage gain of the circuit would roll off at a rate of *6 dB per octave.* Using the 6 dB per octave roll-off rate, we can easily derive the Bode plot for the base circuit of Figure 14.9. This plot is shown in Figure 14.12. Note that the $\Delta A_{v(dB)}$ intervals are now equal to 6 dB and that the frequency is measured in octaves instead of decades. Other than these two points, the plot shown is read in the same manner as the one in Figure 14.11.

Roll-off rates can also be measured at *octave* intervals.

Before we go on to look at the effect of having three cutoff frequencies in an amplifier ($f_{1B}, f_{1C},$ and $f_{1E}$), let's summarize the points that have been made in this section:

**1.** The low-frequency roll-off rate for an *RC* circuit is *20 dB per decade.* This is equal to a rate of *6 dB per octave.*

Summary of *RC* roll-off rates.

**2.** The roll-off rate is independent of the values of *R* and *C* in the terminal circuit. Thus, the rates given in statement 1 apply to all *RC* terminal circuits.

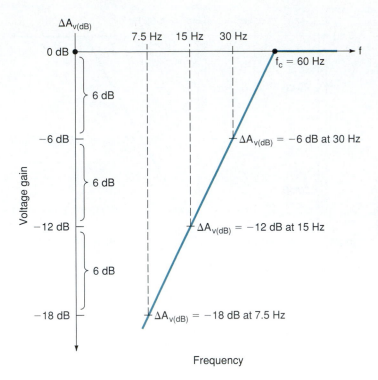

FIGURE 14.12    Roll-off measured in decibels per octave.

3. Equation (14.17) can be used to determine the value of $A_{v(dB)}$ *relative to* $A_{v(mid)}$ at a given frequency.

## The Combined Effects of $f_{1B}$, $f_{1C}$, and $f_{1E}$

You have seen that a circuit like the one in Figure 14.9 has three $RC$ terminal circuits, each with its own cutoff frequency. Next, we will take a look at what happens when more than one cutoff frequency is passed. The overall concept is easy to understand with the help of Figure 14.13. Here we have three separate $RC$ circuits, each having the indicated value of $f_1$. The cutoff frequencies were chosen at one-decade intervals, which simplifies a complex analysis.

In Figure 14.13a, the input frequency is shown to be 500 Hz. Since this frequency is below the value of $f_1$ for $RC_1$, the circuit has cut off and is causing a voltage gain roll-off of 20 dB/decade. At this point, the input frequency is still above $RC_2$ and $RC_3$, so these two circuits are still operating at their respective values of $A_{v(mid)}$. Thus, at this point, the overall gain is dropping by 20 dB/decade, the roll-off rate of $RC_1$.

In Figure 14.13b, the input frequency has dropped to 50 Hz, and the value of $f_c$ for $RC_2$ has also been passed. Now we have $RC_1$ introducing a 20-dB/decade drop and $RC_2$ introducing a 20-dB/decade drop. *The overall roll-off is equal to the sum of these drops, 40 dB/decade.* This should not seem strange to you. As you recall, the total decibel gain of a multistage circuit is equal to the *sum* of the individual decibel gain values. The same principle holds true for roll-off rates.

When the input frequency decreases to 5 Hz (Figure 14.13c), all three of the $RC$ circuits are operated below their respective values of $f_c$. Thus, all three circuits are introducing a 20-dB/decade roll-off. The sum of these roll-off rates is 60 dB/decade.

The circuit action just described is represented by the Bode plot in Figure 14.14. As you can see, the roll-off rate is 20 dB/decade for input frequencies between 1 kHz and 100 Hz. When the input frequency drops past 100 Hz, the roll-off increases to 40 dB/decade. When the input frequency drops past 10 Hz, all three circuits are cut off, and the roll-off rate is shown to be 60 dB/decade.

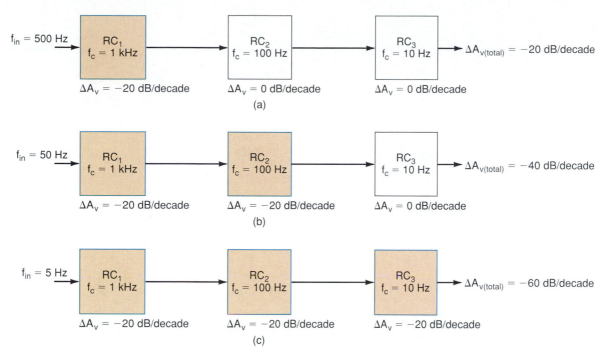

**FIGURE 14.13** The effect of cascaded *RC* circuits on roll-off rates.

Now let's relate this principle to the circuit in Figure 14.9. The low-frequency Bode plot for this circuit is shown in Figure 14.15. Note that we are using the more convenient rate of *6 dB/octave* in this figure. This is acceptable since a roll-off of 6 dB/octave is equal to a roll-off of 20 dB/decade. We have also rounded off the calculated values of $f_{1B}$, $f_{1C}$, and $f_{1E}$, as shown in the illustration.

At 600 Hz, the *RC* circuit in the emitter cuts off. As the operating frequency drops below this value, we get a roll-off of 6 dB/octave. This continues until the 120-Hz cutoff frequency of the collector is reached. Once this is reached, the roll-off rate increases to 12 dB/octave. This rate continues until the operating frequency reaches the 60-Hz cutoff frequency of the base. At this point, all three of the terminal circuits are introducing a 6-dB/octave drop, and the total roll-off rate is 18 dB/octave.

## A Practical Consideration

The Bode plot represents the *ideal* amplifier response to a change in frequency. For example, the Bode plot in Figure 14.15 idealizes the low-frequency response of the circuit in Figure 14.9. That is:

1. It assumes that all cutoff frequencies are exactly as calculated. In other words, there are no resistor tolerances that will affect the actual cutoff frequency values.

2. It assumes that each transistor terminal circuit will have a 0-dB value of $A_{v(dB)}$ until the cutoff frequency is reached.

3. The rolloff is constant at the given roll-off rate.

In practice, none of the above assumptions is valid. Circuit component tolerances and stray capacitance values *will* affect the actual cutoff frequencies. You will find that, in practice, measured cutoff frequencies will vary from the calculated cutoff frequencies.

A more practical approach to the whole situation is as follows:

1. An *RC* circuit will noticeably begin to reduce the gain of an amplifier at approximately *twice* its value of $f_1$. The gain of the amplifier will drop by approximately 3 dB by the time the lower cutoff frequency is reached.

Component tolerances will affect actual cutoff frequencies.

Practical frequency-response characteristics.

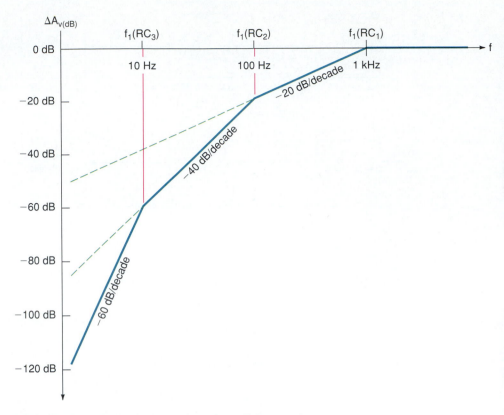

**FIGURE 14.14   The effects of multiple cutoff frequencies.**

2. The roll-off rate for a given *RC* circuit is 20 dB/decade (6 dB/octave).

3. When multiple *RC* circuits are involved (as in the BJT amplifier), the effects of each will be felt well before their actual cutoff frequencies are reached.

How much will these considerations affect the frequency-response curve of a given amplifier? Figure 14.16 shows the original Bode plot for the amplifier in Figure 14.15 and

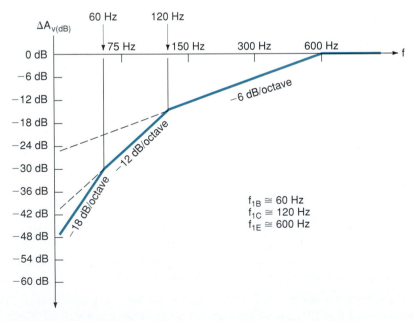

**FIGURE 14.15   The low-frequency Bode plot for the circuit in Figure 14.9.**

**FIGURE 14.16**

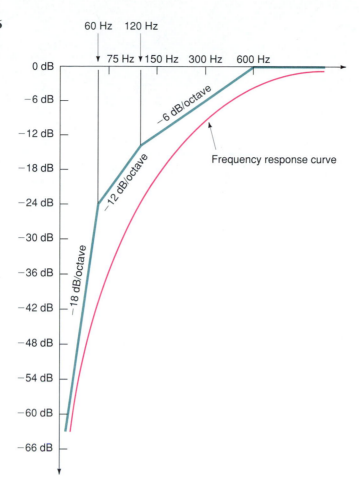

the more realistic frequency-response curve. As you can see, the curve does resemble the Bode plot. However, the combined effects of the *RC* circuits cause the actual curve to have more rounded turning points that occur at higher frequencies than those calculated.

The Bode plot provides a means of predicting the response of a circuit to a change in frequency. However, when we are making these predictions, we need to keep in mind the fact that they are *approximations* based on assuming a number of ideal conditions. Therefore, we should not be surprised when the percentage of error between our predicted and measured frequency-response values approaches 10 percent or 20 percent.

When using Bode plots, keep in mind that they provide approximate values. This is because they are plotted assuming a number of ideal conditions.

One point of interest: The greater the separation between cutoff frequencies, the closer the actual frequency-response curve will come to the Bode plot for an amplifier. For example, an amplifier with cutoff frequencies of 10 kHz, 1 kHz, and 10 Hz would have a frequency-response curve that is closer to the ideal (the Bode plot) than an amplifier with cutoff frequencies of 10 kHz, 5 kHz, and 1 kHz. The proof of this concept is included as a "Brain Drain" problem at the end of the chapter.

## BJT Amplifier High-Frequency Response

In many ways, the high-frequency operation of the BJT amplifier is similar to the low-frequency operation. You have a cutoff frequency for each terminal circuit. Once a given value of $f_2$ is passed, the circuit will cut off and will have a roll-off rate that is approximately equal to 20 dB/decade (6 dB/octave).

The primary differences between low- and high-frequency responses are as follows:

How $f_2$ calculations differ from $f_1$ calculations.

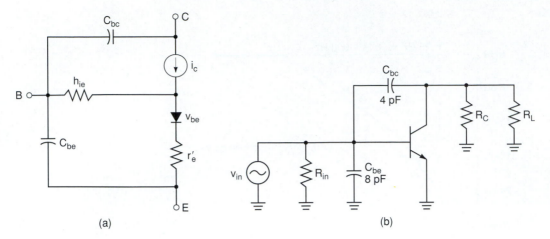

FIGURE 14.17    BJT internal capacitance.

**1.** The values of capacitance used in $f_2$ calculations

**2.** The methods used to determine the total resistance in each terminal circuit

These differences will now be discussed in detail.

## BJT Internal Capacitance

You may recall from Chapter 2 that a *pn* junction has some measurable amount of capacitance. Since the BJT has two internal *pn* junctions, it has two internal capacitances, as shown in Figure 14.17. As you will be shown in this section, these internal capacitances control the high-frequency response of the device.

The capacitance of the collector–base junction ($C_{bc}$) is usually listed on the specification sheet of a given transistor. Note that it may be listed as $C_{bc}$, $C_c$, $C_{ob}$, or $C_{obo}$. If you refer back to the specification sheet for the 2N3904 (Figure 7.24), you will see that it lists a value of collector–base capacitance ($C_{ob}$) equal to 4 pF. The value of $C_{be}$ may or may not be given on the specification sheet. When given, it is usually listed as $C_{in}$, $C_{be}$, $C_{ib}$, or $C_b$. The specification sheet for the 2N3904 lists this value as $C_{ib} = 8$ pF.

When the value of $C_{be}$ is *not* listed on the spec sheet of a given transistor, its value can be approximated using

$$C_{be} \cong \frac{1}{2\pi f_T r'_e} \tag{14.18}$$

**Current–gain bandwidth product**
The frequency at which BJT current gain drops to unity.

where $f_T$ is the **current gain–bandwidth product** of the transistor, as listed on the spec sheet. The current gain–bandwidth product of a transistor is the frequency at which $C_{be}$ will have a low enough reactance to cause the transistor current gain to drop to *unity*. Example 14.12 demonstrates the use of equation (14.18) in determining the value of $C_{be}$.

---

*EXAMPLE 14.12*

The spec sheet for the 2N3904 lists a value of $f_T = 300$ MHz at $I_C = 10$ mA. Determine the value of $C_{be}$ at this current.

***Solution:***    Assuming that $I_C = I_E$, we can determine the value of $r'_e$ to equal 2.9 Ω. Using this value in equation (14.18), we can approximate the value of $C_{be}$ as

---

$$C_{be} \cong \frac{1}{2\pi f_T r_e'}$$

$$= \frac{1}{2\pi (300 \text{ MHz})(2.9 \text{ }\Omega)}$$

$$= 183 \text{ pF}$$

**PRACTICE PROBLEM 14.12**

A transistor has a current gain–bandwidth product of $f_T = 200$ MHz, $h_{ie} = 650$ $\Omega$, and $h_{fe} = 250$ at $I_C = 10$ mA. Determine the value of $C_{be}$ for the device.

The value of $r_e'$ used in this example was found using

$$r_e' = \frac{h_{ie}}{h_{fe}}$$

The operating curves for the 2N3904 (not shown) provide the following values at $I_C = 10$mA:

$$h_{ie} = 500 \text{ }\Omega$$
$$h_{fe} = 170$$

Using these values, $r_e'$ was found to be 2.9 $\Omega$.

As you can see, the value of $C_{be} = 212$ pF in the example is significantly different from the spec sheet rating of 8 pF. The difference between the two values is due to the fact that they were determined for two different values of $I_E$. The spec sheet rating indicates that $C_{be}$ will be no greater than 8 pF when $I_E = 0$ mA. However, as $I_E$ changes, so does the value of $r_e'$ and the width of the base–emitter junction, as shown in the following table.

| $I_E$ | $r_e'$ | Junction Width |
|---|---|---|
| Increase | Decrease | Decrease |
| Decrease | Increase | Increase |

Since the value of $C_{be}$ is *inversely* proportional to the width of the emitter–base junction, it will vary directly with variations in $I_E$. In other words, as $I_E$ increases, so will $C_{be}$. As $I_E$ decreases, so will $C_{be}$.

So which value of $C_{be}$ do you use for a given circuit, the calculated value or the spec sheet value? In most cases, you should use the *calculated* value of $C_{be}$ because it will provide much more accurate results in later calculations. However, remember that even the calculated value of $C_{be}$ will give you only an approximate figure. Since the calculation of $r_e'$ is only an approximation, the value obtained by using equation (14.18) will be an approximation. However, in most cases, this approximated value of $C_{be}$ will be much closer to the actual emitter–base capacitance than the spec sheet rating.

Why have we ignored these capacitance values until now? As you can see, $C_{be}$ and $C_{bc}$ are very small capacitance values. Because of this, their low-frequency reactance is extremely high. For example, the reactance of $C_{bc}$ is approximately 3.98 M$\Omega$ at $f = 10$ kHz. This reactance is so high that it can be considered to be open for all practical purposes. But what happens when the operating frequency of the transistor increases? Since reactance is inversely proportional to frequency for a given capacitor, the value of $X_C$ for $C_{bc}$ will decrease. At 100 MHz, the reactance of $C_{bc}$ is only 398 $\Omega$. This reactance *cannot* be ignored because it will have a definite impact on the operation of the transistor.

◄ *OBJECTIVE 7*

## Miller's Theorem

Before we can get on to some actual circuit calculations, we have to solve one more problem that involves the value of $C_{bc}$. As you saw in Figure 14.17, $C_{bc}$ is the capacitance between the collector and base terminals. We can represent this capacitance as shown in Figure 14.18.

In Figure 14.18a, the transistor has been shown as a block having a certain value of voltage gain, $A_v$. Since $C_{bc}$ exists between the collector and base terminals, it is shown as

◄ *OBJECTIVE 8*

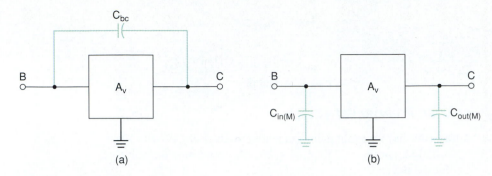

FIGURE 14.18   Miller equivalent circuit for a feedback capacitor.

an external capacitor. The question now is this: Is $C_{bc}$ part of the input (base) circuit or part of the output (collector) circuit? The answer is: *both*.

**Miller's theorem**

Allows a feedback capacitor to be represented as separate input and output capacitances.

**Miller's theorem** allows $C_{bc}$ to be represented as *two* capacitors, one in the base circuit and one in the collector circuit. These two capacitors are shown in Figure 14.18b. For this circuit the *Miller input capacitance* is given as

$$C_{\text{in(M)}} = C_{bc}(A_v + 1) \tag{14.19}$$

*A Practical Consideration:* Miller's theorem applies only to inverting amplifiers (those with a 180° voltage phase shift).

and the *Miller output capacitance* is given as

$$C_{\text{out(M)}} = C_{bc}\frac{A_v + 1}{A_v} \tag{14.20}$$

In most cases, $A_v > 10$. When this is the case, the above equations can be simplified as

$$C_{\text{in(M)}} \cong A_v C_{bc} \tag{14.21}$$

and

$$C_{\text{out(M)}} \cong C_{bc} \tag{14.22}$$

Why we use Miller's theorem.

Note that the above equations can be used whenever the value of $A_v$ is greater than 10.

So, why go to all this trouble? The *feedback* capacitor shown in Figure 14.18a can make circuit calculations extremely complex. A much simpler approach is to represent $C_{bc}$ as two separate capacitors. This way, one of the capacitors is used strictly for input (base) circuit calculations, and the other is used strictly for output (collector) circuit calculations. Example 14.13 demonstrates the method for determining the Miller input/output capacitance values for a given amplifier.

**EXAMPLE 14.13**

An amplifier has values of $C_{bc} = 6$ pF and $A_v = 120$. Determine the Miller input and output capacitance values for the circuit.

**Solution:**   The Miller input capacitance is determined as

$$
\begin{aligned}
C_{\text{in(M)}} &\cong A_v C_{bc}\\
&= (120)(6 \text{ pF})\\
&= 720 \text{ pF}
\end{aligned}
$$

The Miller output capacitance is determined as

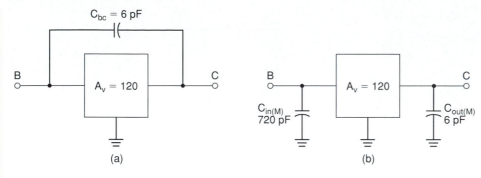

FIGURE 14.19

$$C_{\text{out}(M)} \cong C_{bc}$$
$$= 6 \text{ pF}$$

Using the values obtained here, we can derive the equivalent of the circuit in Figure 14.19a. This circuit is shown in Figure 14.19b.

***PRACTICE PROBLEM 14.13***

A BJT with a value of $C_{bc} = 4$ pF is used in an amplifier that has a value of $A_v = 240$. Determine the Miller input and output equivalent capacitances for $C_{bc}$.

After the Miller capacitance values for an amplifier are determined, we are ready to determine the upper cutoff frequencies for a given BJT amplifier.

## High-Frequency Operation

When we discussed the low-frequency operation of the common-emitter amplifier, you were shown that the value of $f_1$ depended on the various values of circuit resistance and capacitance. The same holds true for the value of $f_2$. However, the resistances are combined in a different fashion, and we use different values of circuit capacitance.

The high-frequency equivalent circuit for the common-emitter amplifier is shown in Figure 14.20. We will reduce this circuit to its simplest *RC* equivalent circuit so that you will be able to see *why* the values of $R$ and $C$ are combined as they are. Before we do this, there are two values shown that should be explained. $C_{\text{in}}$ is the *input capacitance of the next stage*. If this amplifier is driving another, $C_{\text{in}}$ will equal the total *high-frequency* input capacitance of the next stage. (*This point is discussed further in our coverage of multistage amplifiers. At this point we will simply assume a value of $C_{\text{in}}$ for any example problems.*) Also, the *base input impedance*, $h_{ie}$, is shown in the circuit. As you will see, this value is needed for the determination of $f_2$.

The equivalent circuit in Figure 14.20b was derived by performing *two* steps:

1. The collector–base capacitor, $C_{bc}$, was replaced by its equivalent Miller input and output capacitors.
2. The source resistance ($R_S$) and the amplifier input resistance ($R_{\text{in}}$) were combined into a single *parallel* equivalent resistance, $R'_{\text{in}}$.

Now, we will simplify the circuit further by taking three additional steps. As Figure 14.20c shows:

3. The parallel capacitors in the base circuit, $C_{be}$ and $C_{\text{in}(M)}$, are combined into a single input capacitance value.

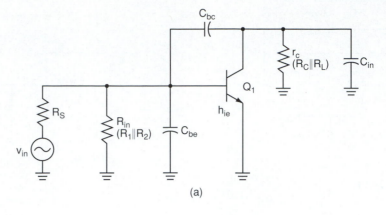

(a)

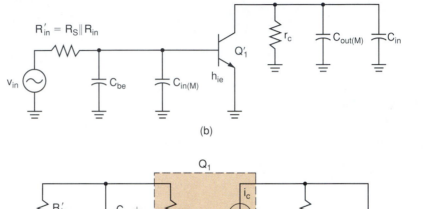

(b)

A Practical Consideration: The value of $C_{in}$ in the collector circuit of Figure 14.20 assumes that the amplifier load has some amount of capacitance. If the load is purely resistive, $C_{in}$ for the load will be zero, and the total capacitance in the output circuit can be assumed to be equal to $C_{out(M)}$ for the transistor.

(c)

**FIGURE 14.20   BJT high-frequency ac equivalent circuit.**

4. The two parallel capacitors in the collector circuit, $C_{in}$ and $C_{out(M)}$, are combined into a single output capacitance value.

5. The transistor has been replaced to show it as a resistance, equal to $h_{ie}$, and a current source.

OBJECTIVE 9 ▶ We can now use this equivalent circuit to analyze the high-frequency response of the original amplifier.

The high-frequency cutoff ($f_2$) is determined using the same basic equation we used to determine $f_1$:

$$f_2 = \frac{1}{2\pi RC}$$

For the base circuit, $R$ is the parallel combination of $R'_{in}$ and $h_{ie}$. Also, $C$ is the combination of $C_{be}$ and $C_{in(M)}$. Thus, for the base circuit,

$$f_{2B} = \frac{1}{2\pi(R'_{in} \parallel h_{ie})(C_{be} + C_{in(M)})} \tag{14.23}$$

where $R'_{in} = R_S \parallel R_1 \parallel R_2$
$\qquad h_{ie}$ = the base input impedance to the transistor

$C_{be}$ = the base–emitter junction capacitance

$C_{in(M)}$ = the Miller input capacitance

Example 14.14 demonstrates the process for determining the value of $f_2$ for the base circuit.

---

Determine the value of $f_2$ for the base circuit of the amplifier shown in Figure 14.21.

EXAMPLE **14.14**

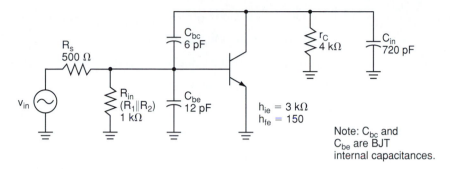

**FIGURE 14.21**

**Solution:** The first step is to determine the value of $A_v$ for the amplifier, as follows:

$$A_v = \frac{h_{fe}r_C}{h_{ie}}$$

$$= 200$$

The value of $A_v$ is now used (along with $C_{bc}$) to determine the value of $C_{in(M)}$:

$$C_{in(M)} = A_v C_{bc}$$
$$= (200)(6 \text{ pF})$$
$$= 1.2 \text{ nF}$$

The value of $R'_{in}$ is found as

$$R'_{in} = R_S \parallel R_{in}$$
$$= 333 \ \Omega$$

Now that we have all the needed values of resistance and capacitance, we can find the value of $f_2$ for the base circuit as follows:

$$f_{2B} = \frac{1}{2\pi(R'_{in} \parallel h_{ie})(C_{be} + C_{in(M)})}$$

$$= \frac{1}{2\pi(333 \ \Omega \parallel 3 \text{ k}\Omega)(12 \text{ pF} + 1.2 \text{ nF})}$$

$$= \frac{1}{2\pi(300 \ \Omega)(1.212 \text{ nF})}$$

$$= \frac{1}{2.283 \times 10^{-6}}$$

$$= 437.94 \text{ kHz}$$

**PRACTICE PROBLEM 14.14**

The amplifier described in Practice Problem 14.8 has values of $h_{fe} = 200$, $f_T = 300$ MHz at $I_C = 10$ mA, and $C_{bc} = 4$ pF. Determine the value of $f_{2B}$ for the circuit.

For the collector circuit, the value of $R$ is equal to $r_C$, and $C$ is the combination of $C_{in}$ for the next stage and $C_{out(M)}$ for the amplifier. Thus, for the collector circuit,

$$f_{2C} = \frac{1}{2\pi r_C(C_{out(M)} + C_{in})} \tag{14.24}$$

The following example demonstrates the use of this equation.

---

**EXAMPLE 14.15**

Determine the value of $f_{2C}$ for the circuit in Figure 14.21.

**Solution:** Since the value of $A_v \gg 10$, the value of $C_{out(M)}$ is equal to $C_{bc}$. Thus,

$$C_{out(M)} = 6 \text{ pF}$$

Since the other circuit values are already known, we can proceed to find $f_2$ for the collector circuit, as follows:

$$
\begin{aligned}
f_{2C} &= \frac{1}{2\pi r_C(C_{out(M)} + C_{in})} \\
&= \frac{1}{2\pi(4 \text{ k}\Omega)(6 \text{ pF} + 720 \text{ pF})} \\
&= \frac{1}{1.825 \times 10^{-5}} \\
&= 54.8 \text{ kHz}
\end{aligned}
$$

**PRACTICE PROBLEM 14.15**

What is the value of $f_{2C}$ for the amplifier described in Practice Problems 14.14 and 14.8?

---

## Gain Roll-Off

OBJECTIVE 10 ▶ The *high-frequency* gain roll-off is found in the same manner as the low-frequency gain roll-off. The exact equation is

$$\Delta A_{v(dB)} = 20 \log \frac{1}{\sqrt{1 + (f/f_2)^2}} \tag{14.25}$$

As you can see from equation (14.25), the gain roll-off for the high-frequency circuit is also independent of the values of $R$ and $C$. We can use equation (14.25) to determine the high-frequency roll-off in *dB per decade* as follows. When $f = f_2$,

$$\Delta A_{v(dB)} = 20 \log \frac{1}{\sqrt{1 + 1^2}}$$
$$= -3 \text{ dB}$$

When $f = 10f_2$,

$$\Delta A_{v(dB)} = 20 \log \frac{1}{\sqrt{1 + 10^2}}$$
$$= -20 \text{ dB}$$

When $f = 100f_2$,

$$\Delta A_{v(dB)} = 20 \log \frac{1}{\sqrt{1 + 100^2}}$$

$$= -40 \text{ dB}$$

The high-frequency roll-off of the common emitter amplifier is equal to the low-frequency roll-off, *20 dB/decade*. Similarly, the high-frequency roll-off can also be computed as equal to 6 dB/octave.

The high-frequency operation of a given common-emitter amplifier is exactly the same as the low-frequency operation. The only difference is in the method used to determine $f_2$.

## Theory Versus Practice

A wide variety of factors can affect the frequency response of a given amplifier. Component tolerances, estimations of BJT internal capacitances, and the effects of test equipment input capacitance can cause the measured values of $f_1$ and $f_2$ to vary significantly from their predicted (calculated) values.

When measuring the values of $f_1$ and $f_2$ for a BJT amplifier, you will find that the largest percentage of error will occur in the $f_2$ measurements. There are two reasons for this:

Why $f_2$ measurements tend to have large percentage of error values.

1. The BJT internal capacitance values are estimated.
2. The capacitance values used in $f_2$ calculations are in the picofarad (pF) range, as are the input capacitance ratings of most pieces of test equipment. Thus, by connecting the test equipment to the circuit, you are significantly altering the total capacitance that determines the values of $f_2$.

When you are predicting the values of $f_1$ and $f_2$ for your standard BJT amplifier, you need to keep in mind that you are dealing with approximated values and, therefore, will get results that are approximations. While this may seem frustrating, it shouldn't be. You see, the purpose of performing the frequency analysis of an amplifier is to ensure that the circuit gain will not be affected by the input frequency. For example, if you want to operate a given amplifier at a frequency of 50 kHz, you want to know that an input frequency of 50 kHz will not cause the amplifier gain to be reduced. Beyond this application, the exact values of $f_1$ and $f_2$ are rarely of any consequence for standard *RC*-coupled BJT amplifiers.

There is one common application where the exact values of $f_1$ and $f_2$ *are* important. This is the case where you are dealing with *tuned amplifiers*. You may recall that tuned amplifiers are designed for specific bandwidths. Since they are designed for specific values of $f_1$ and $f_2$, these values are important when you are dealing with tuned amplifiers. This point is discussed further in Chapter 17.

*A Practical Consideration:* When you connect an oscilloscope or frequency counter to the output of an amplifier, the value of $f_{2C}$ will decrease. The reason is the added input capacitance of the particular piece of test equipment. For example, if an oscilloscope has an input capacitance of 30 pF, you are adding 30 pF when you connect the oscilloscope to the output of the circuit. This added 30 pF may significantly reduce the value of $f_2$ for the circuit.

One way of reducing the percentage of error introduced by an oscilloscope is to use a × *10 probe* when making high-frequency measurments. A × 10 probe decreases the input capacitance of the oscilloscope and thus reduces the error introduced by the device.

**Section Review**

1. Explain why the power gain of a given amplifier drops by 50 percent when the reactance of the input coupling capacitor equals the input resistance of the amplifier.
2. Define the term *roll-off rate*.
3. What is the relationship between the roll-off rate for a given *RC* circuit and the circuit values of *R* and *C*? Explain your answer.
4. What are the standard low-frequency roll-off rates?
5. Describe the BJT amplifier low-frequency Bode plot.

6. Contrast the Bode plot with the actual low-frequency response curve of a BJT amplifier.

7. What are the primary differences between BJT $f_1$ and $f_2$ calculations?

8. Why aren't the BJT internal capacitances considered in the low-frequency analysis of a given amplifier?

9. What does Miller's theorem state? Why is it used?

10. List, in order, the steps taken to perform the high-frequency analysis of a BJT amplifier.

11. Compare high-frequency roll-off rates to low-frequency roll-off rates.

12. Explain why $f_2$ measurements tend to have larger percentage of error values than $f_1$ measurements.

## 14.3 FET AMPLIFIER FREQUENCY RESPONSE

The transition from BJT circuits to FET circuits is actually very simple. All we have to do is come up with the equations for $f_1$ and $f_2$ for a given FET amplifier. Everything else is identical to what we have been doing up to this point.

In this section, we will concentrate on the voltage-divider biased common-source amplifier. This circuit, along with its low-frequency ac equivalent, is shown in Figure 14.22.

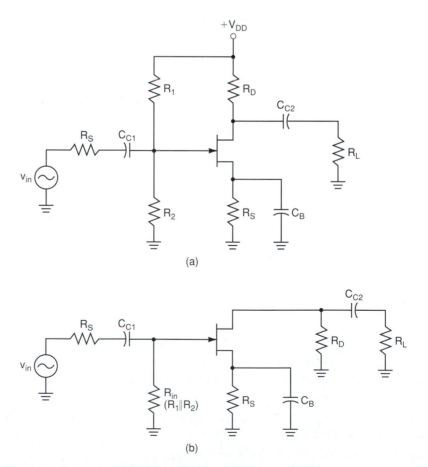

**FIGURE 14.22   FET amplifier low-frequency ac equivalent circuit.**

# Low-Frequency Response

◀ *OBJECTIVE 11*

As you can see, there is very little difference between the FET amplifier and the BJT amplifier as far as the low-frequency equivalent circuit is concerned. In fact, the gate and drain current equations are exactly the same as those used for the BJT amplifier. Thus,

$$f_{1G} = \frac{1}{2\pi(R_S + R_{in})C_{C1}} \qquad (14.26)$$

and

$$f_{1D} = \frac{1}{2\pi(R_D + R_L)C_{C2}} \qquad (14.27)$$

Because of the extremely high input impedance of the FET, the value of $R_{in}$ for the amplifier is found as

$$R_{in} = R_1 \| R_2 \qquad (14.28)$$

As Example 14.16 shows, the high input impedance of the FET means that the FET amplifier will normally have a much lower input cutoff frequency than that of a similar BJT amplifier.

---

**EXAMPLE 14.16**

Determine which of the two circuits shown in Figure 14.23 has the higher input circuit cutoff frequency.

*Solution:* Figure 14.23a is the ac circuit that we analyzed in Example 14.8. At that point, we determined the value of $f_{1B}$ for the circuit to be 60.81 Hz (review Example 14.8 if necessary).

The circuit in Figure 14.23b is identical to the one in Figure 14.23a except that the BJT has been replaced by an FET. Because of the high input impedance of the FET,

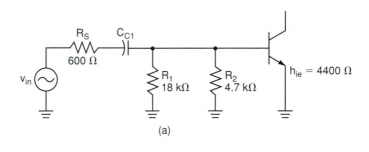

(a)

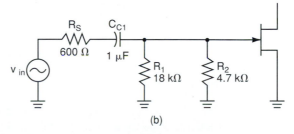

(b)

**FIGURE 14.23**

$$R_{in} = R_1 \| R_2$$
$$= 18 \text{ k}\Omega \| 4.7 \text{ k}\Omega$$
$$= 3727 \ \Omega$$

Using this value for $R_{in}$, the value of $f_{1G}$ is found as

$$f_{1G} = \frac{1}{2\pi(R_S + R_{in})C_{C1}}$$

$$= \frac{1}{2\pi(4327 \ \Omega)(1 \ \mu F)}$$

$$= 36.78 \text{ Hz}$$

Thus, by replacing the BJT with an FET, we have decreased the value of $f_1$ for the input circuit by nearly 50 percent.

In practice, FET amplifiers normally have much higher input resistance values than do BJT amplifiers. As Example 14.17 shows, this has the effect of lowering the value of $f_{1G}$ even more.

EXAMPLE 14.17

Determine the value of $f_{1G}$ for the amplifier shown in Figure 14.24.

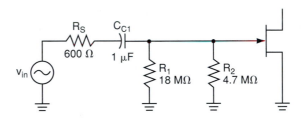

**FIGURE 14.24**

**Solution:** For this circuit,

$$R_{in} = R_1 \| R_2$$
$$= 18 \text{ M}\Omega \| 4.7 \text{ M}\Omega$$
$$= 3.727 \text{ M}\Omega$$

Using this value of $R_{in}$, the gate cutoff frequency is found as

$$f_{1G} = \frac{1}{2\pi(R_S + R_{in})C_{C1}}$$

$$= \frac{1}{2\pi(3.727 \text{ M}\Omega)(1 \ \mu F)}$$

$$= 0.043 \text{ Hz}$$

**PRACTICE PROBLEM 14.17**

An FET amplifier has values of $R_1 = R_2 = 10 \text{ M}\Omega$, $R_S = 100 \ \Omega$, and $C_{C1} = 3.3 \ \mu F$. Determine the value of $f_{1G}$ for the circuit.

The low-frequency analysis of the drain circuit is just like that for the collector circuit of a BJT amplifier. For this reason, it needs no further explanation. The entire low-frequency analysis of a basic FET amplifier is demonstrated in Example 14.18.

Determine the overall value of $f_1$ for the circuit shown in Figure 14.25.

EXAMPLE 14.18

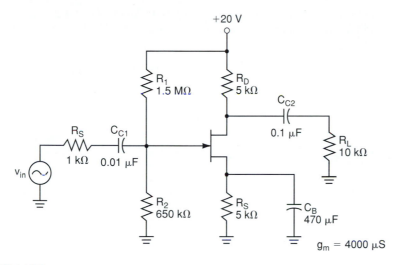

**FIGURE 14.25**

**Solution:** The input resistance to the amplifier is found as

$$R_{in} = R_1 \| R_2$$
$$= 453.5 \text{ k}\Omega$$

Now $f_{1G}$ is found as

$$f_{1G} = \frac{1}{2\pi(R_S + R_{in})C_{C1}}$$

$$= \frac{1}{2\pi(454.5 \text{ k}\Omega)(0.01 \text{ }\mu\text{F})}$$

$$= 35 \text{ Hz}$$

The value of $f_{1D}$ is found as

$$f_{1D} = \frac{1}{2\pi(R_D + R_L)C_{C2}}$$

$$= \frac{1}{2\pi(15 \text{ k}\Omega)(0.1 \text{ }\mu\text{F})}$$

$$= 106 \text{ Hz}$$

**PRACTICE PROBLEM 14.18**

An FET amplifier has values of $R_1 = 10 \text{ M}\Omega$, $R_2 = 1 \text{ M}\Omega$, $R_D = 1.1 \text{ k}\Omega$, $R_S = 820 \text{ }\Omega$, $R_L = 5 \text{ k}\Omega$, $C_{C1} = 0.01 \text{ }\mu\text{F}$, $C_{C2} = 0.1 \text{ }\mu\text{F}$, and $g_m = 2200 \text{ }\mu\text{S}$. Determine the overall value of $f_1$ for the amplifier.

The value of $f_1$ for the amplifier would be assumed to be 106 Hz, the higher of the two cutoff frequencies. Just as with the BJT amplifier, the gain would drop by 3 dB by the time the frequency had dropped to 106 Hz. At this point, the gain would continue to drop (with a continual decrease in frequency) at a rate of *6 dB/octave* until 35 Hz is reached. Then the roll-off rate would increase to *12 dB/octave* again, exactly the same as with the BJT amplifier.

You may be wondering why we aren't considering the cutoff frequency of the source circuit in our analysis. There are two reasons:

1. Determining the cutoff frequency of the source circuit is an extremely complex procedure.

2. The cutoff frequency of the source circuit is normally the lowest of the three terminal cutoff frequencies. Therefore, calculating its value isn't worth the effort involved.

Since the value of $f_1$ for the source circuit will not affect the overall value of $f_1$ for the amplifier, we will consider our analysis complete with finding the gate and drain values of $f_1$.

## High-Frequency Analysis

*OBJECTIVE 12* ▶  The high-frequency response of the FET is limited by values of *internal* capacitance, just as for the BJT. These capacitances are shown in Figure 14.26. As you can see, we again have a situation very similar to that of the BJT. There is a measurable amount of capacitance between each terminal *pair* of the FET. These capacitances each have a reactance that decreases as frequency increases. As the reactance of a given terminal capacitance decreases, more and more of the signal at the terminal is shorted through the component. This is the same problem that we ran into with the BJT.

The high-frequency equivalent circuit of the basic FET amplifier is analyzed in the same fashion as the BJT amplifier. The high-frequency equivalent for the amplifier in Figure 14.22a is shown in Figure 14.27. As you can see, all the terminal capacitance values are included, with the exception of $C_{gd}$. This capacitor has been replaced with the Miller equivalent input and output capacitance values. This is the same thing that we did with $C_{bc}$ in the BJT amplifier.

The value of $C_{\text{in(M)}}$ is found using a form of equation (14.19) as follows:

$$C_{\text{in(M)}} = C_{gd}(A_v + 1)$$

Since the value of $A_v$ for the FET amplifier is equal to $g_m r_D$, the equation above can be rewritten as

**FIGURE 14.26  FET internal capacitances.**

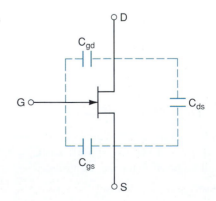

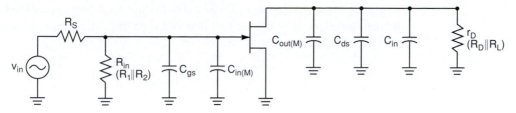

**FIGURE 14.27** FET amplifier high-frequency ac equivalent circuit.

$$C_{\text{in(M)}} = C_{gd}(g_m r_D + 1) \qquad \textbf{(14.29)}$$

The calculation of $C_{\text{in(M)}}$ is demonstrated in Example 14.19.

*A Practical Consideration:* When the value of $g_m r_D$ for the amplifier is greater than (or equal to) 10, equation (14.29) can be simplified as

$$C_{\text{in(M)}} \cong C_{gd} g_m r_D$$

---

A given FET amplifier has values of $C_{gd} = 4$ pF, $g_m = 2500$ μS, and $r_D = 5.6$ kΩ. Determine the value of $C_{\text{in(M)}}$ for the amplifier.

**EXAMPLE 14.19**

*Solution:* The value of $C_{\text{in(M)}}$ is found as

$$
\begin{aligned}
C_{\text{in(M)}} &\cong C_{gd}(g_m r_D + 1) \\
&= (4 \text{ pF})[(2500 \text{ μS})(5.6 \text{ kΩ}) + 1] \\
&= (4 \text{ pF})(15) \\
&= 60 \text{ pF}
\end{aligned}
$$

**PRACTICE PROBLEM 14.19**

A given FET amplifier has values of $C_{gd} = 3$ pF, $g_m = 3200$ μS, and $r_D = 1.8$ kΩ. Determine the value of $C_{\text{in(M)}}$ for the circuit.

---

Equation (14.20) defined $C_{\text{out(M)}}$ as being

$$C_{\text{out(M)}} = C \frac{A_v + 1}{A_v}$$

If we replace $C$ with $C_{gd}$, and $A_v$ with $g_m r_D$, we obtain the following equation for the FET amplifier value of $C_{\text{out(M)}}$:

$$C_{\text{out(M)}} = C_{gd} \frac{g_m r_D + 1}{g_m r_D} \qquad \textbf{(14.30)}$$

Again, when $g_m r_D > 10$, the equation for $C_{\text{out(M)}}$ can be simplified to

$$C_{\text{out(M)}} \cong C_{gd} \qquad \textbf{(14.31)}$$

From the circuit in Figure 14.27, you can see where the following equations come from:

$$C_G = C_{gs} + C_{\text{in(M)}} \qquad \textbf{(14.32)}$$

and

$$C_D = C_{\text{out(M)}} + C_{ds} + C_{\text{in}} \qquad \textbf{(14.33)}$$

where $C_{in}$ is the input capacitance of the following stage. We can now use these values to define the values of $f_2$ for the gate and drain circuits. These equations, which will look very familiar at this point, are as follows:

$$f_{2G} = \frac{1}{2\pi R'_{in}C_G} \qquad (14.34)$$

where

$$R'_{in} = R_S \| R_{in}$$

and

$$f_{2D} = \frac{1}{2\pi r_D C_D} \qquad (14.35)$$

Example 14.20 demonstrates the process for determining the upper cutoff frequency for an FET amplifier.

---

**EXAMPLE 14.20**

Determine the values of $f_{2G}$ and $f_{2D}$ for the amplifier in Figure 14.25. Assume that the FET has values of $C_{gd} = 4$ pF, $C_{gs} = 5$ pF, and $C_{ds} = 2$ pF. Also assume that the input capacitance to the load is 1 pF.

**Solution:** The first step is to draw the high-frequency equivalent of the circuit. This equivalent circuit is shown in Figure 14.28.

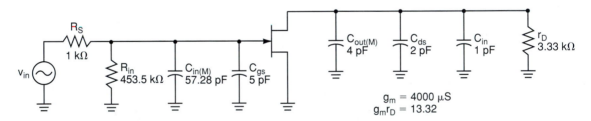

**FIGURE 14.28**

The value of $R_{in}$ was found by taking $R_1$ in parallel with $R_2$. Combining this value with $R_S$, we obtain the total ac resistance in the gate circuit, as follows:

$$R'_{in} = R_S \| R_{in}$$
$$= 998 \ \Omega$$

Now the Miller input capacitance is found as

$$C_{in(M)} \cong C_{gd}(g_m r_D + 1)$$
$$= (4 \text{ pF})(14.32)$$
$$= 57.28 \text{ pF}$$

Using this value of $C_{in(M)}$ and the value of $C_{gs}$, the total gate circuit capacitance is found as

$$C_G = C_{gs} + C_{in(M)}$$
$$= 5 \text{ pF} + 57.28 \text{ pF}$$
$$= 62.28 \text{ pF}$$

We can now use this value along with the value of $R'_{in}$ to find $f_{2G}$, as follows:

$$f_{2G} = \frac{1}{2\pi R'_{in} C_G}$$

$$= \frac{1}{2\pi(998 \ \Omega)(62.28 \ \text{pF})}$$

$$= 2.56 \ \text{MHz}$$

Since $g_m r_D > 10$ for this circuit, the Miller output capacitance is equal to $C_{gd}$, 4 pF. Combining this value with $C_{ds}$ and $C_{in}$, we get the total capacitance in the drain circuit, as follows:

$$C_D = C_{out(M)} + C_{ds} + C_{in}$$
$$= 4 \ \text{pF} + 2 \ \text{pF} + 1 \ \text{pF}$$
$$= 7 \ \text{pF}$$

Using the values of $C_D$ and $r_D$, we can find the upper cutoff frequency of the drain circuit as follows:

$$f_{2D} = \frac{1}{2\pi r_D C_D}$$

$$= \frac{1}{2\pi(3.33 \ \text{k}\Omega)(7 \ \text{pF})}$$

$$= 6.83 \ \text{MHz}$$

***PRACTICE PROBLEM 14.20***

The amplifier described in Practice Problem 14.18 has values of $C_{gd} = 4$ pF, $C_{gs} = 5$ pF, and $C_{ds} = 2$ pF. Determine the values of $f_{2G}$ and $f_{2D}$ for the circuit. Assume a load capacitance of 0 F.

Since the value of $f_{2G}$ is lower than $f_{2D}$, the gate circuit determines the overall value of $f_2$ for the amplifier. Thus, the upper cutoff frequency is 2.56 MHz. If the input frequency reaches this value, the value of $A_v$ will be 3 dB lower than $A_{v(mid)}$. Also, further increases in input frequency would cause the value of $A_v$ to continue to drop at a rate of 6 dB/octave until 6.83 MHz is reached. If frequency continues to increase, the value of $A_v$ will drop at a rate of 12 dB/octave since both the gate and drain circuits will be cut off. The Bode plot representing this circuit action is shown in Figure 14.29.

## Capacitance Specifications

The only catch with analyzing the high-frequency operation of an FET amplifier is the fact that the values of $C_{ds}$, $C_{gd}$, and $C_{gs}$ are not usually listed on the specification sheet for a given FET. Rather, the spec sheet will list the following: $C_{iss}$, $C_{oss}$, and $C_{rss}$.

$C_{iss}$ is the input capacitance to the FET, measured with the output shorted. The measurement of $C_{iss}$ is illustrated in Figure 14.30a. When the output is shorted and the input capacitance is measured, the measurement is equal to the parallel combination of $C_{gs}$ and $C_{gd}$. Thus,

$$C_{iss} = C_{gs} + C_{gd} \tag{14.36}$$

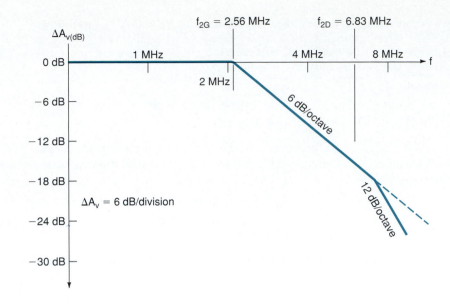

**FIGURE 14.29**

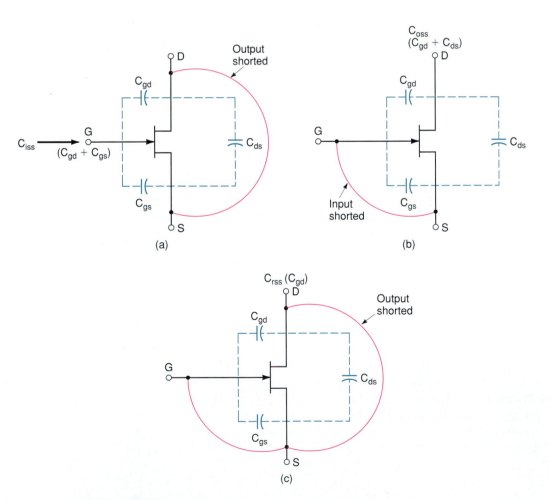

**FIGURE 14.30** Measuring $C_{iss}$, $C_{rss}$, and $C_{oss}$.

$C_{oss}$ is the output capacitance, measured with the gate–source junction shorted. The measurement of $C_{oss}$ is shown in Figure 14.30b. When the gate–source junction is shorted, the total capacitance is the parallel combination of $C_{ds}$ and $C_{gd}$. Thus,

$$C_{oss} = C_{ds} + C_{gd} \qquad \textbf{(14.37)}$$

Finally, $C_{rss}$ is the capacitance measured with both the drain–source and gate–source junctions shorted. With these two junctions shorted, the only capacitance present is $C_{gd}$, as shown in Figure 14.30a. Thus,

$$C_{rss} = C_{gd} \qquad \textbf{(14.38)}$$

Equations (14.36), (14.37), and (14.38) can be used to derive the following useful equations:

$$C_{gd} = C_{rss} \qquad \textbf{(14.39)}$$

$$C_{ds} = C_{oss} - C_{rss} \qquad \textbf{(14.40)}$$

and

$$C_{gs} = C_{iss} - C_{rss} \qquad \textbf{(14.41)}$$

Example 14.21 demonstrates the use of these equations.

---

A given FET has values of $C_{oss} = 6$ pF, $C_{rss} = 2$ pF, and $C_{iss} = 10$ pF. Determine the values of $C_{gs}$, $C_{ds}$, and $C_{gd}$.

**EXAMPLE 14.21**

*Solution:*   The gate–drain capacitance is found as

$$C_{gd} = C_{rss}$$
$$= 2 \text{ pF}$$

The gate–source capacitance is found as

$$C_{gs} = C_{iss} - C_{rss}$$
$$= 10 \text{ pF} - 2 \text{ pF}$$
$$= 8 \text{ pF}$$

Finally, the drain–source capacitance is found as

$$C_{ds} = C_{oss} - C_{rss}$$
$$= 6 \text{ pF} - 2 \text{ pF}$$
$$= 4 \text{ pF}$$

**PRACTICE PROBLEM 14.21**

A given FET has values of $C_{oss} = 4$ pF, $C_{rss} = 3$ pF, and $C_{iss} = 12$ pF. Determine the values of $C_{gs}$, $C_{ds}$, and $C_{gd}$.

## Theory Versus Practice

In our discussion on BJT amplifiers, you were shown that there tends to be a high percentage of error in the circuit $f_2$ calculations. For the FET amplifier, the largest percentage of error tends to show up in the $f_{2G}$ calculation. The reason for this is the fact that $g_m$ is involved in the calculation of $C_{in(M)}$.

In Chapter 12, you were shown that a JFET has two transconductance curves and, therefore, a range in values of $g_m$. The large possible range in $g_m$ values can cause a wide range in the value of $A_v$ for the amplifier. The range of $A_v$ values for an FET amplifier will show up in the calculation of $C_{in(M)}$ and, therefore, in the calculation of $f_{2G}$.

With the wide possible variation in the value of $f_{2G}$, you should not be surprised when the expected value of $f_{2G}$ is significantly lower (or higher) than the measured value. Again, the exact value of $f_2$ for an amplifier is critical only when you are dealing with tuned amplifiers. In any standard FET amplifier, you are concerned only with the approximate value of $f_2$ for the circuit.

---

**Section Review**

1. List, in order, the steps taken to perform the low-frequency analysis of an FET amplifier.

2. List, in order, the steps taken to perform the high-frequency analysis of an FET amplifier.

3. List the typical FET capacitance ratings and the equations used to convert them into usable terminal capacitances.

---

## 14.4  MULTISTAGE AMPLIFIERS

OBJECTIVE 13 ▶ When you cascade amplifier circuits with identical values of $f_2$, the overall value of $f_2$ is found as

$$f_{2T} = f_2 \sqrt{2^{1/n} - 1} \tag{14.42}$$

where $f_{2T}$ = the overall value of $f_2$ for the circuit
$f_2$ = the value of $f_2$ for one stage
$n$ = the total number of stages

The derivation of equation (14.42) is included in Appendix D.

When you cascade several amplifier stages, the overall value of $f_2$ will be *lower* than the value of $f_2$ for a single stage. This point is illustrated in Example 14.22.

---

**EXAMPLE 14.22**

Two amplifier circuits, each having a value of $f_2 = 500$ kHz, are cascaded. Determine the overall value of $f_2$ for the two-stage amplifier.

**Solution:**  The overall value of $f_2$ is found as

$$
\begin{aligned}
f_{2T} &= f_2 \sqrt{2^{1/n} - 1} \\
&= (500 \text{ kHz}) \sqrt{2^{1/2} - 1} \\
&= (500 \text{ kHz})(0.643) \\
&= 321.8 \text{ kHz}
\end{aligned}
$$

---

## PRACTICE PROBLEM 14.22

Four amplifier stages, each having an upper cutoff frequency of $f_2 = 800$ kHz, are cascaded. Determine the overall value of $f_2$ for the amplifier.

Most BJT and FET amplifiers have values of $f_1$, as you have seen in this chapter. When a value of $f_1$ exists for identical cascaded amplifiers, the overall value of $f_1$ changes as follows:

$$f_{1T} = \frac{f_1}{\sqrt{2^{1/n} - 1}} \qquad (14.43)$$

This equation is also derived in Appendix D.

When you cascade several amplifiers with values of $f_1$, the overall value of $f_1$ will be *higher* than the value of $f_1$ for a single stage. This point is illustrated in Example 14.23.

Two amplifiers with values of $f_1 = 5$ kHz are cascaded. Determine the overall value of $f_1$ for the two-stage amplifier.

**EXAMPLE 14.23**

**Solution:**  The overall value of $f_1$ is found as

$$f_{1T} = \frac{f_1}{\sqrt{2^{1/n} - 1}}$$

$$= \frac{5 \text{ kHz}}{\sqrt{2^{1/2} - 1}}$$

$$= \frac{5 \text{ kHz}}{0.0643}$$

$$= 7.78 \text{ kHz}$$

## PRACTICE PROBLEM 14.23

Three amplifier stages, each having a value of $f_1 = 800$ Hz, are cascaded. Determine the overall value of $f_1$ for the three-stage amplifier.

For any multistage amplifier, the total bandwidth is found as

$$BW_T = f_{2T} - f_{1T} \qquad (14.44)$$

When you cascade two or more amplifiers that are *not* identical, the overall circuit is analyzed in the same fashion as we used to analyze the combined effects of $f_{1B}, f_{1C}$, and $f_{1E}$ for the BJT amplifier. For example, consider the Bode plot shown in Figure 14.31. This plot represents the overall response of a two-stage amplifier with the following characteristics:

| Stage | $f_1$ | $f_2$ |
|-------|--------|--------|
| 1 | 1 kHz | 100 kHz |
| 2 | 10 kHz | 1 MHz |

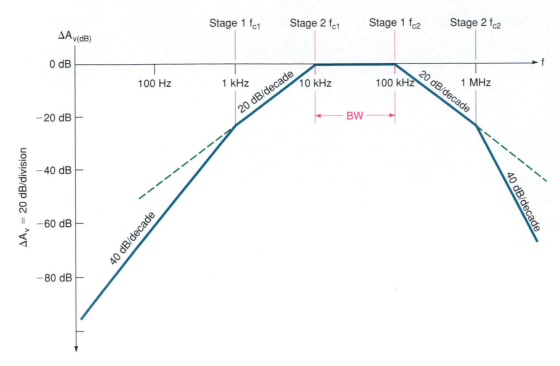

**FIGURE 14.31**   Overall amplifier frequency response.

*Note*: This discussion assumes (for simplicity) that each amplifier stage has only one lower and one upper cutoff frequency. In practice, each may have more (as demonstrated in earlier examples).

On the low end, the two-stage amplifier cuts off at 10 kHz, the value of $f_1$ for stage 2. The gain of the amplifier decreases at a *20-dB/decade* rate until the value of $f_1$ for stage 1 is reached, 1 kHz. At this point, the overall gain decreases at a rate of *40 dB/decade*. The same principles apply to the high-frequency operation. As you can see, the two-stage gain drops by 3 dB at the cutoff frequency of stage 1, 100 kHz. The *20-dB/decade* drop continues until $f_2$ of stage 2 (1 MHz) is reached. At that point, the roll-off rate increases to *40 dB/decade*.

Note that the bandwidth of the overall amplifier is determined by the *highest* $f_1$ value and the *lowest* $f_2$ value. Also, these principles apply to circuits with more than two stages. As each amplifier cuts off, the drop in gain will increase by another 20 dB/decade.

## Section Review

1. As you cascade identical amplifier stages, what happens to the overall value of $f_2$?
2. As you cascade identical amplifier stages, what happens to the overall value of $f_1$?
3. As you cascade identical amplifier stages, what happens to the overall value of BW?
4. How do you plot the frequency response of a cascaded amplifier that does not contain identical stages?

## ■ Chapter Summary

The *bandwidth* of an amplifier is the range of frequencies over which gain remains relatively constant. The bandwidth of a given amplifier depends on the value of the circuit components and on the type of active component(s) used. When an amplifier is operated within its bandwidth, the gain values for the amplifier are calculated as shown earlier in the text. For clarity, these values of $A_i$, $A_v$, and $A_p$ are referred to as *midband* gain values.

The *Bode plot* is a graphic representation of the frequency response of a given amplifier. The bandwidth of the amplifier is bound by the lower and upper *cutoff frequencies*, $f_1$ and $f_2$. At these frequencies, the gain of the amplifier is 3 dB down from the value of midband gain. As the Bode plot shows, gain continues to drop at a rate of 20 dB/decade (6 dB/octave) when the frequency continues beyond $f_1$ or $f_2$.

The low-frequency response of an amplifier is a function of the circuit resistance values and the values of the bypass and coupling capacitors. The high-frequency response is a function of the circuit resistance values and the *internal* capacitance values of the active component.

When making high-frequency calculations for an inverting circuit, you must take into account the *Miller input/output capacitance values*. These values are used in place of the feedback capacitance to make circuit calculations easier. When a given amplifier cuts off, the *gain roll-off rate* is independent of the circuit values of $R$ and $C$. In fact, these values are used only to determine the exact values of $f_1$ and $f_2$. Once these frequencies are passed, the decrease in gain is proportional to the actual frequency.

Since the power gain ($A_p$) of an amplifier drops by 3 dB (50 percent) at $f_1$ and $f_2$, these two points are commonly referred to as the *lower and upper 3-dB points*. At $f_1$ and $f_2$, $A_v$ will equal $0.707A_{v(\text{mid})}$. This relationship gives us a convenient method of measuring the lower and upper 3-dB points with the oscilloscope.

When more than one circuit is in cutoff, the roll-off rate will equal 20 dB *for each cutoff circuit*. Thus, if three circuits are in cutoff, the roll-off rate will equal three times 20 dB/decade. This holds true for any number of cascaded stages.

---

The following terms were introduced and defined in this chapter:

bandwidth (BW)
Bode plot
center frequency ($f_p$)
current–gain bandwidth
   product

cutoff frequencies
decade
frequency-response
   curve

gain–bandwidth product
Miller's theorem
octave
roll-off rate

---

| Equation Number | Equation | Section Number |
|---|---|---|
| (14.1) | $BW = f_2 - f_1$ | 14.1 |
| (14.2) | $f_0 = \sqrt{f_1 f_2}$ | 14.1 |
| (14.3) | $\dfrac{f_0}{f_1} = \dfrac{f_2}{f_0}$ | 14.1 |
| (14.4) | $f_1 = \dfrac{f_0^2}{f_2}$ | 14.1 |
| (14.5) | $f_2 = \dfrac{f_0^2}{f_1}$ | 14.1 |
| (14.6) | $A_{p(\text{dB})} = 10 \log A_p$ | 14.1 |
| (14.7) | $A_{p(\text{dB})} = 10 \log \dfrac{P_{\text{out}}}{P_{\text{in}}}$ | 14.1 |

| Equation Number | Equation | Section Number |
|---|---|---|
| (14.8) | $v_b = v_{in}\left(\dfrac{R_{in}}{\sqrt{X_C^2 + R_{in}^2}}\right)$ | 14.2 |
| (14.9) | $v_b = 0.707v_{in}$ (when $X_C = R_{in}$) | 14.2 |
| (14.10) | $f = \dfrac{1}{2\pi X_C C}$ | 14.2 |
| (14.11) | $f_{1B} = \dfrac{1}{2\pi RC}$ | 14.2 |
| (14.12) | $f_{1B} = \dfrac{1}{2\pi (R_S + R_{in})C_{C1}}$ | 14.2 |
| (14.13) | $f_{1C} = \dfrac{1}{2\pi (R_C + R_L)C}$ | 14.2 |
| (14.14) | $Z_{out} = R_E \parallel \left(r'_e + \dfrac{R'_{in}}{h_{fe}}\right)$ | 14.2 |
| (14.15) | $R_{out} \cong r'_e + \dfrac{R'_{in}}{h_{fe}}$ | 14.2 |
| (14.16) | $f_{1E} = \dfrac{1}{2\pi R_{out}C_E}$ | 14.2 |
| (14.17) | $\Delta A_{v(dB)} = 20\log\dfrac{1}{\sqrt{1 + (f_1/f)^2}}$ | 14.2 |
| (14.18) | $C_{be} \cong \dfrac{1}{2\pi f_T r'_e}$ | 14.2 |
| (14.19) | $C_{in(M)} = C_{bc}(A_v + 1)$ | 14.2 |
| (14.20) | $C_{out(M)} = C_{bc}\dfrac{A_v + 1}{A_v}$ | 14.2 |
| (14.21) | $C_{in(M)} \cong A_v C_{bc}$ | 14.2 |
| (14.22) | $C_{out(M)} \cong C_{bc}$ | 14.2 |
| (14.23) | $f_{2B} = \dfrac{1}{2\pi (R'_{in} \parallel h_{ie})(C_{be} + C_{in(M)})}$ | 14.2 |
| (14.24) | $f_{2C} = \dfrac{1}{2\pi r_C(C_{out(M)} + C_{in})}$ | 14.2 |
| (14.25) | $\Delta A_{v(dB)} = 20\log\dfrac{1}{\sqrt{1 + (f/f_2)^2}}$ | 14.2 |
| (14.26) | $f_{1G} = \dfrac{1}{2\pi (R_S + R_{in})C_{C1}}$ | 14.3 |
| (14.27) | $f_{1D} = \dfrac{1}{2\pi (R_D + R_L)C_{C2}}$ | 14.3 |

| Equation Number | Equation | Section Number |
|---|---|---|
| (14.28) | $R_{in} = R_1 \parallel R_2$ | 14.3 |
| (14.29) | $C_{in(M)} = C_{gd}(g_m r_D + 1)$ | 14.3 |
| (14.30) | $C_{out(M)} = C_{gd} \dfrac{g_m r_D + 1}{g_m r_D}$ | 14.3 |
| (14.31) | $C_{out(M)} \cong C_{gd}$ | 14.3 |
| (14.32) | $C_G = C_{gs} + C_{in(M)}$ | 14.3 |
| (14.33) | $C_D = C_{out(M)} + C_{ds} + C_{in}$ | 14.3 |
| (14.34) | $f_{2G} = \dfrac{1}{2\pi R'_{in} C_G}$ | 14.3 |
| (14.35) | $f_{2D} = \dfrac{1}{2\pi r_D C_D}$ | 14.3 |
| (14.36) | $C_{iss} = C_{gs} + C_{gd}$ | 14.3 |
| (14.37) | $C_{oss} = C_{ds} + C_{gd}$ | 14.3 |
| (14.38) | $C_{rss} = C_{gd}$ | 14.3 |
| (14.39) | $C_{gd} = C_{rss}$ | 14.3 |
| (14.40) | $C_{ds} = C_{oss} - C_{rss}$ | 14.3 |
| (14.41) | $C_{gs} = C_{iss} - C_{rss}$ | 14.3 |
| (14.42) | $f_{2T} = f_2 \sqrt{2^{1/n} - 1}$ | 14.4 |
| (14.43) | $f_{1T} = \dfrac{f_1}{\sqrt{2^{1/n} - 1}}$ | 14.4 |
| (14.44) | $BW_T = f_{2T} - f_{1T}$ | 14.4 |

**14.1.** 1.0995 MHz
**14.2.** 23.45 kHz
**14.3.** Both ratios equal 46.9.
**14.4.** 297 kHz
**14.5.** 120 kHz
**14.6.** 41.55 dB
**14.7.** $A_{pT} = 55.88$ dB
**14.8.** 45.2 Hz
**14.9.** 14.2 Hz
**14.10.** 54.5 Hz
**14.11.** 16.01 dB
**14.12.** 306 pF
**14.13.** 960 pF, 4 pF
**14.14.** 1.65 MHz

*Answers to the Example Practice Problems*

**14.15.** 37.54 MHz

**14.17.** 0.01 Hz

**14.18.** 260.9 Hz

**14.19.** 20.3 pF

**14.20.** $f_{2G} = 11.5$ MHz, $f_{2D} = 22.01$ MHz

**14.21.** $C_{gd} = 3$ pF, $C_{gs} = 9$ pF, $C_{ds} = 1$ pF

**14.22.** 348 kHz

**14.23.** 1.57 KHz

---

## Practice Problems

### §14.1

1. An amplifier has cutoff frequencies of 1.2 kHz and 640 kHz. Calculate the bandwidth and center frequency for the circuit.

2. An amplifier has cutoff frequencies of 3.4 kHz and 748 kHz. Calculate the bandwidth and center frequency for the circuit.

3. An amplifier has cutoff frequencies of 5.2 kHz and 489.6 kHz. Calculate the bandwidth and center frequency for the circuit.

4. An amplifier has cutoff frequencies of 1.4 kHz and 822.7 kHz. Calculate the bandwidth and center frequency for the circuit.

5. An amplifier has cutoff frequencies of 2.6 kHz and 483.6 kHz. Show that the ratio of $f_0$ to $f_1$ is equal to the ratio of $f_2$ to $f_0$ for the circuit.

6. An amplifier has cutoff frequencies of 1 kHz and 345 kHz. Show that the ratio of $f_0$ to $f_1$ is equal to the ratio of $f_2$ to $f_0$ for the circuit.

7. The values of $f_0$ and $f_2$ for an amplifier are measured at 72 kHz and 548 kHz, respectively. Calculate the values of $f_1$ and bandwidth for the amplifier.

8. The values of $f_0$ and $f_2$ for an amplifier are measured at 22.8 kHz and 321 kHz, respectively. Calculate the values of $f_1$ and bandwidth for the circuit.

9. The values of $f_0$ and $f_1$ for an amplifier are measured at 48 kHz and 2 kHz, respectively. Calculate the values of $f_2$ and bandwidth for the circuit.

10. The values of $f_0$ and $f_1$ for an amplifier are measured at 36 kHz and 4.3 kHz, respectively. Calculate the values of $f_2$ and bandwidth for the circuit.

### §14.2

11. Calculate the value of $f_{1B}$ for the amplifier in Figure 14.32.

12. Calculate the value of $f_{1B}$ for the amplifier in Figure 14.33.

13. Calculate the value of $f_{1C}$ for the amplifier in Figure 14.32.

14. Calculate the value of $f_{1C}$ for the amplifier in Figure 14.33.

15. Calculate the value of $f_{1E}$ for the amplifier in Figure 14.32.

16. Calculate the value of $f_{1E}$ for the amplifier in Figure 14.33.

17. A given amplifier has values of $A_{v(mid)} = 32$ dB and $f_1 = 8$ kHz. Calculate the dB voltage gain of the amplifier at operating frequencies of 7, 5, 4, and 1 kHz.

18. A given amplifier has values of $A_{v(mid)} = 16$ dB and $f_1 = 12$ kHz. Calculate the dB voltage gain of the amplifier at operating frequencies of 18, 12, 10, and 2 kHz.

19. Compare the results of Problems 11, 13, and 15. What is the approximate value of $f_1$ for the amplifier described in these three problems?

20. Compare the results of Problems 12, 14, and 16. What is the approximate value of $f_1$ for the amplifier described in these three problems?

21. A transistor has a value of $f_T = 200$ MHz at $I_C = 10$ mA. Determine the value of $C_{be}$ at this current.

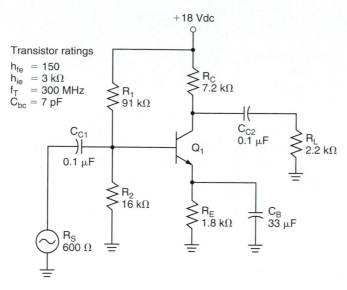

**FIGURE 14.32**

22. A transistor has a value of $f_T = 400$ MHz at $I_C = 1$ mA. Determine the value of $C_{be}$ at this current.

23. An inverting amplifier has values of $C_{bc} = 7$ pF and $A_v = 100$. Determine the Miller input and output capacitance values for the circuit.

24. An inverting amplifier has values of $C_{bc} = 3$ pF and $A_v = 4$. Determine the Miller input and output capacitance values for the circuit.

25. Determine the value of $f_{2B}$ for the amplifier in Figure 14.32.

26. Determine the value of $f_{2B}$ for the amplifier in Figure 14.33.

27. Determine the value of $f_{2C}$ for the amplifier in Figure 14.32.

28. Determine the value of $f_{2C}$ for the amplifier in Figure 14.33.

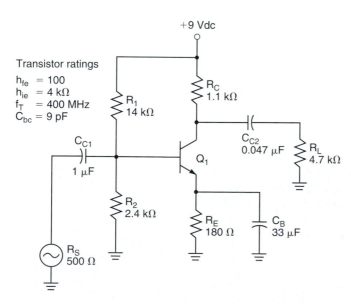

**FIGURE 14.33**

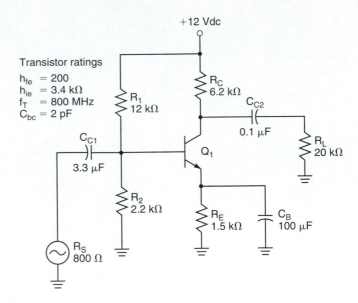

+12 Vdc

Transistor ratings
$h_{fe}$ = 200
$h_{ie}$ = 3.4 k$\Omega$
$f_T$ = 800 MHz
$C_{bc}$ = 2 pF

$R_1$
12 k$\Omega$

$R_C$
6.2 k$\Omega$

$C_{C2}$
0.1 $\mu$F

$C_{C1}$
3.3 $\mu$F

$Q_1$

$R_L$
20 k$\Omega$

$R_2$
2.2 k$\Omega$

$R_E$
1.5 k$\Omega$

$C_B$
100 $\mu$F

$R_S$
800 $\Omega$

**FIGURE 14.34**

**29.** Calculate the overall values of $f_1$ and $f_2$ for the amplifier in Figure 14.34.

**30.** Calculate the overall values of $f_1$ and $f_2$ for the amplifier in Figure 14.35.

*§14.3*

**31.** Calculate the value of $f_{1G}$ for the amplifier in Figure 14.36.

**32.** Calculate the value of $f_{1G}$ for the amplifier in Figure 14.37.

**33.** Calculate the value of $f_{1D}$ for the amplifier in Figure 14.36.

**34.** Calculate the value of $f_{1D}$ for the amplifier in Figure 14.37.

**35.** Calculate the values of $f_{2G}$ and $f_{2D}$ for the amplifier in Figure 14.36.

**36.** Calculate the values of $f_{2G}$ and $f_{2D}$ for the amplifier in Figure 14.37.

**37.** An FET has ratings of $C_{iss}$ = 14 pF, $C_{rss}$ = 9 pF, and $C_{oss}$ = 10 pF. Determine the values of $C_{gd}$, $C_{gs}$, and $C_{ds}$ for the device.

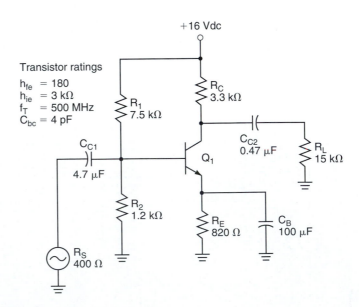

+16 Vdc

Transistor ratings
$h_{fe}$ = 180
$h_{ie}$ = 3 k$\Omega$
$f_T$ = 500 MHz
$C_{bc}$ = 4 pF

$R_1$
7.5 k$\Omega$

$R_C$
3.3 k$\Omega$

$C_{C1}$
4.7 $\mu$F

$C_{C2}$
0.47 $\mu$F

$Q_1$

$R_L$
15 k$\Omega$

$R_2$
1.2 k$\Omega$

$R_E$
820 $\Omega$

$C_B$
100 $\mu$F

$R_S$
400 $\Omega$

**FIGURE 14.35**

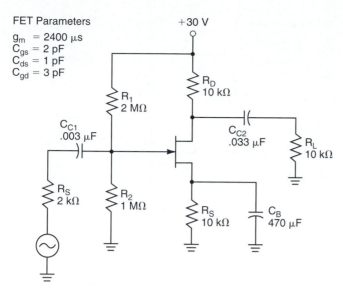

FET Parameters

$g_m$ = 2400 μs
$C_{gs}$ = 2 pF
$C_{ds}$ = 1 pF
$C_{gd}$ = 3 pF

+30 V

$R_1$
2 MΩ

$R_D$
10 kΩ

$C_{C1}$
.003 μF

$C_{C2}$
.033 μF

$R_L$
10 kΩ

$R_S$
2 kΩ

$R_2$
1 MΩ

$R_S$
10 kΩ

$C_B$
470 μF

**FIGURE 14.36**

38. An FET has ratings of $C_{oss}$ = 16 pF, $C_{rss}$ = 9 pF, and $C_{iss}$ = 12 pF. Determine the values of $C_{gd}$, $C_{gs}$, and $C_{ds}$ for the device.

39. Calculate the overall values of $f_1$ and $f_2$ for the amplifier in Figure 14.38.

40. Calculate the overall values of $f_1$ and $f_2$ for the amplifier in Figure 14.39.

41. Two amplifiers, each having a value of $f_2$ = 120 kHz, are cascaded. Determine the overall value of $f_2$ for the circuit.

42. Two amplifiers, each having a value of $f_1$ = 3 kHz, are cascaded. Determine the overall value of $f_1$ for the circuit.

43. Two amplifiers, each having values of $f_1$ = 2 kHz and $f_2$ = 840 kHz, are cascaded. Determine the bandwidth of the circuit.

44. Four amplifiers, each having values of $f_1$ = 1.5 kHz and $f_2$ = 620 kHz, are cascaded. Determine the bandwidth of the circuit.

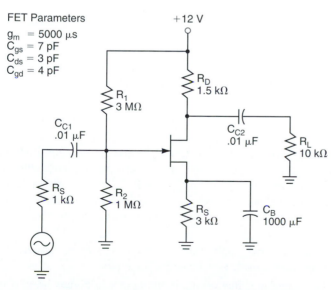

FET Parameters

$g_m$ = 5000 μs
$C_{gs}$ = 7 pF
$C_{ds}$ = 3 pF
$C_{gd}$ = 4 pF

+12 V

$R_1$
3 MΩ

$R_D$
1.5 kΩ

$C_{C1}$
.01 μF

$C_{C2}$
.01 μF

$R_L$
10 kΩ

$R_S$
1 kΩ

$R_2$
1 MΩ

$R_S$
3 kΩ

$C_B$
1000 μF

**FIGURE 14.37**

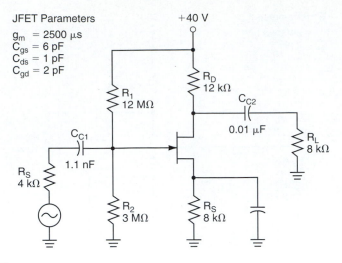

**FIGURE 14.38**

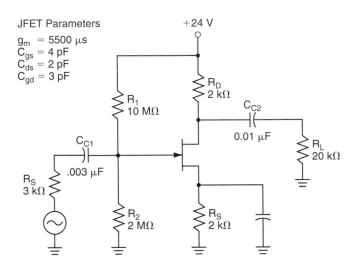

**FIGURE 14.39**

**The Brain Drain**

45. We have two amplifiers. The first amplifier has values of $f_{1B} = 2$ kHz, $f_{1C} = 16$ kHz, and $f_{1E} = 32$ kHz. The second amplifier has values of $f_{1B} = 8$ kHz, $f_{1C} = 12$ kHz, and $f_{1E} = 4$ kHz. For each of these amplifiers:

   **a.** Draw the Bode plot, including the appropriate roll-off rates. The Bode plots should have an octave frequency scale that starts at 2 kHz and continues up to 32 kHz.

   **b.** Plot the frequency-response curve on the same graph as the Bode plot. The frequency-response curve must take into account the combined effects of $f_{1B}, f_{1C}$, and $f_{1E}$ at each major division on the graph. In other words, at each major division, calculate the values of $\Delta A_{v(\text{dB})}$ for each of the terminal circuits, determine the total value of $\Delta A_{v(\text{dB})}$, and plot the point that corresponds to the total value of $\Delta A_{v(\text{dB})}$.

   After completing the two graphs, compare the results to see which frequency-response curve more closely resembles its Bode plot.

46. Calculate the change in $f_o$ that occurs in the circuit in Figure 14.40 if the load resistance opens.

**618** CHAPTER 14 / AMPLIFIER FREQUENCY RESPONSE

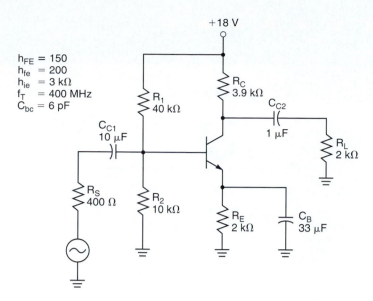

$h_{FE} = 150$
$h_{fe} = 200$
$h_{ie} = 3\ k\Omega$
$f_T = 400\ MHz$
$C_{bc} = 6\ pF$

+18 V

$R_1$
40 kΩ

$R_C$
3.9 kΩ

$C_{C1}$
10 μF

$C_{C2}$
1 μF

$R_L$
2 kΩ

$R_S$
400 Ω

$R_2$
10 kΩ

$R_E$
2 kΩ

$C_B$
33 μF

**FIGURE 14.40**

47. Write a program that will determine the lower cutoff frequencies of a BJT amplifier, given the proper input values.

48. Write a program that will determine the upper cutoff frequencies of a BJT amplifier, given the proper input values.

49. Combine the programs from Problems 47 and 48. Then add program steps to determine the overall values of $f_1, f_2, f_0$, and bandwidth for the circuit.

50. Repeat Problems 47 through 49 for a JFET amplifier.

51. Repeat Problem 50 for an inverting amplifier.

*Suggested Computer Applications Problems*

# 15

# OPERATIONAL AMPLIFIERS

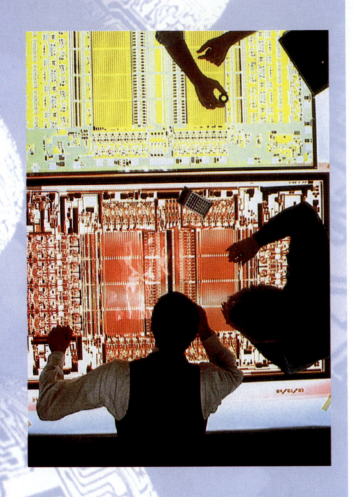

The production of integrated circuits starts with the design of the circuit. Here, several engineers check the final circuit design to ensure its accuracy.

# OUTLINE

# OBJECTIVES

*After studying the material in this chapter, you should be able to:*

1. Describe the *operational amplifier*, or *op-amp*.

2. Define the term *differential amplifier*.

3. Define open-loop voltage gain and state its typical range of values.

4. Describe the operation of a discrete differential amplifier.

5. Describe the various offset voltage and current values for an op-amp.

6. Discuss the effects of common-mode rejection ratio on op-amp operation.

7. Discuss the relationship between slew-rate and op-amp operating frequency.

8. Discuss slew-rate distortion and the means by which it can be reduced.

9. Describe and analyze the operation of the inverting amplifier.

10. Compare and contrast the operating characteristics of the inverting and noninverting amplifiers.

11. List the common op-amp faults and the symptoms of each.

12. Define gain–bandwidth product and explain its significance.

13. Describe the various types of feedback.

14. Describe the effects of negative feedback on inverting amplifier gain and bandwidth.

15. Calculate the *attenuation factor* and *feedback factor* for a given feedback amplifier.

16. Calculate the input and output impedance values for inverting and noninverting amplifiers.

# SHRINKING CIRCUITS

The impact that integrated circuit technology has had on the field of electronics can easily be understood if you consider the history of the operational amplifier, or op-amp. At one time, op-amps were made using vacuum tubes. A typical vacuum tube op-amp contained approximately six vacuum tubes that were housed in a single module. The typical module measured $3 \times 5 \times 12$ in. (approximately $8 \times 13 \times 30$ cm).

In the early 1960s, the semiconductor op-amp became available. This component (whose *equivalent* circuit is shown in Figure 15.1) typically contains close to 200 internal active devices in a case that is approximately $0.4 \times 0.26$ in. ($10.2 \times 6.6$ mm)!

---

**Discrete components**
Components housed in individual packages; that is, one package—one component.

**Integrated circuit (IC)**
A single package that contains any number of active and/or passive components, all constructed on a single piece of silicon.

Individual components, such as the 2N3904 BJT and the 2N5459 FET, are classified as **discrete components**. The term *discrete* indicates that each physical package contains only one component. For example, when you purchase a 2N3904, you are buying a single component housed in its own casing.

Over the years, advances in manufacturing technology have made it possible to produce entire *circuits* on a single piece of semiconductor material. This type of circuit, which is housed in a single casing, is referred to as an **integrated circuit**, or *IC*. ICs range in complexity from simple circuits containing a few active and/or passive components to complex circuits containing hundreds of thousands of components. The more complex the internal circuitry of an IC, the more complex its function.

The major impacts of ICs all relate to their internal operation and relatively low manufacturing cost. Circuit operations that once took hundreds of discrete components to perform can now be accomplished with a single IC. This has made circuits easier to design and troubleshoot. At the same time, the cost of an IC is generally lower than the cost of a comparable discrete-component circuit. This has made electronic systems less expensive to manufacture.

It would be impossible for a single book to cover *every* type of IC available. As your study of electronics continues, you will be introduced to more and more types of ICs. In this book, we will concentrate on the most commonly used *linear* IC, the *operational amplifier,* or *op-amp.* We will discuss the operating principles and basic amplifier applications of op-amps, including op-amp circuit troubleshooting.

## 15.1 OP-AMPS: AN OVERVIEW

OBJECTIVE 1 ▶

**Operational amplifier (op-amp)**
A high-gain dc amplifier that has high input impedance and low output impedance.

The op-amp is a *high gain* dc amplifier that has *high input impedance* and *low output impedance*. The internal circuitry, schematic symbol, and pin diagram for the 741 general-purpose **operational amplifier** are shown in Figure 15.1.

Check out that circuit! How would you like to have to troubleshoot *that* on your average Monday morning? Fortunately, the circuitry of the 741 is all contained in a single component. Since we're dealing with a single component, all we need to be concerned with are the input/output relationships and characteristics of the component. You cannot get into a 741 to repair the internal circuitry, so its complexity is of no consequence.

The signal inputs to the op-amp are labeled *inverting* and *noninverting*. Normally, the input to the amplifier will be applied to one of these two inputs. The other is usually used to control the operating characteristics of the component. The particular application determines which input pin is used as the active input.

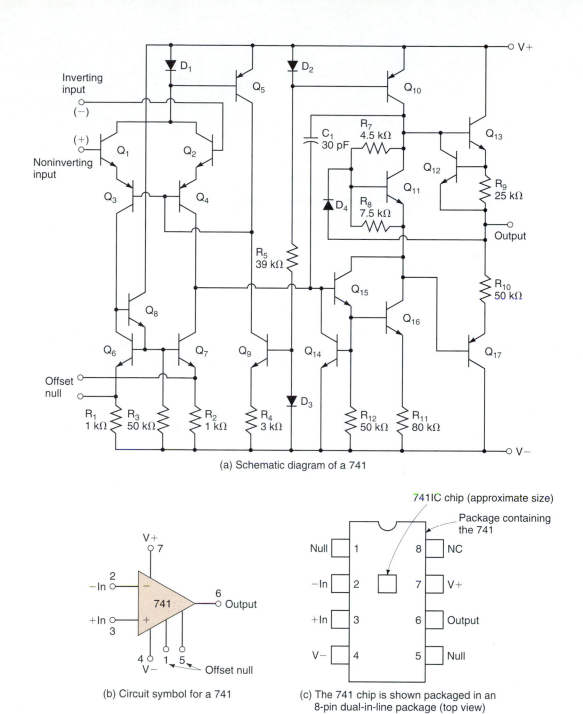

(a) Schematic diagram of a 741

(b) Circuit symbol for a 741

(c) The 741 chip is shown packaged in an 8-pin dual-in-line package (top view)

**FIGURE 15.1    A 741 operational amplifier (op-amp). (Courtesy of Fairchild, a division of National Semiconductor.)***

The op-amp has *two* power supply voltage inputs, labeled $+V$ and $-V$. These two supply pins are normally connected in one of two ways, as illustrated in Figure 15.2. One way is to have $+V$ and $-V$ set to equal voltages that are of opposite polarity, as shown in Figure 15.2a. The other way is to provide a single supply voltage to one of the supply pins while grounding the other. This connection is shown in Figure 15.2b and c. Again, the specific wiring of these two pins depends on the particular application. The *offset null* pins (Figure 15.1) are discussed later in this chapter.

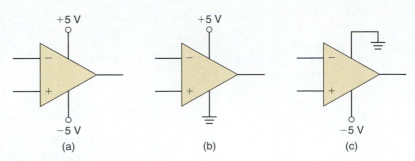

**FIGURE 15.2   Op-amp supply voltages.**

## IC Identification

Hundreds of types of op-amps are produced by various manufacturers. Many op-amps can be identified using a seven-character ID code. An example is shown in Figure 15.3. The *prefix* is used to identify the particular manufacturer. A listing of the most common prefixes is provided in Table 15.1. The **designator code** indicates two things:

1. The three-digit number indicates the specific type of op-amp.
2. The final letter indicates the operating temperature range.

Examples of commonly used temperature codes are shown in Table 15.2. The designator code is used not only to determine the specific type of op-amp you are dealing with; it is also used to help you determine which op-amps can be substituted for each other. This point is discussed further in the section on op-amp circuit troubleshooting. The *suffix* in the ID code indicates the type of package in which the op-amp is housed. The commonly used suffix codes are listed in Table 15.3.

**Designator code**
An IC code that indicates the type of circuit and its operating temperature range.

| Prefix | Designator | Suffix |
|--------|-----------|--------|
| MC | 741C | N |

**FIGURE 15.3   Op-amp ID code.**

TABLE 15.1   **Manufacturers' Prefixes**

| Prefix | Manufacturer |
|--------|--------------|
| AD | Analog Devices |
| CA | RCA |
| LM | National Semiconductor |
| MC | Motorola |
| NE/SE | Signetics |
| OP | Precision Monolithics |
| RC/RM | Raytheon |
| SG | Silicon General |
| TL | Texas Instruments |
| UA | Fairchild[a] |

[a]Fairchild is a division of National Semiconductor.

TABLE 15.2   **Temperature Codes**

| Code | Application | Temperature Range (°C) |
|------|-------------|------------------------|
| C | Commercial | 0 to 70 |
| I | Industrial | −25 to 85 |
| M | Military | −55 to 125 |

**TABLE 15.3    Suffix Codes**

| Code | Package Type |
|------|-------------|
| D | Plastic dual-in-line (DIP) |
| J | Ceramic DIP |
| N, P | Plastic DIP with longer lead |

## Op-Amp Packages

Op-amps are available in *dual-in-line* (DIP) packages, metal cans, and surface-mount packages (SMPs). The DIP and metal can packages are illustrated in Figure 15.4. Metal cans (type TO-5) are available with 8, 10, or 12 leads. DIPs for op-amps commonly have 8 or 14 pins, as shown. Of the three types of packages, DIPs are still the most commonly

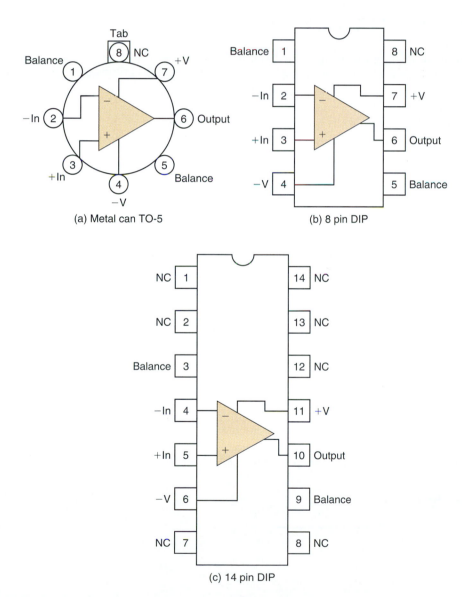

**FIGURE 15.4    Op-amp packages. (Courtesy of Prentice Hall.)**

used. Note that the type of package is extremely important when considering whether or not one op-amp may be substituted for another.

**Section Review**

1. What is a *discrete component*?
2. What is an *integrated circuit*?
3. What is an *operational amplifier*, or *op-amp*?
4. Draw the schematic symbol for an op-amp and identify the signal and supply voltage inputs.
5. What are the three parts of the op-amp identification code? What does each part of the code tell you?
6. What are the three types of op-amp packages?

## 15.2 OPERATION OVERVIEW

*OBJECTIVE 2* ▶

**Differential amplifier**
A circuit that amplifies the difference between two input voltages.

The input stage of the op-amp is a **differential amplifier.** This type of circuit amplifies the *difference between* inputs $v_1$ and $v_2$. This point is illustrated in Figure 15.5. Note that the amplifier sees the input as being a *difference voltage* between the two input voltages. By formula,

$$v_d = v_2 - v_1 \qquad (15.1)$$

where $v_d$ = the difference voltage that will be amplified
$v_1$ = the voltage applied to the *inverting* input
$v_2$ = the voltage applied to the *noninverting* input

*It is important for you to remember that the op-amp is amplifying the difference between the input terminal voltages.*

The output from the amplifier for a given pair of input voltages depends on several factors:

1. The *gain* of the amplifier.
2. The polarity relationship between $v_1$ and $v_2$.
3. The values of the supply voltage, $+V$ and $-V$.

We will now look at these three factors in detail.

### Op-Amp Gain

*OBJECTIVE 3* ▶

The *maximum* possible gain from a given op-amp is referred to as the **open-loop voltage gain,** $A_{OL}$. The value of $A_{OL}$ is generally greater than 10,000. For example, the Fairchild μA741 op-amp has an open-loop voltage gain of 200,000 (typical).

**FIGURE 15.5**

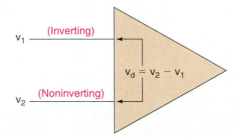

FIGURE 15.6   Op-amp feedback path.

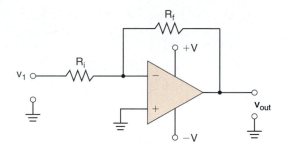

The term *open-loop* indicates a circuit condition where *there is no feedback path from the output to the op-amp input.* You may recall that a feedback path is a connection used to "feed" a portion of the output signal back to the input. Op-amp circuits usually contain one or more feedback paths. In Figure 15.6, the circuit containing $R_f$ is the feedback path. When part of the output signal is fed back to the input, the overall gain of the op-amp is reduced. As we cover specific circuits, you will be shown how to determine the gain of each. In the meantime, you should remember:

**Open-loop voltage gain, $A_{OL}$**
The maximum possible gain of a given op-amp.

What the term *open loop* means.

1. The maximum gain of a given op-amp is $A_{OL}$ (typically, 10,000 or greater).
2. The actual gain of a specific op-amp circuit is reduced when a negative feedback path is added between the component output and input.

The high gain of the op-amp is another advantage that this component has over BJTs and FETs. Values of $A_v$ near the 200,000 mark (which are common for op-amps) are virtually impossible for discrete components.

## Input/Output Polarity

The polarity relationship between $v_1$ and $v_2$ will determine whether the op-amp output voltage swings toward $+V$ or $-V$. If $v_1$ is *more negative* than $v_2$, the op-amp output voltage will swing toward $+V$. This is illustrated in Figure 15.7a. If $v_1$ is *more positive* than $v_2$, the op-amp output voltage will swing toward $-V$. This is illustrated in Figure 15.7b. Note the relationships between the input voltage polarities and the input signs in the schematic symbol. In Figure 15.7a, the polarity symbols of $v_1$ and $v_2$ match the polarity signs in the schematic symbol, and the output is positive. In Figure 15.7b, the polarity signs for $v_1$ and $v_2$ do not match the polarity signs in the schematic symbol, and the output is negative. This leads us to the following statement regarding the input/output polarity relationships for the op-amp:

> *When the input voltage polarities, with respect to each other, match the polarity signs in the schematic symbol, the output voltage is positive. When they do not match, the output is negative.*

Op-amp input/output polarity relationship.

This point is illustrated further in Figure 15.8.

**FIGURE 15.7   Op-amp input polarities.**

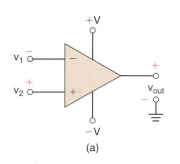

(a)

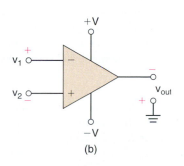

(b)

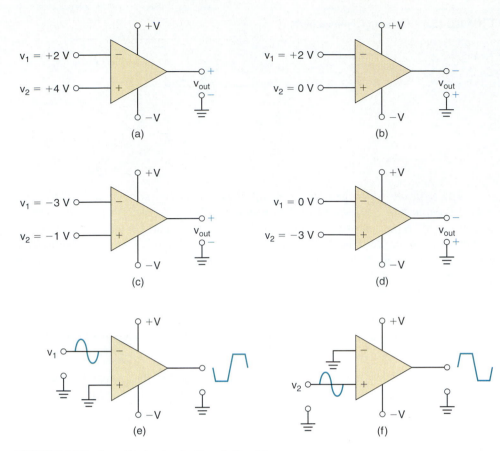

**FIGURE 15.8   Input/output polarity relationships.**

It is important to note that the output polarity is determined by the *relationship between* the polarities of $v_1$ and $v_2$, not by their polarities with respect to ground. For example, look at circuits (a) and (c) in Figure 15.8. In both of these cases, $v_1$ is *more negative* than $v_2$ (the voltage polarities match the polarity symbols), and the output is *positive*. It did not matter whether both inputs were positive or negative, only that $v_1$ was *negative with respect to $v_2$*. If you take a moment to look at circuits (b) and (d), you'll see that $v_1$ is *more positive* than $v_2$ (the voltage polarities do not match the polarity signs), and the output is *negative*. Again, the *relationship between $v_1$ and $v_2$* has determined the output polarity.

Now take a look at circuit (e). Note that the noninverting (+) input is *grounded*. When $v_1$ goes more positive than ground, the voltage polarities do not match signs, and the output goes negative. When $v_1$ goes more negative than ground, the voltage polarities match the polarity signs, and the output goes positive. Note the 180° phase shift that occurs between the op-amp input and output signals. This is where the term **inverting input** comes from. If you apply the same reasoning to circuit (f), you will see why the (+) input is referred to as the **noninverting input.**

There is another method you can use to determine the polarity of the op-amp output voltage for a given set of input voltages. You may remember that equation (15.1) defined the input difference voltage as

**Inverting input**
The op-amp input that produces an output 180° voltage phase shift when used as a signal input.

**Noninverting input**
The op-amp input that does not produce an output voltage phase shift when used as a signal input.

$$v_d = v_2 - v_1$$

When the result of this equation is *positive,* the op-amp output voltage will be *positive*. When the result of this equation is *negative,* the output voltage will be *negative*. This is illustrated in Example 15.1.

EXAMPLE 15.1

Determine the output voltage polarity for circuits (a) and (b) in Figure 15.8 using equation (15.1).

**Solution:**  For circuit (a),

$$v_d = v_2 - v_1$$
$$= 4\text{ V} - 2\text{ V}$$
$$= 2\text{ V}$$

Since $v_d$ is *positive,* the op-amp output voltage will be *positive.* For circuit (b),

$$v_d = v_2 - v_1$$
$$= 0\text{ V} - 2\text{ V}$$
$$= -2\text{ V}$$

Since $v_d$ is *negative,* the op-amp output voltage will be *negative.*

**PRACTICE PROBLEM 15.1**

Using equation (15.1), determine the output voltage polarities for circuits (c) and (d) in Figure 15.8.

---

If you compare the polarities obtained in the example with those determined in the discussion, you will see that we reached the same conclusions in both cases.

You may be wondering why we went to all the trouble of developing an *observation method* of analysis (comparing the voltage polarities to the polarity signs in the schematic symbol) when the *mathematical method* of analysis [using equation (15.1)] seems so much simpler. The reason for establishing two methods of analysis is because each is simpler to use under different circumstances. When you are analyzing the schematic diagram of an op-amp-based circuit or system, it is easier to use the mathematical method of analysis. When you are troubleshooting an op-amp circuit or system with an oscilloscope, the observation method is easier to use because you do not need to determine exact voltage values. You only need to quickly determine the polarity relationship between the inputs, and then you can predict and observe the output voltage polarity.

Up to this point, we have simplified matters by not considering the effects of the $+V$ and $-V$ values on the op-amp output. As you were shown earlier, these supply pins can be set to different values. We will now take a look at the effects of $+V$ and $-V$ on the output voltage from the op-amp.

## Supply Voltages

The supply voltages ($+V$ and $-V$) determine the *limits* of the output voltage swing. No matter what the gain or input signal strength, the output cannot exceed some value *slightly less than* $+V$ or $-V$. For example, consider the circuits shown in Figure 15.9. Circuit (a) has supply voltages of $\pm 15$ V. Assuming that the output can make the full transition between $+V$ and $-V$, the output cannot go any more positive than $+15$ V or any more negative than $-15$ V. For circuit (b), the output would be limited to $+5$ V on the positive transition and $-5$ V on the negative transition. The output from circuit (c) would have limits of $+10$ V and ground, while the output from circuit (d) would have limits of ground and $-10$ V.

In practice, the peak output voltage will not reach either $+V$ or $-V$. The reason for this is the fact that some voltage is dropped across the components in the op-amp output

*The supply voltages limit the output $V_{PP}$.*

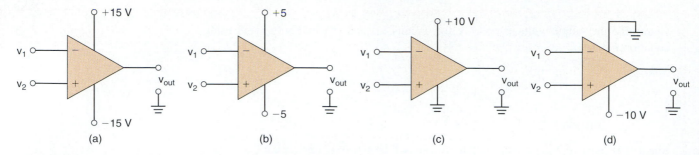

**FIGURE 15.9** Op-amp supply voltages.

circuit (refer to Figure 15.1a). The actual limits on $v_{out}$ depend on the op-amp being used and the value of the load resistance. For example, the spec sheet for the μA741 op-amp lists the following parameters:

| Parameter | Condition | Typical Value[a] |
|---|---|---|
| Output voltage swing | $R_L \geq 10\ k\Omega$ | $\pm 14$ V |
| | $R_L \geq 2\ k\Omega$ | $\pm 13$ V |

[a]$V_S = \pm 15$ V, $T_A = 25°C$.

For this op-amp, with source voltages of $\pm 15$ V, the output will typically be limited to $\pm 14$ V if the load is 10 kΩ or more, and to $\pm 13$ V if the load is between 2 and 10 kΩ. Note that $\pm 14$ V represents a peak-to-peak voltage of 28 V and $\pm 13$ V represents a peak-to-peak voltage of 26 V.

What happens if the load resistance is less than 2 kΩ? The spec sheet for the μA741 contains a graph showing maximum output voltage as a function of load resistance. This graph is shown in Figure 15.10. Note that the output voltages are given as peak-to-peak values. Also note that the curve was derived using supply voltages of $\pm 15$ V. You determine the maximum output voltage for a given load resistance by finding that resistance on the x-axis and determining the corresponding value of maximum output voltage. For example, a load resistance of 200 Ω would limit the output to a peak-to-peak voltage of 10 V, or $\pm 5$ V.

Now another question arises. How do you determine the maximum output swing when the source voltages are set to values other than $\pm 15$ V? The graph of output volt-

**FIGURE 15.10** **Output voltage as a function of load resistance.**

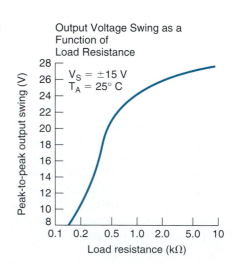

FIGURE 15.11 **Output voltage as a function of supply voltage.**

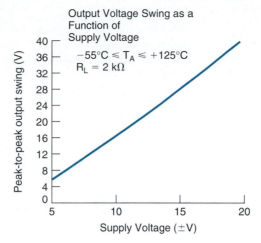

Output Voltage Swing as a Function of Supply Voltage

$-55°C \leqslant T_A \leqslant +125°C$
$R_L = 2 \text{ k}\Omega$

age swing versus supply voltage for the μA741 is shown in Figure 15.11. As you can see, the maximum output voltage increases at a linear rate as the supply voltage values increase. Generally, the following guidelines can be used when dealing with the 741:

| Value of Load Resistance (kΩ) | Max. (+) Output | Max. (−) Output |
|---|---|---|
| Greater than 10 | $(+V) - 1 \text{ V}$ | $(-V) + 1 \text{ V}$ |
| 2 to 10 | $(+V) - 2 \text{ V}$ | $(-V) + 2 \text{ V}$ |

For example, consider the circuit shown in Figure 15.12. The supply voltages are +10 V and ground. If the value of $R_L$ is 10 kΩ or more, the maximum output transition is from +9 to +1 V. If the load resistance is between 2 and 10 kΩ, the maximum output voltage transition is from +8 to +2 V.

## Putting It All Together

We have considered the factors of gain, $v_2$–$v_1$ polarity relationships, load resistance, and supply voltages on the operation of an op-amp. In this section, we will go through a series of examples that put all these factors together. In each example, a gain value is assumed to simplify the problem. Later in the chapter, you will be shown how to determine the gain of each of these circuits.

FIGURE 15.12

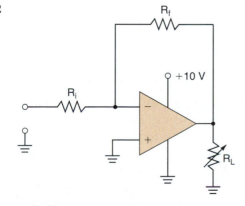

## EXAMPLE 15.2

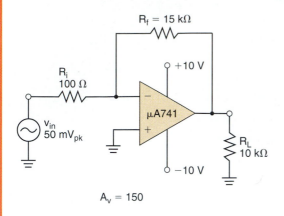

R_f = 15 kΩ

R_i
100 Ω

+10 V

μA741

−10 V

V_in
50 mV_pk

R_L
10 kΩ

A_v = 150

**FIGURE 15.13**

In this example, we converted $v_{in}$ to a peak-to-peak form. When you convert a negative peak voltage to peak-to-peak form, you can drop the minus sign as we did here.

Determine the peak-to-peak output voltage value for the circuit shown in Figure 15.13. Also determine the maximum possible output values for the amplifier. Assume that the gain of the amplifier is 150.

***Solution:*** Since the noninverting input to the amplifier is grounded, the value of $v_2$ is 0 V. The difference voltage is therefore found as

$$v_d = v_2 - v_1$$
$$= -v_1$$

Converting the peak input value of 50 mV$_{pk}$ to 100 mV$_{PP}$, the peak-to-peak output voltage is determined as

$$v_{out} = A_v v_d$$
$$= (150)(100 \text{ mV}_{pp})$$
$$= 15 \text{ V}_{pp}$$

Since the load resistance is 10 kΩ, we can determine the maximum possible output peak values using the general guidelines established for the 741 earlier in the chapter. Thus, the maximum positive and negative peak values are found as

$$V_{pk(+)} = (+V) - 1 \text{ V} = +9 \text{ V} \qquad \text{(maximum)}$$

and

$$V_{pk(-)} = (-V) + 1 \text{ V} = -9 \text{ V} \qquad \text{(maximum)}$$

### PRACTICE PROBLEM 15.2

An op-amp circuit has the following values: $+V = 12$ V, $-V = -12$ V, $v_1 = 20$ mV$_{pk}$, $v_2 = 0$ V (ground), and $A_v = 140$. Determine the value of $v_{out}$ for the circuit.

## EXAMPLE 15.3

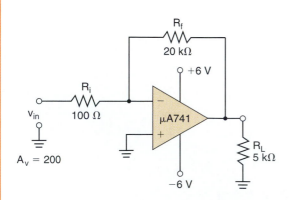

R_f

20 kΩ

+6 V

R_i

V_in    100 Ω

μA741

R_L
5 kΩ

A_v = 200

−6 V

**FIGURE 15.14**

Determine the maximum allowable value of $v_{in}$ for the circuit shown in Figure 15.14. Assume that the gain of the amplifier is 200.

***Solution:*** The first step is to determine the maximum allowable peak output voltage values. Since the load resistance is between 2 and 10 kΩ, the peak output values are found as

$$V_{pk(+)} = (+V) - 2 \text{ V} = +4 \qquad \text{(maximum)}$$

and

$$V_{pk(-)} = (-V) + 2 \text{ V} = -4 \text{ V} \qquad \text{(maximum)}$$

The maximum value of peak-to-peak output voltage would be the difference between these two values, 8 V. Dividing this value by the gain of the amplifier gives us the maximum allowable peak-to-peak value of $v_{in}$, as follows:

$$v_{in(PP)} = \frac{V_{PP(max)}}{A_v}$$

$$= \frac{8\text{ V}}{200}$$

$$= 40\text{ mV}_{PP}$$

*Note*: When one op-amp input is grounded

$$v_d = v_{in}$$

## PRACTICE PROBLEM 15.3

Determine the maximum allowable peak-to-peak input voltage for the amplifier described in Practice Problem 15–2. Assume that the load resistance for the circuit is 20 kΩ.

---

Determine the maximum peak output values for the circuit in Figure 15.15. Also, determine the maximum allowable value of $v_{in}$ for the circuit. Assume that the gain of the amplifier is 121.

*Solution:*  Since the load resistance is less than 2 kΩ, we must make sure that the output voltage swing will not be limited by the load. Checking the graph in Figure 15.10, we see that the maximum possible output swing with a load of 1.5 kΩ is 26 $V_{PP}$, or ±13 $V_{pk}$. Since the supply voltages are well below this value, we can safely assume that the load will not affect the output voltage swing. With this in mind, we can determine the maximum output values using the guidelines established for loads *under* 10 kΩ. Thus,

$$V_{pk(+)} = (+V) - 2\text{ V} = +2\text{ V} \qquad \text{(maximum)}$$

and

$$V_{pk(-)} = (-V) + 2\text{ V} = -2\text{ V} \qquad \text{(maximum)}$$

and our maximum $V_{PP}$ for the circuit is 4 V. Using this value, the maximum $V_{PP}$ of the input is found as

$$v_{in(PP)} = \frac{V_{PP(max)}}{A_v}$$

$$= 33.06\text{ mV}_{PP}$$

**EXAMPLE 15.4**

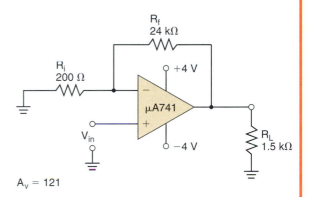

$A_v = 121$

**FIGURE 15.15**

## PRACTICE PROBLEM 15.4

Determine the maximum allowable peak-to-peak input voltage for the amplifier described in Practice Problem 15.2. Assume that the load resistance for the circuit is 1 kΩ. Use the curve in Figure 15.10 to determine the effect of the load resistance on the op-amp output voltage limits.

---

Incidentally, the circuit in Figure 15.15 had the input applied to the *noninverting* terminal, while the circuits in Figures 15.13 and 15.14 had the inputs applied to the *inverting* terminals. The main difference between the circuit in Figure 15.15 and the other two is that there will be no voltage phase shift from input to output. The circuits in Figures 15.13 and 15.14 will both have a 180° voltage phase shift from input to output.

Another point can be made at this time. The circuit configuration shown in Figures 15.13 and 15.14 are the op-amp equivalents of the *common-emitter* amplifier. If you compare these circuits to the voltage-divider biased amplifier, you will see that the op-amp circuits contain fewer components. This is another advantage of using op-amp circuits rather than discrete component circuits. Op-amp circuits require fewer external components to establish the desired operation.

---

**Section Review**

1. What is a *differential amplifier*?
2. What determines the output from an op-amp?
3. What is *open-loop voltage gain* ($A_{OL}$)?
4. What are the typical values of $A_{OL}$ for op-amps?
5. Describe the *observation method* for determining the output voltage polarity for an op-amp.
6. Describe the *mathematical method* for determining the output voltage polarity for an op-amp.
7. What effect do the supply voltages of an op-amp have on the maximum peak-to-peak output voltage?
8. What effect does the load resistance of an op-amp have on the maximum peak-to-peak output voltage?

---

## 15.3 THE DIFFERENTIAL AMPLIFIER AND OP-AMP SPECIFICATIONS

Diodes, BJTs, and FETs all have parameters and electrical characteristics that affect their operation. The op-amp is no exception. Up to this point, we have ignored the op-amp parameters and electrical characteristics to give you a chance to grasp the fundamental concepts of component operation without being bogged down with detail. Now it is time to bring the op-amp parameters and electrical characteristics into the picture.

To understand many of the op-amp electrical characteristics, you need to understand the operation of its input circuit. This input circuit, which is driven by the inverting and noninverting inputs, is called a **differential amplifier.** In this section, we will take a look at the operation of a discrete differential amplifier. We will then cover most of the op-amp electrical characteristics.

**Differential amplifier**
The input circuit of the op-amp driven by the inverting and noninverting inputs. This circuit produces an output that is proportional to the difference between its inputs.

### The Basic Differential Amplifier

A *differential amplifier* is a circuit that accepts two inputs and produces an output that is proportional to the *difference between* those inputs. The basic differential amplifier is shown in Figure 15.16. Note that the input to $Q_2$ is identified as the *noninverting input* (*NI*), and the input to $Q_1$ is identified as the *inverting input* (*I*).

OBJECTIVE 4 ▶

Ideally, the $Q_1$ and $Q_2$ circuits have identical characteristics. Since they are etched on a single piece of silicon, the transistors are identical. By design, the values of the two collector resistors are also equal. With identical characteristics, the quiescent values of $I_E$ for the transistors are equal. When the two inputs are grounded as shown in Figure 15.16:

$$I_{E1} = I_{E2}$$

(15.2)

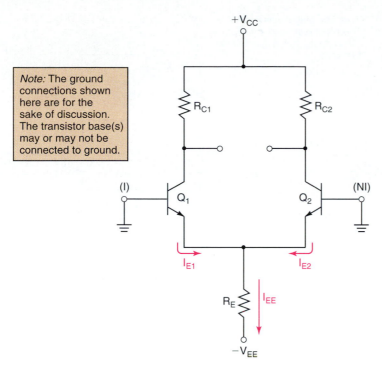

Note: The ground connections shown here are for the sake of discussion. The transistor base(s) may or may not be connected to ground.

**FIGURE 15.16    The basic differential amplifier.**

Since both emitter currents pass through $R_E$,

$$I_{E1} = I_{E2} = \frac{I_{EE}}{2} \qquad\qquad (15.3)$$

where

$$I_{EE} = \frac{V_E - V_{EE}}{R_E} \qquad\qquad (15.4)$$

Assuming that base currents are negligible,

$$I_C \cong I_E$$

and

$$I_{C1} = I_{C2} \cong \frac{I_{EE}}{2}$$

When both collector currents and both collector resistors are equal, then

$$V_{C1} = V_{C2} = V_{CC} - I_C R_C \qquad\qquad (15.5)$$

and since $V_{C1} = V_{C2}$, the output voltage of the differential amplifier is also 0 V ($V_{out1} - V_{out2}$).

Now consider what happens when we apply a signal to the inverting input, as shown in Figure 15.17a. During the positive alternation of the input, the current through $Q_1$ increases. Assuming that the value of $I_{EE}$ is relatively constant, the increase in $I_{E1}$ causes $I_{E2}$ to *decrease*. The increase in $I_{E1}$ causes the voltage drop across $R_{C1}$ to increase, and $V_{C1}$ *decreases*. At the same time, the decrease in $I_{E2}$ causes the voltage drop across

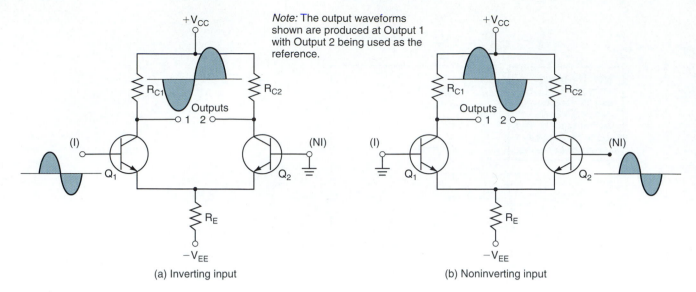

Note: The output waveforms shown are produced at Output 1 with Output 2 being used as the reference.

(a) Inverting input

(b) Noninverting input

FIGURE 15.17

$R_{C2}$ to decrease, and $V_{C2}$ *increases*. Assuming that the output signal ($v_{out}$) is taken from output 1 (with output 2 being the reference), we obtain the negative output alternation shown in the figure.

When the input in Figure 15.17a goes *negative*, $I_{E1}$ decreases and $I_{E2}$ increases. This time, the current changes cause $V_{C1}$ to increase and $V_{C2}$ to decrease. These changes cause $v_{out}$ to *increase*, as shown in the figure. Thus, the output voltage is 180° out of phase with the input voltage.

In Figure 15.17b, a positive input causes $I_{E2}$ to increase and $I_{E1}$ to decrease. As you have been shown, these changes cause a positive alternation to be produced at output 1. A negative-going input causes $I_{E2}$ to decrease and $I_{E1}$ to increase. These changes cause a negative alternation to be produced at output 1. Thus, the input and output voltages (in this case) are in phase.

## Modes of Operation

There are three basic modes of operation for a differential amplifier:

1. *Single-ended mode.* A differential amplifier is operating in a single-ended mode when an active signal is applied to only one input, as shown in Figure 15.17. The inactive is normally connected directly (or via a resistor) to ground. Depending on which input is active, the amplifier is classified as either an *inverting amplifie*r (Figure 15.17a) or a *noninverting amplifier* (Figure 15.17b).

2. *Differential mode.* For differential operation, two active signals are applied to the amplifier. The output magnitude and polarity will reflect the relationship between the two inputs, as described earlier in this chapter.

3. *Common-mode* operation occurs when two signals of the same amplitude, frequency, and phase are applied to the differential amplifier. In this mode, the output of the amplifier is ideally zero when measured between outputs 1 and 2. The outputs at $Q_1$ and $Q_2$ effectively cancel each other due to the inverting action of $Q_1$ and the noninverting action of $Q_2$. The advantage of this type of operation is that any noise or unwanted signal appearing at both inputs of the diff-amp is canceled and will not appear at the output to cause distortion of the signal. (This point is discussed in detail later in this section.)

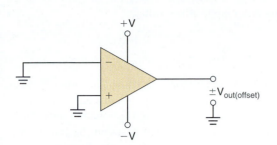

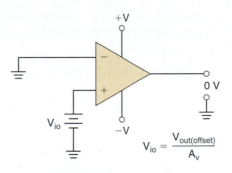

**FIGURE 15.18**

**FIGURE 15.19    Applying an input offset voltage.**

## Output Offset Voltage

Even though the transistors in the differential amplifier are very closely matched, there are some differences in their electrical characteristics. One of these differences is found in the values of $V_{BE}$ for the two transistors. When $V_{BE1} \neq V_{BE2}$, an imbalance is created in the differential amplifier that may show up as an **output offset voltage.** This condition is illustrated in Figure 15.18. Note that with the op-amp inputs grounded, the output shows a measurable voltage. This voltage is a result of the imbalance in the differential amplifier, which causes one of the transistors to conduct harder than the other.

There are several methods that may be used to eliminate output offset voltage. One of these is to apply an **input offset voltage,** $V_{io}$, between the input terminals of the op-amp, as shown in Figure 15.19. The value of $V_{io}$ required to eliminate the output offset voltage is found as

$$V_{io} = \frac{V_{out(offset)}}{A_v} \qquad (15.6)$$

Another method of eliminating output offset voltage is to connect the op-amp's **offset null** pins as shown in Figure 15.20. When the offset null is used, power is applied to the circuit and the potentiometer is adjusted to eliminate the output offset.

Figure 15.1 shows that the offset null pins are connected (indirectly) to the input differential amplifier. When properly adjusted, the offset null circuitry corrects the imbalance in the differential amplifier, causing the output of the op-amp to go to 0 V. As you will be shown later in the chapter, the offset null pins are rarely used. In some applica-

◄  *OBJECTIVE 5*

**Output offset voltage [$V_{out(offset)}$]**
A voltage that may appear at the output of an op-amp; caused by an imbalance in the differential amplifier.

**Input offset voltage ($V_{io}$)**
Voltage applied between the input terminals of the op-amp to eliminate output offset voltage.

**Offset null**
Pins connected to the inputs of the op-amp to eliminate output offset voltage.

**FIGURE 15.20    Offset null connection for the 741 op-amp.**

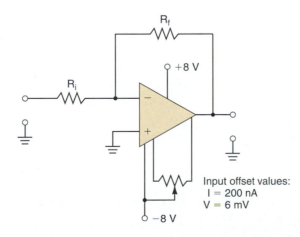

Input offset values:
I = 200 nA
V = 6 mV

tions, critical operation of the op-amp is not required. In other applications, there are simpler methods that can be used to eliminate any output offset voltage.

## Input Offset Current

**Input offset current**
A slight difference in the input currents, caused by differences in the transistor beta ratings.

When the output offset voltage of an op-amp is eliminated (as shown in either Figure 15.19 or 15.20), there could be a slight difference between the input currents to the noninverting and inverting inputs of the device ($i_1$ and $i_2$). This slight difference in input currents, called **input offset current,** is caused by a beta mismatch between the transistors in the differential amplifier. Note that there is no way of predicting which of the two input currents will be greater when the output offset voltage is eliminated.

## Input Bias Current

**Input bias current**
The average quiescent value of dc biasing current drawn by the signal inputs of an op-amp.

The inputs to an op-amp require some amount of dc biasing current for the BJTs in the differential amplifier. The average quiescent value of dc biasing current drawn by the signal inputs of the op-amp is the **input bias current** *rating* of the amplifier. For the μA741, this rating is between 80 nA (typical) and 500 nA (maximum). This means that the op-amp signal inputs draw between 80 nA an 500 nA from the external circuitry when no active signal is applied to the device.

The fact that both transistors in the differential amplifier require an input biasing current leads to the following operating restriction: *An op-amp will not produce the expected output if either of its inputs is open.* For example, look at the circuit shown in Figure 15.21. The noninverting input is shown to have an open between the op-amp and ground. This open circuit does not allow the dc biasing current required for the operation of the differential amplifier. (*The transistor associated with the inverting input will work, but not the one associated with the noninverting input.*) Since the differential amplifier will not work, the overall op-amp circuit will not work either. Thus, an input bias current path must always be provided for *both* op-amp inputs.

## Common-Mode Rejection Ratio (CMMR)

OBJECTIVE 6 ▶     **Common-mode signals** are *identical signals that appear simultaneously at the two inputs of an op-amp.* For example, the two signals shown in Figure 15.22 are common-mode signals.

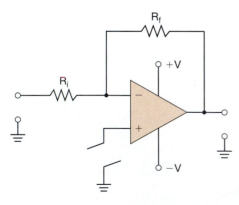

FIGURE 15.21   An open input prevents the op-amp from working.

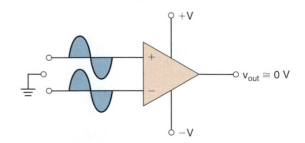

FIGURE 15.22   Common-mode signals.

If the two signals shown *occur at the same time and have the same amplitude,* the *ideal* op-amp will not respond to their presence. Remember that the op-amp is designed to respond to *the difference between* two input signals. If there is no difference between the two inputs, the op-amp won't produce an output.

A measure of the ability of an op-amp to "ignore" common-mode signals is the **common-mode rejection ratio (CMMR).** Technically, it is *the ratio of differential-mode gain to common-mode gain.* For example, let's assume for a moment that the op-amp in Figure 15.22 has a differential-mode voltage gain of 1500. Let's also assume that the op-amp has a common-mode gain of 0.01. The common-mode rejection ratio for the op-amp would be found as

$$CMRR = 1500 : 0.01$$
$$= 150,000 : 1$$

This number means that the gain provided to a difference between the inputs would be 150,000 times as great as the gain provided to two common-mode signals.

The ability of the op-amp to reject common-mode signals is one of its strong points. Common-mode signals are usually *undesired* signals, caused by external interference. For example, any RF signals picked up by the op-amp inputs would be considered undesirable. The common-mode rejection ratio indicates the op-amp's ability to reject such unwanted signals.

The common-mode rejection ratio for a given op-amp is usually measured in *decibels,* or *dB*. For example, the μA741 op-amp has a common-mode rejection ratio of 70 dB. This means that a differential input signal will have a gain that is at least 3163 times as great as the gain of a common-mode signal.

## Power Supply Rejection Ratio

The **power supply rejection ratio** is a rating that indicates *how much the output from an op-amp will change when the supply voltages change.* The μA741 has a power supply rejection ratio of 150 μV/V (maximum). This means that the dc output voltage from the op-amp will change by no more than 150 μV when the power supply voltage changes by 1 V. If the supply voltage for the op-amp was to change by 3 V, the dc output voltage would change by no more than 3 × 150 μV = 450 μV.

## Output Short-Circuit Current

Op-amps are protected internally from excessive current caused by a shorted load. The **output short-circuit current** rating of an op-amp is the maximum value of output current when the load is shorted. For the μA741, this rating is 25 mA. Thus, with a load of 0 Ω, the output current from this op-amp will be no more than 25 mA. Note that this short-circuit protection is provided by $R_9$ and $R_{10}$ (shown in Figure 15.1) in the 741 op-amp.

The short-circuit current rating helps to explain why the output voltage from an op-amp drops when the load resistance decreases. For example, consider the circuit shown in Figure 15.23. The load resistance is shown to be 50 Ω. Assuming that the op-amp is producing the maximum output current of 25 mA, the maximum load voltage is found as

$$V_L = I_{out}R_L$$
$$= (25 \text{ mA})(50 \text{ Ω})$$
$$= 1.25 \text{ V}$$

This is considerably less than the ±10 V supply that normally limits the output voltage. Whenever you use a load resistance that is less than the output resistance of the op-amp, the value of the short-circuit current will drop the maximum output voltage to a very low value.

**Common-mode signals**
Identical signals that appear simultaneously at the two inputs of an op-amp.

**Common-mode rejection ratio (CMRR)**
The ratio of differential gain to common-mode gain.

Common-mode signals are usually undesired signals.

**Power supply rejection ratio**
The ratio of a change in op-amp output voltage to a change in supply voltage.

**Output short-circuit current**
The maximum output current for an op-amp, measured under shorted load conditions.

FIGURE 15.23

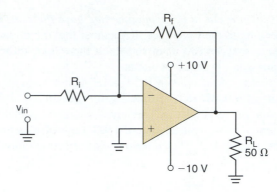

## Slew Rate

OBJECTIVE 7 ▶

**Slew rate**
How fast the op-amp output voltage can change.

**Lab Reference:** The slew rate of an op-amp is measured in Exercise 33.

The **slew rate** of an op-amp is a measure of *how fast the output voltage can change in response to an input signal.* The slew rate of the μA741 is 0.5 V/μs (typical). This means that the output from this amplifier can change by 0.5 V every microsecond. Since frequency is related to time, the *slew rate can be used to determine the maximum operating frequency of the op-amp,* as follows:

$$f_{max} = \frac{\text{slew rate}}{2\pi V_{pk}} \qquad (15.7)$$

Example 15.5 demonstrates the use of this formula.

---

## EXAMPLE 15.5

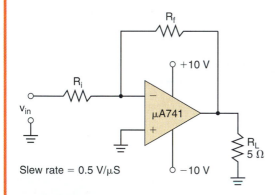

**FIGURE 15.24**

*Note*: The slew rate in the fraction was converted to 500 kHz as follows:

$$\frac{0.5}{1\,\mu s} = 500\text{ kHz}$$

Determine the maximum operating frequency for the circuit shown in Figure 15.24.

**Solution:** The maximum peak output voltage for this circuit is approximately 8 V. Using this value as $V_{pk}$ in equation (15.7), the maximum operating frequency for the amplifier is determined as

$$f_{max} = \frac{\text{slew rate}}{2\pi V_{pk}}$$

$$= \frac{0.5\text{ V/}\mu s}{6.28(8\text{ V})}$$

$$= \frac{500\text{ kHz}}{50.24}$$

$$= 9.95\text{ kHz}$$

### PRACTICE PROBLEM 15.5

An op-amp with a slew rate of 0.4 V/μs has a 10 $V_{pk}$ output. Determine the maximum operating frequency for the component.

---

EXAMPLE 15.6

The amplifier in Figure 15.24 is used to amplify an input signal to a peak output voltage of 100 mV. What is the maximum operating frequency of the amplifier?

*Solution:* Using 100 mV as the peak output voltage, $f_{max}$ is determined as

$$f_{max} = \frac{\text{slew rate}}{2\pi V_{pk}}$$

$$= \frac{0.5 \text{ V}/\mu s}{6.28(100 \text{ mV})}$$

$$= \frac{500 \text{ kHz}}{0.628}$$

$$= 796 \text{ kHz}$$

**PRACTICE PROBLEM 15.6**

The output peak voltage for the op-amp in Practice Problem 15.5 is reduced to 2 $V_{pk}$. Determine the new maximum operating frequency for the device.

---

Examples 15.5 and 15.6 have served to show that the op-amp can be operated at a much higher frequency when it is used as a small-signal amplifier than when it is used as a large-signal amplifier.

The effects of operating an op-amp circuit above the value of $f_{max}$ can be seen in ◄ *OBJECTIVE 8* Figure 15.25. This figure shows the *distortion* caused to the output signal by the op-amp. In both cases shown, the input is changing at a rate that is higher than the op-amp can handle. When this happens, the output from the op-amp changes more slowly than does the input signal, and the distortion shown is the result. Note that the distortion can be eliminated by reducing the input frequency or by using an op-amp with a higher-frequency capability. The distortion can also be eliminated by reducing the peak output voltage of the op-amp. If the gain of the component is reduced, the peak output voltage is also reduced for a given input signal. As you were shown in Example 15.6, the reduction in peak output voltage results in an increase in maximum operating frequency.

## Input/Output Resistance

As was stated earlier in the chapter, op-amps typically have high input resistance and low output resistance ratings. These ratings for the μA741 are 2 MΩ and 75 Ω, respectively. Note that the high input resistance and low output resistance of the op-amp closely resem-

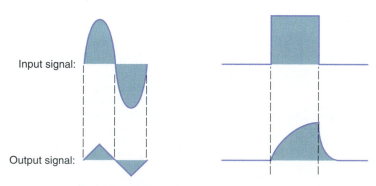

**FIGURE 15.25** Slew-rate distortion.

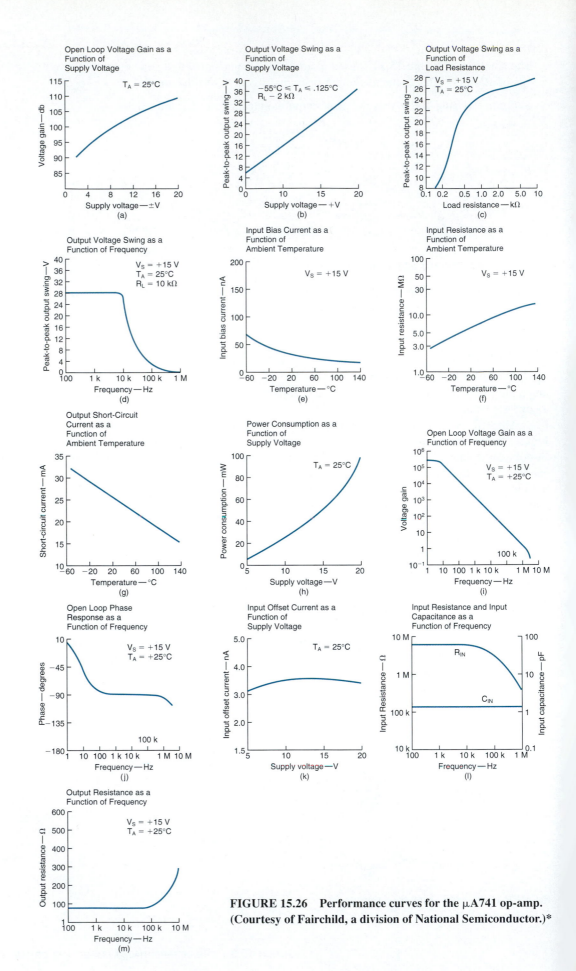

**FIGURE 15.26** Performance curves for the μA741 op-amp. (Courtesy of Fairchild, a division of National Semiconductor.)*

ble the characteristics of the *ideal* voltage amplifier, which has infinite input resistance and zero output resistance.

## Putting It All Together

You have been exposed to many op-amp operating characteristics in this section. Now we will take a look at how they relate to each other. As the graphs in Figure 15.26 illustrate, the basic op-amp characteristics are not independent of each other. Table 15.4 has been derived using the graphs in Figure 15.26. As you read the comments on each characteristic, refer back to the indicated graph to verify the components for yourself. Graph references are given in parentheses after each statement.

The following series of statements will provide a *cause-and-effect* viewpoint of the information from the 741 graphs:

1. As you *increase supply voltage,* you also:
   —increase the open-loop voltage gain.
   —increase the possible output voltage swing.
   —increase the component power consumption.
   —vary the input offset current very slightly.
2. As you *increase the operating frequency,* you also:
   —decrease the maximum output voltage swing ($f > 10$ kHz).
   —decrease the open-loop voltage gain.
   —decrease the input resistance ($f > 10$ kHz).
   —increase the output resistance ($f > 100$ kHz).
3. As the *ambient temperature increases:*
   —input bias current decreases.
   —input resistance increases.
   —output short-circuit current decreases.

As you will see, these cause-and-effect statements come in very handy when you are trying to build a circuit in the lab or when you are troubleshooting a given op-amp circuit.

**TABLE 15.4  Op-Amp Characteristics**

| Characteristic | Comments |
|---|---|
| Open-loop voltage gain | Directly proportional to supply voltage (a); decreases at a linear rate as frequency increases from 10 Hz to 1 MHz (i). |
| Output voltage swing | Directly proportional to $R_L$; drop rate increases significantly at values of $R_L < 500\ \Omega$ (c); directly proportional to supply voltage (b); constant for frequencies up to 10 kHz, then drops off rapidly (d). |
| Input resistance | Directly proportional to ambient temperature (f); constant for frequencies up to 10 kHz, then starts to drop off (1). |
| Input bias current | Inversely proportional to temperature (e); this follows the increase in input resistance. |
| Output short-circuit | Rated for temperature of 25°C; increases below this temperature, and decreases above this temperature (g). |
| Power consumption | The power used by the device is directly proportional to the supply voltage (h). |
| Output resistance | Remains constant for frequencies up to 100 kHz, then increases rapidly (m). |

**Electrical Characteristics** $V_S = \pm 15$ V, $T_A = 25°C$ unless otherwise specified

| Characteristic | Condition | $\mu$A741 | | | $\mu$A741C | | | Unit |
|---|---|---|---|---|---|---|---|---|
| | | Min | Typ | Max | Min | Typ | Max | |
| Input Offset Voltage | $R_S \leq 10$ k$\Omega$ | | 1.0 | 5.0 | | 2.0 | 6.0 | mV |
| Input Offset Current | | | 20 | 200 | | 20 | 200 | nA |
| Input Bias Current | | | 80 | 500 | | 80 | 500 | nA |
| Power Supply Rejection Ratio | $V_S = +10, -20$ $V_S = +20, -10$ V, $R_S = 50$ $\Omega$ | | 30 | 150 | | 30 | 150 | $\mu$V/V |
| Input Resistance | | .3 | 2.0 | | 3 | 2.0 | | M$\Omega$ |
| Input Capacitance | | | 1.4 | | | 1.4 | | pF |
| Offset Voltage Adjustment Range | | | $\pm 15$ | | | $\pm 15$ | | mV |
| Input Voltage Range | | | | | $\pm 12$ | $\pm 13$ | | V |
| Common Mode Rejection Ratio | $R_S \leq 10$ k$\Omega$ | | | | 70 | 90 | | dB |
| Output Short Circuit Current | | | 25 | | | 25 | | mA |
| Large Signal Voltage Gain | $R_L \geq 2$ k$\Omega$, $V_{OUT} = \pm 10$ V | 50k | 200k | | 20k | 200k | | |
| Output Resistance | | | 75 | | | 75 | | $\Omega$ |
| Output Voltage Swing | $R_L \geq 10$ k$\Omega$ | | | | $\pm 12$ | $\pm 14$ | | V |
| | $R_L \geq 2$ k$\Omega$ | | | | $\pm 10$ | $\pm 13$ | | V |
| Supply Current | | | 1.7 | 2.8 | | 1.7 | 2.8 | mA |
| Power Consumption | | | 50 | 85 | | 50 | 85 | mW |
| Transient Response (Unity Gain) — Rise Time | $V_{IN} = 20$ mV, $R_L = 2$ k$\Omega$ $C_L \leq 100$ pF | | .3 | | | .3 | | $\mu$S |
| Transient Response (Unity Gain) — Overshoot | | | 5.0 | | | 5.0 | | % |
| Bandwidth (Note 4) | | | 1.0 | | | 1.0 | | MHz |
| Stew Rate | $R_L \geq 2$ k$\Omega$ | | .5 | | | .5 | | V/$\mu$S |

Notes

4  Calculated value from BW (MHz) = $\dfrac{0.35}{\text{Rise Time }(\mu s)}$

5  All $V_{CC} = 15$ V for $\mu$A741 and $\mu$A741C

6  Maximum supply current for all devices

   25°C = 2.8 mA

   125°C = 2.5 mA

   −55°C = 3.3 mA

**FIGURE 15.27**    $\mu$A741 specification sheet. (Courtesy of Fairchild, a division of National Semiconductor.)*

## *Other Op-Amp Characteristics*

Figure 15.27 shows the electrical characteristics portion of the specification sheet for the $\mu$A741 op-amp.* As you can see, the spec sheet contains all the electrical characteristics we have discussed in this section. Now we will take a brief look at four additional electrical characteristics listed on the spec sheet.

The *input voltage range* indicates *the maximum value that the op-amp can accept without risking damage to the internal differential amplifier.* The minimum value for the $\mu$A741 is $\pm 12$ V. If either input is driven beyond this range, the result could be damage to the amplifier.

*Reprinted with permission of National Semiconductor. National bears no responsibility for any circuitry described. National reserves the right at any time to change without notice said circuitry.

National products are not authorized for use as *critical components in life-support devices or systems* without the expressed written approval of the president of National Semiconductors.

1. *Life-support devices or systems* are devices or systems which

   a. are intended for surgical implant into the body, or

   b. support or sustain life and whose failure to perform when used in accordance with instructions for use provided in the labeling, can be reasonably expected to result in a significant injury to the user.

2. A *critical component* is any component of a life-support device or system whose failure to perform can be reasonably expected to cause the failure of the life-support device or system.

The *large-signal voltage gain* is the open-loop voltage gain for the op-amp. As you can see, the value of $A_{OL}$ for the µA741 is typically around 200,000. This agrees with the earlier statement that the op-amp has extremely high voltage gain capabilities.

The *supply current* rating indicates *the quiescent current that is drawn from the power supply.* When the µA741 does not have an input signal, it will draw a maximum of 2.8 mA from the $+V$ and $-V$ power supplies.

The *power consumption* rating indicates *the amount of power that the op-amp will dissipate when it is in its quiescent state.* As you can see, the power dissipation rating of the µA741 has a maximum value of 85 mW.

There are two more ratings that we have not covered. These are *bandwidth* and *transient response*. Bandwidth was covered in Chapter 14. Transient response will be covered in Chapter 19. In the next section, we will discuss the operation of the two most basic op-amp circuits, the *inverting amplifier* and the *noninverting amplifier.*

**Section Review**

1. Describe the differential amplifier's response to an input signal at the inverting input.

2. Describe the differential amplifier's response to an input signal at the noninverting input.

3. What is *input offset voltage*?

4. What is *output offset voltage*?

5. What effect does the *offset null* control have on the input offset voltage and output offset voltage? Explain your answer.

6. What is *input offset current*?

7. What is *input bias current*? What restriction does input bias current place on the wiring of an op-amp?

8. What are *common-mode signals*?

9. What is the *common-mode rejection ratio (CMRR)*? Why is a high CMRR considered desirable?

10. What is the *power supply rejection ratio*?

11. What is the *output short-circuit current* rating?

12. Discuss the relationship between load resistance and maximum output voltage swing in terms of the output short-circuit current rating.

13. What is *slew rate*? What circuit parameter is limited by the op-amp slew rate?

14. What is the result of trying to operate an op-amp beyond its slew rate limit?

15. Define each of the following op-amp characteristics:
    a. Input voltage range
    b. Large-signal voltage range
    c. Supply current rating
    d. Power consumption rating

# 15.4 INVERTING AMPLIFIERS

◄ *OBJECTIVE 9*

**Inverting amplifier**
A basic op-amp circuit that produces a 180° signal phase shift. The op-amp equivalent of the common-emitter and common-source circuits.

In most of our examples, we have used an op-amp circuit that contains a single input resistor ($R_i$) and a single feedback resistor ($R_f$). This amplifier is the basic **inverting amplifier,** which is the op-amp equivalent of the common-emitter and common-source circuits. The operation of the inverting amplifier is illustrated in Figure 15.28.

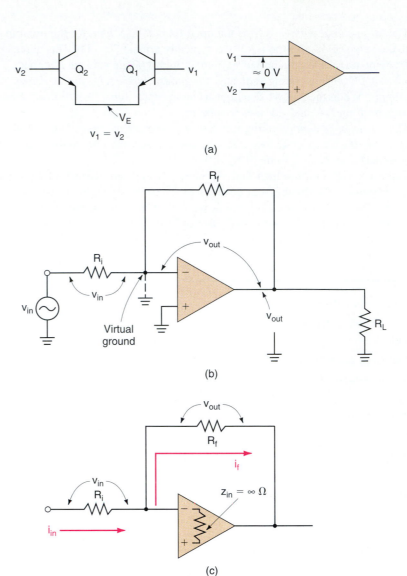

**FIGURE 15.28** Inverting amplifier operation.

The key to the operation of the inverting amplifier lies in the differential input circuit of the op-amp. For reference purposes, this input circuit is shown again in Figure 15.28a. Assuming that the two transistors are perfectly matched, the values of $v_1$ and $v_2$ will always be *approximately* equal to each other. They may differ by a few millivolts, but that is all.

Now, for a moment, let's idealize the differential circuit. For the sake of discussion, we will assume that $v_1$ and $v_2$ are *exactly* equal. With this in mind, look at the circuit in Figure 15.28b. If the voltages at the two inputs are equal, and the noninverting input is grounded, the inverting input to the op-amp is also at ground. Remember, this *virtual ground* is caused by the fact that $v_1$ and $v_2$ are approximately equal, and the noninverting input is grounded.

With the inverting input of the op-amp at virtual ground, the total input voltage can be measured across $R_i$. Also, the total output voltage can be measured across $R_f$. These relationships are illustrated in Figure 15.28b.

Since $v_{out}$ appears across the feedback resistor, its value can be found as:

$$v_{out} = i_f R_f \qquad (15.8)$$

Since $v_{in}$ appears across the input resistor, its value can be found as

$$v_{in} = i_{in}R_i \qquad \textbf{(15.9)}$$

Since an op-amp has extremely high input impedance, it has almost zero input current (as shown in Figure 15.28c). Therefore, $i_{in} \cong i_f$, and equation (15.8) can be rewritten as

$$v_{out} = i_{in}R_f \qquad \textbf{(15.10)}$$

As you know, voltage gain ($A_v$) is found as the ratio of output voltage to input voltage. By formula,

$$A_v = \frac{v_{out}}{v_{in}}$$

Since $v_{out} = i_{in}R_f$ and $v_{in} = i_{in}R_i$, the above equation can be rewritten as

$$A_v = \frac{i_{in}R_f}{i_{in}R_i}$$

or

$$A_v = \frac{R_f}{R_i} \qquad \textbf{(15.11)}$$

For the inverting amplifier, you only have to divide the value of $R_f$ by the value of $R_i$ to determine the approximate gain of the amplifier. In fact, if you go back to look at the examples earlier in this chapter, you'll see where the "assumed" values of voltage gain came from. In every case, the value of $A_v$ was approximately equal to the value of $R_f/R_i$.

Now that we have discussed a basic op-amp circuit, we can give practical meaning to a few terms. The **open-loop voltage gain**, $A_{OL}$, of an op-amp is the gain that is measured when there is no **feedback path** (physical connection) between the output and the input of the circuit. When a feedback path is present, such as the $R_f$ connection in the inverting amplifier, the resulting circuit gain is referred to as the **closed-loop voltage gain**, $A_{CL}$.

## Amplifier Input Impedance

While the op-amp has an extremely high input impedance, the inverting amplifier does not. The reason for this can be seen by referring back to Figure 15.28b. As this figure shows, the voltage source "sees" an input resistance ($R_i$) that is going to (virtual) ground. Thus, the input impedance for the inverting amplifier is found as

$$Z_{in} \cong R_i \qquad \textbf{(15.12)}$$

The value of $R_i$ will always be much less than the input impedance of the op-amp. Therefore, the overall input impedance of an inverting amplifier will also be much lower than the op-amp input impedance.

## Amplifier Output Impedance

If you look at the circuit in Figure 15.29, you can see that the output impedance of the inverting amplifier is the parallel combination of $R_f$ and the output impedance of the op-amp itself. Since $R_f$ is normally much greater than the value of $Z_{out}$ for the op-

Note: The value of $A_v$ for an inverting amplifier is often written as a negative value to indicate the 180° phase shift.

**Lab Reference:** The relationship between the resistor values and voltage gain is demonstrated in Exercise 29.

**Open-loop voltage gain ($A_{OL}$)** The gain of an op-amp with no feedback path.

**Closed-loop voltage gain ($A_{CL}$)** The gain of an op-amp with a feedback path; always lower than the value of $A_{OL}$.

*A Practical Consideration:* The gain of an inverting amplifier is determined by the ratio of $R_f$ to $R_i$. It would follow that a low value of $R_i$ would be desirable to provide a high voltage gain. However, lowering the value of $R_i$ also lowers the amplifier input impedance. Thus, increasing voltage gain is normally done by increasing the value of $R_f$ rather than decreasing the value of $R_i$.

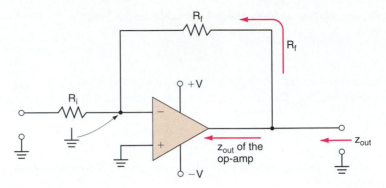

**FIGURE 15.29**

amp, the output impedance of the circuit is usually assumed to equal the $Z_{out}$ rating of the op-amp.

## Amplifier CMRR

In the last section, the *common-mode rejection ratio* (CMRR) of an op-amp was defined as the *ratio of differential gain to* **common-mode gain ($A_{CM}$).** Since the differential gain of the inverting amplifier ($A_{CL}$) is much less than the differential gain of the op-amp ($A_{OL}$), the CMRR of the inverting amplifier will be much lower than that of the op-amp.

The CMRR of the inverting amplifier is found as the ratio of closed-loop gain to common-mode gain. By formula,

*A Practical Consideration:* Most op-amp spec sheets do not provide a value of common-mode gain ($A_{CM}$). Rather, they list values of $A_{OL}$ and CMRR. When this is the case, the value of $A_{CM}$ can be obtained using

$$\text{CMRR} = \frac{A_{CL}}{A_{CM}} \qquad \textbf{(15.13)}$$

where $A_{CL}$ = the closed loop voltage gain of the inverting amplifier
$A_{CM}$ = the *common-mode* gain of the op-amp

$$A_{CM} = \frac{A_{OL}}{\text{CMRR}}$$

If the CMRR is listed in decibel (dB) form, it must be converted to standard numeric form before using the above equation. To convert the CMRR to standard form, use

Even though the CMRR of an inverting amplifier will be lower than that of the op-amp, it will still be extremely high in most cases. This point is illustrated in the upcoming examples.

## Inverting Amplifier Analysis

$$\text{CMRR} = \log^{-1}\left(\frac{\text{CMRR (dB)}}{20}\right)$$

The complete analysis of an inverting amplifier involves determining the circuit values of $A_{CL}$, $Z_{in}$, $Z_{out}$, CMRR, and $f_{max}$. The complete analysis of an inverting amplifier is illustrated in Example 15.7.

---

**EXAMPLE 15.7**

Perform the complete analysis of the circuit shown in Figure 15.30.

**Solution:** First, the closed-loop voltage gain ($A_{CL}$) of the circuit is found as

$$A_{CL} = \frac{R_f}{R_i}$$

$$= \frac{100 \text{ k}\Omega}{10 \text{ k}\Omega}$$

$$= 10$$

The input impedance of the circuit is found as

---

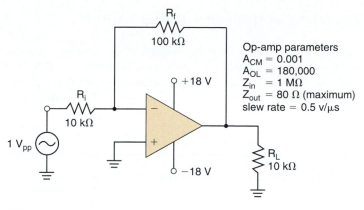

**FIGURE 15.30**

$$Z_{\text{in}} \cong R_i$$
$$= 10 \text{ k}\Omega$$

The circuit output impedance is lower than the output impedance of the op-amp, 80 $\Omega$ (maximum).

The CMRR of the circuit is found as

$$\text{CMRR} = \frac{A_{\text{CL}}}{A_{\text{CM}}}$$

$$= \frac{10}{0.001}$$

$$= 10,000$$

To calculate the maximum operating frequency for the inverting amplifier, we need to determine its peak output voltage. With values of $V_{\text{in}} = 1$ V$_{\text{PP}}$ and $A_{\text{CL}} = 10$, the peak-to-peak output voltage is found as

$$V_{\text{out}} = (1 \text{ V}_{\text{PP}})(10)$$
$$= 10 \text{ V}_{\text{PP}}$$

Thus, the peak output voltage is 5 V$_{\text{pk}}$, and the maximum operating frequency is found as

$$f_{\text{max}} = \frac{\text{slew rate}}{2\pi V_{\text{pk}}}$$

$$= \frac{0.5 \text{ V/}\mu\text{s}}{31.42}$$

$$= \frac{500 \text{ kHz}}{31.42}$$

$$= 15.9 \text{ kHz}$$

## PRACTICE PROBLEM 15.7

An op-amp has the following parameters: $A_{\text{CM}} = 0.02$, $A_{\text{OL}} = 150,000$, $Z_{\text{in}} = 1.5$ M$\Omega$, $Z_{\text{out}} = 50$ $\Omega$ (maximum), and slew rate = 0.75 V/$\mu$s. The op-amp is used in an inverting amplifier with $\pm12$ V$_{\text{dc}}$ supply voltages and values of $V_{\text{in}} = 50$ mV$_{\text{PP}}$, $R_f = 250$ k$\Omega$, and $R_i = 1$ k$\Omega$. Perform the complete analysis of the circuit.

## Other Inverting Amplifier Configurations

In this section, we have covered only the simplest of the inverting amplifiers. While the calculations of gain, input impedance, and output impedance will vary with different configurations, the calculations of $f_{max}$ and CMRR will not. You will also see that the various inverting amplifiers all have the characteristic 180° voltage phase shift from input to output.

---

**Section Review**

1. Explain why the voltage gain of an inverting amplifier ($A_{CL}$) is equal to the ratio of $R_f$ to $R_i$.

2. Why is the input impedance of an inverting amplifier lower than the input impedance of the op-amp?

3. Why isn't the value of $R_f$ normally considered in the output impedance calculation for an inverting amplifier?

4. Why is the CMRR ratio of an inverting amplifier lower than that of its op-amp?

5. List, in order, the steps involved in performing the complete analysis of an inverting amplifier.

---

## 15.5 NONINVERTING AMPLIFIERS

**Noninverting amplifier**
An op-amp circuit with no signal phase shift.

The **noninverting amplifier** has most of the characteristics of the inverting amplifier, with two exceptions:

1. The noninverting amplifier has much higher circuit input impedance.

2. The noninverting amplifier does not produce a 180° voltage phase shift from input to output. Thus, the input and output signals are in phase.

The basic noninverting amplifier is shown in Figure 15.31. Note that the input is applied to the noninverting op-amp input, and the input resistance ($R_i$) is returned to ground.

Because the input signal is applied to the noninverting terminal, we must calculate the gain of this circuit differently. We will start our discussion of circuit gain by again assuming that $v_1 = v_2$ and that the currents through $R_i$ and $R_f$ are equal. Since $v_1 = v_2$ and $v_2 = v_{in}$, we can state the following relationship:

$$v_1 = v_{in} \tag{15.14}$$

**FIGURE 15.31   Noninverting amplifier.**

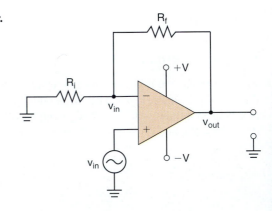

**Lab Reference:** The operation of the noninverting amplifier is demonstrated in Exercise 30.

Note that this is different from the relationship of $v_1 = 0$ V stated for the inverting amplifier with a grounded ($+$) input. Once again, we have $v_{in}$ measured across $R_i$, and $i_{in}$ can be found as

$$i_{in} = \frac{v_{in}}{R_i}$$

or

$$v_{in} = i_{in}R_i$$

Since the voltage across $R_f$ is equal to the difference between $v_{in}$ and $v_{out}$ and $i_f = i_{in}$, we can state that the current through the resistor is found as

$$i_f = \frac{v_{out} - v_{in}}{R_f} \qquad (15.15)$$

and

$$v_{out} = i_{in}R_f + v_{in}$$

Now we can use the equations established to define $A_{CL}$ as

$$A_{CL} = \frac{v_{out}}{v_{in}}$$

$$= \frac{i_{in}R_f + i_{in}R_i}{i_{in}R_i}$$

$$= \frac{i_{in}R_f}{i_{in}R_i} + \frac{i_{in}R_i}{i_{in}R_i}$$

or

$$A_{CL} = \frac{R_f}{R_i} + 1 \qquad (15.16)$$

where $A_{CL}$ is the *closed-loop* voltage gain of the amplifier. Thus, the gain of a noninverting amplifier will always be greater than the gain of an equivalent inverting amplifier by a value of 1. If an inverting amplifier has a gain of 150, the equivalent noninverting amplifier will have a gain of 151.

## Amplifier Input and Output Impedance

Since the input signal is applied directly to the op-amp, the noninverting amplifier has extremely high input impedance. In fact, the presence of the feedback network causes the amplifier input impedance to be even greater than the input impedance of the op-amp in most cases.

The output impedance of the noninverting amplifier is approximately equal to the output impedance of the op-amp, as is the case with the inverting amplifier.

The extremely high input impedance and extremely low output impedance of the noninverting amplifier make the circuit very useful as a *buffer*. You may recall that buffers (like the *emitter follower* and the *source follower*) are circuits that can be used to match a high-impedance source to a low-impedance load. The noninverting amplifier can be used for the same purpose. The primary difference is that the basic noninverting amplifier is capable of high voltage gain values, while the emitter follower and the source follower have $A_v$ values that are less than 1.

## Noninverting Amplifier Analysis

The complete analysis of the noninverting amplifier is almost identical to that of the inverting amplifier. The values of $Z_{out}$, CMRR, and $f_{max}$ are found using the same equations we used for the inverting amplifier. The values of $A_{CL}$ and $Z_{in}$ are found using the equations and principles established in this section. The complete analysis of the noninverting amplifier is illustrated in Example 15.8.

---

**EXAMPLE 15.8**

Perform the complete analysis of the noninverting amplifier shown in Figure 15.32.

**FIGURE 15.32**

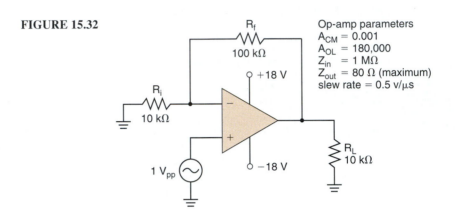

Op-amp parameters
$A_{CM} = 0.001$
$A_{OL} = 180,000$
$Z_{in} = 1\ M\Omega$
$Z_{out} = 80\ \Omega$ (maximum)
slew rate = 0.5 v/µs

***Solution:*** The closed-loop voltage gain ($A_{CL}$) of the circuit is found as

$$A_{CL} = \frac{R_f}{R_i} + 1$$

$$= \frac{100\ k\Omega}{10\ k\Omega} + 1$$

$$= 11$$

The input impedance of the circuit is at least 1 MΩ and the output impedance of the circuit is approximately 80 Ω.

The CMRR for the circuit is found as

$$CMRR = \frac{A_{CL}}{A_{CM}}$$

$$= \frac{11}{0.001}$$

$$= 11,000$$

To determine the value of $f_{max}$, we need to calculate the peak output voltage for the amplifier. The peak-to-peak output voltage is found as

$$V_{out} = A_{CL}V_{in}$$
$$= (11)(1\ V_{PP})$$
$$= 11\ V_{PP}$$

The peak output voltage would be one-half of this value, 5.5 $V_{pk}$. Using this value and the slew rate shown in the figure, the value of $f_{max}$ is calculated as

$$f_{max} = \frac{\text{slew rate}}{2\pi V_{pk}}$$

$$= \frac{0.5 \text{ V/}\mu s}{34.56}$$

$$= \frac{500 \text{ kHz}}{34.56}$$

$$= 14.47 \text{ kHz}$$

### PRACTICE PROBLEM 15.8

The circuit described in Practice Problem 15.7 is wired as a noninverting amplifier. Perform a complete analysis on the new circuit.

The circuit values and op-amp parameters used in Example 15.7 were identical to those used for the inverting amplifier in Example 15.8. The results found in the two examples are summarized below.

◀ OBJECTIVE 10

| Value | Inverting Amplifier | Noninverting Amplifier |
|---|---|---|
| $A_{CL}$ | 10 | 11 |
| $Z_{in}$ | 1 kΩ | 1 MΩ (minimum) |
| $Z_{out}$ | 80 Ω (maximum) | 80 Ω (maximum) |
| CMRR | 10,000 | 11,000 |
| $f_{max}$ | 15.9 kHz | 14.47 kHz |

As you can see, the noninverting amplifier has slightly greater values of $A_{CL}$ and CMRR. It also has a much greater value of input impedance. At the same time, the inverting amplifier has a slightly greater maximum operating frequency. This is due to the higher peak output voltage of the comparable noninverting amplifier. Figure 15.33 summarizes the operation of the noninverting and inverting amplifiers.

## The Voltage Follower

If we remove $R_i$ and $R_f$ from the noninverting amplifier and short the output of the amplifier to the inverting input, we have the **voltage follower.** This circuit, which is the op-amp equivalent of the emitter follower and source follower, is shown in Figure 15.34. You may recall that the characteristics of the emitter and source followers are as follows:

**Voltage follower**
The op-amp equivalent of the emitter follower and the source follower.

1. High $Z_{in}$ and low $Z_{out}$

2. $A_v$ that is approximately equal to 1

3. Input and output signals that are in phase

Characteristics 1 and 3 are accomplished by using an op-amp in a noninverting circuit configuration. The voltage gain for the voltage follower is calculated as

$$A_{CL} = \frac{R_f}{R_i} + 1$$

$$= \frac{0 \text{ }\Omega}{R_i} + 1 \quad (\text{since } R_f = 0 \text{ }\Omega)$$

$$= 1$$

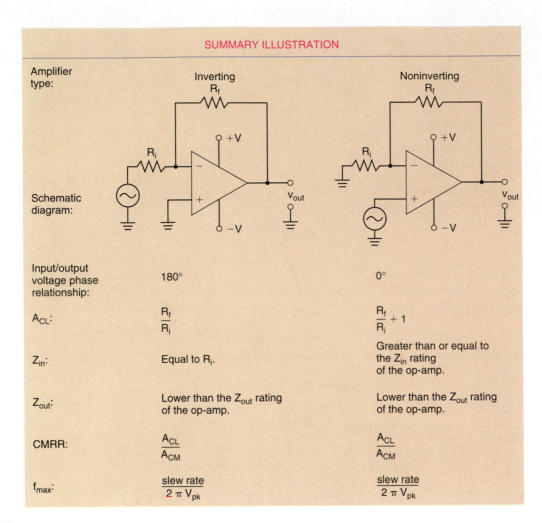

| Amplifier type: | Inverting | Noninverting |
|---|---|---|
| Schematic diagram: | | |
| Input/output voltage phase relationship: | $180°$ | $0°$ |
| $A_{CL}$: | $\dfrac{R_f}{R_i}$ | $\dfrac{R_f}{R_i} + 1$ |
| $Z_{in}$: | Equal to $R_i$. | Greater than or equal to the $Z_{in}$ rating of the op-amp. |
| $Z_{out}$: | Lower than the $Z_{out}$ rating of the op-amp. | Lower than the $Z_{out}$ rating of the op-amp. |
| CMRR: | $\dfrac{A_{CL}}{A_{CM}}$ | $\dfrac{A_{CL}}{A_{CM}}$ |
| $f_{max}$: | $\dfrac{\text{slew rate}}{2\pi V_{pk}}$ | $\dfrac{\text{slew rate}}{2\pi V_{pk}}$ |

**FIGURE 15.33**

Simple enough? (*If you think that calculating $A_{CL}$ for this circuit is easy, wait until you try troubleshooting it!*)

The values of $Z_{in}$, $Z_{out}$, and $f_{max}$ for the voltage follower are calculated using the same equations that we used for the basic noninverting amplifier. Since $A_{CL} = 1$ for the voltage follower, the circuit CMRR is found using

$$CMRR = \frac{1}{A_{CM}}$$

(15.17)

As Example 15.9 demonstrates, the voltage follower is by far the easiest op-amp circuit to analyze.

**FIGURE 15.34    Voltage follower.**

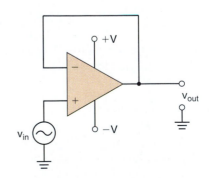

**Lab Reference:** The operation of the voltage follower is demonstrated in Exercise 31.

Perform the complete analysis of the voltage follower in Figure 15.35.

EXAMPLE 15.9

*Solution:* For the voltage follower,

$$A_{CL} = 1$$

The values of $Z_{in}$ and $Z_{out}$ are equal to the rated values for the op-amp, 1 M$\Omega$ and 80 $\Omega$ (maximum), respectively. The CMRR of the circuit is found as

$$CMRR = \frac{1}{A_{CM}}$$

$$= \frac{1}{0.001}$$

$$= 1000$$

Since $A_{CL} = 1$ for the circuit, $V_{out} = V_{in}$. Thus, the peak output voltage is one-half of 6 V$_{PP}$, or 3 V$_{pk}$. Now $f_{max}$ is found as

$$f_{max} = \frac{\text{slew rate}}{2\pi V_{pk}}$$

$$= \frac{500 \text{ kHz}}{18.85}$$

$$= 26.53 \text{ kHz}$$

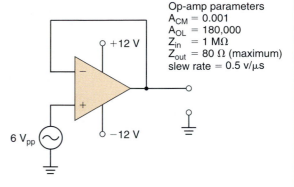

Op-amp parameters
$A_{CM} = 0.001$
$A_{OL} = 180,000$
$Z_{in} = 1$ M$\Omega$
$Z_{out} = 80$ $\Omega$ (maximum)
slew rate = 0.5 v/$\mu$s

**FIGURE 15.35**

**PRACTICE PROBLEM 15.9**

The op-amp described in Practice Problem 15.7 is used in a voltage follower with $\pm 14$ V supply voltages and a 12 V$_{PP}$ input. Perform the complete analysis of the amplifier.

As you can see, a voltage follower will have significantly lower $A_{CL}$ and CMRR values and a higher maximum operating frequency than a comparable "standard" noninverting amplifier.

1. How do you calculate the value of $A_{CL}$ for a noninverting amplifier?
2. How do you determine the values of $Z_{in}$ and $Z_{out}$ for a noninverting amplifier?
3. How do you determine the values of CMRR and $f_{max}$ for a noninverting amplifier?
4. Compare and contrast the inverting amplifier with the noninverting amplifier.
5. Describe the gain and impedance characteristics of the voltage follower.

# 15.6 TROUBLESHOOTING BASIC OP-AMP CIRCUITS

From the technician's viewpoint, op-amp circuits are a dream come true. The basic op-amp circuit has only three or four components that can become faulty, and each fault has distinct symptoms. For example, consider the circuit shown in Figure 15.36. Assuming that the load resistor, the input signal, and the supply voltages are all good, there are only

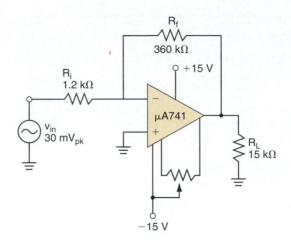

FIGURE 15.36

four components that could cause a given fault: $R_i$, $R_f$, the offset resistor, and the op-amp itself. Let's take a look at what happens when one of the resistors goes bad.

## $R_f$ Open

*OBJECTIVE 11* ▶ When the feedback resistor opens, the entire feedback loop is effectively removed from the circuit. This causes the gain of the amplifier to increase from $A_{CL}$ to the open-loop voltage gain of the op-amp, $A_{OL}$. Now consider what this would do to the circuit in Figure 15.36. The value of $A_{CL}$ for this circuit is 300. The value of $A_{OL}$ for the μA741 is 200,000. With the amplifier gain going to $A_{OL}$, the output would try to go to $\pm 6000$ V, which is clearly impossible. The result of this circuit action is shown in Figure 15.37.

Figure 15.37a shows the normal output from the circuit in Figure 15.36. As you can see, the signal is a clear, unclipped sine wave. The waveform shown in Figure 15.37b is the result of $R_f$ opening. The gain of the amplifier has become so great that the output waveform is clipped on both the positive and negative alternations. This output waveform is classic for an open feedback loop.

**Lab Reference:** Typical fault symptoms for the inverting and noninverting amplifiers are demonstrated in Exercises 29 and 30.

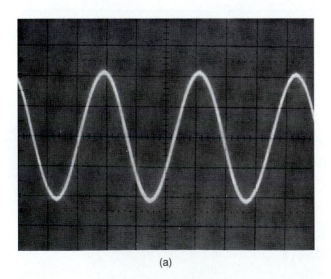

(a)

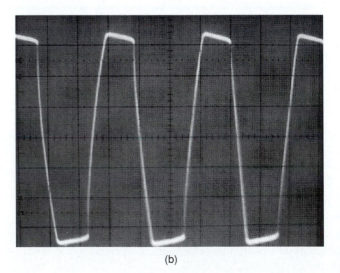

(b)

**FIGURE 15.37    Effect of an open $R_f$ on the inverting amplifier output.**

# $R_i$ Open

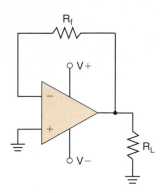

**FIGURE 15.38**

This fault is an interesting one. You would think that an open $R_i$ would simply block the input signal, and that the output would therefore go to 0 V. However, this may not be the case. Let's take a look at the circuit you have when $R_i$ opens. This circuit is shown in Figure 15.38.

Now we will assume for a moment that the output from this circuit is equal to $+V$ at the moment when $R_i$ opens. Here's what can happen:

1. A *positive* signal is fed back to the inverting input from the output via $R_f$.
2. The positive inverting input causes the output from the amplifier to go negative toward $-V$.
3. A *negative* signal is now fed back to the inverting input from the output.
4. As the inverting input goes negative, the output again goes positive. This takes the amplifier back to step 1.

The process above repeats over and over, *causing the amplifier to produce an ac output signal with no input signal*. This signal will be in the low-millivolt range. Incidentally, there are circuits that are *designed* to work in the fashion described above. These circuits, called *oscillators*, are discussed in Chapter 18.

# $R_{offset}$ Open

If the offset adjust resistor opens, the output from the op-amp will be offset from its normal level by an amount equal to the offset voltage times the closed-loop gain of the amplifier, $A_{CL}$. One possible result of an open offset resistor is shown in the photograph in Figure 15.39.

Whether the op-amp offsets in the positive direction or the negative direction depends on the circuit. Just remember, if the output goes to a dc level that is above or below the proper output when there is no input signal, the offset resistor is either open or needs to be adjusted.

# What Happens If the Op-Amp Is Bad?

The answer to this question depends on what goes wrong with the op-amp. If you refer back to Figure 15.1, you will see that there are a lot of components in the op-amp that could go bad.

*Component Substitution:* Most common op-amps are available at any electronics parts store. If an op-amp is bad, you can generally replace it with an equivalent from any manufacturer. Equivalent op-amps will have the same package, designator code, and suffix code (see Figure 15.3).

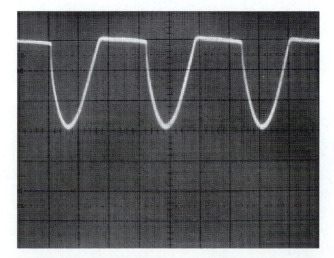

**FIGURE 15.39** Effect of an open offset null resistor on the inverting amplifier output.

**FIGURE 15.40**

The best way to determine that the op-amp is faulty is to determine that everything else is okay. If all the resistors in the amplifier are good and all supply voltages are as they should be, and the amplifier still does not work, the op-amp is the source of the problem and must be replaced.

## Working with ICs

ICs are often placed into circuits using IC sockets, such as those shown in Figure 15.40. These sockets are used to allow you to easily remove and replace the ICs.

When IC sockets are used, a problem can develop. Sometimes an IC will develop an *oxide* layer on its pins. If this oxidation becomes severe enough, it can cause the circuit to have an erratic output. The circuit may work one minute and not the next. When this happens, remove the IC and clean the pins and the socket with contact cleaner. This will eliminate the problem.

One more point: When you replace an IC, make sure that you put it into the circuit properly. It is very simple to put an IC in backward if you are not paying attention. If you *do* place an IC in a socket backward and apply power to the circuit, odds are that you will have to replace that IC with a new one.

---

**Section Review**

1. What is the primary symptom of an open feedback resistor?
2. What is the primary symptom of an open input resistor?
3. What is the primary symptom of an open offset null resistor?
4. How can you tell when an op-amp is faulty?

---

## *15.7* OP-AMP FREQUENCY RESPONSE

You have already been introduced to some of the frequency considerations involved in dealing with op-amps. As a review, here are some of the major points that were made regarding the frequency response of an op-amp:

1. The *slew rate* of an op-amp is a measure of *how fast the output voltage can change,* measured in volts per microsecond (V/μs).
2. The maximum operating frequency of an op-amp is found as

$$f_{max} = \frac{\text{slew rate}}{2\pi \, V_{pk}}$$

Thus, the *peak output voltage limits the maximum operating frequency.*

3. When the maximum output frequency of an op-amp is exceeded, the result is a *distorted* output waveform.

4. Increasing the operating frequency of an op-amp beyond a certain point will:

   **a.** decrease the maximum output voltage swing.

   **b.** decrease the open-loop voltage gain.

In this section we will take a closer look at how frequency affects the operation of an op-amp.

## Frequency Versus Gain

The gain of an op-amp will remain stable from 0 Hz up to some upper cutoff frequency, $f_2$. Then the gain will drop at the standard rate of *20 dB/decade*. This operating characteristic is represented by the Bode plot in Figure 15.41. Since the op-amp is a *dc amplifier,* it will exhibit midband voltage gain at 0 Hz. As the frequency of operation increases from 0 Hz, a point will be reached where the gain starts to drop. This drop in gain is due to internal values of capacitance.

As with any other circuit, $f_2$ is the upper cutoff frequency for the op-amp. When this frequency is reached, the gain of the op-amp will have dropped by 3 dB. As frequency continues to increase, the gain of the op-amp will continue to drop at the standard 20 dB/decade rate. Thus, *increasing the operating frequency decreases the component gain.*

*Don't Forget:* The Bode plot doesn't show the 3 dB drop at $f_2$.

There is another way that we can look at this frequency—gain relationship; that is, *decreasing the gain of an op-amp will increase the maximum operating frequency.* This point is easy to see by taking a look at the Bode plot shown in Figure 15.42. This plot represents the operating characteristics of the μA741 op-amp.

The maximum voltage gain on the Bode plot is shown to be equal to the open-loop voltage gain of the component, $A_{OL}$. For this op-amp, $A_{OL}$ is shown to be approximately

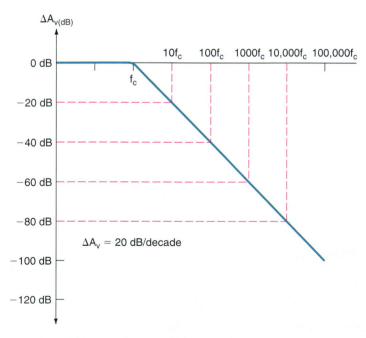

**FIGURE 15.41  Op-amp frequency response.**

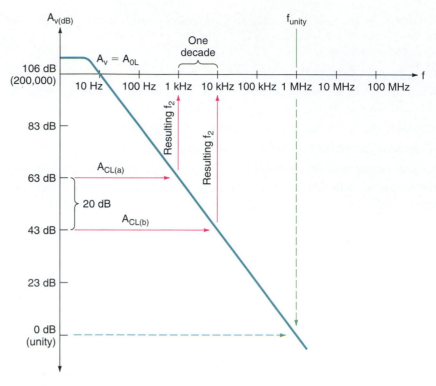

**FIGURE 15.42**

106 dB (200,000). If you want to operate the μA741 so that $A_{v(dB)}$ is equal to the maximum possible value of $A_{OL}$, you are limited to a maximum operating frequency of about 10 Hz. Above this frequency, the maximum gain of the op-amp drops by more than 3 dB, and the device is considered to be beyond the cutoff frequency.

Now, what if we were to use a *feedback path* such that the closed-loop gain of the amplifier was equal to 63 dB? What would your maximum operating frequency be now? Figure 15.42 shows that a value of $A_{CL}$ = 63 dB has a corresponding value of $f_2$ = 1 kHz. In other words, by intentionally decreasing the gain of the amplifier, we have increased the value of $f_2$, and thus the *bandwidth* of the device. If we further reduce the value of $A_{CL}$ to 43 dB, $f_2$ increases to 10 kHz, as does the bandwidth of the amplifier. In fact, each time we decrease the value of $A_{CL}$ by 20 dB, we increase the bandwidth of the amplifier by one decade. Eventually, a point is reached where the value of $A_{CL}$ = 0 dB (unity). The frequency that corresponds to this $A_{CL}$ is designated as $f_{unity}$. For the μA741 op-amp, $f_{unity}$ is approximately equal to 1 MHz, as shown in Figure 15.42.

The relationship between op-amp gain and bandwidth.

Based on the fact that gain and bandwidth are *inversely proportional,* we can make the following statements:

**1.** *The higher the gain of an op-amp, the narrower its bandwidth.*

**2.** *The lower the gain of an op-amp, the wider its bandwidth.*

Thus, we have a *gain–bandwidth trade-off.* If you want a wide bandwidth, you have to settle for less gain. If you want high gain, you have to settle for a narrower bandwidth.

OBJECTIVE 12 ▶

## *Gain–Bandwidth Product*

**Gain–bandwidth product**
A constant, equal to the unity-gain frequency of an op-amp. The product of $A_{CL}$ and bandwidth will always be approximately equal to this constant.

A *figure of merit* for a given op-amp is called the **gain–bandwidth product.** The gain–bandwidth product can be used to find:

**1.** The maximum value of $A_{CL}$ at a given value of $f_2$

**2.** The value of $f_2$ for a given value of $A_{CL}$

For example, let's say that you want to know the value of $f_2$ when $A_{CL} = 45$ dB or the value of $A_{CL}$ that will allow an $f_2$ of 200 kHz. The gain–bandwidth product can be used to solve either of these problems.

The gain–bandwidth product is always equal to the value of $f_{unity}$ for an op-amp. By formula,

$$A_{CL}f_2 = f_{unity} \qquad \textbf{(15.18)}$$

At any frequency, the product of $A_{CL}$ and $f_2$ must equal the unity-gain frequency of the op-amp. For example, refer back to Figure 15.42. As you were shown, $f_{unity} = 1$ MHz for the μA741 op-amp. At 10 Hz,

$$A_{CL}f_2 = (100{,}000)(10 \text{ Hz})$$
$$= 1 \text{ MHz}$$

At 100 Hz,

$$A_{CL}f_2 = (10{,}000)(100 \text{ Hz})$$
$$= 1 \text{ MHz}$$

At 1 kHz,

$$A_{CL}f_2 = (1000)(1 \text{ kHz})$$
$$= 1 \text{ MHz}$$

and so on. By rearranging equation (15.18), we can derive the following useful equations:

$$A_{CL} = \frac{f_{unity}}{f_2} \qquad \textbf{(15.19)}$$

and

$$f_2 = \frac{f_{unity}}{A_{CL}} \qquad \textbf{(15.20)}$$

Examples 15.10 and 15.11 demonstrate the usefulness of these equations.

---

The LM318 op-amp has a gain–bandwidth product of 15 MHz. Determine the bandwidth of the LM318 when $A_{CL} = 500$, and the maximum value of $A_{CL}$ when $f_2 = 200$ kHz.

**EXAMPLE 15.10**

***Solution:*** When $A_{CL} = 500$, the value of $f_2$ is found as

$$f_2 = \frac{f_{unity}}{A_{CL}}$$

$$= \frac{15 \text{ MHz}}{500}$$

$$= 30 \text{ kHz}$$

Since the op-amp is capable of operating as a dc amplifier,

$$\text{BW} = f_2$$
$$= 30 \text{ kHz}$$

When $f_2 = 200$ kHz, the maximum value of $A_{CL}$ is found as

$$A_{CL} = \frac{f_{unity}}{f_2}$$

$$= \frac{15 \text{ MHz}}{200 \text{ kHz}}$$

$$= 75 \ (37.5 \text{ dB})$$

**PRACTICE PROBLEM 15.10**

An op-amp has a gain–bandwidth product of 25 MHz. What is the bandwidth of the device when $A_{CL} = 200$?

The greatest thing about using the gain–bandwidth product is that it allows us to solve various gain–bandwidth problems without the use of a Bode plot. Here's another type of problem that can be solved using the gain–bandwidth product.

**EXAMPLE 15.11**

We need to construct an amplifier that has values of $A_{CL} = 500$ and BW = 80 kHz. Can the μA741 op-amp be used in this application?

**Solution:** The μA741 has a value of $f_{unity} = 1$ MHz. Therefore, any product of $A_{CL}f_2$ must be *less than or equal to* this value. In other words, if $A_{CL}f_2$ is *greater than* $f_{unity}$, the op-amp cannot be used. For our application,

$$A_{CL}f_2 = (500)(80 \text{ kHz})$$
$$= 40 \text{ MHz}$$

Since this value of $A_{CL}f_2$ is greater than the μA741 $f_{unity}$ rating, the μA741 *cannot* be used for this application.

**PRACTICE PROBLEM 15.11**

We need to construct an amplifier with values of $A_{CL} = 52$ dB and BW = 10 kHz. We have an op-amp with a gain–bandwidth product of 5 MHz. Determine whether or not the op-amp can be used in this application.

## Op-Amp Internal Capacitance

If you refer back to the internal diagram of the μA741 op-amp (Figure 15.1), you'll see that the circuit contains an internal *compensating capacitor*, $C_1$. This capacitor, which is used to improve the internal frequency response of the device, limits the high-frequency operation of the component.

As frequency increases, the reactance of $C_1$ decreases. As this reactance decreases, the capacitor acts more and more like a short circuit. Eventually, a point is reached where a portion of the op-amp internal circuitry is short circuited, effectively reducing the gain of the amplifier to unity. The frequency at which this occurs is the unity-gain frequency of the device.

## Determining the Value of $f_{unity}$

The unity-gain frequency of an op-amp can be determined in many ways. Some op-amp spec sheets will list an $f_{unity}$ rating. Others will simply list a *bandwidth* rating. For example, the spec sheet for the μA741 op-amp (Figure 15.27) lists a bandwidth rating of 1 MHz. For this device, 1 MHz is the unity-gain frequency.

When the operating curves for an op-amp are available, the *voltage gain versus operating frequency* curve can be used to determine the value of $f_{unity}$. For example, refer to the curve for the µA741 op-amp shown in Figure 15.43. By taking any frequency value and multiplying it by the corresponding limit on voltage gain, you can obtain the value of $f_{unity}$. From the curve in Figure 15.43, we can approximate the value of $f_{unity}$ as

$$f_{unity} = (100 \text{ kHz})(10)$$
$$= 1 \text{ MHz}$$

Note that the value of $f = 100$ kHz and the corresponding value of $A_{OL} = 10$ were obtained from the curve.

The value of $f_{unity}$ for an op-amp can be measured using the following procedure:

1. Set up an inverting amplifier with a closed-loop gain of 100. (We use this value because it is easy to construct with resistor values of $R_f = 100$ kΩ and $R_i = 1$ kΩ.)

2. Apply an input signal to the amplifier and increase the operating frequency until $f_2$ is reached; that is, until the peak-to-peak output voltage drops to 0.707 times the mid-band value.

3. Take the measured value of $f_2$ and plug it into the following equation:

$$f_{unity} = 100 f_2$$

Note that this equation is simply a form of equation (15.18) that assumes a voltage gain of 100.

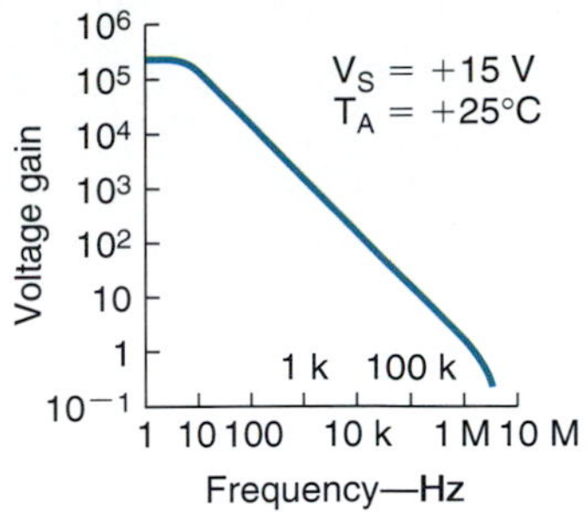

**FIGURE 15.43**

## One Final Point

As you can see, the bandwidth calculations for the op-amp are much simpler than those for the BJT or FET amplifier. This is another of the many advantages that op-amp circuits have over discrete amplifiers.

**Section Review**

1. What is the relationship between slew rate and maximum operating frequency?
2. What is the relationship between peak output voltage and maximum operating frequency?
3. What is the relationship between op-amp operating frequency and voltage gain?
4. What is *gain–bandwidth product*?
5. How can the *voltage gain versus operating frequency* curve for an op-amp be used to determine the value of $f_{unity}$ for the device?
6. How can the value of $f_{unity}$ for an op-amp be measured?

# 15.8 NEGATIVE FEEDBACK

As you know, *feedback* is a term that describes the process of providing a signal path from the output of a circuit back to its input. For example, take a look at the op-amp circuits shown in Figure 15.44. The *feedback resistor* ($R_f$) used in each of the amplifiers shown provides a signal path from the output of the op-amp back to its input. The effect that feedback has on the operation of the circuit depends on a number of factors, as you will be shown in this section.

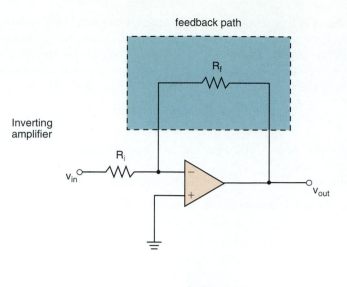

feedback path

Inverting
amplifier

$R_f$

$R_i$

$v_{in}$

$v_{out}$

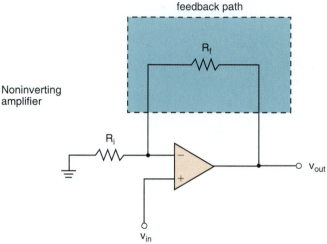

feedback path

Noninverting
amplifier

$R_f$

$R_i$

$v_{out}$

$v_{in}$

**FIGURE 15.44**

## Negative Versus Positive Feedback

*OBJECTIVE 13* ▶

**Negative feedback**
The feedback signal is 180° out of
phase with the input signal.

**Positive feedback**
The feedback signal is in phase with
the input signal.

**Oscillator**
A circuit that converts dc to ac.

Feedback is generally classified as either *negative feedback* or *positive feedback*. **Negative feedback** provides a feedback signal that is 180° out of phase with the input signal. One method of obtaining negative feedback is illustrated in Figure 15.45a. In the circuit shown, the amplifier is providing a 180° voltage phase shift, but the feedback network is not. The result is that that total voltage phase shift around the loop is 180°, and the feedback signal is out of phase with the input signal. The same result can be achieved by using an amplifier with a 0° phase shift and a feedback network with a 180° phase shift, as shown in Figure 15.45b.

Positive feedback provides a feedback signal that is in phase with the circuit input. One method of obtaining positive feedback is represented in Figure 15.46a. In this case, the amplifier and the feedback network *each* introduce a 180° voltage phase shift into the loop. This results in a total voltage phase shift of 360° (or 0°), and the feedback signal is in phase with the circuit input. The same result can be achieved by using the circuit configuration shown in Figure 15.46b. In either case, the feedback and input signals are in phase.

Positive feedback is used in a special type of amplifier called an **oscillator**. An oscillator is a circuit that converts dc to a sinusoidal (or some other varying) output. We

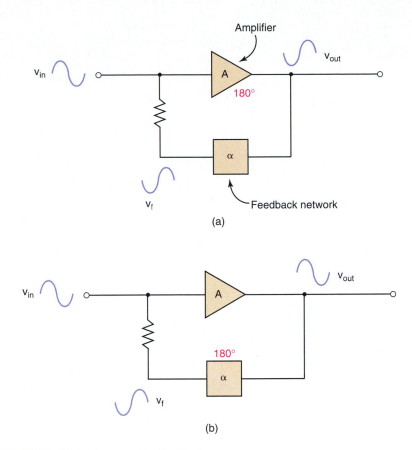

**FIGURE 15.45  Obtaining negative feedback.**

will discuss the operation and applications of oscillators in Chapter 18. In this section, we concentrate on the effects of *negative* feedback on amplifier operation.

## Voltage Versus Current Feedback

Negative feedback can be divided into two types: **voltage feedback** and **current feedback**. Both of these feedback configurations can be represented as shown in Figure 15.47. When *voltage feedback* is used, the inputs of the feedback network are in parallel with the load. Thus, the input to the feedback network is a voltage that is equal to the amplifier output voltage. The output from the feedback network is an *attenuated* (reduced) voltage that is 180° out of phase with the amplifier input voltage.

When *current feedback* is used, the input to the feedback network is in *series* with the load, as shown in Figure 15.47b. As you can see, a portion of the source current bypasses the amplifier through the feedback network. Thus, the amplifier input *current* is reduced.

Why use two different types of negative feedback? As shown in the following table, the effects of negative voltage feedback are significantly different from those of negative current feedback.

| Feedback Type | *Effect On:* | | | |
| | $A_v$ | $A_i$ | $Z_{in}$ | $Z_{out}$ |
| --- | --- | --- | --- | --- |
| Voltage | Decreases | None | Varies[a] | Decreases |
| Current | None | Decreases | Varies[a] | Increases |

[a]Depending on the type of circuit.

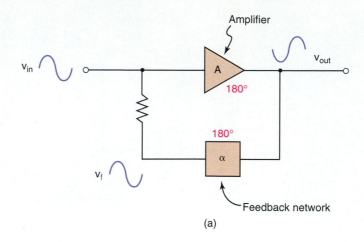

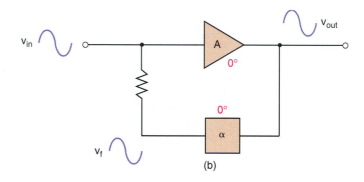

**FIGURE 15.46   Obtaining positive feedback.**

The effects of negative-voltage and negative-current feedback.

Of the two types of negative feedback, *negative voltage feedback* is by far the most commonly used. For this reason, this section will deal exclusively with this type of feedback. Whenever you see a reference to *negative feedback*, remember that we are actually talking about negative *voltage* feedback. The effects of negative current feedback will be covered as needed later in the text.

## Inverting Amplifier Operation

*OBJECTIVE 14* ▶ The feedback resistor ($R_f$) in an inverting amplifier forms a *negative feedback* network. When a negative feedback network is connected to an op-amp, it results in a *decrease in voltage gain* and an *increase in operating bandwidth*. This point is illustrated in Figure 15.48.

Figure 15.48a shows an inverting amplifier with no feedback path. Without the feedback path, the voltage gain of the circuit is equal to the open-loop voltage gain ($A_{OL}$) of the op-amp. In this case, $A_V = A_{OL} = 200,000$. With a unity-gain frequency rating of 3 MHz, the bandwidth of the circuit is found as

$$\text{BW} = \frac{f_{\text{unity}}}{A_{OL}}$$

$$= \frac{3 \text{ MHz}}{200,000}$$

$$= 15 \text{ Hz}$$

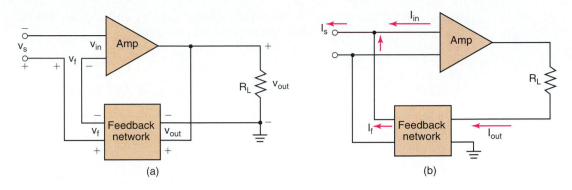

FIGURE 15.47   Voltage feedback versus current feedback.

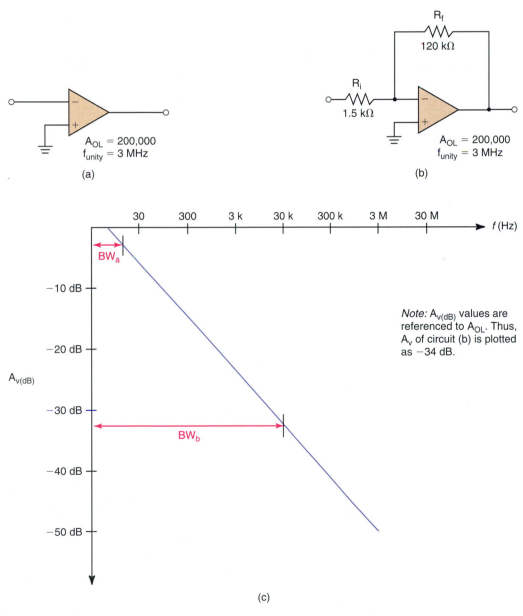

Note: $A_{v(dB)}$ values are referenced to $A_{OL}$. Thus, $A_v$ of circuit (b) is plotted as $-34$ dB.

(c)

FIGURE 15.48

When a negative feedback path is added to the op-amp (as shown in Figure 15.48b), the voltage gain and bandwidth of the circuit are found as

$$A_{CL} = \frac{R_f}{R_i}$$

$$= \frac{120 \text{ k}\Omega}{1.5 \text{ k}\Omega}$$

$$= 80$$

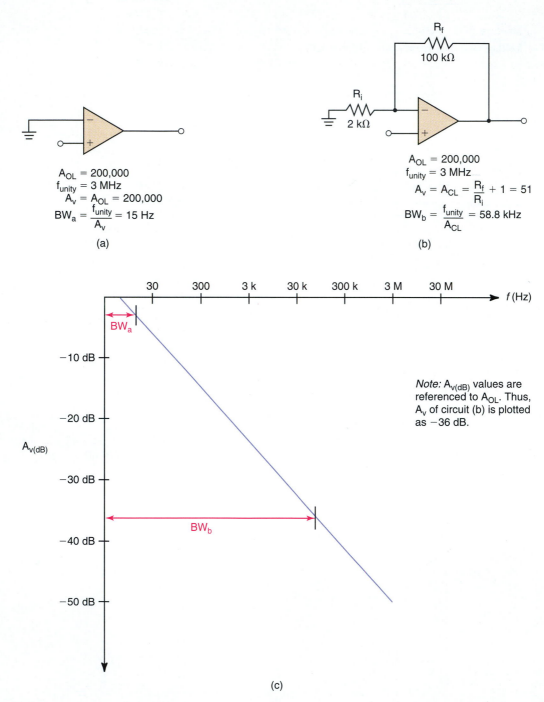

$A_{OL} = 200,000$
$f_{unity} = 3 \text{ MHz}$
$A_v = A_{OL} = 200,000$
$BW_a = \dfrac{f_{unity}}{A_v} = 15 \text{ Hz}$

(a)

$A_{OL} = 200,000$
$f_{unity} = 3 \text{ MHz}$
$A_v = A_{CL} = \dfrac{R_f}{R_i} + 1 = 51$
$BW_b = \dfrac{f_{unity}}{A_{CL}} = 58.8 \text{ kHz}$

(b)

*Note:* $A_{v(dB)}$ values are referenced to $A_{OL}$. Thus, $A_v$ of circuit (b) is plotted as $-36$ dB.

(c)

**FIGURE 15.49**

and

$$BW = \frac{f_{unity}}{A_{CL}}$$

$$= \frac{3 \text{ MHz}}{80}$$

$$= 37.5 \text{ kHz}$$

As you can see, the addition of a negative feedback path results in:

**1.** a *decrease* in voltage gain.

**2.** an *increase* in bandwidth.

These changes are further illustrated in Figure 15.48c. As you can see, the frequency response curve for the circuit in Figure 15.48b shows that it has lower voltage gain and a wider bandwidth than the circuit in Figure 15.48a. This is always the case when negative feedback is used.

## Noninverting Amplifier Operation

In terms of voltage gain and bandwidth, negative feedback has the same effect on the operation of a noninverting amplifier as it has on an inverting amplifier. This is illustrated by the circuits, calculations, and frequency response curve shown in Figure 15.49.

## Mathematical Analysis

Now that you have seen how negative feedback affects the voltage gain and operating bandwidth of an op-amp, it is time to get into its mathematical analysis. For this discussion, we will refer to the circuit shown in Figure 15.50.  ◄ *OBJECTIVE 15*

First, we will consider the feedback amplifier to be a pair of gain values. The amplifier itself is shown to be a voltage gain, defined as

**FIGURE 15.50**

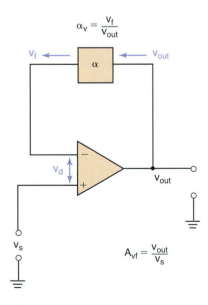

$$\alpha_v = \frac{v_f}{v_{out}}$$

$$A_{vf} = \frac{v_{out}}{v_s}$$

$$A_{\text{OL}} = \frac{v_{\text{out}}}{v_d} \qquad\qquad \textbf{(15.21)}$$

where $v_d$ = the difference between the voltages at the inverting and noninverting inputs to the op-amp

Remember that $A_{\text{OL}}$ is an op-amp parameter and does not necessarily equal the overall gain of the feedback amplifier.

The feedback network is represented as an *attenuation*, $\alpha$. Note that the Greek letter $\alpha$ is commonly used to represent a voltage and/or power *loss*. Don't confuse this use of $\alpha$ with the collector-emitter current ratio discussed earlier in the text.

The **attenuation factor** $(\alpha_v)$ of the feedback network is found as

**Attenuation factor** $(\alpha_v)$
The ratio of feedback voltage to output voltage. The value of $\alpha_v$ is always less than 1.

$$\alpha_v = \frac{v_f}{v_{\text{out}}} \qquad\qquad \textbf{(15.22)}$$

Note that $v_f$ will always be less than $v_{\text{out}}$. For this reason, $\alpha_v$ is always less than 1.

As with any amplifier, the *effective* gain of the feedback amplifier is found as

$$A_{vf} = \frac{v_{\text{out}}}{v_s} \qquad\qquad \textbf{(15.23)}$$

In Appendix D, we use equations (15.21), (15.22), and (15.23) to derive the following useful equation:

$$A_{vf} = \frac{A_v}{1 + \alpha_v A_v} \qquad\qquad \textbf{(15.24)}$$

where   $A_{vf}$ = the *effective* gain of the voltage feedback amplifier
$A_v$ = the *open-loop* voltage gain of the amplifier (that is, the voltage gain that the amplifier would exhibit if there were no feedback path present)

Note that equation (15.24) applies to discrete feedback amplifiers as well as those using op-amps. For an inverting (or noninverting) amplifier, equation (15.24) is often written as

$$A_{\text{CL}} = \frac{A_{\text{OL}}}{1 + \alpha_v A_{\text{OL}}} \qquad\qquad \textbf{(15.25)}$$

As you will see, the value that appears in the denominator of equation (15.25), $1 + \alpha_v A_{\text{OL}}$, appears in almost every equation for a given negative feedback amplifier. To simplify things, we will call this value the **feedback factor** for a given amplifier. In most cases, the circuit gain and impedance values (as well as the cutoff frequencies and bandwidth) are changed from their open-loop values by the feedback factor of the circuit. In other words, the factor by which a circuit characteristic increases or decreases equals the feedback factor of the circuit. Example 15.12 demonstrates the effect of the feedback factor on the voltage gain of an inverting amplifier.

**Feedback factor** $(1 + \alpha_v A_{\text{OL}})$
A value used in the impedance calculations for a given feedback amplifier.

---

**EXAMPLE 15.12**

The inverting amplifier in Figure 15.51 uses an op-amp with a rating of $A_{\text{OL}} = 150{,}000$. Assuming that the circuit has a value of $\alpha_v = 0.005$, determine its value of $A_{\text{CL}}$.

*Solution:*   Using the given values of $A_{\text{OL}}$ and $\alpha_v$, the closed-loop voltage gain of the circuit is found as

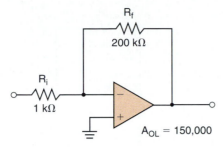

**FIGURE 15.51**

$$A_{CL} = \frac{A_{OL}}{1 + \alpha_v A_{OL}}$$

$$= \frac{150{,}000}{1 + (0.005)(150{,}000)}$$

$$= 199.7$$

***PRACTICE PROBLEM 15.12***

An amplifier like the one in Figure 15.51 has values of $A_{OL} = 200{,}000$ and $\alpha_v = 0.01$. Calculate the value of $A_{CL}$ for the circuit.

The value of $\alpha_v A_v$ (in equation 15.24) is always *much greater than 1*. For this reason, you can approximate the closed-loop voltage gain of a feedback amplifier as

$$A_{vf} \cong \frac{1}{\alpha_v} \tag{15.26}$$

or

$$A_{CL} \cong \frac{1}{\alpha_v} \tag{15.27}$$

for a given inverting (or noninverting) amplifier.

Equation (15.27) is important because it shows how we can calculate the value of $\alpha_v$ for an inverting or noninverting amplifier. First, the value of $A_{CL}$ is found using the equations established earlier in this chapter. Then, the value of $\alpha_v$ is found as

$$\alpha_v \cong \frac{1}{A_{CL}} \tag{15.28}$$

Once the value of $\alpha_v$ is known, we can calculate the value of the feedback factor for the circuit. The feedback factor is then used in the circuit impedance calculations, as you will be shown later in this section. Example 15.13 demonstrates the procedure for calculating the value of the feedback factor for a noninverting amplifier.

Calculate the value of the feedback factor for the noninverting amplifier shown in Figure 15.52.

***EXAMPLE 15.13***

***Solution:*** Using the relationship established earlier in the chapter, the closed-loop voltage gain of the amplifier is found as

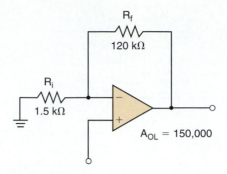

**FIGURE 15.52**

$$A_{CM} = \frac{R_f}{R_i} + 1$$

$$= \frac{120 \text{ k}\Omega}{1.5 \text{ k}\Omega} + 1$$

$$= 81$$

Now, using the values of $A_{OL} = 150{,}000$ and $A_{CL} = 81$, the attenuation factor $(\alpha_v)$ of the circuit is found as

$$\alpha_v = \frac{1}{A_{CL}}$$

$$= \frac{1}{81}$$

$$= 0.0123$$

Finally, the feedback factor of the circuit is found as

$$1 + \alpha_v A_{OL} = 1 + (0.0123)(150{,}000)$$
$$= 1853$$

***PRACTICE PROBLEM 15.13***

An *inverting* amplifier has the following values: $R_f = 220 \text{ k}\Omega$ and $R_i = 2 \text{ k}\Omega$. The op-amp in the circuit has a rating of $A_{OL} = 180{,}000$. Calculate the value of the feedback factor for the circuit.

## The Effects of Negative Feedback on Circuit Impedance Values

OBJECTIVE 16 ▶ The input and output impedance values for inverting and noninverting amplifiers are calculated as shown in Figure 15.53. For the inverting amplifier, the presence of the *virtual ground* at the inverting input causes the amplifier input impedance to be approximately equal to the value of the input resistor $(R_i)$. This point was discussed earlier in the chapter.

For the noninverting amplifier, the presence of the feedback signal at the inverting input reduces the input differential voltage $(v_d)$, and therefore, the amount of current that the amplifier draws from the source. Since

$$Z_{in} = \frac{v_s}{i_s}$$

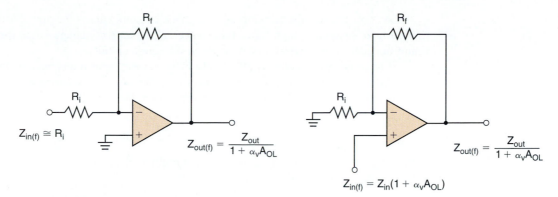

**FIGURE 15.53**

the reduction in source current causes an effective *increase* in amplifier input impedance. The amount by which the input impedance is increased is equal to the feedback factor of the circuit. By formula,

$$Z_{in(f)} = Z_{in} (1 + \alpha_v A_{OL}) \tag{15.29}$$

where $Z_{in(f)}$ = the input impedance to the noninverting amplifier
$\quad\quad Z_{in}$ = the op-amp input impedance

As the following example demonstrates, the input impedance to a noninverting amplifier is significantly higher than that of its op-amp.

---

Refer to the noninverting amplifier in Figure 15.53b. Assume that the op-amp has ratings of $Z_{in}$ = 5 MΩ and $A_{OL}$ = 180,000. If $R_i$ = 1.2 kΩ and $R_f$ = 180 kΩ, what is the value of the amplifier input impedance?

*Example 15.14*

**Solution:** We have to start our calculations by finding the value of the attenuation factor. Using the method established earlier in the text, the closed loop voltage gain of the circuit is found to be 151. Thus,

$$\alpha_v = \frac{1}{A_{CL}}$$
$$= 0.0066$$

Now, using $\alpha_v$ = 0.0066, the value of the feedback factor is found as

$$1 + \alpha_v A_{OL} = 1 + (0.0066)(180,000)$$
$$= 1189$$

Finally, the amplifier input impedance is found as

$$Z_{in(f)} = Z_{in}(1 + \alpha_v A_{OL})$$
$$= (5 \text{ M}\Omega)(1189)$$
$$= 5.95 \text{ G}\Omega$$

**PRACTICE PROBLEM 15.14**

A given op-amp has the following ratings: $Z_{in}$ = 2 MΩ and $A_{OL}$ = 200,000. The op-amp is used in a noninverting amplifier with values of $R_f$ = 220 kΩ and $R_i$ = 1 kΩ. Calculate the value of the amplifier input impedance.

As you can see, the addition of a feedback network causes the noninverting amplifier to have extremely high input impedance. This is an advantage since the amplifier input impedance presents almost no load on its source circuit.

Just as negative feedback effectively increases the input impedance of the op-amp, *it also decreases the op-amp's output impedance*. The amount by which $Z_{out}$ is reduced is also determined by the feedback factor of the circuit. By formula,

$$Z_{out(f)} = \frac{Z_{out}}{1 + \alpha_v A_{OL}} \qquad (15.30)$$

where $Z_{out(f)}$ = the outut impedance of the amplifier
$\quad\quad\ Z_{out}$ = the op-amp output impedance

The following example demonstrates the effect of negative feedback on the output impedance of a noninverting amplifier.

---

**EXAMPLE 15.15**

Refer to Example 15.14. If the op-amp has a rating of $Z_{out} = 80\ \Omega$, what is the value of the amplifier output impedance?

*Solution:* In Example 15.14, the feedback factor was found to have a value of 1189. Using this value and the rated output impedance of the op-amp, the output impedance of the noninverting amplifier is found as

$$Z_{out(f)} = \frac{Z_{out}}{1 + \alpha_v A_{OL}}$$

$$= \frac{80\ \Omega}{1189}$$

$$= 67.3\ m\Omega$$

**PRACTICE PROBLEM 15.15**

Refer to Practice Problem 15.14. If the op-amp has a rating of $Z_{out} = 75\ \Omega$, what is the output impedance of the amplifier?

---

As you can see, the feedback network has greatly reduced the effective output impedance of the op-amp. Again, this is an added benefit of using negative feedback. With the lower output impedance, the circuit is much better suited to driving low-impedance loads. If you refer back to Figure 15.53, you'll see that the use of negative feedback has the same effect on the *inverting* amplifier as it has on the noninverting amplifier.

---

**Section Review**

1. How does negative voltage feedback reduce the effective voltage gain of an amplifier?
2. List and describe the various types of commonly used feedback.
3. What is the *attenuation factor* of a feedback network?
4. What is the *feedback factor* of a negative feedback amplifier?
5. What effect does negative feedback have on the input impedance of an inverting amplifier? A noninverting amplifier?
6. What effect does negative feedback have on the output impedance of an inverting or a noninverting amplifier?

The following terms were introduced and defined in this chapter:

attenuation factor
closed-loop voltage gain
  ($A_{CL}$)
common-mode gain ($A_{CM}$)
common-mode rejection
  ratio (CMRR)
common-mode signals
current feedback
designator code
differential amplifier
feedback factor
feedback path
gain–bandwidth product

input bias current
input offset current
input offset voltage ($V_{io}$)
integrated circuit (IC)
inverting amplifier
inverting input
negative feedback
noninverting amplifier
noninverting input
offset null
open-loop voltage gain
  ($A_{OL}$)

operational amplifier
  (op-amp)
oscillator
output offset voltage
  ($V_{out(offset)}$)
output short-circuit current
positive feedback
power supply rejection
  ratio
slew rate
voltage feedback
voltage follower

| Equation Number | Equation | Section Number | **Equation Summary** |
|---|:---:|:---:|---|
| (15.1) | $v_d = v_2 - v_1$ | 15.2 | |
| (15.2) | $I_{E1} = I_{E2}$ | 15.3 | |
| (15.3) | $I_{E1} = I_{E2} = \dfrac{I_{RE}}{2}$ | 15.3 | |
| (15.4) | $I_{EE} = \dfrac{V_E - V_{EE}}{R_E}$ | 15.3 | |
| (15.5) | $V_{C1} = V_{C2} = V_{CC} - I_C R_C$ | 15.3 | |
| (15.6) | $V_{io} = \dfrac{V_{out(offset)}}{A_v}$ | 15.3 | |
| (15.7) | $f_{max} = \dfrac{\text{slew rate}}{2\pi V_{pk}}$ | 15.3 | |
| (15.8) | $v_{out} = i_f R_f$ | 15.4 | |
| (15.9) | $v_{in} = i_{in} R_i$ | 15.4 | |
| (15.10) | $v_{out} = i_{in} R_f$ | 15.4 | |
| (15.11) | $A_v = \dfrac{R_f}{R_{in}}$ | 15.4 | |
| (15.12) | $Z_{in} \cong R_i$ | 15.4 | |
| (15.13) | $\text{CMRR} = \dfrac{A_{CL}}{A_{CM}}$ | 15.4 | |
| (15.14) | $v_1 = v_{in}$ | 15.5 | |
| (15.15) | $i_f = \dfrac{v_{out} - v_{in}}{R_f}$ | 15.5 | |

| Equation Number | Equation | Section Number |
|---|---|---|
| (15.16) | $A_{CL} = \dfrac{R_f}{R_i} + 1$ | 15.5 |
| (15.17) | $CMRR = \dfrac{1}{A_{CM}}$ | 15.5 |
| (15.18) | $A_{CL}f_2 = f_{unity}$ | 15.7 |
| (15.19) | $A_{CL} = \dfrac{f_{unity}}{f_2}$ | 15.7 |
| (15.20) | $f_2 = \dfrac{f_{unity}}{A_{CL}}$ | 15.7 |
| (15.21) | $A_{OL} = \dfrac{v_{out}}{v_d}$ | 15.8 |
| (15.22) | $\alpha_v = \dfrac{v_f}{v_{out}}$ | 15.8 |
| (15.23) | $A_{vf} = \dfrac{v_{out}}{v_s}$ | 15.8 |
| (15.24) | $A_{vf} = \dfrac{A_v}{1 + \alpha_v A_v}$ | 15.8 |
| (15.25) | $A_{CL} = \dfrac{A_{OL}}{1 + \alpha_v A_{OL}}$ | 15.8 |
| (15.26) | $A_{vf} \cong \dfrac{1}{\alpha_v}$ | 15.8 |
| (15.27) | $A_{CL} \cong \dfrac{1}{\alpha_v}$ | 15.8 |
| (15.28) | $\alpha_v \cong \dfrac{1}{A_{CL}}$ | 15.8 |
| (15.29) | $Z_{in(f)} = Z_{in}(1 + \alpha_v A_{OL})$ | 15.8 |
| (15.30) | $Z_{out(f)} = \dfrac{Z_{out}}{1 + \alpha_v A_{OL}}$ | 15.8 |

**Answers to the Example Practice Problems**

**15.1.** 15.8c: positive; and 15.8d: negative
**15.2.** 5.6 $V_{PP}$
**15.3.** 157 $mV_{PP}$
**15.4.** 142.9 $mV_{PP}$
**15.5.** 6.4 kHz
**15.6.** 31.8 kHz
**15.7.** $A_{CL} = 250$, $Z_{in} \cong 1$ k$\Omega$, $Z_{out} < 50$ $\Omega$, CMRR = 12,500, $f_{max} = 19.1$ kHz
**15.8.** $A_{CL} = 251$, $Z_{in} \geqslant 1.5$ M$\Omega$, $Z_{out} \leqslant 50$ $\Omega$, CMRR = 12,550, $f_{max} = 19.02$ kHz
**15.9.** $A_{CL} = 1$, $Z_{in} = 1.5$ M$\Omega$, $Z_{out} = 50$ $\Omega$, CMRR = 50, $f_{max} = 19.9$ kHz
**15.10.** 125 kHz
**15.11.** Yes, it can be used.

**15.12.** 99.95
**15.13.** 1637.4
**15.14.** 1.82 GΩ
**15.15.** 82.8 mΩ

---

## §15.2

1. Determine the output polarity for each of the op-amps in Figure 15.54.
2. Determine the output polarity for each of the op-amps in Figure 15.55.
3. Determine the maximum peak-to-peak output voltage for the amplifier in Figure 15.56.
4. Determine the maximum peak-to-peak output voltage for the amplifier in Figure 15.57.
5. The amplifier in Problem 3 has a voltage gain of 120. Determine the maximum allowable peak-to-peak input voltage for the circuit.
6. The amplifier in Problem 4 has a voltage gain of 220. Determine the maximum allowable peak-to-peak input voltage for the circuit.
7. Determine the maximum allowable peak-to-peak input voltage for the amplifier in Figure 15.58.

**FIGURE 15.54**

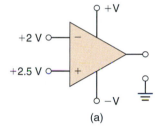
(a)

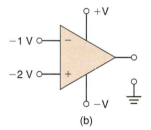
(b)

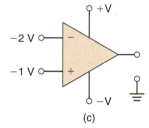
(c)

**FIGURE 15.55**

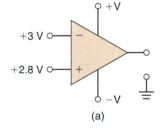
(a)

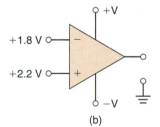
(b)

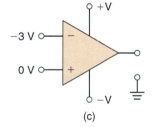
(c)

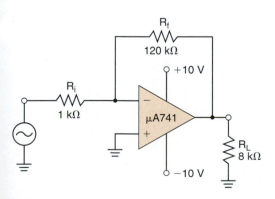

**FIGURE 15.56**

**FIGURE 15.57**

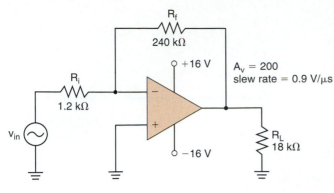

FIGURE 15.58

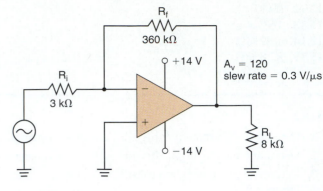

FIGURE 15.59

8. Determine the maximum allowable peak-to-peak input voltage for the amplifier in Figure 15.59.

§15.3

9. The amplifier in Figure 15.58 has an output offset voltage of 2.4 V. Determine the input offset voltage for the circuit.

10. The amplifier in Figure 15.59 has an output offset voltage of 960 mV. Determine the input offset voltage for the circuit.

11. Determine the maximum operating frequency for the amplifier in Figure 15.58. Assume that the circuit has a 10 mV$_{PP}$ input signal.

12. Determine the maximum operating frequency for the amplifier in Figure 15.59. Assume that the circuit has a 40 mV$_{PP}$ input signal.

13. Determine the maximum operating frequency for the amplifier in Figure 15.60.

14. Determine the maximum operating frequency for the amplifier in Figure 15.61.

§15.4

15. Perform the complete analysis of the amplifier in Figure 15.62.

16. Perform the complete analysis of the amplifier in Figure 15.63.

17. Perform the complete analysis of the amplifier in Figure 15.64.

18. Perform the complete analysis of the amplifier in Figure 15.65.

19. Perform the complete analysis of the amplifier in Figure 15.66.

20. Perform the complete analysis of the amplifier in Figure 15.67.

21. Perform the complete analysis of the amplifier in Figure 15.68.

22. Perform the complete analysis of the amplifier in Figure 15.69.

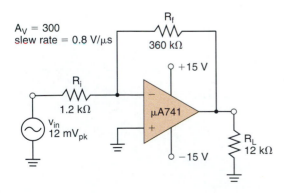

FIGURE 15.60

FIGURE 15.61

**FIGURE 15.62**

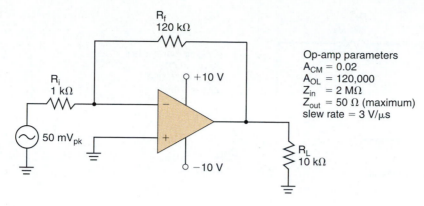

$R_f$
120 kΩ

$R_i$
1 kΩ

50 mV$_{pk}$

+10 V

−10 V

$R_L$
10 kΩ

Op-amp parameters
$A_{CM}$ = 0.02
$A_{OL}$ = 120,000
$Z_{in}$ = 2 MΩ
$Z_{out}$ = 50 Ω (maximum)
slew rate = 3 V/μs

**FIGURE 15.63**

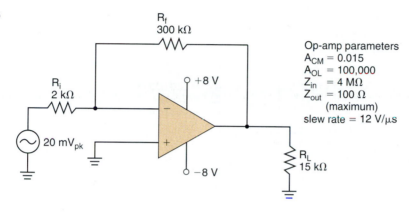

$R_f$
300 kΩ

$R_i$
2 kΩ

20 mV$_{pk}$

+8 V

−8 V

$R_L$
15 kΩ

Op-amp parameters
$A_{CM}$ = 0.015
$A_{OL}$ = 100,000
$Z_{in}$ = 4 MΩ
$Z_{out}$ = 100 Ω
(maximum)
slew rate = 12 V/μs

**FIGURE 15.64**

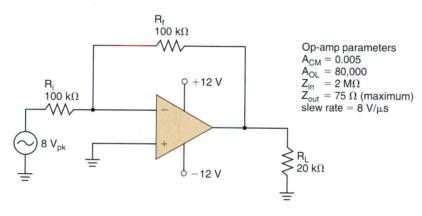

$R_f$
100 kΩ

$R_i$
100 kΩ

8 V$_{pk}$

+12 V

−12 V

$R_L$
20 kΩ

Op-amp parameters
$A_{CM}$ = 0.005
$A_{OL}$ = 80,000
$Z_{in}$ = 2 MΩ
$Z_{out}$ = 75 Ω (maximum)
slew rate = 8 V/μs

**FIGURE 15.65**

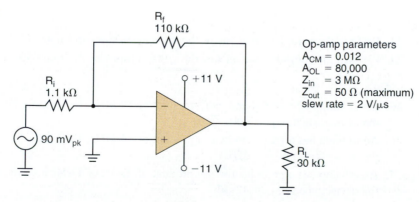

$R_f$
110 kΩ

$R_i$
1.1 kΩ

90 mV$_{pk}$

+11 V

−11 V

$R_L$
30 kΩ

Op-amp parameters
$A_{CM}$ = 0.012
$A_{OL}$ = 80,000
$Z_{in}$ = 3 MΩ
$Z_{out}$ = 50 Ω (maximum)
slew rate = 2 V/μs

FIGURE 15.66

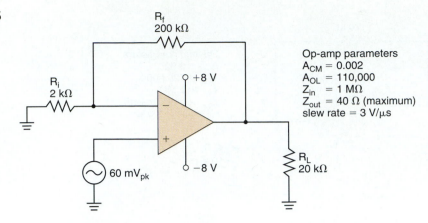

Op-amp parameters
$A_{CM} = 0.002$
$A_{OL} = 110,000$
$Z_{in} = 1\ M\Omega$
$Z_{out} = 40\ \Omega$ (maximum)
slew rate = 3 V/μs

FIGURE 15.67

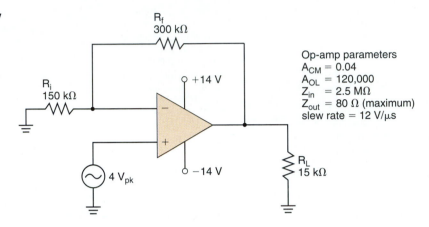

Op-amp parameters
$A_{CM} = 0.04$
$A_{OL} = 120,000$
$Z_{in} = 2.5\ M\Omega$
$Z_{out} = 80\ \Omega$ (maximum)
slew rate = 12 V/μs

FIGURE 15.68

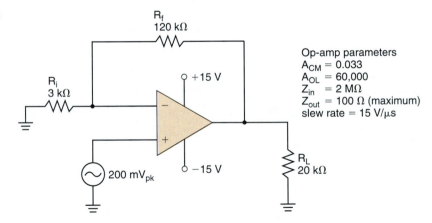

Op-amp parameters
$A_{CM} = 0.033$
$A_{OL} = 60,000$
$Z_{in} = 2\ M\Omega$
$Z_{out} = 100\ \Omega$ (maximum)
slew rate = 15 V/μs

23. Perform the complete analysis of the voltage follower in Figure 15.70.

24. Perform the complete analysis of the voltage follower in Figure 15.71.

*§15.6*

25. An op-amp has a gain–bandwidth product of 12 MHz. Determine the bandwidth of the device when $A_{CL} = 400$.

26. An op-amp has a gain–bandwidth product of 14 MHz. Determine the bandwidth of the device when $A_{CL} = 320$.

27. An op-amp has a gain–bandwidth product of 25 MHz. Determine the bandwidth of the device when $A_{CL} = 42$ dB.

FIGURE 15.69

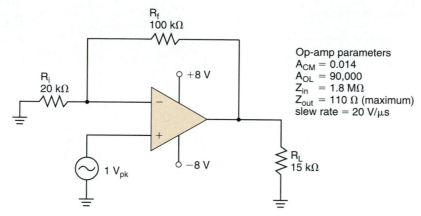

Op-amp parameters
$A_{CM} = 0.014$
$A_{OL} = 90,000$
$Z_{in} = 1.8\ M\Omega$
$Z_{out} = 110\ \Omega$ (maximum)
slew rate = 20 V/μs

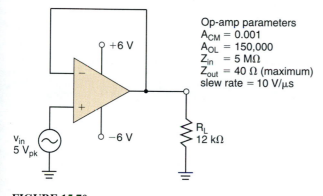

Op-amp parameters
$A_{CM} = 0.001$
$A_{OL} = 150,000$
$Z_{in} = 5\ M\Omega$
$Z_{out} = 40\ \Omega$ (maximum)
slew rate = 10 V/μs

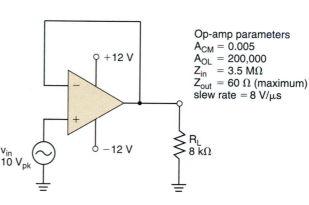

Op-amp parameters
$A_{CM} = 0.005$
$A_{OL} = 200,000$
$Z_{in} = 3.5\ M\Omega$
$Z_{out} = 60\ \Omega$ (maximum)
slew rate = 8 V/μs

FIGURE 15.70                    FIGURE 15.71

28. An op-amp has a gain–bandwidth product of 1 MHz. Determine the bandwidth of the device when $A_{CL} = 20$ dB.

29. We need to construct an amplifier that has values of $A_{CL} = 200$ and $f_2 = 120$ kHz. Can we use an op-amp with $f_{unity} = 28$ MHz for this application?

30. We need to construct an amplifier that has values of $A_{CL} = 24$ dB and $f_2 = 40$ kHz. Can we use an op-amp with $f_{unity} = 1$ MHz for this application?

31. An op-amp circuit with $A_{CL} = 120$ has a measured value of $f_2 = 100$ kHz. What is the gain–bandwidth product of the op-amp?

32. An op-amp circuit with $A_{CL} = 300$ has a measured value of $f_2 = 88$ kHz. What is the gain–bandwidth product of the op-amp?

33. The circuit in Figure 15.72 has a measured $f_2$ of 250 kHz. What is the value of $f_{unity}$ for the op-amp?

FIGURE 15.72

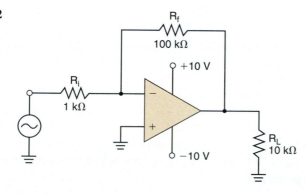

FIGURE 15.73

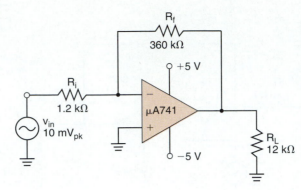

34. The circuit in Figure 15.73 has a measured $f_2$ of 100 kHz. What is the value of $f_{unity}$ for the op-amp?

35. An inverting amplifier has values of $A_v = 1000$ and $\alpha_v = 0.22$. Determine the value of $A_{vf}$ for the circuit.

36. An inverting amplifier has values of $A_v = 588$ and $\alpha_v = 0.092$. Determine the value of $A_{vf}$ for the circuit.

37. An inverting amplifier has values of $A_{OL} = 150,000$ and $\alpha_v = 0.008$. Determine the value of $A_{CL}$ for the circuit.

38. An inverting amplifier has values of $A_{OL} = 200,000$ and $\alpha_v = 0.0015$. Determine the value of $A_{CL}$ for the circuit.

39. The amplifier in Problem 35 has values of $Z_{in} = 48$ k$\Omega$ and $Z_{out} = 220$ $\Omega$. Calculate the values of $Z_{in(f)}$ and $Z_{out(f)}$ for the circuit.

40. The op-amp in Problem 37 has values of $Z_{in} = 2$ M$\Omega$ and $Z_{out} = 80$ $\Omega$. Assume the amplifier has values of $R_i = 1.2$ k$\Omega$ and $R_f = 150$ k$\Omega$. Calculate the values of $Z_{in(f)}$ and $Z_{out(f)}$ for the circuit.

41. Calculate the values of $A_{CL}$, $Z_{in(f)}$, and $Z_{out(f)}$ for the circuit in Figure 15.72. Assume that the op-amp has the following ratings: $A_{OL} = 200,000$, $Z_{in} = 5$M$\Omega$, and $Z_{out} = 100$ $\Omega$.

42. Calculate the values of $A_{CL}$, $Z_{in(f)}$, and $Z_{out(f)}$ for the circuit in Figure 15.73. Assume that the op-amp has the following ratings: $A_{OL} = 180,000$, $Z_{in} = 3$ M$\Omega$, and $Z_{out} = 75$ $\Omega$.

## Troubleshooting Practice Problems

43. The circuit in Figure 15.74 has the waveforms shown. Discuss the possible cause(s) of the problem.

44. The circuit in Figure 15.75 has the waveforms shown. Determine the cause of the problem. (*Hint:* The measured value of $R_f$ is exactly as shown; that is, $R_f$ is not the problem.)

45. The circuit in Figure 15.76 has the following readings: TP-1 = +10 mV, TP-2 = + 20 mV, TP-3 = +120 mV, TP-4 = 0 V, and TP-5 = −240 mV. Determine the possible cause(s) of the problem.

46. The circuit in Figure 15.76 has the following readings: TP-1 = −5 mV, TP-2 = 0 V, TP-3 = +4 V, TP-4 = 0 V, and TP-5 = −4 V. Determine the possible cause(s) of the problem.

47. The circuit in Figure 15.76 has the following readings: TP-1 = +2 mV, TP-2 = +1.8 mV, TP-3 = +4 V, TP-4 = 0 V, and TP-5 = 0 V. Determine the possible cause(s) of the problem.

**FIGURE 15.74**

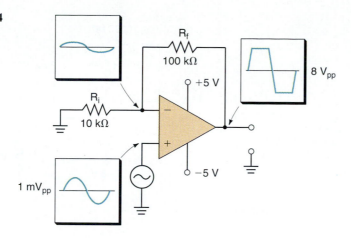

**FIGURE 15.75**

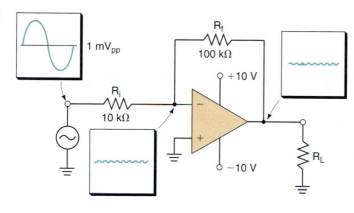

**FIGURE 15.76**

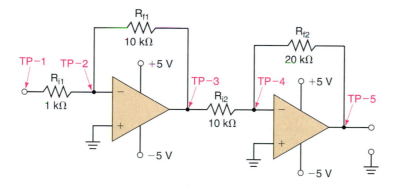

**48.** The circuit in Figure 15.76 has the following readings: TP-1 = −1 V, TP-2 = 0 V, TP-3 = 0 V, TP-4 = 0 V, and TP-5 = 0 V. Determine the possible cause(s) of the problem.

**49.** The op-amp in Figure 15.77 has the input and output voltages shown. Determine the CMRR of the device.

**50.** In Chapter 8, you were shown how to calculate the efficiency of an amplifier. Using those principles, calculate the efficiency of the amplifier in Figure 15.78. (*Hint:* Don't forget to consider the average output current from the op-amp in your current calculations.)

*The Brain Drain*

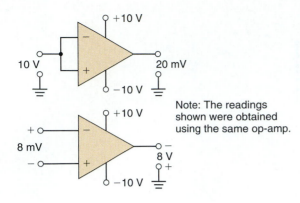

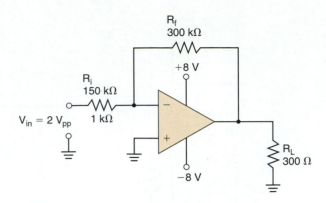

**FIGURE 15.77**

**FIGURE 15.78**

**FIGURE 15.79**

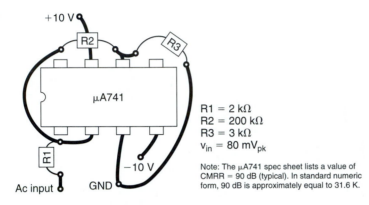

R1 = 2 kΩ
R2 = 200 kΩ
R3 = 3 kΩ
$V_{in}$ = 80 mV$_{pk}$

Note: The μA741 spec sheet lists a value of
CMRR = 90 dB (typical). In standard numeric
form, 90 dB is approximately equal to 31.6 K.

51. Using the μA741, design a noninverting amplifier that will deliver a 25 V$_{PP}$ output to a 20 kΩ load resistance with a 100 mV$_{pk}$ input signal. The available supply voltages are ±18 V$_{dc}$. Include the 8-pin DIP pin numbers in your schematic diagram.

52. Figure 15.79 is the parts placement diagram for an op-amp circuit. Analyze the circuit and draw its schematic diagram. Then perform a complete analysis of the circuit to determine its values of $A_{CL}$, $Z_{in}$, $Z_{out}$, CMRR, and $f_{max}$.

***Suggested Computer Applications Problems***

53. Write a program that will determine the maximum allowable input voltage for a given inverting amplifier when provided with the values of $R_f$, $R_i$, $R_1$, $+V$, and $-V$.

54. Write a program like the one described in Problem 53 for the noninverting amplifier.

# 16

# ADDITIONAL OP-AMP APPLICATIONS

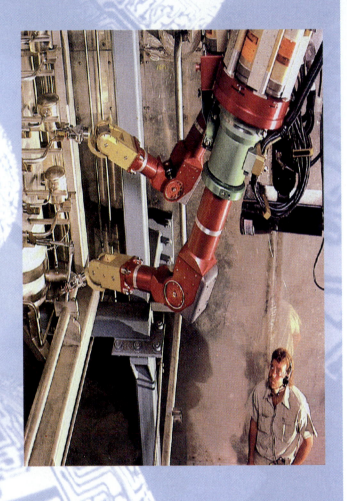

Industrial systems like the one shown require a wide variety of circuits to perform the operations for which they are designed. Op-amp circuits, like some of those covered in this chapter, are often used to aid in these operations.

# OUTLINE

# OBJECTIVES

*After studying the material in this chapter, you should be able to:*

1. State the purpose served by a *comparator*.
2. Describe the operation of a basic comparator circuit.
3. List the typical comparator fault symptoms and the possible causes of each.
4. State the purpose of the *integrator* and describe its operation.
5. State the purpose of a *differentiator* and describe its operation.
6. State the purpose of a *summing amplifier* and describe its operation.
7. Describe the use of the summing amplifier in a *digital-to-analog converter*.
8. Describe the operation of the *averaging amplifier* and the *subtractor*.
9. State several applications of the *instrumentation amplifier* and describe its operation.

# LOOKING TO THE FUTURE

Through the course of this book, you have been provided with glimpses of the history of solid-state devices and their development. The next question is simple: *Where will it go from here?*

While there is no clear picture of the future of solid-state electronics, one thing seems certain: The future of discrete devices is rather bleak. Most of the problems that are inherent to discrete devices are not present in integrated circuits. Most IC amplifiers and switching circuits are faster, cheaper, and easier to work with than discrete component circuits. It would make sense, then, that discrete devices will eventually be extremely limited in use. In fact, future devices textbooks will probably have two or three chapters on discrete component circuits, and then will move on to integrated circuits.

I n this chapter, we take a look at some op-amp circuits that are widely used for a variety of applications. These circuits do not necessarily relate to each other, except for the fact that they are all constructed using one or more op-amps. Most of the chapter will deal with five extremely common circuits: *comparators, integrators, differentiators, summing amplifiers,* and *instrumentation amplifiers.* Each of these circuits will be given extensive coverage. Then we will take a brief look at several other typical op-amp circuits. Remember as you go through the chapter that, other than the use of one or more op-amps, the circuits covered are not necessarily related to each other. Each section should be approached as a separate entity.

## 16.1 COMPARATORS

OBJECTIVE 1 ►

**Comparator**
A circuit used to compare two voltages.

The **comparator** is a relatively simple circuit used to compare two voltages and provide an output indicating the relationship between those two voltages. Generally, comparators are used to compare either:

1. Two changing voltages to each other (as in comparing two sine waves)
2. A changing voltage to a set dc reference voltage

We will look at the second type of comparator in this section. However, you may find it helpful if we take a moment to discuss the applications of comparators before going into any circuit detail.

### Applications

**Digital circuits**
Circuits designed to respond to specific alternating dc voltage levels.

Comparators are most commonly used in *digital* applications. **Digital circuits** are circuits that respond to *alternating dc voltage levels.* The signals that typically appear in a digital system consist of alternating dc voltage levels rather than sinusoidal waveforms. Typical digital signals are shown in Figure 16.1. You will spend a great deal of time studying digital systems, such as personal computers, in the future. For now, we need establish only

**FIGURE 16.1  Digital waveform characteristics.**

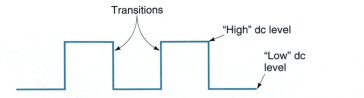

**FIGURE 16.2**

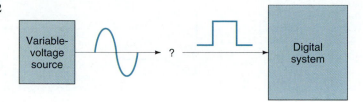

that digital systems tend to respond only to alternating dc levels. These alternating levels almost always take the form of *rectangular waves* or *square waves*.

Now consider the problem illustrated in Figure 16.2. Here we see a digital system and a sine-wave source. Let's assume for a moment that the digital system is to perform some function whenever the sine wave reaches a peak value greater than 10 V. The nature of the function is not important at this point; what *is* important is that the digital system must have a way of knowing whether or not $V_{pk}$ of the sine wave is greater than 10 V. From the illustration, the problem is obvious. The variable source circuit is generating a sine wave, and the digital system requires a varying dc-level input. The problem, then, is how to inform the digital system whether or not the output from the source is greater than 10 V in a manner to which the digital system can respond. A comparator would be used for this purpose. The solution to the problem is shown in Figure 16.3. In Figure 16.3a, a comparator is shown placed between the voltage source and the digital system. The inverting input of the comparator is connected to the reference voltage, +10 V. When the noninverting input is greater than this value, the output will go to the high-voltage level. It remains here until the noninverting input decreases below +10 V. At that time, the output of the comparator goes back to the low-voltage level. This input/output action is illustrated in Figure 16.3b.

Since the digital system is designed to respond to high and low dc levels, we have successfully converted information about the sine wave into a form that the digital system can deal with. This is the primary function served by a comparator.

**FIGURE 16.3   Basic comparator operation.**

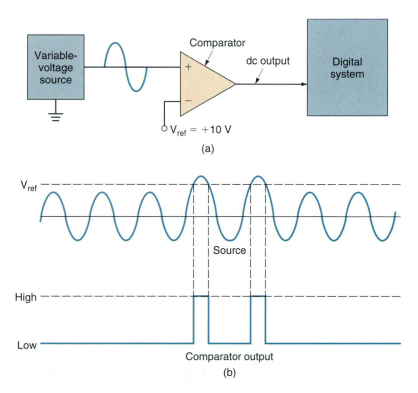

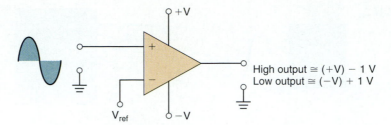

**FIGURE 16.4    A typical comparator circuit.**

Several points can be made at this time:

1. The output from a comparator is normally a dc voltage that indicates the polarity (or magnitude) relationship between the two input voltages.

2. A comparator does not normally convert a sine wave to a square wave. This function is normally performed by a circuit called a *Schmitt trigger*, which is discussed in Chapter 19. Instead, the comparator will merely change its dc output voltage to indicate a specific input voltage condition.

3. When a comparator is used to compare a signal amplitude to a fixed dc level (as was the case in Figure 16.3), the circuit is referred to as a **level detector.**

**Level detector**
Another name for a comparator used to compare an input voltage to a fixed dc reference voltage.

Now that you have an idea of what a comparator is used for, we will take a look at some common circuit configurations.

## Comparator Circuits

Comparator circuit recognition.

The most noticeable circuit recognition feature of the comparator is *the lack of any feedback path* in the circuit. A typical comparator circuit is shown in Figure 16.4. Without a feedback path, the voltage gain of the circuit is equal to the open loop gain of the op-amp.

*OBJECTIVE 2* ▶

When we discussed linear op-amp circuits in Chapter 15, you were shown that an open in the feedback path of an amplifier causes the output to clip at the positive and negative extremes. This clipping is caused by the high gain of the op-amp, $A_{OL}$, which is normally limited by the feedback path. In the case of the comparator, the clipping caused by the high gain of the op-amp is a desired feature. Since the gain of a comparator is equal to $A_{OL}$, virtually any difference voltage at the input will cause the output to go to one of the voltage extremes and stay there until the difference voltage is removed. The polarity of the input difference voltage will determine if the comparator output goes positive or negative. This point is illustrated in the following example.

---

**EXAMPLE 16.1**

The inverting input of the op-amp in Figure 16.5 is connected to a +5 V reference. With the values of $+V$, $-V$, and $A_{OL}$ shown, determine the output voltage from the circuit when the inverting input is at +4.9 and +5.1 V.

***Solution:***    When the input voltage is at +5.1 V, the difference voltage is found as

$$v_d = V_{in} - V_{ref}$$
$$= 5.1 \text{ V} - 5 \text{ V}$$
$$= +0.1 \text{ V}$$

The output voltage is then found as

---

$$V_{out} = A_{OL}v_d$$
$$= (150,000)(0.1 \text{ V})$$
$$= 1500 \text{ V}$$

Since this is clearly beyond the output limits of the circuit, the output voltage is found as

$$V_{out} \cong (V+) - 1 \text{ V}$$
$$= +9 \text{ V}$$

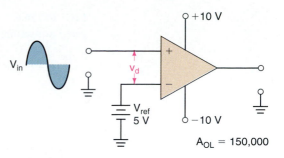

**FIGURE 16.5**

When the input voltage is at +4.9 V, the difference voltage is found as

$$v_d = V_{in} - V_{ref}$$
$$= +4.9 \text{ V} - 5 \text{ V}$$
$$= -0.1 \text{ V}$$

Using this value, we obtain the following output voltage:

$$V_{out} = A_{OL}v_d$$
$$= (150,000)(-0.1 \text{ V})$$
$$= -1500 \text{ V}$$

Again, this output is clearly impossible for the circuit shown. Thus, the output is found as

$$V_{out} \cong (V-) + 1 \text{ V}$$
$$= -9 \text{ V}$$

***PRACTICE PROBLEM 16.1***

A noninverting comparator has its inverting input connected to a +2 $V_{dc}$ reference. The op-amp is connected to ±12 V supplies and has a value of $A_{OL} = 70,000$. Determine the output voltage for input voltages of +2.001 and +1.999 $V_{dc}$.

As Example 16.1 showed, the comparator output voltage will go to one of the extremes even when there is very little difference between the two inputs. In fact, if you divide the 9 V maximum output voltage by the value of $A_{OL}$, you will see that the circuit needs a difference voltage of only 60 μV to cause the output to saturate at +V or −V. Also, as you were shown, the polarity of the input determines to which extreme the output will go.

## Setting the Reference Level

The circuit shown in Figure 16.5 is not very practical simply because most systems do not have a wide variety of power supply values available. In many cases, an electronic system will have one or two different supply voltages, and the reference voltage for the comparator would have to come from one (or both) of them. Most often, a voltage-divider circuit is used to set the reference voltage for a given level detector. Such a circuit is shown in Figure 16.6. For the circuit shown, the reference voltage would be found using the standard voltage-divider formula as follows:

$$V_{ref} = +V \frac{R_2}{R_1 + R_2} \qquad (16.1)$$

Example 16.2 demonstrates the use of this formula in a practical analysis situation.

**FIGURE 16.6 Comparator with a reference-setting circuit.**

**Lab Reference:** The operation of a comparator like the one in Figure 16.6 is demonstrated in Exercise 34.

---

## EXAMPLE 16.2

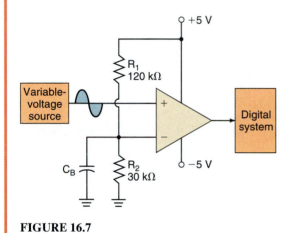

**FIGURE 16.7**

The digital system in Figure 16.7 is used to perform some predetermined function when the variable-voltage source reaches a certain output level. What is this output level?

**Solution:**   The reference voltage is set by the voltage-divider circuit. Thus, $V_{ref}$ is found as

$$V_{ref} = V \frac{R_2}{R_1 + R_2}$$
$$= (+5\ V)(0.2)$$
$$= +1\ V$$

The digital system is "looking" for a change either above or below +1 V.

**PRACTICE PROBLEM 16.2**

A comparator like the one in Figure 16.7 has $\pm 8$ V supplies and resistor values of $R_1 = 3$ kΩ and $R_2 = 1.8$ kΩ. Determine the input reference voltage for the circuit.

---

The circuits shown in Figures 16.6 and 16.7 both contain a *bypass capacitor* in the voltage-divider circuit. This bypass capacitor is included to prevent the variations in $v_{in}$ from being coupled to the voltage-divider circuit through the op-amp. The result is that the output of the circuit will be much more reliable.

Next, we will take a look at some other comparator circuit configurations. As you will see, each comparator circuit configuration is used for a specific application. However, they all work according to the same basic principles.

## Circuit Variations

Up to this point, we have assumed that the comparator is used to provide a *positive* output when the input voltage is more *positive* than some *positive* reference voltage. However, this is not always the case. In fact, you could substitute any combination of the words "positive" and "negative" into the following statement, and you would be correct:

*A comparator is a circuit used to provide a _____ output when the input voltage is more _____ than some _____ reference voltage.*

For example, take a look at the circuits shown in Figure 16.8. In Figure 16.8a, the circuit will have a *positive* output when the input voltage is more *negative* than some *positive* reference voltage. When $v_{in}$ is more positive than this reference voltage, the output will be negative. The circuit in Figure 16.8b will have a *positive* output whenever the input voltage is more *positive* than a *negative* reference voltage. When $v_{in}$ is more negative than this reference voltage, the output will be negative. The circuit in Figure 16.8c will have a *positive* output whenever the input voltage is *positive,* and a negative output otherwise. The circuit in Figure 16.8d will have a *positive* output whenever $v_{in}$ is *negative,* and a negative output otherwise. If you apply your knowledge of the relationship between the inputs of an op-amp and its output, the statements above are easy to visualize.

Another circuit variation is the **variable comparator.** The variable comparator allows you to change the dc reference voltage. Such a circuit is shown in Figure 16.9. By adjusting the value of $R_2$ in the circuit, the reference voltage can be set to any value desired. Note that any of the comparators in Figure 16.8 can be made variable in the same way, with the exception of the ground reference comparators, of course.

*A Practical Consideration:* It is easy to determine the input/output relationship for a comparator. If the input signal is applied to the inverting (−) terminal of the op-amp, the output will be *negative* when $v_{in}$ is more positive than $V_{ref}$. If the input signal is applied to the noninverting (+) terminal of the op-amp, the output will be *positive* when $v_{in}$ is more positive than $V_{ref}$.

**Variable comparator**
A comparator with an adjustable reference voltage.

## Troubleshooting Comparators

One of the really nice things about comparators is that they're extremely easy to troubleshoot. For example, refer back to the circuit in Figure 16.9. There really isn't a whole lot here that can go wrong. In fact, there are only three common problems that could develop. These problems and their possible causes are listed in Table 16.1. Now, let's go through a troubleshooting problem so that you can see how to apply the table to some practical situations.

◄ *OBJECTIVE 3*

**Lab Reference:** The operation of a comparator like the one in Figure 16.8(b) is demonstrated in Exercise 34.

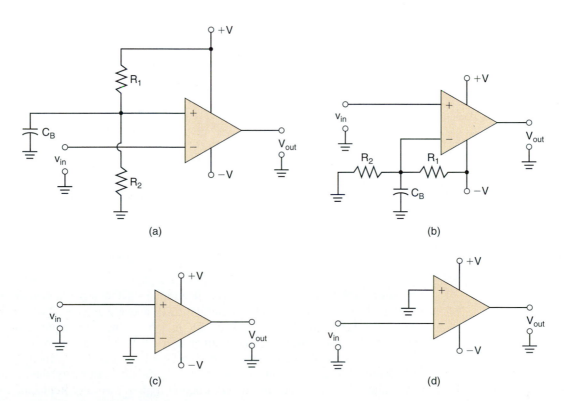

(a)

(b)

(c)

(d)

**FIGURE 16.8** **Comparator circuit variations.**

FIGURE 16.9   Comparator with a variable reference circuit.

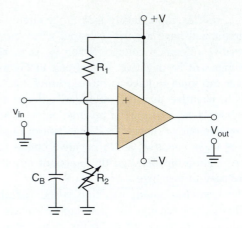

TABLE 16.1   Comparator Troubleshooting

| Problem | Possible Causes |
|---|---|
| 1. No output at all. | No input signal<br>One or both supply voltages out<br>A bad op-amp |
| 2. The output changes at an input voltage that is too *high*. | $R_2$ is out of adjustment<br>$R_1$ is shorted (not likely) |
| 3. The output changes at an input voltage that is too *low*. | $R_1$ is open<br>$R_2$ is out of adjustment<br>$C_B$ is shorted (not likely) |

**EXAMPLE 16.3**

Determine the cause of the problem in the circuit shown in Figure 16.10.

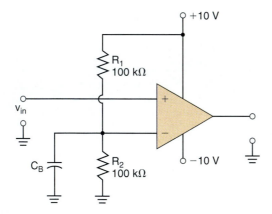

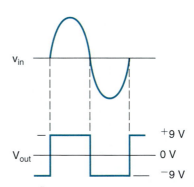

FIGURE 16.10

**Solution:**   The first step, of course, is to determine whether any problem exists at all. If you look at the voltage-divider circuit, you will notice that $R_1 = R_2$. Therefore, the reference voltage should be one-half of $+V$, or $+5$ V. Since the output is changing at the point when the input is at ground rather than $+5$ V, there *is* a problem.

Table 16.1 indicates that this problem could be caused by $R_1$ open, $R_2$ shorted, or $C_B$ shorted. Since the problem of $R_1$ open is by far the most likely, the resistance of this component is checked and found to be too high to measure. Replacing the component solves the problem.

## PRACTICE PROBLEM 16.3

Determine the cause of the problem in the circuit shown in Figure 16.11.

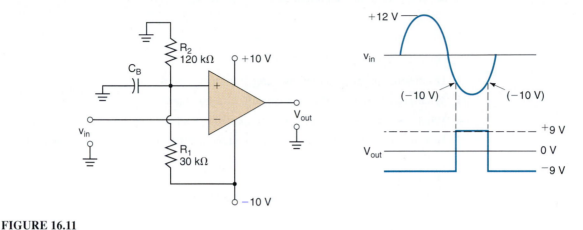

**FIGURE 16.11**

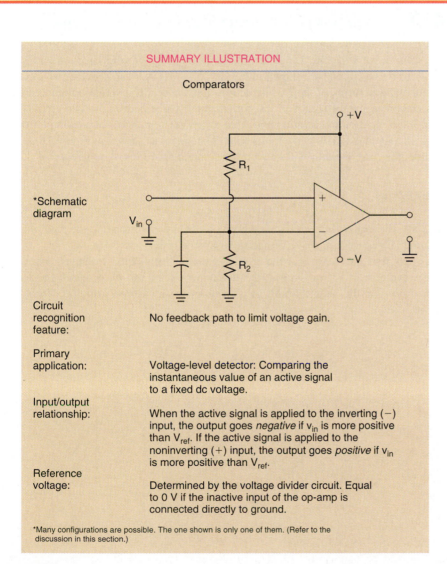

Comparators

*Schematic diagram

Circuit recognition feature: No feedback path to limit voltage gain.

Primary application: Voltage-level detector: Comparing the instantaneous value of an active signal to a fixed dc voltage.

Input/output relationship: When the active signal is applied to the inverting (−) input, the output goes *negative* if $v_{in}$ is more positive than $V_{ref}$. If the active signal is applied to the noninverting (+) input, the output goes *positive* if $v_{in}$ is more positive than $V_{ref}$.

Reference voltage: Determined by the voltage divider circuit. Equal to 0 V if the inactive input of the op-amp is connected directly to ground.

*Many configurations are possible. The one shown is only one of them. (Refer to the discussion in this section.)

**FIGURE 16.12**

As you can see, the comparator is an easy circuit to analyze and troubleshoot. No matter which circuit configuration is used, you will have no problem dealing with the circuit if you remember the basic operating principles of the comparator. The characteristics of the comparator are summarized in Figure 16.12.

## Section Review

1. What is a *comparator*?
2. Discuss the purpose served by comparators in digital systems.
3. What is the circuit recognition feature of a comparator?
4. What does the voltage gain of a comparator equal?
5. How do you determine the reference voltage and the input/output phase relationship of a comparator?
6. What is a *variable comparator*?
7. List the common comparator faults and their causes.

## 16.2 INTEGRATORS AND DIFFERENTIATORS

In this section, we will take a look at the operation of op-amp *integrators* and *differentiators*. These two circuits are shown in Figure 16.13. As you can see, the two circuits are nearly identical in terms of their construction. Each contains a single op-amp and an *RC* circuit. However, the difference in resistor/capacitor placement in the two circuits causes them to have input/output relationships that are exact opposites. For example, the integrator will convert a square wave into a triangular wave, as shown in Figure 16.13. At the same time, the differentiator will convert a triangular wave into a square wave.

### Integrators

OBJECTIVE 4 ▶

**Integrator**
A circuit whose output is proportional to the area of the input waveform.

Technically, the **integrator** is a circuit whose output is proportional to the area of its input waveform. The concept of waveform area is illustrated in Figure 16.14a.

In geometry, we learn that the area of a rectangle is equal to length times height. For the waveform in Figure 16.14a, the length would be the *pulse width* (measured in units of time) and the height would be its *amplitude* (measured in volts). Thus, for the waveform shown, area is found as

$$A = Vt$$

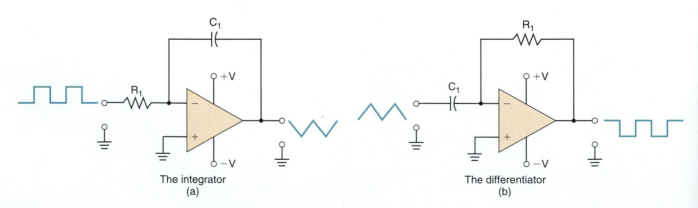

**FIGURE 16.13   Op-amp integrator and differentiator.**

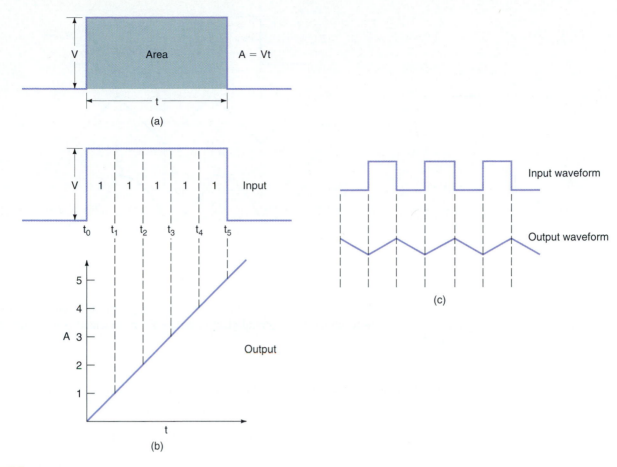

**FIGURE 16.14**

where $A$ = the area of the waveform
   $V$ = the peak voltage of the waveform
   $t$ = the pulse width of the waveform

Figure 16.14b shows how the output of the integrator varies with the area of the input waveform. As you can see, the input square wave has been divided into five equal sections. For ease of discussion, each section is referred to as *1 unit of area*. From $t_0$ to $t_1$, we have one unit of area. In response to this input condition, the output waveform of the integrator goes to 1. When $t_2$ is reached, the area of the waveform has increased to 2 units of area, and the integrator output goes to 2, and so on. As you can see, the output of the integrator indicates the total number of area units at any point on the input waveform.

With continuing cycles of the input waveform, the integrator output will produce the corresponding output waveform shown in Figure 16.14c. Thus, the integrator can be viewed as a square wave–to–triangular wave converter.

The integrator can be viewed as a square wave–to–triangular wave converter.

## Integrator Operation

The basic integrator is simply an *RC* circuit like the one shown in Figure 16.15. With the square-wave input shown, the *RC* integrator would *ideally* have a linear triangular output. This ideal output waveform is shown in the figure. The only problem is that the capacitor does not charge/discharge at a linear rate, but rather at an exponential rate. This produces the actual waveform shown in Figure 16.15.

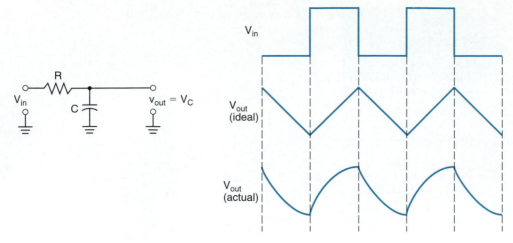

**FIGURE 16.15**  *RC integrator.*

In Chapter 3, you were shown that the capacitor in a given *RC* circuit requires five *time constants* to reach its full potential charge, where a time constant is the time period found as

$$\tau = RC$$

Thus, the time required for the capacitor to reach full charge can be calculated as

$$T_C = 5RC$$

When the value of $T_C$ is equal to the pulse width of the input, the *RC* integrator produces the nonlinear output shown in Figure 16.15.

Unfortunately, the nonlinear output of the *RC* integrator is not nearly close enough to the ideal integrator output for most applications. However, by adding an op-amp to the circuit, we can obtain a linear output from the integrator that is much closer to the ideal.

The key to eliminating the nonlinear output from the integrator is to provide a *constant-current* charge path for the capacitor. In other words, if we can make the capacitor charging current constant, the *rate* of charge for the component will become constant and the output will be linear. Keeping this in mind, let's take a look at the circuit in Figure 16.16. The *constant-current* characteristic of this circuit is based on two well-known points:

1. The inverting input to the op-amp is held at *virtual ground* by the differential amplifier in the component's input circuit.

2. The input impedance of the op-amp is so high that virtually all of $I_1$ is shunted to the resistor.

These two points were established in Chapter 15.

*How the op-amp integrator produces a linear output.*

**FIGURE 16.16   Op-amp integrator constant-current characteristics.**

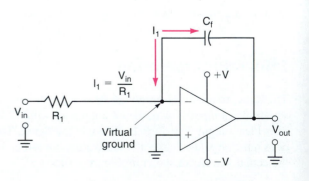

Since the inverting input is held at virtual ground, the value of input current ($I_1$) is found as

$$I_1 = \frac{V_{in}}{R_1}$$

Assuming that $V_{in}$ is constant for a given period of time and $R_1$ is a fixed value, $I_1$ can be assumed also to be a constant value. Since almost all of $I_1$ is drawn from the capacitor, $C$ is charged by a constant-current source. Thus, as long as $V_{in}$ is constant, the capacitor will charge/discharge at a linear rate. This produces the **ramp** output shown in Figure 16.13.

Since the input to the integrator is applied to the inverting input, the output of the circuit is 180° out of phase with the input. Thus, when the input goes positive, the output is a negative ramp. When the input is negative, the output is a positive ramp. This relationship is shown in Figure 16.17.

The waveforms shown in Figure 16.17 lead us to a potential problem with the op-amp integrator. Note that the output is centered around 0 V. In practice, this may not be the case. If the op-amp has not been properly compensated for the input offset voltage, the output may not center itself around 0 V. Thus, we must add a parallel resistor in the feedback path to ensure that any input offset voltage will be compensated for, and the output will be centered around 0 V. This circuit is shown in Figure 16.18. The addition of the feedback resistor has the benefit of centering the output around 0 V, but it also has a disadvantage. The $RC$ feedback circuit will, like any $RC$ circuit, have a cutoff frequency. In this case, it is a *lower* cutoff frequency with which we are concerned. As frequency decreases, the reactance of the capacitor will increase. At some point, the reactance of the capacitor will be greater than the value of $R_f$, and the integrating action of the circuit will stop. As usual, the cutoff frequency is found using

$$f_1 = \frac{1}{2\pi R_f C_f} \qquad \textbf{(16.2)}$$

**Ramp**
Another name for a voltage that changes at a constant (or *linear*) rate.

The feedback resistor (Figure 16.18) eliminates any output offset voltage.

The integrator will *start* to lose its linear output characteristics before the frequency found in equation (16.2) is reached. For an optimum output, the value of $X_C$ should always be less than $0.1R_f$. Therefore, the circuit should not be operated below the frequency found by

$$f_{min} = \frac{10}{2\pi R_f C_f} \qquad \textbf{(16.3)}$$

Example 16.4 illustrates the frequency limits of the integrator.

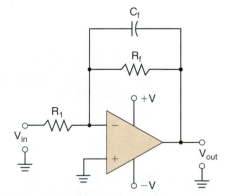

FIGURE 16.17    Op-amp integrator phase relationship.

FIGURE 16.18

**Lab Reference:** The phase relationship shown in Figure 16.17 is observed in Exercise 35.

<table>
<tr><td>EXAMPLE 16.4</td><td>Determine the cutoff frequency for the circuit shown in Figure 16.19. Also, determine the frequency at which the output starts to lose its linear characteristics.</td></tr>
</table>

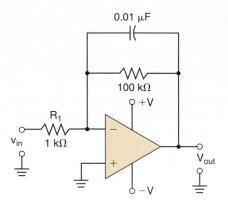

**FIGURE 16.19**

*Solution:*  The cutoff frequency for the circuit is found as

$$f_1 = \frac{1}{2\pi R_f C_f}$$

$$= \frac{1}{2\pi(100 \text{ k}\Omega)(0.01 \text{ }\mu\text{F})}$$

$$= 159 \text{ Hz}$$

If you take a close look at equations (16.2) and (16.3), you will see that the linearity will start to go when the operating frequency is 10 times the cutoff frequency. Therefore, we can save some time and trouble by using

$$f_{min} = 10f_1$$
$$= 1.59 \text{ kHz}$$

**PRACTICE PROBLEM 16.4**

An integrator has values of $C_f = 0.1$ $\mu$F and $R_f = 51$ k$\Omega$. Determine the cutoff frequency for the circuit. Also, determine the frequency at which it will start to lose its linear characteristics.

As you may recall, a feedback capacitor may be broken into two capacitors, one at the input and one at the output, as shown in Figure 16.20. Obviously, this circuit will have an upper cutoff frequency. As operating frequency increases, the feedback capacitor will act more and more like a short circuit. Eventually, if the process were allowed to continue, the capacitor would short out the feedback resistor completely. In this case, the closed-loop gain of the circuit would drop to zero.

## Integrator Troubleshooting

The integrator is another circuit that is relatively easy to troubleshoot. In this section, we will take a look at the potential problems that may develop and the symptoms of each (Table 16.2). Any other problems that could develop with the circuit would be diagnosed

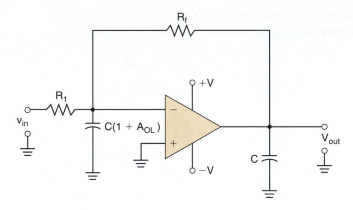

**FIGURE 16.20**
**ac equivalent circuit.**

**TABLE 16.2   Integrator Troubleshooting**

| Fault | Symptoms |
|---|---|
| $R_f$ open | If $R_f$ opens, the resistor is effectively removed from the circuit. Since this resistor is used to keep the output referenced around 0 V, the major symptom is the loss of this output reference. Also, the circuit cutoff frequency will drop to approximately zero. |
| $R_f$ shorted | As you know, resistors do not usually short. However, if something were to short out $R_f$, integration would be lost, as would the gain of the circuit. In other words, the circuit would have little, if any, output signal. |
| $C_f$ open | As you know, a capacitor is an open circuit by nature. However, if the component were to open in such a way as to prevent coupling, the circuit would start to act as a common inverting amplifier. The gain of the circuit would become $$A_{CL} = \frac{R_f}{R_i}$$ and the output would be a square wave. |
| $C_f$ shorted | This would have the same effect as $R_f$ shorting. In fact, if either of these components shorts, the result is that the other component will also be shorted. Therefore, you must test $C_f$ and $R_f$ individually to see which component is actually shorted. |

as discussed earlier. In other words, a problem with $R_1$, the op-amp, or the supply voltages would cause the same symptoms that we described earlier.

## The Differentiator

If we reverse the integrator capacitor and resistor (as shown in Figure 16.13b), we have a circuit called a **differentiator.** Technically, the differentiator is a circuit whose output is proportional to the *rate of change* of its input signal. The input/output relationship of the differentiator is illustrated in Figure 16.21.

◄ *OBJECTIVE 5*

**Differentiator**
A circuit whose output is proportional to the rate of change of its input signal.

The input signal for the differentiator is shown to be a triangular waveform. Between times $t_0$ and $t_2$, the rate of change of the input is constant. Since the change is in a *positive* direction, the rate of change is a *positive constant*. Thus, the output from the circuit is a constant positive voltage. Between times $t_2$ and $t_4$, the rate of change is a *negative constant*. As a result, the output from the circuit switches to a constant negative voltage. Thus, with the triangular input waveform, we get a square-wave output.

Equation (16.2) gave us a limit on the low-frequency operation of the integrator. The same equation can be used to provide a limit on the *high-frequency* operation of the differentiator. Once again, the operating characteristics of the circuits are opposites.

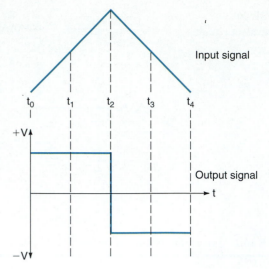

**FIGURE 16.21   The input/output relationship of the differentiator.**

**Lab Reference:** The input/output relationship for the differentiator is demonstrated in Exercise 35.

Just as the integrator will start to lose its operating characteristics at 10 times the cutoff frequency, the differentiator will start to lose its operating characteristics at one-tenth the value of $f_2$. By formula,

$$f_{max} = \frac{1}{20\pi RC} \tag{16.4}$$

If the input frequency to the differentiator exceeds the limit determined by equation (16.4), the output square wave will start to become distorted.

## Differentiator Troubleshooting

Like the integrator, the differentiator is relatively easy to troubleshoot. The primary faults that can develop (other than a bad op-amp) are listed in Table 16.3, along with the symptoms of each. If the differentiator fails to operate and none of the symptoms in Table 16.3 appears, the op-amp is the source of the fault.

## Summary

While they are also used for other applications, the integrator and differentiator are used primarily for waveform conversion. The characteristics of these two circuits are summarized for you in Figure 16.22.

**TABLE 16.3   Differentiator Troubleshooting**

| Fault | Symptoms |
|---|---|
| $R_1$ open | The input will be removed from the op-amp, and the output of the circuit will remain at zero. |
| $R_1$ shorted | If something were to short out $R_1$, the gain of the circuit would increase drastically and the differentiating action of the circuit would be lost. |
| $C_1$ open | In this case, the gain of the circuit equals the value of $A_{OL}$ for the op-amp. Also, differentiating action is lost. |
| $C_1$ shorted | The gain of the circuit will drop to zero, and there will be little (if any) output signal. |

**FIGURE 16.22**

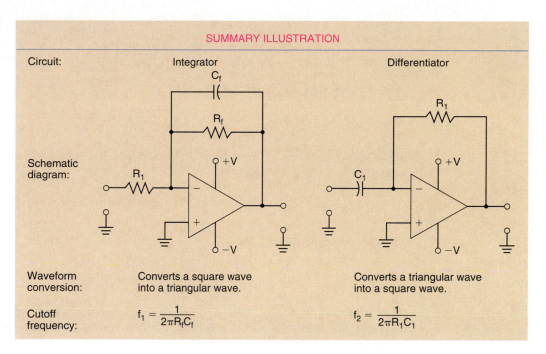

SUMMARY ILLUSTRATION

Circuit:      Integrator                 Differentiator

Schematic diagram:

Waveform conversion:  Converts a square wave into a triangular wave.   Converts a triangular wave into a square wave.

Cutoff frequency:  $f_1 = \dfrac{1}{2\pi R_f C_f}$      $f_2 = \dfrac{1}{2\pi R_1 C_1}$

1. What is an *integrator*?

2. Describe the circuit operation of the op-amp integrator.

3. What is a *ramp*?

4. List the common faults that occur in op-amp integrators and the symptoms of each.

5. What is a *differentiator*?

6. Describe the circuit operation of the differentiator.

7. List the common differentiator faults and the symptoms of each.

# 16.3 SUMMING AMPLIFIERS

The **summing amplifier** is an op-amp circuit that accepts several inputs and provides an output proportional to the *sum* of the inputs. The basic summing amplifier is shown in Figure 16.23. The key to understanding this circuit is to start by considering each input as an individual circuit. If we had an input voltage only at $V_1$, the output would be found as

◄ *OBJECTIVE 6*

**Summing amplifier**
An op-amp circuit that produces an output proportional to the sum of the input voltages.

$$V_{out} = -V_1 \frac{R_f}{R_1}$$

**FIGURE 16.23  Summing amplifier.**

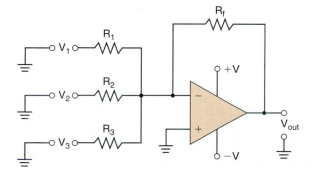

**Lab Reference:** Summing amplifier operation is demonstrated in Exercise 36.

Similarly, if $V_2$ were the only input, we would have

$$V_{\text{out}} = -V_2 \frac{R_f}{R_2}$$

And if $V_3$ were the only input, we would have

$$V_{\text{out}} = -V_3 \frac{R_f}{R_3}$$

If we have voltage inputs at all three of the terminals, the output will be the sum of the equations for each, as follows:

$$V_{\text{out}} = \frac{-V_1 R_f}{R_1} + \frac{-V_2 R_f}{R_2} + \frac{-V_3 R_f}{R_3}$$

Or, after factoring out the value of $-R_f$,

$$V_{\text{out}} = -R_f \left( \frac{V_1}{R_1} + \frac{V_2}{R_2} + \frac{V_3}{R_3} \right) \tag{16.5}$$

Example 16.5 demonstrates the procedure for determining the output from a summing amplifier.

---

## EXAMPLE 16.5

Determine the output voltage from the summing amplifier in Figure 16.24.

**Solution:** Using the values shown in the circuit and equation (16.5), the output voltage is found as

$$
\begin{aligned}
V_{\text{out}} &= -R_f \left( \frac{V_1}{R_1} + \frac{V_2}{R_2} + \frac{V_3}{R_3} \right) \\
&= -10\ \text{k}\Omega \left( \frac{+3\ \text{V}}{10\ \text{k}\Omega} + \frac{+6\ \text{V}}{10\ \text{k}\Omega} + \frac{+4\ \text{V}}{10\ \text{k}\Omega} \right) \\
&= -13\ \text{V}
\end{aligned}
$$

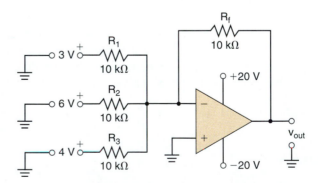

**FIGURE 16.24**

### PRACTICE PROBLEM 16.5

A summing amplifier like the one in Figure 16.24 has the following values: $R_f = 2\ \text{k}\Omega$, $R_1 = R_2 = R_3 = 1\ \text{k}\Omega$, $V_1 = 1\ \text{V}$, $V_2 = 500\ \text{mV}$, and $V_3 = 1.5\ \text{V}$. The supply voltages for the circuit are $\pm 18\ \text{V}$. Determine the value of $V_{\text{out}}$ for the circuit.

If you compare the result in Example 16.5 with the original input voltages, you will see that the amplifier output is equal to the sum of the input voltages. Thus, the term *summing amplifier* is appropriate for the circuit.

In practice, it is not always possible to have an output that is equal to the sum of the input voltages. For example, what if the inputs to Figure 16.24 had been +10 V, +8 V, and +7 V? Clearly, the sum of these voltages is +25 V. However, the maximum possible output voltage from the circuit is ±19 V. Thus, an output of 25 V would be impossible. To solve this problem, the value of $R_f$ is usually set up so that the output voltage is *proportional* to the sum of the inputs. This point is illustrated in Example 16.6.

**EXAMPLE 16.6**

Determine the output voltage from the circuit shown in Figure 16.25.

**Solution:** Using the values shown in the circuit and equation (16.5), the output voltage is found as

$$V_{out} = -R_f\left(\frac{V_1}{R_1} + \frac{V_2}{R_2} + \frac{V_3}{R_3}\right)$$

$$= -1\ \text{k}\Omega\left(\frac{+10\ \text{V}}{10\ \text{k}\Omega} + \frac{+8\ \text{V}}{10\ \text{k}\Omega} + \frac{+7\ \text{V}}{10\ \text{k}\Omega}\right)$$

$$= -2.5\ \text{V}$$

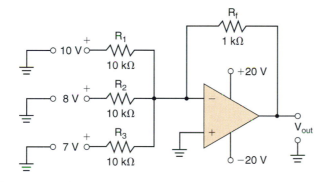

**FIGURE 16.25**

***PRACTICE PROBLEM 16.6***

A summing amplifier like the one in Figure 16.25 has the following values: $R_f = 1\ \text{k}\Omega$, $R_1 = R_2 = R_3 = 3\ \text{k}\Omega$, $V_1 = 2\ \text{V}$, $V_2 = 1\ \text{V}$, and $V_3 = 6\ \text{V}$. The supply voltages for the circuit are ±12 V. Determine the value of $V_{out}$ for the circuit.

In Example 16.6, we solved the problem of dealing with +10, +8, and +7 V inputs by *decreasing* the value of $R_f$. Now the output of the amplifier is not equal to the sum of the inputs, but rather it is *proportional to* the sum of the inputs. In this case, it is equal to one-tenth of the input sum. Picking random values of $V_1$, $V_2$, and $V_3$ and using them in equation (16.5) will show that the circuit always provides an output that is one-tenth of the sum of the inputs. This assumes, of course, that the value obtained by the equation does not exceed ±19 V.

At this time, there are several points that should be made regarding the summing amplifier:

1. There is no practical limit on the number of summing inputs. As long as the amplifier is capable of providing the correct output voltage, the circuit can have any number of inputs. In this case, equation (16.5) would be expanded (or reduced) to include all the amplifier inputs.

2. The resistors used at the amplifier input do not necessarily have the same value. In other words, the inputs can be *weighted* so that certain inputs have more of an effect on the output.

**Lab Reference:** The effect of weighting the inputs to a summing amplifier is demonstrated in Exercise 36.

The second point will be discussed in detail when we look at a very common application for the summing amplifier later in this section. First, we will look at how you can develop a general formula for any summing amplifier that will make the analysis of that circuit easier.

## Circuit Analysis

Often, you need to be able to predict quickly the output from a summing amplifier under a variety of input combinations. This task is made easier if you derive a *general-class* output equation for the amplifier.

**Lab Reference:** The general-class equation for a summing amplifier is derived and used to predict output values in Exercise 36.

The general-class equation for a given summing amplifier is derived as follows:

1. Determine the $R_f/R$ ratio for each input branch.

2. Represent each branch as a product of its resistance ratio times its input voltage.

3. Add the products found in step 2.

Let's apply this process to the circuit shown in Figure 16.26. For the first branch in the circuit, the resistance ratio is

$$\frac{R_f}{R_1} = 1$$

and the branch would be represented as this ratio times $V_1$, or simply $V_1$. The resistance ratio of the second branch would be

$$\frac{R_f}{R_2} = 0.2$$

and the branch would be represented as $0.2V_2$. Performing this procedure on the third and fourth branches would yield results of 0.1 and 0.05, respectively. The output of the circuit can now be found as

**General-class equation**
An equation, derived for a summing amplifier, that is used to predict the output voltage for any combination of input voltages.

$$-V_{out} = V_1 + 0.2V_2 + 0.1V_3 + 0.05V_4$$

This is the **general-class equation** for the circuit and can be used to determine the output voltage for any combination of input voltages, as shown in Example 16.7.

**FIGURE 16.26**

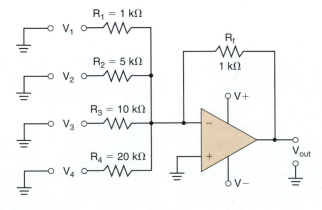

Determine the output from the circuit shown in Figure 16.27 for each of the following input combinations:

EXAMPLE *16.7*

| $V_1(v)$ | $V_2(v)$ | $V_3(v)$ |
|---|---|---|
| +10 | 0 | +10 |
| 0 | +10 | +10 |
| +10 | +10 | +10 |

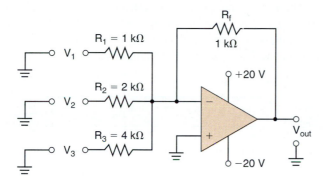

**FIGURE 16.27**

***Solution:*** First, the circuit can be represented by its general-class equation. For this circuit, the equation would be derived as

$$-V_{out} = \frac{R_f}{R_1}V_1 + \frac{R_f}{R_2}V_2 + \frac{R_f}{R_3}V_3$$
$$= V_1 + 0.5V_2 + 0.25V_3$$

Now, using this equation, we can determine the output for the first set of inputs as

$$-V_{out} = 10\text{ V} + 0.5(0\text{ V}) + 0.25(10\text{ V})$$
$$= 12.5\text{ V}$$

For the second set of inputs,

$$-V_{out} = 0\text{ V} + 0.5(10\text{ V}) + 0.25(10\text{ V})$$
$$= 7.5\text{ V}$$

Finally, for the third set of inputs,

$$-V_{out} = 10\text{ V} + 0.5(10\text{ V}) + 0.25(10\text{ V})$$
$$= 17.5\text{ V}$$

***PRACTICE PROBLEM 16.7***

A summing amplifier like the one in Figure 16.27 has the following values: $R_f = 2\text{ k}\Omega$, $R_1 = 200\ \Omega$, $R_2 = 400\ \Omega$, and $R_3 = 2\text{ k}\Omega$. Derive the general-class equation for the circuit. Then use that equation to determine the value of $V_{out}$ when $V_1 = V_2 = V_3 = 1\text{ V}$.

As you can see, deriving the general-class equation for a given summing amplifier makes the analysis of the circuit simpler, especially when you are trying to determine the output for several different input conditions.

## Summing Amplifier Applications

OBJECTIVE 7 ▶

**Digital-to-analog (D/A) converter**
A circuit that converts digital circuit outputs to equivalent analog voltages.

The most common use of the summing amplifier is as a **digital-to-analog (D/A) converter,** or D/A converter. This application is illustrated in Figure 16.28.

First, a word or two about the *digital circuit*. This circuit has only two possible output values on any given line. For our circuit, we will assume these two voltages to be +5 and 0 V. Thus, each of the output lines, labeled 8, 4, 2, and 1, will always be at either +5 V or at 0 V. The exact value on a given line depends on the value that the circuit is presenting at the output. When a given line has +5 V on it, that value is part of the output. For example, if the digital circuit was putting out the equivalent of the decimal number 10, it would have the following output conditions:

| $V_1(8)$ | $V_2(4)$ | $V_3(2)$ | $V_4(1)$ |
|----------|----------|----------|----------|
| +5 V | 0 V | +5 V | 0 V |

With the +5 V outputs at (8) and (2), the output is representing the value $8 + 2 = 10$. If the digital circuit were outputting the value 6, the output voltages, from top to bottom, would be 0, +5, +5, and 0 V.

The D/A converter is used to convert the output value from the digital circuit into a single voltage level. This level is proportional to the numerical output of the circuit. To see how this works, we start by deriving the general-class formula for the circuit. This formula is

$$-V_{out} = V_1 + 0.5V_2 + 0.25V_3 + 0.125V_4$$

Using this equation, the output values in the following table were derived:

| Decimal Value | $V_1(8)$ | $V_2(4)$ | $V_3(2)$ | $V_4(1)$ | $V_{out}$ |
|---------------|----------|----------|----------|----------|-----------|
| 0 | 0 V | 0 V | 0 V | 0 V | −0 V |
| 1 | 0 V | 0 V | 0 V | +5 V | −0.625 V |
| 2 | 0 V | 0 V | +5 V | 0 V | −1.25 V |
| 3 | 0 V | 0 V | +5 V | +5 V | −1.875 V |
| 4 | 0 V | +5 V | 0 V | 0 V | −2.5 V |
| 5 | 0 V | +5 V | 0 V | +5 V | −3.125 V |
| . | | | | | |
| . | | | | | |
| . | | | | | |
| 15 | +5 V | +5 V | +5 V | +5 V | −9.375 V |

It is not important for you to understand the exact workings of the digital circuit output. It is hoped that you have seen a pattern in the +5 V outputs from the circuit. The main point is this: Digital circuits use two-voltage inputs and outputs to represent numerical values. Often, it is necessary to convert the output from a group of digital outputs into a single voltage. This is accomplished using a process called D/A conversion. The conversion is performed by a summing amplifier whose inputs are weighted proportionally to

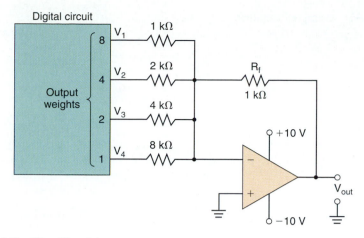

**FIGURE 16.28**  **Simplified D/A converter.**

the weights of the digital outputs. Thus, as the digital circuit output increases in value, the output voltage from the D/A converter increases.

## Circuit Troubleshooting

The biggest problem with troubleshooting a summing amplifier is the fact that one of the input resistors can go bad and the circuit will still work for the other inputs. For example, look at the circuit shown in Figure 16.28. If the resistor on the output 4 line were to open, the rest of the input lines to the summing amplifier would still work, and the amplifier would still be providing an output. Granted, this output may be incorrect for a given set of digital input values, but this is not always easy to see with an oscilloscope.

If you isolate a problem in a given circuit down to a summing amplifier, the best procedure is simply to measure the resistance of each component in the circuit. Since the only components contained in the circuit are resistors (other than the op-amp itself), a series of resistance checks is the fastest way to isolate a problem. If all the resistors prove to be good, replace the op-amp.

How do you tell when a given summing amplifier is faulty? To answer this question, let's take a look at the block diagram shown in Figure 16.29. Let's say that circuit B in the figure is not working properly. Tests indicate that it is not receiving the proper inputs. Testing circuit A shows that it *is* working properly. With these conditions, the problem must be the summing amplifier. In other words, in this case *the summing amplifier was shown to be faulty by proving that the problem wasn't anywhere else*. As depressing as this may sound, it is often the case since summing amplifiers tend to be used in relatively complex digital circuits.

## Circuit Variations

By using the proper input and feedback resistor values, a summing or **averaging amplifier** can be designed to provide an output that is equal to the *average* of any number of inputs. For example, consider the averaging amplifier shown in Figure 16.30

◄ *OBJECTIVE 8*

**Averaging amplifier**
A summing amplifier that provides an output proportional to the *average* of the input voltage.

**FIGURE 16.29**

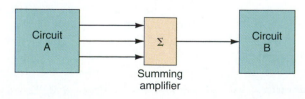

Circuit A → Σ Summing amplifier → Circuit B

FIGURE 16.30 Averaging amplifier.

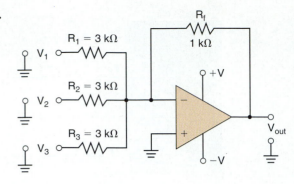

If we were to derive the general-class formula for the circuit in Figure 16.30, we would obtain

$$-V_{out} = \frac{V_1}{3} + \frac{V_2}{3} + \frac{V_3}{3}$$

or

$$-V_{out} = \frac{V_1 + V_2 + V_3}{3}$$

This, by definition, is the average of the three input voltages.

A summing amplifier will act as an averaging circuit whenever *both* of the following conditions are met:

1. All input resistors ($R_1$, $R_2$, and so on) are *equal in value*.

2. The ratio of any input resistor to the feedback resistor is equal to the number of circuit inputs.

For example, in Figure 16.30, all input resistors are equal in value (3 k$\Omega$). If we take the ratio of any input resistor to the feedback resistor, we get 3 k$\Omega$/1 k$\Omega$ = 3. This is equal to the number of inputs to the circuit.

Another variation on the summing amplifier is shown in Figure 16.31. This circuit, called a **subtractor**, will provide an output that is equal to the difference between $V_1$ and $V_2$.

$V_1$ is applied to a standard inverting amplifier that has unity gain. Because of this, the output from the inverting amplifier will equal $-V_1$. This output is then applied to a

**Subtractor**
A summing amplifier that provides an output proportional to the difference between the input voltages.

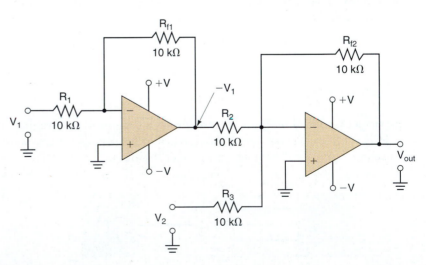

**FIGURE 16.31** Subtractor.

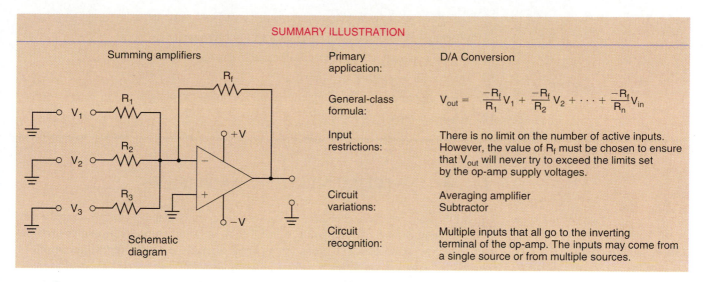

Summing amplifiers

Schematic diagram

Primary application: D/A Conversion

General-class formula:

$$V_{out} = \frac{-R_f}{R_1} V_1 + \frac{-R_f}{R_2} V_2 + \cdots + \frac{-R_f}{R_n} V_{in}$$

Input restrictions: There is no limit on the number of active inputs. However, the value of $R_f$ must be chosen to ensure that $V_{out}$ will never try to exceed the limits set by the op-amp supply voltages.

Circuit variations: Averaging amplifier
Subtractor

Circuit recognition: Multiple inputs that all go to the inverting terminal of the op-amp. The inputs may come from a single source or from multiple sources.

**FIGURE 16.32**

summing amplifier (also having unity gain) along with $V_2$. Thus, the output from the second op-amp is found as

$$-V_{out} = V_2 + (-V_1)$$
$$= V_2 - V_1$$

It should be noted that the gain of the second stage in the subtractor can be varied to provide an output that is proportional to (rather than equal to) the difference between the input voltages. However, if the circuit is to act as a subtractor, the input inverting amplifier *must* have unity gain. Otherwise, the output will not be proportional to the true difference between $V_1$ and $V_2$.

## Summary

The summing amplifier can be designed to provide an output voltage that is equal to the sum of any number of input voltages. Two variations on the basic summing amplifier are the averaging amplifier and the subtractor. The characteristics of the basic summing amplifier are summarized in Figure 16.32.

**Section Review**

1. What is a *summing amplifier*?
2. Why is the output from a summing amplifier often proportional to (rather than equal to) the sum of its input voltages?
3. Is there a limit on the number of inputs to a summing amplifier? Explain your answer.
4. What is the *general-class equation* of a summing amplifier?
5. List the steps used to derive the general-class equation for a summing amplifier.
6. What is a *digital-to-analog (D/A) converter*?

7. Describe the use of a summing amplifier as a D/A converter.

8. Why are summing amplifiers difficult to troubleshoot?

9. How is a summing amplifier usually determined to be faulty?

10. What is an *averaging amplifier*?

11. What are the resistor requirements of an averaging amplifier?

12. What is a *subtractor*?

## *16.4* INSTRUMENTATION AMPLIFIERS

OBJECTIVE 9 ▶

**Instrumentation amplifier**
An amplifier of low-level signals used in process control or measurement applications.

Up to this point, we have concentrated on op-amp circuits that are most commonly used in digital applications. At this point, we will look at an op-amp circuit that is used primarily in process control and measurement applications: the **instrumentation amplifier**. The instrumentation amplifier is a *high-gain, high-CMRR circuit that is used to detect and amplify low-level signals.* The inputs to an instrumentation amplifier are typically in the microvolt and low millivolt range.

A typical application for an instrumentation amplifier is illustrated in Figure 16.33. In an *electroencephalopgraph*, or *EEG*, brain waves are measured and recorded for observation. The sensors shown in the illustration detect brain waves that are in the microvolt range and transmit those signals through the wires to the EEG unit.

Since the outputs from the sensors are in the microvolt range, they must be applied to a high-gain amplifier for the EEG unit to respond to them. The amplifier used must also have a high *common-mode rejection ratio* (CMRR) since the wires between the sensors and the unit are highly susceptible to noise. This noise, which is picked up by all of the wires, can have an amplitude greater than the signal level. Therefore, it must be eliminated if the sensor inputs are to be detected.

The gain and CMRR requirements of the amplifier described are easily provided by an instrumentation amplifier like the one shown in Figure 16.34. Circuits A and B are

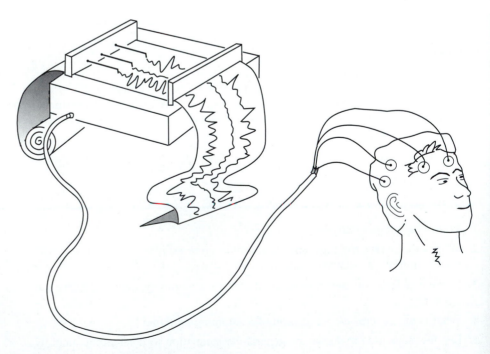

**FIGURE 16.33  Electroencephalograph (EEG).**

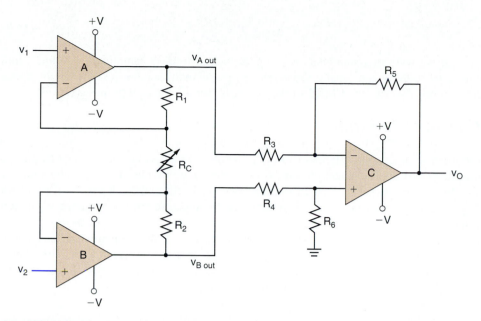

**FIGURE 16.34    Instrumentation amplifier.**

*noninverting amplifiers* and circuit C acts as a *differential amplifier*. We'll start our analysis of the circuit operation by defining the output from circuit A.

Since A is a noninverting amplifier, $v_1$ is amplified by a factor of $(1 + R_1/R_C)$. The signal at the inverting input to A is provided by the output of B. The amplitude of this signal at the inverting input can be shown to equal

$$v_2 \frac{R_2}{R_C}$$

Since the output of A is equal to *the difference between its input voltages*, $v_{A(\text{out})}$ can be found as

$$v_{A(\text{out})} = \left(1 + \frac{R_1}{R_C}\right) v_1 - \frac{R_2}{R_C} v_2$$

Circuit B is identical to circuit A. Therefore, the output from op-amp B can be found as

$$v_{B(\text{out})} = \left(1 + \frac{R_2}{R_C}\right) v_2 - \frac{R_1}{R_C} v_1$$

Assuming that $R_3 = R_4$, the differential input to C is equal to the difference between the above outputs. This differential input is found as

$$v_{B(\text{out})} - v_{A(\text{out})} = \left[\left(1 + \frac{R}{R_C}\right) v_2 - \frac{R}{R_C} v_1\right] - \left[\left(1 + \frac{R}{R_C}\right) v_1 - \frac{R}{R_C} v_2\right]$$

Note that $R_1 = R_2$ (by design), so we don't distinguish between the two in the above equation. Simplifying this equation, we get

$$v_{B(\text{out})} - v_{A(\text{out})} = \left(1 + \frac{2R}{R_C}\right)(v_2 - v_1)$$

The differential amplifier is designed for unity gain. Therefore, the final circuit output is found as

$$v_{(\text{out})} = \left(1 + \frac{2R}{R_C}\right)(v_2 - v_1) \tag{16.6}$$

Since the output from an op-amp circuit is equal to the product of its voltage gain and differential input, we can define the closed-loop voltage gain of the circuit as

$$A_{CL} = 1 + \frac{2R}{R_C} \qquad \textbf{(16.7)}$$

If we were to assume that $v_2 = v_1$, the output from the amplifier [as found using equation (16.6)] would equal 0 V. Therefore, the common-mode gain of the circuit equals *zero*. In other words, the CMRR of the circuit is (for all practical purposes) *infinite*.

## Circuit Calibration

For the instrumentation amplifier to work properly, the gain values of the input amplifiers must be identical. The potentiometer ($R_C$) is included to provide a means of adjusting for any difference between the gain of A and the gain of B. Since the value of $R_C$ affects both $v_{A(out)}$ and $v_{B(out)}$, we can adjust it to compensate for any difference between the input gain values.

The simplest way to set $R_C$ to its proper value is to apply a common-mode signal to the amplifier inputs. Any difference in the input gain values will result in the circuit producing an output voltage. So with the common-mode signals applied, $R_C$ is varied until the output drops to 0 V. When this occurs, the circuit is calibrated properly.

---

**Section Review**

1. What purpose is served by an instrumentation amplifier?
2. Why is it important for an instrumentation amplifier to have extremely high gain and CMRR values?
3. Describe the calibration process for a basic instrumentation amplifier.

---

# 16.5 OTHER OP-AMP CIRCUITS

In this section we will take a *very brief* look at some other op-amp circuits. These circuits are presented merely to give you an idea of the wide variety of applications for the op-amp. We will not go into the details of circuit analysis or troubleshooting for the circuits in this section, it is strictly an applications section.

## Audio Amplifier

**Audio amplifier**
The final audio stage in communications receivers; used to drive the speakers.

The final output stage of most communications receivers is the **audio amplifier**. The audio amplifier is the circuit that drives the system speakers. The ideal audio amplifier has the following characteristics:

1. High gain
2. Minimum distortion in the audio-frequency range (approximately 20 Hz to 20 kHz)
3. High input impedance
4. Extremely low output impedance (to provide maximum coupling to the speakers)

In a *low-power audio system*, an op-amp audio amplifier fulfills the requirements listed very nicely. An op-amp audio amplifier is shown in Figure 16.35. The first thing you will probably notice about this circuit is the fact that the $-V$ input to the op-amp is connected to *ground*. This means that the op-amp output always has some *positive* value. Note that the coupling capacitor between the output of the op-amp and the speaker is used to reference the speaker voltage around 0 V; that is, it removes the positive dc reference from the op-amp output. Also, $C_5$ is included in the $V_{CC}$ line to prevent any transient current caused by the operation of the op-amp from being coupled back to $Q_1$ through the power supply.

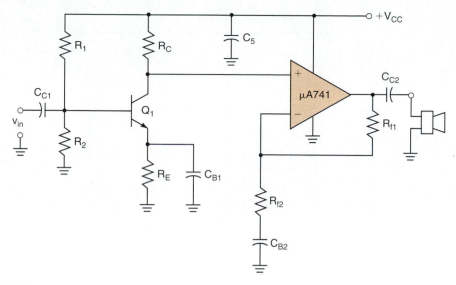

**FIGURE 16.35    Low-power audio amplifier.**

The high-gain requirement is accomplished by the combination of the two amplifier stages. The low $Z_{out}$ of the audio amplifier is accomplished by the op-amp itself, as is the low distortion characteristic.

## High-Impedance Voltmeter

The op-amp can be used in conjunction with a current meter to produce a relatively high $Z_{in}$ voltmeter. Such a circuit is shown in Figure 16.36. In the circuit shown, the closed-loop gain depends on the internal resistance of the meter (represented in the figure as $R_M$). The input voltage is amplified, and the output voltage generates a proportional current through the meter. By adding a small series potentiometer in the feedback loop, the meter can be calibrated to produce a more accurate reading.

## Voltage-Controlled Current Source

As you know, a *current source* is a circuit that produces a constant-value output current. A **voltage-controlled current source** uses an input voltage to set the value of its output current. Such a circuit is shown in Figure 16.37.

**Voltage-controlled current source**
A circuit with a constant-current output controlled by the circuit input voltage.

**FIGURE 16.36    Simple high-impedance voltmeter.**

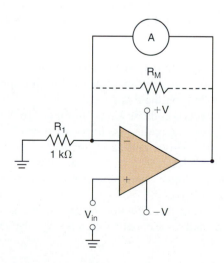

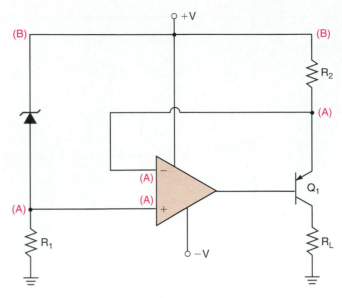

**FIGURE 16.37**   **Voltage-controlled current source.**

To give you an idea of how the circuit works, the two sides of the zender diode have been labeled (A) and (B). Side (B) of the diode is common to the $+V$ side of $R_2$, so this point is also labeled as (B). The lower side of the zener and the input to the op-amp are both labeled (A) since these points are common to each other. The virtual-ground principle allows us to also label the inverting input to the op-amp as (A). Continuing the process, the lower side of $R_2$ is also labeled as (A).

Since the zener diode and $R_2$ have the same (A) and (B) labels, the voltage across the two components are always equal. The zener voltage is constant (and equal to the $V_Z$ rating of the component), and so is the voltage across $R_2$. Therefore, the current through $R_2$ has a fixed value. Because this current is emitter current of the *pnp* transistor, the $Q_1$ collector current is constant, as is the load current. Note that the load current stays constant as long as ($I_C R_L$) does not cause the transistor to saturate. If the transistor goes into saturation, current regulation is lost.

## Precision Diode

**Precision diode**
A clipper that consists of a diode and an op-amp. The circuit is characterized by the ability to clip extremely low-level input signals.

The **precision diode** is an op-amp/diode circuit characterized by its ability to conduct at extremely low diode forward voltages. This relatively simple circuit can be used in any clipper application. For example, consider the positive clipper shown in Figure 16.38a. You may recall that this circuit is used to eliminate the positive portion of its input signal. There is just one problem with the circuit. As shown, the input signal is clipped at approximately 700 mV. What if you want to clip a 100 mV$_{pk}$ input signal at 0 V? The standard diode clipper cannot be used in this application because of the forward voltage required to turn on the diode.

The precision diode circuit in Figure 16.38b can be used to clip a low-level signal. When $V_{in}$ to the circuit is more positive than 0 V by even a few millivolts, the op-amp output goes positive, cutting $D_1$ off. When the input goes negative, so does the output, which turns $D_1$ on. Since there is no voltage divider in the feedback path, the circuit acts as a voltage follower, and the output is identical to the input. The 0.7 V drop across $D_1$ is offset by the op-amp.

Reversing the direction of $D_1$ produces a *negative* clipper. This circuit acts as described above. The only difference is that the *negative* alternation of the input is clipped.

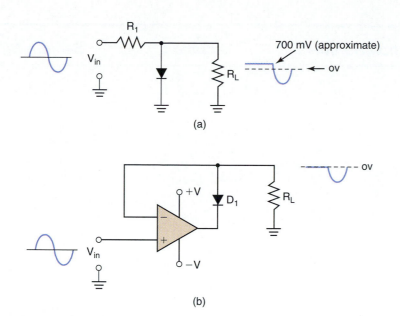

**FIGURE 16.38** Precision diode circuit.

## One Final Note

Obviously, there is a nearly endless list of applications for the op-amp. We could not hope to cover them all in this chapter. In the upcoming chapters, you will see quite a few more op-amp circuits, such as *active filters*, *waveform generators*, and so on.

1. What is an *audio amplifier*?
2. What characteristics of the op-amp make it ideal for audio-amplifier applications?
3. What characteristics of the op-amp make it ideal for use as the input circuit for a voltmeter?
4. Describe the operation of the *voltage-controlled current source*.
5. Describe the operation of the *precision diode*.

The following terms were introduced and defined in this chapter:

**Key Terms**

audio amplifier
averaging amplifier
calibration resistor
comparator
differentiator
digital circuits
digital-to-analog (D/A)
   converter

general-class equation
instrumentation amplifier
integrator
level detector
precision diode
ramp

subtractor
summing amplifier
variable comparator
voltage-controlled current
   source

| Equation Summary | Equation Number | Equation | Section Number |
|---|---|---|---|
| | (16.1) | $V_{ref} = V \dfrac{R_2}{R_1 + R_2}$ | 16.1 |
| | (16.2) | $f_1 = \dfrac{1}{2\pi R_f C_f}$ | 16.2 |
| | (16.3) | $f_{min} = \dfrac{10}{2\pi R_f C_f}$ | 16.2 |
| | (16.4) | $f_{max} = \dfrac{1}{20\pi RC}$ | 16.2 |
| | (16.5) | $V_{out} = -R_f \left( \dfrac{V_1}{R_1} + \dfrac{V_2}{R_2} + \dfrac{V_3}{R_3} \right)$ | 16.3 |
| | (16.6) | $v_{out} = \left( 1 + \dfrac{2R}{R_C} \right)(v_2 - v_1)$ | 16.4 |

## Answers to the Example Practice Problems

**16.1.** $V_{out} = \pm 11$ V
**16.2.** $+3$ V
**16.3.** $R_1$ is shorted or $R_2$ is open.
**16.4.** $f_1 = 31.21$ Hz, $f_{min} = 312.1$ Hz
**16.5.** $-6$ V
**16.6.** $-3$ V
**16.7.** $-V_{out} = 10V_1 + 5V_2 + V_3$, $V_{out} = -16$ V

## Practice Problems

### §16.1

1. Determine the value of $V_{ref}$ for the comparator in Figure 16.39.
2. Determine the value of $V_{ref}$ for the comparator in Figure 16.40.
3. Determine the value of $V_{ref}$ for the comparator in Figure 16.41.
4. Determine the value of $V_{ref}$ for the comparator in Figure 16.42.

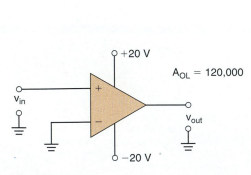

**FIGURE 16.39**

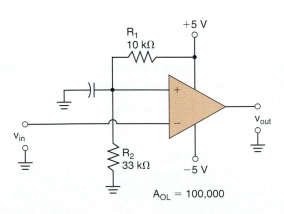

**FIGURE 16.40**

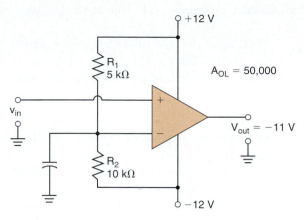

FIGURE 16.41

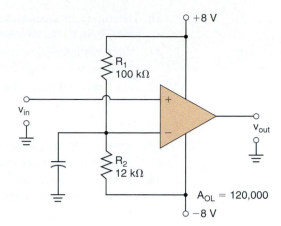

FIGURE 16.42

### §16.2

5. Determine the cutoff frequency for the circuit in Figure 16.43.
6. Determine the frequency at which the circuit in Figure 16.43 will start to lose its linear output characteristics.
7. Determine the cutoff frequency of the circuit in Figure 16.44.
8. Determine the frequency at which the circuit in Figure 16.44 will start to lose its linear output characteristics.
9. Determine the cutoff frequency of the circuit in Figure 16.45.

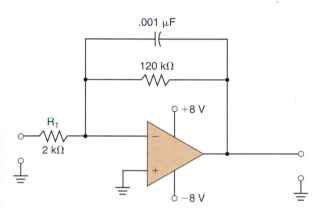

FIGURE 16.43

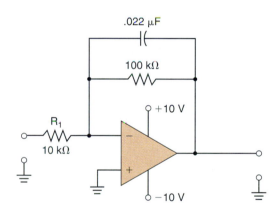

FIGURE 16.44

FIGURE 16.45

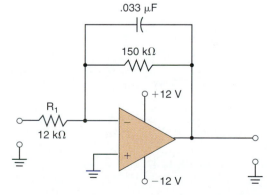

10. Determine the frequency at which the circuit in Figure 16.45 will start to lose its linear output characteristics.

11. A differentiator has values of $R_1 = 10$ kΩ and $C_1 = 0.01$ μF. Determine its maximum linear operating frequency and its cutoff frequency.

12. A differentiator has values of $R_1 = 2.2$ kΩ and $C_1 = 0.15$ μF. Determine its maximum linear operating frequency and its cutoff frequency.

### §16.3

13. Derive the general-class equation for the summing amplifier in Figure 16.46a. Then use the equation to determine the output voltage for the circuit.

14. Derive the general-class equation for the summing amplifier in Figure 16.46b. Then use the equation to determine the output voltage for the circuit.

15. Derive the general-class equation for the summing amplifier in Figure 16.47. Then use the equation to determine the output voltage for the circuit.

16. The feedback resistor in Figure 16.47 is changed to 24 kΩ. Derive the new general-class equation for the circuit and determine its new output voltage.

17. A five-input summing amplifier has values of $R_1 = R_2 = R_3 = R_4 = R_5 = 15$ kΩ. What value of feedback resistor is required to produce an averaging amplifier?

18. A four-input summing amplifier has values of $R_1 = R_2 = R_3 = R_4 = 20$ kΩ. What value of feedback resistor is required to produce an averaging amplifier?

19. Refer back to Figure 16.31. Assume that $V_1 = 4$ V and $V_2 = 6$ V. Determine the output voltage from the circuit.

20. Refer back to Figure 16.31. Assume that $R_{f2}$ in the circuit has been changed to 30 kΩ. Also, assume that the input voltages are $V_1 = 1$ V and $V_2 = 3$ V. Determine the output voltage from the circuit.

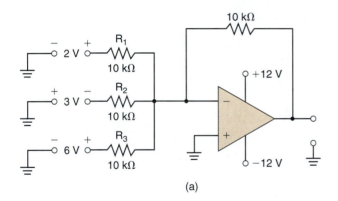

(a)

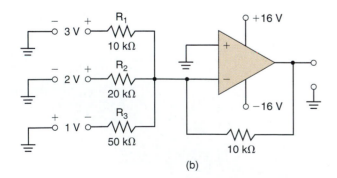

(b)

**FIGURE 16.46**

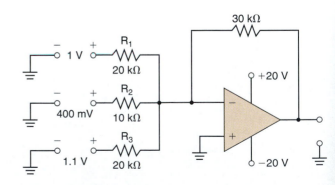

**FIGURE 16.47**

**21.** The circuit in Figure 16.48 has the readings indicated. Determine the possible cause(s) of the problem.

**22.** The circuit in Figure 16.49 has the readings indicated. Determine the possible cause(s) of the problem.

**23.** The circuit in Figure 16.50 has the readings indicated. Determine the possible cause(s) of the problem.

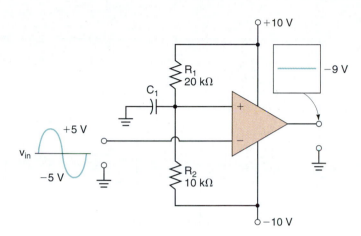

**FIGURE 16.48**

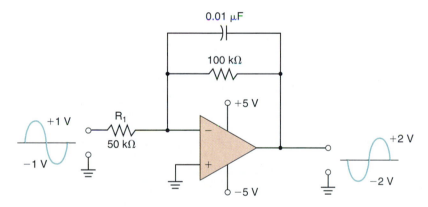

**FIGURE 16.49**

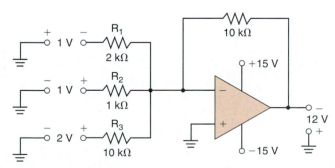

**FIGURE 16.50**

**24.** Calculate the voltage gain of the audio amplifier in Figure 16.51.

**25.** The circuit in Figure 16.52 has the component values shown. Determine the value of $V_{CE}$ for $R_L = 1 \text{ k}\Omega$ and $R_L = 3 \text{ k}\Omega$.

**26.** Design a circuit to solve the following equation:

$$-V_{out} = \frac{V_1 - V_2}{2} + \frac{V_3}{3}$$

The circuit is to contain only two op-amps.

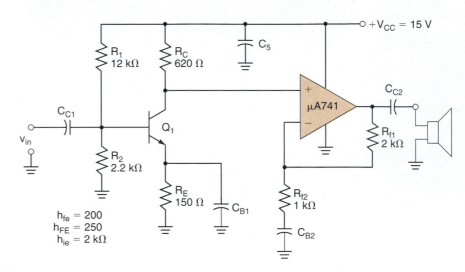

**FIGURE 16.51**

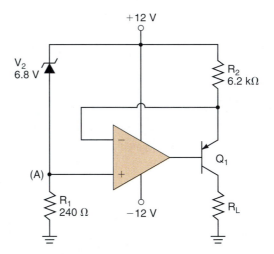

**FIGURE 16.52**

# 17

# TUNED AMPLIFIERS

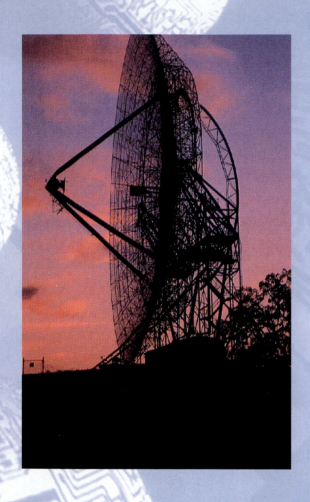

The heart of any communications system is the tuned amplifier. Communications receivers, like the one shown, use tuned circuits to select one out of the many hundreds of signals they receive constantly.

# OUTLINE

# OBJECTIVES

*After studying the material in this chapter, you should be able to:*

1. Compare and contrast the frequency-response characteristics of the *ideal* and *practical* tuned amplifiers.

2. Discuss the *quality (Q) factor* of a tuned amplifier, the factors that affect its value, and its relationship to amplifier bandwidth.

3. Describe the frequency response curves of the *low-pass*, *high-pass*, *band-pass*, and *band-stop (notch)* filters.

4. Perform the gain and frequency analysis of *low-pass* and *high-pass* active filters.

5. Perform the gain and frequency analysis of the *two-stage band-pass*, *multiple-feedback band-pass*, and *notch* filters.

6. Describe active filter fault symptoms and troubleshooting.

7. Describe the frequency-response characteristics of discrete tuned amplifiers.

8. Perform the complete frequency analysis of a discrete tuned amplifier.

9. Discuss the tuning and troubleshooting of common discrete tuned amplifiers.

10. Describe the operation of the basic *class C* amplifier.

**Tuned amplifier**
An amplifier designed for a specific bandwidth.

The heart of any communications system is the **tuned amplifier**. A tuned amplifier is designed for a specific bandwidth. For example, a given tuned amplifier may be designed to amplify only those frequencies that are within $\pm 20$ kHz of 1000 kHz; that is, between 980 and 1020 kHz. As long as the input signal is within these frequencies, it will be amplified. If it goes outside this frequency range, amplification will be drastically reduced.

*A Practical Consideration:* For standard amplifiers, bandwidth is viewed as a limitation. For tuned amplifiers, bandwidth is a desired characteristic that is achieved through circuit design.

In Chapter 14, you were introduced to the concept of bandwidth. You were also shown how amplifier resistance and capacitance values limit the frequency response of a given amplifier. You may be wondering, then, what distinguishes a *tuned* amplifier from any other type of amplifier. After all, don't *all* amplifiers provide gain over a limited band of frequencies? The difference is that tuned amplifiers are designed to have specific, usually narrow, bandwidths. This point is illustrated in Figure 17.1.

In this chapter, we will look at how discrete and op-amp circuits are tuned. We will also take a brief look at some of the typical applications for tuned amplifiers.

## 17.1 TUNED AMPLIFIER CHARACTERISTICS

OBJECTIVE 1 ▶

Before we get into any specific circuits, let's take a moment to discuss some of the bandwidth characteristics of tuned amplifiers. The *ideal* characteristics of such an amplifier are illustrated in Figure 17.2.

Characteristics of the *ideal* tuned amplifier.

The ideal tuned amplifier would have zero ($-\infty$ dB) gain for all frequencies from 0 Hz up to the lower cutoff frequency, $f_1$. At that point, the gain would *instantly* jump to $A_{v(mid)}$. The gain would stay at $A_{v(mid)}$ until $f_2$ was reached. At that time, the gain would *instantly* drop back to zero. Thus, all the frequencies within the bandwidth of the amplifier would be *passed* by the circuit, while all others would be effectively *stopped*. This is where the terms **pass band** and **stop band** come from.

**Pass band**
The range of frequencies passed (amplified) by a tuned amplifier.

**Stop band**
The range of frequencies outside an amplifier's pass band.

In practice, the ideal characteristics of the tuned amplifier cannot be achieved. Figure 17.3 compares the characteristics of the ideal tuned amplifier to those of a more practical circuit. Note that the gain roll-off of the practical circuit curve is not instantaneous. Rather, the gain rolls off from $A_{v(mid)}$ gradually as frequency passes outside the pass band. As you will now be shown, the roll-off rate is an important factor in the overall operation of a tuned amplifier.

**FIGURE 17.1  Tuned amplifier frequency response.**

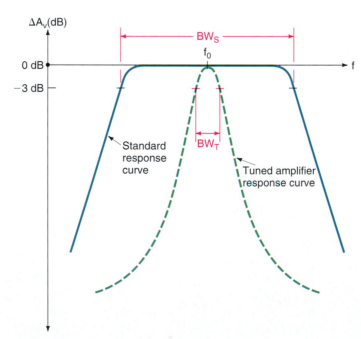

FIGURE 17.2   The ideal
characteristics of a tuned
amplifier.

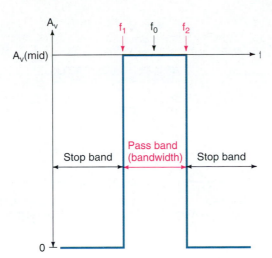

FIGURE 17.3   Ideal versus
practical pass band
characteristics.

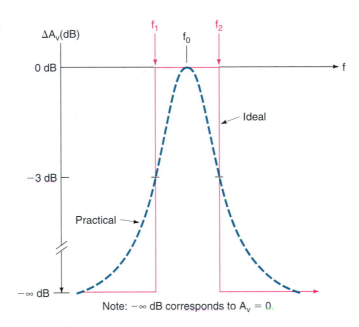

Note: $-\infty$ dB corresponds to $A_v = 0$.

## Roll-Off Rate Versus Bandwidth

We need to start this discussion by establishing two "bottom line" principles:

1. The closer the frequency response curve of a given tuned amplifier to that of the ideal, the better.

2. *In some applications,* the narrower the bandwidth of a tuned amplifier, the better.

The first statement almost goes without saying. As with any circuit, the closer we come to the ideal characteristics of that circuit, the better off we are. This has been the case with every circuit and component we have discussed in this text. The second statement warrants discussion. In Chapter 14, we looked at the frequency response of an amplifier as being a *limitation*. In other words, we considered an *infinite* bandwidth to be the ideal. This is not the case with tuned amplifiers. *With tuned amplifiers, we want a specific, usually narrow, bandwidth.* For example, take a look at Figure 17.4. Here we see a *pass band* (shaded) with two *stop bands*, A and B. If a tuned amplifier were designed to have a center frequency of 1000 kHz and a bandwidth of 40 kHz, that amplifier would be able to pass all the frequencies within the pass band, while rejecting all those in either of the stop

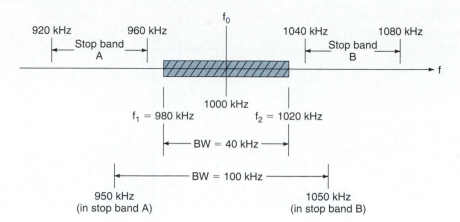

*An Example:* The frequency ranges shown in Figure 17.4 could represent the frequency ranges of three AM radio stations. If the tuned amplifier has a bandwidth of 40 kHz, it will pick up only the radio station to which it is tuned (the center station). If the tuned amplifier has a bandwidth of 100 kHz, it will not only pick up the center station, it will pick up portions of the other two station signals as well.

**FIGURE 17.4**

bands. However, if we used a tuned amplifier with the same center frequency and a bandwidth of 100 kHz, that amplifier would pass a portion of the frequencies in both stop bands. This is not an acceptable situation. A tuned amplifier must pass all the frequencies within the pass band, while stopping all others. Therefore, we want a given tuned amplifier to have a specific, usually narrow, bandwidth. With this in mind, let's take a look at the relationship between roll-off rate and bandwidth.

The relationship between roll-off rates and bandwidth.

If we *decrease* the roll-off rate of a given amplifier, we *increase* its bandwidth. This point is illustrated in Figure 17.5. Here we see the frequency-response curves of four different circuits, (a), (b), (c), and (d). All the circuits are shown to have the same center frequency, $f_0$. Also, the roll-off rates of the circuits become longer as we go from circuit (a) to (b), from (b) to (c), and from (c) to (d). As you can see, the lower the roll-off rate, the wider the bandwidth of the circuit.

OBJECTIVE 2 ▶

Now, apply the principle illustrated in Figure 17.5 to the situation shown in Figure 17.4. It is easy to see that we must limit the roll-off rate of a tuned amplifier if it is to be used for a specific application. The roll-off rate (and, therefore, the bandwidth) of an amplifier is controlled by what is called the $Q$ of the circuit.

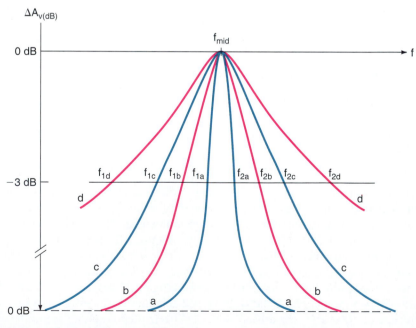

**FIGURE 17.5** **Bandwidth versus roll-off rate.**

The **quality** (**Q**) of a tuned amplifier is a *figure of merit* that is equal to the ratio of center frequency ($f_0$) to bandwidth. By formula,

**Quality (Q)**
A figure of merit for a tuned amplifier that is equal to the ratio of center frequency to bandwidth.

$$Q = \frac{f_0}{BW}$$

$$(17.1)$$

---

**EXAMPLE 17.1**

What is the required $Q$ of a tuned amplifier used for the application shown in Figure 17.4?

*Solution:* The pass band in Figure 17.4 has a center frequency of 1000 kHz and a bandwidth of 40 kHz. Using these two values, the required $Q$ is found as

$$Q = \frac{f_0}{BW}$$

$$= \frac{1000 \text{ kHz}}{40 \text{ kHz}}$$

$$= 25$$

Thus, any tuned amplifier used in this application would require a $Q$ of approximately 25.

**PRACTICE PROBLEM 17.1**

An amplifier with a center frequency of 1200 kHz has a bandwidth of 20 kHz. Determine the $Q$ of the amplifier.

---

It should be noted that the value of $Q$ for a tuned circuit determines its bandwidth. The $Q$ of a given amplifier depends on component values that are within the amplifier. Once the $Q$ is determined, the bandwidth is found as

The $Q$ of an amplifier is determined by circuit component values.

$$BW = \frac{f_0}{Q}$$

$$(17.2)$$

The following example illustrates this point.

---

**EXAMPLE 17.2**

Using the proper circuit calculations, the $Q$ of a given amplifier is found to be 60. If the center frequency for the amplifier is 1860 kHz, what is the bandwidth of the amplifier?

*Solution:* With a center frequency of 1860 kHz and a $Q$ of 60, the amplifier bandwidth is found as

$$BW = \frac{f_0}{Q}$$

$$= \frac{1860 \text{ kHz}}{60}$$

$$= 31 \text{ kHz}$$

**PRACTICE PROBLEM 17.2**

The $Q$ of an amplifier is determined to be 25. If the center frequency of the circuit is 1400 kHz, what is its bandwidth?

With a center frequency of 1860 kHz and a bandwidth of 31 kHz, the cutoff frequencies would be found as $1860 \pm 15.5$ kHz; that is, 1844.5 and 1875.5 kHz.

As we discuss the various tuned circuits, you will be shown how to determine the value of $Q$ for each. Once the $Q$ of an amplifier is known, determining the bandwidth of the amplifier is simple. For now, just remember that the $Q$ of an amplifier is a figure of merit that depends on circuit component values. The bandwidth of the amplifier depends on the value of $Q$.

## Center Frequency

As you may recall, center frequency is found as the geometric average of $f_1$ and $f_2$. By formula,

$$f_0 = \sqrt{f_1 f_2} \tag{17.3}$$

When the $Q$ of an amplifier is greater than (or equal to) 2, the value of $f_0$ approaches the *algebraic average* of $f_1$ and $f_2$ (designated $f_{ave}$); that is, $f_0$ approaches a value that is exactly halfway between $f_1$ and $f_2$. By formula,

$$f_0 \cong f_{ave} \quad \text{(when } Q \geqslant 2\text{)}$$

where $f_{ave}$ is the algebraic average of $f_1$ and $f_2$, found as

$$f_{ave} = \frac{f_1 + f_2}{2} \tag{17.4}$$

The following example illustrates the relationship between $f_0$ and $f_{ave}$ when $Q \geqslant 2$.

---

**EXAMPLE 17.3**

*A Practical Consideration:* The cutoff frequencies of a tuned amplifier are measured using the technique shown in Chapter 14.

The cutoff frequencies for a tuned amplifier are measured as $f_1 = 1160$ kHz and $f_2 = 1240$ kHz. Using equations (17.3) and (17.4), determine the values of $f_0$ and $f_{ave}$ for the circuit. Then verify that the value of $Q$ for the circuit is greater than 2.

**Solution:** Using equation (17.3), the value of $f_0$ is found as

$$f_0 = \sqrt{f_1 f_2}$$
$$= \sqrt{(1160 \text{ kHz})(1240 \text{ kHz})}$$
$$= 1199 \text{ kHz}$$

Using equation (17.4), $f_{ave}$ is found as

$$f_{ave} = \frac{f_1 + f_2}{2}$$
$$= \frac{1160 \text{ kHz} + 1240 \text{ kHz}}{2}$$
$$= 1200 \text{ kHz}$$

The value of $Q$ for the circuit is now found as

$$Q = \frac{f_0}{\text{BW}}$$
$$= \frac{1199 \text{ kHz}}{80 \text{ kHz}}$$
$$= 14.99$$

As you can see, the value of $f_0$ approaches the value of $f_{ave}$ when the $Q$ of an amplifier is greater than (or equal to) 2.

**PRACTICE PROBLEM 17.3**

A tuned amplifier has measured cutoff frequencies of 980 kHz and 1080 kHz. Show that $f_0 \cong f_{mid}$ for the circuit and that the circuit value of $Q$ is greater than 2.

You may be wondering how you can tell when the value of $Q$ for an amplifier is greater than 2. You can assume that the value of $Q$ is greater than (or equal to) 2 *when the value of $f_0$ for the circuit is at least two times its bandwidth.* As long as this condition is met, the $Q$ of the amplifier will be greater than (or equal to) 2, and $f_0$ will be approximately equal to $f_{ave}$. As you will learn, this fact simplifies the frequency analysis of many tuned amplifiers.

## Integrated Versus Discrete Tuned Circuits

Discrete circuits, such as BJT and FET amplifiers, are normally tuned using inductive–capacitive (*LC*) circuits. An example of a discrete tuned amplifier is shown in Figure 17.6a. The frequency response of the circuit is determined primarily by the values of $C_T$ and $L_T$.

An op-amp (IC) equivalent of the discrete tuned amplifier is shown in Figure 17.6b. As you can see, the op-amp tuned circuit does not require the use of inductors. The frequency response of the circuit is determined by its resistor and capacitor values.

In the upcoming sections, we will cover the operation of various op-amp tuned circuits. Then we will cover the operation of the basic discrete tuned amplifier. As you will see, the exact equations used to analyze the operation of these amplifiers will vary from one circuit to another. However, the relationships among bandwidth, $Q$, and $f_0$ discussed in this section will not vary significantly among the various circuits.

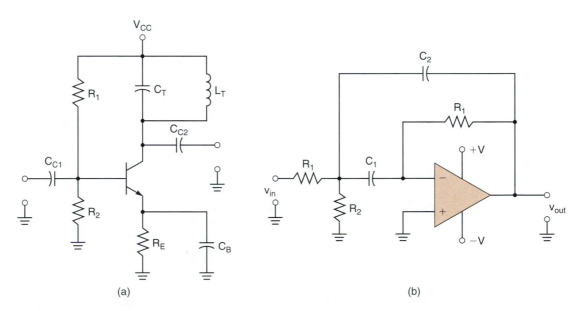

FIGURE 17.6   **A discrete tuned amplifier and its op-amp equivalent.**

1.  What are the characteristics of the *ideal* tuned amplifier?
2.  How does the frequency response of the practical tuned amplifier vary from that of the ideal tuned amplifier?
3.  What is the relationship between the roll-off rate and the bandwidth of a tuned amplifier?
4.  What is the *quality* ($Q$) rating of a tuned amplifier?
5.  What is the relationship between the values of $Q$ and BW for a tuned amplifier?
6.  What determines the value of $Q$ for a tuned amplifier?
7.  What is the relationship between the values of $Q$ and $f_0$ for a tuned amplifier?

# 17.2 ACTIVE FILTERS: AN OVERVIEW

**Active filter**
A tuned op-amp circuit.

Tuned op-amp circuits are generally referred to as **active filters.** There are four basic types of active filters, each with its own circuit configuration and frequency response curve.

## Overview

◄ OBJECTIVE 3 ►

**Low-pass filter**
A filter designed to pass all frequencies *below* a given cutoff frequency.

**High-pass filter**
A filter designed to pass all frequencies *above* a given cutoff frequency.

**Band-pass filter**
One designed to pass all frequencies that fall between $f_1$ and $f_2$.

**Band-stop (notch) filter**
One designed to block all frequencies that fall between $f_1$ and $f_2$.

The frequency-response curves for the four types of active filters are shown in Figure 17.7. The **low-pass filter,** whose frequency-response curve is shown in Figure 17.7a, is designed to pass all frequencies from dc up to an upper cutoff frequency, $f_2$. The **high-pass filter,** represented in Figure 17.7b, is designed to pass all frequencies that are above a lower cutoff frequency, $f_1$. All frequencies normally applied to this circuit that are above $f_1$ are passed by the circuit. Of course, the op-amp does have a unity-gain frequency. Therefore, it is impossible to design a high-pass filter that does not have a value of $f_2$. However, the value of $f_{unity}$ for the op-amp will be well above the range of frequencies that the circuit is being used to pass.

The **band-pass filter,** represented by the curve in Figure 17.7c, is designed to pass only the frequencies that fall between its values of $f_1$ and $f_2$. The **band-stop filter,** or **notch filter,** is used to eliminate all signals within the stop band (between $f_1$ and $f_2$) while passing all frequencies outside this band. The frequency-response curve for the notch filter is shown in Figure 17.7d. Note that the function of the notch filter is the opposite of that of the band-pass filter.

There is one critical point that needs to be made at this time. The concepts of $Q$, center frequency, and bandwidth are related primarily to the band-pass and notch filters. When we are dealing with band-pass and notch filters, we are generally concerned with the bandwidths of such amplifiers, along with their respective values of $Q$ and $f_0$. However, when we are dealing with low-pass and high-pass filters, we are concerned only with one value, $f_2$ or $f_1$, respectively. The concepts of bandwidth and center frequency generally are not applied to these circuits.

## General Terminology

**Pole**
A single *RC* circuit.

Before we get into any circuits, we need to establish some of the terms commonly used to describe active filters. The first of these is the term **pole.** A *pole* is nothing more than an *RC circuit*. Thus, a one-pole filter is one that contains a single *RC* circuit. A two-pole filter contains two *RC* circuits, and so on. The most commonly used type of active filter,

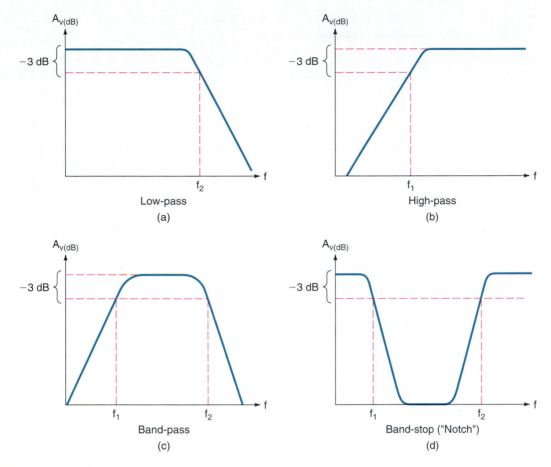

**FIGURE 17.7** Active filter frequency response curves.

the *Butterworth filter,* has a 20-dB/decade roll-off for each pole contained in the circuit. This is shown in the following table:

| Filter Type | Number of RC Circuits | Roll-Off Rate (dB/Decade) |
| --- | --- | --- |
| One pole | 1 | 20 |
| Two pole | 2 | 40 |
| Three pole | 3 | 60 |

*Remember:* Any amplifier with more than one *RC* circuit will have roll-off rates that are additive. This point was discussed in detail in Chapter 14.

The **Butterworth filter** is one that has relatively constant gain across the pass band of the circuit. The term *flat response* is used to describe this constant-gain characteristic. When we talk of a flat response, we mean that the value of $A_{v(dB)}$ is relatively constant across the pass band of the circuit. Since Butterworth filters have the best flat-response characteristics, they are often referred to as *maximally flat,* or *flat-flat,* filters.

Another type of active filter, the **Chebyshev filter,** has a higher roll-off rate than the Butterworth filter containing the same number of poles. A Chebyshev filter has a roll-off rate of 40 dB/decade per pole. However, the frequency-response curve of the Chebyshev filter is *not* flat across the pass band. The typical frequency response of this filter type is compared to that of the Butterworth filter in Figure 17.8. As you can see, there are variations in the pass band gain of the Chebyshev filter. In contrast, the response of the Butterworth filter is relatively flat as shown (dashed line).

**Butterworth filter**
An active filter characterized by flat pass-band response and 20 dB/decade roll-off rates.

**Chebyshev filter**
An active filter characterized by an irregular pass-band response and 40 dB/decade roll-off rates.

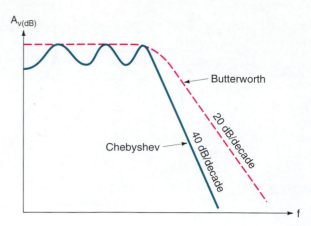

**FIGURE 17.8** Butterworth and Chebyshev response.

Even though the Chebyshev filter has a roll-off rate that is closer to the ideal tuned circuit than the Butterworth filter, it is not used as often. There are two reasons for this:

1. The inconsistent gain of the Chebyshev filter is undesirable.
2. By using a two-pole Butterworth filter, the roll-off rate of the Chebyshev filter can be achieved while still having a flat response.

Since the Butterworth filter is by far the most common active filter, we will concentrate on this circuit type in our discussions on active filters.

## Section Review

1. What is an *active filter*? What are the four types of active filters?
2. Describe the frequency-response characteristics of each type of active filter.
3. What is a *pole*? Why is the number of poles in an active filter important?
4. Contrast the Butterworth response curve with the Chebyshev response curve.

# *17.3* HIGH-PASS AND LOW-PASS FILTERS

## *The Single-Pole Low-Pass Filter*

*OBJECTIVE 4* ▶ The single-pole low-pass Butterworth filter is constructed in one of two ways, as either a *high-gain* circuit or a *voltage follower*. (Recall that a voltage follower has a voltage gain of 0 dB, or unity.) The high-gain circuit is shown in Figure 17.9. As you can see, this cir-

**FIGURE 17.9** Single-pole low-pass active filter.

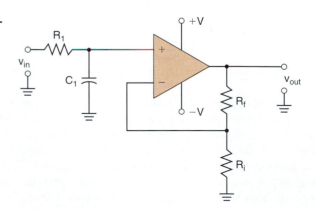

cuit is simply a noninverting amplifier with an added input shunt capacitor. Since the reactance of the capacitor decreases as frequency increases, the high-frequency response is limited. The value of $f_2$ for this type of circuit is found as

$$f_2 = \frac{1}{2\pi RC}$$

(17.5)

This is the same basic relationship that we used to find $f_2$ for the circuits we discussed in Chapter 14. Example 17.4 demonstrates the process for determining the bandwidth of a single-pole low-pass filter.

Determine the bandwidth of the single-pole low-pass filter in Figure 17.10. Also (as a review) determine the value of $A_{CL}$ for the circuit.

*EXAMPLE 17.4*

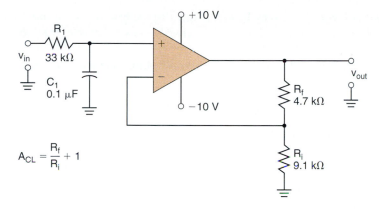

**FIGURE 17.10**

**Solution:** Using the circuit values of $R_1$ and $C_1$, the upper cutoff frequency is found as

$$f_2 = \frac{1}{2\pi R_1 C_1}$$

$$= \frac{1}{2\pi(33 \text{ k}\Omega)(0.1 \text{ }\mu\text{F})}$$

$$= 48.23 \text{ Hz}$$

Since the amplifier is capable of working as a dc amplifier (0 Hz), there is no value of $f_1$ for the circuit. Thus, the bandwidth is equal to $f_2$, or 48.23 Hz.

Because the circuit is a noninverting amplifier, its value of $A_{CL}$ is found using equation (15.16):

$$A_{CL} = \frac{R_f}{R_i} + 1$$

In this case, $R_f = 4.7 \text{ k}\Omega$ and $R_i = 9.1 \text{ k}\Omega$. Therefore, the value of $A_{CL}$ for the circuit is found as

$$A_{CL} = \frac{4.7 \text{ k}\Omega}{9.1 \text{ k}\Omega} + 1$$

$$= 1.52$$

## PRACTICE PROBLEM 17.4

A single-pole low-pass filter like the one in Figure 17.10 has values of $R_1 = 47 \text{ k}\Omega$, $C_1 = 0.033 \text{ }\mu\text{F}$, $R_f = 10 \text{ k}\Omega$, and $R_i = 10 \text{ k}\Omega$. Determine the values of $f_2$ and $A_{CL}$ for the circuit.

*Don't Forget:*
$A_{CL(dB)} = 20 \log A_{CL}$.

The frequency-response curve for the circuit in Example 17.4 is shown in Figure 17.11. As you can see, the midband gain of the amplifier is 1.52, or 3.637 dB. At 48.23 Hz, the gain of the amplifier will be down to 1.08, or 0.637 dB. If the input frequency increases to one octave above the cutoff frequency (96.46 Hz), the gain of the circuit will drop to 0.637 dB − 6 dB = −5.363 dB. This means that the filter can actually *attenuate* the input signal. In other words, at some point, the output voltage will actually be less than the input voltage.

Although the value of $A_{CL}$ in Example 17.4 was relatively low, it can be made much higher by changing the $R_f{:}R_i$ ratio. When unity (0 dB) gain is desired, the unity-gain circuit shown in Figure 17.12 may be used. Note that the bandwidth for this circuit is determined in the same manner as described previously. As you can see, this circuit is nothing more than a *voltage follower*. Thus, the voltage gain of the circuit will be approximately 0 dB. Note that $R_f$ is included for op-amp compensation, and its value does not affect any

**FIGURE 17.11**

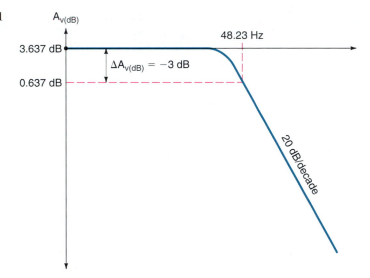

**FIGURE 17.12   Unity-gain circuit.**

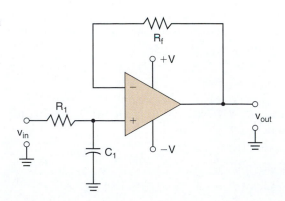

**Lab Reference:** The operation of a unity-gain filter is demonstrated in Exercise 37.

**736**   CHAPTER 17 / TUNED AMPLIFIERS

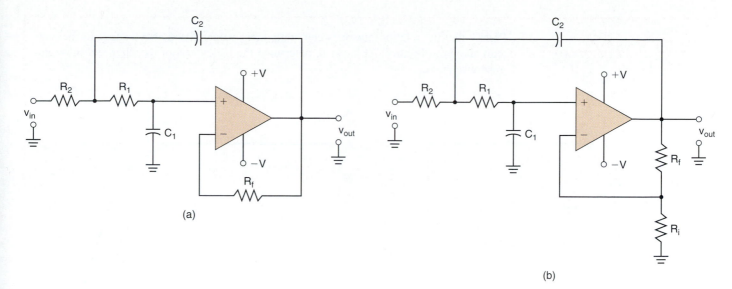

FIGURE 17.13   Common two-pole low-pass filter configurations.

**Lab Reference:** The operation of a two-pole, unity-gain filter is demonstrated in Exercise 37.

circuit calculations. All circuit calculations for this amplifier are the same as described in previous chapters.

## The Two-Pole Low-Pass Filter

As stated earlier, the two-pole low-pass filter has a roll-off rate of 40 dB/decade. Two commonly used two-pole low-pass filter configurations are shown in Figure 17.13. As you can see, each circuit has two $RC$ circuits, $R_1$–$C_1$ and $R_2$–$C_2$. As the operating frequency increases beyond $f_2$, each $RC$ circuit will be dropping $A_{CL}$ by 20 dB, giving a total roll-off rate of 40 dB/decade when operated above $f_2$.

The cutoff frequency for each of the circuits in Figure 17.13 is found as

$$f_2 = \frac{1}{2\pi\sqrt{R_1 R_2 C_1 C_2}}$$

Here's where things get weird. For the two-pole low-pass filter to have a flat-response curve, its value of $A_{CL}$ can be no greater than **1.586 (4 dB)**. This means that you cannot have a high-gain two-pole low-pass filter. The derivation of the value 1.586 involves calculus and is not covered here. In this case, we will simply accept its value as being valid.

Fulfilling the requirements of $A_{CL} < 1.586$ is no problem for the *unity-gain filter* in Figure 17.13a. This circuit, which is actually a voltage-follower, has a standard value of $A_{CL} = 1$. Again, $R_f$ in the circuit is included for offset compensation and its value is unimportant.

The circuit in Figure 17.13b can be designed for any value of $A_{CL}$ between 1 and the upper limit of 1.586. Note that this circuit is normally designed according to the following guidelines:

1. $R_2 = R_1$
2. $C_2 = 2C_1$
3. $R_f \leq 0.586 R_i$

Example 17.5 illustrates these relationships, along with the process for determining the bandwidth of a two-pole low-pass filter.

**EXAMPLE 17.5**

Verify that the relationships between component values listed above apply to the circuit shown in Figure 17.14a. Also, determine the bandwidth of the filter and draw its response curve.

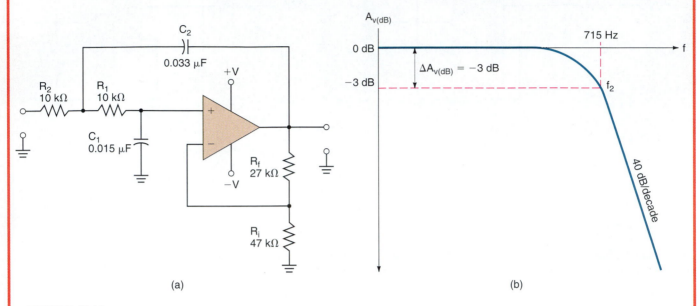

(a)                                          (b)

**FIGURE 17.14**

**Solution:** First, let's verify the relationships listed above. $R_1$ and $R_2$ are equal in value, while $C_2$ is twice the value of $C_1$. Also note that the value of $R_f$ is approximately equal to $0.586R_i$.

The cutoff frequency (and thus the bandwidth) of the filter is found as

$$f_2 = \frac{1}{2\pi\sqrt{R_1 R_2 C_1 C_2}}$$

$$= \frac{1}{2\pi\sqrt{(10\ k\Omega)(10\ k\Omega)(0.015\ \mu F)(0.033\ \mu F)}}$$

$$= 715\ Hz$$

Using the value of $f_2$ for the circuit, the response curve is drawn as shown in Figure 17.14b.

**PRACTICE PROBLEM 17.5**

A filter like the one in Figure 17.14 has the following values: $C_1 = 10$ nF, $C_2 = 22$ nF, and $R_1 = R_2 = 12$ k$\Omega$. Calculate the value of $f_2$ for the circuit.

## The Three-Pole Low-Pass Filter

In Chapter 14, you were shown that the roll-off rates of cascaded stages *add* to form the total roll-off rate for the amplifier. It would follow that a three-pole filter, which has a roll-off rate of 60 dB/decade, could be formed by cascading a single-pole filter with a two-pole filter. This circuit would have the combined roll-off rate of 20 dB + 40 dB = 60 dB/decade. A three-pole low-pass filter is shown in Figure 17.15.

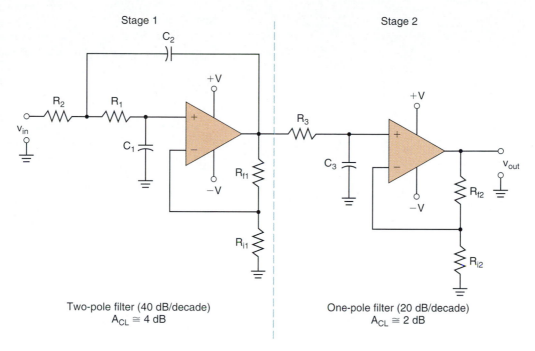

**FIGURE 17.15** **Three-pole low-pass filter.**

As you can see, we again have requirements for the values of $A_{CL}$. The two-pole circuit must have a closed-loop gain of approximately 4 dB, while the single-pole filter must have a closed-loop gain of approximately 2 dB. As long as both of these requirements are fulfilled, the filter will have a Butterworth response curve. Also note that the two filters would be tuned to the same value of $f_2$. All circuit calculations for this amplifier are the same as those performed earlier.

Can active filters have more than three poles? Yes. When you see an active filter with more than three poles, you will have no problem analyzing the circuit if you remember the following points:

*How to analyze a multi-pole active filter.*

1. The filter will have a 20-dB/decade roll-off *for each pole*. For example, a five-pole filter would have a roll-off rate of $5 \times 20 = 100$ dB/decade.

2. All stages will be tuned to the same cutoff frequency. Thus, to find the overall cutoff frequency, simply calculate the value of $f_2$ for any stage.

## High-Pass Filters

Figure 17.16 shows several typical high-pass filters. The high-pass filter differs from the low-pass filter in two respects: First, and most obvious, is the fact that the resistors and capacitors have swapped positions. Second, the multi-pole circuits are designed to fulfill the following conditions:

*The differences between low-pass and high-pass active filters.*

1. $C_1 = C_2$
2. $R_2 = 0.5R_1$

Since the capacitors are in series with the amplifier input, they limit the *low-frequency* operation of the circuit. Note that the value of $f_1$ for each circuit is found using the same equation that we used to find $f_2$ for the equivalent low-pass filter.

As frequency decreases, the reactance of a given series capacitor increases. This causes a larger portion of the input signal to be dropped across the capacitor. When the

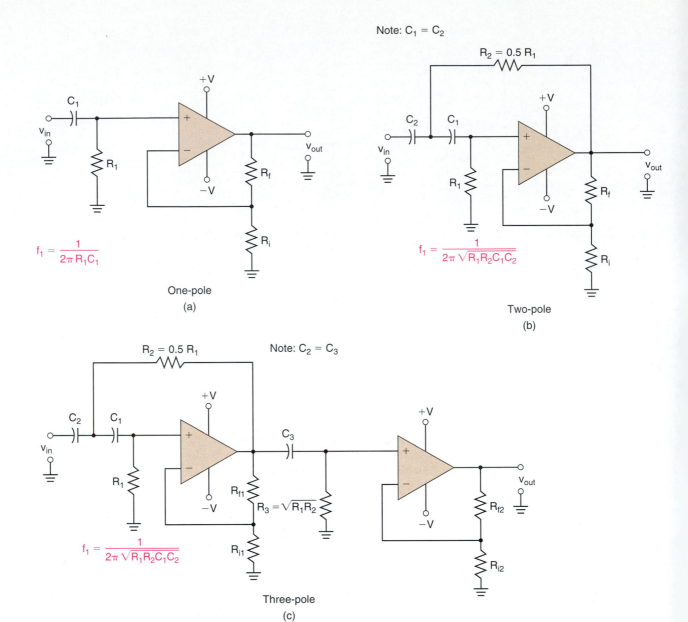

Note: $C_1 = C_2$

$$f_1 = \frac{1}{2\pi R_1 C_1}$$

One-pole

(a)

$$f_1 = \frac{1}{2\pi \sqrt{R_1 R_2 C_1 C_2}}$$

Two-pole

(b)

$R_2 = 0.5 R_1$          Note: $C_2 = C_3$

$$f_1 = \frac{1}{2\pi \sqrt{R_1 R_2 C_1 C_2}}$$

$R_3 = \sqrt{R_1 R_2}$

Three-pole

(c)

**FIGURE 17.16    Typical high-pass active filters.**

**Lab Reference:** The operation of one-pole and two-pole high-pass filters are demonstrated in Exercise 38.

operating frequency has dropped to $f_1$, the series capacitors cause the gain of the amplifier to drop by 3 dB.

We will not spend any time on these circuits since their operating principles should be familiar to you by now. Just remember, these circuits abide by all the rules established up to this point. The only difference is that they will pass frequencies *above* their cutoff frequencies.

## Filter Gain Requirements

In this section, you have been provided with the gain requirements for several low-pass and high-pass active filters. Fulfilling these requirements provides a Butterworth response

**TABLE 17.1** Butterworth Filter Gain Requirements

| Number of Poles | Maximum Overall dc Gain[a] | Approximate Roll-Off Rate (dB/Decade) |
|:---:|:---:|:---:|
| 2 | 1.586  (4 dB) | 40 |
| 3 | 2      (6 dB) | 60 |
| 4 | 2.58   (8 dB) | 80 |
| 5 | 3.29   (10 dB) | 100 |
| 6 | 4.21   (12 dB) | 120 |
| 7 | 5.37   (14 dB) | 140 |

[a]Decibel values are rounded off to nearest whole number.

Note that the gain requirement for low-pass and high-pass active filters increases by 2 dB for each pole added to the circuit. Also note that the approximated roll-off rate is equal to 20 dB/decade per pole.

curve; that is, each filter has relatively constant gain until its cutoff frequency is reached. Then the gain rolls off at an approximate rate of 20 dB/decade per pole.

The gain requirements for a variety of Butterworth active low-pass and high-pass filters are summarized in Table 17.1. The derivations of these gain requirements are way beyond the scope of this text. However, you should be aware that almost every type of low-pass or high-pass active filter has gain requirements that must be fulfilled if the circuit is to have a Butterworth response curve.

Any of the multiple filters described in Table 17.1 could be constructed using the appropriate number of two-pole and one-pole cascaded stages. For example, a five-pole

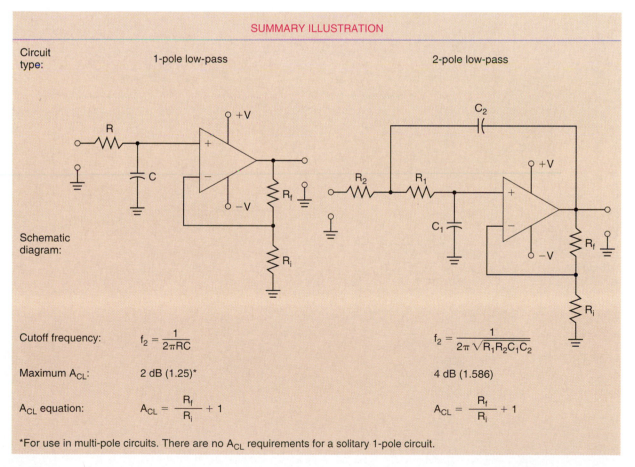

SUMMARY ILLUSTRATION

Circuit type: 1-pole low-pass / 2-pole low-pass

Schematic diagram:

Cutoff frequency:
$$f_2 = \frac{1}{2\pi RC}$$
$$f_2 = \frac{1}{2\pi \sqrt{R_1 R_2 C_1 C_2}}$$

Maximum $A_{CL}$: 2 dB (1.25)* / 4 dB (1.586)

$A_{CL}$ equation:
$$A_{CL} = \frac{R_f}{R_i} + 1$$
$$A_{CL} = \frac{R_f}{R_i} + 1$$

*For use in multi-pole circuits. There are no $A_{CL}$ requirements for a solitary 1-pole circuit.

**FIGURE 17.17**

circuit could be constructed by cascading two two-pole circuits and a one-pole circuit. To obtain the appropriate frequency-response characteristics for a five-pole filter, the following requirements would have to be met by the circuit:

1. Each of the two-pole circuits would require a closed-loop voltage gain of 4 dB and the one-pole circuit would require a closed-loop voltage gain of 2 dB. This would provide an overall voltage gain of 4 dB + 4 dB + 2 dB = 10 dB, the value given in Table 17.1 for a five-pole active filter.

2. The resistor and capacitor values would have to be chosen so that all three stages have identical cutoff frequencies.

As long as these requirements are met, the circuit will operate as a Butterworth filter.

The component requirements for one-pole and two-pole Butterworth low-pass and high-pass active filters are summarized in Figures 17.17 and 17.18. Note that the values of $R$ and $C$ determine the cutoff frequencies, according to the proper form of the following equation:

$$f_c = \frac{1}{2\pi RC}$$

Also note that any of the circuits shown can be converted to unity-gain (0 dB) circuits by removing the lower (grounded) resistor in the feedback path, $R_i$.

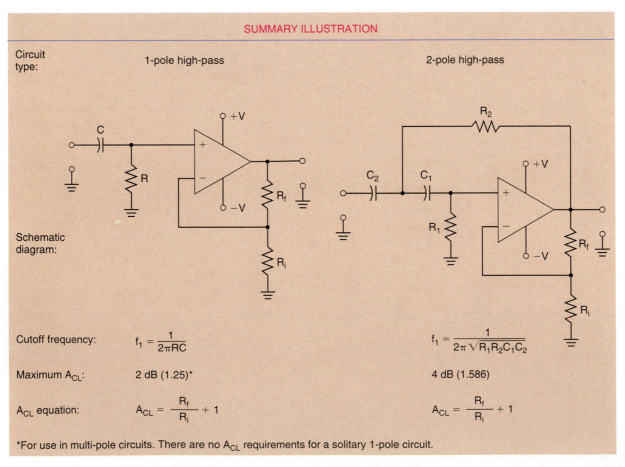

**FIGURE 17.18**

1. What steps are involved in analyzing active one-pole, two-pole, and three-pole low-pass filters?

2. What is the difference between low-pass and high-pass active filters?

3. Describe the dB gain and roll-off characteristics of low-pass and high-pass filters.

## 17.4 BAND-PASS AND NOTCH CIRCUITS

You may recall that *band-pass filters* are designed to *pass* all frequencies within their bandwidths, while *notch filters* (*band-stop filters*) are designed to *block* all frequencies within their bandwidths. In this section, we will take a look at several band-pass and notch filters.

◄ *OBJECTIVE 5*

### The Two-Stage Band-Pass Filter

It is common for a band-pass filter to be constructed by cascading a high-pass filter and a low-pass filter. Such a circuit is shown in Figure 17.19a. The first stage of the amplifier will pass all frequencies that are below its value of $f_2$. All the frequencies passed by the

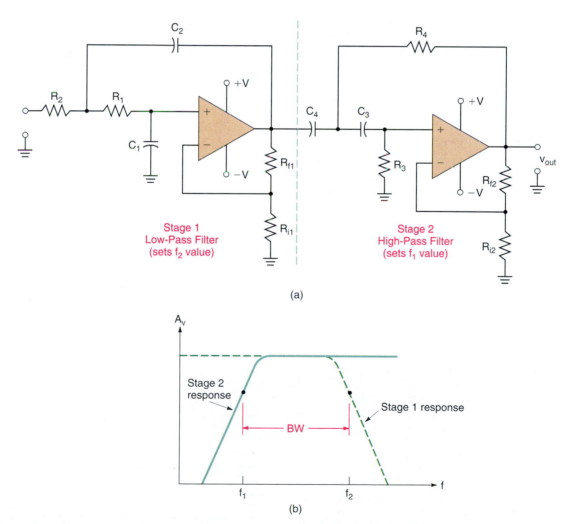

(a)

(b)

**FIGURE 17.19**

*A Practical Consideration:* The order of the low-pass and high-pass filters is unimportant. We could have a high-pass first stage and a low-pass second stage, and the results would be the same.

first stage are coupled to the second stage, which passes all frequencies above its value of $f_1$. The result of this circuit action is shown in Figure 17.19b. Note that the only frequencies that pass through the amplifier are those that fall within the pass band of *both* amplifiers.

The frequency analysis of a circuit like the one in Figure 17.19a is relatively simple. The values of $f_1$ and $f_2$ are found as shown in Section 17.3. Once the values of $f_1$ and $f_2$ are known, the circuit values of bandwidth, center frequency, and $Q$ are found as follows:

$$BW = f_2 - f_1$$
$$f_0 = \sqrt{f_1 f_2}$$
$$Q = \frac{f_0}{BW}$$

The following example demonstrates the complete frequency analysis of the two-stage band-pass filter.

---

**EXAMPLE 17.6**    Perform the complete frequency analysis of the amplifier in Figure 17.20.

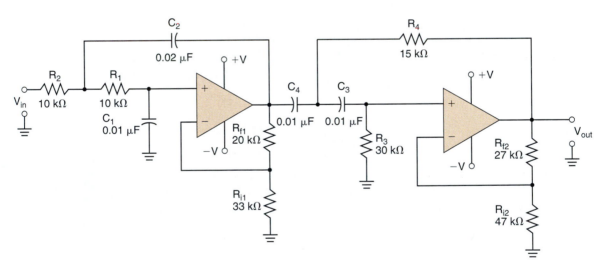

**FIGURE 17.20**

***Solution:***    The value of $f_2$ is determined by the first stage of the circuit (the *low-pass* filter) as follows:

$$f_2 = \frac{1}{2\pi\sqrt{R_1 R_2 C_1 C_2}}$$

$$= \frac{1}{2\pi\sqrt{(10\ k\Omega)(10\ k\Omega)(0.01\ \mu F)(0.02\ \mu F)}}$$

$$= 1.13\ kHz$$

The value of $f_1$ is determined by the second stage of the circuit (the *high-pass* filter) as follows:

---

$$f_1 = \frac{1}{2\pi\sqrt{R_3 R_4 C_3 C_4}}$$

$$= \frac{1}{2\pi\sqrt{(30\ \text{k}\Omega)(15\ \text{k}\Omega)(0.01\ \mu\text{F})(0.01\ \mu\text{F})}}$$

$$= 750\ \text{Hz}$$

Thus, the circuit would pass all frequencies between 750 Hz and 1.13 kHz. All others would effectively be blocked by the band-pass filter.

Now that the values of $f_1$ and $f_2$ are known, we can solve for the amplifier bandwidth as follows:

$$\text{BW} = f_2 - f_1$$
$$= 1.13\ \text{kHz} - 750\ \text{Hz}$$
$$= 380\ \text{Hz}$$

The center frequency is found as

$$f_0 = \sqrt{f_1 f_2}$$
$$= \sqrt{(750\ \text{Hz})(1.13\ \text{kHz})}$$
$$= 921\ \text{Hz}$$

Finally, the value of $Q$ is found as

$$Q = \frac{f_0}{\text{BW}}$$
$$= \frac{921\ \text{Hz}}{380\ \text{Hz}}$$
$$= 2.42$$

**PRACTICE PROBLEM 17.6**

A filter like the one in Figure 17.20 has values of $R_1 = R_2 = 12\ \text{k}\Omega$, $C_1 = C_3 = C_4 = 0.01\ \mu\text{F}$, $C_2 = 0.02\ \mu\text{F}$, $R_3 = 39\ \text{k}\Omega$, and $R_4 = 20\ \text{k}\Omega$. Perform the frequency analysis of the filter.

The two-stage band-pass filter is the easiest of the band-pass filters to analyze. However, it has the disadvantage of requiring two op-amps and a relatively large number of resistors and capacitors. As you will see, the physical construction of the *multiple-feedback band-pass filter* is much simpler than that of the two-stage filter. At the same time, the frequency analysis of the multiple-feedback band-pass filter is a bit more difficult than that of the two-stage filter.

## Multiple-Feedback Band-Pass Filters

The **multiple-feedback band-pass filter,** which is shown in Figure 17.21, derives its name from the fact that it has two feedback networks, one capacitive and one resistive. Note the presence of the input series capacitor ($C_1$) and the input shunt capacitor ($C_2$). The series capacitor affects the low-frequency response of the filter, while the shunt capacitor affects its high-frequency response. This point is illustrated in Figure 17.22.

Let's begin our discussion of the circuit operation by establishing some ground rules. To simplify our discussion, we will assume that:

1. $C_2 < C_1$.
2. A given capacitor will act as an open circuit until a *short-circuit frequency* is reached. At that point, the capacitor can be represented as a short circuit.

**Multiple-feedback band-pass filter**

A band-pass (or band-stop) filter that has a single op-amp and two feedback paths, one resistive and one capacitive.

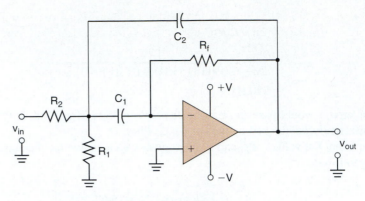

**FIGURE 17.21** A multiple-feedback band-pass filter.

Granted, the operation of the capacitors is more complicated than this, but we want to start by getting the overall picture of how the circuit works.

When the input frequency is below the short-circuit frequency for $C_1$, *both capacitors will act as open circuits*. Since $C_1 > C_2$, we know that the short-circuit frequency for $C_1$ is *lower* than that of $C_2$. Therefore, as long as $C_1$ is an open, $C_2$ is also an open. Having both capacitors acting as opens gives us the equivalent circuit in Figure 17.22a. As you can see, the input signal is completely isolated from the op-amp. Therefore, the circuit will have no output.

Now assume that the lower cutoff frequency for the filter ($f_1$) is equal to the short-circuit frequency for $C_1$. When $f_1$ is reached, $C_1$ becomes a short circuit, while $C_2$ remains

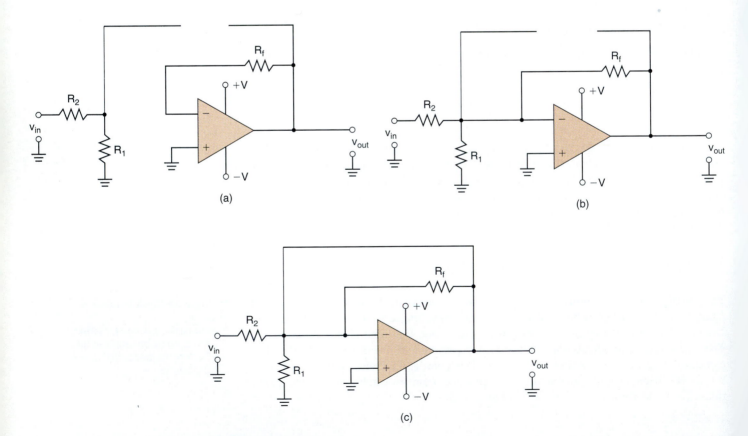

**FIGURE 17.22** Band-pass active filter circuit operation.

open. This gives us the equivalent circuit in Figure 17.22b. For this circuit, $v_{in}$ has no problem making it to the input of the op-amp, and the filter acts as an inverting amplifier. Thus, at frequencies above $f_1$, the filter has an output.

Assume now that $f_2$ is equal to the short-circuit frequency of $C_2$. When $f_2$ is reached, we have the equivalent circuit shown in Figure 17.22c. Now both capacitors are acting as short circuits. The input signal has no problem making it to the op-amp, but now the output is shorted back to the input. Since the gain of an inverting amplifier is found as

$$A_{CL} = \frac{R_f}{R_i}$$

and $R_f$ is effectively shorted, it is easy to see that the closed-loop voltage gain of the circuit is zero at frequencies above $f_2$. If we put these three equivalent circuits together, we have a circuit with zero voltage gain when operated at frequencies below $f_1$, relatively high voltage gain when operated at frequencies between $f_1$ and $f_2$, and zero voltage gain when operated at frequencies above $f_2$. This, by definition, is a band-pass filter.

*Note:* The actual equation used to find $A_{CL}$ for the multiple-feedback filter is a variation on the equation shown here. However, the equation shown here is valid for the point being made.

## Circuit Frequency Analysis

As you know, the cutoff frequency for a given low-pass or high-pass active filter is found using

$$f_c = \frac{1}{2\pi\sqrt{R_1 R_2 C_1 C_2}}$$

This equation must be modified to provide us with the following equation for the center frequency of a multiple-feedback band-pass filter:

$$f_0 = \frac{1}{2\pi\sqrt{(R_1 \parallel R_2)R_f C_1 C_2}} \qquad (17.6)$$

Example 17.7 demonstrates the use of this equation in determining the center frequency of a multiple-feedback filter.

---

Determine the value of $f_0$ for the filter in Figure 17.23.

**Solution:** The center frequency of the circuit is found as

EXAMPLE 17.7

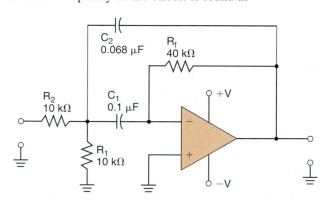

**FIGURE 17.23**

$$f_0 = \frac{1}{2\pi\sqrt{(R_1 \parallel R_2)R_f C_1 C_2}}$$

$$= \frac{1}{2\pi\sqrt{(10 \text{ k}\Omega \parallel 10 \text{ k}\Omega)(40 \text{ k}\Omega)(0.1 \text{ }\mu\text{F})(0.068 \text{ }\mu\text{F})}}$$

$$= \frac{1}{2\pi(0.001166)}$$

$$= 136.5 \text{ Hz}$$

### PRACTICE PROBLEM 17.7

A multiple-feedback filter has the following values: $R_1 = R_2 = 30$ k$\Omega$, $R_f = 68$ k$\Omega$, $C_1 = 0.22$ $\mu$F, and $C_2 = 0.1$ $\mu$F. Determine the center frequency of the circuit.

Once the center frequency of a multiple-feedback filter is known, we can calculate the value of $Q$ for the circuit using the following equations:

$$C = \sqrt{C_1 C_2} \qquad \textbf{(17.7)}$$

and

$$Q = \pi f_0 R_f C \qquad \textbf{(17.8)}$$

Example 17.8 shows how equations (17.7) and (17.8) are used to determine the $Q$ of a multiple-feedback filter.

### EXAMPLE 17.8

Determine the values of $Q$ and bandwidth for the filter in Figure 17.23.

**Solution:** In Example 17.7, the center frequency of the filter was found to be 136.5 Hz. Using this value, the $Q$ of the circuit is found as

$$C = \sqrt{C_1 C_2}$$
$$= \sqrt{(0.1 \text{ }\mu\text{F})(0.068 \text{ }\mu\text{F})}$$
$$= 0.082 \text{ }\mu\text{F}$$

and

$$Q = \pi f_0 R_f C$$
$$= \pi(136.5 \text{ Hz})(40 \text{ k}\Omega)(0.082 \text{ }\mu\text{F})$$
$$= 1.41$$

The bandwidth of the circuit is found as

$$\text{BW} = \frac{f_0}{Q}$$
$$= \frac{136.5 \text{ Hz}}{1.41}$$
$$= 97 \text{ Hz}$$

### PRACTICE PROBLEM 17.8

Determine the values of $Q$ and bandwidth for the filter described in Practice Problem 17.7.

Once the values of center frequency, $Q$, and bandwidth are known, the values of $f_1$ and $f_2$ for the multiple-feedback filter can be determined. As you will see, the equations used to determine the cutoff frequencies of the filter will vary with the $Q$ of the circuit; that is, we will use one set of equations to find $f_1$ and $f_2$ when $Q > 2$ and another set to find $f_1$ and $f_2$ when $Q < 2$.

You may recall that the value of $f_0$ approaches the *algebraic* average ($f_{ave}$) of $f_1$ and $f_2$ when $Q \geqslant 2$, as was discussed in Section 17.1. Since $f_{ave}$ is halfway between the cutoff frequencies, we can use the following equations to approximate the values of $f_1$ and $f_2$ when $Q \geqslant 2$:

$$f_1 \cong f_0 - \frac{BW}{2} \quad \text{(when } Q \geqslant 2) \tag{17.9}$$

and

$$f_2 \cong f_0 + \frac{BW}{2} \quad \text{(when } Q \geqslant 2) \tag{17.10}$$

Remember, the above equations may be used *only if $Q$ is greater than or equal to 2*. If $Q$ is less than 2, we have to use two equations that are a bit more complex than equations (17.9) and (17.10).

The actual relationship between $f_{ave}$ and $f_0$ is given as

$$f_{ave} = f_0 \sqrt{1 + \left(\frac{1}{2Q}\right)^2} \tag{17.11}$$

When the $Q$ of a multiple-feedback filter is less than 2, equations (17.9) and (17.10) must be modified to take into account the difference between $f_{ave}$ and $f_0$, as follows:

$$f_1 = f_0 \sqrt{1 + \left(\frac{1}{2Q}\right)^2} - \frac{BW}{2} \quad \text{(when } Q < 2) \tag{17.12}$$

$$f_2 = f_0 \sqrt{1 + \left(\frac{1}{2Q}\right)^2} + \frac{BW}{2} \quad \text{(when } Q < 2) \tag{17.13}$$

Example 17.9 demonstrates the fact that equations (17.9) and (17.10) can be used to approximate the values of $f_1$ and $f_2$ when $Q$ is greater than (or equal to) 2.

---

A multiple-feedback filter has values of $f_0 = 12$ kHz and BW = 6 kHz. Approximate the values of $f_1$ and $f_2$. Then determine the percentage of error in the approximated values.

*EXAMPLE 17.9*

**Solution:** With the values of $f_0$ and BW given, we know that the $Q$ of the circuit is 2. Thus, the values of $f_1$ and $f_2$ can be approximated as

$$f_1 = f_0 - \frac{BW}{2}$$
$$= 12 \text{ kHz} - 3 \text{ kHz}$$
$$= 9 \text{ kHz}$$

and

$$f_2 = f_0 + \frac{BW}{2}$$

$$= 12 \text{ kHz} + 3 \text{ kHz}$$

$$= 15 \text{ kHz}$$

Now we'll use equations (17.12) and (17.13) to validate the approximated values of $f_1$ and $f_2$ as follows:

$$f_1 = f_0 \sqrt{1 + \left(\frac{1}{2Q}\right)^2} - \frac{BW}{2}$$

$$= (12 \text{ kHz}) \sqrt{1 + \left(\frac{1}{4}\right)^2} - 3 \text{ kHz}$$

$$= (12 \text{ kHz})(1.031) - 3 \text{ kHz}$$

$$= 9.37 \text{ kHz}$$

and

$$f_2 = f_0 \sqrt{1 + \left(\frac{1}{2Q}\right)^2} + \frac{BW}{2}$$

$$= (12 \text{ kHz}) \sqrt{1 + \left(\frac{1}{4}\right)^2} + 3 \text{ kHz}$$

$$= (12 \text{ kHz})(1.031) + 3 \text{ kHz}$$

$$= 15.37 \text{ kHz}$$

Finally, the percentages of error in the approximated values of $f_1$ and $f_2$ are found as

$$f_1: \quad \% \text{ of error} = \frac{|9.347 \text{ kHz} - 9 \text{ kHz}|}{9.37 \text{ kHz}} \times 100$$

$$= 4.1\%$$

and

$$f_2: \quad \% \text{ of error} = \frac{|15.37 \text{ kHz} - 15 \text{ kHz}|}{15.37 \text{ kHz}} \times 100$$

$$= 2.5\%$$

These percentages of error are well below the acceptable limit of 10 percent, so we can use the approximations with confidence.

### PRACTICE PROBLEM 17.9

A multiple-feedback filter has values of $f_0 = 20$ kHz and BW = 5 kHz. Approximate the values of $f_1$ and $f_2$ for the circuit, and then determine the percentages of error in those approximated values.

For the majority of active filters, the value of $Q$ will be greater than 2. However, in those cases when $Q < 2$, you should be aware that the approximated values of $f_1$ and $f_2$ may be too inaccurate. This point is illustrated in Example 17.10.

An active filter has values of $f_0 = 100$ Hz and $Q = 1.02$. Approximate the values of $f_1$ and $f_2$. Then determine the percentages of error in those approximated values.

**EXAMPLE 17.10**

**Solution:** First, the bandwidth of the filter is found as

$$\text{BW} = \frac{f_0}{Q}$$

$$= \frac{100 \text{ Hz}}{1.02}$$

$$\cong 98 \text{ Hz}$$

The values of $f_1$ and $f_2$ are approximated as

$$f_1 \cong f_0 - \frac{\text{BW}}{2}$$

$$= 100 \text{ Hz} - 49 \text{ Hz}$$

$$= 51 \text{ Hz}$$

and

$$f_2 \cong f_0 + \frac{\text{BW}}{2}$$

$$= 100 \text{ Hz} + 49 \text{ Hz}$$

$$= 149 \text{ Hz}$$

Using equations (17.12) and (17.13), the cutoff frequencies for the circuit are found as

$$f_1 = f_0 \sqrt{1 + \left(\frac{1}{2Q}\right)^2} - \frac{\text{BW}}{2}$$

$$= (100 \text{ Hz}) \sqrt{1 + \left(\frac{1}{2.04}\right)^2} - 49 \text{ Hz}$$

$$= (100 \text{ Hz})(1.113) - 49 \text{ Hz}$$

$$= 62 \text{ Hz}$$

and

$$f_2 = f_0 \sqrt{1 + \left(\frac{1}{2Q}\right)^2} + \frac{\text{BW}}{2}$$

$$= (100 \text{ Hz}) \sqrt{1 + \left(\frac{1}{2.04}\right)^2} + 49 \text{ Hz}$$

$$= (100 \text{ Hz})(1.113) + 49 \text{ Hz}$$

$$= 160 \text{ Hz}$$

Finally, the percentages of error are found as

$$f_1: \quad \% \text{ of error} = \frac{62 \text{ Hz} - 51 \text{ Hz}}{62 \text{ Hz}} \times 100$$

$$= 17.74\%$$

$$f_2: \quad \% \text{ of error} = \frac{160 \text{ Hz} - 149 \text{ Hz}}{160 \text{ Hz}} \times 100$$

$$= 6.88\%$$

While the percentage of error in the approximated value of $f_2$ in the example is acceptable, the percentage of error in the $f_1$ approximation is not. Therefore, we cannot approximate the cutoff frequencies for a filter with a $Q$ that is less than 2.

## Filter Gain

In most cases, the analysis of a multiple-feedback filter begins and ends with calculating the frequency response of the circuit. However, being able to predict the closed-loop voltage gain of the circuit provides us with a very valuable troubleshooting tool, as you will see in our discussion on troubleshooting active filters.

The following equation is used to find the closed-loop voltage gain of a multiple-feedback filter:

$$A_{CL} = \frac{R_f}{2R_i} \qquad (17.14)$$

where $R_i$ is the circuit series input resistor. Example 17.11 demonstrates the calculation of $A_{CL}$ for a multiple-feedback filter.

---

**EXAMPLE 17.11**

Determine the closed-loop voltage gain for the filter in Figure 17.23.

**Solution:** For this circuit, $R_1$ is the circuit series input resistor. Using the values of $R_f$ and $R_1$, the value of $A_{CL}$ is found as

$$A_{CL} = \frac{R_f}{2R_1}$$
$$= \frac{40 \text{ k}\Omega}{20 \text{ k}\Omega}$$
$$= 2$$

**PRACTICE PROBLEM 17.11**

Determine the closed-loop voltage gain for the filter described in Practice Problem 17.7. Assume that $R_1$ is the series input resistor for the circuit.

---

This completes the analysis of the multiple-feedback band-pass filter. At this point, we will move on to discuss the operation of two notch filters. The first will be a cascaded notch filter that is similar to the two-stage band-pass filter that we covered earlier. Then we will take a brief look at the multiple-feedback notch filter.

## Notch Filters

As you were told earlier in this section, the notch filter is designed to block all frequencies that fall within its bandwidth. A *multistage notch filter* block diagram and frequency-response curve are shown in Figure 17.24. As you can see, the circuit is made up of a *high-pass filter*, a *low-pass filter*, and a *summing amplifier*. In Chapter 16, you were shown that a *summing amplifier* is a circuit designed to procuce an output that is *proportional* to the sum of its input voltages. In the case of the multistage notch filter, the output of the summing amplifier *equals* the sum of its input voltages.

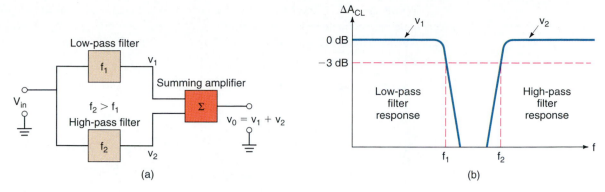

**FIGURE 17.24** Multistage notch filter block diagram and its frequency-response curve.

The multistage notch filter is designed so that $f_1$ (which is set by the low-pass filter) is lower in value than $f_2$ (which is set by the high-pass filter). The gap between the values of $f_1$ and $f_2$ is the bandwidth of the filter. This point is illustrated in the frequency-response curve in Figure 17.24b.

When the circuit input frequency is lower than $f_1$, the input signal passes through the low-pass filter to the summing amplifier. Since the input frequency is below the cutoff frequency of the high-pass filter, $v_2$ is zero. Thus, the output from the summing amplifier equals the output from the low-pass filter.

When the circuit input frequency is higher than $f_2$, the input signal passes through the high-pass filter to the summing amplifier. Since the input frequency is above the cutoff frequency of the low-pass filter, $v_1$ is zero. Now the summing amplifier output equals the output from the high-pass filter. Note that the frequencies below $f_1$ and above $f_2$ have been passed by the notch filter.

When the circuit input frequency is between $f_1$ and $f_2$, neither of the filters produce an output (*ideally*). Thus, $v_1$ and $v_2$ are both equal to zero, and the output from the summing amplifier also equals zero. In practice, of course, the exact output from the notch filter depends on how close the input frequency is to either $f_1$ or $f_2$. But, in any case, the output from the notch filter is greatly reduced when it is operated within its bandwidth.

The circuit represented in Figure 17.24 is constructed as shown in Figure 17.25. While the circuit appears confusing at first, closer inspection shows that it is made up of circuits that we have already discussed. The low-pass filter consists of the op-amp labeled IC1 and all the components with the $L$ subscript. This low-pass filter, though drawn differently than you are used to seeing, is identical to the low-pass input stage of the band-pass filter in Figure 17.19. The high-pass filter consists of the op-amp labeled IC2 and all the components with the $H$ subscript. This circuit is identical to the high-pass output stage of the band-pass filter in Figure 17.19. Finally, the summing amplifier is made up of the op-amp labeled IC3 and all the components with the $S$ subscript. Despite the complex appearance of the circuit, it operates exactly as described in our discussion of the notch filter block diagram.

The frequency analysis of the notch filter shown in Figure 17.25 is identical to that of the two-stage band-pass filter. First, the cutoff frequencies of the low-pass and high-pass filters are determined. Then, using the calculated cutoff frequencies, the bandwidth, center frequency, and $Q$ values for the circuit are determined.

## The Multiple-Feedback Notch Filter

The multiple-feedback notch filter, which is shown in Figure 17.26, is very similar to its band-pass counterpart. The capacitors in the notch filter ($C_1$ and $C_2$) react to a change in frequency exactly as described in our discussion on multiple-feedback band-pass filters.

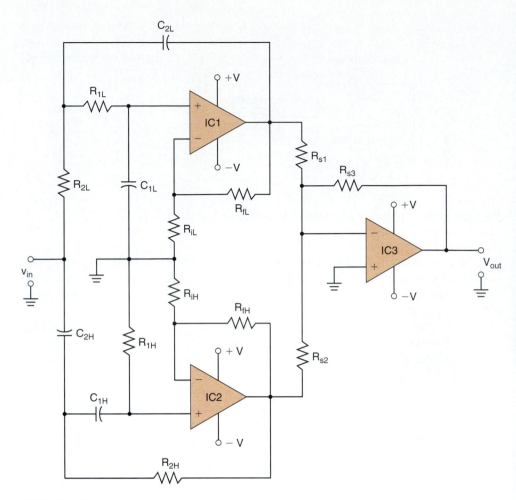

**FIGURE 17.25**

However, the connection of $v_{in}$ to the noninverting input of the op-amp (via the $R_2$–$R_3$ voltage divider) radically alters the overall circuit response, as follows:

1. When $f_{in} < f_1$, $C_1$ prevents the input voltage from reaching the inverting input of the op-amp. However, the signal is still allowed to reach the noninverting input, and a noninverted output is produced by the circuit.

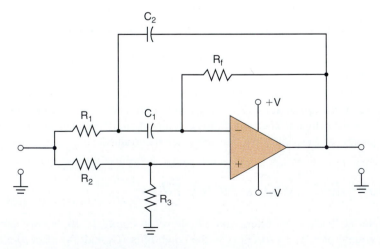

**FIGURE 17.26   A multiple-feedback notch filter.**

2. When $f_{in} > f_1$, the input signal is applied to both the inverting and noninverting inputs of the op-amp. In other words, $v_{in}$ starts to be seen by the op-amp as a common-mode signal. As you recall, op-amps reject common-mode input signals. Therefore, there is little or no output from the filter when the input signal is within the bandwidth of the op-amp.

3. When $f_{in} > f_2$ both capacitors will act as short circuits. While $C_2$ partially shorts out the signal at the inverting input of the op-amp (reducing its amplitude), $v_{in}$ is still applied to the noninverting op-amp input. In this case, the op-amp will amplify the difference between the two input signals. Again, the circuit will have a measurable output.

*A Practical Consideration:* $C_1$ will always produce a phase shift. Even though this phase shift may be extremely small, it will still prevent $v_{in}$ from appearing at both inputs of the op-amp at the same time. Therefore, the circuit will still have a small output signal when operated within its bandwidth.

In a summary, the circuit will act as a noninverting amplifier when the input frequency is lower than $f_1$ or higher than $f_2$. When the input frequency is within the bandwidth of the filter, the op-amp will, to one degree or another, see $v_{in}$ as a pair of common-mode signals and will reject those signals accordingly.

The frequency analysis of the notch filter is very similar to that of the band-pass filter. The only equation modification is as follows:

$$f_0 = \frac{1}{2\pi\sqrt{R_1 R_f C_1 C_2}} \tag{17.15}$$

The final difference lies in the $A_{CL}$ characteristic of the circuit. When operated above $f_2$, $C_2$ acts as a short circuit, and the filter acts as a *voltage follower*. To have equal gain values on both sides of the bandwidth (that is, below $f_1$ and above $f_2$), the values of $R_f$ and $R_1$ are normally selected to provide a low-frequency gain that is as close to unity (1) as possible. Thus, the closed-loop voltage gain can be approximated as

$$A_{CL} \cong 1 \quad (0\ dB) \tag{17.16}$$

## One Final Note

There are far too many types of active filters to cover in a textbook of this type. In this chapter, you have been introduced to several active filters. It should not surprise (or discourage) you to know that there are literally hundreds of active filter configurations. However, the most commonly used are still Butterworth filters. Although we have not been able to cover completely the topic of active filters, you should now have the knowledge needed to pursue the study of active filters further.

**Section Review**

1. Describe how cascaded low-pass and high-pass filters can form a band-pass filter.

2. List, in order, the steps taken to perform the frequency analysis of a two-stage band-pass filter.

3. In terms of circuit construction and analysis, contrast the two-stage band-pass filter with the multiple-feedback band-pass filter.

4. Which capacitor in the multiple-feedback band-pass filter controls the high-frequency response of the circuit? Which one controls its low-frequency response?

5. Briefly describe the operation of the multiple-feedback band-pass filter.

6. List, in order, the steps taken to analyze the frequency response of a multiple-feedback band-pass filter.

7. How does the analysis of a $Q < 2$ multiple-feedback filter differ from that of a $Q \geq 2$ filter?

SECTION 17.4 / BAND-PASS AND NOTCH CIRCUITS  **755**

8. Describe the operation of the multistage notch filter shown in Figure 17.25.

9. How would you analyze the frequency response of a notch filter like the one in Figure 17.25?

10. Briefly describe the operation of the multiple-feedback notch filter.

## 17.5 ACTIVE FILTER APPLICATIONS AND TROUBLESHOOTING

**Biomedical electronics**
The area of electronics that deals with medical test and treatment equipment.

It was stated at the beginning of this chapter that tuned circuits are used primarily in communications electronics. Audio and video circuits make extensive use of tuned circuits. At the same time, tuned circuits are also used in other areas of electronics, such as **biomedical electronics,** the area of electronics that deals with medical test and treatment equipment.

In this section, we will take a look at several audio applications for active filters. We will also take a look at troubleshooting basic active filters.

### An Audio Crossover Network

**Crossover network**
A circuit designed to separate high-frequency audio from low-frequency audio.

A **crossover network** is a circuit that is designed to split the audio signal from a stereo, television, or other communications system so that the high-frequency portion of the audio goes to a small high-frequency speaker (called a *tweeter*) and the low-frequency portion goes to a relatively large low-frequency speaker (called a *woofer*). A crossover network may be placed in an audio system, as shown in Figure 17.27.

Let's assume that the block diagram represents the output of a stereo. In this stereo, the audio signal is applied to a pre–amp power amplifier to boost the power level of the audio signal. Then the high-power signal is applied to the crossover network. This network splits the audio, sending the high-frequency portion of the audio to the tweeter and the low-frequency portion of the audio to the woofer.

The crossover network in Figure 17.27 could be a circuit similar to the one shown in Figure 17.28. This crossover network consists of an input buffer, followed by a two-pole low-pass filter and a two-pole high-pass filter.

The input buffer is a voltage-follower that consists of IC1 and $R_{in}$. $R_{in}$ is used to match the input impedance of the crossover network to the output impedance of the pre-amp.

The output signal from the buffer is applied to both of the active filters. IC2 and its associated circuitry pass the low-frequency audio to the woofer power amplifier while blocking the high-frequency audio. IC3 and its associated circuitry pass the high-frequency audio to the tweeter power amplifier while blocking the low-frequency audio.

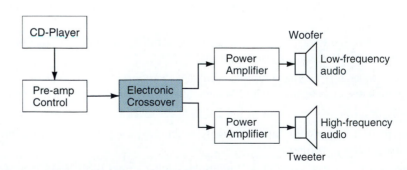

**FIGURE 17.27  Crossover network in an audio system.**

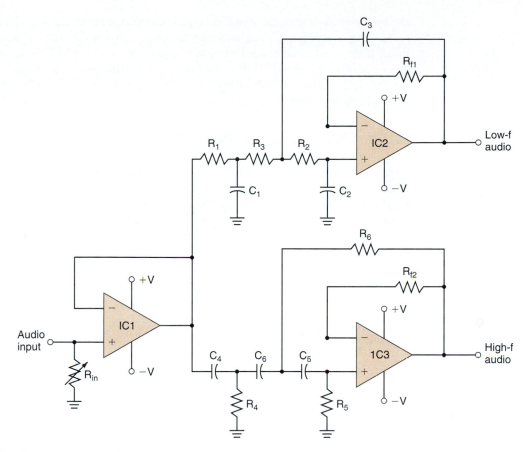

**FIGURE 17.28    A crossover network.**

## A Simple Graphic Equalizer

A **graphic equalizer** is a circuit or system that is designed to allow you to control the amplitude of different audio-frequency ranges. A simplified graphic equalizer block diagram is shown in Figure 17.29. This graphic equalizer is made up of a series of low-$Q$ active filters and a summing amplifier. The number of active filters used increases with the sophistication of the system.

The summing amplifier is made up of the op-amp and the resistors labeled $R_4$, $R_5$, and $R_6$. The audio output from the summing amplifier is fed back to the band-pass filters via their input control potentiometers. When the potentiometer of a given filter is adjusted

**Graphic equalizer**
A circuit or system designed to allow you to control the amplitude of different audio-frequency ranges.

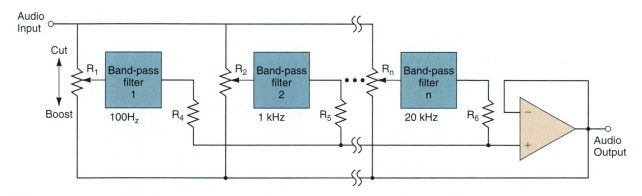

**FIGURE 17.29    Simple graphic equalizer.**

toward the audio output, the signal strength at the filter input *increases*. This has the effect of *boosting* (amplifying) that particular frequency. When the input potentiometer of a given filter is adjusted toward the audio input, the signals strength at the filter input *decreases*. This has the effect of *cutting* (attenuating) that particular frequency. By adjusting the input controls to the various active filters, you are able to boost the frequencies you want to hear while cutting others.

## Some Other Active Filter Applications

High-pass filters can be used to eliminate the low-frequency noise that can be generated in many audio systems. For example, a high-pass filter can be used to eliminate any 60-Hz power line noise. By tuning the high-pass filter so that it has a lower cutoff frequency above 60 Hz, the power line noise is eliminated.

There are far more applications for active filters than could possibly be covered here. However, you should now have a good idea of how versatile these circuits are.

## Active Filter Fault Symptoms

OBJECTIVE 6 ▶ Active filter troubleshooting is relatively simple when you keep in mind the fact that some of the circuit components are used to determine the circuit's frequency-response characteristics while others are used to determine its gain characteristics. For example, consider the two-pole low-pass filter shown in Figure 17.30a. The frequency response of the circuit is controlled by the combination of $R_1$, $R_2$, $C_1$, and $C_2$. The gain of the circuit is controlled primarily by $R_f$ and $R_i$.

If either of the gain components opens, the amplitude of the filter output will change. For example, if $R_i$ opens, the circuit configuration changes to that of a voltage follower, and the $A_{CL}$ of the circuit will drop to 1 (0 dB). If $R_f$ opens, the value of $A_{CL}$ will jump to the open-loop voltage gain of the op-amp, and the output signal will be clipped.

While $R_1$ and $R_2$ are a part of the frequency-response circuit, the output of the filter will drop to zero if either of these components opens. This is because the input signal will be isolated from the op-amp itself. The same thing will occur if $C_1$ is shorted.

If either $C_1$ or $C_2$ opens, there will be two results:

1. The roll-off rate of the circuit will decrease.

2. The cutoff frequency of the circuit will shift.

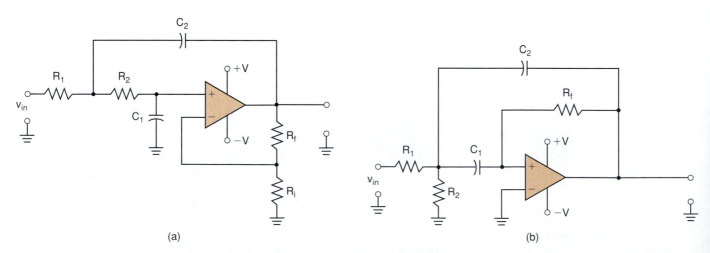

(a)            (b)

**FIGURE 17.30**

**TABLE 17.2  Low-Pass Filter Fault Symptoms and Their Causes**

| Symptom | Possible Cause(s) |
|---------|-------------------|
| $A_{CL}$ drops to 1 | $R_f$ open, $R_i$ shorted[a] |
| $A_{CL} = A_{OL}$ | $R_f C$ open |
| $A_{CL} = 0$ (no output) | $R_1$ open, $R_2$ open, $C_1$ shorted, $C_2$ shorted |
| Higher $f_c$ and lower roll-off rate | $C_1$ or $C_2$ open |

[a] Highly unlikely.

The first point is easy to see when you consider the case of $C_2$ opening. If this capacitor opens, we go from a two-pole filter to a single-pole filter. As you recall, the filter will have a roll-off rate of 20 dB/decade per pole. Thus, the roll-off rate for the filter will drop from 40 to 20 dB/decade (or from 12 to 6 dB/octave).

If either capacitor in the filter opens, the cutoff frequency for the circuit will *increase*. You see, $C_1$ and $C_2$ are both shunt capacitors when viewed from the op-amp. Thus, from the op-amp's point of view, they can be considered (loosely) to be parallel components. If either opens, the total shunt capacitance decreases, causing a decrease in the cutoff frequency of the circuit.

Table 17.2 summarizes the faults that can occur in the low-pass two-pole filter and the symptoms of each. The fault symptoms that can develop in the two-pole high-pass filter are very similar to those listed for the low-pass filter, as can be seen by comparing Table 17.3 with Table 17.2.

*A Practical Consideration:* Don't forget that the circuit won't work if there is no input signal or if either of the power supply connections is faulty.

The two-stage band-pass filter (see Figure 17.20) can develop a fault in either of the two stages. A loss of gain in either stage will direct you to the appropriate components. If $f_1$ is low and the low-frequency roll-off rate is low, the fault is located in the high-pass filter stage. If $f_2$ is low and the high-frequency roll-off rate is low, the fault is located in the low-pass filter stage.

The multiple-feedback band-pass filter is relatively easy to troubleshoot if you keep in mind the equations for the circuit center frequency and gain. These equations were given earlier as

$$f_0 = \frac{1}{2\pi\sqrt{(R_1 \parallel R_2)R_f C_1 C_2}}$$

and

$$A_{CL} = \frac{R_f}{2R_i}$$

For the circuit in Figure 17.30b, $R_i$ is the resistor labeled $R_1$. If this resistor opens, the output of the filter will drop to zero. If $R_2$ opens, the circuit will act as an inverting amplifier. You may recall that the gain of an inverting amplifier is found as

$$A_{CL} = \frac{R_f}{R_i}$$

**TABLE 17.3  High-Pass Filter Fault Symptoms and Their Causes**

| Symptom | Possible Cause(s) |
|---------|-------------------|
| $A_{CL}$ drops to 1 | $R_i$ open, $R_f$ shorted[a] |
| $A_{CL} = A_{OL}$ | $R_f$ open |
| $A_{CL} = 0$ (no output) | $R_1$ open, $R_2$ shorted,[a] $C_1$ open, $C_2$ open |
| Low $f_c$ and roll-off rate | $C_1$ or $C_2$ shorted |

[a] Highly unlikely.

*Don't Forget:* A leaky capacitor will act (to some degree) like a shorted capacitor.

**TABLE 17.4   Band-Pass Filter Fault Symptoms and Their Causes**

| Symptom | Possible Causes |
|---|---|
| $A_{CL} = A_{OL}$ | $R_f$ open, $R_1$ shorted[a] |
| $A_{CL}$ doubles | $R_2$ open |
| $A_{CL} = 0$ (no output) | $R_1$ open, $R_2$ shorted,[a] $C_1$ open, $C_2$ shorted, $R_f$ shorted[a] |
| Circuit acts as a low-pass filter | $C_1$ shorted |
| Circuit acts as a high-pass filter | $C_2$ open |

[a] Highly unlikely.

Thus, if $R_2$ in the filter opens, the value of $A_{CL}$ for the circuit will *double*. Also, the value of $f_0$ for the circuit will decrease since $R_2$ will no longer be in parallel with $R_1$ (which increases the denominator value in the $f_0$ equation).

If $R_f$ opens, the value of $A_{CL}$ will increase to the value of $A_{OL}$ for the op-amp. At the same time, the value of $f_0$ will dramatically decrease.

If $C_1$ shorts, the circuit will act as a single-pole low-pass filter. This is due to the fact that $C_1$ determines the lower cutoff frequency for the circuit. If $C_2$ shorts, the output of the op-amp will be shorted back to the input, and the circuit will not have an output. If $C_1$ opens, the input signal will be isolated from the filter, and the output will drop to zero. If $C_2$ opens, the circuit will act as a single-pole low-pass filter since $C_2$ controls the value of $f_2$. The fault symptoms for the multiple-feedback filter are listed in Table 17.4.

## Active Filter Troubleshooting

Once you know the common fault symptoms for active filters, the troubleshooting procedure for these circuits is easy:

1. Determine the type of filter you are dealing with.
2. Verify that the filter has an input signal.
3. Verify that the load is not the cause of the problem by isolating the filter output from the load.
4. Verify that both supply inputs to the filter are working properly.
5. Using your observations and Tables 17.2 through 17.4, determine the source of the fault within the filter.
6. If none of the passive components is faulty, replace the op-amp.

**Section Review**

1. What is a *crossover network*?
2. Describe the operation of the crossover network in Figure 17.28.
3. What is a *graphic equalizer*?
4. Describe the operation of the graphic equalizer in Figure 17.29.
5. Describe how high-pass filters can be used to eliminate low-frequency noise in audio systems.
6. List the fault symptoms that can occur in active filters (those described in the text) and the possible cause(s) of each.
7. List the steps involved in troubleshooting an active filter.

# 17.6 DISCRETE TUNED AMPLIFIERS

While many tuned circuit applications can be filled by active filters, there are some applications that exceed the power-handling and/or high-frequency limits of active filters. In these applications, *discrete tuned amplifiers* are still commonly used.

Discrete component circuits are tuned using parallel LC (inductive-capacitive) circuits in place of a collector (or drain) resistor. A typical BJT tuned amplifier is shown in Figure 17.31. The collector of the transistor is coupled to $V_{CC}$ via the parallel LC circuit ($C_T$ and $L_T$). The parallel LC circuit determines the frequency response characteristics of the amplifier. This being the case, we will start our discussion of the circuit by briefly reviewing the operating characteristics of parallel LC circuits.

## Parallel LC Circuits

If we were to graph the *reactance versus frequency* characteristics of a capacitor and an inductor on the same chart, the plot would look like the one shown in Figure 17.32. The *origin* of the graph represents 0-$\Omega$ impedance on the *y*-axis and 0-Hz frequency on the *x*-axis. Note that, at 0 Hz, the impedance of the inductor is shown to be 0 $\Omega$, while the value of $X_C$ is shown to approach $\infty$ $\Omega$. As frequency increases, the value of $X_L$ *increases*, while the value of $X_C$ *decreases*. At some frequency, the values of $X_C$ and $X_L$ are *equal* for a given capacitor–inductor pair. This frequency, called the *resonant frequency,* is found as

$$f_r = \frac{1}{2\pi\sqrt{LC}}$$

(17.17)

You may recall from your study of basic electronics that:

**1.** The current in a capacitive branch *leads* the voltage across that branch by 90°.

**2.** The current in an inductive branch *lags* the voltage across that branch by 90°.

In a parallel *LC* circuit, the voltage across the capacitor is equal to the voltage across the inductor. Combining this fact with the statements above, it is easy to see that $I_C$ and $I_L$ are 180° out of phase in a parallel *LC* tank. Thus,

**FIGURE 17.31   Typical BJT discrete tuned amplifier.**

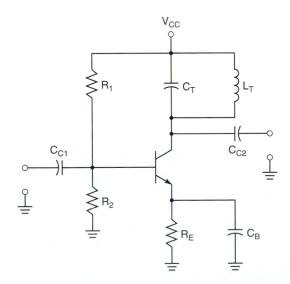

**Lab Reference:** The operation of a tuned *BJT* amplifier is demonstrated in Exercise 39.

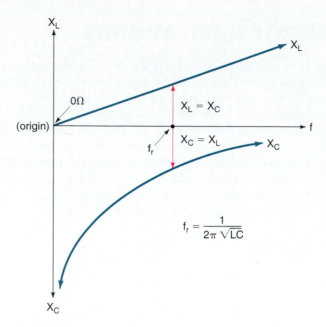

**FIGURE 17.32** Plot of the reactance versus frequency characteristics.

$$I_{net} = |I_C - I_L| \tag{17.18}$$

where $I_{net}$ is the net current entering (and leaving) the tank circuit.

At resonance, $X_C = X_L$.

At resonance, the values of $X_C$ and $X_L$ are equal. Since the voltages across the parallel components are equal, $I_C$ and $I_L$ are ideally equal.
Thus, at resonance,

$$I_{net} = |I_C - I_L|$$
$$= 0 \text{ A}$$

Below resonance, a tank circuit acts inductive. Above resonance, it acts capacitive.

Since the net current through the *LC* tank circuit is 0 A at resonance, the overall impedance of the tank circuit is infinite for all practical purposes. At frequencies below resonance, $X_L < X_C$. Thus, $I_L > I_C$, and the circuit acts inductive. At frequencies above resonance, $X_C < X_L$. Thus, $I_C > I_L$, and the circuit acts capacitive. These points are illustrated in Figure 17.33. Now let's apply these principles to the tuned amplifier shown in Figure 17.31.

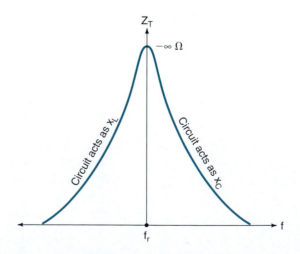

**FIGURE 17.33** Frequency response of a parallel LC tank circuit.

# ac *Circuit Conditions*

The ac equivalent of Figure 17.31 is derived in the usual fashion. The only exception is the fact that the tank circuit components are not shorted. This ac equivalent circuit is shown in Figure 17.34. The load resistance is included to aid our discussion. ◄ *OBJECTIVE 7*

To understand the circuitry completely, we need to look at three frequency conditions:

1. $f_{in} = f_r$
2. $f_{in} < f_r$
3. $f_{in} > f_r$

When $f_{in} = f_r$, the tank circuit acts as an open, as was discussed earlier in this section. Since $R_L$ represents the only path to ground in the collector circuit, it provides the only path for the ac output current. Thus, at $f_{in} = f_r$, the ac load current and amplifier efficiency reach their maximum possible values.  *When $f_{in} = f_r$.*

As $f_{in}$ begins to drop below the value of $f_r$, $I_L$ *increases* and $I_C$ *decreases*. As a result, $I_{net}$ increases. As the input frequency continues to decrease, $I_{net}$ also continues to increase. At some point, the transistor loading caused by the increase in $I_{net}$ causes the load voltage to drop by 3 dB. The frequency at which this occurs is the value of $f_1$ for the circuit.  *When $f_{in} < f_r$.*

As $f_{in}$ begins to climb above the value of $f_r$, $I_C$ *increases* and $I_L$ *decreases*. Again, this results in an increase in $I_{net}$. As the input frequency continues to increase, another point is reached where the transistor loading caused by the increase in $I_{net}$ causes the load voltage to drop by 3 dB. The frequency at which this occurs is the value of $f_2$ for the circuit.  *When $f_{in} > f_r$.*

As you may have figured out by now, the resonant frequency of the tank circuit is the center frequency of the tuned amplifier. By formula, ◄ *OBJECTIVE 8*

$$f_0 = f_r \qquad \textbf{(17.19)}$$

or

$$f_0 = \frac{1}{2\pi\sqrt{LC}} \qquad \textbf{(17.20)}$$

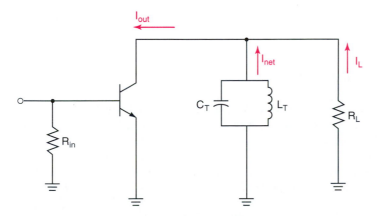

**FIGURE 17.34   An equivalent circuit for a discrete tuned amplifier.**

**EXAMPLE 17.12**

Determine the center frequency of the tuned amplifier shown in Figure 17.35.

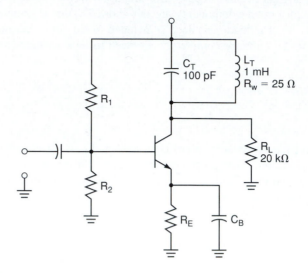

**FIGURE 17.35**

**Solution:** The center frequency is equal to the resonant frequency of the *LC* tank circuit. Using equation (17.20) and the circuit values shown, $f_0$ is found as

$$f_0 = \frac{1}{2\pi\sqrt{LC}}$$

$$= \frac{1}{2\pi\sqrt{(1\text{ mH})(100\text{ pF})}}$$

$$= \frac{1}{1.987 \times 10^{-6}}$$

$$= 503.29\text{ kHz}$$

**PRACTICE PROBLEM 17.12**

A tuned BJT amplifier has values of $L = 33$ mH and $C = 0.1$ μF. Calculate the value of $f_0$ for the circuit.

Once you have determined the value of $f_0$ for a tuned amplifier, the next step is to find the value of $Q$ for the circuit. We take a look at the process for calculating $Q$ now.

## The Figure of Merit, Q

Earlier in this chapter, we stated that the $Q$ of a tuned amplifier is equal to the ratio of center frequency to bandwidth. Although this statement helps to explain the use of $Q$ in bandwidth calculations, it does not really tell you what $Q$ is or where it comes from.

The $Q$ of a discrete tuned amplifier is a property of the tank circuit; more specifically, it is a property of the inductor. Strictly speaking, the **$Q$ of a tank circuit** is *the ratio of energy stored in the circuit to energy lost per cycle in the circuit.* By formula,

**Q (tank circuit)**
The ratio of energy stored in the circuit to energy lost per cycle by the circuit.

$$Q = \frac{\text{energy stored}}{\text{energy lost}}$$

Another way of writing this is

$$Q = \frac{\text{reactive power}}{\text{resistive power}}$$

If you were to take the voltage across an inductor and multiply it by the current through the component, you would obtain an **apparent power** value. We say *apparent* because only a small portion of the power value is actually dissipated by the component. Most of it is stored in the electromagnetic field that surrounds the component.

**Apparent power**
Power stored in the field of a reactive component.

Apparent power is made up of *reactive power* and *resistive (or "true") power.* Reactive power, which is actually *energy stored* by the component, is found as

$$P_X = (I_L^2)(X_L)$$

While most of the apparent power is stored in the electromagnetic field, *some* of the power is actually dissipated across the small amount of winding resistance in the coil. This *resistive power,* or *energy lost,* is found as

$$P_{Rw} = (I_L^2)(R_w)$$

Now, the $Q$ of the coil is found as the ratio of apparent (reactive) power to resistive power. By formula,

$$Q = \frac{P_X}{P_{Rw}}$$

$$= \frac{(I_L^2)(X_L)}{(I_L^2)(R_w)}$$

or, simply,

$$Q = \frac{X_L}{R_w} \qquad\qquad (17.21)$$

where   $Q$ = the quality of the coil, measured at the center frequency
$X_L$ = the reactance of the coil, measured at the center frequency
$R_w$ = the resistance of the winding

The following example demonstrates the process for determining the $Q$ of a tank circuit.

---

Determine the $Q$ of the tank circuit in Figure 17.35.

**EXAMPLE 17.13**

**Solution:**   In Example 17.12, we determined the center frequency of the amplifier to be 503.29 kHz. Using this value, $X_L$ for the circuit is determined as

$$X_L = 2\pi f L$$
$$= 2\pi(503.29 \text{ kHz})(1 \text{ mH})$$
$$= 3.16 \text{ k}\Omega$$

Using this value and the value of $R_w = 25 \ \Omega$ shown in the figure, the $Q$ of the tank circuit is found as

$$Q = \frac{X_L}{R_w}$$

$$= \frac{3.16 \text{ k}\Omega}{25 \ \Omega}$$

$$= 126.5$$

The value of $Q$ obtained in Example 17.13 is accurate for the tank circuit. However, before we can use this value for bandwidth calculations, we have to take into account the effects of *circuit loading*; that is, the effect that $R_L$ will have on the $Q$ of the tank.

## Loaded-Q

To see the effects of loading on circuit $Q$, we have to take another look at the ac equivalent of the collector circuit. The ac equivalent is shown in Figure 17.36a. In this figure, we have included the tank circuit capacitor, $C_T$. However, since it does not really weigh into this discussion, we will not refer to it.

If you look closely at Figure 17.36a, you'll notice that the load resistance is in parallel with the inductor and its winding resistance ($R_w$). Therefore, we must include the resistance of the load in our calculations if we want an accurate value of $Q$.

The first step in determining the loaded-$Q$ of the circuit is to replace $R_w$ with an equivalent parallel resistance, $R_p$. This resistance is shown in Figure 17.36b. The value of $R_p$ is derived by determining the values of $X_L$ and $R_w$ and then solving for the equivalent parallel resistance value. If we were to do this here, we would eventually end up with the following relation:

$$R_P = Q^2 R_w \qquad (17.22)$$

The derivation of equation (17.22) is shown in Appendix D. The total ac resistance of the collector circuit can now be found as

$$r_C = R_P \| R_L \qquad (17.23)$$

and the loaded-$Q$ can be found as

$$Q_L = \frac{r_C}{X_L} \qquad (17.24)$$

Equation (17.24) is also derived in Appendix D. The following example demonstrates the determination of the value of loaded-$Q$ for a tuned amplifier.

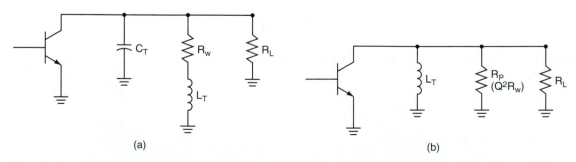

(a)                                                                                          (b)

**FIGURE 17.36    Effects of loading on circuit $Q$.**

Determine the loaded-$Q$ for the circuit shown in Figure 17.35.

EXAMPLE 17.14

**Solution:** The $Q$ of the tank circuit was found in Example 17.13 to be 126.5. Using this value of $Q$ and the value of $R_w = 25\ \Omega$, the value of $R_P$ is found as

$$
\begin{aligned}
R_P &= Q^2 R_w \\
&= (126.5)^2(25\ \Omega) \\
&= 400\ \text{k}\Omega
\end{aligned}
$$

Since $R_P \gg R_L$, we can approximate the parallel combination of the two resistors ($r_C$) to be equal to $R_L$, 20 k$\Omega$. Using this value in equation (17.24), along with the value of $X_L = 3.16$ k$\Omega$ (obtained in Example 17.13), we can find $Q_L$ as follows:

$$
\begin{aligned}
Q_L &= \frac{r_C}{X_L} \\
&= \frac{20\ \text{k}\Omega}{3.16\ \text{k}\Omega} \\
&= 6.33
\end{aligned}
$$

**PRACTICE PROBLEM 17.14**

The circuit described in Practice Problems 17.12 and 17.13 has a 10-k$\Omega$ load. Determine the value of $Q_L$ for the circuit.

As you can see, the loaded-$Q$ value for the circuit is considerably lower than the original value of $Q$ found for the tank circuit. Now that we have determined the loaded-$Q$ value for the circuit, the bandwidth of the amplifier can be found as shown in Example 17.15.

Determine the bandwidth for the amplifier shown in Figure 17.35.

EXAMPLE 17.15

**Solution:** Using the values of $f_0 = 503.29$ kHz and $Q_L = 6.33$, the bandwidth of the amplifier is found as

$$
\begin{aligned}
\text{BW} &= \frac{f_0}{Q_L} \\
&= \frac{503.29\ \text{kHz}}{6.33} \\
&= 79.5\ \text{kHz}
\end{aligned}
$$

**PRACTICE PROBLEM 17.15**

Determine the bandwidth of the amplifier in Practice Problem 17.14.

Once the values of $f_0$ and bandwidth are known, we can calculate the cutoff frequencies of the amplifier using the same guidelines established in our discussion on active filters. For circuits with $Q_L > 2$, the cutoff frequencies can be approximated as

$$f_1 \cong f_0 - \frac{BW}{2}$$

and

$$f_2 \cong f_0 + \frac{BW}{2}$$

When $Q_L < 2$, we need to use the more exact equations:

$$f_1 = f_0 \sqrt{1 + \left(\frac{1}{2Q_L}\right)^2} - \frac{BW}{2}$$

and

$$f_2 = f_0 \sqrt{1 + \left(\frac{1}{2Q_L}\right)^2} + \frac{BW}{2}$$

The only difference between these equations and those presented earlier in the text is the use of the *loaded-Q, $Q_L$.*

---

## Section Review

1. What is the circuit recognition feature of the discrete tuned amplifier?
2. Why is the net current in a parallel resonant tank circuit approximately equal to zero?
3. Describe the tuned discrete amplifier response to frequencies that are less than, equal to, and greater than the value of $f_r$.
4. What is the $Q$ of a tank circuit?
5. How does the presence of a load affect the $Q$ of the tank circuit?

---

## 17.7 DISCRETE TUNED AMPLIFIERS: PRACTICAL CONSIDERATIONS AND TROUBLESHOOTING

*OBJECTIVE 9* ▶ The primary concern here is this: It is common for a technician to build a tuned amplifier and find that the center frequency is not what was expected. There are several reasons for this:

Why variable components are used.

1. Capacitors and inductors usually have large tolerances. Thus, there can be a significant difference between the *rated* value of an inductor (or capacitor) and its actual value. This can throw your calculations off by a significant margin.

2. Amplifiers tend to have many "natural" capacitances that are not accounted for in the center–frequency equation. Values of stray capacitance and junction capacitance (for example) are always present in a discrete amplifier. If the capacitor used in the tank circuit is in the picofarad range, the other circuit capacitances can have a drastic effect on the actual center frequency of the circuit.

Tuning an amplifier.

To overcome these problems, discrete tuned amplifiers are usually built using either a variable capacitor or a variable inductor. Both circuit types are shown in Figure 17.37. In the circuit shown in Figure 17.37a, a variable inductor is used for tuning the amplifier. If the value of $L_T$ is *decreased,* the value of $f_0$ will *increase,* and vice versa. The same principle holds true for the circuit in Figure 17.37b. Of the two configurations, the variable-inductor circuit is used more commonly. This is due to the fact that variable inductors tend to be less expensive than variable capacitors.

**Electronic tuning**
Using voltages to control the tuning of a circuit.

Another method of adjusting the tuning of a circuit is shown in Figure 17.38. This type of tuning is referred to as **electronic tuning.** This name comes from the fact that the

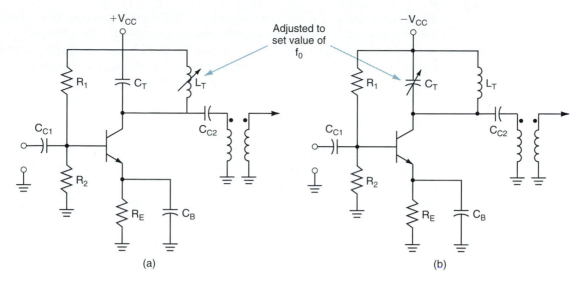

**FIGURE 17.37   Amplifier tuning.**

tuning of the amplifier is *voltage controlled*. The reverse bias on each of the *varactor diodes* ($D_1$ and $D_2$) is adjusted to set the capacitance of the component to the desired value. You may recall (from Chapter 5) that the varactor is used as an electronically variable capacitance when it is reverse biased. The circuit in Figure 17.38 shows exactly this type of application for the component.

When the biasing of $D_1$ is adjusted, its junction capacitance changes. This causes the total capacitance in the tuned circuit to change. Changing the total capacitance in the

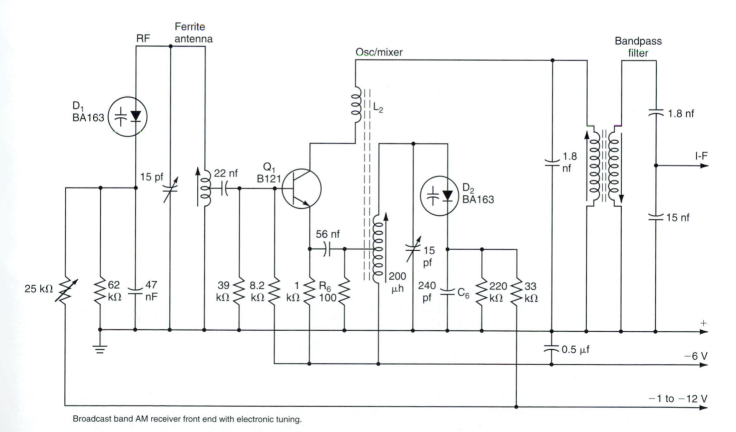

Broadcast band AM receiver front end with electronic tuning.

**FIGURE 17.38   Electronic (varactor) tuning. (Courtesy of Prentice Hall.)**

tuned circuit causes the resonant frequency of the circuit to change. Thus, the tuning of the amplifier is changed by varying the reverse bias applied to the varactor.

Don't get caught up in trying to figure out all of Figure 17.38. This circuit, or one like it, will be covered in detail when you study communications electronics. It is introduced here solely to show you the concept of electronic tuning.

## Tuning a Circuit

The next question is, How do you go about tuning an amplifier? What test procedure is used for adjusting the variable component to obtain the right center frequency? One method is to adjust the variable component using an *impedance meter,* or *Z-meter.* By connecting the meter across the variable component, you can adjust it for the proper value of $L$ or $C$.

There are a couple of problems with using a Z-meter for tuning purposes:

1. The Z-meter cannot account for any other values of $L$ and $C$ in the circuit, and thus will usually only come close to giving you the needed value of $L$ or $C$ for the desired $f_r$.

2. Z-meters cost a *lot* of money. (This problem can tend to outweigh problem 1!)

A more practical approach to the problem is illustrated in Figure 17.39. Here, a *dc current meter* is connected to the tuned amplifier between the tank circuit and $V_{CC}$. With no signal applied, this meter would read the dc value of $I_C$. However, when $f_r$ is applied to the circuit, the meter reading will vary *slightly* around $I_C$.

The procedure for tuning the circuit is as follows:

1. Using an oscilloscope (or a frequency meter), adjust $f_r$ to the desired value.

2. *Slowly* adjust $L_T$ while keeping an eye on the ammeter.

3. When the ammeter shows a *rapid change* in value, called a "null," the circuit is properly tuned.

This test is based on the fact that the net current in the tank circuit is 0 A when it is operated at resonance. When it is operated either above or below $f_r$, a small amount of "alternating" current will start to pass through the tank. This change in current will cause a *momentary* change in the reading on the dc current meter. By holding the input frequency

**FIGURE 17.39**

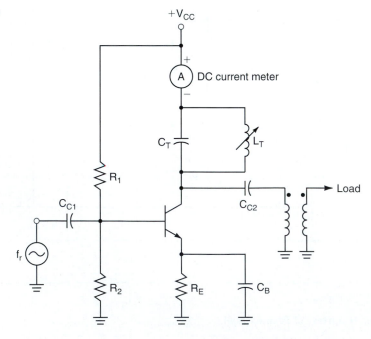

**TABLE 17.5   Tuned Amplifier Troubleshooting**

| Fault | Symptoms |
|---|---|
| $L_T$ open | $V_{CC}$ is effectively removed from the transistor collector circuit (capacitors block dc voltages). Therefore, $I_C$ and $V_C$ both drop to zero. |
| $L_T$ shorted | The ac output from the amplifier is developed across the tank circuit. Shorting $L_T$ effectively removes the tank circuit from the picture. The circuit will now act as an emitter follower, and $V_C$ will stay at $V_{CC}$. |
| $C_T$ open | Capacitors are "open" circuits by nature. However, the capacitor *connections* can open, effectively removing the capacitor from the circuit. When this happens, the amplifier will act inductive, and all tuning is lost (the amplifier will pass *all* high frequencies). |
| $C_T$ shorted | The symptoms for this condition are the same as those that occur when $L_T$ shorts. |

*Don't Forget:* A leaky capacitor will act (to a degree) like a shorted component.

at the desired value of $f_r$ and adjusting $L_T$, the jump in the meter reading will indicate that the tank circuit has been tuned to the value of $f_r$.

## Circuit Troubleshooting

The most common problem that develops in a tuned amplifier is frequency **drift.** Component aging can cause the tuning of an amplifier to change. When this happens, the amplifier may start to pass frequencies that may otherwise have been stopped. A simple cure for this problem is to retune the amplifier. If this does not solve the problem, one or both of the tank circuit components will probably have to be replaced. (*Do not neglect the possibility that the coupling transformer could cause the drift.*)

**Drift**
A change in tuning caused by component aging.

If either the capacitor or the inductor *fails* (opens or shorts), the effects will be much more drastic than those of a drift problem. Possible "fatal" faults that can occur in the tank circuit and their respective symptoms are listed in Table 17.5.

Since either component shorting will cause the same symptoms, further testing is required when $V_C$ stays at $V_{CC}$. The simplest thing to do is to measure the resistance of the capacitor. If the capacitor shows signs of charging, the inductor is the problem. If the capacitor resistance reading stays at a low value, the capacitor is the cause of the problem.

As with any circuit troubleshooting, you should start by making sure that your tuned amplifier is the cause of the problem. Once you have narrowed the problem down to a given tuned amplifier, the table in this section should help you to establish whether or not the fault is in the tank circuit. If it is not, troubleshoot the rest of the circuit components using the procedures taught earlier in the book.

## Summary

Discrete tuned amplifiers use *LC* circuits to provide amplifier tuning. The characteristics of the BJT tuned amplifier are summarized in Figure 17.40.

1. Why is amplifier tuning needed?
2. What is *electronic tuning*?
3. Describe the methods normally used to adjust the tuning of an amplifier.
4. What is *drift*?
5. How is drift corrected?
6. What are the symptoms of a major fault in the tank circuit of a tuned amplifier?

**Section Review**

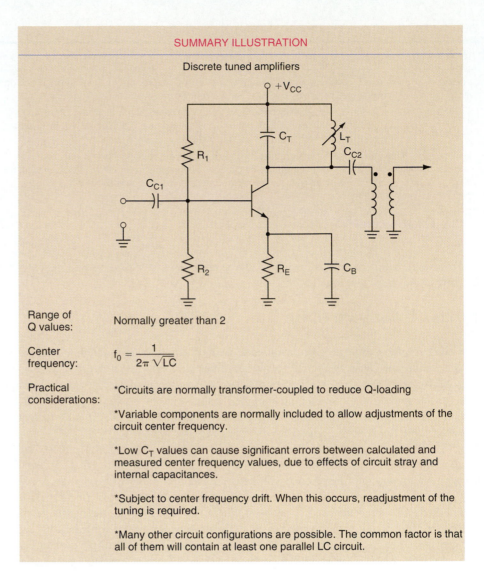

Discrete tuned amplifiers

| Range of Q values: | Normally greater than 2 |
|---|---|

Center frequency:

$$f_0 = \frac{1}{2\pi \sqrt{LC}}$$

Practical considerations:

*Circuits are normally transformer-coupled to reduce Q-loading

*Variable components are normally included to allow adjustments of the circuit center frequency.

*Low $C_T$ values can cause significant errors between calculated and measured center frequency values, due to effects of circuit stray and internal capacitances.

*Subject to center frequency drift. When this occurs, readjustment of the tuning is required.

*Many other circuit configurations are possible. The common factor is that all of them will contain at least one parallel LC circuit.

**FIGURE 17.40**

## *17.8* CLASS C AMPLIFIERS

**Class C amplifiers**
Circuits containing transistors that conduct for less than 180° of the input cycle.

Earlier in the book, you were exposed to class A, class B, and class AB amplifiers. In this section, we will take a brief look at **class C amplifiers.** Class C amplifiers were described briefly in Chapter 11 as *tuned amplifiers.*

Class C amplifiers are circuits containing transistors that conduct for *less than 180° of the input cycle.* A basic class C amplifier is shown in Figure 17.41.

### *dc Operation*

OBJECTIVE 10 ▶

The transistor in a class C amplifier is biased deeply into *cutoff*. This is often done by connecting the base resistor to a negative supply voltage (for *npn* circuits) as shown in Figure 17.41.

The dc load line for the class C amplifier is misleading.

Since a class C amplifier is biased in cutoff, the value of $V_{CEQ}$ is approximately equal to $V_{CC}$. This results in a dc load line like the one shown in Figure 17.42. Note that this dc load line can be somewhat misleading. You see, with the amplifier biased in cut-

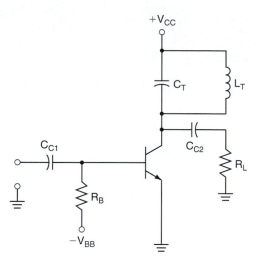

FIGURE 17.41   Class C amplifier.

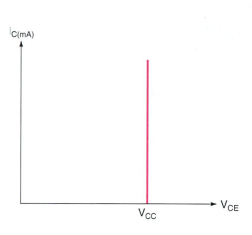

FIGURE 17.42   Class C amplifier dc load line.

Lab Reference: The dc characteristics of a Class C amplifier are observed in Exercise 40.

off, $I_C = 0$ A. The dc load line for the circuit would lead you to believe that the value of $I_C$ can be varied, as long as $V_{CEQ} = V_{CC}$. This is not the case. As long as there is no input signal to the transistor, $V_{CEQ} = V_{CC}$ and $I_{CQ} = 0$ A. Perhaps it would be more appropriate to have a dc load *point* for this circuit, rather than a dc load *line*. However, the dc operation of the circuit is traditionally represented by a load line, so. . . .

When a negative supply voltage is used for $V_{BB}$, it is normally set at a value that fulfills the following relation:

$$V_{in(pk)} + (-V_{BB}) = 1 \text{ V}$$

or

$$-V_{BB} = 1 \text{ V} - V_{in(pk)} \tag{17.25}$$

Thus, if the amplifier in Figure 17.41 had an input peak value of $+4$ V, the value of $V_{BB}$ would be approximately $-3$ V. This would ensure that the transistor would turn on only at the positive peaks of the input cycle. The purpose served by this type of biasing will be discussed later in this section.

## ac Operation

The ac operation of this circuit is based on the characteristics of the parallel resonant tank circuit. These characteristics are illustrated in Figure 17.43. In Figure 17.43a, we have a parallel tank circuit. The waveform shown on the right is produced by the tank circuit if $SW_1$ is closed for an instant and then opened again. The source of the signal is easy to understand if we consider the circuit operation during each time interval shown (see Table 17.6).

If the process described were to continue, we would get a waveform like that shown in Figure 17.43c. Each cycle would be caused by the charge/discharge cycle set up by the capacitor and the inductor. Since the tank circuit would lose some power with each cycle, the waveform would eventually die out.

The charge/discharge cycle described here is known as the **flywheel effect.** Note that the waveform produced by the flywheel effect will have a frequency equal to the resonant frequency of the tank circuit.

**Flywheel effect**
A term used to describe the ability of a parallel *LC* circuit to self-oscillate for a brief period of time.

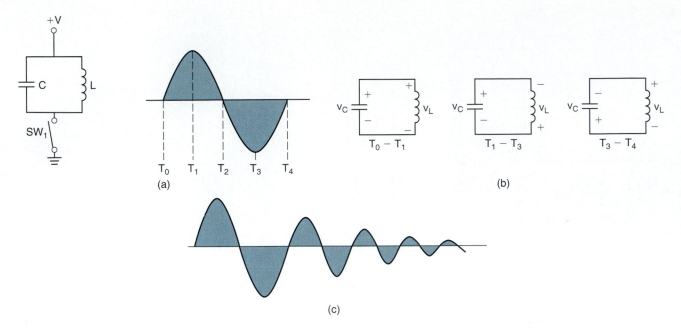

FIGURE 17.43   LC tank operation.

If we wanted to keep the flywheel effect from dying out, we could do so by closing $SW_1$ for an instant on each positive transition from $T_0$ to $T_1$. This point is illustrated in Figure 17.44.

If we could close the switch in Figure 17.43a for just an instant during each cycle produced by the tank circuit, the waveform produced would not die out. This is due to the fact that we would be returning the power lost by the circuit on each cycle. For example, from $T_1$ to $T_2$, the tank circuit will lose a small amount of power. Remember that this power lost is the result of the current through the $R_w$ of the coil. At $T_2$, the closing of the switch would add enough power to the circuit to make up for the loss. If we could continue this cycle of closing and opening the switch, we could get the waveform produced by the tank circuit to continue indefinitely. With this in mind, let's take another look at the original class C amplifier circuit. For convenience, this circuit is shown again in Figure 17.45.

Now, consider the transistor in this circuit to be the switch we talked about earlier. The positive peaks of the input signal cause the transistor to saturate, while the

TABLE 17.6   Circuit Operation

| Time Interval | Circuit Action |
|---|---|
| $T_0$–$T_1$ | During this time interval, the switch is closed. While the switch is closed, current passes through the circuit, charging both components to the polarities shown in Figure 17.43b. Note that, at the end of $T_1$, the electromagnetic field around the inductor is at its *maximum* strength. |
| $T_1$–$T_3$ | At the end of $T_1$, the switch is reopened. At this time, the field around the inductor starts to collapse. Recall that this collapsing field causes a *counter emf*, thus the reversed polarity of $v_L$ shown in Figure 17.43b. With the polarities shown, the current in the tank circuit is reversed. The current will continue in this direction until the capacitor has charged fully and $v_L$ has changed to the value represented at $T_3$ in Figure 17.43a. |
| $T_3$–$T_4$ | At $T_3$, the capacitor has charged fully and $v_L$ has again reversed polarity. Now the capacitor discharges through the inductor, causing the inductor voltage to go toward the value indicated at $T_4$. |

FIGURE 17.44

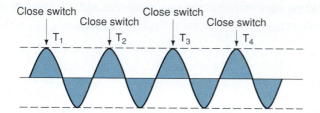

**Lab Reference:** Class C circuit action is demonstrated in Exercise 40.

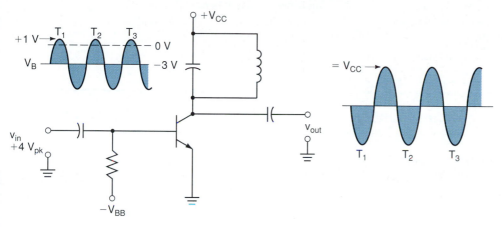

**FIGURE 17.45** Class C circuit action.

transistor is in cutoff the rest of the time. Each time the transistor saturates, it is like closing the switch in Figure 17.43. In other words, on each positive peak of the input, the transistor switch is closed, and the tank circuit gets the current pulse it requires to continue producing the output waveform.

Do not forget that the common-emitter transistor produces a 180° phase shift. Thus, the output signal is labeled on the negative peaks. If we had shown current waveforms, the input and output would have been shown to be in phase, as is the case with the common-emitter amplifier.

Another important point is this: For the class C amplifier to work properly, the tank circuit must be tuned to either the same frequency as $v_{in}$ or to some *harmonic* of that frequency. A **harmonic** is a whole number multiple of a given frequency. For example, a 2-kHz signal would have harmonics of 4 kHz, 6 kHz, 8 kHz, and so on.

When a class C amplifier is tuned to a harmonic of $v_{in}$, the output frequency of the circuit will be equal to that harmonic. For example, a class C amplifier that has an input frequency of 3 kHz and a tank circuit that is tuned to 6 kHz will produce a 6-kHz output. Thus, the class C amplifier can be used as a type of *frequency multiplier.*

Since the bandwidth, $Q$, and $Q_L$ characteristics of the class C amplifier are the same as those described earlier in this chapter, we will not elaborate on them further.

**Harmonic**
A whole number multiple of a given frequency.

---

1. What is the transistor conduction characteristic of the class C amplifier?

2. Why is the dc load line of the class C amplifier misleading?

3. Describe *flywheel effect.*

4. How is the transistor used to maintain the flywheel effect in a class C amplifier?

5. What is the input frequency requirement for proper class C operation?

*Section Review*

The following terms were introduced and defined in this chapter:

| | | |
|---|---|---|
| active filter | electronic tuning | pass band |
| apparent power | flywheel effect | pole |
| band-pass filter | graphic equalizer | Q (tank circuit) |
| band-stop (notch) filter | harmonic | quality (Q) |
| biomedical electronics | high-pass filter | stop band |
| Butterworth filter | low-pass filter | tuned amplifier |
| Chebyshev filter | multiple-feedback band- | |
| crossover network | pass filter | |
| drift | | |

## Equation Summary

| Equation Number | Equation | Section Number |
|---|---|---|
| **(17.1)** | $$Q = \frac{f_0}{\text{BW}}$$ | 17.1 |
| **(17.2)** | $$\text{BW} = \frac{f_0}{Q}$$ | 17.1 |
| **(17.3)** | $$f_0 = \sqrt{f_1 f_2}$$ | 17.1 |
| **(17.4)** | $$f_{\text{ave}} = \frac{f_1 + f_2}{2}$$ | 17.1 |
| **(17.5)** | $$f_2 = \frac{1}{2\pi RC}$$ | 17.3 |
| **(17.6)** | $$f_0 = \frac{1}{2\pi\sqrt{(R_1 \| R_2)R_f C_1 C_2}}$$ | 17.4 |
| **(17.7)** | $$C = \sqrt{C_1 C_2}$$ | 17.4 |
| **(17.8)** | $$Q = \pi f_0 R_f C$$ | 17.4 |
| **(17.9)** | $$f_1 \cong f_0 - \frac{\text{BW}}{2} \quad (\text{when } Q \geq 2)$$ | 17.4 |
| **(17.10)** | $$f_2 \cong f_0 + \frac{\text{BW}}{2} \quad (\text{when } Q \geq 2)$$ | 17.4 |
| **(17.11)** | $$f_{\text{ave}} = f_0 \sqrt{1 + \left(\frac{1}{2Q}\right)^2}$$ | 17.4 |
| **(17.12)** | $$f_1 = f_0 \sqrt{1 + \left(\frac{1}{2Q}\right)^2} - \frac{\text{BW}}{2} \quad (\text{when } Q < 2)$$ | 17.4 |
| **(17.13)** | $$f_2 = f_0 \sqrt{1 + \left(\frac{1}{2Q}\right)^2} + \frac{\text{BW}}{2} \quad (\text{when } Q < 2)$$ | 17.4 |
| **(17.14)** | $$A_{\text{CL}} = \frac{R_f}{2R_i}$$ | 17.4 |
| **(17.15)** | $$f_0 = \frac{1}{2\pi\sqrt{R_1 R_f C_1 C_2}}$$ | 17.4 |

| Equation Number | Equation | Section Number |
|---|---|---|
| **(17.16)** | $A_{CL} \cong 1 \ (0 \ dB)$ | 17.4 |
| **(17.17)** | $f_r = \dfrac{1}{2\pi\sqrt{LC}}$ | 17.6 |
| **(17.18)** | $I_{net} = |I_C - I_L|$ | 17.6 |
| **(17.19)** | $f_0 = f_r$ | 17.6 |
| **(17.20)** | $f_0 = \dfrac{1}{2\pi\sqrt{LC}}$ | 17.6 |
| **(17.21)** | $Q = \dfrac{X_L}{R_w}$ | 17.6 |
| **(17.22)** | $R_P = Q^2 R_w$ | 17.6 |
| **(17.23)** | $r_C = R_P \| R_L$ | 17.6 |
| **(17.24)** | $Q_L = \dfrac{r_C}{X_L}$ | 17.6 |
| **(17.25)** | $-V_{BB} = 1 \ V - V_{in(pk)}$ | 17.8 |

**Answers to the Example Practice Problems**

**17.1.** 60
**17.2.** 56 kHz
**17.3.** $f_0 = 1029$ kHz, $f_{ave} = 1030$ kHz, $Q = 10.29$
**17.4.** $f_2 = 102.6$ Hz, $A_{CL} = 2$
**17.5.** 894.2 Hz
**17.6.** $f_1 = 570$ Hz, $f_2 = 938$ Hz, BW $= 368$ Hz, $f_0 = 731$ Hz, $Q = 1.99$
**17.7.** 33.6 Hz
**17.8.** $Q = 1.06$, BW $= 31.6$ Hz
**17.9.** % of error $(f_1) = 0.91\%$, 0.91% of error $(f_2) = 0.71\%$
**17.11.** 1.13
**17.12.** 2.77 kHz
**17.13.** 31.9
**17.14.** 11.3
**17.15.** 246 kHz

**Practice Problems**

**§17.1**

1. An amplifier has values of $f_0 = 14$ kHz and BW $= 2$ kHz. Calculate the value of $Q$ for the circuit.
2. An amplifier has values of $f_0 = 1200$ kHz and BW $= 300$ kHz. Calculate the value of $Q$ for the circuit.
3. An amplifier with a center frequency of 800 kHz has a $Q$ of 6.2. Calculate the bandwidth of the circuit.
4. An amplifier with a center frequency of 1100 kHz has a $Q$ of 25. Calculate the bandwidth of the circuit.

**5.** Complete the following table.

| $f_0$ (kHz) | BW | Q |
|---|---|---|
| 740 | _____ | 2.4 |
| 388 | 40 kHz | _____ |
| 1050 | _____ | 5.6 |
| 920 | 600 kHz | _____ |

**6.** An amplifier has cutoff frequencies of 1180 and 1300 kHz. Show that $f_0 = f_{ave}$ for the circuit. Also, determine the $Q$ of the circuit.

**§17.3**

**7.** Calculate the bandwidth and closed-loop voltage gain for the filter in Figure 17.46.

**8.** A filter like the one in Figure 17.46 has values of $R_1 = 82$ kΩ, $C_1 = 0.015$ μF, $R_f = 150$ kΩ, and $R_i = 20$ kΩ. Calculate the bandwidth and closed-loop voltage gain of the circuit.

**9.** Calculate the bandwidth and closed-loop voltage gain of the filter in Figure 17.47.

**10.** A filter like the one in Figure 17.13a has values of $R_1 = R_2 = 33$ kΩ, $C_1 = 100$ pF, $C_2 = 200$ pF, and $R_f = 220$ kΩ. Calculate the bandwidth and closed-loop voltage gain of the circuit.

**11.** Calculate the bandwidth and closed-loop voltage gain of the filter in Figure 17.48.

**12.** Calculate the bandwidth and closed-loop voltage gain of the filter in Figure 17.49.

**13.** Calculate the lower cutoff frequency and closed-loop voltage gain of the filter in Figure 17.50.

**14.** Calculate the lower cutoff frequency and closed-loop voltage gain of the filter in Figure 17.51.

**§17.4**

**15.** Calculate the values of $f_1, f_2$, bandwidth, center frequency, and $Q$ for the band-pass filter in Figure 17.52.

**16.** Calculate the values of $f_1, f_2$, bandwidth, center frequency, and $Q$ for the band-pass filter in Figure 17.53.

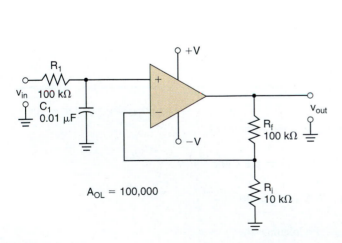

**FIGURE 17.46**

**FIGURE 17.47**

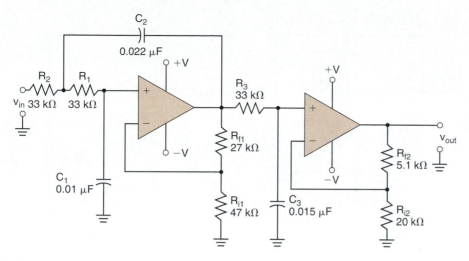

**FIGURE 17.48**

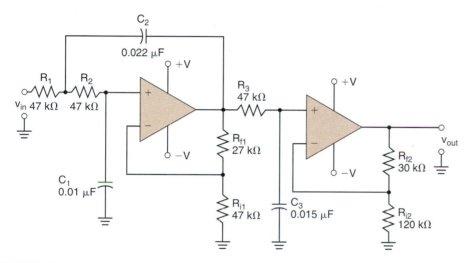

**FIGURE 17.49**

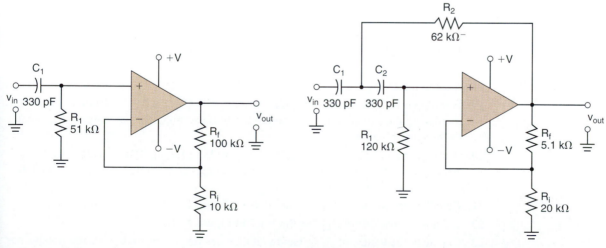

**FIGURE 17.50**

**FIGURE 17.51**

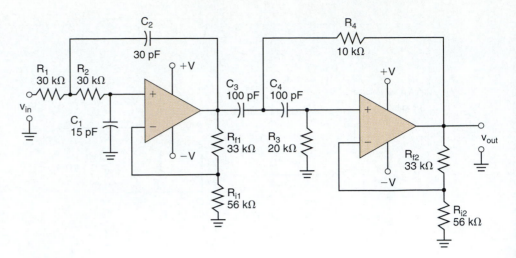

**FIGURE 17.52**

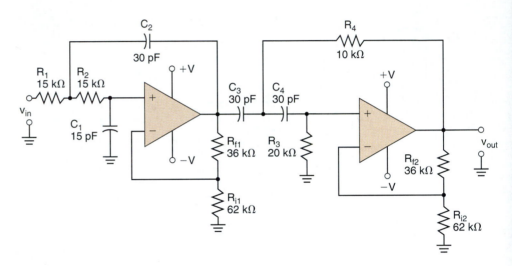

**FIGURE 17.53**

17. Calculate the values of $f_1, f_2$, bandwidth, center frequency, and $Q$ for the band-pass filter in Figure 17.54.

18. Calculate the values of $f_1, f_2$, bandwidth, center frequency, and $Q$ for the band-pass filter in Figure 17.55.

19. A multiple-feedback filter has values of $f_0 = 62$ kHz and BW $= 40$ kHz. Approximate the values of $f_1$ and $f_2$ for the circuit and determine the percentage of error in those approximations.

20. A multiple-feedback filter has values of $f_0 = 482$ kHz and BW $= 200$ kHz. Approximate the values of $f_1$ and $f_2$ for the circuit and determine the percentage of error in those approximations.

21. Calculate the value of $A_{CL}$ for the filter in Figure 17.54.

22. Calculate the value of $A_{CL}$ for the filter in Figure 17.55.

23. Calculate the values of $f_1, f_2$, bandwidth, center frequency, and $Q$ for the notch filter in Figure 17.56.

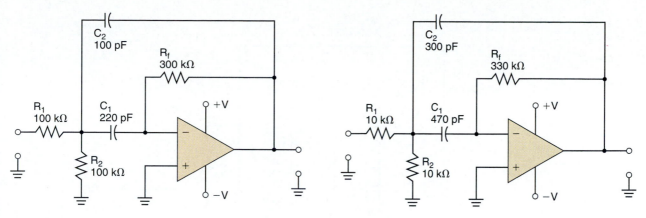

**FIGURE 17.54**                    **FIGURE 17.55**

**24.** Calculate the values of $f_1$, $f_2$, bandwidth, center frequency, and $Q$ for the notch filter in Figure 17.57.

**25.** Calculate the values of $f_1$, $f_2$, bandwidth, center frequency, and $Q$ for the notch filter in Figure 17.58.

**26.** Calculate the values of $f_1$, $f_2$, bandwidth, center frequency, and $Q$ for the notch filter in Figure 17.59.

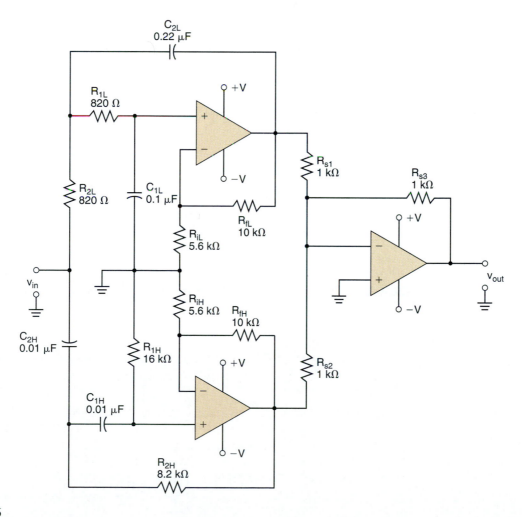

**FIGURE 17.56**

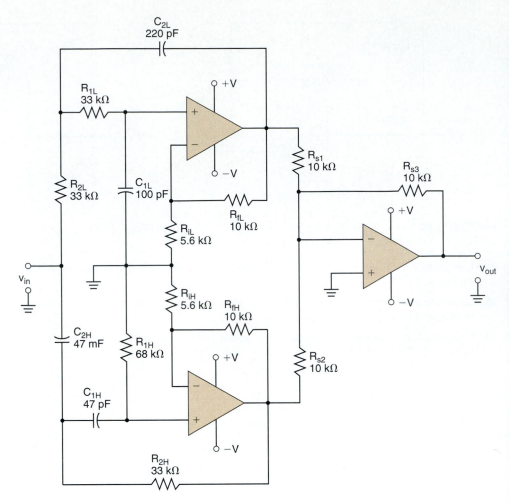

**FIGURE 17.57**

**FIGURE 17.58**

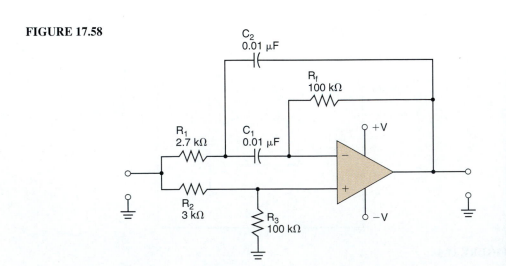

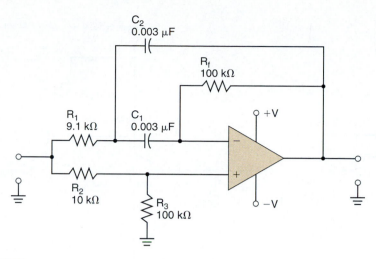

**FIGURE 17.59**

**27.** Calculate the values of $Q$, $Q_L$, center frequency, bandwidth, $f_1$, and $f_2$ for the circuit in Figure 17.60a.

**28.** Calculate the values of $Q$, $Q_L$, center frequency, bandwidth, $f_1$, and $f_2$ for the circuit in Figure 17.60b.

**§17.9**

**29.** Calculate the values of $V_{CEQ}$ and $I_{CQ}$ for the class C amplifier in Figure 17.61.

**30.** Calculate the values of $V_{CEQ}$ and $I_{CQ}$ for the amplifier in Figure 17.62.

**31.** Calculate the minimum acceptable peak input voltage for the amplifier in Figure 17.61.

**32.** Calculate the minimum acceptable peak input voltage for the amplifier in Figure 17.62.

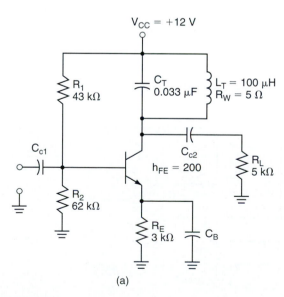

(a)

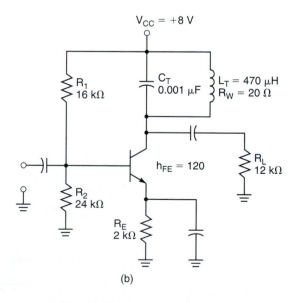

(b)

**FIGURE 17.60**

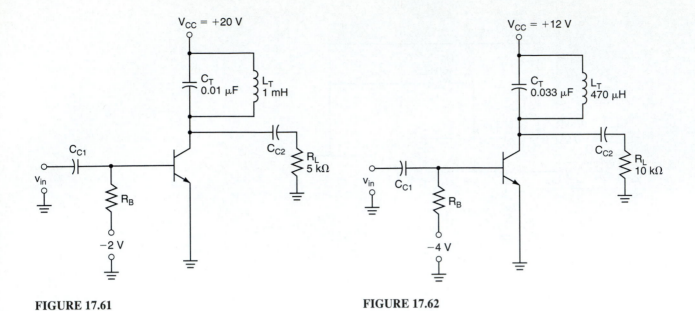

FIGURE 17.61                    FIGURE 17.62

**Troubleshooting Practice Problems**

33. What fault is indicated by the readings in Figure 17.63? Explain your answer.
34. What fault is indicated by the readings in Figure 17.64? Explain your answer.

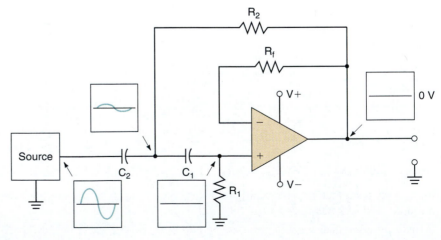

FIGURE 17.63

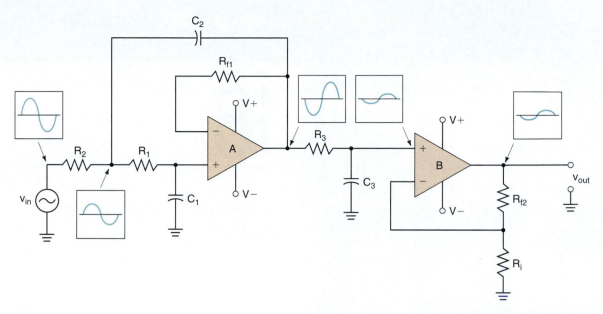

**FIGURE 17.64**

---

**35.** The inductor in Figure 17.65 is replaced by one with a winding resistance of 17 Ω. Calculate the shift that occurs in each of the cutoff frequencies as a result of this change.

*The Brain Drain*

**FIGURE 17.65**

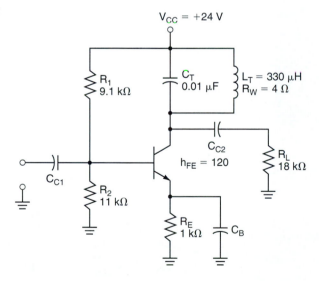

$V_{CC} = +24$ V

$R_1$
9.1 kΩ

$C_T$
0.01 μF

$L_T = 330$ μH
$R_W = 4$ Ω

$C_{C2}$

$h_{FE} = 120$

$R_L$
18 kΩ

$C_{C1}$

$R_2$
11 kΩ

$R_E$
1 kΩ

$C_B$

---

**36.** Write a program that will determine the values of $f_0$, $Q$, and bandwidth for a multiple-feedback band-pass active filter.

**37.** Write a program that will determine the values of $f_0$, $Q$, $Q_L$, and BW for a discrete tuned amplifier.

*Suggested Computer Applications Problems*

# 18

# OSCILLATORS

An important quality of all electronic circuits, including oscillators, is reliability, especially when those circuits are intended for use in remote locations.

## OUTLINE

## OBJECTIVES

*After studying the material in this chapter, you should be able to:*

1. State the function of the oscillator.

2. Describe *positive feedback,* how it is produced, and how it maintains oscillations after an oscillator is triggered.

3. Discuss the *Barkhausen criterion* and its effect on oscillator operation.

4. List the three requirements for proper oscillator operation.

5. Describe the operating characteristics of the *phase-shift* and *Wien-bridge RC* oscillators.

6. Describe the operation and perform the complete frequency analysis of a *Colpitts oscillator.*

7. Describe the operation of the *Hartley, Clapp,* and *Armstrong LC* oscillators.

8. Describe the operating principles of *crystals* and *crystal-controlled oscillators.*

9. Describe the overall approach to (and difficulties involved in) oscillator troubleshooting.

# DC TO AC?

In Chapter 3, we went through the analysis of a dc power supply. As you recall, these circuits are used to convert ac to dc.

In this chapter, we will discuss the operation of *oscillators,* circuits used to convert dc to ac. It would be reasonable to wonder why we want to convert dc to ac after going through the trouble of having a power supply convert ac to dc. Why don't we just use the ac supplied by the wall outlet to begin with?

The line frequency in the United States is 60 Hz standard. However, electronic systems depend on inter-nally generated frequencies that range from a few hertz up to the megahertz range. Microwave systems depend on internally generated frequencies that are in the giga-hertz region. All these frequencies must somehow be generated from the dc voltages present in the system. Generating these frequencies is the function of the oscil-lators in the system.

How common are oscillators? You will see throughout your career that there are very few electronic systems that do not contain one or more oscillators.

OBJECTIVE 1 ▶

**Oscillator**
An ac signal generator.

**A**n **oscillator** is a circuit that produces an output waveform without any external sig-nal source. The only input to an oscillator is the dc power supply. As such, the oscil-lator can be viewed as a *signal generator.*

There are several types of oscillators, each classified according to the type of out-put waveform it produces. In this chapter, we will cover the operation of *sine-wave oscil-lators;* those that have sinusoidal outputs. Then, in Chapter 19, we will discuss (along with a variety of circuits) the operation of a *square-wave oscillator.*

## 18.1 INTRODUCTION

OBJECTIVE 2 ▶

**Positive feedback**
When a feedback signal is in phase with the original amplifier input signal.

In Chapter 16, you were introduced to **positive feedback.** Positive feedback is the key to the operation of oscillators. As you recall, a positive feedback amplifier produces a feed-back voltage, $v_f$, that is *in phase* with the original input signal. This is illustrated in Fig-ure 18.1. In this circuit, we show an input signal, $v_{in}$. This signal is applied to the ampli-fier, which introduces a 180° phase shift. The output signal is applied to the input of the feedback network, which introduces another 180° phase shift. The result is that the signal has been shifted 360° as it has traveled around the loop. As you know, shifting a signal 360° is the same as not shifting it at all. Therefore, the feedback signal is in phase with the original input signal.

**FIGURE 18.1  Positive feedback.**

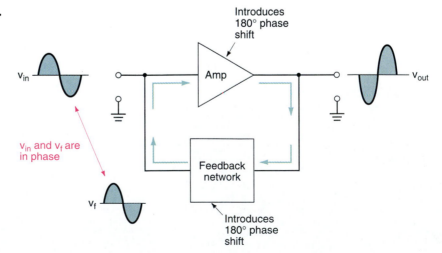

**788**    CHAPTER 18 / OSCILLATORS

# Oscillators: The Basic Idea

Even though the circuit in Figure 18.1 is useful in explaining positive feedback, it has an input signal. This is inconsistent with our definition of an oscillator. However, by modifying Figure 18.1, we can develop a circuit that is very useful for showing you the basic operating principle of the oscillator. This modified circuit is shown in Figure 18.2.

In Figure 18.2, we have added a switch in series with the amplifier input. When the switch is closed, the circuit waveforms are as shown in the figure. Now, assume that the switch is opened while the circuit is in operation. If this happens, $v_{in}$ is removed from the circuit. However, $v_f$ (which is in phase with the original input) is still applied to the amplifier input. The amplifier will respond to this signal in the same way that it did to $v_{in}$. In other words, $v_f$ will be amplified and sent to the output. Since the feedback network sends a portion of the output back to the input, the amplifier receives another input cycle, and another output cycle is produced. As you can see, this process will continue as long as the amplifier is turned on, and the amplifier will produce a sinusoidal output with no external signal source.

In any oscillator, the feedback network is used to *generate* an input to the amplifier, which, in turn, is used to *generate* an input to the feedback network. Since positive feedback produces this circuit action, it is often referred to as **regenerative feedback.** Regenerative feedback is the basis of operation for all oscillators. This point will be demonstrated throughout this chapter.

**Regenerative feedback**
Another name for positive feedback.

It should be noted that an oscillator needs only a quick *trigger* signal to start the oscillating circuit action. In other words, anything that causes a slight signal variation at

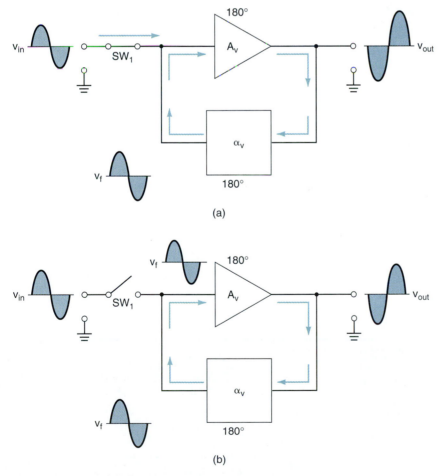

(a)

(b)

**FIGURE 18.2**

any point in the circuit will start the oscillator. It is not necessary for us to provide a complete input cycle from an external source. In fact, most oscillators will provide their own trigger signals. The sources of these trigger signals will be made clear in the next section. For now, just remember the basic requirements for an oscillator to work:

1. The circuit must have a *positive* (regenerative) feedback loop. This means that the amplifier and its feedback circuitry must combine to produce a 360° (or 0°) voltage phase shift.
2. The circuit must receive some trigger, either internally or externally generated, to start the oscillating action.

There is one other requirement that must be fulfilled for an oscillator to work. We will look at this requirement now.

## The Barkhausen Criterion

In Chapter 16, you were shown that the active component in a feedback amplifier introduces a voltage gain, while the feedback network itself introduces a voltage loss. In other words, the $A_v$ of the amplifier is greater than 1, while the feedback factor, $\alpha_v$, of the feedback network is less than 1.

OBJECTIVE 3 ▶ For an oscillator to operate properly, the following relationship must be fulfilled:

$$\alpha_v A_v = 1 \qquad\qquad (18.1)$$

**Barkhausen criterion**
The relationship between the circuit feedback factor ($\alpha_v$) and voltage gain ($A_v$) for proper oscillator operation.

This relationship is called the **Barkhausen criterion.** The results of *not* following this criterion are as follows:

1. If $\alpha_v A_v < 1$, the oscillations will fade out within a few cycles.
2. If $\alpha_v A_v > 1$, the oscillator will drive itself into saturation and cutoff clipping.

These points are easy to understand if we apply them to a couple of circuits using several different gain–value combinations. As a reference, we will use the circuits shown in Figure 18.3. In Figure 18.3a, the output voltage is shown to be the product of the amplifier gain ($A_v$) and the input voltage ($v_f$). By formula,

$$v_{out} = A_v v_f$$

The value of $v_f$ depends on the values of $\alpha_v$ and $v_{out}$, as follows:

$$v_f = \alpha_v v_{out}$$

Now we will use these equations to see what is happening in Figure 18.3a as the circuit progresses through several cycles of operation. We will assume that the initial input to the amplifier is a 0.1 $V_{pk}$ signal. Starting with this value, and using $A_v = 100$ and $\alpha_v = 0.005$ (as shown), the circuit progresses as follows:

| Cycle | $v_{in}$ | $v_{out}$ | $v_f$ |
|-------|----------|-----------|-------|
| 1 | 0.1 $V_{pk}$ | 10 $V_{pk}$ | 0.05 $V_{pk}$ |
| 2 | 0.05 $V_{pk}$ | 5 $V_{pk}$ | 0.025 $V_{pk}$ |
| 3 | 0.025 $V_{pk}$ | 2.5 $V_{pk}$ | 0.0125 $V_{pk}$ |
| 4 | 0.0125 $V_{pk}$ | 1.25 $V_{pk}$ | 0.00625 $V_{pk}$ |

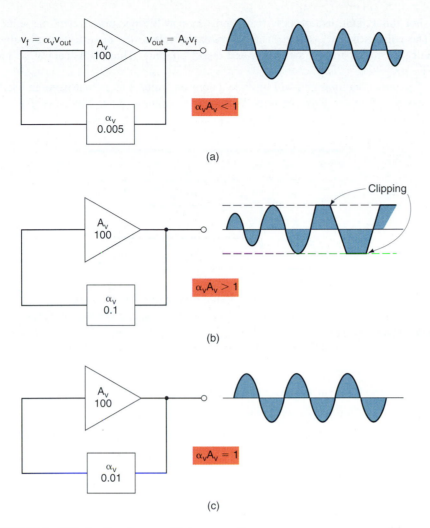

FIGURE 18.3   Effects of $\alpha_v A_v$ on oscillator operation.

Note how the value of $v_f$ produced by each cycle is used as $v_{in}$ for the next. This follows the basic operating principle of the oscillator. Now take a look at the progression of $v_{out}$ values. As you can see, $v_{out}$ decreases from each cycle to the next. If this progression were to continue, $v_{out}$ would eventually reach 0 V for all practical purposes. This output deterioration is illustrated in Figure 18.3a. Note that this circuit has an $\alpha_v A_v$ product that is less than 1. As you can see, when $\alpha_v A_v < 1$, the oscillations will lose amplitude on each progressive cycle and will eventually fade out. This loss of signal is called **damping**.

When $\alpha_v A_v$ is greater than 1, the opposite situation will occur. This situation is shown in Figure 18.3b. Using the same initial 0.1 $V_{pk}$ input as before and the circuit values shown, the circuit in Figure 18.3b would progress as follows:

**Damping**
The fading and loss of oscillations that occurs when $\alpha_v A_v < 1$.

| Cycle | $v_{in}$ | $v_{out}$ | $v_f$ |
|-------|----------|-----------|-------|
| 1 | 0.1 $V_{pk}$ | 10 $V_{pk}$ | 1 $V_{pk}$ |
| 2 | 1 $V_{pk}$ | 100 $V_{pk}$ | 10 $V_{pk}$ |
| 3 | 10 $V_{pk}$ | 1000 $V_{pk}$ | 100 $V_{pk}$ |

For this circuit, it took only two cycles to start hitting some ridiculous values for $v_{out}$. The point is that $v_{out}$ is increasing from each cycle to the next. Eventually, the output will start to experience saturation and cutoff clipping. Thus, $\alpha_v A_v$ cannot be greater than 1.

The only condition that will provide a constant sinusoidal output from an oscillator is to have $\alpha_v A_v = 1$. This condition is shown in Figure 18.3c. Now, using the same 0.1 $V_{pk}$ initial input, let's construct the cycle chart for Figure 18.3c.

| Cycle | $v_{in}$ | $v_{out}$ | $v_f$ |
|-------|----------|-----------|-------|
| 1 | 0.1 $V_{pk}$ | 10 $V_{pk}$ | 0.1 $V_{pk}$ |
| 2 | 0.1 $V_{pk}$ | 10 $V_{pk}$ | 0.1 $V_{pk}$ |
| 3 | 0.1 $V_{pk}$ | 10 $V_{pk}$ | 0.1 $V_{pk}$ |

It is obvious that the cycle will continue over and over. Thus, having a product of $\alpha_v A_v = 1$ will cause the oscillator to have a consistent sinusoidal output, which is the only acceptable situation.

◄ OBJECTIVE 4

Oscillator circuit requirements.

Now we have established the three requirements for an oscillator:

1. Regenerative feedback

2. An initial input trigger to start the oscillations

3. $\alpha_v A_v = 1$ (the Barkhausen criterion)

As long as all these conditions are fulfilled, we have an oscillator. Now, we will look at a variety of circuits to see how these requirements are fulfilled.

## Section Review

1. What is an *oscillator*?

2. What is *positive feedback*?

3. In terms of phase shifts, how is positive feedback usually produced?

4. How does positive feedback maintain the oscillations started by a trigger?

5. What is the *Barkhausen criterion*?

6. What happens when $\alpha_v A_v > 1$?

7. What happens when $\alpha_v A_v < 1$?

8. What is *damping*?

9. List the three requirements for proper oscillator operation.

# 18.2 PHASE-SHIFT OSCILLATORS

◄ OBJECTIVE 5

Probably the easiest oscillator to understand is the *phase-shift oscillator*. This circuit contains *three RC* circuits in its feedback network, as shown in Figure 18.4.

In your study of basic electronics, you were introduced to the concept of the *phase shift* that is produced by an RC circuit at a given frequency. As you may recall, the phase shift of a given RC circuit is found as

$$\theta = \tan^{-1} \frac{-X_C}{R} \qquad (18.2)$$

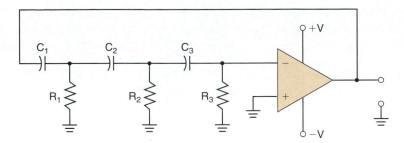

**FIGURE 18.4** Basic phase-shift oscillator.

*A Circuit Variation:* Figure 18.4 shows a phase-shift oscillator made up of series resistors and shunt capacitors. You can also construct a phase-shift oscillator using shunt resistors and series capacitors; that is, by reversing the capacitor and resistor locations.

where    $\theta$ = the phase angle of the circuit
         $\tan^{-1}$ = the *inverse* tangent of the fraction

At this point, we are not interested in the exact value of $\theta$. What we *are* interested in is the fact that $\theta$ changes with $X_C$, and thus with frequency. In other words, a given $RC$ circuit can be designed to produce a specific phase shift at a given frequency.

Now assume that the **phase-shift oscillator** is designed so that the three $RC$ circuits produce a combined phase shift of $180°$ at a resonant frequency, $f_r$. This will cause the circuit to oscillate at that frequency, provided that the Barkhausen criterion has been met. For example, let's assume that the circuit in Figure 18.4 has been designed to produce a $180°$ phase shift at 10 kHz and that, at this frequency, the Barkhausen criterion is met. Then the circuit will oscillate, producing a sinusoidal output waveform.

You would think that each $RC$ circuit in the phase-shift oscillator would be designed to produce a $60°$ phase shift, with the three $60°$ shifts combining to produce the $180°$ shift needed for regenerative feedback. However, this is not the case. Each $RC$ circuit in the phase-shift oscillator acts as a *load* on the previous $RC$ circuit. Just as the loaded-$Q$ of an amplifier was different than the unloaded-$Q$, the phase shift of a loaded $RC$ circuit changes from that of an unloaded $RC$ circuit. Thus, the exact phase shift of each $RC$ circuit in the phase-shift oscillator will be different from the next. However, the overall phase shift of the three will still total $180°$.

**Phase-shift oscillator**
An oscillator that uses three $RC$ circuits in its feedback network to produce a $180°$ phase shift.

## *Practical Considerations*

You should be aware that phase-shift oscillators are rarely used because they are extremely unstable. **Oscillator stability** is a measure of its ability to maintain an output that is constant in *frequency* and *amplitude*. Phase-shift oscillators are extremely difficult to stabilize in terms of *frequency*. Thus, they cannot be used in any application where timing is critical.

Even though the phase-shift oscillator is rarely used, it is introduced here for two reasons. First, as was stated, the phase-shift oscillator is easy to understand. Thus, it is a valuable learning tool. The second reason is that it is very easy to *accidentally* build a phase-shift oscillator. For example, consider the circuit shown in Figure 18.5. In this circuit, we have three $RC$ networks and a combined transistor phase shift of $180°$ (the first two $180°$ shifts cancel out). The feedback path is provided by the dc power supply, $V_{CC}$. Remember that $V_{CC}$ is simply a dc source connected between the "high" side of the circuit and all the ground connections. If the internal impedance of the dc supply is high enough, any oscillations fed back can cause a significant amount of alternating current to be developed across the supply's internal impedance. This current, fed back to the first stage through its biasing circuit, can make a first-rate (although unintentional) oscil-

**Oscillator stability**
A measure of an oscillator's ability to maintain constant output amplitude and frequency.

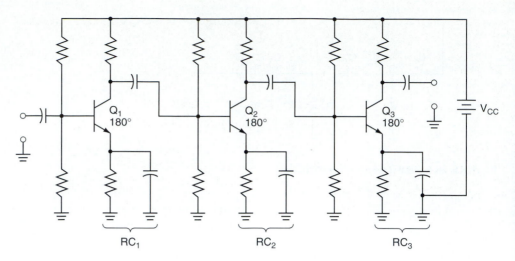

**FIGURE 18.5**

lator out of the circuit. What's the solution? Using a *regulated* power supply. Regulated power supplies have extremely low internal impedance values. This low impedance keeps the ac current caused by the oscillations at a minimum. Thus, while the oscillating current *will* be present, it will be too small to cause any problem. It will simply appear as a small ripple current in the power supply. Another solution is to connect a low-value bypass capacitor across the dc power supply to short any alternating current around the internal resistance of the supply. We discussed the use of a bypass capacitor for this purpose in Chapter 10.

There are several other possible sources of unwanted oscillations. *Stray capacitance* between the first and last stages of an amplifier can cause high-frequency oscillations at the output. Oscillations caused by stray capacitance can be eliminated by increasing the distance between stages or by placing metal shielding over each stage. High-frequency oscillations, called *loop oscillations,* can also be caused by improper ground connections between amplifier stages. Loop oscillations can be prevented by connecting all the stages to the same ground point (if possible) and by keeping the lengths of the ground connections as short as possible.

Another practical consideration involves the Barkhausen criterion. The relationship $\alpha_v A_v = 1$ holds true only for *ideal* circuits. In any practical circuit, the product of $\alpha_v A_v$ must be *slightly greater than 1.* Since each resistive component in an oscillator dissipates *some* amount of power, there is a power loss in the overall circuit. By making $\alpha_v A_v$ slightly greater than 1, the power lost during each cycle is returned to the circuit. How much greater than 1? Just enough to sustain oscillations.

Finally, we must address the question of how the oscillations start to begin with. Consider what happens when power is first applied to the circuit. Since the output is fed back to the *inverting* input, there must exist a 180° phase shift between the two ends of the feedback network. When the power is applied, the circuit *will* establish this phase shift in one of two ways:

1. The input will remain stable and the output will change to the opposite polarity extreme.

2. The output will remain stable, and the input offset voltage will change to the appropriate opposite-polarity voltage.

Which happens is of no consequence. The fact is that, either way, a *transition* occurs between the output and input. This transition is enough to trigger the oscillating process.

*In practice, $\alpha_v A_v$ must be slightly greater than 1.*

*How practical oscillators are triggered.*

1. What is a *phase-shift oscillator*?
2. How is regenerative feedback produced by the phase-shift oscillator?
3. What is *oscillator stability*?
4. Why are phase-shift oscillators rarely used?
5. Explain how a three-stage BJT amplifier can become a phase-shift oscillator.
6. Explain how the problem in Question 5 is prevented.
7. Why must $\alpha_v A_v$ be slightly greater than 1 in a practical oscillator?
8. How is a trigger produced in a practical oscillator?

# 18.3 THE WIEN-BRIDGE OSCILLATOR

The **Wien-bridge oscillator** is one of the more commonly used RC oscillators, up to frequencies of around 1 MHz. This circuit achieves regenerative feedback by producing *no phase shift* at the resonant frequency. In other words, neither the amplifier nor the feedback network will produce a phase shift. This has the same effect as using two 180° phase-shift circuits to produce oscillations. The basic Wien-bridge oscillator is shown in Figure 18.6.

> **Wien-bridge oscillator**
> An oscillator that achieves regenerative feedback by producing no phase shift at $f_r$.

The Wien-bridge oscillator has *two* feedback paths: a positive feedback path (to the noninverting input) and a negative feedback path (to the inverting input). The positive feedback path is used to produce oscillations, while the negative feedback path is used to control the $A_{CL}$ of the circuit.

> Why there are *two* feedback paths.

## The Positive Feedback Path

The positive feedback path consists of $R_1$, $C_1$, $R_2$, and $C_2$. $R_1$ and $C_1$ form a low-pass filter, while $R_2$ and $C_2$ form a high-pass filter. As you were shown in Chapter 17, the series combination of a low-pass filter and a high-pass filter forms a band-pass filter. The circuit will oscillate at the resonant frequency of this band-pass filter.

> The positive feedback circuit is a band-pass filter.

In a typical Wien-bridge oscillator, $R_1 = R_2$ and $C_1 = C_2$. This means that the two circuits have the same cutoff frequency. The result of this condition is illustrated in

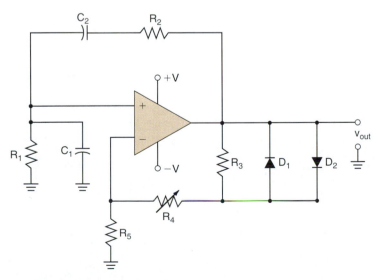

**FIGURE 18.6** Wien-bridge oscillator.

> **Lab Reference:** The operation of the oscillator in Figure 18.6 is demonstrated in Exercise 41.

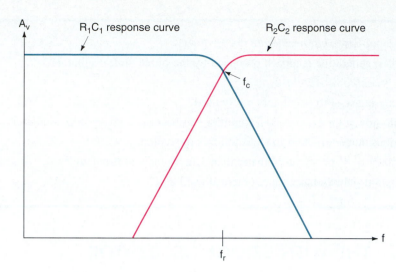

**FIGURE 18.7    Frequency response of the positive feedback path.**

Figure 18.7. Note that the frequency-response curves cross at the $f_c$ of each circuit. This frequency is the frequency of oscillation for the circuit.

As you know, a band-pass filter does not introduce a phase shift when operated at its resonant frequency. Also, there is no phase shift between the noninverting input of the op-amp and the component output terminal. Therefore, the output and input signals are in phase, and the feedback is regenerative.

*Why trimmer pots are used.*    One final note on this circuit: You will often see *trimmer potentiometers* added in series with $R_1$ and $R_2$. These trimmers allow the feedback circuit (and thus the resonant frequency) to be "fine-tuned." The circuit in Figure 18.8 shows the added trimmer pots. The pots are both adjusted to set the frequency of oscillations to the precise value desired.

Why add two potentiometers? Why not just make $R_1$ and $R_2$ potentiometers and save the extra components? The answer to these questions is based on the fact that $R_1$ and $R_2$ are commonly selected to be in the neighborhood of 10 to 200 k$\Omega$. For the circuit to be fine tuned, you want the potentiometers to have a range that is equal to approximately 10 percent of the total series resistance or less. For example, if $R_1 = R_2 = 100$ k$\Omega$, you would want the potentiometers to have a maximum range of 10 k$\Omega$ or less. In all proba-

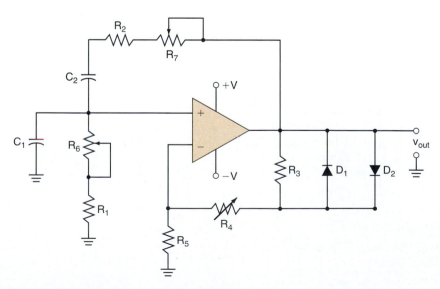

**FIGURE 18.8    Trimmer pots are used for oscillator tuning.**

bility, you would want them to be as low as 1 kΩ. This gives you the ability to vary $f_c$ easily through a relatively small frequency range.

## The Negative Feedback Circuit

The negative feedback circuit is used to set the closed-loop voltage gain ($A_{CL}$) of the circuit. You may recall from earlier discussions that voltage-divider circuits are commonly used for this purpose. The only differences between this feedback path and the other negative feedback circuits covered up until now are:

1. The added potentiometer, $R_4$
2. The diodes in parallel with $R_3$

The potentiometer is added to the circuit to allow adjustment of the circuit $A_{CL}$. The diodes in the feedback network are used to limit the gain of the oscillator. If the oscillator output tries to exceed ($V_{R4} + V_{R5}$) by more than 0.7 V, one of the diodes will turn on, shorting out $R_3$. This results in an increase in the feedback current and a reduction of the overall gain of the op-amp. Thus, the output is prevented from exceeding a specified range. Note that the diodes are essentially acting as *clippers* in this circuit. (And you thought you wouldn't see clippers again.)

**Lab Reference:** The effects of the negative feedback path on circuit operation are observed in Exercise 41.

## Frequency Limits

As you increase the operating frequency of an op-amp, you will eventually reach the point where a phase shift will start to occur between the noninverting input and the output terminal. This phase shift is caused by the **propagation delay** of the op-amp. Propagation delay is the time required for a signal to pass through the component.

When the op-amp starts to introduce a phase shift, the oscillating action of the circuit will start to lose its stability. Remember, the operation of the circuit depends heavily on the phase relationship between the input and output. When you change this relationship, you affect the operation of the oscillator.

For most Wien-bridge oscillators, the upper frequency limit is in the range of 1 MHz. Above this frequency, the stability of the oscillator starts to drop. For any oscillator application, this is an unacceptable situation. Thus, for high-frequency applications, we must use *LC* oscillators. As you will see in the next two sections, *LC* oscillators are much better suited to high-frequency operation.

**Propagation delay**
The time required for a signal to pass through a component.

## Summary

The Wien-bridge oscillator contains an op-amp and two feedback networks. The positive feedback network is used to control the operating frequency of the circuit, while the negative feedback network is used to control its gain. By including variable components in the positive and negative feedback networks, both the gain and the frequency of operation for the circuit can be adjusted.

Because of the propagation delay of the op-amp, Wien-bridge oscillators are usually restricted to operating frequencies of 1 MHz or less. The Wien-bridge oscillator characteristics are summarized in Figure 18.9.

1. Describe the construction of the Wien-bridge oscillator.
2. How does the positive feedback network of the Wien-bridge oscillator control its operating frequency?
3. Explain why potentiometers are normally included in the positive feedback network of a Wien-bridge oscillator.

*Section Review*

The Wien-Bridge Oscillator

Schematic diagram:

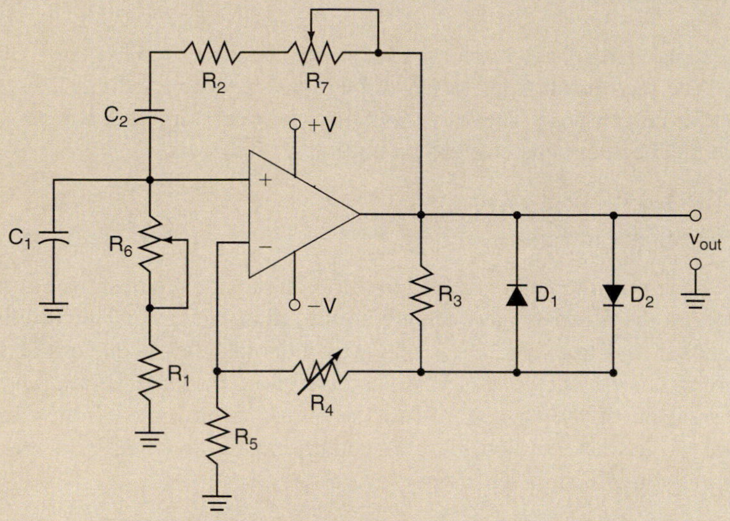

*A Practical Consideration:* The frequency stability of the Wien-bridge oscillator is much higher than that of the phase-shift oscillator. However, it will still experience some frequency drift due to component tolerances and heating. Thus, it is normally used in low-frequency applications where the *exact* operating frequency isn't critical.

| | |
|---|---|
| Circuit recognition: | The dual-feedback networks with the band-pass filter in the positive feedback network. |
| Gain: | Controlled by the negative feedback network. A potentiometer (R₄) is included to allow adjustment. |
| Operating frequency: | • Controlled by the band-pass circuit in the positive feedback network. Added potentiometers (R₆ and R₇) allow fine tuning of the output frequency. |
| | • Operating frequency is limited by the propagation delay of the op-amp; normally restricted to 1 MHz or less. |
| Applications: | Relatively low-frequency systems where the oscillator output frequency is not critical; that is, where some amount of frequency drift can be tolerated by the system. |

**FIGURE 18.9**

4. Explain why a potentiometer is normally included in the negative feedback network of a Wien-bridge oscillator.

5. What is *propagation delay*?

6. How does propagation delay limit the operating frequency of a Wien-bridge oscillator?

## 18.4  THE COLPITTS OSCILLATOR

◄ OBJECTIVE 6

**Colpitts oscillator**
An oscillator that uses a pair of tapped capacitors and an inductor to produce the 180° phase shift in the feedback network.

The **Colpitts oscillator** is a discrete *LC* amplifier that uses a pair of *tapped capacitors* and an inductor to produce the regenerative feedback necessary for oscillations. The Colpitts oscillator is shown in Figure 18.10. As you can see, the transistor configuration is a common emitter. There is a 180° phase shift from the transistor base to the collector. This means that the feedback network must produce a 180° phase shift if the feedback is to be regenerative.

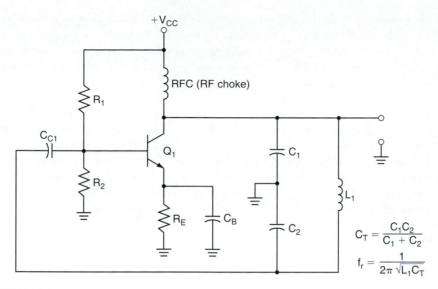

FIGURE 18.10 Colpitts oscillator.

$$C_T = \frac{C_1 C_2}{C_1 + C_2}$$

$$f_r = \frac{1}{2\pi \sqrt{L_1 C_T}}$$

Lab Reference: The operation of an oscillator like the one in Figure 18.10 is demonstrated in Exercise 42.

## The Feedback Network

The key to understanding the Colpitts oscillator is knowing how the feedback network achieves a 180° phase shift. The feedback network for the circuit in Figure 18.10 consists of $C_1$, $C_2$, and $L_1$.

There are several key points that must be made clear so you can get the picture of the overall circuit operation:

1. The output voltage from the amplifier is felt across $C_1$.

2. The feedback voltage is developed across $C_2$.

3. The voltage across $C_2$ is 180° out of phase with the voltage across $C_1$. Therefore, the feedback voltage is 180° out of phase with the output voltage.

Points 1 and 2 are easy to see if we simplify the circuit as shown in Figure 18.11. In Figure 18.11a, we have dropped the inductor $L_1$ from the circuit. Then, in Figure 18.11b, we

Circuit operation key points.

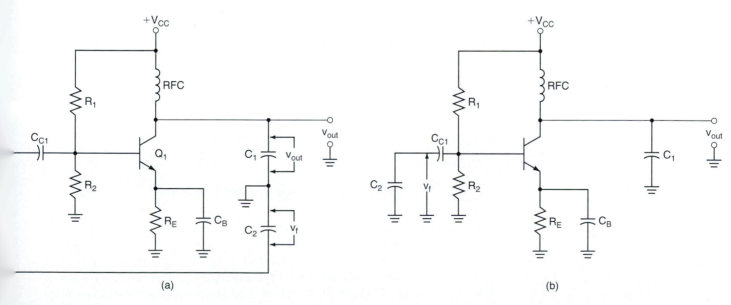

(a)                                                                (b)

FIGURE 18.11

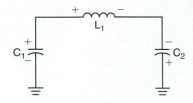

**FIGURE 18.12**

How the feedback circuit produces a 180° phase shift.

split the capacitors at the ground connection. As you can see, the input voltage to the amplifier is developed across $C_2$. Therefore, $C_2$ must be the source of the feedback voltage. $C_1$ is obviously across the output of the amplifier. Therefore, $v_{out}$ is measured across this component. Now, the trick is to see how the voltages across these two components are always 180° out of phase.

The phase relationship between $v_{out}$ and $v_f$ is caused by the *current* action in the feedback network. This action is easy to see if we redraw the feedback network as shown in Figure 18.12. By comparing the circuit in Figure 18.12 with the feedback network in Figure 18.10, you can see that we are dealing with exactly the same circuit in both illustrations. Now, let's assume for a moment that $L_1$ has a voltage across it with the polarities shown. If we view $L_1$ as the voltage source, it is easy to see that it will produce a current in the circuit. With this current in the circuit, voltages are developed across $C_1$ and $C_2$ with the polarities shown. As you can see, these voltages are 180° out of phase. If you reverse the voltage polarity across $L_1$, all polarity signs and current directions reverse, but the voltages across $C_1$ and $C_2$ are still 180° out of phase.

The amount of feedback voltage in the Colpitts oscillator depends on the $\alpha_v$ of the circuit. For this oscillator, $\alpha_v$ is the ratio of $X_{C2}$ to $X_{C1}$. By formula,

$$\alpha_v = \frac{X_{C2}}{X_{C1}} \tag{18.3}$$

Since $X_{C2}$ and $X_{C1}$ are both inversely proportional to the values of $C_2$ and $C_1$ at a given frequency, equation (18.3) can be rewritten as

$$\alpha_v = \frac{C_1}{C_2} \tag{18.4}$$

The validity of equation (18.4) is demonstrated in Example 18.1.

---

**EXAMPLE 18.1**

Determine the value of $\alpha_v$ for the circuit in Figure 18.13 using both equations (18.3) and (18.4).

*Solution:* Before we can determine $\alpha_v$ with equation (18.3), we have to determine the reactance of the two capacitors. Before we can determine $X_{C1}$ and $X_{C2}$, we have to find the operating frequency of the circuit. As always, the operating frequency is equal to the resonant frequency of the feedback network. To find $f_r$, we need to determine the total capacitance in the $C_1$–$C_2$–$L_1$ circuit.

Figure 18.11 demonstrated the fact that $C_1$ and $C_2$ are in *series*. Therefore,

$$C_T = \frac{C_1 C_2}{C_1 + C_2}$$

$$= 3.19 \text{ nF}$$

Now, we can use this value to find $f_r$ as follows:

$$fr = \frac{1}{2\pi \sqrt{L_1 C_T}}$$

$$= 411 \text{ kHz}$$

Using the value of $f = 411$ kHz, we can now calculate values of $X_{C1} = 117.34 \ \Omega$ and $X_{C2} = 3.87 \ \Omega$. Using these values in equation (18.3), we get

$$\alpha_v = \frac{X_{C2}}{X_{C1}}$$

$$= \frac{3.87 \ \Omega}{117.34 \ \Omega}$$

$$= 0.03298$$

$$\cong 0.033$$

Now, if we determine the value of $\alpha_v$ using equation (18.4), we obtain

$$\alpha_v = \frac{C_1}{C_2}$$

$$= \frac{3.3 \ \text{nF}}{0.1 \ \mu\text{F}}$$

$$= 0.033$$

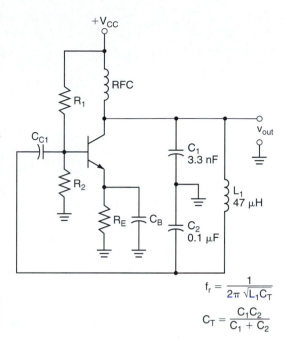

$$f_r = \frac{1}{2\pi \ \sqrt{L_1 C_T}}$$

$$C_T = \frac{C_1 C_2}{C_1 + C_2}$$

**FIGURE 18.13**

### PRACTICE PROBLEM 18.1

A Colpitts oscillator like the one in Figure 18.13 has the following values: $C_1 = 10$ nF, $C_2 = 1.5$ μF, and $L = 10$ μH. Calculate the value of $\alpha_v$ for the circuit using both equations (18.3) and (18.4).

Given the fact that none of us enjoy doing things the hard way, equation (18.4) is definitely the way to go when you want to determine the value of $\alpha_v$.

## Circuit Gain

As with any other oscillator, the product of $\alpha_v A_v$ for the Colpitts oscillator must be slightly greater than 1. This prevents loss of oscillations due to power losses within the amplifier and, at the same time, prevents the circuit from driving itself into saturation and cutoff clipping.

Since the feedback network in the Colpitts oscillator is a parallel resonant tank circuit, it will draw very little current from the transistor collector circuit. Therefore, it can be ignored in determining the value of $A_v$ for the circuit. The value of $A_v$ is found as

$$A_v = \frac{v_{out}}{v_f} \cong \frac{C_2}{C_1}$$

## Amplifier Coupling

As with any parallel resonant tank circuit, the feedback network loses some efficiency when loaded down. To reduce the loading effects of $R_L$, Colpitts oscillators are commonly transformer coupled to the load. A transformer-coupled Colpitts oscillator is shown in Figure 18.14. For this circuit, the primary winding of $T_1$ is the feedback network inductance. Remember that the transformer reduces circuit loading because of the effects of the turns ratio on the reflected load impedance. This point was discussed in Chapter 17.

Also, it is acceptable to use capacitive coupling for the Colpitts oscillator provided that the following relationship is fulfilled:

$$C_C \ll C_T \qquad\qquad \textbf{(18.5)}$$

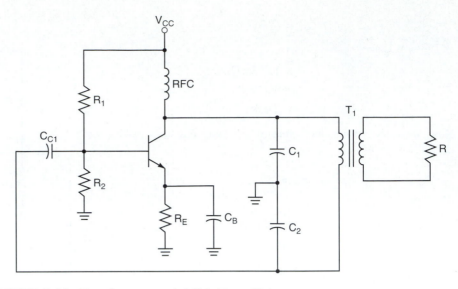

**FIGURE 18.14   Transformer-coupled Colpitts oscillator.**

where $C_T$ is the total series capacitance of the feedback network. By fulfilling this relationship, you ensure that the reactance of the coupling capacitor is much greater than that of the series combination of $C_1$ and $C_2$. This prevents circuit loading almost as well as using transformer coupling.

## Summary

The Colpitts oscillator uses a pair of tapped capacitors and an inductor (all in its feedback network) to produce the 180° feedback phase shift required for oscillations. The frequency of operation for the circuit is approximately equal to the resonant frequency of the feedback circuit.

**Section Review**

1. What is the *Colpitts oscillator*?
2. Explain how the feedback network in the Colpitts oscillator produces a 180° voltage phase shift.
3. List the key points to remember about the operation of the Colpitts oscillator.
4. How do you calculate the value of $A_v$ for a Colpitts oscillator?
5. Why is transformer coupling used in Colpitts oscillators?

## 18.5 OTHER *LC* OSCILLATORS

*OBJECTIVE 7* ▶  Although the Colpitts is the most commonly used *LC* oscillator, several others are worth mentioning here. In this section, we will take a brief look at three other *LC* oscillator circuits, starting with the Hartley oscillator.

## Hartley Oscillators

The *Hartley oscillator* is almost identical to the Colpitts oscillator. The only difference is that the Hartley oscillator uses *tapped inductors* and a single capacitor. The Hartley oscillator is shown in Figure 18.15.

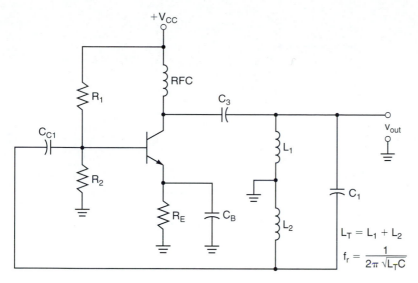

**FIGURE 18.15   Hartley oscillator.**

*A Practical Consideration:* $C_3$ in Figure 18.15 is a *blocking capacitor.* It is included to prevent the RF choke (RFC) and $L_1$ from shorting $V_{CC}$ to ground. The value of $C_3$ is too high (by design) to consider in any circuit frequency calculations.

For the **Hartley oscillator,** the feedback voltage is measured across $L_2$ and the output voltage is measured across $L_1$. Thus, the value of $\alpha_v$ is found as

$$\alpha_v = \frac{X_{L2}}{X_{L1}} \qquad (18.6)$$

**Hartley oscillator**
An oscillator that uses a pair of tapped inductors and a parallel capacitor in its feedback network to produce the 180° voltage phase shift required for oscillation.

or

$$\alpha_v = \frac{L_2}{L_1} \qquad (18.7)$$

Since the inductors are in series in the Hartley oscillator, the total inductance is found as

$$L_T = L_1 + L_2$$

This value of $L_T$ must be used whenever you are calculating the value of $f_r$ for the circuit. Except for these two differences, all calculations and circuit principles are the same for this circuit as for the Colpitts oscillator.

## The Clapp Oscillator

The **Clapp oscillator** is a variation on the Colpitts, as is the Hartley oscillator. The Clapp oscillator was designed to eliminate the effects of *stray capacitance* on the operation of the basic Colpitts. This is done by adding another capacitor, as is shown in Figure 18.16. Just for a moment, ignore the presence of $C_3$ in the Clapp oscillator. As you can see, all you have without $C_3$ is a standard Colpitts oscillator.

**Clapp oscillator**
A Colpitts oscillator with an added capacitor (in series with the feedback inductor) that is used to reduce the effects of stray capacitance.

If you refer back to Figure 18.11, you will see that $C_1$ and $C_2$ in the feedback network of the Colpitts oscillator are in parallel with the rest of the transistor circuit. As you have seen countless times, the transistor has junction capacitance values. These capacitance values are in parallel with $C_1$ and $C_2$ and thus affect the total amount of capacitance in the feedback network. This affects the operating frequency of the circuit.

The Clapp oscillator eliminates the effects of transistor capacitance by adding the capacitor $C_3$. This capacitor is normally much lower in value than $C_1$ or $C_2$. Because of this, it becomes the dominant component in all frequency calculations for the circuit. For the Clapp oscillator,

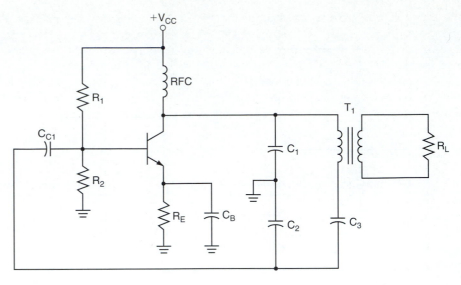

**FIGURE 18.16    Clapp oscillator.**

$$f_r = \frac{1}{2\pi \sqrt{LC_3}}$$

(18.8)

You may be wondering why $C_1$ and $C_2$ are even included in the Clapp oscillator if $C_3$ is the dominant component. The answer to this is simple. Even though the operating frequency of the circuit depends on the value of $C_3$, $C_1$ and $C_2$ are still needed to provide the 180° phase shift required for regenerative feedback. $C_3$ has not replaced $C_1$ and $C_2$; it has just eliminated their values from any frequency considerations. In fact, except for the $f_r$ formula, all other calculations for the Clapp oscillator are exactly the same as those for the Colpitts oscillator.

## The Armstrong Oscillator

**Armstrong oscillator**
An oscillator that uses a transformer in its feedback network to achieve the required 180° voltage phase shift.

The **Armstrong oscillator** is a simple (but effective) oscillator circuit. The basic circuit is shown in Figure 18.17. In the Armstrong oscillator, the required 180° phase shift in the

*Note:* $C_2$ is a blocking capacitor used to prevent the RFC and the primary of $T_1$ from shorting out the dc supply.

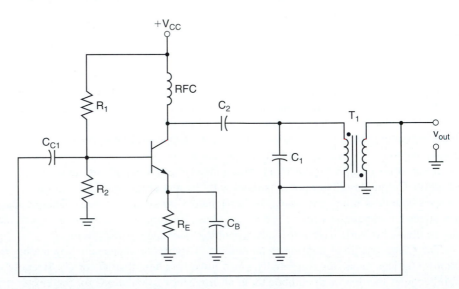

**FIGURE 18.17    Armstrong oscillator.**

feedback network is provided by the transformer, $T_1$. As the polarity dots indicate, the output of the amplifier will be inverted from the transformer primary to its secondary. This provides the 180° phase shift of $v_f$. Note that $v_f$ is developed across the secondary of the oscillator.

The capacitor in the output circuit ($C_1$) is there to provide the tuning of the oscillator. The resonant frequency of the circuit will be determined by the value of $C_1$ and the transformer primary.

## One Final Note

You have been introduced to several *LC* oscillators in the preceding two sections. The BJT oscillators make up only part of the entire picture. Any *LC* oscillator can be constructed using either an FET or an op-amp. For example, consider the circuits shown in Figure 18.18. Figure 18.18a shows a Colpitts oscillator using an op-amp. Figure 18.18b shows an FET Hartley oscillator. These oscillators work according to the same basic principles as their BJT counterparts, so don't let the different active components confuse you.

Another situation that you will commonly see is an oscillator that has the feedback network going from the output of one stage back to the emitter, source, or noninverting input of a previous stage. Again, the overall principles of such a circuit are the same as those discussed earlier.

Recognizing oscillator circuits.

The key to being able to deal with the various oscillators is to learn the circuit recognition features of the feedback networks. For example, the Colpitts oscillator has two tapped capacitors and an inductor in the feedback network. Anytime you see an oscillator with this configuration, it is a Colpitts. It really does not matter what the active component is or to which terminal the feedback network leads. If the circuit is an oscillator with tapped capacitors and an inductor in the feedback network, it is a Colpitts. The recognition features of the other common *LC* oscillators are as follows:

| Oscillator Type | Recognition Feature(s) |
|---|---|
| Hartley | Tapped inductors or a center-tapped transformer with a single parallel capacitor |
| Clapp | Looks like a Colpitts, with a capacitor added in series with the inductor |
| Armstrong | A single transformer with a parallel capacitor |

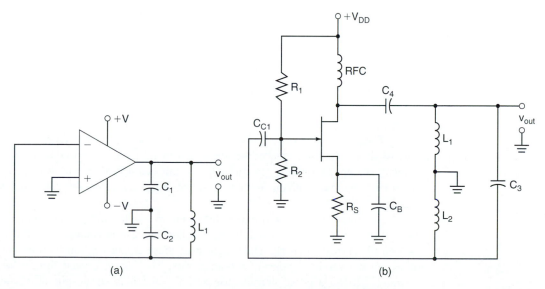

(a)                                        (b)

**FIGURE 18.18**

Once you have established that a given circuit is an oscillator, just remember the recognition features listed. They will tell you which type of circuit you are dealing with.

---

**Section Review**

1. Where are the feedback and output voltages measured in the feedback network of a Hartley oscillator?
2. How is the construction of the Clapp oscillator different from that of the Colpitts oscillator? What purpose does this difference serve?
3. How is the required 180° voltage phase shift accomplished in the feedback network of an Armstrong oscillator?
4. List the circuit recognition features of the *LC* oscillators covered in this section.

---

# *18.6* CRYSTAL-CONTROLLED OSCILLATORS

OBJECTIVE 8 ►

Most communications applications require the use of oscillators that have *extremely* stable output signals. This can pose a problem for any "conventional" oscillator since there are so many things that can cause the output from such an oscillator to change. For example, refer back to the circuit shown in Figure 18.10. The output from this circuit could change if any of the following were to occur:

1. The transistor is replaced. This could change $r_e'$, causing $A_v$ to change, which could change the $\alpha_v A_v$ product. As you know, that can cause you all sorts of grief.
2. Changing any component in the feedback network could change the resonant frequency.
3. The circuit will warm up. This can change resistance values, which can change the load on the feedback network, and so on.

In any system where the stability of the oscillator is critical, the foregoing problems are intolerable. For these types of applications, a **crystal-controlled oscillator** is normally used. Crystal-controlled oscillators have a *quartz crystal* that is used to control the frequency of operation. To help you understand the importance of the crystal, we will take a brief look at what it is and how it works.

**Crystal-controlled oscillators**
Oscillators that use quartz crystals to produce extremely stable output frequencies.

## *Crystals*

**Piezoelectric effect**
The tendency of a crystal to vibrate at a fixed frequency when subjected to an electric field.

The key to the operation of a crystal is called the **piezoelectric effect.** This means that *the crystal vibrates at a constant rate when it is subjected to an electric field.* The frequency of the vibrations depends on the physical dimensions of the crystal. Thus, it is possible to produce crystals with very exact frequency ratings by simply cutting them to the right dimensions.

There are three commonly used crystals that exhibit a piezoelectric effect. When one of these crystals is placed between two metal plates, it vibrates at its *resonant frequency* as long as a voltage is applied. Crystals can also be made to vibrate by applying a signal to them at a given frequency. When this is done, the crystal vibrates at that frequency.

The three crystals used in oscillators are *Rochelle salt, quartz,* and *tourmaline.* The best of the three is Rochelle salt because it has the best piezoelectric activity. However, it is also the easiest to break. The toughest of these crystals is tourmaline, but this crystal does not have a very constant vibration rate. The quartz crystal falls between the two extremes: It has a good piezoelectric activity, and it is strong enough to withstand the vibrations. Quartz is also the least expensive of the three to use.

## Quartz Crystals

A quartz crystal is made of silicon dioxide, $SiO_2$. This is the same compound used as the insulation layer in the gate of a MOSFET. Quartz crystals are very common in nature. They develop as six-sided crystals, as shown in Figure 18.19. When used in an electronic component, a thin slice of crystal is placed between two conducting plates, like those of a capacitor.

As you know, the physical dimensions of a crystal determine its operating frequency. The only factor that will normally alter the physical dimensions of a crystal (and thus its operating frequency) is *temperature*. However, using cooling methods to hold the crystal temperature constant eliminates the effects of temperature.

The electrical operation of the crystal is based on its mechanical properties. However, this does not prevent us from representing the crystal as an equivalent circuit, as shown in Figure 18.20. Figure 18.20a shows the schematic symbol for the crystal. The equivalent circuit for the device is shown in Figure 18.20b. The components shown represent specific characteristics of the crystal, as follows:

$C_C$ = the capacitance of the crystal itself

$C_M$ = the *mounting capacitance* (this is the capacitance between the crystal and the parallel conducting plates that hold it)

$L$ = the inductance of the crystal

$R$ = the resistance of the crystal

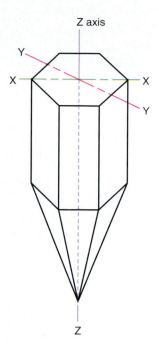

**FIGURE 18.19  Quartz crystal.**

That strange-looking graph in Figure 18.20c tells us quite a bit about the crystal. At very low frequencies, the reactance of the crystal is controlled by the extremely high reactance of $C_M$ and $C_C$. As the frequency of operation increases, the circuit composed of $C_C$ and $L$ approaches its resonant frequency. At the frequency $f_s$, the crystal acts like a *series* resonant circuit. At this frequency, the reactances of $C_C$ and $L$ cancel, and the reactance of the component drops to zero. If the frequency continues to climb, the reactance of the parallel capacitor, $C_M$, starts to drop to a low enough value that it starts to affect the circuit. When $f_p$ is reached, $C_M$ and $L$ resonate, and the circuit acts as a *parallel* resonant circuit. At frequencies above $f_p$, the continuing drop in $C_M$ eventually causes the quartz to act as a short circuit. The bottom line of all this is as follows:

1. At $f_s$, the crystal will act as a series resonant circuit.

2. At $f_p$, the crystal will act as a parallel resonant circuit.

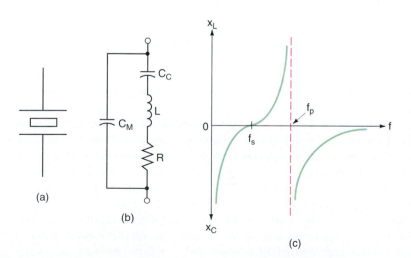

**FIGURE 18.20  Crystal symbol, equivalent circuit, and frequency response.**

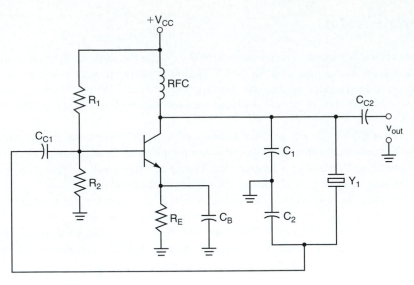

**FIGURE 18.21** Crystal-controlled Colpitts oscillator.

This means that we can use a crystal in place of a series *LC* circuit or in place of a parallel *LC* circuit. If we use it in place of a series *LC* circuit, the oscillator will operate at $f_s$. If we use it in place of a parallel *LC* circuit, the oscillator will operate at $f_p$.

## Overtone Mode

**Fundamental frequency**
The resonant frequency of a crystal or circuit.

**Overtone**
Another word for *harmonic*.

**Overtone mode**
An operating mode where a circuit is tuned to a harmonic of its crystal for the purpose of producing a higher output frequency.

A given crystal produces outputs at its *resonant frequency* and *harmonics* of that frequency. The *resonant frequency* itself is sometimes referred to as the **fundamental frequency,** while the *harmonics* are sometimes referred to as **overtones.** (You were first introduced to the concept of harmonics in our discussion on Class C amplifiers.)

Crystals are limited by their physical dimensions to fundamental outputs of 10 MHz and less. However, by tuning circuits to the overtones, we can produce usable signals at much higher frequencies. For example, tuning a circuit to the tenth harmonic of a 10 MHz crystal produces a relatively stable 100MHz signal. This **overtone mode** of operation is commonly used in digital and communications electronics.

## CCO Circuits

The Colpitts oscillator can be modified into a crystal-controlled oscillator (CCO), as shown in Figure 18.21. As you can see, the only difference here is the addition of the crystal ($Y_1$) in the feedback network. The crystal will act as a parallel-resonant circuit at $f_p$, allowing the maximum allowable energy transfer through the feedback network at that frequency.

To understand the effectiveness of this oscillator, you must realize that the parallel resonant characteristics of the filter will exist only for an extremely limited band of frequencies. Even the smallest deviation from $f_p$ will cause the crystal to act as an effective short. Therefore, the oscillator is capable of oscillating only at $f_p$, and we have an extremely stable oscillator. Note that placing the crystal in the same relative position in either a Hartley or Clapp oscillator will have the same effect.

The simplest of the crystal oscillators is the *Pierce oscillator,* shown in Figure 18.22. It has only four components and a crystal. This oscillator is not highly stable, but its simplicity has kept it around for years, and it will probably be around for years to come.

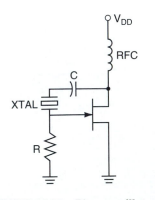

**FIGURE 18.22** Pierce oscillator.

**Section Review**

1. What is a *crystal-controlled oscillator*?
2. What is the *piezoelectric effect*?
3. Explain how a crystal will act as a series resonant circuit at one frequency ($f_s$) and as a parallel resonant component at another ($f_p$).
4. Explain how the addition of a crystal to an *LC* oscillator stabilizes the operating frequency of the circuit.

# 18.7 OSCILLATOR TROUBLESHOOTING

Of all the circuits that we have discussed in this text, oscillators are by far the most challenging when it comes to troubleshooting. They can also be the most time-consuming circuits to troubleshoot. The basic problem is this: Every component in the oscillator, with the exception of the biasing resistors, is involved in the production of the output signal. If any one of them goes bad, nothing will work. For example, refer back to Figure 18.14. Whether a failure of the circuit is caused by $C_1$, $C_2$, $T_1$, or $C_{C1}$, the results are exactly the same—the oscillator does not have an output.

◄ *OBJECTIVE 9*

As with any circuit, you start to troubleshoot an oscillator by making sure that it is the source of the problem. If the circuit does not have an output, check the power supply connections in the circuit. If there is no problem here, the oscillator has a fault. (You do not need to check the oscillator's signal source—it doesn't have one.)

When you have verified that an oscillator is bad, there are a few points to remember that may make the troubleshooting process go a little more smoothly:

1. Remember that, for dc analysis purposes, the oscillator is biased as an emitter, source, or voltage follower. Thus, your dc troubleshooting procedure should be the same as it is for these circuit types.
2. In the case of the oscillator, it is often faster (and therefore less costly) to go ahead and replace the reactive components in the feedback network.

If circumstances require you to test the components in the feedback network, use an analog VOM (set for resistance measurements) to test them and remember the following points:

*A Practical Consideration:* Most oscillators contain one or more variable components for output tuning. When you replace any component in the circuit, you must check the output frequency and, if necessary, retune the circuit.

1. A good capacitor will initially show a low resistance. Then, as the capacitor is charged by the ohmmeter, the resistance reading will rise.
2. An inductor will always have a low resistance reading. When a transformer is used, as in Figure 18.14, remember that you should measure a low resistance between the two primary connections and the two secondary connections. However, you should read a very high resistance when measuring from either primary connection to either secondary connection.

If you decide to go the component-replacement (or *swapping*) route, replace the reactive components one at a time. After you replace each component, apply power to the circuit and retest its operation. When the circuit starts to operate at the proper frequency, your task is complete.

**Section Review**

1. Why are oscillators difficult to troubleshoot?
2. What points should be remembered when troubleshooting an oscillator?
3. When testing the reactive components in an oscillator, what points should be kept in mind?

## Key Terms

The following terms were introduced and defined in this chapter:

| | | |
|---|---|---|
| Armstrong oscillator | damping | phase-shift oscillator |
| Barkhausen criterion | fundamental frequency | piezoelectric effect |
| Clapp oscillator | Hartley oscillator | positive feedback |
| Colpitts oscillator | oscillator | propagation delay |
| crystal-controlled oscillator | overtone | regenerative feedback |
| | overtone mode | Wien-bridge oscillator |

## Equation Summary

| Equation Number | Equation | Section Number |
|---|---|---|
| (18.1) | $\alpha_v A_v = 1$ | 18.1 |
| (18.2) | $\theta = \tan^{-1} \dfrac{-X_C}{R}$ | 18.2 |
| (18.3) | $\alpha_v = \dfrac{X_{C2}}{X_{C1}}$ | 18.4 |
| (18.4) | $\alpha_v = \dfrac{C_1}{C_2}$ | 18.4 |
| (18.5) | $C_C \ll C_T$ | 18.5 |
| (18.6) | $\alpha_v = \dfrac{X_{L2}}{X_{L1}}$ | 18.5 |
| (18.7) | $\alpha_v = \dfrac{L_2}{L_1}$ | 18.5 |
| (18.8) | $f_r = \dfrac{1}{2\pi\sqrt{LC_3}}$ | 18.5 |

## Answer to the Example Practice Problem

**18.1.** $\alpha_v = 6.67 \times 10^{-3}$ (for both solutions)

## Practice Problems

### §18.3

1. Calculate the operating frequency of the circuit in Figure 18.23.
2. Calculate the operating frequency of the circuit in Figure 18.24.

### §18.4

3. Calculate the operating frequency for the circuit in Figure 18.25.
4. Calculate the value of $\alpha_v$ for the circuit in Figure 18.26.
5. A circuit like Figure 18.25 has $C_1 = 10$ nF, $C_2 = 1$ µF, and $L_1 = 3.3$ mH. Calculate the operating frequency, the value of $\alpha_v$, and the value of $A_v$ for the circuit.
6. A circuit like Figure 18.25 has $C_1 = 1$ µF, $C_2 = 10$ µF, and $L_1 = 3.3$ mH. Calculate the operating frequency, the value of $\alpha_v$, and the value of $A_v$ for the circuit.

FIGURE 18.23

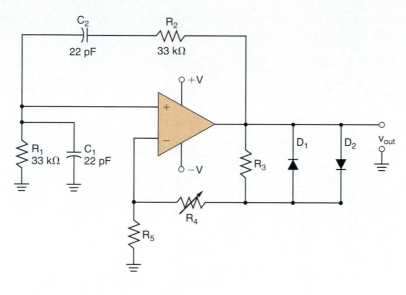

FIGURE 18.24

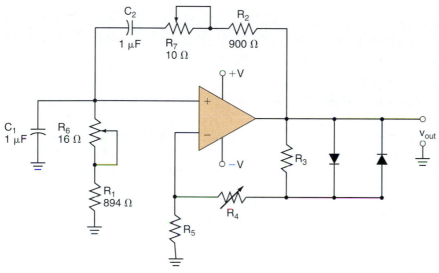

FIGURE 18.25

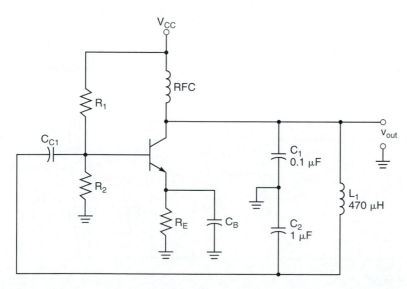

FIGURE 18.26

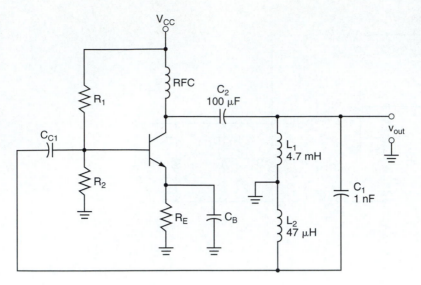

7. A circuit like Figure 18.25 has $C_1 = 0.1$ μF, $C_2 = 1$ μF, and $L_1 = 470$ μH. Calculate the operating frequency, the value of $\alpha_v$, and the value of $A_v$ for the circuit.

### §18.5

8. Calculate the operating frequency, the value of $\alpha_v$, and the value of $A_v$ for the circuit in Figure 18.26.

9. A circuit like Figure 18.26 has $L_1 = 0.1$ H, $L_2 = 1$ mH, and $C_1 = 22$ nF. Calculate the operating frequency, the value of $\alpha_v$, and the value of $A_v$ for the circuit.

10. For the circuit shown in Figure 18.27, calculate the operating frequency, the value of $\alpha_v$, and the value of $A_v$.

11. A circuit like the one in Figure 18.27 has $C_1 = 0.1$ μF, $C_2 = 1$ μF, $C_3 = 100$ pF, and $L_{T1} = 3.3$ mH. Calculate the operating frequency, the value of $\alpha_v$, and the value of $A_v$ for the circuit.

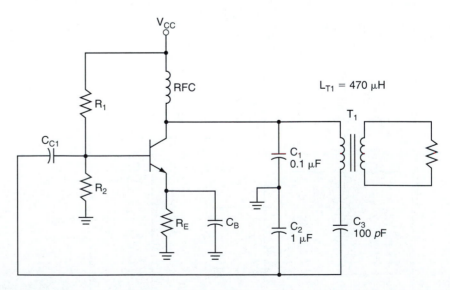

**FIGURE 18.27**

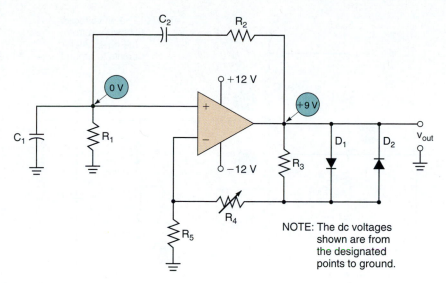

**FIGURE 18.28**

---

**12.** The circuit in Figure 18.28 does not oscillate. A check of the dc voltages provides the readings indicated. Discuss the possible cause(s) of the problem.

**13.** The circuit in Figure 18.29 does not oscillate. A check of the dc voltages provides the readings indicated. Discuss the possible cause(s) of the problem.

**14.** The circuit in Figure 18.30 does not oscillate. A check of the dc voltages provides the readings indicated. Discuss the possible cause(s) of the problem.

*Troubleshooting*
*Practice Problems*

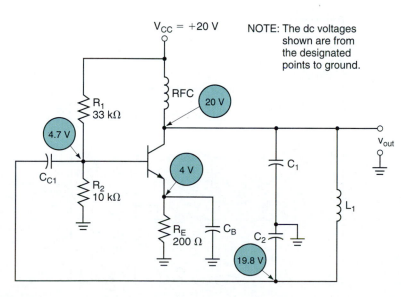

**FIGURE 18.29**

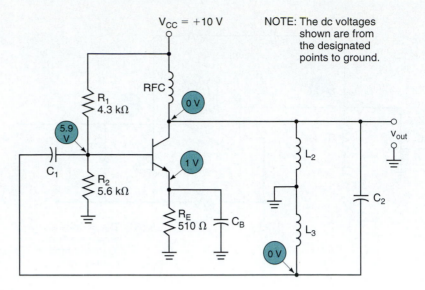

FIGURE 18.30

The Brain Drain

15. Refer to Figure 18.26. Use circuit calculations to show that the circuit fulfills the Barkhausen criterion.

16. Despite its appearance, you've seen the dc biasing circuit in Figure 18.31 before. Determine the type of biasing used, and perform a complete dc analysis of the oscillator.

17. The feedback inductor in Figure 18.31 is shown to be a variable inductor. Determine the adjusted value of $L$ that would be required for an operating frequency of 22 kHz.

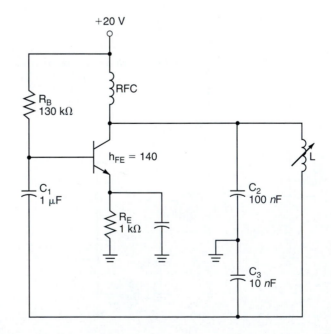

FIGURE 18.31

**18.** Write a program to determine the operating frequency of a Wien-bridge oscillator, given the positive feedback values of $R$ and $C$.

**19.** Write a program to determine the operating frequency and $\alpha_v$ values for a Colpitts oscillator, given the feedback network values of $L$ and $C$.

# 19

# SOLID-STATE SWITCHING CIRCUITS

The study of discrete switching circuits will help you to prepare for the study of modern digital circuits, most of which are made up almost entirely of integrated circuits. The digital circuit shown here is an erasable programmable read-only memory (EPROM).

# OUTLINE

# OBJECTIVES

*After studying the material in this chapter, you should be able to:*

1. Describe and analyze the operation of the basic BJT, JFET, and MOSFET switches.

2. List the common time and frequency characteristics of rectangular waveforms and describe how each is measured.

3. List and describe the factors that affect BJT and FET switching time.

4. List and discuss the methods used to improve the switching times of basic BJT and FET circuits.

5. Compare and contrast the time measurement techniques for *inverters* and *buffers.*

6. Describe and analyze the operation of *inverting Schmitt triggers* and *noninverting Schmitt triggers.*

7. List and describe the output characteristics of the three types of *multivibrators.*

8. Describe the internal construction and operation of the 555 timer.

9. Describe, analyze, and troubleshoot 555 timer *astable* and *monostable* multivibrators.

10. Describe the operation of the 555 timer *voltage-controlled oscillator (VCO).*

817 ■

# COMPUTERS ANYONE?

Solid-state switching circuits are the fundamental components of modern computer systems. It seems appropriate then that we should take a minute to look at the effects that development in solid-state electronics have had on the digital computer.

The first general-purpose digital computer, called ENIAC, was developed in the 1940s at the University of Pennsylvania. This computer was just about as large as a small building, weighed about 50 tons, and contained over 18,000 vacuum tubes. As the story is told, several college students spent their days running around the computer shop with shopping carts full of vacuum tubes since ENIAC burned out several tubes each hour. Amazingly, ENIAC was capable of doing less than your calculator.

When the first transistor computer circuits were developed, the *minicomputer* was born. Minicomputers were capable of performing more advanced functions and were a lot smaller than any vacuum tube computer.

The development of the solid-state *microprocessor,* a complete data-processing unit constructed on a single chip, in the 1970s led to the birth of the *microcomputer.* Since the birth of the microcomputer, computers have been accessible and affordable for almost everyone.

One interesting point: As computer chips became smaller and more powerful, they also became much less expensive. It has been said that, if automobiles had gone through the same evolution as computers, you could now buy a Rolls Royce for less than $35.00. Of course, the car would be smaller than a pencil eraser!

---

**T**hroughout the text, we have concentrated almost entirely on the *linear* operation of solid-state devices and circuits. Another important aspect of solid-state electronics is the way in which devices and circuits respond to *nonlinear* waveforms, such as *square waves.* Circuits designed to respond to nonlinear waveforms (or generate them) are referred to as **switching circuits.** In this chapter, we will discuss the most basic types of switching circuits. As you will see, these circuits are viewed quite differently from linear circuits. They are also easier to understand and troubleshoot.

**Switching circuits**
Circuits designed to respond to (or generate) nonlinear waveforms, such as square waves.

## *19.1* INTRODUCTORY CONCEPTS

For you to understand some of the more complex switching circuits, you must be comfortable with the idea of using a BJT, FET, or MOSFET as a switch. In this section, we will discuss some of the basic principles involved in using a solid-state device as a switch.

### *The BJT As a Switch*

OBJECTIVE 1 ▶

A BJT can be used as a switch simply by driving the component back and forth between *saturation* and *cutoff.* A basic transistor switching circuit is shown in Figure 19.1a. As the figure shows, a rectangular input produces a rectangular output. The way this is accomplished is easy to see if you refer to the dc load line shown in Figure 19.1b.

When the input to the transistor is at $-V$, the emitter–base junction of the transistor is biased off. When the transistor is biased off, the following (ideal) conditions exist:

$$V_{CE} = V_{CC} \qquad \text{and} \qquad I_C = 0$$

Ideal cutoff conditions are similar to those of an *open* switch.

As Figure 19.2a shows, these conditions are the same as those caused by an *open* switch.

When the input to the transistor is at $+V$, the transistor will saturate. This assumes, of course, that the input current is sufficient to cause saturation. When the transistor saturates, the following (ideal) conditions exist:

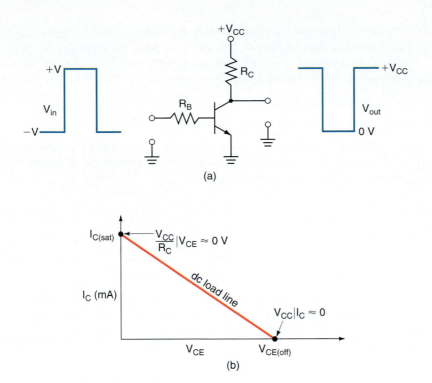

**FIGURE 19.1**  A basic transistor switch.

$$V_{CE} = 0 \text{ V} \qquad \text{and} \qquad I_C = \frac{V_{CC}}{R_C}$$

As Figure 19.2b shows, these conditions are the same as those caused by a *closed* switch. Ideal saturation conditions are similar to those of a *closed* switch.

For the circuit in Figure 19.1, we made several assumptions to simplify the discussion:

1. The $-V_{in}$ is low enough to prevent $Q_1$ from conducting.
2. The $+V_{in}$ is high enough to produce the base current required to saturate $Q_1$.
3. The transistor is an ideal component.

In a practical transistor switching circuit, the value of $-V_{in}$ will usually be equal to the emitter supply connection. For example, in Figure 19.1, $-V_{in}$ would be 0 V, since the transistor emitter is returned to ground. Had the emitter been tied to a negative voltage,

**FIGURE 19.2**  The "open" and "closed" transistor switch.

(a) Open switch

(b) Closed switch

say, $-10$ V, then $-V_{in}$ would have been $-10$ V. By making the low input voltage equal to $V_{EE}$, you ensure that the transistor will be in cutoff when the input equals $-V_{in}$.

Practical input signal characteristics.

The high input voltage, $+V_{in}$, must ensure that the transistor saturates. In a practical transistor switch, this is accomplished by two things:

1. $+V_{in}$ is usually equal to $+V_{CC}$.

2. $R_B$ is made small enough so that the calculated value of $I_B \gg I_{C(sat)}/h_{FE}$.

The result of fulfilling both these requirements can be seen in Example 19.1. Here, we use the relationships established in Chapter 7 to determine the minimum value of $+V_{in}$ required to saturate a transistor switch.

---

## EXAMPLE 19.1

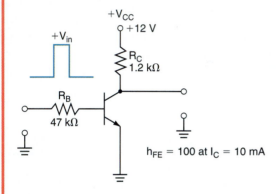

**FIGURE 19.3**

Determine the minimum high input voltage $(+V_{in})$ required to saturate the transistor switch shown in Figure 19.3.

**Solution:** Assuming that the transistor is an ideal component ($V_{CE} = 0$ V at saturation), the value of $I_{C(sat)}$ is found as

$$I_{C(sat)} = \frac{V_{CC}}{R_C}$$
$$= 10 \text{ mA}$$

As shown, the value of $h_{FE}$ for the transistor is 100 when $I_C$ is 10 mA. Using this value of $h_{FE}$, the value of base current required to cause $I_{C(sat)}$ to equal 10 mA is

$$I_B = \frac{I_{C(sat)}}{h_{FE}}$$
$$= 100 \text{ } \mu A$$

Now recall that the base current for a base-biased circuit like the one shown is found as

$$I_B = \frac{+V_{in} - V_{BE}}{R_B}$$

This equation can be rewritten to give us an equation for $V_{in}$, as follows:

$$V_{in} = I_B R_B + V_{BE}$$

Using the values of $I_B = 100$ $\mu A$, $R_B = 47$ k$\Omega$, and $V_{BE} = 0.7$ V, the minimum value of $+V_{in}$ required to saturate the transistor is found as

$$+V_{in} = I_B R_B + V_{BE}$$
$$= 4.7 \text{ V} + 0.7 \text{ V}$$
$$= 5.4 \text{ V}$$

Since we need $+V_{in}$ to equal only 5.4 V to saturate the transistor, a value of $+V_{in} = V_{CC}$ will definitely do the trick.

### PRACTICE PROBLEM 19.1

A BJT switch like the one in Figure 19.3 has the following values: $V_{CC} = +10$ V, $R_C = 1$ k$\Omega$, $R_B = 51$ k$\Omega$, and $h_{FE} = 70$ at $I_C = 100$ mA. Determine the minimum high input voltage required to saturate the transistor.

In our discussion, we have assumed that the transistor is an *ideal* component; that is, we have assumed that $V_{CE} = V_{CC}$ when the transistor is in cutoff, and $V_{CE} = 0$ V when the transistor is saturated. In practice, the outputs from a transistor switch will fall in the following ranges:

Practical output values for a BJT switch.

| Condition | Output Voltage ($V_{CE}$) Range |
|---|---|
| Cutoff | Within 1 V of $V_{CC}$ |
| Saturation | Between 0.2 and 0.4 V |

When a transistor is in cutoff, there is still some leakage current through the component. This leakage current will cause some voltage to be dropped across the collector resistor. Thus, the output voltage may be a bit lower than $V_{CC}$. When the transistor is saturated, you still have some voltage developed across the internal resistance of the component. Thus, the output will never actually be 0 V. However, the range of output voltages for a saturated transistor is so low that we can easily idealize the situation and say that the output equals 0 V.

## The JFET As a Switch

The basic JFET switch differs from the BJT switch in a couple of ways:

The differences between BJT and JFET switches.

1. The JFET switch has a much higher input impedance.

2. A *negative* input square wave (or *pulse*) to an *n*-channel is used to produce a *positive* output pulse.

Both of these points can be seen by looking at the basic JFET switch shown in Figure 19.4. The high input impedance of the JFET switch is caused by the relatively high value of $R_G$ (typically in megohms) and the extremely high input impedance of the JFET gate. For the circuit shown in Figure 19.4, the input impedance of the switch would be equal to the value of $R_G$.

To understand the input/output voltage relationship of the circuit, we need to refer to the transconductance curve shown in Figure 19.4. Recall that $I_D$ is at its maximum value, $I_{DSS}$, when $V_{GS} = 0$ V. Thus, when the input is at 0 V, the JFET is saturated and the output is found as

JFET switch input/output relationships.

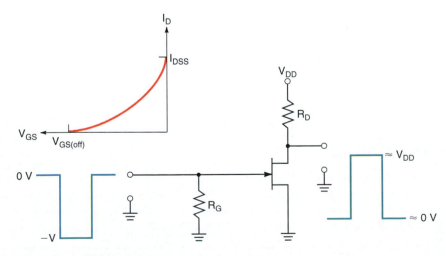

**FIGURE 19.4   A JFET switch.**

$$V_{out} = V_{DD} - I_{DSS}R_D$$

Assuming that the value of $R_D$ has been selected properly, the output from the JFET switch will be close to 0 V when the input is at 0 V.

When the input goes to $-V$, the current through the JFET drops nearly to zero, assuming that $-V$ is greater than or equal to $V_{GS(off)}$. With $I_D$ nearly zero, little voltage is dropped across $R_D$, and the output voltage is close to $V_{DD}$. By formula,

$$V_{out} \cong V_{DD}$$

when $V_{in}$ is more negative than $V_{GS(off)}$. Using these two relationships for $V_{out}$, it is easy to see that the input pulse shown in Figure 19.4 produces the output pulse shown. Example 19.2 demonstrates the analysis of a basic JFET switch.

---

## EXAMPLE 19.2

*A Practical Consideration:.* This example assumes that the JFET has only one value of $I_{DSS}$ and $V_{GS(off)}$. In practice, JFETs typically have minimum and maximum values of $I_{DSS}$ and $V_{GS(off)}$. When this is the case, you must use both values to determine the output voltage *ranges.*

Determine the high and low output voltage values for the circuit shown in Figure 19.5.

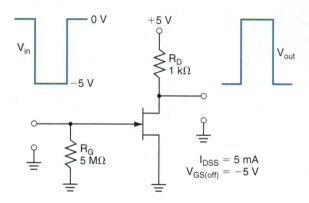

**FIGURE 19.5**

**Solution:** When the input is at 0 V, the JFET has minimum resistance between the source and drain, and $I_D = I_{DSS}$. Thus,

$$V_{out} = V_{DD} - I_{DSS}R_D$$
$$= 5\,V - 5\,V$$
$$= 0\,V$$

When the input is at $-5$ V, it is equal to $V_{GS(off)}$, and thus $I_D$ is zero. Using this value in place of $I_{DSS}$ in the above equation yields

$$V_{out} = V_{DD} - I_D R_D$$
$$= +5\,V - 0\,V$$
$$= +5\,V$$

### PRACTICE PROBLEM 19.2

A JFET switch like the one in Figure 19.5 has the following values: $V_{DD} = +10$ V, $R_D = 1.8$ k$\Omega$, $I_{DSS} = 5$ mA, and $V_{GS(off)} = -3$ V. The input signal to the circuit is a square wave that has low and high output voltages of $-4$ and 0 V, respectively. Determine the high and low output voltage values for the circuit.

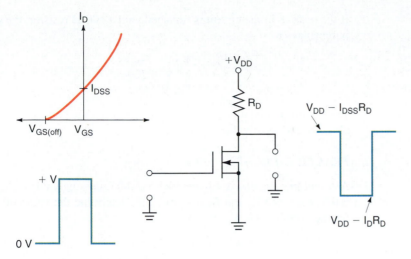

FIGURE 19.6   The MOSFET switch.

In Example 19.2, we have once again assumed that the active component is *ideal*. In practice, the output values would have been slightly off from the ideal values of 0 V and $V_{DD}$. These discrepancies would be caused by the *drain cutoff current*, $I_{D(\text{off})}$ and the *drain–source on-state voltage*, $V_{DS(\text{on})}$, of the JFET.

*Practical output values for a JFET switch.*

## The MOSFET As a Switch

The MOSFET switch has the input impedance advantage of the JFET switch, and, like the BJT, can have a positive or negative input signal. The basic MOSFET switch is shown in Figure 19.6.

Since the depletion-enhancement type MOSFET can be operated with its gate–source junction in enhancement mode we are able to produce a positive output pulse with an input signal that is also a positive voltage. While there is still a 180° phase shift from input to output, both the input and output voltages from the MOSFET switch have a positive polarity. In contrast, the JFET switch used a negative polarity input to produce a positive polarity output. This polarity relationship can be seen by referring back to Figure 19.4.

*The difference between MOSFET and JFET switches.*

When the input to the circuit in Figure 19.6 is at 0 V, $I_D = I_{DSS}$, and

$$V_{\text{out}} = V_{DD} - I_{DSS}R_D$$

For the MOSFET switch shown, lower values of $R_D$ will produce an output that is closer to the value of $V_{DD}$. This point is illustrated in Example 19.3.

---

Determine the output voltage for the circuit shown in Figure 19.7 when the input is at 0 V. Repeat the procedure for the same circuit using $R_D = 100\ \Omega$.

***Solution:***   When the input is at 0 V, the current through the MOSFET is equal to $I_{DSS}$. Therefore, the output voltage is found as

$$\begin{aligned}
V_{\text{out}} &= V_{DD} - I_{DSS}R_D \\
&= 10\text{ V} - 2\text{ V} \\
&= 8\text{ V}
\end{aligned}$$

*EXAMPLE 19.3*

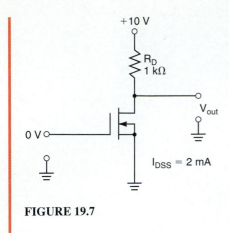

**FIGURE 19.7**

$I_{DSS} = 2$ mA

+10 V

$R_D$
1 kΩ

$V_{out}$

0 V

If the resistor in the circuit is replaced with a 100 Ω resistor, the output changes to

$$V_{out} = V_{DD} - I_{DSS}R_D$$
$$= 10 \text{ V} - 200 \text{ mV}$$
$$= 9.8 \text{ V}$$

Thus, with a smaller value of $R_D$, the high output voltage is much closer to the ideal value, $V_{DD}$.

### PRACTICE PROBLEM 19.3

A circuit like the one in Figure 19.7 has the following values: $V_{DD} = +8$ V, $R_D = 2$ kΩ, and $I_{DSS} = 500$ μA. Determine the value of $V_{out}$ for the circuit when $V_{in} = 0$ V.

*The effect of $R_D$ on the output voltage of a MOSFET switch.*

A smaller $R_D$ may help to produce a high output that is closer to $V_{DD}$, but it will also cause the low output voltage to be farther from ground. With a smaller value of $R_D$, the value of $I_D R_D$ will be lower for a given value of $I_D$. This means that when the MOSFET is conducting to its maximum capability, $R_D$ may not drop enough voltage to produce an output that is near 0 V. Thus, we have a possible trade-off. If having an output close to $V_{DD}$ is important, a low value of $R_D$ *may* be required. This would depend on the current capability of the particular MOSFET.

The CMOS switch discussed in Chapter 13 eliminates the potential trade-off problem that can be caused by $R_D$. As a review, the CMOS switch is shown in Figure 19.8. Recall from our previous discussion that the two MOSFETs shown in the switch are always in opposite operating states. Thus, when $Q_1$ is *on*, $Q_2$ is *off*, and vice versa. The following table is based on the relationship between the operating states of $Q_1$ and $Q_2$.

| $Q_1$ State | $Q_2$ State | $Q_2$ Resistance |
|---|---|---|
| On | Off | High |
| Off | On | Low |

When $Q_2$ is on, its resistance is very low. Thus, the voltage dropped across $Q_2$ is very low, and the output voltage is very close to $V_{DD}$. When $Q_1$ is on, the resistance of $Q_2$

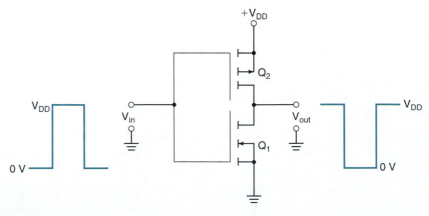

**FIGURE 19.8   The CMOS switch.**

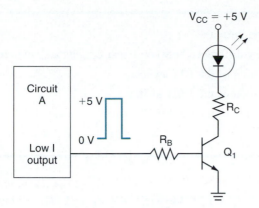

**FIGURE 19.9** A typical application for the simple BJT switch.

is very high. Therefore, the voltage drop across $Q_2$ is very high, and the output voltage is very close to 0 V. This is one of the reasons that CMOS switches are almost always used in place of other MOS switches.

## Basic Switching Applications

A typical application for the simple BJT switch can be seen in Figure 19.9. Here, we see the BJT switch used as an LED *driver*. A **driver** is a circuit used to couple a low-current output to a relatively high current device. In Figure 19.9, the block labeled *circuit A* is assumed to be a switching circuit with a low-current output. Assuming that the output current from circuit A would be insufficient to drive the LED (which typically requires at least 1 mA of forward current to light), a driver would be required to provide the current needed to light the LED.

How does the circuit work? When the output from circuit A is at 0 V, $Q_1$ is in cut-off (acting as an open switch). In this case, the transistor collector current is approximately equal to zero, and the LED doesn't light. When the output from circuit A goes to +5 V, $Q_1$ saturates (acts as a closed switch). In this case, $I_C$ rises to a value that is sufficient to light the LED. Note that the exact value of $I_C$ depends on the forward voltage drop of the LED and the value of $R_C$ (which is acting as a series current-limiting resistor), as suggested by equation (2.10). In this case, the source voltage ($V_S$) is the +5 V value of $V_{CC}$ for the transistor, and $R_S$ is the collector resistor.

Even though the driver shown in Figure 19.9 uses a BJT, drivers are commonly made with each of the active devices covered in this section. The device used in any driver application is determined by the system engineer, and no single active device seems to dominate this application.

**Driver**
A circuit used to couple a low-current output to a relatively high-current device.

A modified version of equation (2.10):

$$I_F = \frac{V_S - V_F}{R_S}$$

## Summary

BJT, JFET, and MOSFET switches are commonly used as *drivers* in switching applications. A driver is a circuit used to couple a low-current circuit output to a relatively high current device, such as an LED.

The basic switching circuits are used to provide high and low dc outputs when driven by an input signal. A high output is one that is within 1 V of the circuit supply voltage (typically). A low voltage is one that is within 1 V of ground.

1. What is a switching circuit?
2. What is the primary difference between linear circuits and switching circuits?
3. Describe the operation of the BJT switch.
4. Describe the operation of the JFET switch.
5. Describe the operation of the MOSFET switch.
6. What is a *driver*?

## 19.2 BASIC SWITCHING CIRCUITS: PRACTICAL CONSIDERATIONS

*OBJECTIVE 2* ▶ In this section, we will take a look at some of the factors that affect the operation and analysis of basic switching circuits. Most of these factors explain the differences between *ideal* and *practical* time and frequency measurements. We will also look at two switching circuit classifications and several devices that are specially designed for switching applications.

**Rectangular waveform**
A waveform made up of alternating (high and low) dc voltages.

**Pulse width (PW)**
The time spent in the active (high) dc voltage state.

**Space width (SW)**
The time spent in the passive (low) dc voltage stage.

**Cycle time ($T_C$)**
The sum of pulse width and space width.

### Practical Measurements

Before we can discuss any practical time or frequency measurements, we must define some of the terms commonly used to describe rectangular waves. We will define these terms with the help of Figure 19.10.

A waveform made up of alternating (high and low) dc voltages is generally classified as a **rectangular waveform.** The time a rectangular waveform spends in the high dc voltage state is generally called the **pulse width (PW)** of the waveform. The time spent in the low dc voltage state is generally called the **space width (SW)** of the waveform. The sum of pulse width and space width gives you the **cycle time ($T_C$)** of the waveform.

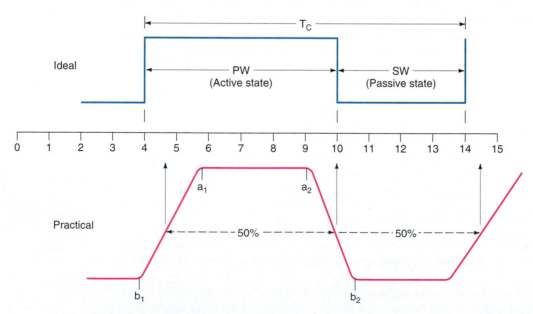

**FIGURE 19.10**  **Measuring pulse width, space width, and cycle time.**

The **square wave** is a special-case rectangular waveform that has equal pulse width and space width values. When PW ≠ SW, the waveform is simply referred to as a rectangular waveform.

**Square wave**
A special-case rectangular waveform that has equal PW and SW values.

When discussing the *ideal* waveform, it is easy to determine the value of pulse width, space width, or cycle time. For example, take a look at the ideal waveform in Figure 19.10. The scale shown between the ideal and practical waveforms is used as a time reference. Assume (for the sake of discussion) that the scale is measuring time in microseconds (μs). For the ideal waveform, it is easy to see that the value of PW is

$$PW = 10 \ \mu s - 4 \ \mu s = 6 \ \mu s$$

Similarly, the value of SW can be found as

$$SW = 14 \ \mu s - 10 \ \mu s = 4 \ \mu s$$

and the cycle time can be found as

$$T_C = 14 \ \mu s - 4 \ \mu s = 10 \ \mu s$$

*A Practical Consideration:* Of course, the value of $T_C$ could have been found as $T_C = PW + SW$.

When we wish to make the time measurements of a practical rectangular waveform, we run into a bit of a problem. You see, the transitions from low to high and from high to low are not perfectly vertical. (The reasons for this are discussed later in this section.) For example, take a look at the *practical* waveform in Figure 19.10. If we measure the PW of this waveform from $a_1$ to $a_2$, its value is found to be approximately 3.5 μs. However, if we measure the PW from $b_1$ to $b_2$, its value is found to be approximately 7 μs. The question here is: Where do we measure the values of PW, SW, and $T_C$ so that everyone gets the same results?

To eliminate the problem of measurement variations, the values of PW, SW, and $T_C$ are *always* measured at the 50 percent points (where the voltage is halfway between the high and low values). For example, if the practical waveform in Figure 19.10 has peak values of 0 V (low) and +5 V (high), we measure the values of PW, SW, and $T_C$ at the +2.5 V points on the waveform. This point is illustrated further in Example 19.4.

Time measurements are always made at the 50% points on the waveform.

Figure 19.11a represents the output from an inverter as seen on an oscilloscope. If the oscilloscope is set to a horizontal calibration of 50 μs/div, what are the values of PW and $T_C$ for the waveform?

*EXAMPLE* **19.4**

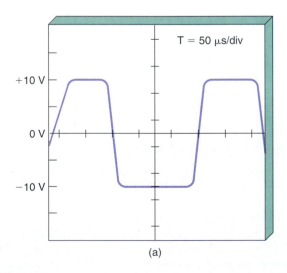

(a)

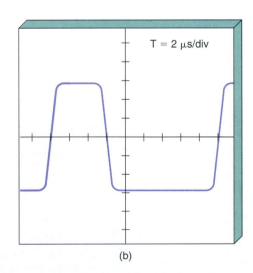

(b)

**FIGURE 19.11**

**Solution:** The waveform has peak values of $\pm 10$ V. Therefore, we will measure PW and $T_C$ at the 0 V points since these are halfway between the peak values.

The PW is measured during the positive transition. There are approximately 2.3 major divisions between the two 0 V points. Since each major division represents 50 $\mu$s, the PW is found as

$$\text{PW} = 2.3 \times 50 \,\mu\text{s}$$
$$= 115 \,\mu\text{s}$$

*A Practical Consideration:* In most circuit analyses, we are interested only in measuring the values of PW and $T_C$. If the value of SW is needed, it is then determined by subtracting PW from $T_C$.

Now, looking at the illustration, we can see that there are approximately 5.3 divisions from the 0 V point on one rising edge to the 0 V point on the next. Since these points correspond to one complete cycle, the cycle time is found as

$$T_C = 5.3 \times 50 \,\mu\text{s}$$
$$= 265 \,\mu\text{s}$$

### PRACTICE PROBLEM 19.4

Determine the values of PW and $T_C$ for the waveform in Figure 19.11b.

**Duty cycle (DC)**
The ratio of pulse width (PW) to cycle time ($T_C$), measured as a percentage.

Another time-relative measurement commonly made with switching circuits is *duty cycle*. **Duty cycle** is the ratio of pulse width to total cycle time, *measured as a percentage*. By formula,

$$\text{duty cycle} = \frac{\text{pulse width}}{\text{cycle time}} \times 100 \qquad \textbf{(19.1)}$$

Example 19.5 demonstrates the use of this equation in determining the duty cycle of a given waveform.

### EXAMPLE 19.5

Determine the duty cycle for the waveform in Figure 19.11a.

**Solution:** In Example 19.4, we determined the waveform to have values of PW = 115 $\mu$s and $T_C$ = 265 $\mu$s. Using these values and equation (19.1), the value of the duty cycle is found as

$$\text{dc} = \frac{\text{PW}}{T_C} \times 100$$

$$= \frac{115 \,\mu\text{s}}{265 \,\mu\text{s}} \times 100$$

$$= 43.4\%$$

### PRACTICE PROBLEM 19.5

Determine the duty cycle of the waveform in Figure 19.11b.

The duty cycle of a waveform indicates *the percentage of each cycle that is taken up by the pulse width*. If a given waveform has a duty cycle of 35 percent, the pulse width takes up 35 percent of each cycle. An ideal square wave would always have a duty cycle of 50 percent since, by definition, the pulse width is one-half of the cycle time in a square wave.

Now that you are familiar with the basic switching circuit measurements, we will examine the measurement characteristics of the BJT and FET individually.

## BJT Switching Time

In an *ideal* BJT switch, the output would change at the *exact* instant that the input changes. The output would also have transitions that would be perfectly vertical. These characteristics of the ideal switch are illustrated in Figure 19.12a. In this figure, we see the ideal input to a BJT. As you can see, the ideal input is shown as changing states *instantly* at $T_0$ and $T_3$. In practice, the BJT would have a response that more closely resembles the *practical* output waveforms shown in Figures 19.12b and c. These waveforms differ from the ideal because:

◄ OBJECTIVE 3

1. There is a delay between each input transition and the time when the output transition starts.

2. The output transitions are not vertical, implying that the transitions require some measurable amount of time to occur.

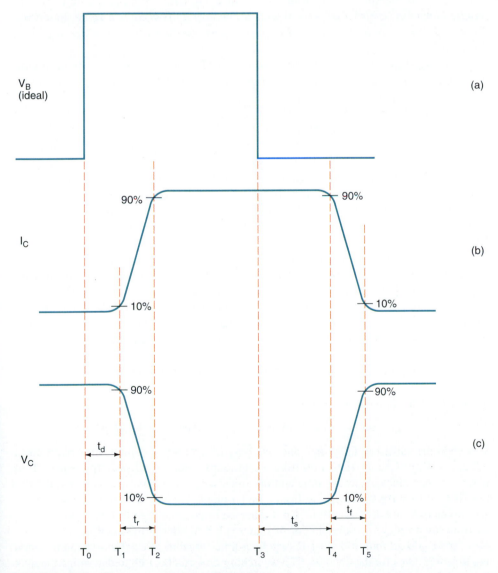

**FIGURE 19.12    The causes of propagation delay.**

**Propagation delay**
The time delay between input and output changes, as measured at the 50 percent point on the two waveforms.

**Delay time ($t_d$)**
The time required for a BJT to come out of cutoff. In terms of $I_C$, it is the time required for $I_C$ to reach 10 percent of its maximum value. In terms of oscilloscope measurements, $t_d$ is the time required for $V_C$ to drop to 90 percent of its maximum value.

**Rise time ($t_r$)**
The time required for the BJT to go from cutoff to saturation. In terms of $I_C$, the time required for the 10 percent to 90 percent transition. In terms of $V_C$, the time required for the 90 percent to 10 percent transition.

**Storage time ($t_s$)**
The time required for a BJT to come out of saturation. In terms of $I_C$, the time required for $I_C$ to drop to 90 percent of its maximum value. In terms of $V_C$, the time required for $V_C$ to rise to 10 percent of its maximum value.

**Fall time ($t_f$)**
The time required for the BJT to make the transition from saturation to cutoff. In terms of $I_C$, the time required for $I_C$ to drop from 90 percent to 10 percent of its maximum value. In terms of $V_C$, the time required to increase from 10 percent to 90 percent of its maximum value.

The overall delay between the time that the input changes and the output changes (as measured at the 50 percent points on the two waveforms) is referred to as **propagation delay.**

There are actually *four* different sources of propagation delay for a BJT. Each of these sources is measured during a specific period, as shown in Figure 19.12. Here, we see the input signal for a BJT, along with the output current ($I_C$) and voltage ($V_C$) waveforms. Note that the input waveform could be used to represent the change in either the input voltage ($V_B$) or input current ($I_B$) since $V_B$ and $I_B$ are always in phase. Also, $V_C$ is shown to be out of phase with $I_C$, as is always the case.

The first contributor to propagation delay is called **delay time ($t_d$)**. This is the time required for the transistor to come out of cutoff. Delay time is measured between times $T_0$ and $T_1$ (Figure 19.12b and c). In terms of $I_C$, delay time is *the time required for $I_C$ to reach 10 percent of its maximum value.*

When you are viewing the output from a BJT switch on an oscilloscope, you are not actually looking at $I_C$. Since the oscilloscope is used to view voltages, we also need to define $t_d$ in terms of collector voltage. In terms of $V_C$, delay time is *the time required for $V_C$ to drop to 90 percent of its maximum value;* in most cases, 90 percent of $V_{CC}$.

**Rise time ($t_r$)** is the time required for the BJT to make the transition from cutoff to saturation. It is measured between times $T_1$ and $T_2$. In terms of $I_C$, rise time is *the time required for $I_C$ to rise from 10 percent up to 90 percent of its maximum value.* In terms of $V_C$, it is *the time required for $V_C$ to drop from 90 percent down to 10 percent of its maximum value.*

**Storage time ($t_s$)** is the time required for a BJT to come out of saturation. In terms of $I_C$, it is *the time required for $I_C$ to drop from 100 percent down to 90 percent of its maximum value,* as can be seen between $T_3$ and $T_4$. In terms of $V_C$, *it is the time required for $V_C$ to rise to 10 percent of its maximum value.* Note that storage time (like delay time) is shown to begin at the point when the input signal changes.

Finally, **fall time ($t_f$)** is the time required for the BJT to make the transition from saturation to cutoff, as shown between $T_4$ and $T_5$. In terms of $I_C$, it is the time required for $I_C$ to drop from 90 percent to 10 percent of its maximum value. In terms of $V_C$, it is the time required for $V_C$ to increase from 10 percent to 90 percent of its maximum value.

An important point needs to be made at this time. Delay time and storage time account for the delay between the input transition and *the start* of the output transition. Rise time and fall time account for the *slope* of the output transitions. These points are illustrated in Figure 19.13. If a BJT had only delay time and storage time, the output would resemble the waveform shown in Figure 19.13b. If a BJT had only rise time and fall time, the output would resemble the waveform shown in Figure 19.13c. As you will see later in this section, we can use external components to greatly reduce delay time and storage time. However, little can be done to change rise time and fall time. Thus, the practical output waveform for a BJT switch will most closely resemble the one in Figure 19.13c.

The values of $t_d$, $t_r$, $t_s$, and $t_f$ are all provided on the specification sheet of a given BJT. For example, the 2N3904 spec sheet (Figure 7.24) lists the following maximum values under *switching characteristics:*

$$t_d = 35 \text{ ns} \qquad t_r = 35 \text{ ns} \qquad t_s = 200 \text{ ns} \qquad t_f = 50 \text{ ns}$$

If we add the values of delay time and rise time, we get the maximum time required for the 2N3904 to make the transition from a high output state to a low output state, 70 ns. If we add the values of storage time and fall time, we get the maximum time required for the 2N3904 to make the transition from a high output state to a low output state, 250 ns. As you can see, it takes more time for the device to go from high to low than it takes to go from low to high. The reasons for this are explained later in this section.

If we add all four of the listed switching times together, we get a maximum switching time of 320 ns for the 2N3904. This switching time can be used to determine the maximum theoretical switching frequency for the device, as follows:

**FIGURE 19.13**

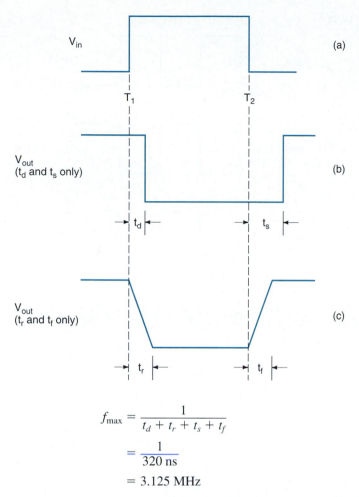

$V_{in}$      (a)

$T_1$      $T_2$

$V_{out}$
($t_d$ and $t_s$ only)      (b)

$t_d$      $t_s$

$V_{out}$
($t_r$ and $t_f$ only)      (c)

$t_r$      $t_f$

*A Practical Consideration:* FETs also experience propagation delay. The sources of FET propagation delay are discussed later in this section.

$$f_{max} = \frac{1}{t_d + t_r + t_s + t_f}$$

$$= \frac{1}{320 \text{ ns}}$$

$$= 3.125 \text{ MHz}$$

An important point needs to be made: While a 2N3904 switching circuit is capable of operating at 3.125 MHz, the output from such a circuit would be an extremely distorted square wave. The reason for this lies in the relationship between the upper cutoff frequency ($f_2$) of the BJT and the frequency of its square-wave input.

In Appendix D, the following equation is derived for calculating the value of $f_2$ for a switching circuit:

$$f_2 = \frac{0.35}{t_r} \tag{19.2}$$

where $t_r$ = the *rise time* of the active device. Using this equation, the value of $f_2$ for a 2N3904 switching circuit would be found as

$$f_2 = \frac{0.35}{35 \text{ ns}} \qquad (t_r = 35 \text{ ns for the 2N3904})$$

$$= 10 \text{ MHz}$$

Thus, the 2N3904 has an upper cutoff frequency of 10 MHz.

To pass a square wave with minimum distortion, *the upper cutoff frequency of a switching circuit must be at least 10 times the frequency of the input square wave.* Thus, the practical limit on the frequency of a switching circuit input signal will be one-tenth the value of $f_2$ for the circuit. By formula,

$$f_{max} = \frac{0.35}{10t_r} \qquad \text{(practical limit)} \tag{19.3}$$

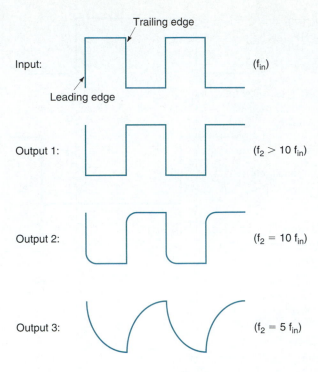

**FIGURE 19.14**

In the case of the 2N3904 switch, this means that the practical limit on $f_{in}$ for a circuit is 1 MHz.

Figure 19.14 shows what happens if the practical limit on the $f_{in}$ for a circuit is exceeded. As you can see, the leading and trailing edges of the circuit output become more and more rounded as the frequency of the input waveform increases. While there is *some* rounding of the square wave corners when $f_2 = 10f_{in}$, this small amount of rounding is considered to be acceptable. Example 19.6 demonstrates the complete frequency analysis of a BJT switching circuit.

---

*EXAMPLE* **19.6**

The Motorola MMBT2907 transistor has the following maximum values listed on its spec sheet: $t_d = 10$ ns, $t_r = 40$ ns, $t_s = 80$ ns, and $t_f = 30$ ns. Determine the theoretical and practical limits on the operating frequency of the device.

***Solution:*** The theoretical limit on $f_{in}$ for the device is found using the same time-to-frequency conversion that we used earlier in this section, as follows:

$$f_{max} = \frac{1}{t_d + t_r + t_s + t_f} \qquad \text{(maximum, theoretical)}$$

$$= \frac{1}{160 \text{ ns}}$$

$$= 6.25 \text{ MHz}$$

The value of $f_2$ for the device is now found as

$$f_2 = \frac{0.35}{t_r}$$

$$= \frac{0.35}{40 \text{ ns}}$$

$$= 8.75 \text{ MHz}$$

The practical limit on $f_{in}$ is one-tenth the value of $f_2$ for the device, 875 kHz. If $f_{in}$ exceeds 875 kHz, the output from the device will be distorted.

**PRACTICE PROBLEM 19.6**

A transistor has the following maximum values listed on its spec sheet: $t_d = 20$ ns, $t_r = 25$ ns, $t_s = 120$ ns, and $t_f = 20$ ns. Determine the theoretical and practical limits on $f_{in}$ for the device.

## Improving BJT Switching Time

Before you can understand the methods used to improve BJT switching time, we need to take a moment to discuss the causes of switching time. The various times are explained easily with the help of an illustration that appeared earlier in the text. Figure 6.5 is repeated as Figure 19.15.

◄ *OBJECTIVE 4*

When the BJT is in cutoff, the depletion layer is at its maximum width, and $I_C$ is essentially at zero. When the input to the BJT goes positive, the depletion layer will start to "dissolve," allowing $I_C$ to begin to increase. Delay time is the time required for the depletion layer to dissolve to the point where 10 percent of the maximum value of $I_C$ is allowed to pass through the component. How long is this period of time? That depends on three factors:

The cause of delay time.

1. The physical characteristics of the particular BJT
2. The amount of reverse bias initially applied to the component
3. The amount of $I_B$ that the input signal generates when it goes positive

We cannot do much about the physical characteristics of the BJT, but we can do several things to improve $t_d$. By keeping the initial reverse bias at a minimum, we can keep the depletion layer at a minimum width. The narrower the depletion layer, the less time it will take for it to dissolve. Also, you should recall that the BJT is a *current-controlled* device; that is, the width of the depletion layer is determined by the amount of $I_B$. Therefore, by providing a very high initial value of $I_B$, delay time is further reduced.

How to reduce delay time.

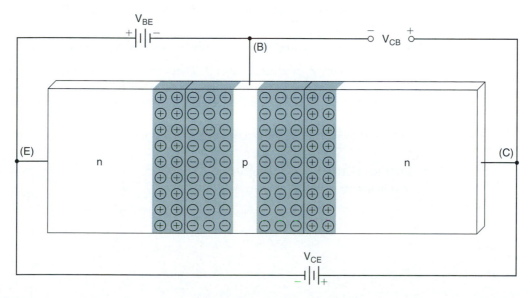

**FIGURE 19.15**

Later in this section, we will discuss the method by which a large initial value of $I_B$ is generated, thus reducing $t_d$.

After delay time has passed, the depletion layer will continue to dissolve, allowing $I_C$ to continue to increase. Rise time is the time required for the depletion layer to dissolve to the point where $I_C$ reaches 90 percent of its maximum value. Rise time is strictly a function of the physical characteristics of the BJT, and nothing can be done to improve it.

The biggest overall delay is *storage time*. Referring back to the specifications of the 2N3904, you can see that storage time is at least four times as long as any of the other switching times. Therefore, reducing $t_s$ will have a major impact on the overall switching time of a given BJT.

*The cause of storage time.*

When a transistor is in saturation, the base region is flooded with current. When the input voltage goes low, it takes a great deal of time for all this current to leave the base region and for a new depletion layer to begin to form. The time required for this to happen is storage time.

The actual value of storage time depends on three factors:

1. The physical characteristics of the BJT

2. The initial value of $I_C$

3. The initial value of reverse bias applied to the base

*How to reduce storage time.*

Again, we cannot do anything about the physical characteristics of the BJT, but we *can* do something about the other two variables. Storage time can be greatly reduced by keeping $I_C$ low enough so that the BJT never actually saturates. In other words, if we keep the BJT operating *just below* saturation, the amount of current in the base region of the BJT is greatly reduced, and storage time is also reduced. Also, if we apply a *very large* initial value of reverse bias, the current is forced out of the base region at a much faster rate, and again storage time is reduced.

Like rise time, fall time is a function of the physical characteristics of the BJT and cannot be reduced by any practical means. Fall time is the time required for the depletion layer to grow to the point where $I_C$ drops to 10 percent of its maximum value.

*A Practical Consideration:* We can reduce propagation delay as described in this section. However, since the practical frequency limit is determined by rise time, improving propagation delay will not improve the operating frequency limit of the switch.

Now, as a summary, let's review the steps we can take to reduce the overall switching time of a given BJT:

1. By applying a high *initial* value of $I_B$, delay time is reduced.

2. By using the *minimum* value of reverse bias required to hold the BJT in cutoff, delay time is further reduced.

3. By limiting $I_B$ to a value lower than that required for the BJT to saturate, storage time is reduced.

4. By applying an *initial* reverse bias that is very large, storage time is further reduced.

Now, look at statements 1 and 3. These statements would indicate that we want $I_B$ to be very high *initially* (to reduce delay time) and then settle down to some level below that required for saturation (to reduce storage time). Statements 2 and 4 can be combined in a similar fashion. We want a very high *initial* value of reverse bias (to reduce storage time) and then a minimum reverse bias (to reduce delay time).

**FIGURE 19.16    A speed-up capacitor ($C_S$) improves switching time.**

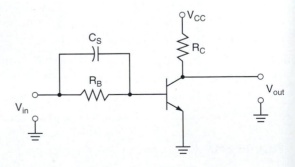

The desired $I_B$ and reverse-bias characteristics described above are both achieved by using what is called a **speed-up capacitor.** A basic BJT switch with an added speed-up capacitor is shown in Figure 19.16. The speed-up capacitor will perform all the required functions when its value is properly selected. The means by which it accomplishes its function are easy to understand if you take a look at the waveforms shown in Figure 19.17.

In Figure 19.16, $R_B$ and $C_S$ act as an *RC differentiator.* If the base–emitter junction of the transistor was open, the base waveform produced by this RC circuit would look like the waveform shown in Figure 19.16b. As indicated, the positive spike produces a very high initial value of $I_B$. Then, as the spike returns to the 0 V level, $I_B$ decreases to some value slightly less than that required for saturation. Thus, we have reduced the delay time while making sure that the value of $I_B$ returns to a value that is low enough to prevent saturation.

When the input signal returns to 0 V, the output of the RC circuit is driven to $-5$ V. This is the high initial value of reverse bias needed to reduce storage time. Since the transistor was prevented from saturating by $I_B$, we have now done all that can be done

**Speed-up capacitor**
A capacitor used to reduce propagation delay by reducing $t_d$ and $t_s$.

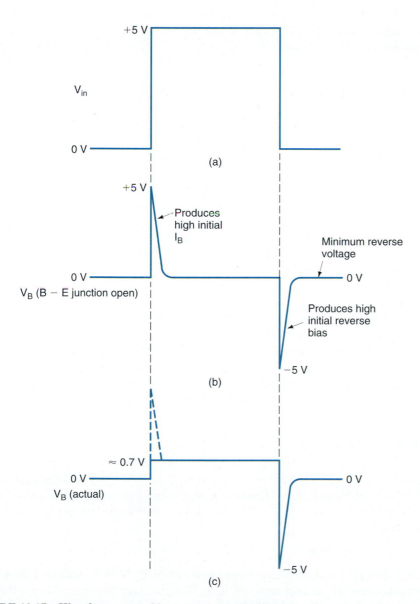

**FIGURE 19.17** **Waveforms caused by a speed-up capacitor.**

to reduce storage time. Also, since $V_B$ returns to 0 V at the end of the negative spike, the final value of reverse bias is at a minimum. This, coupled with the high initial $I_B$ spike, ensures that delay time is held to a minimum.

The value of $C_S$ is normally chosen so that the time constant of the $RC$ circuit is very short at the maximum operating frequency. This requirement is fulfilled by using the following equation to select $C_S$:

The value of $f_{max}$ used in equation (19.4) is the practical $f_{max}$ of the circuit, found as $f_2/10$.

$$C_S < \frac{1}{20R_B f_{max}}$$

(19.4)

As long as the speed-up capacitor fulfills the requirement of equation (19.4), the switching times of the BJT will be greatly reduced.

In Figure 19.17b, we assumed the transistor had an open emitter–base junction to simplify things. The actual waveform that you would see under working conditions would be similar to the one shown in Figure 19.19c. Here, we see the effects of the BJT on the base waveform. On the positive transition, the base–emitter junction will turn on and clip off a majority of the positive spike produced by the $RC$ circuit. Also, since the transistor stays on during the entire positive half-cycle of the input, the output from the $RC$ circuit will not actually return to 0 V. Rather, it will stay at the value of $V_{BE}$ for the transistor, approximately 0.7 V.

## JFET Switching Time

JFET spec sheets usually list values of $t_d$, $t_r$, $t_s$, and $t_f$. The causes of these times in the JFET are similar to those of the BJT. When the value of $t_r$ is listed on the JFET spec sheet, the practical limit on $f_{in}$ can be determined in the exact same fashion as it is for the BJT.

**Turn-on time ($t_{on}$)**
The sum of $t_d$ and $t_r$.

**Turn-off time ($t_{off}$)**
The sum of $t_s$ and $t_f$.

Some JFET spec sheets list only **turn-on time** and **turn-off time.** Turn-on time is the sum of delay time and rise time. Turn-off time is the sum of storage time and fall time. When these are the only values listed in the JFET spec sheet, it is difficult to determine the exact practical limit on $f_{in}$ for the device. This is because you cannot determine the portion of turn-on time that is actually taken up by the rise time of the device. In other words, there is no conversion formula for calculating rise time when turn-on time is known.

*A Practical Consideration:* In some cases, the spec sheet of a BJT will list $t_{on}$ and $t_{off}$ in place of the standard parameters. When this is the case, you must follow the guideline discussed here for approximating the practical limit on $f_{in}$.

When only $t_{on}$ is given, use this value in place of $t_r$ in equation (19.3) to determine the practical operating frequency limit. While the actual limit on $f_{in}$ will be higher than your calculated value, you will still have a good approximation of this limit.

Since JFETs are voltage-controlled devices, speed-up capacitors will only improve the turn-off time of the component. By initially supplying a gate–source voltage that is greater than $V_{GS(off)}$, the speed-up capacitor will decrease the time required to turn off the JFET. However, the capacitor will do little to improve the turn-on time of the component.

## Switching Devices

**Switching transistors**
Devices with extremely low switching times.

Groups of BJTs and JFETs have been developed especially for switching applications. These devices, called **switching transistors,** are designed to have extremely short values of $t_d$, $t_r$, $t_s$, and $t_f$. For example, the 2N2369 has the following maximum values listed on its data sheet:

$$t_d = 5 \text{ ns} \qquad t_r = 18 \text{ ns} \qquad t_s = 13 \text{ ns} \qquad t_f = 15 \text{ ns}$$

With the extremely small delay and storage time values, a 2N2369 switching circuit probably wouldn't need a speed-up capacitor. Also, with the given rise time of 18 ns, the practical limit on $f_{in}$ for the device would be found as

$$f_{max} = \frac{0.35}{10t_r}$$

$$= \frac{0.35}{180 \text{ ns}}$$

$$= 1.94 \text{ MHz}$$

This is almost twice the acceptable $f_{in}$ limit of the 2N3904. Thus, the 2N2369 can be used at much higher switching frequencies than the 2N3904.

## Switching Circuit Classifications

Our discussion on basic switching circuits has been limited to a specific type of switching circuit called an *inverter*. An **inverter** has a 180° voltage phase shift from input to output. Looking back at the various illustrations of BJT and FET circuits, you will see that every circuit had a high output voltage when the input voltage was low, and vice versa.

◄ OBJECTIVE 5

**Inverter**
A basic switching circuit that produces a 180° voltage phase shift.

Another basic switching circuit is the *buffer*. A **buffer** is a switching circuit that does not introduce a 180° voltage phase shift. The BJT and FET buffers are shown in Figure 19.18. As you can see, these circuits are simply the emitter follower and source follower. When a square wave is applied to any of these circuits, the output will be a square wave that is in phase with the input signal. Because of this, we must redefine the $V_{out}$ descriptions of $t_d$, $t_r$, $t_s$, and $t_f$. For the *buffer*:

**Buffer**
A switching circuit that does not produce a voltage phase shift.

1. $t_d$ is the time required for $V_{out}$ to reach 10 percent of its maximum value.

Buffer time measurements.

2. $t_r$ is the time required for $V_{out}$ to rise from 10 percent to 90 percent of its maximum value.

3. $t_s$ is the time required for $V_{out}$ to drop to 90 percent of its maximum value.

4. $t_f$ is the time required for $V_{out}$ to drop from 90 percent to 10 percent of its maximum value.

The changes in the $V_{out}$ definitions for the buffer are necessary because it does not produce a 180° voltage phase shift. Except for this difference, the operation of the buffer and inverter circuits are nearly identical.

## Summary

A waveform made up of alternating (high and low) dc voltages is classified as a *rectangular* waveform. The square wave is a special-wave rectangular waveform that has equal *pulse-width* and *space-width* values.

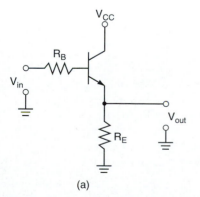
(a)

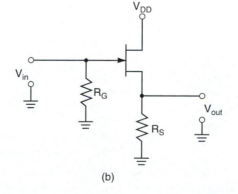
(b)

**FIGURE 19.18** Some basic buffer circuits.

The ideal rectangular waveform would have instantaneous transitions from high to low and from low to high. However, these transitions are not perfectly vertical in practice. To ensure that pulse-width, space-width, and cycle-time measurements are consistent, these values are always measured at the 50 percent points on rectangular waveforms; that is, at the points where the amplitude of the waveform is halfway between the high and low output voltage levels.

The duty cycle of a rectangular waveform indicates the percentage of the waveform taken up by the pulse width. Since pulse width and space width are always equal for a square wave, the duty cycle of a square wave will always be 50 percent.

The time between the input and output changes of a switching circuit is referred to as *propagation delay*. For discrete devices, four switching times contribute to propagation delay:

**1.** Delay time ($t_d$): The time required for the device to come out of cutoff.

**2.** Rise time ($t_r$): The time required for the device to make the transition from cutoff to saturation.

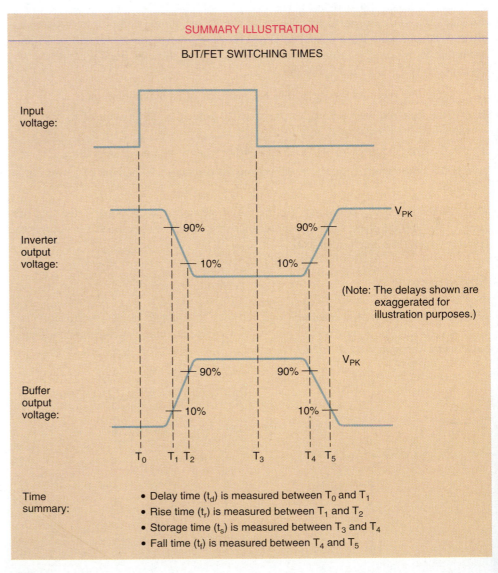

**FIGURE 19.19**

**3. Storage time ($t_s$):** The time required for the device to come out of saturation.

**4. Fall time ($t_f$):** The time required for the device to make the transition from saturation to cutoff.

The times listed are measured using an oscilloscope. Figure 19.19 summarizes these times and shows where they are measured on the output voltage waveforms for the inverter and the buffer. As you can see, the inverter has a 180° voltage phase shift from input to output, while the buffer does not.

The propagation delay of a discrete switch can be reduced by the use of a *speed-up capacitor*. The speed-up capacitor will reduce delay and storage time in the BJT and storage time (only) in an FET switch. The rise time of a device determines the maximum practical operating frequency of a switching circuit. Since a speed-up capacitor does not affect rise time, the practical limit on $f_{in}$ is not increased when a speed-up capacitor is added.

---

1. What is a *rectangular waveform*?
2. Define each of the following terms: *pulse width*, *space width*, and *cycle time*.
3. What is a *square wave*?
4. Where are each of the values in Question 2 measured on a practical rectangular waveform?
5. Why do we need to have a standard for rectangular waveform time measurements?
6. What is *duty cycle*?
7. Why do all square waves have the same duty cycle value? What is this value?
8. What is *propagation delay*?
9. What is delay time?
10. Describe delay time in terms of $I_C$ and $V_C$.
11. What is *rise time*?
12. Describe rise time in terms of $I_C$ and $V_C$.
13. What is *storage time*?
14. Describe storage time in terms of $I_C$ and $V_C$.
15. What is *fall time*?
16. Describe fall time in terms of $I_C$ and $V_C$.
17. Which of the switching times determines the practical limit on $f_{in}$ for the device?
18. What causes delay time?
19. What causes storage time?
20. How does the use of a *speed-up capacitor* reduce $t_d$ and $t_s$?
21. What effect does a speed-up capacitor have on the practical frequency limit of a BJT? Explain your answer.
22. Define *turn-on time* ($t_{on}$) and *turn-off time* ($t_{off}$).
23. What is a *switching transistor*?
24. What is the primary difference between the *inverter* and the *buffer*?
25. List the points where $t_d$, $t_r$, $t_s$, and $t_f$ are measured on the output voltage of an inverter.
26. List the points where $t_d$, $t_r$, $t_s$, and $t_f$ are measured on the output voltage of a buffer.

# 19.3  SCHMITT TRIGGERS

OBJECTIVE 6 ▶

Now that we have covered the basic principles associated with switching circuits, it is time to start discussing some special-purpose switching circuits. As you will see, these more complicated circuits are used to perform functions that cannot be accomplished with the basic inverter or buffer. The first circuit we will cover is the *Schmitt trigger.*

**Schmitt trigger**
A voltage-level detector.

The **Schmitt trigger** is a *voltage-level detector.* When the input to a Schmitt trigger reaches a specified level, the output will change states. When the input then goes below another specified level, the output will return to its original state. This input/output relationship is illustrated in Figure 19.20.

In Figure 19.20, the sine wave applied to the Schmitt trigger passes the **upper trigger point (UTP)** on its positive-going transition. When the UTP is reached, the output from the Schmitt trigger goes to $+V$. Once triggered, the output from the Schmitt trigger stays at $+V$. When the input goes negative again, it passes the **lower trigger point (LTP)**. When this occurs, the output from the Schmitt trigger returns to $-V$. The output now stays at $-V$ until the sine-wave input passes the UTP again.

**Upper trigger point (UTP)**
A reference voltage. When a positive-going input reaches the UTP, the Schmitt trigger output goes high.

**Lower trigger point (LTP)**
A reference voltage. When a negative-going input reaches the LTP, the Schmitt trigger output goes low.

Several points should be made at this time:

1. The UTP and LTP values are determined by the component values in the Schmitt trigger.

2. The Schmitt trigger can be designed so that the UTP and LTP values are not equal. However, the LTP can never be more positive than the UTP.

3. The Schmitt trigger can process any type of input waveform.

The relationship among the LTP, UTP, and circuit component values will be seen when we start to discuss some specific circuits. The effect of UTP and LTP values that are not equal to each other is illustrated in Figure 19.21. When the Schmitt trigger has different UTP and LTP values, the output will not go to $-V$ until the LTP is reached. Therefore, it is possible for the input to go below the UTP without the circuit output changing. In Figure 19.21, the sine wave passes the UTP at $v_1$ and again at $v_2$. At $v_1$, the output goes to $+V$ since the UTP has been reached on a *positive-going transition* of the input signal. When the input passes point $v_2$, the output does not change. Even though the input voltage is now below the UTP, the output is still at $+V$ and will remain there until the LTP is reached. This happens at $v_3$, and the output then returns to $-V$.

It is obvious from the waveforms in Figure 19.21 that the output from a Schmitt trigger will not be affected by any voltage between the UTP and the LTP. This is illustrated further in Figure 19.22. Even though the input is going through some significant changes, the output changes only when:

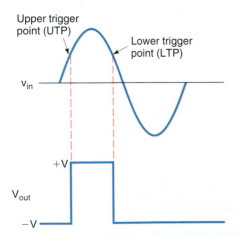

**FIGURE 19.20**  Schmitt trigger input and output signals.

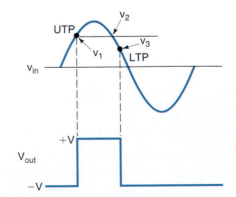

**FIGURE 19.21**  The UTP and LTP.

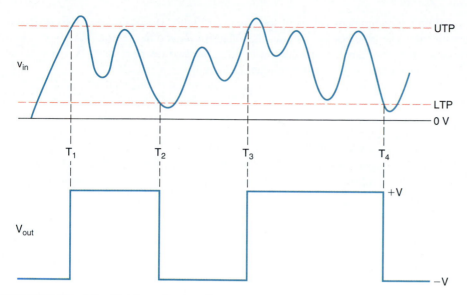

FIGURE 19.22   Schmitt trigger hysteresis.

1. The UTP is reached by a positive-going transition.
2. The LTP is reached by a negative-going transition.

No input voltage between the UTP and the LTP will have any effect on the output of the circuit. Note that the range of voltages between the UTP and LTP that will not affect the output can be referred to as **hysteresis**. For example, if Figure 19.22 had values of UTP = +10 V and LTP = +2 V, the hysteresis of the circuit would be the input voltage range of $+2\text{ V} < V_{in} < +10\text{ V}$. No voltage in this range could ever cause the output of the circuit to change states.

**Hysteresis**
A term that is (sometimes) used to describe the range of voltages between the UTP and the LTP.

## Noninverting Schmitt Triggers

The noninverting Schmitt trigger uses a simple feedback resistor, as shown in Figure 19.23a. Note the similarity between this circuit and the linear inverting amplifiers described earlier. The only difference is the reversal of the op-amp inputs.

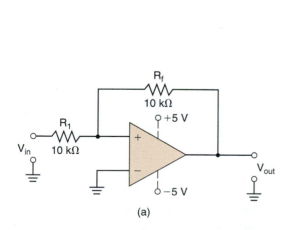

FIGURE 19.23   Noninverting Schmitt trigger operation.

With the component values of the circuit in Figure 19.23a, it is easy to understand how the circuit works. For this circuit, the values of $R_1$ and $R_f$ are equal. Also, you know from earlier discussions that the current through the two resistors must be equal. Therefore, the voltage across $R_1$ is always equal to the voltage across $R_f$.

We will start to discuss the circuit operation by assuming that $V_{out}$ is at $-4$ V (its negative limit) and $V_{in}$ is at $+4$ V. Since 4 V is being dropped across each resistor, the input to the op-amp is at 0 V. As soon as the input starts to go more positive than $+4$ V, the noninverting input to the op-amp becomes more positive than the inverting input, and the output goes to $+4$ V ($+V - 1$ V). This is illustrated in Figure 19.23b. At $T_0$, the input to the Schmitt trigger goes above $+4$ V, and the output responds by changing to $+4$ V. Remember, this happens because the noninverting input is now more positive than the inverting input.

The same circuit action would explain what happens on the negative transition of the input. The only difference is that the polarities are reversed. With the output of the circuit at $+4$ V, the input must pass $-4$ V before the noninverting input goes to a value that is more negative than the inverting input. When this happens, the output returns to $-4$ V and the cycle repeats itself.

As it is drawn, the Schmitt trigger in Figure 19.23a has UTP and LTP values that are equal to $(+V - 1$ V$)$ and $(-V + 1$ V$)$. By changing the ratio of $R_f$ to $R_1$, we can set the circuit up for other trigger point values. Generally, the UTP and LTP values can be determined using the following relationships:

$$\text{UTP} = -\frac{R_1}{R_f}(-V + 1\text{ V}) \qquad \textbf{(19.5)}$$

and

$$\text{LTP} = -\frac{R_1}{R_f}(+V - 1\text{ V}) \qquad \textbf{(19.6)}$$

The following example demonstrates the use of these two equations.

---

**EXAMPLE 19.7**

Determine the UTP and LTP values for the circuit shown in Figure 19.24. Also, verify that the output will change states when the UTP and LTP values are exceeded.

**Solution:** Using equation (19.5), the UTP is found as

$$\text{UTP} = -\frac{R_1}{R_f}(-V + 1\text{ V})$$

$$= -0.2(-12\text{ V})$$

$$= 2.4\text{ V}$$

The LTP is found using equation (19.6), as follows:

$$\text{LTP} = -\frac{R_1}{R_f}(+V - 1\text{ V})$$

$$= -0.2(+12\text{ V})$$

$$= -2.4\text{ V}$$

Now the UTP value can be verified as follows. When the input is at $+2.4$ V, the output is still at $-12$ V. Using the standard voltage-divider formula, we can obtain the voltage drop across $R_1$. This voltage drop is found as

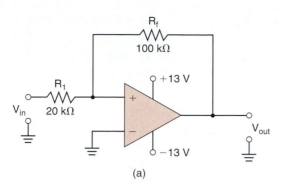

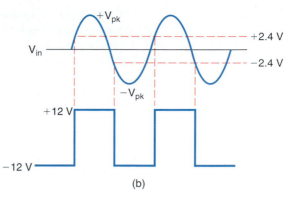

(a)                                    (b)

**FIGURE 19.24**

$$V_1 = (V_{in} - V_{out}) \frac{R_1}{R_1 + R_f}$$

$$= (14.4 \text{ V})(0.167)$$

$$= 2.4 \text{ V}$$

With 2.4 V applied to the circuit and a 2.4 V drop across $R_1$, it is clear that the voltage at the op-amp input is 0 V. Therefore, if the input goes any higher than $+2.4$ V, the noninverting input to the op-amp will be more positive than the inverting input, and the output will go to $+12$ V. The same method can be used to validate the value of LTP $= -2.4$ V. This is left as an exercise for you to do on your own. With the UTP and LTP values found, the input signals would appear as shown in Figure 19.24b.

*PRACTICE PROBLEM 19.7*

A noninverting Schmitt trigger like the one in Figure 19.24 has values of $R_f = 33$ k$\Omega$ and $R_1 = 11$ k$\Omega$. Determine the UTP and LTP values for the circuit.

The noninverting Schmitt trigger in Figure 19.24 is restricted because the magnitudes of the UTP and the LTP must be equal. For example, if the UTP is $+2.4$ V, the LTP must be $-2.4$ V. If the UTP were 4 V, the LTP would have to be $-4$ V. However, the non-inverting Schmitt trigger can be modified so that the magnitudes of the UTP and the LTP are not equal. A circuit that contains this modification is shown in Figure 19.25.

The key to the operation of this circuit is the fact that $D_1$ and $D_2$ will conduct on opposite transitions of the output. When the output from the Schmitt trigger is *negative,* $D_1$ will conduct and $D_2$ will be off. This means that $R_{f2}$ is effectively removed from the circuit, and the UTP is found as

$$\text{UTP} = -\frac{R_1}{R_{f1}}(-V + 1 \text{ V}) \qquad (19.7)$$

When the output is *positive,* $D_1$ will be off and $D_2$ will conduct. This means that $R_{f1}$ is effectively removed from the circuit, and the LTP is found as

$$\text{LTP} = -\frac{R_1}{R_{f2}}(+V - 1 \text{ V}) \qquad (19.8)$$

FIGURE 19.25   A noninverting
Schmitt trigger designed for unequal
UTP and LTP values.

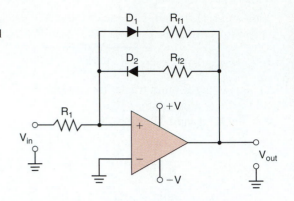

Since the UTP and LTP are determined by separate feedback resistors, these values are completely independent of each other. This point is illustrated in Example 19.8.

**EXAMPLE 19.8**

Determine the UTP and LTP values for the noninverting Schmitt trigger shown in Figure 19.26a.

*Solution:*   The UTP is found as

$$UTP = -\frac{R_1}{R_{f1}}(-V + 1\text{ V})$$

$$= -\frac{3.3\text{ k}\Omega}{11\text{ k}\Omega}(-11\text{ V})$$

$$= 3.3\text{ V}$$

The LTP is found as

$$LTP = -\frac{R_1}{R_{f2}}(+V - 1\text{ V})$$

$$= -\frac{3.3\text{ k}\Omega}{33\text{ k}\Omega}(11\text{ V})$$

$$= -1.1\text{ V}$$

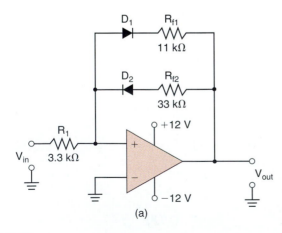

(a)

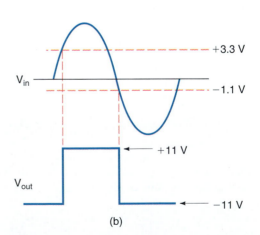

(b)

**FIGURE 19.26**

With these UTP and LTP values, the circuit in Figure 19.26a would have the input/output relationship shown in Figure 19.26b.

**PRACTICE PROBLEM 19.18**

A noninverting Schmitt trigger like the one in Figure 19.26 has $\pm 15$ V supply voltages and the following resistor values: $R_1 = 2$ k$\Omega$, $R_{f1} = 20$ k$\Omega$, and $R_{f2} = 14$ k$\Omega$. Determine the UTP and LTP values for the circuit.

The circuits in this section are all noninverting circuits; that is, their outputs go *high* when the UTP is passed by a *positive-going* input signal and negative when the LTP is passed by a *negative-going* input signal. It is possible to wire an op-amp as an *inverting Schmitt trigger* that will have the opposite input/output relationship from the one described here; that is, the output will go *negative* when the UTP is passed and *positive* when the LTP is passed. We will take a look at the operation of this circuit now.

## Inverting Schmitt Triggers

The basic inverting Schmitt trigger is shown in Figure 19.27. In this circuit, the input ($V_{in}$) is applied to the inverting terminal. A feedback signal is applied to the noninverting terminal from the junction of $R_1$ and $R_2$.

To understand the operation of the circuit, we must first establish the equations for the UTP and LTP. The UTP for the circuit in Figure 19.27 is found as

$$\text{UTP} = \frac{R_2}{R_1 + R_2}(+V - 1\text{ V}) \qquad (19.9)$$

and the LTP is found as

$$\text{LTP} = \frac{R_2}{R_1 + R_2}(-V + 1\text{ V}) \qquad (19.10)$$

Example 19.9 demonstrates the procedure for determining the UTP and LTP values for an inverting Schmitt trigger circuit.

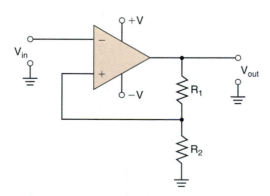

**FIGURE 19.27  An inverting op-amp Schmitt trigger.**

**Lab Reference:** The operation of an inverting Schmitt trigger is demonstrated in Exercise 43.

## EXAMPLE 19.9

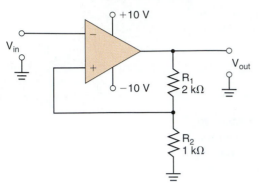

**FIGURE 19.28**

Determine the UTP and LTP values for the circuit shown in Figure 19.28.

**Solution:** The quickest approach is to determine the resistance ratio in the circuit first. This ratio is found to be

$$\frac{R_2}{R_1 + R_2} \cong 0.33$$

Now the UTP is found as

$$UTP = 0.33(10 \text{ V} - 1 \text{ V})$$
$$= 3 \text{ V}$$

and the LTP is now found as

$$LTP = 0.33(-10 \text{ V} + 1 \text{ V})$$
$$= -3 \text{ V}$$

### PRACTICE PROBLEM 19.9

An inverting Schmitt trigger has $\pm 10$ V supply voltages and values of $R_1 = 3$ k$\Omega$ and $R_2 = 1$ k$\Omega$. Determine the UTP and LTP values for the circuit.

We will use the values obtained in Example 19.9 to explain the operation of the circuit. First, remember that we are dealing with an *inverting* Schmitt trigger. Therefore, the output will go *low* when the UTP is passed and *high* when the LTP is passed. This point is illustrated in Figure 19.29. The waveforms shown here are the input and output waveforms for the circuit in Figure 19.28. As you can see, the input and output waveforms are out of phase by 180°.

When the output of the circuit is at +9 V, the voltage across $R_2$ is +3 V. As the input passes +3 V (at $T_0$), the output switches to −9 V. Now, the voltage across $R_2$ is −3 V. With this voltage applied to the noninverting input of the op-amp, the output does not switch again until the input passes −3 V (at $T_1$).

The basic inverting Schmitt trigger has one limitation. If you look at Figure 19.28, it is not very difficult to figure out that the UTP and LTP values must always be equal in magnitude for this circuit. This is due to the constant resistance ratio and the equal magnitudes of the two output voltages. In other words, you could not have a UTP of, say,

**Note:** UTP and LTP values that are equal in magnitude are called **symmetrical** trigger points. Trigger-point values that are *not* equal in magnitude are called **non symmetrical** trigger points.

**FIGURE 19.29   Input/output waveforms for an inverting Schmitt trigger.**

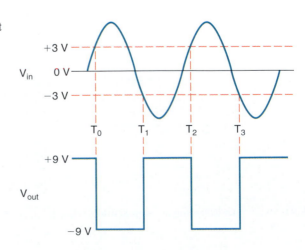

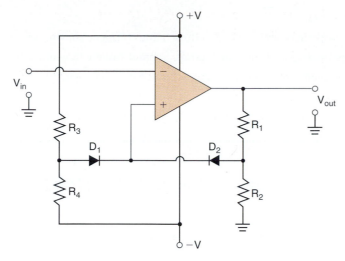

**FIGURE 19.30** An inverting Schmitt trigger designed for unequal (nonsymmetrical) UTP and LTP values.

+5 V and an LTP of −3 V for the circuit. The UTP and LTP values, while opposite in polarity, would have to be equal in value.

The inverting Schmitt trigger can be modified to allow different values of UTP and LTP, as shown in Figure 19.30. In this circuit, the UTP would be determined by $R_1$ and $R_2$, as before. The LTP would be determined by $R_3$, $R_4$, and $D_1$. By adjusting $R_3$ and $R_4$, the anode voltage of $D_1$ can be set to any value between $+V$ and $-V$. The LTP would then be 0.7 V more negative than the anode voltage. For example, let's say that $R_3$ and $R_4$ are set so that the anode voltage of $D_1$ is +2 V. The LTP for the circuit is then +1.3 V. If $R_3$ and $R_4$ are set to provide −4 V at the anode of $D_1$, the LTP is −4.7 V, and so on. Again, the values of $R_1$ and $R_2$ are selected to provide the desired UTP, and $R_3$ and $R_4$ are selected to provide the desired LTP.

## One Final Note

You have seen that the Schmitt trigger is a voltage-level detector that can be constructed in noninverting or inverting form. When the input signal to a Schmitt trigger makes a positive-going transition past the UTP, the output of the circuit goes to one dc output level. The output will stay at that level until the input signal makes a negative-going transition that passes the LTP. At that time, the output goes to the other dc output level. The dc output level for a specific input condition depends on whether the Schmitt trigger is an inverting or a noninverting circuit.

In Chapter 16, you were introduced to the *comparator.* This simple op-amp circuit is also a voltage-level detector that can be constructed in noninverting or inverting form. You may be wondering, then, what the difference is between the comparator and the Schmitt trigger.

The comparator is limited because it has a single reference voltage ($V_{ref}$) that acts as both its UTP and its LTP. The Schmitt trigger, on the other hand, can be designed for totally independent UTP and LTP values, as you have seen in this section. Otherwise, the two circuits serve the same basic purpose.

**Lab Reference:** The operation of a nonsymmetrical Schmitt trigger is demonstrated in Exercise 43.

---

1. What is a *Schmitt trigger?*
2. What is the *upper trigger point (UTP)?*
3. What is the *lower trigger point (LTP)?*

*Section Review*

4. What is *hysteresis*?

5. Explain the operation of the noninverting Schmitt trigger shown in Figure 19.23.

6. Explain the operation of the noninverting Schmitt trigger shown in Figure 19.25.

7. Explain the operation of the inverting Schmitt trigger in Figure 19.27.

8. Explain the operation of the inverting Schmitt trigger in Figure 19.30.

9. Compare and contrast the Schmitt trigger and the comparator.

## *19.4* MULTIVIBRATORS: THE 555 TIMER

OBJECTIVE 7 ▶

**Multivibrators**
Circuits designed to have zero, one, or two stable output states.

**Astable multivibrator**
A switching circuit that has no stable output state. A square-wave oscillator. Also called a **free-running multivibrator.**

**Monostable multivibrator**
A switching circuit with one stable output state. Also called a **one-shot.**

**Bistable multivibrator**
A switching circuit with two stable output states. Also called a **flip-flop.**

**Multivibrators** are circuits designed to have zero, one, or two *stable* output states. The concept of *stable* output states is illustrated in Figure 19.31.

The **astable multivibrator** is a switching circuit that has *no stable output state.* In its own way, this circuit can be viewed as a square-wave generator. The circuit will produce a constant square-wave output with only a supply-voltage input. Because it continually produces this square-wave output, it is often referred to as a **free-running multivibrator.**

The **monostable multivibrator** has a single stable output state. For the circuit represented in Figure 19.31, the stable output would be $-V$. The circuit will stay at this output level as long as no input signal is applied. When the circuit receives an input, generally referred to as a *trigger,* it will produce a single output pulse. The duration of the output pulse depends on the component values used in the circuit. Since the monostable multivibrator produces a single output pulse for each input trigger, it is generally referred to as a **one-shot.**

The **bistable multivibrator** has two output states. When the proper input trigger is received, it will switch from one output state to the other. It will then stay at the new output state until another trigger is received, returning the output to its original output state. The bistable multivibrator is generally referred to as a **flip-flop.**

As you have probably figured out by now, the term *astable* means *not stable,* the term *monostable* means *one stable state,* and the term *bistable* means *two stable states.* If

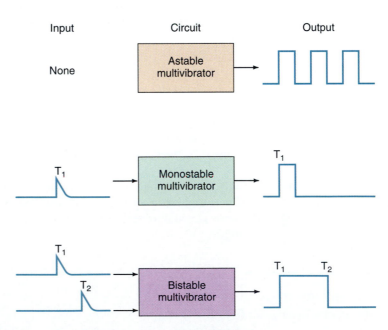

**FIGURE 19.31** **Multivibrator input/output relationships.**

you look closely at Figure 19.31, you might notice something else that is appropriate about these names. The astable multivibrator has no active input. The one-shot generally has one active input. The flip-flop generally has two active inputs. Although there are plenty of exceptions to this, the input relationships shown generally hold true.

Flip-flops are covered extensively in any digital electronics course and thus are not covered in depth here. Although the astable and monostable multivibrators are also covered in digital electronics, they are discussed here to give you an introductory idea of how these circuits operate.

At one time, multivibrators were commonly constructed using discrete components. However, this is no longer the case. More often than any type of circuit we have discussed thus far, multivibrators are constructed using ICs. One of these ICs is the *555 timer,* which is an eight-pin (lead) IC that can be used for a variety of switching applications. Among its many applications are the astable and monostable multivibrator.

The advantage in using the 555 timer is the fact that all the active components required to produce a multivibrator are within a single chip. Simply by adding the proper passive components to the IC, you can use the 555 timer as either a free-running multivibrator or a one-shot. In this section, we will look at the 555 timer internal circuitry and its use as a free-running circuit or a one-shot.

## The 555 Timer

The **555 timer** is a switching circuit that comes in an eight-pin IC. The IC contains two comparators, a flip-flop, an inverter, and several resistors and transistors. The block diagram of the 555 timer is shown in Figure 19.32.

Since you may never have worked with an IC of this type before, there are a few points that should be made. In Figure 19.32, the block diagram is shown inside the IC casing. The small numbered blocks that border the case are the IC pins. Looking at the left side of the case, you will see that there is a small indentation. This indentation is used to identify pin 1. When you hold a 555 timer, hold it so that the indentation is on the

◄ *OBJECTIVE 8*

**555 Timer**
An eight-pin IC designed for use in a variety of switching applications.

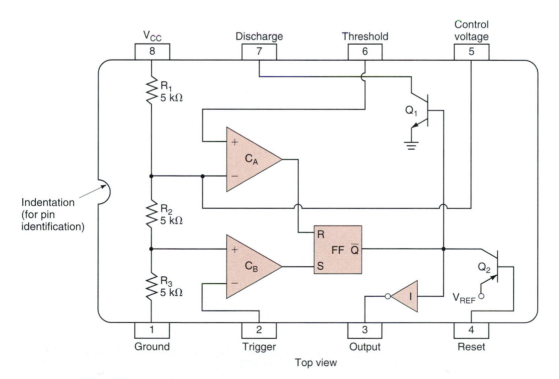

**FIGURE 19.32   The 555 timer.**

left. Then pin 1 is the lower-left pin. The rest of the pins are set in counterclockwise order around the case. This is the same as the pin-numbering scheme for the 741 op-amp.

Inside the chip are two comparators ($C_A$ and $C_B$), a flip-flop (FF), an inverter ($I$), two transistors, and a voltage divider ($R_1$ through $R_3$). The voltage divider is used to set the comparator reference voltages. Since $R_1 = R_2 = R_3$, the reference voltages for $C_A$ and $C_B$ are normally equal to $\frac{2}{3} V_{CC}$ and $\frac{1}{3} V_{CC}$, respectively. The only exception to this condition occurs when the *control voltage* input to the timer (pin 5) is used. This point is demonstrated later in this section.

If the input to pin 6 is greater than the reference voltage for $C_A$, the output from that comparator goes *high*. Otherwise, it remains at its low output state. If the input at pin 2 is greater than the reference voltage for $C_B$, the output from that comparator goes *low*. Otherwise, it remains at its high output state.

Under normal operating conditions, the comparators will be operating in one of three input/output states. These states are illustrated in Figure 19.33. Each of the output state combinations depends on the inputs at pins 2 and 6 of the timer. Note that it *is* possible to drive the outputs from both comparators *high* at the same time, but this is normally not done because of the effect it will have on the flip-flop.

The flip-flop responds to its inputs as shown in Figure 19.34. As you can see, having two *high* inputs is considered to be "invalid." This is due to the fact that there is no way to predict how the flip-flop will respond to that condition. If it were to receive two *high* inputs, the output from the flip-flop (labeled $\overline{Q}$) could be high, low, or anywhere in between the two. Thus, the input combination of *pin 6 = high and pin 2 = low* is never allowed. By disallowing this input combination, the flip-flop never receives two high inputs at the same time.

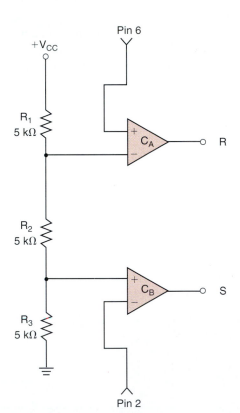

| Inputs | | Outputs | |
|---|---|---|---|
| Pin 6 | Pin 2 | R | S |
| High | High | High | Low |
| Low | Low | Low | High |
| * High | Low | High | High |
| Low | High | Low | Low |

* invalid input state (due to effect on flip-flop operation).

*Note:* For each comparator, the inputs are designated as *high* or *low* with respect to the reference voltage.

**FIGURE 19.33   Comparator input/output combinations.**

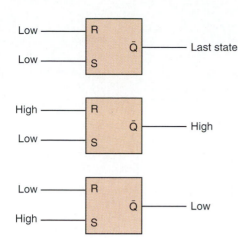

| S | R | $\bar{Q}$ |
|---|---|---|
| Low | Low | L.S.* |
| Low | High | High |
| High | Low | Low |
| High | High | Invalid |

* Last state. The $\bar{Q}$ output will remain in the state it was in before the input combination occurred.

*Note:* When the flip-flop inputs are not equal, the output follows the input at R.

**FIGURE 19.34    Flip-flop input/output combinations.**

We need to clear up only a few more points and we will be ready to look at some circuits. First, the flip-flop output is tied to the *discharge* output (pin 7) via an *npn* transistor. This transistor is merely acting as a current-controlled switch. When the output from the flip-flop is *high,* pin 7 is grounded through the transistor. When the output from the flip-flop is *low,* pin 7 will appear as an open to any external circuitry. The purpose served by this pin will be made clear when we discuss the basic 555 timer circuits. The inverter (pin 3) and the transistor (pin 7) both invert the flip-flop output. The reason that different symbols are used for the two circuits is that the inverter is used to provide some gain for the flip-flop output, whereas the transistor is used only as a switch.

The *reset* input (pin 4) is used to disable the 555 timer. Even though the timer will function (internally) as usual, the output (pin 3) will be held *low* whenever pin 4 is grounded. This input is used only in special applications and is usually held at an inactive level by being tied to $V_{CC}$. When pin 4 is tied to $V_{CC}$, the *pnp* transistor is biased *off* and the 555 timer output will operate normally. When the *reset* input is *low,* the *pnp* transistor is biased *on,* and the output from the flip-flop is shorted out before it reaches the inverter. In all our circuits, pin 4 will be tied to $V_{CC}$, thus disabling the reset circuit.

The purpose of the *reset input.*

When the reset input isn't used, it should be connected to $V_{CC}$.

## The Monostable Multivibrator

The 555 timer can be converted into a one-shot with the addition of a single resistor and a single capacitor. The 555 timer one-shot is shown in Figure 19.35. Before we get into the operation of this circuit, a few practical observations should be made:

◄ *OBJECTIVE 9*

Initial circuit conditions.

1. The *reset* input to the 555 timer (pin 4) is shown to be connected to $V_{CC}$. This disables the *reset* circuit.

2. Pins 8 and 1 are tied to $V_{CC}$ and ground, respectively, because they must be for the IC to operate.

3. The *control voltage* input (pin 5) is shown as *not connected* (N.C.). This means that this pin is not connected to anything. This is acceptable when the control voltage input is not used.

4. Pins 6 and 7 are tied together between *R* and *C*. Thus, the voltage at *both* of these pins will always equal the voltage across the capacitor, $V_C$.

Now let's look at the operation of the circuit. The appropriate waveforms are shown in Figure 19.36. To start with, we will have to make an assumption. We will assume that something caused the output of the flip-flop to be *high* before $T_0$. You will see what caused this condition in a minute. However, we need to assume this high flip-flop output to understand the waveforms at $T_0$.

**Lab Reference:** The operation of a 555-timer one-shot is demonstrated in Exercise 44.

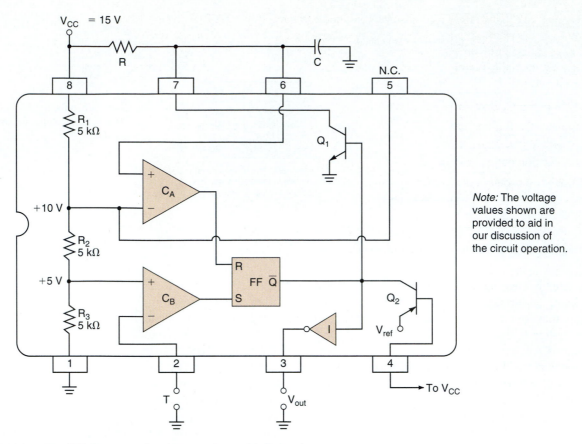

**FIGURE 19.35   The 555 timer one-shot (monostable multivibrator).**

Between $T_0$ and $T_2$, the inactive circuit voltages are shown. Normally, the trigger input ($T$) will be approximately equal to $V_{CC}$ when in its inactive state. This ensures that the output from $C_B$ is low.

Assuming that the output from the flip-flop is high, we know that pins 3 and 7 are both tied *low*. The low output at pin 7 ensures that the voltage across the capacitor ($V_C$) is equal to 0 V, as shown in Figure 19.36. This low voltage is applied to $C_A$ via pin 6.

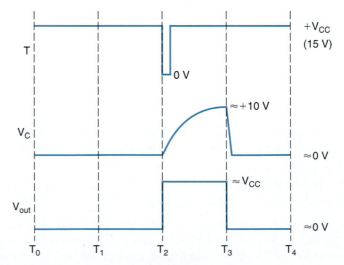

**FIGURE 19.36   The 555 timer one-shot waveforms.**

With a low input to $C_A$, its output is low. Since both inputs to the flip-flop are low, its output does not change from the *high* voltage that was already present. The circuit voltages will remain as shown until something happens to change them.

At $T_2$, the *trigger* (*T*) input is driven *low*. When this occurs, the following sequence of events takes place:

What happens when (T) goes low.

1. The low input to $C_B$ drives the output of the comparator *high*.

2. The inputs to the flip-flop are now unequal, and $\overline{Q}$ will go to the same level as the $R$ input, which is *low*.

3. The low flip-flop output is inverted and pin 3 goes *high,* as shown in Figure 19.36.

4. The low flip-flop output biases the *npn* transistor *off*. Thus, the capacitor will now start to charge toward $V_{CC}$. The charge time of the capacitor depends on the values of $R$ and $C$.

5. Eventually, the capacitor charges to the point where $V_C > +10$ V. At this point, the input to pin 6 is *high*.

6. The high input to $C_A$ causes the comparator's output to go *high*. Since the trigger signal is no longer present, the output from $C_B$ has returned to a low.

7. Since the inputs to the flip-flop are unequal again, $\overline{Q}$ will switch to the level of $R$, which is *high*.

8. The *high* output from the flip-flop causes pin 3 to go low again (shown at $T_3$).

9. The high flip-flop output turns the *npn* transistor back on, which in turn discharges the capacitor. This returns all circuit potentials to their original values (shown at $T_4$).

*Note:* Whenever the inputs to the flip-flop are *not equal*, $\overline{Q}$ goes to the same state as the $R$ input. (This is verified in Figure 19.34.)

As you can see, our initial assumption of $\overline{Q}$ = high is based on the fact that the flip-flop is always set to this value at the end of a trigger cycle.

At this point, you may be thinking that this circuit is very complex. In a way it is, but in a far more practical sense, it is not. In this discussion, we have considered all the details of the internal 555 timer response to a trigger input. If you were working with this circuit in a practical situation, you would be concerned only with the input/output function of the timer. In fact, look at the diagram shown in Figure 19.37. This is what you would see if you were working with the circuit we just discussed. All you have is one IC, one resistor, and one capacitor. If the circuit does not work, all you need to do is make a few basic measurements. If the passive components check out, you replace the 555 timer. How is *that* for simple? (We will discuss the entire troubleshooting process later in this section.)

**FIGURE 19.37**

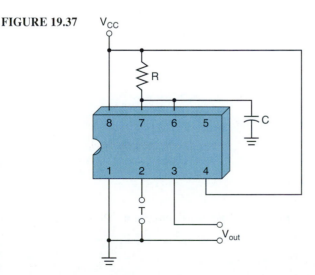

The pulse width of a 555 timer one-shot is determined by the values of $R$ and $C$ used in the circuit. The charge time of the capacitor in an $RC$ circuit is found as

*Note:* The equation shown here is derived in a basic electronics course on $RC$ circuits. In our discussion, we will accept its validity without going through its derivation.

$$t = RC \left[ \ln \frac{V_S - V_I}{V_S - v_C} \right] \tag{19.11}$$

where   $t$ = the time for the capacitor to charge to $v_C$
   $V_S$ = the source voltage used to charge the capacitor
   $V_I$ = the initial charge on the capacitor, usually 0 V
   $v_C$ = the capacitor voltage of interest

Note that "ln" indicates that we are interested in the *natural log* of the fraction. For the 555 timer one-shot, the charging voltage ($V_S$) for the capacitor is $V_{CC}$ and the initial charge ($V_I$) on the capacitor is 0 V. Since the pulse width is determined by the time required for the capacitor to charge to $\frac{2}{3} V_{CC}$, we use this value as $v_C$ in the charge time equation. With these values, we can derive a useful equation that allows us to determine the pulse width of the one-shot for given values of $R$ and $C$, as follows:

$$
\begin{aligned}
t &= RC \left[ \ln \frac{V_S - V_I}{V_S - v_C} \right] \\
&= RC \left[ \ln \frac{V_{CC} - 0\ \text{V}}{V_{CC} - \frac{2}{3} V_{CC}} \right] \\
&= RC \left[ \ln \frac{V_{CC}}{\frac{1}{3} V_{CC}} \right] \\
&= RC \left[ \ln(3) \right]
\end{aligned}
$$

or

$$PW = 1.1RC \tag{19.12}$$

Example 19.10 demonstrates the use of this equation.

---

**EXAMPLE 19.10**

**Lab Reference:** The effect of changing $C$ on the PW of a one-shot is demonstrated in Exercise 44.

A one-shot like the one in Figure 19.37 has values of $R = 1.2\ \text{k}\Omega$ and $C = 0.1\ \mu\text{F}$. Determine the pulse width of the output.

**Solution:**   The pulse width is found as

$$
\begin{aligned}
PW &= 1.1RC \\
&= 1.1(1.2\ \text{k}\Omega)(0.1\ \mu\text{F}) \\
&= 132\ \mu\text{s}
\end{aligned}
$$

**PRACTICE PROBLEM 19.10**

A one-shot like the one in Figure 19.37 has values of $R = 18\ \text{k}\Omega$ and $C = 1.5\ \mu\text{F}$. Determine the pulse width of the time output signal.

---

## Circuit Troubleshooting

The one-shot in Figure 19.37 is extremely simple to troubleshoot. The process involves making the following checks:

Circuit checks.

- Are the $V_{CC}$ and ground connections good?
- Is the *reset* input held at an inactive level?

- Is the circuit receiving a *valid* trigger signal?
- Is the resistor good?
- Is the capacitor good?

If the answer to all these questions is *yes,* replace the 555 timer. If any question has a *no,* you have found the problem.

The troubleshooting procedure shown involves checking for a *valid* trigger signal. The question at this point is this: Just what is a *valid* trigger signal? A valid trigger signal is an input to pin 2 that is *low enough* to cause $C_B$ to have a high output. The value of input voltage required to do this depends on whether or not the *control voltage* input is used. When the control voltage input is *not* used, the maximum input trigger value that will work can be found as

What is a *valid* trigger signal?

$$V_{T(max)} < \tfrac{1}{3} V_{CC} \qquad \text{(19.13)}$$

This equation is based on the fact that the (+) input of $C_B$ is at $\tfrac{1}{3}V_{CC}$ when the control voltage input is not used. Therefore, the ($T$) input must go *below* this voltage level to cause the output of $C_B$ to go high.

When the control voltage input is used, it will be dropped equally across $R_2$ and $R_3$ in the resistive ladder. Therefore, the voltage at the (+) input of $C_B$ will be equal to one-half of the control voltage, and the maximum trigger voltage is found as

$$V_{T(max)} < \tfrac{1}{2} V_{con} \qquad \text{(19.14)}$$

where $V_{con}$ is the voltage at the *control voltage* input. Example 19.11 demonstrates the process for determining the maximum acceptable input trigger voltage for a given one-shot.

---

Determine the value of $V_{T(max)}$ for each of the one-shot circuits shown in Figure 19.38.

**EXAMPLE 19.11**

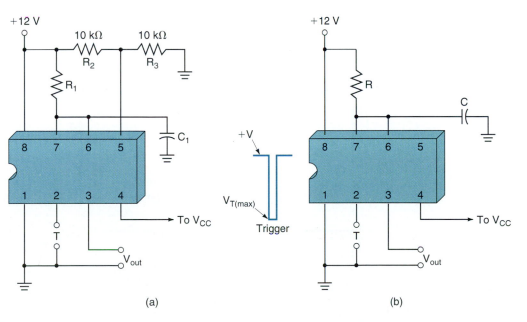

**FIGURE 19.38**

*Solution:* The circuit in Figure 19.38a has a voltage-divider circuit providing a voltage to the control voltage input. Using the standard voltage-divider formula, the control voltage is found to be $V_{con} = +6$ V. Therefore, the maximum allowable value for a low trigger input is

$$V_{T(max)} < \tfrac{1}{2}V_{con}$$
$$< +3 \text{ V}$$

For the circuit in Figure 19.38a, the *low* trigger voltage would have to be lower than +3 V for the circuit to trigger properly.

Since the circuit in Figure 19.38b has an open control voltage input, equation (19.13) is used to determine $V_{T(max)}$, as follows:

$$V_{T(max)} < \tfrac{1}{3}V_{CC}$$
$$< +4 \text{ V}$$

### PRACTICE PROBLEM 19.11

A circuit like the one in Figure 19.38a has values of $R_1 = 2$ kΩ, $R_2 = 10$ kΩ, and $R_3 = 30$ kΩ. If $V_{CC}$ for the circuit is +15 V, what is the value of $V_{T(max)}$ for the circuit?

The advantage of using the *control input* (pin 5).

The circuits used in Example 19.11 demonstrate the advantage of using the control voltage input to the 555 timer. In both of these circuits, the output would be a pulse that has peak voltages of 0 V and +12 V (approximately). However, the circuit in Figure 19.38a can be triggered by a lower voltage than the one in Figure 19.38b. If we had wanted a +3 V trigger without using the control voltage input, we would have had to use a +9 V supply. In other words, by using the control voltage input, we can select any trigger signal level for the circuit without having to change the $V_{CC}$ of the circuit and thus the maximum possible peak output value from the timer.

Using the values obtained in Example 19.11, we can determine which trigger signals are valid and which are not. For example, consider the trigger signals shown in Figure 19.39. All three of the trigger signals in Figure 19.39a are valid for either circuit in Figure 19.38. Since they are all equal to or less than the values of $V_{T(max)}$, they will all cause the two circuits to trigger properly.

The first trigger signal in Figure 19.39b [$V_{T(low)} = 3.2$ V] will work very well for the circuit in Figure 19.38b, as will the second trigger signal [$V_{T(low)} = +4$ V]. However, none of these trigger signals will work for the circuit in Figure 19.38a because their

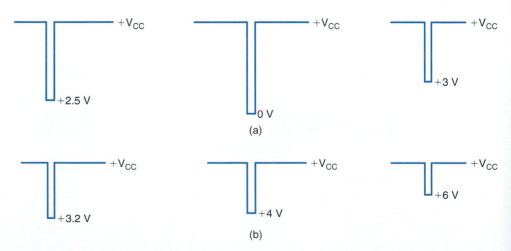

**FIGURE 19.39** One-shot trigger signals.

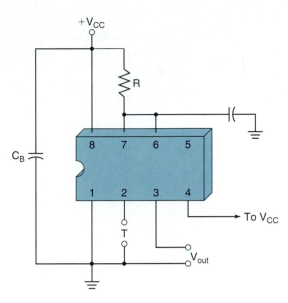

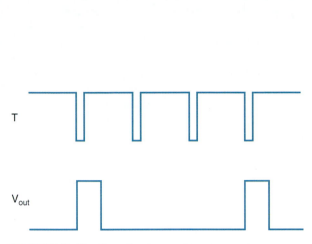

FIGURE 19.40    One-shot intermittent operation.

FIGURE 19.41    Use of a bypass capacitor ($C_B$) can eliminate intermittent output problems.

low levels are greater than the value of $V_{T(max)}$ for that circuit. The third waveform shown will not work for either circuit.

A special note needs to be made at this time. The third waveform in Figure 19.39b may cause the circuit in Figure 19.38a to operate *intermittently*. This means that it is close enough to the value of $V_{T(max)}$ for the circuit that it may work one time and not work another. An example of intermittent operation is illustrated in Figure 19.40. Note that the output triggers on two of the input signals but not on the other two. Any time that you have an intermittent problem in a switching circuit, the trigger signal should be suspect. In most cases, the intermittent operation is caused by a trigger signal that is not going low (or high) enough.

Sometimes an intermittent problem can develop when a one-shot is operated at a relatively high frequency. In this case, the problem is the switching rate of the 555, and not the trigger signal. You see, when the 555 timer switches from one output state to another, there is an instant when an effective short circuit *may* develop inside the chip between the $V_{CC}$ and ground pins. The cause of this short is the active components switching through their active region. The cure for this problem is to use a *bypass capacitor* between the $V_{CC}$ and ground pins of the chip, as shown in Figure 19.41.

If a one-shot circuit has a frequency-related intermittent problem, the bypass capacitor will eliminate the problem by holding the supply current ($I_{CC}$) relatively constant during the switching time. Note that $C_B$ will typically be around 0.1 μF.

How do you know when an intermittent problem is caused by $f_{in}$? Check the trigger signal. If the trigger signal is definitely valid, odds are that the problem is related to the input frequency of the circuit.

## The Astable Multivibrator

The 555 timer can be wired for free-running operation as shown in Figure 19.42. Note that the circuit contains *two resistors and one capacitor* and *does not have an input trigger from any other circuit*. The lack of a trigger signal from an external source is the circuit recognition feature of the free-running multivibrator.

Circuit recognition.

The key to the operation of this circuit is the fact that the capacitor is connected to *both* the trigger input (pin 2) and the threshold input (pin 6). Thus, if the capacitor is caused to charge and discharge between the threshold voltage ($\frac{2}{3}V_{CC}$) and the trigger voltage

Circuit operation.

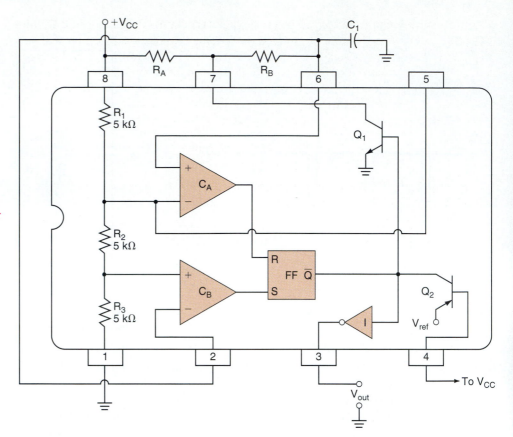

**FIGURE 19.42**  The 555 timer astable (free-running) multivibrator.

Lab Reference: The operation of a 555 timer free-running multivibrator is demonstrated in Exercise 45.

($\frac{1}{3}V_{CC}$), the output will be a steady train of pulses. This point is illustrated in Figure 19.43. Since pins 6 and 2 are tied together, their voltage values will always be equal. When $V_C$ charges to $\frac{2}{3}V_{CC}$, both inputs are high and the output from the timer goes *low*. When $V_C$ discharges to $\frac{1}{3}V_{CC}$, both inputs are low and the output from the timer goes *high*.

The charge–discharge action of $C_1$ is caused by the 555 timer internal circuitry. To see how this is accomplished, we will start by making a few assumptions:

1.  The capacitor is just starting to charge toward the threshold voltage, $V_{th}$.

2.  The output from the flip-flop is *low,* causing pin 3 to be high and pin 7 to be *open.*

With pin 7 open, the capacitor will be charged by $V_{CC}$ via $R_A$ and $R_B$. As the capacitor charges, it eventually reaches $V_{th}$. At this point, both inputs to the 555 timer are *high.*

**FIGURE 19.43**  **Free-running multivibrator waveforms.**

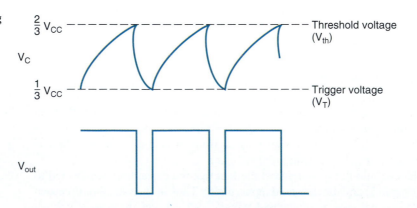

This causes the output from the flip-flop to go high. The high voltage out of the flip-flop causes the output (pin 3) to go low and biases the *npn* transistor *on*. Thus, pin 7 is now acting as a short to ground and will allow the capacitor to discharge. When it discharges down to $V_T$, both inputs will again be low, driving the output from the flip-flop low. This, in turn, causes the output to go high again and causes pin 7 to return to the *open* condition, and the cycle starts again.

The output cycle time for the free-running multivibrator is found as

$$T = 0.693[(R_A + 2R_B)C_1] \qquad \textbf{(19.15)}$$

Converting equation (19.15) into a frequency form yields the following equation for the operating frequency of the circuit:

$$f_0 = \frac{1}{0.693[(R_A + 2R_B)C_1]}$$

or

$$f_0 = \frac{1.44}{(R_A + 2R_B)C_1} \qquad \textbf{(19.16)}$$

The duty cycle of the free-running multivibrator is the ratio of the pulse width to the cycle time. If you look at the waveforms in Figure 19.43, you'll see that the pulse width and cycle time of the circuit are determined by the charge and discharge times of $C_1$. We already know from equation (19.15) that the cycle time is equal to $0.693[(R_A + 2R_B)C_1]$. The pulse width of the circuit is found as

Lab Reference: The values defined in equations (19.15) through (19.18) are measured as part of Exercise 45.

$$PW = 0.693[(R_A + R_B)C_1] \qquad \textbf{(19.17)}$$

If we use the charge/discharge equations in place of PW and cycle time in the duty cycle formula, we obtain the following useful equation:

$$\text{duty cycle} = \frac{0.693[(R_A + R_B)C_1]}{0.693[(R_A + 2R_B)C_1]} \times 100$$

or

$$\text{duty cycle} = \frac{R_A + R_B}{R_A + 2R_B} \times 100 \qquad \textbf{(19.18)}$$

Example 19.12 demonstrates the basic analysis of a free-running multivibrator.

---

Determine the values of $f_0$ and duty cycle of the circuit shown in Figure 19.44. Also, determine the pulse width of the circuit.

**Solution:** The $f_0$ for the circuit is found as

$$f_0 = \frac{1.44}{(R_A + 2R_B)C_1}$$

$$= \frac{1.44}{(3\ k\Omega + 5.4\ k\Omega)(0.033\ \mu F)}$$

$$= 5.19\ kHz$$

*EXAMPLE 19.12*

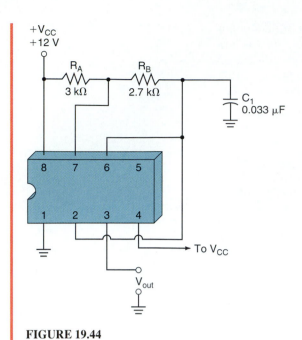

+V_CC
+12 V

R_A 3 kΩ  R_B 2.7 kΩ

C_1 0.033 μF

To V_CC

V_out

**FIGURE 19.44**

The duty cycle of the circuit is found as

$$dc = \frac{R_A + R_B}{R_A + 2R_B} \times 100$$

$$= 67.9\%$$

Finally, the pulse width of the circuit is found as

$$PW = 0.693[(R_A + R_B)C_1]$$
$$= 0.693(5.7 \text{ k}\Omega)(0.033 \text{ μF})$$
$$= 130 \text{ μs}$$

### PRACTICE PROBLEM 19.12

A circuit like the one in Figure 19.44 has the following values: $V_{CC}$ = +10 V, $R_A$ = 5.1 kΩ, $R_B$ = 2.2 kΩ, and $C_1$ = 0.022 μF. Determine the values of $f_0$, duty cycle, and output pulse width for the circuit.

---

**Why two resistors are used in the RC circuit.**

At this point, you might be wondering why the free-running multivibrator requires two resistors in its $RC$ network. After all, we required only a single resistor for the one-shot. To answer this question, let's take a look at what would happen if either $R_A$ or $R_B$ were shorted.

If we short $R_A$, we have a single resistor, $R_B$, between pins 6 and 7. This situation is unacceptable because the 555 timer will short circuit the power supply when pin 7 goes low. Recall that pin 7 goes low during every discharge cycle. If $R_A$ is shorted, this low output is connected directly to $V_{CC}$. The result would be a "well-done" 555 timer. Thus, we must have a resistor between pin 7 and $V_{CC}$.

If $R_B$ were shorted, there would be no $RC$ circuit during the discharge cycle of $C_1$. As you have been shown, $C_1$ charges through $(R_A + R_B)$ and discharges through $R_B$. If $R_B$ is replaced with a short circuit, the discharge cycle time will be 0, and the duty cycle will be 100 percent, an unacceptable situation. Thus, $R_B$ is also required for proper operation.

## Circuit Troubleshooting

The process for troubleshooting the free-running circuit is nearly the same as the procedure used on the one-shot. The only difference is that you have two resistors that must be verified as good. All other tests remain the same.

## Voltage-Controlled Oscillators

*OBJECTIVE 10* ▶

**Voltage-controlled oscillator (VCO)**
A free-running oscillator whose output frequency is controlled by a dc input voltage.

The 555 can be modified to form a circuit called a **voltage-controlled oscillator (VCO).** A VCO (shown in Figure 19.45) is a free-running oscillator whose output frequency depends on a given dc input control voltage, $V_{con}$. When applied to pin 5 of the timer, $V_{con}$ becomes the reference voltage ($V_{ref}$) for $C_A$. The $V_{ref}$ of $C_B$ is equal to ½ of $V_{con}$. When the potentiometer is set to provide a high value of $V_{con}$, the output frequency of the 555 is relatively low. When the value of $V_{con}$ is lower, the output frequency increases. This is because the value of $V_{con}$ directly affects the values of $V_{th}$ and $V_T$. For example, when $V_{con}$

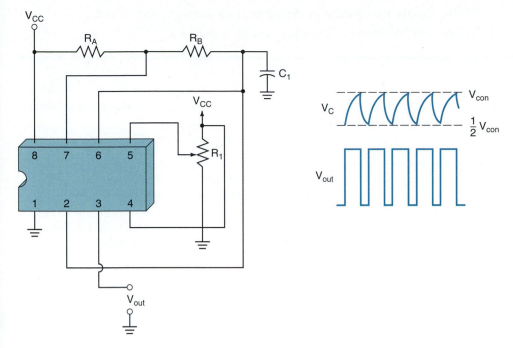

**FIGURE 19.45   A voltage-controlled oscillator (VCO).**

is set to $+10$ V, the value of $V_{th}$ also becomes $+10$ V, and $V_T$ becomes one-half of $V_{con}$, or $+5$ V. Capacitor $C_1$ must charge to the value of $V_{con}$ ($+10$ V) to change the output of the comparator, $C_A$, and discharge to the value of $V_T$ ($+5$ V), to cause the output of $C_B$ to change states.

*Don't forget: $V_{th}$ is the voltage that triggers comparator A. $V_T$ is the input that triggers comparator B.*

When the value of $V_{con}$ is changed, the values of $V_{th}$ and $V_T$ change also. If the value of $V_{con}$ is changed to $+7$ V, $C_1$ only needs to change to $+7$ V and discharge to $+3.5$ V to produce an output cycle. This reduces the time required to charge and discharge the capacitor, which causes the output state of both comparators to switch at a faster rate. The faster the comparators' switch rates, the higher the output frequency for the circuit.

## One Final Note

In this chapter, we have barely scratched the surface of the entire subject of switching circuits. The study of digital electronics provides complete coverage of the subject. However, this chapter has provided you with the fundamental principles of switching circuits in such a way as to allow you to continue the study of digital electronics on your own.

**Section Review**

1. What is a *multivibrator?*
2. Describe the output characteristics of each of the following circuits:
    a. *Astable multivibrator (free-running multivibrator)*
    b. *Monostable multivibrator (one-shot)*
    c. *Bistable multivibrator (flip-flop)*
3. What is the *555 timer?*
4. List the input/output relationships for the 555 timer.
5. Briefly describe the operation of the monostable multivibrator (one-shot).
6. What checks are made when troubleshooting a one-shot?

7. Briefly describe the operation of the astable (free-running) multivibrator.

8. Describe the operation of the voltage-controlled oscillator (VCO).

## Key Terms

The following terms were introduced and defined in this chapter:

| | | |
|---|---|---|
| astable multivibrator | hysteresis | space width (SW) |
| bistable multivibrator | inverter | speed-up capacitor |
| buffer | lower trigger point (LTP) | square wave |
| cycle time ($T_C$) | monostable multivibrator | storage time ($t_s$) |
| delay time ($t_d$) | multivibrator | switching circuits |
| driver | one-shot | switching transistors |
| duty cycle (DC) | propagation delay | turn-off time ($t_{off}$) |
| fall time ($t_f$) | pulse width (PW) | turn-on time ($t_{on}$) |
| 555 timer | rectangular waveform | upper trigger point (UTP) |
| flip-flop | rise time ($t_r$) | voltage-controlled |
| free-running multivibrator | Schmitt trigger | oscillator (VCO) |

## Equation Summary

| Equation Number | Equation | Section Number |
|---|---|---|
| **(19.1)** | $\text{duty cycle} = \dfrac{\text{pulse cycle}}{\text{cycle time}} \times 100$ | 19.2 |
| **(19.2)** | $f_2 = \dfrac{0.35}{t_r}$ | 19.2 |
| **(19.3)** | $f_{max} = \dfrac{0.35}{10 t_r}$ (practical limit) | 19.2 |
| **(19.4)** | $C_S < \dfrac{1}{20 R_B f_{max}}$ | 19.2 |
| **(19.5)** | $\text{UTP} = -\dfrac{R_1}{R_f}(-V + 1\text{ V})$ | 19.3 |
| **(19.6)** | $\text{LTP} = -\dfrac{R_1}{R_f}(+V - 1\text{ V})$ | 19.3 |
| **(19.7)** | $\text{UTP} = -\dfrac{R_1}{R_{f1}}(-V + 1\text{ V})$ | 19.3 |
| **(19.8)** | $\text{LTP} = -\dfrac{R_1}{R_{f2}}(+V - 1\text{ V})$ | 19.3 |
| **(19.9)** | $\text{UTP} = \dfrac{R_2}{R_1 + R_2}(+V - 1\text{ V})$ | 19.3 |
| **(19.10)** | $\text{LTP} = \dfrac{R_2}{R_1 + R_2}(-V + 1\text{ V})$ | 19.3 |
| **(19.11)** | $t = RC\left[\ln \dfrac{V_S - V_I}{V_S - v_c}\right]$ | 19.4 |
| **(19.12)** | $\text{PW} = 1.1RC$ | 19.4 |

| Equation Number | Equation | Section Number |
|---|---|---|
| (19.13) | $V_{T(max)} < \frac{1}{3}V_{CC}$ | 19.4 |
| (19.14) | $V_{T(max)} < \frac{1}{2}V_{con}$ | 19.4 |
| (19.15) | $T = 0.693[(R_A + 2R_B)C_1]$ | 19.4 |
| (19.16) | $f_0 = \dfrac{1.44}{(R_A + 2R_B)C_1}$ | 19.4 |
| (19.17) | $PW = 0.693[(R_A + R_B)C_1]$ | 19.4 |
| (19.18) | $\text{duty cycle} = \dfrac{R_A + R_B}{R_A + 2R_B} \times 100$ | 19.4 |

**19.1.** 7.99 V
**19.2.** When $V_{GS} = 0$ V, $V_{out} = 1$ V. When $V_{GS} = -4$ V, $V_{out} = 10$ V.
**19.3.** 7 V
**19.4.** PW = 6 μs, $T_C = 18$ μs
**19.5.** 33.3%
**19.6.** Theoretical: 5.41 MHz, Practical: 1.4 MHz
**19.7.** UTP = +4 V, LTP = −4 V
**19.8.** UTP = +1.4 V, LTP = −2 V
**19.9.** UTP = +2.25 V, LTP = −2.25 V
**19.10.** 29.7 μs
**19.11.** 5.63 V
**19.12.** $f_0 = 6.9$ kHz, dc = 76.8%, PW = 111.3 μs

## §19.1

1. Determine the minimum high voltage required to saturate the transistor in Figure 19.46.

2. A switch like the one in Figure 19.46 has the following values: $V_{CC} = +12$ V, $h_{FE} = 150$, $R_C = 1.1$ kΩ, and $R_B = 47$ kΩ. Determine the minimum high input voltage required to saturate the transistor.

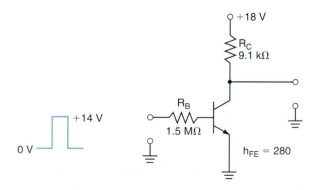

**FIGURE 19.46**

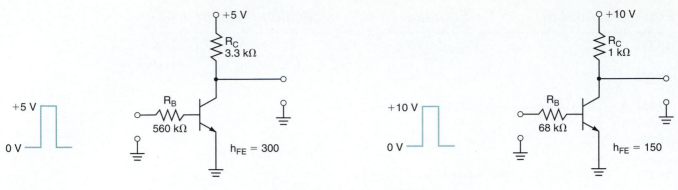

**FIGURE 19.47**

**FIGURE 19.48**

3. Determine the minimum high input voltage required to saturate the transistor in Figure 19.47.
4. Determine the minimum high input voltage required to saturate the transistor in Figure 19.48.
5. Determine the high and low output voltages for the circuit in Figure 19.49.
6. Determine the high and low output voltages for the circuit in Figure 19.50.

*§19.2*

7. Determine the values of PW and $T_C$ for the waveform shown in Figure 19.51.
8. Determine the values of PW and $T_C$ for the waveform shown in Figure 19.52.

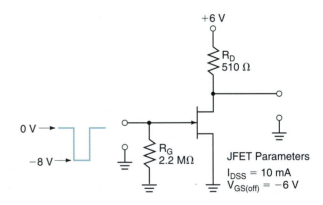

**FIGURE 19.49**

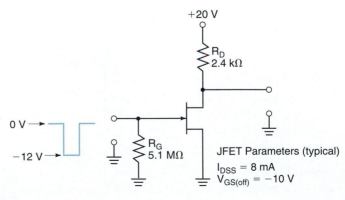

**FIGURE 19.50**

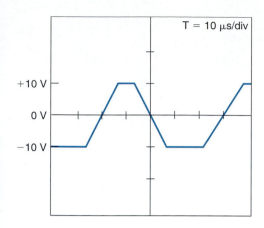

FIGURE 19.51

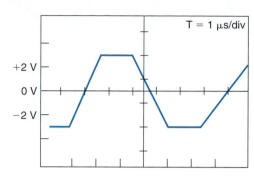

FIGURE 19.52

9. Calculate the duty cycle of the waveform in Figure 19.51.

10. Calculate the duty cycle of the waveform in Figure 19.52.

11. The 2N4264 BJT has the following parameters: $t_d = 8$ ns, $t_r = 15$ ns, $t_s = 20$ ns, and $t_f = 15$ ns. Determine the theoretical and practical limits on $f_{in}$ for the device.

12. The 2N4400 BJT has the following parameters: $t_d = 15$ ns, $t_r = 20$ ns, $t_s = 225$ ns, and $t_f = 30$ ns. Determine the theoretical and practical limits on $f_{in}$ for the device.

13. The BCX587 BJT has the following parameters: $t_d = 16$ ns, $t_r = 29$ ns, $t_s = 475$ ns, and $t_f = 40$ ns. Determine the theoretical and practical limits on $f_{in}$ for the device.

14. The MPS2222 BJT has the following parameters: $t_d = 10$ ns, $t_r = 25$ ns, $t_s = 225$ ns, and $t_f = 60$ ns. Determine the theoretical and practical limits on $f_{in}$ for the device.

15. The 2N3971 JFET has the following parameters: $t_d = 15$ ns, $t_r = 15$ ns, and $t_{off} = 60$ ns. Determine the theoretical and practical limits on $f_{in}$ for the device.

16. The 2N4093 JFET has the following parameters: $t_d = 20$ ns, $t_r = 40$ ns, and $t_{off} = 80$ ns. Determine the theoretical and practical limits on $f_{in}$ for the device.

17. The 2N4351 MOSFET has the following parameters: $t_d = 45$ ns, $t_r = 65$ ns, $t_s = 60$ ns, and $t_f = 100$ ns. Determine the theoretical and practical limits on $f_{in}$ for the device.

18. The 2N4393 JFET has the following parameters: $t_{on} = 15$ ns and $t_{off} = 50$ ns. Determine the theoretical and practical limits on $f_{in}$ for the device.

19. Determine the values of $t_d$, $t_r$, $t_s$, and $t_f$ for the waveform in Figure 19.53.

20. Determine the values of $t_d$, $t_r$, $t_s$, and $t_f$ for the waveform in Figure 19.54.

§19.3

21. Determine the UTP and LTP values for the Schmitt trigger in Figure 19.55.

22. Determine the UTP and LTP values for the Schmitt trigger in Figure 19.56.

23. Determine the UTP and LTP values for the Schmitt trigger in Figure 19.57.

24. Determine the UTP and LTP values for the Schmitt trigger in Figure 19.58.

25. Determine the UTP and LTP values for the Schmitt trigger in Figure 19.59.

26. Determine the UTP and LTP values for the Schmitt trigger in Figure 19.60.

27. Determine the UTP and LTP values for the Schmitt trigger in Figure 19.61.

28. Determine the UTP and LTP values for the Schmitt trigger in Figure 19.62.

FIGURE 19.53

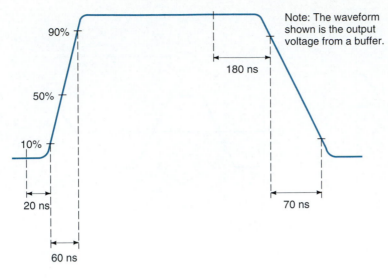

Note: The waveform shown is the output voltage from a buffer.

FIGURE 19.54

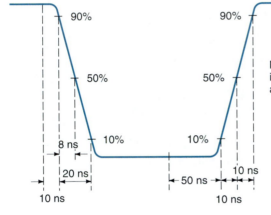

Note: The waveform shown is the output voltage from an inverter.

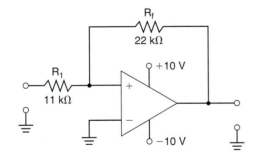

FIGURE 19.55

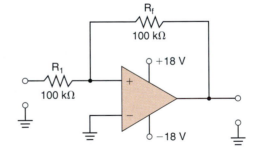

FIGURE 19.56

FIGURE 19.57

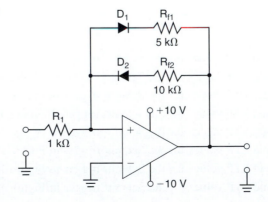

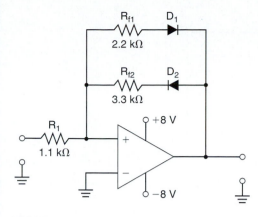

FIGURE 19.58

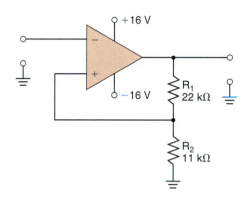

FIGURE 19.59

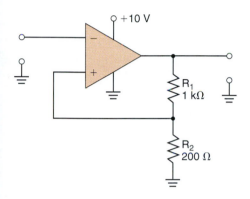

FIGURE 19.60

FIGURE 19.61

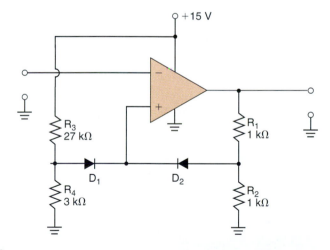

FIGURE 19.62

§19.4

29. Calculate the output pulse width of the one-shot in Figure 19.63.

30. Calculate the output pulse width of the one-shot in Figure 19.64.

31. Calculate the value of $V_{T(\text{max})}$ for the switch in Figure 19.65.

32. The resistor values in Figure 19.65 are changed to $R_1 = 33$ k$\Omega$ and $R_2 = 11$ k$\Omega$. Recalculate the value of $V_{T(\text{max})}$ for the circuit.

FIGURE 19.63

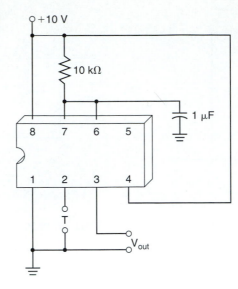

+10 V

10 kΩ

1 μF

8    7    6    5

1    2    3    4

T

V<sub>out</sub>

FIGURE 19.64

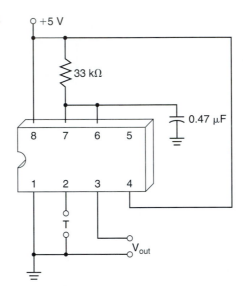

+5 V

33 kΩ

0.47 μF

8    7    6    5

1    2    3    4

T

V<sub>out</sub>

FIGURE 19.65

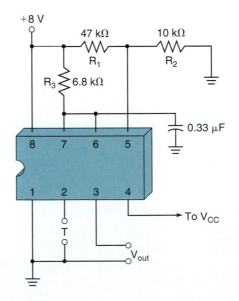

+8 V

47 kΩ    R₁    10 kΩ    R₂

R₃    6.8 kΩ

0.33 μF

8    7    6    5

1    2    3    4

T

To V<sub>CC</sub>

V<sub>out</sub>

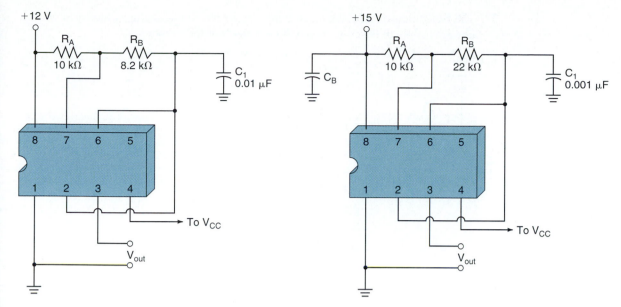

FIGURE 19.66                                    FIGURE 19.67

33. Calculate the operating frequency, duty cycle, and pulse-width values for the free-running multivibrator in Figure 19.66.

34. Calculate the operating frequency, duty cycle, and pulse-width values for the free-running multivibrator in Figure 19.67.

35. The value of $R_B$ in Figure 19.66 is increased to 10 kΩ. Recalculate the operating frequency, duty cycle, and pulse-width values for the circuit.

36. The value of $C_1$ in Figure 19.67 is increased to 0.033 μF. Recalculate the values of operating frequency and pulse width for the circuit.

37. The free-running multivibrator in Figure 19.68 has the dc readings indicated when power is applied. Discuss the possible cause(s) of the problem.

*Troubleshooting Practice Problems*

FIGURE 19.68

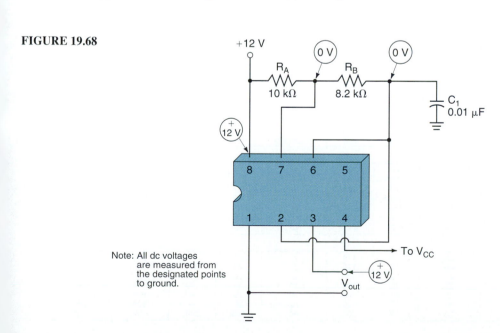

Note: All dc voltages are measured from the designated points to ground.

**38.** The free-running multivibrator in Figure 19.69 has the dc readings indicated when power is applied. Discuss the possible cause(s) of the problem.

FIGURE 19.69

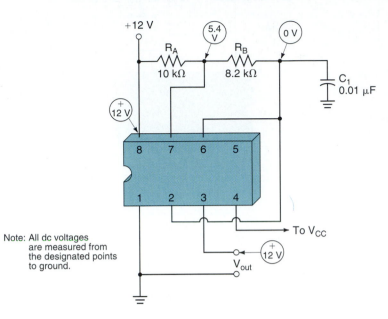

Note: All dc voltages are measured from the designated points to ground.

## The Brain Drain

**39.** The BJT in Figure 19.70a is faulty. Determine which (if any) of the transistors listed in the selector guide can be used as a substitute component. (Be careful; this one is more difficult than it appears!)

FIGURE 19.70a

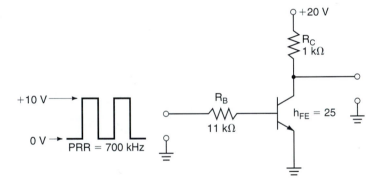

FIGURE 19.70b   **Pinout: 1-Base, 2-Emitter, 3-Collector**

Devices are listed in order of descending $f_T$.

| Device | Marking | Switching Time (ns) | | $V_{(BR)CEO}$ | $h_{FE}$ | | | $f_T$ |
| | | $t_{on}$ | $t_{off}$ | | Min | Max | @ $I_C$ (mA) | Min (MHz) |
|---|---|---|---|---|---|---|---|---|
| **NPN** | | | | | | | | |
| MMBT2369 | 1J | 12 | 18 | 15 | 20 | — | 100 | — |
| BSV52 | B2 | 12 | 18 | 12 | 40 | 120 | 10 | 400 |
| MMBT2222 | 1B | 35 | 385 | 30 | 30 | — | 500 | 250 |
| MMBT2222A | 1P | 35 | 385 | 40 | 40 | — | 500 | 200 |
| MMBT4401 | 2X | 35 | 255 | 40 | 40 | — | 500 | 250 |
| MMBT3903 | 1Y | 70 | 225 | 40 | 15 | — | 100 | 250 |
| MMBT3904 | 1A | 70 | 250 | 40 | 30 | — | 100 | 200 |
| **PNP** | | | | | | | | |
| MMBT3638A | BN | 75 | 170 | 25 | 20 | — | 300 | — |
| MMBT3638 | AM | 75 | 170 | 25 | 20 | — | 300 | — |
| MMBT3640 | 2J | 25 | 35 | 12 | 20 | — | 50 | 500 |
| MMBT4403 | 2T | 35 | 225 | 40 | 90 | 180 | 1 | 150 |
| MMBT2907 | 2B | 45 | 100 | 40 | 30 | — | 500 | 200 |
| MMBT2907A | 2F | 45 | 100 | 60 | 50 | — | 500 | 200 |
| MMBT3906 | 2A | 70 | 300 | 40 | 100 | 300 | 10 | 250 |

FIGURE 19.71

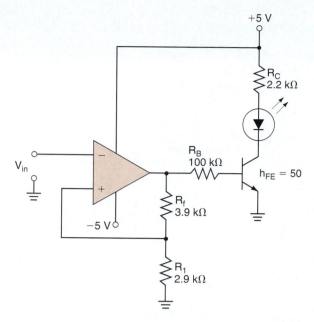

**40.** Using a 555 timer, design a free-running multivibrator that produces a 50 kHz output with a 60 percent duty cycle. The supply voltage for the timer is +10 V, and it is to have an inactive reset input pin.

**41.** Refer to Figure 19.71. The circuit was intended (by design) to fulfill the following requirements:

   **a.** The LED should light when a positive-going input signal passes a UTP value of +1.7 V.

   **b.** The LED should turn off when a negative-going input signal passes an LTP of −1.7 V.

There is a flaw in the circuit design. Find the flaw and suggest a correction that will make the circuit work properly.

**42.** Write a program that will determine the UTP and LTP values for both of the following:

   **a.** A noninverting Schmitt trigger like the one in Figure 19.25.

   **b.** An inverting Schmitt trigger like the one in Figure 19.27.

**43.** Write a program that will determine the PRR, duty cycle, and pulse-width values for a free-running multivibrator.

*Suggested Computer Applications Problems*

# 20

# THYRISTORS AND OPTOELECTRONIC DEVICES

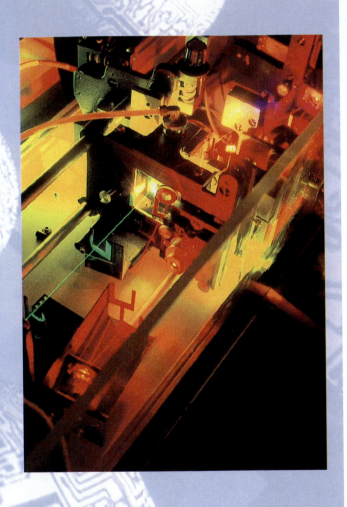

Industrial-optical systems, like the one shown, use both thyristors and optoelectronic devices to perform the functions for which they are designed.

# OUTLINE

# OBJECTIVES

*After studying the material in this chapter, you should be able to:*

1. Describe the operation of the *silicon unilateral switch (SUS)*.

2. List and describe the methods commonly used to drive an SUS into cutoff.

3. Describe the construction of the *silicon-controlled rectifier (SCR)* and the methods used to trigger the device into conduction.

4. Discuss the problem of SCR *false triggering* and the methods commonly used to prevent it.

5. Discuss the use of the SCR as a *crowbar circuit* and a *phase controller*.

6. Discuss the similarities and differences between *diacs* and *triacs*.

7. List and describe the circuits commonly used to control triac triggering.

8. Describe the construction and operation of *unijunction transistors (UJTs)*.

9. Compare and contrast *light emitters* and *light detectors*.

10. Discuss *wavelength* and *light intensity*.

11. Describe the operation and critical parameters of the basic discrete photodetectors.

12. Discuss the operation and applications of *optoisolators* and *optointerrupters*.

**Thyristors**
Devices designed specifically for high-power switching applications.

**Optoelectronic devices**
Devices that are controlled by light and/or emit light.

I n this chapter, we will take a look at two independent groups of electronic devices. First, we will discuss a group of devices called *thyristors:* **Thyristors** are devices designed specifically for high-power switching applications. Unlike BJTs, FETs, and op-amps (which can also be used as switches), most thyristors are not designed to be used as linear amplifying devices.

Second, we wil discuss a group known as **optoelectronic devices.** These devices are controlled by light and/or emit (generate) light. In Chapter 2, we discussed the most basic of the optoelectronic devices, the *light-emitting diode,* or *LED.* In this chapter, we will discuss some other devices that belong to this group.

No central theme ties thyristors and optoelectronic devices together. These two groups of components are covered in the same chapter simply because it is convenient to do so.

# 20.1 INTRODUCTION TO THYRISTORS: THE SILICON UNILATERAL SWITCH (SUS)

OBJECTIVE 1 ▶

**Silicon unilateral switch (SUS)**
A two-terminal, four-layer device that can be triggered into conduction by applying a specified forward voltage across its terminals.

The simplest of the thyristors is the **silicon unilateral switch (SUS)**. The SUS is a two-terminal, four-layer device that can be triggered into conduction by applying a specified forward voltage across its terminals. The construction of the SUS (along with its schematic symbol) is shown in Figure 20.1. As you can see, the device contains two *n*-type regions and two *p*-type regions. Because of its construction, the SUS is often referred to as a *pnpn diode* or a *four-layer diode.* Note that the term *diode* is considered appropriate because the device has two terminals: an *n-type cathode* (K) and a *p-type anode* (A).

The SUS is actually known by a variety of names. In addition to those listed above, the device is sometimes referred to as a *Schockley diode,* a *current latch,* or a *reverse blocking diode thyristor* (the JEDEC designated name).

## SUS Operation

The overall operation of the SUS is easy to understand if we discuss it in terms of a two-transistor equivalent circuit. The derivation of this equivalent circuit is illustrated in Figure 20.2. If we split the SUS as shown in Figure 20.2a, we can see that the device is effectively made up of two transistors: one *pnp* ($Q_1$), and one *npn* ($Q_2$). Note that the collector of each transistor is connected to the base of the other.

Figure 20.3a illustrates the response of an *off* (nonconducting) SUS to an increase in supply voltage. Assuming that both transistors are in cutoff, the total forward current through the device ($I_F$) is approximately equal to zero, as is the total current through the circuit. Since $I_T$ is approximately equal to zero, there is no voltage dropped across the series resistor, and $V_{AK}$ is approximately equal to $V_S$. Note that these circuit conditions

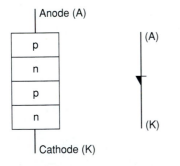

Construction          Schematic Symbol

**FIGURE 20.1   SUS construction and schematic symbol.**

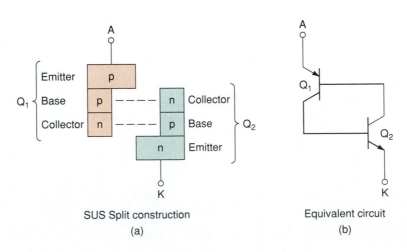

SUS Split construction
(a)

Equivalent circuit
(b)

**FIGURE 20.2   SUS split construction and equivalent circuit.**

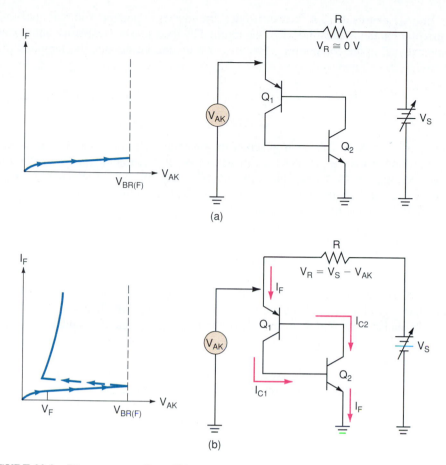

(a)

(b)

**FIGURE 20.3** The response of an off (nonconducting) SUS to an increase in supply voltage.

will exist as long as $V_{AK}$ is less than the **forward breakover voltage ($V_{BR(F)}$)** rating of the SUS. The $V_{BR(F)}$ rating of an SUS is the value of forward voltage that will force the device into conduction.

Figure 20.3b shows what happens when the value of $V_{AK}$ reaches the forward breakover voltage of the device. When this occurs, the two transistors break down and begin to conduct. As soon as conduction starts, both transistors in the SUS equivalent circuit are driven into *saturation*. This causes the voltage across the SUS to drop, as shown in the operating curve. Thus, when the SUS is forced into conduction, two things happen:

**1.** The device current ($I_F$) rapidly increases as the device is driven into saturation.

**2.** The value of $V_{AK}$ rapidly decreases because of the low resistance of the saturated device materials. (The relatively low value of $V_{AK}$ is identified as $V_F$ in the operating curve.)

As you can see, the SUS acts as an open circuit, dropping the applied voltage until $V_{BR(F)}$ is reached. At that time, the device breaks down and becomes a low-resistance conductor.

You may be wondering why the start of conduction causes both transistors in the SUS equivalent circuit to go into saturation. The answer is simple. If you look closely at the SUS equivalent circuit, you'll see that two conditions exist:

**1.** The collector of $Q_2$ is tied to the base of $Q_1$. Thus, $I_{C2} = I_{B1}$.

**2.** The base of $Q_2$ is tied to the collector of $Q_1$. Thus, $I_{B2} = I_{C1}$.

Now, consider what happens when $I_{B1}$ starts to increase. An increase in $I_{B1}$ causes an increase in $I_{C1}$. This, in turn, causes an increase in $I_{B2}$, which causes an additional increase in $I_{B1}$. As this cycle continues, the device is rapidly forced into saturation.

**Forward breakover voltage ($V_{BR(F)}$)**
The value of forward voltage that forces an SUS into conduction.

The fact that the SUS forces itself into saturation is important. You may recall that the current through a standard device is controlled by the values of resistance and voltage that are external to the component. Thus, the forward current through the saturated SUS in Figure 20.3b is found as

$$I_F = \frac{V_S - V_F}{R_S} \qquad \text{(20.1)}$$

where $V_F$ is the forward voltage drop across the saturated SUS. Equation (20.1) is important for several reasons. It provides us with the means of determining the forward current through a saturated SUS. It also shows that the SUS requires the use of a *series current-limiting resistor* ($R_S$). Without such a resistor, there will be no limit on the device current, and the SUS will most likely be destroyed.

## Driving the SUS into Cutoff

OBJECTIVE 2 ▶

**Holding current ($I_H$)**
The minimum value of $I_F$ required to maintain conduction.

**Anode current interruption**
A method of driving an SUS into cutoff by breaking the diode current path or shorting the circuit current around the diode.

**Forced commutation**
Driving an SUS into cutoff by forcing a reverse current through the device. The SUS turns off when $I_F - I_R < I_H$.

Once an SUS is driven into conduction, the device continues to conduct as long as $I_F$ is greater than a specified value, called the **holding current ($I_H$)**. The holding current rating of an SUS indicates the minimum forward current required to maintain conduction. Once $I_F$ drops below $I_H$, the device is driven back into cutoff and remains there until $V_{AK}$ reaches $V_{BR(F)}$ again.

Several methods are used to drive $I_F$ below the value of $I_H$. The first of these methods, called **anode current interruption,** is illustrated in Figure 20.4. In Figure 20.4a, a series switch is opened to interrupt the path for $I_F$, causing the device current to drop to zero. In Figure 20.4b, a shunt switch is closed to divert the circuit current ($I_T$) around the SUS, again causing the device current to drop to zero. In both cases, $I_F$ is forced to a value that is less than the $I_H$ rating of the SUS, and the device is driven into cutoff.

In practice, *series interruption* is generally caused by the *opening of a fuse*. As you'll see later in this chapter, thyristors can be used to protect voltage-sensitive loads from overvoltage conditions. When such a circuit is activated, the thyristor will stay on and protect the load just long enough for a fuse to blow. The blown fuse then acts as the open switch in Figure 20.4a. *Shunt interruption* is also used for load protection. However, the switch in the shunt interruption circuit would normally be a *reset* switch that is pressed when the overvoltage situation has passed. By closing the reset switch, the SUS is driven into cutoff, and normal circuit operation is restored.

Another method for driving an SUS into cutoff is called **forced commutation.** In this method (shown in Figure 20.5), a positive pulse is applied to the cathode of the device, forcing a *reverse current* ($I_R$) through the SUS. Since this current is in the opposite direction of $I_F$, the net current through the SUS is equal to the difference between $I_F$ and $I_R$. The device turns off when $I_F - I_R < I_H$.

**FIGURE 20.4    Anode current interruption.**

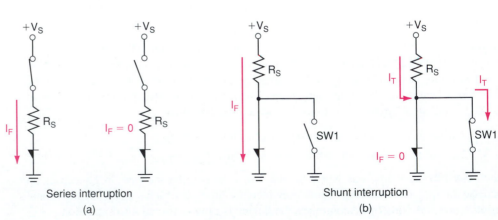

Series interruption
(a)

Shunt interruption
(b)

## SUS Specifications

Figure 20.6 shows the operating curve of the SUS. As you can see, the reverse region of the curve is nearly identical to that of the *pn*-junction diode and the zener diode. The part of the curve that falls between 0 V and the reverse breakdown voltage ($V_{BR(R)}$) is called the *reverse blocking region*. When the SUS is operating in this region, it has the same characteristics as a reverse-biased *pn* junction; that is, reverse current ($I_R$) is in the low $\mu$A range and the reverse voltage across the device ($V_R$) equals the applied voltage. If the reverse voltage exceeds the value of $V_{BR(R)}$, the device breaks down and conducts in the reverse direction. Like the *pn*-junction diode, the SUS is normally damaged or destroyed by this reverse conduction.

The forward operating curve of the SUS is enlarged in Figure 20.6b. Forward operation is divided into two regions. The **forward blocking region** is the area of the curve that illustrates the *off-state* (nonconducting) operation of the device. The **forward operating region** is the area of the curve that illustrates the *on-state* (conducting) operation of the device.

The forward blocking region is defined by two parameters: $V_{BR(F)}$ and $I_{BR(F)}$. As you know, $V_{BR(F)}$ is the value of forward voltage that causes the SUS to break down and conduct in the forward direction. The **forward breakover current ($I_{BR(F)}$)** rating of the SUS is the value of $I_F$ at the point where breakover occurs. For example, the ECG6404 has ratings of $V_{BR(F)} = 10$ V (maximum) and $I_{BR(F)} = 500$ $\mu$A. This means that the forward current through the device is approximately 500 $\mu$A when $V_F$ reaches the breakover rating of 10 V.

The importance of the $I_{BR(F)}$ rating is illustrated in Figure 20.7. The curve shown is for an SUS with an anode-to-cathode voltage ($V_{AK}$) that is lower than the forward breakover rating of the device. As temperature increases, the leakage current through the device also increases. When the temperature reaches 150°C, the leakage current reaches the value of $I_{BR(F)}$ and the device is triggered into its on-state. Thus, *an SUS can be triggered into conduction by a significant increase in operating temperature.*

Once an SUS turns on, the device enters the forward operating region of the curve. The holding current ($I_H$) is the minimum forward current required to maintain the

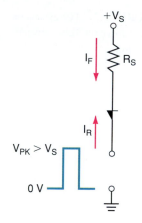

**FIGURE 20.5   Forced commutation.**

**Forward blocking region**
The forward *off-state* (nonconducting) region of operation.

**Forward operating region**
The forward *on-state* (conducting) region of operation.

**Forward breakover current ($I_{BR(F)}$)**
The value of $I_F$ at the point where breakover occurs.

The effects of temperature on SUS breakover.

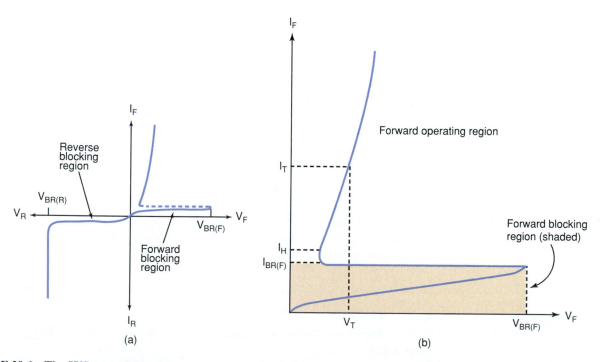

**FIGURE 20.6   The SUS operating curve.**

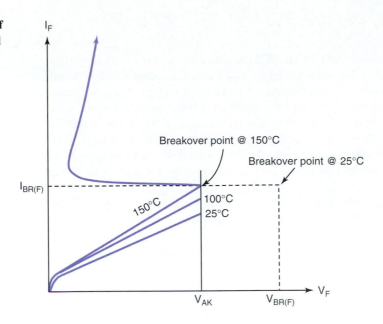

**FIGURE 20.7** The effects of temperature on SUS forward operation.

**Average on-state current ($I_T$)**
The maximum average (dc) forward current.

**Average on-state voltage ($V_T$)**
The value of $V_F$ when $I_F = I_T$.

on-state condition. When $I_F$ falls below the value of $I_H$, the device returns to its off-state, as was stated earlier in this section.

The **average on-state current ($I_T$)** is the maximum average forward current for the SUS. The series current-limiting resistor for an SUS is selected to ensure that the dc forward current through the device is held below this value. The **average on-state voltage ($V_T$)** rating is the value of $V_F$ when $I_F = I_T$.

In addition to the ratings discussed here, the SUS has the standard maximum current, power dissipation, and voltage ratings. Since we have covered these parameters for other components, we will not discuss them here.

## Summary

The *silicon unilateral switch (SUS)* is a thyristor that is forced into conduction when the forward voltage across the device reaches a specified *forward breakover voltage ($V_{BR(F)}$)*. Once triggered, the device becomes a low-impedance conductor. It remains in this *on-state* until its forward current ($I_F$) drops below the holding current ($I_H$) rating for the device. At that time, the component returns to the *off-state* (nonconducting) region of operation.

The SUS can be driven from the on-state to the off-state by one of two methods. *Anode current interruption* involves blocking or diverting the diode current so that $I_F$ falls below the value of $I_H$. *Forced commutation* involves forcing a reverse current through the device to drop the value of $I_F$ below the value of $I_H$. In either case, the device is driven into its off-state and remains there until triggered again.

The SUS is rarely used in modern circuit design. However, you will see in the next two sections that many of the commonly used thyristors are nothing more than variations on the basic SUS. For this reason, it is important that you understand the SUS operating principles and ratings.

## Section Review

1. What are *thyristors*?

2. What distinguishes thyristors from other switching devices, such as BJTs, FETs, and op-amps?

3. What are *optoelectronic* devices?

4. What is the *silicon unilateral switch (SUS)*?

5. Describe the construction of the SUS.

6. What is forward breakover voltage $(V_{BR(F)})$?

7. Describe what happens when the voltage applied to an SUS reaches the $V_{BR(F)}$ of the device.

8. Describe the two methods of *anode current interruption*.

9. Explain how *forced commutation* drives an SUS into cutoff.

10. What is the *forward blocking region* of an SUS?

11. What is the *forward operating region* of an SUS?

12. Explain how temperature can drive an SUS into its forward operating region.

## 20.2 SILICON-CONTROLLED RECTIFIER (SCR)

The **silicon-controlled rectifier (SCR)** is a three-terminal device that is very similar in construction and operation to the SUS. The construction of the SCR is shown in Figure 20.8, along with its schematic symbol. As you can see, the construction of the SCR is identical to that of the SUS, except for the addition of a third terminal called the *gate*. As you will see, the gate provides an additional means of triggering the device into the on-state. Except for that, the operation of the SCR is identical to that of the SUS.

◄ *OBJECTIVE 3*

**Silicon-controlled rectifier (SCR)**
A three-terminal device that is very similar in construction and operation to the SUS. The third terminal, called the *gate*, provides an additional method for triggering the device.

### SCR Triggering

Figure 20.9a shows the two-transistor equivalent circuit for the SCR. Again, note the similarities between the SCR and the SUS. Like the SUS, the SCR can be triggered into conduction by applying an anode-to-cathode voltage $(V_{AK})$ that is equal to the $V_{BR(F)}$ rating of the device.

The addition of the gate terminal provides an additional means of triggering the device, as is shown in Figure 20.9b. Here, we see a positive pulse applied to the gate terminal. When the gate pulse goes positive $(T_1)$, $Q_2$ is forced into conduction. At that point, the device takes itself into saturation, just like the SUS.

When the gate pulse is removed $(T_2)$, the device continues its forward conduction. The forward conduction continues until the *forward current* $(I_F)$ drops below the *holding current* $(I_H)$ rating of the device. Note that the SCR is driven into the off-state using the same two methods we use for the SUS: *anode current interruption* and *forced commutation*.

**FIGURE 20.8   SCR construction and schematic symbol.**

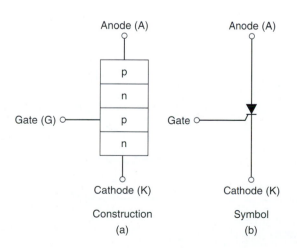

Anode (A)

| p |
| n |
| p |
| n |

Gate (G)

Cathode (K)

Construction

(a)

Anode (A)

Gate

Cathode (K)

Symbol

(b)

FIGURE 20.9

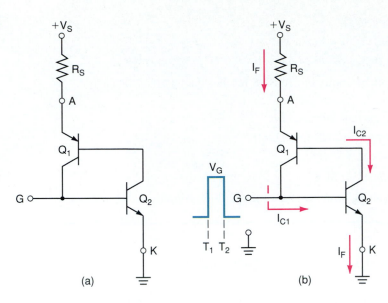

(a)

(b)

## SCR Operating Curve

Figure 20.10a shows the operating curve of the SCR. The enlarged forward operating curve (Figure 20.10b) shows the effects of gate current on forward conduction. As the value of $I_G$ is increased, the device breaks over into the on-state at decreasing values of $V_F$. For example, let's say that we have an applied SCR voltage equal to $V_{AK}$ in Figure 20.10b. If $I_G = I_{G1}$, the device will be in the off-state. However, if we increase $I_G$ to the value of $I_{G2}$, the value of $V_{AK}$ is more than high enough to drive the device into the on-state. Again, when the device turns on, the forward current must be driven below the component's $I_H$ rating to return it to the off-state.

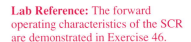

**Lab Reference:** The forward operating characteristics of the SCR are demonstrated in Exercise 46.

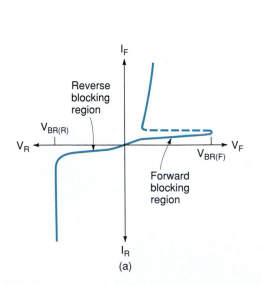

(a)

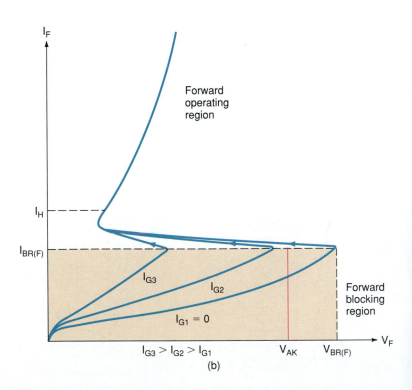

(b)

**FIGURE 20.10   The SCR operating curve.**

## SCR Specifications

Many of the specifications for the SCR are identical to those for the SUS. For example, take a look at the spec sheet for the 2N682–92 series SCRs in Figure 20.11. Table 20.1 lists the specifications contained in this spec sheet that have already been introduced. One

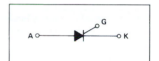

**2N682
thru
2N692**

# Silicon Controlled Rectifiers
## Reverse Blocking Triode Thyristors

. . . designed primarily for half-wave ac control applications, such as motor controls, heating controls and power supplies; or wherever half-wave silicon gate-controlled, solid-state devices are needed.

- Glass Passivated Junctions and Center Gate Fire for Greater Parameter Uniformity and Stability
- Blocking Voltage to 800 Volts

**SCRs
25 AMPERES RMS
50 thru 800 VOLTS**

**CASE 263-04
STYLE 1**

**MAXIMUM RATINGS** ($T_J$ = 125°C unless otherwise noted.)

| Rating | Symbol | Value | Unit |
|---|---|---|---|
| *Peak Repetitive Off-State Blocking Voltage, Note 1 | $V_{RRM}$ or $V_{DRM}$ | | Volts |
| 2N682 | | 50 | |
| 2N683 | | 100 | |
| 2N685 | | 200 | |
| 2N688 | | 400 | |
| 2N690 | | 600 | |
| 2N692 | | 800 | |
| *Peak Non-Repetitive Reverse Voltage | $V_{RSM}$ | | Volts |
| 2N682 | | 75 | |
| 2N683 | | 150 | |
| 2N685 | | 300 | |
| 2N688 | | 500 | |
| 2N690 | | 720 | |
| 2N692 | | 960 | |
| *RMS On-State Current (All Conduction Angles) | $I_{T(RMS)}$ | 25 | Amps |
| *Average On-State Current ($T_C$ = 65°C) | $I_{T(AV)}$ | 16 | Amps |
| *Peak Non-Repetitive Surge Current (One cycle, 60 Hz, preceded and followed by rated current and voltage) | $I_{TSM}$ | 150 | Amps |
| Circuit Fusing Considerations (t = 8.3 ms) | $I^2t$ | 93 | $A^2s$ |
| *Peak Gate Power | $P_{GM}$ | 5 | Watts |
| *Average Gate Power | $P_{G(AV)}$ | 0.5 | Watt |

*Indicates JEDEC Registered Data

(cont.)

Note 1. $V_{DRM}$ and $V_{RRM}$ for all types can be applied on a continuous dc basis without incurring damage. Ratings apply for zero or negative gate voltage. Devices should not be tested for blocking capability in a manner such that the voltage supplied exceeds the rated blocking voltage.

**FIGURE 20.11    The 2N682–92 SCR specification sheet. (Copyright of Motorola. Used by permission.)**

**MAXIMUM RATINGS — continued** ($T_J$ = 125°C unless otherwise noted.)

| Rating | Symbol | Value | Unit |
|---|---|---|---|
| *Peak Forward Gate Current   2N682–2N688<br>2N690, 2N692 | $I_{GM}$ | 2<br>1.2 | Amps |
| *Peak Gate Voltage — Forward<br>Reverse | $V_{FGM}$<br>$V_{RGM}$ | 10<br>5 | Volts |
| *Operating Junction Temperature Range | $T_J$ | −65 to +125 | °C |
| *Storage Temperature Range | $T_{stg}$ | −65 to +150 | °C |
| Stud Torque | — | 30 | in. lb. |

**THERMAL CHARACTERISTICS**

| Characteristic | Symbol | Max | Unit |
|---|---|---|---|
| Thermal Resistance, Junction to Case | $R_{\theta JC}$ | 2 | °C/W |

**ELECTRICAL CHARACTERISTICS** ($T_J$ = 25°C unless otherwise noted.)

| Characteristic | Symbol | Min | Typ | Max | Unit |
|---|---|---|---|---|---|
| *Average Forward or Reverse Blocking Current<br>(Rated $V_{DRM}$ or $V_{RRM}$, gate open, $T_J$ = 125°C)  2N682–2N683<br>2N685<br>2N688<br>2N690<br>2N692 | $I_{D(AV)}$, $I_{R(AV)}$ | —<br>—<br>—<br>—<br>— | —<br>—<br>—<br>—<br>— | 6.5<br>6<br>4<br>2.5<br>2 | mA |
| Peak Forward or Reverse Blocking Current<br>(Rated $V_{DRM}$ or $V_{RRM}$, gate open)  $T_J$ = 25°C<br>$T_J$ = 125°C | $I_{DRM}$, $I_{RRM}$ | —<br>— | —<br>— | 10<br>20 | µA<br>mA |
| *Peak On-State Voltage<br>($I_{TM}$ = 50.3 A peak, Pulse Width ≤ 1 ms, Duty Cycle ≤ 2%) | $V_{TM}$ | — | — | 2 | Volts |
| Gate Trigger Current (Continuous dc)<br>($V_{AK}$ = 12 Vdc, $R_L$ = 50 Ω)<br>*($V_{AK}$ = 12 Vdc, $R_L$ = 50 Ω, $T_C$ = −65°C) | $I_{GT}$ | —<br>— | —<br>— | 40<br>80 | mA |
| Gate Trigger Voltage (Continuous dc)<br>($V_{AK}$ = 12 Vdc, $R_L$ = 50 Ω)<br>*($V_{AK}$ = 12 Vdc, $R_L$ = 50 Ω, $T_J$ = −65°C) | $V_{GT}$ | —<br>— | 0.65<br>— | 2<br>3 | Volts |
| *Gate Non-Trigger Voltage<br>(Rated $V_{DRM}$, $R_L$ = 50 Ω, $T_J$ = 125°C) | $V_{GD}$ | 0.25 | — | — | Volts |
| Holding Current<br>($V_{AK}$ = 12 Vdc, Gate Open) | $I_H$ | — | 7.3 | 50 | mA |
| Critical Rate of Rise of Off-State Voltage<br>(Rated $V_{DRM}$, Exponential Waveform, $T_J$ = 125°C, Gate Open) | dv/dt | — | 30 | — | V/µs |

*Indicates JEDEC Registered Data.

**FIGURE 20.11**   **(Continued)**

**Circuit fusing ($I^2t$) rating**
The maximum forward surge current capability of an SCR.

important point: As shown in Table 20.1, the forward and reverse blocking voltages and currents are equal in magnitude. The same holds true for the SUS.

In addition to the specifications listed in Table 20.1, several other parameters warrant discussion. The first of these is the **circuit fusing ($I^2t$) rating.** This rating indicates the *maximum forward surge current capability* of the device. For example, the circuit fus-

**TABLE 20.1**

| Specification | Symbol | Definition |
|---|---|---|
| Peak repetitive off-state blocking voltage | $V_{RRM}$ or $V_{DRM}$ | The maximum forward or reverse blocking voltage. The same thing as $V_{BR(F)}$ and $V_{BR(R)}$. |
| Peak forward or reverse blocking current | $I_{DRM}$ or $I_{RRM}$ | The maximum forward or reverse blocking current. The same thing as $I_{BR(F)}$ and $I_{BR(R)}$. |
| Holding current | $I_H$ | The minimum forward current required to maintain conduction. |

ing ($I^2t$) rating of the 2N682–92 series SCRs is 93 ampere-square-seconds ($A^2s$). If the product of the square of surge current times the duration (in seconds) of the surge exceeds 93 $A^2s$, the device will be destroyed by excessive power dissipation. Example 20.1 demonstrates the determination of $I^2t$ for a given circuit.

---

**EXAMPLE 20.1**

In a given circuit, the 2N682 is subjected to a 50 A current surge that lasts for 10 ms. Determine whether or not this surge will destroy the device.

**Solution:** The $I^2t$ value for the circuit is found as

$$I^2t = (50\ A)^2(10\ ms)$$
$$= (2500\ A)(10\ ms)$$
$$= 25\ A^2s$$

Since this value is well below the maximum rating of 93 $A^2s$, the device will not be destroyed.

**PRACTICE PROBLEM 20.1**

The 2N6394 has a circuit fusing rating of 40 $A^2s$. Determine whether or not the device will survive a 60 A surge that lasts for 15 ms.

---

When the circuit fusing rating for an SCR is known, we can determine the maximum allowable duration of a surge with a known current value, as follows:

$$t_{max} = \frac{I^2t\ (\text{rated})}{I_S^2} \tag{20.2}$$

where $I_S$ = the known value of surge current. Example 20.2 demonstrates the use of equation (20.2) in determining the maximum allowable duration (time) of a surge.

---

**EXAMPLE 20.2**

The 2N1843A has a circuit fusing rating of 60 $A^2s$. The device is used in a circuit where it could be subjected to a 100 A surge. Determine the limit on the duration of such a surge.

**Solution:** The maximum allowable duration of the surge is found as

$$t_{max} = \frac{I^2t\ (\text{rated})}{I_S^2}$$
$$= \frac{60\ A^2s}{(100\ A)^2}$$
$$= 6\ ms$$

Thus, if the 100 A surge is present for more than 6 ms, the 2N1843A will not survive the current.

**PRACTICE PROBLEM 20.2**

The C35 series SCRs have a circuit fusing rating of 75 $A^2s$. Determine the maximum allowable duration of a 150 A surge that passes through one of these devices.

SECTION 20.2 / SILICON-CONTROLLED RECTIFIER (SCR)   **883**

We can also use a variation on equation (20.2) to determine the maximum allowable surge current value for a given period of time, as follows:

$$I_{S(max)} = \sqrt{\frac{I^2 t \text{ (rated)}}{t_S}}$$

(20.3)

where $t$ = the time duration of the current surge. Example 20.3 demonstrates the use of equation (20.3) in determining the maximum allowable value of a current surge.

---

## EXAMPLE 20.3

*Note:* The value of 60 A²s for the 2N1843A was given in Example 20.2.

Determine the highest surge current value that the 2N1843A can withstand for a period of 20 ms.

**Solution:** The peak value for the surge is found as

$$I_{S(max)} = \sqrt{\frac{I^2 t \text{ (rated)}}{t_s}}$$

$$= \sqrt{\frac{60 \text{ A}^2\text{s}}{20 \text{ ms}}}$$

$$= 54.77 \text{ A}$$

Thus, a 20 ms current surge would have to exceed 54.77 A to damage the 2N1843A.

### PRACTICE PROBLEM 20.3

Refer to Practice Problem 20.2. Determine the highest surge current value that a C35 series SCR can withstand for a duration of 50 ms.

---

**Nonrepetitive surge current ($I_{TSM}$)**
The absolute limit on the forward surge current through an SCR.

**Gate nontrigger voltage ($V_{GD}$)**
The maximum gate voltage that can be applied without triggering the SCR into conduction.

OBJECTIVE 4 ▶

**False triggering**
When a noise signal triggers an SCR into conduction.

**Critical rise rating (*dv/dt*)**
The maximum rate of increase in $V_{AK}$ without causing false triggering.

Later in this section, we will look at several SCR applications. In one of these applications, the SCR is used as a surge-protection device. When the SCR is used in a surge-protection application, the circuit fusing rating becomes critical.

The SCR has another maximum current rating called the **nonrepetitive surge current ($I_{TSM}$)** rating. This is the absolute limit on the surge current through the device. If this rated value is exceeded for *any length of time*, the device will be destroyed.

If you look under the *electrical characteristics* heading in Figure 20.11, you'll see the **gate nontrigger voltage ($V_{GD}$)** rating. This rating indicates the maximum gate voltage that can be applied without triggering the SCR into conduction. If $V_G$ exceeds this rating, the SCR may be triggered into the on-state.

The $V_{GD}$ rating is important because it points out one of the potential causes of **false triggering.** False triggering is a situation where the SCR is accidentally triggered into conduction, usually by some type of noise. For example, the 2N682–92 series of SCRs has a $V_{GD}$ rating of 250 mV. If a noise signal with a peak value greater than 250 mV appears at the gate, the device may be triggered into conduction.

Another common cause of false triggering is a rise in anode voltage that exceeds the **critical rise rating (*dv/dt*)** of the SCR. This rating indicates the *maximum rate of increase in anode-to-cathode voltage that the SCR can handle without false triggering occurring*. For example, the *dv/dt* rating of the 2N682–92 series SCRs is 30 V/μs. This means that a noise signal in $V_{AK}$ with a rate of rise equal to 30 V/μs may cause false triggering.

Obviously, 30 V of noise would never occur under normal circumstances. However, even a relatively small amount of noise can have a sufficient *rate of increase* to cause false triggering. To determine the rate of increase of a given noise signal in V/μs, the following conversion formula can be used:

$$\Delta V = \frac{dv}{dt}\,\Delta t \qquad (20.4)$$

where $\frac{dv}{dt}$ = the critical rise rating of the component

$\Delta t$ = the rise time of the increase in $V_{AK}$
$\Delta V$ = the amount of change in $V_{AK}$

As Example 20.4 demonstrates, even a small amount of noise in $V_{AK}$ can cause false triggering if the rise time of the noise signal is short enough.

---

A 2N682 SCR is used in a circuit where 2 ns (rise time) noise signals occur randomly. Determine the noise amplitude required to cause false triggering.

**EXAMPLE 20.4**

*Solution:* With a critical rise rating of 30 V/μs, the amplitude that will cause false triggering is found as

$$\Delta V = \frac{dv}{dt}\Delta t$$
$$= (30 \text{ V/}\mu\text{s})(2 \text{ ns})$$
$$= 60 \text{ mV}$$

As you can see, with a 2 ns rise time, only 60 mV of noise is required to cause false triggering.

**PRACTICE PROBLEM 20.4**

The C35D SCR has a critical rise rating of 25 V/μs. Determine the amplitude of noise needed to cause false triggering when the noise signal rise time is 100 ps.

---

As you can see, many of the critical SCR parameters deal with the current limits and false triggering limits of the device. At this point, we'll move on to look at the methods commonly used to prevent false triggering.

## Preventing False Triggering

You have been shown that false triggering is usually caused by one of two conditions:

**1.** A noise signal at the gate terminal
**2.** A short rise-time noise signal in $V_{AK}$

Noise from the gate signal source is generally reduced using one of the two methods shown in Figure 20.12. In Figure 20.12a, the gate is connected (via a gate resistor) to a negative dc power supply. For the sake of discussion, let's assume that the gate supply voltage is −2 V. If this is the case, the noise at the gate would have to be greater than $(2V+V_{GD})$ to cause false triggering. This is not likely to occur under any normal circumstances.

*A Practical Consideration:* False triggering can also be caused by temperature, as is the case with the SUS.

The circuit in Figure 20.12b uses a bypass capacitor connected between the gate terminal and the terminal and ground. This capacitor, which would short any gate noise signals to ground, is normally selected at a value between 0.1 and 0.01 μF.

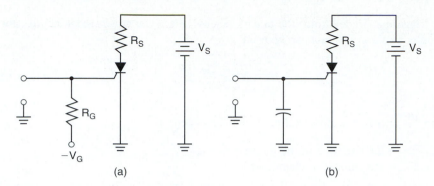

**FIGURE 20.12    Preventing false triggering.**

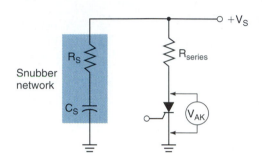

**FIGURE 20.13    An *RC* snubber network.**

Of the two methods shown in Figure 20.12, the bypass capacitor in Figure 20.12b is by far the most commonly used. This is because the capacitor provides a very inexpensive solution to the gate noise problem.

Noise in $V_{AK}$ is normally eliminated by the use of a **snubber network.** A snubber is an *RC* circuit connected between the anode and cathode terminals of the SCR. A snubber network is shown in Figure 20.13. The capacitor ($C_S$) effectively shorts any noise in $V_{AK}$ to ground. The resistor ($R_S$) is included to limit the capacitor discharge current when the SCR turns on.

It should be noted that noise in $V_{AK}$ (Figure 20.13) would originate in the dc voltage source, $V_S$. Assuming that $V_S$ comes from a standard dc power supply, any noise generated in the supply will be coupled to the SCR circuit. This noise will then be reduced by the snubber to a level that prevents it from causing false triggering.

Snubbers are used in most SCR circuits since exceeding the critical rise (*dv/dt*) rating is the most common cause of false triggering. As you will see later in this chapter, snubbers are also used in circuits that contain some of the other thyristors. In other words, SCRs are not the only components that require the use of snubber networks.

**Snubber network**
An *RC* circuit that is connected between the SCR anode and cathode to eliminate false triggering.

## SCR Applications

OBJECTIVE 5 ▶

The SCR is the most commonly used of the thyristors. Most SCR applications are found in the area of **industrial electronics,** which deals with the devices, circuits, and systems used in manufacturing. Although we cannot cover all the possible applications for the SCR, we will touch briefly on a couple of them in this section.

**Industrial electronics**
The area of electronics that deals with the devices, circuits, and systems used in manufacturing.

## The SCR Crowbar

**Crowbar**
A circuit used to protect a voltage-sensitive load from excessive dc power supply output voltages.

One common use for the SCR is in a type of circuit called a **crowbar.** A crowbar is a circuit used to protect a voltage-sensitive load from excessive dc power supply output volt-

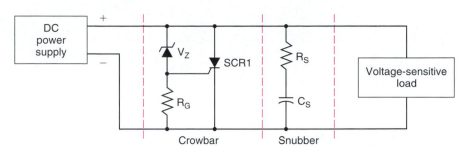

FIGURE 20.14   The SCR crowbar.

DC power supply

$V_Z$

SCR1

$R_S$

$R_G$

$C_S$

Voltage-sensitive load

Crowbar          Snubber

ages. The crowbar circuit (shown in Figure 20.14) consists of a zener diode, a gate resistor ($R_G$), and an SCR. The circuit in Figure 20.14 also contains a snubber to prevent false triggering.

In the crowbar shown, the zener diode and the SCR are normally off. With the zener diode in cutoff, there is no path for current through $R_G$, and no voltage is developed across the resistor. With no voltage developed across $R_G$, the gate of the SCR is at 0 V, holding the device in the off-state. Thus, as long as the zener diode is off, the SCR is acting as an open and does not affect either the dc power supply or the load.

If the power supply output voltage *increases* beyond a certain point, the zener diode breaks down and conducts. The current through the zener diode causes a voltage to be developed across $R_G$. This voltage then forces the SCR into conduction. When the SCR conducts, the output from the dc power supply is shorted through the device and the load is protected.

Several points should be made regarding this circuit:

1. The SCR would be activated only by some major fault within the dc power supply that causes its output voltage to increase drastically. Once activated, the SCR will continue to conduct until the power supply fuse opens.

2. The *circuit fusing* ($I^2t$) rating of the SCR must be high enough to ensure that the SCR isn't destroyed before the power supply fuse has time to blow. If the SCR gives out before the power supply fuse, the power supply will continue to operate and the load will be damaged after all.

3. The SCR in the crowbar is considered to be expendable. This is why there is no series current-limiting resistor in the circuit. The object of the circuit is to protect the load, not the SCR.

4. Most crowbar circuits are self-resetting after the high voltage ceases. However, if not self-resetting, the SCR in a crowbar should be replaced after activation of the circuit as a precaution against future component failure from possible internal damage.

Circuit protection is not the only application for the SCR. A completely different type of application can be found in the *SCR phase controller*.

The SCR in a crowbar is turned off by a blown primary fuse in the dc power supply. This is an example of *series anode current interruption.*

The $I^2t$ rating of the SCR is especially critical when the dc power supply contains a slow-blow fuse.

## The SCR Phase Controller

A **phase controller** is a circuit that is used to control the conduction angle through a load, and thus the average load voltage. For example, by varying the setting of the potentiometer ($R_2$) in the phase controller in Figure 20.15a, we can vary the conduction angle ($\theta$) through the load as shown in Figure 20.15b. Since the *average* (dc equivalent) load voltage varies with the conduction angle, adjusting the setting of $R_2$ also varies the average load voltage. This point is discussed later in this section.

Conduction through the load in Figure 20.15a is controlled by the SCR. When $v_{in}$ causes the voltage at point A to reach a specified level, called the **trigger-point voltage** ($V_{TP}$), the SCR is triggered into conduction. Once triggered, the SCR allows $v_{in}$ to pass through to the load. Note that the SCR then continues to conduct for the remainder of the

**Phase controller**
A circuit used to control the conduction phase angle through a load, and thus the average load voltage.

**Trigger-point voltage ($V_{TP}$)**
The voltage required at the anode of the gate diode to trigger the SCR.

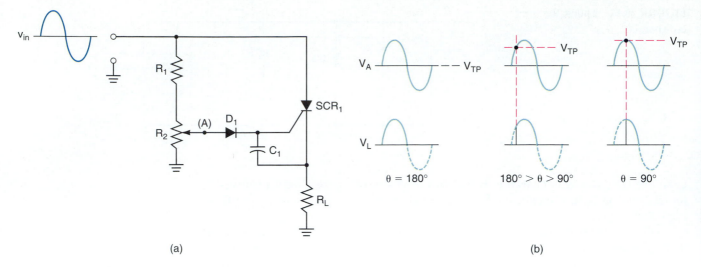

**FIGURE 20.15** The SCR phase controller.

**Lab Reference:** The operation of an SCR phase controller is demonstrated in Exercise 46.

positive alternation of $v_{in}$. The relationship between the voltage at point A (caused by $v_{in}$), the value of $V_{TP}$, and load conduction can be seen in the three waveforms in Figure 20.15b.

For the SCR to be triggered into conduction, the sine wave at point A must reach the level specified by the following equation:

$$V_{TP} = V_F + V_{GT}$$

where $V_F$ = the forward voltage drop across the gate diode ($D_1$)

$V_{GT}$ = the **gate trigger voltage** rating of the SCR

**Gate trigger voltage ($V_{GT}$)**
The value of $V_{GK}$ that will cause $I_G$ to become great enough to trigger the SCR into conduction.

The gate trigger voltage of the SCR is the value of $V_{GK}$ that will cause $I_G$ to become great enough to trigger the SCR into conduction.

When the sine wave at point A reaches the value of $V_{TP}$, the voltage is sufficient to overcome the forward voltage drop across $D_1$ and the value of $V_{GT}$ for the SCR. This triggers the SCR into conduction for the remainder of the positive alternation of $v_{in}$. When $v_{in}$ goes negative, the anode voltage of the SCR also goes negative. This causes the SCR current to drop below the holding current ($I_H$) value of the device, and it returns to the off-state.

Turning an SCR off by reversing the anode voltage polarity is an example of *forced commutation*.

An important point needs to be made regarding the circuit and waveforms shown in Figure 20.15: The waveforms shown in the figure are somewhat misleading. You see, the value of $V_{TP}$ for a given phase controller is a fixed value, while the amplitude of the sine wave at point A is variable. For example, let's say that the value of $V_{GT}$ for the SCR in Figure 20.15a is 1.5 V. This gives the circuit a trigger-point voltage found as

$$\begin{aligned} V_{TP} &= V_F + V_{GT} \\ &= 0.7\ V + 1.5\ V \\ &= 2.2\ V \end{aligned}$$

This is the voltage required at point A to trigger the SCR into conduction. The voltage divider ($R_1$ and $R_2$) applies a sine wave to point A that is *proportional to* $v_{in}$. As the setting of $R_2$ is increased, the amplitude of the sine wave at point A is also increased. Thus, by increasing the setting of $R_2$, we decrease the value of $v_{in}$ required to cause the voltage at point A to reach the value of $V_{TP}$. In the same fashion, decreasing the setting of $R_2$ increases the value of $v_{in}$ required to cause the voltage at point A to reach the value of $V_{TP}$. From this discussion, we can see that adjusting $R_2$ changes the amplitude of $V_A$, not the value of $V_{TP}$ for the circuit.

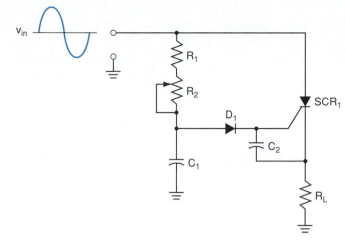

FIGURE 20.16   Modified phase-angle controller.

A few more points should be made with regard to the circuit in Figure 20.15a:

1. The capacitor between the SCR gate and cathode terminals ($C_1$) is included to prevent false triggering.

2. SCRs tend to have relatively low gate breakdown voltage ratings, typically 10 V or less. For this reason, $D_1$ is used in the circuit to prevent the negative alternation of $v_{in}$ from being applied to the SCR gate–cathode junction.

3. The load resistance is in series with the SCR and thus acts as the series current-limiting resistor for the device.

4. The mathematical relationship between the load conduction angle ($\theta$) and the average (dc equivalent) load voltage ($V_{AVE}$) is defined using integral calculus and thus is not given here. However, it can be stated that the value of $V_{AVE}$ for a waveform varies with *the area under the waveform* (shown as shaded areas in Figure 20.15b). As the area under the waveform decreases, so does its value of $V_{AVE}$. Therefore, by reducing the $R_2$ setting in Figure 20.15a, we reduce the load values of $\theta$ and $V_{AVE}$.

Adding a capacitor to Figure 20.15 between the anode of $D_1$ and the bottom of resistor $R_2$ gives us the circuit shown in Figure 20.16. This modification is used to further delay the firing angle of the $V_{TP}$. Setting $R_2$ toward its minimum value will decrease the firing angle, bringing it closer to 0°. This will cause the average load voltage to decrease. Setting the value of $R_2$ to midrange will cause the firing angle to be nearer to 90°, thus causing an increase in average load voltage. The maximum setting of $R_2$ will extend the firing angle beyond 90° and bring it closer to 180°. The added capacitor will allow a range of firing angles from approximately 15° to around 170°. This allows better control of the circuit over a larger range of input phase angles.

The fact that the value of $V_{AVE}$ for a phase controller is variable makes the circuit very useful as a *motor-speed controller*.

Lab Reference: A circuit like the one in Figure 20.17 is used to test the operation of an SCR in Exercise 46.

## SCR Testing

Whenever possible, in-circuit testing on an SCR is performed. An open SCR will remain in the nonconducting state when triggered. A shorted SCR will usually conduct regardless of the polarity of the anode–cathode voltage ($V_{AK}$). In either case, the SCR fault can be diagnosed by checking the gate, anode, and cathode voltages of the device with a voltmeter or an oscilloscope.

In some circuits (like crowbars), it is impractical to perform any type of in-circuit tests on the SCR. In this case, the device must be tested using a circuit like the one shown in Figure 20.17. The circuit is initially set up so that $V_{GK}$ is approximately 0 V. The value

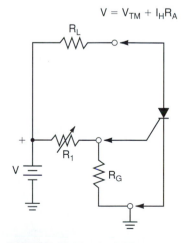

$$V = V_{TM} + I_H R_A$$

FIGURE 20.17   An SCR test circuit.

*A practical consideration*: SCR phase controllers are often used to control the rotating speed of a *dc motor*. In this application, a shorted SCR will cause the motor to *vibrate rapidly* (rather than rotate).

of $R_L$ is selected to equal the value used in the $V_{GT}$ listing on the spec sheet of the device. For example, the 2N682–92 series SCRs list a value of $R_L = 50$ Ω for the testing of $V_{GT}$ (see the spec sheet in Figure 20.11). Thus, when testing one of these SCRs, $R_L$ would be a 50 Ω resistor. The value of $+V$ is set to equal the $V_{BR(F)}$ rating of the SCR under test.

To test the SCR, the setting of $R_1$ is decreased until $V_G$ equals the value of $V_{GT}$. At this point, the SCR should trigger, and $V_{AK}$ should drop to a relatively low value. If the device fails to trigger, $R_1$ is varied across its entire range. If the device continues in the nonconducting state, it is open and must be replaced.

## SCR Summary

The SCR is a three-terminal device that acts as a gated SUS. The device is normally triggered into conduction by applying a gate–cathode voltage ($V_{GK}$) that causes a specific level of gate current ($I_G$). This gate current triggers the device into conduction.

Like the SUS, the SCR is returned to its nonconducting state by either *anode current interruption* or *forced commutation*. When turned off, the device remains in its nonconducting state until another trigger is received.

Two critical parameters for the SCR are the *circuit fusing* and *critical rise* ratings. If the *circuit fusing* ($I^2t$) *rating* is exceeded, the device is destroyed by excessive power dissipation. If the *critical rise* ($dv/dt$) rating is exceeded by any noise in the anode volt-

**FIGURE 20.18**

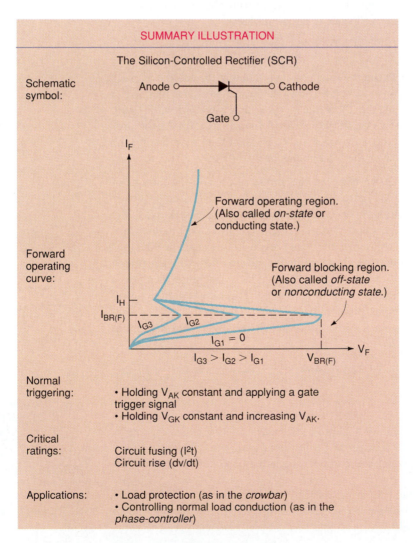

age, the device may experience *false triggering*. False triggering, a condition where the device is unintentionally triggered into conduction, can also be caused by excessive noise in the gate circuit biasing voltage. False triggering is normally prevented by the use of *snubber networks* and *gate bypass capacitors.*

SCR applications generally fall into one of two categories: load overvoltage protection, and limiting the conduction through a load. The characteristics of the SCR are summarized in Figure 20.18.

1. Describe the construction of the SCR.
2. What is the *forward blocking region* of the SCR?
3. What is the *forward operating region* of the SCR?
4. List the methods used to trigger an SCR.
5. List the methods used to force an SCR into cutoff.
6. Using the operating curve in Figure 20.10, describe the forward operation of the SCR.
7. What is the *circuit fusing* ($I^2t$) *rating* of an SCR? Why is it important?
8. What is *false triggering*?
9. Which SCR ratings are tied directly to false triggering problems?
10. How does a gate bypass capacitor prevent false triggering?
11. What is a *snubber network*? What is it used for?
12. What is the function of the SCR *crowbar*?
13. Describe the operation of the crowbar.
14. What is the function of the SCR *phase controller*?
15. Describe the operation of the SCR phase controller.
16. What are the typical in-circuit symptoms of *open* and *shorted* SCRs?

# 20.3 DIACS AND TRIACS

*Diacs* and *triacs* are components that are classified as **bidirectional thyristors.** The term *bidirectional* means that they are capable of conducting in two directions. For all practical purposes, the *diac* can be viewed as a bidirectional SUS, and the *triac* can be viewed as a bidirectional SCR. Since these components are extremely similar in operation to the thyristors we have discussed, most of the material in this section will seem more like a review than new material.

◄ *OBJECTIVE 6*

**Bidirectional thyristor**
A thyristor capable of conducting in two directions.

## Diacs

The **diac** is a two-terminal, three-layer device whose forward and reverse operating characteristics are identical to the forward characteristics of the SUS. The diac symbol and construction are shown in Figure 20.19.

As you can see, the construction of the diac is extremely similar to that of the *npn* (or *pnp*) transistor. However, there are several crucial differences:

1. There is no base connection.
2. The three regions are nearly identical in size.
3. The three regions have nearly identical doping levels.

**Diac**
A two-terminal, three-layer device whose forward and reverse characteristics are identical to the forward characteristics of the SUS. Note: Diacs are often referred to as *bidirectional diodes.*

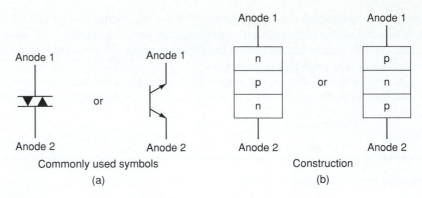

FIGURE 20.19    Diac construction and schematic symbols.

In contrast, the BJT has an extremely narrow and lightly doped base (center) region, as you were shown in Chapter 6. Note that the diac can be constructed in either *npn* or *pnp* form. The operating principles and symbols are identical for both *npn* and *pnp* diacs.

The operating curve of the diac is shown in Figure 20.20. As you can see, the forward and reverse operating characteristics are identical to the forward characteristics of the SUS. In either direction, the device will act as an open until the breakover voltage ($V_{BR}$) is reached. At that time, the device is triggered and conducts in the appropriate direction. Conduction then continues until the device current drops below its holding current ($I_H$) value. Note that the breakover voltage and holding current values are identical for the forward and reverse regions of operation.

Note that the diac can be triggered into conduction by an increase in temperature. The most common application for the diac is as a triggering device for a triac in ac control circuits. This application will be discussed later in this section.

Both *anode current interruption* and *forced commutation* are used to drive a diac into cutoff.

## Triacs

**Triac**
A bidirectional thyristor whose forward and reverse characteristics are identical to the forward characteristics of the SCR.

Triacs are also referred to as *triodes* and *bidirectional triode thyristors* (the JEDEC designated name).

The **triac** is a three-terminal, five-layer device whose forward and reverse characteristics are identical to the forward characteristics of the SCR. The triac symbol and construction are shown in Figure 20.21.

FIGURE 20.20    The
diac operating curve.

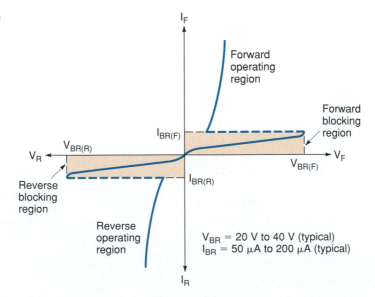

$V_{BR} = 20$ V to $40$ V (typical)
$I_{BR} = 50$ μA to $200$ μA (typical)

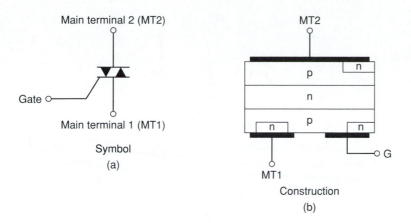

FIGURE 20.21 Triac construction and schematic symbol.

The primary conducting terminals are identified as *main terminal 1 (MT1)* and *main terminal 2 (MT2)*. Note that MT1 is the terminal on the same side of the device as the gate.

If we split the triac as shown in Figure 20.22, it is easy to see that we have complementary SCRs that are connected in parallel. When MT2 is *positive* and MT1 is *negative*, a triggered triac will have the conduction path shown on the left in Figure 20.22b. When MT2 is *negative* and MT1 is *positive*, a triggered triac will have the conduction path shown on the right. Note that the triggered triac will conduct in the appropriate direction as long as the current through the device is greater than its *holding current* $(I_H)$ rating.

Because of its unique structure, the operating curve of the triac is a bit more complicated than any we have seen thus far. The triac operating curve is shown in Figure 20.23. The graph in Figure 20.23 is divided into four *quadrants* labeled *I, II, III,* and *IV*. Each quadrant has the MT2 and G (gate) polarities shown. For example, quadrant I indicates the operation of the device when MT2 is positive (with respect to MT1) and a positive potential is applied to the gate. Quadrant II indicates the operation of the device when MT2 is positive (with respect to MT1) and a negative potential is applied to the gate, and so on.

◄ *OBJECTIVE 7*

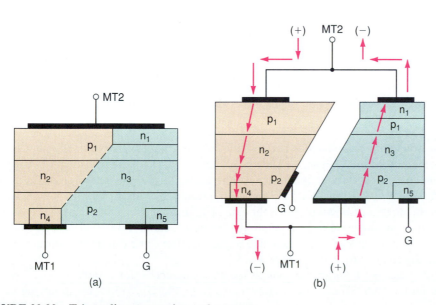

FIGURE 20.22 Triac split construction and currents.

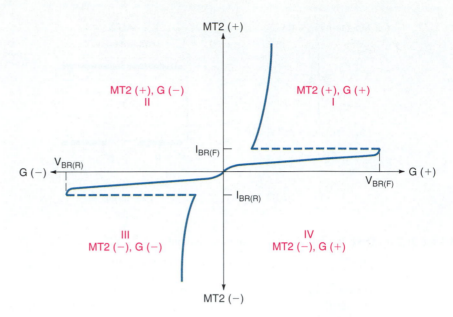

**FIGURE 20.23   The triac operating curve.**

Operation in quadrant I is identical to the operation of the SCR. All the SCR characteristics apply to the triac operated in the first quadrant. For example:

1. The device remains in the off-state until $V_{BR(F)}$ is reached or a positive gate trigger voltage ($V_{GT}$) generates a gate current ($I_G$) sufficient to trigger the device.

2. The device may be false triggered by increases in temperature, gate noise, or anode (MT2) noise.

3. Once triggered, the device continues to conduct until the forward current drops below its holding current ($I_H$) rating.

Note that, when operated in quadrant I, the current through the triac is in the direction shown on the left side of Figure 20.22b.

Quadrant III operation is simply a reflection of quadrant I operation. In other words, when MT2 is negative (with respect to MT1) and a negative gate trigger voltage is applied to the triac, it is triggered into reverse conduction, as shown on the right side of Figure 20.22b. Again, all the SCR characteristics apply to the triac; the direction of conduction has simply been reversed.

Quadrants II and IV illustrate the unique triggering combinations of the triac. The triac can be triggered into conduction by either of the following conditions:

1. Applying a negative gate trigger while MT2 is positive (quadrant II)

2. Applying a positive gate trigger while MT2 is negative (quadrant IV)

Quadrants I and II have the same MT2 polarities. The same is true for quadrants III and IV.

These triggering combinations are possible because the gate is physically connected to both the $p_2$ and $n_5$ regions, as shown in Figure 20.22a. When quadrant II triggering occurs, the current through the device will be in the direction indicated on the left side of Figure 20.22b. When quadrant IV triggering occurs, the current through the device will be in the direction shown on the right side of Figure 20.22b. Since the polarity of MT2 determines the direction of current through the triac, quadrant II triggering will cause current in the same direction as quadrant I triggering. The same can be said for quadrant III triggering and quadrant IV triggering.

Confusing? Let's take a look at the $V_{GT}$ parameters for a typical triac. The spec sheet for the 2N6157–65 series triacs (shown in Figure 20.24) lists the following values:

| $V_{GT}$ (Maximum Value) | Listed Conditions | Indicated Quadrant |
|---|---|---|
| 2 V | MT2(+), G(+) | I |
| 2.1 V | MT2(+), G(−) | II |
| 2.1 V | MT2(−), G(−) | III |
| 2.5 V | MT2(−), G(+) | IV |

# Triacs
## Silicon Bidirectional Triode Thyristors

. . . designed primarily for industrial and military applications for the control of ac loads in applications such as light dimmers, power supplies, heating controls, motor controls, welding equipment and power switching systems; or wherever full-wave, silicon gate controlled solid-state devices are needed.

- Glass Passivated Junctions and Center Gate Fire
- Isolated Stud for Ease of Assembly
- Gate Triggering Guaranteed In All 4 Quadrants

2N6145–47
(See 2N5571)

2N6157
thru
2N6165

TRIACs
30 AMPERES RMS
200 thru 600 VOLTS

MT2 O ─ ▷|◁ ─ O MT1
       O G

CASE 174-04
(TO-203AA)
STYLE 3
2N6157-59

CASE 263-04
STYLE 2
2N6160-62

CASE 311-02
STYLE 2
2N6163-65

### MAXIMUM RATINGS

| Rating | Symbol | Value | Unit |
|---|---|---|---|
| *Peak Repetitive Off-State Voltage <br> ($T_J$ = −65 to +125°C) <br> 1/2 Sine Wave 50 to 60 Hz, Gate Open <br> *Peak Principal Voltage <br>    2N6157, 2N6160, 2N6163 <br>    2N6158, 2N6161, 2N6164 <br>    2N6159, 2N6162, 2N6165 | $V_{DRM}$ | <br><br><br><br>200<br>400<br>600 | Volts |
| *Peak Gate Voltage | $V_{GM}$ | 10 | Volts |
| *RMS On-State Current <br> ($T_C$ = −65 to +85°C) <br> ($T_C$ = +100°C) <br> Full Sine Wave, 50 to 60 Hz | $I_{T(RMS)}$ | <br>30<br>20 | Amps |
| *Peak Non-Repetitive Surge Current <br> (One Full Cycle of surge current at 60 Hz, preceded <br> and followed by a 30 A RMS current, $T_C$ = 85°C) | $I_{TSM}$ | 250 | Amps |
| Circuit Fusing Considerations <br> (t = 8.3 ms) | $I^2t$ | 260 | $A^2s$ |
| *Peak Gate Power <br> ($T_J$ = +80°C, Pulse Width = 2 μs) | $P_{GM}$ | 20 | Watts |
| *Average Gate Power <br> ($T_J$ = +80°C, t = 8.3 ms) | $P_{G(AV)}$ | 0.5 | Watt |
| *Peak Gate Current | $I_{GM}$ | 2 | Amps |
| *Operating Junction Temperature Range | $T_J$ | −65 to +125 | °C |
| *Storage Temperature Range | $T_{stg}$ | −65 to +150 | °C |
| *Stud Torque <br>    2N6160 thru 2N6165 | — | 30 | in. lb. |

### THERMAL CHARACTERISTICS

| Characteristic | Symbol | Max | Unit |
|---|---|---|---|
| *Thermal Resistance, Junction to Case | $R_{\theta JC}$ | 1 | °C/W |

*Indicates JEDEC Registered Data.

**FIGURE 20.24** The 2N6157–65 triac specification sheet. (Copyright of Motorola. Used by permission.)

**ELECTRICAL CHARACTERISTICS** (T$_C$ = 25°C unless otherwise noted.)

| Characteristic | Symbol | Min | Typ | Max | Unit |
|---|---|---|---|---|---|
| *Peak Forward or Reverse Blocking Current<br>(Rated V$_{DRM}$ or V$_{RRM}$)  T$_J$ = 25°C<br>T$_J$ = 125°C | I$_{DRM}$, I$_{RRM}$ | —<br>— | —<br>— | 10<br>2 | μA<br>mA |
| *Peak On-State Voltage (Either Direction)<br>(I$_{TM}$ = 42 A Peak, Pulse Width = 1 to 2 ms, Duty Cycle ≤ 2%) | V$_{TM}$ | — | 1.5 | 2 | Volts |
| Gate Trigger Current (Continuous dc), Note 1<br>(Main Terminal Voltage = 12 Vdc, R$_L$ = 50 Ohms)<br>MT2(+), G(+)<br>MT2(+), G(−)<br>MT2(−), G(−)<br>MT2(−), G(+)<br>*MT2(+), G(+); MT2(−), G(−) T$_C$ = −65°C<br>*MT2(+), G(−); MT2(−), G(+) T$_C$ = −65°C | I$_{GT}$ | <br><br>—<br>—<br>—<br>—<br>—<br>— | <br><br>15<br>20<br>20<br>30<br>—<br>— | <br><br>60<br>70<br>70<br>100<br>200<br>250 | mA |
| Gate Trigger Voltage (Continuous dc)<br>(Main Terminal Voltage = 12 Vdc, R$_L$ = 50 Ohms)<br>MT2(+), G(+)<br>MT2(+), G(−)<br>MT2(−), G(−)<br>MT2(−), G(+)<br>*All Quadrants, T$_C$ = −65°C<br>*Main Terminal Voltage = Rated V$_{DRM}$, R$_L$ = 10 k ohms, T$_J$ = +125°C | V$_{GT}$ | <br><br>—<br>—<br>—<br>—<br>—<br>0.2 | <br><br>0.8<br>0.7<br>0.85<br>1.1<br>—<br>— | <br><br>2<br>2.1<br>2.1<br>2.5<br>3.4<br>— | Volts |
| Holding Current<br>(Main Terminal Voltage = 12 Vdc, Gate Open)<br>(Initiating Current = 500 mA)<br>MT2(+)<br>MT2(−)<br>*Either Direction, T$_C$ = −65°C | I$_H$ | <br><br><br>—<br>—<br>— | <br><br><br>8<br>10<br>— | <br><br><br>70<br>80<br>200 | mA |
| *Turn-On Time<br>(Main Terminal Voltage = Rated V$_{DRM}$, I$_{TM}$ = 42 A,<br>Gate Source Voltage = 12 V, R$_S$ = 50 Ohms, Rise Time = 0.1 μs,<br>Pulse Width = 2 μs) | t$_{gt}$ | — | 1 | 2 | μs |
| Blocking Voltage Application Rate at Commutation,<br>f = 60 Hz, T$_C$ = 85°C<br>On-State Conditions:<br>(I$_{TM}$ = 42 A, Pulse Width = 4 ms, di/dt = 17.5 A/ms)<br>Off-State Conditions:<br>(Main Terminal Voltage = Rated V$_{DRM}$ (200 μs min),<br>Gate Source Voltage = 0 V, R$_S$ = 50 Ω) | dv/dt(c) | — | 5 | — | V/μs |

*Indicates JEDEC Registered Data.
Note 1. All voltage polarities referenced to main terminal 1.

**FIGURE 20.24** **(Continued)**

Now let's apply the values listed to the triacs shown in Figure 20.25. Figure 20.25a shows the quadrant I triggering scheme. MT2 is positive, and a +2 V gate input triggers the device into conduction. Quadrant II triggering (Figure 20.25b) occurs when MT2 is positive and a −2.1 V input is applied to the gate. Note that the triac conduction is the same for quadrants I and II triggering. Quadrant III triggering (Figure 20.25c) occurs when MT2 is negative and a −2.1 V input is applied to the gate. Quadrant IV triggering occurs when MT2 is negative, and a +2.5 V input is applied to the gate. Note that the triac conduction is in the same direction for quadrants III and IV triggering.

Two points should be made at this time:

1. Regardless of the triggering method used, triacs are forced into cutoff in the same fashion as the SCR. Either *device current interruption* or *forced commutation* must be used to decrease the device current below its holding current ($I_H$) rating. At that point, the device enters its nonconducting state and remains in cutoff until another trigger is received.

2. Except for the quadrant triggering considerations, the spec sheet for a triac is nearly identical to that of an SCR. This can be seen by comparing the triac spec sheet in Figure 20.24 with the SCR spec sheet in Figure 20.11.

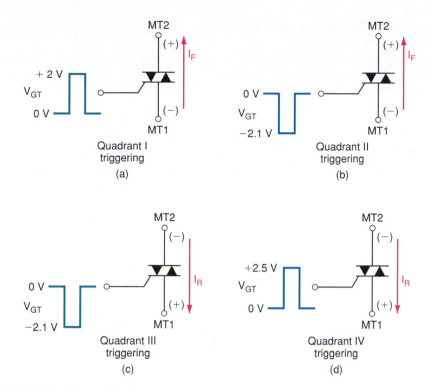

**FIGURE 20.25**   **Triac triggering.**

## Controlling Triac Triggering

The low gate voltages required to trigger the triac can cause problems in many circuit applications. For example, consider the 2N6157 triac shown in Figure 20.26a. Here, we see the approximate triggering voltages required for the device (quadrants I and II triggering). Three questions arise when you think about these gate triggering voltages:

1. What if we want the 2N6157 to be triggered *only* by a *positive* gate voltage?

2. What if we want the 2N6157 to be triggered *only* by a *negative* gate voltage?

3. What if we want positive *and* negative triggering, but at much higher gate voltage values?

The circuit in Figure 20.26b illustrates the method commonly used to allow *single-polarity* triggering. In the circuit shown, $D_1$ is forward biased when $V_{GT}$ is *positive* and reverse biased when $V_{GT}$ is *negative*. Since $D_1$ will conduct only when $V_{GT}$ is positive, the triac can be triggered only by a positive gate signal. The potential required to trigger the device would be equal to the sum of $V_F$ for $D_1$ and the required gate triggering voltage

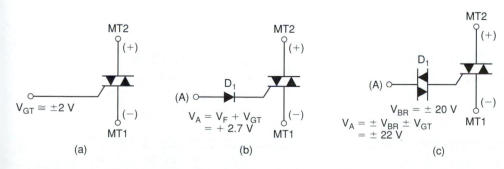

**FIGURE 20.26**   **Controlling triac triggering.**

for the 2N6157. By reversing the direction of the diode, we limit the triggering of the triac to *negative* gate voltages.

In many applications, the dual-polarity triggering characteristic of the triac is desirable, but the rated values of $V_{GT}$ for the device are too low. When this is the case, the triggering levels can be raised by using a *diac* in the gate circuit, as shown in Figure 20.26c. With the diac present, the triac can still be triggered by both positive and negative gate voltages. However, to get conduction in the gate terminal (which is required for triggering), the gate trigger signal must overcome the $V_{BR}$ values for the diac. For example, in Figure 20.26c, we have placed a diac with $V_{BR}$ values of $\pm 20$ V in the gate circuit of the triac. For the 2N6157 to be triggered into conduction, $V_A$ will have to reach a value of

$$V_A = \pm V_{BR} \pm V_{GT}$$
$$= \pm 20 \text{ V} \pm 2 \text{ V}$$
$$= \pm 22 \text{ V}$$

We still have the dual-polarity triggering that we want, but now the gate trigger signal ($V_A$) must reach $\pm 22$ V before the triac will be triggered into conduction. Note that *thyristor triggering control is the primary application for the diac*. It is also one of the applications for a device we will discuss in the next section, the *unijunction transistor (UJT)*.

## Triac Applications

A typical triac application can be seen in the *phase controller* in Figure 20.27. The primary difference between the triac phase controller and the SCR phase controller is the triac's ability to conduct during both alternations of the input signal.

With the input polarity indicated in Figure 20.27, $D_1$ is forward biased and $D_2$ is reverse biased. This puts a positive voltage at point A in the circuit. When $V_A$ is high enough to overcome the $V_{BR}$ rating of the diac and the $V_{GT}$ rating of the triac, the triac is triggered into conduction, providing the positive output cycle shown. When the input polarity reverses, $D_1$ becomes reverse biased and $D_2$ becomes forward biased. In this case, point A is still positive, but MT2 is now negative. When the triac is triggered, the current direction through the load is reversed, and the negative output cycle is produced. Note that the triac in Figure 20.27 is quadrant I triggered during the positive alternation of the input and quadrant IV triggered during the negative alternation of the input. Also, note the presence of the diac and triac *snubber networks*. These are included because the diac and triac are subject to the same false triggering problems as the SCR.

Figure 20.28 shows an improved circuit that can extend the phase control beyond 90°. It uses a variable *RC* network to replace the diodes and diac. The capacitor also eliminates false triggering. $R_3$ is included to prevent a surge current from destroying the triac gate. This control circuit is similar in operation and construction to the SCR circuit of

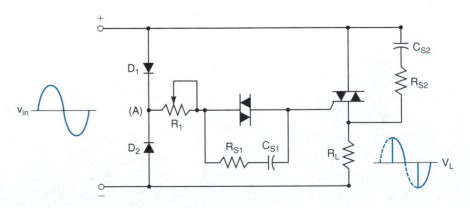

**FIGURE 20.27  A triac phase controller.**

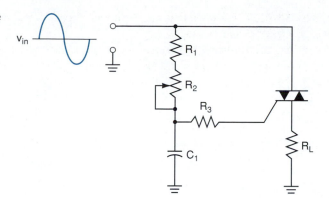

**FIGURE 20.28  Large-angle triac phase controller.**

Figure 20.16, except that this circuit can be used for phase control from approximately 30° to around 320°.

## Component Testing

Diacs and triacs are almost always in-circuit tested. An oscilloscope can be used to check the operation of each of these components. When shorted, both components will act as low-value resistors. When open, the components will not conduct, regardless of the presence of a triggering input.

## Summary

The *diac* acts as a bidirectional SUS. The device conducts in either direction (*anode 1 to anode 2*, or *anode 2 to anode 1*) when the terminal voltage reaches the breakover potential of the device. The device then continues to conduct until its current is driven below the holding current ($I_H$) rating by either anode current interruption or forced commutation.

The *triac* acts as a bidirectional SCR. The triac can be triggered into conduction by any one of four biasing conditions. Because of the ease of triac triggering, external components are often required in triac circuits to restrict the conditions under which triggering will occur. Like all the devices we have covered so far, the diac and triac are subject to false triggering from increases in temperature, gate noise, and anode (or main terminal) noise. The characteristics of the diac and triac are summarized in Figure 20.29

**Section Review**

1. What is a *bidirectional thyristor*?
2. What is a *diac*?
3. What are the two terminals of the diac called?
4. How is a diac triggered into conduction?
5. How is a diac driven into cutoff?
6. What is the most common diac application?
7. What is a *triac*?
8. Draw the schematic symbol for a triac and identify the component terminals.
9. Describe the four-quadrant triggering characteristics of the triac.
10. What methods are used to drive a triac into cutoff?
11. Describe the operation of each of the trigger control circuits in Figure 20.26.
12. Describe the operation of the triac phase controller in Figure 20.27.

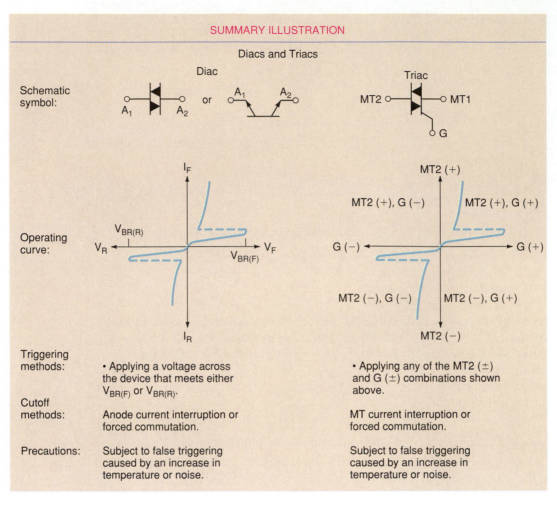

SUMMARY ILLUSTRATION

Diacs and Triacs

Diac

Triac

Schematic symbol:

Operating curve:

Triggering methods:
- Applying a voltage across the device that meets either $V_{BR(F)}$ or $V_{BR(R)}$.

Cutoff methods:
Anode current interruption or forced commutation.

Precautions:
Subject to false triggering caused by an increase in temperature or noise.

Triggering methods:
- Applying any of the MT2 ($\pm$) and G ($\pm$) combinations shown above.

Cutoff methods:
MT current interruption or forced commutation.

Precautions:
Subject to false triggering caused by an increase in temperature or noise.

**FIGURE 20.29**

13. What are the symptoms of a *shorted* diac or triac?
14. What are the symptoms of an *open* diac or triac?

## 20.4 UNIJUNCTION TRANSISTORS (UJTs)

*OBJECTIVE 8* ▶

**Unijunction transistor (UJT)**
A three-terminal device whose trigger voltage is proportional to its applied biasing voltage.

**Peak voltage ($V_P$)**
The value of $V_{EB1}$ that triggers the UJT into conduction.

**Peak current ($I_P$)**
The minimum value of $I_E$ required to maintain emitter conduction.

The **unijunction transistor (UJT)** is a three-terminal switching device whose trigger voltage is proportional to its applied biasing voltage. The schematic symbol for the UJT is shown in Figure 20.30a. As you can see, the terminals of the UJT are called the *emitter, base 1* (B1), and *base 2* (B2). The switching action of the UJT is explained with the help of the circuit in Figure 20.30b. A biasing voltage ($V_{BB}$) is applied across the two base terminals. The emitter–base 1 junction acts as an open until the voltage across the terminals ($V_{EB1}$) reaches a specified value, called the **peak voltage ($V_P$)**. When $V_{EB1} = V_P$, the emitter–base 1 junction triggers into conduction and will continue to conduct until the emitter current drops below a specified value, called the **peak current ($I_P$)**. At that time, the device returns to its nonconducting state. Note that the peak voltage (emitter–base 1 triggering voltage) is proportional to the value of $V_{BB}$. Thus, when we change $V_{BB}$, we change the value of $V_{EB1}$ needed to trigger the device.

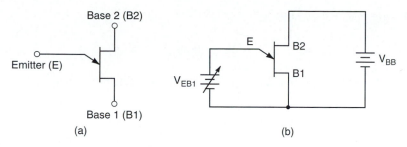

(a)                                                    (b)

**FIGURE 20.30    UJT schematic symbol and biasing.**

## UJT Construction and Operation

Figure 20.31 shows the construction and equivalent circuit of the UJT. As you can see, the UJT structure is very similar to that of the *n*-channel JFET. In fact, the only difference between the two components is the fact that the *p*-type (gate) material of the JFET *surrounds* the *n*-type (channel) material, as was shown in Chapter 12. The similarity in the construction of the UJT and the *n*-channel JFET is the reason that their schematic symbols are so similar.

Looking at the UJT equivalent circuit (Figure 20.31b), we see that the emitter terminal is connected (in effect) to a *pn*-junction diode. The cathode of the diode is connected to a voltage-divider circuit made up of $R_{B1}$ and $R_{B2}$. Note that these resistors represent the resistance of the *n*-type material that lies between each of the terminals and the *pn* junction.

The overall principle of operation is simple. For the emitter–base diode to conduct, its anode voltage must be approximately 0.7 V more positive than its cathode. The cathode potential of the diode ($V_k$) is determined by the combination of $V_{BB}$, $R_{B1}$, and $R_{B2}$. Using the standard voltage-divider equation, the value of $V_k$ is found as

$$V_k = V_{BB} \frac{R_{B1}}{R_{B1} + R_{B2}} \qquad (20.5)$$

The resistance ratio in equation (20.5) is called the **intrinsic standoff ratio, η** (Greek letter *eta*) of the UJT. Since

$$\eta = \frac{R_{B1}}{R_{B1} + R_{B2}}$$

**Intrinsic standoff ratio (η)**
The ratio of emitter–base 1 resistance to the total *interbase resistance* in a UJT. The **interbase resistance** is the total resistance between base 1 and base 2 when the device is not conducting.

**FIGURE 20.31    UJT construction and equivalent circuit.**

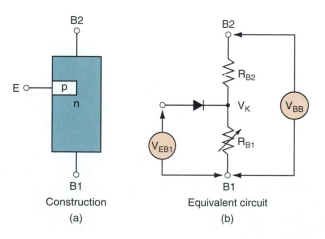

Construction                Equivalent circuit
(a)                              (b)

equation (20.5) can be rewritten as

$$V_k = \eta V_{BB} \qquad (20.6)$$

Adding 0.7 V to the value found in equation (20.6), we obtain an equation for the triggering voltage ($V_P$) for the UJT, as follows:

$$V_P = \eta V_{BB} + 0.7 \text{ V} \qquad (20.7)$$

The intrinsic standoff ratio for a given UJT is listed on its spec sheet. For example, the spec sheet for a 2N5431 UJT lists an intrinsic standoff ratio of $\eta = 0.8$ (maximum). Example 20.5 shows how this value is used to calculate the peak voltage for the 2N5431 at a given value of $V_{BB}$.

---

## EXAMPLE 20.5

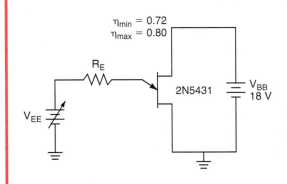

$\eta_{min} = 0.72$
$\eta_{max} = 0.80$

$R_E$

2N5431

$V_{EE}$

$V_{BB}$ 18 V

**FIGURE 20.32**

Determine the peak voltage ($V_P$) value for the 2N5431 UJT in Figure 20.32.

**Solution:** With $\eta_{max} = 0.8$ and $V_{BB} = +18$ V, the peak voltage of the circuit is found as

$$\begin{aligned} V_P &= \eta V_{BB} + 0.7 \text{ V} \\ &= (0.8)(+18 \text{ V}) + 0.7 \text{ V} \\ &= 15.1 \text{ V} \end{aligned}$$

### PRACTICE PROBLEM 20.5

The 2N4870 UJT has a rating of $\eta_{max} = 0.75$. Determine the maximum value of $V_P$ for the device when it is used in a circuit with $V_{BB} = +12$ V.

---

The peak voltage is applied across the emitter–base 1 terminals. Because of this, almost all of $I_E$ is drawn through $R_{B1}$ when the device is triggered, as is shown in Figure 20.33. Since $I_E$ is drawn through $R_{B1}$, this resistance value drops off as $I_E$ increases. The value of $R_{B2}$ is not affected very much by the presence of $I_E$. The results of this component operation are shown in the UJT characteristic curve in Figure 20.34. As you can see,

**FIGURE 20.33   UJT currents.**

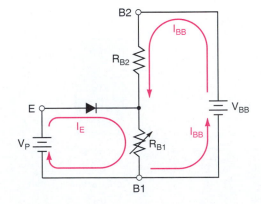

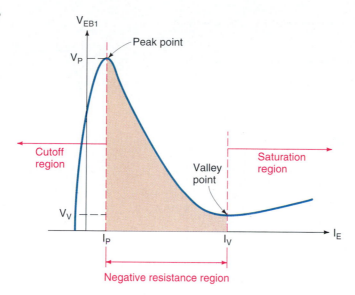

**FIGURE 20.34** The UJT operating curve.

$V_{EB1}$ can be increased until the *peak point* is reached. When the peak point is reached, $V_{EB1} = V_P$, and the emitter diode begins to conduct. The initial value of $I_E$ is referred to as the *peak emitter current* ($I_P$). This term can be confusing since we tend to think of a peak current as a maximum value. However, for the UJT, the peak current ($I_P$) is the *minimum* rating, because it is associated with the first point of conduction. Note that when $I_E < I_P$ the UJT is said to be in its *cutoff* region of operation.

When $I_E$ increases above the value of $I_P$, the value of $R_{B1}$ drops drastically. As a result, the value of $V_{EB1}$ actually *decreases* as $I_E$ increases. The decrease in $V_{EB1}$ continues until $I_E$ reaches the *valley current* ($I_V$) rating of the UJT. When $I_E$ is increased beyond $I_V$, the device goes into *saturation*.

As the characteristic curve in Figure 20.34 shows, $V_{EB1}$ and $I_E$ are inversely proportional between the peak point and valley point of the curve. The region of operation between these two points is called the **negative resistance region.** Note that the term **negative resistance** is used to describe any device with current and voltage values that are *inversely proportional*.

Once the UJT is triggered, the device continues to conduct as long as $I_E$ is greater than $I_P$. When $I_E$ drops below $I_P$, the device returns to the cutoff region of operation and remains there until another trigger is received.

**Negative resistance region**
The region of operation that lies between the peak and valley points of the UJT curve.

**Negative resistance**
A term used to describe any device with current and voltage values that are *inversely proportional*.

## UJT Applications

UJTs are used almost exclusively as *thyristor triggering devices*. An example of this application can be seen in Figure 20.35. The circuit in Figure 20.35a is called a **relaxation oscillator.** A relaxation oscillator is a circuit that uses the charge/discharge characteristics of a capacitor or inductor to produce a pulse output. This pulse output is normally used as the triggering signal for a thyristor, such as an SCR or a triac. In a practical relaxation oscillator, the dc supply voltage ($V_S$) is derived from the SCR anode voltage to control the timing of the oscillator. An example of this type of circuit is shown in Figure 20.35.

The ac input to the circuit is applied to a bridge rectifier. The full-wave rectified signal is then applied to the rest of the circuit. The zener diode ($D_1$) clips the rectified signal, producing the $V_S$ waveform shown in Figure 20.35b. Note that $V_S$ is the biasing voltage for the UJT and its trigger circuit ($R_T$ and $C_T$).

**Relaxation oscillator**
A circuit that uses the charge/discharge characteristics of a capacitor or inductor to produce a pulse output.

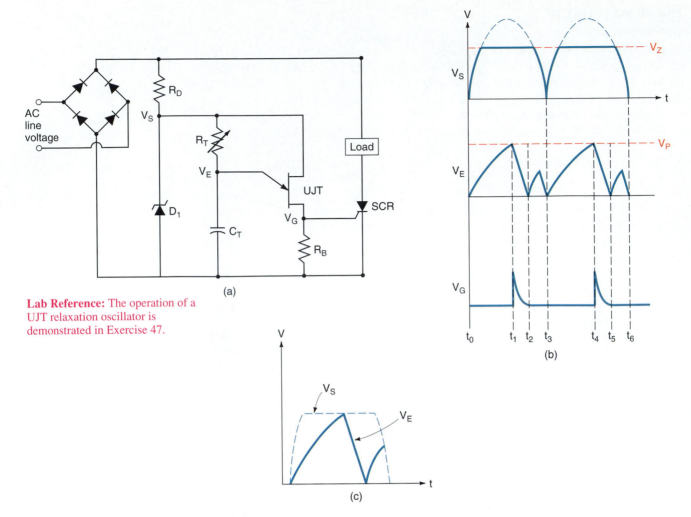

(a)

**Lab Reference:** The operation of a UJT relaxation oscillator is demonstrated in Exercise 47.

(c)

**FIGURE 20.35** A UJT relaxation oscillator.

*Don't Forget:* The time required for a capacitor to charge or discharge is found as $T = 5RC$. Since $R_{B1}$ (during conduction) is much less than $R_T$, the capacitor takes less time to discharge than it does to charge.

From $t_0$ to $t_1$, $C_T$ charges through $R_T$. When the emitter voltage of the UJT ($V_E$) reaches $V_P$, the UJT fires, causing conduction through $R_B$. The current through $R_B$ then produces the gate trigger pulse ($V_G$) required to trigger the SCR into conduction.

During the time period from $t_1$ to $t_2$, $C_T$ is discharging. When the UJT returns to its cutoff state (at $t_2$), the capacitor begins to charge again through $R_T$. However, this time the capacitor charge doesn't make it to $V_P$. The reason for this is shown in Figure 20.35c. As $V_E$ is starting to increase for the second time in the alternation, $V_S$ is decreasing in value. Before $V_E$ can reach the original value of $V_P$, the supply voltage is effectively removed from the UJT. Thus, the second charge cycle of $C_T$ is cut short, producing the $V_E$ waveform shown.

Even if $C_T$ was able to completely recharge during a given alternation, it wouldn't have any effect on the load. The reason for this is simple: The SCR is triggered into conduction during the discharge cycle of $C_T$ and continues to conduct for the remainder of the alternation. Since the SCR is already conducting when the second $C_T$ charge cycle occurs, the production of another trigger signal by the UJT would have no effect on the SCR, and thus none on the load.

Since the initial charge cycle of $C_T$ is controlled by the cycles of $V_S$, the relaxation oscillator is synchronized to the ac load cycle. This method of synchronizing the oscillator to the ac load cycle is used extensively in practice.

## One Final Note

The UJT is technically classified as a *thyristor-triggering device,* rather than as a thyristor. Since triggering thyristors such as SCRs and triacs is the only common application for the UJT, the device is normally covered along with thyristors. Also, you will normally find the spec sheets for UJTs in the same section of a parts manual as the thyristors.

Another thyristor trigger that is commonly used is the *programmable unijunction transistor* (PUT). We will conclude this section with a brief introduction to this device.

## The Programmable UJT (PUT)

The PUT is a four-layer device that is very similar to the SCR. In this case, however, the gate is used as a *reference* terminal, rather than as a triggering terminal. The schematic symbol for the PUT is shown, along with an application circuit, in Figure 20.36. Note the similarity between the PUT schematic symbol and the SCR schematic symbol. The similarity in schematic symbols between the two devices indicates the similarity in their operation.

The PUT blocks anode current until the value of $V_{AK}$ reaches the value of $V_{GK}$. For example, if $V_{GK}$ is +5 V, the device will not permit anode conduction until $V_{AK}$ reaches +5 V. At that time, the device triggers and anode current increases rapidly. Once triggered, the PUT continues to conduct until the anode current drops below the $I_P$ rating of the device. At that point, the device returns to cutoff and will block anode current until triggered again.

The circuit shown in Figure 20.36b demonstrates the use of the PUT in a relaxation oscillator. $R_1$ and $R_2$ are used to establish the value of $V_{GK}$ for the device. $C_T$ is charged through $R_T$, and $V_{AK}$ reaches the value of $V_{GK}$. At that point, the PUT is triggered, causing conduction through $R_3$. The conduction through $R_3$ produces the voltage waveform shown in Figure 20.36c. This pulse then triggers the SCR into conduction.

When the PUT is triggered, $C_T$ discharges through the device. When the current supplied by the capacitor drops below the $I_P$ rating of the PUT, the device returns to cutoff, and the charge cycle begins again.

It should be noted that the PUT relaxation oscillator would be synchronized to the SCR anode signal, as was the case with the UJT relaxation oscillator. Thus, the practical PUT oscillator is actually a bit more complex than the one shown in Figure 20.36b.

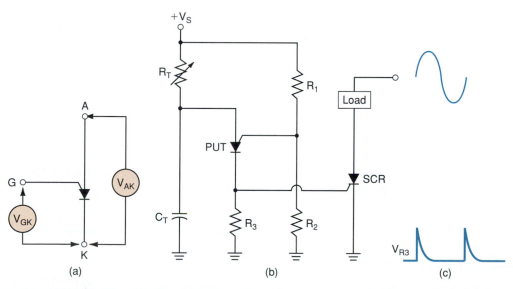

**FIGURE 20.36    A simple PUT relaxation oscillator.**

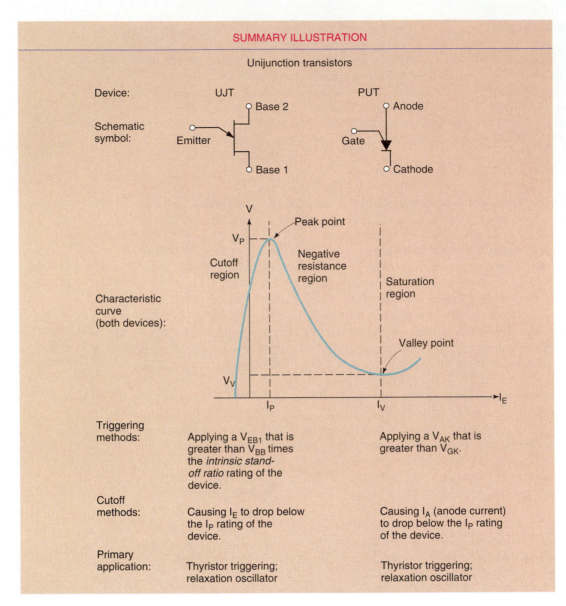

**FIGURE 20.37**

## Summary

The UJT and PUT are devices that are not classified as thyristors, but rather as *thyristor triggers*. These devices are usually used to control the triggering of SCRs and triacs.

The most common UJT and PUT triggering circuits are *relaxation oscillators*. These oscillators are usually synchronized to the ac signal that is applied to the circuit thyristor (SCR or triac) so that proper phase control is provided. The characteristics of the UJT and PUT are summarized in Figure 20.37.

---

**Section Review**

1. What is the *unijunction transistor* (UJT)?
2. What is the *peak voltage* ($V_P$) value of a UJT?
3. What is the *peak current* ($I_P$) value of a UJT?
4. What is the *intrinsic standoff ratio* ($\eta$) rating of a UJT?

---

5. How does the value of $\eta$ for a UJT relate to the value of $V_P$?

6. What is meant by the term *negative resistance*?

7. What current and voltage values define the limits of the negative resistance region of UJT operation?

8. What is the relationship between $I_E$ and $V_{EB1}$ in the negative resistance region of UJT operation?

9. Describe the operation of the *relaxation oscillator* in Figure 20.35.

10. Explain the operation of the *relaxation oscillator* in Figure 20.36.

11. What is the primary difference between the PUT and the SCR?

# 20.5 DISCRETE PHOTODETECTORS

In Chapter 2, we discussed the operation of the *light-emitting diode* (LED). The LED is classified as a **light emitter** because it gives off light when biased properly. In this section, we will take a look at the operation of several **light detectors**. Light detectors are components whose electrical output characteristics are controlled by the amount of light they receive. In other words, while *emitters* produce light, *detectors* respond to light.

When we discuss photodetectors, we need to consider a variety of parameters related to light. For this reason, we need to start this section with a brief discussion on light and its properties.

◄ *OBJECTIVE 9*

**Light emitters**
Optoelectronic devices that produce light.

**Light detectors**
Optoelectronic devices that respond to light.

## Characteristics of Light

**Light** is *electromagnetic energy* that falls within a specific range of frequencies, as shown in Figure 20.38. The entire light spectrum falls within the range of 30 THz to 3 PHz and is further broken down as shown in Table 20.2.

Two parameters are commonly used to describe light. The first of these is **wavelength** (symbolized by $\lambda$, the Greek letter *lambda*). Wavelength is the physical length of one cycle of a transmitted electromagnetic wave.

The concept of wavelength is illustrated in Figure 20.39. The wavelength for a signal at a given frequency is found as

◄ *OBJECTIVE 10*

**Light**
Electromagnetic energy that falls within a specific range of frequencies.

**Wavelength ($\lambda$)**
The physical length of one cycle of a transmitted electromagnetic wave.

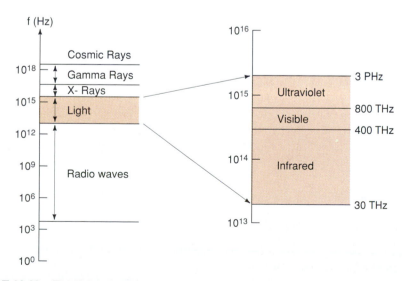

**FIGURE 20.38   The light spectrum.**

**TABLE 20.2   Light Frequency Ranges**

| Type of Light | Frequency Range (Approximate)[a] |
|---|---|
| Infrared | 30 THz to 400 THz |
| Visible | 400 THz to 800 THz |
| Ultraviolet | 800 THz to 3 PHz |

[a]P stands for *peta* $(10^{15})$ and T for *tera* $(10^{12})$.

**FIGURE 20.39   Wavelength.**

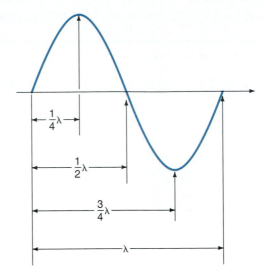

$$\lambda = \frac{c}{f} \qquad\qquad (20.8)$$

*A Practical Consideration:* The wavelength parameters of many optoelectronic devices are given in *nanometers* (nm).

where $\lambda$ = the wavelength of the signal (in *nanometers*)
   $c$ = the *speed of light*, given as $3 \times 10^{17}$ nm/s
   $f$ = the frequency of the transmitted signal

The following example demonstrates the calculation of wavelength for a given signal.

---

**EXAMPLE 20.6**

Determine the wavelength of a 150 THz light signal.

*Solution:*   The wavelength of the signal is found as

$$\lambda = \frac{c}{f}$$

$$= \frac{3 \times 10^{17} \text{ nm/s}}{150 \times 10^{12} \text{ Hz}}$$

$$= 2000 \text{ nm}$$

Thus, the physical length of one cycle of a 150 THz signal is 2000 nm, or 2 $\mu$m.

**PRACTICE PROBLEM 20.6**

The infrared frequency spectrum falls between 30 and 400 THz. Determine the range of wavelengths for this range of frequencies.

---

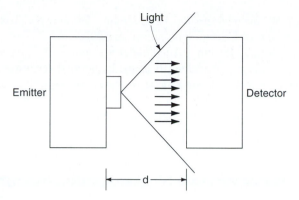

**FIGURE 20.40    Emitter–detector.**

Wavelength is an important concept because photoemitters and photodetectors are rated for specific wavelengths. You see, photoemitters are used to drive the various photodetectors. For a given photoemitter to be used with a given photodetector, the two devices must be rated for the same approximate wavelength. This point will be discussed further later in this section.

The second important light parameter is *intensity*. **Light intensity** is the amount of light per unit area received by a given photodetector. For example, consider the emitter–detector combination shown in Figure 20.40. The light intensity is a measure of the amount of light measured at the detector. Note that the intensity decreases as the distance between the emitter and detector (*d*) increases. As you will see, the output characteristics of most photodetectors are controlled by the light intensity at their inputs, as well as the input wavelength.

## *The Photodiode*

The **photodiode** is a diode whose reverse conduction is light intensity controlled. When the light intensity at the optical input to a photodiode increases, the reverse current through the device also increases. This point is illustrated in Figure 20.41.

In Figure 20.41, a photodiode is enclosed with an LED in an *opaque case*. The term **opaque** is used to describe anything that blocks light. When in the opaque case, the photodiode will receive a light input only when the LED is turned on by its control circuit. When the LED is off, the photodiode reverse current is shown to be 10 nA. When the LED lights, the photodiode reverse current increases to 50 μA. Note that the **light current ($I_L$)** is shown to be approximately 5000 times as great as the **dark current ($I_D$).** This ratio of light current to dark current is not at all uncommon.

The spec sheet for a photodiode lists its values of $I_D$ and $I_L$, as can be seen in Figure 20.42. Note that the value of $I_D$ for the MRD500 photodiode is 2 nA (maximum),

**Light intensity**
The amount of light per unit area received by a given photodetector; also called **irradiance**.

**Photodiode**
A diode whose reverse conduction is light intensity controlled.

**Opaque**
A term used to describe anything that blocks light.

**Light current ($I_L$)**
The reverse current through a photodiode with an active light input.

**Dark current ($I_D$)**
The reverse current through a photodiode with no active light input.

**FIGURE 20.41**

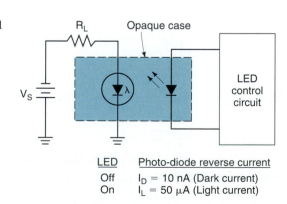

| LED | Photo-diode reverse current |
|-----|------------------------------|
| Off | $I_D$ = 10 nA (Dark current) |
| On  | $I_L$ = 50 μA (Light current) |

while the value of $I_L$ for the device is 6 μA (minimum). These two values show the device to have a minimum ratio of light-to-dark current of 3000.

Several other photodiode ratings listed under the *optical characteristics* heading are important. The first of these is **wavelength of peak spectral response ($\lambda_S$).** This rating tells you the wavelength that will cause the strongest response in the photodiode. For the MRD500, the optimum input wavelength is 0.8 μm (800 nm). As Example 20.7 shows, this value can be used to determine the optimum operating frequency for the device.

**Wavelength of peak spectral response ($\lambda_s$)**
A rating that indicates the wavelength that will cause the strongest response in a photodetector.

# Photo Detectors
## Diode Output

**MRD500**
**MRD510**

**PHOTO DETECTORS
DIODE OUTPUT
PIN SILICON
250 MILLIWATTS
100 VOLTS**

... designed for application in laser detection, light demodulation, detection of visible and near infrared light-emitting diodes, shaft or position encoders, switching and logic circuits, or any design requiring radiation sensitivity, ultra high-speed, and stable characteristics.

- Ultra Fast Response — (<1 ns Typ)
- High Sensitivity — MRD500 (1.2 μA/mW/cm² Min)
  MRD510 (0.3 μA/mW/cm² Min)
- Available With Convex Lens (MRD500) or Flat Glass (MRD510) for Design Flexibility
- Popular TO-18 Type Package for Easy Handling and Mounting
- Sensitive Throughout Visible and Near Infrared Spectral Range for Wide Application
- Annular Passivated Structure for Stability and Reliability

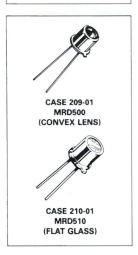

CASE 209-01
MRD500
(CONVEX LENS)

CASE 210-01
MRD510
(FLAT GLASS)

**MAXIMUM RATINGS** ($T_A$ = 25°C unless otherwise noted)

| Rating | Symbol | Value | Unit |
|---|---|---|---|
| Reverse Voltage | $V_R$ | 100 | Volts |
| Total Power Dissipation @ $T_A$ = 25°C<br>Derate above 25°C | $P_D$ | 250<br>2.27 | mW<br>mW/°C |
| Operating Temperature Range | $T_A$ | −55 to +125 | °C |
| Storage Temperature Range | $T_{stg}$ | −65 to +150 | °C |

**STATIC ELECTRICAL CHARACTERISTICS** ($T_A$ = 25°C unless otherwise noted)

| Characteristic | Fig. No. | Symbol | Min | Typ | Max | Unit |
|---|---|---|---|---|---|---|
| Dark Current ($V_R$ = 20 V, $R_L$ = 1 megohm) Note 2<br>$T_A$ = 25°C<br>$T_A$ = 100°C | 2 and 3 | $I_D$ | —<br>— | —<br>14 | 2<br>— | nA |
| Reverse Breakdown Voltage ($I_R$ = 10 μA) | — | $V_{(BR)R}$ | 100 | 200 | — | Volts |
| Forward Voltage ($I_F$ = 50 mA) | — | $V_F$ | — | — | 1.1 | Volts |
| Series Resistance ($I_F$ = 50 mA) | — | $R_S$ | — | — | 10 | Ohms |
| Total Capacitance ($V_R$ = 20 V, f = 1 MHz) | 5 | $C_T$ | — | — | 4 | pF |

**OPTICAL CHARACTERISTICS** ($T_A$ = 25°C unless otherwise noted)

| Characteristic | | Fig. No. | Symbol | Min | Typ | Max | Unit |
|---|---|---|---|---|---|---|---|
| Light Current<br>($V_R$ = 20 V) Note 1 | MRD500<br>MRD510 | 1 | $I_L$ | 6<br>1.5 | 9<br>2.1 | —<br>— | μA |
| Sensitivity at 0.8 μm<br>($V_R$ = 20 V) Note 3 | MRD500<br>MRD510 | — | $S_{(\lambda = 0.8\,\mu m)}$ | —<br>— | 6.6<br>1.5 | —<br>— | μA/mW/<br>cm² |
| Response Time ($V_R$ = 20 V, $R_L$ = 50 Ohms) | | — | $t_{(resp)}$ | — | 1 | — | ns |
| Wavelength of Peak Spectral Response | | 5 | $\lambda_S$ | — | 0.8 | — | μm |

NOTES: 1. Radiation Flux Density (H) equal to 5 mW/cm² emitted from a tungsten source at a color temperature of 2870 K.
2. Measured under dark conditions. (H ≈ 0).
3. Radiation Flux Density (H) equal to 0.5 mW/cm² at 0.8 μm.

**FIGURE 20.42   The MRD500-510 photodetector specification sheet. (Copyright of Motorola. Used by permission.)**

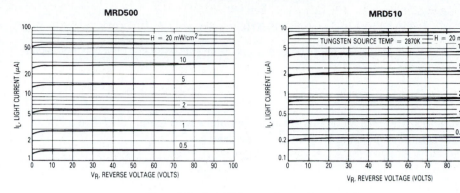

Figure 1. Irradiated Voltage — Current Characteristic

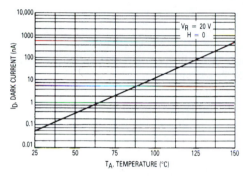

Figure 2. Dark Current versus Temperature

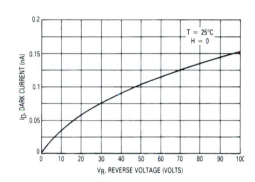

Figure 3. Dark Current versus Reverse Voltage

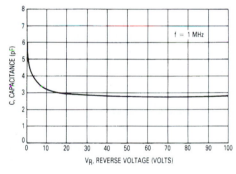

Figure 4. Capacitance versus Voltage

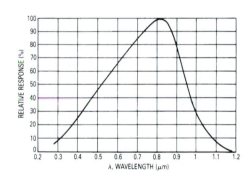

Figure 5. Relative Spectral Response

FIGURE 20.42   (Continued)

---

Determine the optimum operating frequency for the MRD500 photodiode.

*EXAMPLE 20.7*

**Solution:**   Rearranging equation (20.8) gives us the following equation for frequency:

$$f = \frac{c}{\lambda}$$

Using the given optimum wavelength of 800 nm, the optimum operating frequency for the MRD500 is found as

$$f = \frac{3 \times 10^{17} \text{ nm/s}}{800 \text{ nm}}$$

$$= 375 \text{ THz}$$

Since 375 THz is in the *infrared* frequency spectrum, the MRD500 is classified as an *infrared detector*.

### PRACTICE PROBLEM 20.7

The Motorola MRD821 photodiode has an optimum wavelength of 940 nm. Determine the optimum operating frequency for the device.

**Sensitivity**
A rating that indicates the response of a photodetector to a specified light intensity.

The **sensitivity** rating of the photodiode indicates the response of the device to a specified light intensity. For the MRD500, the sensitivity rating (listed under *optical characterics*) is given as 6.6 $\mu A/mW/cm^2$. This means that the reverse current of the device will increase by 6.6 $\mu A$ for every 1 $mW/cm^2$ of light applied to the device. For example, refer back to Figure 20.42. If the light intensity at the photodiode is 1 $mW/cm^2$, the value of $I_R$ for the device is approximately 6.6 $\mu A$. If the light intensity at the input increases to 2 $mW/cm^2$, $I_R$ increases to approximately 13.2 $\mu A$, and so on.

**Spectral response**
A measure of a photodetector's response to a change in input wavelength. Note that response is measured in terms of the device *sensitivity*.

The **spectral response** of a photodiode is a measure of the device's response to a change in input wavelength. The spectral response of the MRD500 is shown in the *relative spectral response* curve in Figure 20.42. As you can see, the response of the photodiode peaks at 800 nm, the value given on the spec sheet. If the input wavelength goes as low as 520 nm or as high as 950 nm, the *sensitivity* of the device drops to 50 percent of its rated value. Thus, at these two frequencies, the sensitivity of the device drops to approximately 3.3 $\mu A/mW/cm^2$.

The *irradiated voltage–current characteristic* curves in Figure 20.42 show that the light current ($I_L$) of the device is relatively independent of the amount of reverse bias across the diode. The dark current ($I_D$) curves show that the amount of dark current is far more dependent on temperature than on the value of $V_R$. This is similar to the reverse current through any *pn*-junction diode, which is primarily temperature dependent.

Several final points need to be made regarding the photodiode:

1. The schematic symbol used in Figure 20.41 is only one of two commonly used symbols. The other is shown in Figure 20.43. Note that the light arrows in the symbol point *toward* the diode, rather than away from it.

2. The photodiode is always operated in its reverse operating region. Since light affects the *reverse conduction* of the device, it wouldn't make any sense to operate it in its forward operating region.

3. Most of the photodiode parameters and ratings we have discussed apply to many other photodetectors. This point will become clear as our coverage of these devices continues.

## The Phototransistor

**Phototransistor**
A three-terminal photodetector whose collector current is controlled by the intensity of the light at its optical input (base).

The **phototransistor** is a three-terminal photodetector whose collector current is controlled by the intensity of the light at its optical input (base). Figure 20.44 shows one possible configuration for a phototransistor amplifier.

In Figure 20.44, the phototransistor is shown to be enclosed with an LED in an opaque casing. When the light from the LED is varied by its control circuit, the change in light causes a proportional change in base current in the phototransistor. This change in base current causes a change in emitter (and collector) current that causes a propor-

**FIGURE 20.43**

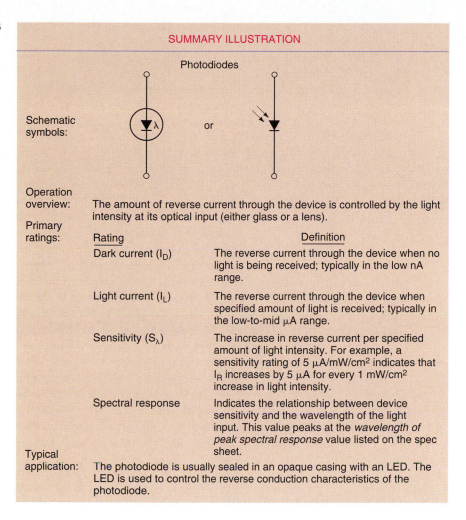

SUMMARY ILLUSTRATION

Photodiodes

Schematic symbols:

or

Operation overview: The amount of reverse current through the device is controlled by the light intensity at its optical input (either glass or a lens).

Primary ratings:

| Rating | Definition |
|---|---|
| Dark current ($I_D$) | The reverse current through the device when no light is being received; typically in the low nA range. |
| Light current ($I_L$) | The reverse current through the device when specified amount of light is received; typically in the low-to-mid μA range. |
| Sensitivity ($S_\lambda$) | The increase in reverse current per specified amount of light intensity. For example, a sensitivity rating of 5 μA/mW/cm² indicates that $I_R$ increases by 5 μA for every 1 mW/cm² increase in light intensity. |
| Spectral response | Indicates the relationship between device sensitivity and the wavelength of the light input. This value peaks at the *wavelength of peak spectral response* value listed on the spec sheet. |

Typical application: The photodiode is usually sealed in an opaque casing with an LED. The LED is used to control the reverse conduction characteristics of the photodiode.

**FIGURE 20.44   Optocoupling.**

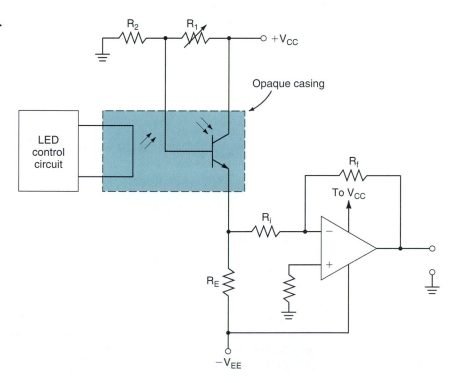

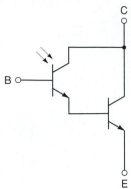

**FIGURE 20.45   The photo-Darlington schematic symbol.**

tional voltage to be developed across $R_E$. This voltage is applied to the inverting amplifier, which responds accordingly.

The phototransistor is identical to the standard BJT in every respect, except for the fact that its collector current is (ultimately) light intensity controlled. Thus, the phototransistor in Figure 20.44 could be a linear amplifier or a switch, depending on the output of the LED control circuit. The biasing circuit ($R_1$ and $R_2$) is used to adjust the dark (quiescent) operation of the phototransistor. When $R_1$ is varied, the dark emitter current of the phototransistor varies. Thus, by adjusting $R_1$, the input voltage to the op-amp is varied, allowing the output of the op-amp to be set to zero. Note that $R_1$ in this circuit is called a *zero adjust* because it is used to set the final circuit output to zero volts.

The arrangement in Figure 20.44 is referred to as *optocoupling* because the output from the LED control circuit is coupled via light to the phototransistor/op-amp circuit. The advantage of optocoupling is that it provides complete dc isolation between the source (the LED control circuit) and the load (the phototransistor/op-amp circuit).

Several points should be made regarding the phototransistor: First, the spec sheet for the phototransistor contains the same basic parameters and ratings as found on the spec sheet for a photodiode. The phototransistor is subject to the same wavelength and sensitivity characteristics as the photodiode. The difference is the fact that the optical characteristics of the phototransistor relate the light input to *collector current* (rather than diode reverse current). Second, the base terminal of the phototransistor is normally returned to ground through some type of biasing circuit, as shown in Figure 20.44. The biasing circuit not only allows for adjusting the output from the device, but it also adds *temperature stability* to the circuit. You see, transistors have the same leakage current considerations as any other device. Without a return path from base to ground, any base leakage current is forced into the emitter terminal of the device. This can cause the phototransistor to be forced into conduction by a change in temperature. By providing a resistive path from the base terminal to ground, you prevent base leakage current from affecting the emitter circuit of the device.

## The Photo-Darlington

**Photo-Darlington**
A phototransistor with a Darlington pair output.

The **photo-Darlington** is a phototransistor constructed in a Darlington configuration, as is shown in Figure 20.45. Except for the increased output current capability (as is always the case with Darlingtons), the operation and characteristics of the device are identical to those of the standard phototransistor.

## The LASCR

**Light-activated SCR (LASCR)**
A three-terminal light-activated SCR.

The **light-activated SCR (LASCR),** or *photo-SCR,* is a three-terminal light-activated SCR. The schematic symbol for the LASCR is shown in Figure 20.46. Again, all the properties of the photodiode can be found in the LASCR. In this case, however, the SCR is *latched* into its conducting state by the light input to the device. Thus, the LASCR could be used in an optically coupled phase-control circuit.

## Summary

Most discrete devices can be obtained in *photodetector* form. The photodetector devices work in a fashion that is very similar to their current-controlled counterparts. However, the photodetector devices are *light-controlled* devices.

When working with photodetectors, you need to remember the *wavelength* and *sensitivity* (intensity) characteristics of the devices. If a light emitter and photodetector are not matched in terms of wavelength and sensitivity, the optocoupling circuit will operate at reduced efficiency or may not work at all.

**FIGURE 20.46   The LASCR schematic symbol.**

The wavelength and sensitivity concerns of discrete photodetector circuits are eliminated by the use of *optoisolators* and *optointerrupters*. These integrated circuits contain both a light emitter and a light detector. As you will see in the next section, these ICs are far easier to deal with than their discrete component counterparts.

**Section Review**

1. What is the difference between a light *emitter* and a light *detector*?

2. What is *wavelength*?

3. Why is the concept of wavelength important when working with optoelectronic devices?

4. What is *light intensity*?

5. What is a *photodiode*?

6. In terms of the photodiode, what is *dark current* ($I_D$)? What is *light current* ($I_L$)?

7. What is the relationship between photodiode light current and dark current?

8. What is the *wavelength of peak spectral response* for a photodiode?

9. What is *sensitivity*?

10. Explain the $\mu A/mW/cm^2$ unit of measure for sensitivity.

11. What is *spectral response*? Which photodiode unit of measure is affected by spectral response?

12. What is a *phototransistor*?

13. What is the relationship between input light and collector current for a phototransistor?

14. What is the difference between the optical ratings of the phototransistor and those of the photodiode?

15. What is the difference between the *photo-Darlington* and the standard phototransistor?

16. What is the *LASCR*? How does it differ from the standard SCR?

17. List the characteristics that must be considered whenever you are working with any discrete photodetector.

## 20.6  OPTOISOLATORS AND OPTOINTERRUPTERS

*Optoisolators* and *optointerrupters* are ICs that contain both a photoemitter and a photodetector. While the photoemitter is always an LED, the photodetector can be any one of the discrete photodetectors we discussed in the last section.

◀ *OBJECTIVE 12*

## *Optoisolators*

The **optoisolator** is an optocoupler; that is, a device that uses light to couple a signal from its input (a photoemitter) to its output (a photodetector). For example, an optoisolator with a transistor output would contain the circuitry shown in the opaque casing in Figure 20.44.

The typical optoisolator comes in a six-pin *dual in-line package* (DIP), as is shown on the spec sheet in Figure 20.47. As the schematic diagram in the figure shows, this transistor-output optoisolator contains an LED and a phototransistor. Thus, the 4N35–7 series optoisolators could be used in the circuit in Figure 20.44.

**Optoisolator**
An optocoupler. A device that uses light to couple a signal from its input (a photoemitter) to its output (a photodetector).

**Lab Reference:** Optoisolator operation is demonstrated in Exercise 48.

# 6-Pin DIP Optoisolators
## Transistor Output

4N35
4N36
4N37

6-PIN DIP
OPTOISOLATORS
TRANSISTOR
OUTPUT

CASE 730A-02
PLASTIC

These devices consist of a gallium arsenide infrared emitting diode optically coupled to a monolithic silicon phototransistor detector.

- Convenient Plastic Dual-In-Line Package
- High Current Transfer Ratio — 100% Minimum at Spec Conditions
- Guaranteed Switching Speeds
- High Input-Output Isolation Guaranteed — 7500 Volts Peak
- UL Recognized. File Number E54915 ℞
- VDE approved per standard 0883/6.80 (Certificate number 41853), with additional approval to DIN IEC380/VDE0806, IEC435/VDE0805, IEC65/VDE0860, VDE0110b, covering all other standards with equal or less stringent requirements, including IEC204/VDE0113, VDE0160, VDE0832, VDE0833, etc. ⟨VDE⟩883
- Meets or Exceeds All JEDEC Registered Specifications
- Special lead form available (add suffix "T" to part number) which satisfies VDE0883/6.80 requirement for 8 mm minimum creepage distance between input and output solder pads.
- Various lead form options available. Consult "Optoisolator Lead Form Options" data sheet for details.

**MAXIMUM RATINGS** ($T_A$ = 25°C unless otherwise noted)

| Rating | Symbol | Value | Unit |
|---|---|---|---|
| **INPUT LED** | | | |
| Reverse Voltage | $V_R$ | 6 | Volts |
| Forward Current — Continuous | $I_F$ | 60 | mA |
| LED Power Dissipation @ $T_A$ = 25°C with Negligible Power in Output Detector Derate above 25°C | $P_D$ | 120 <br> 1.41 | mW <br> mW/°C |
| **OUTPUT TRANSISTOR** | | | |
| Collector-Emitter Voltage | $V_{CEO}$ | 30 | Volts |
| Emitter-Base Voltage | $V_{EBO}$ | 7 | Volts |
| Collector-Base Voltage | $V_{CBO}$ | 70 | Volts |
| Collector Current — Continuous | $I_C$ | 150 | mA |
| Detector Power Dissipation @ $T_A$ = 25°C with Negligible Power in Input LED Derate above 25°C | $P_D$ | 150 <br> 1.76 | mW <br> mW/°C |
| **TOTAL DEVICE** | | | |
| Isolation Source Voltage (1) (Peak ac Voltage, 60 Hz, 1 sec Duration) | $V_{ISO}$ | 7500 | Vac |
| Total Device Power Dissipation @ $T_A$ = 25°C Derate above 25°C | $P_D$ | 250 <br> 2.94 | mW <br> mW/°C |
| Ambient Operating Temperature Range | $T_A$ | −55 to +100 | °C |
| Storage Temperature Range | $T_{stg}$ | −55 to +150 | °C |
| Soldering Temperature (10 seconds, 1/16" from case) | — | 260 | °C |

(1) Isolation surge voltage is an internal device dielectric breakdown rating.
For this test, Pins 1 and 2 are common, and Pins 4, 5 and 6 are common.

**SCHEMATIC**

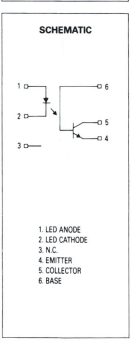

1. LED ANODE
2. LED CATHODE
3. N.C.
4. EMITTER
5. COLLECTOR
6. BASE

**FIGURE 20.47   The 4N35–7 optoisolator specification sheet. (Copyright of Motorola. Used by permission.)**

**Isolation source voltage ($V_{ISO}$)**
The voltage that, if applied across the input and output pins, will destroy an optoisolator.

**Soldering temperature**
The maximum amount of heat that can be applied to any pin of the IC without causing damage to the chip.

Most optoisolator parameters and ratings are the standard LED and phototransistor specifications. However, several new ratings warrant discussion.

The **isolation source voltage ($V_{ISO}$)** rating indicates the input-to-output voltage that will cause the optoisolator to break down and conduct. In other words, it is the voltage that will destroy the device if applied across the input and output pins. For the 4N35–7 series chips, the value of $V_{ISO}$ is 7500 $V_{ac}$.

The **soldering temperature** is the maximum amount of heat that can be applied to any pin of the IC without causing internal damage to the chip. If the total heat

**ELECTRICAL CHARACTERISTICS** ($T_A$ = 25°C unless otherwise noted)

| Characteristic | | Symbol | Min | Typ | Max | Unit |
|---|---|---|---|---|---|---|
| **INPUT LED** | | | | | | |
| Forward Voltage ($I_F$ = 10 mA) | $T_A$ = 25°C | $V_F$ | 0.8 | 1.15 | 1.5 | V |
| | $T_A$ = −55°C | | 0.9 | 1.3 | 1.7 | |
| | $T_A$ = 100°C | | 0.7 | 1.05 | 1.4 | |
| Reverse Leakage Current ($V_R$ = 6 V) | | $I_R$ | — | — | 10 | µA |
| Capacitance (V = 0 V, f = 1 MHz) | | $C_J$ | — | 18 | — | pF |
| **OUTPUT TRANSISTOR** | | | | | | |
| Collector-Emitter Dark Current ($V_{CE}$ = 10 V, $T_A$ = 25°C) | | $I_{CEO}$ | — | 1 | 50 | nA |
| ($V_{CE}$ = 30 V, $T_A$ = 100°C) | | | — | — | 500 | µA |
| Collector-Base Dark Current ($V_{CB}$ = 10 V) | $T_A$ = 25°C | $I_{CBO}$ | — | 0.2 | 20 | nA |
| | $T_A$ = 100°C | | — | 100 | — | |
| Collector-Emitter Breakdown Voltage ($I_C$ = 1 mA) | | $V_{(BR)CEO}$ | 30 | 45 | — | V |
| Collector-Base Breakdown Voltage ($I_C$ = 100 µA) | | $V_{(BR)CBO}$ | 70 | 100 | — | V |
| Emitter-Base Breakdown Voltage ($I_E$ = 100 µA) | | $V_{(BR)EBO}$ | 7 | 7.8 | — | V |
| DC Current Gain ($I_C$ = 2 mA, $V_{CE}$ = 5 V) | | $h_{FE}$ | — | 400 | — | — |
| Collector-Emitter Capacitance (f = 1 MHz, $V_{CE}$ = 0) | | $C_{CE}$ | -- | 7 | — | pF |
| Collector-Base Capacitance (f = 1 MHz, $V_{CB}$ = 0) | | $C_{CB}$ | — | 19 | — | pF |
| Emitter-Base Capacitance (f = 1 MHz, $V_{EB}$ = 0) | | $C_{EB}$ | — | 9 | — | pF |
| **COUPLED** | | | | | | |
| Output Collector Current ($I_F$ = 10 mA, $V_{CE}$ = 10 V) | $T_A$ = 25°C | $I_C$ | 10 | 30 | — | mA |
| | $T_A$ = −55°C | | 4 | — | — | |
| | $T_A$ = 100°C | | 4 | — | — | |
| Collector-Emitter Saturation Voltage ($I_C$ = 0.5 mA, $I_F$ = 10 mA) | | $V_{CE(sat)}$ | — | 0.14 | 0.3 | V |
| Turn-On Time | ($I_C$ = 2 mA, $V_{CC}$ = 10 V, $R_L$ = 100 Ω, Figure 11) | $t_{on}$ | — | 7.5 | 10 | µs |
| Turn-Off Time | | $t_{off}$ | — | 5.7 | 10 | |
| Rise Time | | $t_r$ | — | 3.2 | — | |
| Fall Time | | $t_f$ | — | 4.7 | — | |
| Isolation Voltage (f = 60 Hz, t = 1 sec) | | $V_{ISO}$ | 7500 | — | — | Vac(pk) |
| Isolation Current ($V_{I-O}$ = 3550 Vpk)  4N35 | | $I_{ISO}$ | — | — | 100 | µA |
| ($V_{I-O}$ = 2500 Vpk)  4N36 | | | — | — | 100 | |
| ($V_{I-O}$ = 1500 Vpk)  4N37 | | | — | 8 | 100 | |
| Isolation Resistance (V = 500 V) | | $R_{ISO}$ | $10^{11}$ | — | — | Ω |
| Isolation Capacitance (V = 0 V, f = 1 MHz) | | $C_{ISO}$ | — | 0.2 | 2 | pF |

**FIGURE 20.47**   (Continued)

applied to any pin of the 4N35 reaches 260°C for a duration of 10 s, internal damage may result.

The **isolation current ($I_{ISO}$)** rating indicates the amount of current that can be forced between the input and output at the rated voltage. For example, a 3550 $V_{pk}$ input applied across the 4N35 will generate 100 µA between the input and output pins of the device. The 4N36 requires a 2500 $V_{pk}$ input to generate 100 µA, and so on.

The **isolation resistance ($R_{ISO}$)** rating indicates the total resistance between the input pins and output pins of the device. For the 4N35–7 series, this rating is $10^{11}$ Ω (100 GΩ).

Finally, the **isolation capacitance ($C_{ISO}$)** rating indicates the total capacitance between the device input and output pins. For the 4N35–7 series, this rating is 2 pF.

Except for the ratings listed here, all the parameters and ratings for the optoisolator are fairly standard. This can be seen by quickly looking through the spec sheet. As you glance through the spec sheet, you'll also notice two standard parameters that are missing: those for *wavelength* and *sensitivity*. Since the photoemitter and photodetector are contained in the same chip, the wavelength and sensitivity values of the devices are unimportant.

As stated earlier, optoisolators come with a variety of output devices. Among these are the transistor output, Darlington output, triac output, and SCR output. For each of these devices, the input/output parameters and ratings vary only to reflect the characteris-

**Isolation current ($I_{ISO}$)**
The amount of current that can be forced between the input and output at the rated voltage.

**Isolation resistance ($R_{ISO}$)**
The total resistance between the device input and output pins.

**Isolation capacitance ($C_{ISO}$)**
The total capacitance between the device input and output pins.

tics of the photodetector used. Other than that, they operate exactly as described here and in Section 20.5.

## An Optoisolator Application: The Solid-State Relay

**Solid-state relay (SSR)**
A circuit that uses a dc input voltage to pass or block an ac signal.

A **solid-state relay (SSR)** is a circuit that uses a dc input voltage to pass or block an ac signal. The diagram in Figure 20.48a illustrates the input/output relationship for the SSR. When the dc input has the polarity shown, the waveform at the input (ac 1) is coupled to the ac output (ac 2), and thus to the load. As long as the dc potential remains, the ac signal will be coupled to the load. If the dc potential is removed, the output goes open circuit, and the ac signal is blocked from the load.

The schematic for the SSR is shown in Figure 20.48b. The dc circuit is coupled to the ac circuit via the triac-output optoisolator. As long as the dc input voltage has the polarity shown, the LED in the optoisolator is forward biased and emits light. This light keeps the triac photodetector conducting. As long as the triac photodetector is conducting, the triac in the ac circuit continues to conduct, passing the ac signal from the input terminal (ac 1) to the output terminal (ac 2).

If the dc input voltage is removed, the LED turns off, and the triac photodetector shuts off. This prevents a gate trigger voltage from being developed across $R_3$, and the output triac shuts off. When the output triac shuts off, the ac input is blocked from the ac output terminal.

The SSR is typically used to allow a digital circuit to control the operation (on or off) of a motor. This application is easy to understand if you picture the input to Figure 20.48a as coming from a digital system and the load as being a motor. Note that SSRs are

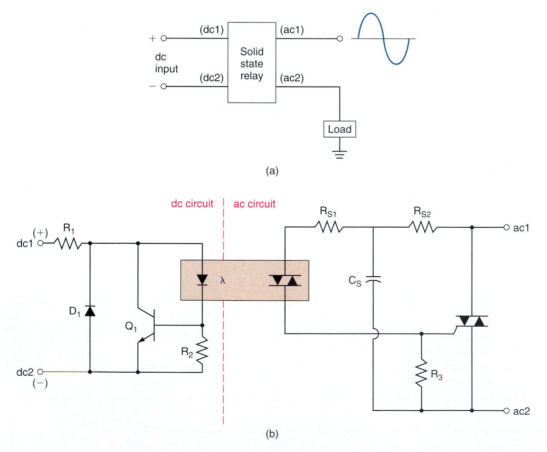

(a)

(b)

**FIGURE 20.48   The solid-state relay.**

| H21A1 |
|---|
| **H21A2** |
| **H21A3** |
| **H22A1** |
| **H22A2** |
| **H22A3** |

## Slotted Optical Switches
### Transistor Output

Each device consists of a gallium arsenide infrared emitting diode facing a silicon NPN phototransistor in a molded plastic housing. A slot in the housing between the emitter and the detector provides the means for mechanically interrupting the infrared beam. These devices are widely used as position sensors in a variety of applications.

- Single Unit for Easy PCB Mounting
- Non-Contact Electrical Switching
- Long-Life Liquid Phase Epi Emitter
- 1 mm Detector Aperture Width

| SLOTTED OPTICAL SWITCHES TRANSISTOR OUTPUT |
|---|

### MAXIMUM RATINGS

| Rating | Symbol | Value | Unit |
|---|---|---|---|
| **INPUT LED** | | | |
| Reverse Voltage | $V_R$ | 6 | Volts |
| Forward Current — Continuous | $I_F$ | 60 | mA |
| Input LED Power Dissipation @ $T_A$ = 25°C Derate above 25°C | $P_D$ | 150 2 | mW mW/°C |
| **OUTPUT TRANSISTOR** | | | |
| Collector-Emitter Voltage | $V_{CEO}$ | 30 | Volts |
| Output Current — Continuous | $I_C$ | 100 | mA |
| Output Transistor Power Dissipation @ $T_A$ = 25°C Derate above 25°C | $P_D$ | 150 2 | mW mW/°C |
| **TOTAL DEVICE** | | | |
| Ambient Operating Temperature Range | $T_A$ | −55 to +100 | °C |
| Storage Temperature | $T_{stg}$ | −55 to +100 | °C |
| Lead Soldering Temperature (5 seconds max) | — | 260 | °C |
| Total Device Power Dissipation @ $T_A$ = 25°C Derate above 25°C | $P_D$ | 300 4 | mW mW/°C |

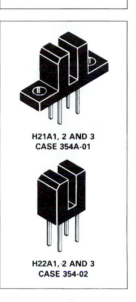

H21A1, 2 AND 3
CASE 354A-01

H22A1, 2 AND 3
CASE 354-02

**FIGURE 20.49** The **H21-2** series optointerruptors specification sheet. (Copyright of Motorola. Used by permission.)

also available in integrated form. While the exact circuitry of the integrated SSR varies from the circuitry shown in Figure 20.48b, the overall principle of operation is the same.

## *Optointerrupters*

The **optointerrupter** or **optical switch** is an IC optocoupler designed to allow an outside object to block the light path between the photoemitter and the photodetector. Two of the many common optointerrupter configurations are shown on the spec sheet in Figure 20.49. Note that the photoemitter is on one side of the slot, while the photodetector is on the other side.

The open gap between the photoemitter and the photodetector is the primary difference between the optointerrupter and the optoisolator. This gap allows the optointerrupter to be activated by some outside object, such as a piece of paper. For example, let's say that the optointerrupter in Figure 20.49 is used in a photocopying machine and that the device is placed so that the edge of a given copy passes through the slot as it (the paper) passes through the machine. When no paper is present in the optointerrupter gap, the light from the emitter reaches the phototransistor, causing it to saturate. Thus, the out-

**Optointerrupter** or **optical switch**
An IC optocoupler designed to allow an outside object to block the light path between the photoemitter and the photodetector.

put from the phototransistor is *low*. When a piece of paper passes through the gap, the emitter light is blocked (by the paper) from the phototransistor. This causes the phototransistor to go into cutoff, causing its collector voltage to go *high*. The low and high output voltages from the phototransistor are used to tell the machine whether or not a piece of paper has passed the point where the optointerrupter is located.

The type of application discussed here is the primary application for the optointerruptor. Any other type of optical-coupling application is usually performed by a discrete optocoupler or an optoisolator.

## Section Review

1. What is an *optoisolator*?
2. List and define the common *isolation* parameters and ratings of the optoisolator.
3. Describe the operation of the *solid-state relay (SSR)*.
4. Contrast the *optointerrupter* with the *optoisolator*.

## Key Terms

The following terms were introduced and defined in this chapter:

anode current interruption
average on-state current ($I_T$)
average on-state voltage ($V_T$)
bidirectional thyristor
circuit fusing rating ($I^2t$)
critical rise rating ($dv/dt$)
crowbar
dark current ($I_D$)
diac
false triggering
forced commutation
forward blocking region
forward breakover current ($I_{BR(F)}$)
forward breakover voltage ($V_{BR(F)}$)
forward operating region
gate nontrigger voltage ($V_{GD}$)
gate trigger voltage ($V_{GT}$)
holding current ($I_H$)
industrial electronics
interbase resistance

intrinsic standoff ratio ($\eta$)
isolation capacitance ($C_{ISO}$)
isolation current ($I_{ISO}$)
isolation resistance ($R_{ISO}$)
isolation source voltage ($V_{ISO}$)
light
light-activated SCR (LASCR)
light current ($I_L$)
light detectors
light emitters
light intensity
negative resistance
negative resistance region
nonrepetitive surge current ($I_{TSM}$)
opaque
optoelectronic devices
optointerruptor
optoisolator
peak current ($I_P$)
peak voltage ($V_P$)

phase controller
photo-Darlington
photodiode
phototransistor
programmable unijunction transistor (PUT)
relaxation oscillator
sensitivity
silicon-controlled rectifier (SCR)
silicon unilateral switch (SUS)
snubber network
soldering temperature
solid-state relay (SSR)
spectral response
thyristors
triac
trigger-point voltage ($V_{TP}$)
unijunction transistor (UJT)
wavelength ($\lambda$)
wavelength of peak spectral response ($\lambda_S$)

| (20.1) | $I_F = \dfrac{V_S - V_F}{R_S}$ | 20.1 |
|---|---|---|
| (20.2) | $t_{max} = \dfrac{I^2t \text{ (rated)}}{I_S^2}$ | 20.2 |
| (20.3) | $I_{S(max)} = \sqrt{\dfrac{I^2t \text{ (rated)}}{t_S}}$ | 20.2 |
| (20.4) | $\Delta V = \dfrac{dv}{dt}\,\Delta t$ | 20.2 |
| (20.5) | $V_k = V_{BB}\dfrac{R_{B1}}{R_{B1} + R_{B2}}$ | 20.4 |
| (20.6) | $V_k = \eta V_{BB}$ | 20.4 |
| (20.7) | $V_P = \eta V_{BB} + 0.7 \text{ V}$ | 20.4 |
| (20.8) | $\lambda = \dfrac{c}{f}$ | 20.5 |

**Answers to the Example Practice Problems**

**20.1.** For the application, $I^2t = 54 \text{ A}^2\text{s}$.
**20.2.** 3.3 ms
**20.3.** 38.73 A
**20.4.** 2.5 mV
**20.5.** 9.7 V
**20.6.** 750 nm to 10,000 nm (10 μm)
**20.7.** 319.1 THz

**Practice Problems**

*§ 20.2*

1. The 2N6237 SCR has a circuit fusing rating of 2.6 $\text{A}^2\text{s}$. Determine whether or not the component can withstand a 25 A surge that lasts for 120 ms.
2. The 2N6342 SCR has a circuit fusing rating of 40 $\text{A}^2\text{s}$. Determine whether or not the device can withstand an 80 A surge that lasts for 8 ms.
3. The 2N6237 SCR has a circuit fusing rating of 2.6 $\text{A}^2\text{s}$. Determine the maximum allowable duration of a 50 A surge through the device.
4. The 2N6342 has a circuit fusing rating of 40 $\text{A}^2\text{s}$. Determine the maximum allowable duration of a 95 A surge through the device.
5. The 2N6237 has a circuit fusing rating of 2.6 $\text{A}^2\text{s}$. Determine the maximum surge current that the device can withstand for 50 ms.
6. The 2N6342 has a circuit fusing rating of 40 $\text{A}^2\text{s}$. Determine the maximum surge current that the device can withstand for 100 ms.
7. The 2N6237 has a critical rise rating of $dv/dt = 10$ V/μs. Determine the anode noise amplitude at $t_r = 10$ ns required to cause false triggering.
8. The 2N6342 has a critical rise rating of $dv/dt = 5$ V/μs. Determine the anode noise amplitude at $t_r = 25$ ns required to cause false triggering.

9. The 2N2646 has a range of $\eta$ = 0.56 to 0.75. Determine the range of $V_P$ values for the device when $V_{BB}$ = +14 V.

10. The 2N4871 UJT has a range of $\eta$ = 0.7 to 0.85. Determine the range of $V_P$ values for the device when $V_{BB}$ = +12 V.

11. The 2N4948 UJT has a range of $\eta$ = 0.55 to 0.82. Determine the range of $V_P$ values for the device when $V_{BB}$ = +16 V.

12. The 2N4949 UJT has a range of $\eta$ = 0.74 to 0.86. Determine the range of $V_P$ values for the device when $V_{BB}$ = +26 V.

### § 20.5

13. Determine the wavelength (in nm) for a 650 THz signal.

14. Determine the wavelength (in nm) for a 220 THz signal.

15. Determine the wavelength (in nm) for a 180 THz signal.

16. The visible light spectrum includes all frequencies between approximately 400 THz and 800 THz. Determine the range of wavelength values for this band of frequencies.

17. A photodetector has a value of $\lambda_S$ = 0.94 nm. Determine the optimum operating frequency for the device.

18. A photodetector has a rating of $\lambda_S$ = 0.84 nm. Determine the optimum operating frequency for the device.

**The Brain Drain**

19. The spec sheet for the SCR in Figure 20.50 is given in Figure 20.11. The output short-circuit current for the dc power supply is 4 A and will remain at that level until its primary fuse blows. For the circuit, determine the following:

   a.  The maximum value of $V_{dc}$ required to trigger the SCR.

   b.  The maximum value of zener current at the instant that the SCR triggers.

   c.  Whether the fuse or the SCR will open first.

20. Refer to Figure 20.51. $R_2$ in the circuit is set to the value shown. The spec sheet for the 2N682 is given in Figure 20.11. Determine the conduction angle for the circuit load. (*Hint:* Start by finding the maximum value of $v_{in}$ required to trigger the SCR.)

21. Refer back to Figure 20.35. The relaxation oscillator shown contains a  51 V zener diode. The UJT has a range of $\eta$ = 0.7 to 0.82. Determine the minimum value of $V_E$ required to trigger the UJT.

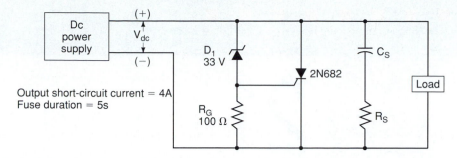

**FIGURE 20.50**

Output short-circuit current = 4A
Fuse duration = 5s

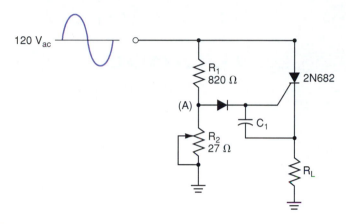

**FIGURE 20.51**

# 21

# DISCRETE AND INTEGRATED VOLTAGE REGULATORS

**A**s you know, a lot of research is now being done in the area of superconductivity. The properties of superconductivity, like the one shown here, may completely change the field of electronics someday.

# OUTLINE

# OBJECTIVES

*After studying the material in this chapter, you should be able to:*

1. List the purposes served by a voltage regulator.

2. Define *line regulation* and its commonly used units of measure.

3. Define *load regulation* and its commonly used units of measure.

4. Describe and analyze the operation of the *pass-transistor regulator.*

5. Describe the means by which *short-circuit protection* is provided for a pass-transistor regulator.

6. Describe and analyze the operation of the *shunt-feedback regulator.*

7. Discuss the need for *overvoltage protection* in a shunt-feedback regulator and the means by which it is provided.

8. List the reasons that the series regulator is preferred over the shunt regulator.

9. List and describe the various types of linear IC voltage regulators.

10. List and describe the commonly used linear IC regulator parameters and ratings.

11. Describe the fundamental operating difference between the linear regulator and the switching regulator.

12. List the four parts of the basic switching regulator and describe the function of each of them.

13. Describe the response of a switching regulator to a change in load demand.

14. Describe the two most commonly used methods for controlling power switch conduction.

*(continued)*

925

**15.** List the circuit recognition features for each of the common switching regulator configurations.

**16.** Describe the functions performed by a typical switching regulator IC.

**17.** In terms of their advantages, disadvantages, and applications, compare and contrast linear and switching regulators.

In Chapter 3, you were introduced to the operation of the basic power supply. At that time, you were shown that a *rectifier* is used to convert the ac output of the transformer to *pulsating dc*. This pulsating dc is then applied to a *filter*, which reduces the variations in the rectifier output. Finally, the filtered dc is applied to a *voltage regulator*. This regulator serves two purposes:

*OBJECTIVE 1* ▶ **1.** It reduces the variations (ripple) in the filtered dc.

**2.** It maintains a constant output voltage regardless of minor changes in the load current demand and/or input voltage.

The block diagram of the basic power supply is shown (along with its waveforms) in Figure 21.1.

The voltage regulator we used in Chapter 3 was simply a *zener diode* connected in parallel with the load. In practice, this type of regulation (known as *brute-force regulation*) is rarely used. The primary problem with the simple zener regulator is the fact that the zener wastes a tremendous amount of power.

Practical voltage regulators contain a number of discrete and/or integrated active devices. In this chapter, we will look at the operation of many of these more practical voltage regulators.

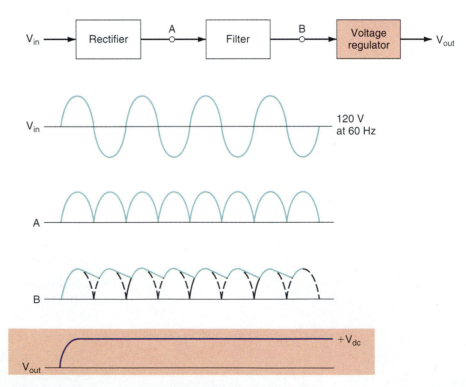

**FIGURE 21.1** **Basic power supply block diagram and waveforms.**

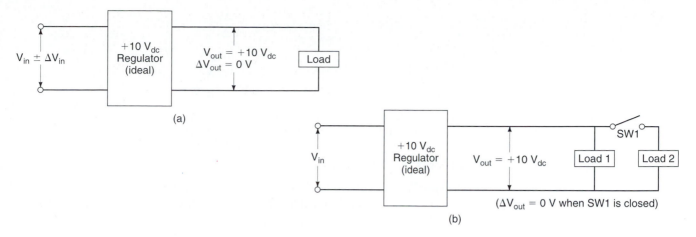

FIGURE 21.2 A +10 $V_{dc}$ regulator.

# 21.1 VOLTAGE REGULATION: AN OVERVIEW

The *ideal* voltage regulator maintains a constant dc output voltage, regardless of changes in either its input voltage or its load current demand. For example, consider the +10 $V_{dc}$ regulator shown in Figure 21.2. A change in the regulator input voltage (shown as $\Delta V_{in}$ in Figure 21.1a) would not be coupled to the regulator output. Note that $\Delta V_{in}$ could be either a change in the steady-state (dc) value of $V_{in}$ or could be some ripple voltage. In either case, the change in $V_{out}$ for the ideal regulator would be 0 V. This assumes, of course, that the value of $V_{in}$ does not decrease below the value required to maintain the operation of the regulator. (A regulator cannot have a 10 $V_{dc}$ output if the input is less than 10 $V_{dc}$.)

Figure 21.2b shows how the ideal voltage regulator would respond to a change in load current demand. Assume that *load 1* and *load 2* have the same resistance. When SW1 is closed, the added load causes the current demand on the regulator to *double*. The ideal voltage regulator maintains a constant output voltage ($\Delta V_{out} = 0$) despite the change in load current demand.

## Line Regulation

In practice, a change in the input voltage to a regulator *does* cause a change in its output voltage. The **line regulation** rating of a voltage regulator indicates the change in output voltage that occurs per unit change in the input voltage. The line regulation of a voltage regulator is found as

◄ *OBJECTIVE 2*

**Line regulation**
A rating that indicates the change in regulator output voltage that will occur per unit change in input voltage.

$$\text{line regulation} = \frac{\Delta V_{out}}{\Delta V_{in}} \qquad (21.1)$$

where $\Delta V_{out}$ = the change in output voltage (usually in microvolts or millivolts)
$\Delta V_{in}$ = the change in input voltage (usually in volts)

Example 21.1 demonstrates the calculation of line regulation for a voltage regulator.

A voltage regulator experiences a 10 μV change in its output voltage when its input voltage changes by 5 V. Determine the value of line regulation for the circuit.

*EXAMPLE 21.1*

*Solution:* The line regulation of the circuit is found as

$$\text{line regulation} = \frac{\Delta V_{\text{out}}}{\Delta V_{\text{in}}}$$

$$= \frac{10\ \mu\text{V}}{5\ \text{V}}$$

$$= 2\ \mu\text{V/V}$$

The 2 $\mu$V/V rating means that the output voltage will change by 2 $\mu$V for every 1 V change in the regulator's input voltage.

### PRACTICE PROBLEM 21.1

The change in output voltage for a voltage regulator is measured at 100 $\mu$V when the input voltage changes by 4 V. Calculate the line regulation rating for the regulator.

As was stated earlier, the *ideal* voltage regulator would have a $\Delta V_{\text{out}}$ of 0 V when the input voltage changes. Thus, for the ideal voltage regulator:

$$\text{line regulation} = \frac{\Delta V_{\text{out}}}{\Delta V_{\text{in}}}$$

$$= \frac{0\ \text{V}}{\Delta V_{\text{in}}}$$

$$= 0$$

Based on the fact that the ideal line regulation rating is *zero,* we can make the following statement: *The lower the line regulation rating of a voltage regulator, the higher the quality of the circuit.*

Line regulation is commonly expressed in a variety of units. The $\mu V/V$ rating used in Example 21.1 is only one of them. The commonly used line regulation units and their meanings are listed in Table 21.1.

## Load Regulation

**Load regulation**
A rating that indicates the change in regulator output voltage per unit change in load current.

The practical voltage regulator also experiences a slight change in output voltage when there is a change in load current demand. The **load regulation** rating indicates the change in regulator output voltage that occurs per unit change in load current. The load regulation of a voltage regulator is found as

*A Practical Consideration:*
Technically, line regulation should have no unit of measure since it is the ratio of one voltage value to another. However, the units listed in Table 21.1 are commonly used on spec sheets, so you need to know what they are.

**TABLE 21.1  Commonly Used Line Regulation Units**

| Unit | Meaning |
|------|---------|
| $\mu$V/V | The change in output voltage (in microvolts) per 1 V change in input voltage |
| ppm/V | *Parts-per-million per volt.* Another way of saying *microvolts per volt* |
| %/V | The percent of change in the output voltage that can occur per 1 V change in the input voltage |
| % | The total percent of change in output voltage that can occur over the rated range of input voltages |
| mV (or $\mu$V) | The actual change in output voltage that can occur over the rated range of input voltages |

$$\text{load regulation} = \frac{V_{NL} - V_{FL}}{\Delta I_L} \qquad \textbf{(21.2)}$$

where $V_{NL}$ = the no-load output voltage (the output voltage when the load is open)
$V_{FL}$ = the full-load output voltage (the output voltage when the load current demand is at its maximum value)
$\Delta I_L$ = the *change in* load current demand

Another way of expressing this is

$$\text{load regulation} = \frac{\Delta V_{out}}{\Delta I_L}$$

Example 21.2 demonstrates the calculation of load regulation for a given voltage regulator.

---

A voltage regulator is rated for an output current of $I_L$ = 0 to 20 mA. Under no-load conditions, the output voltage from the circuit is 5 V. Under full-load conditions, the output voltage from the circuit is 4.9998 V. Determine the value of load regulation for the circuit.

*Solution:* The load regulation of the regulator is found as

$$\text{load regulation} = \frac{V_{NL} - V_{FL}}{\Delta I_L}$$

$$= \frac{5\ V - 4.9998\ V}{20\ mA}$$

$$= 10\ \mu V/mA$$

The 10 $\mu$V/mA rating indicates that the output voltage changes by 10 $\mu$V for each 1-mA change in load current.

**EXAMPLE 21.2**

*PRACTICE PROBLEM 21.2*

A voltage regulator is rated for an output current of $I_L$ = 0 to 40 mA. Under no-load conditions, the output voltage from the circuit is 8 V. Under full-load conditions, the output voltage from the circuit is 7.996 V. Determine the value of load regulation for the circuit.

---

Like the line regulation rating, load regulation is commonly expressed in a variety of units. The commonly used load regulation units and their meanings are listed in Table 21.2.

The ideal voltage regulator would not experience a change in output voltage when the load current demand increases. Thus, for the ideal regulator, $V_{NL} = V_{FL}$, and load regulation equals *zero*. Based on this fact, we can state that *the lower the load regulation rating of a voltage regulator, the higher the quality of the circuit.*

## Line and Load Regulation: Some Practical Considerations

Some manufacturers combine the *line regulation* and *load regulation* ratings into a single **regulation** rating. This rating indicates the *maximum* change in output voltage that can occur when input voltage *and* load current are varied over their entire rated ranges. For example, a given voltage regulator has the following ratings:

**Regulation**
A rating that indicates the maximum change in regulator output voltage that can occur when input voltage *and* load current are varied over their entire rated ranges.

**TABLE 21.2   Commonly Used Load Regulation Units**

| Unit | Meaning |
|---|---|
| $\mu V/mA$ | The change in output voltage (in microvolts) per 1 mA change in load current |
| %/mA | The percent of change in the output voltage per 1 mA change in load current |
| % | The total percent of change in output voltage over the rated range of load current values |
| mV (or $\mu V$) | The actual change in output voltage that can occur over the rated range of load current values |
| $\Omega$ | V/mA, expressed as a resistance value; the rating times 1 mA gives you the V/mA rating of the regulator |

$$V_{in} = 12 \text{ to } 24 \text{ V}_{dc}$$
$$I_L = 40 \text{ mA (maximum)}$$
$$\text{regulation} = 0.33\%$$

These ratings indicate that the output voltage of the regulator will vary by no more than 0.33 percent as long as $V_{in}$ remains between 12 and 24 $V_{dc}$ and load current does not exceed 40 mA. A single regulation rating is always given either as a percentage, or in millivolts or microvolts.

There is one other important consideration we need to discuss. The unit of measure used for line regulation and/or load regulation is a good indicator of the quality of that regulator. For example, let's say that we need to choose between two voltage regulators. These regulators have the following ratings:

| Regulator | dc Output Voltage | Line Regulation |
|---|---|---|
| A | +10 V | 0.12 %/V |
| B | +10 V | 40 $\mu V/V$ |

If the input voltage to regulator A increases by 5 V, the output voltage of the circuit will be

$$V_{out} = V_{dc} + (V_{dc})(0.12\%)(\Delta V_{in})$$
$$= 10 \text{ V} + 0.06 \text{ V}$$
$$= 10.06 \text{ V}$$

If the input voltage to regulator B increases by 5 V, the output voltage of the circuit will be

$$V_{out} = 10 \text{ V} + (40 \text{ } \mu V)(\Delta V_{in})$$
$$= 10 \text{ V} + 200 \text{ } \mu V$$
$$= 10.0002 \text{ V}$$

As you can see, the regulator with the $\mu V/V$ rating had a smaller change in output voltage than the regulator with the %/V rating. Since the ideal voltage regulator would have no change in output voltage when the input voltage changed, regulator B came much closer to the ideal regulator than regulator A.

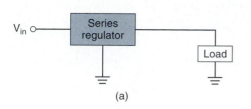

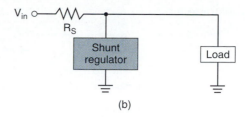

FIGURE 21.3    Series and shunt regulators.

## Types of Regulators

There are two basic types of discrete voltage regulators: the *series regulator* and the *shunt regulator*. These two types of regulator are represented by the blocks in Figure 21.3. The **series regulator** is placed in *series* with the load, as shown in Figure 21.3a. The **shunt regulator** is placed in *parallel* with the load, as shown in Figure 21.3b. As you will see, series and shunt regulators each have their own advantages and disadvantages.

**Series regulator**
A voltage regulator in series with the load.

**Shunt regulator**
A voltage regulator in parallel with the load.

*Section Review*

1. What two purposes are served by a voltage regulator?
2. How would the *ideal* voltage regulator respond to a change in input voltage?
3. How would the *ideal* voltage regulator respond to a change in load current demand?
4. What is *line regulation*?
5. What is the line regulation value of an ideal voltage regulator?
6. What is the relationship between the quality of a voltage regulator and its line regulation rating?
7. List and define the commonly used line regulation ratings.
8. What is *load regulation*?
9. List and define the commonly used load regulation ratings.
10. What is the load regulation value of an ideal voltage regulator?
11. What is the relationship between the quality of a voltage regulator and its load regulation rating?
12. Explain the single *regulation* rating.
13. What is a *series regulator*?
14. What is a *shunt regulator*?

## 21.2  SERIES VOLTAGE REGULATORS

Series voltage regulators can take a variety of forms. However, they all have one or more active devices placed in series with the load. In this section, we will take a look at several of the commonly used series voltage regulators.

## Pass-Transistor Regulator

The **pass-transistor regulator** uses a series transistor, called a *pass-transistor,* to regulate load voltage. The term *pass transistor* comes from the fact that the load current passes through the series transistor, $Q_1$, as shown in Figure 21.4.

◄  *OBJECTIVE 4*

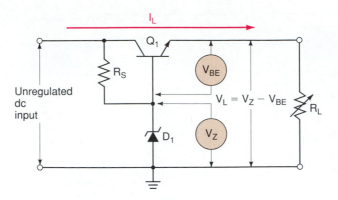

**FIGURE 21.4  Pass-transistor regulator.**

**Pass-transistor regulator**
A regulator that uses a series transistor to maintain a constant load voltage.

The key to the operation of the pass-transistor regulator is the fact that the base voltage is held to the relatively constant voltage across the zener diode. For example, if a 9.1-V zener diode is used, the base voltage of $Q_1$ is held at approximately 9.1 V. Since $Q_1$ is an *npn* transistor, $V_L$ is found as

$$V_L = V_Z - V_{BE} \tag{21.3}$$

*Don't Forget:* The emitter voltage of a conducting *npn* transistor is *less than* the base voltage.

Equation (21.3) is important because it provides the basis for explaining the response of the pass-transistor regulator to a change in load resistance. If load resistance increases, load voltage starts to increase. The increase in load voltage (and thus $V_E$) results in a decrease in $V_{BE}$ (since $V_B$ is constant). The decrease in $V_{BE}$ reduces the conduction of the pass transistor, causing its value of $V_{CE}$ to increase. The increase in $V_{CE}$ offsets the initial increase in $V_E$, and a relatively constant load voltage is maintained.

In a similar fashion, a decrease in load resistance causes $V_E$ to start to decrease. The decrease in $V_E$ causes $V_{BE}$ to increase, increasing conduction through the pass transistor. This increased conduction causes $V_{CE}$ to decrease, offsetting the initial decrease in $V_E$. Again, a relatively constant load voltage is maintained by the regulator.

*What Purpose Is Served by $R_S$?* In many cases, the zener requires more current (to maintain regulation) than the transistor base can allow. $R_S$ provides an alternate path for this additional zener current.

The pass-transistor regulator in Figure 21.4 has relatively good line and load regulation characteristics, but there is a problem with the circuit. If the input voltage or load current values increase, the zener diode may have to dissipate a relatively high amount of power. You see, increases in $V_{in}$ or $I_L$ result in an increase in zener conduction since more current is drawn through $R_S$. The increased zener conduction results in higher power dissipation by the device. This problem, however, is reduced by the use of the *Darlington pass-transistor regulator.*

## Darlington Pass-Transistor Regulator

**Darlington pass-transistor regulator**
A series regulator that uses a Darlington pair in place of a single pass transistor.

The **Darlington pass-transistor regulator** uses a Darlington pair, $Q_1$ and $Q_2$, in place of the single pass transistor, as shown in Figure 21.5. The load voltage for the Darlington circuit is found as

$$V_L = V_Z - 2V_{BE} \tag{21.4}$$

Since the total current gain of a Darlington pair is equal to $h_{FE1}h_{FE2}$, an increase in load current causes very little, if any, increase in zener current. Therefore, the regulator in Figure 21.5 is not subject to the same power concerns as the regulator in Figure 21.4. However, since zener conduction *is* affected by temperature, and the zener current is applied to the high-current-gain Darlington pair, the load current in the Darlington pass-transistor regulator can be severely affected by significant increases in operating temperature. For this reason, the Darlington pass-transistor regulator must be kept relatively cool.

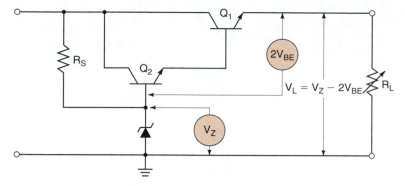

**FIGURE 21.5** Darlington pass-transistor regulator.

## Series Feedback Regulator

The **series feedback regulator** uses an *error detector* to improve on the line and load regulation characteristics of the other pass-transistor regulators. The block diagram for the series feedback regulator is shown in Figure 21.6.

    The error detector receives two inputs: a *reference voltage* derived from the unregulated dc input voltage and a *sample voltage* from the regulated output voltage. The error detector compares the reference and sample voltages and provides an output voltage that is proportional to the difference between the two. This output voltage is amplified and used to drive the pass-transistor regulator.

    The series feedback regulator is capable of responding very quickly to differences between its sample and reference input voltages. This gives the circuit much better line and load regulation characteristics than the other circuits we have discussed in this section.

    The schematic diagram for the series feedback regulator is shown in Figure 21.7. The *sample and adjust circuit* is the voltage divider that consists of $R_3$, $R_4$, and $R_5$. The reference voltage is set by the zener diode. $Q_2$ detects and amplifies the difference between the reference and sample voltages and adjusts the conduction of the pass transistor accordingly.

**Series feedback regulator**
A series regulator that uses an error-detection circuit to improve the line and load regulation characteristics of other pass-transistor regulators.

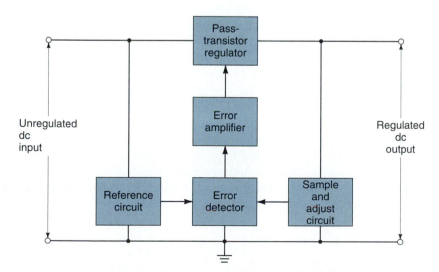

**FIGURE 21.6** Block diagram for the series feedback regulator.

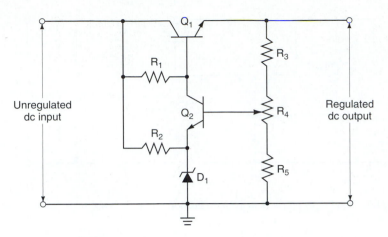

**FIGURE 21.7** Schematic diagram for a basic series feedback regulator.

How the circuit responds to an increase in load resistance.

If the load resistance of the feedback regulator increases, $V_L$ starts to increase. This causes the voltage at the base of $Q_2$ to increase. Since the emitter voltage of $Q_2$ is clamped to the value of $V_Z$, the increase in $V_{B(Q2)}$ increases the conduction through $Q_2$ and its collector resistor ($R_1$). The increased conduction causes $V_{C(Q2)}$ to decrease, which reduces the value of $V_{B(Q1)}$. The reduction in $V_{B(Q1)}$ reduces the conduction through the pass transistor, causing $V_E$ to decrease. The decrease in $V_E$ offsets the initial increase caused by the load.

How the circuit responds to a decrease in load resistance.

If the load resistance decreases, $V_L$ starts to decrease. The decrease in $V_L$ causes $V_{B(Q2)}$ to decrease, which reduces the conduction through the transistor. The reduced conduction through $Q_2$ causes $V_{C(Q2)}$ to increase, increasing $V_{B(Q1)}$. The increase in $V_{B(Q1)}$ causes the pass-transistor conduction to increase. This increases the value of $V_E$, offsetting the initial decrease in load voltage. Again, load regulation is maintained.

## Short-Circuit Protection

*OBJECTIVE 5* ▶ The primary drawback of any series regulator is the fact that the pass transistor can be destroyed by excessive load current if the load is shorted. To prevent a shorted load from destroying the pass transistor, a *current-limiting circuit* is usually added to the regulator, as shown in Figure 21.8.

The current-limiting circuit consists of a transistor ($Q_3$) and a series resistor ($R_S$) that is connected between its base and emitter terminals. In order for $Q_3$ to conduct, the voltage across $R_S$ must reach approximately 0.7 V. This happens when

$$I_L = \frac{0.7\ \text{V}}{1\ \Omega}$$

$$= 700\ \text{mA}$$

Thus, if load current is less than 700 mA, $Q_3$ is in cutoff and the circuit acts exactly as described earlier. If the load current increases above 700 mA, $Q_3$ conducts, decreasing the voltage at the base of $Q_1$. The decreased $V_{B(Q1)}$ reduces the conduction of the pass transistor, preventing any further increases in load current. Thus, the load current for the circuit is limited to approximately 700 mA. In fact, for the type of current-limited regulator shown, the maximum load current can be found as

$$I_{L(\text{max})} \cong \frac{V_{BE(Q3)}}{R_S} \tag{21.5}$$

Thus, the maximum allowable load current for the regulator can be set to any value by using the appropriate value of $R_S$.

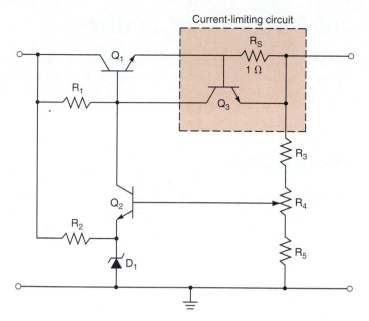

Current-limiting circuit

FIGURE 21.8   Series regulator current-limiting circuit.

You may be wondering why we don't simply use a single current-limiting resistor to protect the pass transistor. If a current-limiting resistor is used, it must have a fairly significant value in order to protect the pass transistor. The voltage drop across such a resistor would be much higher than that caused by a 1-$\Omega$ resistor. The current-limiting circuit is used to provide short-circuit protection without causing a significant drop in output voltage from the regulator.

## One Final Note

The fact that a shorted load can damage a pass transistor is not the only drawback to the series regulator. For one thing, the fact that $I_L$ passes through the pass transistor means that this transistor dissipates a fairly significant amount of power. Also, there is a voltage drop across the pass transistor. This voltage drop reduces the maximum possible output voltage.

Even with its power dissipation and short-circuit protection problems, the series regulator is used more commonly than the shunt regulator. As you will see in the next section, the shunt regulator has several problems that make the series regulator the more desirable of the two.

*Section Review*

1. Describe the basic *pass-transistor regulator.*
2. Describe how the pass-transistor regulator in Figure 21.4 responds to a change in load resistance.
3. Explain how the *Darlington pass-transistor regulator* reduces the problem of excessive zener diode power dissipation.
4. Describe the response of the *series feedback regulator* to a change in load resistance.
5. Describe the operation of the current-limiting circuit in Figure 21.8.
6. List the disadvantages of using series voltage regulators.

# 21.3 SHUNT VOLTAGE REGULATORS

The most basic shunt regulator is the simple zener regulator that we discussed in Chapter 3. As you know, this regulator wastes far too much power for most applications and thus is rarely used. In this section, we will discuss the *shunt feedback regulator*. As you will see, this regulator is very similar to the series feedback regulator we covered in the last section.

## Shunt Feedback Regulator

OBJECTIVE 6 ▶

**Shunt feedback regulator**
A circuit that uses an error detector to control the conduction of a *shunt* transistor.

The **shunt feedback regulator** uses an error detector to control the conduction of a *shunt transistor*. This shunt transistor is shown as $Q_1$ in the shunt feedback regulator in Figure 21.9.

The *sample circuit* is (once again) a simple voltage-divider circuit. The *reference circuit* is made up of $D_1$ and $R_1$. The outputs from the sample and reference circuits are applied to the *error detector/amplifier*, $Q_2$. The output from $Q_2$ is then used to control the conduction of the shunt transistor, $Q_1$.

The operation of the shunt feedback regulator is easiest to understand if we view the shunt transistor ($Q_1$) as a variable resistor, $r_{ce}$. When $Q_1$ is not conducting, $r_{ce}$ is at its maximum value. When $Q_1$ is saturated, $r_{ce}$ is at its minimum value. Thus, we can say that *the value of $r_{ce}$ is inversely proportional to the conduction of $Q_1$.*

Under normal circumstances, $R_3$ is set so that $Q_2$ is conducting. Since $I_{B1} = I_{C2}$, $Q_1$ is biased somewhere in its active region of operation. The conduction of the two-transistor circuit is set so that $r_{ce}$ is approximately midway between its two extremes in value.

How the circuit responds to a decrease in load resistance.

If the load resistance decreases, $V_L$ also starts to decrease. This decrease causes $V_{B(Q2)}$ to decrease. Since $V_{E(Q2)}$ is set to a fixed value by the zener diode, $V_{BE(Q2)}$ decreases. The reduction in $V_{BE(Q2)}$ reduces the conduction through the component, decreasing $I_{B1}$. This causes $r_{ce}$ to increase, increasing the value of $V_{C(Q1)}$. Since $V_L = V_{C(Q1)}$, the increase in $V_{C(Q1)}$ offsets the initial decrease in $V_L$.

How the circuit responds to an increase in load resistance.

If $R_L$ increases in value, $V_L$ starts to increase. The increase in $V_L$ increases $V_{B(Q2)}$, and thus $V_{BE(Q2)}$. This increases the conduction through $Q_2$, causing an increase in $I_{B(Q1)}$. The increase in $I_{B(Q1)}$ causes the shunt transistor's conduction to increase, reducing the value of $r_{ce}$. When $r_{ce}$ decreases, $V_{C(Q1)}$ decreases, offsetting the initial increase in $V_L$.

The purpose served by $R_S$.

The series resistor ($R_S$) combines with $r_{ce}$ to form a variable voltage-divider that provides regulation control. In addition, $R_S$ provides shorted-load protection for the power

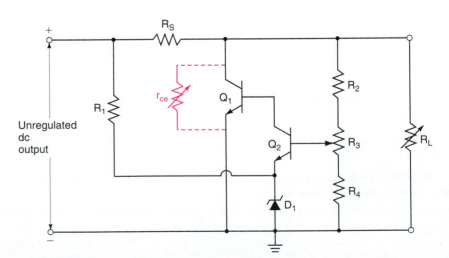

**FIGURE 21.9   Shunt transistor in a shunt feedback regulator.**

supply circuitry. Normally, this resistor must be a high-wattage resistor, since the regulator current and $I_L$ must pass through it.

## Overvoltage Protection

◄ OBJECTIVE 7

Just as the series regulator must be protected against shorted-load conditions, the shunt regulator must be protected from input overvoltage conditions. If the unregulated dc input voltage to the regulator increases, the conduction of the shunt transistor increases to maintain the constant output voltage. Assuming that $V_{CE}$ of $Q_1$ remains relatively constant, the increased conduction through the transistor will cause an increase in its power dissipation.

Several things can be done to ensure that any increase in unregulated dc input voltage will not destroy the shunt transistor. First, you can use a transistor whose $P_{D(max)}$ rating is far greater than the maximum power dissipation that would ever be required in the circuit. For example, assume that the regulator in Figure 21.9 has an input voltage that can go as high as 20 V and that the regulator is designed to provide a regulated 10 $V_{dc}$ output. The worst-case value of transistor current could be approximated as

$$I_{C(max)} = \frac{V_{in} - V_{dc}}{R_s}$$

Assuming that a 100-Ω resistor is used, $I_{C(max)}$ would be

$$I_{C(max)} = \frac{20\ V - 10\ V}{100}$$
$$= 100\ mA$$

Since the transistor is dropping 10 V across its collector–emitter terminals, the maximum power dissipation would be found as

$$P_D = V_{CE}I_{C(max)}$$
$$= (10\ V)(100\ mA)$$
$$= 1\ W$$

Thus, for the circuit described, any transistor with a $P_{D(max)}$ rating that is *greater than* 1 W could be used.

Of course, there are always circumstances where the unregulated dc input voltage can exceed its maximum *rated* value. To protect the circuit from such a circumstance, a *crowbar* circuit can be added to the input of the regulator. You may recall (from Chapter 20) that a *crowbar* is a circuit that uses an SCR to protect its load from an overvoltage condition. If a crowbar is added to the regulator input, the circuit is protected from any extreme input overvoltage condition that may arise.

Another practical consideration involves the potentiometer, $R_3$. This potentiometer is included in the circuit to provide an adjustment for the dc output voltage. Adjusting $R_3$ varies the condition of $Q_1$, and thus the value of $r_{ce}$. Since this resistance forms a voltage divider with $R_S$, varying $R_3$ sets the value of the regulated dc output voltage.

## One Final Note

◄ OBJECTIVE 8

The series voltage regulator is preferred over the shunt regulator for several reasons. The primary problem with the shunt regulator is the fact that a fairly significant portion of the total current through $R_S$ goes through the shunt transistor, rather than to the load. Another problem involves the voltage drop across $R_S$ and the resulting power dissipation. Finally, realize that an input overvoltage condition is far more likely to occur than the shorted-load condition of the series regulator. In other words, the fault that can damage the shunt regulator is more likely to occur than the fault that can damage the series regulator. When

*Don't Forget:* High-voltage problems are far more common than shorted-component problems.

you consider all these drawbacks, it is easy to understand why the series regulator is preferred over the shunt regulator.

1. Describe the response of the shunt feedback regulator to a decrease in load resistance.
2. Describe the response of the shunt feedback regulator to an increase in load resistance.
3. Explain how an input overvoltage condition can destroy the shunt transistor in a shunt feedback regulator.
4. Discuss the means by which a shunt regulator can be protected from an input overvoltage problem.
5. Why are series regulators preferred over shunt regulators?

## 21.4 LINEAR IC VOLTAGE REGULATORS

*OBJECTIVE 9* ▶

**Linear IC voltage regulator**
A device used to hold the output voltage from a dc power supply relatively constant over a wide range of line and load variations.

**Fixed-positive regulator**
A regulator with a predetermined $+V_{dc}$ output.

**Fixed-negative regulator**
A regulator with a predetermined $-V_{dc}$ output.

**Adjustable regulator**
A regulator whose output $V_{dc}$ can be set to any value within specified limits.

**Dual-tracking regulator**
A regulator that provides equal $+V_{dc}$ and $-V_{dc}$ outputs.

The **linear IC voltage regulator** is a device used to hold the output voltage from a dc power supply relatively constant over a wide range of line and load variations. Most of the commonly used IC voltage regulators are three-terminal devices, though some types require more than three terminals. The schematic symbol for a three-terminal regulator is shown in Figure 21.10.

There are basically four types of IC voltage regulators: *fixed positive, fixed negative, adjustable,* and *dual tracking*. The **fixed-positive** and **fixed-negative** IC voltage regulators are designed to provide specific output voltages. For example, the LM309 (fixed positive) provides a $+5$ $V_{dc}$ output (as long as the regulator input voltages and load demand stay within their specified ranges). The **adjustable regulator** can be adjusted to provide any dc output voltage within its two specified limits. For example, the LM317 output can be adjusted to any value between its limits of $+1.2$ and $+32$ $V_{dc}$. Both positive and negative variable regulators are available. The **dual-tracking regulator** provides equal positive and negative output voltages. For example, the RC4195 provides outputs of $+15$ and $-15$ $V_{dc}$. Adjustable dual-tracking regulators are also available. These regulators have outputs that can be varied between their two rated limits. A single adjustment controls both outputs so that they are always equal in value. For example, if an adjustable dual-tracking regulator is adjusted for a positive output of $+2$ $V_{dc}$, the negative output will automatically be adjusted to $-2$ $V_{dc}$.

Regardless of the type of regulator used, the regulator input-voltage polarity must match the device's rated output polarity. In other words, positive regulators must have positive input voltages and negative regulators must have negative input voltages. Dual-tracking regulators require *both* positive and negative input voltages.

IC voltage regulators are *series regulators*. They contain internal pass transistors and transistor control components. Generally, the internal circuitry of an IC voltage regulator resembles that of the series feedback regulator.

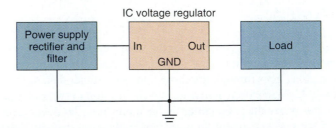

**FIGURE 21.10   Schematic symbol for a three-terminal regulator.**

# IC Regulator Specifications

For our discussion on common IC regulator specifications, we'll use the spec sheet for the LM317 voltage regulator. As was stated earlier, this device is an adjustable regulator whose output can be varied between $+1.2$ and $+32$ $V_{dc}$. The spec sheet for the LM317 is shown in Figure 21.11.

◄ *OBJECTIVE 10*

The **input/output voltage differential** rating (which is listed in the absolute maximum ratings section of the spec sheet) indicates the maximum difference between $V_{in}$ and $V_{out}$ that can occur without damaging the device. For the LM317, this rating is 40 V. The differential voltage rating can be used to determine the maximum allowable value of $V_{in}$ as follows:

**Input/output voltage differential**
The maximum allowable difference between $V_{in}$ and $V_{out}$ for an IC voltage regulator.

$$V_{in(max)} = V_{out(adj)} + V_d \qquad (21.6)$$

where $V_{in(max)}$ = the maximum allowable unrectified dc input voltage
$V_{out(adj)}$ = the *adjusted* output voltage of the regulator
$V_d$ = the input/output voltage differential rating of the regulator

---

The LM317 is adjusted to provide a $+8$ $V_{dc}$ regulated output voltage. Determine the maximum allowable input voltage to the device.

**EXAMPLE 21.3**

***Solution:*** With a $V_d$ rating of 40 V, the maximum allowable value of $V_{in}$ is found as

$$\begin{aligned} V_{in(max)} &= V_{out(adj)} + V_d \\ &= +8\ V_{dc} + 40\ V \\ &= +48\ V \end{aligned}$$

**PRACTICE PROBLEM 21.3**

An adjustable IC voltage regulator has a $V_d$ rating of 32 V. Determine the maximum allowable input voltage to the device when it is adjusted to $+6$ $V_{dc}$.

---

The *line regulation* rating of the LM317 is 0.04%/V (maximum). This rating is measured under the following conditions: $T_A = 25°C$ and $3\ V \le V_{in} - V_{out} \le 40\ V$. This means that the difference between the input and output voltages can be no less than 3 V and no greater than 40 V. For example, if we set the LM317 for a $+10$ $V_{dc}$ output, the input voltage to the device must be between $+13$ and $+50$ V. As long as these conditions are met, the output will vary by no more than $10\ V \times 0.04\% = 4\ mV$.

The *load regulation* rating for the LM317 depends on whether the device is operated for a $V_{out}$ less than or greater than $+5$ $V_{dc}$. If $V_{out}$ is less than $+5$ $V_{dc}$, the output voltage will not vary by more than 25 mV when the load current is varied within the given range of values. If the output voltage is set to a value greater than or equal to $+5$ $V_{dc}$, the same change in load current will cause a maximum output voltage change of 0.5 percent. Note that the rated range of output currents for the device is 10 mA to approximately 1.5 A.

Looking further down the spec sheet, you'll notice another set of line/load regulation ratings. This lower set of ratings indicates the maximum line/load regulation ratings of the device *over the entire range of allowable operating junction temperatures*. For the LM317, this range is 0 to 125°C, as is listed in the absolute maximum ratings section of the spec sheet. Note that the ratings we discussed earlier assumed an operating temperature of 25°C. Thus, the lower set of ratings indicates the *worst-case* operation of the device.

## Absolute Maximum Ratings

| | |
|---|---|
| Power Dissipation | Internally limited |
| Input—Output Voltage Differential | 40V |
| Operating Junction Temperature Range | |
| LM117 | $-55°C$ to $+150°C$ |
| LM217 | $-25°C$ to $+150°C$ |
| LM317 | $0°C$ to $+125°C$ |
| Storage Temperature | $-65°C$ to $+150°C$ |
| Lead Temperature (Soldering, 10 seconds) | $300°C$ |

## Preconditioning

| | |
|---|---|
| **Burn-In in Thermal Limit** | **100% All Devices** |

## Electrical Characteristics (Note 1)

| PARAMETER | CONDITIONS | LM117/217 | | | LM317 | | | UNITS |
|---|---|---|---|---|---|---|---|---|
| | | MIN | TYP | MAX | MIN | TYP | MAX | |
| Line Regulation | $T_A = 25°C$, $3V \leq V_{IN} - V_{OUT} \leq 40V$ (Note 2) | | 0.01 | 0.02 | | 0.01 | 0.04 | %/V |
| Load Regulation | $T_A = 25°C$, $10 mA \leq I_{OUT} \leq I_{MAX}$ | | | | | | | |
| | $V_{OUT} \leq 5V$, (Note 2) | | 5 | 15 | | 5 | 25 | mV |
| | $V_{OUT} \geq 5V$, (Note 2) | | 0.1 | 0.3 | | 0.1 | 0.5 | % |
| Thermal Regulation | $T_Z = 25°C$, 20 ms Pulse | | 0.03 | 0.07 | | 0.04 | 0.07 | %/W |
| Adjustment Pin Current | | | 50 | 100 | | 50 | 100 | µA |
| Adjustment Pin Current Change | $10 mA \leq I_L \leq I_{MAX}$ $3V \leq (V_{IN} - V_{OUT}) \leq 40V$ | | 0.2 | 5 | | 0.2 | 5 | µA |
| Reference Voltage | $3V \leq (V_{IN} - V_{OUT}) \leq 40V$, (Note 3) $10 mA \leq I_{OUT} \leq I_{MAX}$, $P \leq P_{MAX}$ | 1.20 | 1.25 | 1.30 | 1.20 | 1.25 | 1.30 | V |
| Line Regulation | $3V \leq V_{IN} - V_{OUT} \leq 40V$, (Note 2) | | 0.02 | 0.05 | | 0.02 | 0.07 | %/V |
| Load Regulation | $10 mA \leq I_{OUT} \leq I_{MAX}$, (Note 2) | | | | | | | |
| | $V_{OUT} \leq 5V$ | | 20 | 50 | | 20 | 70 | mV |
| | $V_{OUT} \geq 5V$ | | 0.3 | 1 | | 0.3 | 1.5 | % |
| Temperature Stability | $T_{MIN} \leq T_j \leq T_{MAX}$ | | 1 | | | 1 | | % |
| Minimum Load Current | $V_{IN} - V_{OUT} = 40V$ | | 3.5 | 5 | | 3.5 | 10 | mA |
| Current Limit | $V_{IN} - V_{OUT} \leq 15V$ | | | | | | | |
| | K and T Package | 1.5 | 2.2 | | 1.5 | 2.2 | | A |
| | H and P Package | 0.5 | 0.8 | | 0.5 | 0.8 | | A |
| | $V_{IN} - V_{OUT} = 40V$, $T_j = +25°C$ | | | | | | | |
| | K and T Package | 0.03 | 0.4 | | 0.15 | 0.4 | | A |
| | H and P Package | 0.15 | 0.07 | | 0.075 | 0.07 | | A |
| RMS Output Noise, % of $V_{OUT}$ | $T_A = 25°C$, $10 Hz \leq f \leq 10 kHz$ | | 0.003 | | | 0.003 | | % |
| Ripple Rejection Ratio | $V_{OUT} = 10V$, $f = 120 kHz$ | | 65 | | | 65 | | dB |
| | $C_{ADJ} = 10µF$ | 66 | 80 | | 66 | 80 | | dB |
| Long-Term Stability | $T_A = 125°C$ | | 0.3 | 1 | | 0.3 | 1 | % |
| Thermal Resistance, Junction to Case | H Package | | 12 | 15 | | 12 | 15 | °C/W |
| | K Package | | 2.3 | 3 | | 2.3 | 3 | °C/W |
| | T Package | | | | | 4 | | °C/W |
| | P Package | | | | | 12 | | °C/W |

**Note 1:** Unless otherwise specified, these specifications apply $-55°C \leq T_j \leq +150°C$ for the LM117, $-25°C \leq T_j \leq +150°C$ for the LM217, and $0°C \leq T_j \leq +125°C$ for the LM317; $V_{IN} - V_{OUT} = 5V$; and $I_{OUT} = 0.1A$ for the TO-39 and TO-202 packages and $I_{OUT} = 0.5A$ for the TO-3 and TO-220 packages. Although power dissipation is internally limited, these specifications are applicable for power dissipations of 2W for the TO-39 and TO-202, and 20W for the TO-3 and TO-220. $I_{MAX}$ is 1.5A for the TO-3 and TO-220 packages and 0.5A for the TO-39 and TO-202 packages.

**Note 2:** Regulation is measured at constant junction temperature, using pulse testing with a low duty cycle. Changes in output voltage due to heating effects are covered under the specification for thermal regulation.

**Note 3:** Selected devices with tightened tolerance reference voltage available.

**FIGURE 21.11    LM317 specification sheet. (Courtesy of Fairchild, a division of National Semiconductor.)**

**FIGURE 21.12** An LM317 voltage-adjust circuit.

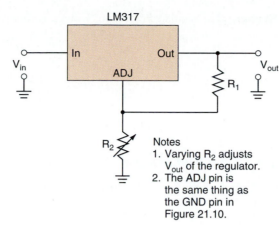

Notes
1. Varying $R_2$ adjusts $V_{out}$ of the regulator.
2. The ADJ pin is the same thing as the GND pin in Figure 21.10.

**Lab Reference:** The operation of the LM317 regulator is demonstrated in Exercise 49.

The **minimum load current** rating indicates the minimum allowable load current demand for the regulator. For the LM317, this rating is 10 mA. If the load current drops below 10 mA (such as when the load is open), regulation of the output voltage is lost.

The **ripple rejection ratio** is the ability of the regulator to drop any ripple voltage at its input. For the LM317, any input ripple is reduced by 65 dB at the output. Using the decibel conversion process discussed in Chapter 8, we find that the input ripple will be reduced by a factor of approximately 1800 by the regulator.

**Minimum load current**
The value of $I_L$ below which regulation is lost.

**Ripple rejection ratio**
The ratio of regulator input ripple to maximum output ripple.

## Output Voltage Adjustment

The output voltage of the LM317 is adjusted using a voltage divider, as is shown in Figure 21.12. The *complete* LM317 spec sheet (not shown) contains the following equation for determining the dc output voltage of the circuit:

$$V_{dc} = 1.25 \left( \frac{R_2}{R_1} + 1 \right) \tag{21.7}$$

where $V_{dc}$ is the regulated dc output voltage of the regulator. Example 21.4 shows how equation (21.7) is used to determine the regulated dc output voltage for an LM317 regulator circuit.

---

$R_2$ in Figure 21.13 is adjusted to 2.4 k$\Omega$. Determine the regulated dc output voltage for the circuit.

**EXAMPLE 21.4**

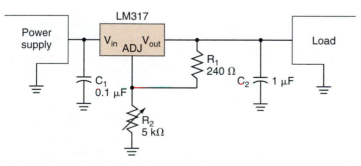

**Lab Reference:** The operation of a complete LM317-regulated power supply (like the one represented in Figure 21.13) is demonstrated in Exercise 50.

**FIGURE 21.13** LM317 circuit connections.

**Solution:** The regulated dc output voltage is found as

$$V_{dc} = 1.25 \left( \frac{R_2}{R_1} + 1 \right)$$

$$= 1.25 \left( \frac{2.4 \text{ k}\Omega}{240 \ \Omega} + 1 \right)$$

$$= (1.25)(11)$$

$$= 13.75 \text{ V}$$

### PRACTICE PROBLEM 21.4

$R_2$ in Figure 21.13 is adjusted to 1.68 k$\Omega$. Determine the regulated dc output voltage for the LM317.

Note that the $V_{dc}$ equation for a given adjustable regulator is always given on its spec sheet.

Figure 21.13 shows two shunt capacitors connected to the regulator input and output pins. The input shunt capacitor is used to prevent the input ripple from driving the regulator into self-oscillations. The output shunt capacitor is used to improve the ripple reduction of the regulator.

## Linear IC Regulator Applications: A Complete Dual-Polarity Power Supply

Figure 21.14 shows a complete dual-polarity power supply. The circuit uses *matched* fixed-positive and fixed-negative regulators to provide equal $+V_{dc}$ and $-V_{dc}$ outputs.

Capacitors $C_1$ and $C_2$ are *filter capacitors*. These capacitors have values in the mid to high microfarad range. Capacitors $C_3$ and $C_4$ are the regulator input shunt capacitors and will be less than 1 $\mu$F in value. Capacitors $C_5$ and $C_6$ are output ripple reduction capacitors and have values in the neighborhood of 1 $\mu$F.

## Circuit Variations

Figure 21.15 shows how a *pass transistor* can be added in parallel with an IC voltage regulator to increase the maximum possible output current. The value of $R_S$ in the circuit is selected to bias $Q_1$ on, as follows:

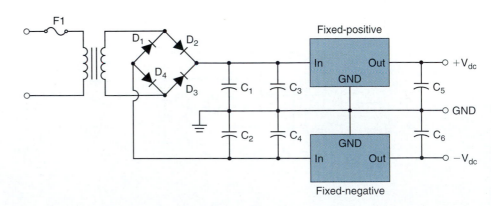

**FIGURE 21.14   Complete dual-polarity power supply.**

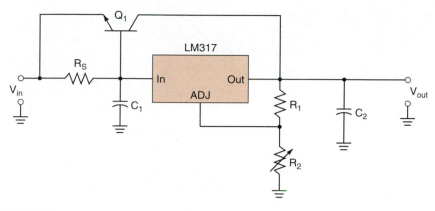

FIGURE 21.15

$$R_S = \frac{V_{BE(Q1)}}{I_{in}}$$

where $I_{in}$ is the regulator input current. With $Q_1$ on, an increase in load current demand (above the capability of the regulator output) results in increased $Q_1$ conduction. This provides the needed load current.

There is one serious drawback to the circuit in Figure 21.15. The input *ripple current* to the circuit is coupled (via $Q_1$) to the load. This problem, however, can be minimized by adding the *current-limiting circuit* ($Q_2$ and $R_{S2}$) shown in Figure 21.16. This current-limiting circuit acts just like the one we discussed for the series feedback regulator in Figure 21.8. When the value of $R_{S2}$ is properly selected, any excessive ripple current at the circuit input is effectively prevented from reaching the load.

## One Final Note

Linear IC voltage regulators are extremely common. While we cannot possibly hope to cover the operation of every type of IC voltage regulator, you should have no problem in dealing with everyday IC voltage regulators. The characteristics of IC voltage regulators are summarized for you in Figure 21.17.

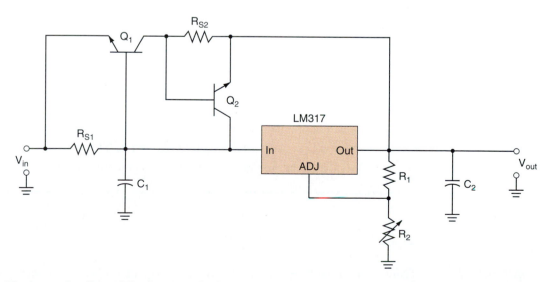

FIGURE 21.16   A complete linear IC voltage regulator.

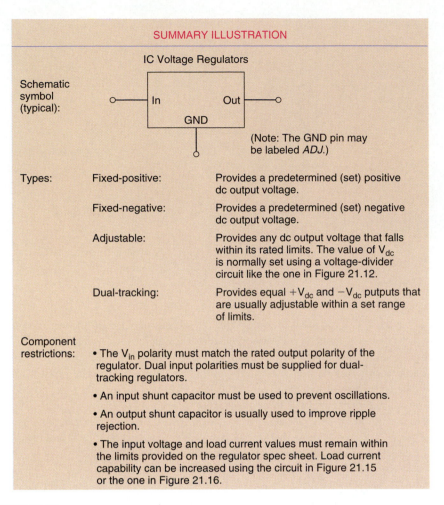

**SUMMARY ILLUSTRATION**

IC Voltage Regulators

Schematic
symbol
(typical):

In        Out
GND

(Note: The GND pin may
be labeled *ADJ.*)

| Types: | | |
|---|---|---|
| Fixed-positive: | Provides a predetermined (set) positive dc output voltage. | |
| Fixed-negative: | Provides a predetermined (set) negative dc output voltage. | |
| Adjustable: | Provides any dc output voltage that falls within its rated limits. The value of $V_{dc}$ is normally set using a voltage-divider circuit like the one in Figure 21.12. | |
| Dual-tracking: | Provides equal $+V_{dc}$ and $-V_{dc}$ putputs that are usually adjustable within a set range of limits. | |

Component
restrictions:

• The $V_{in}$ polarity must match the rated output polarity of the regulator. Dual input polarities must be supplied for dual-tracking regulators.

• An input shunt capacitor must be used to prevent oscillations.

• An output shunt capacitor is usually used to improve ripple rejection.

• The input voltage and load current values must remain within the limits provided on the regulator spec sheet. Load current capability can be increased using the circuit in Figure 21.15 or the one in Figure 21.16.

**FIGURE 21.17**

---

**Section Review**

1. What is an *IC voltage regulator*?

2. List and describe the four types of IC voltage regulators.

3. What input polarity (or polarities) is/are required for each type of voltage regulator?

4. What is the *input/output differential* rating?

5. What is the *minimum load current* rating?

6. What is the *ripple rejection ratio* rating?

7. What type of circuit is normally used to provide the adjustment of the dc output voltage from an adjustable regulator?

8. Where can you find the $V_{dc}$ equation for a given adjustable regulator?

9. What purpose is served by the circuit in Figure 21.15?

10. Why does the circuit in Figure 21.15 need to be modified as shown in Figure 21.16?

---

## *21.5* SWITCHING REGULATORS

*OBJECTIVE 11* ▶  There are two fundamental types of voltage regulators: *linear regulators* and *switching regulators*. The basic block diagram for each type of regulator is shown in Figure 21.18.

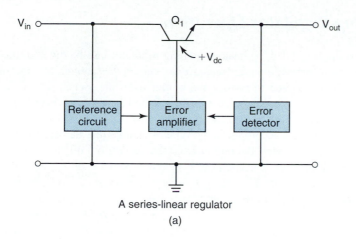

A series-linear regulator

(a)

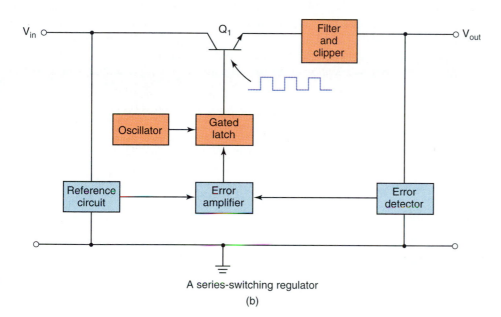

A series-switching regulator

(b)

**FIGURE 21.18   Typical linear and switching regulators.**

The *linear regulator* is designed to provide a continuous path for current between the regulator input and the load. For example, the linear regulator in Figure 21.18a contains a *pass transistor* ($Q_1$) *that is always conducting.* All of the regulators we have covered up to this point have been linear regulators.

The *switching regulator* is designed so that the current path between the regulator input and the load is *not* continuous. For example, the switching regulator in Figure 21.18b contains a pass transistor ($Q_1$) that is rapidly switched back and forth between saturation and cutoff. Note that the pass transistor in a series switching regulator is often referred to as a **power switch.**

When saturated, the power switch provides a current path between the regulator's input and output. When in cutoff, the power switch breaks the conduction path between the input and the load. As you will see later in this section, this circuit action results in:

**Power switch**
A term used to describe the pass transistor in a switching regulator.

1. Higher regulator efficiency
2. Higher power-handling capability

## Switching Regulator Operation

OBJECTIVE 12 ▶ As shown in Figure 21.19, the basic switching regulator can be divided into four circuit groups: the *power switch*, the *filter and clipper*, the *control circuit*, and the *switch driver*. The *control circuit* is used to control the output from the switch driver. In its simplest form, the control circuit acts as a *comparator*. When it senses a change in the regulator output voltage, it sends a signal to the switch driver. This signal causes the output from the switch driver to vary according to the type of change that has occurred. For example, if the load voltage *decreases*, the output from the control circuit causes the switch driver to *increase conduction* through the power switch. Likewise, an *increase* in load voltage results in the switch driver *decreasing conduction* through the power switch. In either case, the load voltage is returned to its proper value.

The *switch driver* contains an oscillator and a *gated latch*. The output from the oscillator is fixed and constant. It is the gated latch that controls the conduction of the power switch. The gated latch accepts inputs from the oscillator and the control circuit. These two inputs are then combined in such a way as to produce a driving signal that either increases or decreases conduction through the power switch. (We will discuss the methods used to produce the driving signal more thoroughly later in this section.)

Since the power switch is constantly changing output states, the voltage at its emitter is (more or less) a *rectangular waveform*. The *filter and clipper* circuit is designed to respond to this waveform as follows:

1. The *capacitor* opposes any change in voltage and, therefore, keeps the load voltage relatively constant.

2. The *inductor* opposes any change in current and, therefore, keeps the load current relatively constant.

3. The *diode* clips the counter emf produced by the *LC* circuit response to a rectangular input. In other words, it provides *transient protection*. (You were first introduced to this application in Chapter 4.)

OBJECTIVE 13 ▶ Now, let's take a look at the overall response of the regulator to a change in load demand. Assume for a moment that the circuit experiences an *increased load*. When a

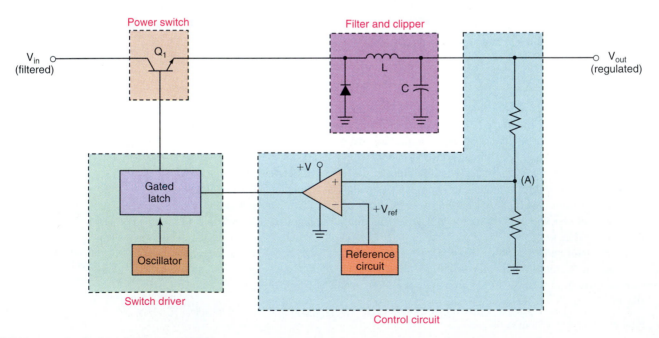

**FIGURE 21.19  Basic switching regulator.**

regulator load is increased, $V_{out}$ starts to *decrease*. The decrease in $V_{out}$ causes the voltage at point (A) in the control circuit to decrease. This, in turn, causes the voltage at the comparator output to *decrease*. The decreased output from the control circuit signals the switch driver to *increase* the conduction of the power switch. The increased power switch conduction causes $V_{out}$ to rise back to its normal value.

*Remember*: The term *increased load* means that the load *current demand* has increased; i.e., the load *resistance* has decreased.

When the circuit experiences a *decreased load*, the results are the opposite of those just given. The *increased load voltage* causes the voltage at point (A) in the control circuit to *increase*. This increase leads to an *increase* in the output from the control circuit. The increased output from the control circuit signals the switch driver to *decrease* the conduction of the power switch. The decrease in power switch conduction causes $V_{out}$ to fall back to its normal value.

At this point, you should have a pretty good idea of how the switching regulator in Figure 21.19 operates. To complete the picture, we need to take a closer look at the operation of the *switch driver* and the means by which it controls the conduction of the power switch.

## Controlling Power Switch Conduction

As you know, the power switch is constantly driven back and forth between saturation and cutoff. When saturated, the transistor couples $V_{in}$ to the load. When in cutoff, the transistor isolates the load from the input.

◄ *OBJECTIVE 14*

The average (dc) value of the waveform produced at the emitter of the power switch can be found as

$$V_{ave} = V_{in}\left(\frac{T_{on}}{T_{on} + T_{off}}\right) \tag{21.8}$$

where $T_{on}$ = the time that the transistor spends in saturation (per cycle)
$\quad$ $T_{off}$ = the time the transistor spends in cutoff (per cycle)

Example 21.5 demonstrates the use of this equation.

---

The regulator shown in Figure 21.19 has the following values: $V_{in} = 24\ V_{dc}$, $T_{on} = 5\ \mu s$, and $T_{off} = 10\ \mu s$. Calculate the dc average of the load voltage.

*EXAMPLE* **21.5**

*Solution:* Using the values given, the average (dc) load voltage can be found as

$$V_{ave} = V_{in}\left(\frac{T_{on}}{T_{on} + T_{off}}\right)$$

$$= (24\ V_{dc})\left(\frac{5\ \mu s}{15\ \mu s}\right)$$

$$= 8\ V_{dc}$$

### PRACTICE PROBLEM 21.5

A regulator like the one shown in Figure 21.20 has the following values: $V_{in} = 36\ V_{dc}$, $T_{on} = 6\ \mu s$, and $T_{off} = 9\ \mu s$. Calculate the value of $V_{ave}$ for the circuit.

---

Equation (21.8) is important because it demonstrates the fact that *the average output voltage from a switching regulator can be controlled by varying the conduction of the power switch*. At this point, we will take a look at two methods commonly used to control the conduction of the power switch and, thus, the output from the switching regulator.

## Pulse-Width Modulation (PWM)

One method commonly used to control the conduction of the power switch (and, therefore, the average emitter voltage) is referred to as **pulse-width modulation**, or **PWM**. In a PWM system, *a signal is used to vary the pulse width of a rectangular waveform while not affecting its total cycle time.*

The control circuit shown in Figure 21.20a is designed to provide PWM for the power switch. The oscillator in this circuit generates a *triangular* waveform ($V_o$). The *error voltage* ($V_{error}$) is a dc voltage that varies directly with the input to the control circuit. The *gated latch* (which actually performs the modulation) *produces a high output whenever the following relationship is fulfilled:*

$$V_o \geq V_{error}$$

For example, take a look at the first set of waveforms in Figure 21.20b. As you can see, the control voltage ($V_c$) goes high as $V_o$ increases beyond the value of $V_{error}$. Then, as $V_o$ drops below the value of $V_{error}$, the control voltage drops back to 0 V. Note that the control voltage shown is nearly a square wave.

Now take a look at the second set of waveforms in the figure. If you compare the voltages shown to those in the first set of waveforms, you'll see that:

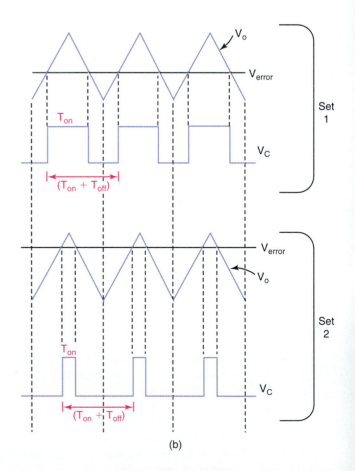

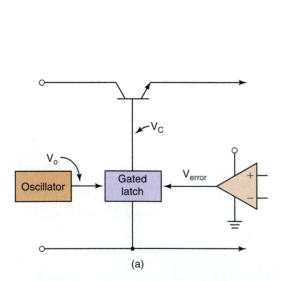

**FIGURE 21.20  Pulse–width modulation (PWM).**

1. The error voltage has increased in value.

2. The *pulse width* ($T_{on}$) of the control voltage has decreased significantly.

3. The total cycle time ($T_{on} + T_{off}$) has not changed.

Since the value of $T_{on}$ has decreased, the average value of $V_C$ has also decreased. Likewise a decrease in the error voltage would cause an *increase* in the value of $V_C$. This is how PWM (as it applies to switching regulators) works.

## Variable Off-Time Modulator

Another method commonly used to control the conduction of a power switch is illustrated in Figure 21.21. In this circuit, *the pulse width of the control voltage is fixed and the total cycle time is variable*. As you can see, the gated latch in this circuit is a *one-shot* and the oscillator is an *astable multivibrator*. As long as the error voltage is high, the *square wave output* from the oscillator is gated to the power switch. When the error voltage goes low, the output from the oscillator is blocked.

The effect that the error voltage has on the cycle time of $V_c$ is shown in the figure. The pulse width ($T_{on}$) has not changed from the first highlighted waveform to the second. However, the total cycle time ($T_{on} + T_{off}$) of the second waveform has been altered by the $V_{error}$ input. As the total cycle time varies, so does the average output from the power switch. Again, the average output voltage from the regulator is controlled.

**Variable off-time modulation**
Using a signal to vary the cycle time of a rectangular wave without affecting its pulse width.

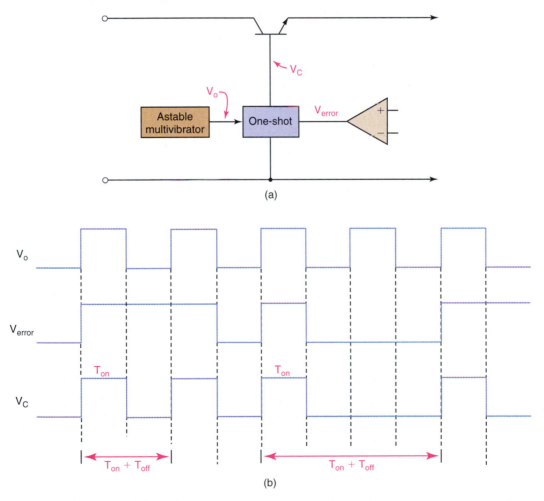

FIGURE 21.21   Variable off-time modulation.

# Switching Regulator Configurations

*OBJECTIVE 15* ▶

**Step-down regulator**
A regulator where $V_{out} \leq V_{in}$.

**Step-up regulator**
A regulator where $V_{out} \geq V_{in}$.

One of the advantages of using a switching regulator is the fact that it can be designed in a variety of configurations. So far, we have discussed only the **step-down regulator**. This configuration, which is represented in Figure 21.22a, produces a dc load voltage that is less than (or equal to) the rectified input voltage.

By modifying the basic configuration shown in Figure 21.22b, we obtain a **step-up regulator**. As you can see, the power switch is now a shunt component and the inductor is placed directly in series with the input source ($V_{in}$).

The step-up switching regulator can provide output voltages greater than the rectified value of $V_{in}$. This is made possible by the positioning of the inductor ($L_1$). In a nutshell, here is how it works: During the *on-time* of the transistor, the current drawn through $L_1$ causes an induced voltage to be developed across its terminals. This voltage adds to the value of $V_{in}$, making it possible for the output voltage to be greater than the input voltage.

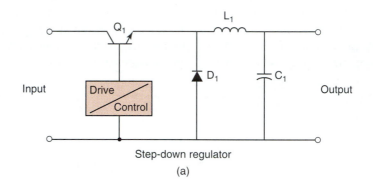

Step-down regulator

(a)

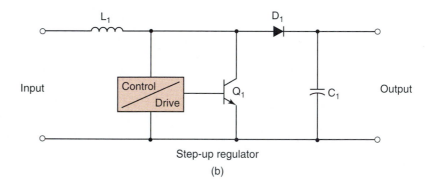

Step-up regulator

(b)

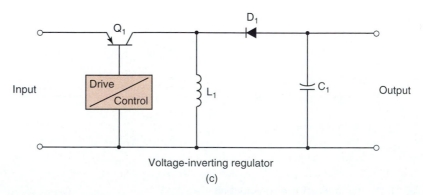

Voltage-inverting regulator

(c)

**FIGURE 21.22** Basic switching regulator configurations.

The **voltage-inverting regulator** shown in Figure 21.22c reverses the polarity of the rectified input voltage. For example, with a rectified input voltage of $+7$ V, the regulator in Figure 21.22c could be designed to be any *negative value* less than, equal to, or greater than $-7$ V (within limits). Note the series power switch and shunt inductor connections characteristic of this configuration.

**Voltage-inverting regulator**
A regulator that reverses the polarity of its regulated dc input.

There are several points that should be made regarding the configurations represented in Figure 21.22. First of all, the *control* and *switch driver* circuits have been lumped into a single block for simplicity. While the control and switch driver circuits may vary from one switching regulator to another, you should always be able to determine the type of regulator you're dealing with by noting the position of the power switch, as follows:

1. If the power switch is in series with the input and the inductor, it is a *step-down* regulator.

2. If the power switch is a *shunt component* placed *after* the inductor, it is a *step-up* regulator.

3. If the power switch is in series with the input and the inductor is a *shunt* component, it is a *voltage-inverting* regulator.

Another important point deals with the fact that there are no "cookbook" equations to help with the analysis of these circuits. Switching regulators are complex circuits that operate on very complex principles. Just as the component placement varies from one switching regulator to another, so do most of the equations used in circuit analysis. This is why the average switching regulator can take up to approximately 6 worker-months to design, build, and debug.

Because switching regulators can be designed for several different input/output voltage relationships, they are sometimes referred to as **dc-to-dc converters**. As the name implies, they can effectively convert one rectified dc voltage to another value.

**dc-to-dc converter**
Another name for a switching voltage regulator.

## IC Switching Regulators

In many cases, the control functions and power switching of a switching regulator are handled by a single integrated circuit. For example, Motorola's MC34063 is an IC that contains the entire control, switch drive, and power switch circuitry. The MC34063 is shown in Figure 21.23. Note that the transistor labeled ($Q_1$) in the IC is the power switch.

◄ *OBJECTIVE 16*

If you compare the internal circuitry of the MC34063 to the block diagram in Figure 21.19, it is easy to see that the IC contains the control, driver, and power switch circuitry. The *control circuit* consists of the 1.25 V reference regulator and the comparator. The control circuit for this IC is used to provide *variable off-time* modulation of the oscillator output.

The MC34063 can be used in any standard switching regulator configuration. For example, the circuit shown in Figure 21.24 is an IC-based *step-down* regulator. This particular circuit was designed to produce a 12 V/750 mA output with a 36 $V_{dc}$ rectified input. Note the positioning of the power switch and the inductor. This is characteristic of a step-down regulator. Also note the excellent efficiency rating of this circuit.

## Switching Regulators: Advantages and Disadvantages

As Figure 21.24 indicates, one of the primary advantages that switching regulators have over linear regulators is *higher efficiency*. While linear regulators are generally limited to efficiency ratings below 60 percent, switching regulators can easily achieve ratings of 90 percent. The higher efficiency of the switching regulator is due to the fact that the power switch is usually in either saturation or cutoff. In either of these operating states, the component dissipates very little power. In contrast, the pass transistor in a linear regulator is

◄ *OBJECTIVE 17*

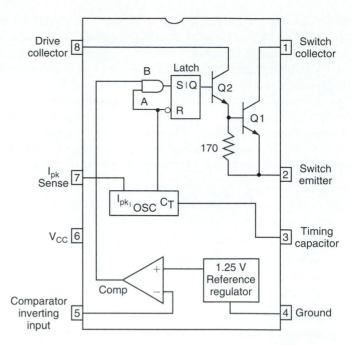

**FIGURE 21.23   The MC34063 switching regulator. (Copyright of Motorola. Used by permission.)**

usually operating within its *active region*. This causes the transistor to dissipate a relatively high amount of power, dropping its efficiency rating.

The power characteristics of switching regulators lead us to another advantage. Since the power switch dissipates very little power, a switching regulator can actually be designed for output power values that are far greater than the maximum power rating of the power switch. For example, a 2 W transistor could easily work in a 20 W switching regulator because the transistor would not be required to dissipate any significant amount of power. The same cannot be said for the linear regulator.

Finally, switching regulators can be built in a variety of configurations (as you were shown earlier in this section). In contrast, linear regulators can be designed only as *step-down* regulators.

In spite of the above advantages, switching regulators *do* have some distinct disadvantages. First, the operation of the power switch and filter can generate a very significant amount of *noise* that can be transmitted into the surrounding environment. Therefore, switching regulators cannot be used in any low-noise application unless shielding is provided.

Another disadvantage is the fact that switching regulators have a longer *transient response time* than linear regulators; that is, they are slower to respond to a change in the load demand. This is due to the time required for the feedback loop (the control and drive circuits) to respond to a change in the error detection circuitry.

Finally, the design of a switching regulator is far more complex and time consuming than that of a similar linear regulator. This adds to the production cost of the switching regulator.

## Regulator Applications

Linear regulators are generally used in low-power applications. For example, most digital systems will have at least one linear regulator for each board-level supply voltage required. These supply voltages are generally not required to handle a significant amount of power.

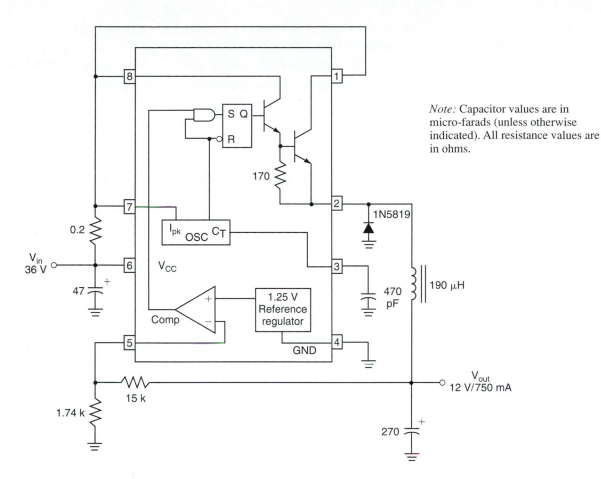

*Note:* Capacitor values are in micro-farads (unless otherwise indicated). All resistance values are in ohms.

FIGURE 21.24   An MC34063-based step-down regulator. (Copyright of Motorola. Used by permission.)

| Test | Conditions | Results |
|---|---|---|
| Line Regulation | $V_{in}$ = 20 to 40 V, $I_{out}$ = 750 mA | Δ = 15 mV or ± 0.063% |
| Load Regulation | $V_{in}$ = 36 V, $I_{out}$ = 100 to 750 mA | Δ = 40 mV or ± 0.17% |
| Output Ripple | $V_{in}$ = 36 V, $I_{out}$ = 750 mA | 60 mV$_{p-p}$ |
| Short Circuit Current | $V_{in}$ = 36 V, $R_L$ = 0.1 Ω | 1.6 A |
| Efficiency | $V_{in}$ = 36 V, $I_{out}$ = 750 mA | 89.5% |

A maximum power transfer of 9.0 watts is possible from an 8-pin dual-in-line package with $V_{in}$ = 36 V and $V_{out}$ = 12 V.

Switching regulators are generally used in *high-power* (above 10 W) applications. They are also used in some lightweight, battery-operated units. This is because of their minimal size requirements and dc-to-dc conversion capabilities.

*Section Review*

1. In terms of conduction characteristics, what is the primary difference between *linear* regulators and *switching* regulators?

2. List the circuits that make up the basic switching regulator and describe the function performed by each.

3. Describe how the switching regulator in Figure 21.19 responds to a change in load demand.

4. Describe the relationship between the duty cycle of a power switch and its average output voltage. (You'll have to stretch on this one!)

5. Compare and contrast *pulse-width modulation* with *variable off-time modulation*.

6. List the circuit recognition features for each of the switching regulators shown in Figure 21.22.

7. Why are switching regulators often referred to as *dc-to-dc converters*?

8. Discuss the advantages, disadvantages, and applications for linear and switching regulators.

## ■ Key Terms

The following terms were introduced and defined in this chapter:

adjustable regulator
Darlington pass-transistor regulator
dc-to-dc converter
dual-tracking regulator
fixed-negative regulator
fixed-positive regulator
input/output voltage differential
linear IC voltage regulator

line regulation
load regulation
minimum load current
pass-transistor regulator
power switch
pulse-width modulation
regulation
ripple rejection ratio
series feedback regulator

series regulator
shunt feedback regulator
shunt regulator
step-down regulator
step-up regulator
variable off-time modulation
voltage-inverting regulator

## Equation Summary

| Equation Number | Equation | Section Number |
|---|---|---|
| (21.1) | $\text{line regulation} = \dfrac{\Delta V_{\text{out}}}{\Delta V_{\text{in}}}$ | 21.1 |
| (21.2) | $\text{load regulation} = \dfrac{V_{NL} - V_{FL}}{\Delta I_L}$ | 21.1 |
| (21.3) | $V_L = V_Z - V_{BE}$ | 21.2 |
| (21.4) | $V_L = V_Z - 2V_{BE}$ | 21.2 |
| (21.5) | $I_{L(\text{max})} \cong \dfrac{V_{BE(Q3)}}{R_S}$ | 21.2 |
| (21.6) | $V_{\text{in(max)}} = V_{\text{out(adj)}} + V_d$ | 21.4 |
| (21.7) | $V_{\text{dc}} = 1.25 \left( \dfrac{R_2}{R_1} + 1 \right)$ | 21.4 |
| (21.8) | $V_{\text{ave}} = V_{\text{in}} \left( \dfrac{T_{\text{on}}}{T_{\text{on}} + T_{\text{off}}} \right)$ | 21.5 |

## Practice Problems

### §21.1

1. A voltage regulator experiences a 20 μV change in its output voltage when its input voltage changes by 4 V. Determine the line regulation rating of the circuit.

2. A voltage regulator experiences a 14 μV change in output voltage when its input voltage changes by 10 V. Determine the line regulation rating of the circuit.

3. A voltage regulator experiences a 15 μV change in its output voltage when its input voltage changes by 5 V. Determine the line regulation rating of the circuit.

4. A voltage regulator experiences a 12 mV change in output voltage when its input voltage changes by 12 V. Determine the line regulation rating of the circuit.

5. A voltage regulator is rated for an output current of 0 to 150 mA. Under no-load conditions, the output voltage of the circuit is 6 V. Under full-load conditions, the output from the circuit is 5.98 V. Determine the load regulation rating of the circuit.

6. A voltage regulator experiences a 20 mV change in output voltage when the load current increases from 0 to 50 mA. Determine the load regulation rating of the circuit.

7. A voltage regulator experiences a 1.5 mV change in output voltage when the load current increases from 0 to 20 mA. Determine the load regulation rating of the circuit.

8. A voltage regulator experiences a 14 mV change in output voltage when the load current increases from 0 to 100 mA. Determine the load regulation rating of the circuit.

### §21.2

9. Determine the approximate value of $V_L$ for the circuit in Figure 21.25.

10. Determine the approximate value of $V_L$ for the circuit in Figure 21.26.

11. Determine the maximum possible value of load current for the circuit in Figure 21.27.

12. $R_S$ in Figure 21.27 is changed to 1.2 Ω. Determine the maximum possible load current for the new circuit.

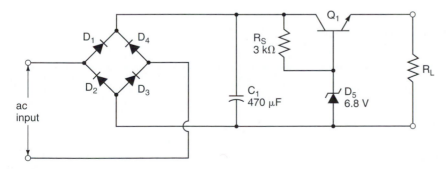

**FIGURE 21.25**

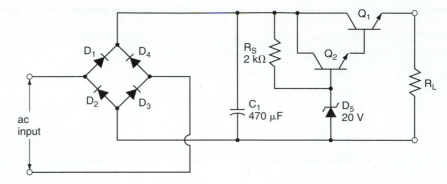

**FIGURE 21.26**

13. An adjustable IC voltage regulator is set for a $+3\ V_{dc}$ output. The $V_d$ rating for the device is 32 V. Determine the maximum allowable input voltage for the device.

14. An adjustable IC voltage regulator is set for a $+6\ V_{dc}$ output. The $V_d$ rating for the device is 24 V. Determine the maximum allowable input voltage for the device.

15. The LM317 is used in a circuit with adjustment resistance values of $R_1 = 330\ \Omega$ and $R_2 = 2.848\ k\Omega$ (adjusted potentiometer value). Determine the output voltage for the circuit.

16. The LM317 is used in a circuit with adjustment resistance values of $R_1 = 510\ \Omega$ and $R_2 = 6.834\ k\Omega$ (adjusted potentiometer value). Determine the output voltage for the circuit.

17. A power switch like the one in Figure 21.19 has the following values: $V_{in} = 36$ V, $T_{on} = 12\ \mu s$, and $T_{off} = 48\ \mu s$. Determine the average output voltage from the power switch.

18. A power switch like the one in Figure 21.19 has the following values: $V_{in} = 24$ V, $T_{on} = 10\ \mu s$, and $T_{off} = 40\ \mu s$. Determine the average output voltage from the power switch.

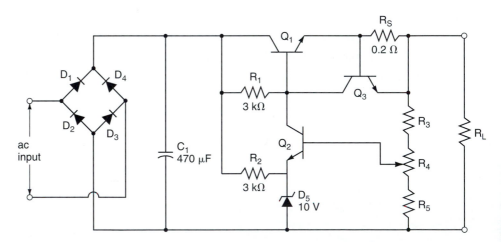

**FIGURE 21.27**

19. A +12 $V_{dc}$ regulator has a line regulation rating of 420 ppm/V. Determine the output voltage for the circuit when $V_{in}$ increases by 10 V.

20. A voltage regulator has the following measured values: $V_{out} = 12.002$ $V_{dc}$ when $V_{in} = +20$ V (rated maximum allowable input), and $V_{out} = 12$ $V_{dc}$ when $V_{in} = +10$ V (rated minimum allowable input). Determine the line regulation rating of the device in mV, %, and %/V.

21. A +5 V regulator has a load regulation rating of 2 $\Omega$ over a range of $I_L = 0$ to 100 mA (maximum). Express the load regulation of the circuit in V/mA, %, and %/mA.

22. A +15 $V_{dc}$ regulator has a 0.02%/mA load regulation rating for a range of $I_L = 0$ to 50 mA (maximum). Assuming that the load current stays within its rated limits, determine the maximum load power that can be delivered by the regulator. (Assume that $V_{out}$ increases as $I_L$ increases.)

# APPENDIX A

# Specification Sheet Guide

## DEVICE LISTING

The following device specification sheets and/or operating curves are located on the pages indicated:

| Part Number | Type of Device | Code* | Page |
|---|---|---|---|
| BB139 | Varactor | | 199 |
| BFW43 | High-voltage *pnp* transistor | MR | 230 |
| MDA2500 | Integrated bridge rectifier | | 103 |
| MLL755 | Zener diode | | 72 |
| MUR 5005 | Power rectifier | | 102 |
| MV209 | Varactor diode | | 178 |
| μA741 | Operational amplifier | OC | 642 |
| IN746–59 | Zener diodes | | 55 |
| IN4001–7 | Rectifier diodes | | 46 |
| IN5283–314 | Current-regulator diodes | | 188 |
| IN5391–9 | Rectifier diodes | | 70 |
| 2N2222 | *npn* transistors | OC | 371 |
| 2N3902 | High-power *npn* transistor | | 231 |
| 2N3903–4 | *npn* transistor | | 225 |
| 2N4237 | High-current *pnp* transistor | MR | 230 |
| 2N5484–6 | N-channel JFET | | 521 |
| 3N169–71 | E-MOSFET | | 548 |
| — | Example zener diode selector guide | | 59 |
| 2N3905–6 | *pnp* transistor | | 959 |
| 2N4013–4 | Switching transistors | | 961 |
| 2N4400 | *npn* transistor | | 358–59 |
| 2N5457–9 | JFETs general purpose | | 962 |
| 2N6426–7 | Darlington transistors | | 963 |

*MR = Maximum ratings only

  OC = Operating curves only

# ADDITIONAL SPECIFICATION SHEETS

## MAXIMUM RATINGS

| Rating | Symbol | Value | Unit |
|---|---|---|---|
| Collector-Emitter Voltage | $V_{CEO}$ | 40 | Vdc |
| Collector-Base Voltage | $V_{CBO}$ | 40 | Vdc |
| Emitter-Base Voltage | $V_{EBO}$ | 5.0 | Vdc |
| Collector Current — Continuous | $I_C$ | 200 | mAdc |
| Total Device Dissipation @ $T_A = 25°C$<br>Derate above 25°C | $P_D$ | 625<br>5.0 | mW<br>mW/°C |
| Total Power Dissipation @ $T_A = 60°C$ | PD | 250 | mW |
| Total Device Dissipation @ $T_C = 25°C$<br>Derate above 25°C | $P_D$ | 1.5<br>12 | Watts<br>mW/°C |
| Operating and Storage Junction Temperature Range | $T_J$, $T_{stg}$ | $-550$ to $+150$ | °C |

## THERMAL CHARACTERISTICS

| Characteristic | Symbol | Max | Unit |
|---|---|---|---|
| Thermal Resistance, Junction to Case | $R_{\theta JC}$ | 83.3 | °C/W |
| Thermal Resistance, Junction to Ambient | $R_{\theta JA}$ | 200 | °C/W |

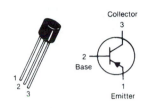

**2N3905**
**2N3906**

**CASE 29-04, STYLE 1**
**TO-92 (TO-226AA)**

Collector 3
Base 2
Emitter 1

**GENERAL PURPOSE TRANSISTOR**

**NPN SILICON**

## ELECTRICAL CHARACTERISTICS ($T_A = 25°C$ unless otherwise noted.)

| Characteristic | | Symbol | Min | Max | Unit |
|---|---|---|---|---|---|
| **OFF CHARACTERISTICS** | | | | | |
| Collector-Emitter Breakdown Voltage(1)<br>($I_C$1.0 mAdc, $I_B = 0$) | | $V_{(BR)CEO}$ | 40 | — | Vdc |
| Collector-Base Breakdown Voltage<br>($I_C$10 μAdc, $I_E = 0$) | | $V_{(BR)CBO}$ | 40 | — | Vdc |
| Emitter-Base Breakdown Voltage<br>($I_E = 10$ μAdc, $I_C = 0$) | | $V_{(BR)EBO}$ | 5.0 | — | Vdc |
| Base Cutoff Current<br>($V_{CE} = 30$ Vdc, $V_{BE} = 3.0$ Vdc) | | $I_{BL}$ | — | 50 | nAdc |
| Collector Cutoff Current<br>($V_{CE} = 30$ Vdc, $V_{BE} = 3.0$ Vdc) | | $I_{CEX}$ | — | 50 | nAdc |
| **ON CHARACTERISTICS(1)** | | | | | |
| DC Current Gain | | $h_{FE}$ | | | — |
| ($I_C = 0.1$ mAdc, $V_{CE} = 1.0$ Vdc) | 2N3905<br>2N3906 | | 30<br>60 | —<br>— | |
| ($I_C \doteq 1.0$ mAdc, $V_{CE} = 1.0$ Vdc) | 2N3905<br>2N3906 | | 40<br>80 | —<br>— | |
| ($I_C = 10$ mAdc, $V_{CE} = 1.0$ Vdc | 2N3905<br>2N3906 | | 50<br>100 | 150<br>300 | |
| ($I_C = 50$ mAdc, $V_{CE} = 1.0$ Vdc) | 2N3905<br>2N3906 | | 30<br>60 | —<br>— | |
| ($I_C = 1]]$ mAdc, $V_{CE} = 1.0$ Vdc) | 2N3905<br>2N3906 | | 15<br>30 | —<br>— | |
| Collector-Emitter Saturation Voltage<br>($I_C = 10$ mAdc, $I_B = 1.0$ mAdc)<br>($I_C = 50$ mAdc, $I_B = 5.0$ mAdc) | | $V_{CE(sat)}$ | —<br>— | 0.25<br>0.4 | Vdc |
| Base-Emitter Saturation Voltage<br>($I_C = 10$ mAdc, $I_B = 1.0$ mAdc)<br>($I_C = 50$ mAdc, $I_B = 5.0$ mAdc) | | $V_{BE(sat)}$ | 0.65<br>— | 0.85<br>0.95 | Vdc |
| **SMALL-SIGNAL CHARACTERISTICS** | | | | | |
| Current-Gain — Bandwidth Product<br>($I_C = 10$ mAdc, $V_{CE} = 20$ Vdc, $f = 100$ MHz) | 2N3905<br>2N3906 | $f_T$ | 200<br>250 | —<br>— | MHz |
| Output Capacitance<br>($V_{CB} = 5.0$ Vdc, $I_E = 0$, $f = 100$ kHz) | | $C_{obo}$ | — | 4.5 | pF |

**(Continued)**

**(Continued)**

**2N3905, 2N3906**

ELECTRICAL CHARACTERISTICS (continued) ($T_A = 25°C$ unless otherwise noted.)

| Characteristic | | Symbol | Min | Max | Unit |
|---|---|---|---|---|---|
| Input Capacitance<br>($V_{CB} = 0.5$ Vdc, $I_C = 0$, $f = 100$ kHz) | | $C_{ibo}$ | — | 10.0 | pF |
| Input Impedance<br>($I_C = 1.0$ mAdc, $V_{CE} = 10$ Vdc, $f = 1.0$ kHz) | 2N3905<br>2N3906 | $h_{ie}$ | 0.5<br>2.0 | 8.0<br>12 | k ohms |
| Voltage Feedback Ratio<br>($I_C = 1.0$ mAdc, $V_{CE} = 10$ Vdc, $f = 1.0$ kHz) | 2N3905<br>2N3906 | $h_{re}$ | 0.1<br>0.1 | 5.0<br>10 | X 10⁻⁴ |
| Small-Signal Current Gain<br>($I_C = 1.0$ mAdc, $V_{CE} = 10$ Vdc, $f = 1.0$ kHz) | 2N3905<br>2N3906 | $h_{fe}$ | 50<br>100 | 200<br>400 | — |
| Output Admittance<br>($I_C = 1.0$ mAdc, $V_{CE} = 10$ Vdc, $f = 1.0$ kHz) | 2N3905<br>2N3906 | $h_{oe}$ | 1.0<br>3.0 | 40<br>60 | $\mu$mhos |
| Noise Figure<br>($I_C = 100\ \mu$Adc, $V_{CE} = 5.0$ Vdc, $R_S = 1.0$ k ohms,<br>$f = 10$ Hz to 15.7 kHz) | 2N3905<br>2N3906 | NF | —<br>— | 5.0<br>4.0 | dB |

**SWITCHING CHARACTERISTICS**

| Delay Time | ($V_{CC} = 3.0$ Vdc, $V_{BE} = 0.5$ Vdc, | | $t_d$ | — | 35 | ns |
|---|---|---|---|---|---|---|
| Rise Time | $I_C = 10$ mAdc, $I_{B1} = 1.0$ mAdc) | | $t_r$ | — | 35 | ns |
| Storage Time | | 2N3905<br>2N3906 | $t_s$ | —<br>— | 200<br>225 | ns |
| Fall Time | ($V_{CC} = 3.0$ Vdc, $I_C = 10$ mAdc,<br>$I_{B1} = I_{B2} = 1.0$ mAdc) | 2N3905<br>2N3904 | $t_f$ | —<br>— | 60<br>75 | ns |

(1) Pulse Test: Pulse Width $\leq 300\ \mu$s, Duty Cycle $\leq 2.0\%$.

FIGURE 1 — DELAY AND RISE TIME
EQUIVALENT TEST CIRCUIT

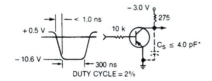

FIGURE 2 — STORAGE AND FALL TIME
EQUIVALENT TEST CIRCUIT

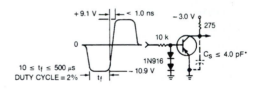

*Total shunt capacitance of test jig and connectors

## MAXIMUM RATINGS

| Rating | Symbol | 2N4013 | 2N4014 | Unit |
|---|---|---|---|---|
| Collector-Emitter Voltage | $V_{CEO}$ | 30 | 50 | Vdc |
| Collector-Base Voltage | $V_{CBO}$ | 50 | 80 | Vdc |
| Emitter-Base Voltage | $V_{EBO}$ | 6.0 | | Vdc |
| Collector Current — Continuous<br>— Peak | $I_C$ | 1.0<br>2.0 | | Adc |
| Total Device Dissipation @ $T_A$ = 25°C<br>Derate above 25°C | $P_D$ | 0.5<br>28.6 | | Watt<br>mW/°C |
| Total Device Dissipation @ $T_C$ = 25°C<br>Derate above 25°C | $P_D$ | 1.4<br>6.8 | | Watts<br>mW/°C |
| Operating and Storage Junction<br>Temperature Range | $T_J$, $T_{stg}$ | −65 to +200 | | °C |

## 2N4013
## 2N4014

**CASE 22-03, STYLE 1**
**TO-18 (TO-206AA)**

3 Collector
2 Base
1 Emitter

## SWITCHING TRANSISTORS

NPN SILICON

## ELECTRICAL CHARACTERISTICS ($T_A$ = 25°C unless otherwise noted.)

| Characteristic | | Symbol | Min | Typ | Max | Unit |
|---|---|---|---|---|---|---|
| **OFF CHARACTERISTICS** | | | | | | |
| Collector-Emitter Breakdown Voltage(1)<br>($I_C$ = 10 mAdc, $I_B$ = 0) | 2N4014<br>2N4013 | $V_{(BR)CEO}$ | 50<br>30 | —<br>— | —<br>— | Vdc |
| Collector-Emitter Breakdown Voltage<br>($I_C$ = 10 μAdc, $V_{BE}$ = 0) | 2N4014<br>2N4013 | $V_{(BR)CES}$ | 80<br>50 | —<br>— | —<br>— | Vdc |
| Collector-Base Breakdown Voltage<br>($I_C$ = 10 μAdc, $I_E$ = 0) | 2N4014<br>2N4013 | $V_{(BR)CBO}$ | 80<br>50 | —<br>— | —<br>— | Vdc |
| Emitter-Base Breakdown Voltage<br>($I_E$ = 10 μAdc, $I_C$ = 0) | | $V_{(BR)EBO}$ | 6.0 | — | — | Vdc |
| Collector Cutoff Current<br>($V_{CB}$ = 60 Vdc, $I_E$ = 0)<br>($V_{CB}$ = 40 Vdc, $I_E$ = 0)<br>($V_{CB}$ = 60 Vdc, $I_E$ = 0, $T_A$ = 100°C)<br>($V_{CB}$ = 40 Vdc, $I_E$ = 0, $T_A$ = 100°C) | 2N4014<br>2N4013<br>2N4014<br>2N4013 | $I_{CBO}$ | —<br>—<br>—<br>— | 0.12<br>0.12<br>—<br>— | 1.7<br>1.7<br>120<br>120 | μAdc |
| Collector Cutoff Current<br>($V_{CE}$ = 80 Vdc, $V_{EB}$ = 0)<br>($V_{CE}$ = 50 Vdc, $V_{EB}$ = 0) | 2N4014<br>2N4013 | $I_{CES}$ | —<br>— | 0.15<br>0.15 | 10<br>10 | μAdc |
| **ON CHARACTERISTICS(1)** | | | | | | |
| DC Current Gain<br>($I_C$ = 10 mAdc, $V_{CE}$ = 1.0 Vdc)<br>($I_C$ = 100 mAdc, $V_{CE}$ = 1.0 Vdc)<br>($I_C$ = 100 mAdc, $V_{CE}$ = 1.0 Vdc, $T_A$ = −55°C)<br>($I_C$ = 300 mAdc, $V_{CE}$ = 1.0 Vdc)<br>($I_C$ = 500 mAdc, $V_{CE}$ = 1.0 Vdc)<br>($I_C$ = 500 mAdc, $V_{CE}$ = 1.0 Vdc, $T_A$ = −55°C)<br>($I_C$ = 800 mAdc, $V_{CE}$ = 2.0 Vdc)<br><br>($I_C$ = 1.0 Adc, $V_{CE}$ = 5.0 Vdc) | <br><br><br><br><br><br>2N4014<br>2N4013<br><br>2N4014<br>2N4013 | $h_{FE}$ | 30<br>60<br>30<br>40<br>35<br>20<br>20<br>25<br><br>25<br>30 | —<br>—<br>—<br>—<br>—<br>—<br>—<br>—<br><br>—<br>— | —<br>150<br>—<br>—<br>—<br>—<br><br><br><br><br> | — |

**(Copyright of Motorola. Used by permission.)**

## MAXIMUM RATINGS

| Rating | Symbol | Value | Unit |
|---|---|---|---|
| Drain-Source Voltage | $V_{DS}$ | 25 | Vdc |
| Drain-Gate Voltage | $V_{DG}$ | 25 | Vdc |
| Reverse Gate-Source Voltage | $V_{GSR}$ | −25 | Vdc |
| Gate Current | $I_G$ | 10 | mAdc |
| Total Device Dissipation @ $T_A = 25°C$<br>Derate above 25°C | $P_D$ | 310<br>2.82 | mW<br>mW/°C |
| Junction Temperature Range | $T_J$ | 125 | °C |
| Storage Channel Temperature Range | $T_{stg}$ | −65 to +150 | °C |

**2N5457
thru
2N5459**

**CASE 29-04, STYLE 5
TO-92 (TO-226AA)**

1 Drain

3 Gate

2 Source

**JFETs
GENERAL PURPOSE**

N-CHANNEL — DEPLETION

Refer to 2N4220 for graphs.

## ELECTRICAL CHARACTERISTICS ($T_A = 25°C$ unless otherwise noted.)

| Characteristic | | Symbol | Min | Typ | Max | Unit |
|---|---|---|---|---|---|---|
| **OFF CHARACTERISTICS** | | | | | | |
| Gate-Source Breakdown Voltage<br>($I_G = -10~\mu Adc$, $V_{DS} = 0$) | | $V_{(BR)GSS}$ | −25 | — | — | Vdc |
| Gate Reverse Current<br>($V_{GS} = -15$ Vdc, $V_{DS} = 0$)<br>($V_{GS} = -15$ Vdc, $V_{DS} = 0$, $T_A = 100°C$) | | $I_{GSS}$ | <br>—<br>— | <br>—<br>— | <br>−1.0<br>−200 | nAdc |
| Gate Source Cutoff Voltage<br>($V_{DS} = 15$ Vdc, $I_D = 10$ nAdc) | 2N5457<br>2N5458<br>2N5459 | $V_{GS(off)}$ | −0.5<br>−1.0<br>−2.0 | —<br>—<br>— | −6.0<br>−7.0<br>−8.0 | Vdc |
| Gate Source Voltage<br>($V_{DS} = 15$ Vdc, $I_D = 100~\mu Adc$)<br>($V_{DS} = 15$ Vdc, $I_D = 200~\mu Adc$)<br>($V_{DS} = 15$ Vdc, $I_D = 400~\mu Adc$) | 2N5457<br>2N5458<br>2N5459 | $V_{GS}$ | —<br>—<br>— | −2.5<br>−3.5<br>−4.5 | —<br>—<br>— | Vdc |
| **ON CHARACTERISTICS** | | | | | | |
| Zero-Gate-Voltage Drain Current*<br>($V_{DS} = 15$ Vdc, $V_{GS} = 0$) | 2N5457<br>2N5458<br>2N5459 | $I_{DSS}$ | 1.0<br>2.0<br>4.0 | 3.0<br>6.0<br>9.0 | 5.0<br>9.0<br>16 | mAdc |
| **SMALL-SIGNAL CHARACTERISTICS** | | | | | | |
| Forward Transfer Admittance Common Source*<br>($V_{DS} = 15$ Vdc, $V_{GS} = 0$, f = 1.0 kHz) | 2N5457<br>2N5458<br>2N5459 | $|y_{fs}|$ | 1000<br>1500<br>2000 | —<br>—<br>— | 5000<br>5500<br>6000 | $\mu$mhos |
| Output Admittance Common Source*<br>($V_{DS} = 15$ Vdc, $V_{GS} = 0$, f = 1.0 kHz) | | $|y_{os}|$ | — | 10 | 50 | $\mu$mhos |
| Input Capacitance<br>($V_{DS} = 15$ Vdc, $V_{GS} = 0$, f = 1.0 MHz) | | $C_{iss}$ | — | 4.5 | 7.0 | pF |
| Reverse Transfer Capacitance<br>($V_{DS} = 15$ Vdc, $V_{GS} = 0$, f = 1.0 MHz) | | $C_{rss}$ | — | 1.5 | 3.0 | pF |

*Pulse Test: Pulse Width ≤ 630 ms; Duty Cycle ≤ 10%.

## MAXIMUM RATINGS

| Rating | Symbol | Value | Unit |
|---|---|---|---|
| Collector-Emitter Voltage | $V_{CEO}$ | 40 | Vdc |
| Collector-Base Voltage | $V_{CBO}$ | 40 | Vdc |
| Emitter-Base Voltage | $V_{EBO}$ | 12 | Vdc |
| Collector Current — Continuous | $I_C$ | 500 | mAdc |
| Total Device Dissipation @ $T_A$ = 25°C<br>Derate above 25°C | $P_D$ | 625<br>5.0 | mW<br>mW/°C |
| Total Device Dissipation @ $T_C$ = 25°C<br>Derate above 25°C | $P_D$ | 1.5<br>12 | Watts<br>mW/°C |
| Operating and Storage Junction<br>Temperature Range | $T_J, T_{stg}$ | − 55 to + 150 | °C |

## THERMAL CHARACTERISTICS

| Characteristic | Symbol | Max | Unit |
|---|---|---|---|
| Thermal Resistance, Junction to Case | $R_{\theta JC}$ | 83.3 | °C/W |
| Thermal Resistance, Junction to Ambient | $R_{\theta JA}$(1) | 200 | °C/W |

(1) $R_{\theta JA}$ is measured with the device soldered into a typical printed circuit board.

## 2N6426
## 2N6427

### CASE 29-04, STYLE 1
### TO-92 (TO-226AA)

Collector 3

Base 2

Emitter 1

### DARLINGTON TRANSISTORS

NPN SILICON

## ELECTRICAL CHARACTERISTICS ($T_A$ = 25°C unless otherwise noted.)

| Characteristic | | Symbol | Min | Typ | Max | Unit |
|---|---|---|---|---|---|---|
| **OFF CHARACTERISTICS** | | | | | | |
| Collector-Emitter Breakdown Voltage(2)<br>($I_C$ = 10 mAdc, $V_{BE}$ = 0) | | $V_{(BR)CES}$ | 40 | — | — | Vdc |
| Collector-Base Breakdown Voltage<br>($I_C$ = 100 µAdc, $I_E$ = 0) | | $V_{(BR)CBO}$ | 40 | — | — | Vdc |
| Emitter-Base Breakdown Voltage<br>($I_E$ = 10 µAdc, $I_C$ = 0) | | $V_{(BR)EBO}$ | 12 | — | — | Vdc |
| Collector Cutoff Current<br>($V_{CE}$ = 25 Vdc, $I_B$ = 0) | | $I_{CEO}$ | — | — | 1.0 | µAdc |
| Collector Cutoff Current<br>($V_{CB}$ = 30 Vdc, $I_E$ = 0) | | $I_{CBO}$ | — | — | 50 | nAdc |
| Emitter Cutoff Current<br>($V_{BE}$ = 10 Vdc, $I_C$ = 0) | | $I_{EBO}$ | — | — | 50 | nAdc |
| **ON CHARACTERISTICS** | | | | | | |
| DC Current Gain(2)<br>($I_C$ = 10 mAdc, $V_{CE}$ = 5.0 Vdc) | 2N6426<br>2N6427 | $h_{FE}$ | 20,000<br>10,000 | —<br>— | 200,000<br>100,000 | — |
| ($I_C$ = 100 mAdc, $V_{CE}$ = 5.0 Vdc) | 2N6426<br>2N6427 | | 30,000<br>20,000 | —<br>— | 300,000<br>200,000 | |
| ($I_C$ = 500 mAdc, $V_{CE}$ = 5.0 Vdc) | 2N6426<br>2N6427 | | 20,000<br>14,000 | —<br>— | 200,000<br>140,000 | |
| Collector-Emitter Saturation Voltage<br>($I_C$ = 50 mAdc, $I_B$ = 0.5 mAdc)<br>($I_C$ = 500 mAdc, $I_B$ = 0.5 mAdc) | | $V_{CE(sat)}$ | —<br>— | 0.71<br>0.9 | 1.2<br>1.5 | Vdc |
| Base-Emitter Saturation Voltage<br>($I_C$ = 500 mAdc, $I_B$ = 0.5 mAdc) | | $V_{BE(sat)}$ | — | 1.52 | 2.0 | Vdc |
| Base-Emitter On Voltage<br>($I_C$ = 50 mAdc, $V_{CE}$ = 5.0 Vdc) | | $V_{BE(on)}$ | — | 1.24 | 1.75 | Vdc |
| **SMALL-SIGNAL CHARACTERISTICS** | | | | | | |
| Output Capacitance<br>($V_{CB}$ = 10 Vdc, $I_E$ = 0, f = 1.0 MHz) | | $C_{obo}$ | — | 5.4 | 7.0 | pF |
| Input Capacitance<br>($V_{BE}$ = 1.0 Vdc, $I_C$ = 0, f = 1.0 MHz) | | $C_{ibo}$ | — | 10 | 15 | pF |

**(Copyright of Motorola. Used by permission.)**

**(Continued)**

(Continued)

## 2N6426, 2N6427

**ELECTRICAL CHARACTERISTICS** (continued) ($T_A$ = 25°C unless otherwise noted.)

| Characteristic | | Symbol | Min | Typ | Max | Unit |
|---|---|---|---|---|---|---|
| Input Impedance ($I_C$ = 10 mAdc, $V_{CE}$ = 5.0 Vdc, f = 1.0 kHz)    2N6426<br>2N6427 | | $h_{ie}$ | 100<br>50 | —<br>— | 2000<br>1000 | k Ω |
| Small-Signal Current Gain ($I_C$ = 10 mAdc, $V_{CE}$ = 5.0 Vdc, f = 1.0 kHz)    2N6426<br>2N6427 | | $h_{fe}$ | 20,000<br>10,000 | —<br>— | —<br>— | — |
| Current Gain — High Frequency ($I_C$ = 10 mAdc, $V_{CE}$ = 5.0 Vdc, f = 100 MHz)    2N6426<br>2N6427 | | $|h_{fe}|$ | 1.5<br>1.3 | 2.4<br>2.4 | —<br>— | — |
| Output Admittance ($I_C$ = 10 mAdc, $V_{CE}$ = 5.0 Vdc, f = 1.0 kHz) | | $h_{oe}$ | — | — | 1000 | $\mu$mhos |
| Noise Figure ($I_C$ = 1.0 mAdc, $V_{CE}$ = 5.0 Vdc, $R_S$ = 100 k$\Omega$, f = 1.0 kHz) | | NF | — | 3.0 | 10 | dB |

(2) Pulse Test: Pulse Width ≤ 300 $\mu$s, Duty Cycle ≤ 2.0%.

# STANDARD RESISTOR VALUES

### 10% Tolerance

| Ω | | | | kΩ | | | MΩ | |
|---|---|---|---|---|---|---|---|---|
| 0.10 | 1.0 | 10 | 100 | 1.0 | 10 | 100 | 1.0 | 10.0 |
| 0.12 | 1.2 | 12 | 120 | 1.2 | 12 | 120 | 1.2 | 12.0 |
| 0.15 | 1.5 | 15 | 150 | 1.5 | 15 | 150 | 1.2 | 15.0 |
| 0.18 | 1.8 | 18 | 180 | 1.8 | 18 | 180 | 1.8 | 18.0 |
| 0.22 | 2.2 | 22 | 220 | 2.2 | 22 | 220 | 2.2 | 22.0 |
| 0.27 | 2.7 | 27 | 270 | 2.7 | 27 | 270 | 2.7 | |
| 0.33 | 3.3 | 33 | 330 | 3.3 | 33 | 330 | 3.3 | |
| 0.39 | 3.9 | 39 | 390 | 3.9 | 39 | 390 | 3.9 | |
| 0.47 | 4.7 | 47 | 470 | 4.7 | 47 | 470 | 4.7 | |
| 0.56 | 5.6 | 56 | 560 | 5.6 | 56 | 560 | 5.6 | |
| 0.68 | 6.8 | 68 | 680 | 6.8 | 68 | 680 | 6.8 | |
| 0.82 | 8.2 | 82 | 820 | 8.2 | 82 | 820 | 8.2 | |

In addition to the values listed in the table, multiples of the following values are available with 5% tolerance only:

| 1.1 | 2.4 | 5.1 |
|---|---|---|
| 1.3 | 3.0 | 6.2 |
| 1.6 | 3.6 | 7.5 |
| 2.0 | 4.3 | 9.1 |

# Approximating Circuit Values

The analysis of a given circuit can often be simplified by approximating many of the resistance and current values in the circuit. In this appendix, we will look at the methods by which you can approximate circuit values and the circumstances that make circuit approximations valid.

## WHAT IS MEANT BY *APPROXIMATING CIRCUIT VALUES?*

When you approximate circuit values, you ignore any resistance and/or current values that will have little impact on the analysis of a given circuit. For example, consider the simple series circuit shown in Figure B.1a. What is total resistance in this circuit? Obviously, it is 10,000,010 $\Omega$. However, *for all practical purposes,* couldn't we just say that we have *approximately* 10 M$\Omega$? In this circuit, the value of $R_1$ will have very little, if any, visible effect on the values of $R_T$, $I_T$, or $V_{R2}$. In other words, these three values would have the same values whether calculated using the value of $R_1$ or assuming that $R_1$ is not in the circuit. Therefore, the value of $R_1$ can be dropped from the circuit with very little loss in the accuracy of our calculations.

Now, take a look at the circuit shown in Figure B.1b. In terms of circuit *current,* which component could be ignored? In this case, it would be valid to ignore the value of $R_2$ since there is very little current through this branch. In fact, if we assume the voltage source to be 10 V, we could use Ohm's law to calculate the current values of $I_1 = 1$ A and $I_2 = 1$ $\mu$A. In this case, the value of $I_2$ has virtually no effect on the value of total circuit current, and thus can be ignored in circuit calculations.

Later, we'll establish the guidelines for approximating circuit values. At this point, we'll take a look at *when* it is valid to use circuit approximations.

**FIGURE B.1**

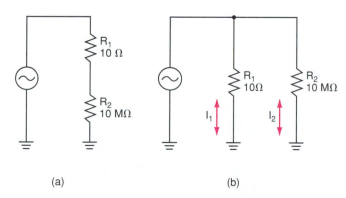

(a)          (b)

## WHEN CAN YOU USE CIRCUIT APPROXIMATIONS?

Any time that you are troubleshooting a given circuit, you can assume that it is all right to use approximated values. Generally, when you are troubleshooting, you are interested only in whether or not a voltage or resistance is *close to* its rated value. For example, consider the circuit shown in Figure B.2. Assume that you are troubleshooting this circuit. As the figure shows, $V_B$ should be a 1 $mV_{PP}$ sine wave, $V_E$ should be a 1 $V_{PP}$ sine wave, $V_C$ should be an 8 $V_{PP}$ sine wave that is 180° out of phase with the other two voltages. When checking this circuit with an oscilloscope, you would not be concerned with whether or not a given value of $V_{PP}$ was off by 0.001V. You would be concerned only with whether or not the value of $V_{PP}$ was *close* to the given value, and whether or not the phase relationships were correct. Therefore, any approximations of circuit $V_{PP}$ values would be completely acceptable.

There are two circumstances when you would not use the circuit approximations covered here. The first is when you are performing an exact circuit analysis involving *h*-parameters. *H*-parameter equations, such as those introduced in Appendix C, lead to very exact values for transistor gain and input/output impedance. It wouldn't make much sense to use these exact equations and then approximate all of the values external to the transistor.

The second instance for not using circuit approximations is when you are designing a circuit for a specific application.

As a summary, the circuit approximation techniques that we are about to discuss can usually be used for general circuit analysis and troubleshooting. They may not be used when an exact analysis is required (such as those involving *h*-parameters) or when designing a circuit.

## APPROXIMATING CIRCUIT RESISTANCE VALUES

*A Practical Consideration:* In this discussion, we are assuming that the tolerance of the resistors used is 10%. For this reason, a percentage of error of 10% in our calculations is considered acceptable. In practice, the acceptable margin of error should be no greater than the tolerance of the components used.

Generally, you can ignore the value of a resistor in a series circuit *when the value of that component is less than 10% of the value of the next smallest resistance value.* For example, take a look at the voltage-dividers shown in Figure B.3.

In Figure B.3a, the value of $R_2$ is equal to $0.1R_1$. Therefore, we can approximate the total resistance in the circuit as being equal to the value of $R_1$, 4.7 kΩ. In Figure B.3b, the value of $R_3$ can be ignored because it is less than $0.1R_1$, and $R_1$ is the next lowest resistance value in the circuit. Note that the value of $R_2$ is not considered because it is greater than the value of $R_1$. In Figure B.3c, none of the resistors can be dropped from the circuit. Why? Even though $R_1$ is less than $0.1R_2$, it is not less than $0.1R_3$. Since it is not less than (or equal to) 10% of the *next lowest individual resistance value,* it must be kept in the circuit.

**FIGURE B.2**

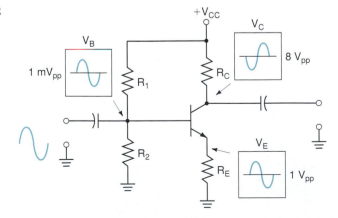

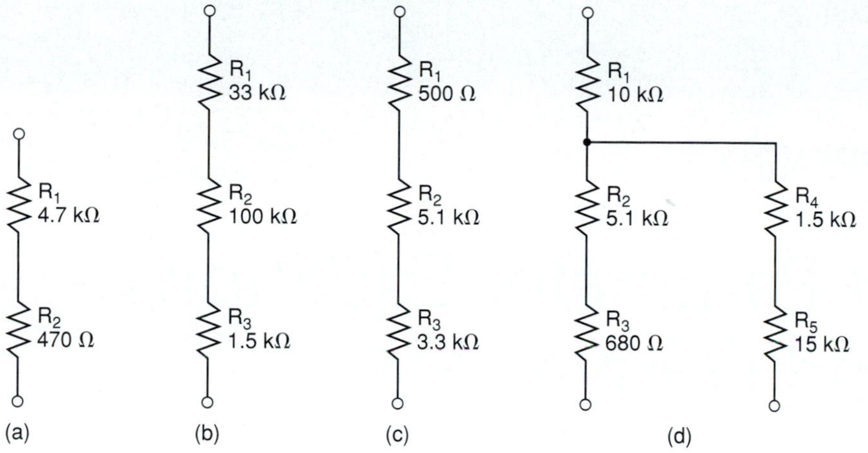

FIGURE B.3

Which resistor (if any) can be dropped from Figure B.3d? If you picked $R_4$, you were correct. The value of $R_4$ is 10% of the value of $R_5$, so it can be dropped from the circuit. Why doesn't the value of $R_1$ (10 k$\Omega$) prevent us from dropping $R_4$? The circuit in Figure B.3d is a *series-parallel* circuit. While $R_1$ is in series with the *combination* of $R_2$, $R_3$, $R_4$, and $R_5$, it is not considered to be in series with $R_4$ directly. This is due to the fact that the current through $R_1$ is not necessarily equal to the current through $R_4$. By definition, two components are in series only when the current through the two components is equal.

For parallel circuits, the 10% rule still holds. In this case, however, *the larger resistance value is the one that can be dropped.* The reason for this can be seen by referring back to Figure B.1b. For the circuit shown, the larger resistor provided less than 10% of the total circuit current. Thus, its value could easily be ignored.

# APPENDIX C

# *H*-Parameter Equations and Derivations

In Chapter 9, you were introduced to the four transistor *h*-parameters and their use in basic circuit analysis. In this appendix, we'll take a more in-depth look at *h*-parameter equations and the transistor hybrid equivalent circuit.

## THE TRANSISTOR HYBRID EQUIVALENT CIRCUIT

Using the *h*-parameter values described in Chapter 9, the *hybrid equivalent circuit* of a transistor is developed. To understand this ac equivalent circuit, the transistor must be viewed as a 4-terminal device. This representation of the transistor is shown in Figure C.1. Since the emitter terminal of the transistor is common to both the input and output circuits in a common emitter amplifier, it is represented as two separate terminals. Realize, however, that both of these terminals represent the single transistor emitter. The input and output voltages are represented as $v_1$ and $v_2$, respectively.

## INPUT CIRCUIT

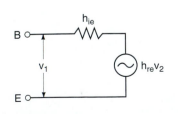

**FIGURE C.1**

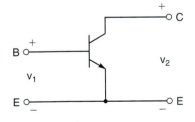

**FIGURE C.2**

The ac equivalent of the input circuit is derived using Thevenin's theorem. Thevenin's theorem states that any circuit can be represented as a single voltage source in series with a single resistance. The hybrid equivalent circuit for the transistor input is derived as illustrated in Figure C.2.

The Thevenin voltage for the input is equal to the value of $v_1$ with the base circuit open. As Figure C.2 shows, this voltage is found as

$$v_1 \cong h_{re}v_2 \qquad \text{(C.1)}$$

This equation is simply another form of equation (9.30). Since $v_{be} = v_1$, and $v_{ce} = v_2$, equation (9.30) can be rewritten as above.

The Thevenin resistance of the transistor input is equal to $h_{ie}$. This value is found as shown in Chapter 9.

So why is the input represented as a Thevenin equivalent circuit? Since the voltage across the base-emitter junction remains fairly constant, it is convenient to define the operation of the input circuit in terms of this voltage. Any changes in $i_b$ or $v_b$ do not change the voltage drop across the emitter-base junction inside the transistor. Therefore, it is a good reference for discussing the ac operation of the input circuit.

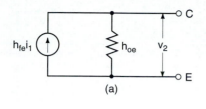

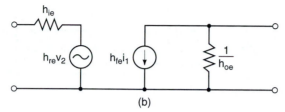

**FIGURE C.3**

# OUTPUT CIRCUIT

The hybrid equivalent of the output circuit is derived using Norton's theorem. Norton's theorem represents a circuit as a current source in parallel with a single resistance. The ac equivalent of the output is represented in Figure C.3.

The Norton current source is represented as the current gain of the transistor ($h_{fe}$) times the base input current ($i_1$). For formula,

$$i_2 = h_{fe}i_1 \tag{C.2}$$

This equation is simply another form of equation (9.28). Since $i_b = i_1$ and $i_c = i_2$, equation (9.28) can be rewritten as above.

The Norton resistance is equal to the collector voltage ($v_2$) divided by the collector current ($i_1$). By formula,

$$r = \frac{v_c}{i_c} = \frac{v_2}{i_2}$$

Since $h_{oe}$ is the reciprocal of the above relationship, the Norton equivalent resistance is found as

$$\frac{1}{h_{oe}} = \frac{v_c}{i_c} = \frac{v_2}{i_2} \tag{C.3}$$

The output circuit is represented as a Norton equivalent circuit because of the fact that $i_c$ is dependent upon factors outside the collector-emitter circuit under normal circumstances. For example, changing the value of $R_C$ will not cause $i_c$ to change significantly. However, changing $i_b$ *will*. Since the transistor output current is relatively independent of the external component values, it is the reference value for output calculations.

The complete hybrid equivalent circuit for the transistor is shown in Figure C.3b. This circuit will be used throughout our discussions on *h*-parameter calculations.

# CALCULATIONS INVOLVING *H*-PARAMETERS

Very exact values of transistor current gain ($A_i$), voltage gain ($A_v$), input impedance ($Z_{in}$), and output impedance ($Z_{out}$) can be obtained using *h*-parameters. *Note that these calculations give you exact values of operation at specific values of $I_C$. If $I_C$ changes, the values of $A_v$, $A_i$, $Z_{in}$, and $Z_{out}$ obtained with the h-parameters will all change.*

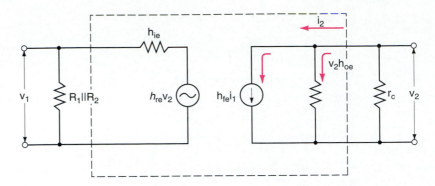

**FIGURE C.4**

## CURRENT GAIN ($A_i$)

*A Practical Consideration:* The discussion here applies *only* to the current gain of the *transistor* in a common-emitter amplifier. The current gain of a common-emitter *amplifier* is found using equation (9.21). The derivation of this equation appears in Appendix D.

The current gain of a transistor is defined as the ratio of output current to input current. By formula,

$$A_i = \frac{i_2}{i_1} \qquad \text{(C.4)}$$

Now, let's apply this equation to the circuit shown in Figure C.4. This circuit is an amplifier ac equivalent circuit. The only difference between this circuit and those discussed earlier is that the transistor has been replaced by its hybrid equivalent circuit. The boundaries of the transistor are represented by the dotted line.

The output current ($i_2$) is equal to the sum of $h_{fe}i_1$ and the current produced by the application of $v_2$ across the Norton resistance. By formula,

$$i_2 = h_{fe}i_1 + v_2 h_{oe} \qquad \text{(C.5)}$$

By substituting this value in place of $i_2$ in equation (9.7), we get

$$A_i = \frac{h_{fe}i_1 + v_2 h_{oe}}{i_1}$$

which can be rewritten as

$$A_i = \frac{h_{fe}i_1}{i_1} + \frac{v_2 h_{oe}}{i_1}$$

or

$$A_i = h_{fe} + \frac{v_2 h_{oe}}{i_1}$$

Since $v_2 = i_2 r_C$,

$$A_i = h_{fe} + \frac{i_2 r_C h_{oe}}{i_1}$$

which can be rewritten as

$$A_i = h_{fe} + \left(\frac{i_2}{i_1}\right) r_C h_{oe}$$

This equation is now rewritten as

$$A_i = h_{fe} + A_i r_C h_{oe}$$

With further algebraic manipulation, this equation becomes

$$A_i = \frac{h_{fe}}{1 + r_C h_{oe}} \qquad \text{(C.6)}$$

Equation (C.6) is our final goal: a formula that defines $A_i$ strictly in terms of the transistor $h$-parameters and the ac resistance of the collector circuit. The following example shows how equation (C.6) is used.

EXAMPLE C.1

Determine the value of $A_i$ for the circuit shown in Figure C.5. Assume that the $h$-parameter values for the transistor are as follows:

$$h_{ie} = 1 \text{ k}\Omega$$
$$h_{re} = 2.5 \times 10^{-4}$$
$$h_{fe} = 50$$
$$h_{oe} = 25 \text{ }\mu\text{S}$$

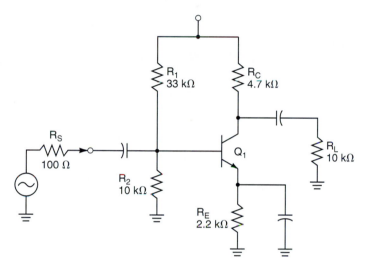

**FIGURE C.5**

**Solution:** To determine the value of $A_i$, we need to determine $r_C$. For the circuit shown,

$$r_C = R_C \| R_L$$
$$= 3.2 \text{ k}\Omega$$

Now, $A_i$ is determined as

$$A_i = \frac{h_{fe}}{1 + r_C h_{oe}}$$

$$= \frac{50}{1 + (32 \text{ k}\Omega)(25 \text{ }\mu\text{S})}$$

$$= 46.3$$

If you take a look at the final result in Example C.1, you'll notice that it is very close to the original value of $h_{fe}$. The reason for this is the fact that $(r_C h_{oe}) \ll 1$. Since this is normally the case, equation (C.6) can be approximated as

$$A_i \cong h_{fe} \qquad \text{(C.7)}$$

# VOLTAGE GAIN $(A_v)$

The voltage gain of an amplifier is defined as the ratio of output voltage to input voltage. By formula,

$$A_v = \frac{v_2}{v_1} \tag{C.8}$$

Now, take a look at the circuit shown in Figure C.6. As is shown, $v_1$ is the sum of $h_{re}v_2$ and the voltage developed across $h_{ie}$. The voltage across $h_{ie}$ is found as $h_{ie}i_1$; therefore,

$$v_1 = h_{re}v_2 + h_{ie}i_1 \tag{C.9}$$

It was already shown that $v_2 = i_2r_C$, and substituting these two equations into equation (C.8) we get

$$A_v = \frac{v_2}{v_1} = \frac{i_2r_C}{h_{re}v_2 + h_{ie}i_1}$$

If we divide both the numerator and the denominator of the fraction by $i_2$, we get

$$A_v = \frac{r_C}{h_{re}r_C + h_{ie}/A_i} \tag{C.10}$$

Confused? When you divide the numerator by $i_2$, you get

$$\frac{i_2r_C}{i_2} = r_C$$

When you divide $h_{re}v_2$ by $i_2$, you get

$$\frac{h_{re}v_2}{i_2} = h_{re}\left(\frac{v_2}{i_2}\right) = h_{re}r_C$$

Finally, when you divide $h_{ie}i_1$ by $i_2$, you get

$$\frac{h_{ie}i_1}{i_2} = h_{ie}\left(\frac{i_1}{i_2}\right) = h_{ie}\left(\frac{1}{A_i}\right) = \frac{h_{ie}}{A_i}$$

These values were substituted into equation (C.10).

The final exact equation for $A_v$ is derived by substituting equation (C.6) for $A_i$ in equation (C.10). This gives us

$$A_v = \frac{h_{fe}r_C}{h_{ie} + (h_{ie}h_{oe} - h_{re}h_{fe})r_C} \tag{C.11}$$

Again, we have defined a gain value strictly in terms of the transistor $h$-parameters and the ac collector resistance. The following example shows the use of this equation.

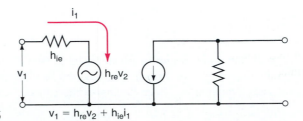

**FIGURE C.6**    $v_1 = h_{re}v_2 + h_{ie}i_1$

EXAMPLE C.2

Determine the voltage gain for the circuit shown in Figure C.5 (Example C.1).

**Solution:** Using the *h*-parameter values listed in Example C.1, we get

$$A_v = \frac{h_{fe}r_C}{h_{ie} + (h_{ie}h_{oe} - h_{re}h_{fe})r_C}$$

$$= \frac{(50)(3.2 \text{ k}\Omega)}{1 \text{ k}\Omega + [(1 \text{ k}\Omega)(25 \text{ }\mu\text{S}) - (2.5 \times 10^{-4})(50)]3.2 \text{ k}\Omega}$$

$$= \frac{1.6 \times 10^5}{1.04 \text{ k}\Omega}$$

$$= 153.85$$

Anyone care for an aspirin? Don't give up yet . . . it gets better. You see, the input impedance of a transistor can be approximated as being equal to $h_{ie}$. When we determined $Z_{\text{in(base)}}$, we defined it as being equal to $h_{fe}r'_e$. Now, based on these two relationships, we'll define $r'_e$ as follows:

$$r'_e = \frac{h_{ie}}{h_{fe}} \qquad \text{(C.12)}$$

Now, let's try attacking the problem in Example C.2 from another angle.

EXAMPLE C.3

Determine the voltage gain for the circuit shown in Figure C.5 (Example C.1).

**Solution:** First, using the *h*-parameters listed in Example C.1, we'll determine the value of $r'_e$ as

$$r'_e = \frac{1 \text{ k}\Omega}{50}$$

$$= 20 \text{ V}$$

The voltage gain can now be calculated as

$$A_v = \frac{r_C}{r'_e}$$

$$= \frac{3.2 \text{ k}\Omega}{20 \text{ }\Omega}$$

$$= 160$$

You will find that the method of determining $A_v$ used in Example C.3 will always give you a result that is well within 10% of the exact value. For almost every analysis and design problem, this value of $A_v$ will work just fine.

## INPUT IMPEDANCE ($Z_{\text{in}}$)

The input impedance to a transistor is defined as the ratio of input voltage ($v_1$) to input current ($i_1$). By formula,

$$Z_{\text{in}} = \frac{v_1}{i_1} \qquad \text{(C.13)}$$

If you refer back to Figure C.6, you'll recall that $v_1 = h_{re}v_2 + h_{ie}i_1$. Substituting this value of $v_1$ into equation (C.13) gives us

$$Z_{\text{in}} = \frac{h_{re}v_2 + h_{ie}i_1}{i_1}$$

which simplifies to

$$Z_{\text{in}} = h_{ie} + \frac{h_{re}v_2}{i_1}$$

This equation can be rewritten as

$$Z_{\text{in}} = h_{ie} + h_{re}\left(\frac{v_2}{i_1}\right)$$

And

$$Z_{\text{in}} = h_{ie} + h_{re}\left(\frac{i_2 r_C}{i_1}\right)$$

Since $\dfrac{i_2}{i_1} = A_1$, the above equation can be rewritten as

$$Z_{\text{in}} = h_{ie} + h_{re}A_i r_C$$

Finally, substituting equation (C.6) into the above equation gives us

$$Z_{\text{in}} = h_{ie} + \frac{h_{re}h_{fe}r_C}{1 + r_C h_{oe}} \tag{C.14}$$

Remember earlier when we assumed that $Z_{\text{in}} \cong h_{ie}$? The following example will show this assumption to be valid.

---

**EXAMPLE C.4**

Determine the exact value of $Z_{\text{in}}$ for the amplifier shown in Example C.1.

**Solution:**  Using the $h$-parameter values from Example C.1, the exact value of $Z_{\text{in}}$ is found as

$$Z_{\text{in}} = h_{ie} + \frac{h_{re}h_{fe}r_C}{1 + r_C h_{oe}}$$

$$= 1\ k\Omega + \frac{(2.5 \times 10^{-4})(50)(3.2\ k\Omega)}{1 + (3.2\ k\Omega)(25\ \mu S)}$$

$$= 1\ k\Omega + \frac{40}{1.08}\ \Omega$$

$$= 1037\ \Omega$$

---

As you can see, the result of the calculation in Example C.4 was only 37 $\Omega$ away from the value of $h_{ie}$.

## OUTPUT IMPEDANCE

The output impedance of a transistor is defined as the ratio of output voltage ($v_2$) to output current ($i_2$). By formula,

$$Z_{\text{out}} = \frac{v_2}{i_2} \tag{C.15}$$

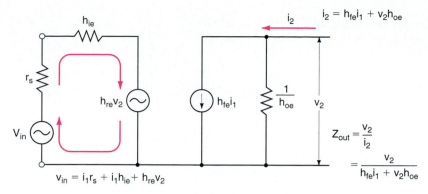

$$i_2 = h_{fe}i_1 + v_2h_{oe}$$

$$v_{in} = i_1r_s + i_1h_{ie} + h_{re}v_2$$

$$Z_{out} = \frac{v_2}{i_2}$$

$$= \frac{v_2}{h_{fe}i_1 + v_2h_{oe}}$$

**FIGURE C.7**

Substituting equation (C.5) in place of $i_2$ gives us

$$Z_{out} = \frac{v_2}{h_{fe}i_1 + v_2h_{oe}} \tag{C.16}$$

Now, take a look at the circuit shown in Figure C.7.

The *Kirchhoff*'s voltage equation for the input is

$$v_{in} = i_1r_s + i_1h_{ie} + h_{re}v_2$$

If we short out the voltage source (a normal step in thevenizing a circuit), we get

$$\begin{aligned}0 &= i_1r_s + i_1h_e + h_{re}v_2\\ &= i_1(r_s + h_{ie}) + h_{re}v_2\end{aligned}$$

Subtracting $(h_{re}v_2)$ from both sides of the equation gives us

$$i_1(r_s + h_{ie}) = -h_{re}v_2$$

And finally,

$$i_1 = \frac{-h_{re}v_2}{r_s + h_{ie}} \tag{C.17}$$

Now, if you substitute equation (C.17) for the value of $i_1$ in equation (C.16), you get

$$Z_{out} = \frac{r_s - h_{ie}}{(r_s + h_{ie})h_{oe} - h_{re}h_{fe}} \tag{C.18}$$

or

$$Z_{out} = \frac{1}{h_{oe} - \left(\dfrac{h_{fe}h_{re}}{h_{ie} + r_s}\right)} \tag{C.19}$$

Determine the value of the output impedance for the transistor used in Example C.1.

*EXAMPLE C.5*

**Solution:**   The source resistance is shown to be 100 Ω. Using this value, and the parameters given in Example C.1, $Z_{out}$ is found as

$$Z_{out} = \cfrac{1}{h_{oe} - \left(\cfrac{h_{fe}h_{re}}{h_{ie} + r_s}\right)}$$

$$= \cfrac{1}{(25 \ \mu S) - \cfrac{(50)(2.5 \times 10^{-4})}{(1 \ k\Omega) + (100 \ \Omega)}}$$

$$= 73.3 \ k\Omega$$

## *H*-PARAMETER APPROXIMATIONS

Most of the *h*-parameter formulas covered in this section can be approximated to a form that is easier to handle. While these formula approximations will not produce results that are as accurate as the original equations, they may be used for most applications.

Equation (C.6) shows *transistor* current gain ($A_i$) to be

$$A_i = \frac{h_{fe}}{1 + r_C h_{oe}}$$

Since $r_C h_{oe} << 1$, the denominator of the equation can be approximated as being equal to one. This leaves us with

$$A_i \cong h_{fe} \tag{C.20}$$

The validity of this approximation is demonstrated in Example C.1. In this example, the exact value of $A_i$ was determined to be 46.3. The value of $h_{fe}$ for the transistor was given as 50. The difference between these two gain values is 3.7; a difference of 7.4%.

The voltage gain ($A_v$) of an amplifier was found in equation (C.11) as

$$A_v = \frac{h_{fe} r_C}{h_{ie} + (h_{ie}h_{oe} - h_{re}h_{fe})r_C}$$

The approximation of this equation is based on several factors. First, we originally defined $A_v$ as

$$A_v = \frac{r_C}{r_e'}$$

This equation can also be written as

$$A_v = r_C\left(\frac{1}{r_e'}\right)$$

Now, equation (C.12) defined $r_e'$ as

$$r_e' = \frac{h_{ie}}{h_{fe}}$$

Therefore,

$$\frac{1}{r_e'} = \frac{h_{fe}}{h_{ie}}$$

Thus, $A_v$ can be approximated as

$$A_v \cong \frac{h_{fe} r_e}{h_{ie}} \tag{C.21}$$

The validity of this approximation was demonstrated in Examples C.2 and C.3. Example C.2 obtained an exact value of $A_v$ equal to 153.85. Example C.3 approximated the value of $A_v$ for the same circuit as 160. The percentage of error for the approximated value was about 4%.

Example C.4 calculated a $Z_{in}$ for an amplifier transistor of 1037 Ω. The value of $h_{ie}$ for the transistor was 1 kΩ. Thus, we can approximate the value of $Z_{in}$ as

$$Z_{in} \cong h_{ie} \tag{C.22}$$

Unfortunately, there is no quick approximation for the output impedance of a transistor. Whenever you need to calculate this value, you need to use equation (C.19). However, there is some good news here. In most cases, there is no reason to calculate the output impedance of a transistor. This is due to the fact that it rarely weighs into any common emitter circuit calculations. When you do need a general idea of how large $Z_{out}$ is for a given amplifier, you can use

$$Z_{out} > \frac{1}{h_{oe}} \tag{C.23}$$

While equation (C.23) will not tell you exactly what the value of $Z_{out}$ is, it will give you an idea of how large the value is.


# OTHER TRANSISTOR CONFIGURATIONS

*H*-parameters apply not only to common emitter circuits, but to common base and common collector circuits as well. However, you must use several conversion factors to obtain the values needed to calculate the $A_i$, $A_v$, $Z_{in}$, and $Z_{out}$ values for these circuit configurations. In this section, you will be introduced to the *h*-parameter conversions, as well as the common base and common collector formulas used for circuit calculations. We will not analyze these formulas and conversions. They are listed simply for future reference.

The common base *h*-parameters are identified by the use of the subscript letter *b* in place of *e*. The following chart illustrates this point.

| Parameter | Common Emitter | Common Base |
|---|---|---|
| Transistor current gain | $h_{fe}$ | $h_{fb}$ |
| Voltage feedback ratio | $h_{re}$ | $h_{rb}$ |
| Input impedance | $h_{ie}$ | $h_{ib}$ |
| Output admittance | $h_{oe}$ | $h_{ob}$ |

To convert the common emitter parameters to common base parameters, you use the following conversion formulas:

$$h_{fb} = \frac{h_{fe}(1 - h_{re}) - h_{ie}h_{oe}}{(1 + h_{fe})(1 - h_{re}) + h_{ie}h_{oe}} \tag{C.24}$$

$$h_{rb} = \frac{h_{ie}h_{oe} - h_{re}(1 + h_{oe})}{(1 + h_{fe})(1 - h_{re}) + h_{ie}h_{oe}} \tag{C.25}$$

$$h_{ib} = \frac{h_{ie}}{(1 + h_{fe})(1 - h_{re}) + h_{ie}h_{oe}} \tag{C.26}$$

$$h_{ob} = \frac{h_{oe}}{(1 + h_{fe})(1 - h_{re}) + h_{ie}h_{oe}} \tag{C.27}$$

The circuit gain and impedance values are now found as

$$A_i = \frac{h_{fb}}{1 + h_{ob}r_C} \tag{C.28}$$

$$A_v = \frac{h_{fb}r_C}{h_{ib} + (h_{ib}h_{ob} - h_{fb}h_{rb})r_C} \tag{C.29}$$

$$Z_{in} = h_{ib} - \frac{h_{rb}h_{fb}r_C}{1 + h_{ob}r_C} \tag{C.30}$$

$$Z_{out} = \frac{1}{h_{ob} - \dfrac{h_{fb}h_{rb}}{h_{ib} + r_s}} \tag{C.31}$$

The *common collector h*-parameters are identified by the letter $c$ in the subscript in place of $e$. The conversions from common emitter parameters to common collector parameters are as follows:

$$h_{fc} = h_{fe} + 1 \tag{C.32}$$

$$h_{rc} = 1 - h_{re} \tag{C.33}$$

$$h_{ic} = h_{ie} \tag{C.34}$$

$$h_{oc} = h_{oe} \tag{C.35}$$

The gains and impedances of the common collector amplifier are now found as

$$A_i = \frac{h_{fc}}{1 + h_{oc}r_E} \quad * \tag{C.36}$$

$$A_v = \frac{h_{fc}r_E}{h_{ic} + (h_{ic}h_{oc} - h_{fc}h_{rc})r_E} \tag{C.37}$$

$$Z_{in} = h_{ic} - \frac{h_{rc}h_{fc}r_E}{1 + h_{oc}r_E} \tag{C.38}$$

$$Z_{out} = \frac{1}{h_{oe} + \dfrac{h_{fc}h_{rc}}{h_{ic} + r_s}} \tag{C.39}$$

The actual gain and impedance equations shown in this section are exactly the same as those used for the common emitter circuit. However, before they can be used, the common emitter *h*-parameters must be converted to the necessary form.

---

* For the common collector circuit, the output is taken from the emitter. Therefore, the output ac resistance is the ac resistance of the emitter circuit, $r_E$.

# APPENDIX D

# Selected Equation Derivations

## EQUATION (9.2)

Schockley's equation for the total current through a *pn* junction is

$$I_T = I_S(\epsilon^{Vq/kT} - 1) \tag{D.1}$$

where

$I_S$ = the reverse saturation current through the diode
$\epsilon$ = the *exponential constant*; approximately 2.71828
$V$ = the voltage across the depletion layer
$q$ = the charge on an electron; approximately $1.6 \times 10^{-19}$ V
$k$ = Boltzmann's constant; approximately $1.38 \times 10^{-23}$ J/°K
$T$ = the temperature of the device, in degrees Kelvin (°K = °C + 273)

We can solve for the value of $q/kT$ at room temperature (approximately 21°C) as

$$q/kT = \frac{1.6 \times 10^{-19} \text{ V}}{(1.38 \times 10^{-23})(294°\text{K})}$$

$$= 40$$

Using this value, the equation for $I_T$ is rewritten as

$$I_T = I_S(\epsilon^{40V} - 1) \tag{D.2}$$

Now, differentiating the above equation gives us

$$\frac{dI}{dV} = 40 I_S \epsilon^{40V} \tag{D.3}$$

If we rearrange equation (D.2) as

$$I_S \epsilon^{40V} = (I_T + I_S)$$

we can use this equation to rewrite equation (D.3) as

$$\frac{dI}{dV} = 40(I_T + I_S) \tag{D.4}$$

Now, if we take the reciprocal of equation (D.4), we will have an equation for the ac resistance of the junction, $r'_e$, as follows:

$$\frac{dV}{dI} = \frac{1 \text{ V}}{40(I_T + I_S)}$$

$$= \frac{1 \text{ V}}{40} \frac{1}{I_T + I_S}$$

$$= 25 \text{ mV} \frac{1}{I_T + I_S}$$

$$= \frac{25 \text{ mV}}{I_T + I_S}$$

In this case, $(I_T + I_S)$ is the current through the emitter-base junction of the transistor, $I_E$. Therefore, we can rewrite the above equation as

$$r'_e = \frac{25 \text{ mV}}{I_E}$$

## EQUATION (9.21)

According to Ohm's law, the input current to a common-emitter amplifier can be found as

$$i_{in} = \frac{v_s}{Z_{in}}$$

As Figure D.1 shows, the input current to a common-emitter amplifier is divided between the resistors in the biasing network and the base of the transistor. Using the current-divider relationship, the value of $i_b$ can be found as

$$i_b = i_{in}\left(\frac{Z_{in}}{Z_{in(base)}}\right) \tag{D.5}$$

where $Z_{in}$ = the parallel combination of the biasing resistors and the base input impedance, $Z_{in(base)}$.

Figure D.1 shows the transistor output current $(i_c)$ to be equal to the product of the transistor ac current gain $(h_{fe})$ and the ac base current. By formula,

$$i_c = h_{fe}i_b$$

Substituting equation (D.5) for the value of $i_b$ in the above equation, we get

$$i_c = h_{fe}i_{in}\left(\frac{Z_{in}}{Z_{in(base)}}\right) \tag{D.6}$$

**FIGURE D.1**

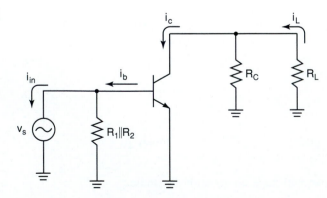

As is the case with the input circuitry, the output circuitry forms a current divider. The output current divider is made up of the collector resistor ($R_C$) and the load. Once we know the value of $i_c$ for a given common-emitter amplifier, the value of the ac load current can be found as

$$i_L = i_c\left(\frac{r_C}{R_L}\right) \tag{D.7}$$

where $r_C$ is the parallel combination of the collector resistor and the load resistance.

The *effective* current gain of any amplifier equals the ratio of *ac load current* to *ac input current*. By equation,

$$A_i = \frac{i_L}{i_{in}}$$

Substituting the relationships we have established for the values of $i_L$ and $i_{in}$, we get

$$A_i = \frac{i_c\dfrac{r_C}{R_L}}{\dfrac{v_s}{Z_{in}}}$$

or

$$A_i = h_{fe}i_b\left(\frac{r_C}{R_L}\right)\left(\frac{Z_{in}}{v_s}\right) \quad \bigg| \quad i_c = h_{fe}i_b$$

Since $\dfrac{Z_{in}}{v_s} = \dfrac{1}{i_{in}}$, the above equation can be rewritten as

$$A_i = h_{fe}\left(\frac{i_b}{i_{in}}\right)\left(\frac{r_C}{R_L}\right) \tag{D.8}$$

According to equation (D.5),

$$\frac{i_b}{i_{in}} = \frac{Z_{in}}{Z_{in(base)}}$$

Substituting this equation for the current ratio in equation (D.8), we get

$$A_i = h_{fe}\left(\frac{Z_{in}r_C}{Z_{in(base)}\,R_L}\right)$$

This is equation (9.21).

# EQUATION (10.15)

The derivation of equation (10.15) is best understood by looking at the circuits shown in Figure D.2. Figure D.2a shows the ac equivalent of the emitter follower. Note that the input circuit consists of the source resistance ($R_S$) in parallel with the combination of $R_1$ and $R_2$. Thus, the total resistance in the base circuit (as seen from the base of the transistor) can be found as

$$R'_{in} = R_1 \parallel R_2 \parallel R_S$$

This resistance is shown as a single resistor in Figure D.2b.

To determine the value of $Z_{out}$, we start by writing the Kirchhoff's voltage equation for the circuit and solving that equation for $i_e$. The voltage equation is

$$v_{in} = i_b R'_{in} + i_e(r'_e + R_E)$$

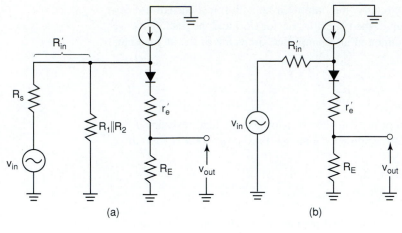

(a)          (b)

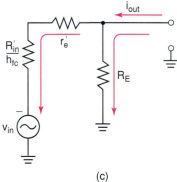

(c)

Since $i_b = \dfrac{i_e}{h_{fc}}$, the voltage equation can be rewritten as

$$v_{in} = \frac{i_e R'_{in}}{h_{fc}} + i_e(r'_e + R_E)$$

Solving for $i_e$ yields

$$i_e = \frac{v_{in}}{r'_e + R_E + \dfrac{R'_{in}}{h_{fc}}}$$

This equation indicates that the emitter "sees" a three-resistor circuit, as shown in Figure D.2c. If we thevenize the circuit, we find that the load sees $R_E$ in parallel with the series combination of the other two resistors. Thus, $Z_{out}$ is found as

$$Z_{out} = R_E \left\| \left(r'_e + \frac{R'_{in}}{h_{fc}}\right)\right.$$

This is equation (10.15).

## EQUATION (10.21)

Equation (10.13) gives the value of the base input impedance as

$$Z_{in(base)} = h_{fc}(r'_e + r_E)$$

where $r_E$ is the parallel combination of $R_E$ and $R_L$. Multiplying on the right side of the equation, we obtain

$$Z_{in(base)} = h_{fc}r'_e + h_{fc}r_E$$

**FIGURE D.3**

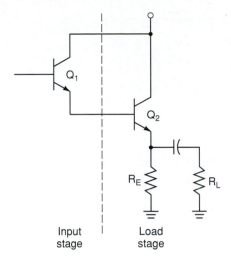

Input stage | Load stage

Now, since $h_{ic} = h_{fc}r'_e$, we can rewrite the above equation as

$$Z_{in(base)} = h_{ic} + h_{fc}r_E$$

Thus, for the Darlington amplifier shown in Figure D.3, the value of $Z_{in(base)}$ for $Q_2$ is

$$Z_{in(base)2} = h_{ic2} + h_{fc2}r_{E2}$$

And the value of $Z_{in(base)}$ for $Q_1$ is

$$Z_{in(base)1} = h_{ic1} + h_{fc1}r_{E1}$$

As Figure D.3 shows, $Q_2$ is the ac load on $Q_1$. Thus, $r_{E1} = Z_{in(base)2}$, thus,

$$Z_{in(base)1} = h_{ic1} + h_{fc1}r_{E1}$$
$$= h_{ic1} + h_{fc1}(h_{ic2} + h_{fc}r_{E2})$$

or

$$Z_{in(base)} = h_{ic1} + h_{fc1}(h_{ic2} + h_{fc}r_E)$$

where

$$Z_{in(base)} = \text{the input impedance of } Q_1$$
$$r_E = \text{the parallel combination of } R_E \text{ and } R_L$$

## RC-COUPLED CLASS A EFFICIENCY

The ideal class A amplifier will have the following characteristics:

$$1. \ V_{CEQ} = \frac{V_{CC}}{2}$$

$$2. \ V_{PP} = V_{CC}$$

$$3. \ I_{CQ} = \frac{I_{C(sat)}}{2}$$

$$4. \ I_{PP} = I_{C(sat)}$$

These ideal characteristics will be used in our derivation of the maximum ideal efficiency rating of 25%.

Equation (11.10) gives us the following equation for load power.

$$P_L = \frac{V_{PP}^2}{8R_L}$$

Since $R_L = V_{PP}/I_{PP}$, we can rewrite the above equation as

$$P_L = \left(\frac{V_{PP}^2}{8}\right)\left(\frac{1}{R_L}\right)$$

$$= \left(\frac{V_{PP}^2}{8}\right)\left(\frac{I_{PP}}{V_{PP}}\right) \tag{D.9}$$

$$= \frac{V_{PP}I_{PP}}{8}$$

Now, we will use the characteristics above and equation (D.9) to determine the ideal maximum load power as follows:

$$P_L = \frac{V_{PP}I_{PP}}{8}$$

$$= \frac{V_{CC}I_{C(\text{sat})}}{8}$$

$$= \frac{(2V_{CEQ})(2I_{CQ})}{8}$$

$$= \frac{4(V_{CEQ}I_{CQ})}{8}$$

$$= \frac{V_{CEQ}I_{CQ}}{2} \quad \text{(maximum, ideal)}$$

Now, recall that the power drawn from the supply of an amplifier is found as

$$P_S = V_{CC}I_{CC}$$

where

$$I_{CC} = I_{CQ} + I_1$$

Since $I_{CQ}$ is normally *much greater than* $I_1$, we can approximate the equation above to

$$P_S \cong V_{CC}I_{CQ}$$

Note that this equation is valid for most amplifier power analyses. We can rewrite the above for the ideal amplifier as follows:

$$P_S = 2V_{CEQ}I_{CQ}$$

We can now use the derived values of $P_L$ and $P_S$ to determine the *maximum ideal value* of $\eta$, as follows:

$$\eta = \frac{P_L}{P_S} \times 100$$

$$= P_L \frac{1}{P_S} \times 100$$

$$= \frac{V_{CEQ}I_{CQ}}{2} \frac{1}{2V_{CEQ}I_{CQ}} \times 100$$

$$= \frac{V_{CEQ}I_{CQ}}{4V_{CEQ}I_{CQ}} \times 100$$

$$= \frac{1}{4} \times 100$$

$$= 25\%$$

# TRANSFORMER-COUPLED CLASS A EFFICIENCY

Assume for a minute that we are dealing with an *ideal* transformer-coupled amplifier. This amplifier will have characteristics of $V_{CEQ} = V_{CC}$ and $V_{CE(max)} = 2V_{CC}$. These characteristics ignore the fact that there will be some voltage dropped across the emitter resistor of the amplifier. Now, recall that the maximum power that can be delivered to a load from a class A amplifier is found as

$$P_{L(max)} = \frac{V_{CEQ}I_{CQ}}{2}$$

This equation was derived in our initial discussion on class A amplifiers. Since the ideal transformer-coupled amplifier has a value of $V_{CEQ} = V_{CC}$, the equation above can be rewritten as

$$P_{L(max)} = \frac{V_{CC}I_{CQ}}{2} \qquad \text{(maximum, ideal)}$$

You may also recall that the power drawn from the supply can be approximated as

$$P_S = V_{CC}I_{CQ}$$

We can use these two equations to calculate the maximum *ideal* efficiency of the transformer-coupled amplifier as follows:

$$\eta = \frac{P_L}{P_S} \times 100$$

$$= P_L \frac{1}{P_S} \times 100$$

$$= \frac{V_{CC}I_{CQ}}{2} \frac{1}{V_{CC}I_{CQ}} \times 100$$

$$= \frac{1}{2} \times 100$$

$$= 50\%$$

# CLASS B AMPLIFIER EFFICIENCY

Equation (11.28) defines $I_{C1(ave)}$ as

$$I_{C1(ave)} = \frac{0.159V_{CC}}{R_L}$$

This equation is based on the relationship

$$I_{C1(ave)} \cong 0.318I_{Pk}$$

The value *0.318* is the "round-off" equivalent of $1/\pi$. A closer approximation of the multiplier used is 0.3183099. Using this value, equation (11.28) becomes

$$I_{C1(ave)} = \frac{0.1591549\, V_{CC}}{R_L}$$

The multiplier shown *is* more accurate, but its length makes it impractical for everyday analysis purposes. However, we will need it when proving that the maximum *ideal* efficiency of a class B amplifier is approximately 78.5%.

In the ideal class B amplifier, $I_{C1(ave)} \gg I_1$. Therefore, we can assume that $I_{CC} = I_{C1(ave)}$ in the ideal class B amplifier. Based on this point, the determination of the ideal maximum efficiency for the class B amplifier proceeds as follows:

$$\eta = \frac{P_L}{P_S} \times 100$$

$$= P_L\left(\frac{1}{P_S}\right) \times 100$$

$$= \left(\frac{V_{CC}^2}{8R_L}\right)\left(\frac{1}{V_{CC}(I_{C1ave})}\right) \times 100$$

$$= \left(\frac{V_{CC}^2}{8R_L}\right)\left(\frac{R_L}{0.1591549V_{CC}^2}\right) \times 100$$

$$= \left(\frac{1}{1.2732495}\right) \times 100$$

$$= 78.54\%$$

The final result can, of course, be assumed to be equal to 78.5%. Remember that this maximum efficiency rating is ideal. Any practical value of efficiency for a class B amplifier would have to be less than this value.

## EQUATION (11.34)

The instantaneous power dissipation in a Class B amplifier is found as

$$p = V_{CEQ}(1 - k \sin \theta)I_{C(\text{sat})}(k \sin \theta) \tag{D.10}$$

where

$V_{CEQ}$ = the quiescent value of $V_{CE}$ for the amplifier
$I_{C(\text{sat})}$ = the saturation current for the amplifier, as determined by the ac load line for the circuit
$\theta$ = the phase angle of the output at the instant that $p$ is measured
$k$ = a constant factor, representing the percentage of the load line that is actually being used by the circuit; $k$ is represented as a decimal value between 0 and 1

The average power dissipation can be found by integrating equation (D.10) for one half-cycle. Note that a half-cycle is represented as being between 0 and $\pi$ *radians* in the integration:

$$P_{\text{ave}} = \frac{1}{2\pi}\int_0^\pi p \, d\theta \tag{D.11}$$

Performing the integration yields

$$P_{\text{ave}} = \left(\frac{V_{CEQ}I_{C(\text{sat})}}{2\pi}\right)\left(2k - \frac{\pi k^2}{2}\right) \tag{D.12}$$

The next step is to find the maximum value of $k$ in equation (D.12). The first step in finding this maximum value is to differentiate $p_{ave}$ with respect to $k$ as follows:

$$\frac{dp_{ave}}{dk} = \frac{V_{CEQ}I_{C(\text{sat})}}{2}(2 - \pi k) \tag{D.13}$$

Next, we set the right-hand side of equation (D.13) equal to zero. This provides us with the following equations:

$$\frac{V_{CEQ}I_{C(\text{sat})}}{2} = 0 \qquad \text{and} \qquad 2 - \pi k = 0$$

Now, we can solve the equation on the right to find the maximum value:

$$k = \frac{2}{\pi} \tag{D.14}$$

At this point, we want to take the value of $k$ obtained in equation (D.14) and plug it into equation (D.11). Reducing this new equation gives us

$$P_{ave} = 0.1\, V_{CEQ} I_{C(sat)} \qquad\qquad \textbf{(D.15)}$$

Finally, we replace the values of $V_{CEQ}$ and $I_{C(sat)}$ in equation (D.15) using the following relationships:

$$V_{CEQ} = \frac{PP}{2} \qquad \text{and} \qquad I_{C(sat)} = \frac{V_{CEQ}}{R_L}$$

This gives us

$$P_D = 0.1 \left( \frac{V_{CEQ} PP}{2 R_L} \right)$$

or

$$P_D = \left( \frac{V_{CEQ} PP}{20 R_L} \right)$$

Now, since $V_{CEQ}$ is equal to $\dfrac{PP}{2}$, we can rewrite the above equation as

$$P_D = (V_{CEQ}) \left( \frac{PP}{20 R_L} \right)$$

$$= \left( \frac{PP}{2} \right)\left( \frac{PP}{20 R_L} \right)$$

or

$$P_D = \frac{PP^2}{40 R_L}$$

This is another form of equation (11.34).

# EQUATION (14.17)

The proof of this equation starts by taking a look at the RC circuit shown in Figure D.4. Since the current through a series circuit is constant, the ratio of $v_{out}$ to $v_{in}$ is equal to the ratio of $X_C$ (the output shunt component) to the total impedance of the circuit. By formula,

$$\frac{v_{out}}{v_{in}} = \frac{X_C}{Z_T} \qquad\qquad \textbf{(D.16)}$$

or

$$\frac{v_{out}}{v_{in}} = \frac{\dfrac{1}{j\omega C}}{\dfrac{1}{j\omega C} + R}$$

**FIGURE D.4**

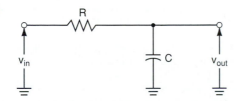

Dividing both the numerator and the denominator in equation (D.16) by $1/j\omega C$, we get

$$\frac{v_{out}}{v_{in}} = \frac{1}{1 + j\omega RC} \quad \text{(D.17)}$$

Now, since $f_c = \dfrac{1}{2\pi RC}$ and $\omega = 2\pi f$, equation (D.17) can be rewritten as

$$\frac{v_{out}}{v_{in}} = \frac{1}{1 + j\dfrac{1}{f_c}} \quad \text{(D.18)}$$

Now, the ratio of $X_C$ to $Z_T$ varies with the ratio of operating frequency of $f_c$. Thus, equation (D.18) may be rewritten as

$$\frac{v_{out}}{v_{in}} = \frac{1}{1 + j(f/f_c)} \quad \text{(D.19)}$$

Finally, we use the relationships

$$A_v = \frac{v_{out}}{v_{in}}$$

and

$$1 + jX = \sqrt{1_1 + X^2}$$

(where $X =$ any quantity) to rewrite equation (D.19) as

$$\Delta A_v = \frac{1}{\sqrt{1 + (f/f_c)^2}} \quad \text{(D.20)}$$

Rewriting equation (D.20) in *dB* form, we obtain

$$\Delta A_{v(dB)} = 20\log\left(\frac{1}{\sqrt{1 + (f/f_c)^{2)}}}\right)$$

## EQUATIONS (14.42) AND (14.43)

In this derivation, we will concentrate on the *low frequency* responses of a multistage amplifier. We will then expand the derivation to include the *high frequency response* of a given multistage amplifier.

In order to minimize any confusion, we now define two variables that will be used extensively in this derivation:

$$A_{v(mid)} = \text{the } \textit{midband} \text{ gain of an amplifier}$$
$$A_v = \text{the instantaneous gain of an amplifier}$$

Note that $A_v = A_{v(mid)}$ when the amplifier is operated above $f_1$, and $A_v < A_{v(mid)}$ when the amplifier is operated below $f_1$.

Earlier in the text, you were shown that the ratio of amplifier gain to its midband gain ($A_v'$) is found as

$$A_v' = \frac{A_v}{A_{v(mid)}} = \frac{1}{\sqrt{1 + (f_1/f)^2}} \quad \text{(D.21)}$$

It follows that two amplifier stages with identical values of $f_1$ and operated at the same value of $f$ will have values of $A_v'$ that are equal. By formula,

$$A_{v1}' = A_{v2}'$$

when $f_{c1} = f_{c2}$ and the two circuits are at the same operating frequency.

Now, recall that the total gain of $n$ cascaded amplifier stages is equal to the product of the individual stage gain values. By formula,

$$A_{vT} = (A_{v1})(A_{v2}) \cdots (A_{vn})$$

The same relationship holds true for the gain ratios of the individual stages. By formula,

$$A'_{vT} = (A'_{v1})(A'_{v2}) \cdots (A'_{vn}) \tag{D.22}$$

Now, assuming that we have $n$ stages with equal values of $A'_v$, the value of $A'_{vT}$ is found as

$$A'_{vT} = (A'_v)^n \tag{D.23}$$

where

$$A'_v = \text{the gain ratio of any individual stage}$$

Substituting equation (D.21) into equation (D.22) gives us

$$A'_{vT} = (A'_v)^n = \left( \frac{1}{\sqrt{1 + (f_1/f)^2}} \right)^n \tag{D.24}$$

$$A'_{vT} = \left( \frac{1}{\sqrt{1 + (f_1/f)^2}} \right)^n \tag{D.25}$$

Now, when $f = f_1$ (the lower 3 dB point), equation (D.24) simplifies as follows:

$$\left( \frac{1}{\sqrt{1 + (f_1/f_1)^2}} \right) = \left( \frac{1}{\sqrt{1 + 1}} \right) = \left( \frac{1}{\sqrt{2}} \right) = (2)^{1/2}$$

If we set $A'_{vT}$ to the value of $(2)^{1/2}$, equation (D.25) can be rewritten as

$$(2)^{1/2} = \left\{ \left[ 1 + (f_1/f)^2 \right]^{1/2} \right\}^n$$

Since the exponent value of $(1/2)$ appears on both sides of the equation, it can be dropped completely, leaving

$$2 = \left[ 1 = (f_1/f)^2 \right]^n$$

or

$$(2)^{1/n} = 1 + (f_1/f)^2 \tag{D.26}$$

Finally, solving equation (D.26) for $f$ yields

$$f = \frac{f_1}{\sqrt{2^{1/n} - 1}}$$

Or, in another form,

$$f_{1T} = \frac{f_1}{\sqrt{2^{1/n} - 1}} \tag{D.27}$$

This is equation (14.42). The same process is then used to derive equation (14.43).

## EQUATION (15.7)

The equation for the instantaneous voltage at any point on a sine wave is given as

$$v = V_{pk}(\sin \omega t)$$

where

$V_{pk} = \text{the peak input voltage}$
$\omega = 2\pi f$
$t = \text{the time from the start of the cycle to the instant that } v \text{ occurs}$

If we differentiate $v$ with respect to $t$, we get

$$\frac{dv}{dt} = \omega V_{pk}(\cos \omega t)$$

Since the slew rate is a maximum rating of a given op-amp, and the rate of change in a sine wave varies, we need to determine the point at which the rate of change in the sine wave is at its maximum value. This point occurs when the sine wave passes the reference voltage, or at $t = 0$. The value of $dv/dt$ at $t = 0$ determines the maximum operating frequency of the op-amp, as follows:

$$\frac{dv}{dt} \, max = \omega_{max} V_{pk}$$

or

$$\text{Slew rate} = 2\pi f_{max} V_{pk}$$

Finally, this equation is rearranged into the form of equation (15.7):

$$f_{max} = \frac{\text{slew rate}}{2\pi V_{pk}}$$

## EQUATION (15.24)

The value of $v_{in}$ for an amplifier with feedback can be given as

$$v_{in} = v_s - v_f$$

According to equation (15.22),

$$v_f = \alpha_v v_{out}$$

Substituting this equation for $v_f$ in the $v_{in}$ equation, we get

$$v_{in} = v_s - \alpha_v v_{out} \tag{D.28}$$

Now, equation (D.28) can be rewritten to solve for $v_s$ as

$$v_s = v_{in} + \alpha_v v_{out} \tag{D.29}$$

Since $v_{out} = A_v V_{in}$, we can rewrite equation (D.29) as

$$v_s = v_{in} + \alpha_v A_v v_{in}$$

or

$$v_s = v_{in}(1 + \alpha_v A_v) \tag{D.30}$$

Now, $A_v$ is the total gain of the feedback amplifier from source to output. By formula,

$$A_{vf} = \frac{v_{out}}{v_s} \tag{D.31}$$

Substituting equation (D.30) for $v_s$ in equation (D.31) gives us

$$A_{vf} = \frac{v_{out}}{v_{in}(1 + \alpha_v A_v)}$$

or

$$A_{vf} = \frac{A_v}{1 + \alpha_v A_v}$$

# EQUATIONS (17.22) AND (17.24)

We start these proofs with an assumption that can be seen intuitively; that is, for any given series reactive–resistive circuit there exists a parallel equivalent reactive–resistive circuit. These two circuits are represented in Figure D.5. All this assumption does is let us assume that it *is* possible to derive a parallel equivalent circuit for a series reactive–resistive circuit.

For the series circuit shown in Figure D.5,

$$Q = \frac{X_S}{R_S} \tag{D.32}$$

and

$$Z_S = R_S \pm jX_S \tag{D.33}$$

For the parallel circuit shown in Figure D.5, the total impedance ($Z_P$) is found as

$$Z_P = \frac{(R_P)(\pm jX_P)}{R_P \pm jX_P} \tag{D.34}$$

If we multiply the right-hand side of equation (D.34) by 1 in the form of

$$\frac{R_P \pm jX_P}{R_P \pm jX_P}$$

we get

$$Z_P = \frac{R_P X_P^2}{R_P^2 + X_P^2} \pm j\frac{X_P R_P^2}{R_P^2 + X_P^2} \tag{D.35}$$

Now, refer back to our original assumption regarding the two circuits in Figure D.5. If these two circuits are equivalent circuits, then the total impedance of the two circuits must be equal. By formula,

$$Z_S = Z_P$$

Substituting this relationship into equation (D.35) gives us

$$Z_S = \frac{R_P X_P^2}{R_P^2 + X_P^2} \pm j\frac{X_P R_P^2}{R_P^2 + X_P^2} \tag{D.36}$$

Now, if we substitute equation (D.33) for $Z_S$ in equation (D.36), we get

$$R_S \pm jX_s = \frac{R_P X_P^2}{R_P^2 + X_P^2} \pm j\frac{X_P R_P^2}{R_P^2 + X_P^2} \tag{D.37}$$

For the two sides of equation (D.37) to be equal, the real components on each side must be equal, and the imaginary components on each side must also be equal. Therefore,

$$R_S = \frac{R_P X_P^2}{R_P^2 + X_P^2} \tag{D.38}$$

**FIGURE D.5**

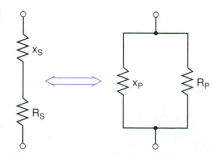

and

$$X_S = \frac{X_P R_P^2}{R_P^2 + X_P^2} \tag{D.39}$$

Using these two equations, we can rewrite equation (D.32) as

$$Q = \frac{\dfrac{X_P R_P^2}{R_P^2 + X_P^2}}{\dfrac{R_P X_P^2}{R_P^2 + X_P^2}}$$

$$= \frac{X_P R_P^2}{R_P X_P^2}$$

and

$$Q = \frac{R_P}{X_P} \tag{D.40}$$

Equation (D.40) will be used to prove equation (17.24) at the end of this proof. For now, we'll concentrate on equation (D.39). Using the relationship

$$Q = \frac{R_P}{X_P}$$

we can rewrite equation (D.39) as

$$X_S = \frac{X_P}{1 + (1/Q)^2} \tag{D.41}$$

Since tuned amplifiers are *high-Q* circuits, it is safe to assume that $(1/Q)^2 \ll 1$. Based on this assumption, equation (D.41) simplifies to

$$X_S \cong X_P \tag{D.42}$$

Now, we can rewrite equation (D.40) as

$$R_P = Q X_P$$

Or, using equation (D.42),

$$R_P = Q X_S$$

Since $X_S = Q R_S$ (from equation D.32), we can solve the above equation as

$$R_P = Q R_S$$
$$= Q(Q R_S)$$

or

$$R_P = Q^2 R_S \tag{D.43}$$

Now, let's relate equations (D.40) and (D.43) to the *LC* tuned circuit. In an *LC* tuned circuit, the series reactance is $X_L$ and the series resistance is $R_W$. Replacing $R_S$ in equation (D.43) with $R_W$ gives us

$$R_P = Q^2 R_W$$

Finally, replacing $X_P$ in equation (D.40) with $X_L$ gives us

$$Q = \frac{R_P}{X_L}$$

or

$$Q = \frac{R_P \| R_L}{X_L}$$

when a load is connected to the equivalent parallel circuit. Of course, this equation is the equivalent of equation (17.24).

# EQUATION (19.2)

The time required for a capacitor to charge to a given value is found as

$$t = RC \left[ \ln \frac{V - V_0}{V - v_c} \right] \tag{D.44}$$

where
$V$ = the charging voltage
$V_0$ = the initial charge on the capacitor
$v_c$ = the capacitor charge of interest
ln = an operator, indicating that we are to take the *natural log* of the fraction

We can use equation (D.44) to find the time required for $v_c$ to reach 0.1 V (10% of $V$) as follows:

$$t = RC \left[ \ln \frac{V - 0\text{ V}}{V - 0.1\text{ V}} \right]$$
$$= RC \left[ \ln \frac{V}{0.9\text{ V}} \right]$$
$$= RC \left[ \ln (1.11) \right]$$
$$= 0.1RC$$

Now, we can use equation (D.44) to find the time required for $v_c$ to reach 0.9 V (90% of $V$) as follows:

$$t = RC \left[ \ln \frac{V - 0\text{ V}}{V - 0.9\text{ V}} \right]$$
$$= RC \left[ \ln \frac{V}{0.1\text{ V}} \right]$$
$$= RC \left[ \ln(10) \right]$$
$$= 2.3RC$$

Since rise time is defined as the time for $v_c$ to fall from 90% of its maximum value to 10% of its maximum value, it can be found as

$$t_r = t_{(90\%)} - t_{(10\%)}$$
$$= 2.3RC - 0.1RC \tag{D.45}$$
$$= 2.2RC$$

Now, the upper cutoff frequency of a given RC circuit is found as

$$f_2 = \frac{1}{2\pi RC}$$

Rewriting the $f_2$ equation, we get

$$RC = \frac{1}{2\pi f_2}$$

Subsituting this equation in equation (D.45), we get

$$t_r = 2.2RC$$

$$= 2.2\left(\frac{1}{2\pi f_2}\right)$$

$$= \frac{2.2}{2\pi f_2}$$

$$= \frac{0.35}{f_2}$$

or

$$f_2 = \frac{0.35}{t_r}$$

# APPENDIX E

# Glossary

**ac beta ($\beta_{ac}$).** The ratio of ac collector current to ac base current; listed on specification sheets as $h_{fe}$.

**Acceptor atoms.** Another name for trivalent atoms.

**ac emitter resistance.** The dynamic resistance of the transistor emitter; used in voltage gain calculations.

**ac equivalent circuit.** A representation of a circuit that shows how the circuit appears to an ac source.

**ac load line.** A graph that represents all possible combinations of $i_c$ and $v_{ce}$.

**Active filter.** A tuned op-amp circuit.

**Active region.** The BJT operating region between saturation (maximum $I_C$) and cutoff (minimum $I_C$).

**Adjustable regulator.** A regulator whose output dc voltage can be set to any value within specified limits.

**AM detector.** A diode clipper that converts a varying amplitude sine wave input to a varying dc level.

**Amplification.** Process of strengthening an ac signal.

**Amplifier.** A circuit used to increase the strength of an ac signal.

**Amplifier input impedance.** The impedance of an amplifier as seen by its source.

**Amplifier output impedance.** The impedance of an amplifier as seen by its load.

**Amplitude.** The maximum value of a changing voltage or current.

**Anode.** The $p$-type terminal of a diode.

**Anode current interruption.** A method of driving an SUS into cutoff by breaking the diode current path or shorting the current around the diode.

**Apparent power.** The sum of power stored in the field of a reactive component plus the power dissipated by its resistance.

**Armstrong oscillator.** An oscillator that uses a transformer in its feedback network to achieve the required 180° voltage phase shift.

**Astable multivibrator.** A switching circuit that has no stable output state. A square-wave oscillator. Also called a *freerunning* multivibrator.

**Attenuation factor ($\alpha_v$).** The ratio of feedback voltage to output voltage. The value of $\alpha_v$ will always be less than one.

**Avalanche current.** The current that occurs when $V_{RRM}$ is reached. Avalanche current can destroy a *pn*-junction diode.

**Average forward current.** The maximum allowable value of dc forward current for a diode.

**Average load current.** The dc equivalent current of a signal.

**Average load voltage.** The dc equivalent voltage of a signal.

**Averaging amplifier.** A summing amplifier that provides an output proportional to the *average* of the input voltage.

**Back-to-back suppressor.** A single package containing two transient suppressors that are connected cathode-to-cathode or anode-to-anode.

**Band.** Another name for an orbital shell (in atomic theory) or a range of frequencies (in frequency analysis).

**Band-pass filter.** A filter designed to pass all frequencies that fall between $f_1$ and $f_2$.

**Band-stop ("notch") filter.** A filter designed to block all frequencies that fall between $f_1$ and $f_2$.

**Bandwidth (BW).** The range of frequencies over which gain is relatively constant.

**Barkhausen criterion.** The relationship between the circuit feedback factor ($\alpha_v$) and voltage gain ($A_v$) for proper oscillator operation.

**Barrier potential.** The natural potential across a *pn* junction.

**Base.** One of three bipolar junction transistor (BJT) terminals. The other two are the *collector* and the *emitter*.

**Base bias.** A BJT biasing circuit that consists of a single base resistor between the base terminal and $V_{CC}$ and no emitter resistor. Also known as *fixed bias*.

**Base current ($I_B$).** Current that can be varied to control the emitter current ($I_E$) and collector current ($I_C$).

**Base curve.** A curve illustrating the relationship between $I_B$ and $V_{BE}$.

**Base cutoff current ($I_{BL}$).** The maximum amount of current through a reverse-biased emitter-base junction.

**Base-emitter junction.** One of two *pn* junctions that make up the bipolar junction transistor. The other is the *collector-base* junction.

**Base-emitter saturation voltage ($V_{BE(sat)}$).** The maximum value of $V_{BE}$ when the transistor is in saturation.

**Beta ($\beta$).** A Greek letter often used to represent the base-to-collector current gain of a bipolar junction transistor.

**Beta curve.** A curve that shows the relationship between beta and temperature and/or collector current.

**Beta-dependent circuit.** A BJT biasing circuit whose $Q$-point values are affected by changes in the dc beta ($h_{FE}$) of the transistor.

**Beta-independent circuit.** A BJT biasing circuit whose $Q$-point values are independent of changes in the dc beta ($h_{FE}$) of the transistor.

**Bias.** A current or voltage applied to a device to obtain a desired mode of operation.

**Biased clamper.** A clamper that allows a waveform to be shifted above or below a dc reference other than 0 V.

**Biased clipper.** A shunt clipper that uses a dc voltage source to bias the diode.

**Bidirectional thyristor.** A thyristor that is capable of conducting in two directions.

**Biomedical electronics.** The area of electronics that deals with medical test and treatment equipment.

**Bipolar junction transistor (BJT).** A three-terminal solid-state device whose output current, voltage, and/or power are normally controlled by its input current.

**Bistable multivibrator.** A switching circuit with two stable output states. Also called a *flip-flop*.

**Bode plot.** A frequency-response curve that assumes $\Delta A_{p(mid)}$ is zero until the cutoff frequency is reached.

**Breakdown voltage ($V_{BR}$).** The value of reverse voltage that will force a *pn* junction to conduct in the reverse direction.

**Bridge rectifier.** A four-diode rectifier used to control the direction of conduction through a load. It is the most commonly used of the various power-supply rectifier circuits.

**Buffer.** Any circuit used for impedance matching. A switching circuit that does not produce a voltage phase shift.

**Bulk resistance.** The natural resistance of a forward-biased diode.

**Butterworth filter.** An active filter characterized by flat pass-band response and 20 dB/decade roll-off rates.

**Bypass capacitor.** A capacitor used to establish an ac ground at a specific point in a circuit.

**Calibration resistor.** A variable resistor used to adjust for minor variations in gain between two input amplifiers.

**Capacitance ratio ($C_R$).** The factor by which the capacitance of a varactor varies from one value of $V_R$ to another.

**Capacitive filter.** A power-supply filter that uses one or more capacitors to oppose any variation in load voltage. *See also* Inductive filter.

**Cascaded amplifiers.** Amplifiers connected in series. Each amplifier in the circuit is referred to as a *stage*.

**Cascode amplifier.** A low-$C_{in}$ amplifier used in high-frequency applications.

**Cathode.** The *n*-type terminal of a diode.

**Center frequency ($f_p$).** Another name for midpoint frequency. Equals the geometric average of $f_1$ and $f_2$.

**Center-tapped transformer.** A transformer with an output lead connected to the center of the secondary winding.

**Channel.** The material that connects the source and drain terminals of an FET.

**Chebyshev filter.** An active filter characterized by an irregular passband response and 40 dB/decade roll-off rates.

**Circuit fusing ($I^2t$) rating.** The maximum forward surge current capability of an SCR.

**Clamper.** Circuits used to change the dc reference of an ac signal. Also known as *dc restorers*.

**Clamping voltage ($V_{RSM}$).** The maximum value of $V_Z$ for a transient suppressor.

**Clapp oscillator.** A Colpitts oscillator with an added capacitor (in series with the feedback inductor) used to reduce the effects of stray capacitance.

**Class A amplifier.** An amplifier that conducts 100% of the input cycle.

**Class AB operation.** When the transistors in an amplifier conduct for slightly more than 180° of the input cycle.

**Class B amplifier.** An amplifier with two transistors that each conduct for 50% (180°) of the input cycle.

**Class C amplifier.** An amplifier that conducts during less than 50% (180°) of the input cycle.

**Clipper.** A circuit used to eliminate a portion of an ac signal. Also called a *limiter*.

**Closed-loop voltage gain ($A_{CL}$).** The gain of an op-amp with a feedback path; always lower than the value of $A_{OL}$.

**Coarse tuning.** A control that allows you to tune a circuit or system over a wide range of frequencies. *See also* Fine tuning.

**Collector.** One of the three terminals of a bipolar junction transistor (BJT). The other two are the *base* and the *emitter*.

**Collector-base junction.** One of two *pn* junctions that make up the bipolar junction transistor (BJT). The other is the *base-emitter* junction.

**Collector biasing voltage ($V_{CC}$).** A dc supply voltage connected to the collector of a bipolar junction transistor (BJT) as part of its biasing circuitry.

**Collector bypass capacitor.** A capacitor connected between $V_{CC}$ and ground parallel with the dc power supply. Often referred to as a *decoupling capacitor*.

**Collector current ($I_C$).** One of the three terminal currents of a bipolar junction transistor (BJT).

**Collector curve.** A curve that relates the values of $I_C$, $I_B$, and $V_{CE}$ for a given transistor.

**Collector cutoff current ($I_{CEX}$).** The maximum value of $I_C$ through a cutoff transistor.

**Collector-emitter saturation voltage ($V_{CE(sat)}$).** The maximum value of $V_{CE}$ when the transistor is in saturation.

**Collector-feedback bias.** A bias circuit constructed so that $V_C$ will directly affect $V_B$.

**Colpitts oscillator.** An oscillator that uses a pair of tapped capacitors and an inductor to produce the 180° phase shift in the feedback network.

**Common-anode display.** A display in which all LED anodes are tied to a single pin.

**Common-base amplifier.** A BJT voltage amplifier that has no current gain.

**Common-cathode display.** A display in which all LED cathodes are tied to a single pin.

**Common-collector amplifier.** A BJT current amplifier that has no voltage gain. Also called an *emitter-follower*.

**Common-drain amplifier.** The JFET equivalent of a common-collector amplifier. It is also referred to as a *source follower*.

**Common-emitter amplifier.** The only BJT amplifier with a 180° voltage phase shift from input to output.

**Common-gate amplifier.** The JFET equivalent of a common-base amplifier.

**Common-mode gain ($A_{CM}$).** The gain a differential amplifier provides to common-mode signals; typically less than 1.

**Common-mode rejection ratio (CMRR).** The ratio of differential gain to common-mode gain.

**Common-mode signals.** Identical signals that appear simultaneously at the two inputs of an op-amp.

**Common-source amplifier.** The JFET equivalent of a common-emitter amplifier.

**Comparator.** A circuit used to compare two voltages.

**Complementary MOS (CMOS).** A logic family made up of MOSFETs.

**Complementary-symmetry amplifier.** A class B circuit configuration using complementary transistors. Also called a *push-pull emitter follower*.

**Complementary transistors.** Two bipolar junction transistors, one *npn* and one *pnp*, that have nearly identical electrical characteristics and ratings.

**Compliance.** The maximum *undistorted* $V_{PP}$ output signal from an amplifier.

**Conduction band.** The band outside the valence shell.

**Constant-current diodes.** Diodes that maintain a constant device current over a wide range of forward operating voltages.

**Counter emf.** A voltage produced by a transformer that has the opposite polarity of the voltage that caused it. Often referred to as *kick emf*.

**Coupling capacitor.** A capacitor connected between amplifier stages to provide dc isolation between the stages while allowing the ac signal to pass without distortion.

**Covalent bonding.** One of the means by which atoms are held together. In a covalent bond, shared electrons hold the atoms together.

**Critical rise rating ($dv/dt$).** The maximum rate of increase in $V_{AK}$ for an SCR that will not cause false triggering.

**Crossover distortion.** Distortion caused by class B transistor biasing. It occurs during the time when neither of the transistors is conducting.

**Crossover network.** A circuit designed to separate high-frequency audio from low-frequency audio.

**Crowbar.** An SCR circuit used to protect a voltage-sensitive load from excessive dc power supply output voltages.

**Crystal-controlled oscillator.** An oscillator that uses a quartz crystal to produce an extremely stable output frequency.

**Current-controlled resistance.** Any resistance whose value is determined by the value of a controlling current.

**Current feedback.** A type of feedback that uses a portion of the output current from a circuit to control the circuit's operating characteristics.

**Current gain.** The factor by which current increases from the base of a transistor (the amplifier input) to the collector of the transistor (the amplifier output), represented as $h_{fe}$, $A_i$, and/or $\beta_{ac}$.

**Current-gain bandwidth product.** The frequency at which BJT current gain drops to unity. Also referred to as *beta-cutoff-frequency*.

**Current-limiting resistor ($R_S$).** A resistor placed in series with an LED to ensure a low $I_F$ through the LED.

**Current regulator diode.** Another name for a constant-current diode.

**Current-source bias.** Configuration that maintains a constant drain current despite the variations in FET parameters, using single or dual supplies.

**Cutoff.** A BJT operating state where $I_C$ is nearly zero.

**Cutoff clipping.** A type of distortion caused by driving a transistor into cutoff.

**Cutoff frequencies.** The frequencies at which the power gain ($A_p$) of an amplifier drops to 50% of its midband value ($A_{p,\text{mid}}$).

**Cycle time ($T_C$).** For a sine wave, the time required for one 360° rotation. For a rectangular wave, the sum of pulse width and space width.

**Damping.** The fading and loss of oscillations that occurs when $\alpha_v A_v < 1$. (*See* Barkhausen criterion.)

**Dark current.** The reverse current through a photodiode with no active light input.

**Darlington amplifier.** A special-case emitter follower that uses two transistors to increase the overall current gain and amplifier input impedance.

**Darlington pair.** Two bipolar junction transistors connected to provide high current gain and input impedance. The emitter of the input transistor is connected to the base of the output transistor, and the collectors of the two are tied together.

**Darlington pass-transistor regulator.** A series regulator that uses a Darlington pair in place of a single pass transistor.

**dB power gain.** The ratio of output to input power, given as ten times the common log of the ratio.

**dB voltage gain.** The ratio of output voltage to input voltage, given as 20 times the common log of the ratio.

**dBm reference.** The ratio of a power value to one milliwatt (1 mW), given as ten times the common log of the ratio.

**dc alpha ($\alpha$).** The ratio of dc collector current to dc emitter current.

**dc beta ($\beta_{dc}$ or $h_{FE}$).** The ratio of dc collector current to dc base current.

**dc blocking voltage.** The maximum allowable value of reverse voltage for a given *pn*-junction diode. It is also referred to as *peak reverse voltage*.

**dc load line.** A graph of all possible combinations of $I_C$ and $V_{CE}$ for a given amplifier.

**dc power dissipation rating.** The maximum allowable value of $P_D$ for a device.

**dc reference voltage.** A dc voltage used to determine the input signal level that will trigger a circuit such as a comparator or Schmitt trigger.

**dc restorer.** Circuits used to change the dc reference of an ac signal. Also called a clamper.

**dc-to-dc converter.** A switching voltage regulator used to change one dc voltage level to another.

**Decade.** A frequency multiplier equal to 10.

**Decibel (dB).** A logarithmic representation of a gain value.

**Decoder-driver.** An IC used to drive a multisegment display.

**Decoupling capacitor.** *See* Collector-bypass capacitor.

**Delay time ($t_d$).** The time required for a BJT to come out of cutoff when provided with a pulse input. In terms of $I_C$, it is the time required for $I_C$ to reach 10% of its maximum value.

**Depletion layer.** The area around a *pn* junction that is depleted of free carriers. Also called the *depletion region*.

**Depletion-mode operation.** Using an input voltage to reduce the channel size of an FET.

**Depletion MOSFET (D-MOSFET).** A MOSFET that can be operated in both the depletion mode and the enhancement mode.

**Designator code.** An IC code that indicates the type of circuit and its operating temperature range.

**Diac.** A two-terminal, three-layer device whose forward and reverse characteristics are identical to the forward characteristics of the silicon unilateral switch (SUS).

**Differential amplifier.** A circuit that amplifies the difference between two input voltages. The input circuit of the op-amp driven by the inverting and noninverting inputs.

**Differentiator.** A circuit whose output is proportional to the rate of change of its input signal.

**Diffusion current.** The $I_F$ below the knee voltage.

**Digital circuits.** Circuits designed to respond to specific alternating dc voltage levels.

**Digital communications.** A method of transmitting and receiving information in digital form.

**Digital-to-analog (D/A) converter.** A circuit that converts digital circuit outputs to equivalent analog voltages.

**Diode.** A one-way conductor.

**Diode bias.** A class AB biasing circuit that uses two diodes in place of the resistor(s) between the bases of the two transistors.

**Diode capacitance ($C_t$).** The rated value (or range) of capacitance for a varactor at a specified value of $V_R$.

**Diode capacitance temperature coefficient ($TC_C$).** A multiplier used to determine the amount by which varactor capacitance changes when temperature changes.

**Discrete.** A term used to describe devices that are packaged in individual casings.

**Distortion.** An unwanted change in the shape of an ac signal.

**Donor atoms.** Another name for *pentavalent atoms.*

**Doping.** Adding impurity elements to intrinsic semiconductors to improve their conductivity.

**Drain.** The JFET equivalent of the BJT collector.

**Drain-feedback bias.** The MOSFET equivalent of collector-feedback bias.

**Drift.** A change in tuning caused by component aging.

**Driver.** A circuit used to couple a low-current output to a relatively high current device.

**Dual-gate MOSFET.** A MOSFET constructed with two gates to reduce gate input capacitance.

**Dual-polarity power supply.** A dc supply that provides both positive and negative dc output voltages.

**Dual-tracking regulator.** A regulator that provides equal +dc and −dc output voltages.

**Duty cycle (DC).** The ratio of pulse width (PW) to cycle time ($T_C$), measured as a percentage.

**Efficiency.** The ratio of ac load power provided by an amplifier to the dc power drawn from its power supply.

**Electrical characteristics.** The guaranteed operating characteristics of the device, listed on the device specification sheet.

**Electron-hole pair.** A free electron and its matching valence band hole.

**Electronic tuning.** Using voltages to control the tuning of a circuit.

**Electron-volt (eV).** The energy absorbed by an electron when it is subjected to a 1-V difference of potential.

**Emitter.** One of the three terminals of a bipolar junction transistor (BJT). The other two are the *base* and the *collector.*

**Emitter bias.** A bias circuit that consists of a split power supply and a grounded base resistor.

**Emitter current ($I_E$).** One of the three terminal currents of a bipolar junction transistor (BJT).

**Emitter-feedback bias.** A bias circuit constructed so that $V_E$ will directly affect $V_B$.

**Emitter follower.** *See* Common-collector amplifier.

**Energy gap.** The space between two orbital shells.

**Enhancement-mode operation.** Using an input voltage to increase the channel size of a MOSFET.

**Enhancement MOSFET (E-MOSFET).** MOSFETs that are restricted to enhancement-mode operation.

**Epicap.** Another name for a *varactor.*

**Extrinsic.** Another word for *impure.*

**Fall time ($t_f$).** The time required for a BJT to make the transition from saturation to cutoff. In terms of $I_C$, the time required for $I_C$ to drop from 90% to 10% of its maximum value.

**False triggering.** When a noise signal triggers an SCR into conduction.

**Feedback bias.** A term used to describe a circuit that "feeds" a portion of the output voltage or current back to the input to control the circuit operation.

**Feedback factor.** A value that appears in almost all the equations for a given negative feedback amplifier.

**Feedback path.** A signal path from the output of an amplifier back to the input.

**Field-effect transistor (FET).** A three-terminal, voltage-controlled device used in amplification and switching applications.

**Filter.** A circuit that reduces the variations in the output of a rectifier.

**Fine tuning.** A control that allows you to tune a circuit or system within a narrow band of frequencies. It is used to select one out of a limited range of frequencies.

**555 timer.** An eight-pin IC that is designed for use in a variety of switching applications.

**Fixed bias.** *See* Base bias.

**Fixed-negative regulator.** A regulator with a predetermined negative dc output voltage.

**Fixed-positive regulator.** A regulator with a predetermined positive dc output voltage.

**Flip-flop.** A bistable multivibrator; a circuit with *two* stable output states.

**Flywheel effect.** A term used to describe the ability of a parallel LC circuit to self-oscillate for a brief time.

**Forced commutation.** Driving an SUS into cutoff by forcing a reverse current through the device.

**Forward bias.** A potential used to reduce the resistance of a *pn* junction.

**Forward blocking region.** The off-state (nonconducting) region of operation for a thyristor.

**Forward breakover current.** The value of $I_F$ at the point where breakover occurs for a given thyristor.

**Forward breakover voltage ($V_{BR(F)}$).** The value of forward voltage that forces an SUS into conduction. The value of $V_F$ at the point where breakover occurs.

**Forward operating region.** The on-state (conducting) region of device operation.

**Forward power dissipation ($P_{D(max)}$).** A diode rating that indicates the maximum possible power dissipation of the forward-biased diode.

**Forward voltage ($V_F$).** The voltage across a forward-biased *pn* junction.

**Free-running multivibrator.** Another name for the *astable multivibrator.*

**Frequency-response curve.** A curve showing the relationship between gain and operating frequency.

**Full load.** A term used to describe a minimum load resistance that draws maximum current.

**Full-wave rectifier.** A power supply circuit that converts ac to pulsating dc. The full-wave rectifier provides two half-cycle outputs for each full cycle input.

**Full-wave voltage doubler.** A circuit that produces a dc output that is twice the peak value of a sinusoidal input. The term *full-wave* indicates that the entire input cycle is used to produce the dc output.

**Fundamental frequency.** In any group of harmonically related signals, it is the lowest frequency signal. The lowest frequency produced by an oscillator or other signal source.

**Gain.** A multiplier that exists between the input and output of an amplifier.

**Gain-bandwidth product.** A constant equal to the unity-gain frequency of an op-amp. The product of $A_{CL}$ and bandwidth will always be approximately equal to this constant.

**Gain-stabilized amplifier.** A term often used to describe a *swamped amplifier.*

**Gate.** The JFET equivalent of the BJT's base.

**Gate bias.** The JFET equivalent of BJT base bias.

**Gate current ($I_G$).** The maximum amount of current that can be drawn through the JFET gate without damaging the device.

**Gate nontrigger voltage ($V_{GD}$).** The maximum gate voltage that can be applied to an SCR without triggering the device into conduction.

**Gate reverse current ($I_{GSS}$).** The maximum amount of gate current that will occur in a JFET when the gate-source junction is reverse biased.

**Gate-source breakdown voltage ($V_{BR(GSS)}$).** The value of $V_{GS}$ that will cause the gate-source junction to break down.

**Gate-source cutoff voltage ($V_{GS(off)}$).** The value of $V_{GS}$ that causes $I_D$ to drop to zero.

**Gate trigger voltage ($V_{GT}$).** The value of $V_{GK}$ that will cause $I_G$ to become great enough to trigger an SCR into conduction.

**General amplifier model.** An amplifier model that represents the circuit as one having gain, input impedance, and output impedance. It is used to show how the circuit is affected by its source and load.

**General-class equation.** An equation derived for a summing amplifier that is used to predict the output voltage for any combination of input voltages.

**Germanium.** One of the semiconductor materials that are commonly used to produce solid-state devices. Another is *silicon.*

**Graphic equalizer.** A circuit or system designed to allow you to control the amplitude of different audio frequency ranges.

**Half-wave rectifier.** A power supply that converts ac to pulsating dc. The half-wave rectifier produces one half-cycle output for each full-cycle input.

**Half-wave voltage doubler.** A half-wave circuit that provides an output like that of a filtered half-wave rectifier.

**Harmonic.** A whole number multiple of a given frequency.

**Hartley oscillator.** An oscillator that uses a pair of tapped inductors and a parallel capacitor in its feedback network to produce the 180° voltage phase shift required for oscillation.

**Heat sink.** A large metallic object that helps to cool components by increasing their surface area.

**Heat-sink compound.** A compound used to aid in the transfer of heat from a component to a heat sink.

**High-current transistors.** BJTs with high maximum $I_C$ ratings.

**High-pass filter.** A filter designed to pass all frequencies *above* a given cutoff frequency.

**High-power transistors.** BJTs with high-power dissipation ratings.

**High-voltage transistors.** BJTs with high reverse breakdown voltage ratings.

**Holding current ($I_H$).** The minimum value of $I_F$ required to maintain conduction through a thyristor.

**Hole.** A gap in a covalent bond.

**Hot-carrier diode.** Another name for the *Schottky diode.*

**h-parameters.** *See* Hybrid parameters.

**Hybrid parameters (h-parameters).** Transistor specifications that describe the operation of small-signal circuits under full-load or no-load conditions.

**Hysteresis.** A term that is sometimes used to describe the range of voltages between the LTP and LTP values of a Schmitt trigger.

**Ideal voltage amplifier.** A theoretical voltage amplifier having infinite gain, input impedance, and bandwidth, and zero output impedance.

**Impedance matching.** A circuit design technique that improves power transfer between source and load.

**Inductive filter.** A power supply filter that uses one or more inductors to oppose any variations in load current. *See also* Capacitive filter.

**Industrial electronics.** The area of electronics that deals with the devices, circuits, and systems used in manufacturing.

**Input bias current.** The average quiescent value of dc biasing current drawn by the signal inputs of an op-amp.

**Input impedance ($h_{ie}$).** The input impedance of the transistor, measured under full-load conditions.

**Input offset current.** A slight difference in op-amp input currents, caused by differences in the transistor beta ratings.

**Input offset voltage ($V_{io}$).** Voltage applied between the input terminals of the op-amp to eliminate output offset voltage.

**Input/output voltage differential.** The maximum allowable difference between $V_{in}$ and $V_{out}$ for a voltage regulator.

**Instrumentation amplifier.** An amplifier of low-level signals used in process control or measurement applications.

**Integrated circuit (IC).** An entire circuit (or group of components) constructed on a single piece of semiconductor material. It contains any number of active and/or passive components.

**Integrated rectifier.** A power supply rectifier that is constructed on a single piece of semiconductor material and housed in a single package.

**Integrated transistors.** A group of transistors having identical characteristics that are constructed on a single piece of semiconductor material and housed in a single package.

**Integrator.** A circuit whose output is proportional to the area of the input waveform.

**Interbase resistance ($\eta$).** The resistance between the two base terminals of a unijunction transistor (UJT).

**Intrinsic.** Another term for *pure*.

**Intrinsic standoff ratio ($\eta$).** The ratio of emitter-base 1 resistance to the total interbase resistance in a UJT. The interbase resistance is the total resistance between base 1 and base 2 when the device is not conducting.

**Inverter.** A basic switching circuit that produces a 180° voltage phase shift. A logic-level converter.

**Inverting amplifier.** A basic op-amp circuit that produces a 180° signal phase shift. The op-amp equivalent of the common-emitter and common-source circuits.

**Inverting input.** The op-amp input that produces an output 180° voltage phase shift.

**Isolation capacitance ($C_{ISO}$).** The total capacitance between the input and output pins of an optoisolator.

**Isolation current ($I_{ISO}$).** The amount of current that can be forced between the input and output pins of an optoisolator at the rated voltage.

**Isolation resistance ($R_{ISO}$).** The total resistance between the input and output pins of an optoisolator.

**Isolation source voltage ($V_{ISO}$).** The voltage that, if applied across the input and output pins, will destroy an optoisolator.

**Isolation transformer.** A transformer with a 1:1 turns ratio that is used to provide physical isolation between two electronic systems.

**Junction capacitance.** The capacitance produced by a reverse biased *pn* junction. This capacitance varies inversely with the amount of reverse bias.

**Kick emf.** *See* Counter emf.

**Knee voltage ($V_k$).** The voltage at which device current suddenly increases or decreases.

**LDMOS.** A high-power MOSFET that uses a narrow channel and a heavily doped *n*-type region to obtain high $I_D$ and low $r_{d(\text{on})}$.

**Leaky.** A term used to describe a component that is partially shorted.

**Level detector.** Another name for a comparator used to compare an input voltage to a fixed dc reference voltage.

**Lifetime.** The time between electron-hole pair generation and recombination.

**Light.** Electromagnetic energy that falls within a specific range of frequencies.

**Light-activated SCR (LASCR).** A three-terminal light-activated SCR.

**Light current.** The reverse current through a photodiode with an active light input.

**Light detector.** An optoelectronic device that responds to light.

**Light emitter.** An optoelectronic device that produces light.

**Light-emitting diode (LED).** A diode designed to produce one or more colors of light under specific biasing conditions.

**Light intensity.** The amount of light per unit area received by a given photodetector; also called *irradiance*.

**Limiter.** A circuit used to eliminate a portion of a signal. Also called a *clipper*.

**Linear.** A change that occurs at a constant rate.

**Linear IC voltage regulator.** A device used to hold the output voltage from a dc power supply relatively constant over a wide range of line voltages and load current demands.

**Linear power supply.** A power supply whose output is a linear function of its input. Any power supply whose output is controlled by a linear voltage regulator.

**Line regulation.** The ability of a voltage regulator to maintain a stable dc output voltage despite variations in line voltage. A rating that indicates the change in regulator output voltage that will occur per unit change in input voltage.

**Liquid-crystal display (LCD).** A display made of segments that can be made to reflect (or not reflect) ambient light.

**Loaded-Q.** The Q (*quality*) of an LC tank circuit when a resistive load is connected to that circuit. The loaded-Q of a tank circuit is always *lower* than its unloaded Q.

**Load regulation.** The ability of a regulator to maintain a constant output voltage despite changes in load current demand. A rating that indicates the change in regulator output voltage that will occur per unit change in load current.

**Logic family.** A group of digital circuits with nearly identical characteristics.

**Logic levels.** A term used to describe the two dc levels used to represent information in a digital circuit or system.

**Lower trigger point (LTP).** A reference voltage. When a negative-going input reaches the LTP, the Schmitt trigger output changes state.

**Low-pass filter.** A filter designed to pass all frequencies *below* a given cutoff frequency.

**Majority carrier.** In a given semiconductor material, the charge carrier that is introduced by doping. The electrons in a pentavalent material. The holes in a trivalent material.

**Matched transistors.** Transistors that have the same operating characteristics.

**Maximum limiting voltage ($V_L$).** The voltage at which a constant-current diode starts to limit current.

**Maximum ratings.** Diode parameters that must not be exceeded under any circumstances.

**Maximum reverse stand-off voltage ($V_{RWM}$).** The maximum allowable peak or dc reverse voltage that will not turn on a transient suppressor.

**Maximum reverse surge current.** In a bidirectional thyristor, the maximum surge current that the device can handle when biased in its reverse operating region.

**Maximum temperature coefficient.** The percentage of change in $V_{BR}$ per 1°C rise in operating temperature.

**Maximum zener current ($I_{ZM}$).** A zener diode rating indicating the maximum allowable value of reverse current through the device.

**Metal-oxide-semiconductor FET (MOSFET).** A field effect transistor designed to operate in the depletion and enhancement modes, or the enhancement mode only. The gate construction allows for enhancement mode operation.

**Midpoint bias.** Having a $Q$-point that is centered on the load line.

**Miller's theorem.** Allows a feedback capacitor to be represented as separate input and output capacitances.

**Minimum load current.** For a voltage regulator, the value of $I_L$ below which regulation is lost.

**Minority carrier.** In a given semiconductor material, the charge carrier that is not introduced by doping. The holes in a pentavalent material. The electrons in a trivalent material.

**Model.** A representation of a component or circuit.

**Modulator.** A circuit that combines two signals of different frequencies into a single signal.

**Monostable multivibrator.** A switching circuit with one stable output state. Also called a *one-shot*.

**MOSFET.** An FET that can be operated in the enhancement mode.

**Multicolor LED.** An LED that emits different colors when the polarity of the supply voltage changes.

**Multiple-feedback filter.** A band-pass (or band-stop) filter that has a single op-amp and two feedback paths, one resistive and one capacitive.

**Multisegment display.** An LED circuit used to display alphanumeric symbols (numbers, letters, and punctuation marks).

**Multistage amplifier.** A term used to describe two or more amplifiers that are connected in series. Each amplifier is called a *stage*.

**Multivibrators.** Switching circuits designed to have zero, one, or two stable output states.

**Negative clamper.** A circuit that shifts an entire input signal *below* a dc reference voltage.

**Negative feedback.** A type of feedback in which the feedback signal is 180° out of phase with the input signal.

**Negative resistance.** Any device whose current and voltage are inversely proportional.

**Negative resistance oscillator.** An oscillator whose operation is based on a negative resistance device, such as a tunnel diode or a unijunction transistor.

**Negative resistance region.** The region of operation that lies between peak and valley points of the UJT curve.

**Noise.** Any unwanted interference with a transmitted signal.

**Nominal zener voltage.** The rated value of $V_Z$ for a given zener diode.

**Noninverting amplifier.** An op-amp circuit with no signal phase shift.

**Noninverting input.** The op-amp input that does not produce an output voltage phase shift.

**Nonlinear distortion.** A type of distortion caused by driving the base-emitter junction of a transistor into its nonlinear operating region.

**Nonrepetitive surge current ($I_{TSM}$).** The absolute limit on the forward surge current through a device, such as a *pn*-junction diode or an SCR.

**Notch filter.** A filter designed to reject (or block) a band of frequencies while allowing the frequencies outside its bandwidth to pass. Also referred to as a *band-stop filter*.

**npn transistor.** A BJT with *n*-type emitter and collector materials and a *p*-type base.

**n-type material.** A semiconductor that has added pentavalent impurities.

**Octave.** A frequency multiplier equal to 2.

**Off characteristics.** The characteristics of a transistor that is in cutoff.

**Offset null.** Pins connected to the inputs of the op-amp to eliminate output offset voltage.

**On characteristics.** The characteristics of a transistor in either the active or saturation regions of operation.

**One-shot.** Another name for the monostable multivibrator.

**Opaque.** A term used to describe anything that blocks light.

**Open-loop voltage gain ($A_{OL}$).** The gain of an op-amp with no feedback path.

**Operational amplifier (op-amp).** A high-gain dc amplifier that has high input impedance and low output impedance.

**Optocoupler.** A device that uses light to couple a signal from its input (a photoemitter) to its output (a photodetector).

**Optoelectronic devices.** Devices that are controlled by and/or emit light.

**Optointerruptor.** An IC optocoupler designed to allow an outside object to block the light path between the photoemitter and the photodetector.

**Optoisolator.** An optocoupler that uses light to couple a signal from its input (a photoemitter) to its output (a photodetector).

**Oscillator.** An ac signal generator. A dc-to-ac converter.

**Oscillator stability.** A measure of an oscillator's ability to maintain constant output amplitude and frequency.

**Output admittance ($h_{oe}$).** The admittance of the collector-emitter circuit, measured under no-load conditions. This parameter is used mainly in amplifier design.

**Output impedance.** The impedance that a circuit presents to its load. The output impedance of a circuit is effectively the source impedance for its load.

**Output offset voltage ($V_{out(offset)}$).** A voltage that may appear at the output of an op-amp; caused by an imbalance in the differential amplifier.

**Output short-circuit current.** The maximum output current for an op-amp, measured under shorted load current.

**Overtone.** Another word for *harmonic*. A whole-number multiple of a given frequency.

**Overtone mode.** A term used to describe the operation of a filter (or other circuit) that is tuned to a harmonic of a given frequency.

**Parameter.** A limit.

**Pass band.** The range of frequencies passed (amplified) by a tuned amplifier.

**Pass-transistor regulator.** A regulator that uses a series transistor to maintain a constant load voltage.

**Peak current ($I_P$).** One of two current values that define the limits of a negative resistance region of operation for a device such as a tunnel diode or unijunction transistor (UJT). The other is the *valley current* ($I_V$).

**Peak inverse voltage (PIV).** The maximum reverse bias that a diode will be exposed to in a rectifier.

**Peak power dissipation rating ($P_{pk}$).** Indicates the amount of surge power that the suppressor can dissipate.

**Peak repetitive reverse voltage ($V_{RRM}$).** The maximum reverse voltage allowable for a diode.

**Peak reverse voltage.** The maximum amount of reverse bias that a *pn* junction can safely handle.

**Peak voltage ($V_P$).** One of two voltage values that define the limits of a negative resistance region of operation for a device such as a tunnel diode or unijunction transistor (UJT). The other is the *valley voltage* ($V_V$).

**Pentavalent elements.** Elements with five valence shell electrons.

**Phase controller.** A circuit used to control the conduction phase angle through a load, and thus the average load voltage.

**Phase-shift oscillator.** An oscillator that uses three RC circuits in its feedback network to produce a 180° phase shift.

**Photo-Darlington.** A phototransistor with a Darlington pair output.

**Photodiode.** A diode whose reverse conduction is light intensity controlled.

**Phototransistor.** A three-terminal photodetector whose collector current is controlled by the intensity of the light at its optical input (base).

**Piezoelectric effect.** The tendency of a crystal to vibrate at a fixed frequency when subjected to an electric field.

**Pinch-off voltage ($V_P$).** With $V_{GS}$ constant, the value of $V_{DS}$ that causes maximum $I_D$.

**PIN diode.** A diode made up of *p*-type and *n*-type materials that are separated by an insulating material.

***pnp* transistor.** A BJT with *p*-type emitter and collector terminals and an *n*-type base.

**Pole.** A single RC circuit.

**Positive clamper.** A circuit that shifts an entire input signal *above* a dc reference voltage.

**Positive feedback.** A type of feedback where the feedback signal is *in phase* with the circuit input signal. The type of feedback used to produce oscillations. Also referred to as *regenerative feedback.*

**Power gain ($Ap$).** The ratio of circuit output power to circuit input power. A power multiplier that exists between the input and output of a circuit.

**Power rectifiers.** Diodes with extremely high forward current and/or power dissipation ratings.

**Power supply.** A group of circuits that combine to convert ac to dc.

**Power supply rejection ratio.** The ratio of a change in op-amp output voltage to a change in supply voltage.

**Power switch.** The power transistor in a switching regulator that controls conduction through the load.

**Precision diode.** An op-amp circuit that provides rectification for extremely low-voltage signals.

**Programmable unijunction transistor (PUT).** A four-layer thyristor trigger whose construction and operation are similar to that of the SCR. However, the PUT gate is used to provide a reference for the triggering voltage rather than as a trigger input.

**Propagation delay.** The time required for a signal to get from the input of a component or circuit, to the output; measured at the 50% point on the two waveforms.

***p*-type material.** Silicon or germanium that has trivalent impurities.

**Pulse width (PW).** The time spent in the active dc output state.

**Pulse-width modulation.** Modulating a rectangular waveform so that its pulse width varies while its overall cycle time remains unchanged. Essentially, it varies the duty cycle of the modulated waveform.

**Push-pull emitter follower.** *See* Complementary-symmetry amplifier.

**Q (tank circuit).** The ratio of energy stored in the circuit to energy lost per cycle by the circuit.

**Q-point.** A point on a load line that indicates the values of $V_{CE}$ and $I_C$ for an amplifier at rest.

**Q-point shift.** A condition where a change in operating temperature (or device substitution) indirectly causes a change in $I_C$ and $V_{CE}$.

**Quality (Q).** A figure of merit for a tuned amplifier that is equal to the ratio of center frequency to bandwidth.

**Quiescent.** At rest.

**Ramp.** Another name for a voltage that changes at a constant rate.

**RC-coupled class A amplifier.** A class A amplifier that uses a capacitor to couple its output signal to a load.

**Recombination.** When a free electron returns to the valence shell.

**Rectangular waveform.** A waveform made up of alternating (high and low) dc voltages.

**Rectifier.** A circuit that converts ac to pulsating dc.

**Regenerative feedback.** Another name for *positive feedback.*

**Regulated dc power supply.** A dc power supply with extremely low internal resistance.

**Regulation.** A rating that indicates the maximum change in regulator output voltage that can occur when input voltage and load current are varied over their entire rated ranges.

**Relaxation oscillator.** A circuit that uses the charge/discharge characteristics of a capacitor or inductor to produce a pulse output.

**Reverse bias.** A potential that causes a *pn* junction to have a high resistance.

**Reverse blocking region.** The off-state (nonconducting) region of operation between 0 V and $V_{BR(R)}$.

**Reverse breakdown voltage ($V_{BR}$).** The minimum reverse voltage that causes a device to break down and conduct in the reverse direction.

**Reverse current ($I_R$).** The current through a reverse-biased diode.

**Reverse saturation current ($I_S$).** A current caused by thermal activity in a reverse-biased diode. $I_S$ is temperature dependent.

**Reverse voltage ($V_R$).** The voltage across a reverse-biased *pn* junction.

**Reverse voltage feedback ratio ($h_{re}$).** The ratio of $v_{be}$ to $v_{ce}$ for a given transistor, measured under no-load conditions.

**RF amplifier.** Radio-frequency amplifier; the input circuit of a radio receiver.

**Ripple rejection ratio.** The ratio of regulator input ripple voltage to maximum output ripple voltage.

**Ripple voltage.** The remaining variation in the output from a filter.

**Rise time ($t_r$).** The time required for a BJT to go from cutoff to saturation. In terms of $I_C$, the time required for the 10% to 90% transition.

**Roll-off rate.** The rate of gain reduction for a circuit when operated beyond the cutoff frequencies.

**Saturation.** A BJT operating region where $I_C$ reaches its maximum value.

**Saturation clipping.** A type of distortion caused by driving a transistor into saturation.

**Schmitt trigger.** A voltage-level detection circuit.

**Schottky diode.** A high-speed diode with very little junction capacitance.

**Selector guide.** A publication that groups components according to their critical ratings and/or characteristics.

**Self-bias.** A JFET biasing circuit that uses a source resistor to establish a negative $V_{GS}$.

**Semiconductor.** An element that is neither an insulator nor a conductor.

**Sensitivity.** A rating that indicates the response of a photodetector to a specified light intensity.

**Series clipper.** A clipper that has an output when the diode is forward biased (conducting).

**Series current regulator.** A circuit that uses a constant-current diode to maintain a constant circuit input current over a wide range of input voltages.

**Series feedback regulator.** A series regulator that uses an error-detection circuit to improve the line and load regulation characteristics of other pass-transistor regulators.

**Series regulator.** A voltage regulator that is in series with the load.

**Seven-segment display.** A display made up of seven LEDs shaped in a figure 8 that is used (primarily) to display numbers.

**Shorted gate-drain current ($I_{DSS}$).** The maximum possible value of $I_D$ for a JFET.

**Shunt clipper.** A clipper that has an output when the diode is reverse biased (not conducting).

**Shunt feedback regulator.** A circuit that uses an error detector to control the conduction of a shunt transistor.

**Shunt regulator.** A voltage regulator that is in parallel with the load.

**Silicon.** One of the semiconductor materials that are commonly used to produce solid-state devices. Another is *germanium*.

**Silicon-controlled rectifier (SCR).** A three-terminal device very similar in construction and operation to the SUS. A third terminal, called the *gate*, provides another method for triggering the device.

**Silicon unilateral switch (SUS).** A two-terminal, four-layer device that can be triggered into conduction by applying a specified forward voltage across its terminals.

**Slew rate.** The maximum rate at which op-amp output voltage can change.

**Snubber network.** An RC circuit that is connected between the SCR anode and cathode to eliminate false triggering.

**Soldering temperature.** The maximum amount of heat that can be applied to any pin of the IC without causing damage to the chip.

**Solid-state relay (SSR).** A circuit that uses a dc input voltage to pass or block an ac signal.

**Source.** The JFET equivalent of the BJT's emitter.

**Source follower (common-drain amplifier).** The JFET equivalent of the emitter follower.

**Space width (SW).** The time spent in the passive dc output stage. The time between pulses in switching circuits.

**Specification sheet.** A listing of all the important parameters and operating characteristics of a device or circuit.

**Spectral response.** A measure of a photodetector's response to a change in input wavelength. Spectral response is measured in terms of the device sensitivity.

**Speed-up capacitor.** A capacitor used in the base circuit of a BJT to allow the device to switch rapidly between saturation and cutoff. It reduces propagation delay by reducing $t_d$ and $t_s$.

**Square wave.** A special-case rectangular waveform that has equal pulse width (PW) and space width (SW) values.

**Stage.** A single amplifier in a cascaded group of amplifiers.

**Standard push-pull amplifier.** A class B circuit that uses two identical transistors and a center-tapped transformer.

**Static reverse current ($I_R$).** The reverse current through a diode when $V_R$ is less than the reverse breakdown value.

**Step-down regulator.** A switching regulator whose dc output voltage is lower than its dc input voltage.

**Step-down transformer.** One with a secondary voltage that is less than the primary voltage.

**Step-recovery diode.** A heavily doped diode with an ultrafast switching time.

**Step-up regulator.** A switching regulator whose dc output voltage is greater than its dc input voltage.

**Step-up transformer.** One that has a secondary voltage greater than the primary voltage.

**Stop band.** The range of frequencies outside an amplifier's pass band.

**Storage time ($t_s$).** The time required for a BJT to come out of saturation. In terms of $I_C$, the time required for $I_C$ to drop to 90% of its maximum value.

**Substrate.** The foundation material of a MOSFET.

**Subtractor.** A summing amplifier that provides an output proportional to the difference between the input voltages.

**Summing amplifier.** A circuit that produces an output voltage proportional to the sum of the input voltages.

**Surface leakage current ($I_{SL}$).** A current along the surface of a reverse-biased diode. $I_{SL}$ is $V_R$ dependent.

**Surface-mount component (SMC).** A term used to describe a type of IC package designed to be mounted on the surface of a PC board. SMCs are smaller than their DIP package counterparts.

**Surge current.** The high initial current in a power supply. Any current caused by a nonrepetitive high-voltage condition.

**Swamped amplifier.** An amplifier that uses a partially bypassed emitter resistance to increase the ac emitter resistance.

**Switching circuits.** Circuits designed to respond to (or generate) nonlinear waveforms, such as square waves.

**Switching power supply.** A power supply that contains a switching regulator.

**Switching time constants.** A term used to describe a condition when a capacitor has different charge and discharge times.

**Switching transistor.** A device with extremely low switching times.

**Switching voltage regulator.** Voltage regulator that operates either fully *on* or fully *off*, creating very high efficiency.

**Thermal contact.** Placing two or more components in physical contact with each other so that their operating temperatures will be equal.

**Threshold voltage ($V_{GS(\text{th})}$).** The value of $V_{GS}$ that turns the E-MOSFET on.

**Thyristors.** Devices designed specifically for high-power switching applications. Also called *breakover devices*.

**Transconductance ($g_m$).** A ratio of a change in drain current ($I_D$) to a change in $V_{GS}$, measured in microsiemens ($\mu$S) or micromhos ($\mu$mhos).

**Transconductance curve.** A plot of all possible combinations of $V_{GS}$ and $I_D$.

**Transducer.** A device that converts energy from one form to another.

**Transformer-coupled class A amplifier.** A class A amplifier that uses a transformer to couple the output signal to the load.

**Transient.** An abrupt current or voltage spike. A surge.

**Transient suppressor.** A zener diode that has extremely high surge-handling capabilities.

**Transistor.** A three-terminal device whose output current, voltage, and/or power are controlled (under normal circumstances) by its input current.

**Transistor checker.** A piece of test equipment designed to test the operation and electrical characteristics of any given transistor.

**Triac.** A bidirectional thyristor whose forward and reverse characteristics are identical to the forward characteristics of the SCR.

**Trigger-point voltage ($V_{TP}$).** The voltage required at the anode of the gate diode to trigger the SCR.

**Trivalent elements.** Elements with three valence electrons.

**Troubleshooting.** The process of locating faults in electronic equipment.

**Tuned amplifier.** A circuit designed to have a specific value of power gain over a specified range of frequencies. An amplifier designed to have a specific bandwidth.

**Tunnel diode.** A heavily doped diode used in high-frequency communications circuits.

**Tunnel diode oscillator.** A signal generator whose operation is based on the negative resistance characteristics of a tunnel diode.

**Turn-off time ($t_{\text{off}}$).** The sum of storage time ($t_s$) and fall time ($t_f$).

**Turn-on time ($t_{\text{on}}$).** The sum of delay time ($t_d$) and rise time ($t_r$).

**Ultrahigh frequency (UHF).** The band of frequencies between 300 MHz and 3 GHz.

**Unijunction transistor (UJT).** A three-terminal device whose trigger voltage is proportional to its applied biasing voltage.

**Upper trigger point (UTP).** A reference voltage. When a positive-going input reaches the UTP, the Schmitt trigger output changes state.

**Valence shell.** The outermost shell of an atom. The number of electrons in the valence shell determines the conductivity of the atom.

**Valley current ($I_V$).** One of two current values that define the limits of a negative resistance region of operation for a device such as a tunnel diode or unijunction transistor (UJT). The other is the *peak current* ($I_P$).

**Valley voltage ($V_V$).** One of two voltage values that define the limits of a negative resistance region of operation for a device such as a tunnel diode or unijunction transistor (UJT). The other is the *peak voltage* ($V_P$).

**Varactor.** A diode with high junction capacitance.

**Variable comparator.** A comparator with an adjustable reference voltage.

**Variable off-time modulation.** A method of modulating a rectangular waveform so that cycle time varies while pulse width remains constant.

**Varicap.** Another name for a *varactor*.

**Vertical MOS (VMOS).** An E-MOSFET designed to handle high values of drain current.

**Very high frequency (VHF).** The band of frequencies between 30 and 300 MHz.

**Voltage-controlled current source.** A current source whose output is determined by an input voltage.

**Voltage-controlled oscillator (VCO).** A free-running oscillator whose output frequency is controlled by a dc input voltage.

**Voltage-divider bias.** A transistor biasing circuit that contains a voltage divider.

**Voltage doubler.** A circuit whose dc output is twice the peak value of its sinusoidal input.

**Voltage feedback.** A type of feedback circuit that uses a portion of the output voltage to control the circuit operation.

**Voltage follower.** The op-amp equivalent of the emitter follower and the source follower.

**Voltage gain ($A_v$).** The factor by which ac signal voltage increases from the input of an amplifier to its output.

**Voltage-inverting regulator.** A switching regulator that produces a negative output when its input voltage is positive, and vice versa.

**Voltage multiplier.** Circuits used to produce a dc output voltage that is some multiple of an ac peak input voltage.

**Voltage quadrupler.** Produces a dc output voltage that is four times the peak input voltage.

**Voltage regulator.** A circuit designed to maintain a constant dc load voltage despite variations in line voltage and load current demand.

**Voltage tripler.** Produces a dc output voltage that is three times the peak input voltage.

**Wavelength ($\lambda$).** The physical length of one cycle of a transmitted electromagnetic wave.

**Wavelength of peak spectral response ($\lambda_s$).** A rating indicating the wavelength that will cause the strongest response in a photodetector.

**Wien-bridge oscillator.** An oscillator that achieves regenerative feedback by producing no phase shift at $f_r$.

**Zener breakdown.** A type of reverse breakdown that occurs at low values of $V_R$.

**Zener clamper.** A clamper that uses a zener diode to determine the dc reference voltage.

**Zener diode.** Diodes designed to work in the reverse operating region.

**Zener impedance ($Z_Z$).** A zener diode's opposition to a change in current.

**Zener knee current ($I_{zk}$).** The minimum value of zener reverse current required to maintain voltage regulation.

**Zener test current ($I_{zt}$).** The value of zener current at which the nominal values of the component are measured.

**Zener voltage ($V_Z$).** The approximate voltage across a zener diode when it is operated in its reverse breakdown region of operation.

**Zero bias.** A term used to describe the biasing of a *pn* junction at room temperature with no potential applied. A D-MOSFET biasing circuit that has quiescent values of $V_{GS} = 0$ V and $I_D = I_{DSS}$.

# Transistor Amplifier Design

The complete common-emitter design process starts with knowing the requirements of the amplifier and the values you have to work with. Normally, you will be provided with the following information:

- The available supply voltage and ac input voltage
- The load resistance or impedance
- The required output voltage or the required value of $A_V$
- The specifications of the transistor being used

The step-by-step process for designing a common-emitter amplifier is as follows:

Equation (11.6):
$$PP = 2V_{CEQ}$$

1. The required output voltage swing is the minimum compliance (PP) of the amplifier. Using this value and equation (11.6), determine the minimum allowable value of $V_{CEQ}$. Whenever possible, set $V_{CEQ}$ to one-half of $V_{CC}$.

2. Determine the desired value of $I_{CQ}$. The exact value is not usually critical in terms of input and output voltages. However, remember that a lower value of $I_{CQ}$ means that the amplifier will dissipate less power. Typical values of $I_{CQ}$ fall between 1 and 10 mA for small-signal amplifiers.

3. For the value of $I_{CQ}$ used, check the specification sheet to determine the value of $h_{FE}$.

4. Once you have determined the proper value of $h_{FE}$, calculate the value of $I_E$ using

$$I_E = I_{CQ}\left(1 + \frac{1}{h_{FE}}\right) \tag{F.1}$$

5. Set $V_E$ to approximately 0.1 $V_{CC}$.

6. Using the values from steps 4 and 5, find $R_E$ as

$$R_E = \frac{V_E}{I_E}$$

7. Determine $V_{R_C}$ as

$$V_{R_C} = V_{CC} - (V_{CEQ} + V_E)$$

8. Find $R_C$ as

$$R_C = \frac{V_{RC}}{I_C}$$

9. Find $V_B$ as

$$V_B = V_E + V_{BE}$$

**10.** Find $I_B$ as

$$I_B = \frac{I_{CQ}}{h_{FE(\min)}}$$

**11.** Set $I_2$ as

$$I_2 \geqslant 10(I_B)$$

**12.** Select the value of $R_2$ as

$$R_2 = \frac{V_B}{I_2}$$

**13.** Now, determine the value of current through $R_1$ as

$$I_1 = I_2 + I_B$$

**14.** Determine the voltage across $R_1$ as

$$V_1 = V_{CC} - V_B$$

**15.** Find the value of $R_1$ as

$$R_1 = \frac{V_1}{I_1}$$

At this point, the dc design process is complete. Now you have to analyze the ac operation of the circuit. Using the value of $I_{CQ}$, you need to determine the $h$-parameter values for the transistor. Then:

**16.** Determine the value of $r_C$.

**17.** Determine the value of $r_e'$ using equation (9.31).

**18.** Determine the value of $A_v$ for the circuit. If the value of $A_v$ is too low, you must redesign the circuit using a higher value of $I_{CQ}$. If it is too high, you may want to consider swamping the emitter circuit to reduce the gain.

The entire process is illustrated in the following example.

Equation (9.31):
$$r_e' = \frac{h_{ie}}{h_{fe}}$$

---

Design an amplifier to the following specifications: The voltage gain is to be 50 with an input voltage of 40 mV. The available supply voltage is +10 V. The transistor used is the 2N3904. Use an $I_{CQ}$ of 1 mA. $R_L = 12\ k\Omega$.

*EXAMPLE F.1*

*Solution:*
A model circuit for this example is shown in Figure F.1. Along with the circuit, this figure contains the $h$-parameters for the 2N3904 at $I_C = 1$ mA. First, the required output compliance from the amplifier is found by converting the value of $v_{out}$ to a peak-to-peak value. The value of $v_{out}$ is found as

$$\begin{aligned} v_{out} &= A_v v_{in} \\ &= 2\ V \end{aligned}$$

The peak-to-peak output voltage is found as

$$\begin{aligned} &= 2.828(v_{out}) \\ &= 5.66\ V_{PP} \end{aligned}$$

Therefore, the minimum value of $V_{CEQ}$ is

$$\begin{aligned} V_{CEQ} &= \frac{V_{PP}}{2} \\ &= 2.83\ V \end{aligned}$$

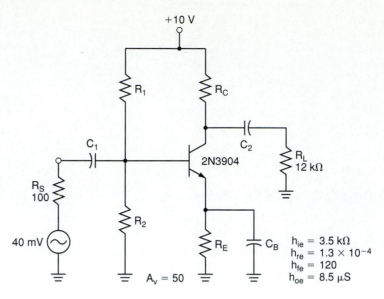

**FIGURE F.1**

Since 2.83 V < ½$V_{CC}$, we will use $V_{CEQ} = 5$ V.

At $I_{CQ} = 1$ mA, the 2N3904 has a value of $h_{FE(min)} = 70$. Using this value, $I_E$ is found as

$$I_E = I_{CQ}\left(1 + \frac{1}{h_{FE}}\right)$$
$$= 1.01 \text{ mA}$$

The value of $V_E$ is found as

$$V_E = 0.1\, V_{CC}$$
$$= 1 \text{ V}$$

and the value of $R_E$ is found as

$$R_E = \frac{V_E}{I_E}$$
$$= 990\ \Omega\ (\text{use } 910\ \Omega\text{-standard})$$

Using $R_E = 910\ \Omega$, the value of $V_E$ is now recalculated as

$$V_E = I_E R_E$$
$$= 920 \text{ mV}$$

Now $V_{R_C}$ is found as

$$V_{R_C} = V_{CC} - (V_{CEQ} + V_E)$$
$$= 4 \text{ V}$$

and $R_C$ is found as

$$R_C = \frac{V_{R_C}}{I_C}$$
$$= 4 \text{ k}\Omega\ (\text{use } 3.9 \text{ k}\Omega\text{-standard})$$

This completes the emitter-collector circuit. Now the first step in designing the base circuit is to find $V_B$ as

$$V_B = V_E + V_{BE}$$
$$= 1.62 \text{ V}$$

Next, $I_B$ is found as

$$I_B = \frac{I_{CQ}}{h_{FE(\text{min})}}$$

$$= 14.3 \ \mu A$$

and $I_2$ is set as

$$I_2 = 10 I_B$$

$$= 143 \ \mu A$$

Now $R_2$ is determined as

$$R_2 = \frac{V_B}{I_2}$$

$$= 11.3 \ k\Omega \ (\text{use } 11 \ k\Omega\text{-standard})$$

Using $R_2 = 11 \ k\Omega$, the value of $I_2$ is now recalculated as

$$I_2 = \frac{V_B}{R_2}$$

$$= 147.3 \ \mu A$$

The value of $I_1$ is found as

$$I_1 = I_2 + I_B$$

$$= 161.6 \ \mu A$$

and the value of $V_1$ is found as

$$V_1 = V_{CC} - V_B$$

$$= 8.4 \ V$$

Finally, the value of $R_1$ is found as

$$R_1 = \frac{V_1}{I_1}$$

$$= 51.9 \ k\Omega \ (\text{use } 51 \ k\Omega\text{-standard})$$

At this point, we need to take a look at the ac operation of our biasing circuit. The first step is to determine the total ac resistance in the collector circuit. This resistance is found as

$$r_C = R_C \parallel R_L$$

$$= 3 \ k\Omega$$

From the specification sheet, $r'_e$ is found as

$$r'_e = \frac{h_{ie}}{h_{fe}}$$

$$= 29 \ \Omega$$

and $A_v$ is found as

$$A_v = \frac{h_{fe} r_C}{h_{ie}}$$

$$= 103$$

The required voltage gain was given as 50. We can swamp the emitter to reduce the gain to the desired value. Recall that

$$A_v = \frac{r_C}{r'_e + r_E}$$

This formula can be rearranged to give us

$$r_E = \frac{r_C - r'_e A_v}{A_v}$$

This equation will give us the method value of $r_E$ when used with the correct value of $r'_e$ and the desired value of $A_v$. For this circuit

$$r_E = 31 \ \Omega$$

The final circuit is shown in Figure F.2. Note that the original value of $R_E$ (910 $\Omega$) has been reduced. The total emitter resistance is now 850 $\Omega$. This change is due to the availability of standard value components.

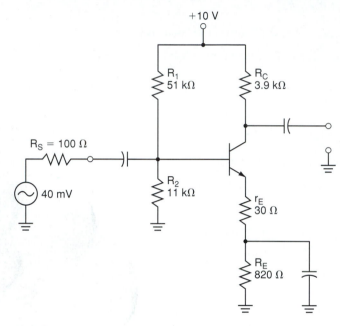

**FIGURE F.2**

# APPENDIX G

# Answers To Selected Odd-Numbered Problems

## Chapter 2:

**7.** $V_{D1} = 10$ V, $V_{D2} = 0$ V    **9.** $V_{D1} = 0.7$ V, $V_{R1} = 0.3$ V, $I_T = 3$ mA
**11.** $V_{D1} = 10$ V, $V_{R1} = 0$ V, $I_1 = 0$ A, $V_{D2} = 0.7$ V, $V_{R2} = 9.3$ V, $I_2 =$ 5.16 mA    **13.** $V_{D1} = V_{D2} = 0.7$ V, $I_T = 25.3$ mA, $V_{R1} = 2.53$ V, $V_{R2} =$ 5.06 V    **15.** 14.8%, not acceptable    **17.** 5.51%    **19.** 120 V
**21.** 250 V    **23.** 8.12 mW    **25.** 120 V    **27.** 0.798 V
**29.** 100 mV    **31.** 2.17 V    **33.** 300 μA    **35.** MR2504
**37.** MR756    **39.** Figures a, c, d, and e    **41.** 147 mA
**43.** 4.44 W    **45.** IN5362A    **47.** IN5349A    **49.** 1.6 kΩ
**51.** good    **53.** open

## Chapter 3:

**1.** 32.5 $V_{pk}$    **3.** 3 A    **5.** 12.5 $V_{rms}$    **7.** 31.8 $V_{pk}$    **9.** 16.27 $V_{pk}$
**11.** 10.12 $V_{dc}$    **13.** 5.18 $V_{dc}$    **15.** $-20.51$ $V_{pk}$, $-6.53$ $V_{dc}$, 1.28 mA
**17.** 21.21 $V_{pk}$    **19.** 12.56 $V_{pk}$, 7.99 $V_{dc}$, 799 μA    **21.** 20.51 $V_{pk}$, 13.1 $V_{dc}$, 2.11 mA    **23.** 33.24 $V_{pk}$    **25.** $-12.56$ $V_{pk}$, $-7.99$ $V_{dc}$, 7.99 μA    **27.** $V_{out(pk)} = 51.63$ $V_{pk}$, $V_{ave} = 32.87$ $V_{dc}$, $I_{pk} = 5.16$ mA, $I_{ave} = 3.29$ mA    **29.** 26.32 $V_{pk}$    **31.** 2.53A    **33.** 16.95 $V_{dc}$, 56.68 $mV_{pp}$    **35.** 520 $mV_{pp}$, 23.8 $V_{dc}$, 29.34 mA    **37.** 33.24 $V_{pk}$
**39.** 13.33 mA    **41.** 10.33 mA    **43.** 487 Ω    **45.** 34.35 mA
**47.** 25 mA, 20 mA, 5 mA    **49.** 9.1 $V_{dc}$, 4.55 mA    **51.** 33 mA, 33 V, 138.99 $mV_{pp}$    **53.** One of the diodes is open.    **55.** Yes
**57.** There is a problem.

## Chapter 4:

**1.** 18.02 $V_{pk}$    **3.** $-11.25$ $V_{pk}$    **5.** $-0.7$ V, 3.77 $V_{pk}$    **7.** 6.36 $V_{pk}$, $-2.7$ $V_{pk}$    **9.** 4.7 V and $-14.67$ $V_{pk}$    **11.** $-12$ $V_{dc}$
**13.** $T_C = 564$ μs, $T_D = 51.7$ ms    **15.** $T_C = 1.32$ ms, $T_D = 198$ ms
**17.** 0 V and $-12$ $V_{pk}$    **19.** $+6$ $V_{pk}$ and $-24$ $V_{pk}$    **21.** $+28$ $V_{pk}$ and $-8$ $V_{pk}$    **23.** 48 $V_{dc}$, 96 $V_{dc}$    **25.** 35.36 $V_{dc}$, 35.36 $V_{dc}$, 70.72 $V_{dc}$
**27.** 20 $V_{dc}$    **29.** 21.21 $V_{dc}$, 42.44 $V_{dc}$, 21.21 $V_{dc}$, 63.64 $V_{dc}$
**31.** $V_{C1} = V_{C3} \cong 50.91$ $V_{dc}$, $V_{C2} = V_{C4} \cong 101.82$ $V_{dc}$, $V_{C5} \cong 203.65$ $V_{dc}$
**33.** 50.91 $V_{dc}$    **35.** $D_1$ is shorted.    **37.** The circuit is operating properly.    **39.** $C_1$ is shorted.    **41.** The circuit is operating properly.
**45.** 1000 $V_{dc}$

## Chapter 5:

**1.** $25 \times 10^{-15}$ F/°C (or 0.025 pF/°C)    **3.** 726.4 kHz, 1.59 MHz
**5.** 4 kW    **7.** 0.4 $V_{dc}$    **9.** 1N6298

## Chapter 6:

**1.** 3.84 mA    **3.** 256.54 mA    **5.** 1.12 mA    **7.** 3.75 mA, 200, 24 mA, 61.54 μA    **9.** (a) 1.025 mA, (b) 180 μA, (c) 2.88 mA, (d) 500 μA, (e) 19.95 mA, (f) 10 mA    **11.** 60 mA, 60.15 mA    **13.** 27.3 μA, 12.03 mA    **15.** 170.6 μA, 64.83 mA    **17.** 0.9977    **19.** 1 mA
**21.** 32 V    **23.** 2 V

## Chapter 7:

**7.** 4 mA, 5 V    **9.** 4.71 mA, $-7.29$ V    **11.** Midpoint biased
**13.** Not midpoint biased    **15.** 6 mA, 2.5 V    **17.** 1.88 mA, 6.35 V
**19.** 2.67 mA, 24 V    **21.** 4.76 mA, 7.37 V, 6.52 mA, 30 V
**23.** 660 kΩ    **25.** 2.03 mA, 17.8 V, 13.44 μA    **27.** 11.75 mA, $-5.89$ V, 64.94 μA    **29.** 75    **31.** 5 mA, 30 V, not midpoint biased
**33.** 16.67 mA, $-20$ V, not midpoint biased    **35.** Not midpoint biased
**37.** 1.83 mA, 4.35 V    **39.** 1.52 mA, $-5.43$ V    **41.** 1 mA, 29.9 V
**43.** 11.27 mA, $-6.32$ V    **45.** $R_B$ is open and/or $Q_1$ is open    **47.** $R_C$ is open.    **49.** $R_1$ is open.    **51.** 9.5 μA

## Chapter 8:

**1.** 250, 90, 91.6, 175    **3.** 129.6 mV, 420 mV, 10.56 V, 576 mV
**5.** 521.7 mV    **7.** 8.46 V    **9.** 342.25    **11.** 233.07
**13.** 13.64%    **15.** 14.29%    **17.** 26.99 dB, 10 dB, 35.05 dB, $-3$ dB
**19.** 63095.7    **21.** 475.47 mW    **23.** 50.48 mW    **25.** 695.7 mW
**27.** 10 W    **29.** 758.6 μW    **31.** 29.54 dB, 84.08 dB, 36.9 dB, 6.02 dB    **33.** 3.98    **35.** 0.2    **37.** 58.8 dB

## Chapter 9:

**1.** 2.08 Ω    **3.** 10.34 Ω    **5.** 12 Ω    **7.** 32.35 Ω    **13.** 33.3
**15.** 96.49    **17.** 108    **19.** 132.48    **21.** 237.24    **23.** 2.85 $V_{ac}$
**25.** 12    **27.** 30.6    **29.** 44 W    **31.** 185.47    **33.** 318.58
**35.** 4.85 kΩ, 1.34 kΩ    **37.** 3.38 kΩ, 2.71 kΩ    **39.** 150.4
**41.** 67.5    **43.** $8 \times 10^3$, $2.31 \times 10^4$, $1.85 \times 10^8$    **45.** $6.22 \times 10^5$, $1.2 \times 10^6$, $7.47 \times 10^{11}$    **47.** 16.15    **49.** 63.32 kΩ, 1.81 kΩ
**51.** 5 kΩ, 41.6 Ω, 76    **53.** 3.5 kΩ, 89.63    **55.** 948.7 Ω, 158.7

## Chapter 10:

**1.** 6 V, 5.3 V, 2.65 mA    **5.** 0.9862, 7.34, 7.24    **7.** 4.83 kΩ
**9.** 20.98 Ω    **11.** $Av = 0.9639$, $A_i = 6.27$, $A_p = 6.05$, $Z_{in} = 3.26$ kΩ
**13.** 85 kΩ    **15.** 10.13 mA, 4.58 MΩ    **17.** 71.3
**19.** 19.89 kΩ, 18.9 Ω    **21.** $A_i = 71.3$, $Av \cong 1$, $Z_{in} = 144$ kΩ, $Z_{out} =$ 18.75 Ω    **23.** 190.3    **25.** 17.5 Ω, 5 kΩ    **27.** $R_1$ is open or $Q_1$ is open.    **29.** $R_6$ is open.    **31.** $Q_1$ is open or $R_1$ is open.
**33.** $Q_1$ is leaky.

## Chapter 11:

**1.** 8.76 $V_{PP}$    **3.** 6.73 $V_{PP}$, saturation    **5.** 52.79 mW    **7.** 187.5 μW    **9.** 799 μW    **11.** 1.51%    **13.** 132.09 mW, 1.89 mW, 1.43%    **15.** 0.71%    **19.** 250 mW    **21.** 37.4%    **25.** 13.5 mW
**27.** 23.13 W    **29.** 18 W    **31.** 77.8%    **33.** 4 V, 4.7 V, 3.3 V
**35.** 68.87%    **37.** 78%    **39.** 33.73 mW    **41.** 529.52 mW
**43.** 133.3 mW    **45.** 312.5 mW    **51.** 7.41 mW    **53.** Yes, cutoff clipping

## Chapter 12:

**1.** 2 mA    **3.** 14 mA    **5.** 4.32 mA    **13.** 1.5 mA, 5.5 V
**15.** 2 mA to 3.3 mA    **17.** 0 μA to 640 μA    **19.** $V_{GS} = -150$ mV
to $-1.4$ V, $I_D = 200$ μA to 1.5 mA, $V_{DS} = 5.55$ V to 11.14 V
**21.** $V_{GS} = -500$ mV to $-3$ V, $I_D = 500$ μA to 3 mA, $V_{DS} = 7$ V to 14.5 V
**23.** 1 mA to 1.5 mA, 13.35 V to 18.9 V    **25.** 2.5 mA to 3.5 mA, 10 V
to 14 V    **27.** 1000 μS to 5600 μS    **29.** 4800 μS, 3000 μS, 1200 μS
**31.** 1.24 to 1.55    **33.** 1.23 to 1.89    **35.** 2.54 to 3.20    **37.** 667 kΩ
**39.** 0.3, 2.55 MΩ, 500Ω    **41.** 0.53, 3.1 MΩ, 216 Ω    **43.** 4.2,
192Ω, 476Ω    **45.** The JFET gate is shorted.    **47.** $C_{C2}$ is shorted.

## Chapter 13:

**5.** 1000 μS, 2000 μS, 3000 μS    **7.** 3 mA    **9.** 0 mA, 8 mA, 14.2
mA    **11.** 8 mA, 16 V    **13.** 10 mA, 7.2 V    **15.** 20 mA, 6 V
**17.** $Q_1$ is open.    **19.** 2.6 to 3.2    **21.** 0 mA to 1.8 mA    **23.** 200 Ω

## Chapter 14:

**1.** 639 kHz, 27.7 kHz    **3.** 484.4 kHz, 50.5 kHz    **5.** 13.64
**7.** 9.5 kHz, 538.5 kHz    **9.** 1.152 MHz, 1.150 MHz    **11.** 520.4 Hz
**13.** 169.3 Hz    **15.** 202.4 Hz    **17.** 28.4 dB, 26.5 dB, 25 dB, 13.9 dB
**19.** 520.4 Hz    **21.** 318.3 pF    **23.** 7 pF    **25.** 507.8 kHz
**27.** 13.49 MHz    **29.** 80.4 MHz, 582.8 kHz    **31.** 79 Hz
**33.** 241 kHz    **35.** 1.95 MHz, 7.49 MHz    **37.** 9 pF, 5 pF, 1 pF
**39.** 795.7 Hz, 1.25 MHz    **41.** 77.23 kHz    **43.** 537.5 kHz

## Chapter 15:

**1.** (+), (−), (+)    **3.** 16 $V_{PP}$    **5.** 133.3 m$V_{PP}$    **7.** 150 m$V_{PP}$
**9.** 12 mV    **11.** 143.24 kHz    **13.** 35.4 kHz    **15.** $A_{CL} = 120$, $Z_{in}$
$\cong 1$ kΩ, $Z_{out} < 50$ Ω, CMRR = 6000, $f_{max} = 79.6$ kHz    **17.** $A_{CL} = 1$,
$Z_{in} \cong 100$ kΩ, $Z_{out} \leq 75$ Ω, CMRR = 200, $f_{max} = 159$ kHz    **19.** $A_{CL}$
$= 101$, $Z_{in} > 1$ MΩ, $Z_{out} < 40$ Ω, CMRR = 50,500, $f_{max} = 78.79$ kHz
**21.** $A_{CL} = 41$, $Z_{in} > 2$ MΩ, $Z_{out} < 100$ Ω, CMRR = 1242, $f_{max} =$
291.14 kHz    **23.** $A_{CL} = 1$, $Z_{in} = 5$ MΩ, $Z_{out} = 40$ Ω, CMRR =
1000, $f_{max} = 318.31$ kHz    **25.** 30 kHz    **27.** 198.6 kHz    **29.** Yes
**31.** 12 MHz    **33.** 25 MHz    **35.** 4.52    **37.** 124.9    **39.** $Z_{in(f)}$
$= 48$ kΩ, $Z_{out(f)} = 1$ Ω    **41.** $A_{CL} = 100$, $Z_{in(f)} = 1$ kΩ, $Z_{out(f)} =$
49.98 mΩ    **43.** $R_f$ is open.    **45.** The input op-amp is faulty.
**47.** The inverting input of the first-stage op-amp is open.    **49.** 500,000

## Chapter 16:

**1.** 0 V    **3.** +4 V    **5.** 1.33 kHz    **7.** 72.34 Hz    **9.** 32.15 Hz
**11.** 159 Hz, 1.59 kHz    **13.** $-V_{out} = V_1 + V_2 + V_3$, $-5$ V

**15.** $-V_{out} = 1.5V_1 + 3V_2 + 1.5V_3$, $-4.35$ V    **17.** 3 kΩ
**19.** $-2$ V    **21.** $R_1$ is open or $R_2$ is shorted    **23.** $R_1$ is open
**25.** 4.1 V, 1.9 V

## Chapter 17:

**1.** 7    **3.** 129 kHz    **5.** 308.3 kHz, 9.7, 187.5 kHz, 1.53
**7.** 159.2 Hz, 11    **9.** 10.7 kHz, 1.5    **11.** 321.5 Hz, 1.98
**13.** 9.46 kHz, 11    **15.** 112.5 kHz, 250.1 kHz, 137.6 kHz, 167.7 kHz,
1.22    **17.** 5.89 kHz, 13.0 kHz, 7.15 kHz, 8.76 kHz, 1.225
**19.** $f_1$: 6.97%, $f_2$: 3.69%.    **21.** 1.5    **23.** 1.31 kHz, 1.39 kHz, 80 Hz,
1.35 kHz, 16.88    **25.** 809.4 Hz, 1127.8 Hz, 318.3 Hz, 968.6 Hz, 3.04
**27.** 11, 9.82, 87.6 kHz, 8.92 kHz, 83.15 kHz, 92.07 kHz    **29.** 20 V, 0 A
**31.** +3 V    **33.** $C_2$ is open.    **35.** $\Delta f_1 = \Delta f_2 = 3.1$ kHz

## Chapter 18:

**1.** 219.2 kHz    **3.** 24.35 kHz    **5.** 27.84 kHz, 0.01, 100    **7.** 24.35
kHz, 0.1, 10    **9.** 3.38 kHz, 0.01, 100    **11.** 277.05 kHz, 0.1, 10
**13.** $Q_1$ is open.    **17.** 5.76 mH

## Chapter 19:

**1.** 11.3 V    **3.** 3.53 V    **5.** 0.9 V, 6 V    **7.** 20 μs, 50 μs
**9.** 40%    **11.** 17.2 MHz, 2.3 MHz    **13.** 1.79 MHz, 1.21 MHz
**15.** 11.1 MHz, 2.33 MHz    **17.** 3.70 MHz, 538.5 kHz    **19.** 20 ns,
60 ns, 180 ns, 70 ns    **21.** 4.5 V, $-4.5$ V    **23.** 1.8 V, $-0.9$ V
**25.** 4.67 V, $-4.67$ V    **27.** 5 V, $-5$ V    **29.** 11 ms    **31.** 0.7 V
**33.** 5.47 kHz, 68.9%, 126 μs    **35.** 4.8 kHz, 66.7%, 138.6 μs
**37.** $R_A$ is open or pin 7 is shorted.    **39.** MMBT2222, MMBT2222A,
MMBT4401

## Chapter 20:

**1.** No    **3.** 1.04 ms    **5.** 7.21 A    **7.** 100 mV    **9.** 8.54 V to
11.20 V    **11.** 9.5 V to 13.8 V    **13.** 461.5 nm    **15.** 1666.7 nm
**17.** 319 PHz (P = $10^{15}$)    **21.** 34.6 V

## Chapter 21:

**1.** 5 μV/V    **3.** 3 μV/V    **5.** 133.3 μV/mA    **7.** 75 μV/mA
**9.** 6.1 V    **11.** 3.5 A    **13.** 35 V    **15.** 12.0 V    **19.** 0.2 mV,
0.017%, 0.0017%/V    **21.** 757.5 mW

# Index

Midband gain values, 568
Midpoint biased circuits, 244, 247–48
Military specs, 56
Miller's theorem, 591–93
Miniaturization of components, 240
Minicomputers, development of, 818
Minimum dynamic impedance ratings, 187
Minimum load current ratings, 941
Minority carriers, 8–9
Mr. Meticulous machine, 324
MMBV109L series varactor diodes, 177–78
Models
    for amplifiers, 286–87, 289–90
    for diodes, 23–32
    for zener diodes, 53–54
Modulators
    PIN diodes for, 195
    variable off-time, 949
Monostable multivibrators, 848–49
    555 timers as, 851–57
    for pulse-width modulation, 949
    troubleshooting, 854–57
MOSFETs, 538
    in cascode amplifiers, 558
    CMOS, 555–57
    construction of, 538–39
    depletion-type, 541–46
    in digital communications, 559–60
    dual-gate, 551–53, 558–59
    enhancement-type, 546–51
    handling, 539–40
    power, 553–55, 559–60
    in RF amplifiers, 559
    specification sheets for, 547–48
    as switches, 823–25
Motor control, 889, 918–19
Mounting capacitance with quartz crystals, 807
MRD500 and MRD510 photodiodes, 910–11
Multicolor light-emitting diodes, 61–62
Multiple-feedback band-pass filters, 745–47
    frequency analysis for, 747–52
    gain of, 752
Multiple-feedback notch filters, 753–54
Multipliers
    frequency, 775
    voltage, 138, 153–58
Multisegment displays, 138, 159–60
Multistage amplifiers, 329
    frequency response of, 608–10
    gain in, 346–48, 575
Multistage notch filters, 752–53
Multivibrators, 848
    555 timers, 849–51
    astable, 857–60
    monostable, 851–57
    troubleshooting, 854–57
MUR50 series rectifiers, 102
MV209 series varactor diodes, 177–79

N-channel JFETs, 475
N-channel MOSFETs, 547
Negative-biased clippers, 144
Negative charge, 4, 10–12
Negative clampers, 147–48, 151
Negative clippers, 138–41
Negative feedback
    and impedance, 672–74
    mathematical analysis of, 669–72
    for operational amplifiers, 663–74
    in Wien-bridge oscillators, 797
Negative full-wave rectifiers, 93
Negative half-wave rectifiers, 82–83, 88–89
Negative resistance

with tunnel diodes, 191–92
with unijunction transistors, 903
Negative resistance oscillators, 192
Net charge, 4
Neutrons, 3
Noise
    active filters for, 758
    common-mode differential amplifier
        operation for, 636
    in instrumentation amplifiers, 712–13
    in JFETs, 525
    in silicon-controlled rectifier triggering,
        884–86
    in switching regulators, 952
    in triac triggering, 894
Nominal zener voltages, 57
Noninverting amplifiers, 650–51
    analysis of, 652–53
    impedance of, 651
    operation of, 669
    voltage followers, 653–55
Noninverting inputs
    in differential amplifiers, 634–35
    in operational amplifiers, 622, 628
Noninverting Schmitt triggers, 841–45
Nonlinear distortion
    in common-emitter amplifiers, 362–64
    from cutoff clipping, 419
Nonlinearity in class B amplifiers, 437
Nonrepetitive surge current ratings, 884
Nonsymmetrical trigger points, 846
Norton's theorem, 969
Notch filters, 732, 752–55
Npn transistors, 204–5, 228
N-type materials
    doping for, 7–8
    forward biasing, 15
    in JFETs, 474–75
    reverse biasing, 16
Nuclei of atoms, 3
Null currents, 770

Observation analysis methods, 629
Octave scales
    in frequency response, 576
    in gain roll-off plots, 585–86
Off characteristics
    of BJTs, 226
    of JFETs, 522
Offset current, 638
Offset null pins, 623, 637–38
Offset resistors, 657
Offset voltages, 637–38
Ohmmeters for testing
    BJTs, 226–27
    capacitors, 126
    diodes, 62–63
On characteristics, 226
One-pole filters, 732
One-shot multivibrators, 848–49
    555 timers as, 851–57
    for pulse-width modulation, 949
    troubleshooting, 854–57
One-way conductors. See Diodes
Op-amps. See Operational amplifiers
Opaque cases, 909
Open filter capacitors, 125, 127
Open JFETs, 517–18
Open-loop voltage gain
    in inverting amplifiers, 647
    of operational amplifiers, 626–27
Open resistors
    in emitter bias circuits, 257–58

in voltage-divider bias circuits, 269–70
Open switches, diodes as, 24
Open transformer windings, 123
Open zener diodes, 125
Operating characteristics. See Specifications
Operating curves
    for diacs, 892
    for silicon-controlled rectifiers, 880
    for triacs, 893–94
Operating regions
    for BJTs, 206
    for diacs, 892
    for silicon-controlled rectifiers, 880
    for silicon unilateral switches, 877
Operating temperature ratings, 47, 50
Operating voltage ratings, 187
Operational amplifiers, 622–24, 631–34
    for audio amplifiers, 714–15
    common-mode rejection ratio for, 638–39
    for comparators, 688–96
    differential amplifiers in, 626, 634–37
    for differentiators, 701–2
    feedback paths in, 627, 656, 660, 663–74
    frequency response of, 658–63
    gain-bandwidth product of, 660–62
    gain of, 626–27, 645, 659–60, 663
    identifying, 624–25
    input bias current for, 638
    input offset current for, 638
    input/output polarity of, 627–29
    input/output resistance of, 641, 643
    input voltage range of, 644
    for integrators, 696–701
    internal capacitance of, 662
    for instrumentation amplifiers, 712–14
    for inverting amplifiers, 645–50, 666–69
    large-signal voltage gain of, 645
    maximum operating frequency of, 988–89
    negative feedback for, 663–74
    for noninverting amplifiers, 650–55, 669
    output offset voltage for, 637–38
    output short-circuit current for, 639–40
    packages for, 625–26
    performance curves of, 642–43
    power consumption rating of, 645
    power supply rejection ratio for, 639
    for precision diodes, 716–17
    slew rate for, 640–41
    sockets for, 658
    specification sheets for, 643–44
    for summing, 703–11
    supply current rating of, 645
    supply voltages for, 623–24, 629–31, 643
    troubleshooting, 655–58
    for tuned amplifiers, 731
    unity-gain frequency of, 662–63
    for voltage-controlled current sources,
        715–16
    for voltmeters, 715
Optical switches, 919–20
Optocoupling, 912–14
Optoelectronic devices, 874. See also Light-
    emitting diodes (LEDs); Photodetectors
Optointerrupters, 919–20
Optoisolators, 915–19
Orbital shells in atoms, 3–4
Oscillators, 788
    Armstrong, 804–5
    Barkhausen criterion for, 790–92
    Clapp, 803–4
    Colpitts, 798–802
    crystal-controlled, 806–8
    Hartley, 802–3

# Selected Equations

## Chapter 2

**Diodes:** $V_R = V_S - 0.7\,V$ $\quad I_T = (V_S - 0.7\,V)/R_T$ $\quad$ % of error $= \dfrac{|X - X'|}{X} \times 100$

$I_{F(max)} = P_{D(max)}/V_F$ $\quad V_F = 0.7\,V + I_F R_B$ $\quad I_R = I_S + I_{SL}$

$I'_R = I_R(2^x)$ $\qquad x = (T - 25^\circ C)/10$

**Zeners:** $Z_Z = \Delta V_Z / \Delta I_Z$ $\quad I_{ZM} = P_{D(max)}/V_Z$

**LEDs:** $R_S = (V_{out(pk)} - V_F)/I_F$

## Chapter 3

**Half-wave:** $V_{ave} = V_{pk}/\pi$ $\quad V_{ave} = 0.318 V_{pk}$ $\quad I_{ave} = I_{pk}/\pi$ $\quad I_{ave} = 0.318 I_{pk}$

$PIV = V_{2(pk)}$

**Full-wave:** $V_{L(pk)} = \dfrac{V_{2(pk)}}{2} - 0.7\,V$ $\quad V_{ave} = 2V_{L(pk)}/\pi$ $\quad V_{ave} = 0.636 V_{L(pk)}$

$PIV = 2V_{L(pk)}$ $\qquad PIV = V_{2(pk)} - 0.7\,V$

**Bridge:** $V_{L(pk)} = V_{2(pk)} - 1.4\,V$

**Filters:** $\tau = RC$ $\quad T = 5(RC)$ $\quad I_{surge} = V_{2(pk)}/(R_w + R_B)$ $\quad C = I(t)/\Delta V_C$

$t = CV/I$ $\quad V_{dc} = V_{pk} - \dfrac{V_r}{2}$ $\quad V_r = I_L t/C$ $\quad PIV = 2V_{2(pk)}$ (half-wave)

**Regulator:** $I_T = (V_S - V_Z)/R_S$ $\quad I_L = V_Z/R_L$ $\quad I_Z = I_T - I_L$ $\quad R_{L(min)} = V_Z/I_{L(max)}$

$V_{r(out)} = \dfrac{(Z_Z \| R_L)}{(Z_Z \| R_L) + R_S} V_r$

## Chapter 4

**Clippers:** $V_L = V_{in} - 0.7\,V$ $\quad V_L = \dfrac{R_L}{R_L + R_S} V_{in}$ $\quad V_L = -V_F$ $\quad V_R = V_{in} + 0.7\,V$

**Clampers:** $T_C = 5R_{D1}C_1$ $\qquad T_D = 5R_L C_1$

## Chapter 5

**Varactors:** $C_T = \varepsilon \dfrac{A}{W_d}$ $\qquad f_r = \dfrac{1}{2\pi\sqrt{LC}}$

## Chapter 6

**Currents:** $I_E = I_B + I_C$ $\quad I_C \cong I_E$ $\quad \beta = I_C/I_B$ $\quad I_C = \beta I_B$ $\quad I_E = I_B(1+\beta)$

$\alpha = I_C/I_E$ $\quad I_C = \alpha I_E$ $\quad I_E = I_C/\alpha$ $\quad I_B = I_E(1-\alpha)$ $\quad \alpha = \dfrac{\beta}{1+\beta}$

**Voltage:** $V_{CE} = V_{CC} - I_C R_C$

## Chapter 7

**Load line:** $I_{C(sat)} = V_{CC}/R_C$ (ideal) $\quad V_{CE(off)} = V_{CC}$

**Base bias:** $I_B = (V_{CC} - V_{BE})/R_B$ $\quad I_{CQ} = h_{FE} I_B$ $\quad V_{CEQ} = V_{CC} - I_C R_C$

**Emitter bias:** $I_{CQ} \cong I_E = (V_{EE} - V_{BE})/R_E$ $\quad V_{CEQ} = V_{CC} - I_{CQ} R_C$

$I_{C(sat)} = 2V_{CC}/(R_C + R_E)$ $\quad V_{CE(off)} = 2V_{CC}$ $\quad R_{IN(base)} = h_{FE} R_E$

**V/D bias:** $V_B = \dfrac{R_2}{R_1 + R_2} V_{CC}$ $\quad V_E = V_B - 0.7\,V$ $\quad I_E = V_E/R_E$

$V_{CEQ} = V_{CC} - I_{CQ}(R_C + R_E)$ $\quad I_B = I_E/(1 + h_{FE})$

$h_{FE(ave)} = \sqrt{h_{FE(min)} \times h_{FE(max)}}$ $\quad I_{C(sat)} = V_{CC}/(R_C + R_E)$ $\quad V_{CE(off)} = V_{CC}$

When $h_{FE} R_E < 10R_2$: $\quad R_{eq} = R_2 \| h_{FE} R_E$ $\quad V_B = V_{CC} \dfrac{R_{eq}}{R_1 + R_{eq}}$

**C/FB bias:** $I_B = \dfrac{V_{CC} - V_{BE}}{R_B + h_{FE} R_C}$ $\quad$ **E/FB bias:** $I_B = \dfrac{V_{CC} - V_{BE}}{R_B + (h_{FE} + 1)R_E}$

## Chapter 8

**Gain:** $A_v = v_{out}/v_{in}$ $\quad v_{out} = A_v v_{in}$ $\quad A_i = i_{out}/i_{in}$ $\quad A_p = P_{out}/P_{in}$

$v_{in} = v_S \dfrac{Z_{in}}{R_S + Z_{in}}$ $\quad v_L = v_{out} \dfrac{R_L}{Z_{out} + R_L}$ $\quad A_{v(eff)} = v_L/v_S$ $\quad A_p = A_v A_i$

**Efficiency:** $\eta = \dfrac{P_L}{P_{dc}} \times 100$

**Decibels:** $A_{p(dB)} = 10 \log A_p$ $\quad A_{p(dB)} = 10 \log (P_{out}/P_{in})$ $\quad A_p = \log^{-1}(A_{p(dB)}/10)$

$P_{dBm} = 10 \log \dfrac{P}{1mW}$ $\quad A_{v(dB)} = 20 \log A_v$ $\quad A_{v(dB)} = 20 \log (v_{out}/v_{in})$

$A_v = \log^{-1}(A_{v(dB)}/20)$ $\quad \Delta A_{p(dB)} = \Delta A_{v(dB)}$

## Chapter 9

**ac concepts:** $r'_e = \dfrac{25\,mV}{I_E}$ $\quad r'_e = \dfrac{\Delta V_{BE}}{\Delta I_B}$ $\quad \beta_{ac} = \dfrac{i_c}{i_b}$

$\beta_{ac} = \dfrac{\Delta I_C}{\Delta I_B}$

$A_i = h_{fe}$ (for the transistor only) $\quad r'_e = h_{ie}/h_{fe}$

**Gain:** $A_v = r_C/r'_e$ $\quad r_C = R_C \| R_L$ $\quad v_{out} = A_v v_{in}$

$A_v = \dfrac{h_{fe} r_C}{h_{ie}}$ $\quad P_{out} = A_p P_{in}$ $\quad A_i = h_{fe} \dfrac{Z_{in} r_C}{Z_{in(base)} R_L}$

$A_p = A_i A_v$

**Impedance:** $Z_{in} = R_1 \| R_2 \| Z_{in(base)}$ $\quad Z_{in(base)} = h_{fe} r'_e$ $\quad Z_{in(base)} = h_{ie}$

$Z_{in(base)} = h_{fe}(r'_e + r_E)$

## Chapter 10

**Emitter follower:** $V_{CEQ} = V_{CC} - V_E$ $\quad I_{C(sat)} = \dfrac{V_{CC}}{R_E}$ $\quad r_E = R_E \| R_L$

$A_v = \dfrac{r_E}{r'_e + r_E}$ $\quad A_i = h_{fe} \dfrac{Z_{in} r_E}{Z_{in(base)} R_L}$

$h_{fc} = h_{fe} + 1 \cong h_{fe}$ $\quad h_{ic} = h_{ie}$

**Impedance:** $Z_{in(base)} = h_{fc}(r'_e + r_E)$ $\quad Z_{in} = R_1 \| R_2 \| Z_{in(base)}$

$Z_{out} = R_E \left\| \left( r'_e + \dfrac{R'_{in}}{h_{fc}} \right) \right.$

**Darlington:** $I_{E2} = \dfrac{V_{B1} - 2V_{BE}}{R_E}$ $\quad R_{IN1} = h_{FC1} h_{FC2} R_E$ $\quad Z_{in(base)} \cong h_{fe1} h_{fe2} r_E$

$Z_{out} \cong r'_{e2} + \dfrac{r'_{e1}}{h_{fc2}}$ $\quad A_i = h_{fc1} h_{fc2} \dfrac{Z_{in} r_E}{Z_{in(base)} R_L}$

**Common base:** $A_v = \dfrac{r_C}{r'_e}$ $\quad Z_{in} = r'_e \| R_E$ $\quad Z_{out} \cong R_C$ $\quad A_i \cong 1$

## Chapter 11

**ac load line:** $i_{c(sat)} = I_{CQ} + (V_{CEQ}/r_C)$ $\quad v_{ce(off)} = V_{CEQ} + I_{CQ} r_C$ $\quad PP = 2I_{CQ} r_C$

$PP = 2V_{CEQ}$

**Class A**

**RC-coupled:** $P_S = V_{CC} I_{CC}$ $\quad P_L = \dfrac{V_L^2}{R_L} = \dfrac{V_{PP}^2}{8R_L}$ $\quad P_{L(max)} = \dfrac{PP^2}{8R_L}$

**Transformer:** $Z_1 = \left( \dfrac{N_1}{N_2} \right)^2 Z_2$ $\quad \Delta I_C = \dfrac{\Delta V_{CE}}{Z_1}$ $\quad I_{C(max)} = I_{CQ} + \Delta I_C$

$P_S = V_{CC} I_{CQ}$

$PP = V_{CE(max)} - V_{CE(min)}$ $\quad V_{PP(max)} = \left( \dfrac{N_2}{N_1} \right) PP$

**Class B:** $V_{CEQ} = V_{CC}/2$ $\quad I_{C(sat)} = \dfrac{V_{CC}}{2R_L}$ $\quad V_{CE(off)} = V_{CC}/2$

$Z_{in(base)} = h_{fe}(r'_e + R_L)$ $\quad Z_{out} = r'_e + \dfrac{R_{in}}{h_{fe}}$ $\quad PP \cong V_{CC}$

$I_{CC} = I_{C1(ave)} + I_1$ $\quad I_{C1(ave)} = (0.159 V_{CC})/R_L$

**Class AB:** $V_{B(Q2)} = \dfrac{R_2}{R_1 + R_2}(V_{CC} - 1.4\,V)$ $\quad I_1 = \dfrac{V_{CC} - 1.4\,V}{R_1 + R_2}$

**Power:** $P_D = V_{CEQ} I_{CQ}$ $\quad P_D = \dfrac{V_{PP}^2}{40R_L}$

## Chapter 12

**JFET:** $\quad I_D = I_{DSS}\left(1 - \dfrac{V_{GS}}{V_{GS(off)}}\right)^2 \qquad I_G \cong 0 \qquad I_D \cong I_S$

**Gate bias:** $\quad V_{GS} = V_{GG} \qquad V_{DS} = V_{DD} - I_D R_D$

**Self-bias:** $\quad V_S = I_D R_S \qquad V_{GS} = -I_D R_S \qquad V_{DS} = V_{DD} - I_D(R_D + R_S)$

**V/D bias:** $\quad V_G = V_{DD}\dfrac{R_2}{R_1 + R_2} \qquad I_D = \dfrac{V_G - V_{GS}}{R_S}$

**Transconductance:** $\quad g_m = g_{m0}\left(1 - \dfrac{V_{GS}}{V_{GS(off)}}\right) \qquad g_{m0} \cong \dfrac{2 I_{DSS}}{V_{GS(off)}}$

**CS:** $\quad A_v = g_m r_D \qquad r_D = R_D \| R_L \qquad A_v = \dfrac{r_D}{r_S + (1/g_m)}$ (swamped)

$\qquad Z_{in} = R_1 \| R_2$

**CD:** $\quad A_v = \dfrac{r_S}{r_S + (1/g_m)} \qquad Z_{out} = R_S \|(1/g_m)$

**CG:** $\quad Z_{in} = R_S \|(1/g_m) \qquad Z_{out} = R_D \| r_d \qquad r_d = 1/y_{os}$

## Chapter 13

**D-MOSFETS:** (All equations are identical to those listed for the JFET)

**E-MOSFETS:** $\quad I_D = k[V_{GS} - V_{GS(th)}]^2 \qquad k = \dfrac{I_{D(on)}}{[V_{GS} - V_{GS(th)}]^2}$

**D/F Bias:** $\quad V_{GS} = V_{DS} \qquad V_{DS} = V_{DD} - R_D I_{D(on)}$

## Chapter 14

**Response:** $\quad BW = f_2 - f_1 \qquad f_0 = \sqrt{f_1 f_2} \qquad f_1 = f_0^2/f_2 \qquad f_2 = f_0^2/f_1$

**BJT:** $\quad f_{1B} = \dfrac{1}{2\pi(R_S + R_{in})C_{C1}} \qquad f_{1C} = \dfrac{1}{2\pi(R_C + R_L)C_{C2}} \qquad R_{out} \cong r'_e + (R'_{in}/h_{fe})$

$\qquad f_{1E} = \dfrac{1}{2\pi R_{out} C_E} \qquad \Delta A_{v(dB)} = 20\log\dfrac{1}{\sqrt{1 + (f_1/f)^2}} \qquad C_{be} \cong \dfrac{1}{2\pi f_T r'_e}$

$\qquad C_{in(M)} \cong A_v C_{be} \qquad C_{out(M)} \cong C_{bc} \qquad f_{2B} = \dfrac{1}{2\pi(R'_{in}\|h_{ie})(C_{be}+C_{in(M)})}$

$\qquad f_{2C} = \dfrac{1}{2\pi_C(C_{out(M)}+C_{in})} \qquad \Delta A_{v(dB)} = 20\log\dfrac{1}{\sqrt{1 + (f/f_2)^2}}$

**JFET:** $\quad f_{1G} = \dfrac{1}{2\pi(R_S + R_{in})C_{C1}} \qquad f_{1D} = \dfrac{1}{2\pi(R_D + R_L)C_{C2}} \qquad C_{in(M)} \cong g_m r_D C_{gd}$

$\qquad C_{out(M)} \cong C_{gd} \qquad C_G = C_{gs} + C_{in(M)} \qquad C_D = C_{ds} + C_{out(M)} + C_{in}$

$\qquad f_{2G} = \dfrac{1}{2\pi R'_{in}C_G} \qquad f_{2D} = \dfrac{1}{2\pi_D C_D} \qquad C_{gd} = C_{rss} \qquad C_{ds} = C_{oss} - C_{rss}$

$\qquad C_{gs} = C_{iss} - C_{rss}$

**Multistage:** $\quad f_{2T} = f_2\sqrt{2^{1/n} - 1} \qquad f_{1T} = f_1/\sqrt{2^{1/n} - 1} \qquad BW_T = f_{2T} - f_{1T}$

## Chapter 15

**Op-amps:** $\quad v_d = v_2 - v_1 \qquad V_{io} = \dfrac{V_{out(offset)}}{A_v} \qquad f_{max} = \dfrac{slew\ rate}{2\pi V_{Pk}}$

**Inverting:** $\quad A_{CL} = R_f/R_i \qquad Z_{in} \cong R_i \qquad CMRR = A_{CL}/A_{CM}$

**Noninverting:** $\quad A_{CL} = (R_f/R_i) + 1 \qquad$ **Follower:** $\quad CMRR = 1/A_{CM}$

**Frequency:** $\quad A_{CL}f_2 = f_{unity}$

**Feedback:** $\quad A_{CL} = \dfrac{A_{OL}}{1 + \alpha_v A_{OL}} \qquad \alpha_v \cong 1/A_{CL}$

$\qquad Z_{in(f)} = Z_{in}(1 + \alpha_v A_{OL}) \qquad Z_{out(f)} = \dfrac{Z_{out}}{1 + \alpha_v A_{OL}}$

## Chapter 16

**Integrator:** $\quad f_1 = \dfrac{1}{2\pi R_f C_f} \qquad f_{min} = \dfrac{10}{2\pi R_f C_f} \qquad$ **Differentiator:** $\quad f_{max} = \dfrac{1}{20\pi RC}$

**Sum Amp:** $\quad V_{out} = -R_f\left(\dfrac{V_1}{R_1} + \dfrac{V_2}{R_2} + ...\right) \qquad$ **Inst. Amp:** $\quad v_{out} = \left(1 + \dfrac{2R}{R_C}\right)(v_2 - v_1)$

## Chapter 17

**Response:** $\quad BW = \dfrac{f_0}{Q} \qquad f_0 = \sqrt{f_1 f_2} \qquad f_{ave} = \dfrac{f_1 + f_2}{2}$

**Low-pass:** $\quad f_2 = \dfrac{1}{2\pi RC} \qquad$ **Two-pole LP:** $\quad f_2 = \dfrac{1}{2\pi\sqrt{R_1 R_2 C_1 C_2}}$

**High-pass:** (The $f_1$ equations are identical to the $f_2$ equations for the low-pass filter)

**Band-pass:** $\quad f_0 = \dfrac{1}{2\pi\sqrt{(R_1\|R_2)R_f C_1 C_2}} \qquad C = \sqrt{C_1 C_2}$

$\qquad Q = \pi f_0 R_f C$

When $Q \geq 2$: $\quad f_1 \cong f_0 - (BW/2) \qquad f_2 \cong f_0 + (BW/2)$

When $Q < 2$: $\quad f_{ave} = f_0\sqrt{1 + \left(\dfrac{1}{2Q}\right)^2} \qquad f_1 = f_{ave} - \dfrac{BW}{2}$

$\qquad f_2 = f_{ave} + \dfrac{BW}{2}$

**Multiple-feedback:** $\quad A_{CL} = \dfrac{R_f}{2R_i} \qquad$ **Notch:** $\quad f_0 = \dfrac{1}{2\pi\sqrt{R_1 R_f C_1 C_2}}$

**Discrete:** $\quad f_0 = \dfrac{1}{2\pi\sqrt{LC}} \qquad Q = X_L/R_w \qquad R_p = Q^2 R_w$

$\qquad r_C = R_p \| R_L$

$\qquad Q_L = r_C/X_L \qquad$ **Class C:** $\quad -V_{BB} = 1V - V_{in(Pk)}$

## Chapter 18

**Barkhausen:** $\quad \alpha_v A_v = 1 \qquad$ **Phase shift:** $\quad \theta = \tan^{-1}(-X_C/R)$

**Colpitts:** $\quad \alpha_v = C_1/C_2 \qquad$ **Hartley:** $\quad \alpha_v = L_2/L_1$

**Clapp:** $\quad f_r = \dfrac{1}{2\pi\sqrt{LC_3}}$

## Chapter 19

**Basic circuits:** $\quad$ duty cycle (%) $= \dfrac{PW}{T_C} \times 100 \qquad f_2 = 0.35/t_r$

$\qquad f_{max} = \dfrac{0.35}{10 t_r}$

**Schmitt triggers:**

**Noninverting:** $\quad UTP = -\dfrac{R_1}{R_f}(-V + 1V) \qquad LTP = -\dfrac{R_1}{R_f}(+V - 1V)$

$\qquad UTP = -\dfrac{R_1}{R_{f1}}(-V + 1V) \qquad LTP = -\dfrac{R_1}{R_{f2}}(+V - 1V)$

**Inverting:** $\quad UTP = \dfrac{R_2}{R_1 + R_2}(+V - 1V) \qquad LTP = \dfrac{R_2}{R_1 + R_2}(-V + 1V)$

**555 One-shot:** $\quad PW = 1.1RC \qquad V_{T(max)} < \dfrac{1}{3}V_{CC} \qquad V_{T(max)} < \dfrac{1}{2}V_{con}$

**555 Astable:** $\quad f_0 = \dfrac{1.44}{(R_A + 2R_B)C_1} \qquad PW = 0.693(R_A + R_B)C_1$

$\qquad$ duty cycle (%) $= \dfrac{R_A + R_B}{R_A + 2R_B} \times 100$

## Chapter 20

**SCR:** $\quad t_{max} = \dfrac{I^2 t\ (rated)}{I_S^2} \qquad I_{S(max)} = \sqrt{\dfrac{I^2 t\ (rated)}{t_S}}$

$\qquad \Delta V = \dfrac{dv}{dt}\Delta t$

**UJT:** $\quad V_k = \eta V_{BB} \qquad V_P = \eta V_{BB} + 0.7V \qquad$ **Light:** $\quad \lambda = \dfrac{c}{f}$

## Chapter 21

**Regulation:** $\quad$ line regulation $= \dfrac{\Delta V_{out}}{\Delta V_{in}} \qquad$ load regulation $= \dfrac{V_{NL} - V_{FL}}{\Delta I_L}$